Modern Techniques in Computational Chemistry: MOTECC-90

Modern Techniques in Computational Chemistry: MOTECC™-90

Edited by

Enrico Clementi

International Business Machines Corporation
Kingston, New York 12401, U.S.A.

ESCOM
Leiden 1990

CIP-Data Koninklijke Bibliotheek, Den Haag

Modern

Modern techniques in computational chemistry:
MOTECC-90 / ed. by Enrico Clementi. – Leiden :
ESCOM – Ill.
With ref.
SISO 540 UDC 681.3:54
Subject heading: computational chemistry.

ISBN-13: 978-94-010-7495-7 e-ISBN-13: 978-94-009-2219-8
DOI: 10.1007/978-94-009-2219-8

Published by:

ESCOM Science Publishers B.V.
P.O. Box 214
2300 AE Leiden
The Netherlands

Foreword to MOTECC-90

MOTECC-89 has been well received both in the academic and industrial worlds. However, it is time to move on, to revise, consolidate, expand, and explore. MOTECC-90 has indeed these characteristics *vis-à-vis* MOTECC-89.† I am pleased to acknowledge the most constructive support and participation from many Universities and from the IBM Research Centers in Almaden, Yorktown and Zurich. Our plans for MOTECC-91 show a continued growth, and, as done for MOTECC-90, we shall revise, correct and expand the chapters presently in MOTECC-90.

We can now state that many of the most well-known and advanced non-commercial codes either are part of MOTECC-90 or will be included in MOTECC-91. We can deal not only with the electronic structure of atoms and molecules, but also of polymers and solid state materials both in the crystalline and amorphous states. The application scope of Newtonian molecular dynamics has been expanded from proteins to solutions of proteins interacting with DNA because of the new interaction potentials obtained from *ab initio* quantum mechanical computations. In addition, quantum molecular dynamics is now present, allowing simulations on clusters and solids. The non-Newtonian dynamics has also been expanded as is also the case for fluid dynamics, which now extends to air pollution and makes use of finite elements techniques to solve structures.

Supercomputing is addressed in two chapters dealing with parallelism for numerical algorithms and with our software for the LCAP experimental supercomputer. However, in MOTECC-90 we have stressed also the use of new and powerful workstations. Graphics and animation not only pervades the application areas, but are also the main tool for interactive input generation.

The goal of making MOTECC a connected set of moduli is still with us but will be implemented starting in MOTECC-91; up to now we have assembled the "raw material" for the interconnected moduli of tomorrow.

In MOTECC-90 we continue to stress *global simulations*, but in an interdisciplinary sense. The necessary tools are finally becoming available, namely organized superstructure of computer programs, like the one offered by MOTECC-90, with easy Input/Output and portable from the powerful and nicely priced work-

---

† Modern Techniques in Computational Chemistry: MOTECC-89, E. Clementi, Editor, ESCOM, Leiden, The Netherlands, 1989.

stations to supercomputers. There are still many obstacles and one must be found in the rigidity of teaching by Universities (beautifully deep, but too narrow).

Yet, interdisciplinarity is absolutely essential if we wish to have some limited understanding of the real world which surrounds us. Below I shall give a simple example of what I mean for "limited understanding of the real world," and to do so let us pretend we are gathered in some pleasant location, for example at a sea-side. We ask ourselves which "model" should we use to "scientifically appreciate" the scenery we are looking. After all, *if we wish to preserve the environment, clearly a global scientific understanding is needed*; keeping in mind that we have only a finite number of neurons, we shall have to compromise for a somewhat shallow but general background understanding, and a deeper specialization in some limited areas. We argue that today's general level of scientific understanding is much too shallow and narrow and has not kept pace with the computational development. With the global simulation approach we "collectively" can develop a deeper and broader scientific understanding, which should be the cornerstone of any rational society.

But now let us return to our example. Sitting on a hillside overlooking the beach, we watch the waves relentlessly crashing against the shore and note the formation of exotic patterns which vary smoothly with space and time. If this, the breaking of the waves on the shore, is what we care to see, then the phenomena can be beautifully recreated by solving the shallow water equations, an adaptation of the Navier-Stokes equations.

If, however, the sea is struck with some gusty winds, then the mosaic of the sea surface changes much more rapidly. The waves, having derived energy from the air, become menacing and we see complex structures pervading the entire expanse of the sea surface. To simulate the energy transfer in the air-sea interaction, we need to supplement the fluid dynamic description with a kinetic description, such as the Boltzmann equation or the Fokker-Planck equations. The time and length scales of the motion in this region do not allow a satisfactory solution with continuum models. We could explore the possibility of modelling the phenomenon directly, without recourse to equations, by cellular automata methods which depend upon simple collision rules incorporating certain conservation laws.

Let us analyze a bit more in detail the new structures generated by adding energy (from the wind) to the water system. Whereas before there were a few well-defined modes of motion of the surface, there is now a collection of modes that span a wide range of spatial and temporal scales. The continuum equations still remain valid in describing the motion of the water; nonlinear interaction creates new modes that transfer energy from the large eddies to the small eddies until it is eventually dissipated in molecular motion. The Reynolds number which indicates the ratio of inertial to viscous forces has exceeded critical values within regions of the water and we witness turbulent behavior characterized by erratic motion at small length scales. The Reynolds number can vary by orders of magnitude depending on the flow situation. For instance, the Reynolds number of the flow inside a common journal bearing is typically on the order of 1. Due to the high viscosity of the lubricant, viscous and inertial forces are on the same order. On the other hand, the Reynolds number of the flow around a tadpole swimming in water is on the order of $10^2 \sim 10^3$. The self-propulsion of a tadpole gives rise to inertial forces which are much larger than the viscous forces. Reynolds numbers of flows around birds in flight tend to be $10^4 \sim 10^5$, mostly due to the fact that the viscosity of air is two orders of magnitude smaller than that of water. If we consider the flow around an automobile cruising at 100 k.p.h., we have Reynolds numbers on the order of a million. Reynolds numbers of flows around the wings of a commercial jet plane at cruising speed are typically on the order of 10^8 , 10^9 . Thus fluid flows are capable of displaying diverse phenomena without violating the assumptions of a continuum.

Then when the wind subsides, and the sky is clear, we see, not far from the shore, corks holding up a fisherman's net. The corks, all in all, stand still. But a closer look with the binoculars reveals that the corks are not standing still after all, but are performing incessant irregular motions. To describe the motion of the corks we may use the techniques of Brownian Dynamics wherein the effect of the surrounding fluid elements is incorporated in the form of a random force term in a Langevin equation, which we consider as the natural extension of Newton equations for systems too complex to allow specific descriptions for all the particles (thus averages are introduced for motion at shorter time scales than those of the phenomenon of interest, like the corks in our example). If we look in the immediate neighborhood of the cork, we discover much irregularity in the water motion: indeed the overall trend set up by the waves is disturbed in the immediate vicinity of a cork , which acts as an obstruction. But if we were really able to see the water, namely the water molecules, then suddenly we would see the translational and rotational motions which can be fairly well approximated by Newtonian mechanics, which, however, does not ask any probing questions on the "true origin" of the force or a torque.

An even more accurate probe would reveal that a water molecule is a pulsating body, with very specific stretching and bending vibrations, which vary little from one molecule of water to another. Furthermore, their description necessitates quantization of the energy for time scales which are shorter than any mentioned phenomena in the bay. Of course the investigation should not stop here and one should look more closely and first realize that there are electrons and harder but very small and massive bodies, the nuclei. Their relative motions cannot be described by classical mechanics and we have to use quantum mechanics, in particular the Schrödinger equation and its many approximations. With this equation we can follow molecular reactions for simple and very complex chemical systems, *i.e.* the present and future goals of quantum chemistry. Notice that hardily any medium would be superior to liquid water for sustaining many chemical reactions including those leading to "living matter" (for this reason there is so much emphasis on water in some of the chapters).

Notice that the Newtonian equations are invalid not only for modeling objects on the atomic scale but also for describing classical bodies moving with velocities comparable with the speed of light; here we need relativistic classical mechanics and for nuclei and electrons we need a relativistic quantum mechanical formulation, the Fock-Dirac equations to start with.

Then the night comes. The sky and the stars remind us that, even if we are infinitesimally small and very fragile, still we can understand, simulate and predict any complexity we select to work on. Incidently, cosmology and astronomy are two choice examples for global simulations.

Here we stop in MOTECC, but the story continues with new forces and time scales characteristic of sub-nuclear particles. Thus natural phenomena can be studied in different time scales, associated with each of these scales is a distinct computational universe which expresses the dynamics of motion at that particular resolution. Clearly it makes little sense to describe the waves in the bay with quantum mechanics, especially if we are interested in an overall description; but we need to know how to connect phenomena in vastly different space-time scales.

The traditional approach to simulating natural phenomena has been to construct models which are applicable within a range of time and length scales; these models are then translated into equations which are solved by suitable methods. When the phenomenon departs from this range of behavior the model breaks down and the equations are no longer valid, even by adding non-linear corrections; it is necessary to construct new models and use new equations within this new regime. Even within regions of validity of the model and its accompanying equations, the behavior may not be predictable in any rigorous manner (*e.g.*, chaotic vibrations and turbulent flow).

It is very tempting to think that collision rules and a mathematical expression of the physical invariance is all that is needed to describe any mechanics at any scale of time and space. Perhaps, someday, cellular automata might develop into this most powerful framework. Until that time we shall keep on with attempts to solve more and more accurately the many equations developed over the last four hundred years to describe moving objects.

Models have limits: the secret is to have an appropriate model which "fully" answers a limited question. MOTECC is a tool which associates models and equations to computer programs to facilitate our computational tasks.

It is my pleasure to acknowledge the most valued help of Dr. Steven Chin in editing MOTECC-90.

Enrico Clementi

Foreword to MOTECC-89

In a way the MOTECC-89 project started in the early sixties at the IBM Research Laboratory in San Jose, California. The six years of post-doctoral research, first with Giulio Natta on conductive polymers, with Michael Kasha on spin-orbit effects, with Kenneth S. Pitzer on high temperature molecules and thermodynamics and with R.S. Mulliken in the quantum chemistry of small molecules had demonstrated pragmatically the importance of a broad-based research and also let me taste some of the excitement to be derived from interdisciplinarity. Thus when I started to gather a department in the newly opened IBM Research Laboratory in San Jose, California, I purposely named it "Large Scale Scientific Computation Department," avoiding reference to chemistry, physics, statistical mechanics or fluid dynamics, which were our main tasks. In the sixties interdisciplinarity was more and more recognized as a most important if not necessary avenue to cope with the technical needs of our society. However, at that time interdisciplinarity was synonymous with "team work," and *true* interdisciplinarity was a formidably difficult objective. Although I headed an excellent group of scientists with different backgrounds and there was *much* progress and creativity, still each one of us was more or less conducting his *own* research in his *own* field with occasional cross-field partnerships and with some of the computational techniques as our common base.

Later, in 1974, I made a second attempt, this time starting a new department at the Donegani Institute, Montedison, in Novara, Italy. Here the department name I selected was "Dipartimento di Calcolo Chimico," and the main stress was computational chemistry in quantum chemistry, drug-design and statistical mechanics with some fluid dynamics in air control study, for example modelling the air quality of the industrial area of Porto Marghera, near Venice.

My third department is now here at IBM-Kingston, New York. In the last twenty years, computing and computers have advanced enormously and this, in itself, brings about new horizons and opportunities. I feel I am getting nearer to the old and original goal; to start with, I am more clear within myself on the direction but — most importantly — the mature and established scientists who have joined me fully understand and support the interdisciplinary style. After all, interdisciplinarity today is an accepted viewpoint; computers and simulations are ubiquitous and gone are the polemics and doubts about *ab initio* chemistry in particular and *computational* chemistry in general. To be sure, the intensity of our involvement in recent years in computer architecture and related system software was underestimated but, on the other hand, "parallelism" is serving as a most useful glue and, for us, is becoming something of a trademark. And it helps to be convinced that we are on the right track at the right moment.

Today, interdisciplinarity essentially should mean that one is *competent, proficient* and *creative in many areas.* The key to this is not some utopian superman, but ample availability of supercomputer time and access to a broad, easy to use library of powerful application programs. Teamwork continues to be important, but is no longer the main focus. Clearly, to reach our goals many challenges will have to be faced and, in particular, our teaching methodology will have to change: *more emphasis on the fundamentals and less on the details,* especially those which can be obtained reliably by pushing the computer "start" button. And MOTECC-89 should help.

The "global simulation" methodology, which we strongly advocate, is expected to become the standard approach to complex engineering problems, especially when the cost of the simulation represents a small fraction of the overall cost and when the availability of simulated data brings about substantial cuts in the time needed to complete the entire research and development process.

Today we are fascinated by the possibility of producing new chemicals obtained with specifically designed proteins and enzymes. At the same time, commercially available proteins are being altered (by adding or removing one or more amino acid residue), not only to learn their mechanism of reaction, but also to obtain *new products.* Another area of dramatic interest is related to DNA sequence alterations obtained by adding or subtracting parts of the code. In both these instances we obtain new chemicals and even new forms of life. In a very different area of endeavor today, we are equally fascinated by the revolutionary proposals and models which characterize modern astronomy and cosmology; at the same time, all at once solar space seems to have shrunk to nearly human dimensions with deep space probes, satellites, shuttles, and space stations.

The implications of these new and very important technologies are not presently well understood at the socio-economic level, and even less at higher levels. It is therefore truly essential to prepare those *instruments* which can be used to simulate some of these processes, keeping in mind that simulations can be used either to avoid unnecessary experimentation or to "optimize" these processes.

In the past man devised tools mainly to enhance his physical dimension, but the computer is a particular tool designed to enhance our intellectual dimension and, hopefully, our ethical and aesthetic standards.

The *intensity* in the use of tools to enhance our mental dimensions does not at all match the pervasive use of the tools to enhance our "physical dimensions"; this is an imbalance which must be reconsidered. At the Seoul Olympic Games in 1988, the high-jump record was won by Mr. Guennadi Avdeenko and by Ms. Louise Ritter, who used their physical and mental ability after years of training to reach nearly incredible strength; the record shows that the height of the jumps were 7 feet, 9.75 inches and 6 feet, 8 inches, respectively. Exceptional people! In *contrast,* millions of people (including infants, the aged and handicapped) can jump to 30 ×

10^3 feet, *routinely*, with an everyday tool — the airplane. We have become accustomed to the latter "incredible" events and take them for granted because we have been using tools for the last few thousand years. On the contrary and most unfortunately, we are still far from even starting *true exploitation* of computing machines which will enhance man's intellectual ability a thousand-fold and not be limited to "number crunching."

The decision to go to press in 1989 is certainly a bit risky, because partly premature, but a delay to 1990 would help only marginally, not substantially, since our front is so broad and extends so far. For example, the last two chapters of this book have been added at the last moment; more will be forthcoming in the future. We have planted a seed. We expect to experience a basic step forward when we open MOTECC to collaborations with universities and research institutions. A possible "danger" could be in too much uncoordinated help which could transform MOTECC into a fragmented collection of unrelated items. Thus we felt that MOTECC-89 was needed to clearly indicate, for better or worse, the texture of a design, which should grow with control and harmony and much joint work with many institutions. This growth should be in extending areas already started, but also in addressing new areas like chemical kinetics, reaction rates and scattering, namely those time-dependent aspects which are basic to the understanding of chemical reactions in gases, liquids and solids.

The main idea is to pull together researchers from different fields, sharing enthusiasm for computing, modelling, and for learning from each other the different ways the scientific method evolves today, stressing logic, mathematics and physics, but also capable of dealing more and more with the complexity of biology, neurology and, eventually, sociology.

I am looking forward to much constructive criticism of MOTECC-89 in order to produce superior revisions: MOTECC-9X. Whatever the outcome, I have no doubt that one of the main ideas, namely the "global simulation" approach, will survive whether or not there will be a MOTECC-2001.

In closing I would like to express my sincere appreciation to the authors of this volume who have joined me in this experiment and are willing to continue with MOTECC-199X. A very special acknowledgment is extended to Dr. Steven Chin, who has accepted the task of overseeing the graphics for the entire volume, and to Ms. Dee Keeling for her incredible endurance in typing, re-typing, and typesetting the finished volume, always with good temper and humor.

Enrico Clementi

Contents

Chapter 3. Kinetically Balanced Geometric Gaussian Basis Set Calculations for Relativistic Many-Electron Atoms

A. Mohanty and E. Clementi
IBM Corporation, Dept. 48B/428, Neighborhood Road
Kingston, NY 12401 USA

Chapter 4. Non-Relativistic Configuration Interaction Calculations for Many-Electron Atoms: ATOMCI

F. Sasaki, M. Sekiya, T. Noro, K. Ohtsuki and Y. Osanai
Department of Chemistry, Faculty of Science, Hokkaido University
Sapporo 060, Japan

Chapter 7. KGNMOL: A Program for Large Molecules and Molecular Interactions

E. Clementi, G. Corongiu and S. Chakravorty
IBM Corporation, Dept. 48B/428, Neighborhood Road
Kingston, NY 12401 USA

Chapter 8. SIRIUS: A General Purpose Direct Second Order MCSCF Program

H.A. Jensen, H. Ågren, and J. Olsen†*

Department of Chemistry, University of Aarhus
DK-8000 Aarhus C, Denmark

**Institute of Quantum Chemistry, University of Uppsala*
Box 518, S-75120 Uppsala, Sweden

†Theoretical Chemistry, Chemistry Center, University of Lund
Box 124, S-22100 Lund, Sweden

Chapter 9. MOLCAS: A General Purpose Quantum Chemistry Program System for Correlated Wavefunctions

*B.O. Roos, G. Karlström, P.-Å. Malmqvist, A.J. Sadlej and P.-O. Widmark**

Department of Theoretical Chemistry, Chemical Centre P.O.B. 124, S-221 00 Lund, Sweden

**S&TC group, ACIS, IBM Sweden*

Chapter 10. MELD: A Many Electron Description

E.R. Davidson
Department of Chemistry, Indiana University
Bloomington, IN 47405 USA

Chapter 11. ALCHEMY II: A Research Tool for Molecular Electronic Structure and Interactions

*A.D. McLean, M.Yoshimine, B.H. Lengsfield**
P.S. Bagus and B. Liu

IBM Research Division, Almaden Research Center, 650 Harry Road
San Jose, CA 95120-6099 USA

**Lawrence Livermore Laboratory, Livermore, CA*

I. Cacelli, V. Carravetta, A. Rizzo* and R. Moccia†*

Scuola Normale Superiore, Piazza dei Cavalieri, 56126 Pisa, Italy

**I. C. Q. E. M. del CNR, Via Risorgimento 35, 56126 Pisa, Italy*

†Dipartimento di Chimica e Chimica Industriale, Università di Pisa Via Risorgimento 35, 56126 Pisa, Italy

Chapter 13. Dirac-Fock Self-Consistent Field Calculations for Closed Shell Molecules with Kinetic Balance and Finite Nuclear Size

A. Mohanty and E. Clementi
IBM Corporation, Dept. 48B/428, Neighborhood Road
Kingston, NY 12401 USA

Chapter 14. First Principles Molecular Dynamics

M. Parrinello
IBM Research Division, Zurich Research Laboratory
CH-8803 Rüschlikon, Switzerland

Chapter 15. LCAO Ab Initio Band Structure Calculations for Polymers

J.M. André, J.L. Brédas, J. Delhalle, D.J. Vanderveken, D.P. Vercauteren and J.G. Fripiat*

Laboratoire de Chimie Théorique Appliquée and
Laboratoire de Chimie Moléculaire Structurale
Facultés Universitaires Notre-Dame de la Paix, 61 rue de Bruxelles
B-5000 Namur, Belgium

**Service de Chimie des Matériaux Nouveaux, Université de Mons-Hainaut, Avenue Maistriau 21, B-7000 MONS Belgium*

Chapter 16. BNDPKG2: A Linear Combination of Gaussian Orbitals (LCGO) Band Structure Program for Cubic Crystals With One Atom Per Unit Cell

*N.E. Brener, J. Callaway, and J.M. Tyler**
Department of Physics and Astronomy and
**Department of Computer Science*
Louisiana State University, Baton Rouge, LA 70803 USA

Chapter 17. Monte Carlo and Molecular Dynamics Simulations

E. Clementi, G. Corongiu, M. Aida, U. Niesar, and G. Kneller
IBM Corporation, Dept. 48B/428, Neighborhood Road
Kingston, NY 12401 USA

Chapter 18. Molecular Dynamics Simulations of Fluid Flows

D.K. Bhattacharya and E. Clementi
IBM Corporation, Dept. 48B/428, Neighborhood Road
Kingston, NY 12401 USA

Chapter 19. Brownian Dynamics Simulations of a Complex Fluid System

D.K. Bhattacharya and E. Clementi
IBM Corporation, Dept. 48B/428, Neighborhood Road
Kingston, NY 12401 USA

Chapter 20. Cellular Automata

R. Panda, V. Sonnad and E. Clementi
IBM Corporation, Dept. 48B/428, Neighborhood Road
Kingston, NY 12401 USA

Chapter 21. The Equations of Fluid Flow and Their Solution by Finite Element Methods

V. Sonnad, S. Hassanzadeh, R. Panda and S. Foresti
IBM Corporation, Dept. 48B/428, Neighborhood Road
Kingston, NY 12401 USA

Chapter 22. The Equations of Elasticity and Their Solution by Finite Element Methods

V. Sonnad, H. Murakami, S. Foresti, S. Hassanzadeh and B. Jiang
IBM Corporation, Dept. 48B/428, Neighborhood Road
Kingston, NY 12401 USA

Chapter 23. Modeling of Atmospheric Pollutant Transport in Shorelines

Z.D. Christidis
IBM Research, T.J. Watson Research Center
Yorktown Heights, NY 10598, U.S.A.

Chapter 24. Visualization Techniques and Windowing Interfaces: KGNGRAF, XWIB and REMOTE

S. Chin, R. Tagliavini, J.P. Prost and M. Martins-Costa
IBM Corporation, Dept. 48B/428, Neighborhood Road
Kingston, NY 12401 USA

Chapter 25. LCAP: Loosely Coupled Array of Processors Parallel Processing Software

D. Folsom
IBM Corporation, Dept. 48B/428, Neighborhood Road
Kingston, NY 12401 USA

Chapter 26. Parallel Algorithms

D. Logan
IBM Corporation, Dept. 48B/428, Neighborhood Road
Kingston, NY 12401 USA

Chapter 1.

Introduction to MOTECC-90†

E. Clementi
IBM Corporation
Scientific and Engineering Computations
Dept. 48B/428
Kingston, New York 12401, U.S.A.

1. Introduction

Developed societies are becoming progressively integrated at the economic and technical level. This trend brings about increasing social-economic constraints whose impact is often difficult to evaluate. As a consequence the society's margins of safety are reduced or, conversely, the society's risk factors are increased. For instance, the degree of air, water and soil pollution could be described as a "failure" — hopefully reversible and temporary — caused by insufficient appreciation of "risk factors." We seem to have an "inborn" trait that makes us strive to improve our material standard of living, thus placing enormous pressure on our social-economic structures. We seem to be ready to accept more and more "risks" — perhaps unconsciously at times — thereby steadily increasing the probability of "failures." The drawback of this behavior is that at times some failures can be very expensive with *irreversible* (if not *terminal*) consequences.

From this scenario it follows the very real and pressing need to decrease risks; we advocate, when feasible, the strategy of realistic simulations where the failures would be experienced by a "virtual society" rather than by the "real society." Obviously, the more thorough the simulation the more reliable the corresponding predictions. It should be possible to eliminate "naked risks," namely those decisions made with inadequate analyses and insufficiently reliable simulations. Therefore, computers, supercomputers, simulations, interdisciplinary approaches, expert systems, artificial intelligence, signal recognition and robotics will be essential aspects of science and technology in the twenty-first century.

---

† This chapter is a revision of Chapter 1 in MOTECC-89.

Notice, however, the rich spectrum of opportunities which derive from this combination of forces and reactions. Indeed, we wish to stress in a balanced fashion both the need to avoid risks and, at the same time, take advantage of the opportunities and move into a more rational and more "human" society. Most unfortunately, "naked risks," biased assumptions, compartmentalization, empiricism and irrationality are still present, not only in today's society but also in our technological goals. In this perspective MOTECC could be considered as a call for "*neo-illuminism*," where once more one looks at rationality and knowledge as a key for social progress.

In this chapter I shall limit myself to a simple facet of the above complexity and I shall address "simple" problems in mechanics and dynamics, particularly for chemistry and chemical engineering.

2. Global Simulations

Science is knowledge, based on the measurement and interpretation of natural processes. The means of measurement has always involved the calculating machine. With the abacus or supercomputer, mankind has always struggled for a truth which implies the rational understanding of many complicated natural phenomena. Such an understanding must be based on a small number of axioms, capable of generating self-consistently numerous experimental observations.

In the science of dynamics of moving bodies, electrons, molecules, or asteroids, there have been many sets of axioms, some of which have survived modification better than others; none of which, however, have remained immutable. The change in the fundamental appreciation of nature has come about as the direct result of calculation, by the use of calculating machines, from insatiable human curiosity. It is this same process that drives us towards the construction of powerful computer architectures in our own times, and it is the purpose of this volume to help our readers form an appreciation of natural dynamical processes from the single perspective of simulations with supercomputers, particularly those of parallel architecture.

It is now possible to build up a "*global*" approach to computer simulation of dynamical processes in complex systems, for example liquid water, as part of a more general global strategy introduced below. We start by constructing the molecules of the liquid from the fundamental entities of matter, electrons and nuclei, using quantum mechanics.[1-5] Next we obtain the potential energy surfaces between interacting water molecules, using the *ab initio* principles of quantum mechanics.[6-10] Then we use Monte Carlo[11-14] and molecular dynamics computer simulation, based at present on classical mechanics, to model the motion and interaction of molecules,[15-20] for example, in liquid water, and to extract from these its measurable macroscopic properties. Finally we use the speed[21-25] of a parallel supercomputer architecture to build from Newton's equations the phenomena[26-29] of hydrodynamics, such as the Rayleigh Benard patterns and flow past objects in two and three dimensions. We are thus able to construct macroscopic properties from the fundamental building blocks of aqueous matter. Now we are in the continuum and we can look at turbulence, a most fascinating aspect of fluid dynamics.

This is only one example of the global approach; it is becoming more and more apparent that this method addresses *any* complex problem.

In the *global simulation*, we decompose a complex problem into *n* sub-problems each corresponding to a sub-model 1,...,i,....,n. The operational rules are that the input needed in the sub-model "i" is fully obtained as the output from sub-model "i-1" and that the input for the *first* sub-model must be very simple. Section 2 of this chapter, for example, illustrates a *model of type 1*: from *ab initio* computations of the intermolecular potential, the input for *model 2*, classical computer simulations of molecular dynamical processes are performed. In turn, model 2 produces data for *model 3*, of flow and convective phenomena which were previously thought to be describable only with the equations of hydrodynamics. In short, "global simulation" and "scientific method" are synonymous, the former however restricted to computational rather than laboratory observations.

In the last forty years we have seen the development which makes this approach possible, the birth and evolution of computers, personal computers, workstations, minicomputers, and supercomputers. However, all the available indicators point towards the beginning of a new social-economic era, *where computing and signal processing machines will play a determining role.* Ages are often classified in terms of some characteristic of their production tools; we have the stone, the bronze and the iron ages. "Tools" enhance the dimension and power of mankind; in the past the *physical dimension*, but today the intellectual and mental dimension. Clearly, our age is the computer age, ("the information society"). The use of supercomputers allows universality in science, the global approach providing quantitative modelling for any specific event, and also in general. Engineering attempts to put to work this understanding in practical applications of ever-increasing complexity and boldness. These goals and attempts have been with mankind since the beginning of our evolution, but today seem to be near realization. A turning point on this learning curve is the ability to perform "*global simulations*", but this *can be accomplished only if supercomputing is a common practice.*

For some time our I.B.M. Kingston Scientific and Engineering Computations Center has been advocating the "global simulation" approach to computer simulations in science and engineering. The idea is borrowed by considering human evolution in history, in particular from the events leading Great Britain to the first industrial revolution, when *human sized assembly lines* were introduced, thus enormously increasing the productivity per worker. In an "assembly line" one starts from simple raw materials, and step by step, without interruption, transforms the raw materials into more and more finished and complex products.

Today we use *robot sized assembly lines*, and this is sometimes referred to as the second industrial revolution. Of course, *the basic concept remains the same,* whether we make use either of "humans" or "robots" or "intelligent robots." In the "global simulation model" we start with the most "simple" assumption, and we build up to more and more complex models. For example, in chemistry, we start with the assumption that molecules are built up of point charge nuclei and electrons and this is one of the bases of quantum chemistry. With quantum mechanics we assemble simple molecules, and increasingly are able to tackle more complicated structures such as proteins. With statistical mechanics, we are able to consider many molecules at a given pressure and temperature, and consider trajectories involving time. Finally, for larger systems, we can use micro-hydrodynamics, cellular automata, and fluid dynamics when viscosity, transport coefficients, convection, turbulence, and so on can be analyzed directly from the fundamental equations of motion. This is all achieved at the operational level in such a way that the entire process is self-consistent and feasible with super-computer simulation.

Considering, now, some details of the relationship between "global simulation modelling" and productivity, the latter correlates both to *goods* and to *information.* The evolution of productivity in human history follows an S-type curve growth and we are still far from the limits of possibility (see Fig. 1). The productivity of a material good such as iron is an arithmetic process, one pound of iron produced and added to another makes exactly two pounds. However, information growth is multiplicative rather than additive, and thus a much more rapid evolution ensues after an initial lag.

Let me elaborate this concept and show that the *productivity of information lags behind the main phases of that of goods in a historical context.* In earliest times goods were produced by hunting, a process which then evolved into agriculture and craftsmanship, reaching a high level by about 1700 A.D. During this time information evolved more slowly through stages involving language, astronomy, geometry and mechanics, until the time of the Renaissance and Illuminism, printing of books helped the diffusion of the scientific method, the dissemination of objectivity. Towards the end of the eighteenth century the first phases of the industrial revolution meant a rapid increase in the *"man-sized" automated production* of goods, leading more and more rapidly to the early stages of the information society of the mid twentieth century (ca. 1960), and a new industrial revolution with the appearance of *"robot-sized" production lines.* We are now entering the third "industrial" revolution where "intelligent robots" are in charge of the new assembly lines, thus we talk of "intelligent robot-sized assembly lines."

The evolution in the production of information evolved contemporaneously with the structure of society, from fragmentation and elitism in the early nineteenth century to interdisciplinarity, mass education, and computers. There have been four generations of electronic computers and we are rapidly moving toward a new generation, either with machine intelligence (fifth generation) or without it but

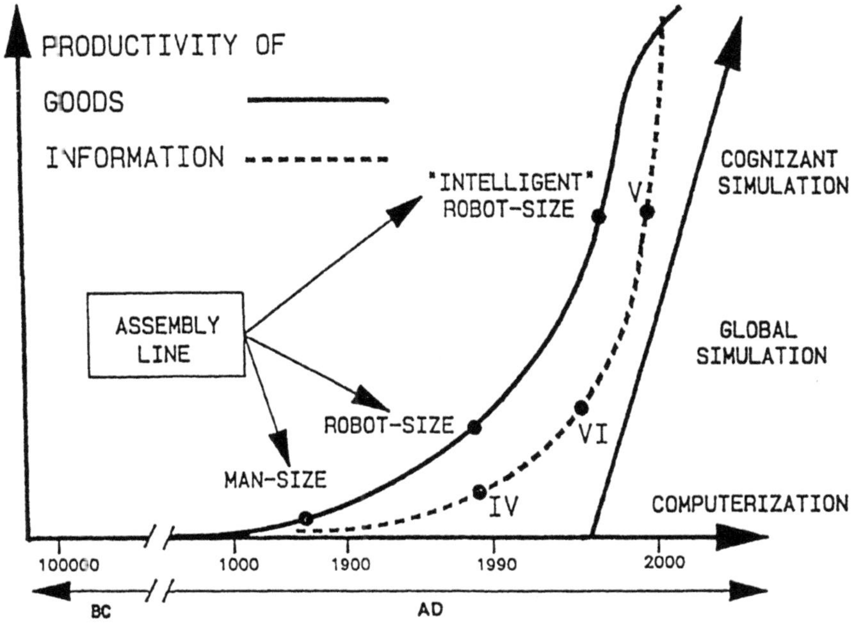

Figure 1. Evolution of computer applications.

with much more power in speed and memory size (sixth generation). Let us discuss the computer's evolution in some detail.

The *first generation* (1946-1955) starts with the ENIAC computer and includes the IBM-650; static memory, vacuum tubes, hand-operated; these are some of its characteristics. The *second generation* (1955-1965) witnesses the transistor logic, magnetized core memory, tapes and punched cards; the IBM-7090 can be taken as a prototype for this generation. The *third generation* (1965-1975) is typified, for example, by the IBM-360 family; among the characteristics we recall integrated circuits (medium to large scale), channels, magnetic disks, system software, virtual memory, and time sharing.

We are now at the end of the *fourth generation* (1975-1989±2), which is characterized by vector processing on one hand and parallelism on the other. Of course, pipelines were invented before (STRECH) and so was parallelism (ILIAC IV). Characteristics of this generation are VLSI, large memories (from mega- to gigabytes), high speed in the top performing machines (from 100 megaflops to a few gigaflops), and a very broad spectrum of products from supercomputers down to PCs. The parallelism is mostly achieved with a few processors (up to 20), seldom massive (up to several thousand).

The beginning of the *fifth generation* is often taken to be 1984, with its MITI's announcement in Japan. However, it has been standard practice to set a generation's birthday at the time a product is introduced into the market; in this sense, since we are only beginning to have "knowledge processing" computers, let me write 1991 ± 1 as the start and 2000 ± 3 as the end of the fifth generation.

In the meantime, we have a more and more clear forecast on the *sixth generation* [1991 ± 2 to 2000 ± 2], with cluster-to-cluster parallelism (four to six or more clusters with 8 to 16 CPUs per cluster) and truly massive parallelism (thousands of CPUs), very large memories with hierarchical organization (CACHE, CPU-PRIVATE, SHARED-INTRA-CLUSTER, SHARED-INTER-CLUSTER), very high speed with peak performance from a few to 500 ± 100 gigaflops, connectivity between systems, workstations, both of medium and very high performance, Personal Computers; networking, distributed computing, and "friendly" systems. Another aspect of the sixth generation will be the very high pick performance achieved on a single processor as already announced, for example, by CRAY, NEC, and Fujitsu. Massive parallelism will become a standard feature for many applications.

Let us now proceed to the *seventh generation* (199X-2010) which will most likely see products based on ideas related to neural networks, the merging of number-crunching machines with signal recognition engines (voice, patterns), full use of optical disks, and an intimate integration of expert-systems and artificial intelligence with system and application softwares. Possibly superconductive inorganic material will replace silicon and gallium, and some preliminary devices with micro-circuits at the molecular level will be tested. The "tool" for producing information will be fully user-friendly due to the progress in graphics and visualization, voice and pattern recognition, and also voice "understanding and simulation" and pattern "rendering" by the computer itself. Special purpose accelerators will replace general purpose supercomputers for specific applications, for example in chemistry and fluid dynamics.

The productivity of information is now benefitting from the introduction of *supercomputers* from what may be termed "phase 1", analogous to the industrial phase in the production of goods, and is moving toward phase 2, that of the *supercomputers* of the sixth generation. This will allow the information society to evolve into "knowledge society" by about the middle of the twenty-first century using the *intelligent supercomputers* of the seventh generation.

At this stage the graph of growth (see Fig. 1) in the productivity of goods will be overtaken by the growth in productivity in information, the former will stagnate as the earth's resources are depleted, but the latter will signal the onset of a more intelligent society, capable of responsible management of the environment.

Whereas the production of goods via robot-sized assembly lines was introduced in the last decade, the "global simulation method", its counterpart in the production of information, is being made (marginally) possible only now, and will blossom with the appearance of the sixth generation supercomputers. Two conditions, market demand and tools (the automated loom and furnace), were essential in initiating a revolution in the textile and smelting industries. Equivalently, the need and demand for interdisciplinarity and information, essential to avoid errors in our highly technical (and thus risk prone) society, had to wait for the evolution of supercomputers in order to be able to set up an "assembly line" for information, that which we call the "global simulation method."

Contemporary advances in materials science, brain research, networking, signal processing, cybernetics, and artificial intelligence are reliable signs of the emergence of intelligent robots for productivity of goods and equivalently intelligent supercomputers for that of information. The former are *already available* on some production lines, while intelligent supercomputers (the fifth generation) are only in the prototype stage. We have noted that it is not unreasonable to suspect that the *sixth* generation of computers might appear before those of the *fifth* generation; indeed the latter is a true conceptual breaking point with the past and thus very difficult to assemble. On the contrary, the sixth generation computers are "simply" engines capable of one to a few dozen gigaflops *sustained* performance. Therefore, they represent "no more" than the "expected" evolution from the fourth generation.

In summary, we are moving towards the third industrial revolution in human productivity, having passed from *man*-sized to *robot*-sized production lines and moving towards *intelligent-robot-sized production lines.* Similarly, we are at the end of the fourth generation of computers, moving toward the sixth and, later, the fifth and seventh generations.

The three revolutions in the productivity of goods and information are mirrored in the evolution of the way to approach the solution of a problem via computer simulations. The first stage can be called "*computerization of a simulation*", and it corresponds to the automation of manual computation. In the early 1920's, Richardson — the father of modern meteorology — envisioned a system of hundreds of humans positioned in a large and specially built room with the task of hand-computing the numerical equation governing weather changes. Well, the first stage was to use a supercomputer for the task envisioned by Richardson! In chemistry, programs such as GaussianXX, HONDO, KGNMOL, CHARMM, and so on, are examples of "computerized simulations." There are equivalent examples in hydrodynamics such as MSC/NASTRAN and FIDAP, and similarly in most other disciplines. However, this is the first stage in the evolutionary process, and it has come to fruition.

The second stage is "*the global simulation.*"[22] This is "modelling within a simulation environment," and Ref. 22 deals with that of *mechanics*, from quantum, to statistical, to fluid. However, most contemporary human endeavor, from space station to genome, to oil exploration at sea are additional examples. The second stage in this respect is equivalent to the evolution from mechanization to robotic assembly lines. Each industry has its own environment and thus each one should have its own modelling environment. This has been realized today partly in the space industry and in a few others where large libraries of programs are collected, updated and maintained and realized to be a key research and essential to any *manufacturing* activity, not just "a service." However, the concept has not yet made enough of an impact in the computational fields of chemistry and physics, especially in the academic world, and MOTECC-89 is likely one of the first attempts to fill this gap by "modelling within a simulation environment."

The third stage will again be *global,* but with "intelligence," capable of dynamical feedback such as error detection and automatic correction, i.e. true knowledge-based simulation. This is equivalent to an assembly line with an intelligent robot and one can refer to this stage as *"cognizant simulation"* (see Fig. 1). In what follows we consider scientific and/or engineering applications as a prototype to measure the impact of such a numerical revolution. MOTECC-89 reaches the second stage, and MOTECC-9X will hopefully mark a timid beginning toward the third stage.

The "tool" to perform global simulations *must be a supercomputer* with super scalar, vector and parallel characteristics. Let us recall that when we "operate on numbers," we can select between either the scalar and sequential -S- or the vector -V- or the parallel -P- modes of operation. Let us consider S, V, P as unit vectors of a three-dimensional space. Any computation can then be associated with a trajectory within the SVP space, as shown in Fig. 2. For complex algorithms, the corresponding trajectories often will occur within a preferential volume in the SVP space, thus defining an *envelope*, characteristic of the given algorithm. The ideal computer for a given *algorithm* is the one which *has high performance for that algorithm's envelope* in the SVP space. For "global simulations" the ideal computer must have super-performance along each one of the three S,V,P dimensions; *it being most unlikely that any simpler choice will suffice.* Alternatively stated, the envelope for global simulations extends over most of the SVP space. It is noted that "performance studies" can be considered as studies which compare different envelopes in SVP space relating it to a given algorithm, and attempts to define an "optimized envelope," and then an optimized architecture. Thus we start with the algorithm and we search for an architecture, and not from the latter to the first as is customary.

In the early 1980s, when we extended our simulations from quantum chemistry and molecular dynamics to micro-dynamics and fluid dynamics, there was no such supercomputer; henceforth the need for us to build one as a feasibility experiment aimed at performance and flexibility. Because of its features it was called lCAP, an acronym for "loosely coupled array of processors."

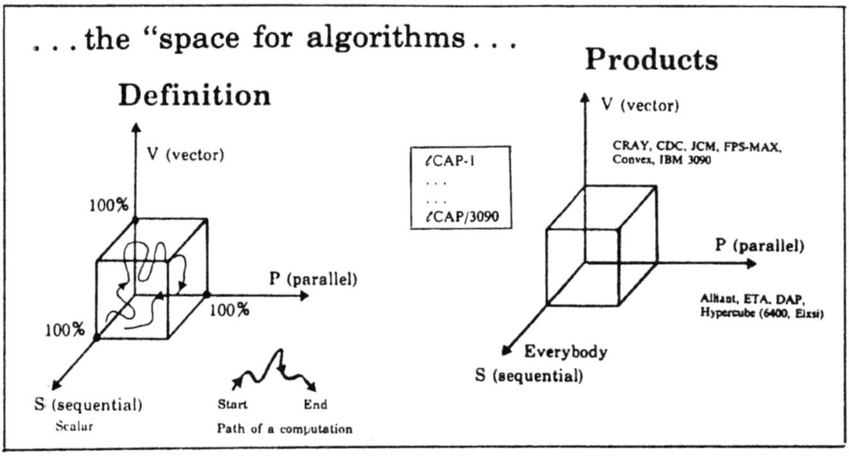

Figure 2. The SVP space.

3. MOTECC: A First "Assembly Line" to Produce Chemical Information

The purpose of the MOTECC initiative is to provide a unified framework for applications in the field of scientific and engineering computing with cross reference from workstations to supercomputer systems. MOTECC-89 collects chapters edited, in general, by senior members of the Department of Scientific Engineering Computations at the I.B.M. site in Kingston, New York, each in a seemingly different field of specialization but each closely inter-related by the ability of "global simulations" viewpoint to jell together the various fundamental axioms into a coherent whole.

MOTECC-90 reports on a collection of algorithms and techniques which have evolved over the past half-century and are now embodied in many application programs in physics, chemistry and engineering. Therefore, we shall discuss topics already present in standard, open reference, advanced, non-proprietary literature. What *is new* is the selection criteria and the overall goal, down to operational concreteness, which is exemplified in the "input" preparation explained in Chapter 24. MOTECC-90 provides the conceptual framework of many specific application programs which are discussed, sometimes in full detail and sometimes only briefly, but are not provided with this MOTECC-90 volume. An Input/Output Documentation Volume, available to the MOTECC-users complements MOTECC-90.

Figure 3 can be useful in explaining the *components* of MOTECC-89. The main three application areas are: 1) quantum mechanics, 2) statistical mechanics, and 3) fluid dynamics. Graphics and parallel computer architectures are ubiquitous.

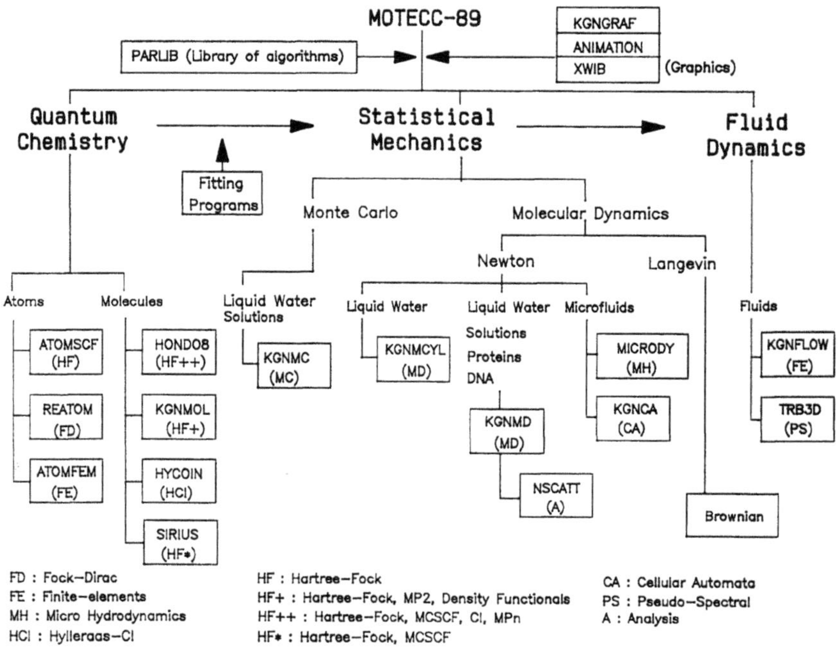

Figure 3. MOTECC-89 overall view of the fields, techniques, and specific application programs.

Fig. 4 provides a summary of the techniques and computer programs of MOTECC-90. Comparing Fig. 4 with Fig. 3 one can note the large increase in the scope of MOTECC-90, resulting from collaborations with Universities in many countries and with two other IBM locations, namely the Research Laboratories in Almanden and Zurich.

Concerning quantum mechanics MOTECC-90 discusses application programs for either atoms or molecules or solids. In atomic computations it deals either with the Hartree-Fock, HF, model[1,2,6-9] or with the Dirac-Fock model.[4,30-33] For the former we use *global functions,* either exponential basis sets of Slater Type Functions (STF) or Gaussian Type Functions (GTF), both spherical and cartesian, or oscillating STOs; *local* functions are used in the finite element method (FEM), also applied to atoms. For the Dirac-Fock we use GTF of spherical type in the specific form of the geometrical basis set.[34] A third atomic chapter has been added in MOTECC-90, namely F. Sasaki[35] computer program, ATOMCI, for atomic CI functions. MOTECC-90 includes also a rather detailed discussion on atomic density functionals (see Chapter 2); it appears that there is much need of accurate CI functions to calibrate the parameterization of the density functionals.

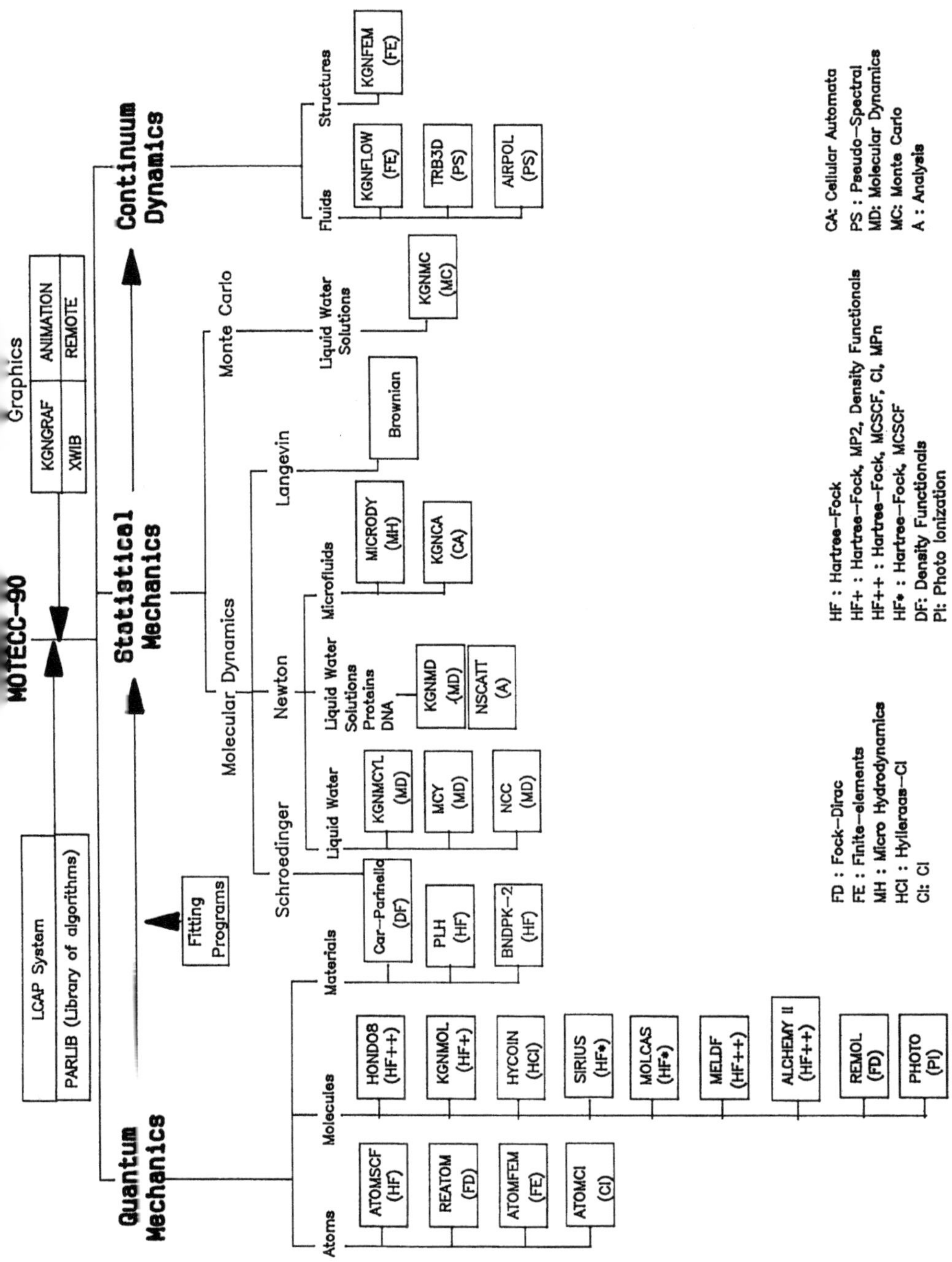

Figure 4. MOTECC-90 overall view of the fields, techniques, and specific application programs.

MOTECC-89 provides four main codes in molecular quantum mechanical computations named HYCOIN, HONDO,[36,37] KGNMOL,[38] and SIRIUS, presented in Chapters 5, 6, 7, and 8, respectively.

Over the years many features have been incorporated into HONDO. These include: calculation of single configuration self-consistent-field (SCF) wavefunctions (closed shell RHF, spin unrestricted UHF, restricted open shell ROHF), of generalized valence bond wavefunctions (GVB) in the perfect pairing approximation, of general multiconfiguration self-consistent-field wavefunctions (MCSCF), and of configuration interaction wavefunctions (CI). The algorithms coded in the program are described in references 39 – 46. Another feature includes the calculation of the electron correlation correction to the energy of closed shell RHF wavefunctions by means of Moller-Plesset perturbation theory (MP2, MP4). Molecular symmetry[47-50] is implemented; effective core potential,[51] geometry optimization,[52] force constant and IR and Raman spectra,[53,54] dipole moment and polarizability,[55] determination of the transition state structure[56] and intrinsic reaction coordinate (IRC) pathway[57] and other expectation values[58-60] can be computed. Therefore, HONDO is a self-connected library of computer programs for atoms and molecules.

KGNMOL is the present name for two previous programs, namely: PP1,[61] developed when our department was located in Poughkeepsie, New York, and IBMOL,[62] one of the first programs in quantum chemistry. The original version, IBMOL-1, was updated a number of times[62-67] to keep up with both computer technology and new aspects of theoretical chemistry. Presently KGNMOL yields Hartree-Fock wave functions for closed or open shell configurations using molecules with contracted or uncontracted functions, GTF; a special integral package is dedicated to GTFs of geometrical type.[34] The feature which has in the past marked IBMOL, and now KGNMOL, is its ability to deal with large molecules since the program was designed in an "open ended" fashion since its first version. Another characteristic feature of the program is the *Add* feature for potential energy functions, whereby the interaction of molecule B with A can be obtained without recomputing the integrals of A. A related feature useful for the computation of surface is the ease in obtaining basis set superposition correction, BSSE, with the counter poised, CP, method of Boys and Bernardi.[68,69] The correlation energy is either estimated using the new Clementi-Chakravorty model (based on recently obtained pair correlation energies for the atoms He to Sr), or is computed using the MP2 approximation, or by using several density functionals proposed over the last fifty years.

The Hylleraas-CI package[70-72] presented in MOTECC-90 uses GTF functions; the package can be used only for few-electrons systems, since it is most expensive both in storage and computer time. The two-electron-integral packages (from one to eight centers) and the *three* and *four* electron integrals have been recently coded at our laboratory.[73-75] This program can be used for computations aiming at spectroscopic accuracy, namely with an error between a few cm^{-1} to a few tenths of one cm^{-1}. In the '89 edition of MOTECC, there was a very short chapter for SIRIUS[76] added the last moment before going to press; in MOTECC-90 revised and extended chapter on SIRIUS is presented. Work is now being pursued to write a general Dirac-Fock relativistic program, REMOL, using GTFs for molecules of any geometry,[77] and hopefully this will be part of MOTECC-90 Input/Output Documentation Manual.

The new methods and programs included in MOTECC-90 cover in depth CI, MCSCF, and photon-ionization cross section techniques. We can state that these new addition to MOTECC represent today's quantum chemistry at its best. Specifically the following codes and techniques are presented: MOLCAS, a general MCSCF program from Sweden by B. Roos and P. Siegbhan *et al.*;[78] MELDF, a general ÇI molecular structure program by E. Davidson *et al.*;[79] the CI and MCSCF program ALCHEMY-II by B. Liu and D. McLean *et al.*,[80] and from Italy a molecular program for one and two photon ionization cross sections, PHOTO by R. Moccia *et al.*.[81] These programs are in Chapters 9 to 12, respectively. (The above survey is much too brief but it is written before obtaining the manuscript of the corresponding chapters; I shall amend with a longer commentary in MOTECC-91).

Solid state, either as crystalline or amorphous or polymeric materials, are the subject of three techniques and corresponding programs, namely the BNDPK-2 pachage by Brener, Callaway and Tyler,[82] the Car-Parrinello code on quantum molecular dynamics[83] and the PL2 code by J.M. André *et al.*[84] Whereas BDNPK-2 and PL2 yield the band structure for cubic lattice crystals and for linear polymers respectively, the Car-Parrinello method expands those limits and includes not only amorphous materials and clusters but also (because it makes use of the computationally very fast algorithms for density functionals[85] and pseudo-potentials) can simulate the dynamics of the system. We note that presently the Car-Parrinello method and code can not be used to simulate the dynamics of macromolecular systems such as enzymes, proteins and nucleic acids. It is our opinion that this limitation can be lifted eventually; indeed it is possible to merge the Car-Parrinello techniques with standard classical molecular dynamics of the type used today to simulate macromolecules. It should be recalled that in many macromolecular reactions, only the active region might be in need of a quantum mechanical description, whereas the rest of the system can be dealt with newtonian mechanics. This approach "quantum mechanics when needed and classical mechanics when and where acceptable" has been proposed and tested some time ago[86] in a study of papain. The advent of much faster computers and the refinement in the algorithms for density functionals and pseudo-potentials makes this suggestion operational. It is noted that standard *ab initio* quantum mechanical computations at each time step can be performed easily even today, but at a very high computer time cost; as a consequence the simulation can not deal with equilibration and are carried out for "statistically not significant" experiments. Some very recent HONDO applications are of this type.

The molecular programs can be used to compute interaction energies needed to fit the parameters in interaction potentials. A special purpose program, called KGNFIT, is available, to obtain the fitting parameters of the interaction energies in *ab initio* analytical potentials which, in turn, are used in Molecular Dynamics, Monte Carlo and Microdynamics simulations of macromolecular systems.

Graphics routines are utilized to display molecular structures either as balls and sticks or as three-dimensional representations of the electronic density obtained from quantum mechanical codes; interactive graphics are also used to prepare user-friendly molecular inputs. Chapter 24 deals with visualization and interfacing in an AIX environment. The main codes in this context are KGNGRAF, XWIB and REMOTE, respectively.

Monte Carlo programs are discussed in Chapter 17 of MOTECC-90 to study either liquid water or a solute molecule considered rigid and surrounded by water molecules, with or without ionic species. Property programs like, for example, the computation of pair correlation functions $g(\alpha, \beta)$, complement the Monte Carlo packages. These programs have been previously described[87-89] mainly for study of liquid water either with or without simple ions. The programs were later extended to study systems like DNA and water[90] either with or without counter ions[91]. The most recent extension of the program was written to study the interaction of water with proteins in crystals, e.g., myoglobin.[92] *Molecular dynamics* is embodied in a number of application packages, also discussed in Chapter 17. These are the KGNMD, KGNNCC, KGNMCYL, and NSCATT computer programs. The main differentiation between the packages is the chemical system for which the specific computer program has been written. Thus we have *one*-component systems – liquid water, *two*-component systems – single ion and water molecules as a solvent, or a single molecule as solute and water molecules as solvent.

Finally, we shall discuss a more refined package which can handle several components. In the latter the molecules (solute or solvent) are assumed either rigid or flexible. A very extensive number of properties can be analyzed using the trajectories obtained in the MD simulation. For example, pair correlation functions, auto correlation functions, static and dynamic structure factors with related properties and many thermodynamics values. Again, extensive graphical outputs of different quality are utilized in MOTECC to represent molecules, macromolecules and liquids or solutions. In addition, routines for animation of MD data, very important in describing time-dependent phenomena, are discussed (Chapter 24).

As it is well known, both Monte Carlo and molecular dynamics simulation are, at best, as reliable as the interaction potential used in the computations.[93,94] Ideally, at each time step or at each random configuration, one should perform a full quantum mechanical simulation of the energetic of the system; this avenue is today computationally too demanding for macromolecules in solution and one must resort to use analytical potentials. The history of attempts to obtain reliable interaction potentials is as long as the history of physical chemistry. For much of the time in the past, theoretical models guided the use of analytical expressions which were supposed to represent the interaction energy of two or more molecules in their trajectories. In general these analytical expressions are parametrized using experimental data. Since, *very many* parameters are needed to obtain a good fit, usually more than the available experimental data, in the past not much care was exercised in differentiating data for a two-body potential vs. data for a many-body potential. We should recall that the underlying assumption was that two-body pair additivity was *sufficient*. In reality, this assumption is not too bad if one thinks in terms of a two-body *effective* potential, namely a many-body potential written essentially in the proper analytical form for a two-body, with parameters coming from two- and n-body experiments; this, however, does not hold for dynamics.

A different viewpoint has been pioneered in our laboratory since early '70s, namely the use of *ab initio* quantum mechanical (data rather than experimental data) and a careful build up either of two-body, three-body, etc., or polarization models yielding a set of analytical potentials of increasing reliability. Clearly, here we have the practical limitations of the available quantum mechanical tools which are presently often not as reliable or practical as the laboratory tools. However, if we truly believe that computational sciences are, or will become, equal partners with experimental and theoretical sciences, then *the use of quantum mechanically derived potentials is the "obvious" path*. Above, the term *ab initio* is used in a strict sense, therefore it excludes the use of semi-empirical methods or of density functionals with empirical parameterization.

For the last twenty years we have been working at obtaining the pair potentials presented in Fig. 5 using *ab initio* quantum-chemical methods. The symmetric matrix is generated by writing a set of chemical species as indices of the rows and columns. Each "matrix element" contains those pair-potentials with which we can discuss the chemical problems reported in Fig. 5. For example, later on we will show how much can be accomplished if one knows the interaction potentials between water molecules (see the first matrix element of Fig. 5). In a similar way the interaction potentials implicit in the first column constitute the basic input needed to understand the solvation of amino acids, proteins, base pairs, nucleic acids, etc. Other "matrix elements" in this figure provide the basic starting point necessary, for example, to determine the structure of proteins or the interaction between proteins and nucleic acids, etc. Thus, Fig. 5 summarizes *the "frontiers" in simulations of bio-engineering problems.*

By now, approximately 4×10^4 *ab initio* computations on the interaction energy between pairs of molecules have been performed and interaction potentials are available for water molecules interacting with: 1) water, 2) simple ions, 3) amino acids and proteins, 4) bases, sugar, phosphate groups and nucleic acids, as well as for 5) protein interacting with protein and, finally, proteins interacting with nucleic acids. In general these potentials are first obtained with *ab initio* computations with "minimal basis sets" (low level accuracy) and are subsequently improved either by including correction for the basis set superposition error or by using "extended basis sets" and finally by including a correlation correction (high level accuracy). The computational cost and the human effort in this enterprise is very high, but *we are motivated to ensure that computational chemistry and physics will become reality more and more and as soon as possible.* MOTECC-90 includes in Chapter 17 a detailed discussion on the derivation of recently obtained *ab initio* potentials for proteins and nucleic acids, to be used in molecular dynamics and Monte Carlo simulations.

Dynamical systems in nature are generally not at equilibrium conditions. The nonequilibrium statistical mechanics we have implemented in MOTECC-90 is often refereed to as "MICRODYNAMICS", which is a connecting bridge between discrete descriptions of matter (statistical mechanics) and continuous descriptions embodied in fluid dynamics. This topic is discussed in Chapters 18-20. The code "MICRODY", detailed in the Chapter-18 is designed for a large scale molecular dynamics simulation of the fluid flow in a three dimensional channel or a flow past an obstacle. It can be used to study the nonequilbrium flow phenomena over a wide range of density, from very dense liquid to a rarefied gas. It can also be applied to explore the transport phenomena in subsonic and supersonic regions, and to examine the critical influence of the imposed temperature gradient and the boundary conditions on the development of the flow profiles. This code can

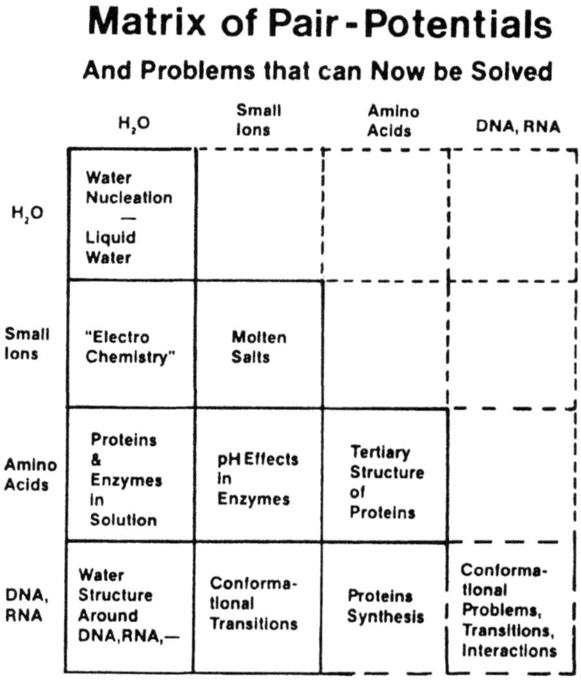

Figure 5. Matrix of interaction potentials between chemical species and model compounds indicated in the top row and left column.

provide information on the behavior of a fluid system in a domain where the continuum hydrodynamics or the transport theory based on the Boltzmann equation breaks down. This is illustrated by studying the phenomena of transition regime gas flows, which has some interesting applications such as the problem of the estimation of thermal heating on the surface of a high speed space vehicle reentering the upper atmosphere.

The code BROWNIAN, which forms the basis of the discussion in Chapter 19, enables one to analyze the rheology of a colloidal suspension. The hydrodynamics of the colloidal system is important not only for analyzing industrial problem like the handling of highly concentrated coal suspensions, but also for understanding numerous biological phenomena such as the passage of drugs through the cell membrane. The code BROWNIAN is based on the generalization of Ermak and McCammon algorithm[95] for a fluid system under a shear, and can be used for examining the shear rate dependence, the shear viscosity and diffusion constant of a colloidal suspension. It also provides a framework for analyzing recently observed shear induced ordering phenomena in colloidal dispersions.

While in Chapters 18 and 19, the macroscopic behavior of a fluid system is obtained by solving the Newton or Langevin equation, Chapter 20 discusses the simulation of fluid flows using the techniques of cellular automata. The fluid particles here have no molecular identity, the macroscopic fluid behavior is the outcome of specific collision rules[96] which satisfy the conservation laws. The computational universe of lattice gas dynamics, and the code KGNCA thus are an appropriate transition from the molecular hydrodynamics to the continuum hydrodynamics.

The motion of a fluid under various conditions is composed of highly complex phenomena which interact at various scales of length and time. There is no single set of equations governing the flow of fluids under all conditions. However, the motion of many fluids under many conditions of practical interest are satisfactorily described by the Navier-Stokes equations. In this introductory section we discuss in brief the Navier-Stokes equations and numerical methods for the solution of these equations. Chapter 21 considers both Navier-Stokes equations and turbolence; the corresponding codes are KGNFLOW and TRB3D. A very useful byproduct of this work is in the field of structural analysis using finite elements techniques. This is taken up in Chapter 22 and its corresponding computer program is called KGNFEM.

The Navier-Stokes equations[97,98] express the principle of conservation of mass and momentum of a fluid in the form of a set of coupled partial differential equations. There are different routes to deriving these equations; the simplest approach is to use the control volume method which is readily described in two dimensions: an imaginary rectangular shaped region (the "control volume") is chosen in a small region of a moving fluid. Conservation of mass is enforced by requiring that the rate of change of mass within the control volume is equivalent to the net flux of fluid across the surface of the control volume. If the sides of the rectangle are reduced to infinitesimals, the above gives a partial differential equation — the continuity equation. Conservation of momentum requires that the rate of change of momentum of the fluid within the control volume in a given direction be balanced by the applied surface stresses and the net flux of momentum across the surfaces of the control volume. The surface stresses are expressed in terms of the velocities by suitable constitutive equations. Once again, reducing the sides of the rectangle to infinitesimals gives the partial differential equations corresponding to the conservation of momentum. The continuity and momentum equations together constitute the Navier-Stokes equations.

The Navier-Stokes equations are a set of coupled nonlinear partial differential equations which are either parabolic (for time dependent problems) or elliptic in nature (for steady state problems).[99] They are capable of describing highly complex fluid behavior such as turbulent flow and shock formation. Analytical solutions are available for a very limited class of problems,[100] and it is very often necessary to use numerical techniques to obtain solutions of practical interest. The numerical solution of the Navier-Stokes equations is a challenging task because of the wide range of complex physical behavior generated by the nonlinearity in the equations. We will examine briefly the major numerical techniques used to solve these equations pointing out the advantages and limitations of each (for details see Chapters 20 to 23).

The *finite difference method* for the solution of partial differential equations consists of replacing differentials by suitably chosen differences. Because of the ease of concept and programming, it is the most widely used method for the solution of the Navier-Stokes equations and a large number of fluid flow problems are routinely solved with this method.[101] However, it is difficult to use this technique when arbitrary boundary conditions are to be imposed on an irregular domain.

The pseudo-spectral method is gaining in importance as a tool for highly accurate solutions of the Navier-Stokes equations,[102] and is often used in direct numerical simulations of turbulence. The key idea in this method is to expand the unknowns in terms of orthogonal polynomials (frequency domain representation) and use a dual representation for each of the unknowns in the grid point domain. At each time step the variables are transformed between these two representations to obtain highly accurate derivative information. If the frequency domain representation is in terms of trigonometric functions, the FFT can be used for efficient transformations between these two representations. This technique is again difficult to use in irregular regions with arbitrary boundary conditions.

The finite element method which, diring the past decade, has been widely used in structural analysis has been extended to a number of different fields including fluid flow.[103] The theoretical framework of the finite element method is now based upon the *method of weighted residuals*, and "weak variational" formulations to allow the Navier-Stokes equations to be discretized with the finite elements. A major advantage of the finite element method is the ease with which it is possible to solve problems in complex geometries with arbitrary boundary conditions. A major disadvantage of the finite element method implemented in the traditional manner is the very large storage and high computing costs involved in the solution of global matrix equations with direct factorization methods.

It is useful to note here that the method of weighted residuals includes each of the above methods (as well as a number of other discretization techniques) as a special case of a single general formulation.[104] For any given operator, the general formulation defines a residual based upon the use of interpolating functions for the unknowns. The residual is then required to be orthogonal to a set of weighting functions; this condition gives rise to a set of linear algebraic equations. The finite difference method can be derived by using relatively low order polynomials as interpolating functions, and spatially distributed delta functions as weighting functions. The pseudo-spectral method is obtained by using sets of globally spanning orthogonal functions as the interpolating functions and delta functions as the weighting functions. The finite element method uses locally spanning polynomials for interpolation and weighting.

While each of the methods has its place in the solution of the Navier-Stokes equations, we favor the use of the finite element method, primarily because of the ability of this method to handle complex geometries and arbitrary boundary conditions. In order to achieve high accuracy, we have used the p-version of the finite element method, which uses high order polynomials of the Legendre type.[105] In order to overcome the problems associated with forming and storing a global method, we have used iterative solution methods which can greatly reduce the storage requirements.[106,107] A delightful and recent application of this technique is provided in Chapter 2, namely highly accurate solutions of Hartree-Fock equations in atomic systems. Work is presently in progress at our laboratory to expand this approach from atoms to molecules of arbitrary geometry. In this context, namely the use of finite element techniques for quantum mechanical problems, I recall that the time-dependent Schroedinger equation represents a nice application for finite elements techniques.

We look forward to the day when these techniques will be fully familiar to the chemist, as familiar as quantum chemistry is today. The distinction between chemistry and chemical engineering is partly due to different training, particularly in applied mathematics. MOTECC should help – hopefully – to reduce this gap.

In the introductory section we mentioned the overall stress on parallel supercomputers throughout this volume. As it is well known, modern numerical analysis is heavily engaged in rewriting as many algorithms as possible for parallel architectures. This topic is the main subject of Chapters 25 and 26, where parallel algorithms and the system software for the LCAP-system are discussed.

A chapter dealing with air pollution, a most important application of fluid dynamics, is now part of MOTECC-90; we feel rather strongly that much of today's problems in the earth's environment are due to ignorance, particularly ignorance of the chemists and chemical engineers in planning experiments and in the design of chemical plants. Therefore, we have included this chapter on air pollution which we hope will have much following in MOTECC-9X. We also hope to include in MOTECC-91 combustion and flame studies, since these provide an excellent example of global simulations very relevant to energy saving, oil industry, and space research.

This concludes our brief overview on the algorithms and applications considered in MOTECC-90.

4. An Example of Global Simulation: From 3 Nuclei and 10 Electrons to a Million Molecules

The global simulation viewpoint is still very young and needs a practical "demonstration" to prove its operational feasibility. Of course, an example — by definition — is of limited interest. The scientist or engineer reading this section should think of his own research area where, most definitively, there is ample need — still untapped — of global simulations.

I do hope to expand MOTECC-9X by including chapters on computational research for flames in particular and combustion in general. There is much need for this extension: for example, to save energy, to control pollution, and to foster space navigation. In such areas one starts with the quantum chemistry of the many reactions related to burning CH_4 in a O_2 atmosphere; the different temperatures in the flames bring about different kinetics and the corresponding transport equations are a basic aspect in the fluid dynamics of the flames.

Today, in MOTECC-90, we shall consider once more the simulation of some chemical in the liquid phase and to be more specific let us consider water, a rather complicated but most interesting liquid. To start with (namely, the "input" for the first submodel) all I need to know is that one molecule of water has the stoichiometric composition of two hydrogen atoms and one oxygen atom, which — to fit our submodel — we translate into the notion that our molecule is made up of ten electrons and has two one-charge point-like nuclei plus one with eight positive charges. Standard quantum chemistry yields a stable molecule; within the Hartree-Fock approximation,[108] the simulated water molecule has an HOH angle slightly too wide relative to experimental data.[109] The correct — within experimental accuracy — value is easily obtained by correlating the electron's motions.[110] Thus now we know the geometry and the energetics of the water molecule from simulations; it is not difficult to obtain also the vibrational motions by mapping the intra-atomic potential surface in the vicinity of the equilibrium configuration.[110]

Let us move from one molecule of water to two, three . . . several[111] and finally many.[112] Computations of the interaction between two molecules even at the near to Hartree-Fock limit,[112] yield the well-known hydrogen-bond dimer. We can improve by using correlated wave functions and energies.[113] In Table IV of Chapter 7 we report on significant computations published in the past twenty years on the water and water dimer. The computed energy points, sampling the potential energy surface, can be fitted to some analytical expression, within the "guidelines" of some simple physical model. In this way we obtained the two-rigid-waters intermolecular potential (two body), known as MCY,[113] or, by relaxing the water "rigidity" and allowing vibrational flexibility, the MCYL two body flexible potential.[114] The MCY potential is by now rather old and at present we are working on its refinement, the NCC potential.[115] A study of small water clusters[111] reminded us that even for few water molecules, the cluster can have many stable geometries, all very near in energy; namely that we must face a difficult minimization problem. It should be noted that the cluster study provides us also with very interesting geometrical arrangements of hydrogen bonded structures, leading to "clathrate-type" complexes. Obviously one should not build up indefinitely the cluster size, not only because it would become computationally impractical and eventually impossible, but mainly because it would be the logically incorrect approach to a problem whose answers are to be found in the understanding of statistical fluctuations.

It is time to move to the *second submodel*, namely equilibrium statistical mechanics, limiting ourself to two main techniques, namely Metropolis Monte Carlo[12] and molecular dynamics.[116] In the former, N molecules are placed in a volume V at a given temperature T and many (millions) geometrical configurations are randomly generated and their energies are computed with the interaction potentials obtained from quantum mechanics. Properties are statistically averaged in the Boltzmann sense. Periodic boundary conditions are imposed on the N particles. In molecular dynamics again we place N molecules in a given volume V and let them interact according to the quantum-mechanically derived potential; for each particle Newton's equations of motion are solved, namely we compute molecular trajectories, i.e., each molecule will modify its translational and rotational motion reacting to the field exerted by the N-1 surrounding molecules. Only motions corresponding to a very small temperature range (ideally *one* temperature value) are considered. The motions are discretized in time steps generally of 0.5×10^{-15} seconds for rigid molecules and 1.25×10^{-16} seconds for vibrating molecules (for details see Chapter 17).

The simulated properties like pair correlation functions, the X-ray and neutron beam scattering spectra, translational and rotational auto-correlation functions, the simulated infrared and Raman spectra, sound frequency, diffusion coefficients and NMR relaxation time, and many thermodynamics data[117-122] show that within the "global simulation spirit" we can indeed progress from nuclei and electrons to a few water molecules, to many, up to the hydrodynamic limit. *Let us now progress into the area hitherto exclusively reserved for fluid dynamics, i.e., for*

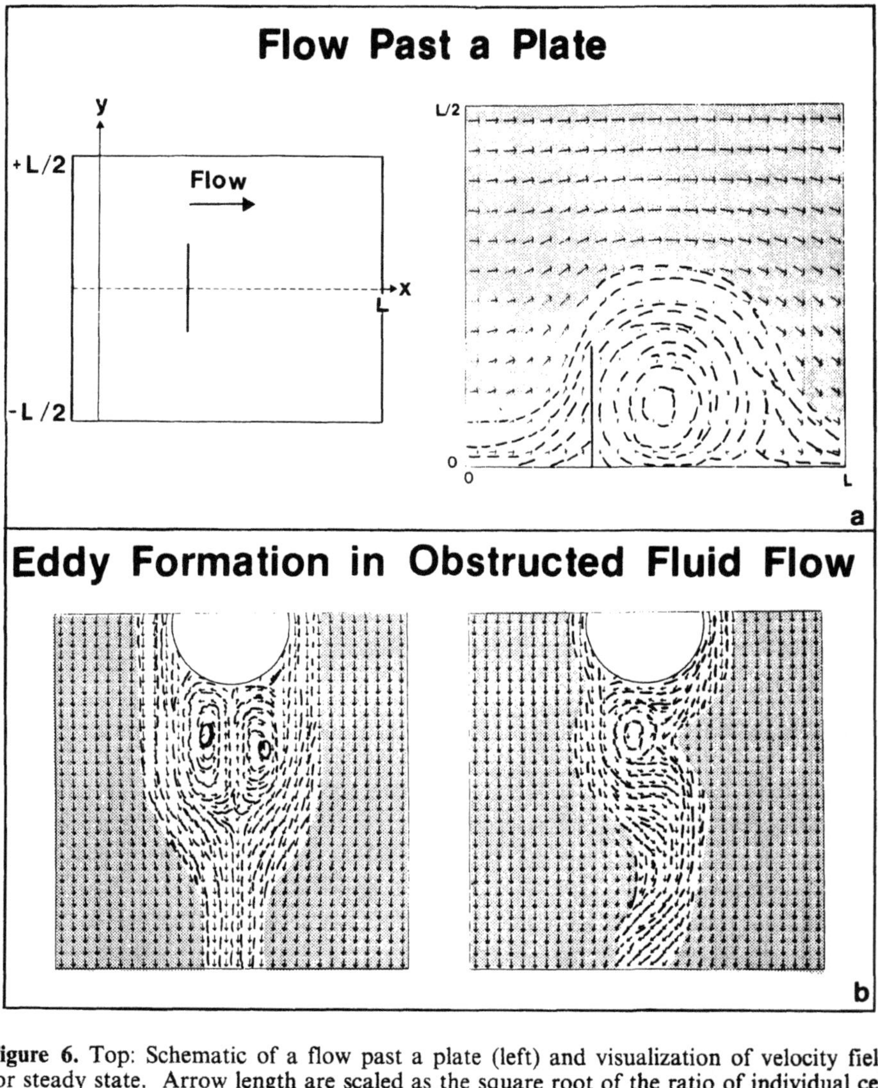

Figure 6. Top: Schematic of a flow past a plate (left) and visualization of velocity field for steady state. Arrow length are scaled as the square root of the ratio of individual cell velocity to maximum cell velocity. Bottom: Velocity field for flow past a cylinder.

solutions of Navier-Stokes equations. We ask the question: can we use the Navier-Stokes equations for a *very small system* of few hundred Å size and for events having a lifetime in the *pico-* or *nano-second range?* Alternatively stated, the question concerns the description of systems generally larger than those analyzed by molecular dynamics but much smaller relative to fluid dynamics. We did answer these basic questions by performing molecular dynamics-type computations with molecules or atoms interacting via intermolecular potentials, but in an open (nonequilibrium) system and with those boundary conditions which are typical in fluid dynamics. For example in Fig. 6 we consider a flow past a plate[27] (upper insert) or past a cylinder[26-28] (bottom insert). Our simulations were performed on $\sim 10^5$ particles for events in the picosecond to nanosecond range over a few hundreds Å scale. We have reported that such systems scale amazingly well and very accurately from results and laboratory experiments of systems at the cm^2 size and with a lifetime of seconds! For example, the Reynolds number at which we observe wakes past the cylinder is about 25 ± 10 , i.e., essentially equal to 35, the Reynolds number for which laboratory data observed the same behavior.[27]

Equally gratifying were preliminary results on Benard type flow obtained by considering 110,000 argon atoms,[27-29] given in Fig. 7. To obtain true Benard patterns, however, we have to consider more carefully the boundary conditions imposed on our system,[29] particularly the vertical boundaries (the horizontal boundaries are associated with temperature T_1, bottom, and T_2, top, with $T_1 > T_2$). After much numerical experimentation we know that the solutions yielding the classical Benard patterns are very unstable to perturbations originating at the vertical boundaries. Stability was obtained by *covering the vertical walls with a layer of square cells,* each one containing a nearly constant number of fluid particles. When a particle of the fluid is reflected at the vertical boundaries, its velocity is constrained to be equal to the average of the velocities of the particle in the cell in which the reflection occurs. In Fig. 8 we report "proper" Benard flow with correct temperature and density profiles. These results are rather recent[29] and have solved the numerical instability previously observed. Notice that the introduction of the square cells, *which cover the vertical wall,* is a step towards replacing the vertical boundaries with a "line of atoms." Indeed, the physical (atomic) boundary would exchange energy with the fluid and eventually thermalize itself with the fluid. The next step in our simulation is the explicit use of "atomic" or "molecular" boundaries, or "membranes." Of particular interest will be the interaction of collective modes of the "membrane" with the fluid, and also the interactions of the atoms in the membrane with those of the fluid. Clearly, both problems are relevant to "slip" or "no-slip" conditions.[123]

These results are exciting, but, in the context of this paper, the most relevant aspect is that indeed our "*global method*" *is verified, starting from systems with few nuclei and electrons up to vastly larger systems, previously described only with fluid dynamics.*

The above study on a liquid is cast into the form of an example of the "Global Simulation" approach; to be sure we did not work in our laboratory on water only to have a *show test-case* for the global simulation methodology. Our interest in water relates to our work in biophysics and biochemistry.

Water is the life bearer and is a solvent, for example, for DNA either in the B or Z conformations with Li, and K counterions.[90,91,124,125] Computer simulation in the global approach has been used to determine the number of water molecules

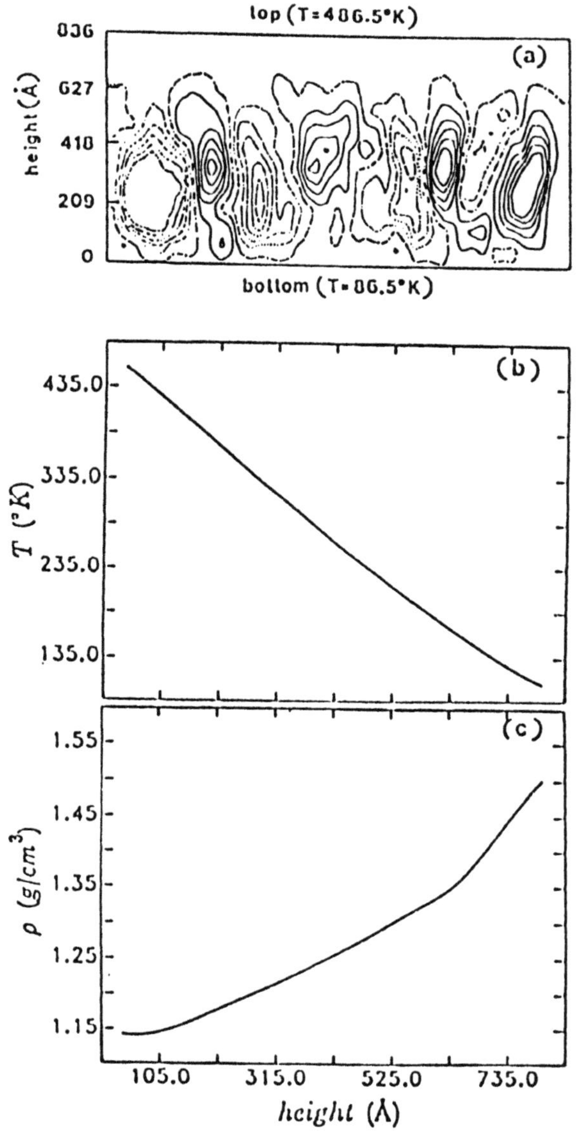

Figure 7. (a) Streamline contours, (b) temperature profiles, and (c) density profiles for Rayleigh-Benard convection. Note clockwise and counter-clockwise rotation in (a) are dotted or full contours, respectively.

hydrating the phosphate and sugar groups as well as the base pairs. It has shown that a filament of water molecules, bound together by hydrogen bonds, runs the length of the minor groove in B-DNA. This differs from the clathrate-like structure of bulk water and for water near a hydrophobic site. The filament length is proportional to the attraction forces of the hydrated site and it has a measurable

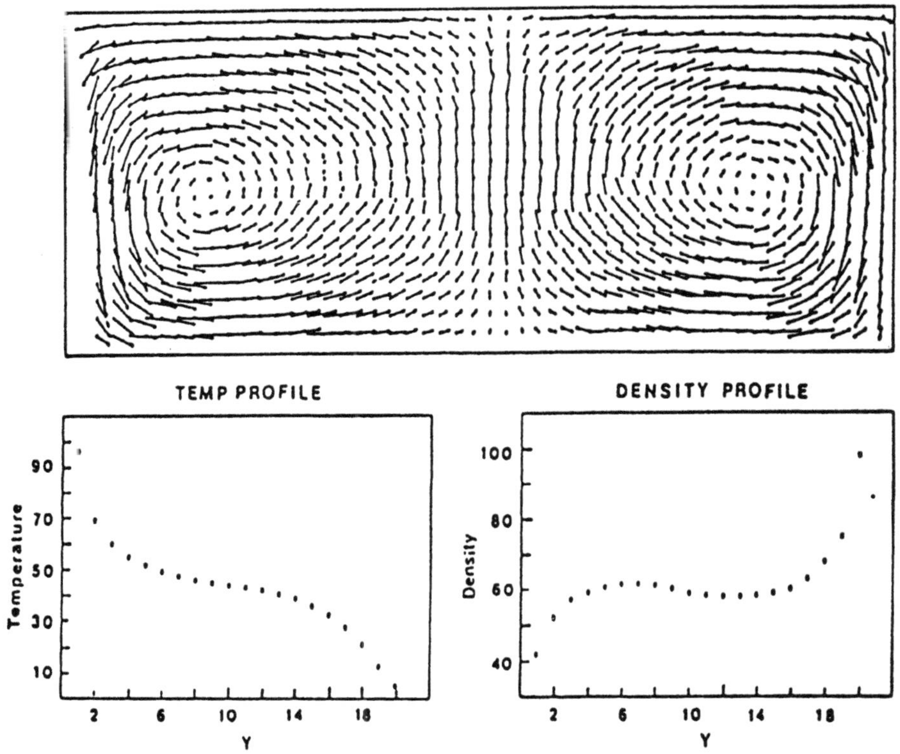

Figure 8. Top: Benard steady state flow. Bottom: Temperature and density profiles.

lifetime exceeding that of the angular correlation time of the diffusing water molecule. In the first few angstroms around DNA the aqueous structure is very different from that of bulk water, and the counter ions are arranged in a well defined pattern for times longer than the picoseconds of the simulation. Similar detail has been obtained from a glycine DNA simulation, with interesting consequences for the breaking of genetic code. This is a situation where a mechanism needs water as a solvent in order to exist. But there are infinitely many and different examples. Let us consider one more where we focus on the interactions between ions and water. We shall not discuss ionic solutions, a too well known case, but a biological case, namely of monovalent ions passing through a gramicidin A transmembrane channel.[126,127] The ion coordination number from Monte Carlo simulation remains essentially the same inside or outside the channel, implying that the interaction energy behaves similarly. The progress of the ion through the channel is, however, only possible when the necessary adjustments have been made to the *ab initio* computed interaction potential for the many amino acid residues forming the channel. In order to correct for this error in interaction energy it is necessary to compute the interaction energy of the ion with the entire transmembrane channel. These are massive and direct computations which need to be repeated hundreds of times for each position of the ion. This was work accomplished on the experimental supercomputer system at IBM, Kingston, as described below in section 5. This computation revealed that the ion only moves easily through the transmembrane channel when it is sandwiched between two water molecules. It rattles at the mouth of the channel when the water carriers are absent. The rotation and libration of the two water molecules conspire to guide the ion dynamically through the channel, in an aqueous pas de trois.

5. ICAP/3090: Parallel Processing for Large-scale Scientific and Engineering Problems

In the last several years, it has been generally agreed that in the future supercomputing and parallel processing will be to a large extent synonymous. This is evidenced by the fact that systems such as the IBM 3090 multiprocessor family and the CRAY-YMP series are now being increasingly employed in a parallel mode to solve problems that cannot be feasibly solved on a single processor. It is also clear that parallel supercomputers must employ a balanced approach, using nodes with high-performance scalar and vector capabilities. In short, as a concept, supercomputing is the union of all approaches that maximize performance.

A distinction between types of parallel systems is whether they purport to be general- or special-purpose systems. The general-purpose systems include most of the products previously mentioned. However, many such systems have been built without a clear focus on the types of applications to which they will be put, and thus the mapping of applications to machine has often become an experiment in parallel processing after the fact, with little guarantee of success.

The alternative approach, i.e., the design of special-purpose parallel machines, is targeted to a very limited class of applications. Included here are a number of systems tailored to solve problems in fundamental particle physics (the IBM GF11 machine[128]) or applications in fluid dynamics (the Navier-Stokes machine[129]). The design process has usually been a bottom-up approach, with specific layouts of chips, hard-wired interconnection networks, and special memory devices. Thus, as with the general-purpose systems, the elapsed time between design and inception is large.

An exception to these approaches has been the development of the ICAP parallel processing system. This system was conceived with a clearly defined set of applications in mind. The goal was to solve these problems by using parallel processing at a rate comparable to, but at a cost less than, that of the fastest commercial supercomputers of the day. The class of applications was broad enough not to be considered as special-purpose usage but, on the other hand, could not be classified as general-purpose usage either. The approach taken in the design of this system was to minimize the hardware and software complexity of the task, and thereby minimize the time needed to build the system, by simply interconnecting commercially available offerings. The objective was therefore not to conduct an experiment in parallel processing, but rather *to use parallel processing to solve real problems.*

Our goal was quickly achieved;[130,131] this was not surprising, because we had realized beforehand the requirements of our applications and their large-grain-size, parallel structures. Thus we considered the design of this system obvious and not worthy of a large treatise on the theory of parallel processing. Indeed, many results have been published from calculations performed on the system, often without mentioning that it was a parallel computer. We were interested in results rather than means.

Because of this success, we began to explore whether we might meet with equal success in areas other than theoretical chemistry. In short, we wanted to see whether our system, which we call lCAP (for loosely coupled array of processors), could be considered a general-purpose, parallel computer. This meant surveying a large spectrum of applications in science, engineering, and mathematics and attempting to adapt them to lCAP. A program for visitors was instituted to solicit professionals, skilled in areas that we were interested in studying, to experiment with parallelism.

The outcome of our experiments culminated in the development of the lCAP/3090 system. A brief description of this system and parallel language issues is given in the next section. We then discuss general principles of parallel processing that apply to all computers of the Multiple Instruction/Multiple Data stream (MIMD) variety. Following this discussion, we describe the experiments in parallel processing that have led to our lCAP/3090 efforts. Finally, we offer our conclusions. For details on parallel algorithms and on the LCAP operating system we refer to Chapters 26 and 27, respectively.

The lCAP/3090 Experimental System

The IBM 3090 vector multiprocessor family encompasses a variety of models ranging from a two-processor system (Model 200) to a six-processor complex (Model 600). These systems can increase the throughput of a large workload by using the multiple processors on independent jobs. This has been the traditional motivation behind the development of such systems. An important corollary of this approach is that memory must be increased proportionately. This contributed to the development of expanded storage on the 3090 to alleviate the effect of paging that occurs when multiple jobs compete for real memory. Overall, the design of this system emphasizes a balance[132,133] of memory access, I/O, and processing capabilities.

The performance potential of a 3090 multiprocessor can also be realized by allowing multiple processors to work on a single problem. To address utilization of the multiple processors of the 3090, a parallel FORTRAN capability, including automatic-compiler-generated parallel code, has been developed, the Parallel FORTRAN (PF) package.[134-135]

The lCAP approach is to increase still further the level of parallelism by coupling readily available commercial processors (in this case, IBM 3090s) to form a system that is not massively parallel, but rather is modular and can be expanded to match the degree of parallelism that a set of applications can support.

lCAP/3090 architecture. The principal idea of lCAP/3090 was to couple together clusters of IBM 3090 multiprocessors; today, in 1990, price/performance and other reasons suggest to cluster RS/6000 workstations.[136] We intend this coupling to be extendable, so that, as faster versions of IBM multiple-processor mainframes are introduced, we may couple them together in a similar fashion. That is, whatever the current power of the fastest machine, we would like to be able to increase this arbitrarily by simple replication and coupling.

The justification for employing more parallel processors than are currently offered in a single integrated system is supported by our findings of the large degree of parallelism that exists in many scientific and engineering applications and their successful implementation on our earlier lCAP systems. This covers, to name a few, calculations in quantum chemistry, statistical mechanics, many engineering applications dealing with heat transfer and computational fluid dynamics, celestial mechanics, and fundamental algorithms used in all areas of science and engineering.

The "degree of parallelism" of a given piece of code can be characterized by three attributes: (1) how much of the computation can be run in parallel; (2) how often communication is required; (3) and how evenly the work can be divided across multiple processes.

First, there are many applications that are almost entirely parallel and—when partitioned across a number of processors—require very infrequent communication. Such applications are said to exhibit *coarse-grain parallelism.* For these we do not need a fast coupling and can use IBM 4.5-MB/S channels and channel-to-channel coupling to support interprocessor communication. Thus, our initial plan will be to couple all systems via channels with full point-to-point connectivity.

The applications are, however, more aptly described as subjects for distributed computing, rather than for more communication-intensive parallel processing. For applications with smaller parallel grain size, i.e., characterized by more frequent communication, we require faster paths, because the overhead associated with communication can degrade performance severely.

The great majority of applications with *small parallel grain size* are characterized by frequent synchronizations and transfers of relatively small amounts of data. Overall, a single IBM channel operating at its peak rate would be satisfactory to serve in this capacity. However, given the large latencies involved, the resulting effective data rate is inadequate. Thus, to achieve faster coupling one can explore the use of two complementary approaches, both of which can have smaller latencies and have been used successfully in our previous lCAP systems. The first approach employs shared memory, and the second utilizes fast message passing. Each serves different purposes.

In general, message-passing and shared-memory communication have the following characteristics. Message passing is most simply realized by a bus interconnect and the specification of synchronous communication protocols. Shared-memory communication is done asynchronously and is implemented in its simplest form by a multiported memory that is mapped into the same address space of the cooperating processors. Both communication methods have advantages and disadvantages. Synchronous communication tends to ensure correctness of parallel execution, because programs that use this mode are rigidly constrained by the data dependencies defined by message passing. The disadvantages are that a bus interconnect is limited by topological constraints, a tendency of the facility to become saturated when many processors attempt to use it, and also by substantial latencies because of the necessity for some type of handshaking protocol. Alternatively, a shared memory tends to have faster data-access rates, smaller latencies, and fewer connectivity restrictions. However, this requires more careful programming of accesses to guarantee program correctness, because implicit synchronization is lacking.

In Fig. 9, an lCAP/3090 system with four IBM-3090/400s (four processor complexes) is illustrated. In this simple system dual-ported, rather than fully shared fourfold-ported, shared memories are used to couple the clusters in a nearest-neighbor ring structure. Also indicated in Fig. 9 are large shared "disk farms" that can be implemented with existing technology. Disk farms serve as a secondary level of shared storage to greatly enhance the size of problems that can be addressed.

In Fig. 10 we report the speedup obtained with an lCAP/3090 configuration composed of two 3090/400 bases, one 3090/600E and one 3090/600S. These examples compare elapsed time and include all the needed I/O related time for the job in consideration. Inserts a, b, c and d report the speedup for a Hylleraas-CI computation, a KGNMOL computation, a Monte Carlo code, and for a graphic package dealing with the ray tracing technique. Once more we recall that the system is somewhat unbalanced since the S engine is faster than the E which, in turn, is faster than those in the 3090/400 *base* systems. In Fig. 10 we have compared all cases assuming that the ideal time (100% parallelization) is obtained for a system composed by S engines, uniformly. Therefore, the measured speedup percent-wise is somewhat better than the one reported in the figure.

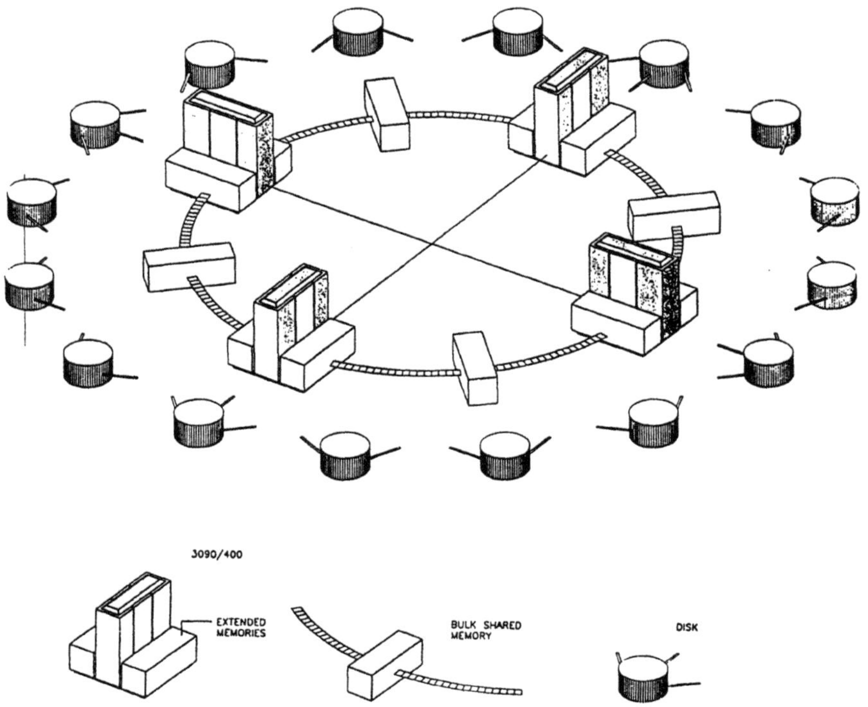

Figure 9. ICAP/3090 model based on four IBM 3090/400s.

Parallel processing performance issues

One of the fundamental reasons for using parallel processing is to speed up the elapsed time of an application. We mentioned in the preceding section that the best performance is bounded by the degree of parallelism that our code exhibits, characterized by three attributes: (1) the fraction of the code that is parallelizable, (2) the amount of communication required, and (3) load imbalancing when we partition the problem across multiple processors. In fact, while we list these as separate factors, they are all interrelated. For simplicity, however, we consider each factor separately, pointing out, where appropriate, the subtleties that exist.

Fraction of parallel code. The most fundamental factor in determining parallel performance is the fraction of the code that is parallel. We define this fraction to be that of the total time required for the job. Consider a problem with that fraction equal to X. The remainder, $1-X$, is sequential. Suppose that we have P processors over which the parallel portion can be distributed equally without any overhead. The sequential portion can be executed by one of the P processors.

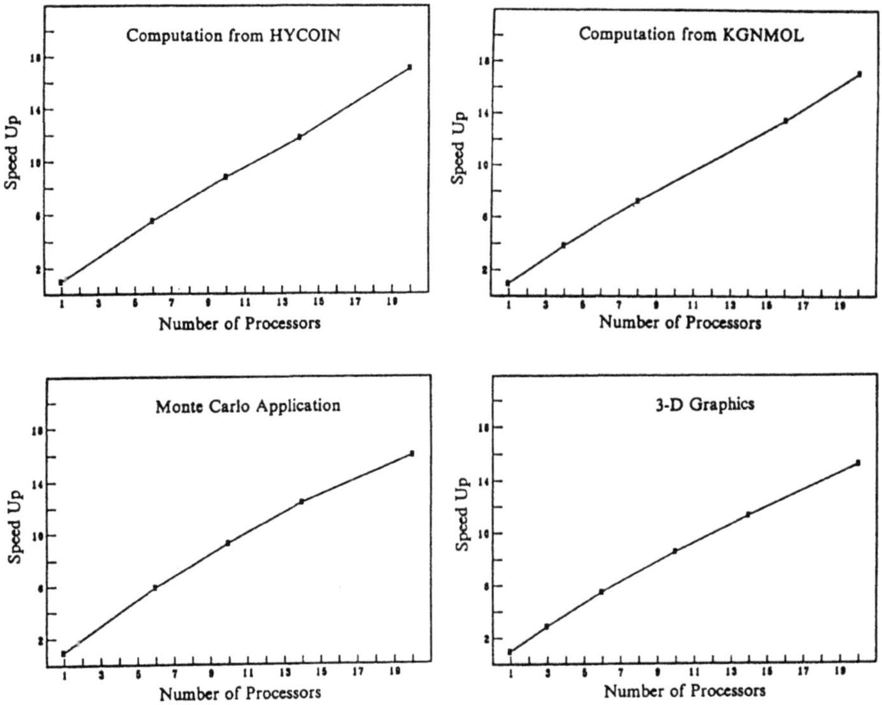

Figure 10. Speed-up measured on ICAP/3090 with 20 nodes.

Then the best we can expect, compared to running the entire problem on only one processor, is realized by a speedup factor S_P of

$$S_P = \frac{1}{1 - X + X/P}.$$

If X is unity, i.e., all of the code is parallel, then the speedup is equal to the number of processors (linear speedup). Otherwise, this limiting expression, known as Amdahl's law, applies.

Communication. While the fraction of parallel code is the primary consideration in determining how profitable it is to use parallel processing, it is by no means the final consideration. This is because, as we have seen previously, we must account for the extra time required for our processors to communicate as this overhead increases the effective sequential fraction of the code.

The degradation that communication can cause may be seen, perhaps in a simpler fashion, by examining the expression for the total time spent in solving our problem with a set of P processors, if the total parallel time is T_p and the sequential time is T_s, the total execution time is given as follows:

$$T_{par} = T_s + \frac{T_p}{P} + f(P).$$

The extra term $f(P)$ represents the additional time spent in communication, and at minimum, can be some constant or a positive function of the number of processors.

Load balancing. The basic goal of parallel processing is to keep all processors busy all of the time. Deviations from this ideal are called load-balancing problems. It is clear that load balancing is faced with the following dilemmas. First, for a given problem size, there is only so much work to be performed, and thus we are limited to some maximum number of processors. Second, because of the discrete nature of any parallel calculation, the division of labor is more evenly balanced for certain numbers of processors than others. These constraints may be compounded if the amount of parallel work changes as the job proceeds and communication is required. A specific example of load balancing is given in Chapter 7.

Early lCAP systems

To introduce the earlier lCAP systems and their influence in defining lCAP/3090, it is appropriate to begin by discussing the applications that led to their design. The motivation behind lCAP was initially limited to solving large problems in theoretical chemistry. This area includes quantum chemistry calculations employing self-consistent field methods, and statistical mechanics calculations using Monte Carlo and molecular dynamics methods.

In view of the characteristics of our applications, it was obvious what type of parallel computer we could assemble to perform these types of calculations. All that we required was an elementary master/slave system in which the slaves would, for example, calculate the two-body contributions at each step and send the results to the host or master. The host would then add them up, predict the new positions, and send these back to the slaves for the next period of computation. The example given has thus far been that for a typical molecular dynamics calculation, but analogous operations also pertain to the remainder of the applications. We now describe the architecture of the initial lCAP system.

Architecture. Because we wanted to assemble such a system in a short period of time and move toward the main goal of solving large problems, the selection of master and slaves was made from those available at the time the project began in early 1983. The slaves or attached processors (APs) had to satisfy the following requirements. Because of cost, space, and complexity of the overall system, the total number of attached processors was limited to approximately 10. This demanded—given our goal of supercomputing performance—that an AP be a powerful processor in its own right and have 64-bit precision floating-point hardware, as was required by our applications. Additionally, each AP needed a real memory of the order of a megaword (64-bit words) and several hundred megawords of disk space. Given these considerations, the optimal choice at the time was a Floating Point Systems Model 164 (FPS-164).

The choice of host was based on the following considerations. First, because data transfer between slaves and host represents overhead in our parallel computations, we required superior transfer rates to the extent possible. Second, we required a host that had superior I/O capability and offered an easy-to-use operating environment. On the basis of these considerations, an IBM 43XX mainframe was selected. For example, a dyadic 4381 can have up to 24 independent channels, any number of which can be coupled to a corresponding AP. Each channel can operate at a maximum transfer rate of 3 megabytes per second (MB/s) and the channels can be driven in parallel with one another. For ease of use, the 43XX offers the Virtual Machine (VM/SP) operating system for which FPS offers all the intersystem support software as a standard product.

With these choices, the system was incremented gradually over approximately two years to include 10 FPS-164s coupled to two IBM hosts, as shown in Fig. 11. As indicated, one host was a dyadic IBM 4381 that could couple to up to all ten APs, and the other host was an IBM 4341 that could couple to up to three APs. The 3088 switching units allowed the latter three APs to be configured to either host. Additionally, a third IBM 4341 was placed in the system and served as a graphics station attached to a number of graphics terminals. All of the 43XX systems had channel-to-channel coupling. Finally, large amounts of external storage in the form of IBM 3350 and 3380 disk systems were included, totaling approximately 25 gigabytes systemwide. This system was called lCAP/1.

In parallel with the development of lCAP/1, we assembled a similar but more powerful lCAP system called lCAP/2. Like the lCAP/1 system, it was a master-slave system, but it differed in that the host was a single dyadic IBM 3081 that coupled to a number of the Floating Point Systems FPS-264s. The FPS-264 is very similar architecturally to the FPS-164, but it is approximately three to four times faster, with a cycle time of 52 nanoseconds. Again, IBM 3-MB/s channels were employed to couple host to slaves. Each slave had its own disk subsystem (each with approximately one gigabyte), and the host a separate large complement of disk storage. Eventually, as with lCAP/1, this system included ten attached processors. One further distinction between the two systems was that the IBM 3081 ran under the IBM MVS operating system rather than the VM/SP operating system.

Performance. The initial goal of achieving supercomputer performance was obtained almost immediately. This is indicated in Tables I and II, where elapsed times for a set of applications are compared on lCAP/1 and lCAP/2, respectively (as a function of the number of slaves used in parallel) versus a single-processor

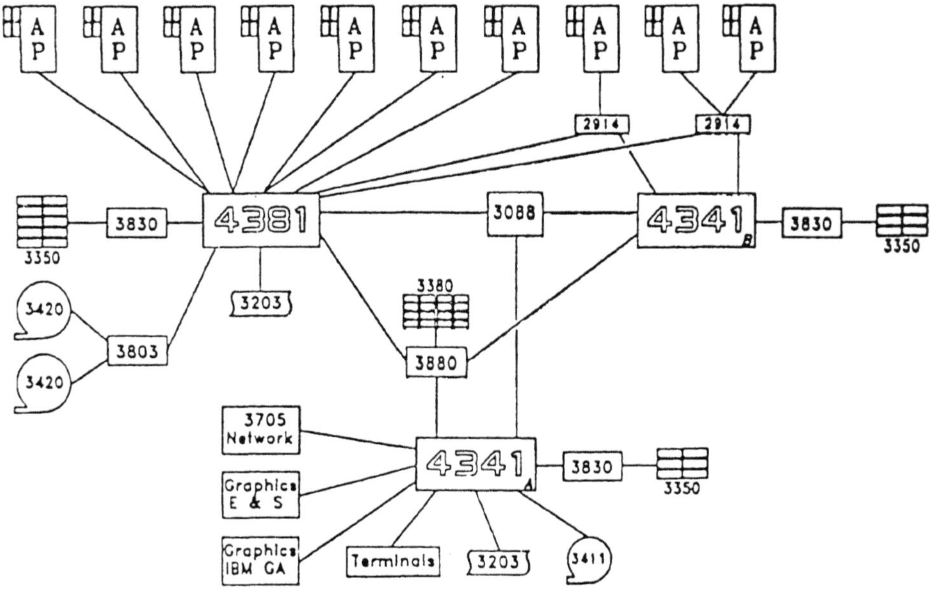

Figure 11. Early, loosely coupled array of processors (lCAP/1).

CRAY-XMP. The applications cited are an integrals-generating program used to construct matrix elements for a self-consistent field (SCF) calculation, for the iterative SCF code, and for illustrative many-body-interaction molecular dynamics and Metropolis Monte Carlo codes. It is important to note that the codes were not optimized for any of the systems, and thus these measurements cannot be used as anything like a definitive comparison. However, they do serve to indicate roughly the aggregate computing power of the machines for these applications.

With this qualification, Tables I and II show that for these applications, lCAP/1, with eight to eleven processors, and lCAP/2, with three to four processors, each roughly equal the computing power of a single CRAY-XMP processor. No data were available for comparison of these jobs running in parallel on an XMP.

The success in applying lCAP to these application areas was attributable to their large parallel grain sizes. The amount of time spent computing was large compared to the amount of interprocessor communication required. A representative but more detailed description of a number of parallel applications in theoretical chemistry executed on the lCAP system may be found in Reference 63. This characteristic was found to apply to many other applications, including the analysis of high-energy physics data, the determination of protein structures, seismic migration analysis, and circuit simulation. Typical performance results are indicated in Fig. 12 for several such applications. Here the speedup factor is plotted versus the number of parallel attached processors used. *Speedup* is defined as the

Table I. Applications performance on lCAP/1.

Job	Elapsed Time for lCAP/1 (minutes)				
	1 AP	3 APs	6 APs	10 APs	CRAY-XMP
Integrals (27 atoms)	71.7	24.0	12.3	7.8	7.6
SCF (27 atoms)	25.2	9.4	5.9	4.9	3.6
Integrals (42 atoms)	203.7	68.9	38.3	21.2	23.2
SCF (42 atoms)	73.0	26.0	14.3	10.6	8.7
Monte Carlo	162.1	57.8	32.0	22.0	20.4
Molecular dynamics	99.6	34.6	19.3	13.7	17.0
Seismic	33.8	11.8	6.6	4.3	5.6

Table II. Applications performance on lCAP/2.

Job	Elapsed Time for lCAP/1 (minutes)				
	1 AP	3 APs	6 APs	10 APs	CRAY-XMP
Integrals (27 atoms)	19.1	6.5	3.3	2.3	7.6
SCF (27 atoms)	10.6	5.2	3.7	3.4	3.6
Integrals (42 atoms)	55.0	18.7	9.3	6.1	23.2
SCF (42 atoms)	24.1	9.1	5.6	4.7	8.7
Monte Carlo	60.0	20.9	11.4	7.7	20.4
Molecular dynamics	29.6	10.6	5.9	4.2	17.0

ratio of the times a given application took to run on *P* processors compared to that running sequentially on one processor.

These examples include (going from left to right, top to bottom, and indicating, where appropriate, collaborators and their affiliations) molecular energy determinations,[130,131] molecular dynamics studies of water, oil reservoir seismic analysis, predictions of protein structure (Professor H. A. Scheraga[137] at Cornell University); determination of the wavefunction for HeH^+ employed in predicting the mass of the neutrino (Professor W. Kolos at the University of Florida); electronic circuit simulation,[138] high-energy physics data analysis (Drs. F. Carminati, R. Mount, H. Newman, and H. Pohl at CERN); chemical reactivity determination through electron density calculations (Professor R. F. W. Bader at McMaster University, Ontario, Canada); stochastic simulations of fluids, study of the liquid-solid silicon interface (Dr. E. Gawlinsky at Temple University), silicon crystal growth (Professor K. Kaski at Tampere University of Technology, Finland); and graphics of biologically active chemicals (Professor G. Richards at Oxford University, England).

However, the lCAP systems did not do as well for applications characterized by medium- or small-grain parallelism. That is, the burden of interprocessor communication through a host intermediary using 3-MB/S channels was often too great. This was the case particularly for engineering applications, where partial differential equations are typically solved following standard discretization techniques

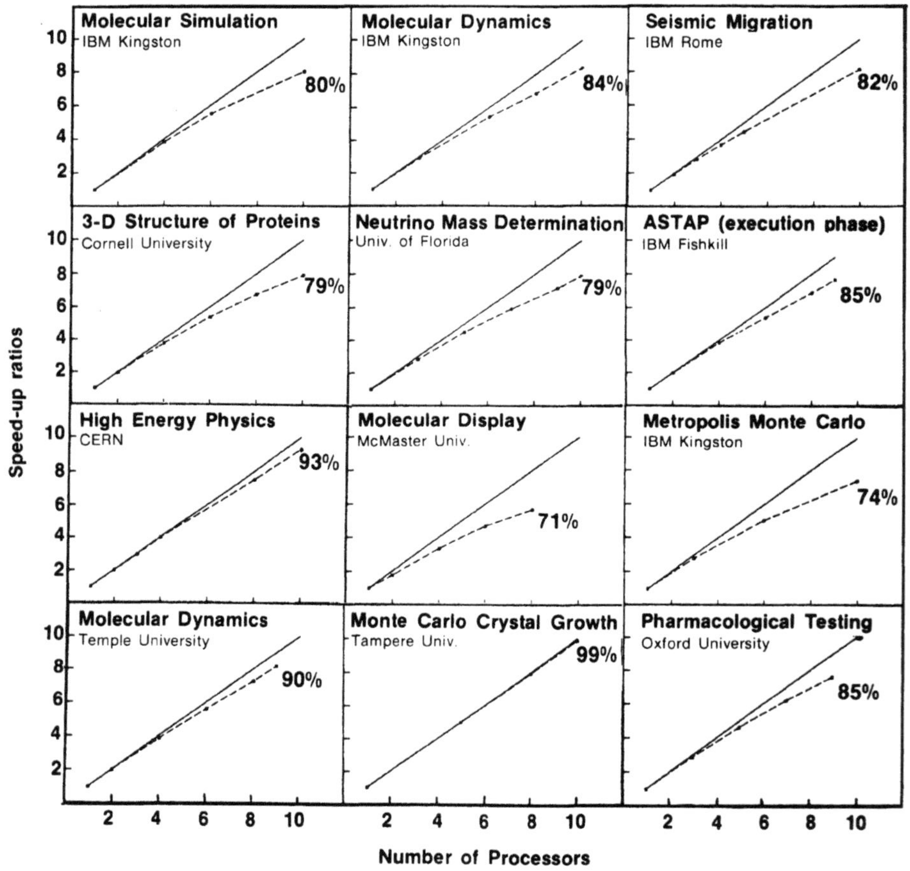

Figure 12. Speedup for typical large parallel grain size applications.

such as finite differences or finite elements. Here, the basic algorithms[139] that are required to solve large linear sets of equations or that perform matrix diagonalization are not characterized by large parallel grain sizes. These algorithms require more efficient interprocessor communication. This concern gave impetus to attempting to improve communication on ICAP.

Communication. The extensions made to the initial lCAP systems dealt primarily with enhancing communication. Two hardware additions sought to achieve this.

The first was the inclusion of a number of shared memories developed by Professor Martin Schultz[140] of Yale University and Scientific Computing Associates (SCA) These memories were fast solid-state disks that were coupled to the I/O ports of the FPS machines and were directly addressable by the attached processors. This configuration is illustrated in Fig. 13 for the lCAP/1 and lCAP/2, which was identically extended. Shown are five memories, each multiplexed four ways and each being 32 MB in size, linked in a double-ring structure around the ten processors. Software to employ these memories was developed by SCA[140] and extended within our laboratory to include functions such as synchronization between processors and locks supporting critical sections.

The second hardware addition to the lCAP clusters was a 32-bit-wide fast bus (built by Floating Point Systems) that linked all attached processors.[141] This is also illustrated in Fig. 13, showing each processor attached to a node on the bus. The nominal peak speeds were 22 MB/S from AP to node and approximately 32 MB/S between nodes, for both lCAP/1 and lCAP/2. Software to use this facility was developed by FPS[141] and included the ability for any non-nearest-neighbor communication, as well as for broadcasting from one to all.

As is also evident from Fig. 13, each lCAP cluster was hosted by a single master—a dyadic IBM 3081 on lCAP/1 and a four-processor IBM 3084 on lCAP/2. The greater computing power of the IBM 3081 eliminated the need for a dual-host system on lCAP/1, which had previously employed IBM 43XX systems. As before, lCAP/1 allocated a number of dedicated attached processors specifically for parallel program development and the remainder for production. However, lCAP/2 was used exclusively for production. Again, the IBM 3081 and the 3084 ran under the VM/SP and MVS operating systems, respectively.

The parallel scheduler. As stated previously, an lCAP cluster was essentially a single-user system. That is, there existed no efficient means of dynamically sharing processors among multiple parallel jobs. To make the system more responsive to a general-user environment, we developed a parallel scheduler to perform the sharing. We refer to Chapter 26 for details on this topic.

Applications. With these improvements in communication, all of the programs that exhibited marginal to poor performance on the early lCAP systems performed adequately. A partial listing of application areas for which codes have been successfully parallelized on either the initial or the extended systems is given in Chapter 26. References are included where more detailed descriptions of implementation may be found. All of these applications obtained from 60 percent to 95 percent of linear speedup, when partitioned on up to ten processors for problem sizes of interest. The same statement may be made with respect to a large number of parallel algorithms. These are of critical importance in being able to parallelize

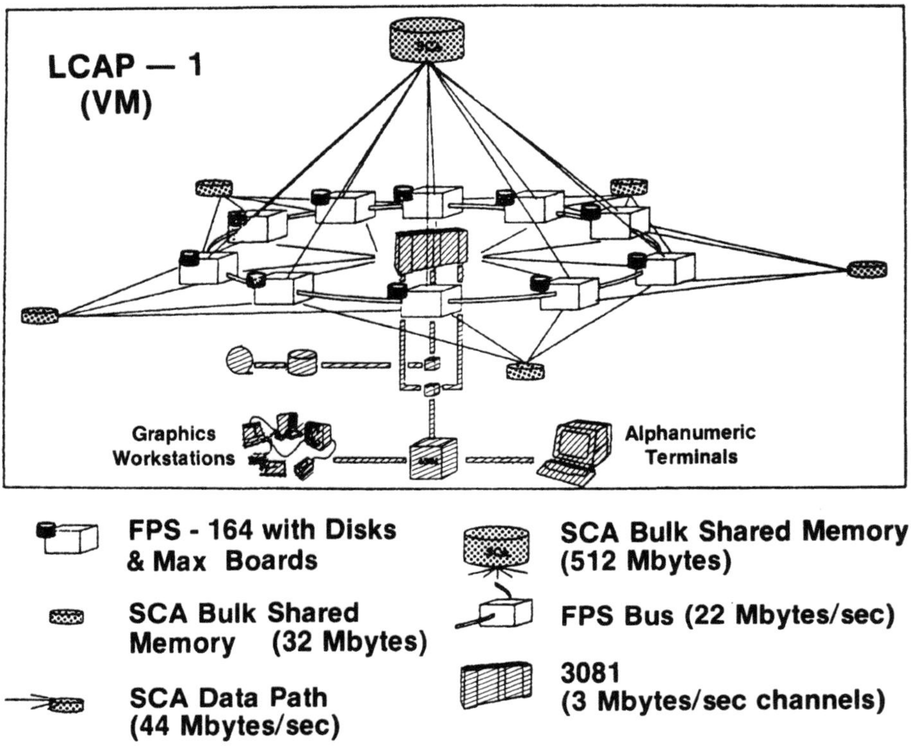

Figure 13. Extended lCAP/1 and lCAP/2.

engineering applications, but they also are of general use in many scientific areas, we refer the interested reader to Chapter 27 for details.

6. Conclusions

We have briefly outlined the evolution of productivity in human history and drawn a strong parallel with the evolution of information-gathering, the latter being conditioned by the evolution of the computer industry. These trends radically affect the way we use computers to perform computations and to write application programs: thus we have commented on the evolution of computer applications.

In section 3 of this chapter we have outlined the content of MOTECC-89, which is seen as an interconnected, modular, and interdisciplinary library of application programs to perform global simulations on fourth generation computers and with the aim of moving to cognizant simulations for fifth and sixth generation computers. An example of global simulation, taken from our work, is discussed in detail.

Finally, we have briefly surveyed an experimental computer assembled in our department in IBM-Kingston, mainly to bring up the issue of parallelism, the necessary way to compute in the near future if we wish to obtain ultra-high performance. We have not discussed an alternative path to ultra-computing, namely the use of special purpose computers each devoted to a single numerical application. This second avenue has notable potential providing that today's barriers at experimenting with computer architectures (hardware computers, not simulated ones) are substantially decreased. Even if rather visionary, we note that if the finite element method would become a really competitive technique in quantum physics and chemistry, it would then emerge as "the special technique" for much of science and engineering, truly deserving the attention of special purpose hardware. In this respect, a Monte Carlo engine, a Newton engine, and a finite element engine connected to a general purpose supercomputer could provide a most useful and integrated tool for scientific information handling and processing.

The algorithms for application programs described in MOTECC-90 have been written for sequential machines, for sequential machines with vector features, for parallel machines with vector features at each node, and for clusters of parallel machines with vector features at each node. This feature, while of large practical importance, is, however, not stressed in MOTECC-90 since our interest is mainly in connecting, within a logical and integrated framework, known concepts and algorithms.

References

1. D. R. Hartree, "The Calculation of Atomic Structures", (Wiley, New York, 1957).
2. E. U. Condon, and G. H. Shortley, "Theory of Atomic Spectra", (Cambridge Univ. Press, 1935).
3. J. C. Slater, "Quantum Theory of Matter", (McGraw Hill, New York, 1951).
4. P. A. M. Dirac, "The Principles of Quantum Mechanics", (4th. ed., Oxford Univ. Press, 1958).
5. C. A. Coulson, "Valence", (Oxford Univ. Press, 1961).
6. D. R. Hartree, Proc. Cambridge Phil. Soc., **24**, 89 (1927).
7. V. Fock, Z. Phys., **61**, 126; *ibid.* **62**, 795 (1930)
8. J. C. Slater, "Quantum Theory of Atomic Structure", (McGraw-Hill, New York, 1960), vols. 1 and 2.
9. C. C. J. Roothaan, Rev. Mod. Phys., **23**, 69 (1951); see also B. Roos, C. Salez, A. Veillard and E. Clementi, "A General Program for Calculation of Atomic SCF Orbitals by the Expansion Method," IBM Research Report **RJ#518** (1968).
10. E. Clementi, J. Chem. Phys., **38**, 996 (1961).
11. N. Metropolis and S. Ulam, J. Am. Chem. Soc., **44**, 335 (1949).

12. N. Metropolis, A. W. Rosenbluth, M. N. Rosenbluth, A. H. Teller, and E. Teller, J. Chem. Phys., **21**, 1087 (1953).
13. J. M. Hamersley and D. C. Handscombe, "Monte Carlo Methods" (Methuen, London, 1964).
14. R. Y. Rubinstein, "Simulation and the Monte Carlo Method", (Wiley, New York, 1981).
15. J. P. Hansen and I. R. McDonald, "Theory of Simple Liquids", (Academic, London, 1976).
16. A. Rahman, Rev. Nuovo Cimento, **1**, 36 (1969).
17. M. W. Evans, G. J. Evans, W. T. Coffey and P. Grigolini, "Molecular Dynamics", (Wiley Interscience, New York, 1982).
18. L. Verlet, Phys. Rev. **69**, 98 (1967).
19. C. W. Gear, "Numerical Initial Value Problems in Ordinary Differential Equations.", (Prentice Hall, Englewood Cliffs, 1971).
20. J. A. Barker and R. O. Watts, Mol. Phys., **26**, 789 (1973).
21. E. Clementi, and S. Chin, G. Corongiu, J. H. Detrich, M. Dupuis, G. C. Lie, D. Logan and V. Sonnad, in "Biological and Artificial Intelligence Systems", eds. E. Clementi and S. Chin, (ESCOM Science Publishers, Leiden, 1988).
22. E. Clementi, Phil. Trans. R. Soc. London, **A326**, 445 (1988); see also Int. J. Quant. Chem., **35**, 3 (1989).
23. E. Clementi, D. Logan, and J. Saarinen, IBM Systems Journal, **27**, 475 (1988).
24. M. Dupuis, ed., "Lecture Notes in Chemistry", No. 44 (Springer-Verlag, 1986).
25. E. Clementi and S. Chin, Eds. "Structure and Dynamics of Nucleic Acids, Proteins and Membranes.", (Plenum, New York, 1986).
26. D. C. Rapaport and E. Clementi, Phys. Rev. Letters, **57**, 695 (1986).
27. L. Hannon, G. C. Lie, and E. Clementi, Phys. Lett. A, **110**, 84 (1986).
28. L. Hannon, G. C. Lie, and E. Clementi, J. Sci. Comp., **1**, 55 (1986).
29. J.A. Given and E. Clementi, J. Chem. Phys., **90**, 7376 (1989).
30. I.P. Grant, J. Phys., **B19**, 3187 (1986) and references therein given.
31. Relativistic Effects in Atoms, Molecules and Solids, G.L. Malli, Ed., NATO ASI Series, Plenum Press, New York (1983).
32. P. Pyykko, Chem. Rev., **88**, 563 (1988).
33. A. Mohanty and E. Clementi, Chem. Phys. Letters, **157**, 384 (1989); see also MOTECC-89, Chapter 3.
34. G. Corongiu and E. Clementi, Chem. Phys. Letters, **90**, 359 (1983).
35. F. Sasaki, Int. J. Quant. Chem., **8**, 605 (1974).
36. HONDO: QCPE Program #544, Univ. of Indiana, Bloomington, Indiana.
37. M. Dupuis, J.D. Watts, H.O. Villar and G.J.B. Hurst, IBM Tech. Rep. KGN-181 (July 1988), and references therein given.
38. KGNMOL: QCPE Program #538, Univ. of Indiana, Bloomington, Indiana.
39. M. Dupuis, J. Rys and H.F. King, J. Chem. Phys., **65**, 111 (1976).
40. M. Dupuis and H.F. King, Int. J. Quant. Chem., **11**, 613 (1977).
41. C.C.J. Roothaan, Rev. Mod. Phys., **32**, 179 (1960).

42. J.A. Pople and R.K. Nesbet, J. Chem. Phys., **22**, 571 (1954).
43. E.R. Davidson, Chem. Phys. Lett., **21**, 565 (1973).
44. F. Bobrowicz and W.A. Goddard, III, in "Modern Theoretical Chemistry," Plenum, New York (1977).
45. B.R. Brooks, W.D. Laidig, P. Saxe, J.D. Goddard, Y. Yamaguchi and H.F. Schaefer, III, J. Chem. Phys., **72**, 4652 (1980).
46. J.A. Pople, J.S. Binkley and R. Seeger, Int. J. Quant. Chem., **S10**, 1 (1976); R. Krishnan, M.J. Frisch and J.A. Pople, J. Chem. Phys., **72**, 4244 (1980).
47. M. Dupuis and H.F. King, J. Chem. Phys., **68**, 3998 (1978).
48. T. Takada, M. Dupuis and H.F. King, J. Chem. Phys., **75**, 332 (1981).
49. T. Takada, M. Dupuis and H.F. King, J. Comp. Chem., **4**, 234 (1983).
50. P. Carsky, B.A. Hess, Jr., and L.S. Schaad, J. Comp. Chem., **5**, 280 (1984).
51. L.R. Kahn, P. Baybutt and D.G. Truhlar, J. Chem. Phys., **65**, 3826 (1976).
52. D. Spangler, I.H. Williams and G. Maggiora, J. Comp. Chem., **4**, 524 (1983).
53. P. Pulay, Mol. Phys., **17**, 197 (1969); **18**, 473 (1970); **21**, 329 (1971).
54. J.A. Pople, R. Krishnan, H.B. Schlegel and J.S. Binkley, Int. J. Quant. Chem., **S13**, 225 (1979).
55. A. Komornicki and J.W. McIver, Jr., J. Chem. Phys., **70**, 2014 (1979).
56. A. Cerjean and W. Miller, J. Chem. Phys., **75**, 2800 (1981).
57. M. Schmid, M. Gordon and M. Dupuis, J. Chem. Phys., **66**, 2153 (1977).
58. R.S. Mulliken, J. Chem. Phys., **23**, 1833 (1955).
59. D. Neumann and J.W. Moskowitz, J. Chem. Phys., **49**, 2056 (1968).
60. C.E. Dykstra and P. Jasien, Chem. Phys. Lett., **109**, 388 (1984).
61. P. Habitz and E. Clementi, Comp. Phys. Comm., **29**, 301 (1983).
62. E. Clementi and D.R. Davies, J. Comp. Phys., **1**, 223 (1966); see also K. Davies and E. Clementi, "IBMOL: Computation of wave Functions of General Geometry," IBM Research Report (December 1965).
63. A. Veillard and E. Clementi, IBM Research Report (August 1966).
64. E. Clementi and J.W. Mehl, IBM System/360 IBMOL-5 Program, General Information Manual, IBM Research Report, **RJ 853** (May 1971).
65. E. Ortoleva, G. Castiglione and E. Clementi, IBMOL-7, Computer Phys. Comm., **12**, 337 (1980).
66. R. Gomperts and E. Clementi, IBM Research Report, KGN-118 (June, 1988).
67. See Chapter 7, Section 16.
68. S.F. Boys and F. Bernardi, Mol. Phys., **19**, 553 (1970).
69. W. Kolos, Theor. Chim. Acta, **54**, 187 (1980).
70. E.A. Hylleraas, Z. Phys., **54**, 347 (1929).
71. J.S. Sims and S. Hagstrom, J. Chem. Phys., **55**, 4699 (1971).
72. W. Kolos and L. Wolniewics, J. Chem. Phys., **49**, 409 (1978); see also A. Preiskorn and W. Woznicki, Mol. Phys., **52**, 1291 (1984).
73. A. Largo-Cabrerizo and E. Clementi, Int. J. Quant. Chem. Symp., **21**, 667 (1987).
74. C. Urdaneta, A. Largo-Cabrerizo, J. Levin, G.C. Lie and E. Clementi, J. Chem. Phys. (in press, 1989).

75. D. Frye, G.C. Lie and E. Clementi, J. Chem. Phys., **91** 2366 (1989).
76. H.J.Aa Jensen and H. Ågren, Chem. Phys. Letters, **110**, 140 (1984); Chem. Phys. **104**, 229 (1986); Chem. Phys. Letters, **137**, 431 (1987); P. Jorgensen, H.J.Aa. Jensen and J. Olsen, J. Chem. Phys., **89**, 3654 (1988); J. Olsen, B.O.Roos, P. Jorgensen, and H.J.Aa. Jensen, J. Chem. Phys., **89**, 2185 (1988); H. Ågren, C. Medina-LLanos, K. Mikkelsen and H.J.Aa Jensen, Chem. Phys. Letters, **152**, 322 (1989).
77. A. Mohanty, and E. Clementi, (J. Chem. Phys., August 1990, in press).
78. U.G. Jorgensen, J. Almolof, B. Gustafsson, M. Larsson, and P. Siegbahn, J. Chem. Phys., **83**, 3034 (1985); M. Larsson and P.E.M. Siegbahn, J. Chem. Phys., **85**, 4208 (1986); C.W. Bauschicher and P.E.M. Siegbahn, J. Chem. Phys., **85**, 2802 (1986); C.W. Bauschicher, Jr., P.S. bagus, C.J. Nelin, and B.O. Roos, J. Chem. Phys., **85**, 354 (1986); J.M.O. Matos, B.O. Roos, and P. Å Malmqvist, J. Chem. Phys., **86**, 1458 (1987); J. Olsen, B.O. Roos, P. Jorgensen, and H.J. Aa. Jensen, J. Chem. Phys., **89**, 2185 (1988); C.M. Marian, M.R.A. Blomberg, and P.E.M. Siegbahn, J. Chem. Phys., **91**, 3589 (1989).
79. L.E. McMurchie and E.R. Davidson, J. Comp. Phys., **26**, 218 (1978); *ibid.* **44**, 289 (1978); S. Langhoff and E.R. Davidson, Int. J. Quant. Chem., **8**, 61 (1974); D. Feller and E.R. Davidson, J. Chem. Phys., **74**, 3977 (1981); R.J. Cave and E.R. Davidson, J. Chem. Phys., **89**, 67 (1988).
80. B. Liu and A.D. McLean, J. Chem. Phys., **72**, 3418 (1980); B.H. Lengsfield, A.D. McLean, M. Yoshimine, and B. Liu, J. Chem. Phys., **79**, 1891 (1983); A.D. McLean, B.H. Lengsfield, III, J. Pacansky, and Y. Ellinger, J. Chem. Phys., **83**, 3567 (1985); R.P. Saxon and B. Liu, J. Chem. Phys., **85**, 2099 (1986); K. Balasubramanian and A.D. McLean, J. Chem. Phys., **85**, 5117 (1986); A. Palma, S. Green, D.J. DeFrees, ans A.D. McLean, J. Chem. Phys., **89**, 1401 (1988); A.D. McLean, B. Liu, and J.A. Barker, J. Chem. Phys., **89**, 6339 (1988); M.R.A. Blomberg and B. Liu, J. Chem. Phys., **89**, 6870 (1988); M. Yoshimine, J. Chem. Phys., **90**, 378 (1989); T. Noro and M. Yoshimine, J. Chem. Phys., **91**, 3012 (1989); B. Liu and A.D. McLean, J. Chem. Phys., **91**, 2348 (1989).
81. I. Cacelli, V. Carravetta, and R. Moccia, J. Chem. Phys., **85**, 7038 (1986); I. Cacelli, V, Carravetta, R. Moccia, and A. Rizzo, J. Phys. Chem., **92**, 979 (1988); Chem. Phys. Letters, **155**, 210 (1989); J. Chem. Phys. **89**, 7301 (1988); *ibid.* **92**, 2883 ((1990); I. Cacelli, R. Moccia, and V. Carravetta, Chem. Phys. **90**, 313 (1984); J. Phys. B **18**, 1375 (1985); Mol. Phys. **59**, 835 (1986).
82. J. Callaway, *Quantum Theory of Solid State*, (Academic press, N.Y. 1974); C.S. Wang and J. Callaway, Comp. Phys. Commun., **14**, 327 (1978); J. Callaway and J.L. Fry, *Computational Methods in Band Theory*, P.M. Marcus, J.F. Jarak, and A.R. Williams, Eds., (plenum Press, N. Y. 1971)
83. R. Car and M. Parrinello, Phys. Letters, **55**, 2471 (1985); Phys. Rev. Letters **60**, 27 (1988); D. Hohl, R.O. Jones, R. Car, and M. Parrinello, J. Am. Chem. Soc., **111**, 825 (1989); G. Galli, R.M. Martins, R. Car, and M. Parrinello, Phys. Rev. Letters, **62**, 555 (1989); *ibid.* **63**, 988 (1989).

84. J.M. André, J. Chem. Phys., **50**, 1536 (1969); J.M. André and G. Leroy, Chem. Phys. Letters, **5**, 71 (1970); J.M. André, J. Delhalle, G.S. Kapsomenos, and G. Leroy, Chem. Phys. Letters, **14**, 485 (1972); L. Piela, J.M. André, J.G. Fripiat, and J. Delhalle, Int. J. Quant. Chem. Symp., **14**, 405 (1980); J.M. André, J.G. Fripiat, C. Demanet, J.L. Brédas, and J. Delhalle, Int. J. Quant. Chem. Symp., **12**, 233 (1978); J. Ladik and J.M. André, Eds., *Quantum Theory of Polymers: Solid State Aspects*, NATO-ASI Series, **C123**, (Reidel Publishing Co., 1984); J.M. André, *Strategy for Computer Chemistry*, C. Tosi, Ed., (Kluver Academic Publ., **45**, 1989).
85. For density functionals see Chapters 2 and 7.
86. G. Bolis, M. Ragazzi, D. Salvaderi, D.R. Ferro, and E. Clementi, Int. J. Quant. Chem., **14**, 815 (1978).
87. H. Kistenmacher, H. Popkie, E. Clementi and R.O. Watts, J. Chem. Phys., **60**, 4455 (1974).
88. R.O. Watts, E. Clementi and J. Fromm, J. Chem. Phys., **61**, 2550 (1974).
89. S. Romano and E. Clementi, Int. J. Quant. Chem., **14**, 839 (1978).
90. E. Clementi and G. Corongiu, Biopolymers, **18**, 2431 (1979); ibid., **20**, 551 (1981).
91. E. Clementi and G. Corongiu, Biopolymers, **20**, 2427 (1981).
92. P. Procacci, G. Corongiu and E. Clementi, IBM Tech. Rep. KGN-195 (1989).
93. E. Clementi, Lecture Notes in Chemistry, Vol. 2, "Determination of Liquid water Structure, Coordination Numbers for Ions and Solvation of Biological Molecules," Springer-Verlag, New York (1976).
94. E. Clementi, Lecture Notes in Chemistry, Vol. 19, "Computational Aspects for Large Chemical Systems," Springer-Verlag, New York (1980).
95. D.L. Ermak and J.A. McCammon, J. Chem. Phys., **69**, 1352 (1978).
96. J.P. Rivet, M. Henon, U. Frisch and D. d'Humieres, Europhys. Lett., **7**, 3 (1988).
97. M. Navier, "Memoire sur les Lois Mouvement des Fluids," Mem. de l'Acad. des Sciences, vi (1882).
98. G.G. Stokes, "On the Theories of Internal Friction of Fluids in Motion," Camb. Trans., viii (1845).
99. O.A. Ladyzhenskaya, The Mathematical Theory of Viscous Incompressible Flow, Gordon and Breach, New York (1963).
100. H. Lamb, Hydrodynamics, Cambridge University Press (1879).
101. P. Roache, Computational Fluid Dynamics, Hermosa Publishers, Albuquerque, New Mexico (1972).
102. S.A. Orszag, "Comparison of pseudospectral and spectral approximation," Stud. App. Math, **51**, 253 (1972).
103. O.C. Zienkiewicz, The Finite Element Method, McGraw Hill, New York (1977).
104. B.A. Finlayson, The Method of Weighted Residuals, Academic Press, New York (1972).
105. I. Babuska, B. Szabo and I.N. Katz, "The P-verson of the Finite Element Method," SIAM J. Num. Anal., **18**, 515 (1981).

106. S. Foresti, G. Brussino and V. Sonnad, "Multilevel Solution Method for the P-version of Finite Elements: Parallel Implementation and Comparison with other Solution Methods," IBM-Kingston Research Report, KGN-137 (1988).
107. A.W. Craig and O.C. Zienkiewicz, "A Multigrid Algorithm using a Hierarchical Finite Element Basis," in Multigrid Methods for Integral and Differential Equations, D.J. Paddon and Holstein, Eds., Clarendon Press, Oxford, 310 (1985).
108. E. Clementi and H. Popkie, J. Chem. Phys., **57**, 1077, (1972).
109. W. S. Benedict, N. Gailer and E. K. Plyler, J. Phys. Chem., **24**, 1139, (1956).
110. R. J. Bartlett, I. Shavitt and G. D. Purvis, J. Chem. Phys., **71**, 281, (1979).
111. K. S. Kim, M. Dupuis, G. C. Lie and E. Clementi, Chem. Phys., Letters, **131**, 451, (1986).
112. H. Popkie, H. Kistenmacher and E. Clementi, J. Chem. Phys., **59**, 1325, (1973).
113. O. Matsuoka, E. Clementi and M. Yoshimine, J. Chem. Phys., **64**, 1351, (1976).
114. G. C. Lie and E. Clementi, Phys. Rev. **33**, No. 4, 2679, (1986).
115. U. Niesar, G. Corongiu, M-J. Huang, M. Dupuis and E. Clementi, Int. J. Quant. Chem. Symp. **23**, 421 (1989); U. Nieser, G. Corngiu, E. Clementi, G. Kneller, and D. Battacharya, (J. Phys. Chem., in press).
116. B. J. Alder, J. Chem. Phys., **40**, 2724, (1964).
117. G. C. Lie, E. Clementi and M. Yoshimine, J. Chem. Phys., **64**, 235, (1976).
118. E. Clementi and G. Corongiu, Int. J. Quant. Chem. Symp., **10**, 31, (1983).
119. J. Detrich, G. Corongiu, and E. Clementi, Int. J. Quant. Chem., **9**, 701, (1984).
120. M. Wojcik and E. Clementi, J. Chem. Phys., **85**, 3544, (1986).
121. M. Wojcik and E. Clementi, J. Chem. Phys., **85**, 6085, (1986).
122. P.A. Madden and P.W. Impey, Chem. Phys. Letters, **123**, 502 (1986).
123. L. Hannon, G.C. Lie and E. Clementi, J. Stat. Physics, **51**, 965 (1988).
124. S. Kandadai and E. Clementi, Biopolymers, **28**, 1497 (1989)
125. G. Corongiu and E. Clementi in "Structure and Dynamics: Nucleic Acids and Proteins," E. Clementi and R.H. Sarma, Eds., Adenine Press (1983).
126. K.S. Kim and E. Clementi, J. Comp. Chem., **8**, 57 (1987).
127. R. Gompers and E. Clementi, IBM Tech. Rep. KGN-107 (February 1987).
128. J. Beetem, M. Denneau and D. Weingarten, in "Special Topics in Super Computing, Vol. 1, Experimental Parallel Computing Architectures," J.J. Dongarra, ed., Elsevier Science Publishers, New York (1987).
129. D.M. Nosenchuck, M.G. Littman and W. Flannery, J. Sci. Comp., **1**, 53 (1986).
130. E. Clementi, G. Corongiu, J. Detrich, S. Chin and L. Domingo, Intl. J. Quant. Chem. Symp., **18**, 601 (1984).
131. J.H. Detrich, G. Corongiu and E. Clementi, Chem. Phys. Lett., **112**, 426 (1984).
132. D.H. Gibson, D.W. Rain and H.F. Walsh, IBM Sys. J., **25**, 36 (1986).
133. Y. Singh, G.M. King and J.W. Anderson, IBM Sys. J., **25**, 20 (1986).

134. D.L. Meck, IBM Tech. Rep. POK-38 (1984).
135. Parallel FORTRAN Language and Library Reference, SC23-0431-0, IBM Corporation, available through IBM branch offices.
136. See the IBM J. of R.&D., **34**, (1990), which is dedicated to the IBM RISC System/6000 Processors.
137. Lecture Notes in Chemistry, **44**, M. Dupuis, Ed., Springer-Verlag, Berlin (1986); Structure and Dynamics of Nucleic Acids, Proteins and Membranes, E. Clementi and S. Chin, Eds., Plenum Publishing Co., New York (1986).
138. ASTAP is an IBM Installed User Program (IUP), Program No. 5796-PBH.
139. J.M. Ortega and R.G. Voigt in Solution of Partial Differential Equations on Vector and Parallel Computers, SIAM Press, Philadelphia (1985).
140. Shared Bulk Memory System Software Manual, Ver. 2.0, Scientific Computing Associates, Yale University, New Haven, CT (1987).
141. FPSBUS Software Manual, Rel. G, Publication No. 860-7313-004A, Floating Point Systems Inc., Beaverton, OR (1986).

Chapter 2.

Independent Electron Models: Hartree-Fock for Many-Electron Atoms†

E. Clementi, S.J. Chakravorty, G. Corongiu, and V. Sonnad
IBM Corporation
Scientific and Engineering Computations
Dept. 48B/428
Kingston, New York 12401, U.S.A.

1. Introduction and Chapter Organization

One of the main problems of quantum chemistry and solid state theory is to develop computational techniques to accurately simulate the interaction of electrons amongst each other in a many electron system. Although the *exact* solution of the time independent, non-relativistic, many electron Schrödinger equation for atoms may be unattainable for complex systems, we can still routinely obtain highly accurate solutions to explain, rationalize, and predict chemical phenomena. Among the various ways of approximately solving the Schrödinger equation the variation and perturbation methods are the most common. In the variational approach, the theorem

$$E = \langle \mathscr{H} \rangle = \frac{\int \overline{\Psi} \mathscr{H} \Psi d\tau}{\int \overline{\Psi} \Psi d\tau} \geq W \tag{1}$$

is the basis of the method. The theorem states that for any trial function Ψ which is a theoretically acceptable wave function, the expectation value $\langle \mathscr{H} \rangle$ is an *upper* bound to the true energy. If a good judgement is exercised in choosing the trial function, Ψ, the value of E obtained may be close to the true energy W. To a first approximation the trial many-electron wave function $\Psi(\vec{r}_1, \vec{r}_2, ..., \vec{r}_N)$ may be

† This chapter is a revision and extension of the chapter by S.J. Chakravorty, G. Corongiu, J.R. Flores, V. Sonnad, V. Carravetta and I. Cacelli, "Independent Electron Models: Hartree-Fock for Many-Electron Atoms", Chapter 3 in MOTECC-89.

written as a product of one-electron functions $u_i(\vec{r})$ as devised by D.R. Hartree. Here it is assumed that each electron is described by a one-electron function, viz. an orbital, and as a consequence one solves the following equation for each orbital function:

$$\left\{ -\frac{1}{2}\nabla^2 - V_{ext}(\vec{r}) + \sum_{j\neq i}\int \frac{\bar{u}_j(\vec{r}_0)u_j(\vec{r}_0)dr_0}{|\vec{r} - \vec{r}_0|} \right\} u_i(\vec{r}) = \varepsilon_i u_i(\vec{r}) \tag{2}$$

From this equation it may be seen that the operator $-\nabla^2/2$ is the usual kinetic energy operator, $V_{ext}(\vec{r})$ is the external potential due to the nucleus and the sum over $j\neq i$ is the potential energy from the probability density of the rest of the electrons besides the one in orbital u_i. Thus, rather than solving a 3N-dimensional differential and eigenvalue equation

$$\mathscr{H}\Psi(\vec{r}_1,\ldots\ldots,\vec{r}_N) = W\Psi(\vec{r}_1,\ldots\ldots\ldots,\vec{r}_N) \tag{3}$$

the Hartree approximation[1], derived on an intuitively base, reduces this equation to a computationally tractable form consisting of N, 3-dimensional pseudo eigenvalue differential equations (Eq. 2). This in essence is the requirement for an **independent particle model** The N equations defined in (2) are really not independent of each other, but constrained such that the knowledge of u_j's all at once is required to find u_i. A procedure of successive approximations is employed for the calculation of u_i's and is described as follows: 1) A guess is made for the set "u_i". 2) The differential equations are constructed with this set. 3) The differential equations in Eq. (2) are solved for a new set of u_i's with constraints of orthonormality 4) This set $\{u_i\}$ is updated. Steps 2 to 4 are repeated until there is a consistency in the process since the effective potential for every orbital changes with every iteration. Ultimately a cycle may be carried through for which the effective potential for an orbital no longer changes. The effective field corresponding to this cycle is called the self consistent field for the orbital, and this powerful method of tackling the many electron problem is called the Self Consistent Field procedure. However, the Hartree solution leaves the motion of electrons uncorrelated and violates the antisymmetry principle which states that the total wave function of a many electron system must change sign when the four coordinates *x, y, z* and spin *s* of any two electrons are interchanged, viz. violates the Pauli exclusion principle.

The simplest form of Ψ for a many electron system (obeying the antisymmetry constraint) is the one proposed by Slater,[2] namely the antisymmetrized product of orbitals, e.g. the Slater determinant:

$$\Psi(r_1,\ldots\ldots,r_N) \simeq \frac{1}{\sqrt{N!}} \begin{vmatrix} \phi_1(1) & \phi_2(1) & \ldots\ldots & \phi_N(1) \\ \phi_1(2) & \phi_2(2) & \ldots\ldots & \phi_N(2) \\ \ldots & \ldots & \ldots & \ldots \\ \ldots & \ldots & \ldots & \ldots \\ \phi_1(N) & \phi_2(N) & \ldots\ldots & \phi_N(N) \end{vmatrix} \tag{4}$$

The above choice of wave function and application of the variation principle yields the possibility to rigorously derive the Hartree equation.[1] This much was done independently by J.C. Slater[3] and by V. Fock,[4] leading to what is today called the "Hartree-Fock approximation," a somewhat questionable designation since "Hartree-Fock-Slater" seems more appropriate. However, not to create confusion, we shall continue with the biased notation HF for Hartree-Fock, and we shall intend "Hartree-Fock-Slater." The revised Hartree-Fock equations are the eigenvalue equations of the form:

$$\left\{ -\frac{1}{2}\nabla^2 - V_{ext}(\vec{r}) + \sum_j \int \frac{\overline{\phi}_j(\vec{r}_0)\phi_j(\vec{r}_0)}{|\vec{r} - \vec{r}_0|} dr_0 \right. \tag{5}$$

$$\left. - \sum_j \delta(m_{si}, m_{sj}) \int \frac{\overline{\phi}_j(\vec{r}_0)\phi_i(\vec{r}_0)\phi_j(\vec{r})\overline{\phi}_i(\vec{r})}{|\vec{r} - \vec{r}_0| \overline{\phi}_i(\vec{r})\phi_i(r)} dr_0 \right\} \phi_i(\vec{r}) = \varepsilon_i \phi_i(\vec{r})$$

This is yet another example of an independent particle model. However in this case the *potential* term is a *non local* one. By a non local potential we mean that one needs the knowledge of the orbital at every point in space to compute the effective potential for the orbital at any point $\vec{r}$. The non local part of the *potential* is called the *exchange* term.

In the original Hartree-Fock (HF) method for atoms, the orbitals were numerically calculated. Notice that, when we use analytical basis sets, then the computation of the exchange terms is no more difficult then the computation of the Coulomb terms. This however, was not at all the case for numerical solutions: from this the "need" to find approximations which in turn led to the origin of the density functionals. However, in molecules the orbitals were described as a linear combination of atomic orbitals, the latter expressed as analytical functions, which were introduced by Zener[5] but are often referred to as Slater type orbitals (STO) or exponential type orbitals (ETO), following Slater. If we think of the STO's as an analytical basis set, then we can construct the Hartree-Fock orbitals directly in terms of STO's namely $\phi_i = \sum c_{ij}\chi_j$ where χ_j are members of the basis set $\chi_1, \chi_2, \ldots$. Roothaan[6,7] reformulated the Hartree-Fock numerical technique in terms of χ_i, a set of basis functions and this approach is popularly known as the *Analytic Hartree-Fock method* or the *expansion method*. In the following we shall be concerned with atomic systems, but it is a simple matter to use the same analyses for molecules simply considering that in the latter there is a different external potential due to more than one nucleus.

To start with, we shall present the detailed formulae and the equations of the Analytic Hartree-Fock method in section 2. In section 3 some of the currently available methods for the open shell Hartree-Fock will be briefly presented. The vector coupling coefficients for the open shell analytic Hartree-Fock method are also presented for atoms. Various types of basis functions, viz the Slater type orbitals, the Gaussian type orbitals and the Oscillatory type orbitals, the latter by Cacelli, Moccia and Carravetta[8] are presented in section 4. Orbital exponent optimization and contraction coefficients are also discussed in detail in section 4. Geometrical basis sets and their advantages are discussed in detail in section 5. In section 6 various integral formulae for each and every matrix element is presented. A discussion on correlation energy and pair correlation energy and its empirical evaluation for atoms via simple models is detailed in section 7 and followed in section 8 with a discussion on density functionals. In section 9 we give expressions for the $< r^n >$ and $< p^n >$ values and their significance. Finally, in section 10 we present a different approach to the HF equations, the *atomic finite element method*, based on the recent work by Flores *et al.*[9] The chapter ends with two appendices. In Appendix A we report the analytical expressions for special integrals used to estimate the Coulomb hole (see section 7). In Appendix B, in a summary way, we deal with the pseudopotential approach, since of interest in molecular computations.

2. The Analytic Hartree-Fock Method

The analytic Hartree-Fock method has been presented by Roothaan for closed shells,[6] by Roothaan for a single open shell,[7] by Huzinaga for two open shells,[10] by Roothaan and Bagus for many open shells of different symmetry.[11] A number of formulations for the expansion method are available in the literature for the general case of many open shells.[11,12]

In what follows we shall adhere closely to the presentation and choice of symbols adopted by Roothaan. The expectation value of the total Hamiltonian for a chosen electronic configuration of a system may be written as

$$E = 2\sum_k H_k + \sum_{kl}(2J_{kl} - K_{kl}) + f\left[2\sum_m H_m + f\sum_{mn}(2aJ_{mn} - bK_{mn}) + 2\sum_{km}(2J_{km} - K_{km})\right] \quad (6)$$

Here, k and l refer to orbitals of closed shells and m and n are open shell orbitals. The constants a and b are the vector coupling coefficients of Roothaan, the number f is in general the fractional occupation of the shell. Employing the variation principle and taking the functional derivative with respect to open and closed shell orbitals, we get the required self-consistent field equations. This open shell treatment permits symmetry orbitals as its solutions of the SCF equations, thus one may additionally impose this constraint. This means that orbitals are grouped in sets, each set transforming under symmetry operations according to an irreducible representation, λ, of the symmetry group. To make the best possible use of symmetry properties of the orbitals, the notation $\phi_{i\lambda\alpha}$ is introduced in lieu of ϕ_i.

The orbitals $\phi_{i\lambda\alpha}$ are characterized by an index λ, which indicates the symmetry species (for atoms, λ corresponds to the quantum number l), by an index α, which indicates the subspecies (the subspecies labels the individual members of the degenerate set that transforms according to the representation λ) and by an index i, which refers to the i th orbital of symmetry λ. Thus for the $2p_0$, $2p_{+1}$, and $2p_{-1}$ orbitals one has $i = 1$ and $\lambda = 1$ and $\alpha = -1$, 0 and $+1$. Indices $i\lambda\alpha$ correspond to the familiar *nlm* quantum numbers of atomic spectra. The orbital $\phi_{i\lambda\alpha}$ is expanded in terms of basis functions according to

$$\phi_{i\lambda\alpha} = \sum_p \chi_{p\lambda\alpha} C_{i\lambda p} \tag{7}$$

The subscript p refers to the pth basis function of symmetry λ and subspecie α.

Let us now give an expression for the total energy. It will turn out to be convenient to subdivide E into two components, $E^{(1)}$ for the one-electron operators and $E^{(2)}$ for the two-electron operators:

$$E = <\Psi| - \sum_i \left(\frac{1}{2}\nabla_i^2 + \frac{Z}{r_i}\right) |\Psi> +$$

$$<\Psi| \sum_{i>j} \frac{1}{r_{ij}} |\Psi> = E^{(1)} + E^{(2)} \tag{8}$$

In Eq. (8) the symbol Ψ refers to the Hartree-Fock function and does not designate the Ψ of Eq. (3); this unfortunate misuse of symbols is due to an insufficient number of type fonts. In the rest of this chapter Ψ will be used to represent a Hartree-Fock wave function.

We shall start by introducing the overlap matrix element $S_{\lambda pq}$ defined as

$$S_{\lambda pq} = \frac{1}{d_\lambda} \sum_\alpha < \chi_{p\lambda\alpha} \mid \chi_{q\lambda\alpha} > \tag{9}$$

where d_λ is the dimension of the irreducible representation λ.

The matrix elements of a given λ can be collected in a square matrix S_λ ; the total collection of all the $S_{\lambda pq}$ for every λ can be collected in a square matrix S. Whenever we shall need to consider the matrix elements of a matrix as a one-dimensional array, we shall call such an array a *supervector*. The expansion coefficients $C_{i\lambda p}$ can be collected in a p component column vector $C_{i\lambda}$. We shall assume that the vectors $C_{i\lambda}$ form an orthonormal set:

$$C_{i\lambda}^{\dagger}\, S_{\lambda} C_{j\lambda} = \delta_{ij} \tag{10}$$

We indicate with $N_{i\lambda}$ the occupation number of the shell $i\lambda$ also, $N_{c\lambda}$ and $N_{o\lambda}$ are the occupation numbers of the closed and open shells of symmetry λ respectively, and we shall define the shell, closed shell, open shell and total-density matrices:

$$D_{i\lambda} = N_{i\lambda}\, C_{i\lambda}\, C_{i\lambda}^{\dagger}\ ; \quad D_{c\lambda} = \sum_{k} D_{k\lambda}\ ; \quad D_{o\lambda} = \sum_{m} D_{m\lambda}\ ; \quad D_{T\lambda} = D_{c\lambda} + D_{o\lambda} \tag{11}$$

where $k = i \subset$ closed and $m = i \subset$ open. (It is noted that $C_{\lambda}^{\dagger}$ is a row vector and C_{λ} is a column vector, hence $C_{\lambda} C_{\lambda}^{\dagger}$ is a matrix).

We are finally in a position to compute $E^{(1)}$. Remembering Eq. (8), let us define the nuclear potential and the kinetic energy matrix elements as

$$U_{\lambda pq} = \frac{1}{d_{\lambda}} \sum_{\alpha} < \chi_{p\lambda\alpha} \mid \frac{1}{r} \mid \chi_{q\lambda\alpha} > \tag{12}$$

$$T_{\lambda pq} = -\frac{1}{d_{\lambda}} \sum_{\alpha} < \chi_{p\lambda\alpha} \mid \frac{1}{2} \nabla^{2} \mid \chi_{q\lambda\alpha} > \tag{13}$$

We can now define the matrices T_{λ}, U_{λ} and the total collection of matrix elements for all symmetries designated by T, U in the same way we have defined the matrices S_{λ} and S. In addition we can define the one-electron Hamiltonian $\boldsymbol{H}$, sometimes referred to as the bare-nuclear Hamiltonian,

$$H = -ZU + T \tag{14}$$

and

$$E^{(1)} = H^{\dagger} D_{T} \tag{15}$$

Following Roothaan,[7] the two-electron part of the energy may be defined in terms of the following integrals:

$$J_{\lambda pq,\,\mu rs} = \frac{1}{d_{\lambda} d_{\mu}} \sum_{\alpha\beta} \int\!\!\int \bar{\chi}_{p\lambda\alpha}(1) \chi_{q\lambda\alpha}(1) \frac{1}{r_{12}} \bar{\chi}_{r\mu\beta}(2) \chi_{s\mu\beta}(2) d\tau_{1} d\tau_{2} \tag{16}$$

$$K_{\lambda pq,\,\mu rs} = \frac{1}{d_{\lambda} d_{\mu}} \sum_{\alpha\beta} \int\!\!\int \bar{\chi}_{p\lambda\alpha}(1) \chi_{s\mu\beta}(1) \frac{1}{r_{12}} \bar{\chi}_{r\mu\beta}(2) \chi_{q\lambda\alpha}(2) d\tau_{1} d\tau_{2} \tag{17}$$

This symmetry factoring may be transparent due to the form of the Coulomb and exchange integrals viz.

$$J_{ij} = < ij \mid ij > \text{ and } K_{ij} = < ij \mid ji > \tag{18}$$

in the Hartree-Fock method. It may be noted that the supermatrices J and K are symmetrical for the exchange $\lambda pq \Leftrightarrow \mu rs$ and Hermitean for the simultaneous exchange $p \Leftrightarrow q$, $r \Leftrightarrow s$. In the case of actual calculations as may be seen from Eq. (16), $J_{\lambda pq,\, \mu rs}$ is symmetrical for the independent exchanges $p \Leftrightarrow q$ and $r \Leftrightarrow s$. This stronger symmetry is due to the fact that basis sets are chosen to consider only real coefficient vectors. Thus the necessary and sufficient set of elements $J_{\lambda pq,\, \mu rs}$ is the triangular part of a real symmetric matrix, the rows and columns of which are labeled with supervector indices λpq and μrs with $q \leq p$ and $r \leq s$. For $K_{\lambda pq,\, \mu rs}$, the symmetrized quantity viz.

$$\frac{1}{2}(K_{\lambda pq,\, \mu rs} + K_{\lambda pq,\, \mu sr}) \tag{19}$$

is used.

The above integrals are computed as supermatrices with indices λpq, μrs. From these two supermatrices two additional supermatrices are defined viz. $\mathbb{P}_{\lambda pq,\, \mu rs}$ and $\mathbb{Q}_{\lambda pq,\, \mu rs}$:

$$\mathbb{P}_{\lambda pq,\, \mu rs} = J_{\lambda pq,\, \mu rs} - \frac{1}{2} K_{\lambda pq,\, \mu rs} \tag{20}$$

and

$$\mathbb{Q}_{\lambda pq,\, \mu rs} = \alpha J_{\lambda pq,\, \mu rs} - \frac{1}{2} \beta K_{\lambda pq,\, \mu rs} \tag{21}$$

With these considerations, the general expression for the two-electron part of the energy, $E^{(2)}$, in supervector and supermatrix form for a general case is as follows:

$$E^{(2)} = \frac{1}{2} D_T^{\dagger} \mathbb{P} D_T - \frac{1}{2} D_o^{\dagger} \mathbb{Q} D_o, \tag{22}$$

where D_c is the sum of the open shell density matrices and $\mathbb{Q}$ is the equivalent to $\mathbb{P}$ generalized to the case of electrons in open shells. Thus the total energy may be simply written down as :

$$E = E^{(1)} + E^{(2)} \tag{23}$$

We shall now sketch how one obtains wave functions which satisfy the above energy expressions.

In order to obtain a Hartree-Fock function in the expansion method, one makes use of the variational method for the density matrices and of "good sense" in the choice of the basis set. Unfortunately "good sense" does not go too far and a process of trial and error in the choice of the *basis set* is needed; this process is generally referred to as *optimization of the orbital exponent* and, nowadays is carried out by the computer, instructed to compromise between efficiency for a specific case and complete generality of the procedure. A method for orbital exponent optimization will be presented later in this chapter. Let us assume we have settled on a given basis set.

We now ask for the *best function* that can be obtained for a given basis set, that is we ask for a variational function. Let us subject every orbital ϕ_i to a variation by an infinitesimal amount $\delta\phi_i$ yielding an energy variation δE; since the basis set is fixed, this is tantamount to an infinitesimal variation of the expansion coefficients $C_{i\lambda}$. The orbital ϕ_i always has to conform to the orthogonality condition. Therefore the variation $\delta C_{i\lambda p}$ is restricted by those conditions which are obtained by subjecting Eq. (10) to an infinitesimal variation.

The standard mathematical technique is to add to the δE the constraints which must be satisfied, multiplied by the Lagrange multipliers. Let us call $\delta E'$ the result of this addition. The problem is now that of finding the condition for $\delta E' = 0$ for any choice of the δC, without restriction, while at the same time giving suitable values to the Lagrange multipliers.

In general, we have Lagrange multipliers between orbitals of closed shells, between orbitals of open shells, and between orbitals of closed and open shells. The latter type of multiplier is eliminated by introducing operators R_c and R_o for closed and open shells, expressed in terms of P and Q, where

$$P = \mathbb{P}D_T \tag{24a}$$

$$Q = \mathbb{Q}D_o \tag{24b}$$

In addition, let us designate by F_c and F_o the quantities

$$F_{c,\,\lambda pq} = H_{\lambda pq} + P_{\lambda pq} + R_{o,\,\lambda pq} \tag{25}$$

and

$$F_{o,\,\lambda pq} = H_{\lambda pq} + P_{\lambda pq} - Q_{\lambda pq} + R_{c,\,\lambda pq} \tag{26}$$

The $R_{c\lambda}$ and $R_{o\lambda}$ are

$$R_{c\lambda} = \left(\frac{1}{N_{c\lambda} - N_{o\lambda}}\right) [S_\lambda D_{c\lambda} Q_\lambda + Q_\lambda D_{c\lambda} S_\lambda] \tag{27a}$$

$$R_{o\lambda} = \left(\frac{1}{N_{c\lambda} - N_{o\lambda}}\right) [S_\lambda D_{o\lambda} Q_\lambda + Q_\lambda D_{o\lambda} S_\lambda] \tag{27b}$$

By designating as $\varepsilon_{m\lambda}$ the diagonal elements we have, in compact notation

$$F_{c\lambda} C_\lambda = \varepsilon_\lambda S_\lambda C_\lambda \tag{28}$$

$$F_{o\lambda}C_{\lambda} = \varepsilon_{\lambda}S_{\lambda}C_{\lambda} \tag{29}$$

These are the Hartree-Fock equations for closed and open shells. They are pseudo eigenvalue equations, because the matrices F_c and F_o depend on the solution **C**. Self-consistency is achieved when the same vectors, from which these matrices are constructed according to Eqs. (25) and (26), are solutions of Eqs. (28) and (29). All this is rather straightforward and even rather simple: once the orbitals are obtained, the energy can be computed using Eqs. (8), (15) and (22). However, from the above derivation one can immediately appreciate the problems implicit in computational quantum chemistry. Generally, for atoms, convergence problems in the SCF procedure may arise. These problems may be avoided by solving the eigenvalue equations for each symmetry separately, and/or by extrapolation procedures for the solution vectors **C** at every iteration step.

3. Open Shell Methods and Roothaan Vector Coupling Coefficients

Generally, the Hartree-Fock wave function for a system consisting of electrons with a definite total azimuthal angular momentum L and total spin angular momentum S needs to be approximated by a linear combination of Slater determinants containing a fixed set of orthonormal orbitals.

$$\Psi_{HF} = \Sigma C_p D_p \tag{30}$$

$$D_p = \det\{\phi_1\alpha, \phi_1\beta, \ldots \phi_k\alpha, \phi_k\beta, \phi_{k+1}\chi_{1,p} \cdots \phi_{k+l}\chi_{l,p}\} \tag{31}$$

Since the D_p form an orthonormal set and $\Sigma|C_p|^2 = 1$, one needs to choose C_p such that Ψ_{HF} is an eigenfunction of L^2, S^2, L_z and S_z. The energy expression for such a wave function may be written in a variety of ways. For instance, Davidson's open shell formulation[12] requires:

$$E = \Sigma n_i h_i + \Sigma \frac{1}{2} n_i n_j (J_{ij} - \frac{1}{2}\mathrm{K}_{ij}) + \frac{1}{4}\sum_{k<i}^{k+l}\sum_{k<j}^{k+l}\beta_{ij}\mathrm{K}_{ij} \tag{32}$$

where

$$h_i = <\phi_i|h|\phi_i> \tag{33}$$

$$J_{ij} = [\overline{\phi}_i(1)\phi_i(1) \mid \frac{1}{r_{12}} \mid \overline{\phi}_j(2)\phi_j(2)] \tag{34}$$

$$\mathrm{K}_{ij} = [\overline{\phi}_i(1)\phi_j(1) \mid \frac{1}{r_{12}} \mid \overline{\phi}_j(2)\phi_i(2)] \tag{35}$$

with $n_i = (0, 1 \text{ or } 2)$ the occupation number of ϕ_i and

$$\beta_{ij} = \beta_{ji} = 1 - 2\sum_{p,q} \overline{C}_p C_q < \chi_{ip} \mid \chi_{jq} > \prod_{l \neq i,j} < \chi_{lp} \mid \chi_{lq} > \quad (36)$$

Another quite generalized form for the open shell theory of Eq. (32) is the approach of Goddard and coworkers[13] derived from their Generalized Valence Bond method. Here the total energy is written as:

$$E = 2\sum_{i=1}^{n} f_i h_i + \sum_{i,j=1}^{n} (a_{ij} J_{ij} + b_{ij} K_{ij}) \quad (37)$$

where n is number of spatial orbitals.

For a closed shell system, the values of f, a and b are given as follows

$$f_1 = f_2 = 1$$

$$a_{11} = a_{22} = a_{12} = 2 \quad (38)$$

$$b_{11} = b_{22} = b_{12} = -1$$

For a general open shell case, the a_{ij} and b_{ij} need to be evaluated case by case by the user of the program, although there may exist some semi-automatic and automatic techniques to obtain the coupling coefficients. For some simple cases, e.g., the high spin open shell cases, we have

$$f_2 = \frac{1}{2} \quad a_{12} = a_{22} = 0 \quad \text{and} \quad b_{12} = b_{22} = -1. \quad (39)$$

Here the subscript 1 refers to the closed shell orbital and 2 to the open shell orbital.

In the open shell formalism of Roothaan one deals directly with shells, density matrices and basis functions. For atoms, the open shell supermatrix is defined as follows[5]

$$\mathbb{Q}_{\lambda pq,\, \mu rs} = \sum_{v=0}^{2(\lambda,\,\mu)_<} \alpha_{\lambda\mu v} J^{v}_{\lambda pq,\, \mu rs} - \frac{1}{2} \sum_{v=|\lambda-\mu|}^{\lambda+\mu} \beta_{\lambda\mu v} K^{v}_{\lambda pq,\, \mu rs} \quad (40)$$

As pointed out by Malli and Olive,[14] one can obtain the vector coupling coefficients from the Slater coefficients $f_v(\lambda, \mu)$ and $g_v(\lambda, \mu)$. They observed that for the $\lambda = \mu$ only the combination $\alpha_{\lambda\lambda v} - \beta_{\lambda\lambda v}/2$ matters; λ and μ refer to symmetry and v is defined in Eq. (40) and refers to the specific radial integral resulting from the angular expansion of $1/r_{12}$ and shall be defined later in Section 6. In addition, making use of this freedom and imposing

$$\alpha_{\lambda\lambda\nu} = 0 \tag{41}$$

the following relations were derived[14]:

$$\alpha_{\lambda\mu o} = 0 \tag{42}$$

$$\alpha_{\lambda\mu\nu} = -f_\nu(\lambda, \mu)/N_\lambda N_\mu, \quad \nu \neq 0 \tag{43}$$

$$\beta_{\lambda\lambda o} = A_{\lambda\lambda o} - 2/N_\lambda \tag{44}$$

$$\beta_{\lambda\lambda\nu} = A_{\lambda\lambda\nu} + 4f_\nu(\lambda, \lambda)/N_\lambda^2, \quad \nu \neq 0 \tag{45}$$

$$\beta_{\lambda\mu\nu} = A_{\lambda\mu\nu} + 2g_\nu(\lambda, \mu)/N_\lambda N_\mu \;\; \lambda \neq \mu. \tag{46}$$

where

$$A_{\lambda\mu\nu} = A_{-\lambda+\mu+\nu}A_{\lambda-\mu+\nu}A_{\lambda+\mu-\nu}/(\lambda + \mu + \nu + 1)A_{\lambda+\mu+\nu},$$

$$A_\rho = \rho!/\left[(\tfrac{1}{2}\rho)!\right]^2, \qquad \rho = 0,\ 2,\ 4,\ \ldots$$

These equations have been then used to calculate the coefficients α and β from the tabulations of Slater coefficients. A detailed tabulation of $\alpha_{\lambda\mu\nu}$ and $\beta_{\lambda\mu\nu}$ can be found in a file of a revised version of the computer program referred to in Ref. 15.

However, it should be remarked that the vector coupling coefficients derived above depend on the form of the basis functions used in representing the atomic orbitals, $\chi(\vec{r}) = R_{n\lambda}(r)Y_{\lambda\alpha}(\theta, \phi)$. Thus, in order to use the cartesian basis functions we have slightly modified the definition for the open shell supermatrix, $\mathbb{Q}$.

$$\mathbb{Q} = J^a_{\lambda pq,\,\mu rs} - \frac{1}{2}K^b_{\lambda pq,\,\mu rs} \tag{47}$$

where

$$J^a_{\lambda pq,\,\mu rs} = (d_\lambda d_\mu)^{-1}\sum_{\alpha,\beta} a^{\alpha\beta}_{\lambda\mu}\int\!\!\int \bar{\chi}_{p\lambda\alpha}(1)\chi_{q\lambda\alpha}(1)\frac{1}{r_{12}}\bar{\chi}_{r\mu\beta}(2)\chi_{s\mu\beta}(2)d\tau_1 d\tau_2 \tag{48}$$

$$K^b_{\lambda pq,\,\mu rs} = (d_\lambda d_\mu)^{-1}\sum_{\alpha,\beta} b^{\alpha\beta}_{\lambda\mu}\int\!\!\int \bar{\chi}_{p\lambda\alpha}(1)\chi_{s\mu\beta}(1)\frac{1}{r_{12}}\bar{\chi}_{r\mu\beta}(2)\chi_{q\lambda\alpha}(2)d\tau_1 d\tau_2. \tag{49}$$

The $a^{\alpha\beta}_{\lambda\mu}$ and $b^{\alpha\beta}_{\lambda\mu}$ are listed for open s and p shells in Table I and some of the ground states for the transition elements are presented in Table II. Note that the vector coupling coefficients for these and many more cases are available in a data set of the Hartree-Fock atomic program in MOTECC-89 and MOTECC-90. For

Table I. Vector coupling coefficients for s, p^n, for s, p^n and sp^n and sp^n configurations.

Config.	State	a_{00}^{00}	b_{00}^{00}	$a_{10}^{\alpha 0}$	$b_{10}^{\alpha 0}$	$a_{11}^{\alpha\beta}$	$b_{11}^{\alpha\beta}$
					$\{\alpha, \beta = -1, 0, 1\}$		
s	^2S	0	−1				
p	^2P					1	1
p^2	^3P					0.25	−0.5
	^1D					0.55	1.3
	^1S					1.0	4.0
p^3	^4S					0	−1
	^2D					0.2	0.2
	^2P					$0.3333\overline{3}$	1
p^4	^3P					0.0625	−0.125
	^1D					0.1375	0.025
	^1S					0.25	1.0
p^5	^2P					0.04	0.04
sp	^3P	0	−1	0	−1	1	1
	^1P	0	−1	0	3	1	1
sp^2	^4P	0	−1	0	−1	0.25	−0.5
	^2D	0	−1	0	0	0.55	1.3
	^2S	0	−1	0	0	1.0	4.0
	^2P	0	−1	0	2	0.25	−0.5
sp^3	^5S	0	−1	0	−1	0	−1
	^3D	0	−1	0	$-0.333\overline{3}$	0.2	0.2
	^3P	0	−1	0	$-0.333\overline{3}$	$0.333\overline{3}$	1
	^1D	0	−1	0	1	0.2	0.2
	^3S	0	−1	0	$-1.666\overline{6}$	0	−1
	^1P	0	−1	0	1	$0.333\overline{3}$	1
sp^4	^4P	0	−1	0	−0.5	0.0625	−0.125
	^2D	0	−1	0	0	0.1375	0.025
	^2S	0	−1	0	0	0.25	1.0
	^2P	0	−1	0	1	0.0625	−0.125
sp^5	^3P	0	−1	0	−0.2	0.04	0.04
	^1P	0	−1	0	0.6	0.04	0.04

(a) $0.333\overline{3}$ means 0.333333....
(b) For p-shell α can take values −1, 0 and 1.

additional information see the MOTECC-89 Input/Output Documentation Manual.[16]

4. Basis Sets for Atomic Computations

An atomic orbital can be expressed as a linear combination of functions of a basis set. Ideally we wish a basis set yielding compact mathematical expressions for the two-electron integrals so as to reduce substantially the computational time. For the Slater type orbitals (STOs) and Gaussian type orbitals (GTOs) the expressions for the various integrals required in an atomic energy calculation are available[11,15] and it may be seen that they have the simple expressions reported in section 6. We first define these basis sets and discuss some of their qualitative aspects.

Slater Type Orbitals. The dominant profile of an atomic orbital is well represented by functions of the form

$$\chi_{\lambda p \alpha} = r^{n_\lambda - 1} e^{-\zeta_{\lambda p} r} \cdot Y_{\lambda \alpha}(\theta, \phi) \cdot N_{\lambda p}$$
$$\text{with } N_{\lambda p} = [(2n_\lambda)!]^{-1/2} (2\zeta_{\lambda p})^{n_\lambda + 1/2} \tag{50}$$

Here $Y_{\lambda\alpha}(\theta, \phi)$'s are normalized spherical harmonics. These functions have been extensively employed in atomic structure calculations. The value of the orbital exponents can be inferred from X-ray data, assuming a single STO per orbital (apart orthogonality corrections); such orbital exponent values can be obtained from the "Slater rules" and from this the origin of the "Slater orbital" designation.[1]

Table II. Vector coupling coefficients for some states of the d^n configurations.

Config.	State	$a_{22}^{\alpha\beta}$	$b_{22}^{\alpha\beta}$	$b_{22}^{\pm 2, \pm 2}$
		$\{\alpha, \beta = \pm 2, \pm 1, 0\}$	$\{\alpha = \pm 2, \pm 1, 0, \beta = \pm 1, 0\}$	
d^1	^2D	1	1	1
d^2	^3F	$0.85714\overline{285714}$	−4.42857142857142816	18.0714285714285694
d^3	^4F	0.380952380952381028	−2.52380952380952350	7.47619047619047561
d^4	^5D	0.0625	−0.875	−0.875
d^5	^6S	0	−1	−1
d^6	^5D	$0.0277\overline{7}$	$-0.388\overline{8}$	$-0.388\overline{8}$
d^7	^4F	0.0699708454810496622	−0.463556851311953233	1.37317784256559783
d^8	^3F	0.0535714285714286864	−0.276785714285713969	1.12946428571428514
d^9	^2D	0.012345679	0.012345679	0.012345679

(a) $0.0277\overline{7}$ means 0.0277777 ...

(b) These coefficients are rational numbers however they have been calculated on a computer. Therefore, only 16 significant figures are correct.

However, the value of the orbital exponent can be derived by optimization of the orbital exponent, as done by Clementi and Raimondi.[17] These rules are still of use today to obtain a "feeling" of an orbital exponent. Note also that a Slater exponent is proportional to the square root of a gaussian exponent. Roothaan and Bagus[11] designed and coded a general and efficient atomic SCF program later rewritten in FORTRAN by Roos *et al.* and eventually documented for general distribution.[15,16]

With these computer programs the ground and excited states for the lowest electronic configurations of the atoms and many ions (iso-electronic series) were computed by Clementi[18] for the first row (He to Ne), second row (Na to Ar) and third row (K to Kr). An equivalent tabulation for negative ions was later provided.[19] All these data were collected into Atomic Tables.[20] Later this work was extended to the forth row atoms and ions by Clementi and Roetti[21] and a new set of Tables of Atomic Functions was published.[22] The main goal was to obtain information on the values of the correlation energy (by subtraction of Hartree-Fock energies from the experimental ones, relativistically corrected) and to provide basis sets for molecular computations.

The following terminology is commonly used to describe the size of the STO basis sets: single zeta, double zeta, and the extended. The single zeta STO basis set is a minimal basis set, where the number of STOs is equal to the number of subshells in an atom. This is to convey that there is a single exponential (zeta) per subshell. In the same way, if there are two exponentials per subshell the collection of STOs is called a double zeta basis and, obviously, this is no longer a minimal basis set since twice the number of STOs are employed to express a set of atomic subshells. We note that stepping from single-zeta to multiple-zeta -the latter being an obvious starting point nowadays- was tantamount to abandon the LCAO approximation and adopt the Hartree-Fock model: This happened at M.I.T., the Universities of Chicago and Cambridge (U.K.) in the late '50s. The extended basis set contains an adequate number of STOs to nearly reach the Hartree-Fock energy limit. It should be noted that — to be semantically correct — one should use the expression "Hartree-Fock energy" and "Hartree-Fock wavefunction" only when such limit is reached; less accurate work should be referred to as "Hartree-Fock *type*" or simply "SCF."

We recall that, for the hydrogen atom a single STO is adequate to express the wave function exactly; moreover, the Kato cusp condition[23] and asymptotic behavior of the exact one-electron density can be easily imposed onto a linear combination of STOs; therefore, one may conclude that STOs are naturally well suited to describe accurately these regions of the wave function.[18f]

Oscillating Slater Type Orbitals. The basis sets presented here have been obtained from the studies in Refs. 8, 24-27. It should be noted that while a single STO is enough to describe the hydrogen atom ground state, a linear combination of

STOs is involved in the exact expression of any discrete excited state. The hydrogenic radial function can be written as

$$R_{nl}(r) = \left[\left(\frac{2\xi}{n} \right)^3 \frac{(n-l-1)!}{2n[(n+l)!]^3} \right]^{1/2} \left(\frac{2\xi r}{n} \right)^l L_{n+l}^{2l+1}\left(\frac{2\xi r}{n} \right) e^{-\xi r/n} \tag{51}$$

where L is a generalized Laguerre polynomial

$$L_{n+l}^{2l+1}(\rho) = \sum_{j=0}^{n-l-1} \frac{[(n+l)!]^2(-1)^{j+1}\rho^j}{(n-l-1-j)!\,(2l+1+j)!j!} \tag{52}$$

The linear combination of several STOs is evidently necessary to obtain the nodes of the radial function whose number increases with the principal quantum number n. The shape of the radial function is even more complicated if we consider the continuum excited states, i.e. those corresponding to positive values of the energy, of the hydrogen atom that are commonly called regular Coulomb waves[28]

$$R_{El}(r) =$$

$$\sqrt{\frac{2k}{\pi}} \frac{e^{\frac{\pi}{2k}}}{(2l+1)!} |\Gamma(l+1-\frac{i}{k})| (2kr)^l e^{-ikr} F\left(\frac{i}{k}+l+1, 2l+2, 2ikr\right) \tag{53}$$

where $k = \sqrt{2E}$, F is the confluent hypergeometric function and normalization per unit energy range is adopted.

A continuum orbital extends to infinity of course, but we can think of giving a reasonable description of it in a region more or less close to the nucleus, by a projection on a basis set of square integrable functions. The number of nodes for unit range of r depends on the continuous energy value E that takes the place of the principal quantum number n for discrete excited states.

A Slater-type basis set does not seem adequate to reproduce the behavior of a continuum orbital; a more suitable function would be

$$R_{n\xi k}(r) = A_{n\xi k} r^{n-1} e^{-\xi r} \cos(kr) \tag{54}$$

This kind of oscillating basis function was actually introduced a few years ago for a better description of atomic continua as well as molecular continua in the one-center approximation. It proved to be a rather efficient basis set in the numerical implementation of L^2 methods for the study of several processes involving continuous spectra like photoionization,[8] Auger emission,[24] shake-off,[25] resonances in photoelectron spectra,[26] as well as properties strongly depending on an adequate description of the electronic continuum like polarizability.[27] It can also be foreseen that the high flexibility of such basis functions may be a valuable advantage in calculations like CI and MCSCF that include electronic correlation effects.

As already observed, an ideal basis set should not only provide a good description of the atomic orbitals, it should also imply compact mathematical expressions for the main integrals, principally the two-electron integrals, that are involved in a numerical application. When the oscillating Slater functions in Eq. (54) have the same origin the basis set is available for atomic and molecular calculations in the one center approximation. Several numerical applications of the formulas that will be presented in next sections have shown that, despite the apparently complicated expression of the two-electron integral, the computational effort going from Slater to oscillating Slater basis sets increases on the average of a factor 3 or 4. It should be noted that expression Eq. (54) reduces to a standard Slater type function when k=0; the inclusion of hydrogenic functions like the one in Eq. (51) leads finally to a quite complete set of square integrable basis functions to be employed in quantum calculations of both bound and scattering states and generally when a satisfactory global description of the electronic spectrum is required.

Gaussian Type Orbitals. Yet another choice of basis functions are the Gaussian type orbitals[29,30] (GTOs), viz.

$$\chi_{\lambda p\alpha} = r^{\lambda} e^{-\alpha_p r^2} Y_{\lambda\alpha}(\theta, \phi) \cdot N_{\lambda p}$$

$$N_{\lambda p} = 2^{\lambda+2}[(2\lambda + 1)!!]^{-1/2}(2\pi)^{-1/4}\alpha^{(2\lambda+3)/2} \tag{55}$$

In the GTOs the power of r is restricted to the lowest for each symmetry. This restriction allows one to naturally take over the atomic basis sets for molecular calculations in which the cartesian representation is widely used.

$$\chi_{lmn}(x, y, z) = N(l, m, n; \alpha)x^l y^m z^n e^{-\alpha r^2} \tag{56}$$

$$N(l, m, n; \alpha) = (2/\pi)^{3/4}[(2l - 1)!!(2m - 1)!!(2n - 1)!!]^{-1/2}\alpha^{[l+m+n+(3/2)]/2}$$

And explicitly they are

$$
\begin{aligned}
&s\text{-type: } l+m+n=0 \quad e^{-\alpha r^2}\\
&p\text{-type: } l+m+n=1 \quad (x, y, z)e^{-\alpha r^2}\\
&d\text{-type: } l+m+n=2 \quad (x^2, xy, xz, y^2, yz, z^2)e^{-\alpha r^2}\\
&f\text{-type: } l+m+n=3 \quad (x^3, x^2y, x^2z, xy^2, xyz, xz^2, y^3, y^2z, yz^2, z^3)e^{-\alpha r^2}\\
&g\text{-type: } l+m+n=4 \quad (x^4, x^3y, x^3z, x^2y^2, x^2yz, x^2z^2, xy^3, xy^2z, xyz^2,\\
&\qquad\qquad xz^3, y^4, y^3z, y^2z^2, yz^3, z^4)e^{-\alpha r^2}\\
&h\text{-type: } l+m+n=5 \quad (x^5, x^4y, x^4z, x^3y^2, x^3yz, x^3z^2, x^2y^3, x^2y^2z,\\
&\qquad\qquad x^2yz^2, x^2z^3, xy^4, xy^3z, xy^2z^2, xyz^3,\\
&\qquad\qquad xz^4, y^5, y^4z, y^3z^2, y^2z^3, yz^4, z^5)e^{-\alpha r^2}
\end{aligned}
\tag{57}
$$

The choice of Cartesian primitive GTOs has been made for the reason that all molecular integrals over them are simple and straightforward. Note that there are redundancies in the d and higher symmetry orbitals for the cartesian representation. For example, the d-functions in Eq. (52) are six-membered. Of course, one may extract six mutually orthogonal functions, viz.

$$
\begin{aligned}
&(2z^2 - x^2 - y^2,\ x^2 - y^2,\ xy,\ yz,\ xz)e^{-\alpha r^2}\\
&\text{and} \quad (x^2 + y^2 + z^2)e^{-\alpha r^2}.
\end{aligned}
\tag{58}
$$

Here the last function is a 3s-type function which needs to be considered along with the s-type functions. To guarantee a one-one correspondence between a calculation performed with spherical gaussian functions and another with cartesian type orbitals, one needs to consider the five orthogonal d-functions and discard the extra 3s function. An identical strategy is necessary for the 10-membered f functions from which the seven $4f$ orthogonal functions are constructed.

In the literature we can find a large number of basis sets which are supposed to help the chemist in performing chemical computations on molecules. This viewpoint has its historical base in the LCAO approximation, which is surviving in a time when we have extended basis sets. It is instructive and sobering to recall that no self-respecting quantum chemist in the sixties would have avoided optimizing the orbital exponents of his molecular basis set (or at least discuss the point), even if atomic basis sets were available.[31] Actually, this was done even in the late fifties (see, for example, B. Ransil, Ref. 32). That was the time of the STO and the beginning of the GTO.

In the seventies a "McDonald's Fast Food" mentality had spread among users of quantum chemistry programs, and the idea of optimizing a basis set (or part of it) was abandoned because of the heavy labor involved. In addition, it should be noted that the chemical literature, as a result, was flooded with computations which essentially compared experiments with basis set superposition error data. In this section we do not wish to review the literature on basis sets since this has already been done,[33-36] but rather we shall recall criteria which could help toward a critical choice among basis sets.

The question is, therefore, "how can we asses the quality of atomic basis sets to be used in molecular computations?" One criterion comes immediately to mind, namely how well the basis set reproduces atoms. However, this is a necessary but not sufficient condition since molecular basis sets must be balanced,[31] that is, have proper polarization functions on all the atoms of the molecule and proper distribution between polarization functions and those functions needed to reproduce the atom being careful to use basis sets which minimize the basis set superposition error (see Chapter 7). Naively one could attempt to simplify the problem by misrepresenting the inner shell electrons using a small basis set; this is actually the "best recipe" for ensuring a large basis set superposition *error!* A large body of papers have fallen into this trap. We shall return on this subject in Chapter 7.

In Table III, we provide the Hartree-Fock energies for the ground state configuration of neutral atoms. These energies are taken from the Atomic Tables by Clementi and Roetti.[22] We also tabulate the single-zeta and double-zeta energies using STFs. We then provide in several tables the atomic energy of a few of the basis sets which are present in the library of MOTECC-89. Energies are given for both the uncontracted and contracted basis sets (almost all molecular computations make use of contracted basis set techniques as introduced and applied originally by Clementi and Davies, see Ref. 37).

It might be of interest to recall that some of the Gaussian basis sets which did, and sometimes still do, enjoy popularity in the seventies and eighties are remarkably poor and represent a notable step backward compared to the single-zeta (Clementi and Raimondi, Ref. 17) and the double-zeta basis sets.[22] Since, in the sixties, the single-zeta STOs were also called "minimal" basis sets, some of the popular GTOs, which yield poorer energies, were referred to as "sub-minimal" (see E. Clementi, Ref. 36).

The basis set superposition error is still with us today, as it was yesterday, and is a very much relevant problem in modern computational chemistry. Thus it is once more important to optimize the basis sets in the molecular computation. This is one of the motivations for using the geometrical basis sets introduced by Clementi and Corongiu.[38] This basis set originates from some early systematic work by F. Van Duijeneveldt at the IBM San Jose Laboratory[39]; in optimizing basis sets for the atoms He to Ne, it was found that the best exponents for large basis sets approaching saturation have the orbital exponent equally spaced if plotted on a

Table III. Hartree-Fock type total energies (in hartrees) for atoms with single zeta, double zeta and extended STO basis sets.(a)

Z	Atom	State	Configuration	Extended	Single Zeta	Double Zeta
2	He	^1S	1s(2)	-2.8616799	-2.8476562	-2.8616726
3	Li	^2S	...2s(1)	-7.4327257	-7.4184820	-7.4327214
4	Be	^1S	...2s(2)	-14.573021	-14.556740	-14.572369
5	B	^2P	...2s(2)2p(1)	-24.529057	-24.498369	-24.527920
6	C	^3P	...2s(2)2p(2)	-37.688612	-37.622389	-37.686749
7	N	^4S	...2s(2)2p(3)	-54.400924	-54.268900	-54.397951
8	O	^3P	...2s(2)2p(4)	-74.809370	-74.540365	-74.804323
9	F	^2P	...2s(2)2p(5)	-99.409300	-98.942113	-99.401309
10	Ne	^1S	...2s(2)2p(6)	-128.54705	-127.81218	-128.53511
11	Na	^2S	...3s(1)	-161.85890	-161.12392	-161.84999
12	Mg	^1S	...3s(2)	-199.61461	-198.85779	-199.60701
13	Al	^2P	...3s(2)3p(1)	-241.87668	-241.15376	-241.87307
14	Si	^3P	...3s(2)3p(2)	-288.85431	-288.08996	-288.85116
15	P	^4S	...3s(2)3p(3)	-340.71869	-339.90988	-340.71595
16	S	^3P	...3s(2)3p(4)	-397.50485	-396.62762	-397.50229
17	Cl	^2P	...3s(2)3p(5)	-459.48187	-458.52369	-459.47960
18	Ar	^1S	...3s(2)3p(6)	-526.81739	-525.76525	-526.81511
19	K	^2S	...4s(1)	-599.16453	-598.08987	-599.16241
20	Ca	^1S	...4s(2)	-676.75803	-675.63390	-676.75594
21	Sc	^2D	...4s(2)3d(1)	-759.73555	-758.40414	-759.72637
22	Ti	^3F	...4s(2)3d(2)	-848.40575	-846.81561	-848.38875
23	V	^4F	...4s(2)3d(3)	-942.88420	-940.97197	-942.85728
24	Cr	^5D	...4s(2)3d(4)	-1043.3095	-1041.0063	-1043.2900

(Continued)

Table III (cont.). Hartree-Fock type total energies (in hartrees) for atoms with single zeta, double zeta and extended STO basis sets.(a)

Z	Atom	State	Configuration	Extended	Single Zeta	Double Zeta
25	Mn	^6S	...4s(2)3d(5)	-1149.8657	-1147.1067	-1149.8140
26	Fe	^5D	...4s(2)3d(6)	-1262.4432	-1259.0855	-1262.3715
27	Co	^4F	...4s(2)3d(7)	-1381.4142	-1377.3744	-1381.3205
28	Ni	^3F	...4s(2)3d(8)	-1506.8705	-1502.0487	-1506.7517
29	Cu	^2S	...4s(2)3d(9)	-1638.9628	-1632.3354	-1638.7496
30	Zn	^1S	...4s(2)3d(10)	-1777.8477	-1771.1509	-1777.6699
31	Ga	^2P	...4s(2)3d(10)4p(1)	-1923.2604	-1916.5167	-1923.1110
32	Ge	^3P	...4s(2)3d(10)4p(2)	-2075.3591	-2068.5139	-2075.2284
33	As	^4S	...4s(2)3d(10)4p(3)	-2234.2382	-2227.2649	-2234.1207
34	Se	^3P	...4s(2)3d(10)4p(4)	-2399.8658	-2392.7274	-2399.7563
35	Br	^2P	...4s(2)3d(10)4p(5)	-2572.4408	-2565.1131	-2572.3415
36	Kr	^1S	...4s(2)3d(10)4p(6)	-2752.0546	-2744.5192	-2751.9613
37	Rb	^2S	...5s(1)	-2938.3470	-2930.6931	-2938.2708
38	Sr	^1S	...5s(2)	-3131.5379	-3123.7176	-3131.4652
39	Y	^2D	...5s(2)4d(1)	-3331.6712	-3324.7806	-3331.6538
40	Zr	^3F	...5s(2)4d(2)	-3538.9821	-3531.3181	-3538.9632
41	Nb	^6D	...5s(1)4d(4)	-3753.5845	-3745.4826	-3753.5591
42	Mo	^5D	...5s(2)4d(4)	-3975.4280	-3967.0398	-3975.4131
43	Tc	^6S	...5s(2)4d(5)	-4204.7753	-4196.0536	-4204.7590
44	Ru	^5D	...5s(2)4d(6)	-4441.4749	-4432.3604	-4441.4569
45	Rh	^4F	...5s(1)4d(8)	-4685.8833	-4676.2637	-4685.8548
46	Pd	^3F	...5s(2)4d(8)	-4937.7712	-4927.8059	-4937.7504
47	Ag	^2D	...5s(2)4d(9)	-5197.5052	-5187.0705	-5197.4836

(Continued)

Table III (cont.). Hartree-Fock type total energies (in hartrees) for atoms with single zeta, double zeta and extended STO basis sets.(a)

Z	Atom	State	Configuration	Extended	Single Zeta	Double Zeta
48	Cd	^1S	...5s(2)4d(10)	-5465.1178	-5454.1908	-5465.0971
49	In	^2P	...5s(2)4d(10)5p(1)	-5740.1570	-5729.0986	-5740.1392
50	Sn	^3P	...5s(2)4d(10)5p(2)	-6022.9220	-6011.6720	-6022.9057
51	Sb	^4S	...5s(2)4d(10)5p(3)	-6313.4755	-6302.0043	-6313.4618
52	Te	^3P	...5s(2)4d(10)5p(4)	-6611.7748	-6600.0387	-6611.7621
53	I	^2P	...5s(2)4d(10)5p(5)	-6917.9727	-6905.9462	-6917.9602
54	Xe	^1S	...5s(2)4d(10)5p(6)	-7232.1302	-7219.7923	-7232.1189

(a) All values in Table III are from E. Clementi and C. Roetti, At. Data and Nucl. Data Tables, **14**, 177 (1974).

semi-log scale.[39] The well known, even-tempered basis set (see Ref. 40) is a restatement of Van Duijeneveldt's findings.

Later some other sets, called universal, were introduced in the literature;[41-42] in these sets the orbital exponents are related by a power law rather than by a semi-log relation. These attempts, however, were only able to yield reasonable results for a very few atoms (three or four) along the periodic table and, thus, the "universality" was mostly a desire. A step forward was accomplished by Clementi and Corongiu[38] when they proposed the so-called "geometrical basis sets" which hold up reasonably well, with the same value of α_0 and C, from Z=2 to Z=38 including some positive and negative ions (see section 5 for definitions). In addition, these sets have been proven to work notably well for relativistic calculations for atoms as shown by Mohanty and Clementi.[43] Thus it appears that geometrical basis sets provide a reasonable alternative to perform atomic computations for the entire periodic system. In addition, they provide a reasonable and systematic avenue for orbital exponent optimization even in polyatomic molecules. Let us now define in detail the optimization of orbital exponents and how contraction coefficients are obtained before returning back to geometrical basis sets.

Optimization of Orbital Exponents

The description of the optimization method is obtained from Ref. 44. The optimization of exponents is performed in two steps. Suppose that $n(\leq 20)$ exponents $\zeta_1, \zeta_2, ... \zeta_n$ are to be optimized. The first step consists of a one-dimensional optimization of each of the exponents in the same order as they are given. If the increment to ζ_i is $\Delta\zeta_i$ (given as input), it is required to go through the SCF procedure for the values

$$\zeta_i + m\Delta\zeta_i \;\; (m = 0,(\pm)1,(\mp)2 \ldots) \tag{59}$$

until three subsequent values

$$\zeta_i + (m-1)\Delta\zeta_i; \;\; \zeta_i + m\Delta\zeta_i; \;\; \zeta_i + (m+1)\Delta\zeta_i, \tag{60}$$

with the total energy $E_m < E_{m-1}$ and $E_m < E_{m+1}$ are found. A parabola $E(\zeta_i)$ is then fitted to these values of ζ_i and E, and the "optimized" value of ζ_i is found at the minimum of this parabola. This procedure is repeated for $i = 1, 2 \ldots n$.

The complete one-dimensional optimization procedure described above is repeated p (given as input) times. Suppose that the exponents obtained in the $(p-1)^{\text{th}}$ and p^{th} optimization are $\zeta_i^{(p-1)}$ and $\zeta_i^{(p)}$ $(i = 1, \ldots n)$ respectively, with the differences $\Delta\zeta_i^{(p)}$.

The second step uses these differences as the increments in a simultaneous variation of the orbital exponents ζ_i. The problem is thus solved for the following exponent values:

$$\zeta_1^{(p)} + m\Delta\zeta_1^{(p)}, \; \zeta_2^{(p)} + m\Delta\zeta_2^{(p)} \ldots, \; \zeta_n^{(p)} + m\Delta\zeta_n^{(p)}, \tag{61}$$

where $m = 1, 2 \ldots$ etc. The procedure is continued until $E_{m+1} > E_m$ and $E_m < E_{m-1}$. A parabola $E(m)$ is fitted to these values on m and E, and the "optimized" value for m is obtained from the minimum of this parabola. This value of m then gives the optimized exponents. A final calculation is made with these exponents.

The complete optimization procedure (step one plus step two) can be repeated as many times as desired. This is controlled by another parameter j. If j = 0, only the first step (one-dimensional optimization) is performed. If j > 0, step one and step two are performed j times. Experience shows that good convergence usually can be obtained with p = 2 and j = 2, at least for up to 5 exponents.

Contraction Coefficients

The computational cost in obtaining, storing and retrieving the multi-center, two-electron integrals employed in a molecular calculation is one of the main reasons for the introduction of contracted gaussian functions. Typically, the number of two-electron integrals which need to be stored increases as N^3 to N^4 for a given basis set of size N. There are many methods of determining contraction coefficients. For instance, one of the most commonly employed techniques is a least squares fitting to the numerical Hartree-Fock orbitals or the STO functions, after which one generally does an energy minimization to obtain atomic vectors. On the other hand, one may obtain the contraction coefficients by a nonlinear optimization technique (the same as used for exponent variation) which was employed earlier. In later versions of the computer program a quite different approach has been employed (see Ref. 44) and is described verbatim in what follows:

The traditional way of selecting the contracted functions from a given set of primitive gaussian functions relies essentially on good sense, namely the primitive functions, with predominant SCF expansion coefficients in a given orbital, are selected as those functions to be used to describe that orbital (with the contracted functions).

This preliminary selection yields very acceptable results, but one can use the variation principle to improve matters, as explained here. Let us indicate with $\phi_{i\lambda}$ the ith orbital of symmetry λ, obtained by a variationally computed linear combination of contracted functions $\gamma(p, \lambda)$, each one in turn constructed by combining $Q(p, \gamma)$ primitive functions (the primitive gaussian functions are here considered as contracted functions with $Q = 1$). In the first step of this method, one decides the number of primitive functions and optimizes the orbital exponents in the standard way.

In the second step one selects the specific contractions schemes to be used (on the basis of good sense, as previously done), *i.e.* the number of contractions and the length, Q, of each contraction. In the third step, the following cyclic method is used, in order to optimize variationally the linear combination coefficients (contraction coefficients) which define a contracted function and which in the older scheme were taken as equal to the SCF expansion coefficients obtained from the uncontracted atomic basis set.

Then the contracted functions are ordered by symmetry, *i.e.* first the s functions, then the p functions, then the d functions and so on. In addition, for each symmetry one starts from the inner to the outer shells.

One may designate this ordered contracted set as $\gamma(1)$, $\gamma(2)$, ... $\gamma(N)$, with $Q(1)$, $Q(2)$, ... $Q(N)$ primitive gaussian functions in each contracted set.

Table IV. Schematic representation of the contraction procedure.

Cycle No.	$Q(1)$	$Q(2)$	$Q(3)$	$Q(4)$	...	$Q(N)$
0	$\gamma(1)$	$\gamma(2)$	$\gamma(3)$	$\gamma(4)$		$\gamma(N)$
	$\gamma(1)$	pr	pr	pr	...	pr
1	$\gamma(1)$	$\gamma'(2)$	pr	pr		pr
2	$\gamma(1)$	$\gamma'(2)$	$\gamma'(3)$	pr		pr
3	$\gamma(1)$	$\gamma'(2)$	$\gamma'(3)$	$\gamma'(4)$		pr
.	.	.	.	.		.
.	.	.	.	.		.
.	.	.	.	.		.
$N-1$	$\gamma(1)$	$\gamma'(2)$	$\gamma'(3)$	$\gamma'(4)$		$\gamma'(N)$
N	pr	$\gamma'(2)$	$\gamma'(3)$	$\gamma'(4)$		$\gamma'(N)$
	$\gamma'(1)$	$\gamma'(2)$	$\gamma'(3)$	$\gamma'(4)$		$\gamma'(N)$

In the *first cycle*, having selected the $\gamma(1)$, ... $\gamma(N)$ in the standard way, all the contracted functions, except $\gamma(1)$, are de-contracted (*i.e.*, reverted to primitive) and the SCF process is used to optimize variationally the primitive function's SCF expansion coefficients. Having done this, one obtains from the SCF coefficients the contraction coefficients for $\gamma(2)$ which – being different from the original one – defines a new contracted function designated as $\gamma'(2)$. The second cycle leaves as contracted $\gamma(1)$ and $\gamma'(2)$ but uncontracts the remaining functions and, after a new SCF process, yields the new contraction coefficients for $\gamma'(3)$. This process is repeated N-1 times, thus yielding $\gamma(1)$, $\gamma'(2)$, $\gamma'(3)$, ... $\gamma'(N)$. In the last (and Nth) cycle, one retains the contracted $\gamma'(2)$, $\gamma'(3)$, ... , $\gamma'(N)$, uncontracts $\gamma(1)$, and repeats the SCF process, obtaining expansion coefficients which define $\gamma'(1)$. In the schematic representation given in Table IV, the N cycles which transform the original N contracted functions $\gamma(1)$, ..., $\gamma(N)$ into the new contracted functions $\gamma'(1)$, ..., $\gamma'(N)$ are indicated. The notation "pr" indicates that the Q primitive functions of γ are *uncontracted* into Q primitive functions.

The technique can be iteratively used more than once, but one may find that one iteration with N cycles is sufficient. Clearly, the energy improvement gained by using the γ' contracted function, rather than the γ contracted functions, becomes smaller and smaller the larger the basis set and the closer Q approaches unity. The input/output specifications for an atomic program that includes the above scheme, are reported elsewhere[16]. This program has a number of special features, given as options, and in addition prepares a considerable part of the input for the molecular SCF program, KGNMOL, described in Chapter 7.

5. Geometrical Basis Sets

The theoretical considerations and the construction scheme of the basis sets presented here has been taken from Clementi and Corongiu.[38] Let us consider an atomic function with orbital symmetries denoted as 1, ..., λ, ..., μ, and let us describe it with standard gaussian functions (primitives) of the form in Eq. (55). These primitives define sets of orbital exponents $\{\alpha\}_\lambda = \alpha_\lambda(1), \ldots \alpha_\lambda(i) \ldots \alpha_\lambda(j), \alpha_\lambda(M)$, where we assume that $\alpha_\lambda(i) < \alpha_\lambda(j)$ for i = 1, ..., M and for any value of $\lambda = 1, \ldots, \mu$. The *geometrical basis set* is obtained by imposing on all the $\{\alpha\}_\lambda$ sets the constraints listed below. To start with, a reference set of orbitals exponents $\alpha(1), \ldots, \alpha(i) \ldots, \alpha(j) \ldots, \alpha(f) \ldots, \alpha(N)$ was constructed such that for any term of the set — say the j-th term — the relation

$$\alpha(j) = \alpha_0 \, C^{j-1} \qquad j = 1, \ldots, N \tag{62}$$

applies; α_0 and C are real numbers. *First constraint*: all the α's of any $\{\alpha\}_\lambda$ set are contained in the reference set. The following symbols, I_λ and F_λ the initial and final indices for the $\{\alpha\}_\lambda$ set, corresponding to the two indices i and f in the reference set, are introduced. *Second constraint*: for any $\{\alpha\}_\lambda$ the inequalities $I_\lambda \leq I'_\lambda \leq F'_\lambda \leq F_\lambda$ hold for any $\lambda' = \lambda + 1$. *Third constraint*: the set defined by integers I_λ and F_λ contains *all* the integer values starting with i and ending with f. In molecular computations, the last two constraints hold also for the polarization functions. *Fourth constraint*: all the atoms composing a molecular system are described with the same geometrical sets, but the length of the basis varies, being *at most* N in value. *Fifth constraint*: since, in general, the precomputed tables (see Chapter 7) are of dimension $N^2/2$ or $N^4/8$, the value of N in Eq. (62) must be kept reasonably small. *Sixth constraint*: the same reference set must be comparably valid to describe all the neutral atoms from Z = 1 up to a given Z value, sufficiently high so as to cover a very large number of molecules.

A version of a molecular computer program based on geometrical basis sets (and the concomitant mathematical improvements and time saving in the two-electron integral computations) is part of KGNMOL (see Chapter 7).

In Table V, the data obtained from three *reference sets*, two with C = 3.0, α_0 = 0.012 and N = 20, the third one with C = 2.6, α_0 = 0.016 and N = 20, are tabulated. The first two sets differ in the number of functions used in the expansion and limited to a maximum value of N = 20. Detailed tabulation of the wave functions for the 55 cases reported in the table are available.[38] In Table V, the ground state energies (in a.u.) for the neutral atoms from Z = 1 to Z = 38 (H to Sr) and for several ions are presented. By different selections of I and F, two geometrical basis sets are developed using the C = 3.0, α_0 = 0.012 and N = 20 reference set.

Table V. Energies (in a.u.) from uncontracted Geometrical Basis sets[a]

Case		STO Hartree-Fock	(L,2.6,0.016,20)				(M,3.0,0.012,20)				(S,3.0,0.012,20)			
				s	p	d		s	p	d		s	p	d
H	^2S	-0.5000	-0.5000	10	0	0	-0.4999	8	0	0	-0.4998	6	0	0
He	^1S	-2.8617	-2.8616	10	0	0	-2.8614	8	0	0	-2.8572	6	0	0
Li^+	^1S	-7.2364	-7.2363	10	0	0	-7.2340	7	0	0	-7.2253	6	0	0
Be^{2+}	^1S	-13.6113	-13.6111	10	0	0	-13.6089	7	0	0	-13.6012	6	0	0
Li	^2S	-7.4327	-7.4326	13	0	0	-7.4321	11	0	0	-7.4303	9	0	0
Be^+	^2S	-14.2774	-14.2772	13	0	0	-14.2764	11	0	0	-14.2748	9	0	0
Be	^1S	-14.5730	-14.5728	13	0	0	-14.5706	10	0	0	-14.5626	9	0	0
B	^2P	-24.5291	-24.5285	13	8	0	-24.5270	11	7	0	-24.5208	9	5	0
C	^3P	-37.6886	-37.6881	13	8	0	-37.6838	10	6	0	-37.6687	9	5	0
N	^4P	-54.4009	-54.4000	13	8	0	-54.3966	11	6	0	-54.3690	9	5	0
O	^3P	-74.8094	-74.8076	13	8	0	-74.8039	11	7	0	-74.7733	9	5	0
F	^2P	-99.4093	-99.4059	13	8	0	-99.3996	11	7	0	-99.3611	9	5	0
Ne	^1S	-128.5470	-128.5440	13	8	0	-128.5281	10	6	0	-128.4647	9	5	0
Na^+	^1S	-161.6769	-161.6742	13	8	0	-161.6508	10	6	0	-161.5305	9	5	0
Mg^{2+}	^1S	-198.8308	-198.8259	13	8	0	-198.8172	11	7	0	-198.7654	9	5	0
F^-	^1S	-99.4594	-99.4554	13	8	0	-99.4494	11	7	0	-99.3602	9	5	0
Na	^2S	-161.8589	-161.8561	16	8	0	-161.8435	14	6	0	-161.7658	12	5	0
Mg^+	^2S	-199.3715	-199.3691	16	8	0	-199.3581	13	7	0	-199.3055	12	5	0
Mg	^1S	-199.6146	-199.6119	16	8	0	-199.6010	13	7	0	-199.5463	12	5	0
Al	^2P	-241.8767	-241.8744	16	11	0	-241.8635	14	10	0	-241.8428	12	8	0
Si	^3P	-288.8543	-288.8511	16	11	0	-288.8371	14	10	0	-288.8229	12	8	0
P	^4P	-340.7187	-340.7150	16	11	0	-340.6951	13	9	0	-340.6793	12	8	0
S	^3P	-397.5048	-397.5009	16	11	0	-397.4729	13	9	0	-397.4493	12	8	0
Cl	^2P	-459.4819	-459.4775	16	11	0	-459.4399	13	9	0	-459.4052	12	8	0
Ar	^1S	-526.8174	-526.8118	16	11	0	-526.7650	13	9	0	-526.7150	12	8	0
Cl^-	^1S	-459.5767	-459.5684	16	11	0	-459.5312	13	9	0	-459.4965	12	8	0
K^+	^1S	-599.0173	-599.0102	16	11	0	-598.9637	14	10	0	-598.9472	12	8	0
Ca^{2+}	^1S	-676.1540	-676.1465	16	11	0	-676.0910	13	9	0	-676.0784	12	8	0
K	^2S	-599.1645	-599.1585	19	11	0	-599.1111	16	10	0	-599.0931	15	8	0
Ca^+	^2S	-676.5699	-676.5621	19	11	0	-676.5095	16	10	0	-676.5035	15	8	0

(Continued)

Table V (cont.). Energies (in a.u.) from uncontracted Geometrical Basis sets(a)

Case		STO Hartree-Fock	(L,2.6,0.016,20)	s	p	d	(M,3.0,0.012,20)	s	p	d	(S,3.0,0.012,20)	s	p	d
Ca	^1S	-676.7580	-676.7500	19	11	0	-676.6979	17	9	0	-676.6918	15	8	0
Sc	^2D	-759.7355	-759.7251	19	11	8	-759.6682	17	10	7	-759.6487	15	8	5
Ti	^3F	-848.4057	-848.3883	19	11	8	-848.3168	17	9	6	-848.3039	15	8	5
V	^4F	-942.8842	-942.8669	19	11	8	-942.7804	16	9	6	-942.7606	15	8	5
Cr	^7S	-1043.3552	-1043.3234	19	11	8	-1043.2403	16	10	7	-1043.1807	15	8	5
Mn	^6S	-1149.8657	-1149.8485	19	11	8	-1149.7241	16	10	7	-1149.6746	15	8	5
Fe	^5D	-1262.4432	-1262.4237	19	11	8	-1262.2764	16	10	7	-1262.2039	15	8	5
Co	^4F	-1381.4142	-1381.3935	19	11	8	-1381.2233	16	10	7	-1381.1179	15	8	5
Ni	^3F	-1506.8705	-1506.8467	19	11	8	-1506.6577	16	10	7	-1506.5053	15	8	5
Zn^{2+}	^1S	-1776.9583	-1776.9288	19	11	8	-1776.5081	13	9	6	-1776.4799	12	8	5
Cu	^2S	-1638.9628	-1638.9310	19	11	8	-1638.7324	16	10	7	-1638.5042	15	8	5
Zn^+	^2S	-1777.5666	-1777.5362	19	11	8	-1777.3327	17	10	7	-1777.0695	15	8	5
Zn	^1S	-1777.8477	-1777.8162	19	11	8	-1777.6099	16	10	7	-1777.3177	15	8	5
Ga	^2P	-1923.2604	-1923.2270	19	14	8	-1922.9990	16	12	7	-1922.8742	15	11	5
Ge	^3P	-2075.3591	-2075.3217	19	14	8	-2075.0909	17	13	7	-2075.0229	15	11	5
As	^4P	-2234.2382	-2234.1966	19	14	8	-2233.9436	16	12	7	-2233.8959	15	11	5
Se	^3P	-2399.8658	-2399.8229	19	14	8	-2399.5395	16	12	7	-2399.4862	15	11	5
Br	^2P	-2572.4408	-2572.3930	19	14	8	-2572.0695	16	12	7	-2571.9999	15	11	5
Kr	^1S	-2752.0546	-2752.0026	18	13	8	-2751.6283	16	12	7	-2751.5356	15	11	5
Br^-	^1S	-2572.5355	-2572.4879	19	14	8	-2572.1649	17	13	7	-2572.0882	15	11	5
Rb^+	^1S	-2938.2072	-2938.1629	18	13	8	-2937.7275	16	12	7	-2937.5792	15	11	5
Sr^{2+}	^1S	-3130.9804	-3130.9355	19	14	8	-3130.4259	16	12	6	-3130.2740	15	11	5
Rb	^2S	-2938.3470	-2938.3005	20	13	8	-2937.8665	19	12	6	-2937.7497	18	11	5
Sr^+	^2S	-3131.3671	-3131.3125	20	13	8	-3130.8084	19	12	7	-3130.6558	18	11	5
Sr	^1S	-3131.5379	-3131.4824	20	13	8	-3130.9812	20	12	7	-3130.8276	18	11	5

(a) All values in Table V are from E. Clementi and G. Corongiu, IBM Tech. Report POK-11 (1982).

The small size basis set — designated as (S, 3.0, 0.012, 20) — is superior to the (7-s, 3-p) and the (9-s, 6-p) basis set used to describe second and third row atoms in computations of large molecules[36] containing 50 to 80 atoms. A third and large size geometrical basis set is obtained using a $C=2.6$, $\alpha_0 = 0.016$ and $N=20$ reference set. This large basis set — designated as (L, 2.6, 0.016, 22) — yields notably good energies, among the best in literature[45,46] for gaussian type basis sets and fairly near to the Hartree-Fock, HF energy limit[22,47] are also reported in Table V. The number of s, p and d primitive gaussian functions used for the three geometrical sets are given in Table V. A pleasant feature of the large basis set is that the outer orbitals are well described with three to four primitive functions.

Clementi and Corongiu[38] concluded that a very reasonable and *systematic* representation of many atoms and ions can be obtained with geometrical basis sets characterized by not too large value of N. *For large molecules*, they proposed the use of (S, 3.0, 0.012, 20) which yields superior wave functions and energies at less computer cost. *For small molecules*, the use of the intermediate and large geometrical sets is recommended.

Let us now return to the orbital exponent optimization. Generally, relatively few different types of atoms are present, even in a large molecule. By relaxing the *sixth* constraint and by considering only those atoms present in a *given molecule*, one can further improve the optimization of the C and α_0 values. One may recall that orbital exponent optimization is a very tedious and computationally expensive task if traditional basis sets are optimized; in the case of the geometrical basis set, the parameters to be optimized are $2(\mu+1)$, namely, I_λ and F_λ for $\lambda = 1, \ldots, \mu$ and the C and α_0 constants. The latter *two* parameters are the most important. For example, for the Sr atom, $2(\mu+1)=8$, whereas the number of different α's (in a standard type but equivalent quality gaussian basis set) ranges from 30 to 40.
In Tables VI to IX, respectively, the contraction schemes employed for the Geosmall (small), Geomedium (medium), and two different schemes for the Geolarge (large) geometrical basis sets are presented along with the respective Hartree-Fock energies. In addition, Clementi and Chakravorty have also relaxed the constraint of "universality" of the basis sets viz. the sixth constraint and have obtained HF quality wave functions for each of the atoms from Z = 1 to Z = 86. These HF energies with extended GTO basis sets are tabulated in Table X along with the size of the basis set, α_0 and C values. Thus all the parameters necessary to construct the basis sets for $Z=1$ through $Z=86$ are supplied in Table X. It is noted that these basis sets are also an excellent starting point for Dirac-Fock functions.[43]

Table VI. Energies (in a.u.) for the Small Geometrical Basis Set Contracted.(a)

Atom		GTO	c-GTO	Contraction Scheme
H	^2S	-0.499788	-0.499788	6
He	^1S	-2.857187	-2.857187	6
Li^+	^1S	-7.225269	-7.225269	6
Be^{2+}	^1S	-13.601205	-13.601205	6
Li	^2S	-7.430252	-7.424900	7,2
Be^+	^2S	-14.274781	-14.260949	7,2
Be	^1S	-14.562620	-14.546788	6,3
B	^2P	-24.520804	-24.495670	7,2/5
C	^3P	-37.668656	-37.631849	6,3/5
N	^4S	-54.358208	-54.308283	7,2/5
O	^3P	-74.773328	-74.712619	7,2/5
F	^2P	-99.361078	-99.272644	6,3/5
Ne	^1S	-128.464724	-128.353501	6,3/5
Na^+	^1S	-161.530490	-161.379567	6,3/5
Mg^{2+}	^1S	-198.765368	-198.587482	7,2/5
F^-	^1S	-99.360184	-99.274026	6,3/5
Mg^+	^2S	-199.305500	-199.114888	7,3,2/5
Na	^2S	-161.765765	-161.607965	7,3,2/5
Mg	^1S	-199.546281	-199.349911	7,3,2/5
Al	^2P	-241.842759	-241.622161	8,2,2/6,2
Si	^3P	-288.822912	-288.569760	8,2,2/6,2
P	^4S	-340.679261	-340.296588	7,3,2/6,2
S	^3P	-397.449270	-396.906805	7,3,2/5,3
Cl	^2P	-459.405178	-458.821051	7,2,3/5,3
Ar	^1S	-526.715022	-526.008439	7,2,3/5,3
Cl^-	^1S	-459.496534	-458.913379	7,2,3/5,3
K^+	^1S	-598.947191	-598.090565	8,2,2/6,2
$Ca^{2\cdot}$	^1S	-676.078418	-675.045797	8,2,2/6,2
K	^2S	-599.093081	-598.230729	8,2,3,2/6,2
Ca^+	^2S	-676.503481	-675.457862	9,2,2,2/6,2

(Continued)

Table VI (cont.). Energies (in a.u.) for the Small Geometrical Basis Set Contracted.(a)

Atom		GTO	c-GTO	Contraction Scheme
Ca	^1S	-676.691844	-675.642001	9,2,2,2/6,2
Sc	^2D	-759.648677	-758.346068	9,2,2,2/6,2/5
Ti	^3F	-848.303880	-846.745132	9,2,2,2/5,3/5
V	^4F	-942.760646	-941.536281	9,2,2,2/5,3/5
Cr	^7S	-1043.180690	-1041.749920	8,3,2,2/5,3/5
Mn	^6S	-1149.674600	-1148.255440	8,2,3,2/5,3/5
Fe	^5D	-1262.203900	-1260.568390	8,2,3,2/5,3/5
Co	^4F	-1381.117900	-1379.134590	8,2,2,3/5,3/5
Ni	^3F	-1506.505250	-1504.089680	8,2,2,3/5,3/5
Zn^{2+}	^1S	-1776.479900	-1773.111520	8,2,2/6,2/5
Cu	^2S	-1638.504210	-1635.633700	8,2,2,3/5,3/5
Zn^+	^2S	-1777.069500	-1773.678280	9,2,2,2/6,2/5
Zn	^1S	-1777.317710	-1773.917340	8,2,2,3/6,2/5
Ga	^2P	-1922.874230	-1918.844270	9,2,2,2/7,2,2/5
Ge	^3P	-2075.022940	-2070.263570	9,2,2,2/7,2,2/5
As	^4S	-2233.895930	-2227.988960	9,2,2,2/6,3,2/5
Se	^3P	-2399.486160	-2394.699320	9,2,2,2/6,3,2/5
Br	^2P	-2571.999920	-2568.032140	9,2,2,2/6,2,3/5
Kr	^1S	-2751.535640	-2748.101610	9,2,1,3/6,2,3/5
Br^-	^1S	-2572.088190	-2568.101540	9,2,2,2/6,2,3/5
Rb^+	^1S	-2937.579240	-2934.350000	10,2,1,2/7,2,2/5
Sr^{2+}	^1S	-3130.273950	-3126.868310	10,2,1,2/7,2,2/5
Rb	^2S	-2937.749710	-2934.509450	10,2,1,3,2/6,2,3/5
Sr^+	^2S	-3130.655760	-3127.239360	11,2,1,2,2/6,2,3/5
Sr	^2S	-3130.827560	-3127.409470	11,2,1,2,2/6,2,3/5

(a) All values in this Table VI are from C. Urdaneta, G. Corongiu and E. Clementi, IBM Tech. Report KGN-133 (1987).

Table VII. Energies (in a.u.) for the Medium Geometrical Basis Set Contracted.(a)

Atom		GTO	c-GTO	Contraction Scheme
H	^2S	-0.499946	-0.499946	6,2
He	^1S	-2.861365	-2.861365	6,2
Li^+	^1S	-7.234039	-7.234039	5,2
Be^{2+}	^1S	-13.608917	-13.608917	5,2
Li	^2S	-7.432100	-7.431527	6,2,1,2
Be^+	^2S	-14.276444	-14.274040	6,2,1,2
Be	^1S	-14.570596	-14.568409	5,2,2,1
B	^2P	-24.527038	-24.526209	6,2,1,2/5,2
C	^3P	-37.683794	-37.678609	5,2,2,1/4,2
N	^4S	-54.396594	-54.394368	6,2,1,2/4,2
O	^3P	-74.803855	-74.799847	6,2,1,2/5,2
F	^2P	-99.399590	-99.393294	6,1,1,3/5,2
Ne	^1S	-128.528123	-128.515137	5,2,1,2/4,2
Na^+	^1S	-161.650845	-161.642158	5,2,1,2/4,2
Mg^{2+}	^1S	-198.817189	-198.813342	6,2,1,2/5,2
F^-	^1S	-99.449399	-99.443516	6,1,1,3/5,2
Mg^+	^2S	-199.358116	-199.354283	6,2,1,2,1,1/5,2
Na	^2S	-161.843462	-161.833514	6,2,1,2,1,2/4,2
Mg	^1S	-199.600958	-199.596875	6,2,1,2,1,1/5,2
Al	^2P	-241.863528	-241.849887	7,2,1,1,1,2/5,2,2,1
Si	^3P	-288.837077	-288.785685	6,2,1,2,1,2/5,2,1,2
P	^4S	-340.695078	-340.636844	6,2,1,2,1,1/5,1,1,2
S	^3P	-397.472932	-397.422919	7,1,1,2,1,1/5,1,1,2
Cl	^2P	-459.439876	-459.373967	6,2,1,1,1,2/4,2,1,2
Ar	^1S	-526.764958	-526.707437	6,2,1,1,1,2/5,1,1,2
Cl^-	^1S	-459.531250	-459.465626	6,2,1,1,1,2/4,2,1,2
K^+	^1S	-598.963695	-598.881131	7,2,1,1,1,2/5,2,1,2
Ca^{2+}	^1S	-676.091020	-676.067702	7,2,1,1,1,1/4,2,1,2
K	^2S	-599.111136	-599.071171	7,2,1,1,1,1,1,2/6,1,1,2
Ca^+	^2S	-676.509536	-676.485817	8,2,1,1,1,1,1,1/4,2,1,3

(Continued)

Table VII (cont.). Energies (in a.u.) for the Medium Geometrical Basis Set Contracted.(a)

Atom		GTO	c-GTO	Contraction Scheme
Ca	^1S	-676.697854	-676.668142	8,2,1,1,1,1,1,2/6,1,1,1
Sc	^2D	-759.668199	-759.621763	8,2,1,1,1,1,1,2/6,1,1,2/5,2
Ti	^3F	-848.316835	-848.208906	8,2,1,1,1,1,1,2/5,1,1,2/4,2
V	^4F	-942.780359	-942.585457	7,2,1,2,1,1,1,1/5,1,1,2/4,2
Cr	^7S	-1043.240320	-1043.124100	8,1,1,2,1,1,1,1/5,1,1,3/4,3
Mn	^6S	-1149.724110	-1149.624250	8,1,1,1,1,2,1,1/5,1,1,3/4,3
Fe	^5D	-1262.276390	-1262.133830	8,1,1,1,1,1,2,1/5,1,1,3/4,3
Co	^4F	-1381.223320	-1381.008130	8,1,1,1,1,1,2,1/5,1,1,3/4,3
Ni	^3F	-1506.657740	-1506.380520	8,1,1,1,1,1,2,1/5,1,1,3/4,3
Zn^{2+}	^1S	-1776.508120	-1776.243590	7,2,1,1,1,1/6,1,1,1/5,1
Cu	^2S	-1638.732370	-1638.424440	7,2,1,1,1,1,2,1/5,1,1,3/4,3
Zn^+	^2S	-1777.332690	-1777.068390	8,2,1,1,1,1,1,2/6,1,1,2/5,2
Zn	^1S	-1777.609900	-1777.340190	7,2,1,1,1,1,2,1/6,1,1,2/4,3
Ga	^2P	-1922.998950	-1922.812120	8,2,1,1,1,1,1,1/5,2,1,2,1,1/5,2
Ge	^3P	-2075.090890	-2074.933710	8,2,1,1,1,1,1,2/7,1,1,1,1,2/4,3
As	^4S	-2233.943590	-2233.872170	8,2,1,1,1,1,1,1/6,1,1,2,1,1/4,3
Se	^3P	-2399.539520	-2399.44864	8,2,1,1,1,1,1,1/6,1,1,2,1,1/4,3
Br	^2P	-2572.069510	-2571.95264	8,2,1,1,1,1,1,1/6,1,1,1,1,2/4,3
Kr	^1S	-2751.628310	-2751.381880	8,2,1,1,1,1,1,1/5,2,1,1,1,2/4,3
Br^-	^1S	-2572.164900	-2572.054140	8,2,1,1,1,1,1,2/5,2,1,1,1,3/4,3
Rb^+	^1S	-2937.727460	-2937.632800	8,1,1,1,1,1,2,1/5,2,1,1,1,2/4,3
Sr^{2+}	^1S	-3130.425850	-3130.278410	9,1,1,1,1,1,1,1/7,1,1,1,1,1/4,2
Rb	^2S	-2937.866550	-2937.771770	9,1,1,1,1,1,1,2,1,1/5,2,1,1,1,2/4,2
Sr^+	^2S	-3130.808440	-3130.659900	10,1,1,1,1,1,1,1,1,1/6,1,1,1,1,2/4,3
Sr	^2S	-3130.981190	-3130.847780	10,1,1,1,1,1,1,1,1,2/5,2,1,1,1,2/4,3

(a) All values in this Table VII are from C. Urdaneta, G. Corongiu and E. Clementi, IBM Tech. Report KGN-133 (1987).

Table VIII. Energies (in a.u.) for the Large Geometrical Basis Set Contracted.(a) First Contraction Scheme

Atom		GTO	c-GTO	Contraction Scheme
H	^2S	-0.499992	-0.499992	7,1,2
He	^1S	-2.861608	-2.861608	7,1,2
Li^+	^1S	-7.236262	-7.236262	7,1,2
Be^{2+}	^1S	-13.611112	-13.611112	7,1,2
Li	^2S	-7.432568	-7.432407	8,2,1,1,1
Be^+	^2S	-14.277190	-14.276677	8,2,1,1,1
Be	^1S	-14.572833	-14.572394	8,2,1,1,1
B	^2P	-24.528486	-24.527247	7,2,1,1,2/5,1,2
C	^3P	-37.688081	-37.687110	8,2,1,1,1/6,1,1
N	^4S	-54.400012	-54.397350	7,3,1,1,1/5,2,1
O	^3P	-74.807580	-74.805087	7,2,1,1,2/5,1,2
F	^2P	-99.405863	-99.403635	7,2,1,1,2/5,1,2
Ne	^1S	-128.543969	-128.535378	8,2,1,1,1/5,1,2
Na^+	^1S	-161.674201	-161.667671	7,2,1,1,2/5,1,2
Mg^{2+}	^1S	-198.825864	-198.818527	8,2,1,1,1/6,1,1
F^-	^1S	-99.455373	-99.453266	7,2,1,1,2/5,1,2
Mg^+	^2S	-199.369086	-199.359369	8,2,2,1,1,1,1/5,3
Na	^2S	-161.856137	-161.847711	7,2,1,3,1,1,1/6,2
Mg	^1S	-199.611888	-199.603315	8,2,1,2,1,1,1/5,3
Al	^2P	-241.874405	-241.865343	8,2,1,2,1,1,1/7,1,1,1,1
Si	^3P	-288.851137	-288.841980	8,2,1,2,1,1,1/7,1,1,1,1
P	^4S	-340.715035	-340.696049	9,2,1,1,1,1,1/6,2,1,1,1
S	^3P	-397.500892	-397.477012	8,3,1,1,1,1,1/7,1,1,1,1
Cl	^2P	-459.477486	-459.444454	8,2,1,2,1,1,1/7,1,1,1,1
Ar	^1S	-526.811846	-526.781167	8,2,1,2,1,1,1/6,1,1,1,2
Cl^-	^1S	-459.568426	-459.536254	8,2,1,2,1,1,1/7,1,1,1,1
K^+	^1S	-599.010196	-598.980197	8,2,1,2,1,1,1/6,1,1,1,2
Ca^{2+}	^1S	-676.146512	-676.115497	9,2,1,1,1,1,1/7,1,1,1,1
K	^2S	-599.158467	-599.129645	9,2,1,2,1,1,1,1,1/5,2,1,3
Ca^+	^2S	-676.562055	-676.527420	9,2,1,2,1,1,1,1,1/7,1,1,2

(Continued)

Table VIII (cont.). Energies (in a.u.) for the Large Geometrical Basis Set Contracted.(a) First Contraction Scheme

Atom		GTO	c-GTO	Contraction Scheme
Ca	^1S	-676.750048	-676.718170	9,2,1,2,1,1,1,1,1/6,2,1,2
Sc	^2D	-759.725081	-759.692117	9,2,1,1,2,1,1,1,1/5,2,1,3/5,1,2
Ti	^3F	-848.388339	-848.331862	9,2,1,1,1,2,1,1,1/6,2,1,2/5,2,1
V	^4F	-942.866850	-942.818647	9,2,1,1,1,2,1,1,1/6,1,1,1,2/5,2,1
Cr	^7S	-1043.323420	-1043.229770	9,2,1,1,1,2,1,1,1/6,1,1,1,2/5,2,1
Mn	^6S	-1149.848450	-1149.775140	8,3,1,1,1,2,1,1,1/5,2,1,1,2/5,1,2
Fe	^5D	-1262.423740	-1262.330200	8,3,1,1,1,1,1,1,2/6,1,1,3/5,1,2
Co	^4F	-1381.393480	-1381.304270	9,2,1,2,1,1,1,1,1/5,2,1,1,2/5,1,2
Ni	^3F	-1506.846690	-1506.743510	9,2,1,2,1,1,1,1,1/5,2,1,1,2/5,1,2
Zn^{2+}	^1S	-1776.928760	-1776.806230	11,2,1,1,1,2,1/6,1,1,1,2/5,1,2
Cu	^2S	-1638.931030	-1638.802310	9,2,1,2,1,1,1,1,1/5,2,1,1,2/5,2,1
Zn^+	^2S	-1777.536180	-1777.411750	9,2,1,1,1,2,1,1,1/6,1,1,3/5,1,2
Zn	^1S	-1777.816250	-1777.688180	9,2,1,1,2,1,1,1,1/6,1,1,3/5,1,2
Ga	^2P	-1923.226990	-1923.099010	9,2,1,1,1,2,1,1,1/7,1,1,2,1,1,1/5,3
Ge	^3P	-2075.321660	-2075.226730	9,2,1,1,1,2,1,1,1/6,1,1,1,2,1,1,1/5,3
As	^4S	-2234.196550	-2234.110760	9,2,1,1,1,2,1,1,1/6,1,1,2,1,1,2/5,3
Se	^3P	-2399.822870	-2399.738710	10,2,1,1,1,1,1,1,1/7,1,1,2,1,1,1/6,2
Br	^2P	-2572.393040	-2572.295870	9,2,1,1,1,1,1,1,2/6,1,1,1,1,1,1,2/5,3
Kr	^1S	-2752.002650	-2751.865790	9,2,1,1,1,1,1,1,1/6,1,1,2,1,1,1/5,3
Br^-	^1S	-2572.487870	-2572.338550	9,2,1,1,1,1,1,1,2/7,1,1,1,1,1,2/6,2
Rb^+	^1S	-2938.162880	-2937.950680	9,2,1,1,1,1,1,1,1/5,2,1,2,1,1,1/5,3
Sr^{2+}	^1S	-3130.935530	-3130.738250	10,2,1,1,1,1,1,1,1/6,2,1,2,1,1,1/5,3
Rb	^2S	-2938.300490	-2938.088220	9,2,1,1,1,1,1,1,1,1,1/5,2,1,2,1,2/5,3
Sr^+	^2S	-3131.312460	-3131.102660	10,1,1,1,1,1,1,1,1,1,1/6,1,1,2,1,2/5,3
Sr	^2S	-3131.482440	-3131.089750	9,2,1,1,1,1,1,1,1,1,1/7,1,1,1,1,2/5,3

(a) All values in this Table VIII are from C. Urdaneta, G. Corongiu and E. Clementi, IBM Tech. Report KGN-133 (1987).

Table IX. Energies (in a.u.) for the Large Geometrical Basis Set Contracted.(a) Second Contraction Scheme

Atom		GTO	c-GTO	Contraction Scheme
H	^2S	-0.499992	-0.499992	7,1,2
He	^1S	-2.861608	-2.861608	7,1,2
Li^+	^1S	-7.236262	-7.236262	7,1,2
Be^{2+}	^1S	-13.611112	-13.611112	7,1,2
Li	^2S	-7.432568	-7.432426	7,1,2,1,1,1
Be^+	^2S	-14.277190	-14.276753	6,2,2,1,1,1
Be	^1S	-14.572833	-14.572522	6,2,2,1,1,1
B	^2P	-24.528486	-24.528228	6,2,1,1,1,2/5,1,2
C	^3P	-37.688081	-37.687156	8,1,1,1,1,1/6,1,1
N	^4S	-54.400012	-54.399763	7,1,1,1,1,2/5,1,2
O	^3P	-74.807580	-74.806798	7,1,1,1,1,2/5,1,2
F	^2P	-99.405863	-99.405270	5,2,2,2,1,1/5,1,2
Ne	^1S	-128.543969	-128.543426	7,1,2,1,1,1/5,1,2
Na^+	^1S	-161.674201	-161.673381	7,1,1,1,1,2/5,1,2
Mg^{2+}	^1S	-198.825864	-198.825736	7,1,1,1,1,2/6,1,1
F^-	^1S	-99.455373	-99.454440	6,1,2,1,1,2/5,1,2
Mg^+	^2S	-199.369086	-199.368980	7,1,1,1,1,1,2,1/5,1,2
Na	^2S	-161.856137	-161.855215	7,1,1,1,2,1,1,1,1/5,2,1
Mg	^1S	-199.611888	-199.610027	8,1,1,1,1,1,1,1,1/5,2,1
Al	^2P	-241.874405	-241.870429	8,1,1,1,1,1,1,1,1/6,1,1,1,1,1
Si	^3P	-288.851137	-288.847989	7,1,2,1,1,1,1,1,1/6,1,1,1,1,1
P	^4S	-340.715035	-340.714087	8,1,2,1,1,1,1,1/6,1,1,1,1,1
S	^3P	-397.500892	-397.495169	8,1,2,1,1,1,1,1/6,1,1,1,1,1
Cl	^2P	-459.477486	-459.474509	8,1,1,1,1,1,1,1,1/6,1,1,1,1,1
Ar	^1S	-526.811846	-526.806304	8,1,1,1,1,1,1,1,1/5,2,1,1,1,1
Cl^-	^1S	-459.568426	-459.565428	8,1,1,1,1,1,1,1,1/6,1,1,1,1,1
K^+	^1S	-599.010196	-599.002274	8,1,1,1,1,1,1,1,1/5,1,1,1,1,2
Ca^{2+}	^1S	-676.146512	-676.144911	7,2,1,1,1,1,1,1,1/6,1,1,1,1,1
K	^2S	-599.158467	-599.150137	9,1,1,1,2,1,1,1,1,1/5,2,1,3
Ca^+	^2S	-676.562055	-676.561191	8,1,1,1,1,1,1,1,1,1,1,1/6,1,1,1,2

(Continued)

Table IX (cont.). Energies (in a.u.) for the Large Geometrical Basis Set Contracted.[(a)] Second Contraction Scheme

Atom		GTO	c-GTO	Contraction Scheme
Ca	^1S	-676.750048	-676.744885	8,1,1,1,1,2,1,1,1,1,1/6,2,1,2
Sc	^2D	-759.725081	-759.720243	8,1,1,1,1,1,2,1,1,1,1/5,1,1,1,1,2/5,1,2
Ti	^3F	-848.388339	-848.385402	8,1,1,1,1,1,1,2,1,1,1/6,1,1,1,1,1/5,2,1
V	^4F	-942.866850	-942.863067	8,1,1,1,1,1,1,2,1,1,1/6,1,1,1,1,1/5,2,1
Cr	^7S	-1043.323420	-1043.316130	8,1,2,1,1,1,2,1,1,1/6,1,1,1,1,1/5,2,1
Mn	^6S	-1149.848450	-1149.840690	8,1,1,1,1,1,1,2,1,1,1/5,2,1,1,1,1/5,1,2
Fe	^5D	-1262.423740	-1262.423060	7,1,1,1,1,1,1,1,1,1,1,2/5,1,1,1,1,2/5,2,1
Co	^4F	-1381.393480	-1381.392640	8,1,1,1,1,1,1,1,1,1,1,1/5,1,1,1,1,2/5,1,2
Ni	^3F	-1506.846690	-1506.844260	7,2,1,1,1,1,1,1,1,1,1/5,1,1,1,1,2/5,2,1
Zn^{2+}	^1S	-1776.928760	-1776.924450	9,2,1,1,1,1,1,2,1/5,1,1,1,1,2/5,1,2
Cu	^2S	-1638.931030	-1638.929580	8,1,1,1,1,1,1,1,1,1,1/5,1,1,1,1,2/5,2,1
Zn^+	^2S	-1777.536180	-1777.532560	8,1,1,1,1,1,1,2,1,1,1/5,1,1,1,1,2/5,1,2
Zn	^1S	-1777.816250	-1777.814250	8,1,1,1,1,1,1,1,1,1,1/5,1,1,1,1,2/5,2,1
Ga	^2P	-1923.226990	-1923.225260	8,1,1,1,1,1,1,1,1,1,1,1/5,1,1,1,1,2,1,1,1/5,3
Ge	^3P	-2075.321660	-2075.317760	8,1,1,1,1,1,1,1,1,1,1,1/5,1,1,1,1,2,1,1,1/5,3
As	^4S	-2234.196550	-2234.190740	8,1,1,1,1,1,1,2,1,1,1/6,1,1,1,1,1,1,2/5,3
Se	^3P	-2399.822870	-2399.815270	9,1,1,1,1,1,1,1,1,1,1/7,1,1,1,1,1,1,1/6,2
Br	^2P	-2572.393040	-2572.388060	8,1,1,1,1,1,1,1,1,1,1,1/5,1,1,1,1,1,1,1,2/5,3
Kr	^1S	-2752.002650	-2751.996010	8,1,1,1,1,1,1,1,1,1,1/5,1,1,1,1,1,1,1,1/5,3
Br^-	^1S	-2572.487870	-2572.478950	8,1,1,1,1,1,1,1,1,1,1,1/5,1,1,1,1,1,1,1,2/5,3
Rb^+	^1S	-2938.162880	-2938.161540	7,1,1,1,1,1,1,1,1,1,2/5,1,1,1,1,1,1,1,1/5,3
Sr^{2+}	^1S	-3130.935530	-3130.934080	9,1,1,1,1,1,1,1,1,1,1/6,1,1,1,1,1,1,1,1/5,3
Rb	^2S	-2938.300490	-2938.291670	8,1,1,1,1,1,1,1,1,1,1,1,1/5,1,1,1,1,1,1,1,1/5,3
Sr^+	^2S	-3131.312460	-3131.310930	8,1,1,1,1,1,1,1,1,1,1,1/5,1,1,1,1,1,1,2/5,3
Sr	^2S	-3131.482440	-3131.480990	7,1,1,1,1,1,1,1,1,1,1,1,1,1/5,1,1,1,1,1,1,1,1/5,3

(a) All values presented in Table IX are from C. Urdaneta, G. Corongiu and E. Clementi, IBM Tech. Report KGN-133 (1987).

Table X. Geometrical Gaussian Basis Sets for atoms Z = 1 through Z = 86. The numerical Hartree-Fock energies are presented when available. (a.u.)

Atom		Numerical[a]	Geo extra large[b]	K	C	N_s	N_p	N_d	N_f	I_s	I_p	I_d	I_f
2 He	^{1}S	2.8616800	2.861679919	0.04850670	2.14774520	20	0	0	0	0	0	0	0
3 Li	^{2}S	7.4327269	7.432726628	0.02031060	2.08932430	23	0	0	0	0	0	0	0
4 Be	^{1}S	14.573023	14.57302275	0.03270700	2.09753060	23	0	0	0	0	0	0	0
5 B	^{2}P	24.529061	24.52905991	0.03591970	2.10343410	23	16	0	0	0	0	0	0
6 C	^{3}P	37.688619	37.68861771	0.05292410	2.10662820	23	16	0	0	0	0	0	0
7 N	^{4}S	54.400934	54.40093197	0.06291970	2.13743840	23	16	0	0	0	0	0	0
8 O	^{3}P	74.809398	74.80939576	0.08350860	2.08687310	23	16	0	0	0	0	0	0
9 F	^{2}P	99.409349	99.40934542	0.09899170	2.12318180	23	16	0	0	0	0	0	0
10 Ne	^{1}S	128.54710	128.5470926	0.11485350	2.13164810	23	16	0	0	0	0	0	0
11 Na	^{2}S	161.85891	161.8588998	0.00922630	2.11413310	26	16	0	0	0	4	0	0
12 Mg	^{1}S	199.61463	199.6146191	0.01215850	2.12183620	26	16	0	0	0	4	0	0
13 Al	^{2}P	241.87671	241.8766983	0.01449350	2.06073230	26	20	0	0	1	0	0	0
14 Si	^{3}P	288.85436	288.8543509	0.01861460	2.08563660	26	20	0	0	1	0	0	0
15 P	^{4}S	340.71878	340.7187678	0.02147790	2.04879270	26	20	0	0	1	0	0	0
16 S	^{3}P	397.50490	397.5048754	0.01978920	2.06216660	26	20	0	0	1	0	0	0
17 Cl	^{2}P	459.48207	459.4820518	0.02451470	2.04628670	26	20	0	0	1	0	0	0
18 Ar	^{1}S	526.81751	526.8174840	0.02635040	2.08675200	26	20	0	0	1	0	0	0
19 K	^{2}S	599.16479	599.1647600	0.01822040	2.02715220	29	20	0	0	0	2	0	0
20 Ca	^{1}S	676.75818	676.7581414	0.01646570	2.01465650	29	20	0	0	0	2	0	0
21 Sc	^{2}D	759.73572	759.7356890	0.01254190	2.01605240	30	20	18	0	0	2	0	0
22 Ti	^{3}F	848.40600	848.4059695	0.01195490	2.01800000	30	21	17	0	0	2	1	0
23 V	^{4}F	942.88433	942.8843114	0.02492140	1.98803560	30	21	17	0	0	2	1	0
24 Cr	^{7}S	---	1043.356355	0.02568400	1.98984000	30	21	17	0	0	2	1	0
25 Mn	^{6}S	1149.8662	1149.866208	0.02664480	2.01694380	30	21	17	0	0	2	1	0
26 Fe	^{5}D	1262.4437	1262.443611	0.01355000	2.01824090	30	21	17	0	0	2	1	0
27 Co	^{4}F	1381.4146	1381.414500	0.01702210	1.98359400	30	21	17	0	0	2	2	0
28 Ni	^{3}F	1506.8709	1506.870851	0.02163180	1.96797340	30	22	17	0	0	3	2	0
29 Cu	^{2}S	---	1638.963683	0.02304480	1.98955180	30	20	17	0	0	3	2	0
30 Zn	^{1}S	1777.8481	1777.848057	0.02754320	1.98074320	30	21	17	0	0	3	2	0
31 Ga	^{2}P	1923.2610	1923.260927	0.02005530	2.00551070	30	23	17	0	1	0	2	0
32 Ge	^{3}P	2075.3597	2075.359672	0.02003000	2.00000030	30	24	17	0	1	0	2	0
33 As	^{4}S	2234.2386	2234.238580	0.02055620	2.00609100	30	23	17	0	1	0	2	0

(Continued)

Table X (cont.) Geometrical Gaussian Basis Sets for atoms Z=1 through Z=86. The numerical Hartree-Fock energies are are presented when available. (a.u)

Atom		Numerical[a]	Geo extra large[b]	K	C	N_s	N_p	N_d	N_f	I_s	I_p	I_d	I_f
34 Se	^3P	2399.8676	2399.867531	0.02230400	2.00701000	30	24	17	0	1	0	2	0
35 Br	^2P	2572.4413	2572.441215	0.02685790	2.01508710	30	24	17	0	1	0	2	0
36 Kr	^1S	2752.0550	2752.054830	0.02960430	2.01960430	30	24	17	0	1	0	2	0
37 Rb	^2S	2938.3574	2938.357090	0.00768400	2.00031000	32	25	17	0	1	1	4	0
38 Sr	^1S	3131.5457	3131.544919	0.01401940	1.99563960	33	24	17	0	0	1	4	0
39 Y	^2D	3331.6842	3331.683006	0.01558470	1.98971210	33	24	20	0	0	1	0	0
40 Zr	^3F	3538.9951	3538.993279	0.01705520	1.97976190	33	24	20	0	0	1	0	0
41 Nb	^6D	---	3753.597436	0.01527040	1.97989290	33	24	20	0	0	1	0	0
42 Mo	^7S	---	3975.549276	0.01879720	1.97909240	32	24	20	0	0	1	0	0
43 Tc	^6S	4204.7887	4204.788515	0.01497550	1.98508430	32	24	20	0	0	2	1	0
44 Ru	^5F	---	4441.539167	0.01856670	1.98515010	32	24	20	0	0	2	1	0
45 Rh	^4F	---	4685.881433	0.01487000	1.98502970	32	24	20	0	0	2	1	0
46 Pd	^1S	---	4937.920781	0.01762500	1.97672850	30	24	20	0	2	2	1	0
47 Ag	^2S	---	5197.698193	0.01863310	1.97862730	32	24	20	0	0	2	1	0
48 Cd	^1S	5465.1331	5465.132700	0.01347150	1.97990020	33	24	20	0	0	2	2	0
49 In	^2P	5740.1691	5740.167812	0.00890265	1.97979410	33	27	20	0	2	0	2	0
50 Sn	^3P	6022.9317	6022.930896	0.00895215	1.98001000	33	27	20	0	2	0	2	0
51 Sb	^4S	6313.4853	6313.484658	0.01490290	1.97979980	33	26	20	0	0	1	2	0
52 Te	^3P	6611.7840	6611.783492	0.01490390	1.98009310	33	26	20	0	0	1	2	0
53 I	^2P	6917.9809	6917.979929	0.01425880	1.97794750	33	26	20	0	2	1	2	0
54 Xe	^1S	7232.1384	7232.137656	0.01430130	1.97784450	33	26	20	0	2	1	2	0
55 Cs	^2S	7553.9336	7553.932485	0.00499318	1.97784450	32	25	17	0	1	3	6	0
56 Ba	^1S	7883.5438	7883.542832	0.00500392	1.97764820	32	25	17	0	1	3	6	0
57 La	^2D	---	8221.065624	0.00557083	1.97765150	32	25	20	0	1	3	3	0
58 Ce	^1G	---	8566.873223	0.00547531	1.97768750	32	25	20	16	1	3	3	3
59 Pr	^4I	8921.1810	8921.180179	0.00813143	1.96986600	32	25	17	16	1	3	6	4
60 Nd	^5I	9283.8829	9283.88173	0.00489201	1.97765720	32	25	17	16	1	3	6	4
61 Pm	^6H	9655.0990	9655.097658	0.00499000	1.97768120	32	25	17	16	1	3	6	4
62 Sm	^7F	10034.953	10034.95126	0.00615587	1.97745700	32	25	17	16	1	3	6	4
63 Eu	^8S	10423.543	10423.54170	0.00769959	1.97570240	32	25	17	16	1	3	6	4
64 Gd	^9D	---	10820.65978	0.00706610	1.97629350	32	25	20	16	1	3	3	4
65 Tb	^6H	11226.568	11226.56732	0.01019150	1.96900000	33	26	18	16	1	3	6	4
66 Dy	^5I	11641.453	11641.45090	0.01334320	1.97350000	33	26	18	16	1	3	6	4
67 Ho	^4I	12065.290	12065.28824	0.01331360	1.97493000	32	24	17	14	0	2	5	4

(Continued)

Table X (cont.) Geometrical Gaussian Basis Sets for atoms Z=1 through Z=86. The numerical Hartree-Fock energies are presented are presented when available. (a.u.)

Atom		Numerical[a]	Geo extra large[b]	K	C	N_s	N_p	N_d	N_f	I_s	I_p	I_d	I_f
68 Er	^3H	12498.153	12498.15112	0.01434040	1.97597670	32	24	17	14	0	2	5	4
69 Tm	^2F	12940.174	12940.17256	0.01529430	1.97809240	32	24	17	14	0	2	5	4
70 Yb	^1S	13391.456	13391.45423	0.01312770	1.97644360	32	24	17	14	0	2	5	4
71 Lu	^2D	13851.808	13851.80597	0.01745150	1.96998000	31	24	20	14	1	2	2	4
72 Hf	^3F	14321.250	14321.24756	0.01639750	1.97223830	31	24	20	14	1	2	2	4
73 Ta	^4F	14799.813	14799.81058	0.01603680	1.97462110	31	24	20	14	1	2	2	4
74 W	^5D	15287.546	15287.54401	0.02030570	1.97756000	31	24	20	14	1	2	2	4
75 Re	^6S	15784.533	15784.53083	0.02057180	1.97645760	31	24	20	14	1	2	2	4
76 Os	^5D	16290.649	16290.64615	0.02058210	1.97725820	32	24	20	15	0	2	2	4
77 Ir	^4F	16806.113	16806.11031	0.02219380	1.97749380	32	24	20	15	0	2	2	4
78 Pt	^3D	---	17331.06739	0.02216280	1.97946280	32	24	20	15	0	2	2	4
79 Au	^2S	---	17865.39726	0.02168500	1.97852130	32	24	20	15	0	2	2	4
80 Hg	^1S	18408.991	18408.98771	0.02177860	1.98045190	32	24	20	15	0	2	2	4
81 Tl	^2P	18961.825	18961.82160	0.02275000	1.97000000	31	25	18	13	1	0	3	6
82 Pb	^3P	19524.008	19524.00458	0.02317030	1.97713580	31	27	18	13	1	0	3	6
83 Bi	^4S	20095.586	20095.58342	0.02672860	1.97537880	32	27	17	13	1	0	3	6
84 Po	^3P	20676.501	20676.49740	0.02745360	1.97545360	31	27	17	13	1	0	3	6
85 At	^2P	21266.882	21266.87870	0.02616310	1.97338370	32	27	19	13	1	0	3	6
86 Rn	^1S	21866.772	21866.76911	0.02429220	1.97294240	32	27	19	13	1	0	3	6

(a) For convenience $-E_{HF}$ is tabulated.
N_s = Number of s functions.
I_s = Index of initial s function in the geometric sequence (see text for details).
(b) Clementi and Chakravorty (unpublished, 1989).

6. Matrix Elements, Spherical Symmetry and Integrals

The matrix elements, $S_{\lambda pq}$, $U_{\lambda pq}$, $T_{\lambda pq}$, $P_{\lambda pq,\mu rs}$, $Q_{\lambda pq,\mu rs}$ and their expression in terms of radial integrals are now considered. We have

$$\chi_{p\lambda\alpha}(r,\vartheta,\varphi) = R_{\lambda p}(r)\, Y_{\lambda\alpha}(\vartheta,\varphi). \tag{63}$$

The matrices which enter the calculation are

	λq					
λp	01	02	03	11	12	21
01	011	021	031			
02		022	032			
03			033			
11				111	121	
12					122	
21						211

Figure 1. Storage and ordering of matrix elements for three *s*-, two *p*- and one *d*-basis functions.

$$S_{\lambda pq} = \int_0^\infty du \cdot u^2 \, R_{\lambda p}(u) R_{\lambda q}(u),$$

$$U_{\lambda pq} = \int_0^\infty du \cdot u \, R_{\lambda p}(u) R_{\lambda q}(u), \tag{64}$$

$$T_{\lambda pq} = \frac{1}{2} \int_0^\infty du[u^2 R'_{\lambda p}(u) R'_{\lambda q}(u) + \lambda(\lambda + 1) R_{\lambda p}(u) R_{\lambda q}(u)],$$

where $R'(u)$ means dR/du. The basis functions are normalized such that $S_{\lambda pp} = 1$. The matrix elements are defined only between radial basis functions which belong to the same symmetry species. The square matrices, which collect the elements for a particular symmetry, are denoted by S_λ, U_λ, T_λ; the total collection of matrix elements for all symmetries will be designated by S, U, T.

The matrices actually contain redundant elements since $S_{\lambda pq} = S_{\lambda qp}$, etc. Accordingly, the triangular arrays which contain only the distinct elements with $q \leq p$, in dictionary order on λ, p and q are required and are stored in a one-dimensional array called a supervector, as shown in Fig. 1.

The supermatrices which enter the iterative SCF process are

$$\mathbb{P}_{\lambda pq,\,\mu rs} = J^{0}_{\lambda pq,\,\mu rs} - \frac{1}{2} \sum_{\nu=|\lambda-\mu|}^{\lambda+\mu} A_{\lambda\mu\nu} K^{\nu}_{\lambda pq,\,\mu rs},$$

$$\mathbb{Q}_{\lambda pq,\,\mu rs} = \sum_{\nu=0}^{2(\lambda,\,\mu)_<} \alpha_{\lambda\mu\nu} J^{\nu}_{\lambda pq,\,\mu rs} - \frac{1}{2} \sum_{\nu=|\lambda-\mu|}^{\lambda+\mu} \beta_{\lambda\mu\nu} K^{\nu}_{\lambda pq,\,\mu rs} \tag{65}$$

In the definition of $\mathbb{Q}$ in Eq. (65) a tacit generalization of the original definition by Roothaan in Eq. (21) to include d and f shells had been introduced for atoms, where

$$J^{\nu}_{\lambda pq,\,\mu rs} = \int_0^{\infty} du \cdot u^{-\nu+1} \int_0^{u} dw \cdot w^{\nu+2}$$

$$\times [R_{\lambda p}(u)R_{\lambda q}(u)R_{\mu r}(w)R_{\mu s}(w) + R_{\mu r}(u)R_{\mu s}(u)R_{\lambda p}(w)R_{\lambda q}(w)],$$

$$K^{\nu}_{\lambda pq,\,\mu rs} = \frac{1}{2}\int_0^{\infty} du \cdot u^{-\nu+1} \int_0^{u} dw \cdot w^{\nu+2} \tag{66}$$

$$\times [R_{\lambda p}(u)R_{\mu r}(u)R_{\lambda q}(w)R_{\mu s}(w) + R_{\lambda q}(u)R_{\mu s}(u)R_{\lambda p}(w)R_{\mu r}(w)$$

$$+ R_{\lambda p}(u)R_{\mu s}(u)R_{\lambda q}(w)R_{\mu r}(w) + R_{\lambda q}(u)R_{\mu r}(u)R_{\lambda p}(w)R_{\mu s}(w)],$$

and $A_{\lambda\mu\nu}$, $\alpha_{\lambda\mu\nu}$, and $\beta_{\lambda\mu\nu}$ are numerical coefficients. The $A_{\lambda\mu\nu}$ are defined by

$$P_{\lambda}(x)P_{\mu}(x) = \sum_{\nu=|\lambda-\mu|}^{\lambda+\mu} (2\nu+1)A_{\lambda\mu\nu}P_{\nu}(x); \tag{67}$$

and $A_{\lambda\mu\nu}$ is defined:

$$A_{\lambda\mu\nu} = A_{-\lambda+\mu+\nu}A_{\lambda-\mu+\nu}A_{\lambda+\mu-\nu}/(\lambda+\mu+\nu+1)A_{\lambda+\mu+\nu},$$

$$A_{\rho} = \rho!/[(\tfrac{1}{2}\rho)!]^2, \quad \rho = 0, 2, 4, \ldots \tag{68}$$

The elements of $\mathbb{P}$ and $\mathbb{Q}$ supermatrices are stored in a one-dimensional array as shown in Fig. 2.

When the basis functions are taken to be simple normalized Slater-type functions, the integrals defined by Eqs. (64, 65) become

$$S_{\lambda pq} = [V_{2n_{\lambda p}}(\zeta_{\lambda p}) V_{2n_{\lambda q}}(\zeta_{\lambda q})]^{-1/2} V_{n_{\lambda p}+n_{\lambda q}}[\tfrac{1}{2}(\zeta_{\lambda p}+\zeta_{\lambda q})],$$

$$U_{\lambda pq} = 2[V_{2n_{\lambda p}}(\zeta_{\lambda p}) V_{2n_{\lambda q}}(\zeta_{\lambda q})]^{-1/2} V_{n_{\lambda p}+n_{\lambda q}-1}[\tfrac{1}{2}(\zeta_{\lambda p}+\zeta_{\lambda q})],$$

$$\begin{aligned} T_{\lambda pq} &= \tfrac{1}{2}\zeta_{\lambda p}\zeta_{\lambda q}[V_{2n_{\lambda p}}(\zeta_{\lambda p}) V_{2n_{\lambda q}}(\zeta_{\lambda q})]^{-1/2} \\ &\times \{V_{n_{\lambda p}+n_{\lambda q}}[\tfrac{1}{2}(\zeta_{\lambda p}+\zeta_{\lambda q})] \\ &- [W_{\lambda, n_{\lambda p}}(\zeta_{\lambda p}) + W_{\lambda, n_{\lambda q}}(\zeta_{\lambda q})] V_{n_{\lambda p}+n_{\lambda q}-1}[\tfrac{1}{2}(\zeta_{\lambda p}+\zeta_{\lambda q})] \\ &+ W_{\lambda, n_{\lambda p}}(\zeta_{\lambda p}) W_{\lambda, n_{\lambda q}}(\zeta_{\lambda q}) V_{n_{\lambda p}+n_{\lambda q}-2}[\tfrac{1}{2}(\zeta_{\lambda p}+\zeta_{\lambda q})]\}, \end{aligned} \tag{69}$$

where

$$V_i(x) = x^{-i-1} i!, \qquad W_{i,j}(x) = 2x^{-1}(j-i-1); \tag{70}$$

and

	μrs									
λpq	011	021	022	031	032	033	111	121	122	211
011	1	2	4	7	11	16	22	29	37	46
021		3	5	8	12	17	23	30	38	47
022			6	9	13	18	24	31	39	48
031				10	14	19	25	32	40	49
032					15	20	26	33	41	50
033						21	27	34	42	51
111							28	35	43	52
121								36	44	53
122									45	54
211										55

Figure 2. Storage and ordering of supermatrix elements for three *s*-, two *p*- and one *d*-basis function.

$$J^{v}_{\lambda pq,\,\mu rs} = 2[V_{2n_{\lambda p}}(\zeta_{\lambda p})V_{2n_{\lambda q}}(\zeta_{\lambda q})V_{2n_{\mu r}}(\zeta_{\mu r})V_{2n_{\mu s}}(\zeta_{\mu s})]^{-1/2}$$

$$\times \Big\{V_{n_{\lambda p}+n_{\lambda q}-v-1}[\tfrac{1}{2}(\zeta_{\lambda p}+\zeta_{\lambda q})]V_{n_{\mu r}+n_{\mu s}+v}[\tfrac{1}{2}(\zeta_{\mu r}+\zeta_{\mu s})]$$

$$\times C_{n_{\lambda p}+n_{\lambda q}-v-1,\,n_{\mu r}+n_{\mu s}+v}[(\zeta_{\lambda p}+\zeta_{\lambda q})/(\zeta_{\mu r}+\zeta_{\mu s})] \tag{71a}$$

$$+ V_{n_{\mu r}+n_{\mu s}-v-1}[\tfrac{1}{2}(\zeta_{\mu r}+\zeta_{\mu s})]V_{n_{\lambda p}+n_{\lambda q}+v}[\tfrac{1}{2}(\zeta_{\lambda p}+\zeta_{\lambda q})]$$

$$\times C_{n_{\mu r}+n_{\mu s}-v-1,\,n_{\lambda p}+n_{\lambda q}+v}[(\zeta_{\mu r}+\zeta_{\mu s})/(\zeta_{\lambda p}+\zeta_{\lambda q})]\Big\},$$

$$K^{v}_{\lambda pq,\,\mu rs} = [V_{2n_{\lambda p}}(\zeta_{\lambda p})V_{2n_{\lambda q}}(\zeta_{\lambda q})V_{2n_{\mu r}}(\zeta_{\mu r})V_{2n_{\mu s}}(\zeta_{\mu s})]^{-1/2}$$

$$\times \{V_{n_{\lambda p}+n_{\mu r}-v-1}[\tfrac{1}{2}(\zeta_{\lambda p}+\zeta_{\mu r})]V_{n_{\lambda q}+n_{\mu s}+v}[\tfrac{1}{2}(\zeta_{\lambda q}+\zeta_{\mu s})]$$

$$\times C_{n_{\lambda p}+n_{\mu r}-v-1,\,n_{\lambda q}+n_{\mu s}+v}[(\zeta_{\lambda p}+\zeta_{\mu r})/(\zeta_{\lambda q}+\zeta_{\mu s})]$$

$$+ V_{n_{\lambda q}+n_{\mu s}-v-1}[\tfrac{1}{2}(\zeta_{\lambda q}+\zeta_{\mu s})]V_{n_{\lambda p}+n_{\mu r}+v}[\tfrac{1}{2}(\zeta_{\lambda p}+\zeta_{\mu r})]$$

$$\times C_{n_{\lambda q}+n_{\mu s}-v-1,\,n_{\lambda p}+n_{\mu r}+v}[(\zeta_{\lambda q}+\zeta_{\mu s})/(\zeta_{\lambda p}+\zeta_{\mu r})] \tag{71b}$$

$$+ V_{n_{\lambda p}+n_{\mu s}-v-1}[\tfrac{1}{2}(\zeta_{\lambda p}+\zeta_{\mu s})]V_{n_{\lambda q}+n_{\mu r}+v}[\tfrac{1}{2}(\zeta_{\lambda q}+\zeta_{\mu r})]$$

$$\times C_{n_{\lambda p}+n_{\mu s}-v-1,\,n_{\lambda q}+n_{\mu r}+v}[(\zeta_{\lambda p}+\zeta_{\mu s})/(\zeta_{\lambda q}+\zeta_{\mu r})]$$

$$+ V_{n_{\lambda q}+n_{\mu r}-v-1}[\tfrac{1}{2}(\zeta_{\lambda q}+\zeta_{\mu r})]V_{n_{\lambda p}+n_{\mu s}+v}[\tfrac{1}{2}(\zeta_{\lambda p}+\zeta_{\mu s})]$$

$$\times C_{n_{\lambda q}+n_{\mu r}-v-1,\,n_{\lambda p}+n_{\mu s}+v}[(\zeta_{\lambda q}+\zeta_{\mu r})/(\zeta_{\lambda p}+\zeta_{\mu s})]\Big\},$$

where

$$C_{\alpha\beta}(t) = (1+t)^{-\alpha-\beta-1}\sum_{\lambda=0}^{\alpha}\binom{\alpha+\beta+1}{\lambda}t^{\lambda}, \tag{72}$$

and $\binom{\alpha}{\lambda}$ is the usual notation for the binomial coefficients.

With GTOs as basis orbitals, the integrals are calculated by the following formulas.[15]

$$
\begin{aligned}
S_{\lambda pq} &= [V_{2n_{\lambda p}}(\zeta_{\lambda p}) \cdot V_{2n_{\lambda q}}(\zeta_{\lambda q})]^{-1/2} V_{n_{\lambda p}+n_{\lambda q}}(\zeta_{\lambda pq}) \\
U_{\lambda pq} &= 2\sqrt{\frac{2}{\pi}}\, S_{\lambda pq} \cdot V_{n_{\lambda p}+n_{\lambda q}-1}(\zeta_{\lambda pq})/V_{n_{\lambda p}+n_{\lambda q}}(\zeta_{\lambda pq}) \\
T_{\lambda pq} &= \frac{1}{2}\zeta_{\lambda p} \cdot \zeta_{\lambda q} \cdot S_{\lambda pq} \cdot \{V_{n_{\lambda p}+n_{\lambda q}+2}(\zeta_{\lambda pq}) \\
&\quad - [W_{\lambda, n_{\lambda p}}(\zeta_{\lambda p}) + W_{\lambda, n_{\lambda q}}(\zeta_{\lambda q})] V_{n_{\lambda p}+n_{\lambda q}}(\zeta_{\lambda pq}) \\
&\quad + W_{\lambda, n_{\lambda p}}(\zeta_{\lambda p}) W_{\lambda, n_{\lambda q}}(\zeta_{\lambda q}) \cdot V_{n_{\lambda q}+n_{\lambda q}-2}(\zeta_{\lambda pq})\}/V_{n_{\lambda p}+n_{\lambda q}}(\zeta_{\lambda pq})
\end{aligned}
\tag{73}
$$

where

$$
\begin{aligned}
H_{\lambda pq} &= -Z \cdot U_{\lambda pq} + T_{\lambda pq} \\
\zeta_{\lambda pq} &= \frac{1}{2}(\zeta_{\lambda p} + \zeta_{\lambda q})
\end{aligned}
\tag{74}
$$

and

$$
V_i(x) = (i-1)!!/x^{(i+1)/2} \tag{75}
$$

and

$$
W_{i,j}(x) = 2x^{-1}(j-i-1) \tag{76}
$$

The results are stored in the matrices S (overlap), U (potential energy), and T (kinetic energy).

The formulae for coulomb and exchange integrals, $J^{\nu}_{\lambda pq,\,\mu rs}$ and $K^{\nu}_{\lambda pq,\,\mu rs}$, for GTOs as basis orbitals are given by:

$$
\begin{aligned}
J^{\nu}_{\lambda pq,\,\mu rs} &= I^{\nu}_{\lambda p\lambda q,\,\mu r\mu s} + I^{\nu}_{\mu r\mu s,\,\lambda p\lambda q} \\
K^{\nu}_{\lambda pq,\,\mu rs} &= \frac{1}{2}\{I^{\nu}_{\lambda p\mu r,\,\lambda q\mu s} + I^{\nu}_{\lambda q\mu s,\,\lambda p\mu r} + I^{\nu}_{\lambda p\mu s,\,\lambda q\mu r} + I^{\nu}_{\lambda q\mu r,\,\lambda p\mu s}\}
\end{aligned}
\tag{77}
$$

where

$$I_{\lambda p\lambda q,\,\mu r\mu s} = 2\sqrt{\frac{2}{\pi}}\,S_{\lambda pq}S_{\mu rs}$$

$$\cdot\frac{V_{n_{\lambda p}+n_{\lambda q}-v-1}(\zeta_{\lambda pq})\cdot V_{n_{\mu r}+n_{\mu s}+v}(\zeta_{\mu rs})}{V_{n_{\lambda p}+n_{\lambda q}}(\zeta_{\lambda pq})\cdot V_{n_{\mu r}+n_{\mu s}}(\zeta_{\mu rs})} \tag{78}$$

$$\cdot\, C_{n_{\lambda p}+n_{\lambda q}-v-1,\,n_{\mu r}+n_{\mu s}+v}(\zeta_{\lambda pq}/\zeta_{\mu rs})$$

and

$$I_{\lambda p\mu r,\,\lambda q\mu s} = 2\sqrt{\frac{2}{\pi}}\cdot S_{\lambda pq}\cdot S_{\mu rs}\frac{V_{n_{\lambda p}+n_{\mu r}-v-1}(\zeta_{\lambda p\mu r})\cdot V_{n_{\lambda q}+n_{\mu s}+v}(\zeta_{\lambda q\mu s})}{V_{n_{\lambda p}+n_{\lambda q}}(\zeta_{\lambda pq})\cdot V_{n_{\mu r}+n_{\mu s}}(\zeta_{\mu rs})} \tag{79}$$

$$\cdot\, C_{n_{\lambda p}+n_{\mu r}-v-1,\,n_{\lambda q}+n_{\mu s}+v}(\zeta_{\lambda p\mu r}/\zeta_{\lambda q\mu s})$$

with

$$C_{\alpha,\,\beta}(t) = (1+t)^{-1/2(\beta+1)}\sum_{\lambda=0}^{(\alpha-1)/2}\frac{(2\lambda+\beta-1)!!}{(2\lambda)!!(\beta-1)!!}\left(\frac{t}{1+t}\right)^{\lambda}. \tag{80}$$

The integrals computed with the Oscillating Slater type orbitals are the following: overlap S, kinetic energy T, nuclear potential U, and two-electron interaction Z. Using the short notation $p\equiv\{n, \xi, k\}$, $q\equiv\{n', \xi', k'\}$, etc., they can be expressed as

$$S_{\lambda pq} = A_{n\xi k}A_{n'\xi'k'}I_1[(n+n'),(\xi+\xi'), k, k'] \tag{81}$$

$$\begin{aligned} T_{\lambda pq} &= -\frac{1}{2}A_{n\xi k}A_{n'\xi'k'}\{(\xi'^2 - k'^2)I_1[(n+n'),(\xi+\xi'), k, k'] + \\ &+ 2\xi'k'I_2[(n+n'),(\xi+\xi'), k, k'] - 2n'\xi'I_1[(n+n'-1),(\xi+\xi'), k, k'] - \\ &- 2n'k'I_2[(n+n'-1),(\xi+\xi'), k, k'] + \\ &+ [n'(n'-1) - l(l+1)]I_1[(n+n'-2),(\xi+\xi'), k, k']\} \end{aligned} \tag{82}$$

$$U_{\lambda pq} = A_{n\xi k}A_{n'\xi'k'}I_1[(n+n'-1),(\xi+\xi'), k, k'] \tag{83}$$

$$Z^{a,b}_{\lambda pq,\,\mu rs} = \int_0^\infty dr_2 \int_0^{r_2} dr_1 R_{n_1\xi_1k_1}(r_1) R_{n_2\xi_2k_2}(r_1) r_1^a\, r_2^b\, R_{n_3\xi_3\kappa_3}(r_2) R_{n_4\xi_4k_4}(r_2) =$$

$$\begin{aligned}
&= \frac{1}{4}\{I_3[(n_3+n_4+b-2),(n_1+n_2+a-2),\,\xi_{34},\,\xi_{12},(k_3+k_4),(k_1+k_2)] + \\
&+ I_3[(n_3+n_4+b-2),(n_1+n_2+a-2),\,\xi_{34},\,\xi_{12},(k_3+k_4),(k_1-k_2)] + \\
&+ I_3[(n_3+n_4+b-2),(n_1+n_2+a-2),\,\xi_{34},\,\xi_{12},(k_3-k_4),(k_1+k_2)] + \\
&+ I_3[(n_3+n_4+b-2),(n_1+n_2+a-2),\,\xi_{34},\,\xi_{12},(k_3-k_4),(k_1-k_2)]\}
\end{aligned} \tag{84}$$

with $\xi_{34} = \xi_3 + \xi_4$ and $\xi_{12} = \xi_1 + \xi_2$ where

$$I_1[n,\,\xi,\,k,\,k'] = \frac{1}{2}\{I_c[n,\,\xi,(k+k')] + I_c[n,\,\xi,(k-k')]\} \tag{85}$$

$$I_2[n,\,\xi,\,k,\,k'] = \frac{1}{2}\{I_s[n,\,\xi,(k+k')] + I_s[n,\,\xi,(k'-k)]\} \tag{86}$$

$$\begin{aligned}
I_3[n,\,n',\,\xi,\,\xi',\,k,\,k'] = \sum_{j=1}^{n'+1} \frac{(-1)^{j+1} n'!}{(n'-j+1)!\,(\xi'^2+k'^2)^{j/2}} \\
\{\cos(jt) I_1[(n+n'-j+1),(\xi+\xi'),\,k,\,k'] + \\
+ \sin(jt) I_2[(n+n'-j+1),(\xi+\xi'),\,k,\,k']\} + \\
+ \frac{(-1)^{n'+1} n'!}{(\xi'^2+k'^2)^{(n'+1)/2}} \cos[(n'+1)t] I_c[n,\,\xi,\,k]
\end{aligned} \tag{87}$$

with $t = \arctan\left(\frac{k'}{\xi'}\right)$.

$$I_c[n,\,\xi,\,k] = \frac{n!}{(\xi^2+k^2)^{(n+1)/2}} \cos[(n+1)t] \tag{88}$$

$$I_s[n,\,\xi,\,k] = \frac{n!}{(\xi^2+k^2)^{(n+1)/2}} \sin[(n+1)t] \tag{89}$$

with $t = \arctan\left(\frac{k}{\xi}\right)$.

7. The Correlation and Pair Correlation Energies

Irrespective of spin and symmetry considerations the electrostatic repulsion alone prevents two electrons from approaching arbitrarily close to one another so that a many particle wave function must correlate the motion of electrons. The term "electronic correlation" and correlation energy was first used by Wigner.[48] Independent particle models neglect this correlation at various levels of approximation, leading to the Hartree and Hartree-Fock Models. In the former there is an unphysical finite probability of finding two electrons in the same place. In the latter, electrons with parallel spin are prevented from overlapping but not those with antiparallel spin. Overlap is prevented in the Hartree-Fock model by the antisymmetry principle, leading to the concept of "Fermi Hole." There is a need, however, for a "Coulomb Hole" around each electron.[48,49] In consequence the one electron models give expectation values of the $< \mathscr{H} >$ higher than the true energy E_{Tot}. The Coulomb hole *correlation energy* is defined by the following equation

$$E_c = E_{exact,\, non\, rel.} - E_{HF} \tag{90}$$

This definition[50] has the advantage of being formally correct but unfortunately the real world is a bit "fuzzy"; indeed the Hartree-Fock function does not always provide a proper zero order approximation in atoms and molecules, as detailed below and in Chapter 7. This drawback then is transferred to the definition of Eq. (90), which requires the knowledge of an unknown (and difficult to obtain) quantity, the exact non relativistic energy. Thus it is necessary to filter out the relativistic effects from the experimental results or, alternatively, make a relativistic computation using the Dirac-Fock procedure inclusive of the Breit terms (DF). In practice we could write the correlation energy as

$$E'_c = E_{experimental} - E_{DF} \tag{91}$$

or, alternatively

$$E_c = E_{experimental,\, relativistically\, corrected} - E_{HF} \tag{92}$$

which now becomes a function of the experimental resolution. The definition of Eqs. (91) and (92), can be challenged in its practical value since it tacitly accepts as physically meaningful an approximate rather than an accurate energy partitioning, namely

$$E_{Tot} = E_{HF} + E_C + E_{Rel} \tag{93}$$

If we choose to defer the definition of E_{Rel}, then we would also defer the definition of E_C. If we equate the relativistic correction to the difference of Dirac-Fock energy and the Hartree-Fock energy

$$E_{DF} - E_{HF} = E'_{Rel} \tag{94}$$

then we have $E_{Tot} = E_{DF} + E'_C$ with $E'_C \neq E_C$,. For reasonable accuracy we shall use the energy partitioning of Eq. (93). We have introduced this discussion in order to point out that different approaches to the correlation problem might benefit from somewhat different definitions as a starting point.

There are a number of ways for dealing with electron correlation, such as the configuration interaction (CI) method, the MCSCF (multi-configuration self-consistent field) method, the Möller-Plesset perturbation approaches (MP2, MP3, MP4), the coupled cluster and the MBPT. All those are discussed in MOTECC-90 by the leaders of today quantum chemistry. No method, however, combines full accuracy with relative simplicity. By "relative" we imply that with such methods we still cannot obtain "total" electron correlation for a "simple" problem for example the electron affinity of iodine or the binding energy of benzene to within an accuracy of a few wavenumbers. Indeed, the old Coulson's dilemma — can we obtain the weight of the ship's captain by weighting the captain and the ship and by subtracting the weight of the ship — is still haunting us. Still worse, it is not clear whether the atom (the ship's captain) in the molecule is much related to the 'free' atom at all. Thus it is useful to have simple even if somewhat rough techniques to estimate the correlation correction. Let us present an extremely simple approximation.

It can be assumed that the correlation energy may be expanded into sums of pair, triple, quartet correlations, viz.

$$E_c = \sum_{i,j} \eta_{ij} + [\sum_{i,j,k} \eta_{ijk} + \sum_{i,j,k,l} \eta_{ijkl}] + \dots\dots\dots\dots \tag{95}$$

where the terms within brackets are the pair-pair interactions and the + ... indicates pair-pair-pair interactions (and higher terms) as in the configuration $2p^6$ or the triple bond of acetylene. In the above partitioning, the indices i,j,k,l refer to spin orbitals. Let us recall that in the independent particle model the Hartree-Fock energy is written as

$$E_{HF} = \sum \eta_i \tag{96}$$

where $\eta_i = n_i(\varepsilon_i + h_i)$ with ε_i, the orbital energy, h_i, the one-electron energy, and n_i, the occupation number. Thus the total non-relativistic energy

$$E = E_{HF} + E_c = \sum \eta_i + \sum_{i,j} \eta_{ij} + \sum_{i,j,k} \eta_{ijk} + \sum_{i,j,k,l} \eta_{ijkl} + \cdots \tag{97}$$

The partitioning in Eq. (97) may be truncated at the second term, if we assume it dominates the contribution to the correlation energy. This is essentially the independent electron pair approximation[51]. To gain insight and to obtain some numerical feeling for the magnitude of η_{ij} semiempirical models may be used.[52,53] Based upon an earlier work due to Clementi[52] we have recently proposed a simple model.[53] In this approach a soft Coulomb hole was introduced around each electron with a smooth exponential cut-off, which replaces an earlier square well cut-off.[52] Essentially, the electrons i and j do repel each other according to the standard operator $(1/r_{ij})$ leading to Eqs. (48), (49) and (22) but are in addition subjected to a repulsive perturbation of the form

$$\frac{-e^{-\omega_{\lambda pq,\mu rs} r_{ij}^2}}{r_{ij}} \tag{98}$$

representing a "soft hole" which keeps electron i away from electron j. The value of $\omega_{\lambda pq,\mu rs}$ was empirically determined and ensures that the electron pair represented by $\phi_{\lambda i}$ is more correlated than those in $\phi_{\lambda i}$ and $\phi_{\lambda j}$ or in $\phi_{\lambda i}$ and $\phi_{\mu j}$. In Appendix 2A, detailed integral formulae for additional integrals due to the term in (98) are presented. Notice that we have used in Eq. (98) the form r_{ij}^2 in order to use Gaussian basis sets and thus extend this approach to molecules. The corresponding four center gaussian integrals with Eq. (98) as operator (needed in molecular computations) are presently being coded.[54] In Fig. 3 we report the computed total correlation energy correction from He (Z = 2) to Xe (Z = 54).

Let us consider the correlation energy of the neon atom in its ground state ^1S ($1s^2 2s^2 2p^6$) and the first positive ion Ne^+ in its ground state (one less p electron); the corresponding empirical values of the pair correlation energies η_{ij} in hartrees are estimated:

Ne	1s	2s	2p	1s	2s	2p	Ne^+
1s	0.043			0.043			
2s	0.004	0.045		0.004	0.046		
2p	0.006	0.060	0.189	0.006	0.057	0.133	

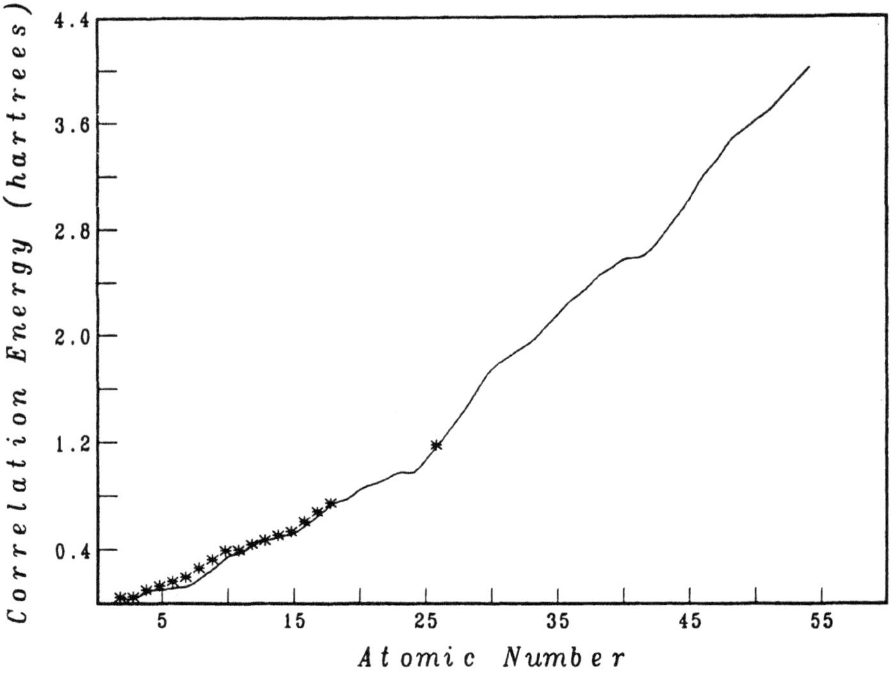

Figure 3. Correlation energy computed by using the "soft coulomb hole" model.

Thus the estimated total correlation energy in Ne and Ne^+ are 0.347 a.u. and 0.283 a.u., respectively, namely the algebraic sums of the above partial contributions. The values obtained by using Eq. (98) are somewhat too small, but this is the best we could initially do in determining a general expression for ω in Eq. (98). The ionization potential is another quantity which allows us to assess the quality of this estimate for the correlation correction. In the ionization of Ne (^1S) to Ne^+ (^2P) there is a loss of a 2p pair. To a first approximation this means that all the interactions involving the 2p orbital, i.e. (2p,2p), (2s,2p) and (1s,2p) should change, the dominant difference being in (2p,2p) (the 2p pair correlation energy). The ionization potential for Ne may be written as

$$I(\mathrm{Ne}) = E_{HF}(\mathrm{Ne}^+) - E_{HF}(\mathrm{Ne}) + E_C(\mathrm{Ne}^+) - E_C(\mathrm{Ne}) = I_{HF}(\mathrm{Ne}) + \Delta E_C(\mathrm{Ne}) \qquad (99)$$

Both the experimental ionization potential and the Hartree-Fock ionization potential are available to sufficient accuracy and their differences, i.e., $I_{expt} - I_{HF} = 1.71$ eV, compares very well with the 1.74 eV (0.064 a.u.) which is the value for ΔE_C from the empirical model. Let us now use some of these concepts in the case of molecular systems. Note that in Eq. (99) we have assumed zero contribution from relativistic effects, because i) we are dealing with the outermost electron and ii) the relativistic effects are relative small in the Ne atom.

In Fig. 4 we present the first ionization potentials from He to Xe and we compare the experimental values to those from Hartree-Fock and from Hartree-Fock with the perturbation correction from Eq. (98). The results are gratifying. We note that the location of the excited state is known to be incorrectly given in the Hartree-Fock approximation. However, using Eq. (98) in computing the excited states of the lowest electronic configuration in the atoms C, O, N, Si, P, and S the obtained spectral differences are in reasonable agreement with experiments.

In Table XI we report the computed pair correlations for the neutral atoms in the ground state from He to Ar; for additional data we refer to the Technical Report of Ref. 52.

In molecules the knowledge of the electron correlation is often essential to determine quantities as dissociation energy,[52-57] bond energy, and potential energy surfaces. For example the Hartree-Fock energies for the ground states of H_2 and LiH in their equilibrium configuration are 1.1336 and 7.9873 a.u. respectively. The corresponding dissociation limits are 2H (^2S) and H(^2S)Li(^2S) for which the correlation energies are 0 for H(^2S) (no correlation energy for a single electron) and 0.046 for Li (with $\eta_{1s,1s} = 0.044$ and $\eta_{1s,2s} = 0.002$). In these molecules the pair correlation scheme is given from the second term of Eq. (97) as:

H_2	LiH
$\eta_{1\sigma,\,1\sigma}$	$\eta_{1\sigma,\,1\sigma} + \eta_{1\sigma,\,2\sigma} + \eta_{2\sigma,\,2\sigma}$

We assume that

$$\eta_{1\sigma,\,1\sigma}\,(LiH) \simeq \eta_{1s,\,1s}(Li)$$

This approximation could be used for the 1σ orbital of H_2 and the 2σ orbital of LiH, leading to 0 pair correlations.

$$D_e(H_2) = E_{HF}(H_2) + E_C(H_2) - 2E_{HF}(H) \tag{100}$$

and for LiH

$$D_e(LiH) = E_{HF}(LiH) + E_C(LiH) - E_{HF}(Li) - E_C(Li) - E_{HF}(H) \tag{101}$$

The term $E_{HF}(H)$ is the exact energy of H; however, neglecting relativistic corrections, the $E_C(H_2)$ and $E_C(LiH)$ (by summing the atomic pair correlation energies) are underestimated because of an extra electron pair is formed in each case. We return to a rather old and abandoned topic, namely the use of *atomic* total and pair-correlation energies to estimate the *molecular* correlation energy, in general, and the "molecular extra correlation energy" in particular, since the latter is needed to compute molecular dissociation energies once the Hartree-Fock energy is known. (We use the term "dissociation energy" rather loosely; more accurately we should talk of "atomization energy," namely dissociation into atomic fragments.)

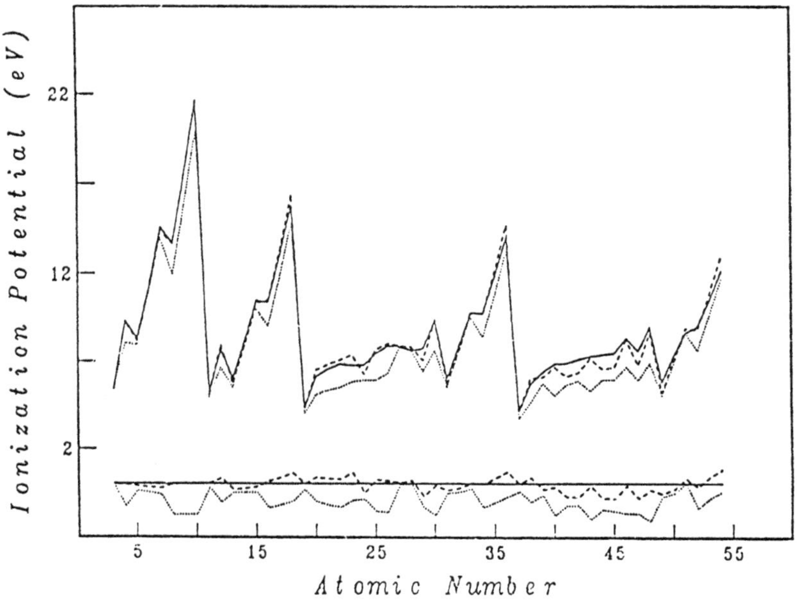

Figure 4. Ionization potentials from He to Xe, obtained either from experiments (solid line) or (thin dots) from the Coulomb Hole model (thick dots) of Eq. (98). The two curves at the bottom report the correlation energy error unaccounted by the CC functional and the much larger Hartree-Fock error.

The energy difference obtained by subtracting from the correlation energy of a molecule the correlation energy of the component atoms (in the correct dissociation state) has been called *molecular extra correlation energy.* [31,52] The name implicitly contains the connotation of an increase in correlation energy above and beyond the sum of the correlation energies from the corresponding atoms. The main reason is that the formation of the molecule always increases the number of electron pairs. This increase is twofold: the most important one is due to the fact that electrons (previously unpaired in the separated atoms) pair during molecule formation. For example, the hydrogen atom has one unpaired spin, the fluorine atom has nine electrons, eight paired and one unpaired, but in the hydrogen fluoride molecule the two unpaired spins (the one of H and the one of F) pair up ($H(^2S) + F(^2P) \rightarrow HF(^1\Sigma^+)$). Secondly, the increase in the number of pairs leads to more interpair correlation energy contributions because of the increase in the number of electron pairs. The molecular extra correlation energy can be a substantial fraction of the total dissociation energy. Let us indicate with *HF, c, r* quantities obtained from the Hartree-Fock methods, due to correlation and relativistic effects, respectively. Let us indicate the component atoms of an N-atom molecule, m, by the subscript a, and finally let us use the symbols D for dissociation energy and E for total energy.

Table XI. Pair correlation energies (a.u.) for neutral atoms in their ground states.

Atom				Atomic Correlation Energy
H				0.000
1s	0.000			
He				0.044
1s	0.044			
Li				0.046
1s	0.044			
2s	0.002	0.000		
Be				0.092
1s	0.044			
2s	0.004	0.044		
B				0.1030
1s	0.044			
2s	0.004	0.045		
2p	0.001	0.009	0.000	
C				0.115
1s	0.044			
2s	0 004	0.045		
2p	0.002	0.019	0.002	
N				0.130
1s	0.044			
2s	0.004	0.046		
2p	0.003	0.029	0.005	
O				.196
1s	0.043			
2s	0.004	0.046		
2p	0.004	0.039	0.060	

(Continued)

Table XI (cont.). Pair correlation energies (a.u.) for neutral atoms in their ground states.

Atom						Atomic Correlation Energy
F						.268
1s	0.043					
2s	0.004	0.046				
2p	0.005	0.049	0.122			
Ne						.347
1s	0.043					
2s	0.004	0.045				
2p	0.006	0.060	0.189			
Na						.375
1s	0.042					
2s	0.004	0.045				
2p	0.007	0.063	0.007	0.205		
3s	0.000	0.002	0.000			
Mg						.442
1s	0.042					
2s	0.004	0.045				
2p	0.008	0.065	0.015	0.218		
3s	0.001	0.004	0.041			
Al						.467
1s	0.042					
2s	0.004	0.045				
2p	0.009	0.066	0.017	0.229		
3s	0.001	0.004	0.042			
3p	0.000	0.000	0.007	0.001	0.000	

(Continued)

Table XI (cont.). Pair correlation energies (a.u.) for neutral atoms in their ground states.

Atom						Atomic Correlation Energy
Si						.492
1s	0.042					
2s	0.004	0.045				
2p	0.010	0.067	0.017	0.239		
3s	0.001	0.004	0.042			
3p	0.000	0.001	0.015	0.002	0.002	
P						.517
1s	0.041					
2s	0.004	0.045				
2p	0.010	0.068	0.018	0.248		
3s	0.001	0.004	0.042			
3p	0.000	0.002	0.024	0.004	0.005	
S						.585
1s	0.041					
2s	0.004	0.044				
2p	0.011	0.069	0.018	0.256		
3s	0.001	0.004	0.042			
3p	0.000	0.003	0.033	0.006	0.052	
Cl						.658
1s	0.041					
2s	0.004	0.044				
2p	0.011	0.070	0.019	0.263		
3s	0.001	0.004	0.042			
3p	0.001	0.004	0.042	0.008	0.105	
Ar						.736
1s	0.040					
2s	0.004	0.044				
2p	0.012	0.070	0.019	0.270		
3s	0.001	0.004	0.042			
3p	0.001	0.005	0.051	0.010	0.163	

The dissociation energy, D, of a molecule is then given by the relation

$$D = D_{HF} + D_c + D_r \tag{102}$$

or

$$D = D_{HF} + M_c + D_r \tag{103a}$$

where

$$D_{HF} = \sum_{a=1}^{N} E_{a,HF} - E_{m,HF} \tag{103b}$$

and where the molecular extra correlation energy is defined as

$$M_c \equiv D_c = \sum_{a=1}^{N} E_{a,c} - E_{m,c} \tag{104}$$

and finally

$$D_r = \sum_{a=1}^{N} E_{a,r} - E_{m,r} \tag{105}$$

In other words, we partition the dissociation energy into a Hartree-Fock component, a correlation component and a relativistic energy component (the latter often neglected for low Z elements). The quantity D_{HF} is obtained directly from the molecular computations of Hartree-Fock type (now routinely available) and from the atomic Hartree-Fock energies.[22] We wish to find an easy way either to obtain or at least to reliably estimate the value of M_c.

With the availability of accurate Hartree-Fock and atomic correlation energies from semiempirical and *ab initio* methods, it was hoped that one could be in a position to estimate the M_c and dissociation energy of a molecule to within a few percent.[55,56] Let us now consider these intuitive corrections in a somewhat more systematic way, even if we shall retain the empirical approach. We could note that it is somewhat unfortunate that this avenue was abandoned in favor of CI computations and other post-Hartree-Fock techniques. The questionable efficiency of most of the present many-body techniques for large molecules will likely compel us to reassess this strategy.

Proposals to use atomic data in order to estimate molecular energies followed immediately after the availability of tabulations of atomic Hartree-Fock energies[18-22] and estimates of the atomic correlation energies.[57,58] For example, after considering diatomic molecules[56], Clementi expanded in 1963 to CH_4 as an example of a polyatomic molecule[59] and, later, Hollister and Sinanoglu systematized algorithms to obtain binding energies using pair correlation data.[55] The theoretical models can be considered as derivations and adaptations of the Bethe-Goldstone model as advanced primarily by Nesbet,[60] Kelly[61] and Sinanoglu.[62]

Recently, we have revisited this problem, and in Chapter 7 we shall report on the Clementi-Chakravorty electron pair functional, where we use the idea of pair correlation transferability from atoms to molecules. Thus, we make use of the pair correlation obtained from Eq. (98) for the ground state atoms (dissociation products of the molecule) and of the pair correlation energies of the "atoms when in the molecule", namely in the electronic cloud of the bonding region. For example, in CH_4, C_2H_2, C_2H_4, etc., the C atom experiences the presence of the electrons of the hydrogen atoms, as long ago pointed out by the "octet" rule in valence considerations. Thus at dissociation the carbon atom has configuration $1s^2(2s, 2p)^4$, but within the CH_4 molecule it has configuration which could be written as $1s^2(2s, 2p)^8$; equivalently, the hydrogen atom in CH_4 experiences a configuration which can be thought of as $1s^2$, different from the one at dissociation, namely $1s^1$. The correlation energies of the "atoms in the molecule" are given in Table XII. We should stress that the pair correlation energies in Tables XI and XII are somewhat underestimated; this calls for a recalibration of the parameters leading to the value of $\omega_{\lambda pq,\, \mu rs}$ of Eq. (98). Alternatively, one might simply scale up the pair correlation values.

8. Density Functionals for Atomic Computations

The perturbation method of the previous section which introduces a *Coulomb hole* around each electron, to be seen by electrons of opposite spin, has its physical origin in Wigner's work[48] and was presented by Clementi[52] at the same time of Hohenberg and Kohn[63] work on density functionals, DF, also pioneered by Wigner[48]. We shall expand on the density functionals in the Chapter 7, section 8, where we report a few DFs and compare their ability to predict molecular dissociation energies. In this section we shall discuss *density functionals for atoms.*

However, before doing this, it might be advisable to have some quantitative "feeling" about the correlation energy. In the early sixties this very question was addressed by Clementi[58] and his approach was to compute an extensive library of Hartree-Fock functions[18-22] (at that time this task was very formidable), estimate relativistic corrections,[64,65] and subtract these quantities from the sum of the ionization potentials (or from corresponding estimates, when not available). In this way it was possible to gain *for the first time a reasonably accurate estimate of correlation energy for any atomic system and to correlate such information to molecular binding.*

From Fig. 3 we saw that the correlation increases with the number of electrons rather uniformly. In Fig. 5 (top and bottom inserts) we plot the correlation energy for atomic and ionic isoelectronic series for the 2 to the 10 electrons (He to Ne).

Table XII. Pair correlation energies for "atoms in molecules" used for the calculation of closed-shell molecular correlation energy. All energies are given in hartrees.

Atom						Atomic Correlation Energy
H						0.0000
1s	0.028					
Li						0.0453
1s	0.044					
2s	0.004	0.042				
Be						0.0944
1s	0.044					
2s	0.004	0.044				
B						0.1240
1s	0.043					
2s	0.004	0.045				
2p	0.002	0.020	0.040			
2p	0.002	0.020	0.035	0.040		
2p	0.002	0.020	0.035	0.035	0.040	
C						0.1551
1s	0.043					
2s	0.004	0.045				
2p	0.002	0.020	0.0411			
2p	0.002	0.020	0.0359	0.0411		
2p	0.002	0.020	0.0359	0.0359	0.0411	
N						0.1861
1s	0.043					
2s	0.004	0.045				
2p	0.002	0.020	0.0446			
2p	0.002	0.020	0.0391	0.0446		
2p	0.002	0.020	0.0391	0.0391	0.0446	

(Continued)

Table XII (cont.). Pair correlation energies for "atoms in molecules" used for the calculation of closed-shell molecular correlation energy. All energies are given in hartrees.

Atom							Atomic Correlation Energy
O							.2339
1s	0.043						
2s	0.004	0.045					
2p	0.002	0.020	0.040				
2p	0.002	0.020	0.035	0.040			
2p	0.002	0.020	0.035	0.035	0.040		
F							.3160
1s	0.043						
2s	0.004	0.045					
2p	0.002	0.020	0.040				
2p	0.002	0.020	0.035	0.040			
2p	0.002	0.020	0.035	0.035	0.040		
Ne							.381
1s	0.043						
2s	0.004	0.045					
2p	0.002	0.020	0.040				
2p	0.002	0.020	0.035	0.040			
2p	0.002	0.020	0.035	0.035	0.040		
Mg							.428
1s	0.040						
2s	0.004	0.044					
2p	0.004	0.023	0.051				
2p	0.004	0.023	0.039	0.051			
2p	0.004	0.023	0.039	0.039	0.051		
3s	0.001	0.004	0.006	0.006	0.006	0.042	

(Continued)

Table XII (cont.). Pair correlation energies for "atoms in molecules" used for the calculation of closed-shell molecular correlation energy. All energies are given in hartrees.

Atom										Atomic Correlation Energy
Si										.494
1s	0.040									
2s	0.004	0.044								
2p	0.004	0.023	0.051							
2p	0.004	0.023	0.039	0.051						
2p	0.004	0.023	0.039	0.039	0.051					
3s	0.001	0.004	0.006	0.006	0.006	0.042				
3p	0.000	0.002	0.002	0.000	0.000	0.017	0.029			
3p	0.000	0.002	0.000	0.002	0.000	0.017	0.026	0.029		
3p	0.000	0.002	0.000	0.000	0.002	0.017	0.026	0.026	0.029	
S										.595
1s	0.040									
2s	0.004	0.044								
2p	0.004	0.023	0.051							
2p	0.004	0.023	0.039	0.051						
2p	0.004	0.023	0.039	0.039	0.051					
3s	0.001	0.004	0.006	0.006	0.006	0.042				
3p	0.000	0.002	0.002	0.000	0.000	0.017	0.029			
3p	0.000	0.002	0.000	0.002	0.000	0.017	0.026	0.029		
3p	0.000	0.002	0.000	0.000	0.002	0.017	0.026	0.026	0.029	
Cl										.667
1s	0.040									
2s	0.004	0.044								
2p	0.004	0.023	0.051							
2p	0.004	0.023	0.039	0.051						
2p	0.004	0.023	0.039	0.039	0.051					
3s	0.001	0.004	0.006	0.006	0.006	0.042				
3p	0.000	0.002	0.002	0.000	0.000	0.017	0.029			
3p	0.000	0.002	0.000	0.002	0.000	0.017	0.026	0.029		
3p	0.000	0.002	0.000	0.000	0.002	0.017	0.026	0.026	0.029	

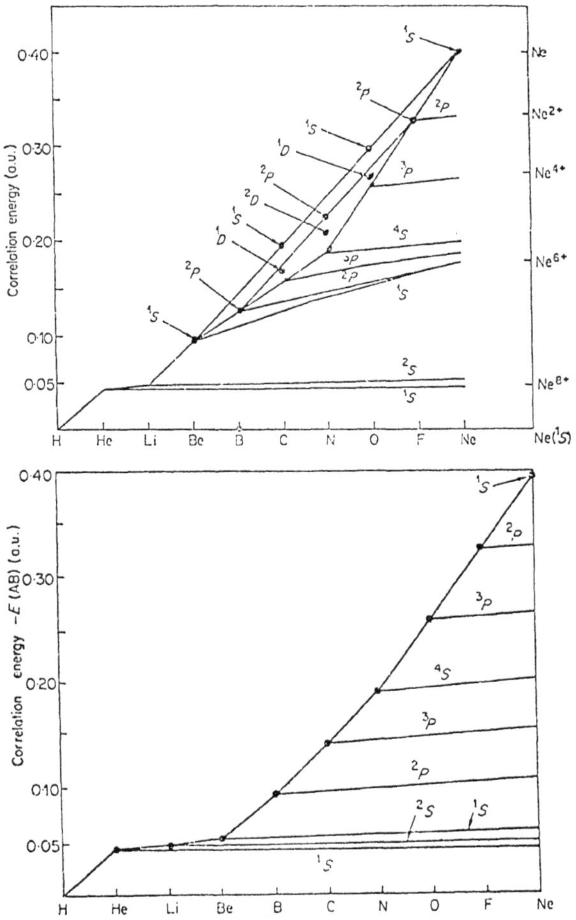

Figure 5. Correlation energies for atoms and ions from He to Ne. *Top*: Excited and ground states of the lowest electronic configuration for neutral atoms and ground states for ionic iso-electronic series. *Bottom*: Same as above, but for ground states and after removal of the 2s-2p degeneracy (E_{AB}).

Fig. 5 is taken from Clementi[58] (top inset) and Clementi-Veillard[66] (bottom inset). From the top inset of Fig. 5 we learn that the correlation energy for Be(^{1}S), C(^{1}S) O(^{1}S) and Ne(^{1}S) is on a straight line (we progress from 4, to 6, to 8 to 10 electrons), that the same holds for the ^{2}P state of B, N, and F (5 to 7 to 9 electrons). By considering now the isoelectronic series with 2,3,4,5,6,7,8 and 9 electrons (^{1}S, ^{2}S, ^{1}S, ^{2}P, ^{3}P, ^{4}S, ^{3}P and ^{2}P) and Ne we see an increase each time we add one electron, *stronger* when the added electron makes a new pair (case of O(^{3}P), F(^{2}P) and Ne(^{1}S)), *less so* when otherwise (case of Li (^{2}S), B(^{2}P), C(^{3}P)). Thus, there is experimental evidence on the amount of correlation energy within each pair, for pair-pair interaction, and also clear indication on the dependency on total angular momentum for the case of a given atom in different states.

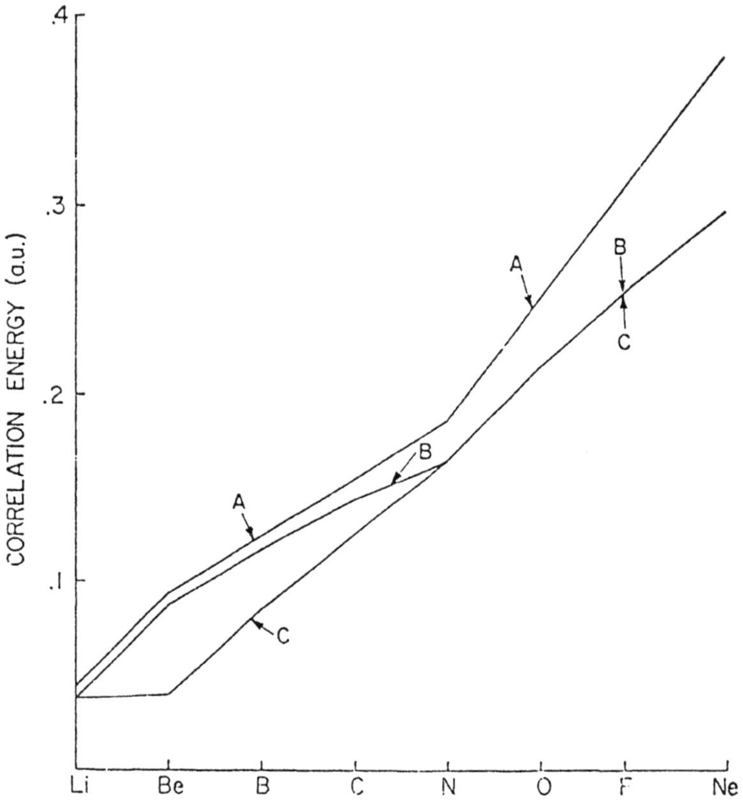

Figure 6. Simplified diagrams of the correlation energy for ground state neutral atoms from He to Ne. Curve A: Same as Fig. 5 (top); curve B: Same as Fig. 5 (bottom); curve C: From a three-determinant MCSCF function.

Fig. 5 (bottom) reports the correlation energies this time only for ground states and after removal of the 2s-2p degeneracy (Z effect) as computed from two MCSCF determinants.[66] Here we learn once more that the intra-pair correlation energy is larger than the inter-pair correlation and that the 2s correlation is minimal after 2s-2p degeneracy removal. All this is well known, and supports the empirical electron pair model of the previous section. At the same time however it shows that *the correlation energy is a very simple function of the number of electrons, thus of "the integral of the electronic density."*

The data of Fig. 5 also open more pressingly the question on the "proper reference" for atoms; indeed the one determinant appears to be not the best choice and two determinants for the case of (ns-np) degeneracy seem needed. Notice that we aim at a correlation energy model such that a "left over correlation error" can be most easily estimated. Along these lines we report in Fig. 6 the correlation energy obtained with three determinants -one being the Hartree-Fock- using non-orthogonal orbitals; this choice simplifies even further the correlation energy diagram for the first row atoms in the ground state.

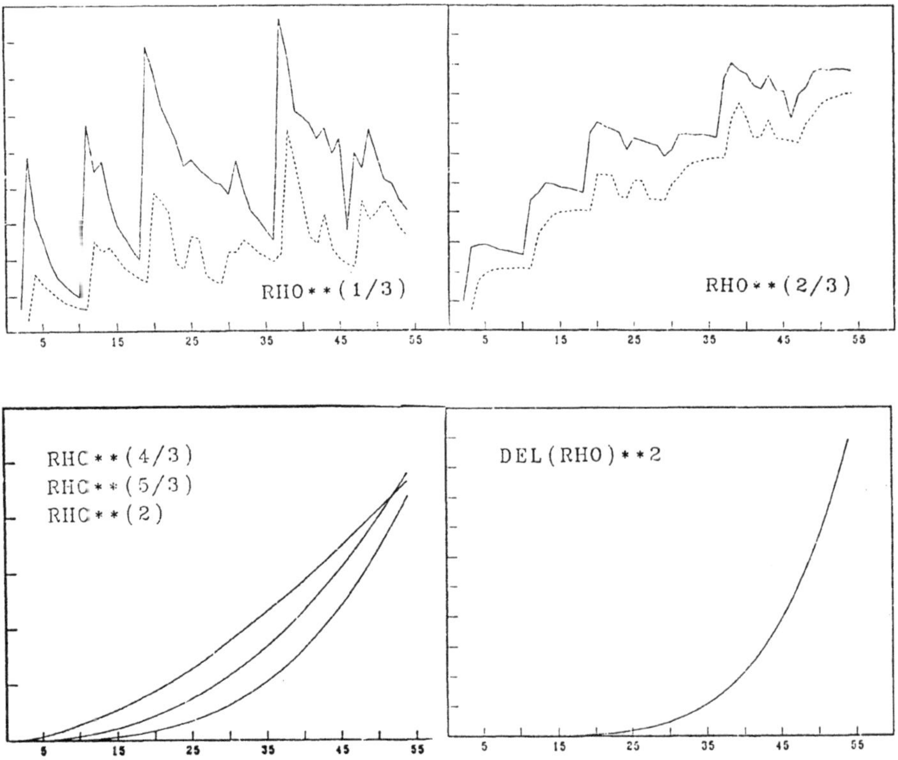

Figure 7. Few integrands of the Hartree-Fock density (from He to Xe).

Let us briefly digress and report the computation of a few integrals in terms of some function of the Hartree-Fock density. Specifically, in Fig. 7 we report for the ground state atoms He to Xe the value of the quantities

$$\int \rho^n d\rho \qquad \int (\nabla \rho)^2 d\rho$$

where $n=$ 1/3, 2/3, 4/3, 5/3, and 2, in order to learn on the curvature of these functions, which will appear in the density functional expressions given below.

Let us now consider several DFs. We refer to section 8 of Chapter 7 for their expression, exception made for the three DFs, Wigner[48] (W), Gombas[67] (G), and Wigner-Clementi[68] (W-SPP), given below.

Wigner:

$$E_c = -\int a_1\rho^{1/3}(a_2 + \rho^{1/3})^{-1}\rho d\rho$$

Gombas:

$$E_c = -\int [a_1\ \rho^{1/3}(a_2 + \rho^{1/3})^{-1} + b_1 \ln(1 + b_2\rho^{1/3})]\rho d\rho$$

Wigner-Clementi:

$$E_c = -\int \{a_1\ (\rho_\alpha^{1/3} + \rho_\beta^{1/3})[a_2 + (\rho_\alpha^{1/3} + \rho_\beta^{1/3})]^{-1}\}\rho d\rho$$

$$+ \int \{a'_1\ \rho_\alpha^{1/3}[a'_2 + \rho_\alpha^{1/3}]^{-1}\}\rho_\alpha d\rho + \int \{a'_1\ \rho_\beta^{1/3}[a'_2 + \rho_\beta^{1/3}]^{-1}\}\rho_\beta d\rho$$

to be used with the self-interaction correction.[69] Note that in the W-SPP functional a logarithmic term (following Gombas) was added to account for high electronic density; however, during optimization of the parameters this term became vanishingly small and therefore is not given in the functional. In addition we consider the Lie-Clementi functional[70] (LC), the Gunnarson and Lundqvist functional[71] with self interaction correction (GL-SPP), the Wosko-Wilk-Nusair functional[72] also with the self interaction correction (WWN-SPP), the Perdew-Zunger functional[73] with the Perdew gradient correction[74] (PZ), the Becke functional[75] (B), and the Lee-Yang-Parr functional[76] (LYP). The parameter values for the above functionals are those reported in Chapter 7.

For all these functionals we have determined a new set of parameter values by fitting to those atomic and ionic correlation energies presented in Clementi and Veillard[66] (but somewhat modified to take care of a sign error in the estimate of the Lamb shift and new data on the relativistic correction). This fitting is done to test the sensitivity of the parameters.

In Table XIII we report the new values of the constants, which can be compared with those of section 8 of Chapter 7. In this table we report also the parameters for the WC potential where however the logarithmic terms are omitted as a result of the fitting. This experimentation, as well the re-fitting of Gombas density functional, seams to strongly point to the conclusion that, at least for the atoms from He to Xe, there is no need of a special correction for intermediate density. In other words, we can state that the Wigner expression, derived for low electronic density, seams to be appropriate especially after a re-fit of its parameters. The

Table XIII. Value of parameters in DF (original = old, refitted = new). For details see also Chapter 7, Eqs. 8-19.

W(old)	$a_1 = 0.0564$	$a_2 = 0.0795$		
W(new)	$a_1 = 0.1890$	$a_2 = 5.8032$		
G(old)	$a_1 = 0.0357$	$a_2 = 0.0562$	$b_1 = 0.03111$	$b_2 = 2.39$
G(new)	$a_1 = 11.7182$	$a_2 = 950.0$	$b_1 = 0.0042$	$b_2 = 39.5494$
W-SPP	$a_1 = 0.6272$	$a_2 = 3.8956$	$a'_1 = 0.4981$	$a'_2 = 1.8400$
LC(old)	$a_1 = 0.02096$	$a_2 = 1.2$	$b_1 = 0.02096$	$b_2 = 2.39$
LC(new)	$a_1 = 10.6121$	$a_2 = 949.9$	$b_1 = 0.0028$	$b_2 = 950.$
GLSPP(old)	$C_I = 0.0666$	$C_{II} = 0.0406$	$x_I = 11.4$	$x_{II} = 15.9$
GLSPP(new)	$C_I = -2.4500$	$C_{II} = -2.5466$	$x_I = 0.9504$	$x_{II} = 1.1953$
VWNSPP(old)	$A_I = 0.06218$	$A_{II} = 0.03109$	$A_{III} = -0.03377$	$b_I = 3.72744$
	$b_{II} = 7.06042$	$b_{III} = 1.13107$	$c_I = 12.9352$	$c_{II} = 18.0578$
	$c_{III} = 13.0045$	$x_{oI} = -0.10498$	$x_{oII} = -0.32500$	$x_{oIII} = -4.7584D-3$
VWNSPP(new)	$A_I = -0.4036$	$A_{II} = -0.4742$	$A_{III} = -0.0901$	$b_I = 3.4976$
	$b_{II} = 5.5495$	$b_{III} = -0.3267$	$c_I = 12.6876$	$c_{II} = 15.7717$
	$c_{III} = 12.7286$	$x_{oI} = 0.2305$	$x_{oII} = 0.1437$	$x_{oIII} = -9.4680$
PZ(old)	$\gamma_I = -0.1423$	$\gamma_{II} = -0.0843$	$\beta_I^1 = 1.0529$	$\beta_I^2 = 0.3334$
	$\beta_{II}^1 = 1.3981$	$\beta_{II}^2 = 0.2611$	$A_I = 0.0311$	$A_{II} = 0.01555$
	$B_I = -0.048$	$B_{II} = -0.0269$	$C_I = 0.0020$	$C_{II} = 0.0007$
	$D_I = -0.0116$	$D_{II} = -0.0048$	$c_1 = 1.667D-3$	$c_2 = 2.568D-3$
	$c_3 = 2.3266D-2$	$c_4 = 7.389D-6$	$c_5 = 8.723$	$c_6 = 0.472$
	$c_7 = 7.389D-2$			
PZ(new)	$\gamma_I = -0.1277$	$\gamma_{II} = -1.6585$	$\beta_I^1 = 1.2253$	$\beta_I^2 = 0.0463$
	$\beta_{II}^1 = 87.2906$	$\beta_{II}^2 = -14.3669$	$A_I = 0.0482$	$A_{II} = -0.4461$
	$B_I = -0.0368$	$B_{II} = -0.4083$	$C_I = -0.0444$	$C_{II} = 0.1572$
	$D_I = -0.0404$	$D_{II} = 0.3807$	$c_1 = 2.721D-3$	$c_2 = 3.848D-3$
	$c_3 = 5.954D-3$	$c_4 = 9.378D-3$	$c_5 = 3.2232$	$c_6 = 0.8385$
	$c_7 = 1.7025$			
B(old)	$c_{\alpha\beta} = 0.63$	$c_{\sigma\sigma} = 0.96$	$a = 0.00375$	$b = 0.007$
B(new)	$c_{\alpha\beta} = 1.2987$	$c_{\sigma\sigma} = 1.7047$	$a = 0.4008$	$b = 0.5180$
LYP(old)	$a = 0.04918$	$b = 0.132$	$c = 0.2533$	$d = 0.349$
LYP(new)	$a = 0.03725$	$b = 0.4975$	$c = 1.1348$	$d = 0.065$

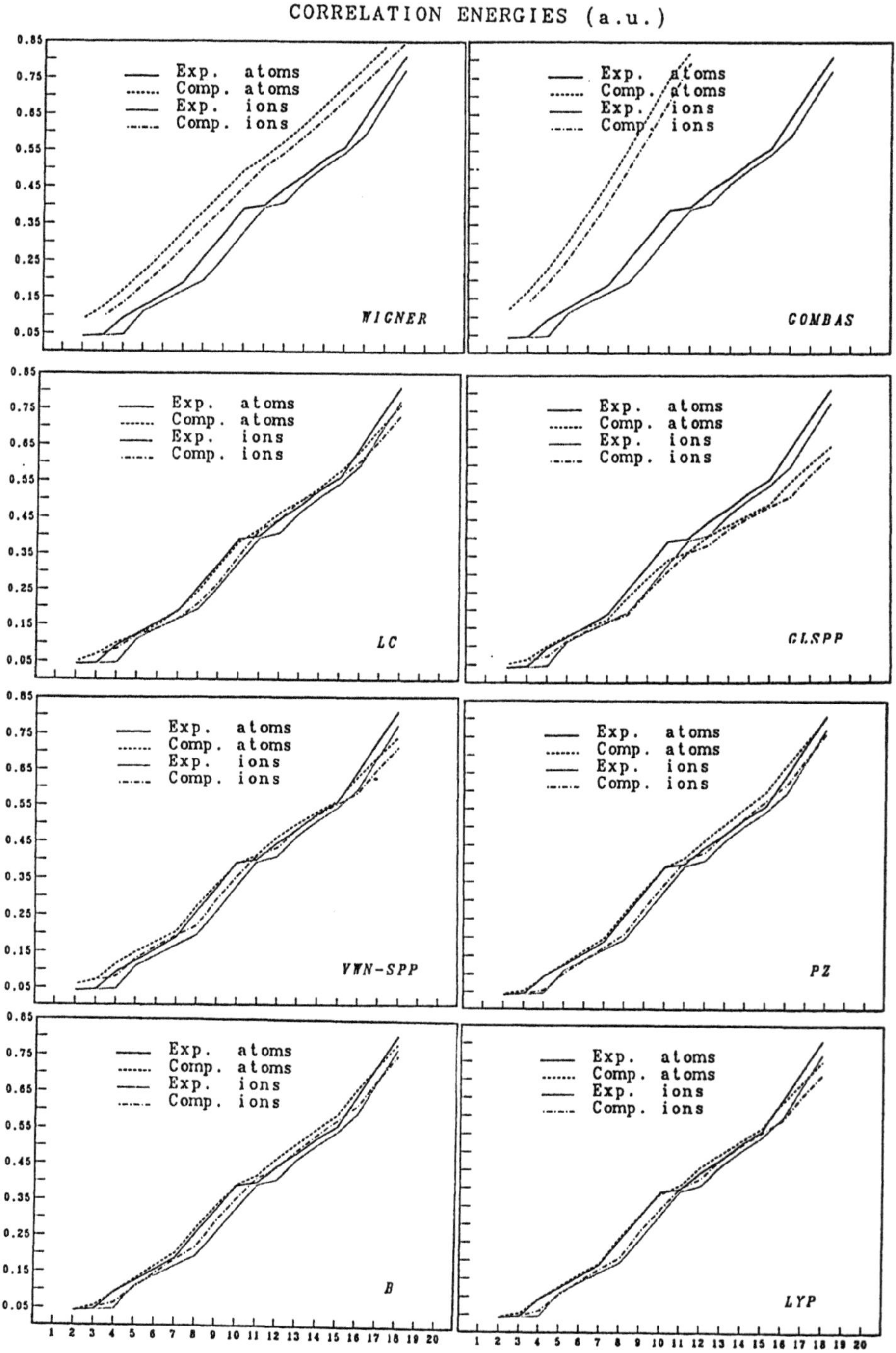

Figure 8. Density functional derived correlation energies from He to Ar (neutral atoms and first positive ions) computed using the original values of the parameters.

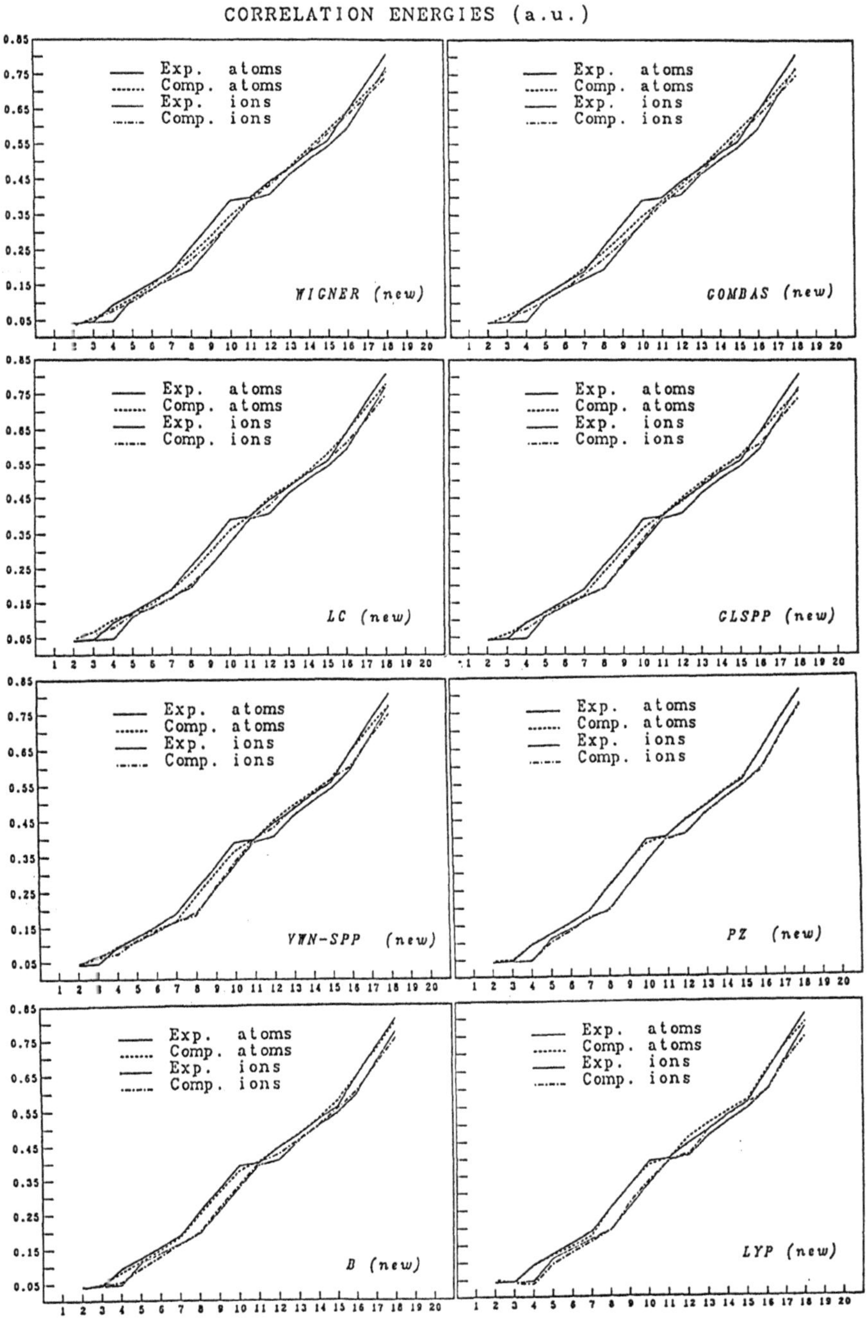

Figure 9. Density functional derived correlation energies from He to Ar (neutral atoms and first positive ions) computed using the constants of Table XIII.

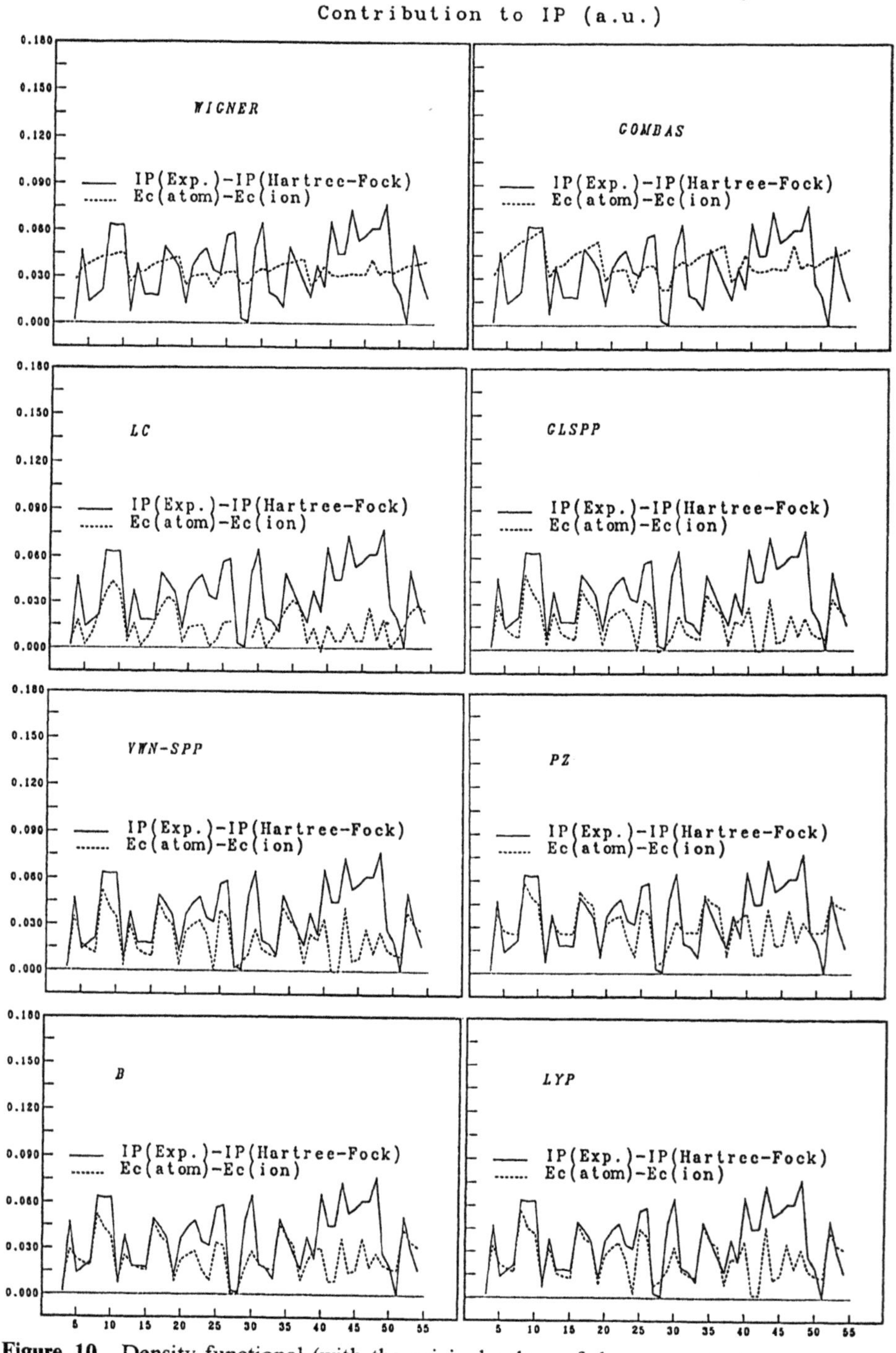

Figure 10. Density functional (with the original values of the parameters) and Hartree-Fock errors in the computation of the ionization potentials from Li to Xe.

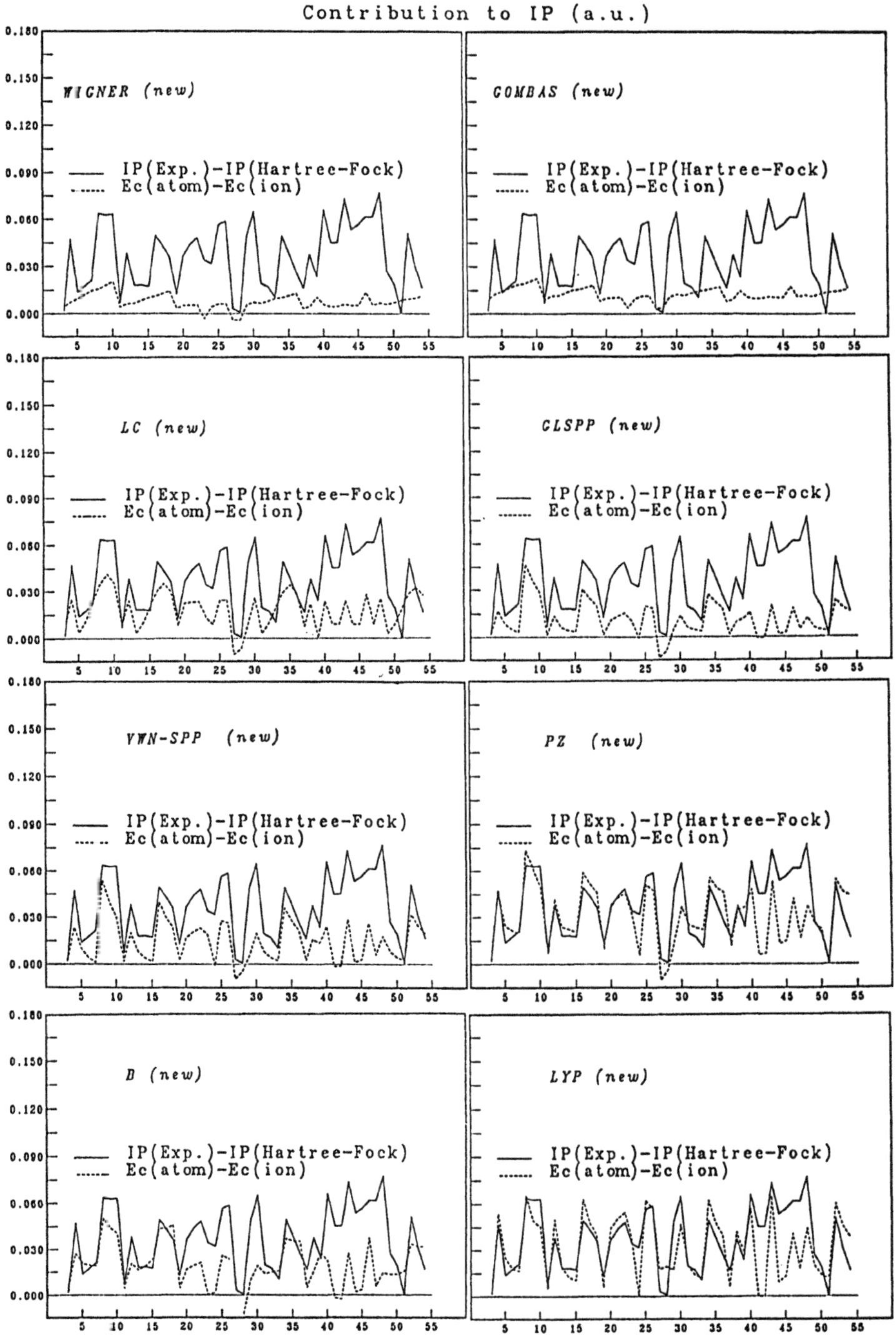

Figure 11. Density functional (with the new values of the parameters) and Hartree-Fock errors in the computation of the ionization potentials from Li to Xe.

new parameters are notably different from the original ones and this points to linear dependency and non uniqueness. In Fig. 8 we report the correlation energies from He to Ar, atoms and ions, as obtained from the original parameters, and in Fig. 9 we report the equivalent quantities using the new parameters of Table XIII. Finally, in Figs. 10 and 11 we compare the Hartree-Fock and density functional errors present in the computation of the ionization potentials, both with the old and new values of the parameters (as previously reported at the bottom of Fig. 4). Note that the detailed analyses of the error in computing the ionization potentials represents a good test to assess the reliability of the DFs.

The first conclusion is that in general the functionals provide a most reasonable estimates of the correlation energy in atoms; however, the error is still too large for chemical accuracy as it can be seen from the diagrams in Figs. 9 and 10. Fitting on a more extensive set of data will likely bring about a sufficiently reliable formulation. We also conclude that it is becoming more and more important to attempt to perform CI or MCSCF benchmark computations on Ar, Kr, and Xe, so as to guide and check the density functional parametrization studies. These studies will also require inclusion of open shell cases and, equally important, of ionic iso-electronic series to insure that the DFs will not contain sistematic errors in the parameterization. The computer program by Sasaki, Noro and Sekiya, called ATOMCI,[77] is being presently used in this laboratory in collaboration with the above colleagues, exactly in the direction indicated.

9. Configuration and Momentum Expectation Values

The physical significance of $< r^n >$ and $< p^n >$ and their relationship to experimental properties is considered here briefly. Remarkable progress in the development of high speed computers and automatic diffractometers has made the experimental determination of charge densities of atoms and molecules of all sizes a routine technique.[78] The quantity that is obtained from experiment is the scattering intensity $I(\vec{k})$ and the form factor $f(\vec{k})$. The form factor $f(\vec{k})$ is the Fourier transform of the electron density $\rho(\vec{r})$

$$f(\vec{k}) = \int \rho(\vec{r}) e^{i\vec{k}\cdot\vec{r}} d\vec{r} \tag{106}$$

and

$$I(\vec{k}) = \bar{f}(\vec{k}) f(\vec{k}). \tag{107}$$

It may be noted that within the HF model the electron density is $\rho(\vec{r}) = \sum_i^{occ} n_i \bar{\phi}_i(\vec{r}) \phi_i(\vec{r})$, where ϕ_i's are the orbitals and n_i's are occupation numbers.

The $< r^n >$-values are defined as

$$< r^n > = \int \rho(\vec{r}) r^n d\vec{r} \tag{108}$$

For atoms, this simplifies to

$$< r^m > = 4\pi \sum n_i \int_0^\infty R_{nl}^2(r) r^{m+2} dr \tag{109}$$

here $R_{nl}(r)$ are the radial part of the atomic orbitals.

The $< r^n >$ may also be obtained directly from the $f(\vec{k})$ (see Ref. 79) for atoms:

$$< r^n > = \begin{cases} (-1)^{n/2}(n+1) f^{(n)}(0) & \text{for even } n \\ (-1)^{(n+1)/2}\left(\frac{2}{\pi}\right)(n+1)\int_0^\infty \frac{f^{(n)}(k)}{k}\, dk & \text{for odd } n \end{cases} \tag{110}$$

For $< r^{-n} >$ one has

$$< r^{-2} > = \int_0^\infty dk\, k\, f(k) \tag{111}$$

and

$$< r^{-1} > = (2/\pi)\int_0^\infty dk\ f(k) \tag{112}$$

The $< r^o >$-value is simply the normalization constant and is equal to N, the number of electrons. The $< r^{-1} >$-value is numerically equal to Z^{-1} times the nuclear-electron attraction energy and the $< r >$-value is related to the atomic size. Also, the $< r^2 >$-value is proportional to the diamagnetic susceptibility which can be experimentally determined. There exist tabulations of various $< r^n >$ -values[80] and can be computed without much numerical labor. These properties also serve well in assessing the quality of the wave function.

In recent years several theoretical studies of electron momentum distributions have been reported. With the advent of X-ray, synchrotron and (e,2e) experimental techniques, very accurate measurements of Compton profiles of atoms and molecules have become possible[81]. One can evaluate the $< p^n >$-values from these experimental Compton profiles using the following equation[82]

$$< p^n > = 2(n+1)\int_0^\infty J(q) q^n dq. \tag{113}$$

for $0 \leq n \leq 4$. Here, $J(q)$ is the spherically averaged Compton profile and q is the projection of the electrons initial momentum upon the direction of the scattering vector,

$$J(q) = \int_{|q|}^{\infty} \frac{I(p)}{2p} dp \tag{114}$$

where

$$I(p) = \iint \gamma(\vec{p}) p^2 \sin \theta_p d\theta_p d\phi_p \tag{115}$$

is the radial momentum density and $\gamma(\vec{p})$ is the corresponding atomic momentum density. There exists an exhaustive tabulation of various $< p^n >$-values of atoms and ions and also detailed studies on the shell structure of $I(p)$ and the non-monotonicity of the electron momentum density in the literature.[83] The importance of some $< p^n >$-values was explicitly pointed out by Epstein.[84] The $< p >$-value could be correlated with the shielding in nuclear magnetic resonance spectroscopy. The electronic energy is numerically equal to $< p^2 > /2$ by the virial theorem. Relativistic corrections to the total energy in the Breit-Pauli perturbation theory are proportional to $< p^4 >$. Another observable directly available from the Compton profile is $J(0)$, which is equal to $< p^{-1} > /2$. The Fourier transform, $\phi(\vec{p})$, of an orbital $\psi(\vec{r})$ is defined by

$$\phi(\vec{p}) = \frac{1}{(2\pi)^{3/2}} \int \psi(\vec{r}) e^{i\vec{p} \cdot \vec{r}} d\vec{r} \tag{116}$$

For a central field problem this reduces to

$$\phi_{nl}(p) = \left(\frac{2}{\pi}\right)^{1/2} \int_0^{\infty} R_{nl}(r) j_l(pr) r^2 dr \tag{117}$$

where $R_{nl}(r)$ is the radial part of the orbital and $j_l(pr)$ is the spherical Bessel function of order l. The above integral is analytic for STOs and GTOs. Thus the atomic electron momentum density $\gamma(p)$ and the radial momentum density $I(p)$ can be readily evaluated employing analytic expressions.

$$\gamma(p) = \sum n_i \phi_{nl}^2(p) \tag{118}$$

and

$$I(p) = 4\pi p^2 \gamma(p) \tag{119}$$

The various $< p^n >$ values are obtained employing

$$< p^n > = 4\pi \int_0^\infty \gamma(p) p^{n+2} dp = \int_0^\infty I(p) p^n dp. \tag{120}$$

A special property package to evaluate $< p^n >$, $< r^n >$ and $J(q)$ for STOs and GTOs is now included in the Atomic package of MOTECC.

10. The Finite Element Method (FEM)

As we mentioned in the previous sections, the most commonly used basis sets are global in nature, viz the STO and GTO functions that, despite their relative success in atomic and molecular calculations, are not completely satisfactory, especially in the latter. Thus it is entirely reasonable to search for alternative avenues, particularly those where much work is pursued by modern numerical analyses.

The finite element method (FEM) offers an alternative to global basis set methods [85] In the FEM, the region of solution is divided into elements, and (local) basis functions are defined within the elements to represent the wavefunction in that particular region of the space. Applications of the FEM to quantum chemistry are relatively recent, and have shown the high degree of accuracy that can be achieved.[86] Flores, Clementi, and Sonnad[9] have recently formulated a FE technique which yields very accurate Hartree-Fock energies for atoms from $Z=1$ to $Z=86$; the use of the p-version of the finite element technique[87] permits systematic improvement in the description of the orbitals, without devoting much effort towards the optimization and choice of basis functions. The atomic program using this technique has been named ATOMFEM and is part of the MOTECC-89 (see Ref.16).

In this context, the simplest way to understand the FEM is to pretend that the local basis functions can be considered global bases which are, however, non-zero only in a particular element. Thus, one can express the orbitals as linear combinations of local basis functions, calculate the Fock and overlap matrices and resolve the eigenvalue problem as in the usual expansion method.

There are two important differences between global and local basis function approaches. In the former, the continuity of the orbitals is guaranteed by the use of continuous global basis functions, but in the latter it is imposed at both ends of each element. The other difference is that in the FEM, because of the local nature of the basis functions, the matrices of local operators have a particular band structure that does not appear in the global approach. These two features must be considered in developing a solution technique, and in interpreting the results of a calculation using the FEM.

In their formulation, Flores *et al.*[9] use spherical coordinates and express the atomic orbitals (AO) as products of a radial and an angular part. The angular part is known to be a spherical harmonic, so only the radial coordinate is divided in elements; i.e. they use a one-dimensional grid. The AO can then be written as follows

$$\phi_{i\lambda\alpha}(r,\theta,\psi) = \sum_{x}^{\mathrm{NE}} \sum_{p}^{n_x} \left(c_{i\lambda p}^{x} \, \chi_p^{x}(r) \right) Y_{\lambda}^{\alpha}(\theta,\psi) \qquad (121)$$

where $\chi_p^x(r)$ is the pth radial basis function of element x, NE is the number of elements and n_x is the number of basis functions defined in element x. Note that the radial basis functions are independent not only of the symmetry subspecies but also of the species, in contrast to Eq. (63). Note also that n_x is independent of λ; in other words, the same set of radial basis functions is used for all symmetries. As a consequence, the n_x terms describing, for example, a ground state wavefunction can be used to describe excited states too, of course with different $c_{i\lambda p}^x$'s. Hereafter, we will simply refer to $\{\chi_p^x(r)\}$ as basis functions, although properly speaking, the basis functions include the spherical harmonics. For linear molecules a two dimensional grid would be necessary and a general molecular program would employ a three dimensional grid. The local nature of the basis function ensures that in molecules there will be no basis set superposition error. In addition, one can think of "prefabricated" element partitioning for functional groups which could speed up the SCF process.

In order to define the local basis functions and calculate matrix elements, we use the following transformation between the global and local radial coordinates

$$r = \bar{r}_x + t\,\delta_x \ ; \quad -1 \leq t \leq 1 \qquad (122)$$

where $\bar{r}_x$ is the mid point and δ_x is the half width of element x.

The basis functions can be written in terms of the variable t as follows

$$I_1(t) = \frac{P_0 - P_1}{2} = \frac{1-t}{2}$$

$$I_2(t) = \frac{P_0 + P_1}{2} = \frac{1+t}{2} \qquad (123)$$

and

$$s_j(t) = (4j+2)^{-1/2} \left[P_{j+1}(t) - P_{j-1}(t) \right]$$

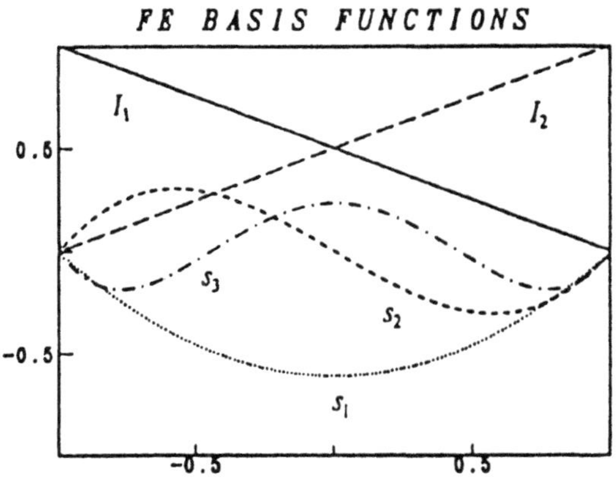

Figure 12. Some of the local basis functions used in this work.

where $I(t)$ and $I_2(t)$ are linear interpolation functions, $s_j(t)$ is the shape function (SF) of order j and $P_j(t)$ is the jth order Legendre polynomial. Fig. 12 shows I_1, I_2 and SFs up to third order.

The coefficients that go together with the interpolation functions represent values of the radial orbital, that are required to be coincident at the nodes. Thus, it is clear that inclusion of both interpolation functions is mandatory in every element. Inclusion of SFs permits the orbital to have non-zero higher order derivatives (s_1 provides a constant non zero second derivative, s_2 makes the second derivative linear in the element and provides a constant non-zero third derivative and so on). An element is described to order i if $s_1, ...s_i,$ and both interpolation functions are used to represent the orbital in that element.

The lack of continuity condition across the elements for the orbital derivatives does not downgrade the quality of the solution. In fact it has been found that, as we will show below, even with relatively few basis functions, the first and often higher derivatives are virtually continuous at the nodes.

The calculation of one-electron integrals (arising from one-electron operators) is straightforward but for the sake of completeness we give the formulas. Integration over the angular coordinates is done separately and involves only the spherical harmonics.

We first define the following functions

$$
\begin{aligned}
P_{ij}(n) &= \int_{-1}^{+1} P_i(t)\, P_j(t)\, t^n\, dt \qquad i,j \geq 0 \\
S_{ij}(n) &= \int_{-1}^{+1} s_i(t)\, s_j(t)\, t^n\, dt \qquad i,j \geq 1 \\
I_{ab}(n) &= \int_{-1}^{+1} I_a(t)\, I_b(t)\, t^n\, dt \qquad (a,b) = \{1,2\} \\
SI_{i,a}(n) &= \int_{-1}^{+1} s_i(t)\, I_a(t)\, t^n\, dt \\
C(i,j) &= \left(\frac{2j+1}{2}\right)^{1/2} \left(\frac{2i+1}{2}\right)^{1/2}
\end{aligned}
\tag{124}
$$

where the integrals S_{ij}, I_{ab} and $SI_{i,a}$ are calculated in terms of P_{ij}, that are computed using recursion relations between Legendre polynomials.

The overlap integrals are given by

$$
\begin{aligned}
\int_{-1}^{+1} I_a(t)\, I_b(t)\, (\bar{r} + t\delta)^2\, \delta\, dt &= \sum_{k=1}^{3} \binom{2}{k-1} \bar{r}^{(3-k)}\, \delta^k\, I_{ab}(k-1) \\
\int_{-1}^{+1} s_i(t)\, s_j(t)\, (\bar{r} + t\delta)^2\, \delta\, dt &= \sum_{k=1}^{3} \binom{2}{k-1} \bar{r}^{(3-k)}\, \delta^k\, S_{ij}(k-1) \\
\int_{-1}^{+1} s_i(t)\, I_a(t)\, (\bar{r} + t\delta)^2\, \delta\, dt &= \sum_{k=1}^{3} \binom{2}{k-1} \bar{r}^{(3-k)}\, \delta^k\, SI_{i,a}(k-1)
\end{aligned}
\tag{125}
$$

For the nuclear attraction operator we have the following integrals

$$
\begin{aligned}
-Z \int_{-1}^{+1} I_a(t)\, I_b(t)\, (\bar{r} + t\delta)\, \delta\, dt &= -Z\,\delta\, [\bar{r}\, I_{ab}(0) + \delta\, I_{ab}(1)] \\
-Z \int_{-1}^{+1} s_i(t)\, s_j(t)\, (\bar{r} + t\delta)\, \delta\, dt &= -Z\,\delta\, [\bar{r}\, S_{ij}(0) + \delta\, S_{ij}(1)] \\
-Z \int_{-1}^{+1} s_i(t)\, I_a(t)\, (\bar{r} + t\delta)\, \delta\, dt &= -Z\,\delta\, [\bar{r}\, SI_{i,a}(0) + \delta\, SI_{i,a}(1)]
\end{aligned}
\tag{126}
$$

For the kinetic energy operator

$$
\hat{T} = \frac{1}{(\bar{r} + t\delta)^2} \left\{ \frac{-1}{2\delta^2} \frac{\partial}{\partial t} \left[(\bar{r} + t\delta)^2 \frac{\partial}{\partial t} \right] + \frac{l(l+1)}{2} \right\} \tag{127}
$$

we have

$$T_{11}^{l} = T_{22}^{l} = \frac{1}{24\delta^2}\left[(\bar{r}+\delta)^3 - (\bar{r}-\delta)^3\right] + \frac{l(l+1)}{3}\,\delta$$

$$T_{12}^{l} = -\frac{1}{24\delta^2}\left[(\bar{r}+\delta)^3 - (\bar{r}-\delta)^3\right] + \frac{l(l+1)}{6}\,\delta$$

$$T_{ij}^{S} = \frac{C(i,j)}{2\delta}\sum_{k=1}^{3}\binom{2}{k-1}\bar{r}^{(3-k)}\,\delta^{(k-1)}\,P_{ij}(k-1) + \frac{l(l+1)}{2}\,S_{ij}(0)\,\delta \tag{128}$$

$$T_{i,1-2}^{SI} = (\mp 1/2)\frac{1}{2\delta}\left(\frac{2i+1}{2}\right)^{1/2}\sum_{k=1}^{3}\binom{2}{k-1}\bar{r}^{(3-k)}\,\delta^{(k-1)}\,P_{i,0}(k-1)$$

$$+\frac{l(l+1)}{2}\,SI_{i,1-2}(0)\,\delta$$

where the $-$ and $+$ signs of the last expression stand for interpolation functions 1 and 2, respectively.

The calculation of electron repulsion integrals (ERI), that are necessary to construct the Coulomb and exchange matrices, presents some interesting features and will be discussed in detail. The simplest way to calculate the ERI is to use the multipole expansion of r_{12}^{-1}. In this way, integration over the angular coordinates can be done separately and then, every ERI is a finite linear combination of radial ERI. The general expression for a radial ERI is

$$(pq|rs)_{x,y}^{l} = \int_x\int_y \chi_p^x(r_1)\,\chi_q^x(r_1)\,\chi_r^y(r_2)\,\chi_s^y(r_2)\,\frac{r_<^l}{r_>^{l+1}}\,r_1^2 r_2^2\,dr_1 dr_2 \tag{129}$$

where l indicates the term of the multipole expansion and x,y stand for the elements where the basis functions are defined. It is clear that for a radial ERI to be non-zero, basis functions corresponding to the *same electron* must belong to the *same element*. Thus, there are two cases:

(a) $x \neq y$. In this case either $r_1 > r_2$ or $r_2 > r_1$. Assuming the first possibility we have

$$(pq|rs)_{x,y}^{l} = \int_x \chi_p^x(r_1)\,\chi_q^x(r_1)\,r_1^{1-l}\,dr_1\int_y \chi_r^y(r_2)\,\chi_s^y(r_2)\,r_2^{2+l}dr_2 = (pq)_x^{1-l}\,(rs)_y^{2+l} \tag{130}$$

(b) $x = y$. In this case the radial ERI may be written as follows

$$(pq|rs)^{l}_{x,x} = \int_{r_x^-}^{r_x^+} \chi_p^x(r_1)\, \chi_q^x(r_1)\, r_1^{1-l} \int_{r_x^-}^{r_1} \chi_r^x(r_2)\, \chi_s^x(r_2)\, r_2^{2+l}\, dr_2 dr_1 + \int_{r_x^-}^{r_x^+} \chi_p^x(r_1)\, \chi_q^x(r_1)\, r_1^{2+l} \int_{r_1}^{r_x^+} \chi_r^x(r_2) \chi_s^x(r_2)\, r_2^{1-l}\, dr_2 dr_1 \qquad (131)$$

In order to calculate expressions (130) and (131) we use transformation (122). It is evident from formulae (123) that the basis functions can be written as linear combinations of powers of t

$$\chi_p(t) = C_p \sum_{\mu=1}^{\max(p)} a_{\mu p}\; t^{\mu-1} \qquad (132)$$

When $l \leq 1$, the expressions for integrals (130) and (131) are very simple

$$(pq)^n_x = \sum_{i=1}^{n+1} \binom{n}{i-1} \bar{r}^{(n-i+1)}\, \delta^i \int_{-1}^{+1} t^{i-1}\, \chi_p^x(t)\, \chi_q^x(t)\, dt \qquad (133)$$

$$(pq|rs)^l_{x,x} = \sum_{i=1}^{3+l} \sum_{j=1}^{2-l} \binom{2+l}{i-1} \binom{1-l}{j-1}\, \delta^{(i+j)}\, \bar{r}^{(5-i-j)} \qquad (134)$$

$$\left[\int_{-1}^{+1} t_1^{j-1}\, \chi_p^x \chi_q^x \int_{-1}^{t_1} t_2^{i-1}\, \chi_r^x \chi_s^x\, dt_2\, dt_1 + \int_{-1}^{1} t_1^{i-1}\, \chi_p^x \chi_q^x \int_{t_1}^{1} t_2^{j-1}\, \chi_r^x \chi_s^x\, dt_2\, dt_1 \right]$$

where

$$\int_{-1}^{+1} t^{i-1}\, \chi_p\, \chi_q\, dt = C_p\, C_q \left[A_{p,q,i-1}(t)\right]_{-1}^{+1} \qquad (136)(135)$$

$$\int_{-1}^{+1} t^{\alpha}\, \chi_p\, \chi_q \int_{t_1}^{(\mp 1)} t_2^{\beta}\, \chi_r\, \chi_s\, dt_2\, dt_1 = C_p\, C_q\, C_r\, C_s \left\{ A_{r,s,\,\beta}(\mp 1)\, [A_{p,q,\,\alpha}(t)]_{-1}^{+1} - [B^{\alpha,\,\beta}_{p,q,r,s}(t)]_{-1}^{+1} \right\}$$

with

$$A_{r,s,\,\beta}(t) = \sum_{\lambda,\,\sigma} \frac{a_{\lambda r}\, a_{\sigma s}}{\lambda + \sigma + \beta - 1}\, t^{\lambda+\sigma+\beta-1} \qquad (137)$$

$$B^{\alpha,\,\beta}_{p,q,r,s}(t) = \sum_{\mu,\,\nu,\,\lambda,\,\sigma} \frac{a_{\mu p}\, a_{\nu q}\, a_{\lambda r}\, a_{\sigma s}}{\lambda + \sigma + \beta - 1}\; \frac{t^{\mu+\nu+\lambda+\sigma+\alpha+\beta-2}}{\mu + \nu + \lambda + \sigma + \alpha + \beta - 2} \qquad (138)$$

Note that integrals (135) and (136) are independent of the grid.

When $l > 1$ a new type of integral appears in expression (133)

$$(pq)_x^{1-l} = C_p\, C_q \sum_{\mu,\,\nu} a_{\mu p}\, a_{\nu q}\, D(\mu + \nu - 2, l - 1, x) \tag{139}$$

where

$$D(\mu + \nu - 2, l - 1, x) = \int_{-1}^{+1} \frac{t^{\mu+\nu-2}}{(\bar{r}_x + t\delta_x)^{l-1}}\, \delta_x\, dt \tag{140}$$

For integral (131) we have

$$(pq|rs)_{x,x}^{l} = C_p\, C_q\, C_r\, C_s \sum_{i=1}^{3+l} \binom{2+l}{i-1} \delta_x^i\, \bar{r}_x^{(3+l-i)}$$

$$\Bigg\{ (-A_{r,s,i-1}(-1)) \sum_{\mu\,\nu} a_{\mu p}\, a_{\nu q}\, D(\mu + \nu - 2, l - 1, x) + \sum_{\mu\,\nu\,\lambda\,\sigma} a_{\mu p}\, a_{\nu q}\, a_{\lambda r}\, a_{\sigma s} \tag{141}$$

$$\left[\frac{D(\mu + \nu + \lambda + \sigma + i - 4, l - 1, x)}{\lambda + \sigma + i - 2} + G(\mu + \nu + i - 3, \lambda + \sigma - 2, l - 1, x) \right] \Bigg\}$$

where

$$G(\alpha, \beta, n, x) = \int_{-1}^{+1} t_1^{\alpha} \int_{t_1}^{1} \frac{t_2^{\beta}}{(\bar{r}_x + t_2\delta_x)^n}\, \delta_x\, dt_2\, dt_1 \tag{142}$$

It is possible to obtain simple formulae to calculate integrals G and D, but direct application causes precision problems even when the computations are carried out in extended precision. We have therefore derived recursion schemes that are free of this problem and also more efficient.

We have used the following recursion relations.

$$D(\alpha, 1, x) = \left[\frac{t^{\alpha}}{\alpha}\right]_{-1}^{+1} - \frac{\bar{r}_x}{\delta_x} D(\alpha -1, 1, x) \qquad \alpha > 0$$

with (143)

$$D(0,1,x) = \ln \frac{\bar{r}_x + \delta_x}{\bar{r}_x - \delta_x}$$

When a complete set of $D(\alpha,1,x)$ integrals has been calculated, we use the following recursion to obtain the rest

$$D(\alpha, n, x) = \frac{\delta_x}{(n-1)\bar{r}_x}\left[\frac{t^{\alpha+1}}{(\bar{r}_x + t\delta_x)^{n-1}}\right]_{-1}^{+1} - \frac{\alpha - n + 2}{(n-1)\bar{r}_x} D(\alpha, n-1, x) \quad n > 1 \tag{144}$$

For the G integrals we apply similar recursion relations in the same way

$$G(\alpha, \beta, n, x) = \frac{\delta_x}{(n-1)\bar{r}_x}\left\{\frac{1}{(\bar{r}_x + \delta_x)^{n-1}}\left[\frac{t^{\alpha+1}}{\alpha+1}\right]_{-1}^{+1} - \frac{1}{\delta_x} D(\alpha + \beta + 1, n-1, x)\right\}$$
$$- \frac{\beta - n + 2}{(n-1)\bar{r}_x} G(\alpha, \beta, n-1, x) \tag{145}$$

$$G(\alpha, \beta, 1, x) = \frac{1}{\beta}\left[\frac{t^{\alpha+1}}{\alpha+1} - \frac{t^{\alpha+\beta+1}}{\alpha+\beta+1}\right]_{-1}^{+1} - \frac{\bar{r}_x}{\delta_x} G(\alpha, \beta - 1, 1, x) \tag{146}$$

and to calculate $G(\alpha, 0, 1, x)$ we use

$$G(\alpha, 0, 1, x) = -\frac{(-1)^{\alpha+1}}{\alpha+1} \ln \frac{\bar{r}_x + \delta_x}{\bar{r}_x - \delta_x} + \frac{1}{\alpha+1} D(\alpha+1, 1, x) \tag{147}$$

Of course, formulas can be derived for integral (131) including only either G or D integrals. All these schemes appear to be free of precision problems.

The number of integrals $(pq|rs)^{l}_{x,y}$ needed (for each value of l) is

$$\frac{1}{8}\sum_{x}^{\mathrm{NE}} (n_x^4 + 2n_x^3 + 3n_x^2 + 2n_x) \tag{148}$$

for case (b) $(x = y)$, where n_x is the number of basis functions of element x. For case (a) $(x \neq y)$ we have

$$\sum_{\substack{x \\ x>y}}^{NE} \sum_{y}^{NE} \left(\frac{n_x(n_x+1)}{2}\right) \left(\frac{n_y(n_y+1)}{2}\right) \tag{149}$$

It is evident from these expressions that the number of ERIs involved in a FE-HF calculation may be quite high, even though it depends only on the second power of NE. However, since the basis functions are independent of the atom, the ERIs are transferable between different atoms, if the same grid is used.

The next step is the construction of the Fock matrix using one and two-electron (ERI) integrals. For elements of the core and overlap matrices we have the following expressions

$$\begin{aligned} h^{\lambda}_{n_p, n_q} &= (\chi^x_p \mid \hat{h}^{\lambda} \mid \chi^x_q) + \ldots\ldots \\ S_{n_p, n_q} &= (\chi^x_p \mid \chi^x_q) + \ldots\ldots \end{aligned} \tag{150}$$

where the periods indicate that integrals involving functions of adjacent spatial elements may contribute to the same matrix elements, because of the local boundary conditions. The indices n_p, n_q take into account these boundary conditions, i.e., the size of the matrices has been reduced to the number of independent coefficients.

Considering for instance a closed shell case the matrix elements of the Coulomb operator may be written as follows

$$J^{\lambda}_{n_p, n_q} = \sum_{\mu} \sum_{y} \sum_{r,s}^{n_y} D^{\mu}_{n_r, n_s} J^{x,y}_{\lambda pq, \mu rs} + \ldots\ldots \tag{151}$$

where $J^{x,y}_{\lambda pq, \mu rs}$ is the same ERI defined in expression (16), but including the indices x and y to indicate the element of basis functions p,q and r,s respectively. For the exchange matrix we have

$$K^{\lambda}_{n_p, n_q} = \sum_{\mu} \sum_{s}^{n_x} \sum_{r}^{n_y} D^{\mu}_{n_r, n_s} K^{x,y}_{\lambda pq, \mu rs} + \ldots\ldots \tag{152}$$

where $K^{x,y}_{\lambda pq, \mu rs}$ is the exchange integral defined according to expression (17), the superindices indicating that the basis functions p and s belong to element x and r,q belong to element y. $D^{\mu}_{n_r n_s}$ is an element of the $D_{c\mu}$ density matrix, as defined in Eq. (11), i.e.

$$D^{\mu}_{n_r, n_s} = \sum_{i\mu} N_{i\mu} \, c^x_{i\mu r} \, c^y_{i\mu s} \tag{153}$$

where $N_{i\mu}$ is the occupation number of the orbital i corresponding to symmetry μ, and elements x and y can be the same or different; in other words, the density matrix is , of course, dense.

The Fock operators corresponding to the restricted open shell case can be handled without difficulty as well. All matrices have been factorized into disconnected symmetry blocks. Note also that the matrices h^λ, S and J^λ, that correspond to local operators, have block structure, each block corresponding to a particular spatial element and having one term in common with the next because of the local connectivity conditions.

```
x                  ---------------> l_1^1
x x                -------------> s_1^1
x x x              -----------> s_2^1
x x x x            ---------> l_2^1 + l_1^2
o o o x x          -------> s_1^2
o o o x x x        -----> l_2^2 + l_1^3
o o o o o x x      ---> s_1^3
o o o o o x x x    -> l_2^3
```

In the K^λ matrix however, all spatial elements are intrinsically connected; hence this matrix, and consequently the Fock matrix, are both dense. As in the global basis function approach, the matrices are symmetric. The above drawing shows the structure of one symmetry block of the Fock matrix. The element distribution of this matrix correspond to the following ordering of basis functions within each spatial element, l_1, $s_1 \ldots s_n$, l_2, if the element is described to order n. The core and Coulomb operators contribute only to the area marked with "x" but the exchange operator contributes to all matrix elements. Note, in this respect, that application of the FEM is particularly promising in combination with methods that involve only local potentials such as SCF-X_α, or density functional approaches.

Having constructed the Fock matrix, the next step is to solve the generalized eigenvalue problem which, of course, can be done using the traditional Housholder diagonalization method. Iterative methods, however, are particularly attractive because the ratio between the number of solutions needed and the dimension of the problem is usually rather small.

The hierarchic nature of the basis functions in the p-version can be used to develop an efficient computational scheme that is relatively easy to implement. A grid is first chosen, and the problem is solved at the lowest polynomial level. The AOs resulting from the computation at this level are used as initial guess for the next higher level, and this approach is repeated until the results are satisfactory. It

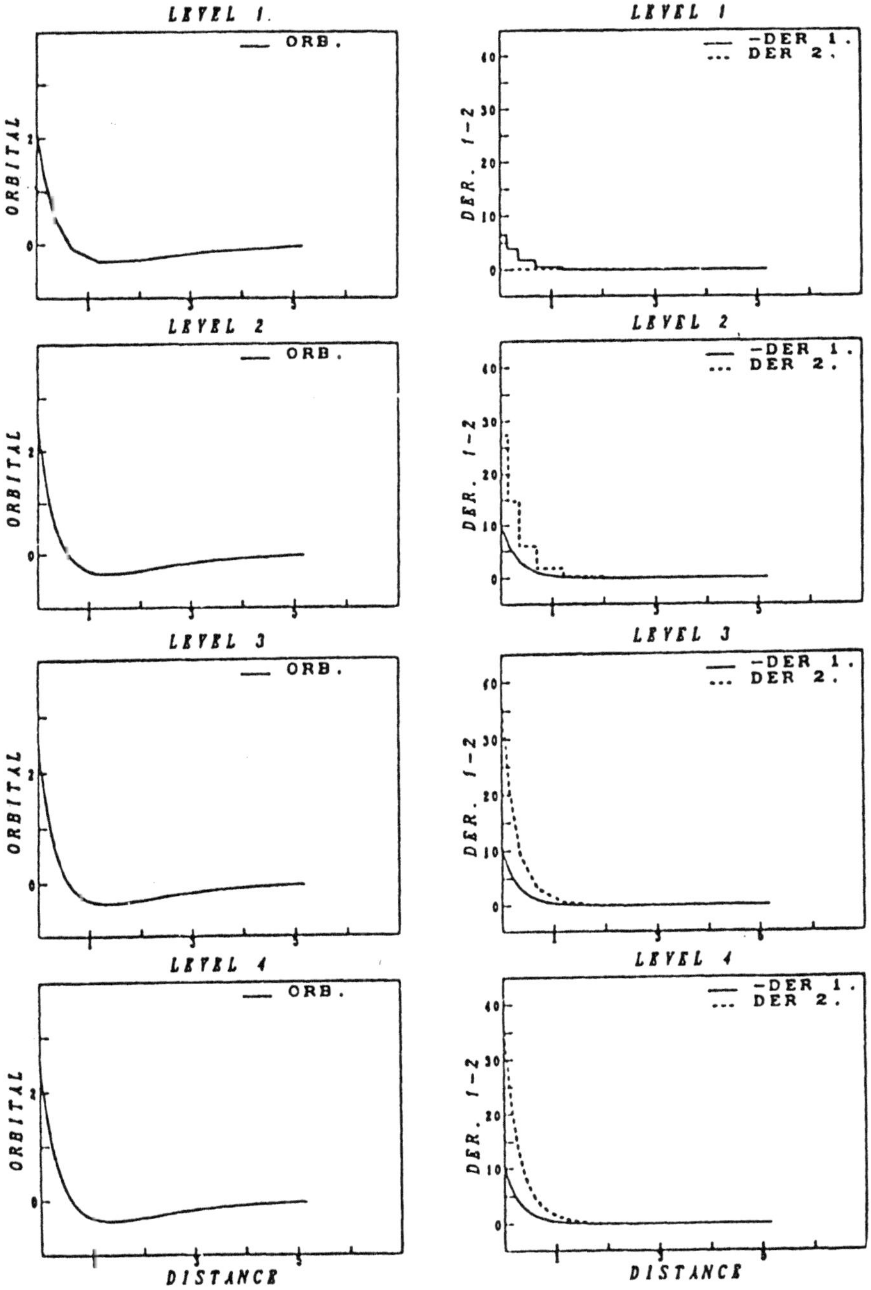

Figure 13. 2s radial orbital of Be and derivatives calculated at four levels.

Table XIV. Element layout and distribution of SF for the different levels. R(i) represents the end of region i (in a.u.) that is divided in NE(i) equal-width elements. The number j in the level's columns indicate that every element of region i is described to jth order. (The only exception is region 1, in which the first element is described with SF up to order 5 at every level). The total number of elements is 25. The numbers in parentheses represent the dimension of a single symmetry block of the Fock matrix.

			Level				
i	R(i)	NE(i)	**1**	**2**	**3**	**4**	**5**
1	0.2	6	2	3	3	4	4
2	0.5	3	2	3	3	4	4
3	1.0	3	2	3	3	4	4
4	2.0	3	1	1	2	2	3
5	6.0	4	1	1	2	2	3
6	12.0	3	1	1	2	2	3
7	20.0	1	1	1	2	2	3
8	34.0	1	0	0	1	1	2
9	60.0	1	0	0	0	0	1
			(64)	(75)	(87)	(98)	(111)

is thus possible to use multigrid concepts without actually constructing multiple grids.[88] Note that there is no need to recalculate the complete set of integrals at successive levels; one simply needs those arising from the new basis functions.

We present two examples of this kind of computation. Figure 13 shows the 2s radial orbital and its first and second derivatives (drawn from 0 to 5.2 a.u.), resulting from calculations on the Beryllium atom. Four levels have been employed, defined as follows: in level m every element is described to order m-1. In the first level only (linear) interpolation functions are used so the resulting orbital consists of linear segments, the vertices of these segments showing the location of the nodes. The first derivative is constant inside the elements and presents large discontinuities at the nodes, the second derivative is, of course, zero. At the second level, the inclusion of first order shape functions, makes the orbital smoother, the first derivative is linear with tiny discontinuities at the nodes that cannot be seen in the figure, and the second derivative is constant inside the elements with large discontinuities at the nodes. At the third level the second derivative is linear inside the elements and at the last level, functions and derivatives are smooth even though a relatively crude mesh is used.

The other example is the calculations reported[9] for rare gas atoms from He to Rn and for various states of Vanadium. A common grid is employed for the calcu-

Table XV. Total energies (in Hartree), relative errors with respect to the Hartree-Fock limits and virial ratios for the He isovalent series. The number in parentheses (x) of the virial ratios indicates that the first x decimal figures are 0 (− 2.00 ...) or 9 (− 1.99 ...). The grid and SF distribution is described in Table XIV.

	Level				
Atom	1	2	3	4	5
He	-2.8608054	-2.8608054	-2.8616657	-2.8616657	-2.8616798
	-1.(3)75	-1.(3)75	-2.(5)14	-2.(5)14	-2.(7)82
	3.1 10^{-4}	3.1 10^{-4}	4.9 10^{-6}	4.9 10^{-6}	6.9 10^{-8}
Ne	-128.54225	-128.54242	-128.54699	-128.54700	-128.54710
	-1.(4)53	-1.(3)76	-1.(6)43	-1.(6)39	-2.(8)75
	3.8 10^{-5}	3.6 10^{-5}	8.6 10^{-7}	7.8 10^{-7}	0.0
Ar	-526.80476	-526.80732	-526.81735	-526.81738	-526.81751
	-1.(4)69	-1.(4)65	-2.(6)14	-1.(7)53	-1.(7)89
	2.4 10^{-5}	1.9 10^{-5}	3.0 10^{-7}	2.5 10^{-7}	0.0
Kr	-2752.0072	-2752.0318	-2752.0544	-2752.0547	-2752.0550
	-2.(4)16	-1.(5)00	-1.(6)87	-1.(6)74	-1.(7)71
	1.7 10^{-5}	8.4 10^{-6}	2.2 10^{-7}	1.1 10^{-7}	0.0
Xe	-7231.8117	-7232.1094	-7232.1339	-7232.1374	-7232.1384
	-2.(3)14	-1.(5)40	-2.(5)17	-1.(6)07	-1.(7)64
	4.5 10^{-5}	4.0 10^{-6}	6.2 10^{-7}	1.4 10^{-7}	0.0
Rn	-21864.181	-21866.645	-21866.702	-21866.769	-21866.772
	-2.(3)13	-1.(5)15	-2.(5)85	-1.(6)58	-2.(6)18
	1.2 10^{-4}	5.8 10^{-6}	3.2 10^{-6}	1.4 10^{-7}	0.0

Table XVI. Total energies (in hartree), and relative energies (in millihartree) with respect to the ^4F state, for several electronic states of V. We have used the grid described in Table XIV.

Term	Level 1	2	3	4	5
^6D	-942.86288	-942.86775	-942.87936	-942.87942	-942.87978
	3.60	3.61	4.52	4.53	4.55
^2G	-942.80222	-942.80710	-942.81958	-942.81965	-942.82004
	64.26	64.26	64.30	64.30	64.29
^2F	-942.71867	-942.72358	-942.73600	-942.73607	-942.73646
	147.81	147.78	147.88	147.88	147.87
^2D	-942.70498	-942.70986	-942.72227	-942.72234	-942.72273
	161.53	161.50	161.61	161.61	161.60
^4P	-942.80239	-942.80730	-942.81976	-942.81983	-942.82021
	64.09	64.06	64.12	64.12	64.12
^2P	-942.78115	-942.78605	-942.79850	-942.79856	-942.79895
	85.33	85.31	85.38	85.39	85.38
^4F	-942.86648	-942.87136	-942.88388	-942.88395	-942.88433
	0.00	0.00	0.00	0.00	0.00

lations on all atoms and states, so that the ERIs need to be calculated only once. Table XIV shows the grid, and defines the SF distribution for the different levels. Table XV shows results for the rare gas atoms. Total energies, their relative errors with respect to the Hartree-Fock limits resulting from numerical calculations,[89] along with the virial ratios, are given for every level. The final energies are in complete agreement with the Hartree-Fock limits, with the exception of He where a very small difference was encountered. Note also that the virial ratios are consistent with the quality of the total energy. Table XVI shows absolute and relative energies (with respect to the ^4F state) for some states of V. It is readily seen that the relative energies are very consistent at the different levels. Even for the ^6D state, whose electron configuration $(..3d^4 4s^1)$ is different from the others $(..3d^3 4s^2)$, the third level already provides a good relative energy. Note that global basis sets are usually optimized (exponents of STO or GTO) for the ground state or at most a few excited states of each atom.[22] In the FE method, optimization of the shape function distribution may be done using adaptive computation techniques, that automatically increase the number of basis functions only in the region where they are most needed, during the course of the calculation.

To conclude it must be said that that the FEM which has long been applied successfully in engineering, represents an alternate strategy for obtaining accurate analytic solutions for many electron problems.

Appendix 2A. Two-Electron Integrals

The integrals for kinetic, nuclear-electron attraction, electron-electron repulsion energies and overlap terms using cartesian gaussian basis sets have not been given here because they are readily available in the literature (see also Chapter 7). However, the expressions for the two electron integrals in cartesian gaussian functions related to the perturbation term of Eq. (98) are presented here.

$$E_2 = \iint g_1(1) g_2(1) \frac{e^{-\omega r_{12}^2}}{r_{12}} g_3(2) g_4(2)\, d\vec{r}_1 d\vec{r}_2 \tag{A.1}$$

This may be written in terms of x_1, y_1, z_1 and x_2, y_2, z_2, the coordinates of the two electrons, and cartesian gaussian basis functions as follows:

$$E_2 = \int_{-\infty}^{\infty}\int_{-\infty}^{\infty}\int_{-\infty}^{\infty}\int_{-\infty}^{\infty}\int_{-\infty}^{\infty}\int_{-\infty}^{\infty} x_1^{l_1+l_2} y_1^{m_1+m_2} z_1^{n_1+n_2} x_2^{l_3+l_4} y_2^{m_3+m_4} z_2^{n_3+n_4} \times e^{-(\gamma_1 r_1^2 + \gamma_2 r_2^2)} \frac{e^{-\omega r_{12}^2}}{r_{12}}\, dx_1 dy_1 dz_1 dx_2 dy_2 dz_2 \tag{A.2}$$

here, $\gamma_1 = \alpha_1 + \alpha_2$ and $\gamma_2 = \alpha_3 + \alpha_4$. The above integral finally factorizes to:

$$E_2 = \frac{(2\pi)^{5/2}}{(\gamma_1 + \gamma_2 y^2)^{3/2}} \left[\frac{1-y^2}{\gamma_2}\right]^{3/2} \frac{1}{\delta^{1/2}} H_x H_y H_z \frac{1}{(\mu+1)} \tag{A.3}$$

where $y^2 = \alpha/(\gamma_2 + \alpha)$, $\mu = \tilde{L} + \bar{L} + \tilde{M} + \bar{M} + \tilde{N} + \bar{N}$, and H_x is given by :

$$\begin{aligned} H_x = &\sum_{\bar{l}=0}^{l'} \binom{l'}{\bar{l}} y^{2(l'-\bar{l})} \frac{\bar{l}!}{2^{\bar{l}}} \sum_{q=0}^{[\bar{l}/2]} \frac{1}{q!\bar{L}!} \left[\frac{1-y^2}{\gamma_2}\right]^{\bar{l}-q} \frac{(-1)^{\tilde{l}}\tilde{l}!}{2^{\tilde{l}}} \\ &\times \sum_{p=0}^{[\tilde{l}/2]} \frac{1}{p!\tilde{L}!} \frac{(1-y^2)^{\tilde{L}}}{(\gamma_1 + \gamma_2 y^2)^{\tilde{l}-p}} (\bar{L} + \tilde{L} - 1)!! \left[\frac{-2}{\delta}\right]^{(\bar{L}+\tilde{L})/2} \end{aligned} \tag{A.4}$$

Here $l = l_1 + l_2$, $l' = l_3 + l_4$ and $\tilde{l} = l + l' - \bar{l}$, and $\tilde{L} = \tilde{l} - 2p$ and $\bar{L} = \bar{l} - 2q$. H_y and H_z are analogously defined. δ is given by the following expression

$$\delta = \frac{1-y^2}{\gamma_2} + \frac{(1-y^2)^2}{\gamma_1 + \gamma_2 y^2} \tag{A.5}$$

Another integral which can also be evaluated is given below

$$S_2 = \iint g_1(1) g_2(1) e^{-\omega r_{12}^2} g_3(2) g_4(2)\, d\vec{r}_1 d\vec{r}_2 \tag{A.6}$$

This may be written in terms of x_1, y_1, z_1 and x_2, y_2, z_2, the coordinates of the two electrons and cartesian gaussian basis functions as follows:

$$\begin{aligned} S_2 = &\int_{-\infty}^{\infty}\int_{-\infty}^{\infty}\int_{-\infty}^{\infty}\int_{-\infty}^{\infty}\int_{-\infty}^{\infty}\int_{-\infty}^{\infty} x_1^{l_1+l_2} y_1^{m_1+m_2} z_1^{n_1+n_2} x_2^{l_3+l_4} y_2^{m_3+m_4} z_2^{n_3+n_4} \\ &\times e^{-(\gamma_1 r_1^2 + \gamma_2 r_2^2)}\, e^{-\omega r_{12}^2}\, dx_1 dy_1 dz_1 dx_2 dy_2 dz_2 \end{aligned} \tag{A.7}$$

The above integral reduces to:

$$S_2 = \frac{\pi^3}{(\gamma_1 + \gamma_2 y^2)^{3/2}} \left[\frac{1-y^2}{\gamma_2}\right]^{3/2} P_x P_y P_z \tag{A.8}$$

where

$$P_x = \sum_{\bar{l}=0}^{l'} \binom{l'}{\bar{l}} y^{2(l'-\bar{l})} \left[\frac{1-y^2}{\gamma_2}\right]^{\bar{l}/2} \frac{1}{(\gamma_1 + \gamma_2 y^2)^{\tilde{l}/2}} \frac{(\bar{l}-1)!!}{2^{\bar{l}/2}} \frac{(\tilde{l}-1)!!}{2^{\tilde{l}/2}} \tag{A.9}$$

P_y and P_z are analogously defined.

Thus, employing the identity $r_{12}^2 = (x_1 - x_2)^2 + (y_1 - y_2)^2 + (z_1 - z_2)^2$, the integrals of the type:

$$D_2 = \iint g_1(1)g_2(1)(a + b\,r_{12})\frac{e^{-\omega r_{12}^2}}{r_{12}}g_3(2)g_4(2)\; d\vec{r}_1 d\vec{r}_2$$

and

$$A_2 = \iint g_1(1)g_2(1)(a + b\,r_{12})e^{-\omega r_{12}^2}g_3(2)g_4(2)\; d\vec{r}_1 d\vec{r}_2$$

may be easily evaluated. The current version of our computer program, ATOMSCF,[16] evaluates the integrals D_2 and A_2 for a given value of a, b and ω over the cartesian gaussian basis functions.

Appendix 2B. Pseudopotentials

A feature currently available in the ATOMSCF program is the possibility of using effective core or pseudopotentials with STOs or GTOs. For STOs a convenient form of the pseudopotential, $V_{c\lambda}(r)$ is

$$V_{c\lambda}(r) = \sum_{K=1}^{m_\lambda} d_{\lambda k} r^{n_{\lambda k}-2} e^{-\zeta'_{\lambda k} r} \tag{B.1}$$

where λ is the symmetry, m_λ is the number of terms in the pseudopotential; the exponents, $\zeta_{\lambda k}$ and the n_k are supplied by the user.

For GTOs the form of the pseudopotential used is

$$V_{c\lambda}(r) = \sum_{K=1}^{m_\lambda} d_{\lambda k} r^{n_{\lambda k}-2} e^{-\alpha'_{\lambda k} r^2} \tag{B.2}$$

The parameters $d_{\lambda k}$, $n_{\lambda k}$, $\zeta_{\lambda k}$, α', and m_λ have been extensively calibrated for various atoms;[90-93] in the molecular code KGNMOL we have used those by W.J. Stevens *et al.*[91]

With these choices of the pseudopotentials the one electron part of the Fock operator (see Eqs. 5, 6, and 8) contains an additional term, $H'_{\lambda pq}$, given below

$$H'_{\lambda pq} = \frac{1}{d_\lambda} \sum_{\alpha} < \chi_{p\lambda\alpha} \mid V_{c\lambda} \mid \chi_{q\lambda\alpha} > \tag{B.3}$$

Exploiting the spherical symmetry of atoms, the $H'_{\lambda pq}$ for STO's is given by

$$H'_{\lambda pq} = [V_{2n\lambda p}(\zeta_{\lambda p}) V_{2n\lambda q}(\zeta_{\lambda q})]^{-1/2} \sum_{k=1}^{m_\lambda} d_{\lambda k} V_{n\lambda p + n\lambda q + n\lambda k - 2}[\frac{1}{2}(\zeta_{\lambda p} + \zeta_{\lambda q} + \zeta_{\lambda k})] \tag{B.4}$$

where $V_i(x) = x^{-i-1} i$

and for GTOs a basis orbitals $H'_{\lambda pq}$ is given by

$$H'_{\lambda pq} = [V_{2n\lambda p}(\alpha_{\lambda p}) V_{2n\lambda q}(\alpha_{\lambda q})]^{-1/2} \sum_{k=1}^{m_\lambda} d_{\lambda k} V_{n\lambda p + n\lambda q + n\lambda k - 2}[\frac{1}{2}(\alpha_{\lambda p} + \alpha_{\lambda q} + \alpha_{\lambda k})] \tag{B.5}$$

where $V_i(x) = \dfrac{(i-1)!!}{x^{(i+1)/2}}$

References

1. D.R. Hartree, Proc. Cambridge Phil. Soc., **24**, 89 (1928).
2. J.C. Slater, Phys. Rev., **34**, 1293 (1929)
3. J.C. Slater, Phys. Rev., **35**, 210 (1930)
4. V. Fock, Z. Physik, **61**, 126 (1930)
5. C. Zener, Phys. Rev., **36**, 51 (1930); J.C. Slater, Phys. Rev., **36**, 57 (1930); C. Eckart, Phys. Rev., **36**, 878 (1930).
6. C.C.J. Roothaan, Revs. Mod. Phys., **23**, 69 (1951).
7. C.C.J. Roothaan, Revs. Mod. Phys., **32**, 179 (1960).
8. I. Cacelli, R. Moccia and V. Carravetta, Chem. Phys., **90**, 313 (1984); I. Cacelli, V. Carravetta and R.Moccia, J. Chem. Phys., **85**, 7038 (1986); I. Cacelli, V. Carravetta, R. Moccia and A. Rizzo, J. Phys. Chem., **92**, 979 (1988); I. Cacelli, V. Carravetta, R. Moccia and A. Rizzo, J. Chem. Phys., **89**, 7301 (1988).
9. J.R. Flores, E. Clementi and V. Sonnad, J. Chem. Phys., **91**, 7030 (1989)
10. S. Huzinaga, Phys. Rev., **120**, 866 (1960); **127**, 131 (1961); also see J. Chem. Phys., **51**, 3971 (1969).
11. C.C.J. Roothaan and P.S. Bagus in *Methods in Computational Physics,* edited by B. Alder, S. Fernbach and M. Rotenberg (Academic, New York, 1963), Vol. 2, p. 47; C.C.J. Roothaan, L.M. Sachs and A.W. Weiss, Rev. Mod. Phys., **32**, 186 (1960).
12. E.R. Davidson, Chem. Phys. Lett., **21**, 565 (1973).
13. W.J. Hunt, P.J. Hay and W.A. Goddard III, J. Chem. Phys., **57**, 738 (1972); W.J. Hunt, W.A. Goddard III and T.H. Dunning, Jr., Chem. Phys. Lett., **6**, 147 (1970); Chem. Phys. Lett., **3**, 606 (1969) and **4**, 231 (1969).
14. G.L. Malli and J.P. Olive, J. Chem. Phys., **43**, 861 (1965); see also Laboratory of Molecular Structure, Tech. Rep. 1962-63, Part 2, The Univ. of Chicago, p. 258.
15. B. Roos, C. Salez, A. Veillard and E. Clementi, IBM Tech. Rep., **RJ-518** (1968).
16. MOTECC-89, Input/Output Documentation Manual, available to the MOTECC users. For additional information contact one of us (E.C.)
17. E. Clementi and D.L. Raimondi, J. Chem. Phys., **38**, 2686 (1963); E. Clementi, D.L. Raimondi, and W. Reinhardt, J. Chem. Phys. **47**, 1300 (1967).
18. a) E. Clementi, J. Chem. Phys., **38**, 996 (1963); b) *ibid.* **38**, 1001 (1963); c) *ibid.* **40**, 1944 (1964); d) *ibid.* **41**, 295 (1964); e) *ibid.* **41**, 303 (1964); f) E. Clementi, C.C.J. Roothaan and M. Yoshimine, Phys. Rev., **127**, 1618 (1962).
19. E. Clementi and A.D. McLean, Phys. Rev., **133**, A419 (1964); E. Clementi, A.D. McLean, D.L. Raimondi, and M. Yoshimine, **133**, A1274 (1964); E. Clementi, Phys. Rev., **135**, A980 (1964).
20. E. Clementi, IBM Journal, Supplement **9**, 2 (1965).
21. C. Roetti and E. Clementi, J. Chem. Phys., **60**, 4725 (1974); *ibid.* **60**, 3342 (1974).
22. E. Clementi and C. Roetti, At. Data Nucl. Data Tables, **14**, 177 (1974).

23. T. Kato, Commun. Pure Apl. Math., **10**, 151 (A57); N.H. March, Phys. Rev. A., **33**, 88 (1986).
24. V. Carravetta and H. Ågren, Phys. Rev. A, **35**, 1022 (1987); A. Cesar, H. Ågren and V. Carravetta, Phys. Rev. A, (1989).
25. H. Ågren and V. Carravetta, J. Chem. Phys., **87**, 370 (1987); H. Ågren and V. Carravetta, Phys. Rev. A, **38**, 2707 (1988).
26. V. Carravetta, J. Phys. B: At. Mol. Opt. Phys., **21**, 1777 (1988)
27. V. Carravetta, H. Ågren, H.J. Jensen, P. Jorgensen, and Jeppe Olsen, J. Phys. B., (1989).
28. L.D. Landau and E.M. Lifshitz, "Quantum Mechanics," Pergamon Press (1977).
29. R.F. McWeeny, Proc. Roy. Soc., **A196**, 215 (1946).
30. S.F. Boys, Proc. Roy. Soc., **A200**, 542 (1950).
31. (a) E. Clementi, J. Chem. Phys., **36**, 33 (1962); (b) E. Clementi, J. Chem. Phys. **46**, 3851 (1967).
32. B. Ransil, Rev. Mod. Phys., **32**, 239, 245 (1960).
33. S. Huzinaga, J.Andzelm, M. Klobukowski, E. Radzio-Andzelm, Y.Sakai and H.Tatewaki, Eds. Gaussian Basis Sets for Molecular Calculations (Elsevier Amsterdam, 1984)
34. See S. Wilson in Ab Initio Methods in Quantum Chemistry Part I edited by K.P.Lawley: Adv. in Chem. Phys. vol LXVII (1987)
35. E.R. Davidson and D. Feller, Chem. Rev., **86** 681 (1986).
36. E. Clementi, J. Phys. Chem., **84**, 2122 (1980).
37. E. Clementi and D.R. Davies, J. Comp. Phys., **1**, 223 (1966); for further details see D.R. Davies and E. Clementi, IBM Tech. Rep. (1955); for a later version refer to E. Ortoleva, G. Castiglione and E. Clementi, Comp. Phys. Comm., **19**, 337 (1980).
38. E. Clementi and G. Corongiu, Chem. Phys. Lett., **90**, 359 (1982); IBM Tech Rep., **POK-11** (1982).
39. F. Van Duijeneveldt, IBM Tech. Rep., **RJ-945** (1971).
40. R.C. Raffenetti, J. Chem. Phys., **59**, 5936 (1973). C.M. Reeves, J. Chem. Phys. **39**, 1 (1963); K. Ruedenberg , R.C. Raffenetti and R. Bardo in Energy, Structure and Reactivity Proc. 1972 Boulder Res. Conf. Theor. Chem., Wiley, New York, 1973.
41. D.M. Silver and W.C. Nieuwpoort, Chem. Phys. Lett., **57**, 421 (1978).
42. P.G. Mezey, Theor. Chim. Acta, **53**, 183 (1979).
43. A. Mohanty and E. Clementi, Chem. Phys. Lett, (1989); J. Chem. Phys. (in press).
44. L. Gianolio, R. Pavani and E. Clementi, Gazz. Chim. Italiana, **108**, 181 (1978).
45. S. Huzinaga and V. Sakai, J. Chem. Phys., **51**, 1371 (1969).
46. H. Tatewaki and S. Huzinaga, J. Comp. Chem., **3**, 205 (1980); R. Ditchfield, W.M. Hehre and J.A. Pople, J. Chem. Phys., **52**, 5001 (A70); A. Veillard, Theor. Chim. Acta, **12** 405 (1968); A.J.H. Wachters, J. Chem. Phys., **52**, 1033 (1970).

47. S. Fraga and Jacek Karwowski, Tables of Hartree-Fock Atomic Data, University of Alberta, Tech. Report (1974).
48. E. Wigner, Phys. Rev., **46**, 1002 (1934). E. Wigner and F. Seitz, Phys. Rev., **43**, 804 (1933).
49. C.A. Coulson and A.H. Neilson, Proc. Phys. Soc. **78**, 831(1961).
50. P.O. Löwdin, Advances in Chemical Physics, **2** (Interscience Publishers, New York, 1959).
51. For a detailed review dealing in electron correlation in atoms and molecules, see *Methods in Computational Physics*, edited by S. Wilson (Plenum, New York), Vol. I (1987); and also see G. Corongiu and E. Clementi, Electronic Structure of Atoms in *Claire Francqui Lecture Series: Part I*
52. E. Clementi, IBM Journal of Research and Development, **9**, 2 (1965).
53. S. Chakravorty and E. Clementi, Phys. Rev. A., **39**, 2290 (1989). In the IBM Tech. Rep. KGN-156 (Oct. 6, 1988) the pair correlation energies for atoms and positive ions from He to Xe are tabulated.
54. E. Clementi and S. Chakravorty, J. Chem. Phys. (in press); IBM Tech. Rep. KGN-215 (April 9, 1990).
55. C. Hollister and O. Sinanoglu, J. Am. Chem. Soc., **88** 13 (1966).
56. E. Clementi, J. Chem. Phys., **38**, 2780 (1963).
57. L.C. Allen, E. Clementi, and H.M. Gladney, Rev. Mod. Phys., **35**, 465 (1963).
58. E. Clementi, J. Chem. Phys., **38**, 2248 (1963); *ibid.* **39**, 175 (1963); *ibid.* **42**, 2783 (1965).
59. E. Clementi, J. Chem. Phys., **39**, 487 (1963).
60. For a review, see R.K. Nesbet, Advances in Chem. Phys., Vol. 9, 321 (1965); R.K. Nesbet, Phys. Rev., **175**, 2 (1968).
61. H.P. Kelly, Phys. Rev., **131**, 684 (1963); H.P. Kelly, Phys. Rev., **136**, B896 (1964).
62. O. Sinanoglu, J. Chem. Phys., **37**, 191 (1962).
63. P.Hohenberg and W. Kohn, Phys. Rev. B **136**,864 (1964)
64. H. Hartmann and E. Clementi, Phys. Rev., **133**, A1295 (1964).
65. E. Clementi, J. Mol. Spec., **12**, 18 (1964).
66. E. Clementi and A. Veillard, J. Chem. Phys., **44**, 3050 (1966); A. Veillard and E. Clementi, J. Chem. Phys., **49**, 2415 (1968).
67. P. Gombas, *Pseudopotentiale*, Springer Verlag, Wien (1967).
68. E. Clementi and G. Corongiu (unpublished).
69. H. Stoll, C.M.E. Pavlidou and H. Preuss, Theor. Chem. Acta, **49**, 143 (1978); A. Savin, H. Stoll and H. Preuss, Theor. Chim. Acta, **70**, 407 (1986).
70. G.C. Lie and E. Clementi, J. Chem. Phys., **60**, 1275 (1974); ibid, **60**, 1288 (1974).
71. O. Gunnarson and B.I. Lundqvist, Phys. Rev. B., **13**, 4274 (1976).
72. S.H. Vosko, L. Wilk and M. Nusair, Can. J. Phys., **58**, 1200 (1980); J.B. Lagowski and S.H. Vosko, J. Phys. B., **21**, 203 (1988).
73. J.P. Perdew and A. Zunger, Phys. Rev. B., **23**, 5048 (1981).
74. J.P. Perdew, Phys. Rev. B., **33**, 8822 (1986).
75. A.D. Becke, J. Chem. Phys., **88**, 1053 (1988).
76. C. Lee, W. Yang and R.G. Parr, Phys. Rev. B. **37**, 785 (1988).

77. a) F. Sasaki and M. Yoshimine, Phys. Rev. A, **9**, 17 (1974); *ibid.* **9**, 26 (1974). b) F. Sasaki, K. Tanaka, T. Noro, M. Togasi, T. Nomura, M. Sekiya, T. Gonoi, and K. Ohno, Theor. Chim. Acta, **72**, 123 (1987).
78. J.D. Dunitz, "X-ray analysis and structure of Organic Molecules" (Cornell Univ. Press, Ithaca, 1979); J.P. Glusker, B.K. Patterson and M. Rossi, Patterson and Pattersons (Oxford, New York, 1986); P. Coppens and M.B. Hall, Eds., Electron Distributions and Chemical Bond, Plenum, New York (1982).
79. A. Thakkar and T. Koga, J. Chem. Phys., **83**, 747 (1985); R. Benesch, S.R. Singh and V.H. Smith, Chem. Phys. Lett., **10** 151 (1971).
80. S. Fraga, J. Karwowski, K.M.S. Saxena, At. Data Nucl. Dat. Tables, **12**, 467 (1973); S.Fraga and K.M.S. Saxena, ibid., **4**, 255, 269 (1972); S. Fraga, K.M.S. Saxena and B.W.N. Lo, **3**, 323 (1971); C. Froese Fischer, ibid., **4**, 301 (1972); **12**, 87 (1973); A.D. McLean and R.S. McLean, ibid., **26**, 197 (1981); for Relativistic $< r^n >$-values refer to J.P. Desclaux, ibid., **12**, 311 (1973); C.C. Lu, T.A. Carlson, F.B. Malik, T.C. Tucker and C.W. Nestor, Jr., ibid., **3**, 1 (1971); **14**, 89 (1974).
81. B.G. Williams (ed.), Compton Scattering, McGraw-Hill, New York (1977) and I.R. Epstein, MTP International Review of Science, Physical Chemistry Series II, Theoretical Chemistry, A.D. Buckingham and C.A. Coulson, Eds., Butterworths, London (1975).
82. R. Benesch and V.H. Smith in *Wave Mechanics — The First Fifty Years*, ed. by W.C. Price, S.S. Chissick and T. Ravensdale, Butterworths, London, 357-377 (1973).
83. W.E. Duncanson and C.A. Coulson, Proc. Phys. Soc., **60**, 175 (1948); S.R. Gadre, S. Chakravorty and R.K. Pathak, J. Chem. Phys., **78**, 4581 (1983); S.R. Gadre, S.P. Gejji and S.J. Chakravorty, At. Data Nucl. Data Tables, **28**, 477 (1983).
84. I.R. Epstein, Phys. Rev. A., **8**, 160 (1973).
85. M. Duff, H. Rabitz, A. Askar, A. Cakmak and M. Ablowitz, J. Chem. Phys., **72**, 1543 (1980).
86. D. Heinemann, B. Fricke and D. Kolb, Phys. Rev. A., **38**, 4994 (1988); J.L. Gazquez and H.J. Silverstone, J. Chem. Phys., **67**, 1887 (1977); S.R. White, J.W. Wilkins, M.P. Teter, Phys. Rev., **B 39**, 5819 (1989); F.S. Levin and J. Shertzer, Phys. Rev., **A 32**, 3285 (1985); N. Sato and S. Iwata, J. Chem. Phys., **89**, 2932 (1988); V. Sethuraman and P.M. Hunt, J. Chem. Phys., **88**, 6938 (1988).
87. I. Babuska, B. Szabo and I.N. Katz, SIAM J. Num. Anal., **18**, 515 (1981).
88. A. W. Craig and O.C. Zienkiwicz, in "Multigrid Methods for Integral and Differential Equations," Clarendon Press, Oxford (1985).
89. C. Froese-Fischer, "The Hartree-Fock Method for Atoms," Wiley, New-York (1977).
90. L.F. Pacios and P.C. Christiansen, J. Chem. Phys., **82**, 2664 (1985).
91. W.J. Stevens, H. Bash, and M. Krauss, J. Chem. Phys., **81**, 6026 (1984).
92. L.E. McMurchie and E.R. Davidson, J. Comp. Phys., **44**, 289 (1981).
93. P.J. Hay and W.R. Wadt, J. Chem. Phys., **82**, 299 (1985).

Chapter 3.

Kinetically Balanced Geometric Gaussian Basis Set Calculations For Relativistic Many-Electron Atoms

A. Mohanty and E. Clementi

IBM Corporation
Scientific and Engineering Computations
Dept. 48B/428
Kingston, New York 12401, U.S.A.

1. Introduction

Recently there has been an increased interest in solving the Dirac equation for many-electron atoms using the method of basis set expansion. In spite of an earlier problem[1] of "variational collapse" or bound failure associated with the basis set methods, it is now generally agreed[2-7] that, with proper boundary conditions imposed upon the basis set, Dirac-Fock (DF) self consistent field (SCF) calculations can be performed on many-electron atoms using techniques similar to nonrelativistic methods. Crucial to the success of basis set approach lies in imposing kinetic balance[2-7] between the large and the small component basis. Grant et al.[5-8] used kinetically balanced Slater type orbitals (STOs) to perform basis set calculations on H-, He-, Be and Ne-like atoms with low to moderately high atomic charge ($Z = 2 - 50$), and obtained results comparable in accuracy (one part in 10^8) to the numerical DF methods.[9,10] The same authors have also reported[8] calculations for mercury ($Z = 80$).

Although STOs have the correct functional behavior to describe the relativistic wave function of atoms at the origin, they are not particularly suitable for analytic SCF molecular calculations. On the other hand, Gaussian type orbitals (GTOs) are useful in the evaluation of multi-center integrals in molecules but they do not possess the correct functional behavior at the origin. However, with a finite nuclear size model,[11] the Gaussian basis sets provide a natural description of the relativistic wavefunction within the nucleus. Aerts et al.[12,13] used Gaussian basis sets to solve DF equations for one- and many-electron systems. Malli[14] proposed the use of GTOs with both integer and non-integer powers for r for the relativistic DF calculation of atomic properties for a wide range of atomic charge, including Xenon, in the point nucleus approximation. Matsuoka et al.[15-18] reported kinetically balanced well-tempered basis set calculations with the finite nucleus size approximation and gave configuration average Dirac-Hartree-Fock (DHF) energies of various atoms, comparable to the numerical DF limit of Desclaux.[19] The present authors have also performed[20] DF many-electron atomic structure calculations using geometric-type Gaussian basis functions,[21] with kinetic balance[2-7] between the large and small component spinors.

In this paper we give the details of our calculations for closed- and open-shell atoms. We also present new numerical results for atoms (He – Rn) which support our claim that the prescription of kinetic balance works well when Gaussian type orbitals with integer powers of radial distance are employed, particularly for the finite nucleus charge distribution. There is no "variational collapse" and the eigenvalues converge from above to the numerical Dirac-Hartree-Fock result[9,10] as the number of basis functions increases.

The organization of the rest of the paper is as follows. In section 2 we describe the Dirac Hamiltonian and wavefunctions for the many-electron systems. In section 3 we give simple formulae for the one- and two-electron integrals which are needed for deriving the Dirac-Fock equation. In section 4 we present details of the basis set used in our calculation. In section 5 we give matrix elements of the unperturbed Hamiltonian in terms of our basis sets. In section 6 we discuss the angular coefficients needed in computing the Fock matrices. In section 7 the evaluation of one- and two-electron integrals, in terms of elementary functions, is presented. In section 8 we give details of our finite nucleus size approximation. In section 9 we present open-shell calculations and in section 10 we give some of the vector-coupling coefficients used in our calculation. In section 11 we present our numerical results and, finally, in section 12 we give our conclusions.

2. Preliminary

The total Dirac-Breit Hamiltonian for a many-electron system has two terms: a one-electron part consisting of kinetic, mass energy and nuclear attraction:

$$\hat{H}_D(i) = c\vec{\alpha}_i \cdot \vec{p}_i + c^2(\beta_i - 1) + V_{nuc}(r_i), \tag{1}$$

and the two-electron part H_{12} consisting of Coulomb repulsion and the Breit interaction

$$H_{12} = \sum_{\substack{\mathrm{ij} \\ i<j}} \frac{1}{r_{\mathrm{ij}}} + \sum_{\substack{\mathrm{ij} \\ i<j}} \mathbf{B}_{\mathrm{ij}}. \tag{2}$$

In the low-frequency limit;[22]

$$B_{12} = -\frac{1}{(2r_{12})}\{\vec{\alpha}_1 \cdot \vec{\alpha}_2 + [(\vec{\alpha}_1 \cdot \vec{r}_{12})(\vec{\alpha}_2 \cdot \vec{r}_{12})/r_{12}^2]\}. \tag{3}$$

The notation and phase choices of Grant et al.[9] will be adhered to as closely as possible and atomic units (e = $\hbar$ = m = 1) will be used throughout unless explicitly mentioned. The value of the speed of light is taken as 137.0373 (for comparison purposes).

In eqn. (1) we have subtracted the rest mass energy to get the binding energy. Dirac operators $\vec{\alpha}$ and β are expressed by the matrices

$$\vec{\alpha} = \begin{pmatrix} 0 & \vec{\sigma} \\ \vec{\sigma} & 0 \end{pmatrix} \text{ and } \beta = \begin{pmatrix} I & 0 \\ 0 & -I \end{pmatrix}, \tag{4}$$

where $\vec{\sigma}$ stands for the three Pauli matrices and I is the 2×2 unit matrix.

In the central field approximation, the solution of the Dirac equation is given in terms of four component spinors

$$\Psi_{n\kappa m}(r, \theta, \phi) = r^{-1}\begin{pmatrix} P_{n\kappa}(r) & \chi_{\kappa m}(\theta, \phi) \\ iQ_{n\kappa}(r) & \chi_{-\kappa m}(\theta, \phi) \end{pmatrix}, \tag{5}$$

where $P_{n\kappa}(r)$ and $Q_{n\kappa}(r)$ are the large and small components of the radial wave function, and they satisfy the orthonormality condition

$$\int_0^\infty [P_{n\kappa}(r)P_{n'\kappa}(r) + Q_{n\kappa}(r)Q_{n'\kappa}(r)]dr = \delta_{nn'}. \tag{6}$$

The spinor $\chi_{\kappa m}(\theta, \phi)$ is given by

$$\chi_{\kappa m} = \sum_\sigma Y_{l,m-\sigma}(\theta, \phi)\phi^\sigma\, c(l\tfrac{1}{2}j;\, m-\sigma,\ \sigma). \tag{7}$$

The angular state is characterized by the quantum number κ given by

$$\kappa = -2(j-l)(j+\tfrac{1}{2}), \tag{8}$$

where l is the orbital quantum number and $j = l \pm \frac{1}{2}$ is the total angular momentum number, the $c(l\frac{1}{2}j;\ m-\sigma, \sigma)$ are the Clebsch-Gordan coefficients and ϕ^σ are the two-component spinors:

$$\phi_{1/2} = \begin{pmatrix} 1 \\ 0 \end{pmatrix}, \quad \phi_{-1/2} = \begin{pmatrix} 0 \\ 1 \end{pmatrix}. \tag{9}$$

We shall use the spectroscopic notation similar to that of Grant et al.,[9] where a state with $j = l - \frac{1}{2}$ is distinguished from a state with $j = l + \frac{1}{2}$ by placing a bar over the spectroscopic label corresponding to l, e.g., $\bar{p} \equiv (l = 1, j = ½, \kappa = +1)$ or $p \equiv (l = 1, j = ^3/_2, \kappa = -2)$.

3. Matrix Elements of the Hamiltonian

Following the nonrelativistic case, the wavefunction for the many-electron system will be taken to consist of the sum of antisymmetrized products of one-electron central field orbitals given by equation (5). The total energy E then reduces to the sum of one electron integral I_A and two types of two electron integrals, the direct integral J_{AB} and the exchange integral K_{AB}, where the subscripts **A** and **B** denote the set of orbitals (n, κ, m), viz,

$$E = \sum_A I_A + 1/2 \sum_{A,B} (J_{AB} - K_{AB}). \tag{10}$$

Here we define

$$I_A \equiv < \psi_A | H_D | \psi_A >, \tag{11}$$

$$J_{AB} \equiv < \psi_A(1)\, \psi_B(2) | r_{12}^{-1} | \psi_A(1) \psi_B(2) >, \tag{12}$$

and

$$K_{AB} \equiv < \psi_A(1)\, \psi_B(2) | r_{12}^{-1} | \psi_B(1)\, \psi_A(2) > . \tag{13}$$

The one electron integrals involve the matrix elements $< n\kappa m | H_D(i) | n'\kappa' m' >$, the angular parts of which can be easily evaluated to yield

$$\begin{aligned} I_A = \int_0^\infty dr \Big\{ &-2c^2 [Q_A(r)]^2 + V_{nuc}(r)\, ([P_A(r)]^2 + [Q_A(r)]^2) \\ &+ cQ_A(r)\, [P'_A(r) + \frac{\kappa_A}{r} P_A(r)] - cP_A(r)\, [Q'_A(r) - \frac{\kappa_A}{r} Q_A(r)] \Big\}, \end{aligned} \tag{14}$$

where the prime denotes the differentiation with respect to radial coordinates.

A general expression for the two-electron integrals is given by[23]

$$\begin{aligned} C_{ABCD} &= < \psi_A(1)\, \psi_B(2) | \frac{1}{r_{12}} | \psi_C(1)\, \psi_D(2) > \\ &= \sum_\nu C_\nu\, (j_C m_C;\ j_A m_A)\, C_\nu\, (j_B m_B;\ j_D m_D) \\ &\times R_\nu\, (AC;BD)\, \delta_{m_A + m_B,\, m_C + m_D} \end{aligned} \tag{15}$$

where $R_\nu(AC; BD)$ is the relativistic radial integral and C_ν are the angular coefficients. The sum over the magnetic quantum numbers m can be carried out explicitly and for the closed shell structure one can write[23]

$$\sum_B J_{AB} = \sum_{n_B\kappa_B} \bar{a}_0^c(j_A, j_B)\,(2j_B+1) \iint dr_1\, dr_2\; U_\nu(1,2) \times \{[P_A(1)]^2 + [Q_A(1)]^2\} \times \{[P_B(2)]^2 + [Q_B(2)]^2\}. \tag{16a}$$

Similarly

$$\sum_B \mathrm{K}_{AB} = \sum_\nu \sum_{n_B\kappa_B} (2j_B+1)\, \bar{b}_\nu^c(j_A j_B) \int_0^\infty \int_0^\infty dr_1 dr_2\; U_\nu(1,2)\, [P_A(1)\, P_B(1) + Q_A(1)\, Q_B(1)] \times [P_A(2)\, P_B(2) + Q_A(2)\, Q_B(2)], \tag{16b}$$

where

$$\begin{aligned} U_\nu\,(1,2) &= r_1^\nu / r_2^{\nu+1}, \quad \text{if } r_1 < r_2 \\ &= r_2^\nu / r_1^{\nu+1}, \quad \text{if } r_1 > r_2. \end{aligned} \tag{17}$$

The analysis of the two-electron integral for the Breit interaction is much more involved. We refer the reader to Kim[23] and Grant[22] for more details. Following Kim[23] we write

$$M_{AB} = \left< \psi_A(1)\, \psi_B(2) \left| \frac{\vec{\alpha}_1 \cdot \vec{\alpha}_2}{r_{12}} \right| \psi_A(1)\, \psi_B(2) \right> \tag{18}$$

and

$$N_{AB} = \left< \psi_A(1)\, \psi_B(2) \left| \frac{\alpha_1 \cdot \alpha_2}{r_{12}} \right| \psi_B(1)\, \psi_A(2) \right> \tag{19}$$

as the direct and exchange matrix elements of the magnetic interaction term. The exchange matrix element of the retardation term (the second term in Eq. 3) can be written as[23]

$$R_{AB} = \tfrac{1}{2} \left< \psi_A(1)\, \psi_B(2) \left| [(\vec{\alpha}_1 \cdot \vec{\nabla}_1)\, (\vec{\alpha}_2 \cdot \vec{\nabla}_2)\, \vec{r}_{12}] \right| \psi_B(1)\, \psi_A(2) \right>. \tag{20}$$

It is worth noting that the direct part of the retardation term vanishes. Therefore, the relativistic correction arising from the Breit interaction can then be written as follows:[23]

$$E_B = -\frac{1}{2}\sum_{A,B}(M_{AB} - N_{AB} - R_{AB}). \tag{21}$$

For a closed-shell system as has been shown by Kim[23]

$$\sum_{AB} M_{AB} = 0. \tag{22}$$

Hence, we can write the combined exchange and retardation terms as[23]

$$\begin{aligned}\sum_B [N_{AB} + R_{AB}] = \sum_B \sum_\nu (2j_B + 1)\\ [\bar{e}_\nu^c\,(\kappa_A\,\kappa_B)\int_0^\infty\int_0^\infty dr_1\,dr_2\,U_\nu(1,\,2)\,P_A(1)\,Q_B(1)\,P_A(2)\,Q_B(2)\\ +\,\bar{e}_\nu^c\,(\kappa_A\,\kappa_B)\int_0^\infty\int_0^\infty dr_1\,dr_2\,U_\nu(1,2)\,P_B(1)\,Q_A(1)\,P_B(2)\,Q_A(2)\\ +\,2\bar{d}_\nu^c\,(\kappa_A\kappa_B)\int_0^\infty\int_0^\infty dr_1\,dr_2\,U_\nu(1,\,2)\,P_A(1)\,Q_B(1)\,P_B(2)\,Q_A(2)\\ +\,2\bar{g}_\nu^c\,(\kappa_A\kappa_B)\int_0^\infty\int_{r_1}^\infty dr_1\,dr_2\,U_\nu(1,\,2)\,P_A(1)\,Q_B(1)\,P_B(2)\,Q_A(2)]\end{aligned} \tag{23}$$

All these basic integrals can be expressed in terms of the basis functions in the form analogous to nonrelativistic calculations. The coefficients $\bar{a}_\nu^c$, $\bar{b}_\nu^c$, $\bar{e}_\nu^c$, $\bar{d}_\nu^c$, $\bar{g}_\nu^c$, can be obtained in terms of 3j- and 9j-symbols using Racah Algebra.[8,23] We shall return to the discussion of these angular factors in section 7.

4. Choice of Basis Sets

The radial functions are expanded in terms of the basis sets as

$$P_{n\kappa}(r) = \sum_p C_{\kappa p}^L\; g_{\kappa p}^L(r) \tag{24}$$

and

$$Q_{n\kappa}(r) = \sum_p C_{\kappa p}^S\; g_{\kappa p}^S(r), \tag{25}$$

where the summation index p runs over the number of basis functions, N, $g_{\kappa p}^L(r)$ and $g_{\kappa i}^S(r)$ are the basis functions belonging to the large and the small components respectively, and $C_{\kappa p}^L$ and $C_{\cdot\kappa p}^S$ are the corresponding expansion coefficients for each type of orbital of symmetry κ. For brevity we have omitted the summation over various types of orbitals of a particular symmetry.

For the large component, we have chosen GTOs of the form:

$$g_{\kappa i}^{L}(r) = C_{N_{\kappa i}}^{L}\, r^{n_\kappa}\, e^{-\alpha_i r^2}, \tag{26}$$

where n_κ takes integer values 1, 2, 3, 4 for $s, \bar{p}, p;\ \bar{d}, d;\ \bar{f}, f$ symmetries respectively, and $C_{N_{\kappa i}}^{L}$ is the normalization factor given by

$$C_{N_{\kappa i}}^{L} = \left[\frac{\sqrt{\pi}}{2^{2n_\kappa + 3/2}} \; \frac{(2n_\kappa - 1)!!}{(\alpha_i)^{\frac{2n_\kappa + 1}{2}}} \right]^{-1/2}. \tag{27}$$

We have constrained our small component radial basis according to the kinetic balance condition[4-7]

$$g_{\kappa i}^{S}(r) = C_{N_{\kappa i}}^{S} \left(\frac{d}{dr} + \frac{\kappa}{r}\right) g_{\kappa i}^{L}(r), \tag{28a}$$

where

$$C_{N_{\kappa i}}^{S} = \left\{ \frac{\alpha_i}{(2n_\kappa - 1)} \left[(4\kappa^2 - 1) + 4(n_\kappa + \kappa) \right] \right\}^{-1/2}. \tag{28b}$$

Hence, the same set of exponents have been used for the large as well as the small component. The exponents α_i's are generated according to

$$\alpha_i = \alpha_0\, \beta^{i-1},\ i = 1,2, \ldots N \tag{29}$$

where α_0 and β are parameters determined by the basis set size N and the particular atomic system being considered. For Xenon (Z = 54) N was taken to be of the order 30 for the $1s_{1/2}$ orbital. These geometric basis sets defined by Eq. (29) were introduced by Clementi et al.[21] for nonrelativistic atomic and molecular computations.

5. Relativistic Hartree-Fock-Roothaan Equation

The derivation of relativistic Hartree-Fock-Roothaan equation in the basis set expansion form is familiar from the work of Kim[23] and Kagawa.[24] The inclusion of the Breit interaction in the SCF equation was studied by Grant et al,[7] whose formulation we shall adopt. For closed shell system, the equation takes the matrix form:

$$FC = \varepsilon SC \tag{30}$$

In Eq. (30) the basis sets are used to parametrize the large- and small-component parts of the atomic spinors. If we choose our basis sets according to the kinetic balance prescription, the Fock matrix will be of $2N\kappa \times 2N\kappa$ dimension, where κ denotes the number of symmetry species being considered. The matrices are block diagonalized according to symmetry type κ:

$$F_\kappa C_\kappa = \varepsilon_\kappa S_\kappa C_\kappa, \tag{31}$$

where

$$C_\kappa = \begin{bmatrix} C_\kappa^L \\ C_\kappa^S \end{bmatrix}. \tag{32}$$

Equation (31) gives a set of coupled equations to be solved for the expansion coefficients C_κ as in the nonrelativistic equations.

The overlap matrix can be partitioned into $N \times N$ blocks of the form

$$S_\kappa = \begin{bmatrix} S_\kappa^{LL} & 0 \\ 0 & S_\kappa^{SS} \end{bmatrix}, \tag{33}$$

where the subscripts LL and SS denote the fact that large and small component wavefunctions are used.

The Fock operator has three contributions in general: viz,

$$F_\kappa = h_\kappa + g_\kappa + b_\kappa, \tag{34}$$

where

$$h_\kappa = \begin{bmatrix} V_\kappa^{LL} & c\Pi_\kappa^{LS} \\ c\Pi_\kappa^{SL} & V_\kappa^{SS} - 2c^2 S_\kappa^{SS} \end{bmatrix} \tag{35}$$

is the one-electron bare nucleus operator,

$$g_\kappa = \begin{bmatrix} J_\kappa^{LL} - \mathrm{K}_\kappa^{LL} & -\mathrm{K}_\kappa^{LS} \\ -\mathrm{K}_\kappa^{SL} & J_\kappa^{SS} - \mathrm{K}_\kappa^{SS} \end{bmatrix} \tag{36}$$

gives the Coulomb repulsion term and

$$b_\kappa = \begin{bmatrix} \mathrm{B}_\kappa^{LL} & \mathrm{B}_\kappa^{LS} \\ \mathrm{B}_\kappa^{SL} & \mathrm{B}_\kappa^{SS} \end{bmatrix} \tag{37}$$

denotes the contribution due to the Breit term. The matrix form of the one- and two-electron operators appearing in Eqs. (35)−(37) can be deduced from the expressions for I_A, J_{AB}, K_{AB} and $N_{AB}+R_{AB}$ given by Eqs. (14), (16a), (16b), and (23) respectively.

If we write the general form of a single element of Gaussian function as

$$\psi_{\kappa i}(r) = \begin{bmatrix} g^L_{\kappa i}(r) \\ g^S_{\kappa i}(r) \end{bmatrix}, \tag{38}$$

the one-electron matrices given in eqn (35) have the elements

$$V^{TT}_{\kappa ij} = \int_0^\infty g^T_{\kappa i}(r)\; V_{nuc}(r)\; g^T_{\kappa j}(r)\,(r)dr, \tag{39}$$

$$S^{TT}_{\kappa ij} = \int_0^\infty g^T_{\kappa i}(r)\; g^T_{\kappa j}\;(r)dr, \tag{40}$$

where the superscript TT indicates LL or SS, and

$$\Pi^{SL}_{\kappa ij} = \int_0^\infty dr\, g^S_{\kappa i}(r)\left[\frac{d}{dr}+\frac{\kappa}{r}\right]g^L_{\kappa j}(r), \tag{41a}$$

$$\Pi^{LS}_{\kappa ij} = \int_0^\infty dr\, g^L_{\kappa i}(r)\left[-\frac{d}{dr}+\frac{\kappa}{r}\right]g^S_{\kappa j}(r). \tag{41b}$$

Integrating by parts we can see that Π^{LS} and Π^{SL} are matrix adjoints of each other, provided that boundary terms vanish which, of course, is true in case of spherical Gaussian functions.

The matrices J^{TT}_κ and K^{TT}_κ are given by

$$J^{TT}_{\kappa pq} = \sum_{\kappa' rs}\left[D^{TT}_{\kappa' rs}\; J^{0TTTT}_{\kappa pq,\,\kappa' rs} + D^{\overline{T}\overline{T}}_{\kappa' rs}\; J^{0TT\overline{T}\overline{T}}_{\kappa pq,\,\kappa' rs}\right] \tag{42}$$

and

$$\mathrm{K}^{TT'}_{\kappa pq} = \sum_{\nu}\sum_{\kappa' rs}\,[\bar{b}^c_\nu(jj')]\; D^{TT'}_{\kappa' rs}\; \mathrm{K}^{\nu TT',TT'}_{\kappa pq,\,\kappa' rs}. \tag{43}$$

The superscripts $T\overline{T}$ represent a pair of LS or SL and the superscripts T and T' are either L or S. The density matrices are defined as

$$D^{TT'}_{\kappa pq} = N_\kappa\, C^T_{\kappa p}\, C^{T'}_{\kappa q}. \tag{44}$$

For a closed shell $N_\kappa = 2j + 1$. The Coulomb integrals $J^{0TT,T'T'}_{\kappa pq,\kappa' rs}$ and $K^{\nu TT',TT'}_{\kappa pq,\kappa' rs}$ are defined as usual

$$J^{\nu TT,T'T'}_{\kappa pq,\kappa' rs} = \int_0^\infty \int_0^\infty g^T_{\kappa p}(r_1)\, g^T_{\kappa q}(r_1)\, U_\nu(r_1, r_2)\, g^{T'}_{\kappa' r}(r_2)\, g^{T'}_{\kappa' s}(r_2)\, dr_1\, dr_2 \quad (45)$$

where

$$U_\nu(r_1, r_2) \equiv U_\nu(1,2) = r_1^\nu / r_2^{\nu+1}, \quad \text{if } r_1 < r_2$$
$$r_2^\nu / r_1^{\nu+1}, \quad \text{if } r_1 > r_2 \quad (46)$$

and

$$\mathrm{K}^{\nu TT',TT'}_{\kappa pq,\kappa' rs} = \int_0^\infty \int_0^\infty g^T_{\kappa p}(r_1) g^T_{\kappa' r}(r_1) U_\nu(r_1, r_2) g^{T'}_{\kappa q}(r_2) g^{T'}_{\kappa' s}(r_2) dr_1 dr_2. \quad (47)$$

The general form of the Breit interaction matrices is:

$$\mathrm{B}^{TT}_{\kappa pq} = \sum_\nu \sum_{\kappa' rs} \bar{e}^c_\nu(\kappa\kappa')\, D^{\bar{T}\bar{T}}_{\kappa rs}\, \mathrm{K}^{\nu TT\bar{T}\bar{T}}_{\kappa pq,\kappa' rs} \quad (48)$$

and

$$\mathrm{B}^{T\bar{T}}_{\kappa pq} = \sum_\nu \sum_{\kappa' rs} (\bar{d}^c_\nu(\kappa\kappa')\, D^{\bar{T}T}_{\kappa' rs}\, \mathrm{K}^{\nu T\bar{T},\bar{T}T}_{\kappa pq,\kappa' rs} + \bar{g}^c_\nu(\kappa\kappa')\, D^{\bar{T}T}_{\kappa rs}\, M^{\nu T\bar{T},\bar{T}T}_{\kappa pq,\kappa' rs}), \quad (49)$$

where

$$\mathrm{K}^{\nu T\bar{T},\bar{T}T}_{\kappa pq,\kappa' rs} = \int_0^\infty \int_0^\infty g^T_{\kappa p}(r_1)\, g^{\bar{T}}_{\kappa' r}(r_1)\, U_\nu(r_1, r_2)\, g^{\bar{T}}_{\kappa q}(r_2)\, g^T_{\kappa' s}(r_2)\, dr_1\, dr_2, \quad (50)$$

and

$$M^{\nu T\bar{T},\bar{T}T}_{\kappa pq,\kappa' rs} = \int_0^\infty \int_{r_1}^\infty g^T_{\kappa p}(r_1)\, g^{\bar{T}}_{\kappa' r}(r_1)\, U_\nu(r_1, r_2)\, g^{\bar{T}}_{\kappa q}(r_2)\, g^T_{\kappa' s}(r_2)\, dr_1\, dr_2 \quad (51)$$

The angular coefficients $\bar{e}^c_\nu(\kappa\kappa')$ $\bar{d}^c_\nu(\kappa\kappa')$ and $\bar{g}^c_\nu(\kappa\kappa')$ can be evaluated explicitly by using Grant's general angular[9] package.

6. Angular Coefficients

To evaluate the angular coefficients $\bar{b}_\nu^c(\kappa\kappa')$, $\bar{e}_\nu^c(\kappa\kappa')$, $\bar{g}_\nu^c(\kappa\kappa')$ and $\bar{d}_\nu^c(\kappa\kappa')$ that appear in the two-electron matrix elements, we follow Quincy et al.[8] and write the general matrix element as

$$< \psi_a(1)\,\psi_b(2)\left|\frac{1}{r_{12}}\right|\psi_c(1)\,\psi_d(2) > = \sum_{\nu q} \begin{Bmatrix} m_a & \nu & j_c \\ j_a & q & m_c \end{Bmatrix} \begin{Bmatrix} m_b & \nu & j_d \\ j_b & q & m_d \end{Bmatrix} X^\nu(abcd), \tag{52}$$

where m is the magnetic quantum number and $X^\nu(abcd)$ is the effective interaction strength. For the Coulomb interaction:

$$\frac{1}{r_{12}} = \sum_\nu U_\nu(r_1, r_2)\; C^\nu(1) \cdot C^\nu(2), \tag{53}$$

$$C_q^\nu(\theta, \phi) = \sqrt{4\pi/(2\nu+1)}\; Y_{\nu q}(\theta, \phi), \tag{54}$$

and the interaction strength can be written as follows:

$$\begin{aligned} X^\nu(abcd) = \{j_a\ j_c\ \nu\}\,\{j_b\ j_d\ \nu\}\,\pi^e(\kappa_a, \kappa_c, \nu)\pi^e(\kappa_b, \kappa_d, \nu) \\ (-1)^\nu < j_a \| C^{(\nu)} \| j_c > \; < j_b \| C^{(\nu)} \| j_d > \; R^\nu(ab;cd) \end{aligned} \tag{55}$$

where

$$R^\nu(ab;cd) = \iint \rho^e_{ac}(r_1)\;\rho^e_{bd}(r_2)\;U^\nu\;(r_1, r_2)\,dr_1\,dr_2, \tag{56}$$

$$\rho^e_{ac}(r) = P_a(r)\,P_c(r) + Q_a(r)\,Q_c(r), \tag{57}$$

$$< j \| c^\nu \| j' > = (-1)^{j+1/2}\;[jj']\begin{bmatrix} j & \nu & j' \\ 1/2 & 0 & -1/2 \end{bmatrix}, \tag{58}$$

$$\pi^e(\kappa, \kappa', \nu) = 1/2[1 - \delta\delta'(-1)^{j+j'+\nu}]\;(\delta = -\,sign(\kappa)), \tag{59}$$

and the triangular factors $\{jj'\nu\}$ express the restriction

$$|j - j'| \le \nu \le j + j'. \tag{60}$$

By summing over the magnetic quantum number, the coefficients $\bar{a}_\nu^c$ and $\bar{b}_\nu^c$ for the closed shell case can be evaluated explicitly to give[9]

$$\bar{a}_\nu^c = \delta_{\nu 0} \tag{61}$$

and

$$\bar{b}_v^c = \begin{bmatrix} j & v & j' \\ ½ & 0 & -½ \end{bmatrix}^2. \tag{62}$$

The matrix element of the Breit interaction has a more complicated structure. Here we follow the derivation of Grant et al.[22] and give the equations for the sake of completeness. The effective interaction strength in the low frequency limit[22] has the form

$$\begin{aligned} X_T^v(ab;cd) =& (-1)^v < j_a \| C^{(v)} \| j_c > \; < j_b \| C^v \| j_d > \{j_a j_c v\} \{j_b j_d v\} \\ & \times \left\{ \sum_{\lambda=v-1}^{v+1} \pi^0 (\kappa_a, \kappa_c, \lambda)\, \pi^0 (\kappa_b, \kappa_d, \lambda) \sum_{\mu=1}^{4} r_\mu^{\lambda,v}(abcd)\; R_\mu^\lambda(ab;cd) \right. \\ & \left. + \pi^0 (\kappa_a, \kappa_c, v-1)\; \pi^0(\kappa_b, \kappa_d, v+1) \sum_{\mu=1}^{8} s_\mu^v(abcd)\; S_\mu^v(ab;cd) \right\} \end{aligned} \tag{63}$$

where $\pi^0(\kappa, \kappa', v) = 1 - \pi^e(\kappa, \kappa', v)$. The radial integrals $R_v^\lambda(ab;cd)$ and $S_\mu^v(ab;cd)$ are defined[22] in terms of standard integrals

$$R^v(ac; bd) = \int\int \rho_{ac}^0 (r_1)\, [U_v (r_1, r_2)]\, \rho_{bd}^0 (r_2)\, dr_1\, dr_2 \tag{64}$$

and

$$S^v(ac; bd) = \int\int \rho_{ac}^0 (r_1)\, W_{v-1,\, v+1,\, v} (r_1, r_2) \rho_{bd}^0 (r_2)\, dr_1\, dr_2. \tag{65}$$

In Eqs. (64) and (65)

$$\rho_{ac}^0 = P_a (r_1)\; Q_c (r_2) \tag{66}$$

$$\rho_{bd}^0 = P_b (r_1)\; Q_d (r_2) \tag{68}$$

and

$$W_{v-1,\, v+1,\, v}(r_1, r_2) = -\, ½[v][\bar{U}_{v-1}(r_1,r_2) - \bar{U}_{v+1}(r_1,r_2)], \tag{69}$$

where

$$\bar{U}_v(r_1, r_2) = \begin{cases} r_1^v / r_2^{v+1} & \text{if } r_1 < r_2 \\ 0 & \text{if } r_1 > r_2 \end{cases} \tag{70}$$

so that

$$U_v(r_1, r_2) = \bar{U}_v(r_1, r_2) + \bar{U}_v(r_2, r_1). \tag{71}$$

The coefficients $r_{\mu}^{\lambda,\nu}$ and s_{μ}^{λ} are given by Grant and McKenzie[21] and can be further combined to give the coefficients $\bar{e}_{\nu}^{c}(\kappa\kappa')$, $\bar{g}_{\nu}^{c}(\kappa\kappa')$ and $\bar{d}_{\nu}^{c}(\kappa\kappa')$ in the matrix representation. It has recently been demonstrated by Quincy et al.[7] that the Breit interaction can be readily incorporated in the SCF iterations. We shall, however, ignore the Breit interaction in the remaining discussion.

7. Evaluation of Matrix Elements

The expression for one-electron integrals is especially simple for Gaussian functions with integer powers of r.

The overlap integral has two parts

$$S_{\kappa ij}^{LL} = [V_{2n_{\kappa i}}(\alpha_{\kappa i})\, V_{2n_{\kappa j}}(\alpha_{\kappa j})]^{-1/2}\; V_{n_{\kappa i}+n_{\kappa j}}(\alpha_{\kappa ij}) \tag{72a}$$

and

$$\begin{aligned} S_{\kappa ij}^{SS} = {} & C_{N_{\kappa i}}^{S}\, C_{N_{\kappa j}}^{S}\, [V_{2n_{\kappa i}}(\alpha_{\kappa i})\, V_{2n_{\kappa j}}(\alpha_{\kappa j})]^{-1/2} \\ & \times [4(n_{\kappa i}+\kappa)(n_{\kappa j}+\kappa)\, V_{n_{\kappa i}+n_{\kappa j}-2}(\alpha_{\kappa ij}) \\ & - 2\{\alpha_{\kappa j}(n_{\kappa i}+\kappa) + \alpha_{\kappa i}(n_{\kappa j}+\kappa)\}\, V_{n_{\kappa i}+n_{\kappa j}}(\alpha_{\kappa ij}) \\ & + \alpha_{\kappa i}\alpha_{\kappa j}\, V_{n_{\kappa i}+n_{\kappa j}+2}(\alpha_{\kappa ij})], \end{aligned} \tag{72b}$$

where

$$\alpha_{\kappa ij} = \frac{1}{2}(\alpha_{\kappa i} + \alpha_{\kappa j}), \tag{73}$$

$$V_i(x) = \frac{(i-1)!!}{x^{\frac{i+1}{2}}}, \tag{74}$$

and $C_{N_{\kappa j}}^{S}$ is the normalization factor given by Eq. (28b).

The kinetic energy integral

$$\begin{aligned} \Pi_{\kappa ij}^{SL} &= \langle g_{\kappa i}^{S} | \vec{\sigma}\cdot\vec{p}\, | g_{\kappa j}^{L} \rangle \\ &= \int_0^{\infty} g_{\kappa i}^{S}(r)\left(\frac{d}{dr} + \frac{\kappa}{r}\right) g_{\kappa j}^{L}(r)\, dr \\ &= S_{\kappa ij}^{SS} / C_{N_{\kappa j}}^{S} = \Pi_{\kappa ji}^{LS}. \end{aligned} \tag{75}$$

The electron-nucleus Coulomb interaction integral for the point nucleus of charge Z also consists of two parts

$$V_{\kappa ij}^{LL} = (-Z)\frac{2\sqrt{2}}{\sqrt{\pi}} [V_{2n_{\kappa i}}(\alpha_{\kappa i})\, V_{2n_{\kappa j}}(\alpha_{\kappa j})]^{-1/2} \times V_{n_{\kappa i}+n_{\kappa j}-1}(\alpha_{\kappa ij}) \tag{76}$$

and

$$\begin{aligned} V_{\kappa ij}^{SS} = (-Z)\frac{2\sqrt{2}}{\sqrt{\pi}} C_{N_{\kappa i}}^{S} C_{N_{\kappa j}}^{S} [V_{2n_{\kappa i}}(\alpha_{\kappa i})\, V_{2nKj}(\alpha_{\kappa j})]^{-1/2} \\ [4(n_{\kappa i}+\kappa)(n_{\kappa j}+\kappa)\, V_{n_{\kappa i}+n_{\kappa j}-3}(\alpha_{\kappa ij}) \\ -2\{\alpha_{\kappa i}(n_{\kappa j}+\kappa)+\alpha_{\kappa j}(n_{\kappa i}+\kappa)\}\, V_{n_{\kappa i}+n_{\kappa j}-1}(\alpha_{\kappa ij}) \\ +\alpha_{\kappa i}\alpha_{\kappa j}\, V_{n_{\kappa i}+n_{\kappa j}+1}(\alpha_{\kappa ij})]. \end{aligned} \tag{77}$$

The expressions for all the required two-electron Coulomb repulsion integrals can be obtained in terms of the incomplete Beta functions as shown by Malli.[14] We have, however, adopted a different algorithm previously developed.[25]

The two-electron integrals (due to Breit and Coulomb) can be evaluated in terms of the primitive integral

$$I_{\kappa p \kappa q,\, \kappa' r \kappa' s}^{\nu TT',\, T''T'''} = \int_0^\infty \int_0^\infty dr_1\, dr_2\, g_{\kappa p}^{T}(r_1)\, g_{\kappa q}^{T'}(r_1)\, U^{\nu}(r_1, r_2)\, g_{\kappa' r}^{T''}(r_2)\, g_{\kappa' s}^{T'''}(r_2) \tag{78}$$

For example, the direct integral for the large component can be written as

$$\begin{aligned} J_{\kappa pq,\, \kappa' rs}^{\nu LL;\, LL} \equiv I_{\kappa pq,\, \kappa' rs}^{\nu LL,LL} = \frac{2\sqrt{2}}{\sqrt{\pi}} \left[S_{\kappa pq}^{LL} S_{\kappa' rs}^{LL} / V_{n_{\kappa p}+n_{\kappa q}}(\alpha_{\kappa pq})\, V_{n_{\kappa' r}+n_{\kappa' s}}(\alpha_{\kappa' rs})\right] \\ \times V_{n_{\kappa p}+n_{\kappa q}-\nu-1}(\alpha_{\kappa pq})\, V_{n_{\kappa' r}+n_{\kappa' s}+\nu}(\alpha_{\kappa' rs}) \\ \times C_{n_{\kappa p}+n_{\kappa q}-\nu-1,\, n_{\kappa' r}+n_{\kappa' s}+\nu}\left(\frac{\alpha_{\kappa pq}}{\alpha_{\kappa' rs}}\right), \end{aligned} \tag{79}$$

where

$$C_{\alpha,\beta}(t) = (1+t)^{-\frac{1}{2}(\beta+1)} \sum_{\lambda=0}^{(\alpha-1)/2} \frac{(2\lambda+\beta-1)!!}{(2\lambda)!!(\beta-1)!!}\left(\frac{t}{1+t}\right)^{\lambda}. \tag{80}$$

Similarly, the basic exchange integral has the following symmetrized form

$$\begin{aligned} K_{\kappa \underline{pq},\, \kappa' \underline{rs}}^{LL,LL} = \tfrac{1}{2}[I_{\kappa p \kappa' r,\, \kappa q \kappa' s}^{\nu LL,LL} + I_{\kappa q \kappa' s,\, \kappa p \kappa' r}^{\nu LL,LL} \\ + I_{\kappa p \kappa' s,\, \kappa q \kappa' r}^{\nu LL,LL} + I_{\kappa q \kappa' r,\, \kappa p \kappa' s}^{\nu LL,LL}], \end{aligned} \tag{81}$$

where

$$I^{\nu LL,LL}_{\kappa p\kappa' r,\,\kappa q\kappa' s} = \frac{2\sqrt{2}}{\sqrt{\pi}} \frac{S_{\kappa pq} S_{\kappa' rs}}{V_{n_{\kappa p}+n_{\kappa' r}}(\alpha_{\kappa pr})\, V_{n_{\kappa q}+n_{\kappa' s}}(\alpha_{\kappa qs})}$$
$$\times\, V_{n_{\kappa p}+n_{\kappa' r}-\nu-1}(\alpha_{\kappa p\kappa' r})\, V_{n_{\kappa q}+n_{\kappa' s}+\nu}(\alpha_{\kappa q\kappa' s}) \tag{82}$$
$$\times\, C_{n_{\kappa p}+n_{\kappa' r}-\nu-1,\; n_{\kappa q}+n_{\kappa' s}+\nu}\left(\frac{\alpha_{\kappa p\kappa' r}}{\alpha_{\kappa q\kappa' s}}\right),$$

and

$$\alpha_{\kappa p\kappa' r} = \tfrac{1}{2}(\alpha_{\kappa p} + \alpha_{\kappa' r}), \tag{83}$$

$$\alpha_{\kappa q\kappa' s} = \tfrac{1}{2}(\alpha_{\kappa q} + \alpha_{\kappa' s}). \tag{84}$$

The M-type of integrals which occur in the matrix elements of Breit interaction are readily obtained from the primitive integral (K-type) given in Eq. (81) by retaining only half the terms. Since these integrals are analytic, they can be quite accurately evaluated which, of course is critical for the success of the analytic calculations.

The radial matrix elements of Coulomb and low frequency Breit interactions for a particular set of indices $\{pqrs\}$ depend upon primitive integrals which differ in tensor order 'v' and on the labels T. The set of integral lists which are required to perform relativistic super matrix calculation have the labels [*LL*,*LL*], [*LL*,*SS*], [*SS*,*LL*], [*SS*,*SS*], [*SL*,*SL*] and [*SL*,*LS*] and hence can be generated easily with proper indexing structure.

8. Finite Nucleus Approximation

It is well known that Gaussians do not have the proper functional form for representing relativistic wavefunctions for atoms near the origin. For example, in the hydrogenic case the relativistic $1s_{1/2}$ wavefunctions behaves as[26]

$$P_{1s}(r) \sim r^{\gamma},\ \gamma = \sqrt{1 - Z^2\alpha^2}\,. \tag{85}$$

Equation (85) clearly shows that the derivative of P_{1s} has a singularity at the origin and such type of cusp behavior is more pronounced for high Z value. In order to accurately represent this cusp behavior even for Xenon (Z=54), one needs to have exponents as high as 9×10^9. This actually corresponds to a wavefunction which is highly localized in the nucleus region and the number of exponents needed for accurate results is dramatically large. However, in the extended nucleus approximation the $1s_{1/2}$ solution at the origin can be represented as[11]

$$P_{1s_{1/2}}(r) = r + g_3^L r^3 + g_5^L r^5 + \cdots \tag{86a}$$

and

$$Q_{1s_{1/2}}(r) = g_2^S r^2 + g_4^S r^4 + \cdots , \tag{86b}$$

so that for the $1s_{1/2}$ case

$$\frac{1}{r}\begin{pmatrix} P_{n\kappa}(r) \\ Q_{n\kappa}(r) \end{pmatrix} \cong \begin{pmatrix} \exp(-\alpha_L r^2) \\ r\exp(-\alpha_S r^2) \end{pmatrix}. \tag{87}$$

Of course, in the matched basis set, $\alpha_L = \alpha_S$, and Eq. (87) shows that in this case Gaussians give a natural description of the wavefunction within the nucleus. Ishikawa et al.[11] had shown that by representing the nucleus by a uniform charge distribution, the cusp condition is automatically relaxed and one needs optimal exponents much smaller in number in those expansions in which finite nucleus size is considered than those in which the point nucleus is used.The present authors have extended [20] the finite-size nucleus calculations to many-electron atoms (Z=2-86) and have shown that a similar convergence pattern is observed in all calculations when geometric -type Gaussian basis functions are used.

The only part of the Fock matrix which needs modification due to the finite nuclear charge distribution is the nuclear potential matrix elements V_κ^{LL} and V_κ^{SS}.

Let $V_{nuc}(r)$ denote nuclear potential. By assuming a uniformly charged sphere for the nucleus, the effect of the finite nuclear size is included in $V_{nuc}(r)$ and is written as[10]

$$V_{nuc}(r) = \begin{cases} -(Z/2R)(3 - r^2/R^2), & \text{if } r < R \\ -Z/r, & \text{if } r \geq R. \end{cases} \tag{88}$$

For the radius of the nucleus we use R = 2.2677×10^{-5} $A^{1/3}$ in atomic units[21] where A is the atomic number.

The modification of the nuclear potential due to finite nuclear size introduces new terms in the electron-nucleus attraction integrals over GTOs. For the basis sets we have chosen, the bare nucleus matrix elements are given by (we are using the same type of notation as the point nucleus matrix element)

$$\begin{aligned} V_{\kappa pq}^{LL} &= \int_0^R g_{\kappa p}^L(r)\left[-\frac{Z}{2R}(3 - r^2/R^2)\right] g_{\kappa q}^L(r)\,dr \\ &+ \int_R^\infty g_{\kappa p}^L(r)(-Z/r)\, g_{\kappa q}^L(r)\,dr \end{aligned} \tag{89}$$

and

$$V^{SS}_{\kappa pq} = \int_0^R g^S_{\kappa p}(r)\,[-Z/2R\,(3 - r^2/R^2)]\,g^S_{\kappa q}\,dr + \int_R^\infty g^S_{\kappa p}(r)\,(-Z/R)\,g^S_{\kappa q}(r). \tag{90}$$

The integrals in Eqs. (89) and (90) can be expressed in a straightforward manner in terms of elementary functions and error functions; viz,

$$V^{LL}_{\kappa pq} = (-Z)\frac{2\sqrt{2}}{\sqrt{\pi}}(2R)^{n_{\kappa pq}}[V_{2n_{\kappa p}}(\alpha_{\kappa p})V_{2n_{\kappa q}}(\alpha_{\kappa q})]^{-1/2} \times \left[I(n_{\kappa pq}/2,t) + 3/2\,F(n_{\kappa pq}/2,t) - 1/2\,F((n_{\kappa pq}+2)/2,t)\right] \tag{91}$$

where

$$t = (\alpha_{\kappa p} + \alpha_{\kappa q})R^2, \tag{92}$$

$$n_{\kappa pq} = n_{\kappa p} + n_{\kappa q}, \tag{93}$$

$$I(v,t) = \int_1^\infty y^{2v-1}\exp(-ty^2)dy = \sum_{n=1}^{v}\frac{1}{(2t)^n}\frac{(2v-2)!!}{(2v-2n)!!}\exp(-t),\quad t>0 \tag{94}$$

and $F_v(t)$ is the incomplete Gamma Function defined by

$$F_v(t) = \int_0^1 dx\,x^{2v}\exp(-tx^2). \tag{95}$$

Similarly,

$$\begin{aligned} V^{SS}_{\kappa pq} = {}& \frac{2\sqrt{2}}{\sqrt{\pi}}(2R)^{n_{\kappa pq}}[V_{2n_{\kappa p}}(\alpha_{\kappa p})V_{2n_{\kappa q}}(\alpha_{\kappa q})]^{-1/2}\,C^S_{N_{\kappa p}}\,C^S_{N_{\kappa q}} \\ &\times\Big[(R^{-2})(n_{\kappa p}+\kappa)(n_{\kappa q}+\kappa)\{I(n_{\kappa pq}-2,t) + 3/2\,F((n_{\kappa pq}-2)/2,t) \\ &- 1/2\,F(n_{\kappa pq}/2,t)\} \\ &- \{2\,\alpha_{\kappa p}(n_{\kappa p}+\kappa) + 2\,\alpha_{\kappa q}(n_{\kappa q}+\kappa)\}\{I(n_{\kappa pq},t) + 3/2F(n_{\kappa pq}/2,t) \\ &- 1/2F((n_{\kappa pq}+2)/2,t)\} \\ &+ 4\alpha_{\kappa p}\alpha_{\kappa q}R^2\{I(n_{\kappa pq}+2,t) + 3/2F((n_{\kappa pq}+2)/2,t) \\ &- 1/2F((n_{\kappa pq}+4)/2,t)\}\Big]. \end{aligned} \tag{96}$$

It is easily seen from the definition of $I(v, t)$ and $F_v(t)$ that the extra terms arising from the integration inside the nucleus go to zero as $R \rightarrow 0$. Thus, the effect of finite nucleus is to add extra easily computable terms in the formulae for nuclear attraction integral. However, as we shall see later, the results differ considerably from point nucleus calculations and the convergence patterns are very much different.

9. Open-Shell Calculations

In section 5 we described the pseudo-eigenvalue equation for a closed-shell system which is to be solved for the expansion coefficients C_κ (Eq. 31). However, it is well known from the work of Roothaan[27] that in the Hartree-Fock theory of open-shell systems one has to deal with two pseudo-eigenvalue equations, one for the closed- and the other one for the open-shell orbitals. The extension of the Relativistic Hartree-Fock (RHF) theory to open-shells was treated by LeClercq[27] and by Kagawa.[24] Here we shall follow Kagawa[24] to derive the Relativistic Hartree-Fock Roothaan (RHFR) theory for open-shell calculations, and we assume the following:

1. The state has no more than one open-shell for each symmetry species defined by κ.

2. The wavefunction for the state is expressed by a single configuration.

3. When the configuration contains only one open-shell, the state can be uniquely specified by the total angular momentum quantum number J and the seniority number V.

4. When the configuration contains more than one open-shell, the state can be uniquely specified by the total angular quantum number J.

The unperturbed energy for a single determinant is given by

$$\begin{aligned} E = & \sum_A N_A I_A + \sum_{A'} N_{A'} I_{A'} \\ & + \frac{1}{2} \sum_A \sum_B N_A N_B \{J_{AB} - K_{AB}\} \\ & + \sum_A \sum_{A'} N_A N_{A'} \{J_{AA'} - K_{AA'}\} \\ & + \frac{1}{2} \sum_{A'} \sum_{B'} N_{A'} N_{B'} \{J_{A',B'} - K_{A'B'}\}. \end{aligned} \tag{97}$$

Here N_A and $N_{A'}$ are the number of electrons in the closed and open shell, respectively. N_A always equals to $2j_A + 1$, whereas $N_{A'} < 2j_{A'} + 1$. It should be noted that in Eq. (97) we have explicitly shown the sum over the number of electrons of each orbital type defined by the set of quantum numbers $A \equiv nkm$ or $A' = n'k'm'$. The interactions between the closed shells can be obtained from Eqs. (16a) and (16b). For the closed-open shell interaction, the sum over the m quantum numbers can be performed[24] to give

$$J_{ss'} = \frac{1}{(2j+1)(2j'+1)} \sum_m \sum_{m'} < \psi_{sm}(1)\psi_{s'm'}(2) \mid \frac{1}{r_{12}} \mid \psi_{sm}(1)\psi_{s'm'}(2) > \\ = \sum_\nu \bar{a}_\nu^c(\kappa\kappa')R^\nu(ss;\ s's') \tag{98}$$

and

$$K_{s,s'} = \frac{1}{(2j+1)(2j'+1)} \sum_m \sum_{m'} < \psi_{sm}(1)\psi_{s'm'}(2) \mid \frac{1}{r_{12}} \mid \psi_{sm}(2)\psi_{s'm'}(1) > \\ = \sum_\nu \bar{b}_\nu^c(\kappa\kappa')R^\nu(ss';\ ss'). \tag{99}$$

Here s and s' stand for either A or A' and denote the quantum numbers of the closed- and open-shell orbitals, respectively.

Similarly, the open-open shell interaction can be written in the form:

$$J_{A'B'} = \sum_\nu a_\nu^o(\kappa_{A'}\kappa_{B'}J)R^\nu(A'A';\ B'B') \tag{100}$$

and

$$K_{A'B'} = \sum_\nu b_\nu^o(\kappa_{A'}\kappa_{B'}J)R^\nu(A'B';\ A'B'). \tag{101}$$

In order to obtain the SCF equation for the open-shell system we add to the the energy functional (97) the orthonormality constraint

$$C_{i\kappa p}SC_{j\kappa q}^\dagger = \delta ij \tag{102}$$

multiplied by a lagrange multiplier θ_{ij}, (the index i or j denote the type of orbital $1s$, $2s$, $3s$, ... , etc., for a particular symmetry κ) and minimize the entire expression with respect to the expansion coefficients $C_{i\kappa p}$.

In the closed shell case the orbitals can always be subjected to a unitary transformation,[27] which brings the matrix of Lagrange multipliers into diagonal form. We then end up with the pseudo-eigenvalue equation given by Eq. (30). However, for the open-shell case, as discussed in Roothaan's paper,[27] there are only unitary transformations available which transform the open and closed shells within themselves. Such transformations can eliminate only the off-diagonal Lagrange multipliers, say θ_{AB} and $\theta_{A'B'}$ but not the multiplier $\theta_{AA'}$ which couple the closed and open shells. Only when closed and open-shell orbitals have no common symmetry, the off-diagonal multipliers vanish automatically and in that case one obtains two pseudo-eigenvalue equations, one for the closed and one for the open shells. We now review the derivations leading to these two pseudo-eigenvalue equations, which is very similar to the nonrelativistic approach.[27]

We shall have Lagrangian multipliers between orbitals of closed-shells, between orbitals of open-shells, and between orbitals of closed- and open-shell. The latter type of multiplier will be eliminated by introducing new operators R_A and $R_{A'}$ for closed- and open-shells, respectively. The introduction of R_A and $R_{A'}$ in order to eliminate the Lagrangian multipliers is convenient because the R operators, called *coupling operators,* are expressed in terms of the orbitals as follows:

$$\hat{R}_s \psi_{s'm'} = \frac{N_s}{N_A} \sum_m [< \psi_{sm} \mid \hat{J}^{A'} - \hat{K}^{A'} \mid \psi_{s'm'} > \psi_{sm} + < \psi_{sm} \mid \psi_{s'm'} > (\hat{J}^{A'} - \hat{K}^{A'}) \psi_{sm}] \tag{103}$$

where, as before, both s and s' denote A or A'. Here $\hat{J}^{A'}$ and $\hat{K}^{A'}$ are the direct and exchange operators defined by Roothaan[27] as

$$\hat{J}^s(1)\psi_{s'}(1) = \int [\psi^\dagger s(2) \frac{1}{r_{12}} \psi_s(2)\, d\vec{r}_2]\, \psi_{s'}(1) \tag{104a}$$

and

$$\hat{K}^s(1)\psi_{s'}(1) = \int [\psi^\dagger_{s'}(2) \frac{1}{r_{12}} \psi_s(2)\, d\vec{r}_2]\, \psi_s(1). \tag{104b}$$

When $\kappa = \kappa'$, the $a^o_\nu(\kappa\kappa' J)$ and $b^o_\nu(\kappa\kappa' J)$ coefficients are not independent of each other because $R^\nu(A'A'; A'A')$ is the same in both $J_{A'A'}$ and $K_{A'A'}$. We assume, following Kagawa,[24]

$$a^o_\nu(\kappa\kappa' J) = \delta_{\nu o}. \tag{105}$$

This leads to the simplification of the coupling operator (103) which we now describe.

By summing over m and m' we obtain

$$\frac{1}{N_A^2}\sum_{m,m'} < \psi_{sm} \mid \hat{J}^{A'} - \hat{K}^{A'} \mid \psi_{s'm'} >$$
$$= \frac{N_A}{N_A - N_{A'}}\left[\sum_{\nu}\{\alpha_\nu(\kappa J)R^\nu(A'A'; ss') - \beta_\nu(\kappa J)R^\nu(A's; A's')\}\right] \quad (106)$$

where

$$\alpha_\nu(\kappa J) = \bar{a}_\nu^c(\kappa J) - a_\nu^0(\kappa J) \quad (107a)$$

$$\beta_\nu(\kappa J) = \bar{b}_\nu^c(\kappa J) - b_\nu^0(\kappa J). \quad (107b)$$

The direct integral involving $\hat{J}^{o\kappa}$ vanishes because of Eq. (105) and Eq. (61).

By using Eq. (106) in Eq. (103) we obtain the two coupling operators R_A and $R_{A'}$ in terms of the supermatrices S, D and Q,

$$R_A = [1/(N_A - N_{A'})][S_\kappa D_A Q_{A'} + Q_{A'} D_A S_\kappa] \quad (108a)$$

and

$$R_{A'} = [1/(N_A - N_{A'})][S_\kappa D_{A'} Q_{A'} + Q_{A'} D_{A'} S_\kappa], \quad (108b)$$

where the closed- and open-shell densities are defined respectively as

$$D_A = \sum_{i \subset \text{closed}} D_{Ai} \quad (109a)$$

and

$$D_{A'} = \sum_{i \subset \text{open}} D_{A'i}. \quad (109b)$$

Further, we define

$$D_T = D_A + D_{A'} \quad (109c)$$

as the total density matrix.

The supermatrix Q in (108) is defined by

$$Q_{\kappa pq,\, \kappa rs}^{TT',\, TT'} = - \sum_{\nu = k_1,\, k_1+2 \ldots}^{k_2} \beta_\nu(\kappa J) K_{\kappa pq,\, \kappa rs}^{\nu TT',\, TT'} \quad \text{when } \kappa = \kappa', \quad (110a)$$

and

$$Q^{TT',TT'}_{\kappa pq,\kappa' rs} = \sum_{\nu=0,2\ldots}^{k_0} \alpha^{\nu}(\kappa\kappa' J) J^{\nu TT',TT'}_{\kappa pq,\kappa' rs} - \sum_{\nu=k_1,k_1+2\ldots}^{k_2} \beta^{\nu}(\kappa\kappa' J) K^{\nu TT',TT'}_{\kappa pq,\kappa' rs} \quad (110b)$$

when $\kappa \neq \kappa'$.

Here the vector-coupling coefficients $\alpha_\nu(\kappa\kappa' J)$ and $\beta_\nu(\kappa\kappa' J)$ for $\kappa \neq \kappa'$ are defined by the relation

$$\alpha_\nu(\kappa\kappa' J) = \left.\begin{matrix} 0, & \nu = 0 \\ -\,a^{o}_{\nu}(\kappa\kappa' J) & \nu \neq 0 \end{matrix}\right\} \quad (111a)$$

$$\beta_\nu(\kappa\kappa' J) = \bar{b}^{c}_{\nu}(\kappa\kappa' J) - b^{o}_{\nu}(\kappa\kappa' J). \quad (111b)$$

Further, we define supermatrix P as

$$P^{TT',TT'}_{\kappa pq,\kappa' rs} = J^{oTT',TT'}_{\kappa pq,\kappa rs} - \sum_{\nu=k_1,k_1+2\ldots}^{k_2} \bar{b}^{c}_{\nu}(\kappa\kappa' J) K^{\nu TT',TT'}_{\kappa pq,\kappa' rs}. \quad (112)$$

In Eqs. (109a)−(109c)

$$\begin{aligned} \kappa_1 &= |j - j'|, \text{ if } \kappa\kappa' > 0, \\ &= |j - j'| + 1, \text{ if } \kappa\kappa' < 0, \\ \kappa_2 &= j + j', \text{ and } k_0 = (2j-1, 2j'-1) \end{aligned} \quad (113)$$

In addition, let us designate by F_c and F_o the quantities

$$F_{c,\kappa pq} = h_{\kappa pq} + P_{\kappa pq} + R_{o,\kappa pq} \text{ and } F_{o,\kappa pq} = H_{\kappa pq} + P_{\kappa pq} - Q_{\kappa pq} + R_{c,\kappa pq}, \quad (114)$$

where we have used the notation $A \equiv c\kappa$ and $A' \equiv o\kappa$ to denote the quantum numbers of closed- and open-shells of symmetry κ and we have defined

$$P^{TT'}_{\kappa pq} = \sum_{\kappa' rs} P^{TT',TT'}_{\kappa pq,\kappa' rs} D^{TT'}_{T\kappa' rs} \quad (115)$$

and

$$Q^{TT'}_{\kappa pq} = \sum_{\kappa' rs} Q^{TT',TT'}_{\kappa pq,\kappa' rs} D^{TT'}_{T\kappa' rs} \quad (116)$$

By making use of a unitary transformation on the closed and open shells separately, we can bring the matrices of Lagrangian multiplier into diagonal form. By designating as $\varepsilon_{s\kappa}$, the diagonal elements of the transformed Lagrangian matrices, we can finally write

$$\Sigma_q F_{c,\,\kappa pq} C_{k\kappa q} = \varepsilon_{c\kappa} \Sigma_q S_{\kappa pq} C_{k\kappa q}, \tag{117}$$

$$\Sigma_q F_{o,\,\kappa pq} C_{m\kappa q} = \varepsilon_{o\kappa} \Sigma_q S_{\kappa pq} C_{m\kappa q}, \tag{118}$$

where $k = i \subset$ closed- and $m = i \subset$ open-shell respectively or, in a more compact notation, we have

$$F_c C = \varepsilon S C, \tag{119}$$

$$F_o C = \varepsilon S C. \tag{120}$$

The Eqs. (119) and (120) are similar to the Hartree-Fock equations for closed and open shells as in the nonrelativistic case.[27] They are pseudo-eigenvalue equations, because the matrices F_c and F_o depend on the solution C.

The expression for DHF total energy is obtained in supervector and supermatrix notation

$$E_{DHF} = \frac{1}{2} D_T^\dagger P D_T - \frac{1}{2} D_o^\dagger Q D_o + h^\dagger D_T. \tag{121}$$

10. Vector Coupling Coefficients

The vector coupling coefficient $\beta_\nu(\kappa\kappa' J)$ given by Eq. (111) are evaluated using the coefficients of fractional parentage (CFP).[29] The computer program by Grant et al.[9] has been used to generate the coefficients $\bar{b}_\nu^c(\kappa\kappa' J)$ and $b_\nu^o(\kappa\kappa' J)$ for the jj-coupled configurations. In Table I we give coefficients $\bar{b}_\nu^c(\kappa\kappa')$ up to J = 7/2. These were obtained by Eq. (62).

For the open-shell coefficients $b_\nu^o(\kappa\kappa' J)$ one can have two types of terms, one consisting of diagonal terms which come from open-shells like p^n, d^m, f^n, etc., and the other consisting of off-diagonal terms resulting from open shells $p^n s$, $d^m f$, etc. General formulae for integrals of two-electron operator V_{12} and the vector coupling coefficients are given in Refs. 29 and 30. Even with more than two open shells involved, the general formula reduces to the following two open-shell cases[24,29]

$$\begin{aligned} < j_1^{n_1}(V_1 J_1) j_2^{n_2}(V_2 J_2) J \mid \sum_{i<\kappa}^{n_1+n_2} V_{i\kappa} \mid j_1^{n_1}(V_1 J_1) j_2^{n_2}(V_2 J_2) \mid J > \\ = A_1 + A_2 + B, \end{aligned} \tag{122}$$

where the diagonal matrix elements A_λ is written as

Table I. $\bar{b}_{\nu}^{c}(\kappa\kappa')$

		$\nu = 0$						
κ	κ'	0	1	2	3	4	5	6
∓ 1	∓ 1	1/2	–	–	–	–	–	–
∓ 1	± 1	–	1/6	–	–	–	–	–
∓ 2	∓ 1	–	1/6	–	–	–	–	–
∓ 2	± 1	–	–	1/10	–	–	–	–
∓ 2	∓ 2	1/4	–	1/20	–	–	–	–
∓ 2	± 2	–	1/60	–	9/140	–	–	–
∓ 3	∓ 1	–	–	1/10	–	–	–	–
∓ 3	± 1	–	–	–	1/14	–	–	–
∓ 3	∓ 2	–	1/10	–	1/35	–	–	–
∓ 3	± 2	–	–	1/70	–	1/21	–	–
∓ 3	∓ 3	1/6	–	4/105	–	1/63	–	–
∓ 3	± 3	–	1/210	–	4/315	–	25/693	–
-4	-1	–	–	–	1/14	–	–	–
-4	$+1$	–	–	–	–	1/18	–	–
-4	-2	–	–	9/140	–	5/252	–	–
-4	$+2$	–	–	–	1/84	–	5/132	–
-4	-3	–	1/14	–	1/42	–	5/462	–
-4	$+3$	–	–	1/210	–	5/462	–	25/858
-4	-4	1/8	–	5/168	–	9/616	–	25/3432

$$A_\lambda = < j_\lambda^{n_\lambda} V_\lambda J_\lambda \mid \sum_{i<\kappa}^{n_\lambda} V_{i\kappa} \mid j_\lambda^{n_\lambda} V_\lambda J_\lambda > \tag{123}$$

and the off-diagonal matrix element as

$$B = < j_1^{n_1}(V_1 J_1) j_2^{n_2}(V_2 J_2) \mid \sum_{i=1}^{n} \sum_{\kappa=n_1+1}^{n_1+n_2} V_{i\kappa} \mid j_1^{n_1}(V_1 J_1) j_2^{n_2}(V_i J_i) \mid J > , \tag{124}$$

where n_λ denotes the number of electrons in the J_λ shell, and V_λ and J_λ are the seniority and total angular momentum quantum numbers.

The diagonal elements A can be expressed[29] in terms of the coefficients of fractional parentage by use of seniority number V yielding

Table II. Values of $\beta_\nu(\kappa\kappa J)$.

κ	j	n	V	J	$\nu = 0$	$\nu = 2$	$\nu = 4$	$\nu = 6$
∓1	1/2	1	1	1/2	−1/2	–	–	–
∓2	3/2	1	1	3/2	−3/4	1/20	–	–
		2	2	2	−1/4	−1/100	–	–
		3	1	3/2	−1/12	1/180	–	–
∓3	5/2	1	1	5/2	−5/6	0.0380952380952	0.015873015873	–
		2	2	4	−1/3	−0.01904761904769	0.0079365079365079	–
		3	3	9/2	−1/6	0.018503401360544	0.00075585789871	–
		4	2	4	−1/12	−0.0047619047619047	0.0019841269841269	–
		5	1	5/2	−1/30	0.0015238095238095	0.00063492063492063	–
±4	7/2	1	1	7/2	−7/8	0.0297619047619	0.01461038961	0.00728438228
		2	2	6	−3/8	−0.00992063492206349	−0.0048701298701298	0.0065372661526507
		3	3	15/2	−5/24	−0.0231481481481	−0.003492719401810	0.003571441508504
		4	4	8	−1/8	−0.0184240362811790	−0.003731657952437	0.0013413413054699 7
		5	3	15/2	−3/40	−0.00833333333	−0.0012573789846516	0.0012857189430616
		6	2	6	−1/24	−0.0011022927689593	−0.000541125411254	0.00072636290585
		7	1	7/2	−1/56	0.00607385811467	0.00029817121653585	0.0001486608629465

$$A = \sum_\nu \bar{b}^o_\nu(\kappa\kappa J) R^\nu(o\kappa, o\kappa; o\kappa, o\kappa), \tag{125a}$$

The expression for the off-diagonal matrix elements B is much more involved. Details of the calculations can be found in the article by Grant.[30] After angular integration and by using CFPs we can write

$$B = \left[\sum_\nu \bar{a}^o_\nu(\kappa_1\kappa_2 J) R^\nu(o\kappa_1, o\kappa_1; o\kappa_2 o\kappa_2) - \bar{b}^o_\nu(\kappa_1\kappa_2 J) R^\nu(o\kappa_1, o\kappa_2; o\kappa_1, o\kappa_2)\right]. \tag{125b}$$

In Eqs. (125a) and (125b) the bar over $b^o_\nu(\kappa\kappa' J)$ and $a^o_\nu(\kappa\kappa' J)$ denotes the coefficient which includes the sum over a distinct pair of electrons.

Numerical values of coefficients $\bar{a}^0_\nu(\kappa_1\kappa_2 J)$ and $\bar{b}^0_\nu(\kappa_1\kappa_2 J)$ can be obtained by using the graphical techniques.[30] Using the general program by Grant et al.[9] these coef-

Table III(a). Vector coupling coefficients for the $\bar{f}-\bar{d}$ interaction in some of the high spin, high total angular momentum states of $f^n d^1$ configuration.

	$\alpha_2(\bar{f}\bar{d}J)$	$\beta_1(\bar{f}\bar{d}J)$	$\beta_3(\bar{f}\bar{d}J)$
$\bar{f}^1_{1;\,5/2})_{5/2}\,\bar{d}^1_{1;3/2})_{3/2})_{J=4}$	0.071428571428571	−0.3	0.016326530612245
$\bar{f}^2_{2;\,4})_4\,\bar{d}^1_{1;3/2})_{3/2})_{J=11/2}$	−0.0228571428527428	−0.18	0.00775510204816
$\bar{f}^3_{3;9/2})_{9/2}\,\bar{d}^1_{1;3/2})_{3/2})_{J=6}$	–	−0.1	–
$\bar{f}^4_{2;4})_4\,\bar{d}^1_{1;3/2})_{3/2})_{J=11/2}$	0.011428571428571	−0.05	−0.005102040816326
$\bar{f}^5_{1;5/2})_{5/2}\,\bar{d}_{1;3/2})_{3/2})_{J=4}$	0.0114285714285714	−0.02	0.005714285714285

ficients have been further combined with $\bar{b}^c_\nu(\kappa_1\kappa_2 J)$ to obtain the vector coupling coefficients(Eqs. (107a) and (107b). For the LS-coupling scheme Malli and Olive[31] have shown that the vector coupling coefficients can be generated from Slater coefficients. We use similar relation in the relativistic case, viz,

$$\left.\begin{aligned}\alpha^\nu(\kappa\kappa' J) &= -\,\bar{a}^0_\nu(\kappa\kappa' J)/N_A N_{A'},\ \nu \neq 0\\ \beta^\nu(\kappa\kappa' J) &= \bar{b}^c_\nu(\kappa\kappa' J) + \bar{b}^0_\nu(\kappa\kappa' J)/N_A N_{A'}\end{aligned}\right\},\ \kappa \neq \kappa' \tag{126}$$

$$\beta^\nu(\kappa\kappa J) = \bar{b}^c_\nu(\kappa\kappa J) + 2\bar{b}^0_\nu(\kappa\kappa J)/N_A^2 \quad \nu \neq 0,\ \kappa = \kappa', \tag{127a}$$

$$\beta^\nu(\kappa\kappa J) = \bar{b}^c_\nu(\kappa\kappa J) - 1/N_A \quad \nu = 0, \quad \kappa = \kappa'. \tag{127b}$$

We have used the above relations with the Slater coefficients from Grant's program.[9] Some of the vector coupling coefficients for the case $\kappa = \kappa'$ are given in Table II and for the case $\kappa \neq \kappa'$ ($\bar{d}-\bar{f}$ interaction and $\bar{d}-f$ interactions) are given in Table III(a) and III(b).

11. Numerical Results

Calculations were performed for the neutral states of all for atoms (He through Rn), using kinetically balanced Gaussian-type functions with integer powers of r. The results are presented in Tables IV through VIII in atomic units (a.u.). It should be noted that all these calculations were performed in the finite-size nucleus approximation, where atomic weights enter as parameters related to the radii of the nuclei(see Eq. (88)). These atomic weights averaged over the abundance of the isotopes are given in Table IV, column 3. The Breit interaction is not included in our calculations.

The jj-coupling scheme we have used is the same as by Grant et al[9], and we have adopted their notation (as explained in footnote (a), Table IV) to describe the

Table III(b). Vector coupling coefficients for the $f-\bar{d}$ interaction in some of the high spin, high total angular momentum states of $f^n\bar{d}^1$ configuration.

	$\alpha_2(f\bar{d}J)$	$\beta_3(f\bar{d}J)$	$\beta_5(f\bar{d}J)$
$f^1_{1;7/2})_{7/2}\,\bar{d}^1_{1;3/2})_{3/2})_{J=5}$	0.0666666666666666	−0.0039682539682539	0.03696051423324
$f^2_{2;6})_6\,\bar{d}^1_{1;3/2})_{3/2})_{J=15/2}$	−0.038095238095238	−0.0073696145124716	0.03558310376492
$f^3_{3;15/2})_{15/2}\,\bar{d}^1_{1;3/2})_{3/2})_{J=9}$	−0.0158730158730158	−0.0085034013605442	0.033287419651056
$f^4_{4;8})_8\,\bar{d}^1_{1;3/2})_{3/2})_{J=19/2}$	−	−0.0079365079365079	0.029843893480257
$f^5_{3;15/2})_{15/2}\,\bar{d}^1_{1;3/2})_{3/2})_{J=9}$	0.0095238095238095	−0.006235827664399	0.0250229568411386
$f^6_{2;6})_6\,\bar{d}^1_{1;3/2})_{3/2})_{J=15/2}$	0.0095238095238095	−0.0017006802721088	0.010330578512396
$f^7_{1;7/2})_{7/2}\,5\bar{d}^1_{1;3/2})_5$	0.0126984126984126	−0.0039682539682539	0.0185950413223140

open-shell states (column a, Table IV). It should be noted that in some cases lowest-energy single configuration states defined in column c differ from the nonrelativistic ground state configurations (see chapter 3) This difference can be understood by taking chromium as an example. As reported in Table IV column c, chromium has the total DHF energy $\simeq$ -1049.676 a.u. corresponding to the single configuration state ($3\bar{d}^4\,4s^2$) with J = 0 (this differs from the lowest-energy nonrelativistic configuration ($3d^5\,4s^1$) with J = 3). Indeed, the single configuration state ($3\bar{d}^4 3d^1\,4s^1$) with J =3, has the total DHF energy $\simeq$-1049. 624 a.u.(which is confirmed by numerical Dirac-Fock (DF) calculation (Average Level) using the program of Grant et al.[9] However, if we take the nonrelativistic configuration ($3d^5\,4s^1$) , and form all the relativistic configurations with J=3 (there are 17 of them), the multiconfiguration calculations (Average Level) using the program by Grant et al.[9] yields for the lowest-energy configuration a DF total energy $\simeq$ -1049.807 a.u., which is now lower than the DF total energy for the state ($3\bar{d}^4\,4s^2$) with J = 0 .

In Table IV we compare the total DHF energies (in a.u.) for all atoms, He through Rn, with the numerical DF energies obtained from the program of Grant et al.[9]

These results confirm our previous claim[20] that the prescription of kinetic balance gives excellent results even for large Z atoms when Gaussian-type functions are employed, particularly in the finite-size nucleus model. There is no "variational collapse" and the eigenvalues converge from above to the numerical Dirac-Fock (DF) limit.[9] shown in column d. Some of the important conclusions about our results are listed below.

1. For all atoms up to Z=86 (Rn), it was possible to find a basis set which gave energies within a few millihartrees of the numerical values. The initial basis

Table IV. Fully relativistic Dirac-Hartree-Fock total energies (in a.u.).

Z	Atom	Atomic Weight	(a) State	(b) Configuration	(c)This work $-\varepsilon_{DHF}$	(d)Grant et al. $-\varepsilon_{DF}$
2	He	4.0026	$1s^2$	$1s^2$	2.86181328	2.86181335
3	Li	6.939	$2s^1_{1;1/2})_{1/2}$	(He) $2s^1$	7.433532996	7.43353322
4	Be	9.0122	$2s^2$	(He) $2s^2$	14.5758915	14.5758919
5	B	10.811	$2\bar{p}^1_{1;1/2})_{1/2}$	(He) $2s^2\ 2\bar{p}^1$	24.5366160	24.5366169
6	C	12.01115	$2\bar{p}^2$	(He) $2s^2\ 2\bar{p}^2$	37.65741796	37.6574196
7	N	14.0067	$2p^1_{1;3/2})_{3/2}$	(He) $2s^2\ 2\bar{p}^2\ 2p^1$	54.3169598	54.3169626
8	O	15.9994	$2p^2_{2;2})_2$	(He) $2s^2\ 2\bar{p}^2\ 2p^2$	74.8393203	74.8393234
9	F	18.9984	$2p^3_{1;3/2})_{3/2}$	(He) $2s^2\ 2\bar{p}^2\ 2p^3$	99.5022976	99.5023027
10	Ne	20.183	$2p^4$	(He) $2s^2\ 2\bar{p}^2\ 2p^4$	128.691927	128.691938
11	Na	22.9898	$3s^1_{1;1/2})_{1/2}$	(Ne) $3s^1$	162.078089	162.078100
12	Mg	24.312	$3s^2$	(Ne) $3s^2$	199.935070	199.935083
13	Al	26.9815	$3\bar{p}^1_{1;1/2})_{1/2}$	(Ne) $3s^2\ 3\bar{p}^1$	242.331126	242.331141
14	Si	28.086	$3\bar{p}^2$	(Ne) $3s^2\ 3\bar{p}^2$	289.449851	289.449869
15	P	30.9738	$3p^1_{1;3/2})_{3/2}$	(Ne) $3s^2\ 3\bar{p}^2\ 3p^1$	341.488940	341.488962
16	S	32.064	$3p^2_{2;2})_2$	(Ne) $3s^2\ 3\bar{p}^2\ 3p^2$	398.608727	398.608749
17	Cl	35.453	$3p^3_{1;3/2})_{3/2}$	(Ne) $3s^2\ 3\bar{p}^2\ 3p^3$	460.939846	460.939870
18	Ar	39.948	$3p^4$	(Ne) $3s^2\ 3\bar{p}^2\ 3p^4$	528.683800	528.683840
19	K	39.102	$4s^1_{1;1/2})_{1/2}$	(Ar) $4s^1$	601.526017	601.526048
20	Ca	40.08	$4s^2$	(Ar) $4s^2$	679.710239	679.710276
21	Sc	44.956	$3\bar{d}^1_{1;3/2})_{3/2}$	(Ar) $3\bar{d}^1\ 4s^2$	763.379306	763.379350
22	Ti	47.90	$3\bar{d}^2_{2;2})_2$	(Ar) $3\bar{d}^2\ 4s^2$	852.841025	852.841097
23	V	50.942	$3\bar{d}^3_{1;3/2})_{3/2}$	(Ar) $3\bar{d}^3\ 4s^2$	948.206618	948.206681
24	Cr	51.996	$3\bar{d}^4$	(Ar) $3\bar{d}^4\ 4s^2$	1049.67637	1049.67644

(Continued)

Table IV (cont.). Fully relativistic Dirac-Hartree-Fock total energies (in a.u.).

Z	Atom	Atomic Weight	(a) State	(b) Configuration	(c)This work $-\varepsilon_{DHF}$	(d)Grant et al. $-\varepsilon_{DF}$
25	Mn	54.938	$3d^{1}_{1;5/2})_{5/2}$	(Ar) $3\bar{d}^4\ 3d^1\ 4s^2$	1157.35190	1157.35199
26	Fe	55.847	$3d^{2}_{2;4})_{4}$	(Ar) $3\bar{d}^4\ 3d^2\ 4s^2$	1271.44029	1271.44038
27	Co	58.9332	$3d^{3}_{3;9/2})_{9/2}$	(Ar) $3\bar{d}^4\ 3d^3\ 4s^2$	1392.06875	1392.06884
28	Ni	58.71	$3d^{4}_{2;4})_{4}$	(Ar) $3\bar{d}^4\ 3d^4\ 4s^2$	1519.34397	1519.34409
29	Cu	63.54	$3d^{5}_{1;5/2})_{5/2}$	(Ar) $3\bar{d}^4\ 3d^5\ 4s^2$	1653.46159	1653.46168
30	Zn	65.37	$4s^2$	(Ar) $3\bar{d}^4\ 3d^6\ 4s^2$	1794.61331	1794.61340
31	Ga	69.72	$4\bar{p}^{1}_{1;1/2})_{1/2}$	(Ar) $3\bar{d}^4\ 3d^6\ 4s^2\ 4\bar{p}^1$	1942.56665	1942.56674
32	Ge	72.59	$4\bar{p}^2$	(Ar) $3\bar{d}^4\ 3d^6\ 4s^2\ 4\bar{p}^2$	2097.46666	2097.46675
33	As	74.9216	$4p^{1}_{1;3/2})_{3/2}$	(Ar) $3\bar{d}^4\ 3d^6\ 4s^2\ 4\bar{p}^2\ 4p^1$	2259.44457	2259.44468
34	Se	78.96	$4p^{2}_{2;2})_{2}$	(Ar) $3\bar{d}^4\ 3d^6\ 4s^2\ 4\bar{p}^2\ 4p^2$	2428.60582	2428.60594
35	Br	79.909	$4p^{3}_{1;3/2})_{3/2}$	(Ar) $3\bar{d}^4\ 3d^6\ 4s^2\ 4\bar{p}^2\ 4p^3$	2605.02986	2605.03002
36	Kr	83.80	$4p^4$	(Ar) $3\bar{d}^4\ 3d^6\ 4s^2\ 4\bar{p}^2\ 4p^4$	2788.86149	2788.86168
37	Rb	85.47	$5s^{1}_{1;\,1/2})_{1/2}$	(Kr) $5s^1$	2979.80592	2979.80718
38	Sr	87.62	$5s^2$	(Kr) $5s^2$	3178.08115	3178.08133
39	Y	88.905	$4\bar{d}^{1}_{1;3/2})_{3/2}$	(Kr) $4\bar{d}^1\ 5s^2$	3383.76421	3383.76461
40	Zr	91.22	$4\bar{d}^{2}_{2;2})_{2}$	(Kr) $4\bar{d}^2\ 5s^2$	3597.10295	3597.10314
41	Nb	92.906	$4\bar{d}^{3}_{1;3/2})_{3/2}$	(Kr) $4\bar{d}^3\ 5s^2$	3818.18956	3818.18977
42	Mo	95.94	$5s^2$	(Kr) $4\bar{d}^4\ 5s^2$	4047.17894	4047.17917
43	Tc	99.0	$4d^{2}_{2;4})_{4},\ 5s^{1}_{1;1/2})_{1/2})_{9/2}$	(Kr) $4\bar{d}^4\ 4d^2\ 5s^1$	4284.19183	4284.19211
44	Ru	101.07	$4d^{3}_{3;9/2})_{9/2},\ 5s^{1}_{1;1/2})_{1/2})_{5}$	(Kr) $4\bar{d}^4\ 4d^3\ 5s^1$	4529.33236	4529.33269
45	Rh	102.95	$4d^{4}_{2;4})_{4},\ 5s^{1}_{1;1/2})_{1/2})_{9/2}$	(Kr) $4\bar{d}^4\ 4d^4\ 5s.ct^1$	4782.68776	4782.68805
46	Pd	106.4	$4d^{5}_{1;5/2})_{5/2},\ 5s^{1}_{1;1/2})_{1/2})_{3}$	(Kr) $4\bar{d}^4\ 4d^5\ 5s^1$	5044.40597	5044.40632

(Continued)

set was taken as the one optimized to the closest nonrelativistic configuration and then augmented to give better numerical accuracy for the relativistic calculation. The agreement with numerical results[9] is excellent (one part in 10^8).

Table IV (cont.). Fully relativistic Dirac-Hartree-Fock total energies (in a.u.).

Z	Atom	Atomic Weight	(a) State	(b) Configuration	(c)This work $-\varepsilon_{DHF}$	(d)Grant et al. $-\varepsilon_{DF}$
47	Ag	107.87	$5s^1_{1;1/2})_{1/2}$	(Kr) $4\bar{d}^4\,4d^6\,5s^1$	5314.63638	5314.63674
48	Cd	112.40	$5s^2$	(Kr) $4\bar{d}^4\,4d^6\,5s^2$	5593.32242	5593.32286
49	In	114.82	$5\bar{p}^1_{1;1/2})_{1/2}$	(Kr) $4\bar{d}^4\,4d^6\,5s^2\,5\bar{p}^1$	5880.44109	5880.44152
50	Sn	118.69	$5\bar{p}^2$	(Kr) $4\bar{d}^4\,4d^6\,5s^2\,5\bar{p}^2$	6176.14055	6176.14105
51	Sb	121.75	$5p^1_{1;3/2})_{3/2}$	(Kr) $4\bar{d}^4\,4d^6\,5s^2\,5\bar{p}^2\,5p^1$	6480.53566	6480.53622
52	Te	127.60	$5p^2_{2;2})_2$	(Kr) $4\bar{d}^4\,4d^6\,5s^2\,5\bar{p}^2\,5p^2$	6793.72960	6793.73021
53	I	126.9044	$5p^3_{1;3/2})_{3/2}$	(Kr) $4\bar{d}^4\,4d^6\,5s^2\,5\bar{p}^2\,5p^3$	7115.80947	7115.81005
54	Xe	131.30	$5p^4$	(Kr) $4\bar{d}^4\,4d^6\,5s^2\,5\bar{p}^2\,5p^4$	7446.89956	7446.90018
55	Cs	132.905	$6s^1_{1;1/2})_{1/2}$	(Xe) $6s^1$	7786.77523	7786.77603
56	Ba	137.34	$6s^2$	(Xe) $6s^2$	8135.64919	8135.65006
57	La	138.91	$5\bar{d}^1_{1;3/2})_{3/2}$	(Xe) $5\bar{d}^1\,6s^2$	8493.65223	8493.65318
58	Cc	140.12	$4\bar{f}^1_{1;5/2})_{5/2},\ 5\bar{d}^1_{1;3/2})_{3/2})_4$	(Xe) $4\bar{f}^1\,5\bar{d}^1\,6s^2$	8861.09775	8861.09872
59	Pr	140.907	$4\bar{f}^2_{2;4})_4,\ 5\bar{d}^1_{1;3/2})_{3/2})_{11/2}$	(Xe) $4\bar{f}^2\,5\bar{d}^1\,6s^2$	9238.27995	9238.28055
60	Nd	144.24	$4f^3_{3;9/2})_{9/2},\ 5\bar{d}^1_{1;3/2})_{3/2})_6$	(Xe) $4\bar{f}^3\,5\bar{d}^1\,6s^2$	9625.30609	9625.30702
61	Pm	145.0	$4\bar{f}^4_{2;4})_4,\ 5\bar{d}^1_{1;3/2})_{3/2})_{11/2}$	(Xe) $4\bar{f}^4\,5\bar{d}^1\,6s^2$	10122.2764	10022.2774
62	Sm	150.35	$4\bar{f}^5_{1;5/2})_{5/2},\ 5\bar{d}_{1;3/2})_{3/2})_4$	(Xe) $4\bar{f}^5\,5\bar{d}^1\,6s^2$	10429.3605	10429.3616
63	Eu	151.96	$5\bar{d}^1_{1;3/2})_{3/2}$	(Xe) $4\bar{f}^6\,5\bar{d}^1\,6s^2$	10846.7569	10846.7582
64	Gd	157.25	$4f^1_{1;7/2})_{7/2},\ 5\bar{d}^1_{1;3/2})_{3/2})_5$	(Xe) $4\bar{f}^6\,4f^1\,5\bar{d}^1\,6s^2$	11274.4393	11274.4405
65	Tb	158.924	$4f^2_{2;6})_6,\ 5\bar{d}^1_{1;3/2})_{3/2})_{15/2}$	(Xe) $4\bar{f}^6\,4f^2\,5\bar{d}^1\,6s^2$	11712.7263	11712.7275
66	Dy	162.50	$4f^3_{3;15/2})_{15/2},\ 5\bar{d}^1_{1;3/2})_{3/2})_9$	(Xe) $4\bar{f}^6\,4f^3\,5\bar{d}^1\,6s^2$	12161.7420	12161.7431
67	Ho	164.930	$4f^4_{4;8})_8,\ 5\bar{d}^1_{1;3/2})_{3/2})_{19/2}$	(Xe) $4\bar{f}^6\,4f^4\,5\bar{d}^1\,6s^2$	12621.6044	12621.6056
68	Er	167.26	$4f^5_{3;15/2})_{15/2},\ 5\bar{d}^1_{1;3/2})_{3/2})_9$	(Xe) $4\bar{f}^6\,4f^5\,5\bar{d}^1\,6s^2$	13092.4173	13092.4186

(Continued)

Table IV (cont.). Fully relativistic Dirac-Hartree-Fock total energies (in a.u.).

Z	Atom	Atomic Weight	(a) State	(b) Configuration	(c)This work $-\varepsilon_{DHF}$	(d)Grant et al. $-\varepsilon_{DF}$
69	Tm	168.934	$4f^{6}_{2;6})_{6}, 5\bar{d}^{1}_{1;3/2})_{3/2})_{15/2}$	(Xe) $4\bar{f}^{6}\ 4f^{6}\ 5\bar{d}^{1}\ 6s^{2}$	13574.3815	13574.3828
70	Yb	173.04	$4f^{7}_{1;7/2})_{7/2}, 5\bar{d}^{1}_{1;3/2})_{3/2})_{5}$	(Xe) $4\bar{f}^{6}\ 4f^{7}\ 5\bar{d}^{1}\ 6s^{2}$	14067.6852	14067.6889
71	Lu	174.97	$5\bar{d}^{1}_{1;3/2})_{3/2}$	(Xe) $4\bar{f}^{6}\ 4f^{8}\ 5\bar{d}^{1}\ 6s^{2}$	14572.5330	14572.5346
72	Hf	178.49	$5\bar{d}^{2}_{2;2})_{2}$	(Xe) $4\bar{f}^{6}\ 4f^{8}\ 5\bar{d}^{2}\ 6s^{2}$	15088.8152	15088.8166
73	Ta	180.948	$5\bar{d}^{3}_{1;3/2})_{3/2}$	(Xe) $4\bar{f}^{6}\ 4f^{8}\ 5\bar{d}^{3}\ 6s^{2}$	15616.6609	15616.6625
74	W	183.85	$5\bar{d}^{4}$	(Xe) $4\bar{f}^{6}\ 4f^{8}\ 5\bar{d}^{4}\ 6s^{2}$	16156.2463	16156.2484
75	Re	186.2	$5d^{1}_{1;5/2})_{5/2}$	(Xe) $4\bar{f}^{6}\ 4f^{8}\ 5\bar{d}^{4}\ 5d^{1}\ 6s^{2}$	16707.6792	16707.6814
76	Os	190.2	$5d^{2}_{2;4})_{4}$	(Xe) $4\bar{f}^{6}\ 4f^{8}\ 5\bar{d}^{4}\ 5d^{2}\ 6s^{2}$	17271.1551	17271.1573
77	Ir	192.2	$5d^{3}_{3;9/2})_{9/2}$	(Xe) $4\bar{f}^{6}\ 4f^{8}\ 5\bar{d}^{4}\ 5d^{3}\ 6s^{2}$	17846.8566	17846.8590
78	Pt	195.09	$5d^{5}_{1;5/2})_{5/2}, 6s^{1}_{1;1/2})_{1/2})_{3}$	(Xe) $4\bar{f}^{6}\ 4f^{8}\ 5\bar{d}^{4}\ 5d^{5}\ 6s^{1}$	18434.9109	18434.9133
79	Au	196.967	$6s^{1}_{1;1/2})_{1/2}$	(Xe) $4\bar{f}^{6}\ 4f^{8}\ 5\bar{d}^{4}\ 5d^{6}\ 6s^{1}$	19035.5586	19035.5612
80	Hg	200.59	$6s^{2}$	(Xe) $4\bar{f}^{6}\ 4f^{8}\ 5\bar{d}^{4}\ 5d^{6}\ 6s^{2}$	19648.8665	19648.8692
81	Ti	204.37	$6\bar{p}^{1}_{1;1/2})_{1/2}$	(Xe) $4\bar{f}^{6}\ 4f^{8}\ 5\bar{d}^{4}\ 5d^{6}\ 6s^{2}\ 6\bar{p}^{1}$	20274.8261	20274.8278
82	Pb	207.19	$6\bar{p}^{2}$	(Xe) $4\bar{f}^{6}\ 4f^{8}\ 5\bar{d}^{4}\ 5d^{6}\ 6s^{2}\ 6\bar{p}^{2}$	20913.7151	20913.7184
83	Bi	208.98	$6p^{1}_{1;3/2})_{3/2}$	(Xe) $4\bar{f}^{6}\ 4f^{8}\ 5\bar{d}^{4}\ 5d^{6}\ 6s^{2}\ 6\bar{p}^{2}\ 6p^{1}$	21565.6930	21565.6960
84	Po	209.0	$6p^{2}_{2;2})_{2}$	(Xe) $4\bar{f}^{6}\ 4f^{8}\ 5\bar{d}^{4}\ 5d^{6}\ 6s^{2}\ 6\bar{p}^{2}\ 6p^{2}$	22230.9946	22230.9978
85	At	210.0	$6p^{3}_{1;3/2})_{3/2}$	(Xe) $4\bar{f}^{6}\ 4f^{8}\ 5\bar{d}^{4}\ 5d^{6}\ 6s^{2}\ 6\bar{p}^{2}\ 6p^{3}$	22909.7492	22909.7526
86	Rn	222.0	$6p^{4}$	(Xe) $4\bar{f}^{6}\ 4f^{8}\ 5\bar{d}^{4}\ 5d^{6}\ 6s^{2}\ 6\bar{p}^{2}\ 6p^{4}$	23601.9706	23601.9742

(a) For open shell cases, we adopt a notation similar to that of Grant et al.[9] For a single open shell case we denote the state by $nl^{q}_{vj})_{x}$ where n is the principal quantum number, l labels the orbital angular momentum quantum number, q denotes the occupation number, v denotes the seniority number, j denotes the resultant angular momentum quantum number of the open shell, and x is the total angular momentum quantum number of the state. Where there are two open shells, the state is defined by its total angular momentum quantum number J. The closed shell structures are denoted by the last shell and its occupation number.

(b) The configurations are defined with respect to the nearest closed shell structure. Calculation was performed by us on several electronic configurations corresponding to each atom, and only the configuration with the lowest energy state among several electronic configurations is given.

(c) This work (using nonrelativistic optimized geometric Gaussian basis sets).

(d) Numerical DF value[9] obtained with the average level (AL) option (Breit interaction not included).

Table V. Radial Expectation Values (in a.u.) for Ar. Numerical DF values[9] obtained with the average level (AL) option are given in parenthesis.

Orbital	$<1/r>$	$<r>$	$<r^2>$
1s	17.70262598	0.08562456	0.009863280
	(17.702650)	(0.08562454)	(0.009863278)
2s	3.58895526	0.40995835	0.19911321
	(3.5889538)	(0.40995755)	(0.19911223)
3s	0.96793412	1.41613721	2.33104982
	(0.96791243)	(1.4161612)	(2.3311181)
$2\bar{p}$	3.48009481	0.37305593	0.17243040
	(3.4801113)	(0.37305400)	(0.17242876)
$3\bar{p}$	0.81876081	1.65570617	3.28238157
	(0.81882635)	(1.6556346)	(3.2821931)
2p	3.45150000	0.37533778	0.17440355
	(3.4514977)	(0.37533790)	(0.17440356)
3p	0.81333962	1.66538020	3.32155645
	(0.81333304)	(1.6653913)	(3.3216004)

Furthermore, our results for the total DHF energy and orbital energies are much closer to the numerical DF[9] limit than the corresponding results for atoms (He-Hg) recently reported by by Okada and Matsuoka.[17]

2. The value of the total DHF energy was always found to approach (as the number of basis functions increases) the numerical DF limit[9] from above.

3. The relativistic virial theorem[17]

$$<T> = -<V>$$

was tested in all point nucleus cases and was satisfied at least to one part in 10^5.

4. We also tried a common geometric basis set for atoms from Z=72 to Z=86 and obtained results within a few millihartrees of the numerical DF limit. However, the results improved consistently when we used for each atom a

Table VI. Convergence Pattern for the Ground State of Xenon.

$N^{a}_{s_{1/2}}$	$N^{b}_{p_{1/2}}$	$-E^{c,e}_{DHF}$	$-E^{d,e}_{DHF}$
28	26	7446.852770	7447.066014
31	26	7446.897507	7447.142336
33	26	7446.897690	7447.151981
33	28	7446.899177	7447.153555
33	30	7446.899510	7447.153975
34	30	7446.899511	7447.155759
35	30	7446.899512	7447.156711
35	31	7446.899548	7447.156787
35	32	7446.899556	7447.156827
35	33	7446.899557	7447.156848
E_{DHF} (by Grant et al.[9])		(7446.90018)	(7447.15856)

(a) $N_{s_{1/2}}$ is the number of geometric basis functions used for the large component of $s_{1/2}$ symmetry.

(b) $N_{p_{1/2}}$ is the number of geometric basis functions used for the large component of $p_{1/2}$ symmetry.

(c) Finite nucleus size approximation used.

(d) Point nucleus calculation.

(e) In this calculation we have kept the following as constant: $N_{p_{3/2}} = 26$; $N_{d_{3/2}} = 20$; $N_{d_{5/2}} = 20$. The exponents were generated by using $\alpha_0 = 0.0143013$ and $\beta = 1.9778445$.

basis set independently optimized to the closest nonrelativistic ground state configuration.

5. We calculated $< r^n >$, $n = -1, 1, 2$ with our atomic orbitals and compared them with the expectation values obtained by numerical methods. One example of these expectation values for argon is given in Table II. There is excellent agreement between the values calculated using the numerical DF wavefunctions[9] and the corresponding expectation values obtained by using our geometric Gaussian basis functions. We should note that a similar calculation for argon has been recently reported by Quieny et al.[33] using a Slater basis.

Table VII. Orbital and Total DHF Energies of Xe Ground State (in a.u.).

Orbital	$-\varepsilon^a$	$-\varepsilon^b$
1*s*	1277.2588019	1277.2586115
2*s*	202.4650992	202.4650254
3*s*	43.0104331	43.0104285
4*s*	8.4298806	8.4298898
5*s*	1.0101234	1.0101354
$2\bar{p}$	189.6795387	189.6794723
$3\bar{p}$	37.6598865	37.6598869
$4\bar{p}$	6.4524449	6.4524608
$5\bar{p}$	0.4925637	0.4925662
2*p*	177.7046897	177.7045060
3*p*	35.3253609	35.3252341
4*p*	5.9828279	5.9827651
5*p*	0.4398111	0.4397988
$3\bar{d}$	26.0234451	26.0233447
$4\bar{d}$	2.7113564	2.7113095
3*d*	25.5372327	25.5370856
4*d*	2.6338201	2.6337404
DHF Total	-7446.900181	-7446.899558

(a) Same as in footnote (d) of Table IV (the DHF energy given by Okada et al.[17] is −7446.8937 a.u. and the corresponding numerical value given by by Desclaux[10] is −7446.9006 a.u.).

(b) This work.

Table VIII. Orbital and Total DHF Energies of Ground State of Radon (in a.u.).

Orbital	$-\varepsilon^a$	$-\varepsilon^b$
$1s$	3641.152594	3641.151683
$2s$	668.803231	668.803240
$3s$	166.831068	166.831296
$4s$	41.312994	41.313174
$5s$	8.408858	8.408980
$6s$	1.071426	1.074159
$2\bar{p}$	642.327556	642.327511
$3\bar{p}$	154.894066	154.894288
$4\bar{p}$	36.019227	36.019414
$5\bar{p}$	6.408880	6.408966
$6\bar{p}$	0.540362	0.540372
$2p$	541.103035	541.102712
$3p$	131.73119	131.731021
$4p$	30.120838	30.120712
$5p$	5.175965	5.175900
$6p$	0.384038	0.384019
$3\bar{d}$	112.567303	112.567168
$4\bar{d}$	21.548307	21.548232
$5\bar{d}$	2.189770	2.189705
$3d$	107.759555	107.759278
$4d$	20.439131	20.438872
$5d$	2.016730	2.016597
$4\bar{f}$	9.194232	9.194121
$4f$	8.928566	8.928415
DHF Total	-23601.97421	-23601.97060

(a) Same as in footnote (d) of Table IV (the DHF energy given by Matsuoka et al.[18] is −23601.732 a.u. and the corresponding numerical value given by Desclaux[10]. is −23601.978 a.u.)

(b) This work.

In Table VI we present an example of our improved basis set for Xenon (this is different from the one given in our earlier work[20]) and show the convergence pattern for Xenon in the finite nucleus case. Indeed, our results for Xenon confirm the conclusion drawn by Ishikawa et al.[11] from their results for H-like Mercury ion that the convergence is much faster in the finite nucleus case than the point nucleus calculation. From Table VI we see that by the time we have reached $N_{s_{1/2}} = 33$ (number of basis functions for the symmetry $s_{1/2}$) and $N_{p_{1/2}} = 28$ (number of basis functions for the $p_{1/2}$ symmetry), the DHF energy has saturated and we have the limiting DHF total energy of -7446.8996 a.u. compared to the numerical value of -7446.9002 a.u. If we go beyond $N_{1s_{1/2}} = 33$ and $N_{2p_{1/2}} = 28$ the result for the DHF energy does not change to any appreciable degree. In contrast to this, even with $N_{1s_{1/2}} = 35$ and $N_{p_{1/2}} = 28$, we cannot get close to the numerical limit of the DHF total energy[9] for Xenon in the point nucleus approximation, which is -7447.15856 a.u. We think that the kinetic balance condition given by (5), along with proper boundary conditions imposed upon the basis set, contribute to the accelerated convergence and to the variational stability, as has been demonstrated by earlier workers.[4-13,16] This type of convergence pattern seems to be a simple generalization of the result of Ishikawa et al.[11] to the many-electron case.

In Table VII we have given the DHF energy and the orbital energies for Radon (the heaviest atom reported by us) with an improved basis set. It is seen that the orbital energies of the inner shell as well as outer shells can be calculated to an accuracy comparable with the numerical method; the maximum discrepancy is ~ 0.0001 a.u. which occurs for $1s_{1/2}$ orbital. We also predict the energy difference between the $4\bar{f}$ and $4f$ orbitals to be within 0.0001 a.u. of numerical DF method. Similar results for other heavy atoms such as gold(Z=79) and mercury(Z=80) also have been reported by us recently.[34] In passing, we should mention that convergence pattern for all heavy atoms is similar to that presented in Table VI for Xenon in the finite nucleus approximation, although we do not present the results explicitly.

12. Conclusions

In conclusion, our results demonstrate that the use of kinetically balanced geometric Gaussian basis sets for Dirac-Fock SCF calculations in many-electron atoms (He − Rn) can yield results comparable in accuracy to those of numerical DF methods, particularly in the finite-size nucleus model. Furthermore, "variational collapse" does not occur if we use a matched basis set satisfying kinetic balance and proper boundary conditions for bound state orbitals − a point which has been emphasized by Grant et al. previously.[5-9] Also, our calculations support the viewpoint that the finite-size nucleus approximation facilitates the use of GTOs in DF calculations for heavy atoms: the DHF total energy converges more rapidly to the true value and converges from above; therefore, one avoids the necessity for a very large basis set. The development of these compact Gaussian basis sets within the DHF approach has potential applications in the study of molecules containing heavy elements (work in progress). The computer program for DHF atomic computations is available as a part of MOTECC ; the wave functions of Table IV are also available upon request.

Acknowledgments

The authors would like to express their sincere thanks to Dr. S. Chakravorty for providing us with his optimized nonrelativistic geometric Gaussian basis sets.[21] Thanks are also due to Prof. G. Malli and Dr. Daniel Frye for their helpful comments and useful suggestions during the course of the preparation of this article.

References

1. For a review, see W. Kutzelnigg, Int. J. Quantum Chem., **25**, 107 (1984).
2. Y. Ishikawa, R.C. Binning and K.M. Sando, Chem. Phys. Letters, **101**, 111 (1983).
3 K.G. Dyall, I.P. Grant and S. Wilson, J. Phys., **B17**, 493 (1984); ibid **L45**; ibid **B17**, 1201.
4. R.E. Stanton and S. Havriliak, J. Chem. Phys., **81**, 1910 (1984).
5. I.P. Grant, J. Phys., **B19**, 3187 (1986).
6. I.P. Grant and H.M. Quiney, Adv. in Atom. and Mol. Phys., **23**, 37 (1988).
7. H.M. Quiney, I.P. Grant and S. Wilson, J. Phys., **B20**, 1413 (1987).
8. H.M. Quiney, I.P. Grant and S. Wilson, "On the Relativistic Many-Body Perturbation Theory of Atomic and Molecular Electronic Structure," to be published in *Lecture Notes in Chemistry*, U. Kaldor, Ed., Springer Verlag, Berlin.
9. I.P. Grant, B.J. McKenzie, P.H. Norrington, D.F. Mayers and N.C. Pyper, Comput. Phys. Communication, **21**, 207 (1980).
10. J.P. Desclaux, Comput. Phys. Communication, **9**, 31 (1975).
11. Y. Ishikawa, R. Baretty and R.C. Binning, Jr., Chem. Phys. Letters, **121**, 130 (1985).
12. P.J.C. Aerts and W.C. Nieuwpoort, Chem. Phys. Letters, **113**, 165 (1985).
13. O. Visser, P.J.C. Aerts, D. Hegarty and W.C. Nieuwpoort, Chem. Phys. Letters, **134**, 34 (1987).
14. G. Malli, Chem. Phys. Letters, **68**, 529 (1979).
15. O. Matsuoka, M. Klobukowski and S. Huzinaga, Chem. Phys. Letters, **113**, 395 (1985).
16. O. Matsuoka and S. Huzinaga, Chem. Phys. Letters, **140**, 567 (1987).
17. S. Okada and O. Matsuoka, J. Chem. Phys., **91(7)**, 4193 (1989).
18. O. Matsuoka and S. Okada, Chem. Phys. Letters, **155**, 547 (1989).
19. J.P. Desclaux, Atomic and Nuclear Data Tables, **12**, 311 (1973).
20. A.K. Mohanty and E. Clementi, Chem. Phys. Letters, **157**, 348 (1989).
21. E. Clementi and G. Corongiu, Chem. Phys. Letters, **90**, 359 (1982).
22. I.P. Grant and N.C. Pyper, J. Phys. B., **9**, 761 (1976); I.P. Grant and B.J. McKenzie, J. Phys. B. Atom. Molec. Phys., **13**, 2671 (1980).

23. Y.K. Kim, Phys. Rev., **154**, 17 (1967); **159**, 190 (1967).
24. T. Kagawa, Phys. Rev. A., **12**, 2245 (1975); **22**, 2340 (1980).
25. B. Roos, C. Salez, A. Veillard and E. Clementi, Tech. Rep. RJ518, IBM Research (August 12, 1968).
26. H.A. Bethe and E.S. Salpeter, *Quantum Mechanics of one- and two-electron atoms,* Springer Verlag, 70 (1957).
27. C.C.J. Roothaan, Rev. Mod. Phys., **23**, 69 (1951); **32**, 179 (1960).
28. J.M. Leclercq, Phys. Rev. A., **1**, 1358 (1970).
29. S. DeShalit and I. Talmi, Nuclear Shell Theory, Academic Press (1961).
30. I. Grant, *Relativistic effects in atoms and molecules,* S. Wilson, Ed., Plenum Press (1988).
31. G.L. Malli and J.P. Olive, "Vector Coupling Coefficients for Atomic Self-Consistent Field Calculations," Laboratory of Molecular Structure, Technical Report 1962-3, Part 2, University of Chicago, 258 (1963).
32. For a discussion of relativistic virial theorem see Ref. 23; Also, P.J.C. Aerts, *Towards Relativistic Quantum Chemistry,* Ph.D. Thesis, University of Groningen (1986). If we scale the coordinates of N-electron wavefunction $\Phi(\vec{r}_1, \vec{r}_2 \ldots \vec{r}_N)$ by a parameter λ, then

$$\Phi(\vec{r}_1, \vec{r}_2, \ldots \vec{r}_N) \rightarrow \Phi(\lambda\vec{r}_1, \lambda\vec{r}_2, \ldots \lambda\vec{R}_N),$$

and

$$< H > = < T > + < V > + < M > \rightarrow \lambda(T + V) + M.$$

This is because $T = c\vec{\alpha} \cdot \vec{p}$ contains the first derivative with respect to the position coordinate r, operator V contains only terms with inverse power r, and the operator $M = \beta' c^2$ does not depend upon r. For any correct eigenfunction the infinitesimal variation of H with respect to scale parameter λ gives the relativistic virial theorem

$$\frac{\partial < H >}{\partial \lambda} = (< T > + < V >) = 0$$

$$\text{or} \quad < T > = - < V > \quad \text{or} \quad < E > = < M >$$

In order to see the correspondence between the relativistic and nonrelativistic virial theorem, we calculate $< T >$ and $< M >$ by using the free particle spinors in the momentum representation, viz,

$$\psi_r = \sqrt{\frac{mc^2}{E\tau}}\, u^r(p) \exp(i\vec{p} \cdot \vec{x} - \frac{iEt}{\hbar})$$

with

$$u^r(p) = \sqrt{\frac{E + mc^2}{2mc}} \begin{pmatrix} \phi^\sigma \\ \dfrac{(\vec{\sigma} \cdot \vec{p})c}{E + mc^2} \phi^\sigma \end{pmatrix}.$$

Here ϕ^σ are the two-component Pauli spinors, p is the momentum, and τ is the volume over which the integration is performed.

And then taking the limit $c \rightarrow \infty$ one obtains the smooth nonrelativistic limit

$$< T > = 2 < T_{nr} >$$

$$\text{or} \quad < M > = - < T_{nr} > = < E > .$$

33. H.M. Quiney, I.P. Grant and S. Wilson, J. Phys., **B22**, L-15 (1989).
34. A.K. Mohanty and E. Clementi, "Kinetically Balanced Geometric Gaussian Basis Sets for Relativistic Atoms", J.Chem. Phys. (in press)

Chapter 4.

Non-Relativistic Configuration Interaction Calculations for Many-Electron Atoms: ATOMCI

F. Sasaki, M. Sekiya,† T. Noro,
K. Ohtsuki and Y. Osanai

Department of Chemistry
Faculty of Science
Hokkaido University
Sapporo 060, Japan

Introduction

In the theory of atomic spectra, the Racah's methods[2,7,8] proved quite useful in dealing with complex structures. Among others, Racah clarified and extended the concepts of "irreducible tensor operator" and "fractional parentage coefficient." Algebraic manipulation of tensor operators provides a systematic and efficient means to make use of the symmetry of the system. The exclusion principle or the antisymmetry of the total wavefunction is taken care of by the fractional parentage. This method is also successfully applied to systems having lower symmetry such as transition metal ions in crystals.[3,13,14] The approach adopted here[10] is based on the irreducible tensor operator method but, unlike the approaches cited above, fractional parentage coefficients do not play any particular role. We work on operators in the quantized space where all vectors satisfy the exclusion principle. Therefore, no additional considerations on the antisymmetry of wavefunctions are required. The fractional parentage coefficient appears just as a particular case of vacuum expectation values in the general reduction formulas.

There are several papers in which the group theory or the group representation theory is applied to quantum theory of elementary particles, nucleons, atoms, molecules, solids, and so on, and reviewing them is not attempted: it is also entirely out of our ability. This attitude may be, to a certain extent, justified since

† Present address: IBM Research Laboratory, San Jose, California 95193, U.S.A.

mathematical foundation of the present approach was established a long time ago by the work of Wigner and Racah. Nevertheless, it may be useful to point out the relationship of the present work to some of the works in the field of atomic and molecular quantum theory because that will help to get the overall view of the present approach.

The first remark is on the use of a group which is not that of the symmetry of the object quantum system possesses. For atoms in LS-scheme, the apparent symmetry group is O(3)×SU(2), where O(3) and SU(2) are for rotations in the three-dimensional space and in the spin space, respectively. Well known examples of the extension of the group are the introduction of the "seniority number" for the classification of d^n (Ref. 8), and the Lie group G_2 for f^n (Ref. 9). In the present approach, a complete set of atomic shell states l^n is generated automatically in a computer, and the procedure is described in Section 3. In order to make core space compact and to save time for generation, atomic shell states are classified in LSQ-scheme, where the last Q stands for "quasi spin".[1,4] Formulation in LSQ-scheme is equivalent to the use of seniority number, but its group theoretical implications can be seen much more clearly. Classification scheme by G_2 is not considered in the generation of shell states since it applies only for f^n.

The second point we would like to mention is on the unitary group approach,[5,6] which is based on the observation that all the n-electron functions having the same total spin angular momentum and its z-component constitute an irreducible representation space of the unitary group U(N), where N stands for the total number of spatial orbitals under consideration. Gel'fand-Tsetlin representation[19] of the infinitesimal generators of the unitary group may be used for the calculation of matrix elements. Based on this idea, Shavitt[11] has succeeded in bringing matrix elements of one- and two-electron operators into a factorized form. Combining this with the use of graphs for state vectors,[12] he showed that quite efficient algorithm could be devised for the evaluation of matrix elements.

The present approach has an advantage over the unitary group approach in the following points.

1. Spatial symmetry properties of the system can be fully taken into account as long as its symmetry group is simple reducible.[15] Besides the rotational group of the atomic system considered here, the majority of molecular symmetry groups have this property.

2. The factorization of matrix elements comes out as a natural consequence of the derivation.

3. Spin dependent properties can be calculated in the same way as other spin singlet operators.

Section 1 introduces tensor operators and their adjoint operators. In the next section, formulas which involve various recoupling transformations of these tensor operators are derived. They form a basis of the analysis of the following sections. Generation schemes of a complete and orthonormal set of shell states are presented in section 3. Representation of CI vectors in terms of tensor operators is described in the last section, and two useful reduction formulas are derived for the evaluation of matrix elements of operators. The properties of the representation of simply reducible groups are fully discussed in Appendix 4A, and explicit formulas for three-dimensional rotation group is given in Appendix 4B.

This article is essentially an enlarged version of Ref. 10, although the contents of Section 3 are only briefly mentioned there. Some of the original proofs of Wigner regarding the properties of the simply reducible groups are replaced by simpler ones in Appendix 4A.

1. Tensor Operator

An orthonormal and complete set of spin orbitals $\{\xi_\alpha(x)\}$ may be introduced as follows Each spin orbital is characterized by its radial, angular and spin indices $n\ell$, m and μ, which will also be denoted by a single symbol α, i.e.,

$$\xi_\alpha(x) = \xi_{n\ell m\mu}(\mathbf{r}, \sigma) = \begin{cases} R_{n\ell}(r) Y_{\ell m}(\theta, \phi)\alpha(\sigma) \\ R_{n\ell}(r) Y_{\ell m}(\theta, \phi)\beta(\sigma) \end{cases} \text{for } u = \begin{cases} 1/2 \\ -1/2. \end{cases} \tag{1.1}$$

Here, x stands for the spatial and spin coordinates $(\mathbf{r}, \sigma)$, and r, θ and ϕ are the spherical coordinates of $\mathbf{r}$. Spherical harmonics $Y_{\ell m}$'s are normalized according to Condon and Shortley.[18]

These spin orbitals experience a linear transformation when the axes of the space and spin coordinates rotate.[a] Let R denote a rotation of the space coordinate axes. These R's form the group of orthogonal transformations O(3), and the spherical harmonics $Y_{\ell m}(m = -\ell, -\ell + 1, \ldots, \ell)$ span an irreducible representation space of the orthogonal group as

---

a Transformational properties of functions or operators can be studied in two quite different ways. We follow the one adopted in Ref. 2 where functions or physical systems are fixed in space, and only the axes of the coordinates are transformed. In the other approach, functions or physical systems are actually rotated and the axes of coordinates are fixed once for all. Although both approaches are closely related, one should be careful not to mix up concepts or formulas derived from different standpoints. For example, although SR is defined as a rotation which is equivalent to two successive rotations of R and S in this order in both of the approaches, the rotations R and S refer to *the fixed common coordinate system* in the latter. The rotation S in this article is defined *relative to the rotated coordinate axes* by R. See page 105 of Ref. 16 for

$$Y_{\ell m}(\theta', \phi') = \sum_p D^{(\ell)}(R)_{mp} Y_{\ell p}(\theta, \phi). \tag{1.2}$$

Here, the symbols θ and ϕ stand for the angular part of the spherical coordinates of a point P in the original frame, and θ' and ϕ', those of the same point after the frame is rotated by R. The matrix $D^{(\ell)}(R)$ is the irreducible representation for the angular momentum ℓ. Its explicit functional form is given in Appendix 4B. Let θ'' and ϕ'' be the angular part of the coordinates of the point P after the frame is further rotated by S. Then, the following equation holds.

$$\begin{aligned} Y_{\ell m}(\theta'', \phi'') &= \sum_p D^{(\ell)}(S)_{mp} Y_{\ell p}(\theta', \phi') \\ &= \sum_p D^{(\ell)}(SR)_{mp} Y_{\ell p}(\theta, \phi) \end{aligned} \tag{1.3}$$

From Eqs. (1.2) and (1.3) it can be seen that the $D^{(\ell)}(R)$'s form representation matrices of the group O(3), i.e.,

$$D^{(\ell)}(SR)_{mp} = \sum_q D^{(\ell)}(S)_{mq} D^{(\ell)}(R)_{qp}.$$

Similarly, the spin functions $\alpha(\sigma)$ and $\beta(\sigma)$, which are fixed to the spin coordinate frame, transform when the spin coordinate axes rotate but, unlike the spatial coordinate r, the spin coordinate σ is quantized to the z-axis of the spin coordinate. Therefore, there exists no one-to-one correspondence between the spin coordinate σ and σ' which belong to different frames of the spin space. Nevertheless, one can express the spin functions α and β, which are fixed to the rotated frame, in terms of those in the original frame since they constitute a complete set of spin functions:

$$\begin{aligned} \alpha(\sigma') &= u_{11}\alpha(\sigma) + u_{12}\beta(\sigma) \\ \beta(\sigma') &= u_{21}\alpha(\sigma) + u_{22}\beta(\sigma). \end{aligned}$$

The notation $\alpha(\sigma')$ and $\beta(\sigma')$ are used symbolically to indicate the spin functions α and β which are fixed to the rotated frame, and the same functions expressed in the original frame are given on the right hand side of the equations. The two-by-two matrix u, which corresponds to a rotation of the spin coordinate axes, is unitary and its determinantal value is 1.

For the set of spin orbitals $\{\xi_\alpha\}$, both kinds of the rotations of the coordinate axes may be applied, namely

$$\xi_{n\ell m\mu}(\mathbf{r}', \sigma') = \sum_{p\pi} D^{(\ell)}(R)_{mp} D^{(1/2)}(u)_{\mu\pi} \xi_{n\ell p\pi}(\mathbf{r}, \sigma), \tag{1.4}$$

which shows that the set of $4\ell+2$ functions

$$\xi_{n\ell p\pi}(\mathbf{r}, \sigma), \ (m = -\ell, -\ell+1, \dots, \ell, \ \mu = -1/2, \ 1/2)$$

spans an irreducible representation space of the direct product group O(3)×SU(2). Its group element is given by the pair R and u, which correspond to rotations of the spatial and spin coordinate axes, and its irreducible representation can be denoted by the usual spectroscopic notation in LS scheme, e.g., ${}^2\ell$ in this case.

Expanding the second quantized field operators in terms of the spin orbitals, we obtain the annihilation operator a_α and its Hermitian adjoint creation operator $a_\alpha\dagger$ as

$$\psi(x) = \sum_\alpha \xi_\alpha(x)\mathbf{a}_\alpha \tag{1.5}$$

$$\psi\dagger(x) = \sum_\alpha \xi_\alpha(x)^*\mathbf{a}_\alpha\dagger, \tag{1.6}$$

where the complex conjugation is indicated by the symbol $*$. They satisfy the anti-commutation relations

$$\begin{aligned} [\mathbf{a}_\alpha, \mathbf{a}_\beta]_+ = [\mathbf{a}_\alpha\dagger, \mathbf{a}_\beta\dagger]_+ &= 0 \\ [\mathbf{a}_\alpha, \mathbf{a}_\beta\dagger]_+ &= \delta_{\alpha,\beta}. \end{aligned} \tag{1.7}$$

The field operator $\psi^\dagger(x)$ creates an electron at the point whose space-spin coordinate is given by x. When the coordinate axes are rotated, the operator $\psi'\dagger$ defined in this frame should have the same physical effect as $\psi\dagger$, because we only rotate the coordinate axes but not the physical system. Therefore, the following equality should hold.

$$\psi'\dagger(x') = \psi\dagger(x). \tag{1.8}$$

The creation operator $\mathbf{a}'_{n\ell m\mu}\dagger$ in the rotated frame is defined as

$$\psi'\dagger(x') = \sum_\alpha \xi_\alpha(x')^*\mathbf{a}'_\alpha\dagger. \tag{1.9}$$

From Eqs. (1.4), (1.8) and (1.9), the following equation holds.

$$\begin{aligned} \psi\dagger(x) &= \sum_{n\ell m\mu} \xi_{n\ell m\mu}(x')^* \, \mathbf{a}'_{n\ell m\mu}\dagger \\ &= \sum_{n\ell m\mu}\sum_{p\pi} D^{(\ell)}(R)^*{}_{mp} D^{(1/2)}(u)^*{}_{\mu\pi} \xi_{n\ell p\pi}(x)^* \mathbf{a}'_{n\ell p\pi}\dagger. \end{aligned}$$

Equating this to Eq. (1.6), and considering the unitarity of the matrices $D^{(j)}$ we obtain

$$\mathbf{a}'_{n\ell m\mu}\dagger = \sum_{p\pi} D^{(\ell)}(R)_{mp} D^{(1/2)}(u)_{\mu\pi} \mathbf{a}_{n\ell p\pi}\dagger, \tag{1.10}$$

which shows that the set $\mathbf{a}_{n\ell}\dagger = \{\mathbf{a}_{n\ell p\pi}\dagger\}$ obeys the same representation as the set of spin orbitals $\xi_{n\ell p\pi}(\mathbf{r}, \sigma)$.

The representation matrix depends on the basis, and we call

$$D^{(\Lambda)}(X)_{\kappa\lambda} = D^{(\ell)}(R)_{mp} D^{(s)}(u)_{\mu\pi} \tag{1.11}$$

the standard representation, and a set of operators $\mathbf{A} = \{\mathbf{A}_\kappa\}$ which forms a basis of a standard representation will be called a tensor operator. Namely, under the rotation of the coordinate axes, a tensor operator $\mathbf{A}$ transforms as

$$\mathbf{A}'_\kappa = \sum_\lambda D^{(\Lambda)}(X)_{\kappa\lambda} \mathbf{A}_\lambda. \tag{1.12}$$

Here, Λ stands for an irreducible representation $^{2s+1}\ell$, X and for a group element R u, κ and λ for the matrix element $m\mu$ and $p\pi$. An example of a tensor operator is the set $\mathbf{a}_{n\ell}\dagger$ as described above.

From a tensor operator $\mathbf{A}$, one can construct a set of Hermitian adjoint operators $\{A_\kappa\dagger\}$. It is not a tensor operator in general, since its representation is the complex conjugate representation of the original tensor operator.

$$\mathbf{A}'_\kappa\dagger = \sum_\lambda (D^{(\Lambda)}(X)_{\kappa\lambda}\mathbf{A}_\lambda)\dagger = \sum_\lambda D^{(\Lambda)}(X)_{\kappa\lambda}{}^* \mathbf{A}_\lambda\dagger \tag{1.13}$$

However, since the character is a real function for the group O(3)×SU(2), any representation is equivalent to its complex conjugate. Therefore, there exists a unitary matrix $U^{(\Lambda)}$ for which

$$\sum_\lambda D^{(\Lambda)}(X)_{\kappa\lambda}{}^* U^{(\Lambda)}_{\lambda\mu} = \sum_\lambda U^{(\Lambda)}_{\kappa\lambda} D^{(\Lambda)}(X)_{\lambda\mu} \tag{1.14}$$

holds for any X. Using the matrix $U^{(\Lambda)}$, we define a set of operators $\mathbf{B} = \{B_\kappa\}$ as follows.

$$\mathbf{B}_\kappa = \sum_\lambda U^{(\Lambda)*}_{\kappa\lambda} \mathbf{A}_\lambda\dagger. \tag{1.15}$$

Under the coordinate rotation, the operator **B** transforms into **B′** as

$$\begin{aligned}\mathbf{B}'_\kappa &= \sum_\lambda U^{(\Lambda)*}_{\kappa\lambda}\mathbf{A}_\lambda{}^{\dagger\prime} = \sum_{\lambda\mu} U^{(\Lambda)*}_{\kappa\lambda} D^{(\Lambda)}(X)_{\lambda\mu}{}^*\mathbf{A}_\mu{}^\dagger \\ &= \sum_{\lambda\mu} D^{(\Lambda)}(X)_{\kappa\lambda} U^{(\Lambda)*}_{\lambda\mu}\mathbf{A}_\mu{}^\dagger = \sum_\lambda D^{(\Lambda)}(X)_{\kappa\lambda}\mathbf{B}_\lambda,\end{aligned}$$

which indicates that the operator set **B** is indeed a tensor operator. Hereafter, the set **B** will be called simply the *adjoint tensor operator of* **A** and will be indicated by a Greek letter upsilon as $\mathbf{A}^\Upsilon$.

The unitary matrix $U^{(\Lambda)}$ is uniquely determined within an overall phase factor once the standard representation is determined. Properties of the U matrix are discussed in detail in Appendix 4A.

2. Recoupling Transformation

Given two tensor operators **A** and **B**, a set of the operators $\{\mathbf{A}_\kappa\mathbf{B}_\lambda\}$ spans a reducible representation space of the transformation group O(3)×SU(2). There exists a unitary transformation of the basis which decomposes the reducible representation into standard irreducible representations. Let **C** be one of the tensor operators thus obtained. It will be denoted as $(\mathbf{AB})^{(C)}$ and its components are expressed as

$$\mathbf{C}_\mu = (\mathbf{AB})^{(C)}_\mu = \sum_{\kappa\lambda} (C\mu\,|\,A\kappa B\lambda)\mathbf{A}_\kappa\mathbf{B}_\lambda, \tag{2.1}$$

where the transformation matrix elements $(C\mu\,|\,A\kappa B\lambda)$ iarecalled generalized Clebsch-Gordan coefficients and they are uniquely determined except for an arbitrary common phase factor. In order to express formulas compactly, we have introduced a convention that a tensor symbol also stands for its irreducible representation. It should be noted that since the group O(3)×SU(2) is *multiplicity free*, i.e., the same irreducible representation never appears more than once in the product representation of any two irreducible representations, no additional index other than the symbol C is required in the generalized Clebsch-Gordan coefficient to identify the irreducible subspace. From the condition that the set **C** transforms as a tensor

$$\mathbf{C}'_\mu = \sum_{\kappa\lambda} (C\mu\,|\,A\kappa B\lambda)\mathbf{A}'_\kappa\mathbf{B}'_\lambda = \sum_\nu D^{(C)}(X)_{\mu\nu}\mathbf{C}_\nu,$$

one can easily see that the following equation holds for any X.

$$\sum_{\kappa\lambda}(C\mu\,|\,A\kappa B\lambda)D^{(A)}(X)_{\kappa\pi}D^{(B)}(X)_{\lambda\rho}=\sum_{\nu}D^{(C)}(X)_{\mu\nu}(C\nu\,|\,A\pi B\nu).$$

Since we are working on the direct product group O(3)×SU(2), the generalized Clebsch-Gordan coefficient can be expressed as a product of the Clebsch-Gordan coefficient of each group

$$(C\mu\,|\,A\kappa B\lambda)=(C_\ell\mu_\ell\,|\,A_\ell\kappa_\ell B_\ell\lambda_\ell)(C_s\mu_s\,|\,A_s\kappa_s B_s\lambda_s),$$

where suffixes ℓ and s indicate the space and spin part of the corresponding parameters. The Clebsch-Gordan coefficients in the right side of the equation are those for the rotation group.

Properties of the Clebsch-Gordan coefficients and other related symbols are discussed in detail in Appendix 4A. We shall use formulas derived there for the manipulation of equations in this and in the following sections. The term "generalized" will be omitted, since it will not cause any confusion for the interpretation of formulas.

Tensor operators may be coupled in different ways, and there exists a relationship among them. The simplest case is about the tensors $(\mathbf{AB})^{(C)}$ and $(\mathbf{BA})^{(C)}$. If tensor components $\mathbf{A}_\kappa$ and $\mathbf{B}_\lambda$ mutually commute, both tensors are defined in the same representation space and, therefore, they differ at most only by a multiplicative factor which will be denoted by $(BA\,|\,AB)^{(C)}$.

$$(\mathbf{AB})^{(C)}=(BA\,|\,AB)^{(C)}(\mathbf{BA})^{(C)}$$

From Eq. (A.24) we have

$$(BA\,|\,AB)^{(C)}=(-1)^{A+B-C}, \tag{2.2}$$

where the tensor index such as A which appears in the exponent should be interpreted as $A_\ell+A_s$.

More than two tensor operators can be coupled one by one. For example, three tensor operators **A**, **B** and **C** may be coupled into $((\mathbf{AB})^{(D)}\mathbf{C})^{(F)}$ whose components are given by

$$((\mathbf{AB})^{(D)}\mathbf{C})^{(F)}_\mu=\sum_{\kappa\lambda\nu\rho}(D\nu\,|\,A\kappa B\lambda)(F\mu\,|\,D\nu C\rho)\mathbf{A}_\kappa\mathbf{B}_\lambda\mathbf{C}_\rho.$$

A different order of coupling gives rise to a different tensor

$$(\mathbf{A}(\mathbf{BC})^{(G)})^{(F)}.$$

Since both kinds of tensors are defined in the same space spanned by the operators $\{A_\kappa B_\lambda C_\rho\}$ and complete if all possible intermediate representations, D or G, are taken, a tensor constructed in one of the coupling schemes can be expressed as a linear combination of the tensors in the other coupling scheme.

$$((\mathbf{AB})^{(D)}\mathbf{C})^{(F)} = \sum_G ((AB)^{(D)}C \,|\, A(BC)^{(G)})^{(F)} (\mathbf{A}(\mathbf{BC})^{(G)})^{(F)}$$

It should be noted that the equality holds for all components since both sides of the tensors obey the same standard irreducible representation.

The coefficient $(A(BC)^{(D)} \,|\, (AB)^{(G)}C)^{(F)}$ is called Racah's coefficient which is typical one of the so-called recoupling coefficients. It can be expressed in terms of the 6j-symbol as (see Appendix 4A)

$$((AB)^{(D)}C \,|\, A(BC)^{(G)})^{(F)} = (-1)^{A+B+C+F}\sqrt{[D][G]} \begin{Bmatrix} ABD \\ CFG \end{Bmatrix}. \tag{2.3}$$

From four tensors, one can construct invariant operators in a different order of coupling. A typical recoupling transformation is given by

$$\begin{aligned} &((\mathbf{AB})^{(E)}(\mathbf{CD})^{(F)})^{(O)} \\ &\quad = \sum_G ((AB)^{(E)}(CD)^{(F)} \,|\, A(BC)^{(G)}D)^{(O)} (\mathbf{A}(\mathbf{BC})^{(G)}\mathbf{D})^{(O)}, \end{aligned}$$

where the recoupling coefficient is expressed as

$$\begin{aligned} ((AB)^{(E)}(CD)^{(F)} \,|\, A(BC)^{(G)}D)^{(O)} &= \delta_{EF}((AB)^{(E)}C \,|\, A(BC)^{(G)})^{(D)} \\ &= (-1)^{A+B+C+D}\delta_{EF}\sqrt{[E][G]} \begin{Bmatrix} ABE \\ CDG \end{Bmatrix}. \end{aligned} \tag{2.4}$$

A more complex recoupling transformation of four tensors is expressed as

$$\begin{aligned} &((\mathbf{AB})^{(E)}(\mathbf{CD})^{(F)})^{(G)} \\ &\quad = \sum_{HJ} ((AB)^{(E)}(CD)^{(F)} \,|\, (AC)^{(H)}(BD)^{(J)})^{(G)} ((\mathbf{AC})^{(H)}(\mathbf{BD})^{(J)})^{(G)}, \end{aligned}$$

where the tensor components $\mathbf{B}_\kappa$ and $\mathbf{C}_\lambda$ are assumed to be commutative. This recoupling coefficient is expressed by using a 9j-symbol as follows.

$$((AB)^{(E)}(CD)^{(F)} \,|\, (AC)^{(H)}(BD)^{(J)})^{(G)} = \sqrt{[E][F][H][J]} \begin{Bmatrix} A & B & E \\ C & D & F \\ H & J & G \end{Bmatrix}. \tag{2.5}$$

The 6j- and 9j-symbols exhibit high symmetry for a permutation of the parameters. Relevant properties of these symbols are summarized in Appendix 4A.

Before concluding this section, expressions which involve adjoint operators will be studied. From the definition of the adjoint operator (Eq. 1.15) and the U matrix (Eq. A.9), we have

$$(\mathbf{A}^{\Upsilon})^{\Upsilon}_{\kappa} = \sum_{\lambda} U^{(A)*}_{\kappa\lambda} \mathbf{A}^{\Upsilon}_{\lambda}{}^{\dagger} = \sum_{\lambda\mu} U^{(A)*}_{\kappa\lambda} (U^{(A)*}_{\lambda\mu} \mathbf{A}_{\mu}{}^{\dagger})^{\dagger}$$
$$= \sum_{\lambda\mu} U^{(A)*}_{\kappa\lambda} U^{(A)}_{\lambda\mu} \mathbf{A}_{\mu} = (-1)^{2A} \mathbf{A}_{\kappa},$$

or in short,

$$(\mathbf{A}^{\Upsilon})^{\Upsilon} = (-1)^{2A} \mathbf{A} \tag{2.6}$$

holds.

The operator $(\mathbf{A}^{\Upsilon}\mathbf{A})^{(O)}_0$ is given by

$$(\mathbf{A}^{\Upsilon}\mathbf{A})^{(O)}_0 = \sum_{\kappa\lambda} (OO \mid A\kappa A\lambda) \mathbf{A}^{\Upsilon}_{\kappa} \mathbf{A}_{\lambda} = [A]^{-1/2} \sum_{\kappa\lambda\mu} U^{(A)}_{\kappa\lambda} U^{(A)*}_{\kappa\mu} \mathbf{A}_{\mu}{}^{\dagger} \mathbf{A}_{\lambda}$$
$$= [A]^{-1/2} \sum_{\lambda} \mathbf{A}_{\lambda}{}^{\dagger} \mathbf{A}_{\lambda}, \tag{2.7}$$

which shows that it is a positive definite operator.

By using Eqs. (1.15), (2.1), (A.23), (A.21) and (A.39), the adjoint operator $\mathbf{C}^{\Upsilon} = (\mathbf{AB})^{(C)\Upsilon}$ can be expressed as follows.

$$
\begin{aligned}
\mathbf{C}_\nu^\Upsilon &= \sum_{\mu} U_{\nu\mu}^{(C)*}\mathbf{C}_\mu\dagger = \sum_{\mu\kappa\lambda} U_{\nu\mu}^{(C)*}(C\mu\,|\,A\kappa B\lambda)^*(\mathbf{A}_\kappa\mathbf{B}_\lambda)\dagger \\
&= (-1)^{2B}\sqrt{[C]}\sum_{\kappa\lambda\mu\tau} U_{\nu\mu}^{(C)*}U_{\mu\tau}^{(C)}\begin{bmatrix} ABC \\ \kappa\lambda\tau \end{bmatrix}^* \mathbf{B}_\lambda\dagger\mathbf{A}_\kappa\dagger \\
&= (-1)^{2B}\sqrt{[C]}\sum_{\kappa\lambda\mu\tau\rho\pi} U_{\nu\mu}^{(C)*}U_{\mu\tau}^{(C)}\begin{bmatrix} ABC \\ \kappa\lambda\tau \end{bmatrix}^* \\
&\quad \times \mathbf{B}_\rho^\Upsilon U_{\rho\lambda}^{(B)}\mathbf{A}_\kappa^\Upsilon U_{\pi\kappa}^{(A)} \\
&= (-1)^{2B}\sqrt{[C]}\sum_{\kappa\lambda\rho\pi\sigma\tau} U_{\nu\mu}^{(C)*}\begin{bmatrix} ABC \\ \kappa\lambda\tau \end{bmatrix}^* \\
&\quad \times U_{\kappa\pi}^{(A)}U_{\lambda\rho}^{(B)}U_{\tau\mu}^{(C)}\mathbf{B}_\rho^\Upsilon\mathbf{A}_\pi^\Upsilon \\
&= (-1)^{2B}\sqrt{[C]}\sum_{\rho\pi\mu} U_{\nu\mu}^{(C)*}\begin{bmatrix} ABC \\ \pi\rho\mu \end{bmatrix}\mathbf{B}_\rho^\Upsilon\mathbf{A}_\pi^\Upsilon \\
&= (-1)^{2B+A+B+C}\sqrt{[C]}\sum_{\rho\pi\mu} U_{\nu\mu}^{(C)*}\begin{bmatrix} BAC \\ \rho\pi\mu \end{bmatrix}\mathbf{B}_\rho^\Upsilon\mathbf{A}_\pi^\Upsilon \\
&= (-1)^{A+B-C}\sum_{\mu\rho\pi}(C\mu\,|\,B\rho A\pi)^*\mathbf{B}_\rho^\Upsilon\mathbf{A}_\pi^\Upsilon \\
&= (-1)^{A+B-C}(\mathbf{B}^\Upsilon\mathbf{A}^\Upsilon)_\nu^{(C)}
\end{aligned}
$$

i.e.,

$$\mathbf{C}^\Upsilon = (-1)^{A+B-C}(\mathbf{B}^\Upsilon\mathbf{A}^\Upsilon)^{(C)} \tag{2.8}$$

Equation (A.30) shows that $((\mathbf{AB})^{(C)}\mathbf{C})^{(O)}$ is equal to $(\mathbf{A}(\mathbf{BC})^{(A)})^{(O)}$. So it may be expressed as $(\mathbf{ABC})^{(O)}$. By using Eq. (2.8), the adjoint operator of $(\mathbf{ABC})^{(O)}$ is expressed as

$$(\mathbf{ABC})^{(O)\Upsilon} = (-1)^{2C}(\mathbf{C}^\Upsilon(\mathbf{AB})^{(C)\Upsilon})^{(O)} = (-1)^{A+B+C}(\mathbf{C}^\Upsilon\mathbf{B}^\Upsilon\mathbf{A}^\Upsilon)^{(O)}. \tag{2.9}$$

3. Complete Set of Shell States

We describe in this section a procedure which generates a complete and orthonormal set of n-electron states arising from a shell of angular momentum ℓ. First, a somewhat simple algorithm will be described for easy understanding of the procedure. The algorithm actually used in ATOMCI is based on the same approach but slightly modified for a more efficient generation and for the reduction of a table size in a computer. This algorithm together with explanation of notations used in an output listing of ATOMCI will be explained in the latter half of this section.

3.1 Shell Creation Tensor Operators in LS-scheme

The vacuum state which will be denoted by the ket symbol > transform as ${}^1S$ or equivalently as the identity representation O. If $\mathbf{a}_{n\ell}\dagger$ is operated from left, all $4\ell + 2$ one-electron wavefunctions as a set are obtained. The normalization of these functions may be checked by calculating the vacuum expectation value of $< ((\mathbf{a}_{n\ell}\dagger)^{\Upsilon}\mathbf{a}_{n\ell}\dagger)^{O} >$ as

$$
\begin{aligned}
< (\mathbf{a}_{n\ell}\dagger)^{\Upsilon}\mathbf{a}_{n\ell}\dagger > &= \sum_{\lambda\mu}[{}^2\ell]^{-1/2}\delta_{\lambda\mu} < \mathbf{a}_{n\ell,\lambda}\mathbf{a}_{n\ell,\mu}\dagger >, \\
&= \sqrt{[{}^2\ell]} = \sqrt{4\ell+2},
\end{aligned}
\tag{3.1}
$$

which immediately follows from Eq. (2.6) and the commutation relations Eq. (1.7). Note that we have omitted the identity coupling specification of the tensor operator in the vacuum expectation value since only an invariant tensor may have non-zero expectation value.

Similarly, for any tensor operator $\mathbf{A}$

$$
< \mathbf{A}^{\Upsilon}\mathbf{A} > = \sum_{\lambda}[A]^{-1/2} < \mathbf{A}_\lambda\dagger\mathbf{A}_\lambda > = \sqrt{[A]}\ < \mathbf{A}_\lambda\dagger\mathbf{A}_\lambda > \tag{3.2}
$$

holds. The Wigner-Eckart theorem, Eq. (A.36), proves the last equality. If we require $\mathbf{A}_\lambda >$ being a normalized wavefunction, we have

$$
< \mathbf{A}^{\Upsilon}\mathbf{A} > = \sqrt{[A]}\,. \tag{3.3}
$$

The equation (3.1) shows that $\mathbf{a}_{n\ell,\mu}\dagger >$ are properly normalized.

One of the anticommutation relations, Eq. (1.7), can be re-expressed as

$$
\begin{aligned}
\sum_{\lambda} U^{({}^2\ell)*}_{\kappa\lambda}[\mathbf{a}_{n\ell,\lambda}\mathbf{a}_{n\ell,\mu}\dagger]_+ &= (\mathbf{a}_{n\ell}\dagger)^{\Upsilon}_{\kappa}\mathbf{a}_{n\ell}\dagger_\mu + \mathbf{a}_{n\ell}\dagger_\mu(\mathbf{a}_{n\ell}\dagger)^{\Upsilon}_{\kappa} \\
&= \sum_{\lambda} U^{({}^2\ell)*}_{\kappa\lambda}\delta_{\lambda\mu} = U^{({}^2\ell)*}_{\kappa\mu} \\
&= \sqrt{[{}^2\ell]}\,(OO\,|\,{}^2\ell\kappa\,{}^2\ell\mu)^*,
\end{aligned}
$$

by using Eqs. (1.15) and (A.9). Multiplying by $(f\nu\,|\,{}^2\ell\kappa\,{}^2\ell\nu)$ and using the unitarity condition, Eq. (A.6), we have

$$
(\mathbf{b}^{\Upsilon}\mathbf{b})^{(f)} + (-1)^{2b-f}(\mathbf{b}\mathbf{b}^{\Upsilon})^{(f)} = \delta_{fO}\sqrt{[{}^2\ell]}\,. \tag{3.4}
$$

The tensor $\mathbf{a}_{n\ell}\dagger$ will be denoted by **b** throughout this section in order to shorten formulas. Its dimension and symmetry index are given by

$$[b] = [^2\ell] = 4\ell + 2$$

and

$$b = \ell + 1/2.$$

The other anticommutation relations of Eq. (1.7) can be obtained in a similar way. They are expressed as

$$(\mathbf{bb})^{(f)} + (-1)^{2b-f}(\mathbf{bb})^{(f)} = O$$

and

$$(\mathbf{b}^{\Upsilon}\mathbf{b}^{\Upsilon})^{(f)} + (-1)^{2b-f}(\mathbf{b}^{\Upsilon}\mathbf{b}^{\Upsilon})^{(f)} = O,$$

or equivalently

$$(\mathbf{bb})^{(f)} = (\mathbf{b}^{\Upsilon}\mathbf{b}^{\Upsilon})^{(f)} = O \quad \text{for } odd\, f \tag{3.5}$$

since $2b$ is always odd.

The tensor operator

$$\mathbf{f} = (\mathbf{bb})^{(f)}$$

creates a two-electron state coupled into the symmetry f. Equation (3.5) shows the well known fact that only even representations such as ^3P, ^1D, ^1S, ... appear as two-electron states. The norm can be calculated as follows.

$$
\begin{aligned}
<\mathbf{f}^{\Upsilon}\mathbf{f}> &= <((\mathbf{bb})^{(f)})^{\Upsilon}(\mathbf{bb})^{(f)}> \\
&=(-1)^{2b-f}<(\mathbf{b}^{\Upsilon}\mathbf{b}^{\Upsilon})^{(f)}(\mathbf{bb})^{(f)}> \\
&=(-1)^{2b-f}\sum_{k}\sqrt{[f][k]}\begin{Bmatrix} b & b & f \\ b & b & k \end{Bmatrix}<\mathbf{b}^{\Upsilon}(\mathbf{b}^{\Upsilon}\mathbf{b})^{(k)}\mathbf{b}> \\
&=(-1)^{2b-f}\sqrt{[f][b]}\begin{Bmatrix} b & b & f \\ b & b & O \end{Bmatrix}<\mathbf{b}^{\Upsilon}\mathbf{b}> \\
&\quad -\sum_{k}(-1)^{k-f}\sqrt{[f][k]}\begin{Bmatrix} b & b & f \\ b & b & k \end{Bmatrix}<\mathbf{b}^{\Upsilon}(\mathbf{bb}^{\Upsilon})^{(k)}\mathbf{b}> \\
&=(-1)^{2b-f+2b+f}\sqrt{[f]}\,\delta(b,b,f)-\sum_{kh}(-1)^{k-f} \\
&\quad \times\sqrt{[f][k]}\begin{Bmatrix} b & b & f \\ b & b & k \end{Bmatrix}\sqrt{[k][h]}\begin{Bmatrix} b & b & h \\ b & b & k \end{Bmatrix}<(\mathbf{b}^{\Upsilon}\mathbf{b})^{(h)}(\mathbf{b}^{\Upsilon}\mathbf{b})^{(h)}> \\
&=(-1)^{2b-f+2b+f}\sqrt{[f]}\,\delta(b,b,f) \\
&\quad -\sum_{h}(-1)^{-f-f-h}\sqrt{[f][h]}\begin{Bmatrix} b & b & f \\ b & b & h \end{Bmatrix}<(\mathbf{b}^{\Upsilon}\mathbf{b})^{(h)}(\mathbf{b}^{\Upsilon}\mathbf{b})^{(h)}> \\
&=\sqrt{[f]}\,\delta(b,b,f) \\
&\quad -\Sigma(-1)^{-2f+2b+f}\sqrt{[f]}\,\delta(b,b,f)<(\mathbf{b}^{\Upsilon}\mathbf{b})^{(O)}><(\mathbf{b}^{\Upsilon}\mathbf{b})^{(O)}>/[b] \\
&=(1+(-1)^{f})\sqrt{[f]}\,\delta(b,b,f).
\end{aligned}
$$

In the above, the summation over k is done by using Eq. (A.47). Note also that the expectation value $<(\mathbf{b}^{\Upsilon}\mathbf{b})^{(h)}(\mathbf{b}^{\Upsilon}\mathbf{b})^{(h)}>$ is non-zero only when the representation h is O, since the state $(\mathbf{b}^{\Upsilon}\mathbf{b})^{(h)}>$ is a zero-electron state and the vacuum is the only state having no electron. This equation again proves that the norm vanishes for odd f, and is 2 for even f. Thus, we have obtained a set of tensor operators

$$
\mathbf{f}=\frac{1}{\sqrt{2}}(\mathbf{bb})^{(f)} \quad \text{for even } f \text{ and } \delta(b,b,f)=1. \tag{3.6}
$$

These operators are complete and normalized in the sense that the $\mathbf{f}>$'s span 2-electron states completely and $<\mathbf{f}^{\Upsilon}\mathbf{f}>=\sqrt{[f]}$.

We now describe a procedure which will inductively generate a complete and orthonormal set of tensor operators of the shell $n\ell$. First, let us assume that we have a set of tensor operators $\{\mathbf{f}_{N,i}\}$ which satisfies the following conditions:

1. A $\mathbf{f}_{N,i}$ creates N electrons in the shell $n\ell$.

2. The set $\{\mathbf{f}_{N,i}\}$ is complete and orthonormalized, i.e., any N-electron states can be expanded on the $\mathbf{f}_{N,i}>$'s and

$$
<\mathbf{f}_{N,i}^{\Upsilon}\mathbf{f}_{N,j}>=\delta_{ij}\sqrt{[f_{N,i}]}\,. \tag{3.7}
$$

3. The expansion of $\mathbf{bf}_{N-1,k}$ in terms of the set $\{\mathbf{f}_{N,i}\}$ is known. Namely,

$$\sqrt{[h]}\,(\mathbf{bf}_{N-1,k})^{(h)} = \sum_i \delta_{h,f_{N,i}} C_{N,ik}\mathbf{f}_{N,i}. \tag{3.8}$$

The identity operator 1, **b** and the **f**'s given by Eq. (3.6) are such sets of operators for $N = 0$, 1 and 2.

The expansion coefficient $C_{N,ik}$ is easily seen to be given by the following vacuum expectation values from Eqs. (3.8) and (2.9).

$$C_{N,ik} = < \mathbf{f}^{\Upsilon}_{N,i}\mathbf{bf}_{N-1,k} > \tag{3.9}$$

$$\begin{aligned} C_{N,ik}{}^{*} &= < (\mathbf{f}^{\Upsilon}_{N,i}\mathbf{bf}_{N-1,k})^{(O)\Upsilon} > \\ &= (-1)^{b+f_{N-1,k}-f_{N,i}} < \mathbf{f}^{\Upsilon}_{N-1,k}\mathbf{b}^{\Upsilon}\mathbf{f}_{N,i} > . \end{aligned} \tag{3.10}$$

An invariant operator

$$\mathbf{P}_N = \sum_i (-1)^{2f_{N,i}} \sqrt{[f_{N,i}]}\,(\mathbf{f}_{N,i}\mathbf{f}^{\Upsilon}_{N,i})^{(O)} \tag{3.11}$$

works as the identity operator in the space of N-electron wavefunctions of the $n\ell$ shell, since

$$\begin{aligned} \mathbf{P}_N\mathbf{f}_{N,j} > &= \sum_i (-1)^{2f_{N,i}} \sqrt{[f_{N,i}]}\,(\mathbf{f}_{N,i}\mathbf{f}^{\Upsilon}_{N,i})^{(O)}\mathbf{f}_{N,j} > \\ &= \sum_i (-1)^{2f_{N,i}} \sqrt{[f_{N,i}]} \begin{Bmatrix} f_{N,i} & f_{N,i} & O \\ f_{N,j} & f_{N,j} & O \end{Bmatrix} \mathbf{f}_{N,i}(\mathbf{f}^{\Upsilon}_{N,i}\mathbf{f}_{N,j})^{(O)} > \\ &= \sum_i \frac{1}{\sqrt{[f_{N,i}]}} \delta_{f_{N,i}f_{N,j}} \mathbf{f}_{N,i} > < \mathbf{f}^{\Upsilon}_{N,i}\mathbf{f}_{N,j} > = \mathbf{f}_{N,j} > \end{aligned} \tag{3.12}$$

Consider a set of tensor operators $\mathbf{g}_i = (\mathbf{bf}_{N,i})^{(g)}$. They create $N+1$ electrons and all belong to the same irreducible representation g. The metric matrix $M_{N+1,ij} = < \mathbf{g}_i^{\Upsilon}\mathbf{g}_j >$ can be calculated as follows.

$$
\begin{aligned}
M_{N+1,\,ij} &= < \mathbf{g}_i^{\Upsilon} \mathbf{g}_j > = < (\mathbf{b}\mathbf{f}_{N,\,i})^{(g)\Upsilon} (\mathbf{b}\mathbf{f}_{N,\,j})^{(g)} > \\
&= (-1)^{f_{N,\,i}+b-g} < (\mathbf{f}_{N,\,i}^{\Upsilon} \mathbf{b}^{\Upsilon})^{(g)} (\mathbf{b}\mathbf{f}_{N,\,j})^{(g)} > \\
&= \sum_k (-1)^{2f_{N,\,i}+b-g+f_{N,j}+2b} \sqrt{[g][k]} \begin{Bmatrix} f_{N,i} & b & g \\ b & f_{N,j} & k \end{Bmatrix} < \mathbf{f}_{N,\,i}^{\Upsilon} (\mathbf{b}^{\Upsilon}\mathbf{b})^{(k)} \mathbf{f}_{N,j} > \\
&= (-1)^{2f_{N,\,i}+3b-g+f_{N,j}} \sqrt{[g][b]} \begin{Bmatrix} f_{N,\,i} & b & g \\ b & f_{N,j} & O \end{Bmatrix} < \mathbf{f}_{N,\,i}^{\Upsilon} \mathbf{f}_{N,j} > \\
&\quad - \sum_k (-1)^{2f_{N,\,i}+b-g+f_{N,j}+k} \sqrt{[g][k]} \begin{Bmatrix} f_{N,i} & b & g \\ b & f_{N,j} & k \end{Bmatrix} < \mathbf{f}_{N,\,i}^{\Upsilon} (\mathbf{b}\mathbf{b}^{\Upsilon})^{(k)} \mathbf{f}_{N,j} > \\
&= \sqrt{[g]}\, \delta_{ij} \delta(f_{N,\,i}, b, g) \\
&\quad - \sum_k (-1)^{2f_{N,\,i}+b-g+f_{N,j}+k} \sqrt{[g][k]} \begin{Bmatrix} f_{N,i} & b & g \\ b & f_{N,j} & k \end{Bmatrix} \\
&\quad \times \sum_h (-1)^{2b+f_{N,\,i}+f_{N,j}} \sqrt{[h][k]} \begin{Bmatrix} f_{N,i} & b & h \\ b & f_{N,j} & k \end{Bmatrix} < (\mathbf{f}_{N,\,i}^{\Upsilon} \mathbf{b})^{(h)} (\mathbf{b}^{\Upsilon} \mathbf{f}_{N,\,j})^{(h)} > \\
&= \sqrt{[g]}\, \delta_{ij} \delta(f_{N,\,i}, b, g) \\
&\quad - \sum_h (-1)^{f_{N,\,i}-b-2g-h} \sqrt{[g][h]} \begin{Bmatrix} f_{N,i} & b & g \\ f_{N,j} & b & h \end{Bmatrix} < (\mathbf{f}_{N,\,i}^{\Upsilon} \mathbf{b})^{(h)} (\mathbf{b}^{\Upsilon} \mathbf{f}_{N,\,j})^{(h)} >
\end{aligned}
$$

Using $\mathbf{P}_{N-1}$, one can express the vacuum expectation value in the last term as

$$
\begin{aligned}
& < (\mathbf{f}_{N,\,i}^{\Upsilon} \mathbf{b})^{(h)} \mathbf{P}_{N-1} (\mathbf{b}^{\Upsilon} \mathbf{f}_{N,\,j})^{(h)} > \\
&\quad = \sum_k [f_{N-1,\,k}]^{-1/2} \delta_{h,f_{N-1,\,k}} < \mathbf{f}_{N,\,i}^{\Upsilon} \mathbf{b} \mathbf{f}_{N-1,\,k} > < \mathbf{f}_{N-1,\,k}^{\Upsilon} \mathbf{b}^{\Upsilon} \mathbf{f}_{N,j} > \\
&\quad = \sum_k [f_{N-1,\,k}]^{-1/2} (-1)^{b+f_{N-1,\,k}-f_{N,j}} \delta_{h,f_{N-1,\,k}} C_{N,\,ik} C_{N,jk}^{*}.
\end{aligned}
$$

Substituting this expression, we obtain the metric as follows.

$$
\begin{aligned}
M_{N+1,\,ij} &= \sqrt{[g]}\, \delta_{ij} \delta(f_{N,\,i}, b, g) \\
&= \sum_k (-1)^{f_{N,\,i}-f_{N,j}-2g} \sqrt{[g]} \begin{Bmatrix} f_{N,i} & b & g \\ f_{N,j} & b & f_{N-1,\,k} \end{Bmatrix} C_{N,\,ik} C_{N,jk}^{*}
\end{aligned}
\tag{3.13}
$$

Since the matrix M is Hermitian and positive semi-definite, there exists a matrix T for which

$$
T^{\dagger} M T = \sqrt{[g]}\, 1_L
$$

holds, where L denotes the rank of the matrix M and 1_L is the L by L unit matrix. The operators

$$\mathbf{h}_p = \sum_i \mathbf{g}_i T_{ip}, \quad \text{for } p = 1, 2, \dots, L \tag{3.14}$$

are orthogonal and properly normalized, since

$$\begin{aligned} < \mathbf{h}_p^{\Upsilon} \mathbf{h}_q > &= \sum_{ij} < (\mathbf{g}_i T_{ip})^{\Upsilon} \mathbf{g}_j T_{jq} > = \sum_{ij} T_{ip}{}^* M_{N+1,\, ij} T_{jq} \\ &= \sqrt{[g]}\, \delta pq \end{aligned}$$

holds. Repeating this procedure for all possible representations **g**, one obtains a set of tensor operators $\{\mathbf{f}_{N+1,p}\}$ which satisfies the requirements 1) and 2) of shell creation tensor operators. The expansion coefficient C_{N+1} can be evaluated as follows.

$$C_{N+1,\, pi} = < \mathbf{h}_p^{\Upsilon} \mathbf{b} \mathbf{f}_{N,\, i} > = \Sigma T_{jp}{}^* < \mathbf{g}_j^{\Upsilon} \mathbf{g}_i > = \Sigma T_{jp}{}^* M_{N+1,\, ji} \tag{3.15}$$

3.2 Complete Shell States in LSQ-scheme

The operators **b** and $\mathbf{b}^{\Upsilon}$ obey the same representation ${}^2\ell$ when the space and spin axes rotate.

$$\begin{aligned} \mathbf{b}'_\mu &= \sum_\nu U^{(b)}(X)_{\mu\nu} \mathbf{b}_\nu \\ \mathbf{b}_\mu^{\Upsilon'} &= \sum_\mu U^{(b)}(X)_{\mu\nu} \mathbf{b}_\nu^{\Upsilon} \end{aligned} \tag{3.16}$$

A new transformation

$$\begin{aligned} \mathbf{b}'_\mu &= v_{11} \mathbf{b}_\mu + v_{12} \mathbf{b}_\mu^{\Upsilon} \\ \mathbf{b}_\mu^{\Upsilon'} &= v_{21} \mathbf{b}_\mu + v_{22} \mathbf{b}_\mu^{\Upsilon} \end{aligned} \tag{3.17}$$

where the matrix v is a unitary matrix with its determinant equal to 1, commutes with the transformation Eq. (3.16). Therefore, the set of operators $\mathbf{b}_\mu$ and $\mathbf{b}_\mu^{\Upsilon}$ serve as a representation space of O(3)×SU(2)×SU(2), where the last SU(2) refers to the transformation Eq. (3.17) by the matrix v. Since these transformations are analogous to rotations in the spin space, we call them rotations in the *quasi-spin space*. The tensor operators $\mathbf{b}_\mu$ and $\mathbf{b}_\mu^{\Upsilon}$ are regarded as the α and β components of the quasi-spin doublet of a new single tensor operator **B** of the enlarged group O(3)×SU(2)×SU(2).

$$\mathbf{B}_\sigma = \mathbf{B}_{\mu\tau} = \begin{cases} \mathbf{b}_{\mu\Upsilon} \\ -\mathbf{b}_\mu^{\Upsilon} \end{cases} \text{for } \tau = \begin{cases} 1/2 \\ -1/2 \end{cases} \tag{3.18}$$

The factor -1 for $\tau = -1/2$ adjusts the phase of some of the equations. This group is simply reducible; all properties derived in Appendix 4A are also applicable to this group. Its irreducible representation is specified by three quantum numbers, the spatial, spin and quasi-spin angular momenta, and denoted ${}^{2s+1}L^{2q+1}$ (${}^2\ell^2$ for **B**), where $2q+1$ gives the quasi-spin multiplicity.

The anticommutator relations Eqs. (3.4) and (3.5), i.e.,

$$\mathbf{b}_\kappa^\Upsilon \mathbf{b}_\lambda + \mathbf{b}_\lambda \mathbf{b}_\kappa^\Upsilon = \sqrt{[b]}\,(OO\,|\,b\kappa b\lambda)^* \tag{3.19}$$

$$\mathbf{b}_\kappa \mathbf{b}_\lambda + \mathbf{b}_\lambda \mathbf{b}_\kappa = O \tag{3.20}$$

and

$$\mathbf{b}_\kappa^\Upsilon \mathbf{b}_\lambda^\Upsilon + \mathbf{b}_\lambda^\Upsilon \mathbf{b}_\kappa^\Upsilon = O \tag{3.21}$$

can be combined into a single equation as

$$\mathbf{B}_\rho \mathbf{B}_\sigma + \mathbf{B}_\sigma \mathbf{B}_\rho = \sqrt{[B]}\,(OO\,|\,B\rho B\sigma)^*. \tag{3.22}$$

This may be proved by separating the tensor component indices ρ and σ of Eq. (3.22) into space-spin and quasi-spin parts.

$$\mathbf{B}_{\kappa\tau}\mathbf{B}_{\lambda\upsilon} + \mathbf{B}_{\lambda\upsilon}\mathbf{B}_{\kappa\tau} = \sqrt{2[b]}\,(OO\,|\,b\kappa b\lambda)^*(OO\,|\,1/2,\tau,1/2,\upsilon)^*.$$

Substitution of $(-1/2, 1/2)$, $(1/2, 1/2)$ and $(-1/2, -1/2)$ for (τ, υ) into the above equation gives Eqs. (3.19), (3.20) and (3.21), respectively, since the Clebsch-Gordan coefficients for these parameter values are given by

$$\sqrt{2}\,(OO\,|\,1/2,\tau,1/2,\upsilon)^* = \begin{bmatrix} 1/2 & 1/2 & O \\ \tau & \upsilon & O \end{bmatrix}^* = (-1)^{1/2-\tau}\delta_{\tau,-\upsilon}.$$

By using the unitarity Eq. (A.6), the anticommutation relation Eq. (3.22) is expressed as

$$(\mathbf{BB})^{(F)} = \delta_{FO}\sqrt{[B]}/2 \qquad \text{for } even\ F.$$

The adjoint operator $\mathbf{B}^\Upsilon$ is expressed as

$$\begin{aligned}
\mathbf{B}_{\kappa\tau}^\Upsilon &= \sum_{\lambda\upsilon} U^{(B)}_{\kappa\tau,\lambda\upsilon}{}^* \mathbf{B}_{\lambda\upsilon}\dagger \\
&= \sum_{\lambda} U^{(b)}_{\kappa\lambda}{}^* (U^{(1/2)}_{\tau,1/2}{}^* \mathbf{b}_\lambda\dagger - U^{(1/2)}_{\tau,-1/2}{}^* \mathbf{b}_\lambda^\Upsilon\dagger) \\
&= U^{(1/2)}_{\tau,1/2}{}^* \mathbf{b}_\kappa^\Upsilon + U^{(1/2)}_{\tau,-1/2}{}^* \mathbf{b}_\kappa = \mathbf{B}_{\kappa\tau}
\end{aligned}$$

i.e.,

$$\mathbf{B}^{\Upsilon} = \mathbf{B} \tag{3.24}$$

Although the equations (3.23) and (3.24) constitute the starting point of the present analysis, the vacuum state needs to be reconsidered. The rotation, Eq. (3.17), in the quasi-spin space mixes creation operators with annihilation operators. As a result, the vacuum state in LS-scheme no longer remains invariant; it belongs to the $2\ell + 2$ fold degenerate ${}^1S^{2\ell+2}$state which consists of ${}^1S$ states of $0, 2, \ldots, 4\ell + 2$ electrons. We denote this multiplet by the symbol >, and the corresponding bra multiplet by <. Each component of a tensor is specified by the z-components of space, spin and quasi-spin, i.e., ℓ_z, s_z and q_z. For the ket or bra multiplet > or <, space and spin z-components are 0, q_z takes the values $-\ell - 1/2,\ -\ell + 1/2, \ldots, \ell + 1/2$, and they are normalized as usual.

$$(<>)^{(O)} = \sqrt{[>]} = \sqrt{[{}^1S^{2\ell+2}]} = \sqrt{2\ell + 2}\,. \tag{3.25}$$

General ket and bra tensors may be introduced as follows. A set of kets $F>_\mu$ which forms a basis of a standard representation is called a ket tensor and denoted by $F>$. From a ket tensor $F> = \{F>_\mu\}$, we obtain a set of bra's

$$\sum_{\mu} U^{(F)*}_{\kappa\mu}(<F)_\mu$$

which obey the same standard representation as $F>$. It will be denoted by $<F^{\Upsilon}$. Note the use of unstressed ket and bra symbols here. Since the stressed ket > and bra < introduced before are degenerate, the use of a symbol like $F>$ for a ket tensor would make its representation ambiguous. The number of electrons and q_z of a ket or bra tensor are related as

$$N = 2q_z + 2\ell + 1, \tag{3.26}$$

since we adopt the convention that the α-component of the tensor $\mathbf{B}$ corresponds to the creation operators of electrons.

The expectation value

$$(<\mathbf{F}^{\Upsilon}(\mathbf{HG}>)^{(F)})^{(O)} = ((<\mathbf{F}^{\Upsilon}\mathbf{H})^{(G)}\mathbf{G}>)^{(O)}$$

will be denoted by $<\mathbf{F}^{\Upsilon}\mathbf{HG}>$, since only the identity representation for the tensor product may have non-zero expectation values. $<\mathbf{G}^{\Upsilon}\mathbf{H}^{\Upsilon}\mathbf{F}>$ is given by

$$
\begin{aligned}
< \mathbf{G}^{\Upsilon}\mathbf{H}^{\Upsilon}\mathbf{F} > &= (-1)^{2H} \sum_{\kappa\lambda\rho\mu\nu} \begin{bmatrix} GHF \\ \kappa\lambda\rho \end{bmatrix} U_{\kappa\mu}^{(G)*} U_{\lambda\nu}^{(H)*} < \mathbf{G}_{\mu}\mathbf{H}_{\nu}\dagger\mathbf{F}_{\rho} > \\
&= (-1)^{2H} \sum_{\rho\mu\nu\sigma} \begin{bmatrix} GHF \\ \mu\nu\sigma \end{bmatrix}^* U_{\rho\sigma}^{(F)} < \mathbf{F}_{\rho}\mathbf{H}_{\nu}\mathbf{G}_{\mu} > * \\
&= (-1)^{2H+2F} \sum_{\rho\mu\nu} \begin{bmatrix} GHF \\ \mu\nu\rho \end{bmatrix}^* < \mathbf{F}_{\rho}^{\Upsilon}\mathbf{H}_{\nu}\mathbf{G}_{\mu} > * \\
&= (-1)^{-F+G+H} < \mathbf{F}^{\Upsilon}\mathbf{H}\mathbf{G} > *
\end{aligned}
\tag{3.27}
$$

Generation of a complete and orthonormal set of shell states in LSQ-scheme proceeds as follows. We assume that a set of ket tensors $\{\mathbf{F}_{p,i} >\}$ which satisfies the following conditions are known for $p = \ell + 1/2,\ \ell,\ \ell - 1/2,\ \ldots,\ q$:

1. The quasi-spin angular momentum of $\mathbf{F}_{p,i} >$ is p.
2. The $F_{p,i} >$'s are complete and orthonormalized, i.e.,

$$< \mathbf{F}_{p,i}^{\Upsilon}\mathbf{F}_{p,j} > = \delta_{ij}\sqrt{[F_{p,i}]} \tag{3.28}$$

and any ket tensor having quasi-spin angular momentum p can be expanded on the $\mathbf{F}_{p,i} >$'s. The projection operator $\mathbf{P}_p$, which works as the identity operator in the space of the quasi-spin angular momentum p can be expressed as

$$\mathbf{P}_p = \sum_i (-1)^{2F_{p,i}}\sqrt{[F_{p,i}]}\,(\mathbf{F}_{p,i} > < \mathbf{F}_{p,i}^{\Upsilon})^{(O)}, \tag{3.29}$$

and the expansion coefficient is expressed as

$$C_{p,ik} = < \mathbf{F}_{p,i}^{\Upsilon}\mathbf{B}\mathbf{F}_{p+1/2,k} > \tag{3.30}$$

and

$$C_{p,ik}^{*} = (-1)^{F_{p+1/2,k} + B - F_{p,i}} < \mathbf{F}_{p+1/2,k}^{\Upsilon}\mathbf{B}\mathbf{F}_{p,i} > . \tag{3.31}$$

These equations correspond to the equations (3.9) - (3.11) in LS-scheme.

The ket > has the largest quasi-spin angular momentum, $\ell + 1/2$, and there are no other kets which possess that momentum, because the closed shell state is the only state with $4\ell + 2$ electrons, the number of electrons being given by equation (3.25). We shall now describe a procedure to generate $\{F_{q-1/2,i} >\}$ from the known set of ket tensors.

Let each of $\mathbf{F} >$ and $\mathbf{G} >$ are one of the $\mathbf{F}_{q,i} >$'s. Using the anticommutation relation, Eq. (3.23), we have

$$< \mathbf{F}^{\Upsilon}(\mathbf{BB})^{(H)}\mathbf{G} > = \delta_{HO}\delta_{FG}\sqrt{[B][F]}/2$$
$$= \sum_J (-1)^{2B+F+G}\sqrt{[H][J]}\begin{Bmatrix} BBH \\ FGJ \end{Bmatrix}(< \mathbf{F}^{\Upsilon}\mathbf{B})^{(J)}(\mathbf{BG} >)^{(J)} \quad \text{for } \textit{even } H.$$

From Eqs. (A.46) and (A.47) we can derive the following equality.

$$2\sum_{\textit{even } C} [C]\begin{Bmatrix} AAC \\ DEF \end{Bmatrix}\begin{Bmatrix} AAC \\ DEG \end{Bmatrix}$$
$$= \frac{\delta_{FG}\delta(A, E, F)\delta(D, A, F)}{[F]} + (-1)^{F+G}\begin{Bmatrix} AEF \\ ADG \end{Bmatrix}.$$

Using this equality, we obtain

$$\delta_{FG}\sqrt{[B][F]}\begin{Bmatrix} BBO \\ FGK \end{Bmatrix} = \delta_{FG}(-1)^{B+G+K}\delta(B, G, K)$$
$$= \frac{(-1)^{2B+F+G}}{\sqrt{[K]}}(< \mathbf{F}^{\Upsilon}\mathbf{B})^{(K)}(\mathbf{BG} >)^{(K)}$$
$$+ \sum_J (-1)^{2B+F+G+J+K}\sqrt{[J]}\begin{Bmatrix} BFJ \\ BGK \end{Bmatrix}(< \mathbf{F}^{\Upsilon}\mathbf{B})^{(J)}(\mathbf{BG} >)^{(J)}.$$

We split the irreducible representations in the above equation into space-spin and quasi-spin parts as $K = \{K_d, K_q\}$. Then multiplying both sides by

$$(-1)^{-B-G-K}[K_q]\begin{Bmatrix} B_q\ F_q\ q+1/2 \\ B_q\ G_q\ \ K_q \end{Bmatrix} = (-1)^{-B-G-K}[K_q]\begin{Bmatrix} 1/2\ q\ q+1/2 \\ 1/2\ q\ \ K_q \end{Bmatrix}$$

and summing over K_q, we obtain

$$\delta_{FG}\delta(B_d, G_d, K_d)\left\{(2q+2)\begin{Bmatrix} 1/2\ q\ q+1/2 \\ 1/2\ q\ q+1/2 \end{Bmatrix} + 2q\begin{Bmatrix} 1/2\ q\ q+1/2 \\ 1/2\ q\ q-1/2 \end{Bmatrix}\right\}$$
$$= (-1)^{2q+1}\delta_{FG}\delta(B_d, G_d, K_d)$$
$$= \frac{(-1)^{B_d - K_d + F_d + 2q+1}}{(2q+1)\sqrt{(2q+2)[K_d]}}(< \mathbf{F}^{\Upsilon}\mathbf{B})^{(K_d,\, q+1/2)}(\mathbf{BG} >)^{(K_d,\, q+1/2)}$$
$$+ \frac{(-1)^{B_d - K_d + F_d + 2q+2}\sqrt{2q}}{(2q+1)\sqrt{[K_d]}}(< \mathbf{F}^{\Upsilon}\mathbf{B})^{(K_d,\, q-1/2)}(\mathbf{BG} >)^{(K_d,\, q-1/2)}$$
$$+ \sum_{J_d}\frac{(-1)^{B_d + F_d + J_d + 2q+1}\sqrt{[J_d]}}{\sqrt{2q+2}}\begin{Bmatrix} B_d\ F_d\ J_d \\ B_d\ G_d\ K_d \end{Bmatrix}$$
$$\times (< \mathbf{F}^{\Upsilon}\mathbf{B})^{(J_d,\, q+1/2)}(\mathbf{BG} >)^{(J_d,\, q+1/2)},$$

where the following values for the 6j-symbols are used.

$$\begin{Bmatrix} 1/2 & q & q+1/2 \\ 1/2 & q & q+1/2 \end{Bmatrix} = \frac{(-1)^{2q+1}}{(2q+1)(2q+2)}$$

$$\begin{Bmatrix} 1/2 & q & q+1/2 \\ 1/2 & q & q-1/2 \end{Bmatrix} = \frac{(-1)^{2q+1}}{2q+1}$$

Equation (3.32) enables one to generate an orthonormal and complete set of ket tensors for the quasi-spin angular momentum $q - 1/2$. We replace $\mathbf{F}>$ and $\mathbf{G}>$ by $\mathbf{F}_{q,i}>$ and $\mathbf{F}_{q,j}>$ and denote a ket tensor

$$(\mathbf{BF}_{q,i}>)^{(K_d,\, q-1/2)}$$

by $\mathbf{K}_i>$. By using Eqs. (3.24) and (3.27) - (3.31), Eq. (3.32) can be expressed as

$$\begin{aligned} <\mathbf{K}_i^{\Upsilon}\,\mathbf{K}_j> &= \frac{\sqrt{[K_d]}}{\sqrt{2q}}\delta_{ij}\delta(B_d,(F_{q,i})_d, K_d) \\ &- \frac{(-1)^{F_{q,i}-F_{q,j}}}{\sqrt{2q}\sqrt{[K_d]}\,(2q+1)(2q+2)}\sum_k \delta_{(F_{q+1/2,k})_d,\,K_d} C_{q,ik}C_{q,jk}^{*} \\ &- \sum_k \frac{(-1)^{F_{q,i}-F_{q,j}+2F_{q+1/2,k}}\sqrt{[K_d]}}{\sqrt{2q}\,(2q+2)} \begin{Bmatrix} B_d & (F_{q,i})_d & (F_{q+1/2,k})_d \\ B_d & (F_{q,j})_d & K_d \end{Bmatrix} \\ &\times C_{q,ik}C_{q,jk}^{*}. \end{aligned} \tag{3.33}$$

This metric matrix is positive semi-definite as Eq. (3.13), so one can find a matrix T for which

$$\sum_{ij} T_{ik}^{*}T_{jm} <\mathbf{K}_i^{\Upsilon}\mathbf{K}_j> = \delta_{km}\sqrt{2q[K_d]}$$

holds. The ket tensors

$$\sum_i T_{ik}\mathbf{K}_i> \tag{3.34}$$

can be used as members of an orthonormalized set for the quasi-spin angular momentum $q - 1/2$. Repeating this procedure for all possible cases of space-spin representations K_d, we obtain a complete set of ket tensors.

Orthonormal ket tensors in LS-scheme are obtained by limiting the group to O(3)×SU(2). A tensor $\mathbf{F}>$ in LSQ-scheme splits into invariant subspaces, each of which consists of tensor components having common z-components of quasi-spin angular momentum $\mathbf{F}_r>$. Namely, tensor components are given by

$$\mathbf{F}_r>_k = \mathbf{F}>_{k,r}, \tag{3.35}$$

where k and r are the space-spin and the quasi-spin indices of $\mathbf{F}>$. The number of electrons is given by Eq. (3.26). For example, the ${}^3F^4$ ket tensor of d-shell splits into ${}^3F$ ket tensors of 2, 4, 6 and 8 electrons in LS-scheme, and their respective q_z values are $-3/2$, $-1/2$, $1/2$ and $3/2$.

The bra tensor was defined by the equation

$$< \mathbf{F}_\kappa^\Upsilon = \sum_\mu U_{\kappa\mu}^{(F)*}(< \mathbf{F})_\mu$$

By separating indices into space-spin and quasi-spin parts, it is expressed as

$$< \mathbf{F}_{k,r}^\Upsilon = \sum_{ms} U_{km}^{(F_d)*} U_{rs}^{(F_q)*}(< \mathbf{F})_{m,s} = \sum_s U_{rs}^{(F_q)*} < \mathbf{F}_{sk}^\Upsilon, \tag{3.36}$$

where the bra tensor which corresponds to $\mathbf{F}_r >$ is denoted by $< \mathbf{F}_r^\Upsilon$.

An operator $\mathbf{A}$ in LSQ-scheme also splits into tensors $\mathbf{A}_r$ in LS-scheme. Their components are given by

$$(\mathbf{A}_r)_k = (\mathbf{A})_{k,r}. \tag{3.37}$$

From these equations, the expectation value $< \mathbf{F}^\Upsilon \mathbf{AG} >$ is expressed as

$$< \mathbf{F}^\Upsilon \mathbf{AG} > = \sum_{ustr} U_{ur}^{(F_q)*} < \mathbf{F}_r^\Upsilon \mathbf{A}_s \mathbf{G}_t > (OO\,|\,F_q u A_q s G_q t).$$

Its reverse is obtained as

$$< \mathbf{F}_r^\Upsilon \mathbf{A}_s \mathbf{G}_t > = < \mathbf{F}^\Upsilon \mathbf{AG} > (-1)^{2A_q} \sum_u U_{ur}^{(F_q)} \begin{bmatrix} F_q & A_q & G_q \\ u & s & t \end{bmatrix}^* \tag{3.38}$$

by using the Wigner-Eckart theorem. In particular, the expansion coefficient $C_{N,ik}$ (Eq. 3.9) is expressed as

$$< \mathbf{f}_{N,i}^\Upsilon b \mathbf{f}_{N-1,k} > = (-1)^{\ell + F_q + (N-1)/2} \times \begin{Bmatrix} F_q & A_q & G_q \\ \frac{2\ell - N + 1}{2} & \frac{1}{2} & \frac{N - 2\ell - 2}{2} \end{Bmatrix}^* < \mathbf{F}^\Upsilon \mathbf{BG} >, \tag{3.39}$$

where $\mathbf{F}>$ and $\mathbf{G}>$ are the tensors in LSQ-scheme to which $\mathbf{f}_{N,i}>$ and $\mathbf{f}_{N-1,k}>$ belong.

3.3 Generation of Shell States in ATOMCI

In ATOMCI, shell states are generated in a computer by using Eqs. (3.33) and (3.34). Users specify the minimum of the quasi-spin angular momentum for each shell. For example, if one sets 2ℓ for q_{min}, states for which $q = 2\ell + 1$, $2\ell + 1/2$ and 2ℓ will be generated. They include all states with zero, one and two electrons. By setting 1 for q_{min}, all states will be generated. A few hundred K-bytes of core allocation are enough for the generation of all states for an f shell. The maximum of the orbital angular momentum is restricted because of the overflow of factorials in some part of the code. The limit is around $\ell = 10 \sim 11$.

If one runs CFPA and CFPB, which generate a complete set of shell states in LSQ-scheme, or uses the DISPLAY command in the step of CSF (configuration state function) generation, a table of shell states will be printed. A shell state is labeled by a four-figure hexadecimal number. The top two hex-digits are sequential, and indicate the spin and quasi-spin angular momenta. The last two figures indicate the total orbital angular momentum of the shell state. An example of such list is shown in Table I. The minimum and the maximum number of electron occupations are given instead of the quasi-spin angular momentum in the table. In a CSF listing, a shell state in LS-scheme is specified by this four-figure label and the number of electrons. For example, a shell specification (*0C01* 4) means $^1G^4$ state of d-shell and its quasi-spin z-component is $-1/2$. This can be seen from the row *0C0* for the digit 1~3, where the spin multiplicity, the minimum and the maximum of the electron occupation numbers are tabulated as 1, 2 and 8, i.e., it is spin singlet and quasi-spin quartet. The third and the fourth digit, *01*, specify the total orbital angular momentum which is G in this case.

4. Matrix Elements

4.1 Orthonormal Tensors for Many Shells

In section 3.1 the shell creation tensor operators were defined, and an algorithm for their generation was presented in section 3.2. Orthonormal ket tensors for many shells can be expressed by using these tensor operators as follows.

Let $\mathbf{f}_i$ be a member of the complete and orthonormal set of the shell creation operators for a shell $n_i\ell_i$. We define ket tensors $\mathbf{F}_k >$ as

$$\begin{aligned} \mathbf{F}_O > &= > \\ \mathbf{F}_k > &= (\mathbf{f}_k \mathbf{F}_{k-1})^{(F_k)} > \qquad \text{for } k = 1, 2, \dots, M, \end{aligned} \tag{4.1}$$

where M is the total number of shells under consideration. Similarly, ket tensors $\mathbf{G}_k >$ are defined from $\mathbf{g}_i$'s as

Table I. List of Shell States in LSQ-scheme.

S TYPE ORBITAL. L = 0

Digit 1-3	Spin Mult.	Elec. Min.	# Max	Digit 4: 0	1	2	3	4	5	6	7	8	9	A	B	C	D	E	F
010	1	0	2	–	S														
020	2	1	1	–	S														

P TYPE ORBITAL. L = 1

Digit 1-3	Spin Mult.	Elec. Min.	# Max	Digit 4: 0	1	2	3	4	5	6	7	8	9	A	B	C	D	E	F
030	1	0	6	–	S														
040	2	1	5	–	P														
050	3	2	4	–	P														
060	1	2	4	–	D														
070	4	3	3	–	S														
080	2	3	3	–	D														

D TYPE ORBITAL. L = 2

Digit 1-3	Spin Mult.	Elec. Min.	# Max	Digit 4: 0	1	2	3	4	5	6	7	8	9	A	B	C	D	E	F
090	1	0	10	–	S														
0A0	2	1	9	–	D														
0B0	3	2	8	–	F	P													
0C0	1	2	8	–	G	D													
0D0	4	3	7	–	F	P													
0E0	2	3	7	–	H	G	F	D	P										
0F0	5	4	6	–	D														
100	3	4	6	–	H	G	F	D	P										
110	1	4	6	–	I	G	F	D	S										
120	6	5	5	–	S														
130	4	5	5	–	G	D													
140	2	5	5	–	I	G	F	D	S										

F TYPE ORBITAL. L = 3

Digit 1-3	Spin Mult.	Elec. Min.	# Max	Digit 4: 0	1	2	3	4	5	6	7	8	9	A	B	C	D	E	F
150	1	0	14	–	S														
160	2	1	13	–	F														

L TYPE ORBITAL. L = 8

Digit 1-3	Spin Mult.	Elec. Min.	# Max	Digit 4: 0	1	2	3	4	5	6	7	8	9	A	B	C	D	E	F
290	1	0	34	–	S														
2A0	2	1	33	–	L														
2B0	3	2	32	–	U	R	O	M	K	H	F	P							
2C0	1	2	32	–	V	T	Q	N	L	I	G	D							

$$\mathbf{G}_O > = >$$
$$\mathbf{G}_k > = (\mathbf{g}_k\mathbf{G}_{k-1})^{(G_k)} > \qquad \text{for } k = 1, 2, \dots, M. \tag{4.2}$$

Then we have

$$
\begin{aligned}
<\mathbf{F}_k^{\Upsilon}\mathbf{G}_k> &= \delta_{F_k, G_k} < (\mathbf{f}_k\mathbf{F}_{k-1})^{(F_k)\Upsilon}(\mathbf{g}_k\mathbf{G}_{k-1})^{(G_k)} > \\
&= \delta_{F_k, G_k}(-1)^{F_{k-1}+f_k-F_k} < (\mathbf{F}_{k-1}^{\Upsilon}\mathbf{f}_k^{\Upsilon})^{(F_k)}(\mathbf{g}_k\mathbf{G}_{k-1})^{(G_k)} > \\
&= \delta_{F_k, G_k}(-1)^{G_{k-1}+g_k+F_k} \\
&\quad \times \sum_j \sqrt{[F_k][j]} \begin{Bmatrix} F_{k-1} & f_k & F_k \\ g_k & G_{k-1} & j \end{Bmatrix} < \mathbf{F}_{k-1}^{\Upsilon}(\mathbf{f}_k^{\Upsilon}\mathbf{g}_k)^{(j)}\mathbf{G}_{k-1} > \\
&= \delta_{F_k, G_k}(-1)^{G_{k-1}+g_k+F_k} \\
&\quad \times \sqrt{[F_k]} \begin{Bmatrix} F_{k-1} & f_k & F_k \\ g_k & G_{k-1} & 0 \end{Bmatrix} < \mathbf{f}_k^{\Upsilon}\mathbf{g}_k > < \mathbf{F}_{k-1}^{\Upsilon}\mathbf{G}_{k-1} > \\
&= \delta_{F_k, G_k}\delta_{f_k, g_k}\sqrt{[F_k]}\,\delta(F_{k-1}, f_k, F_k)\frac{< \mathbf{F}_{k-1}^{\Upsilon}\mathbf{G}_{k-1} >}{\sqrt{[F_{k-1}]}}
\end{aligned}
\tag{4.3}
$$

Here, the equality

$$< \mathbf{F}_{k-1}^{\Upsilon}(\mathbf{f}_k^{\Upsilon}\mathbf{g}_k)^{(j)}\mathbf{G}_{k-1} > = \delta_{j0} < \mathbf{f}_k^{\Upsilon}\mathbf{g}_k > < \mathbf{F}_{k-1}^{\Upsilon}\mathbf{G}_{k-1} >$$

can be proved by observing that the tensors $\mathbf{F}_{k-1}$ or $\mathbf{G}_{k-1}$ never create or annihilate electrons in the $n_k\ell_k$ shell. Therefore, the above expectation value can be non-zero only when the vacuum component is restored in $\mathbf{f}_k^{\Upsilon}\mathbf{g}_k >$.

The equation (4.3) shows that the ket tensors, Eq. (4.1) or (4.2), are orthogonal and normalized, i.e.,

$$< \mathbf{F}_M^{\Upsilon}\mathbf{G}_M > = \sqrt{[F_M]}\prod_{k=1}^{M}\delta_{F_k, G_k}\delta_{f_k, g_k}\delta(F_{k-1}, f_k, F_k)$$

holds. In ATOMCI, CSF's (configuration state functions) are expressed in tensor form as Eq. (4.1).

Other types of orthonormal CSF's can be easily constructed. For example, let $\mathbf{F} >$ and $\mathbf{G} >$ be

$$\mathbf{F} > = ((\mathbf{f}_4\mathbf{f}_3)^{(f_{43})}(\mathbf{f}_2\mathbf{f}_1)^{(f_{21})})^{(F)} >$$

and

$$\mathbf{G} > = ((\mathbf{g}_4\mathbf{g}_3)^{(g_{43})}(\mathbf{g}_2\mathbf{g}_1)^{(g_{21})})^{(G)} > .$$

Then $< \mathbf{F}^{\Upsilon}\mathbf{G} >$ is $\sqrt{[F]}$ only when $\mathbf{F} >$ and $\mathbf{G} >$ are identical and zero in all other cases. One can easily show that shell creation operators can be coupled in any order. The resultant tensor operators will be an orthonormal set as long as they are ccupled in the same order. In many cases, this general coupling scheme would render a more appropriate set of CSF's. The reason for the adoption of Eq. (4.1) in ATOMCI is simply to save effort for coding.

4.2 Hamiltonian Operator

The non-relativistic Hamiltonian operator for an atom is given by

$$\mathbf{H} = \mathbf{h}_1 + \mathbf{h}_2, \tag{4.4}$$

where

$$\mathbf{h}_1 = \int \psi\dagger(x)\left\{-\frac{1}{2}\Delta - \frac{Z}{r}\right\}\psi(x)dx \tag{4.5}$$

and

$$\mathbf{h}_2 = \frac{1}{2}\int\int :\psi\dagger(x)\psi(x)\frac{1}{|r-r'|}\psi\dagger(x')\psi(x'):dxdx', \tag{4.6}$$

assuming infinite mass for the nucleus. The symbol : : means that the normal product is taken for the enclosed operator. Substituting Eqs. (1.5) and (1.6) into Eqs. (4.5) and (4.6), we have

$$\mathbf{h}_1 = -\sum_{nn'\ell}\sqrt{2\ell}\, h(n\ell, n'\ell)(\mathbf{b}_{n\ell}\mathbf{b}^{\Upsilon}_{n'\ell})^{(^1S)} \tag{4.7}$$

and

$$\begin{aligned}\mathbf{h}_2 = &\sum_{kn_1\ell_1n_2\ell_2n_3\ell_3n_4\ell_4} H^k(\ell_1, \ell_2, \ell_3, \ell_4)R^k(n_1\ell_1, n_2\ell_2, n_3\ell_3, n_4\ell_4)\\ &\times :\left((\mathbf{b}_{n_1\ell_1}\mathbf{b}^{\Upsilon}_{n_2\ell_2})^{(^1k)}(\mathbf{b}_{n_3\ell_3}\mathbf{b}^{\Upsilon}_{n_4\ell_4})^{(^1k)}\right)^{(^1S)}:\end{aligned} \tag{4.8}$$

where

$$h(n\ell, n'\ell) = \int_0^\infty R_{n\ell}(r)^*\left\{-\frac{d^2}{2dr^2} + \frac{\ell(\ell+1)}{r^2} - \frac{Z}{r}\right\}R_{n'\ell}(r)dr,$$

$$\begin{aligned}&R^k(n_1\ell_1, n_2\ell_2, n_3\ell_3, n_4\ell_4)\\ &= \int_0^\infty r^2dr\int_0^\infty r'^2dr'\frac{r_<^k}{r_>^{k+1}}R_{n_1\ell_1}(r)^*R_{n_2\ell_2}(r)R_{n_3\ell_3}(r')^*R_{n_4\ell_4}(r')\end{aligned}$$

and

$$
\begin{aligned}
&H^k(\ell_1,\ell_2,\ell_3,\ell_4)\\
&\quad=(-1)^{(\ell_1-\ell_2+\ell_3-\ell_4)/2}\sqrt{[\ell_1][\ell_2][\ell_3][\ell_4]/[k]}\,F(k,\ell_1,\ell_2)F(k,\ell_3,\ell_4).
\end{aligned}
$$

The function $F\,(a,b,c)$ is defined as

$$
F(a,b,c)=\begin{cases} 0, & \text{for } a+b+c \text{ odd} \\ \dfrac{\delta(a,b,c)g!\sqrt{(2b-2a)!(2g-2b)!(2g-2c)!}}{\sqrt{(2g+1)!}\,(g-a)!(g-b)!(g-c)!} & \text{for } 2g=a+b+c \text{ even.} \end{cases}
$$

The normal product term in Eq. (4.8) can be evaluated from Eq. (3.4) as follows.

$$
\begin{aligned}
&:\left((\mathbf{b}_{n_1\ell_1}\mathbf{b}^{\Upsilon}_{n_2\ell_2})^{({}^1k)}(\mathbf{b}_{n_3\ell_3}\mathbf{b}^{\Upsilon}_{n_4\ell_4})^{({}^1k)}\right)^{({}^1S)}:\\
&=\begin{Bmatrix} {}^2\ell_1 & {}^2\ell_2 & {}^1k \\ {}^2\ell_3 & {}^2\ell_4 & h \end{Bmatrix}(\mathbf{b}_{n_1\ell_1}:(\mathbf{b}^{\Upsilon}_{n_2\ell_2}\mathbf{b}_{n_3\ell_3})^{(h)}:\mathbf{b}^{\Upsilon}_{n_4\ell_4})^{({}^1S)}\\
&=\begin{Bmatrix} {}^2\ell_1 & {}^2\ell_2 & {}^1k \\ {}^2\ell_3 & {}^2\ell_4 & h \end{Bmatrix}(\mathbf{b}_{n_1\ell_1}(\mathbf{b}^{\Upsilon}_{n_2\ell_2}\mathbf{b}_{n_3\ell_3})^{(h)}\mathbf{b}^{\Upsilon}_{n_4\ell_4})^{({}^1S)}\\
&\quad-\delta_{n_2\ell_2,\,n_3\ell_3}\sqrt{{}^2\ell_2}\begin{Bmatrix} {}^2\ell_1 & {}^2\ell_2 & {}^1k \\ {}^2\ell_3 & {}^2\ell_4 & {}^1S \end{Bmatrix}(\mathbf{b}_{n_1\ell_1}\mathbf{b}^{\Upsilon}_{n_4\ell_4})^{({}^1S)}\\
&=\left((\mathbf{b}_{n_1\ell_1}\mathbf{b}^{\Upsilon}_{n_2\ell_2})^{({}^1k)}(\mathbf{b}_{n_3\ell_3}\mathbf{b}^{\Upsilon}_{n_4\ell_4})^{({}^1k)}\right)^{({}^1S)}\\
&\quad-\frac{(-1)^{{}^2\ell_1+{}^2\ell_2+{}^1k}}{\sqrt{{}^2\ell_1}}\delta_{n_2\ell_2,\,n_3\ell_3}\delta_{\ell_1,\,\ell_4}(\mathbf{b}_{n_1\ell_1}\mathbf{b}^{\Upsilon}_{n_4\ell_4})^{({}^1S)}
\end{aligned}
\tag{4.9}
$$

4.3 Reduction Formulas for Matrix Elements

Both the wavefunctions, Eq. (4.1), and the Hamiltonian operator, Eqs. (4.7) - (4.9), are expressed in terms of tensor operators. The recoupling transformation technique which was described in section 2 may be used in various ways for the evaluation of matrix elements. We give some examples of the use of this technique in this subsection.

Let the shells be grouped into two sets. Tensor operators which consist solely of operators from one of the sets will be denoted by **f′**, **g′** and **h′**, and from the other set, by **f″**, **g″** and **h″**. We further assume that these operators will create or annihilate a definite number of electrons, i.e.,

$$
\mathbf{N}\mathbf{j}-\mathbf{j}\mathbf{N}=n(\mathbf{j})\mathbf{j}
$$

where **N** is the number operator which transforms as the identity representation, and $n(\mathbf{j})$ is the number of electrons created by the tensor operator **j**. If two tensors, e.g., $\mathbf{g}'$ and $\mathbf{h}''$, are taken from the different groups, their components anticommute when both operators create or annihilate odd numbers of electrons and commute otherwise. Therefore, we have

$$(\mathbf{h}''\mathbf{g}')^{(j)} = (-1)^{n(\mathbf{g}')n(\mathbf{h}'')+g'+h''-j}(\mathbf{g}'\mathbf{h}'')^{(j)}. \tag{4.10}$$

A matrix element

$$< (\mathbf{f}'\mathbf{f}'')^{(f)\Upsilon}(\mathbf{h}'\mathbf{h}'')^{(h)}(\mathbf{g}'\mathbf{g}'')^{(g)} >$$

can be evaluated by recoupling the last four tensors.

$$\begin{aligned}
&< (\mathbf{f}'\mathbf{f}'')^{(f)\Upsilon}(\mathbf{h}'\mathbf{h}'')^{(h)}(\mathbf{g}'\mathbf{g}'')^{(g)} > \\
&=\sum_{jk}(-1)^{n(\mathbf{g}')n(\mathbf{h}'')+f'+f''-f}\sqrt{[h][g][j][k]}\begin{Bmatrix} h' & h'' & h \\ g' & g'' & g \\ j & k & f \end{Bmatrix} \\
&\quad\times < (\mathbf{f}''^{\Upsilon}\mathbf{f}'^{\Upsilon})^{(f)}(\mathbf{h}'\mathbf{g}')^{(j)}(\mathbf{h}''\mathbf{g}'')^{(k)} > \\
&=\sum_{jk}(-1)^{n(\mathbf{g}')n(\mathbf{h}'')+f'+f''-f}\sqrt{[h][g][j][k]}\begin{Bmatrix} h' & h'' & h \\ g' & g'' & g \\ j & k & f \end{Bmatrix} \\
&\quad\times (-1)^{f'+f''+j+k}\sqrt{[f]}\begin{Bmatrix} f'' & f' & f \\ j & k & O \end{Bmatrix} < \mathbf{f}'^{\Upsilon}\mathbf{h}'\mathbf{g}' >< \mathbf{f}''^{\Upsilon}\mathbf{h}''\mathbf{g}'' > \\
&=(-1)^{n(\mathbf{g}')n(\mathbf{h}'')}\sqrt{[h][g][f]}\begin{Bmatrix} h' & h'' & h \\ g' & g'' & g \\ f' & f'' & f \end{Bmatrix} < \mathbf{f}'^{\Upsilon}\mathbf{h}'\mathbf{g}' >< \mathbf{f}''^{\Upsilon}\mathbf{h}''\mathbf{g}'' >
\end{aligned} \tag{4.11}$$

Another reduction formula which is useful for the evaluation of a product of tensor operators is

$$\begin{aligned}
&< \mathbf{f}^{\Upsilon}(\mathbf{h}_1\mathbf{h}_2)^{(h)}\mathbf{g} > \\
&=\sqrt{[h]}\sum_{k}(-1)^{h_1+h_2+f+g}\begin{Bmatrix} h_1 & h_2 & h \\ g & f & k \end{Bmatrix} < \mathbf{f}^{\Upsilon}\mathbf{h}_1\mathbf{k} >< \mathbf{k}^{\Upsilon}\mathbf{h}_2\mathbf{g} >
\end{aligned} \tag{4.12}$$

where the sum is taken over a complete orthonormal set of tensor operators **k**. This formula may be derived as follows. First, the left-hand side of Eq. (4.12) is recoupled into

$$\sum_{k}(-1)^{h_1+h_2+f+g}\sqrt{[h][k]}\begin{Bmatrix} h_1 & h_2 & h \\ g & f & k \end{Bmatrix} < (\mathbf{f}^{\Upsilon}\mathbf{h}_1)^{(k)}(\mathbf{h}_2\mathbf{g})^{(k)} >.$$

Inserting the identity operator (cf. equations (3.11) and (3.12))

$$\sum_{\mathbf{k}}(-1)^{2k}\sqrt{[k]}\,(\mathbf{k}><\mathbf{k}^{\Upsilon})^{(O)}$$

in the middle of the expectation value $<(\mathbf{f}^{\Upsilon}\mathbf{h}_1)^{(k)}(\mathbf{h}_2\mathbf{g})^{(k)}>$, we obtain Eq. (4.12).

Both equations (4.11) and (4.12) may be used in various ways for the evaluation of operators. Although the latter equation may be used to express expectation values of any operator in terms of $C_{N,ik}$'s (cf. Eqs. (3.9) and (3.10)), the need for the summation over the intermediate states **k** makes the use of it less favorable when **f** and **g** are complex many-shell operators. In the original version of ATOMCI, Eq. (4.12) is utilized in order to reduce the matrix elements of two-electron operators into those of one-electron operators, but the present speeded-up version limits the use of it only to operators within a shell.

On the other hand, the equation (4.11) enables one to express matrix elements in a factorized form. For this purpose we write down tensor operators $\mathbf{H}_k>$ in a similar form as Eq. (4.1):

$$\begin{aligned}\mathbf{H}_O &= 1\\ \mathbf{H}_k &= (\mathbf{h}_k\mathbf{H}_{k-1}) \qquad \text{for } k=1,2,\ldots,M\end{aligned} \tag{4.13}$$

Then using Eqs. (4.1), (4.2), (4.11) and (4.13), we obtain

$$<\mathbf{F}_M^{\Upsilon}\mathbf{H}_M\mathbf{G}_M> = \prod_{k=1}^{M} L_k, \tag{4.14}$$

where

$$L_k = (-1)^{n(\mathbf{g}_k)n(\mathbf{H}_{k-1})}\sqrt{[H_k][G_k][F_k]}\begin{Bmatrix} h_k & H_{k-1} & H_k\\ g_k & G_{k-1} & G_k\\ f_k & F_{k-1} & F_k\end{Bmatrix}<\mathbf{f}_k^{\Upsilon}\mathbf{h}_k\mathbf{g}_k>. \tag{4.15}$$

Any operator can be brought into a linear combination of the form Eq. (4.13) by appropriate recoupling transformations. For example, an operator

$$:((\mathbf{b}_j\mathbf{b}_k^{\Upsilon})^{(p)}(\mathbf{b}_i\mathbf{b}_\ell^{\Upsilon})^{(p)})^{(O)}: \tag{4.16}$$

which appears in the electron repulsion term $\mathbf{h}_2$, can be expressed as

$$\begin{aligned}:((\mathbf{b}_j\mathbf{b}_k^{\Upsilon})^{(p)}(\mathbf{b}_i\mathbf{b}_\ell^{\Upsilon})^{(p)})^{(O)}: &= (-1)^{b_j+b_k-p+2b_\ell}:(\mathbf{b}_\ell^{\Upsilon}(\mathbf{b}_k^{\Upsilon}\mathbf{b}_j)^{(p)}\mathbf{b}_i)^{(O)}:\\ &= \sum_q(-1)^{b_\ell-b_i-p}\sqrt{[p][q]}\begin{Bmatrix} b_k & b_j & p\\ b_i & b_\ell & q\end{Bmatrix}(\mathbf{b}_\ell^{\Upsilon}(\mathbf{b}_k^{\Upsilon}(\mathbf{b}_j\mathbf{b}_i)^{(q)})^{(b_\ell)})^{(O)}\end{aligned} \tag{4.17}$$

where $\mathbf{b}_i$'s stand for $\mathbf{b}_{n_i\ell_i}$'s and an inequality $i<j<k<\ell$ is assumed. The reordered term

$$(\mathbf{b}_\ell^\Upsilon(\mathbf{b}_k^\Upsilon(\mathbf{b}_j\mathbf{b}_i)^{(q)})^{(b_\ell)})^{(O)}$$

is expressed in the form of Eq. (4.13) by defining $\mathbf{h}_r$ and H_r as follows.

$$\mathbf{h}_r = \begin{cases} 1 \\ \mathbf{b}_r \\ \mathbf{b}_r^\Upsilon \end{cases} \text{for} \begin{cases} r \neq i, j, k \text{ and } \ell \\ r = i \text{ or } j \\ r = k \text{ or } \ell \end{cases} \tag{4.18}$$

$$H_r = \begin{cases} {}^1S \\ b_i \\ q \\ b_\ell \end{cases} \text{for} \begin{cases} r < i \text{ or } r \geq \ell \\ i \leq r < j \\ j \leq r < k \\ k \leq r < \ell \end{cases} \tag{4.19}$$

By stubstituting $\mathbf{h}_r$ and H_r's into Eq. (4.15), the factor L_k is finally expressed in terms of a 9j-symbol and $<\mathbf{f}_k^\Upsilon\mathbf{g}_k>$, $<\mathbf{f}_k^\Upsilon\mathbf{b}_k^\Upsilon\mathbf{g}_k>$ or $<\mathbf{f}_k^\Upsilon\mathbf{b}_k\mathbf{g}_k>$, whose derivation is discussed in section 3.

The recoupling transformation Eq. (4.17) involves a summation over the intermediate representation q which satisfies the triangular conditions, but it should be noted that these expectation values for different q's can be used for the calculation of Eq. (4.16) for various p's. For the Coulomb interaction $\mathbf{h}_2$, the spin part of the representations has to be included for q's. Therefore, about twice as many number of the reordered expectation values are enough for the evaluation of the expectation values of the original operators.

A factorized expression like Eq. (4.14) for matrix elements was rediscovered by Shavitt[12] in conjunction with the graphical unitary group approach, and he showed its usefulness for extremely efficient coding of a configuration interaction procedure since many of these factors can be shared for the calculation of different matrix elements.

The present approach differs from his approach in that a) not only the spin but also the spatial symmetry of the system is fully taken into account, b) any matrix elements such as spin-orbit, spin-other-orbit or spin-spin interactions can be brought into factorized form, and c) the factorized expression is derived straightforwardly with no mathematical trick. The reason for such difference may be understood as follows. The graphical unitary group approach is based on the representation theory of the unitary group and the operator $\mathbf{h}_2$ is expressed as a linear combination of products of *two* infinitesimal generators of the unitary group. Factorization is possible, only this is further decomposed into products of four by an intricate mathematical manipulation of formulas. It should also be noted that only spin-singlet operators are expressible in terms of the generators of the unitary group.

In the present approach, the basic tensor operators **b** and $\mathbf{b}^{\Upsilon}$ are constructed from creation and annihilation operators. Any kind of operator is expressible in terms of these basic tensor operators. In particular, the coulomb operator $\mathbf{h}_2$ is expressed as a linear combination of products of *four* tensor operators from the beginning, and the factorization for each shell is a natural consequence of this operator decomposition.

Appendix 4A. Simply Reducible Group

Wigner[15,17] defined "simply reducible groups" in his research on group representations. The concept is quite useful because many of the symmetry groups we encounter in atomic and molecular systems belong to that category, and algebraic manipulations of tensor operators become much easier for those groups.

A finite or compact group which satisfy the following two conditions is called simply reducible: a) The Kronecker product of any two irreducible representations of the group does not contain any irreducible representation more than once. b) A group element and its reciprocal element belong to the same class.

Besides the rotational groups O(3) or SU(2), molecular symmetry groups such as $D_{\infty h}$, $C_{\infty v}$, C_{2v}, C_{3v}, C_{2h}, D_{3h}, D_{3d}, D_{6h}, T_d and O_h have this property. From the definition, it is obvious that the direct product group of two simply reducible groups is also simply reducible.

Although the properties of SR groups are fully discussed in the above mentioned references, we derive properties of simply reducible groups (hereafter abbreviated as SR groups) which are relevant for applications in ATOMCI. Slightly modified phase conventions of Ref. 2 are adopted for some symbols since they seem to be more appropriate.

A.1 Integer Representation and Half Integer Representation

Due to the compactness of the group, any representation of a SR group is completely reducible and can be brought into unitary form by a similarity transformation. We choose a "standard unitary irreducible representation" for each of the irreducible representations. Let

$$D^{(A)}(X)_{\kappa\lambda}$$

and

$$\chi^{(A)}(X) = \sum_{\kappa} D^{(A)}(X)_{\kappa\kappa}$$

be $\kappa\lambda$ matrix element and the character of the standard irreducible representation A of a group element X of the SR group G under consideration. The dimension of the irreducible representation A will be denoted by $[A]$. Since it is a representation,

$$\sum_{\lambda} D^{(A)}(X)_{\kappa\lambda} D^{(A)}(Y)_{\lambda\mu} = D^{(A)}(XY)_{\kappa\mu} \tag{A.1}$$

holds.

In the following, the group under consideration is assumed to be a finite group and its order will be denoted by

$$\sum_{X} 1 = h,$$

but the derivations can be extended for compact groups by replacing summation over group elements by invariant integration on the space of group elements.

Since an irreducible representation of a SR group is equivalent to its complex conjugate representation due to the requirement b), there exists a unitary matrix which transforms the representation into its complex conjugate form as

$$D^{(A)}(X)^* U = U D^{(A)}(X), \tag{A.2}$$

where the symbol * stands for the complex conjugation. Multiplying by U^* from the left, we have

$$U^* D^{(A)}(X)^* U = U^* U D^{(A)}(X) = D^{(A)}(X) U^* U.$$

Because of the irreducibility of the representation and the unitarity of the matrix U, the matrix $U^* U$ is a constant c of modulus 1 times the unit matrix. The constant is either 1 or -1 since

$$c^* 1 = \{U^* U\}^* = UU^* = UU^* UU^{-1} = UcU^{-1} = c1.$$

From the identity

$$U^* U = c1 = \pm 1 \tag{A.3}$$

it follows the relation

$$U^* = cU^{-1} = c\{U^*\}^t,$$

where the superscript t means the transposition. This shows that the matrix U is symmetric when $c = 1$ and skew symmetric when $c = -1$.

For any equivalent representation $D'^{(A)}(X) = TD^{(A)}(X)T^{-1}$, the matrix $U' = T^*UT^{-1}$ satisfies the equations corresponding to Eqs. (A.2) and (A.3) as

$$D'^{(A)}(X)^* U' = U' D'^{(A)}(X)$$

and

$$U'^* U' = U^* U,$$

because

$$\begin{aligned} D'^{(A)}(X)^* T^* U T^{-1} &= T^* D^{(A)}(X)^* U W^{-1} \{T^{-1}\}^* T^* U T^{-1} \\ &= T^* U D^{(A)}(X) T^{-1} = T^* U T^{-1} D'^{(A)}(X) \end{aligned}$$

and

$$U'^* U' = \{T^* U T^{-1}\}^* T^* U T^{-1} = T c T^{-1} = c1 = U^* U \qquad (A.4)$$

holds. The identity Eq. (A.4) shows that the property $c = \pm 1$ does not depend on a particular choice of a representation. Following Wigner[15], irreducible representations for which $c^{(A)} = 1$ holds will be called *integer representations*, and those for which $c^{(A)} = -1$ holds will be called *half integer representations*. For SR groups, any irreducible representation is either integer or half integer because of the requirement b). The name integer or half integer comes from the analogy of that property for the rotational group.

The Kronecker product of two standard representations A and B can be brought by a unitary matrix into completely reduced form in which each block diagonal matrix takes the standard form. Namely, there exists a unitary matrix $(C\mu \,|\, A\kappa B\lambda)$ for which

$$\sum_{\kappa\lambda} (C\mu \,|\, A\kappa B\lambda) D^{(A)}(X)_{\kappa\pi} D^{(B)}(X)_{\lambda\rho} = \sum_{\nu} D^{(C)}(X)_{\mu\nu} (C\nu \,|\, A\pi B\rho) \qquad (A.5)$$

holds for any element X in G. Since an irreducible representation C appears no more than once due to the requirement a) of the SR group, no additional index is required to specify the block. The matrix elements $(C\mu \,|\, A\kappa B\lambda)$ are called Clebsch-Gordan coefficients and the unitarity of the matrix can be expressed as

$$\sum_{\kappa\lambda} (C\mu \,|\, A\kappa B\lambda)^* \, (D\nu \,|\, A\kappa B\lambda) = \delta_{CD}\delta_{\mu\nu}, \qquad (A.6)$$

$$\sum_{C\mu} (C\mu \,|\, A\kappa B\lambda)^* \, (C\mu \,|\, A\pi B\rho) = \delta_{\kappa\pi}\delta_{\lambda\rho} \qquad (A.7)$$

and

$$[A][B] = \sum_{C} \delta(A, B, C)[C].$$

The function $\delta(A, B, C)$ denotes the number of occurrences of the representation C in the Kronecker product representation $A \times B$. The requirement a) of SR group asserts that $\delta(A, B, C)$ is either 0 or 1.

The identity representation in the Kronecker product $A \times B$ deserves special attention. The Clebsch-Gordan coefficient $(OO|A\kappa B\lambda)$, where both the identity representation and its unique component are denoted by the symbol O, is seen to satisfy the following equation owing to Eq. (A.5).

$$\sum_{\kappa\lambda} (OO|A\kappa B\lambda)\, D^{(A)}(X)_{\kappa\pi} D^{(B)}(X)_{\lambda\rho} = (OO|A\pi B\rho)$$

Multiplying this equation by $D^{(A)}(X)_{\mu\pi}{}^*$ and summing over π, we obtain

$$\sum_{\lambda} (OO|A\mu B\lambda) D^{(B)}(X)_{\lambda\rho} = \sum_{\pi} D^{(A)}(X)_{\mu\pi}{}^* \,(OO|A\pi B\rho), \qquad (A.8)$$

which shows that the coefficient $(OO|A\kappa B\lambda)$ can be non-zero only when A and the complex conjugate representation of B, hence B itself, are equivalent. Comparing Eq. (A.8) with (A.2) and from the unitarity of the Clebsch-Gordan coefficient Eqs. (A.6) and (A.7), one can see that the matrix U in Eq. (A.2) can be expressed as

$$U^{(A)}_{\kappa\lambda} = c^{(A)} U^{(A)}_{\lambda\kappa} = \sqrt{[A]}\,(OO|A\kappa A\lambda). \qquad (A.9)$$

The Kronecker product representation of two irreducible representations A and B is given by

$$D(X)_{\kappa\lambda,\, \pi\rho} = D^{(A)}(X)_{\kappa\pi} D^{(B)}(X)_{\lambda\rho}.$$

The matrix

$$W_{\kappa\lambda,\, \pi\rho} = U^{(A)}_{\kappa\pi} U^{(B)}_{\lambda\rho}, \qquad (A.10)$$

satisfies the following equation

$$D(X)^* W = W D(X),$$

since

$$\{D(X)^*W\}_{\kappa\lambda,\,\pi\rho} = \sum_{\mu\nu} D^{(A)}(X)_{\kappa\mu}{}^* D^{(B)}(X)_{\lambda\nu}{}^* W_{\mu\nu,\,\pi\rho}$$
$$= \sum_{\mu\nu} U^{(A)}_{\kappa\mu} D^{(A)}(X)_{\mu\pi} U^{(B)}_{\lambda\nu} D^{(B)}(X)_{\nu\rho} - \{WD(X)\}_{\kappa\lambda,\,\pi\rho}$$

From the definition Eq. (A.10) of the matrix W,

$$W^* W = c^{(A)} c^{(B)} 1$$

holds. By repeating the manipulation which leads to Eq. (A.4), one can easily see that the above property of W is independent of the representation. Thus, if the transformed representation matrices $D'(X)$ take completely reduced form, the matrix W' which brings them into complex conjugate representation also becomes block-diagonal, and each block satisfies Eq. (A.2). This proves that all irreducible representations C which appear in the Kronecker product $A \times B$ have a common value for c, and it is given by

$$c^{(C)} = c^{(A)} c^{(B)}. \qquad (A.11)$$

In other words, the Kronecker product of two integer or two half-integer representations contains only integer representations, and that of an integer and a half integer representations contains only half integer representations.

A.2 Even and Odd Representation

The Kronecker product of an irreducible representation with itself, i.e., $A \times A$, decomposes into the symmetric and antisymmetric parts

$$D_{sym}(X)_{\kappa\lambda,\,\pi\rho} = \frac{1}{2}\{D^{(A)}(X)_{\kappa\pi} D^{(A)}(X)_{\lambda\rho} + D^{(A)}(X)_{\kappa\rho} D^{(A)}(X)_{\lambda\pi}\}$$

and

$$D_{asym}(X)_{\kappa\lambda,\,\pi\rho} = \frac{1}{2}\{D^{(A)}(X)_{\kappa\pi} D^{(A)}(X)_{\lambda\rho} - D^{(A)}(X)_{\kappa\rho} D^{(A)}(X)_{\lambda\pi}\}.$$

They are both invariant subspaces and their characters are given by

$$\chi_{sym}(X) = \frac{1}{2}\{\chi^{(A)\,2}(X) + \chi^{(A)}(X^2)\} \qquad (A.12)$$

and

$$\chi_{asym}(X) = \frac{1}{2}\{\chi^{(A)\,2}(X) - \chi^{(A)}(X^2)\}. \qquad (A.13)$$

An irreducible representation which is contained in the symmetric part of the Kronecker product of an integer representation with itself, or in the antisymmetric part of that of a half integer representation with itself is called an *even representation,* and an irreducible representation which is contained in the antisymmetric part of the Kronecker product of an integer representation with itself, or in the symmetric part of that of a half integer representation with itself is called an *odd representation.* They are both integer representations because their c is given by $\{c^{(A)}\}^2 = 1$ from Eq. (A.11).

Wigner has shown that no irreducible representation can be both even and odd. This may be proved if

$$\sum_X \{\chi^{(A)\,2}(X) + c^{(A)}\chi^{(A)}(X^2)\}\{\chi^{(B)\,2}(X) - c^{(B)}\chi^{(B)}(X^2)\} = O \qquad (A.14)$$

holds for any A and B, since Eq. (A.14) is, from the orthogonality relation and the realness of the characters of a SR group, equal to $4hn$, where n is the total number of irreducible representations which appear as even representations in $A \times A$ and simultaneously as odd ones in $B \times B$. Because of the positive semi-definiteness of the left side of Eq. (A.14), one can replace Eq. (A.14) by the equation in which Eq. (A.14) is summed over A and B. Thus, instead of Eq. (A.14), we have the following equality to be proved.

$$\sum_{AB}\left\{\sum_X \chi^{(A)\,2}(X)\chi^{(B)\,2}(X) - \sum_X c^{(A)}c^{(B)}\chi^{(A)}(X^2)\chi^{(B)}(X^2)\right\} = O \qquad (A.15)$$

From Eq. (A.5) and the orthogonality relation Eq. (A.7), we have

$$D^{(A)}(X)_{\kappa\pi}D^{(B)}(X)_{\lambda\rho} = \sum_C{}' \sum_{\mu\nu}(C\mu\,|\,A\kappa B\lambda)^* D^{(C)}(X)_{\mu\nu}(C\nu\,|\,A\pi B\rho), \qquad (A.16)$$

where the symbol Σ' indicates that the summation is to be extended only for those C's which are contained in $A \times B$. By using Eqs. (A.16) and (A.6), the right side term in Eq. (A.15) can be evaluated as follows.

$$\begin{aligned}\sum_X \chi^{(A)}(X^2)\chi^{(B)}(X^2) &= \sum_{X\kappa\pi\lambda\rho} D^{(A)}(X)_{\kappa\pi} D^{(A)}(X)_{\pi\kappa} D^{(B)}(X)_{\lambda\rho} D^{(B)}(X)_{\rho\lambda} \\ &= \sum_{X\kappa\pi\lambda\rho} \sum_{CD}{}' \sum_{\mu\nu\sigma\tau} (C\mu \,|\, A\kappa B\lambda)^* D^{(C)}(X)_{\mu\nu} (C\nu \,|\, A\pi B\rho) \\ &\quad \times (D\sigma \,|\, A\pi B\rho)^* D^{(D)}(X)_{\sigma\tau} (D\tau \,|\, A\kappa B\lambda) \\ &= \sum_X \sum_{CD}{}' \sum_{\mu\nu\sigma\tau} \delta_{CD}\delta_{\mu\tau}\delta_{DC}\delta_{\nu\sigma} D^{(C)}(X)_{\mu\nu} D^{(D)}(X)_{\sigma\tau} \\ &= \sum_X \sum_C{}' \sum_{\mu\nu} D^{(C)}(X)_{\mu\nu} D^{(C)}(X)_{\nu\mu} \\ &= \sum_X \sum_C{}' \chi^{(C)}(X^2)\end{aligned} \tag{A.17}$$

From Eq. (A.2) and the orthogonality and completeness relationship of the representation

$$\sum_X D^{(C)}(X)_{\kappa\lambda}{}^* D^{(D)}(LX)_{\mu\nu} = h[C]^{-1}\delta_{CD}\delta_{\kappa\mu}\delta_{\lambda\nu},$$

we obtain

$$\begin{aligned}\sum_X \chi^{(C)}(X^2) &= \sum_{X\kappa\lambda} D^{(C)}(X)_{\kappa\lambda} D^{(C)}(X)_{\lambda\kappa} \\ &= \sum_{X\kappa\lambda\mu\nu} U^{(C)-1}_{\kappa\mu} D^{(C)}(X)^*{}_{\mu\nu} U^{(C)}_{\nu\lambda} D^{(C)}(X)_{\lambda\kappa} \\ &= h[C]^{-1} \sum_{\kappa\lambda\mu\nu} \delta_{\lambda\mu}\delta_{\kappa\nu} U^{(C)-1}_{\kappa\mu} U^{(C)}_{\nu\lambda} \\ &= h[C]^{-1} \sum_{\kappa\lambda} U^{(C)-1}_{\kappa\lambda} U^{(C)}_{\kappa\lambda} \\ &= h[C]^{-1} \sum_{\kappa\lambda} U^{(C)*}_{\lambda\kappa} U^{(C)}_{\kappa\lambda} = c^{(C)}h.\end{aligned} \tag{A.18}$$

Substituting Eq. (A.18) into (A.17), we have

$$\sum_X \chi^{(A)}(X^2)\chi^{(B)}(X^2) = h\sum_C{}' c^{(C)} = hc^{(A)}c^{(B)}\sum_C{}' 1. \tag{A.19}$$

The left side of Eq. (A.15) is much easier to evaluate. It can be expressed as

$$\sum_X \chi^{(A)\,2}(X)\chi^{(B)\,2}(X) = \sum_X \sum_{CD}{}' \delta(A, B, C)\chi^{(C)}(X)\delta(A, B, D)\chi^{(D)}(X)$$
$$= h\sum_C{}' \delta(A, B, C) = h\sum_C{}' 1.$$

This equation together with Eq. (A.19) completes the proof that Eq. (A.15) is valid.

Thus, the integer representations may be further classified into even representations and odd representations. The present analysis does not exclude the possibility of the existence of an integer representation which is neither even nor odd. Although the rotational group does not contain such representations, one can show that some SR group has such a representation[16].

A.3 Three-j Symbols

A triple Kronecker product representation $A \times B \times C$ contains the identity representation $\delta(A, B, C)$ times. When $\delta(A, B, C) = 1$, the corresponding row vector of the unitary transformation may be expressed as

$$(OO\,|\,A\kappa B\lambda C\mu) = \sum_\nu (C\nu\,|\,A\kappa B\lambda)(OO\,|\,C\nu C\mu) = c^{(B)}\begin{bmatrix} A & B & C \\ \kappa & \lambda & \mu \end{bmatrix}. \qquad (A.20)$$

Although the last symbol in Eq. (A.20) differs from the 3j-symbol introduced by Wigner[2,16] by a phase factor, we call it 3j-symbol.

Any permutation of columns leaves the 3j-symbol invariant except for a phase factor of modulus 1, since there exists only one one-dimensional invariant subspace which corresponds to the identity representation in $A \times B \times A$. Therefore, $(OO\,|\,B\lambda A\kappa C\mu)$, $(OO\,|\,C\mu A\kappa B\lambda)$, etc., are all proportional to $(OO\,|\,A\kappa B\lambda C\mu)$. In order to express the phase factor, the following convention will be introduced.

We have seen that all irreducible representations of a SR group can be classified into four categories: 1) even integer, 2) odd integer, 3) integer but neither even nor odd, and 4) half integer. We assign an even number, an odd number, an integer (even number can be chosen) and a half integer for each representation in the respective categories, and call them *symmetry indices* of irreducible representations. We introduce a convention that whenever symbols of irreducible representations appear in the exponent part of (-1), they stand for these indices. For example, using this convention one can express $c^{(A)}$ as $(-1)^{2A}$, since it is 1 for an integer representation and -1 for a half integer representation. Different numbers in modulo 2 in the case of integer representations, or modulo 1 in the case of half integer representations may be equally assigned. For the three-dimensional rotational group, the angular momentum can be used for those numbers.

By choosing appropriate phases for Clebsch-Gordan coefficients, it is possible to normalize 3j-symbols as follows.

$$\sum_{\kappa\lambda\mu}\begin{bmatrix} A & B & C \\ \kappa & \lambda & \mu \end{bmatrix} = (-1)^{A+B+C}\begin{bmatrix} B & A & C \\ \lambda & \kappa & \mu \end{bmatrix} = (-1)^{A+B+C}\begin{bmatrix} C & B & A \\ \mu & \lambda & \kappa \end{bmatrix} \qquad (A.21)$$

Note that Eq. (A.11) can be expressed as

$$(-1)^{2A+2B+2C} = c^{(A)}c^{(B)}c^{(C)} = 1.$$

Therefore, $(-1)^{A+B+C}$ is either $+1$ or -1, and from Eq. (A.21) one can see that for even permutation of columns, 3j-symbols remain unaltered, and for odd permutation they change by a factor of $(-1)^{A+B+C}$. Since this factor is just one of the one-dimensional representations of the symmetric group S_3, the assigned symmetry property Eq. (A.21) is at least internally consistent.

In the case where the representations A, B and C are all different from each other, permuted 3j-symbols correspond to a different set of Clebsch-Gordan coefficients. Since we are free to choose any phase factor for them, internally consistent Eq. (A.21) can be adopted.

When the first two representations of a 3j-symbol are the same, i.e.,

$$\begin{bmatrix} A & A & C \\ \kappa & \lambda & \mu \end{bmatrix} = c^{(A)}\sum_{\nu}(C\nu\,|\,A\kappa A\lambda)(OO\,|\,C\nu C\mu), \qquad (A.22)$$

the permutation of the first two columns induces only the exchange of κ and λ in the Clebsch-Gordan coefficient $(C\nu\,|\,A\kappa A\lambda)$. From Eqs. (A.21) and (A.22) we obtain

$$(C\nu\,|\,A\lambda A\kappa) = (-1)^{2A+C}(C\nu\,|\,A\kappa A\lambda).$$

The above equation means that the representation of C is in the symmetric or in the antisymmetric part of $A \times A$ depending on withether $(-1)^{2A+C}$ is $+1$ or -1, but this is nothing but the symmetry property of SR groups discussed in this section. C can never be a representation which is integer but neither even nor odd so that the indeterminancy of $(-1)^{C}$ for such representations does not cause any problem. The consistency for other permutations can easily be checked.

Having established the property of the 3j-symbols, we now derive some important properties of them and the Clebsch-Gordan coefficients.

Using the unitarity Eqs. (A.9) and (A.20) we can express the Clebsch-Gordan coefficients as follows.

$$\sum_{\mu} c^{(B)}[C](OO|C\rho C\mu)^* \begin{bmatrix} A & B & C \\ \kappa & \lambda & \mu \end{bmatrix} = \sum_{\nu\mu}[C](C\nu|A\kappa B\lambda)(OO|C\nu C\mu)(OO|C\rho C\mu)^*$$
$$= (C\rho|A\kappa B\lambda)$$

i.e.,

$$\begin{aligned}(C\mu|A\kappa B\lambda) &= (-1)^{2B}[C]\sum_{\nu}(OO|C\mu C\nu)^* \begin{bmatrix} A & B & C \\ \kappa & \lambda & \nu \end{bmatrix} \\ &= (-1)^{2B}\sqrt{[C]}\sum_{\nu} U_{\mu\nu}^{(C)*} \begin{bmatrix} A & B & C \\ \kappa & \lambda & \nu \end{bmatrix}\end{aligned} \tag{A.23}$$

holds. From Eq. (A.23) we have

$$\begin{aligned}(C\mu|B\lambda Ak) &= (-1)^{2A}[C]\sum_{\nu}(OO|C\mu C\nu)^* \begin{bmatrix} B & A & C \\ \lambda & \kappa & \nu \end{bmatrix} \\ &= (-1)^{2A-2B+A+B+C}(C\mu|A\kappa B\lambda) = (-1)^{A+B-C}(C\mu|A\kappa B\lambda),\end{aligned} \tag{A.24}$$

where the identity $(-1)^{4A} = 1$ is used.

From the definition Eq. (A.20) and the unitarity we have

$$\sum_{\kappa\lambda\mu} \begin{bmatrix} A & B & C \\ \kappa & \lambda & \mu \end{bmatrix} \begin{bmatrix} A & B & C \\ \kappa & \lambda & \mu \end{bmatrix}^* = 1. \tag{A.25}$$

From Eqs. (A.5) and (A.8) it follows the equation

$$\begin{aligned}&\sum_{\kappa\lambda\mu} \begin{bmatrix} A & B & C \\ \kappa & \lambda & \mu \end{bmatrix} D^{(A)}(X)_{\kappa\pi} D^{(B)}(X)_{\lambda\rho} D^{(C)}(X)_{\mu\nu} \\ &= c^{(B)} \sum_{\kappa\lambda\mu\sigma} (C\sigma|A\kappa B\lambda)(OO|C\sigma C\mu) D^{(A)}(X)_{\kappa\pi} D^{(B)}(X)_{\lambda\rho} D^{(C)}(X)_{\mu\nu} \\ &= c^{(B)} \sum_{\mu\sigma\tau} D^{(C)}(X)_{\mu\tau}(C\tau|A\pi B\rho)(OO|C\sigma C\mu) D^{(C)}(X)_{\mu\nu} \\ &= c^{(B)} \sum_{\sigma\tau\upsilon} D^{(C)}(X)_{\upsilon\tau}(C\tau|A\pi B\rho)(OO|C\upsilon C\nu) D^{(C)}(X)_{\sigma\upsilon}^* \\ &= c^{(B)} \sum_{\tau\upsilon} \delta_{\tau\upsilon}(C\tau|A\pi B\rho)(OO|C\upsilon C\nu) = \begin{bmatrix} A & B & C \\ \pi & \rho & \nu \end{bmatrix},\end{aligned} \tag{A.26}$$

which shows the invariance of the 3j-symbol as it should be because it is the component of the invariant vector in the space of $A \times B \times C$. The same vector may be obtained by reducing the space $B \times C$ first as

$$\sum_{\nu}(OO|A\kappa A\nu)(A\nu|B\lambda C\mu). \tag{A.27}$$

By using Eq. (A.23), Eq. (A.27) can be expressed as

$$\begin{aligned}\sum_{\nu}(OO|A\kappa A\nu)(A\nu|B\lambda C\mu) &= (-1)^{2C}[A]\sum_{\nu\rho}(OO|A\kappa A\nu)(OO|A\nu A\rho)^{*}\begin{bmatrix} B & C & A \\ \lambda & \mu & \rho \end{bmatrix} \\ &= (-1)^{2C}\sum_{\nu\rho}U^{(A)}_{\kappa\nu}U^{(A)*}_{\nu\rho}\begin{bmatrix} B & C & A \\ \lambda & \mu & \rho \end{bmatrix} \\ &= (-1)^{2C+2A}\sum_{\rho}\{U^{(A)*-1}U^{(A)*}U^{(A)}U^{(A)*}\}_{\kappa\rho}\begin{bmatrix} B & C & A \\ \lambda & \mu & \rho \end{bmatrix} \\ &= (-1)^{2B}\sum_{\rho}\delta_{\kappa\rho}\begin{bmatrix} B & C & A \\ \lambda & \mu & \rho \end{bmatrix} \\ &= \sum(C\nu|A\kappa B\lambda)(OO|C\nu C\mu).\end{aligned} \tag{A.28}$$

This equality, as expressed as a recoupling coefficient, leads to

$$((AB)^{(C)}C|A(BC)^{(A)})^{(0)} = 1, \tag{A.29}$$

and for tensor operators to the following equality of expectation values.

$$< (\mathbf{AB})^{(C)}\mathbf{C} > = < \mathbf{A}(\mathbf{BC})^{(A)} > \tag{A.30}$$

In fact, if the equality Eq. (A.28) were not true, the notation $(OO|A\kappa B\lambda C\mu)$ in Eq. (A.20) would have been ambiguous since it does not explicitly specify whether $A \times B \times C$ is reduced $A \times B$ first or $B \times C$ first. The same reasoning may apply for Eq. (A.30); the expectation value Eq. (A.30) can be denoted by

$$< \mathbf{ABC} > \tag{A.31}$$

with no risk of confusion.

Another important conclusion one can derive from Eq. (A.26) is that any invariant vector in the space $A \times B \times C$ is proportional to the 3j-symbol; i.e., if $\{f_{\kappa\lambda\mu}\}$ satisfies the following equation for any X,

$$\sum_{\kappa\lambda\mu} f_{\kappa\lambda\mu}D^{(A)}(X)_{\kappa\pi}D^{(B)}(X)_{\lambda\rho}D^{(C)}(X)_{\mu\nu} = f_{\pi\rho\nu} \tag{A.32}$$

it can be expressed as

$$f_{\kappa\lambda\mu} = f\begin{bmatrix} A & B & C \\ \kappa & \lambda & \mu \end{bmatrix}, \tag{A.33}$$

where f is given by

$$f = \sum_{\kappa\lambda\mu} f_{\kappa\lambda\mu} \begin{bmatrix} A & B & C \\ \kappa & \lambda & \mu \end{bmatrix}^* . \tag{A.34}$$

The vacuum expectation value < **ABC** > is given by

$$< \mathbf{ABC} > = \sum_{\kappa\lambda\mu} (OO \mid A\kappa B\lambda C\mu) < \mathbf{A}_\kappa \mathbf{B}_\lambda \mathbf{C}_\mu > . \tag{A.35}$$

Under the group transformation, a tensor operator transforms as

$$\mathbf{A'}_\kappa = \sum_\lambda D^{(A)}(X)_{\kappa\lambda} \mathbf{A}_\lambda .$$

The expectation value $< \mathbf{A}_\kappa \mathbf{B}_\lambda \mathbf{C}_\mu >$, therefore, transforms as

$$\begin{aligned} < \mathbf{A'}_\kappa \mathbf{B'}_\lambda \mathbf{C'}_\mu > &= \sum_{\pi\rho\nu} D^{(A)}(X)_{\kappa\pi} D^{(B)}(X)_{\lambda\rho} D^{(C)}(X)_{\mu\nu} < \mathbf{A}_\pi \mathbf{B}_\rho \mathbf{C}_\nu > \\ &= < \mathbf{A}_\kappa \mathbf{B}_\lambda \mathbf{C}_\mu > , \end{aligned}$$

where the last equality results from the invariance of the vacuum state. Multiplying by $D^{(A)}(X)^*{}_{\kappa\pi} \cdot D^{(B)}(X)^*{}_{\lambda\rho} \cdot D^{(C)}(X)^*{}_{\mu\nu}$ and adding over κ, λ and μ, we obtain

$$\sum_{\kappa\lambda\mu} < \mathbf{A}_\kappa \mathbf{B}_\lambda \mathbf{C}_\mu > D^{(A)}(X)^*{}_{\kappa\pi} D^{(B)}(X)^*{}_{\lambda\rho} D^{(C)}(X)^*{}_{\mu\nu} = < \mathbf{A}_\pi \mathbf{B}_\rho \mathbf{C}_\nu > .$$

Since this equation is just the complex conjugate of Eq. (A.32), one sees that $< \mathbf{A}_\kappa \mathbf{B}_\lambda \mathbf{C}_\mu >$ is proportional to the complex conjugate of the 3j-symbol. The factor can be evaluated from Eq. (A.35), and finally we obtain

$$< \mathbf{A}_\kappa \mathbf{B}_\lambda \mathbf{C}_\mu > = (-1)^{2B} \begin{bmatrix} A & B & C \\ \kappa & \lambda & \mu \end{bmatrix}^* < \mathbf{ABC} > . \tag{A.36}$$

This equation is known as the Wigner-Eckart theorem.

Substituting Eq. (A.2) into the complex conjugate of Eq. (A.26), we have

$$\sum_{\kappa\lambda\mu\pi\rho\nu} \begin{bmatrix} A & B & C \\ \kappa & \lambda & \mu \end{bmatrix}^* D^{(A)}(X)^*_{\kappa\pi} D^{(B)}(X)^*_{\lambda\rho} D^{(C)}(X)^*_{\mu\nu} U^{(A)}_{\pi\sigma} U^{(B)}_{\rho\tau} U^{(C)}_{\nu\upsilon}$$
$$= \sum_{\kappa\lambda\mu\pi\rho\nu} \begin{bmatrix} A & B & C \\ \kappa & \lambda & \mu \end{bmatrix}^* U^{(A)}_{\kappa\pi} U^{(B)}_{\lambda\rho} U^{(C)}_{\mu\nu} D^{(A)}(X)_{\pi\sigma} D^{(B)}(X)_{\rho\tau} D^{(C)}(X)_{\nu\upsilon}$$
$$= \sum_{\pi\rho\nu} \begin{bmatrix} A & B & C \\ \pi & \rho & \nu \end{bmatrix}^* U^{(A)}_{\pi\sigma} U^{(B)}_{\rho\tau} U^{(C)}_{\nu\upsilon}.$$

Comparison of this equation with Eq. (A.32) gives the following equality

$$\sum_{\kappa\lambda\mu} \begin{bmatrix} A & B & C \\ \kappa & \lambda & \mu \end{bmatrix}^* U^{(A)}_{\kappa\pi} U^{(B)}_{\lambda\rho} U^{(C)}_{\mu\nu} = c \begin{bmatrix} A & B & C \\ \pi & \rho & \nu \end{bmatrix}. \tag{A.37}$$

From the unitarity of U matrices, one can show that the absolute values of c is 1. If A, B, and C are all different from the identity representation, one can multiply a common phase factor to 3j-symbols so that c becomes 1 in Eq. (A.37), since the condition Eq. (A.21) remains valid for such a common phase factor. When one of the representations, say C, is the identity representation, Eq. (A.37) reduces to

$$\sum_{\kappa\lambda} \begin{bmatrix} A & A & O \\ \kappa & \lambda & O \end{bmatrix}^* U^{(A)}_{\kappa'} U^{(A)}_{\lambda\rho} U^{(O)}_{OO} = c \begin{bmatrix} A & A & O \\ \pi & \rho & O \end{bmatrix}. \tag{A.38}$$

By using Eqs. (A.9) and (A.23), the 3j-symbols in Eq. (A.38) can be expressed as

$$\begin{bmatrix} A & A & O \\ \kappa & \lambda & O \end{bmatrix} = (-1)^{2A} [A]^{-1/2} (OO \mid OOOO) U^{(A)}_{\kappa\lambda}.$$

Substituting this into Eq. (A.38), we have

$$U^{(A)}_{\pi\rho} = c (OO \mid OOOO) U^{(A)}_{\pi\rho}.$$

Natural choice for $(OO \mid OOOO)$ is of course 1, so c should be 1 in this case as well. Thus we can impose the condition $c = 1$ in all cases and we have

$$\sum_{\kappa\lambda\mu} \begin{bmatrix} A & B & C \\ \kappa & \lambda & \mu \end{bmatrix}^* U^{(A)}_{\kappa\pi} U^{(B)}_{\lambda\rho} U^{(C)}_{\mu\nu} = \begin{bmatrix} A & B & C \\ \pi & \rho & \nu \end{bmatrix} \tag{A.39}$$

From the unitarity of U matrices Eq. (A.39) can also be expressed as

$$\begin{bmatrix} A & B & C \\ \kappa & \lambda & \mu \end{bmatrix}^* = \sum_{\pi\rho\nu} U^{(A)*}_{\kappa\pi} U^{(B)*}_{\lambda\rho} U^{(C)*}_{\mu\nu} \begin{bmatrix} A & B & C \\ \pi & \rho & \nu \end{bmatrix}. \tag{A.40}$$

A.4 Six-j Symbol and Nine-j Symbol

The recoupling coefficient

$$((AB)^{(D)}C\,|\,A(BC)^{(G)})^{(F)}$$

can be evaluated by comparing typical row vectors of the unitary transformation matrices for both the coupling schemes. The matrix elements of the νth row of the representation F and (κ, λ, μ)th column of $A \times B \times C$ are given by

$$\sum_{\sigma}(F\nu\,|\,D\sigma C\mu)(D\sigma\,|\,A\kappa B\lambda)$$

and

$$\sum_{\rho}(F\nu\,|\,A\kappa G\rho)(G\rho\,|\,B\lambda C\mu)$$

for the left and right sides of the coupling schemes.

Since both are unitary, the recoupling coefficient is expressed as

$$
\begin{aligned}
&((AB)^{(D)}C\,|\,A(BC)^{(G)})^{(F)} \\
&= \sum_{\nu}[F]^{-1}\sum_{\kappa\lambda\mu\rho\sigma}(F\nu\,|\,D\sigma C\mu)^*(D\sigma\,|\,A\kappa B\lambda)^*(F\nu\,|\,A\kappa G\rho)(G\rho\,|\,B\lambda C\mu) \\
&= (-1)^{2B+2G}\sqrt{[D][G]}\sum_{\kappa\lambda\mu\rho\sigma\nu\tau\upsilon\phi\chi}U^{(F)}_{\nu\tau}U^{(D)}_{\sigma\upsilon}U^{(F)*}_{\nu\phi}U^{(G)*}_{\rho\chi} \\
&\quad\times\begin{bmatrix}D & C & F\\ \sigma & \mu & \tau\end{bmatrix}^*\begin{bmatrix}A & B & D\\ \kappa & \lambda & \upsilon\end{bmatrix}^*\begin{bmatrix}A & G & F\\ \kappa & \rho & \phi\end{bmatrix}\begin{bmatrix}B & C & G\\ \lambda & \mu & \chi\end{bmatrix} \\
&= (-1)^{2B+2G}\sqrt{[D][G]}\sum_{\kappa\lambda\mu\rho\sigma\tau\upsilon\chi\alpha\beta\gamma\delta\varepsilon\zeta}U^{(D)}{}_{\sigma\upsilon}U^{(G)}{}_{\rho\chi^*} \\
&\quad\times U^{(D)*}_{\sigma\alpha}U^{(C)*}_{\mu\beta}U^{(F)*}_{\tau\gamma}U^{(A)*}_{\kappa\delta}U^{(B)*}_{\lambda\varepsilon}U^{(D)*}_{\nu\zeta} \\
&\quad\times\begin{bmatrix}D & C & F\\ \alpha & \beta & \gamma\end{bmatrix}^*\begin{bmatrix}A & B & D\\ \delta & \varepsilon & \zeta\end{bmatrix}^*\begin{bmatrix}A & G & F\\ \kappa & \rho & \tau\end{bmatrix}\begin{bmatrix}B & C & G\\ \lambda & \mu & \chi\end{bmatrix} \\
&= (-1)^{2B+2G}\sqrt{[D][G]} \\
&\quad\times\sum_{\kappa\lambda\mu\rho\tau\upsilon\chi\beta\gamma\delta\varepsilon\zeta}U^{(G)*}_{\rho\chi}U^{(C)*}_{\mu\beta}U^{(F)*}_{\tau\gamma}U^{(A)*}_{\kappa\delta}U^{(B)*}_{\lambda\varepsilon}U^{(D)*}_{\upsilon\zeta} \\
&\quad\times\begin{bmatrix}D & C & F\\ \upsilon & \beta & \gamma\end{bmatrix}^*\begin{bmatrix}A & B & D\\ \delta & \varepsilon & \zeta\end{bmatrix}^*\begin{bmatrix}A & G & F\\ \kappa & \rho & \tau\end{bmatrix}\begin{bmatrix}B & C & G\\ \lambda & \mu & \chi\end{bmatrix} \\
&= (-1)^{2B+2G+A+F+G+C+B+G+2B}\sqrt{[D][G]} \\
&\quad\times\sum_{\kappa\lambda\mu\rho\tau\upsilon\chi\beta\gamma\delta\varepsilon\zeta}U^{(A)*}_{\kappa\delta}U^{(B)*}_{\varepsilon\lambda}U^{(D)*}_{\upsilon\zeta}U^{(C)*}_{\mu\beta}U^{(F)*}_{\tau\gamma}U^{(G)*}_{\rho\chi} \\
&\quad\times\begin{bmatrix}A & B & D\\ \delta & \varepsilon & \zeta\end{bmatrix}\begin{bmatrix}A & F & G\\ \kappa & \tau & \rho\end{bmatrix}\begin{bmatrix}C & B & G\\ \mu & \lambda & \chi\end{bmatrix}\begin{bmatrix}C & F & D\\ \beta & \gamma & \upsilon\end{bmatrix} \\
&= (-1)^{A+B+C+F}\sqrt{[D][G]}\begin{Bmatrix}A & B & D\\ C & F & G\end{Bmatrix},
\end{aligned}
\tag{A.41}
$$

where Eqs. (A.9), (A.21) and (A.23) are used for manipulation. It should also be noted that the average over ν is evaluated instead of a single term, since each term gives the same value.

The last symbol is called Racah's coefficient or Wigner's six-j symbol and defined by

$$
\begin{aligned}
\begin{Bmatrix}A & B & D\\ C & F & G\end{Bmatrix} = &\sum_{\kappa\lambda\mu\rho\tau\upsilon\chi\beta\gamma\delta\varepsilon\zeta}U^{(A)*}_{\kappa\delta}U^{(B)*}_{\varepsilon\lambda}U^{(D)*}_{\upsilon\zeta}U^{(C)*}_{\mu\beta}U^{(F)*}_{\tau\gamma}U^{(G)*}_{\rho\chi} \\
&\times\begin{bmatrix}A & B & D\\ \delta & \varepsilon & \zeta\end{bmatrix}\begin{bmatrix}A & F & G\\ \kappa & \tau & \rho\end{bmatrix}\begin{bmatrix}C & B & G\\ \mu & \lambda & \chi\end{bmatrix}\begin{bmatrix}C & F & D\\ \beta & \gamma & \upsilon\end{bmatrix}.
\end{aligned}
\tag{A.42}
$$

The 6j-symbols are all real.

$$
\begin{Bmatrix}A & B & D\\ C & F & G\end{Bmatrix}^* = \begin{Bmatrix}A & B & D\\ C & F & G\end{Bmatrix}
\tag{A.43}
$$

This can be proved by substituting Eq. (A.39) into Eq. (A.42) and using the symmetry property of U matrices (Eq. A.9).

The symmetry property of the 6j-symbol may also be deduced from the definition Eq. (A.42). It remains invariant for 1) any permutations of its three columns, and 2) exchange of rows on any *two* of its columns. For example, if the row elements of the first two columns are exchanged, we have

$$\begin{aligned}\begin{Bmatrix} C & F & D \\ A & B & G \end{Bmatrix} &= \sum_{\kappa\lambda\mu\rho\tau\upsilon\chi\beta\gamma\delta\varepsilon\zeta} U_{\kappa\delta}^{(C)*} U_{\varepsilon\lambda}^{(F)*} U_{\upsilon\zeta}^{(D)*} U_{\mu\beta}^{(A)*} U_{\tau\gamma}^{(B)*} U_{\rho\chi}^{(G)*} \\ &\quad \times \begin{bmatrix} C & F & D \\ \delta & \varepsilon & \zeta \end{bmatrix} \begin{bmatrix} C & B & G \\ \kappa & \tau & \rho \end{bmatrix} \begin{bmatrix} A & F & G \\ \mu & \lambda & \chi \end{bmatrix} \begin{bmatrix} A & B & D \\ \beta & \gamma & \upsilon \end{bmatrix} \\ &= (-1)^{2B+2D+2F+2G} \\ &\quad \times \sum_{\kappa\lambda\mu\rho\tau\upsilon\chi\beta\gamma\delta\varepsilon\zeta} U_{\mu\beta}^{(A)*} U_{\gamma\tau}^{(B)*} U_{\zeta\upsilon}^{(D)*} U_{\kappa\delta}^{(C)*} U_{\lambda\varepsilon}^{(F)*} U_{\chi\rho}^{(G)*} \\ &\quad \times \begin{bmatrix} A & B & D \\ \beta & \gamma & \upsilon \end{bmatrix} \begin{bmatrix} A & F & G \\ \mu & \lambda & \chi \end{bmatrix} \begin{bmatrix} C & B & G \\ \kappa & \tau & \rho \end{bmatrix} \begin{bmatrix} C & F & D \\ \delta & \varepsilon & \zeta \end{bmatrix} \\ &= (-1)^{2B+2D+2F+2G} \begin{Bmatrix} A & B & D \\ C & F & G \end{Bmatrix} = \begin{Bmatrix} A & B & D \\ C & F & G \end{Bmatrix},\end{aligned}$$

where the last equality holds because of

$$1 = (-1)^{2A+2B+2D} = (-1)^{2A+2F+2G}.$$

Altogether, 24 permutations of its parameters leave the value of the 6j-symbol invariant. Examples are

$$\begin{Bmatrix} A & B & C \\ D & E & F \end{Bmatrix} = \begin{Bmatrix} D & E & C \\ A & B & F \end{Bmatrix} = \begin{Bmatrix} B & A & C \\ E & D & F \end{Bmatrix} = \begin{Bmatrix} B & C & A \\ E & F & D \end{Bmatrix}. \tag{A.44}$$

The symmetry of the 6j-symbol is that of a regular tetrahedron with A, B, ... as its edges and opposing edges in one column: the second, the third and the fourth equalities in the above formula correspond to a rotation about a 2-fold and a 3-fold axis and a reflection of a regular tetrhedron. Three edges of each triangle should satisfy the trinagular condition

$$\delta(A, B, C) = \delta(D, E, C) = \delta(A, E, F) = \delta(D, B, F) = 1$$

in order not to have a vanishing value.

When one of the representations is the identity representation, the 6j-symbol is expressed as

$$\begin{Bmatrix} A & B & C \\ D & E & O \end{Bmatrix} = \frac{(-1)^{A+B+C} \delta_{AE} \delta_{BD} \delta(A,B,C)}{\sqrt{[A][B]}}. \tag{A.45}$$

This can be seen by substituting F in Eq. (A.41) with the identity representation and from Eq. (A.29) as

$$((AB)^{(D)}C\,|\,A(BC)^{(G)})^{(0)} = (-1)^{A+B+C}\sqrt{[D][G]}\begin{Bmatrix} A & B & D \\ C & 0 & G \end{Bmatrix}$$
$$= \delta_{CD}\delta_{AG}((AB)^{(C)}C\,|\,A(BC)^{(A)})^{(0)} = 1.$$

Important sum rules are

$$\sum_C [C]\begin{Bmatrix} A & B & C \\ D & E & F \end{Bmatrix}\begin{Bmatrix} A & B & C \\ D & E & G \end{Bmatrix} = \frac{\delta_{FG}\delta(A,E,F)\delta(D,B,F)}{[F]} \tag{A.46}$$

and

$$\sum_C (-1)^{C+F+G}[C]\begin{Bmatrix} A & B & C \\ D & E & F \end{Bmatrix}\begin{Bmatrix} A & B & C \\ E & D & G \end{Bmatrix} = \begin{Bmatrix} A & E & F \\ B & D & G \end{Bmatrix}. \tag{A.47}$$

The equation (A.46) can be proved by considering the following recoupling coefficient and the realness of 6j-symbols.

$$((AE)^{(F)}D\,|\,(AE)^{(G)}D)^{(B)} = \delta_{FG}\delta(A,E,F)\delta(D,B,F)$$
$$= \sum_C ((AE)^{(F)}D\,|\,A(ED)^{(C)})^{(B)}(A(ED)^{(C)}\,|\,(AE)^{(G)}D)^{(B)}$$
$$= \sum_C [C]\sqrt{[F][G]}\begin{Bmatrix} A & B & C \\ D & E & F \end{Bmatrix}\begin{Bmatrix} A & B & C \\ D & E & G \end{Bmatrix}^{*}.$$

Similar relations on the recoupling coefficient lead to Eq. (A.47).

$$((EA)^{(F)}D\,|\,E(AD)^{(G)})^{(B)} = (-1)^{E+A+D+B}\sqrt{[F][G]}\begin{Bmatrix} E & A & F \\ D & B & G \end{Bmatrix}$$
$$= \sum_C (EA\,|\,AE)^{(F)}((AE)^{(F)}D\,|\,A(ED)^{(C)})^{(B)}(ED\,|\,DE)^{(C)}$$
$$\times (A(DE)^{(C)}\,|\,(AD)^{(G)}E)^{(B)}(GE\,|\,EG)^{(B)}$$
$$= \sum_C (-1)^{E+A-F+E+D-C+G+E-B+2(A+E+D+B)}[C]\sqrt{[F][G]}$$
$$\times \begin{Bmatrix} A & E & F \\ D & B & C \end{Bmatrix}\begin{Bmatrix} A & D & G \\ E & B & C \end{Bmatrix}^{*}$$

Note that the recoupling coefficient $(AB\,|\,BA)^{(C)}$ is expressed as

$$(AB\,|\,BA)^{(C)} = (-1)^{A+B-C}$$

from Eq. (A.24), and the following equality

$$(-1)^{E+A-F+E+D-C+G+E-B+2(A+E+D+B)-(E+A+D+B)}$$
$$=(-1)^{4E+2A-F+2D-C+G}=(-1)^{2A+2D+2G-F-C-G}=(-1)^{F+C+G}$$

Consider a recoupling coefficient

$$((AB)^{(E)}(CD)^{(F)}\,|\,(AC)^{(H)}(BD)^{(J)})^{(G)}.$$

It is expressed as

$$\begin{aligned}((AB)^{(E)}(CD)^{(F)}\,|\,(AC)^{(H)}(BD)^{(J)})^{(G)} &= \sum_K ((AB)^{(E)}F\,|\,(A(BF)^{(K)})^{(G)}(CD\,|\,DC)^{(F)} \\ &\times (B(DC)^{(F)}\,|\,(BD)^{(J)}C)^{(K)}(JC\,|\,CJ)^{(K)}(A(CJ)^{(K)}\,|\,(AC)^{(H)}J)^{(G)} \\ &= \sum_K (-1)^{A+B+F+G+C+D-F-(B+D+C+K)+J+C-K-(A+C+J+G)} \\ &\times \sqrt{[E][K][F][J][K][H]}\begin{Bmatrix} A & B & E \\ F & G & K\end{Bmatrix}\begin{Bmatrix} B & D & J \\ C & K & F\end{Bmatrix}^{*}\begin{Bmatrix} A & C & H \\ J & G & K\end{Bmatrix} \\ &= \sqrt{[E][F][J][H]}\begin{Bmatrix} A & B & E \\ C & D & F \\ H & J & G\end{Bmatrix},\end{aligned} \tag{A.48}$$

Where the last symbol with nine parameters of irreducible representation is called 9j-symbol which is expressed by 6j-symbols as

$$\begin{Bmatrix} A & B & E \\ C & D & F \\ H & J & G\end{Bmatrix} = \sum_K (-1)^{2K}[K]\begin{Bmatrix} A & B & E \\ F & G & K\end{Bmatrix}\begin{Bmatrix} B & D & J \\ C & K & F\end{Bmatrix}\begin{Bmatrix} A & C & H \\ J & G & K\end{Bmatrix}. \tag{A.49}$$

An interchange of any two rows or columns of parameters of 9j-symbols multiplies it by a factor $(-1)^{A+B+C+D+E+F+G+H+J}$, and transposition of its 9 parameters along diagonal leaves the value invariant.

$$\begin{Bmatrix} A & B & E \\ C & D & F \\ H & J & G\end{Bmatrix} = (-1)^{A+B+C+D+E+F+G+H+J}\begin{Bmatrix} C & D & F \\ A & B & E \\ H & J & G\end{Bmatrix} = \begin{Bmatrix} A & C & H \\ B & D & J \\ E & F & G\end{Bmatrix} \tag{A.50}$$

Note that three representations in any one row or column must satisfy the triangular condition in order to have non-vanishing value.

When one of the parameters is the identity representation, it reduces to a 6j-symbol because

$$\begin{aligned}\sqrt{[E][F][J][H]}\begin{Bmatrix} A & B & E \\ C & D & F \\ H & J & O\end{Bmatrix} &= ((AB)^{(E)}(CD)^{(F)}\,|\,(AC)^{(H)}(BD)^{(J)})^{(O)} \\ &= \delta_{EF}\delta_{HJ}(AB\,|\,BA)^{(E)}(A(CD)^{(E)}\,|\,(AC)^{(H)}D)^{(B)}(BH\,|\,HB)^{(D)} \\ &= \delta_{EF}\delta_{HJ}(-1)^{-(A+B-E)+A+C+D+B+B+H-D}\sqrt{[E][H]}\begin{Bmatrix} A & C & H \\ D & B & E\end{Bmatrix}^{*},\end{aligned}$$

or a renamed equivalent expression

$$\begin{Bmatrix} A & B & C \\ D & E & F \\ G & H & O \end{Bmatrix} = \frac{(-1)^{B+C+D+G}\delta_{CF}\delta_{GH}}{\sqrt{[C][G]}} \begin{Bmatrix} A & B & C \\ E & D & G \end{Bmatrix} \tag{A.51}$$

holds.

Appendix 4B.

Rotation Group

Since the Lie algebra O(3) and SU(2) are isomorphic, one can employ the same irreducible representations for them. Inversion symmetry for O(3) is known as parity, but this symmetry can be accounted for simply by counting the number of electrons in $p, f, h, \ldots$ orbitals. Therefore we limit the group O(3) only for proper rotations. In this section, explicit formulas of the standard representation matrices, 3j- and 6j-symbols will be presented. Derivation of these formulas may be found in Ref. 2.

A rotation R of the coordinate axes may be specified by the Euler angles (ψ, θ, ϕ). An irreducible representation is labeled by the angular momentum ℓ. The standard representation matrix is expressed as

$$D^{(\ell)}(\psi, \theta, \phi) = D_z^{(\ell)}(\psi) D_y^{(\ell)}(\theta) D_z^{(\ell)}(\phi),$$

and its dimension is given by

$$[\ell] = \sqrt{2\ell + 1}$$

where D_z and D_y are the transformation matrices of the corresponding Euler angles about the z and y axes. They are given by

$$D_z^{(\ell)}(\zeta)_{m\mu} = e^{i\zeta m}\delta_{m\mu}$$

and

$$D_y^{(\ell)}(\theta)_{m\mu} = \sum_r (-1)^r \frac{\sqrt{(\ell + m)!(\ell - m)!(\ell + \mu)!(\ell - \mu)!}}{(\ell - m - r)!(\ell + \mu - r)!r!(r + m - \mu)!} (\cos\frac{1}{2}\theta)^{2\ell - m + \mu - 2r} (\sin\frac{1}{2}\theta)^{2r + m - \mu} \qquad (B.1)$$

The summation over r extends only for those integers for which the factorials in the denominator are defined. The 3j-symbol is given by the following formula.

$$\begin{bmatrix} a & b & c \\ \alpha & \beta & \gamma \end{bmatrix} = \delta_{\alpha+\beta+\gamma,\,0}\Delta(a, b, c)\sum_q(-1)^{q+2b+c+\gamma}$$
$$\times \frac{\sqrt{(a+\alpha)!(a-\alpha)!(b+\beta)!(b-\beta)!(c+\gamma)!(c-\gamma)!}}{q!(a+b-c-q)!(a-\alpha-q)!(c-b+\alpha+q)!(b+\beta-q)!(c-a-\beta+q)!} \tag{B.2}$$

where a, b, ... and α, β, ... are the angular momenta and their corresponding z-components, and $\Delta(a, b, c)$ is given by

$$\Delta(a, b, c) = \delta(a, b, c)\sqrt{(a+b-c)!(b+c-a)!(c+a-b)!/(a+b+c+1)!} \tag{B.3}$$

The summation over q extends only for those values for which factorials are defined, and $\delta(a, b, c)$ is the number of the occurrence of the identity representation in the direct product representation $a \times b \times c$. For the rotation group, it takes the value of 1 when the angular momenta a, b and c satisfy the triangular condition and zero otherwise.

By using Eqs. (B.2), (A.9) and (A.23), the U matrix is expressed as

$$U^{(\ell)}_{m\mu} = (-1)^{2\ell}\sqrt{[\ell]}\begin{bmatrix} \ell & \ell & O \\ m & \mu & O \end{bmatrix} = (-1)^{\ell-m}\delta_{m,\,-\mu}. \tag{B.4}$$

The 6j-symbol is given by the following formula.

$$\begin{Bmatrix} a & b & c \\ d & e & f \end{Bmatrix} = \Delta(a, b, c)\Delta(d, e, c)\Delta(b, d, f)\Delta(a, e, f)\sum_q(-1)^q(q+1)!$$
$$\times \{(q-a-b-c)!(q-d-e-c)!(q-d-b-f)!(q-a-e-f)!$$
$$\times (a+b+d+e-q)!(a+c+d+f-q)!(b+c+e+f-q)!\}^{-1} \tag{B.5}$$

The 9j-symbol can be calculated by using Eq. (A.49) from 6j-symbols Eq. (B.5).

References

1. B.H. Flowers and S. Szpikowski, Proc. Roy. Soc. (London), **A222**, 84 (1964).
2. U. Fano and G. Racah, "Irreducible Tensorial Sets," Academic Press Inc., New York (1959).
3. J.S. Griffith, "The Irreducible Tensor Method for Symmetry Groups," Prentice-Hall Inc., New Jersey (1962).
4. B.R. Judd, " Second Quantization and Atomic Spectroscopy," The Johns Hopkins Press, Maryland (1967).
5. J. Paldus, J. Chem. Phys., **61**, 5321 (1974).
6. J. Paldus, Intern. J. Quantum Chem., **S9**, 165 (1975).
7. G. Racah, Phys. Rev., **62**, 438 (1942).
8. G. Racah, Phys. Rev., **63**, 367 (1943).
9. G. Racah, Phys. Rev., **76**, 1352 (1949).
10. F. Sasaki, Intern. J. Quantum Chem., **8**, 605 (1974).
11. I. Shavitt, Intern. J. Quantum Chem., **S11**, 131 (1977).
12. I. Shavitt, Intern. J. Quantum Chem., **S12**, 5 (1978).
13. S. Sugano, Y. Tanabe and H. Kamimura, "Multiplets of Transition-Metal Ions in Crystals," Academic Press, New York (1970).
14. Y. Tanabe and S. Sugano, J. Phys. Soc. Japan, **9**, 753 (1954).
15. E.P. Wigner, Am. J. Math., **63**, 57 (1941).
16. E.P. Wigner, "Group Theory and its Application to the Quantum Mechanics of Atomic Spectra," Academic Press, New York (1959).
17. E.P. Wigner, "On the Matrices which reduce the Kronecker Products of Representations of S.R. Groups," in *Quantum Theory of Angular Momentum,* L.C. Biedenharn and H. van Dam, Eds., Academic Press, New York (1965).
18. E.U. Condon and G.H. Shortley, "The Theory of Atomic Spectra," Cambridge Univ. Press, Cambridge (1963).
19. I.M. Gel′fand and M.L. Tsetlin, Dokl. Akad. Nauk. SSSR, **71**, 825, 1070, (1950).

Chapter 5.

HYCOIN: Hylleraas Configuration Interaction Method Using Gaussian Functions

D. Frye, A. Preiskorn, G. C. Lie, and E. Clementi

IBM Corporation
Scientific and Engineering Computations
Dept. 48B/428
Kingston, New York 12401, U.S.A.

"The underlying physical laws necessary for the mathematical theory of ... the whole of chemistry are thus completely known, and the difficulty is only that the exact application of these laws leads to equations much too complicated to be soluble."

— Paul A. M. Dirac (1929)

1. Introduction

The fundamental difficulty in accurately describing atomic and molecular structures is that the time-independent Schrödinger equation,

$$H\Psi = E\Psi, \tag{1}$$

cannot be solved exactly, except for the trivial cases of the hydrogen atom and the hydrogen molecular ion H_2^+. Devising various approximations and struggling to improve them are thus the daily chores that every quantum chemist has to face.

The simplest description of any multi-electron system generally starts with the independent-particle model to describe the wavefunction (for a full explanation see Chapters 2 and 6). In the so-called Hartree-Fock (HF) approximation,[1] the wavefunction Ψ is written as a determinantal product of the individual orbitals to which the electrons in a given system are assigned. In molecular calculations, these orbitals are referred to as molecular orbitals (MOs) and are found independently in the average potential formed by the other electrons. Since MOs in this model have to be solved iteratively until self consistency is reached, the HF approximation is very often also referred to as the self-consistent field (SCF) method.[1]

The energy difference between the Hartree-Fock energy and the exact energy obtained from solving the Schrödinger equation is termed the *correlation* energy. Since the exact solution of the Schrödinger equation is not known, experimental non-relativistic energies or estimates are usually employed to determine the correlation energy. Although Hartree-Fock calculations usually give very good *total* energies (very often over 99.5% of the experimental energies), it is, unfortunately, in the correlation energies that one is invariably interested. A well known example is the binding energy of the F_2 molecule: The total energy calculated from the Hartree-Fock approximation is better than 99.6% of the experimental energy, but the molecule is predicted *not* to be bound! Since Hartree-Fock calculations are usually first carried out in solving Eq. (1), recovering the correlation energies is therefore the major problem of quantum chemistry calculations. In the following we will discuss some of the standard methods available and, in turn, our proposal for the calculation of correlation energies.

One of the earliest and still most widely employed methods used to recover correlation energies is the configuration interaction (CI) method (see Chapter 10 for details).[2] It is well known that the method of configuration interaction is an easy-to-use and powerful technique for introducing correlation into the wavefunction. CI normally introduces correlation by including into the wavefunction configuration state functions (CSFs) of the atom or molecule other than the one of interest. These CSFs, configurations with the proper spin and orbital symmetry, are usually constructed from orbitals obtained from the solutions of the HF equations. The complete wavefunction Ψ is then a linear combination of these configurations. For example, in calculating the ground state of a given molecule, excited configurations of the same symmetry can be included in the wavefunction. The ground configuration plus these excited configurations then together form an N-particle basis set for the wavefunction in an expanded Hilbert space. This process of including static configurations with electron densities that are different from that of the ground configuration approximates the effect of the dynamic correlation that the moving electrons actually see. The effect of the potential produced by the occupation of these configurations on the ground configuration is then included into the energy through the interaction among these states. In a matrix solution of Eq. (1), this would be seen in non-zero off-diagonal matrix elements between different configurations of the system. The inclusion of all possible states of a given symmetry of an atom or molecule as basis functions of the wavefunction allows the exact solution of the problem to the limit of the accuracy of the basis set. However, the convergence of Eq. (1) to the exact solution is very slow as the number of basis functions increases. Furthermore, for almost all types of calculations the necessary truncation in the CI expansion due to limitations in computer memory and speed is made with significant amounts of correlation not included in the expansion.

Extensions to this bare-bones CI approach have emerged which attempt to include correlation in other ways that complement the correlation included by the CI method. In one approach, called the multi-configuration Hartree-Fock (MCHF) method[3] or multi-configuration self-consistent field (MCSCF) method,[4] the orbitals that comprise the limited CI basis states are allowed to vary in a self-consistent fashion in a potential composed of all the configurations included. These new orbitals (often called *pseudo-orbitals*) then evolve into forms such that correlation contributions are felt from configurations that are not explicitly included in the CI expansion. This method therefore *implicitly* brings in further electronic correlation and has been shown to be very successful in calculating atomic and molecular structures. A special version of the MCSCF method is the Complete Active Space, SCF, or CASSCF, where all the configuration state functions that can be generated from a given set of active MOs are included.[5] An even more powerful approach to the correlation problem is the combination of the MCSCF and the CI.[6,7] In this approach MCSCF calculations are first performed to obtain MOs, which are then used in the CI calculations.

All of the previously mentioned methods share a common feature, *i.e.*, the correlation is introduced *implicitly* into the wavefunctions. They thus suffer a common shortcoming of converging rather slowly to the true solution of Eq. (1). Instead of the implicit approach, one can argue that the most powerful approach should be the one where inter-electronic coordinates are built *explicitly* into the wavefunction. Hylleraas was the first to develop this approach and he used it to calculate the energy and wavefunction of the helium atom with great success.[8] A number of different techniques expanding on this idea for atoms were later tried.[9-16] James and Coolidge[17] extended this method to the hydrogen molecule and Kolos and Wolniewicz[18] have shown that, to date, it is the most efficient method available for calculating accurate potential energy curves in the hydrogen molecule. The results of the latter authors even challenged the experimental spectroscopic data. The verification of their prediction was hailed as "a great triumph of *ab initio* calculations" by Nobel laureate R. S. Mulliken.[19] However, due to the numerical complications introduced by the inclusion of inter-electronic coordinates in the wavefunctions, Hylleraas' approach was limited to simple cases such as the helium atom and the linear hydrogen molecule.

With the advent of supercomputers, a more general extension of the principle of explicit inclusion of the inter-electronic separation into the wavefunction started to flourish.[20,21] The method has become known as the Hylleraas-CI (HCI) method in the literature. In these calculations, Gaussian lobe functions are employed as the basis functions due to the speed with which the greatly increased number of electron integrals (in comparison with conventional CI) can be evaluated. However, use of Gaussian lobe functions is known to present some disadvantages, such as serious round-off errors and lack of rotational invariance,[22] and may suffer slower convergence compared with other type of functions. This led us to explore the use of cartesian gaussian functions in the HCI calculations. The required 2-, 3-, and 4-electron integral formulae have all been derived,[23,24] and a program called HyCoIn (*Hy*lleraas-*Co*nfiguration-*In*teraction) that performs HCI calculations for 2-electron molecular systems has been written and documented.[25]

Using HyCoIn, it has been shown that the HCI method employing general cartesian gaussian functions as basis sets for the MO expansions can work successfully in calculating the potential surfaces of the ground and excited states of molecules.[26-32] Section 2 of this chapter describes the general theoretical HCI framework from the calculation of the atomic integrals through the solution of the HCI equivalent of Eq. (1). Section 3 gives a brief outline of the applications possible with this theory and gives some examples of recent HCI calculations. The explicit form that the integrals take when using cartesian gaussians are given in the Appendices. Also included in the Appendices are the explicit forms that integrals take for the three- and four-electron cases. The theory for this more general use of the HCI method has not been included in the description in section 2, but all the information in this section can easily be extended to the multi-electron problem. Furthermore, Appendix 5E contains the integral equations necessary for a possible future expansion of HYCOIN to include correlation in an exponential form rather than as a linear algebraic factor.

2. Theory

The standard CI approach consists of expanding the many-electron wavefunction $\Psi(1,2, \ldots N)$ in a linear combination of configuration state functions (CSFs) Φ_k such that

$$\Psi_{CI}(1,2, \ldots N) = \sum_{k=1}^{NCSF} C_k \Phi_k(1,2, \ldots N), \tag{2}$$

where Φ_k is a linear combination of totally anti-symmetric determinantal states each consisting of N independent molecular orbital functions φ_i and NCSF is the total number of CSFs included. The HCI approach modifies Eq. (2) by expanding the wavefunction in powers of the inter-electronic distances r_{ij} weighted by the original CSFs Φ_k, *i.e.*

$$\Psi_{HCI}(1,2, \ldots N) = \sum_{\nu} \sum_{i<j=1}^{N} \sum_{k=1}^{NCSF} C_{\nu,k}\, r_{ij}^{\nu}\, \Phi_k(1,2, \ldots N). \tag{3}$$

The effect of the r_{ij} terms is to explicitly introduce correlation into the wavefunction. The simple interpretation for devising this construct is that now we have terms in the wavefunction that *look* like they take correlation into account. This expansion in the inter-electronic distance is generally assumed to be a power series expansion,[33] thereby limiting ν to non-negative integers, so that the terms corresponding to large r_{ij} (therefore lessening electron repulsion and decreasing the energy) are given the most weight. It has been shown by others[34,35] that of the possible non-zero ν values, $\nu = 1$ is the most important and our discussion will be limited to the $\nu = 0$ (normal CI) and $\nu = 1$ (what we term HCI) terms. It has also been shown repeatedly that inclusion of the HCI terms does speed the convergence of configuration expansions and that the HCI method is a viable alternative when high-accuracy calculations are needed. In this discussion, we will restrict ourselves to the two-electron problem so that Eq. (3) becomes

$$\Psi_{HCI} = \sum_{k=1}^{NCSF} C_{0,k}\Phi_k + r_{12}\sum_{k=1}^{NCSF} C_{1,k}\Phi_k. \tag{4}$$

The addition of the explicit correlation term in the wavefunction given in Eq. (4) means that our HCI wavefunction is no longer simply an expansion in orthonormal basis functions, therefore the equivalent form of Eq. (1) becomes, in matrix notation,

$$(\tilde{H} - E\tilde{S})C = 0, \tag{5}$$

where $\tilde{S}$ is the HCI overlap matrix and C represents the set of vectors that function as the weight coefficients in the HCI expansion. Since we have restricted ourselves to the two-electron case we can write $\tilde{H}$ and $\tilde{S}$ in Eq. (5) as

$$\tilde{H} = \begin{pmatrix} \tilde{H}^{00} & \tilde{H}^{01} \\ \tilde{H}^{10} & \tilde{H}^{11} \end{pmatrix}, \tag{6}$$

where

$$H_{KL}^{\nu_1\nu_2} = \iint d\tau_1 d\tau_2 \Phi_K(1,2)\, r_{12}^{\nu_1} H r_{12}^{\nu_2}\, \Phi_L(1,2), \tag{7}$$

and

$$\tilde{S} = \begin{pmatrix} \tilde{S}^{00} & \tilde{S}^{01} \\ \tilde{S}^{10} & \tilde{S}^{11} \end{pmatrix}, \tag{8}$$

where

$$S_{KL}^{\nu_1\nu_2} = \iint d\tau_1 d\tau_2 \Phi_K(1,2)\, r_{12}^{\nu_1+\nu_2}\, \Phi_L(1,2). \tag{9}$$

For our purposes, we have assumed that the $\Phi_K(1,2)$ form an orthonormal set so that the submatrix $\tilde{S}^{00}$ will be the unit matrix. Furthermore, inspection of Eqs. (7) and (9) will show that $\tilde{H}$ and $\tilde{S}$ are symmetric matrices. Additional inspection reveals that the submatrices $\tilde{S}^{10}$, $\tilde{S}^{11}$, $\tilde{H}^{00}$, and $\tilde{H}^{11}$ are themselves symmetric. Unfortunately, the $\tilde{H}^{10}$ submatrix is not symmetric due to the commutation properties of r_{i_j} and the kinetic operator in H. However, the symmetric properties of the matrices taken as a whole greatly reduce the computational scope of the problem.

To evaluate Eqs. (7) and (9) we need to define the actual form of the wavefunctions. The *molecular* orbital functions φ_i, mentioned in the introduction as composing the Φ_Ks, are assumed here to be made up of linear combinations of primitive cartesian gaussians, η_{aA}s,

$$\varphi_i(r) = \sum_{aA} c_{i,a}\, \eta_{aA}(r_A)\,, \tag{10}$$

where A is the origin about which the gaussian function η_{aA} is defined and a runs through the number of basis functions centered at A. The unnormalized gaussian basis functions are written

$$\eta_{aA}(r_A) = x_A^{l_a} y_A^{m_a} z_A^{n_a} \exp(-\alpha_{aA}\, r_A^2), \tag{11}$$

where α_{aA} is a positive real number, the l_a, m_a, n_a are non-negative integers with $j = l_a + m_a + n_a$, and r_A represents the distance to the center located at A. It is customary to label the gaussians according to the index j. All gaussians where j is equal to zero are called s, equal to one are p, *etc.*

The matrix elements of $\tilde{H}$ and $\tilde{S}$ can be broken naturally into 3 general groups when using gaussian orbitals, these are:

a) the one-electron integrals

$$S_1 = \int d\tau_1 \eta_{aA}(1) \eta_{bB}(1), \tag{12}$$

$$K_1 = \int d\tau_1 \eta_{aA}(1) \left(-\frac{1}{2} \nabla_1^2 \right) \eta_{bB}(1), \tag{13}$$

$$N_1 = \sum_V \int d\tau_1 \eta_{aA}(1) \left(-\frac{Z_V}{r_{1V}} \right) \eta_{bB}(1), \tag{14}$$

b) the two-electron repulsion integral

$$E_2 = \iint d\tau_1 d\tau_2 \eta_{aA}(1) \eta_{cC}(2) (r_{12}^{-1}) \eta_{bB}(1) \eta_{dD}(2), \tag{15}$$

and c) the two-electron HCI integrals

$$S_2^{\nu} = \iint d\tau_1 d\tau_2 \eta_{aA}(1) \eta_{cC}(2) r_{12}^{\nu} \eta_{bB}(1) \eta_{dD}(2), \tag{16}$$

$$K_2^{\nu_1\nu_2} = \iint d\tau_1 d\tau_2 \eta_{aA}(1) \eta_{cC}(2) r_{12}^{\nu_1} \left(-\frac{1}{2} \nabla_1^2 - \frac{1}{2} \nabla_2^2 \right) r_{12}^{\nu_2} \eta_{bB}(1) \eta_{dD}(2), \tag{17}$$

$$N_2^v = \sum_V \iint d\tau_1 d\tau_2 \eta_{aA}(1)\eta_{cC}(2) r_{12}^v \left(-\frac{Z_V}{r_{1V}} - \frac{Z_V}{r_{2V}} \right) \eta_{bB}(1)\eta_{dD}(2). \tag{18}$$

The sum over V in Eqs. (14) and (18) represents the sum over all the nuclear centers (a restriction to the two-electron case does not restrict the number of nuclear centers). The expanded versions of Eqs. (16) to (18) which make use of the specific properties of the cartesian gaussians are reproduced in Appendix 5A. The full expanded versions of the other integrals can be seen in Chapter 7. Each of the integrals in Eqs. (12) to (18) presents different computational problems, but only the N_2^1 integrals must be approximated by a numerical integration.

The next stage of the general problem is the transformation from atomic orbitals to molecular orbitals. Since this procedure is standard and is treated extensively in many works, it will not be repeated here.

The final theoretical step in the calculation is the actual solution of the HCI secular equation as given by Eq. (5). This is accomplished in two stages. The first step is to diagonalize the overlap matrix $\tilde{S}$ to obtain $\tilde{U}$ which transforms $\tilde{S}$ to the diagonal representation D. Then the new matrix

$$\tilde{H}' = \tilde{D}^{-1/2}\tilde{U}^{-1}\tilde{H}\tilde{U}\tilde{D}^{-1/2} \tag{19}$$

is diagonalized to obtain the HCI eigenvalues.

3. Applications and Specific Examples

The HCI theory described in this chapter is, at the moment, limited to two-electron, N-center molecular problems. However, within this group of problems there are no significant limitations. Virtually any ground or excited state of any two-electron system can be examined with this method. To date, this package has been applied to the ground ${}^1\Sigma_g^+$ potential curve in H_2, various levels of the ${}^3\Sigma_u^+$ potential curves in H_2, the ground and first excited bound potential surfaces of H_3^+, and (in an earlier version) the ground ${}^1\Sigma$ state of HeH^+. It has not yet been applied to states of non-Σ total symmetry, but in principle there is no difficulty in doing so. In each of these examples, significant improvements in energy over results from all forms of general variational calculation methods have been realized. In the case of the ground state of H_2, the method of Kolos *et al.*[36] seems still to converge faster, but that method is limited to diatomic molecules.

There are many interesting two-electron problems that can be studied. In the following we will discuss some of the results that have been obtained in the past two years in this laboratory.

Table I. The effects of the basis size in H_2 at an inter-nuclear distance of 1.4011 bohrs (total energies are given in hartrees). Note that all the basis sets, except the last one, are composed of gaussian functions whose exponents form a geometric series (see Refs. 28 and 29 and Chapters 3 and 7 for details).

Basis Set	SCF	Conventional-CI	Hylleraas-CI
(13*s*)	-1.128532	-1.154881	-1.168870
(13*s*7*p*)/[13*s*1*p*]	-1.133561	-1.170495	-1.174334
(13*s*7*p*)/[13*s*2*p*]	-1.133561	-1.171378	-1.174380
(13*s*7*p*1*d*)/[13*s*1*p*1*d*]	-1.133610	-1.171661	-1.174399
(13*s*7*p*1*d*)/[13*s*2*p*1*d*]	-1.133618	-1.172596	-1.174456
(13*s*7*p*1*d*)	-1.133618	-1.173306	-1.174467
(13*s*7*p*2*d*)	-1.133619	-1.173858	-1.174473
(14*s*8*p*3*d*)	-1.133621	-1.173894	-1.174474
(14*s*7*p*2*d*1*f*)	-1.133622	-1.173987	-1.174474
(15*s*7*p*2*d*1*f*)	-1.133622	-1.173987	-1.174475

We will first look at the effect of the basis set sizes on the HCI calculations for the ground state of the hydrogen molecule.[26,28,29] The results are summarized in Table I. The "exact" variational limit for this state was established by Kolos *et al.*,[36] who used a specialized elliptical coordinate basis set with an explicitly correlated wavefunction, at -1.1744757 hartrees. The largest basis set of our calculation, (15*s*7*p*2*d*1*f*) , produces an HCI energy that is less than 0.3 cm^{-1} above this variational limit. If spectroscopic accuracy is defined as having an error of less than ~1 cm^{-1}, Table I indicates that it can be achieved with a basis set of size as small as (13*s*7*p*2*d*).

Compared with conventional CI, the size of the basis set seems to have much less effect on the recovering of the correlation energy. For example, addition of the first contracted *p* -gaussian lowers the conventional CI energy by 0.016614 hartree, but brings down the HCI energy by only 0.005464 hartree. Similarly, one *d*-type function added to the (13*s*7*p*)/[13*s*1*p*] basis set has an effect of 0.00116 hartree in conventional CI, compared with 0.00065 hartree in HCI. All of these seem to be due to the fact that most of the correlation energy has already been picked up in the *s*-function basis set, 86% with HCI compared to 52% with conventional CI. Thus the HCI method should be much superior to conventional CI in terms of convergence to a given accuracy in energy. That is indeed the case. For example, the HCI energy obtained with (13*s*7*p*) contracted to [13*s*1*p*], -1.174334 hartree, is already much better than CI employing 5*s*3*p*3*d*2*f* *Slater* functions, -1.174142 hartree.[37] Addition of one *d*-function lowers our calculated energy to -1.174399 hartree, which is slightly better than the calculation employing uncontracted Gaussians with angular momenta up to *g*: (12*s*6*p*5*d*4*f*2*g*).[38]

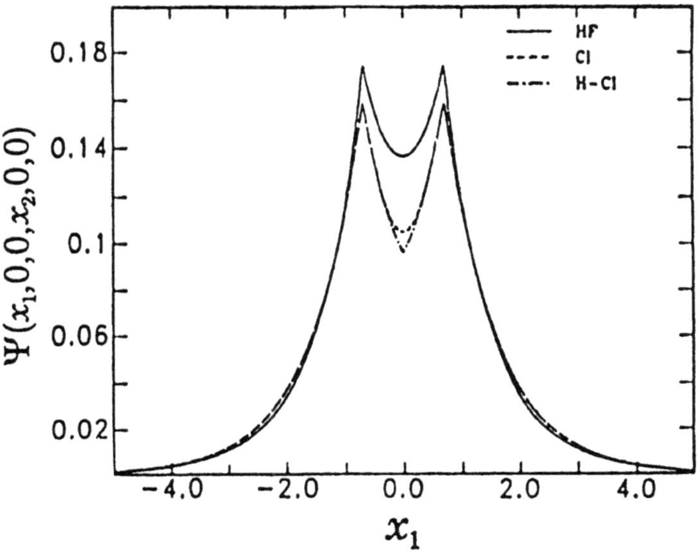

Figure 1. Comparison of the ground state wavefunctions, $\Psi(x_1, y_1, z_1, x_2, y_2, z_2)$, of the hydrogen molecule obtained from the Hartree-Fock (HF), the conventional CI, and the HCI calculations. The nuclei are located along the x-axis at -0.70055 and 0.70055 bohr, whereas one of the two electrons is located at the origin (*i.e.*, $x_2, y_2,$ and $z_2=0$).

One of the most interesting results we have obtained can be seen in the comparison of the HCI wavefunction with the conventional CI wavefunction, Fig. 1 shows one such comparison.[28] We see from the figure that although the CI wavefunction agrees almost everywhere with the HCI wavefunction, it lacks the inter-electronic cusp as r_{12} goes to zero. Thus the reason that the CI energy is higher than the HCI energy by $\sim$100 cm^{-1} seems to be this missing inter-electronic cusp. Since it is difficult to obtain this type of cusp in the wavefunction without explicit correlation, we can see that there is probably an effective limit to how much correlation can be recovered by conventional CI, no matter how many basis functions are employed. It is difficult to estimate this limit, but judging from the fact that the (12*s*6*p*5*d*4*f*2*g*) calculation just mentioned still leaves more than 16 cm^{-1} unrecovered, one would be surprised if spectroscopic accuracy can be achieved by the conventional CI approach.

The HCI method described here has also been used to study the potential curves of the excited $b^3\Sigma_u^+$, $e^3\Sigma_u^+$, and $f^3\Sigma_u^+$ states of H_2.[27,30] The calculated energies at small internuclear distances for the repulsive $b^3\Sigma_u^+$ state are the best in the literature, even surpassing the calculations employing confocal elliptical functions. The same is true for the attractive *e* and *f* states. The calculated HCI curves for these states were superior to both the most extensive CI calculation, that of Borondo *et al.*,[39] and to a recent correlated wavefunction calculation using confocal elliptical coordinates, that of Rychlewski[40] on the *e* state only. For example, not only is our curve lower than that of Rychlewski everywhere for the e state, our curve seems to have a higher quality also. The potential barrier predicted by Rychlewski, 860 cm^{-1}, is definitely much too high compared with our value of 230 cm^{-1}. The potential curves were used to predict an array of vibrational levels and spectroscopic constants in H_2 and D_2, resulting in excellent agreement with experiment.[30]

We have also recently extended the HCI calculations to the study of the non-linear H_3^+ molecule, a problem that cannot be studied by the less general elliptical coordinate methods. We examined both the equilibrium energy and calculated a number of points on the potential energy surfaces of both the ground $^1A'_1$ state and first excited $^3\Sigma_u^+$ state of H_3^+.[31,32] The lowest energy obtained by us for the equilateral triangle geometry of the ground state is -1.3438279 hartree at the internuclear distance of 1.6500 bohrs with the use of a $13s5p3d$ basis set on each site (a total of 138 basis functions). This energy is compared in Table II with other recent *ab initio* calculations on the equilibrium energy. From Table II we can see that this energy is significantly lower than the previous best published variational calculation of -1.343500 hartree,[26] and is in excellent agreement with the results of the latest quantum Monte Carlo calculations, $-1.34387 \pm .00005$ hartree[48] and $-1.3433 \pm .0005$ hartree.[49] The energy is also below the soon-to-be published result of Alexander *et al.*,[50] who used a Random-Tempered Optimization method with Gaussian-Type Geminals (also a variational calculation with an explicitly correlated wavefunction), and obtained -1.3438220 hartree at an equilibrium separation of 1.6504 bohrs.

Furthermore, a smaller basis set of $13s3p1d$ orbitals at each site (84 total basis functions) has been used to calculate 70 points of the H_3^+ hypersurface. This new calculation is everywhere more than 170 cm^{-1} below the calculation that has been the standard to-date[43] and in some regions of the surface is more than 350 cm^{-1} below the earlier work. Meyer *et al.*[46] have recently completed a new potential surface for H_3^+ employing full CI with a $(10s4p2d)/[7s4p2d]$ Gaussian basis set. Although it is difficult to compare our surface directly with theirs due to the use of different grid points, it is found that our value at R_e is about 160 cm^{-1} lower than that obtained by Meyer *et al.*[46]

In addition to the ground state, the first excited bound state of H_3^+ has been examined. This symmetric linear ($D_{\infty h}$) state, which has $^3\Sigma_u^+$ symmetry, was found to have an electronic energy of -1.1161027 hartree with a basis set of $14s5p1d$ cartesian gaussians at each site (105 total functions). The potential surface was calculated at 18 points with a smaller $14s3p1d$ basis set. Both the equilibrium energy and the points on the potential surface were significantly lower in total energy than previous calculations.[51,52] The computed vibrational frequencies (in the harmonic oscillator approximation) of 1235 cm^{-1} (symmetric stretch), 823 cm^{-1} (asymmetric stretch), and 720 cm^{-1} (doubly degenerate bend), give the reaction

$$H_2^+(^2\Sigma_g^+) + H(^2\Sigma) = H_3^+(^3\Sigma_u^+) \tag{20}$$

Table II. Selected *ab initio* calculations for the ground state equilibrium energy of H_3^+.

Authors	Method	$R_{eq}(b)$	$E(h)$
Salmon, *et al.*[41] (1973)	CI, 18 SP	1.6500	−1.34335
Carney, *et al.*[42] (1976)	CI, 21 FSGO	1.6500	−1.33936
Dykstra, *et al.*[43] (1979)	CI, 63 CGTO	1.6504	−1.342784
Mentch, *et al.*[44] (1981)	Random Walk	1.6500	−1.3439 ± .0002
Preiskorn, *et al.*[21] (1984)	SCC, 24 CGLO	1.6504	−1.343422
Burton *et al.*[45] (1985)	CI, 81 PNO	1.6525	−1.34188
	CI, 108 PNO	1.6525	−1.34272
Meyer, *et al.*[46] (1986)	CI, 87 CGTO	1.6504	−1.34309
	CI, 104 CGTO	1.6504	−1.34340
Anderson[47] (1987)	Random Walk	1.6500	−1.34376 ± .00003
Traynor, *et al.*[48] (1988)	Random Walk	1.6500	−1.34387 ± .00005
Urdaneta, *et al.*[26] (1988)	HCI, 48 CGTO	1.6504	−1.343500
Huang, *et al.*[49] (1990)	Random Walk	1.6500	−1.3433 ± .0005
Alexander, *et al.*[50] (1990)	RTO, 700 GTG	1.6504	−1.3438220
Present results	HCI, 84 GTO	1.6499	−1.3437933
	HCI, 138 GTO	1.6500	−1.3438279

Note: SP = Singer Polynomial, FSGO = Floating Singer Gaussian Orbitals, CGTO = Contracted Gaussian Type Orbitals, CGLO = Contracted Gaussian Lobe Orbitals, PNO = Pseudo Natural Orbitals, MP = Many-Body Perturbation, RTO = Random-Tempered Optimization, GTG = Gaussian Type Geminal, GTO = Gaussian Type Orbitals.

a negative enthalpy of formation of $\Delta H_0^0 = -6.75$ kcal/mol including zero-point vibration energies. Therefore, this state is "bound," barely, within the harmonic oscillator approximation. Similar analysis is currently underway for the D_3^+ isotopic relative.

As mentioned previously, the present version of the HyCoIn program is restricted to the calculations of energies and wavefunctions of 2-electron molecular systems only. Work in progress includes extending the program to many-electron molecules and the calculation of electronic properties, such as dipole moments.

Appendix 5A.
Two-electron Integral Formulas

In this appendix, the formulas found by expanding Eqs. (16) to (18) and using the properties of the cartesian gaussian basis sets are given. Throughout this appendix we will follow the convention that a GTF labeled η_{aA} describes an orbital a centered at A with angular momentum quantum numbers l_a, m_a, and n_a and orbital exponent α_a. The electron number labels (1 & 2) of the GTFs are not written explicitly in the formulas but we will use the convention that GTFs a and b are always associated with electron 1 and GTFs c and d are similarly associated with electron 2. Furthermore, in order to avoid confusion between electron number and number of electrons whenever possible, we will associate electron 1 with unprimed indices and electron 2 with primed indices. We will retain the electron number explicitly only for the HCI term r_{12} itself, the related R_{12} operator and in the distance from the electron to the various nuclei, i.e., $r_{i\nu}$. Lastly, we will extensively use the Gaussian Product Theorem (GPT) as derived in Chapter 7, which states that the product of two gaussians can be written

$$\eta_{aA}\eta_{bB} = W \sum_{l=0}^{l_a+l_b} f_l(l_a, l_b, \overline{PA}_x, \overline{PB}_x) x_P^l \exp(-\gamma x_P^2)$$
$$\cdot \sum_{m=0}^{m_a+m_b} f_m(m_a, m_b, \overline{PA}_y, \overline{PB}_y) y_P^m \exp(-\gamma y_P^2) \tag{A.1}$$
$$\cdot \sum_{n=0}^{n_a+n_b} f_n(n_a, n_b, \overline{PA}_z, \overline{PB}_z) z_P^n \exp(-\gamma z_P^2),$$

where we use

$$\vec{P} = \frac{\alpha_a \vec{A} + \alpha_b \vec{B}}{\gamma}, \tag{A.2}$$

$$W = \exp(-\alpha_a \alpha_b \overline{AB}^2 / \gamma), \tag{A.3}$$

$$\gamma = \alpha_a + \alpha_b, \tag{A.4}$$

and where f (often called the *newton* function) is

$$f_l(l_a, l_b, A, B) = \sum_{j=\max(0,l-l_b)}^{\min(l,l_a)} \frac{l_a! l_b! A^{l_a-j} B^{l_b-l+j}}{j!(l_a-j)!(l-j)!(l_b-l+j)!}. \tag{A.5}$$

For the product of gaussians η_{cC} and η_{dD} we similarly get

$$\eta_{cC}\eta_{dD} = W' \sum_{l'=0}^{l_c+l_d} f_{l'}(l_c, l_d, \overline{QC}_x, \overline{QD}_x) x_Q^{l'} \exp(-\gamma' x_Q^2) \cdot \sum_{m'=0}^{m_c+m_d} f_{m'}(m_c, m_d, \overline{QC}_y, \overline{QD}_y) y_Q^{m'} \exp(-\gamma' y_Q^2) \cdot \sum_{n'=0}^{n_c+n_d} f_{n'}(n_c, n_d, \overline{QC}_z, \overline{QD}_z) z_Q^{n'} \exp(-\gamma' z_Q^2), \tag{A.6}$$

where we use

$$\vec{Q} = \frac{\alpha_c \vec{C} + \alpha_d \vec{D}}{\gamma'}, \tag{A.7}$$

$$W' = \exp(-\alpha_c \alpha_d \overline{CD}^2/\gamma'), \tag{A.8}$$

and

$$\gamma' = \alpha_c + \alpha_d. \tag{A.9}$$

Furthermore, we will extensively use the one-electron and two-electron integral formulas derived in Chapter 7. To wit, OV in our notation represents the product of the integral portions of I_x, I_y, and I_z, Eq. (7.28), found in the derivation of the one-electron overlap integral. N_1^2 represents the integral portion of the one-electron nuclear attraction integral in Eq. (7.45), and F_2 is the integral portion of the electron repulsion integral in Eq. (7.51).

1. The S Integral

Upon application of the Gaussian Product Theorem to Eq. (16), the general four-center integral is reduced to a two-center integral

$$S_2^\nu = WW' \sum_{l=0}^{l_a+l_b} f_l \sum_{m=0}^{m_a+m_b} f_m \sum_{n=0}^{n_a+n_b} f_n \sum_{l'=0}^{l_c+l_d} f_{l'} \sum_{m'=0}^{m_c+m_d} f_{m'} \sum_{n'=0}^{n_c+n_d} f_{n'} \cdot I_2^\nu, \tag{A.10}$$

where

$$I_2^v = \int\int x_P^l y_P^m z_P^n x_Q^{l'} y_Q^{m'} z_Q^{n'} \exp(-\gamma r_P^2 - \gamma' r_Q^2) r_{12}^v d\tau\, d\tau', \tag{A.11}$$

and to save space we have shortened $f_l(l_a, l_b, \overline{PA_x}, \overline{PA_y})$ to f_l. The solution of Eq. (A.11) is different, of course, for v = odd or v = even, but the solution is found in both cases by applying the R_{ij}^v operator detailed in Appendix 5B.

For the case of v = even, we can rewrite Eq. (A.11) as

$$I_2^v = R_{12}^v \int\int x_P^l y_P^m z_P^n x_Q^{l'} y_Q^{m'} z_Q^{n'} \exp(-\gamma r_P^2 - \gamma' r_Q^2) d\tau\, d\tau'. \tag{A.12}$$

This integral then separates into a linear combination of products of simple one-electron overlap integrals,

$$I_2^v = R_{12}^v \cdot OV \cdot OV'. \tag{A.13}$$

For the case of v = odd, use of the R_{12} operator leaves an inverse power of r_{12} or

$$I_2^v = \int\int x_P^l y_P^m z_P^n x_Q^{l'} y_Q^{m'} z_Q^{n'} \exp(-\gamma r_P^2 - \gamma' r_Q^2) \frac{r_{12}^{v+1}}{r_{12}} d\tau\, d\tau', \tag{A.14}$$

which becomes

$$I_2^v = R_{12}^{v+1} \int\int x_P^l y_P^m z_P^n x_Q^{l'} y_Q^{m'} z_Q^{n'} \exp(-\gamma r_P^2 - \gamma' r_Q^2) \frac{1}{r_{12}} d\tau\, d\tau' \tag{A.15}$$

or

$$I_2^v = R_{12}^{v+1} F_2(l, m, n;\ l', m', n'). \tag{A.16}$$

2. The K Integral

The K_2 integral can be split into cartesian components, such that

$$K_2^{v_1, v_2} = -\frac{1}{2}\left(K_{2x}^{v_1, v_2} + K_{2y}^{v_1, v_2} + K_{2z}^{v_1, v_2}\right), \tag{A.17}$$

where

$$K_{2x}^{v_1, v_2} = \int\int \eta_{aA}\eta_{cC} r_{12}^{v_1} \left(\frac{\partial^2}{\partial x^2} + \frac{\partial^2}{\partial x'^2}\right) r_{12}^{v_2} \eta_{bB}\eta_{dD}\, d\tau\, d\tau'. \tag{A.18}$$

After performing the partial differentiation and examining the individual terms, the components of K_2 can be expanded in terms of the S_2^ν type of integral. For example, the x component is

$$\begin{aligned} K_{2x}^{\nu_1,\nu_2} &= l_b(l_b-1)S_2^\nu(l_b-2) - 2\alpha_b(2l_b+1)S_2^\nu + 4\alpha_b^2 S_2^\nu(l_b+2) \\ &+ l_d(l_d-1)S_2^\nu(l_d-2) - 2\alpha_d(2l_d+1)S_2^\nu + 4\alpha_d^2 S_2^\nu(l_d+2) \\ &+ 2\nu_2\Big\{[l_b+\frac{1}{6}\nu_2+\frac{1}{6}]S_2^{\nu-2} + 2\alpha_b[S_2^{\nu-2}(l_b+1,l_d+1) - S_2^{\nu-2}(l_b+2)] \\ &- l_bS_2^{\nu-2}(l_b-1,l_d+1) + \overline{BD}_x[l_bS_2^{\nu-2}(l_b-1) - 2\alpha_bS_2^{\nu-2}(l_b+1)] \\ &+ [l_d+\frac{1}{6}\nu_2+\frac{1}{6}]S_2^{\nu-2} + 2\alpha_d[S_2^{\nu-2}(l_b+1,l_d+1) - S_2^{\nu-2}(l_d+2)] \\ &- l_dS_2^{\nu-2}(l_b+1,l_d-1) + \overline{DB}_x[l_dS_2^{\nu-2}(l_d-1) - 2\alpha_dS_2^{\nu-2}(l_d+1)]\Big\}, \end{aligned} \tag{A.19}$$

where $\nu = \nu_1 + \nu_2$ and the variations of the quantum numbers caused by the differentiation are denoted inside the parentheses. One should note that the $+ 1/6$ term is partly due to a combination of terms from each of the components that together greatly simplify the expression.

3. The N Integral

Applying the Gaussian Product Theorem as was done for S_2^ν, the N_2^ν integral becomes

$$\begin{aligned} N_2^\nu = -WW'\sum_V Z_V \sum_{l=0}^{l_a+l_b} f_l \sum_{m=0}^{m_a+m_b} f_m \sum_{n=0}^{n_a+n_b} f_n \\ \cdot \sum_{l'=0}^{l_c+l_d} f_{l'} \sum_{m'=0}^{m_c+m_d} f_{m'} \sum_{n'=0}^{n_c+n_d} f_{n'} \cdot (J_{21}^\nu + J_{22}^\nu), \end{aligned} \tag{A.20}$$

where

$$J_{21}^\nu = \iint x_P^l y_P^m z_P^n x_Q^{l'} y_Q^{m'} z_Q^{n'} \exp(-\gamma r_P^2 - \gamma' r_Q^2) r_{12}^\nu \left(\frac{1}{r_{1V}}\right) d\tau\, d\tau', \tag{A.21}$$

and similarly for J_{22}^ν. For even ν, the operation is similar to that performed on the S_2 integral, *i.e.*,

$$J_{21}^\nu = R_{12}^\nu \iint x_P^l y_P^m z_P^n x_Q^{l'} y_Q^{m'} z_Q^{n'} \exp(-\gamma r_P^2 - \gamma' r_Q^2) \frac{1}{r_{1V}} d\tau\, d\tau'. \tag{A.22}$$

This expression then separates simply into products of one-electron overlap and one-electron nuclear attraction integrals

$$J_{21}^{\nu} = R_{12}^{\nu} \cdot N_1 \cdot OV'. \tag{A.23}$$

It is the case where ν is odd that produces the most complicated of the two-electron HCI integrals, namely one that must be performed numerically. To begin with we can write

$$J_{21}^{\nu} = R_{12}^{\nu+1} \iint x_P^l y_P^m z_P^n x_Q^{l'} y_Q^{m'} z_Q^{n'} \exp(-\gamma r_P^2 - \gamma' r_Q^2) \frac{1}{r_{1V}} \frac{1}{r_{12}} d\tau\, d\tau'. \tag{A.24}$$

After some involved manipulations, Eq. (A.24) can be rewritten as

$$J_{21}^{\nu} = R_{12}^{\nu+1} \frac{4\pi^2}{\gamma\gamma'^{3/2}} \int_0^1 \frac{1}{\delta^{1/2}} H_x H_y H_z F_\mu\left(\frac{\Theta^2}{\delta}\right) d\kappa, \tag{A.25}$$

where the H function is defined by

$$\begin{aligned} H_x = {} & \frac{(-1)^l\, l!\, l'!}{2^{l'}} \exp(-\gamma\kappa^2 \overline{PV}_x^2) \sum_{\bar{l}=0}^{l} \frac{(\kappa^2 \overline{PV}_x)^{l-\bar{l}}}{(l-\bar{l})!\,2^{\bar{l}}} \sum_{\bar{q}=0}^{[\bar{l}/2]} \frac{1}{\bar{q}!\,\bar{L}!} \left(\frac{1-\kappa^2}{\gamma}\right)^{\bar{l}-\bar{q}} \\ & \cdot \sum_{q'=0}^{[l'/2]} \frac{1}{q'!\,L'!\,(\gamma')^{l'-q'}}\, \mathscr{L}! \sum_{u=0}^{[\mathscr{L}/2]} \frac{(-1)^u}{u!\,(\mathscr{L}-2u)!} \frac{(2\Theta_x)^{\mathscr{L}-2u}}{\delta^{\mathscr{L}-u}}, \end{aligned} \tag{A.26}$$

and all the symbols are as previously defined except that

$$L = l - 2q, \quad L' = l' - 2q', \quad \text{and} \quad \mathscr{L} = L + L'. \tag{A.27}$$

Also, δ, Θ_x, and Θ are given by

$$\delta = \frac{1}{\gamma'} + \frac{1-\kappa^2}{\gamma}, \tag{A.28}$$

$$\Theta_x = \overline{PQ}_x + \kappa^2 \overline{VP}_x, \quad \text{and} \quad \Theta^2 = \Theta_x^2 + \Theta_y^2 + \Theta_z^2. \tag{A.29}$$

In addition, $[l/2]$ is the largest integer less than or equal to $l/2$, and $F_\mu(T)$ denotes the incomplete Gamma function as described in Chapter 7 and μ is defined by

$$\mu = \mu_x + \mu_y + \mu_z \quad \text{and} \quad \mu_x = \mathscr{L} - u. \tag{A.30}$$

Appendix 5B.
The $\mathbf{R}_{ij}$ Operator

Consider the basic integral for the S_2^2 problem:

$$I_2^2 = \int\int x_P^l y_P^m z_P^n x_Q^{l'} y_Q^{m'} z_Q^{n'} \exp(-\gamma r_P^2 - \gamma' r_Q^2) r_{12}^2 d\tau\, d\tau'. \tag{B.1}$$

Now expand r_{12}^2 by

$$r_{12}^2 = r_x^2 + r_y^2 + r_z^2, \tag{B.2}$$

where

$$r_x^2 = (x_P - x_Q)^2 + 2\overline{PQ}_x(x_P - x_Q) + \overline{PQ}_x^2 \tag{B.3}$$

and $\overline{PQ}_x$ is the x component of $\vec{P} - \vec{Q}$. Using similar expansions for the y and z coordinates the integral I_2^2 now becomes

$$\begin{aligned} I_2^2 = {} & \overline{PQ}^2 OV \cdot OV' + OV'[OV(l+2) + OV(m+2) + OV(n+2) \\ & + 2\overline{PQ}_x OV(l+1) + 2\overline{PQ}_y OV(m+1) + 2\overline{PQ}_z OV(n+1)] \\ & + OV[OV'(l'+2) + OV'(m'+2) + OV'(n'+2) \\ & - 2\overline{PQ}_x OV'(l'+1) - 2\overline{PQ}_y OV'(m'+1) - 2\overline{PQ}_z OV'(n'+1)] \\ & - 2[OV(l+1)OV'(l'+1) + OV(m+1)OV'(m'+1) + OV(n+1)OV'(n'+1)], \end{aligned} \tag{B.4}$$

where, following the notation in Appendix 5A, OV and OV' are the integral portions of the one-electron overlaps over electron 1 and electron 2, respectively.

Eq. (B.1) can now be rewritten in a simpler form using the operator R_{12}^2 as

$$I_2^2 = R_{12}^2 \int\int x_P^l y_P^m z_P^n x_Q^{l'} y_Q^{m'} z_Q^{n'} \exp(-\gamma r_P^2 - \gamma' r_Q^2) d\tau\, d\tau', \tag{B.5}$$

where R_{12}^2 implies the linear combination of terms given in Eq. (B.4).

Appendix 5C.
Three-electron Integral Formulas

In the main text of this chapter, the future expansion to the three- and four-electron theory was discussed. In the next two appendices, we will give the integral formulas that will be necessary for this expansion. The three-electron integrals will be in this appendix and the four-electron integrals will be in the following appendix. The three-electron integrals which one has to evaluate when applying the HCI method are of the following types

$$S_3^{\nu_1,\nu_2} = \iiint d\tau d\tau' d\tau'' \eta_{aA}\eta_{cC}\eta_{eE} r_{12}^{\nu_1} r_{23}^{\nu_2} \eta_{bB}\eta_{dD}\eta_{fF}, \tag{C.1}$$

$$T_3^{\nu_1,\nu_2} = \iiint d\tau d\tau' d\tau'' \eta_{aA}\eta_{cC}\eta_{eE} r_{12}^{\nu_1} r_{23}^{\nu_2} r_{13}^{-1} \eta_{bB}\eta_{dD}\eta_{fF}, \tag{C.2}$$

$$K_3^{\nu_1,\nu_2} = \iiint d\tau d\tau' d\tau'' \eta_{aA}\eta_{cC}\eta_{eE} r_{12}^{\nu_1} \left(-\frac{\nabla}{2} - \frac{\nabla'}{2} - \frac{\nabla''}{2} \right) r_{23}^{\nu_2} \eta_{bB}\eta_{dD}\eta_{fF}, \tag{C.3}$$

$$N_3^{\nu_1,\nu_2} = \sum_V \iiint d\tau d\tau' d\tau'' \eta_{aA}\eta_{cC}\eta_{eE} r_{12}^{\nu_1} \left(-\frac{Z_V}{r_{1V}} - \frac{Z_V}{r_{2V}} - \frac{Z_V}{r_{3V}} \right) r_{23}^{\nu_2} \eta_{bB}\eta_{dD}\eta_{fF}. \tag{C.4}$$

In the preceding equations, the value of ν_1 and ν_2 can be any non-negative integer except in Eq. (C.1) where one of the two exponents can take the value -1.

Throughout this appendix we will follow the conventions stated in Appendix 5A. In order to prevent confusion in differentiating between electron number and number of electrons for a given type of integral we will associate electron 1 with unprimed indices, electron 2 with primed indices and electron 3 with double primed indices. We will, however, leave the reference to electron number explicitly for the HCI factor r_{ij}, the related R_{ij} operator and in the distance between the electron and the nuclear centers, r_{iV}. Furthermore, we will exclusively associate GTFs e and f with electron 3. In addition to the formulas given in Appendix 5A, we get for the product of GTFs e and f,

$$\eta_{eE}\eta_{fF} = W'' \sum_{l''=0}^{l_e+l_f} f_{l''} x_R^{l''} \sum_{m''=0}^{m_e+m_f} f_{m''} y_R^{m''} \sum_{n''=0}^{n_e+n_f} f_{n''} z_R^{n''} \exp(-\gamma'' r_R^2), \tag{C.5}$$

where we use

$$\vec{R} = \frac{\alpha_e \vec{E} + \alpha_f \vec{F}}{\gamma''}, \tag{C.6}$$

$$W'' = \exp(-\alpha_e \alpha_f \overline{EF}^2 / \gamma''), \tag{C.7}$$

and

$$\gamma'' = \alpha_e + \alpha_f. \tag{C.8}$$

Recall that in Appendix 5A we related OV, N_1, and F_2 to integrals derived in Chapter 7.

1. The S Integral

Upon application of the Gaussian Product Theorem to Eq. (C.1), the general six-center integral is reduced to a three-center integral

$$\begin{aligned} S_3^{\nu_1, \nu_2} = WW'W'' &\sum_{l=0}^{l_a+l_b} f_l \sum_{m=0}^{m_a+m_b} f_m \sum_{n=0}^{n_a+n_b} f_n \sum_{l'=0}^{l_c+l_d} f_{l'} \sum_{m'=0}^{m_c+m_d} f_{m'} \sum_{n'=0}^{n_c+n_d} f_{n'} \\ &\cdot \sum_{l''=0}^{l_e+l_f} f_{l''} \sum_{m''=0}^{m_e+m_f} f_{m''} \sum_{n''=0}^{n_e+n_f} f_{n''} \cdot I_3^{\nu_1, \nu_2}, \end{aligned} \tag{C.9}$$

where

$$\begin{aligned} I_3^{\nu_1, \nu_2} &= \iiint x_P^l y_P^m z_P^n x_Q^{l'} y_Q^{m'} z_Q^{n'} x_R^{l''} y_R^{m''} z_R^{n''} r_{12}^{\nu_1} r_{23}^{\nu_2} \exp(-\gamma r_P^2 - \gamma' r_Q^2 - \gamma'' r_R^2) d\tau d\tau' d\tau'' \\ &= I_3^{\nu_1, \nu_2}(l, m, n;\ l', m', n';\ l'', m'', n''). \end{aligned} \tag{C.10}$$

The solution of Eq. (C.9) is different, of course, depending on the values of ν_1 and ν_2 and we will treat each case separately. Recall that the R_{ij}^{ν} operator is detailed in Appendix 5B. The following sections are labelled explicitly by the values of ν_1 and ν_2.

even, −1

In this case the problem separates into a product of one- and two-electron integrals,

$$I_3^{\nu_1, -1} = R_{12}^{\nu_1} \cdot OV \cdot F_2(l', m', n';\ l'', m'', n''). \tag{C.11}$$

odd, − 1

Using the R_{12} operator we obtain

$$I_3^{\nu_1, -1} = R_{12}^{\nu_1+1} \iiint x_P^l y_P^m z_P^n x_Q^{l'} y_Q^{m'} z_Q^{n'} x_R^{l''} y_R^{m''} z_R^{n''} \frac{1}{r_{12}} \frac{1}{r_{23}} \cdot \exp(-\gamma r_P^2 - \gamma' r_Q^2 - \gamma'' r_R^2) d\tau d\tau' d\tau''. \tag{C.12}$$

This integral can be solved using integral transform methods and after some manipulation one obtains

$$I_3^{\nu_1, -1} = R_{12}^{\nu_1+1} \frac{4\pi^{7/2}}{\gamma^{3/2}\gamma'} \int_0^1 \frac{\varepsilon^{3/2}}{\delta^{1/2}} H_x H_y H_z F_\mu\left(\frac{\Theta^2}{\delta}\right) d\kappa, \tag{C.13}$$

where

$$\begin{aligned} H_x = & \frac{(-1)^l\, l!\, l'!}{2^l} \exp\left[-\gamma'\gamma''\varepsilon\kappa^2 \overline{QR}_x^2\right] \sum_{q=0}^{[l/2]} \frac{1}{q!L!}\left(\frac{1}{\gamma}\right)^{l-q} \\ & \cdot \sum_{\bar{l}=0}^{l'} \frac{(\kappa^2)^{l'-\bar{l}}}{(l'-\bar{l})!\, 2^{\bar{l}}} \sum_{\bar{q}=0}^{[\bar{l}/2]} \frac{1}{\bar{q}!\bar{L}!}\left(\frac{1-\kappa^2}{\gamma'}\right)^{\bar{l}-\bar{q}} \\ & \cdot \sum_{\underline{l}=0}^{l''+l'-\bar{l}} \frac{f_{\underline{l}}\, \underline{l}!}{2^{\underline{l}}} \sum_{\underline{q}=0}^{[\underline{l}/2]} \frac{(\kappa^2)^{\underline{L}}}{\underline{q}!\underline{L}!} \varepsilon^{\underline{l}-\underline{q}} \\ & \cdot \mathscr{L}! \sum_{u=0}^{[\mathscr{L}/2]} \frac{(-1)^u}{u!(\mathscr{L}-2u)!} \frac{(2\Theta_x)^{\mathscr{L}-2u}}{\delta^{\mathscr{L}-u}}. \end{aligned} \tag{C.14}$$

Also,

$$L = l - 2q, \qquad \bar{L} = \bar{l} - 2\bar{q}, \qquad \underline{L} = \underline{l} - 2\underline{q}, \tag{C.15}$$

$$\mathscr{L} = L + \bar{L} + \underline{L}, \tag{C.16}$$

$$\delta = \frac{1}{\gamma} + \frac{1-\kappa^2}{\gamma'} + \kappa^4\varepsilon \quad \text{with} \quad \varepsilon = \frac{1}{\gamma'' + \kappa^2\gamma'}, \tag{C.17}$$

$$\Theta_x = (\overline{PQ}_x + \gamma''\kappa^2\varepsilon\, \overline{QR}_x), \tag{C.18}$$

and

$$\mu = \mu_x + \mu_y + \mu_z \quad \text{with} \quad \mu_x = \mathscr{L} - u. \tag{C.19}$$

In addition, we have used $f_{\underline{l}} = f_{\underline{l}}(l'', l' - \bar{l}, \Xi_x, \Upsilon_x)$, where

$$\overline{\Xi}_x = \gamma'\kappa^2\varepsilon\,\overline{QR}_x \quad \text{and} \quad \overline{\Upsilon}_x = \gamma''\varepsilon\,\overline{RQ}_x. \tag{C.20}$$

even, even

In this case the problem completely separates into a product of one-electron integrals,

$$I_3^{\nu_1,\nu_2} = R_{12}^{\nu_1} R_{23}^{\nu_2} \cdot OV \cdot OV' \cdot OV''. \tag{C.21}$$

even, odd

As in the (even, − 1) case the problem separates into a product of one- and two-electron integrals,

$$I_3^{\nu_1,\nu_2} = R_{12}^{\nu_1} R_{23}^{\nu_2+1} \cdot OV \cdot F_2(l', m', n'; l'', m'', n''). \tag{C.22}$$

odd, odd

This can be obtained from the (odd, − 1) case,

$$I_3^{\nu_1,\nu_2} = R_{23}^{\nu_2+1} I_3^{\nu_1,-1}. \tag{C.23}$$

2. The T Integral

Upon application of the Gaussian Product Theorem to Eq. (C.2), the general six-center integral is reduced to a three-center integral

$$\begin{aligned} T_3^{\nu_1,\nu_2} = WW'W'' \sum_{l=0}^{l_a+l_b} f_l \sum_{m=0}^{m_a+m_b} f_m \sum_{n=0}^{n_a+n_b} f_n \sum_{l'=0}^{l_c+l_d} f_{l'} \sum_{m'=0}^{m_c+m_d} f_{m'} \sum_{n'=0}^{n_c+n_d} f_{n'} \\ \cdot \sum_{l''=0}^{l_e+l_f} f_{l''} \sum_{m''=0}^{m_e+m_f} f_{m''} \sum_{n''=0}^{n_e+n_f} f_{n''} \cdot A_3^{\nu_1,\nu_2}, \end{aligned} \tag{C.24}$$

where

$$\begin{aligned} A_3^{\nu_1,\nu_2} = \int\!\!\int\!\!\int x_P^l y_P^m z_P^n x_Q^{l'} y_Q^{m'} z_Q^{n'} x_R^{l''} y_R^{m''} z_R^{n''} r_{12}^{\nu_1} r_{23}^{\nu_2} r_{13}^{-1} \\ \cdot \exp(-\gamma r_P^2 - \gamma' r_Q^2 - \gamma'' r_R^2) d\tau d\tau' d\tau''. \end{aligned} \tag{C.25}$$

Proceeding as before and analyzing each case depending on whether v_1 and v_2 are even or odd, we get the following equations.

even, even

Here the basic integral separates into a combination of one-electron overlap and two-electron electron repulsion integrals,

$$A_3^{v_1, v_2} = R_{12}^{v_1} R_{23}^{v_2} \cdot OV' \cdot F_2(l, m, n; l'', m'', n''). \tag{C.26}$$

even, odd

In this case, if we permute the electron indices cyclicly twice we can write

$$A_3^{v_1, v_2} = R_{13}^{v_1} \cdot I_3^{v_2, -1}, \tag{C.27}$$

where I_3 was defined in Eq. (C.23).

odd, odd

In this case,

$$A_3^{v_1, v_2} = \frac{8\pi^3}{\gamma\gamma'} R_{12}^{v_1+1} R_{23}^{v_2+1} \int_0^1 \int_0^1 \frac{\varepsilon^{3/2}}{\delta^{1/2}} H_x H_y H_z F_\mu\left(\frac{\Theta^2}{\delta}\right) d\kappa \, d\lambda, \tag{C.28}$$

where

$$H_x = \exp\left(\frac{-\varepsilon\gamma\lambda^2}{\gamma'' + \gamma'\kappa^2}(\gamma''\overline{PR}_x + \gamma'\kappa^2\overline{PQ}_x)^2\right) \exp\left(\frac{-\gamma\gamma''\kappa^2\overline{QR}_x^2}{\gamma'' + \gamma'\kappa^2}\right)$$

$$\cdot \sum_{\hat{l}=0}^{l} \frac{(-1)^{\hat{l}} l!(\lambda^2)^{l-\hat{l}}}{2^{\hat{l}}(l-\hat{l})!} \sum_{\hat{q}=0}^{[\hat{l}/2]} \frac{1}{\hat{q}!\hat{L}!} \left(\frac{1-\lambda^2}{\gamma}\right)^{\hat{l}-\hat{q}}$$

$$
\begin{aligned}
&\cdot \sum_{\bar{l}=0}^{l'} \frac{l'!(\kappa^2)^{l'-\bar{l}}}{2^{\bar{l}}(l'-\bar{l})!} \sum_{\bar{q}=0}^{[\bar{l}/2]} \frac{1}{\bar{q}!\bar{L}!}\left(\frac{1-\kappa^2}{\gamma'}\right)^{\bar{l}-\bar{q}} \\
&\cdot \sum_{\dot{l}=0}^{l''+l'-\bar{l}} f_{\dot{l}} \sum_{\underline{l}=0}^{\dot{l}+l-\hat{l}} \frac{f_{\underline{l}}(-1)^{\underline{l}}\underline{l}!}{2^{\underline{l}}} \sum_{\underline{q}=0}^{[\underline{l}/2]} \frac{(\lambda^2-\kappa^2)^{\underline{L}}}{\underline{q}!\underline{L}!}\varepsilon^{\underline{l}-\underline{q}} \\
&\cdot \mathscr{L}! \sum_{u=0}^{[\mathscr{L}/2]} \frac{(-1)^u}{u!(\mathscr{L}-2u)!}\frac{(2\Theta_x)^{\mathscr{L}-2u}}{\delta^{\mathscr{L}-u}},
\end{aligned}
\tag{C.29}
$$

and where everything is as previously defined except

$$
\mathscr{L} = \hat{L} + \bar{L} + \underline{L},
\tag{C.30}
$$

$$
\delta = \frac{1-\lambda^2}{\gamma} + \frac{1-\kappa^2}{\gamma'} + \varepsilon(\lambda^2-\kappa^2)^2,
\tag{C.31}
$$

$$
\varepsilon = \frac{1}{\gamma'' + \kappa^2\gamma' + \lambda^2\gamma},
\tag{C.32}
$$

and

$$
\Theta_x = \overline{PQ}_x + \lambda^2\varepsilon(\gamma''\overline{RP}_x + \gamma'\kappa^2\overline{QP}_x) + \kappa^2\varepsilon(\gamma''\overline{QR}_x + \gamma\lambda^2\overline{QP}_x).
\tag{C.33}
$$

In addition, we have used $f_{\underline{l}} = f_{\underline{l}}(\dot{l}, l-\hat{l}, \Xi_x, \Upsilon_x)$, where

$$
\Xi_x = \frac{\gamma\kappa^2\varepsilon}{\gamma''+\kappa^2\gamma'}(\gamma''\overline{PR}_x + \gamma'\kappa^2\overline{PQ}_x)
\tag{C.34}
$$

and

$$
\bar{\Upsilon}_x = \varepsilon(\gamma''\overline{RP}_x + \gamma'\kappa^2\overline{QP}_x),
\tag{C.35}
$$

and $f_{\dot{l}} = f_{\dot{l}}(l'', l'-\bar{l}, \bar{\Omega}_x, \bar{\Lambda}_x)$, where

$$
\bar{\Omega}_x = \frac{\gamma'\kappa^2}{\gamma''+\kappa^2\gamma'}\overline{QR}_x
\tag{C.36}
$$

and

$$
\bar{\Lambda}_x = \frac{\gamma''}{\gamma''+\kappa^2\gamma'}\overline{RQ}_x.
\tag{C.37}
$$

3. The K Integral

The K_3 integral can be split into cartesian components

$$K_3^{\nu_1,\nu_2} = -\frac{1}{2}\left(K_{3x}^{\nu_1,\nu_2} + K_{3y}^{\nu_1,\nu_2} + K_{3z}^{\nu_1,\nu_2}\right), \tag{C.38}$$

where

$$K_{3x}^{\nu_1,\nu_2} = \iiint \eta_{aA}\eta_{cC}\eta_{eE} r_{12}^{\nu_1}\left(\frac{\partial^2}{\partial x^2} + \frac{\partial^2}{\partial x'^2} + \frac{\partial^2}{\partial x''^2}\right) \cdot r_{23}^{\nu_2}\eta_{bB}\eta_{dD}\eta_{fF} d\tau d\tau' d\tau'' \tag{C.39}$$

and similarly for K_{3y} and K_{3z}.

After performing the partial differentiation and examining the individual terms, the components of K_3 can be expanded in terms of the $S_3^{\nu_1,\nu_2}$ type of integral, defined in Eq. (C.9), or

$$\begin{aligned} K_{3x}^{\nu_1,\nu_2} &= l_b(l_b-1)S_3^{\nu_1,\nu_2}(l_b-2) - 2\alpha_b(2l_b+1)S_3^{\nu_1,\nu_2} + 4\alpha_b^2 S_3^{\nu_1,\nu_2}(l_b+2) \\ &+ l_d(l_d-1)S_3^{\nu_1,\nu_2}(l_d-2) - 2\alpha_d(2l_d+1)S_3^{\nu_1,\nu_2} + 4\alpha_d^2 S_3^{\nu_1,\nu_2}(l_d+2) \\ &+ l_e(l_e-1)S_3^{\nu_1,\nu_2}(l_e-2) - 2\alpha_e(2l_e+1)S_3^{\nu_1,\nu_2} + 4\alpha_e^2 S_3^{\nu_1,\nu_2}(l_e+2) \\ &+ 2\nu_2\Big\{(l_d + \frac{\nu_2}{6} + \frac{1}{6})S_3^{\nu_1,\nu_2-2} - l_d S_3^{\nu_1,\nu_2-2}(l_d-1, l_f+1) \\ &+ 2\alpha_d[S_3^{\nu_1,\nu_2-2}(l_d+1, l_f+1) - S_3^{\nu_1,\nu_2-2}(l_d+2)] \\ &+ \overline{DF}_x[l_d S_3^{\nu_1,\nu_2-2}(l_d-1) - 2\alpha_d S_3^{\nu_1,\nu_2-2}(l_d+1)]\Big\} \end{aligned} \tag{C.40}$$

where the variations of the quantum numbers caused by the differentiation are denoted inside the parentheses.

4. The N Integral

Upon application of the Gaussian Product Theorem to Eq. (C.4), the general six-center integral is reduced to a three-center integral

$$N_3^{\nu_1,\nu_2} = -WW'W''\sum_V Z_V \sum_{l=0}^{l_a+l_b} f_l \sum_{m=0}^{m_a+m_b} f_m \sum_{n=0}^{n_a+n_b} f_n \sum_{l'=0}^{l_c+l_d} f_{l'} \sum_{m'=0}^{m_c+m_d} f_{m'} \cdot \sum_{n'=0}^{n_c+n_d} f_{n'} \sum_{l''=0}^{l_e+l_f} f_{l''} \sum_{m''=0}^{m_e+m_f} f_{m''} \sum_{n''=0}^{n_e+n_f} f_{n''} \cdot J_3^{\nu_1,\nu_2}, \tag{C.41}$$

where

$$J_3^{\nu_1,\nu_2} = \iiint x_P^l y_P^m z_P^n x_Q^{l'} y_Q^{m'} z_Q^{n'} x_R^{l''} y_R^{m''} z_R^{n''} r_{12}^{\nu_1} \left(\frac{1}{r_{1V}} + \frac{1}{r_{2V}} + \frac{1}{r_{3V}} \right) \cdot r_{23}^{\nu_2} \exp(-\gamma r_P^2 - \gamma' r_Q^2 - \gamma'' r_R^2) d\tau d\tau' d\tau''. \quad (C.42)$$

As we did for the S_3 integral, we will consider separately the cases where ν_1 and ν_2 are even or odd and will label the cases as such. Furthermore, we will need to consider as unique the three separate terms corresponding to each of the $1/r_{iV}$ operators.

even, even

Here the basic integral completely separates into one-electron overlap and nuclear attraction integrals,

$$J_3^{\nu_1,\nu_2} = R_{12}^{\nu_1} R_{23}^{\nu_2} [N_1 \cdot OV' \cdot OV'' + OV \cdot N_1' \cdot OV'' + OV \cdot OV' \cdot N_1'']. \quad (C.43)$$

even, odd

Here again the basic integral separates into one-electron and two-electron integrals,

$$J_3^{\nu_1,\nu_2} = R_{12}^{\nu_1} R_{23}^{\nu_2+1} [N_1(l,m,n) \cdot F_2(l',m',n';l'',m'',n'') + OV(l,m,n) \cdot [J_{21}^{\nu_2}(l',m',n';\ l'',m'',n'') + J_{22}^{\nu_2}(l',m',n';\ l'',m'',n'')]]. \quad (C.44)$$

odd, odd

The basic integral can written here in terms of two types of fairly complicated three-electron integrals,

$$J_3^{\nu_1,\nu_2} = J_{31}^{\nu_1,\nu_2} + J_{32}^{\nu_1,\nu_2} + J_{33}^{\nu_1,\nu_2}. \quad (C.45)$$

To begin, J_{31} can be written as

$$J_{31}^{\nu_1,\nu_2} = \frac{8\pi^3}{\gamma\gamma'} R_{12}^{\nu_1+1} R_{23}^{\nu_2+1} \int_0^1 \int_0^1 \frac{\varepsilon^{3/2}}{\delta^{1/2}} H_x H_y H_z F_\mu \left(\frac{\Theta^2}{\delta} \right) d\lambda d\kappa, \quad (C.46)$$

where

$$H_x = (-1)^{l} l!\, l'!\, \exp(-\gamma\lambda^2 \overline{PV}_x^2)\ \exp(-\gamma''\gamma'\kappa^2\varepsilon \overline{QR}_x^2)$$
$$\cdot \sum_{\hat{l}=0}^{l} \frac{(\lambda^2)^{l-\hat{l}} (\overline{VP}_x)^{l-\hat{l}}}{2^{\hat{l}}(l-\hat{l})!} \sum_{\hat{q}=0}^{[\hat{l}/2]} \frac{1}{\hat{q}!\hat{L}!}\left(\frac{1-\lambda^2}{\gamma}\right)^{\hat{l}-\hat{q}}$$
$$\cdot \sum_{\bar{l}=0}^{l'} \frac{(\kappa^2)^{l-\bar{l}}}{2^{\bar{l}}(l'-\bar{l})!} \sum_{\bar{q}=0}^{[\bar{l}/2]} \frac{1}{\bar{q}!\bar{L}!}\left(\frac{1-\kappa^2}{\gamma'}\right)^{\bar{l}-\bar{q}} \tag{C.47}$$
$$\cdot \sum_{\underline{l}=0}^{l''+l'-\bar{l}} \frac{f_{\underline{l}}\underline{l}!}{2^{\underline{l}}} \sum_{\underline{q}=0}^{[\underline{l}/2]} \frac{1}{\underline{q}!\underline{L}!} (\kappa^2)^{\underline{L}} \varepsilon^{\underline{l}-\underline{q}}$$
$$\cdot\ \mathscr{L}! \sum_{u=0}^{[\mathscr{L}/2]} \frac{(-1)^u}{u!(\mathscr{L}-2u)!} \frac{(2\Theta_x)^{\mathscr{L}-2u}}{\delta^{\mathscr{L}-u}}$$

and everything is as previously defined except

$$\mathscr{L} = \hat{L} + \bar{L} + \underline{L}, \tag{C.48}$$

$$\delta = \left(\frac{1-\lambda^2}{\gamma} + \frac{1-\kappa^2}{\gamma'} + \kappa^4\varepsilon\right) \quad \text{with} \quad \varepsilon = \frac{1}{\gamma'' + \kappa^2\gamma'}, \tag{C.49}$$

$$\Theta_x = (\overline{PQ}_x + \lambda^2 \overline{VP}_x + \gamma''\kappa^2\varepsilon\, \overline{QR}_x), \tag{C.50}$$

and $f_{\underline{l}} = f_{\underline{l}}(l'', l' - \bar{l}, \Xi_x, \Upsilon_x)$, where

$$\Xi_x = \gamma'\kappa^2\varepsilon\, \overline{QR}_x \quad \text{and} \quad \Upsilon_x = \gamma''\varepsilon\, \overline{RQ}_x. \tag{C.51}$$

Note that, in all these formulae, V signifies the nuclear centers.

Furthermore, J_{32} can be written as

$$J_{32}^{\nu_1,\nu_2} = \frac{8\pi^3\gamma'^{1/2}}{\gamma^{3/2}\gamma''} R_{12}^{\nu_1+1} R_{23}^{\nu_2+1}$$
$$\cdot \int_0^1\int_0^1 \frac{\varepsilon^{3/2}}{\delta^{1/2}}(1-\lambda^2)^{-3/2} H_x H_y H_z F_\mu\left(\frac{\Theta^2}{\delta}\right) d\lambda d\kappa \tag{C.52}$$

where

$$H_x = \frac{(-1)^l l! l'! \, l''!}{2^l} \exp(-\gamma' \lambda^2 \overline{VQ}_x^2) \exp\left(\frac{-\gamma'\gamma''\varepsilon\kappa^2}{1-\lambda^2} (\overline{QR}_x + \lambda^2 \overline{VQ}_x)^2 \right)$$

$$\cdot \sum_{q=0}^{[l/2]} \frac{1}{q!L!} \left(\frac{1}{\gamma}\right)^{l-q} \sum_{\hat{l}=0}^{l'} \frac{1(\lambda^2 \overline{VQ}_x)^{l'-\hat{l}}}{\hat{l}!(l'-\hat{l})!}$$

$$\cdot \sum_{\bar{l}=0}^{l''} \frac{(\kappa^2)^{l''-\bar{l}}}{\bar{l}!(l''-\bar{l})!} \left(\frac{1-\kappa^2}{\gamma''}\right)^{\bar{l}/2} \frac{(\bar{l}-1)!!}{2^{\bar{l}/2}} \tag{C.53}$$

$$\cdot \sum_{\underline{l}=0}^{\hat{l}+l''-\bar{l}} \frac{f_{\underline{l}} \underline{l}!}{2^{\underline{l}}} \sum_{\underline{q}=0}^{[\underline{l}/2]} \frac{1}{\underline{q}!\underline{L}!} \varepsilon^{\underline{l}-\underline{q}} \mathscr{L}! \sum_{u=0}^{[\mathscr{L}/2]} \frac{(-1)^u}{u!(\mathscr{L}-2u)!} \frac{(2\Theta_x)^{\mathscr{L}-2u}}{\delta^{\mathscr{L}-u}},$$

and we have

$$\mathscr{L} = L + \underline{L}, \quad \delta = \frac{1}{\gamma} + \varepsilon, \tag{C.54}$$

$$\varepsilon = \frac{1-\lambda^2}{\gamma' + \gamma''\kappa^2(1-\lambda^2)}, \tag{C.55}$$

and

$$\Theta_x = \overline{PQ}_x + \varepsilon\left(\frac{\gamma'\lambda^2}{1-\lambda^2} \overline{QV}_x + \gamma''\kappa^2 \overline{QR}_x \right). \tag{C.56}$$

Also we use $f_{\underline{l}} = f_{\underline{l}}(\hat{l}, l''-\bar{l}, \Xi_x, \Upsilon_x)$, where

$$\Xi_x = \gamma''\varepsilon\kappa^2(\overline{RQ}_x + \lambda^2 \overline{QV}_x) \quad \text{and} \quad \Upsilon_x = \frac{\gamma'\varepsilon}{1-\lambda^2}(\overline{QR}_x + \lambda^2 \overline{VQ}_x). \tag{C.57}$$

J_{33} is the same as J_{31} with rotated indices.

Appendix 5D.
Four-electron Integral Formulas

Following the lead of the previous Appendix, we give here the formulas necessary for expanding the HCI method to four-electron problems. The four-electron integrals which one has to evaluate when applying the HCI method are of the following types

$$S_4^{\nu_1,\nu_2} = \iiiint d\tau d\tau' d\tau'' d\tau''' \eta_{aA}\eta_{cC}\eta_{eE}\eta_{gG} r_{12}^{\nu_1} r_{23}^{\nu_2} r_{34}^{-1} \eta_{bB}\eta_{dD}\eta_{fF}\eta_{hH}, \quad (D.1)$$

$$T_4^{\nu_1,\nu_2} = \iiiint d\tau d\tau' d\tau'' d\tau''' \eta_{aA}\eta_{cC}\eta_{eE}\eta_{gG} r_{12}^{\nu_1} r_{23}^{-1} r_{34}^{\nu_2} \eta_{bB}\eta_{dD}\eta_{fF}\eta_{hH}, \quad (D.2)$$

$$U_4^{\nu_1,\nu_2} = \iiiint d\tau d\tau' d\tau'' d\tau''' \eta_{aA}\eta_{cC}\eta_{eE}\eta_{gG} r_{12}^{\nu_1} r_{13}^{\nu_2} r_{14}^{-1} \eta_{bB}\eta_{dD}\eta_{fF}\eta_{hH}. \quad (D.3)$$

In the preceding equations, the value of ν_1 and ν_2 are limited to positive integers.

Throughout this appendix we will follow the conventions stated in Appendices 5A and 5C. In addition to the rules laid down therein, we will associate electron 4 with triply primed indices. Furthermore, we will exclusively associate GTFs g and h with electron 4. We get for the product of GTFs g and h,

$$\eta_{gG}\eta_{hH} = W''' \sum_{l'''=0}^{l_g+l_h} f_{l'''} x_S^{l'''} \sum_{m'''=0}^{m_g+m_h} f_{m'''} y_S^{m'''} \sum_{n''=0}^{n_g+n_h} f_{n'''} z_R^{n'''} \exp(-\gamma'' r_S^2), \quad (D.4)$$

where we use

$$\vec{S} = \frac{\alpha_g \vec{G} + \alpha_h \vec{H}}{\gamma'''}, \quad (D.5)$$

$$W''' = \exp(-\alpha_g\alpha_h \overline{GH}^2/\gamma'''), \quad (D.6)$$

and

$$\gamma''' = \alpha_g + \alpha_h. \quad (D.7)$$

1. The S Integral

Upon application of the Gaussian Product Theorem to Eq. (D.1), the general eight-center integral is reduced to a four-center integral

$$S_4^{\nu_1,\nu_2} = WW'W''W''' \sum_{l=0}^{l_a+l_b} f_l \sum_{m=0}^{m_a+m_b} f_m \sum_{n=0}^{n_a+n_b} f_n \sum_{l'=0}^{l_c+l_d} f_{l'} \sum_{m'=0}^{m_c+m_d} f_{m'} \sum_{n'=0}^{n_c+n_d} f_{n'}$$
$$\cdot \sum_{l''=0}^{l_e+l_f} f_{l''} \sum_{m''=0}^{m_e+m_f} f_{m''} \sum_{n''=0}^{n_e+n_f} f_{n''} \sum_{l'''=0}^{l_g+l_h} f_{l'''} \sum_{m'''=0}^{m_g+m_h} f_{m'''} \sum_{n'''=0}^{n_g+n_h} f_{n'''} \cdot I_4^{\nu_1,\nu_2}, \qquad (D.8)$$

where

$$I_4^{\nu_1,\nu_2} = \iiiint x_P^l y_P^m z_P^n x_Q^{l'} y_Q^{m'} z_Q^{n'} x_R^{l''} y_R^{m''} z_R^{n''} x_S^{l'''} y_S^{m'''} z_S^{n'''}$$
$$\cdot\ r_{12}^{\nu_1} r_{23}^{\nu_2} r_{34}^{-1} \exp(-\gamma r_P^2 - \gamma' r_Q^2 - \gamma'' r_R^2 - \gamma''' r_S^2) d\tau d\tau' d\tau'' d\tau'''. \qquad (D.9)$$

The solution of Eq. (D.9) is different, of course, depending on the values of ν_1 and ν_2 and we will treat each case separately. Recall that the R_{ij}^{ν} operator is detailed in Appendix 5B. The following sections are labelled explicitly by the values of ν_1 and ν_2.

even,even

In this case the problem separates into a product of one- and two-electron integrals,

$$I_4^{\nu_1,\nu_2} = R_{12}^{\nu_1} R_{23}^{\nu_2} \cdot OV \cdot OV' \cdot F_2(l'', m'', n''; l''', m''', n'''). \qquad (D.10)$$

even,odd

Here we find a linear combination of three-electron integrals,

$$I_4^{\nu_1,\nu_2} = R_{12}^{\nu_1} \cdot OV \cdot I_3^{\nu_2,-1}(l', m', n'; l'', m'', n''; l''', m''', n'''), \qquad (D.11)$$

where $I_3^{\nu_2,-1}$ is defined in Eq. (C.11).

odd,even

For this problem we get a combination of two-electron integrals,

$$I_4^{\nu_1,\nu_2} = R_{12}^{\nu_1+1} R_{23}^{\nu_2} \cdot F_2(l,m,n;l'm'n') \cdot F_2(l'',m'',n'';l''',m''',n'''). \qquad (D.12)$$

odd,odd

The difficult case for the S_4 integral is the odd-odd problem where we obtain

$$I_4^{\nu_1,\nu_2} = R_{12}^{\nu_1+1} R_{23}^{\nu_2+1} \cdot D_4^{\nu_1,\nu_2}, \qquad (D.13)$$

and

$$D_4^{\nu_1,\nu_2} = \frac{8\pi^{9/2}}{\gamma^{3/2}\gamma'\gamma'''} \int_0^1 \int_0^1 \frac{\varepsilon^{3/2}}{\delta^{1/2}} H_x H_y H_z F_\mu\left(\frac{\Theta^2}{\delta}\right) d\kappa\, d\lambda \qquad (D.14)$$

where

$$\begin{aligned} H_x = {} & \frac{(-1)^l l! l'! l'''!}{2^l} \exp\left[\frac{-\gamma'''\lambda^2\varepsilon(\gamma''\overline{RS}_x + \gamma'\kappa^2\overline{QS}_x)^2}{(\gamma''+\kappa^2\gamma')}\right] \exp\left[\frac{-\gamma''\gamma'\kappa^2}{\gamma''+\kappa^2\gamma'}\overline{QR}_x^2\right] \\ & \cdot \sum_{q=0}^{[l/2]} \frac{1}{q!L!}\left(\frac{1}{\gamma}\right)^{l-q} \sum_{\bar{l}=0}^{l'} \frac{(\kappa^2)^{l'-\bar{l}}}{(l'-\bar{l})!2^{\bar{l}}} \sum_{\bar{q}=0}^{[\bar{l}/2]} \frac{1}{\bar{q}!\bar{L}!}\left(\frac{1-\kappa^2}{\gamma'}\right)^{\bar{l}-\bar{q}} \\ & \cdot \sum_{\tilde{l}=0}^{l''+l'-\bar{l}} f_{\tilde{l}} \sum_{l^*=0}^{l'''} \frac{(\lambda^2)^{l'''-l^*}}{l^*!(l'''-l^*)!}\left(\frac{1-\lambda^2}{\gamma'''}\right)^{l^*/2} \frac{(l^*-1)!!}{2^{l^*/2}} \\ & \cdot \sum_{\underline{l}=0}^{\tilde{l}+l'''-l^*} \frac{f_{\underline{l}}\underline{l}!}{2^{\underline{l}}} \sum_{\underline{q}=0}^{[\underline{l}/2]} \frac{(\kappa^2)^{\underline{L}}}{\underline{q}!\underline{L}!} \varepsilon^{\underline{l}-\underline{q}} \mathscr{L}! \sum_{u=0}^{[\mathscr{L}/2]} \frac{(-1)^u}{u!(\mathscr{L}-2u)!} \frac{(2\Theta_x)^{\mathscr{L}-2u}}{\delta^{\mathscr{L}-u}}. \end{aligned} \qquad (D.15)$$

Also,

$$L = l - 2p, \quad \bar{L} = \bar{l} - 2\bar{q}, \quad \underline{L} = \underline{l} - 2\underline{q}, \qquad (D.16)$$

$$\mathscr{L} = L + \bar{L} + \underline{L}, \qquad (D.17)$$

$$\delta = \frac{1}{\gamma} + \frac{1-\kappa^2}{\gamma'} + \kappa^4\varepsilon \text{ with } \varepsilon = \frac{1}{\gamma''+\kappa^2\gamma'+\lambda^2\gamma'''}, \qquad (D.18)$$

$$\Theta_x = \overline{PQ}_x + \kappa^2\varepsilon(\gamma''\overline{QR}_x + \lambda^2\gamma'''\overline{QS}_x), \qquad (D.19)$$

and

$$\mu = \mu_x + \mu_y + \mu_z \quad \text{with} \quad \mu_x = \mathscr{L} - u. \tag{D.20}$$

In addition, we have used $f_{\bar{l}} = f_{\bar{l}}(l'' - \bar{l}, l'', \Xi_x, \Upsilon_x)$, where

$$\Xi_x = \frac{\gamma''}{\gamma'' + \kappa^2\gamma'} \overline{RQ}_x, \tag{D.21}$$

and

$$\Upsilon_x = \frac{\gamma'\kappa^2}{\gamma'' + \kappa^2\gamma'} \overline{QR}_x. \tag{D.22}$$

Also, $f_{\underline{l}} = f_{\underline{l}}(\tilde{l}, l''' - l^*, \overline{\Omega}_x, \overline{\Lambda}_x)$, where

$$\overline{\Omega}_x = \frac{\gamma'''\lambda^2\varepsilon(\gamma''\overline{SR}_x + \gamma'\kappa^2\overline{SQ}_x)}{(\gamma'' + \kappa^2\gamma')} \tag{D.23}$$

and

$$\overline{\Lambda}_x = \varepsilon(\gamma''\overline{RS}_x + \gamma'\kappa^2\overline{QS}_x). \tag{D.24}$$

2. The T Integral

We now turn to the second type of four-electron integral, Eq. (D.2). Applying the Gaussian Product Theorem we get

$$T_4^{\nu_1,\nu_2} = WW'W''W''' \sum_{l=0}^{l_a+l_b} f_l \sum_{m=0}^{m_a+m_b} f_m \sum_{n=0}^{n_a+n_b} f_n \sum_{l'=0}^{l_c+l_d} f_{l'} \sum_{m'=0}^{m_c+m_d} f_{m'} \sum_{n'=0}^{n_c+n_d} f_{n'}$$
$$\cdot \sum_{l''=0}^{l_e+l_f} f_{l''} \sum_{m''=0}^{m_e+m_f} f_{m''} \sum_{n''=0}^{n_e+n_f} f_{n''} \sum_{l'''=0}^{l_g+l_h} f_{l'''} \sum_{m'''=0}^{m_g+m_h} f_{m'''} \sum_{n'''=0}^{n_g+n_h} f_{n'''} \cdot A_4^{\nu_1,\nu_2}, \tag{D.25}$$

where

$$A_4^{\nu_1,\nu_2} = \iiiint x_P^l y_P^m z_P^n x_Q^{l'} y_Q^{m'} z_Q^{n'} x_R^{l''} y_R^{m''} z_R^{n''} x_S^{l'''} y_S^{m'''} z_S^{n'''}$$
$$\cdot r_{12}^{\nu_1} r_{23}^{-1} r_{34}^{\nu_2} \exp(-\gamma r_P^2 - \gamma' r_Q^2 - \gamma'' r_R^2 - \gamma''' r_S^2) d\tau d\tau' d\tau'' d\tau'''. \tag{D.26}$$

Once again we label the various forms depending on the values of ν_1 and ν_2.

even,even

In this case the problem separates into a product of one- and two-electron integrals,

$$A_4^{\nu_1,\nu_2} = R_{12}^{\nu_1} R_{34}^{\nu_2} \cdot OV \cdot F_2(l', m', n'; l'', m'', n'') \cdot OV'''. \qquad (D.27)$$

even,odd

Here we find a linear combination of three-electron integrals,

$$A_4^{\nu_1,\nu_2} = R_{12}^{\nu_1} \cdot OV(l, m, n) \cdot I_3^{-1,\nu_2}(l', m', n'; l'', m'', n''; l''', m''', n'''). \quad (D.28)$$

odd,even

For this problem we again get a combination of three-electron integrals,

$$A_4^{\nu_1,\nu_2} = R_{34}^{\nu_2} \cdot OV'''(l''', m''', n''') \cdot I_3^{\nu_2,-1}(l, m, n; l', m', n'; l'', m'', n''). \quad (D.29)$$

odd,odd

This case can be written in terms of the D_4 integral detailed in Eq. (D.14) or

$$A_4^{\nu_1,\nu_2} = R_{12}^{\nu_1+1} R_{34}^{\nu_2+1} \cdot D_4^{\nu_1,\nu_2}. \qquad (D.30)$$

3. The U Integral

Upon application of the Gaussian Product Theorem to Eq. (D.3), the general eight-center integral is reduced to a four-center integral

$$U_4^{\nu_1,\nu_2} = WW'W''W''' \sum_{l=0}^{l_a+l_b} f_l \sum_{m=0}^{m_a+m_b} f_m \sum_{n=0}^{n_a+n_b} f_n \sum_{l'=0}^{l_c+l_d} f_{l'} \sum_{m'=0}^{m_c+m_d} f_{m'} \sum_{n'=0}^{n_c+n_d} f_{n'}$$
$$\cdot \sum_{l''=0}^{l_e+l_f} f_{l''} \sum_{m''=0}^{m_e+m_f} f_{m''} \sum_{n''=0}^{n_e+n_f} f_{n''} \sum_{l'''=0}^{l_g+l_h} f_{l'''} \sum_{m'''=0}^{m_g+m_h} f_{m'''} \sum_{n'''=0}^{n_g+n_h} f_{n'''} \cdot B_4^{\nu_1,\nu_2}, \qquad (D.31)$$

where

$$B_4^{\nu_1,\nu_2} = \iiiint x_P^l y_P^m z_P^n x_Q^{l'} y_Q^{m'} z_Q^{n'} x_R^{l''} y_R^{m''} z_R^{n''} x_S^{l'''} y_S^{m'''} z_S^{n'''} \cdot r_{12}^{\nu_1} r_{13}^{\nu_2} r_{14}^{-1} \exp(-\gamma r_P^2 - \gamma' r_Q^2 - \gamma'' r_R^2 - \gamma''' r_S^2) d\tau d\tau' d\tau'' d\tau'''. \quad (D.32)$$

even,even

In this case the problem separates into a product of one- and two-electron integrals,

$$B_4^{\nu_1,\nu_2} = R_{12}^{\nu_1} R_{13}^{\nu_2} \cdot OV \cdot F_2(l, m, n; l''', m''', n''') \cdot OV''. \quad (D.33)$$

even,odd

Here we find a linear combination of one-electron and three-electron integrals,

$$B_4^{\nu_1,\nu_2} = R_{12}^{\nu_1} \cdot OV'(l', m', n') \cdot I_3^{\nu_2,-1}(l'', m'', n''; l, m, n; l''', m''', n'''). \quad (D.34)$$

odd,even

Similar to the (even, odd) case, we obtain

$$B_4^{\nu_1,\nu_2} = R_{13}^{\nu_1} \cdot OV'(l'', m'', n'') \cdot I_3^{\nu_2,-1}(l', m', n'; l, m, n; l''', m''', n'''). \quad (D.35)$$

odd,odd

The difficult case for the U_4 integral is once again the odd-odd problem where we obtain

$$B_4^{\nu_1,\nu_2} = R_{12}^{\nu_1+1} R_{13}^{\nu_2+1} \frac{8\pi^{9/2}}{\gamma^{3/2}\gamma'\gamma'''} \int_0^1 \int_0^1 \frac{\varepsilon^{3/2}}{\delta^{1/2}} H_x H_y H_z F_\mu\left(\frac{\Theta^2}{\delta}\right) d\kappa d\lambda, \quad (D.36)$$

with

$$
\begin{aligned}
H_x = {} & \frac{l!l'!l'''!}{2^l} \exp\left[\frac{-\gamma'''\lambda^2\varepsilon(\gamma''\overline{RS}_x + \gamma'\kappa^2\overline{QS}_x)^2}{(\gamma'' + \kappa^2\gamma')}\right] \exp\left[\frac{-\gamma''\gamma'\kappa^2}{\gamma'' + \kappa^2\gamma'}\overline{QR}_x^2\right] \\
& \cdot \sum_{q=0}^{[l/2]} \frac{1}{q!L!}\left(\frac{1}{\gamma}\right)^{l-q} \sum_{\bar{l}=0}^{l'} \frac{(\kappa^2)^{l'-\bar{l}}(1-\kappa^2)^{\bar{l}/2}(\bar{l}-1)!!}{\bar{l}!(l'-\bar{l})!(2\gamma')^{\bar{l}/2}} \\
& \cdot \sum_{\tilde{l}=0}^{l''+l'-\bar{l}} f_{\bar{l}} \sum_{l^*=0}^{l'''} \frac{(\lambda^2)^{l'''-l^*}}{l^*!(l'''-l^*)!}\left(\frac{1-\kappa^2}{\gamma'''}\right)^{l^*/2} \frac{(l^*-1)!!}{2^{l^*/2}} \\
& \cdot \sum_{\underline{l}=0}^{\tilde{l}+l'''-l^*} \frac{f_{\underline{l}}(-1)^{\underline{l}}\underline{l}!}{2^{\underline{l}}} \sum_{\underline{q}=0}^{[\underline{l}/2]} \frac{\varepsilon^{\underline{l}-\underline{q}}}{\underline{q}!\underline{L}!} \mathscr{L}! \sum_{u=0}^{[\mathscr{L}/2]} \frac{(-1)^u}{u!(\mathscr{L}-2u)!} \frac{(2\Theta_x)^{\mathscr{L}-2u}}{\delta^{\mathscr{L}-u}} .
\end{aligned}
\tag{D.37}
$$

Also,

$$L = l - 2p, \qquad \underline{L} = \underline{l} - 2\underline{q}\,, \tag{D.38}$$

$$\mathscr{L} = L + \underline{L}\,, \tag{D.39}$$

$$\delta = \frac{1}{\gamma} + \varepsilon \quad \text{with} \quad \varepsilon = \frac{1}{\gamma'' + \kappa^2\gamma' + \lambda^2\gamma'''}\,, \tag{D.40}$$

$$\Theta_x = \varepsilon(\gamma'\kappa^2\overline{PQ}_x + \gamma''\overline{RP}_x + \lambda^2\gamma'''\overline{SP}_x)\,, \tag{D.41}$$

and

$$\mu = \mu_x + \mu_y + \mu_z \quad \text{with} \quad \mu_x = \mathscr{L} - u. \tag{D.42}$$

In addition, we have used $f_{\bar{l}} = f_{\bar{l}}(l' - \bar{l}, l''', \Xi_x, \Upsilon_x)$, where

$$\Xi_x = \frac{\gamma''}{\gamma'' + \kappa^2\gamma'}\overline{RQ}_x \quad \text{and} \quad \Upsilon_x = \frac{\gamma'\kappa^2}{\gamma'' + \kappa^2\gamma'}\overline{QR}_x. \tag{D.43}$$

Also, $f_{\underline{l}} = f_{\underline{l}}(\tilde{l}, l''' - l^*, \overline{\Omega}_x, \overline{\Lambda}_x)$, where

$$\overline{\Omega}_x = \frac{\gamma''\lambda^2\varepsilon(\gamma''\overline{SR}_x + \gamma'\kappa^2\overline{SQ}_x)}{(\gamma'' + \kappa^2\gamma')} \tag{D.44}$$

and

$$\overline{\Lambda}_x = \varepsilon(\gamma''\overline{RS}_x + \gamma'\kappa^2\overline{QS}_x). \tag{D.45}$$

Appendix 5E.
Exponential Integral Formulas

In the main text of this chapter, the future extension of this theory to explicit correlation in the form of an exponential argument was mentioned. In this appendix, we will detail the integrals necessary for such an expansion. The form the correlation takes in this section is $\exp(-\rho r_{12}^2)$, where ρ is an arbitrary parameter. Notice that higher powers of the correlation only change ρ, therefore this type of correlation, unlike the power series expansion of r_{12} discussed in the main chapter, gives only one set of integrals, that is

$$SE_2^\nu = \int\int d\tau d\tau' \eta_{aA}\eta_{cC} \exp(-\nu\rho r_{12}^2)\eta_{bB}\eta_{dD}, \tag{E.1}$$

$$EE_2^\nu = \int\int d\tau d\tau' \eta_{aA}\eta_{cC} \exp(-\nu\rho r_{12}^2) r_{12}^{-1}\ \eta_{bB}\eta_{dD}, \tag{E.2}$$

$$KE_2^{\nu_1,\nu_2} = \int\int d\tau d\tau' \eta_{aA}\eta_{cC} \exp(-\nu_1\rho r_{12}^2)\left(-\frac{\nabla}{2}-\frac{\nabla'}{2}\right)\exp(-\nu_2\rho r_{12}^2)\eta_{bB}\eta_{dD}, \tag{E.3}$$

$$NE_2^\nu = \sum_V\int\int d\tau d\tau' \eta_{aA}\eta \exp(-\nu\rho r_{12}^2)\left(-\frac{Z_V}{r_{1V}}-\frac{Z_V}{r_{2V}}\right)\eta_{bB}\eta_{dD}. \tag{E.4}$$

Throughout this appendix we will follow the conventions put forth in Appendix 5A. We will associate electron 1 with unprimed indices and electron 2 with primed indices. Explicit reference to electron number will only remain in the correlation factor r_{12} and in the distance between the electrons and the nuclear centers, r_{iV}. We will continue to associate GTFs a and c with electron 1 and GTFs b and d with electron 2. We will also use the results of Eqs.(A1-A9) without explicit reference.

1. The SE Integral

Upon application of the Gaussian Product Theorem to Eq. (E1), the general four-center integral is reduced to a two-center integral

$$SE_2^\nu = WW' \sum_{l=0}^{l_a+l_b} f_l \sum_{m=0}^{m_a+m_b} f_m \sum_{n=0}^{n_a+n_b} f_n \sum_{l'=0}^{l_c+l_d} f_{l'} \sum_{m'=0}^{m_c+m_d} f_{m'} \sum_{n'=0}^{n_c+n_d} f_{n'} \cdot ISE_2^\nu, \tag{E.5}$$

where

$$ISE_2^{\nu} = \int\int x_P^l y_P^m z_P^n x_Q^{l'} y_Q^{m'} z_Q^{n'} \exp(-\gamma r_P^2 - \gamma' r_Q^2 - \nu\rho r_{12}^2) d\tau\, d\tau'. \tag{E.6}$$

Notice that we continue to use f_l to represent $f_l(l_a, l_b, \overline{PA}_x, \overline{PB}_x)$. We will now restrict ourselves to $\nu = 1$ and after some straightforward manipulations we obtain

$$ISE_2 = \frac{\pi^3}{\varepsilon^{3/2}(\gamma' + \rho)^{3/2}} H_x H_y H_z, \tag{E.7}$$

where

$$H_x = l'! \exp\left(\frac{-\gamma\gamma'\rho\overline{PQ}_x^2}{\varepsilon(\gamma' + \rho)}\right) \sum_{\bar{l}=0}^{l'} \frac{1}{\bar{l}!(l' - \bar{l})!} \left(\frac{\rho}{\gamma' + \rho}\right)^{l' - \bar{l}} \cdot \frac{(\bar{l} - 1)!!}{2^{\bar{l}/2}(\gamma' + \rho)^{\bar{l}/2}} \sum_{\tilde{l}=0}^{l+l'-\bar{l}} f_{\tilde{l}} \frac{(\tilde{l} - 1)!!}{2^{\tilde{l}/2}\varepsilon^{\tilde{l}/2}} \tag{E.8}$$

and

$$\varepsilon = \gamma + \frac{\gamma'\rho}{\gamma' + \rho}. \tag{E.9}$$

In addition we have used

$$f_{\tilde{l}} = f_{\tilde{l}}(l, l' - \bar{l}, \Xi_x, \Upsilon_x), \tag{E.10}$$

where

$$\Xi_x = \left(\frac{\gamma'\rho}{\varepsilon(\gamma' + \rho)}\right)\overline{QP}_x \quad \text{and} \quad \Upsilon_x = \gamma\varepsilon^{-1}\overline{PQ}_x. \tag{E.11}$$

The $\nu = 2$ integrals can be found by multiplying ρ by 2 in the above equations.

2. The EE Integral

Applying the Gaussian Product Theorem to Eq. (E2), the EE_2 integral becomes

$$EE_2^{\nu} = WW' \sum_{l=0}^{l_a+l_b} f_l \sum_{m=0}^{m_a+m_b} f_m \sum_{n=0}^{n_a+n_b} f_n \sum_{l'=0}^{l_c+l_d} f_{l'} \sum_{m'=0}^{m_c+m_d} f_{m'} \sum_{n'=0}^{n_c+n_d} f_{n'} \cdot IEE_2^{\nu}, \tag{E.12}$$

where

$$IEE_2^{\nu} = \iint x_P^l y_P^m z_P^n x_Q^{l'} y_Q^{m'} z_Q^{n'} \exp(-\gamma r_P^2 - \gamma' r_Q^2 - \nu\rho r_{12}^2) r_{12}^{-1} d\tau\, d\tau'. \qquad (E.13)$$

Proceeding as before, we find that

$$IEE_2 = \frac{2\pi^{5/2}}{\delta^{1/2}\varepsilon^{3/2}(\gamma' + \rho)^{3/2}} H_x H_y H_z F_\mu\left(\frac{\Theta^2}{\delta}\right), \qquad (E.14)$$

where

$$H_x = l'! \exp\left(\frac{-\gamma\gamma'\rho\overline{PQ}_x^2}{\varepsilon(\gamma' + \rho)}\right) \sum_{\bar{l}=0}^{l'} \frac{1}{2^{\bar{l}}(l' - \bar{l})!}\left(\frac{\rho}{\gamma' + \rho}\right)^{l' - \bar{l}}$$
$$\cdot \sum_{\bar{q}=0}^{[\bar{l}/2]} \frac{1}{\bar{q}!\bar{L}!(\gamma' + \rho)^{\bar{l} - \bar{q}}} \sum_{\tilde{l}=0}^{l + l' - \bar{l}} \frac{f_{\tilde{l}}(-1)^{\tilde{l}}\tilde{l}!}{2^{\tilde{l}}}$$
$$\cdot \sum_{\tilde{q}=0}^{[\tilde{l}/2]} \frac{1}{\tilde{q}!\tilde{L}!\varepsilon^{\tilde{l} - \tilde{q}}}\left(\frac{\gamma'}{\gamma' + \rho}\right)^{\tilde{L}} \mathscr{L}! \sum_{u=0}^{[\mathscr{L}/2]} \frac{(-1)^u (2\Theta_x)^{\mathscr{L} - 2u}}{u![\mathscr{L} - 2u]!\delta^{\mathscr{L} - u}} \qquad (E.15)$$

with

$$\delta = \frac{1}{\gamma' + \rho} + \frac{1}{\varepsilon}\left(\frac{\gamma'}{\gamma' + \rho}\right)^2, \qquad (E.16)$$

$$\Theta_x = \frac{\gamma}{\varepsilon}\left(\frac{\gamma'}{\gamma' + \rho}\right)\overline{PQ}_x, \qquad (E.17)$$

and ε and $f_{\tilde{l}}$ are the the same as in Eqs. (E.9 & E.10), respectively. In addition, we have used

$$\bar{L} = \bar{l} - 2\bar{q}, \quad \tilde{L} = \tilde{l} - 2\tilde{q}, \quad \text{and} \quad \mathscr{L} = \bar{L} + \tilde{L}. \qquad (E.18)$$

3. The KE Integral

The KE integral is only one among this set in which the formulation is different depending on whether one or two correlation factors are present in the integral. Therefore, the $KE_2^{1,0}$ and $KE_2^{1,1}$ cases will be handled separately. As before however, higher powers of the correlation factor will produce the same forms for the integral and hence we can ignore them. The KE_2 integral can be split into cartesian components, such that

$$KE_2^{\nu_1, \nu_2} = -\frac{1}{2}(KE_{2x}^{\nu_1, \nu_2} + KE_{2y}^{\nu_1, \nu_2} + KE_{2z}^{\nu_1, \nu_2}), \qquad (E.19)$$

where

$$K_{2x}^{\nu_1,\nu_2} = \iint \eta_{aA}\eta_{cC} \exp(-\nu_1 \rho r_{12}^2)\left(\frac{\partial^2}{\partial x^2} + \frac{\partial^2}{\partial x'^2}\right) \exp(-\nu_2 \rho r_{12}^2)\eta_{bB}\eta_{dD} d\tau d\tau'. \quad (E.20)$$

After performing the partial differentiation, we can see that the components can be expanded in terms of the SE_2^ν type of integral where ν can be either 1 or 2. The x component of the 1,0 case can be written simply as

$$\begin{aligned} KE_{2x}^{1,0} = {} & l_b(l_b - 1)SE_2^1(l_b - 2) - 2\alpha_b(2l_b + 1)SE_2^1 + 4\alpha_b^2 SE_2^1(l_b + 2) \\ & + l_d(l_d - 1)SE_2^1(l_d - 2) - 2\alpha_d(2l_d + 1)SE_2^1 + 4\alpha_d^2 SE_2^1(l_d + 2) \end{aligned} \quad (E.21)$$

and similarly for the 1,1 case we obtain

$$\begin{aligned} K_{2x}^{1,1} = {} & l_b(l_b - 1)SE_2^2(l_b - 2) - 4\rho l_b\overline{BD}_x SE_2^2(l_b - 2) + 4\rho l_b SE_2^2(l_b - 1, l_d + 1) \\ & + l_d(l_d - 1)SE_2^2(l_d - 2) - 4\rho l_d\overline{DB}_x SE_2^2(l_d - 2) + 4\rho l_d SE_2^2(l_b + 1, l_d - 1) \\ & + 8\rho(2\rho + \alpha_b)\overline{BD}_x SE_2^2(l_b + 1) + 4((\rho + \alpha_b)^2 + \rho^2)SE_2^2(l_b + 2) \\ & + 8\rho(2\rho + \alpha_d)\overline{DB}_x SE_2^2(l_d + 1) + 4((\rho + \alpha_d)^2 + \rho^2)SE_2^2(l_d + 2) \\ & (-4\rho(l_b + l_d + 1) + 8\rho^2(\overline{BD}_x^2) - 2\alpha_b(2l_b + 1) - 2\alpha_d(2l_d + 1))SE_2^2 \\ & - 8(\rho + \alpha_b + \rho + \alpha_d)SE_2^2(l_b + 1, l_d + 1). \end{aligned} \quad (E.22)$$

4. The NE Integral

Application of the Gaussian Product Theorem to Eq. (E.4) produces a two-center integral of the form

$$\begin{aligned} NE_2^\nu = {} & -WW'\sum_V Z_V \sum_{l=0}^{l_a+l_b} f_l \sum_{m=0}^{m_a+m_b} f_m \sum_{n=0}^{n_a+n_b} f_n \\ & \cdot \sum_{l'=0}^{l_c+l_d} f_{l'} \sum_{m'=0}^{m_c+m_d} f_{m'} \sum_{n'=0}^{n_c+n_d} f_{n'} \cdot (JE_{21}^\nu + JE_{22}^\nu), \end{aligned} \quad (E.23)$$

where

$$JE_{21}^\nu = \iint x_{P}^{l} y_{P}^{m} z_{P}^{n} x_{Q}^{l'} y_{Q}^{m'} z_{Q}^{n'} \exp(-\gamma r_P^2 - \gamma' r_Q^2 - \nu\rho r_{12}^2)\left(\frac{1}{r_{1V}}\right) d\tau\, d\tau', \quad (E.24)$$

and similarly for JE_{22}^ν. Eq. (E.24) can be simplified, with $\nu = 1$, to

$$JE_{21} = \frac{2\pi^{5/2}\delta}{(\gamma' + \rho)^{3/2}} H_x H_y H_z F_\mu\left(\frac{\Theta^2}{\delta}\right), \quad (E.25)$$

where

$$H_x = l'! \exp\left(- \frac{\gamma\gamma'\delta}{\gamma' + \rho} \overline{PQ}_x^2 \right) \sum_{\bar{l}=0}^{l'} \frac{1}{\bar{l}!(l' - \bar{l})!} \left(\frac{\rho}{\gamma' + \rho} \right)^{l' - \bar{l}} \frac{(\bar{l} - 1)!!}{2^{\bar{l}/2}(\gamma' + \rho)^{\bar{l}/2}}$$
$$\cdot \sum_{\tilde{l}=0}^{l+l'-\bar{l}} \frac{f_{\tilde{l}}(-1)^{\tilde{l}}\tilde{l}!}{2^{\tilde{l}}} \sum_{\tilde{q}=0}^{[\tilde{l}/2]} \frac{\delta^{\tilde{l}-\tilde{q}}}{\tilde{q}!} \tilde{L}! \sum_{u=0}^{[\tilde{L}/2]} \frac{(-1)^u (2\Theta_x)^{\tilde{L}-2u}}{u![\tilde{L} - 2u]!\delta^{\tilde{L}-u}}. \tag{E.26}$$

In addition to ε and $f_{\tilde{l}}$ defined in Eqs. (E.9 and E.10), we have used

$$\delta = \frac{1}{\varepsilon} \quad \text{and} \quad \Theta_x = \gamma\delta\overline{PV}_x + \frac{\gamma'\rho}{\gamma' + \rho}\overline{QV}_x. \tag{E.27}$$

References

1. C. C. J. Roothaan, Rev. Mod. Phys. **23**, 69 (1951); *ibid.* **32**, 179 (1960).
2. For an extensive review of the CI method, see I. Shavitt, in *Methods of Electronic Structure Theory,* H. F. Schaefer III, ed. (Plenum Press, New York, 1977), pp.189-275.
3. C. F. Fischer, *The Hartree-Fock Method for Atoms* (John Wiley & Sons, New York, 1977), 308 pp.
4. For an extensive review of the MCSCF method, see A. C. Wahl and G. Das, in *Methods of Electronic Structure Theory,* H. F. Schaefer III, ed. (Plenum Press, New York, 1977), pp.51-78.
5. B. O. Roos, P. R. Taylor, and P. E. M. Siegbahn, Chem. Phys., **48**, 157 (1980).
6. G. C. Lie, J. Hinze, and B. Liu, J. Chem. Phys. **59**, 1872 (1973); *ibid.* 1887 (1973).
7. J. W. Liu, G. C. Lie, and B. Liu, J. Phys. B **20**, 2853 (1987).
8. E. A. Hylleraas, Z. Physik **54**, 347 (1929).
9. T. Kinoshita, Phys. Rev. **105**, 1490 (1957); *ibid.* **115**, 366 (1959).
10. C. L. Pekeris, Phys. Rev. **112**, 1649 (1958); *ibid.* **115**, 1216 (1959).
11. K. Frankowski and C. L. Pekeris, Phys. Rev. **146**, 46 (1966).
12. H. M. James and A. S. Coolidge, Phys. Rev. **49**, 688 (1936).
13. S. Larsson, Phys. Rev. **169**, 49 (1968).
14. J. F. Perkins, Phys. Rev. A **5**, 514 (1972); *ibid.* **13**, 915 (1976).
15. Y. K. Ho, Int. J. Quant. Chem. **20**, 1077 (1981).
16. F. W. King and V. Shoup, Phys. Rev. A **33**, 2940 (1986).
17. H. M. James and A. S. Coolidge, J. Chem. Phys. **1**, 825 (1933).
18. W. Kolos and L. Wolniewicz, J. Chem. Phys. **41**, 3663 (1964); *ibid.* **43**, 2429 (1965); Phys. Rev. Lett. **20**, 243 (1968).
19. R. S. Mulliken and W. C. Ermler, *Diatomic Molecules: Results of Ab Initio Calculations* (Academic, New York, 1977).
20. J. S. Sims and S. Hagstrom, Phys. Rev. A, **4**, 908 (1971).

21. A. Preiskorn and W. Woznicki, Chem. Phys. Lett. **86**, 369 (1982); *ibid.*, Mol. Phys. **52**, 1291 (1984).
22. V. R. Saunders, in *Methods in Computational Physics,* G. H. F. Diercksen and S. Wilson ed. (Reidel, Dordrecht, 1983), Vol. 2, p. 1.
23. A. Largo-Cabrerizo and E. Clementi, J. Comp. Chem. **8**, 1191 (1987).
24. A. Largo-Cabrerizo, C. Urdaneta, G. C. Lie, and E. Clementi, Int. J. Quant. Chem.: Quant. Chem. Symp. **21**, 677 (1987).
25. D. Frye, G. C. Lie, and E. Clementi, *HyCoIn: Documentation and User's Guide*, (IBM Kingston Technical Report KGN-180).
26. C. Urdaneta, A. Largo-Cabrerizo, J. Lievin, G. C. Lie, and E. Clementi, J. Chem. Phys. **88**, 2091 (1988).
27. D. Frye, G. C. Lie, and E. Clementi, J. Chem. Phys. **91**, 2366 (1989).
28. D. Frye, G. C. Lie, and E. Clementi, J. Chem. Phys. **91**, 2369 (1989).
29. D. Frye, G. C. Lie, and E. Clementi, *HCI Calculations on the Potential Energy Curve for the* $X^1\Sigma_g^+$ *State of* H_2, (IBM Kingston Technical Report KGN-179).
30. A. Preiskorn, D. Frye, G. C. Lie, and E. Clementi, J. Chem. Phys. **92, 4941** (1990).
31. D. Frye, A. Preiskorn, G.C. Lie, and E. Clementi, J. Chem. Phys. **92**, 4948 (1990).
32. A. Preiskorn, D. Frye, and E. Clementi, *Gaussian Functions in Hylleraas-CI Calculations. VI. The* $^3\Sigma_u^+$ *bound excited state of* H_3^+ *and* D_3^+, (IBM Kingston Technical Report KGN-220), in preparation.
33. H. A. Bethe, E. E. Salpeter, *Quantum Mechanics of One- and Two-Electron Atoms* (Plenum/Rosetta, New York, 1977), 370 pp.
34. N. C. Handy, in *Computational Techniques in Quantum Chemistry and Molecular Physics*, G. H. F. Diercksen, B. T. Sutcliffe, and A. Veillard, ed. (Reidel, Dordrecht, 1975), p.425.
35. W. Kutzelnigg, Theor. Chim. Acta **68**, 445 (1985).
36. W. Kolos, K. Szalewicz, and H. Monkhorst, J. Chem. Phys. **84**, 3259 (1986).
37. B. Liu, J. Chem. Phys. **80**, 581 (1984).
38. D. Feller, private communication (E. R. Davidson and D. Feller, to appear in Rev. Comp. Chem.).
39. F. Borondo, F. Martín, and M. Yáñez, J. Chem. Phys. **86**, 4982 (1987).
40. J. Rychlewski, Chem Phys. Lett. **151**, 553 (1988).
41. L. Salmon and R. D. Poshusta, J. Chem. Phys. **59**, 3497 (1973).
42. G. D. Carney and R. N. Porter, J. Chem. Phys. **65**, 3547 (1976).
43. C. E. Dykstra and W. C. Swope, J. Chem. Phys. **70**, 1 (1979).
44. F. Mentch and J. B. Anderson, J. Chem. Phys. **74**, 6307 (1981).
45. P. G. Burton, E. Von N.-Felsobuki, G. Doherty, and M. Hamilton, Mol. Phys. **55**, 527 (1985).
46. W. Meyer, P. Botschwina, and P. Burton, J. Chem. Phys. **84**, 891 (1986).
47. J. B. Anderson, J. Chem. Phys. **86**, 2839 (1987).
48. C. A. Traynor and J. B. Anderson, Chem. Phys. Lett. **147**, 389 (1988).
49. S. Huang, Z. Sun, and W. A. Lester, Jr., J. Chem. Phys. **92**, 597 (1990).

50. S. A. Alexander, H. J. Monkhorst, R. Roeland, and K. Szalewicz, to be published.
51. L. J. Schaad and W. V. Hicks, J. Chem. Phys. **61**, 1934 (1974).
52. R. Ahlrichs, C. Votava, and C. Zirz, J. Chem. Phys. **66**, 2771 (1977).

Chapter 6.

HONDO: A General Atomic and Molecular Electronic Structure System†

M. Dupuis, A. Farazdel,
S.P. Karna‡ and S.A. Maluendes

IBM Corporation
Scientific and Engineering Computations
Dept. 48B/428
Kingston, New York 12401, U.S.A.

A. Introduction

In this chapter we are concerned with the determination of molecular wavefunctions and properties in the framework of the Born-Oppenheimer approximation[1] whereby the molecular wavefunction Φ is written as the product of a nuclear wavefunction X and an electronic wavefunction Ψ which depends on the nuclear coordinates $\underset{\sim}{R}$ as parameters, i.e. $\Phi(\underset{\sim}{r}; \underset{\sim}{R}) = \Psi(\underset{\sim}{r}; \underset{\sim}{R})X(\underset{\sim}{R})$. At all levels of theory the electronic wavefunction $\Psi(\underset{\sim}{r}; \underset{\sim}{R})$ is built from molecular orbitals $\underset{\sim}{\phi}$ which are functions of the coordinates of one electron only and are expressed as linear combinations of atomic orbitals $\underset{\sim}{\chi}$. The many-electron wavefunction is obtained by solving the time-independent electronic Schrödinger equation,

$$H\,\Psi(\underset{\sim}{r}; \underset{\sim}{R}) = E_{elec}(\underset{\sim}{R})\,\Psi(\underset{\sim}{r}; \underset{\sim}{R}) \qquad (A.1)$$

for a given $\underset{\sim}{R}$ where H is the electronic Hamiltonian and E_{elec} is the electronic energy. Throughout the years a great deal of research has been concerned with various approximations for expressing and calculating the electronic wavefunction. More recently equal attention has been directed toward determining the nuclear wavefunction and obtaining a more dynamical picture of molecular systems in the context of molecular spectroscopy or chemical reactivity.

---

† HONDO has been enhanced since last year and hence this chapter has been revised accordingly (compared to its predecessor, Chapter 7 in MOTECC-89). Essentially, we have added the new sections C.1.c, D.7, E.1.a, and E.1.b and have made a few minor modifications.

‡ Present address: Photonics Laboratory, Department of Chemistry, S.U.N.Y., Buffalo.

The development of a computer program with sufficiently general and comprehensive capabilities so that it can be used for very accurate studies of small molecules or for less accurate studies of larger molecules, or for the determination of reaction pathways or the calculation of electronic properties, is by no means an easy task and certainly requires several man-years of work. Over the years we have developed such a program,[2] HONDO, in collaboration with many other researchers, starting from the doctoral work of one of us (MD) and of John Rys, under the supervision of Professor Harry F. King. The capabilities of the present computer code constitute the underlying backbone of this chapter.

The goal of this chapter is to provide an overview of HONDO, and of the methodological developments that we have contributed to this program. Thus this text gathers in one place several items which have been described in more detail elsewhere. We feel that this collection of methodological presentations will be useful to the researcher interested in expanding the capabilities of the code as part of his own research program. Throughout the chapter many other methods and algorithms are mentioned and discussed in general terms, but are not presented in detail. Appropriate references are given which the interested reader should turn to for further study of the problem.

The chapter is organized as follows: first we outline various approximations used for the determination of the electronic wavefunction. Second we discuss the calculation of molecular properties which can be extracted from the electronic density. Third we focus our attention on algorithms which provide a representation of the motion of the nuclei on the potential surface of the electronic energy. Fourth we discuss miscellaneous aspects which form an integral part of an efficient computer implementation of wavefunction calculations. We finish the chapter with an overview of some studies carried out with HONDO. These were selected to highlight the capabilities of the code.

B. Wavefunctions and Energies

1. Closed Shell Hartree-Fock (SCF) Wavefunction

(a) Theory and Energy expression: The electronic wavefunction is written as a Slater determinant, an antisymmetrized product of spin-orbitals, an ansatz which guarantees that the Pauli exclusion principle[3] is satisfied:

$$\Psi(r_1, r_2, \ldots, r_n) = (n!)^{-1/2} \sum_{P} (-1)^p \hat{P}[\phi_1(1)\phi_2(2)\ldots\phi_n(n)] \qquad (B.1)$$

where $\hat{P}$ is the permutation operator of the electrons among the orbitals, and p is the signature of the permutation. For a closed shell system in which each spatial orbital ϕ is associated with two electrons with spin α and β, the electronic energy is given by:

$$E_{elec} = \sum_i 2(i|h|i) + \sum_{i,j} [2(ii|jj) - (ij|ij)] \qquad (B.2)$$

The total energy of the system is equal to the electronic energy plus the electrostatic nuclear repulsion

$$E_{total} = E_{elec} + E_{nuc} \qquad (B.3)$$

$$E_{nuc} = \sum_{A,B}^{Nuc} \frac{Z_A Z_B}{R_{AB}} \qquad (B.4)$$

In Eq. (B.2), $(i|h|j)$ is the matrix element over the orbital ϕ_i and ϕ_j of the bare nucleus operator

$$h(1) = -\frac{1}{2}\nabla^2(1) - \sum_{A=1}^{Nuc} \frac{Z_A}{r_{1A}} \qquad (B.5)$$

$$(i|h|j) = \int \phi_i(1)\, h(1)\, \phi_j(1)\, d\tau_1 \qquad (B.6)$$

and $(ij|ij)$ is the electrostatic self-interaction of the $\phi_i(1)\phi_j(1)$ charge distribution:

$$(ij|ij) = \iint \phi_i(1)\phi_j(1) \frac{1}{r_{12}} \phi_i(2)\phi_j(2)\, d\tau_1\, d\tau_2 \qquad (B.7)$$

The molecular orbitals ϕ are expressed as linear combinations of the atomic orbitals χ according to:

$$\phi = \chi C \qquad (B.8)$$

and are subject to the orthonormality condition:

$$C^\dagger S C = 1 \qquad (B.9)$$

where

$$S_{\mu\nu} = \int \chi_\mu(1)\chi_\nu(1) d\tau_1 \qquad (B.10)$$

From the application of the variational principle,[4] the C coefficients corresponding to the lowest energy are solutions of the Roothaan-Hall[5] equations:

$$F C = S C \varepsilon \qquad (B.11)$$

with

$$F_{\mu\nu} = h_{\mu\nu} + \sum_{\rho,\sigma} D_{\rho\sigma}[(\mu\nu|\rho\sigma) - \frac{1}{2}(\mu\rho|\nu\sigma)] \qquad (B.12)$$

in which

$$D_{\mu\nu} = \sum_{i}^{occ} 2\, C_{\mu i}\, C_{\nu i} \qquad (B.13)$$

The matrix of orbital coefficients C is obtained by solving Eq. (B.11) which in turn depends on C through F and D. This is the basis of the self-consistent-field (SCF) method. Starting from an initial guess for the molecular orbitals (the bare nucleus hamiltonian may be used, although better guesses can be generated), the density matrix is constructed and combined with the electron repulsion integrals to form the Fock matrix F.

(b) Solving the SCF equations: The generalized eigenvalue problem of Eq. (B.11) is converted to a true eigenvalue problem by means of any orthonormalizing transformation. Indeed if Q is a matrix such that:

$$Q^{\dagger} S\, Q = I \qquad (B.14)$$

It follows that:

$$\tilde{F}\, \tilde{C} = \tilde{C}\, \varepsilon \qquad (B.15)$$

where

$$\tilde{F} = Q^{\dagger} F\, Q \qquad (B.16)$$

and

$$\tilde{C} = Q^{-1}\, C \qquad (B.17)$$

In practice Q is calculated, and so is $\tilde{F}$, and $\tilde{C}$ is obtained from the diagonalization of $\tilde{F}$. Actually since we are interested in C, there is no need to calculate Q^{-1}, but instead we use the equation:

$$C = Q\, \tilde{C} \qquad (B.18)$$

There are several choices for Q. One possibility is to use the C matrix from the previous iteration. This approach is not desirable since numerical inaccuracies accumulate during the SCF process, and a reorthonormalization of the molecular orbital coefficients should be performed every few cycles. Another possibility for Q is to use the matrix $S^{-1/2}$, where S is the overlap matrix defined in Eq. (B.10). If we write the eigenvalue equation for S:

$$S\, U = U\, s \qquad (B.19)$$

then we find that

$$S^{-1/2} = U\, s^{-1/2}\, U^{\dagger} \qquad (B.20)$$

The disadvantage associated with this choice of Q lies in the fact that it may happen that some of the eigenvalues of S become very small when the basis set is nearly redundant, thus leading to numerical difficulties when calculating the inverse square root of these eigenvalues.

The approach implemented in HONDO is to use the so-called canonical orthonormal orbitals, defined as follows:

$$Q = \tilde{U}\tilde{s}^{-1/2} \tag{B.21}$$

where the $\sim$ in Eq. (B.21) indicates that the eigenvectors corresponding to the small eigenvalues s are excluded from the transformation and from the matrix Q. The numerical difficulties associated with near-linear dependencies of the basis set are thus eliminated. In practice, the program makes use of a user specified parameter which defines the threshold value at which the basis set is considered linearly-dependent, and accordingly discards the eigenvalues (and associated eigenvectors) smaller than the threshold.

(c) Construction of the Fock matrix: The construction of the Fock matrix is an integral driven procedure. The electron repulsion integrals are read from external storage into memory along with the four indices of the basis functions, while the density matrix resides in memory. In computing the electron repulsion integrals, the permutational symmetry of the integrals is taken into account so that only $(\mu\nu|\rho\sigma)$ is computed and stored. None of the equivalent integrals $(\mu\nu|\sigma\rho)$, $(\nu\mu|\rho\sigma)$, $(\nu\mu|\sigma\rho)$, $(\rho\sigma|\mu\nu)$, $(\rho\sigma|\nu\mu)$, $(\sigma\rho|\mu\nu)$, $(\sigma\rho|\nu\mu)$ is computed. To compensate, the integral $(\mu\nu|\rho\sigma)$ is used to make six different contributions to the Fock matrix in general:

$$\begin{array}{rl} D_{\mu\nu}\,(\mu\nu|\rho\sigma) & \rightarrow F_{\rho\sigma} \\ D_{\rho\sigma}\,(\mu\nu|\rho\sigma) & \rightarrow F_{\mu\nu} \\ -\frac{1}{4}D_{\mu\rho}\,(\mu\nu|\rho\sigma) & \rightarrow F_{\nu\sigma} \\ -\frac{1}{4}D_{\mu\sigma}\,(\mu\nu|\rho\sigma) & \rightarrow F_{\nu\rho} \\ -\frac{1}{4}D_{\nu\rho}\,(\mu\nu|\rho\sigma) & \rightarrow F_{\mu\sigma} \\ -\frac{1}{4}D_{\nu\sigma}\,(\mu\nu|\rho\sigma) & \rightarrow F_{\mu\rho} \end{array} \tag{B.22}$$

When there are coincidences among the four indices of an integral, fewer contributions should be made, and for this reason the processing of the eleven possible different types of integrals ought to proceed through different parts of the computer code An alternative approach is used in HONDO whereby all the integrals are processed in an identical manner, regardless of possible coincidences among indices. To make up for too large contributions to the Fock matrix when coincidences exist, the integrals written on external storage differ by a factor of 2, 4, or 8 from the electron repulsion integrals. The quantities stored are:

$$\{\mu\nu|\rho\sigma\} = (\mu\nu|\rho\sigma) \times \frac{1}{(1+\delta_{\mu,\nu})} \times \frac{1}{(1+\delta_{\rho,\sigma})} \times \frac{1}{(1+\delta_{\mu\nu,\rho\sigma})} \qquad (B.23)$$

where δ represents a Kronecker symbol. The simplicity of the computer code for processing the modified integrals makes up for the slightly increased number of floating point operations, and results in a highly efficient code.

Raffenetti[6] suggested that if supermatrix integrals were stored instead of integrals, a significantly reduced number of operations would result. These supermatrix integrals are defined as follows:

$$P_{\mu\nu\rho\sigma} \equiv [\mu\nu|\rho\sigma] = (\mu\nu|\rho\sigma) - \frac{1}{4}[(\mu\rho|\nu\sigma) + (\mu\sigma|\nu\rho)] \qquad (B.24)$$

Each supermatrix integral makes two contributions to the Fock matrix according to:

$$D_{\mu\nu}[\mu\nu|\rho\sigma] \rightarrow F_{\rho\sigma}$$
$$D_{\rho\sigma}[\mu\nu|\rho\sigma] \rightarrow F_{\mu\nu} \qquad (B.25)$$

Just as discussed above these two contributions are made for all supermatrix integrals regardless of coincidences between $\mu\nu$ and $\rho\sigma$. To make up for too large contributions in cases of coincidences, the supermatrix integrals are divided by the factor $(1+\delta_{\mu\nu,\rho\sigma})$. As with the integrals, the simplicity of the code more than makes up for the slightly greater number of floating point operations, and the processing of the supermatrix integrals is highly efficient.

It can be seen from the contributions described above that in theory the processing of the supermatrix integrals requires three times less floating point operations than the processing of the normal electron repulsion integrals. In practice a factor of two or more is observed. Indeed only integrals greater than a cutoff value are written on external storage. Experience shows that the same cutoff value results in more supermatrix integrals being written out than integrals.

(d) Symmetry Blocking: The transformation step from the generalized eigenvalue problem to the true eigenvalue problem, may be used additionally to block the Fock matrix $\tilde{F}$ according to the irreducible representations of the symmetry point group to which the molecule belongs, if any. To this end the program generates, for any symmetry point group, a matrix T defining the symmetry adapted basis functions χ':

$$\chi' = \chi T \qquad (B.26)$$

In the symmetry adapted basis, the overlap matrix is S' obtained from S according to:

$$S' = T^{\dagger} S T \qquad (B.27)$$

The matrix Q' is derived from S' according to Eq. (B.21), and the orthonormalizing transformation becomes:

$$Q = T\, Q' \qquad (B.28)$$

It is easy to show that the Q matrix just defined does convert the Fock equation to a true eigenvalue problem. Furthermore, because of the definition of T, the matrices S' and $\tilde{F}$ are blocked by irreducible representations, and the columns of Q' transform like the irreducible representations, and the partners of the representations when those are degenerate. If N is the size of the atomic basis set, the blocking of matrices by irreducible representations reduces the computational task from $N^3 = (\sum n_{irrep})^3$ dependence to a $\sum n_{irrep}^3$ where n_{irrep} is the dimension of the subspace that transforms as the representation *irrep*.

Other advantages obtained by taking point symmetry into account are described later. These involve the need to deal only with symmetry unique electron repulsion integrals and their derivatives when calculating energies and energy gradients and hessians.[7-9]

(e) Extrapolation, Damping, and Level Shifting: The iterative process of the SCF method is not guaranteed to converge. It is almost always advantageous to help control the convergence process, and to accelerate it. This is done using a variety of methods such as extrapolation, damping, and level shifting. The DIIS method[10] devised by Pulay is generally believed to improve significantly the convergence of the SCF process, and has become very popular in the past few years. It is not currently implemented in HONDO, which relies on damping and level shifting techniques to stabilize and accelerate the SCF process. The damping algorithm in HONDO is borrowed from the code developed in the group of Davidson and is summarized below.

Let us denote D_0, D_1, ...D_i the density matrices at iterations 0, 1, ...i. We want to devise a modified density matrix $\tilde{D}_i$. This modified matrix is the damped density matrix defined as follows:

$$\tilde{D}_i = \frac{(D_i + \lambda D_{i-1})}{(1.0 + \lambda)} \qquad (B.29)$$

When $\lambda = 0.0$, there is no damping, and the iterative process progresses unaltered. When $\lambda = \infty$, the SCF process is stopped, as the new density matrix is equal to the old one. A small value of λ generally suffices to stabilize the SCF process, and leads to convergence. Davidson devised an improved scheme, however, according to which the value of the damping parameter λ is updated automatically by the program at each iteration. The value of λ becomes large when the convergence behavior of the energy is not smooth, and it tends to zero as the energy is well behaved. The algorithm used for updating λ and devised by Davidson,[11] is empirical. Under the right circumstances (smooth convergence), the damping parameter ends up being set to zero and the program switches to an extrapolation procedure first implemented in the GAUSSIAN-70 program.[12] We have found this damping technique to be very robust, and it is the method of choice in HONDO.

In difficult circumstances, damping is not sufficient. This is particularly true for molecules containing transition metal atoms. In this case the option exists to use the technique of level shifting devised by Saunders and Hillier.[13] According to this technique, if we use the previous orbitals as the $\tilde{Q}$ matrix (as described above), then $\tilde{F}$ defines how much mixing of the virtual orbitals and the occupied orbitals should occur to lower the energy. As the SCF process progresses the $\tilde{F}$ matrix thus defined becomes nearly diagonal, with the diagonal elements equal to the orbital energies. Adding a constant factor, the level shift, to the diagonal elements corresponding to the virtual orbitals, artificially restricts the mixing of the occupied and virtual orbitals. A large level shift prevents the SCF process from progressing, while a small level shift usually results in a stable process. Because we use this technique mostly only when convergence is difficult, no attempt is made to reduce the level shift parameter automatically. A possible scenario involves running say ten iterations with some value of the level shift parameter, and then restarting the calculation with a smaller level shift for another ten iterations, and so on. Once the process is well under control, it is then best to return to the damping option described above.

2. Spin Unrestricted Open Shell Hartree-Fock (UHF) Wavefunction

(a) Theory And Energy Expression: If we consider now a wavefunction made up of n_α spin-orbitals with spin α, and n_β spin-orbitals with spin β without the requirement that a given molecular orbital be occupied with two electrons of opposite spin, but rather that spin up electrons occupy spatial orbitals different than the spatial orbitals occupied by the spin down electrons, we have what is called a spin unrestricted Hartree-Fock wavefunction.[14] The energy expression for such a wavefunction is:

$$E_{elec} = \sum_{i}^{\alpha+\beta} (i|h|i) + \frac{1}{2}\sum_{i,j}^{\alpha+\beta} (ii|jj) - \frac{1}{2}\sum_{i,j}^{\alpha} (ij|ij) - \frac{1}{2}\sum_{i,j}^{\beta} (ij|ij) \quad (B.30)$$

Minimizing the energy with respect to the molecular orbital coefficients C^α and C^β gives rise to the following two equations:

$$F^\alpha C^\alpha = S C^\alpha \varepsilon^\alpha \quad (B.31)$$

$$F^\beta C^\beta = S C^\beta \varepsilon^\beta \quad (B.32)$$

$$F^\alpha_{\mu\nu} = h_{\mu\nu} + \sum_{\rho,\sigma} [(D^\alpha_{\rho\sigma} + D^\beta_{\rho\sigma})(\mu\nu|\rho\sigma) - D^\alpha_{\rho\sigma}(\mu\rho|\nu\sigma)] \quad (B.33)$$

$$F^{\beta}_{\mu\nu} = h_{\mu\nu} + \sum_{\rho,\sigma} [(D^{\alpha}_{\rho\sigma} + D^{\beta}_{\rho\sigma})(\mu\nu|\rho\sigma) - D^{\beta}_{\rho\sigma}(\mu\rho|\nu\sigma)] \qquad (B.34)$$

in which

$$D^{\alpha}_{\mu,\nu} = \sum_{i}^{\alpha occ} C^{\alpha}_{\mu,i} C^{\alpha}_{\nu,i} \qquad (B.35)$$

$$D^{\beta}_{\mu,\nu} = \sum_{i}^{\beta occ} C^{\beta}_{\mu,i} C^{\beta}_{\nu,i} \qquad (B.36)$$

The algorithms for solving the SCF equations described in detail for closed shell SCF calculations are easily extended to the UHF case. The orthonormalizing transformation to reduce the SCF equations to true eigenvalue problems is valid in the present case with the choice of the Q matrix that blocks the Fock matrix by symmetry species. The processing of the electron repulsion integrals proceeds in a similar manner, just as the processing of the supermatrix integrals. We note that in addition to the integral combination $P_{\mu\nu\rho\sigma}$ given in Eq. (B.24), the exchange combination by itself is needed:

$$K_{\mu\nu\rho\sigma} = \frac{1}{4} [(\mu\rho|\nu\sigma) + (\mu\sigma|\nu\rho)] \qquad (B.37)$$

The extrapolation, damping, and level shifting procedures used for closed shell SCF calculations apply equally well to UHF calculations, with the added feature that one may want to use different level shift parameters for the open shells and for the virtual orbitals.

3. High Spin Restricted Open Shell Hartree-Fock (ROHF) Wavefunction

(a) Theory and Energy Expression: The energy expression for a wavefunction made up of closed shells and half-filled open shells can be written:

$$E_{elec} = \sum_{i}^{cl,op} n_i (i|h|i) + \sum_{i,j}^{cl,op} \frac{n_i n_j}{2} [(ii|jj) - \frac{1}{2}(ij|ij)] + \sum_{i,j}^{op,op} \frac{1}{4} \beta_{i,j} (ij|ij) \qquad (B.38)$$

In Eq. (B.38) n_i is the occupation number of the shell i, $n = 2$ for closed shells and $n = 1$ for open shells. The β_{ij} are the coupling coefficients. This expression, used by Davidson,[15] is valid for any single configuration with open shells. In such cases the open shells are grouped into sets, and each set corresponds to a different Fock operator. It can be shown that two half occupied shells, whose spins are coupled to form a triplet, belong to the same set, i.e. the molecular orbitals are eigenvectors of the same operator. In that case $\beta_{ij} = -1.0$. It is clear then that the wavefunction is totally determined by only a few parameters. These are the parameters required as input to HONDO. The number of closed shells is denoted *nco*, the number of sets of open shells is denoted *nseto*, and the number of open shells in the $i - th$ set is denoted no_i. Finally the spin multiplicity of the wavefunction is denoted *nspin*.

In HONDO, Davidson's algorithm is implemented for high spin cases only, such as doublet, triplet, quartet, quintet ... spins for one, two, three, four ... unpaired electrons outside the closed shells. In these cases the β_{ij} values are always the same and therefore need not be input. For more general wavefunctions where not all open shells are high spin coupled, the algorithm outlined below is used in HONDO. Extrapolation, damping and differential level shifting (see UHF) are available to help convergence.

4. General Restricted Open Shell Hartree-Fock and Generalized Valence Bond Wavefunctions (ROHF-GVB)

(a) Theory and Energy Expression: The energy expression has the following form:

$$E_{elec} = \sum_{i}^{cl,\,op} 2f_i\,(i|h|i) + \sum_{i,j}^{cl,\,op} a_{ij}\,(ii|jj) + \sum_{i,j}^{cl,\,op} b_{ij}\,(ij|ij) \qquad (B.39)$$

The f_i are the shell occupation numbers and the a_{ij} and b_{ij} are coupling coefficients. Such an energy expression applies to any open shell single configuration as well as to wavefunctions of the Generalized Valence Bond type.[16] In addition to the parameters described above, which fully determine the wavefunction, GVB wavefunctions are characterized by pairs of electrons which participate in valence bond mixing. Each such pair is characterized by two molecular orbitals, a bonding orbital and a correlating (anti-bonding) orbital. The wavefunction has the form of a linear combination of doubly excited configurations. The additional input parameter needed to fully specify a GVB wavefunction is denoted (quite naturally) *npair*. For ease of use, default values of the coupling coefficients are given for the most often encountered molecular wavefunctions. For more complex cases the user may input these parameters. This option is most useful when calculating average state energies in the case of atoms or linear molecules, as mentioned briefly below.

The computer code for GVB calculations found in HONDO was adapted from the early version of Goddard and Hunt's program,[17] into which the damping and level shifting schemes described earlier were incorporated to help stabilize and accelerate convergence. We note a slight difference between this code and the SCF codes described earlier. The difference has to do with the criterion used to decide whether the SCF process has converged. For the closed shell program, the UHF program, and the ROHF program convergence is based on the largest change in the density matrix. A convergence threshold of 1.0E-05 is generally sufficient. For the GVB program convergence is based on the magnitude of the elements of the orbital rotation matrix. This is a more stringent convergence condition, and here the convergence threshold of 1.0E-04 is generally sufficient.

The generality of the energy expression is beneficial when calculating the energy of configurations in which one or more open shells belong to a degenerate irreducible representation. More precisely the average energy of the degenerate members of the wavefunction must be calculated if one wants to obtain a symmetric charge density. As an example consider an oxygen atom in its ground $O(^3P)$ state. A fully spin and symmetry restricted wavefunction is obtained by averaging the energy of the $O(^3P_{+1})$ state, of the $O(^3P_0)$ state, and of the $O(^3P_{-1})$ state.

5. Configuration Interaction (CI) wavefunction

(a) Theory and Energy Expression: A CI wavefunction is a linear combination of configuration state functions (CSF) which are spin and symmetry adapted linear combinations of Slater determinants. Thus we write:

$$\Psi_{CI} = \sum_I C_I \Phi_I \qquad (B.40)$$

In Eq. (B.40) the Φ_I represent the configuration state functions, which form an orthonormal basis of the n electrons wavefunction space, and the C_I are the CI coefficients. These coefficients are obtained by applying the variational principle which yields the secular equation:

$$\mathbf{H}\,\mathbf{C} = E_{elec}\,\mathbf{C} \qquad (B.41)$$

The **H** matrix is the matrix of CI interaction elements and has the general form:

$$\mathbf{H}_{IJ} = \sum_{i,j} \gamma_{IJ}^{ij}\,(i|h|j) + \frac{1}{2}\sum_{i,j,k,l} \Gamma_{IJ}^{ijkl}\,(ij|kl) \qquad (B.42)$$

The elements γ_{IJ}^{ij} and Γ_{IJ}^{ijkl} are called structure constants and their values depend on the occupation numbers of the molecular orbitals used to construct the configuration state functions and on the spin couplings of the electrons. In contrast with the multiconfiguration SCF (MCSCF) method described later, the molecular orbitals which enter the configuration state functions are fixed. They are generally chosen as the orbitals optimized for a single configuration wavefunction, or for a MCSCF wavefunction calculated prior to carrying out the CI calculation.

The efficient evaluation of these structure constants has been the subject of extensive research. In particular the graphical unitary group approach developed by Shavitt and others[18] has proven to be extremely efficient. The code incorporated in HONDO is the original code developed by Schaefer and coworkers.[19] Very large CI expansions are routinely handled by codes based on more recent versions of the unitary group approach. The existing code in HONDO is in practice limited to a CI expansion of the order of a few tens of thousands configuration state functions.

At this point it is appropriate to introduce a few more quantities. They are the one-particle reduced density matrix (DM1) and the two-particle reduced density matrix elements (DM2). Given the CI coefficients $\mathbf{C}_I$ for a CI state (an eigenvector of the Hamiltonian matrix **H**), these elements are given by:

$$\gamma^{ij} = \sum_{I,J} \mathbf{C}_I \gamma_{IJ}^{ij} \mathbf{C}_J \tag{B.43}$$

and

$$\Gamma^{ijkl} = \sum_{I,J} \mathbf{C}_I \Gamma_{IJ}^{ijkl} \mathbf{C}_J \tag{B.44}$$

The electronic energy is thus equal to:

$$E_{elec} = \sum_{I,J} \mathbf{C}_I \mathbf{H}_{IJ} \mathbf{C}_J = \sum_{i,j} \gamma^{ij} (i|h|j) + \frac{1}{2} \sum_{i,j,k,l} \Gamma^{ijkl} (ij|kl) \tag{B.45}$$

The order of the DM1 matrix γ defined according to Eq. (B.43) is equal to the total number of orbitals which are occupied in at least one CSF. The eigenvectors of γ are the natural orbitals of the CI wavefunction, and the eigenvalues are the corresponding occupation numbers.

(b) Diagonalization: While the CI expansion may be as large as several thousands of configurations, we are only interested in obtaining the first few eigenvectors of the CI matrix corresponding to the lowest few eigenvalues. The method proposed by Davidson[20] is generally accepted as being the one best adapted to solving the CI problem in the framework of the unitary group approach to the calculation of CI matrix elements. It is an iterative method whereby, given an approximate solution to the eigenvalue problem, a correction vector is determined which, when added to the original vector, yields an improved approximate solution. If $\mathbf{C}^{(i)}$ is the approximate solution at the $i-th$ iteration, the new correction vector is obtained after calculating the product:

$$\sigma^{(i+1)} = \mathbf{H} \cdot \mathbf{C}^{(i)} \tag{B.46}$$

a step which represents the time consuming part of the computation. The process is repeated till convergence is reached, i.e., when the eigenvalue no longer changes within some threshold. When interested in electronic excited states, we must obtain the few lowest eigenvalues and corresponding eigenvectors of the CI matrix. Liu[21] extended Davidson's method to extract several eigenvectors simultaneously. This variant of Davidson's method is implemented in HONDO.

6. Multiconfiguration Hartree-Fock (MCSCF) wavefunction

(a) Theory and Energy Expression: The MCSCF theory uses the CI theory just outlined as a starting point, and determines the molecular orbitals which yield the lowest possible energy for the CI state of interest. The problem at hand is to rotate the molecular orbitals among themselves so as to decrease the electronic energy. To this end we make use of the Newton-Raphson technique to minimize a function of many variables.

Let us consider a function $E(\Theta)$ where Θ represents a set of parameters $(\theta_1, \theta_2, ..., \theta_n)$. For $\Theta = \Theta_0$, we have $E = E_0$. For a value Θ not too far from Θ_0 we can write:

$$E = E_0 + g\,\Delta\Theta + \frac{1}{2}\Delta\Theta^\dagger H\,\Delta\Theta + ... \qquad (B.47)$$

g and H are the gradient and hessian of the function E with respect to the parameters Θ. If the function to be minimized is truly a quadratic function of the parameters, then the optimum change in Θ leading to the minimum of the function is given by:

$$\Delta\Theta = -H^{-1}g \qquad (B.48)$$

If the function E is not exactly quadratic in Θ then one expects the change defined in Eq. (B.48) to lead to a lower value of the function. The process is repeated till convergence is reached, i.e., $g = 0$. This is the Newton-Raphson procedure. In practice it is known that the process does not converge unless it is started close to the minimum (the radius of convergence or trust radius is very small). It is then necessary to apply some sort of damping to stabilize the process. The Augmented Hessian Matrix approach[22] is one available technique. It was shown to be equivalent to a dynamic damping technique in which the damping factor goes to zero as convergence is reached. In this approach the augmented hessian matrix is constructed as shown in Eq. (B.49), for which we require the lowest eigenvector which gives the optimum change for this iteration:

$$\begin{bmatrix} 0 & g^\dagger \\ g & H \end{bmatrix} \Delta\Theta = \lambda\,\Delta\Theta \qquad (B.49)$$

Given the energy expression of Eq. (B.45), we want to find a unitary transformation U which yields a lower energy when applied to the present molecular orbitals. The orbital change is written as follows:

$$C' = C\,U \qquad (B.50)$$

As an approximation we assume that the CI coefficients stay unchanged, and so do the DM1 and DM2 density elements. Upon rotation of the orbitals the new energy is E'. In second-order we write:

$$U = 1 + dW + \frac{1}{2} d^2W + \ldots \tag{B.51}$$

It follows that:

$$\begin{aligned} E' = E &+ \sum_{r,i} 2\,\varepsilon^{ri}\, dW_{ri} + \frac{1}{2}\sum_{r,j} 2\,\varepsilon^{rj}\, d^2W_{rj} \\ &+ \frac{1}{2}\sum_{r,i}\sum_{s,j} dW_{ri}\, 2\,(r|F^{ij}|s)\, dW_{sj} + \frac{1}{2}\sum_{r,i}\sum_{s,k} dW_{ri}\, 2\,\Gamma^{ijkl}\,(rj|sl)\, dW_{sk} \end{aligned} \tag{B.52}$$

with the operator F^{ij} defined as:

$$F^{ij} = h\,\gamma^{ij} + \sum_{k,l} |kl)\,\Gamma^{ijkl} \tag{B.53}$$

and

$$\varepsilon^{ri} = \sum_{j} (r|F^{ij}|j) \tag{B.54}$$

In Eq. (B.52) we note that the Lagrangian terms defined by Eq. (B.54) contribute to the energy gradient and to the energy hessian: these contributions are the first-order changes in the energy, and the second-order changes in the energy due to second-order changes in the parameters (here the orbital rotations). We call the fourth contribution in Eq. (B.52) the coulomb terms, and the fifth contribution the exchange terms. These two contributions correspond to second-order changes in the energy due to the product of two first-order changes in the parameters.

Let us suppose that in Eq. (B.51) we can write d^2W formally as the square of dW, i.e., $d^2W = dW \cdot dW$. It follows that:

$$\sum_{r,j} 2\,\varepsilon^{rj}\, d^2W_{rj} = \sum_{r,i}\sum_{s,j} dW_{ri}\, 2\,\varepsilon^{rj}\,\delta_{i,s}\, dW_{sj} \tag{B.55}$$

In these conditions we can regroup the first- and second-order terms and write:

$$E' = E + g\,dW + \frac{1}{2} dW^{\dagger} H\, dW \tag{B.56}$$

with

$$g_{ri} = 2\,\varepsilon^{ri} \tag{B.57}$$

and

$$H_{ri,sj} = h'_{ri,sj} + h''_{ri,sj} \tag{B.58}$$

where $h'_{ri,sj}$ originates from Eq. (B.55) and $h''_{ri,sj}$ is given by:

$$h''_{ri,sj} = \gamma^{ij}(r|h|s) + \sum_{k,l} \Gamma^{ijkl}(rs|kl) + \sum_{k,l} 2\,\Gamma^{ikjl}(rk|sl) \qquad (B.59)$$

To preserve the orthonormality of the molecular orbitals we use Levy's proposal[23] that U be the exponential of an antisymmetric matrix:

$$U = e^{X} = 1 + X + \frac{1}{2} X \cdot X + \dots \qquad (B.60)$$

d^2W has the form for which Eq. (B.55) was derived. Therefore we can substitute dW_{ri} by dX_{ri} in the equations above. Further, the sums in these equations can now be modified to take into account the antisymmetry of X. In Eq. (B.55) r runs over the whole space (core, fractionally occupied, and virtual orbitals) and i runs over the occupied space (core, fractionally occupied orbitals) independently, as there are no restrictions on dW_{ri}. When substituting the X matrix we now have the restriction that *r be greater than i*. Thus we substitute:

$$dW_{ri} \equiv +X_{ri} \quad \text{when } r > i$$

$$dW_{ri} \equiv -X_{ri} \quad \text{when } r < i \qquad (B.61)$$

The determination of the 'best' orbital rotations is thus done by constructing the augmented hessian matrix in the space of the rotation parameters X, by extracting the lowest eigenvector X of the augmented hessian matrix, and by constructing U according to:

$$U = 1 + X + \frac{1}{2} X \cdot X \qquad (B.62)$$

followed by orthonormalization of U. With the new orbitals, the CI problem is solved again, and the process is repeated until it converges.

The dimension of the orbital rotation space grows roughly as the product *nocc* × *nvir* where *nocc* is the number of orbitals occupied in at least one configuration, and *nvir* is the number of virtual orbitals. Thus the size of the augmented hessian matrix grows rapidly, so that we use Davidson's iterative method to extract the eigenvector of that matrix.

We now give specific expressions for the orbital gradient and hessian matrix elements according to the type of orbitals involved. Generally there are three types of orbitals (as mentioned briefly above): the *core* orbitals, denoted $a, b...$, are doubly occupied in all configurations, they have an occupation number equal to 2; the *valence* orbitals, denoted $i, j...$, are variably occupied with an occupation number between 0 and 2; the *virtual* orbitals, denoted $x, y...$, are never occupied. The density matrix elements DM1 and DM2 have the following properties:

$$\gamma^{ab} = 2\,\delta_{a,b}$$

$$\gamma^{ai} = \gamma^{ax} = \gamma^{ix} = 0$$

$$\Gamma^{pqrs} = \gamma^{pq}\gamma^{rs} - \frac{1}{4}(\gamma^{pr}\gamma^{qs} + \gamma^{ps}\gamma^{qr}) + \Gamma'^{pqrs} \qquad (B.63)$$

with $\Gamma'^{pqrs} \neq 0$ only if p, q, r, s all belong to the valence space. In addition it can be shown that the energy is invariant with respect to rotations among core orbitals, invariant with respect to rotations among the valence orbitals if the CI expansion is complete in the valence space (CASSCF)[24] , and invariant with respect to rotations among virtual orbitals. Therefore we need consider only some of the rotations: core-valence, core-virtual, valence-virtual, and valence-valence rotations (these latter rotations being ignored in CASSCF calculations). To this end it is convenient to define two Fock-like matrices F^{core} and F^{val} constructed from the following density matrices:

$$D^{core}_{\rho\sigma} = \sum_{k}^{core} 2\, C_{\rho k}\, C_{\sigma k} \qquad (B.64)$$

$$D^{val}_{\rho\sigma} = \sum_{k,l}^{val} \gamma^{kl}\, C_{\rho k}\, C_{\sigma l} \qquad (B.65)$$

We define:

$$F^{core}_{\mu\nu} = (\mu|h|\nu) + \sum_{\rho,\sigma} D^{core}_{\rho\sigma}\,[(\mu\nu|\rho\sigma) - \frac{1}{2}(\mu\rho|\nu\sigma)] \qquad (B.66)$$

$$F^{val}_{\mu\nu} = \sum_{\rho,\sigma} D^{val}_{\rho\sigma}\,[(\mu\nu|\rho\sigma) - \frac{1}{2}(\mu\rho|\nu\sigma)] \qquad (B.67)$$

Transformation to the molecular orbital basis yields:

$$F^{core}_{pq} = \sum_{\mu,\nu} C_{\mu p}\, F^{core}_{\mu\nu}\, C_{\nu q} \qquad (B.68)$$

$$F^{val}_{pq} = \sum_{\mu,\nu} C_{\mu p}\, F^{val}_{\mu\nu}\, C_{\nu q} \qquad (B.69)$$

In terms of these quantities we have the following expressions for the Lagrangian and hessian terms, for any r and s orbital:

- Lagrangian, *i core*

$$\varepsilon^{ri} = 2\,(F^{core}_{ri} + F^{val}_{ri}) \qquad (B.70)$$

- Lagrangian, *i valence*

$$\varepsilon^{ri} = 2\, F_{ri}^{core} + \sum_{j,k,l}^{val} \Gamma^{ijkl}\, (rj\,|\,kl) \qquad (B.71)$$

■ hessian, *i core, j core*

$$h'_{ri,\,sj} = \gamma^{ij}\,(F_{rs}^{core} + F_{rs}^{val}) + 8\,(ri\,|\,sj) - 2\,(rj\,|\,si) - 2\,(rs\,|\,ij) \qquad (B.72)$$

where $\gamma^{ij} = 2\,\delta_{i,j}$ in this case.

■ hessian, *i core, j valence*

$$h'_{ri,\,sj} = \sum_{l}^{val} \gamma^{jl}\,[\,4\,(ri\,|\,sl) - (rl\,|\,si) - (rs\,|\,il)] \qquad (B.73)$$

■ hessian, *i valence, j core*

$$h'_{ri,\,sj} = \sum_{l}^{val} \gamma^{il}\,[\,4\,(rl\,|\,sj) - (rj\,|\,sl) - (rs\,|\,lj)] \qquad (B.74)$$

■ hessian, *i valence, j valence*

$$h'_{ri,\,sj} = \gamma^{ij}\,F_{rs}^{core} + \sum_{k,l}^{val} [\,\Gamma^{ijkl}\,(rs\,|\,kl) + 2\,\Gamma^{ikjl}\,(rk\,|\,sl)\,] \qquad (B.75)$$

The formulae given above have been implemented in HONDO. Our experience points to the robustness of the augmented hessian matrix algorithm, which converges often rapidly. The program has been used to performed calculations involving several thousand configurations, with basis sets including up to 120 basis functions. In certain situations, the augmented hessian matrix approach is known to fail to converge. In this case HONDO provides the option of using a fully quadratic approach[25] in which the rotations between configurations is taken into account at the same time as the orbital rotations. This approach improves the convergence, and in most cases achieves the convergence that the less sophisticated approach failed to reach. The present implementation in HONDO, however, restricts the use of the fully quadratic algorithm to no more than one hundred configurations. Finally it is important to recall the redundancy among the orbital rotations for the valence orbitals of a CASSCF wavefunction. In practice it means that these orbitals are not uniquely defined. A useful unique definition of the valence orbitals is in effect if one uses the natural orbitals of the CASSCF wavefunction (see above). Further the doubly occupied orbitals can be chosen to be the canonical Hartree-Fock orbitals, i.e., the eigenvectors of the core Fock operator given by Eq. (B.68). In that case all the orbitals of the CASSCF wavefunction are well defined.

7. Moller-Plesset Perturbation (MP2, MP3, MP4)

(a) Theory and Energy Expression: In this section we give the equations on which the MP4 program in HONDO is based. These have been adapted from a series of papers by Pople and coworkers.[26-29] We give the general spin orbital equations that are applicable both to closed shell restricted Hartree-Fock (RHF) and open shell unrestricted Hartree-Fock (UHF) wavefunctions. Of course, in a computer program the energies are evaluated from integrals over *spatial*, rather than *spin*, orbitals. Each spin orbital equation leads to several spatial orbital equations, one for every allowed combination of spins (spin cases). Presently the code in HONDO is limited to the case of closed shell SCF reference wavefunctions. Additionally we note that a special program exists in HONDO to perform MP2 calculations only, also for the case of a closed shell SCF reference. Indeed, owing to the much simpler nature of the problem, a different computational strategy is more appropriate than what is described below.

The following notation conventions have been adopted in this section. The "double-bar" integrals $(pq||rs)$ are defined by

$$(pq||rs) = (pr|qs) - (ps|qr) \tag{B.76}$$

where we have used charge-cloud notation for the "single-bar" integrals $(pr|qs)$ and $(ps|qr)$. Under index exchange the double-bar integrals have the following properties:

$$(pq||rs) = -(qp||rs) = -(pq||sr) = (rs||pq) \tag{B.77}$$

For double-bar integrals divided by excitation energies we use the a symbols:

$$a_{rs}^{pq} = \frac{(pq||rs)}{\varepsilon_p + \varepsilon_q - \varepsilon_r - \varepsilon_s} \tag{B.78}$$

It follows from Eqs. (B.77 and B.78) that these symbols have the following index exchange properties:

$$a_{rs}^{pq} = -a_{rs}^{qp} = -a_{sr}^{pq} = -a_{pq}^{rs} \tag{B.79}$$

Since in the later sections we wish to refer to the individual components of a_{ij}^{ab}, we introduce the notation:

$$a_{ij}^{ab} = [ab||ij] = [ai|bj] - [aj|bi] \tag{B.80}$$

We use $a, b, c, d,$ and e to denote virtual spin orbitals and $i, j, k, l,$ and m to denote occupied spin orbitals. The number of active occupied spin orbitals is denoted *nocc* or *n* and the number of virtual spin orbitals *nvir* or *N*. Later on we use these letters and abbreviations to refer to spatial orbitals.

Moller-Plesset perturbation theory[30] is standard Rayleigh-Schrödinger perturbation theory using the Fock operator as the zeroth-order Hamiltonian and a Slater determinant of spin orbitals as the zeroth-order wavefunction. The total electronic energy is given by the sum

$$E = E^{(0)} + E^{(1)} + E^{(2)} + E^{(3)} + \dots \tag{B.81}$$

The zeroth-order energy is the sum of the eigenvalues of the occupied spin-orbitals, and the sum of the zeroth- and first-order energies is the Hartree-Fock energy. The sum of the remaining terms in the series is the correlation energy.

The second-order energy $E^{(2)}$ is given by:

$$E^{(2)} = -\frac{1}{4}\sum_{a,\,b=1}^{nvir}\sum_{i,j=1}^{nocc} a_{ij}^{ab} a_{ij}^{ab}(\varepsilon_a + \varepsilon_b - \varepsilon_i - \varepsilon_j) \tag{B.82}$$

The second-order energy accounts for the major effect of doubly excited configurations. It is necessarily negative, an energy lowering.

The third-order energy also accounts for some of the effects of double excitations, though it is generally much smaller in absolute value than the second-order energy. It may be positive or negative and is given by:

$$E^{(3)} = \frac{1}{4}\sum_{a,\,b=1}^{nvir}\sum_{i,j=1}^{nocc} (W1_{ij}^{ab} + W2_{ij}^{ab} + W3_{ij}^{ab})a_{ij}^{ab} \tag{B.83}$$

where

$$W1_{ij}^{ab} = \frac{1}{2}\sum_{c,\,d=1}^{nvir} (ab\,||\,cd\,)a_{ij}^{cd} \tag{B.84}$$

$$W2_{ij}^{ab} = \frac{1}{2}\sum_{k,\,l=1}^{nocc} (ij\,||\,kl\,)a_{kl}^{ab} \tag{B.85}$$

$$W3_{ij}^{ab} = -\sum_{k=1}^{nocc}\sum_{c=1}^{nvir}[(kb\,||\,jc)a_{ik}^{ac} - (kb\,||\,ic)a_{jk}^{ac} - (ka\,||\,jc)a_{ik}^{bc} + (ka\,||\,ic)a_{jk}^{bc}] \tag{B.86}$$

At the fourth-order of perturbation, single, double, triple, and quadruple excitations contribute to the energy. The fourth-order energy is the sum of several components:

$$E^{(4)} = E_R^{(4)} + E_S^{(4)} + E_D^{(4)} + E_T^{(4)} + E_Q^{(4)} \qquad (B.87)$$

The first term in Eq. (B.87) is known as the renormalization term and is given by:

$$E_R^{(4)} = -E^{(2)} \times \frac{1}{4} \sum_{a,b=1}^{nvir} \sum_{i,j=1}^{nocc} a_{ij}^{ab} a_{ij}^{ab} \qquad (B.88)$$

The contributions from the single, double, triple, and quadruple excitations are as follows:

$$E_S^{(4)} = -\sum_{a=1}^{nvir} \sum_{i=1}^{nocc} \frac{(W4_{ai} + W5_{ai})^2}{(\varepsilon_a - \varepsilon_i)} \qquad (B.89)$$

with

$$W4_{ai} = \frac{1}{2} \sum_{l=1}^{nocc} \sum_{c,d=1}^{nvir} (al||cd)\, a_{il}^{cd} \qquad (B.90)$$

and

$$W5_{ai} = -\frac{1}{2} \sum_{k,l=1}^{nocc} \sum_{d=1}^{nvir} (kl||id)\, a_{kl}^{ad} \qquad (B.91)$$

$$E_D^{(4)} = -\frac{1}{4} \sum_{a,b=1}^{nvir} \sum_{i,j=1}^{nocc} \frac{(W1_{ij}^{ab} + W2_{ij}^{ab} + W3_{ij}^{ab})^2}{(\varepsilon_a + \varepsilon_b - \varepsilon_i - \varepsilon_j)} \qquad (B.92)$$

$$E_T^{(4)} = -\frac{1}{36} \sum_{a,b,c=1}^{nvir} \sum_{i,j,k=1}^{nocc} \frac{W_{ijkabc}^2}{(\varepsilon_a + \varepsilon_b + \varepsilon_c - \varepsilon_i - \varepsilon_j - \varepsilon_k)} \qquad (B.93)$$

with

$$\begin{aligned} W_{ijkabc} = \sum_{e=1}^{nvir} & a_{ij}^{ae}(bc||ek) - a_{ij}^{be}(ac||ek) - a_{ij}^{ce}(ba||ek) + \\ & a_{ik}^{ae}(bc||je) - a_{ik}^{be}(ac||je) - a_{ik}^{ce}(ba||je) - \\ & a_{jk}^{ae}(bc||ie) + a_{jk}^{be}(ac||ie) + a_{jk}^{ce}(ba||ie) \end{aligned}$$

$$+\sum_{m=1}^{nocc} a_{im}^{ab}(cm||jk) + a_{im}^{bc}(am||jk) - a_{im}^{ac}(bm||jk) -$$
$$a_{jm}^{ab}(cm||ik) - a_{jm}^{bc}(am||ik) + a_{jm}^{ac}(bm||ik) + \tag{B.94}$$
$$a_{km}^{ab}(cm||ij) + a_{km}^{bc}(am||ij) - a_{km}^{ac}(bm||ij)$$

$$E_Q^{(4)} = -\frac{1}{4}\sum_{a,b=1}^{nvir}\sum_{i,j=1}^{nocc} a_{ij}^{ab} y_{ij}^{ab} \tag{B.95}$$

with

$$y_{ij}^{ab} = \frac{1}{4}\sum_{c,d=1}^{nvir}\sum_{k,l=1}^{nocc}(kl||cd)\Big[\ (a_{ij}^{ab}a_{kl}^{cd} + a_{ij}^{cd}a_{kl}^{ab}) - 2(a_{ij}^{ac}a_{kl}^{bd} + a_{ij}^{bd}a_{kl}^{ac})$$
$$- 2(a_{ik}^{ab}a_{jl}^{cd} + a_{ik}^{cd}a_{jl}^{ab}) + 4(a_{ik}^{ac}a_{jl}^{bd} + a_{ik}^{bd}a_{jl}^{ac})\Big] \tag{B.96}$$

Noting that the renormalization term cancels the first term in Eq. (B.96) and defining the arrays:

$$X1_{ijkl} = \sum_{a,b=1}^{nvir} a_{kl}^{ab}a_{ij}^{ab} \times (\varepsilon_a + \varepsilon_b - \varepsilon_k - \varepsilon_l) \tag{B.97}$$

$$Y1_{ijkl} = \sum_{a,b=1}^{nvir} a_{kl}^{ab}a_{ij}^{ab} \tag{B.98}$$

$$X2_{bc} = \sum_{d=1}^{nvir}\sum_{k,l=1}^{nocc}(kl||cd)a_{kl}^{bd} \tag{B.99}$$

$$Y2_{bc} = \sum_{d=1}^{nvir}\sum_{k,l=1}^{nocc} a_{kl}^{cd}a_{kl}^{bd} \tag{B.100}$$

$$X3_{jk} = \sum_{c,d=1}^{nvir}\sum_{l=1}^{nocc} a_{jl}^{cd}(kl||cd) \tag{B.101}$$

$$Y3_{jk} = \sum_{c,d=1}^{nvir}\sum_{l=1}^{nocc} a_{jl}^{cd}a_{kl}^{cd} \tag{B.102}$$

$$X4_{bcjk} = \sum_{a=1}^{nvir}\sum_{i=1}^{nocc} a_{ik}^{ac}a_{ij}^{ab} \times (\varepsilon_a + \varepsilon_c - \varepsilon_i - \varepsilon_k) \tag{B.103}$$

$$Y4_{bcjk} = \sum_{a=1}^{nvir} \sum_{i=1}^{nocc} a_{ik}^{ac} a_{ij}^{ab} \qquad (B.104)$$

The quadruples contribution to the energy can be written more compactly as:

$$E_Q^{(4)} = -E_R^{(4)} + \frac{1}{16} \sum_{i,j,k,l=1}^{nocc} X1_{ijkl} Y1_{ijkl} - \frac{1}{4} \sum_{b,c=1}^{nvir} X2_{bc} Y2_{bc} - \frac{1}{4} \sum_{j,k=1}^{nocc} X3_{jk} Y3_{jk} + \frac{1}{2} \sum_{a,c=1}^{nvir} \sum_{i,k=1}^{nocc} X4_{acik} Y4_{acik} \qquad (B.105)$$

We note that $E_D^{(4)}$ can be calculated at the same time as $E^{(3)}$ and that the matrices $X2$ and $X3$ are easily obtained from $X4$ and $X1$, and $Y2$ and $Y3$ from $Y4$ and $Y1$, as follows:

$$X2_{bc} = \sum_{l=1}^{nocc} X4_{bcll} \qquad (B.106)$$

$$X3_{jk} = -\sum_{l=1}^{nocc} X1_{ljkl} \qquad (B.107)$$

$$Y2_{bc} = \sum_{l=1}^{nocc} Y4_{bcll} \qquad (B.108)$$

$$Y3_{jk} = -\sum_{l=1}^{nocc} Y1_{ljkl} \qquad (B.109)$$

It is evident that the most time consuming step in an MP4 calculation is the evaluation of the contribution of the triple substitutions, $E_T^{(4)}$. This step requires $o(n^3N^4)$ operations. Of the other steps, the most costly are the evaluation of the third-order energy, $E^{(3)}$, and the contribution of quadruple substitutions, $E_Q^{(4)}$. Thus, the evaluation of $W1$ and $W3$ require $o(n^2N^4)$ and $o(n^3N^3)$ operations, while the evaluation of the contribution of $X4$ to $E_Q^{(4)}$ requires $o(n^3N^3)$ operations. The calculation of the second-order energy, $E^{(2)}$, and the contribution of single substitutions, $E_S^{(4)}$, are somewhat less costly, requiring respectively only $o(n^2N^2)$ and $o(n^3N^2)$ operations.

Prior to the evaluation of the various energy expressions given above, it is necessary to transform the two-electron integrals from the atomic orbital (AO) basis to the molecular orbital (MO) basis. The task of the two-electron integral transformation is to carry out the fourth-rank tensor transformation given by:

$$(pq|rs) = \sum_{\mu=1}^{m} C_{\mu p} \sum_{\nu=1}^{m} C_{\nu q} \sum_{\rho=1}^{m} C_{\rho r} \sum_{\sigma=1}^{m} C_{\sigma s} (\mu\nu|\rho\sigma) \tag{B.110}$$

In this equation m is the number of basis functions, the $C_{\mu p}$ are MO coefficients, and the $(\mu\nu|\rho\sigma)$ and $(pq|rs)$ are the two-electron integrals in the AO and MO bases. The transformation requires $o(m^5)$ operations and $o(m^4)$ words of storage. In an MP4 calculation the full transformation [Eq. (B.109)] is not required since the largest set of transformed integrals, $(ab|cd)$, is not needed. $W1$ can in fact be calculated directly from the AO integrals as follows:

$$\tilde{a}_{ij\rho\sigma} = \sum_{c,\, d=1}^{N} c_{\rho c} c_{\sigma d} a_{ij}^{cd} \tag{B.111}$$

$$\tilde{W}_{ij\mu\nu} = \sum_{\rho,\, \sigma=1}^{m} (\mu\nu||\rho\sigma) \tilde{a}_{ij\rho\sigma} \tag{B.112}$$

$$W1_{abij} = \sum_{\mu,\, \nu=1}^{m} c_{\mu a} c_{\nu b} \tilde{W}_{ij\mu\nu} \tag{B.113}$$

By avoiding the calculation of the $(ab|cd)$ integrals, the time for the transformation is reduced, only $o(nm^4)$ operations now being required. The AO approach makes the calculation of $W1$ an $o(n^2m^4)$ process, rather than $o(n^2N^4)$ of Eq. (B.84). Since the number of virtual orbitals is usually not much less than the number of basis functions, this extra work is expected to be outweighed by the reduction in the time required for the integral transformation. This may not be true for systems with several heavy atoms and a lot of frozen core orbitals.

(b) MP4 Implementation: In this section we discuss the organization of the MP4 program, giving particular attention to integral storage, and using a terminology that is closely related to the structure of HONDO.

A. Pre-MP4 Steps

The first stage in an MP4 calculation is an SCF calculation. In this procedure one- and two-electron integrals are calculated and stored, Fock matrices are formed and diagonalized, and the SCF wave function and energy are obtained. The data from this step that are needed in the remaining steps of the MP4 calculation are the two-electron integrals, the SCF wavefunction, which is in the form of a matrix of MO coefficients, and molecular orbital energies. The two-electron integrals are stored along with labels in a fixed-length record sequential file on unit 8. The MO coefficients and energies are stored on the general scratch direct access file on unit 10.

Following the SCF step, the two-electron integrals are put into square canonical order in preparation for the two-electron integral transformation. The integral sorting is accomplished by standard bin-sorting techniques. The ordered two-electron integrals are stored as rows of the square matrix in packed form on unit 9. These integrals are used in the calculation of $\tilde{W}$ for which square canonical order is convenient.

The next step is the two-electron integral transformation. This takes ordered AO integrals from file 9 and writes out transformed integrals on unit 8. A direct access file on unit 20 is also used if the transformation cannot be done in core. Only a partial transformation is actually needed for an MP4 calculation. After the integral transformation the transformed integrals are sorted into square canonical order and the ordered transformed integrals are written to file 11.

B. Calculation of The MP4 Energy

The MP4 calculation proper now begins. The first step is the formation of the a_{ij}^{ab} pseudointegrals. This is done by reading all the $(ai|bj)$ transformed integrals from file 11 and dividing them by the energy denominator $(\varepsilon_a + \varepsilon_b - \varepsilon_i - \varepsilon_j)$. The a_{ij}^{ab} are written to a direct access file on unit 30. Since they are needed in all remaining steps of the computation, file 30 may not be overwritten. At this point the second-order energy $E^{(2)}$ and the contribution of quadruple excitations to the fourth-order energy $E_Q^{(4)}$ are calculated; these quantities can be calculated from the a_{ij}^{ab} and MO energies ε_i alone from Eqs. (B.82) and (B.97) to (B.105).

The next step is the calculation of the single and triple excitation contributions to the fourth-order energy, $E_S^{(4)}$ and $E_T^{(4)}$ [Eqs. (B.89) to (B.91) and Eqs. (B.93) and (B.94)]. For these terms the $(ai||jk)$ and $(ab||ci)$ integrals are needed. These integrals are read from the transformed integral list on file 11 and written to the direct access file on unit 20. With these integrals, the a_{ij}^{ab}, and the MO energies $E_S^{(4)}$ and $E_T^{(4)}$ may be calculated. We note that the $(ai||jk)$ and $(ab||ci)$ are not needed again and can be overwritten.

The final step in the calculation is the evaluation of the third-order energy $E^{(3)}$ and the fourth-order contribution of double excitations $E_D^{(4)}$. These terms are evaluated from $W1$, $W2$, $W3$, and the a_{ij}^{ab}. To form $W1$, the AO integrals are read from file 9 and put on file 20, the a_{ij}^{ab} are transformed according to Eq. (B.111), the $\tilde{a}_{ij\rho\sigma}$ are combined with the AO integrals to form the $\tilde{W}$ [Eq. (B.112)], and $W1$ is obtained from the transformation [Eq. (B.105)]. $W2$ is formed from the $(ij||kl)$ and the a_{ij}^{ab} [Eq. (B.85)]. The former integrals are read from file 11 and held in core and the a_{ij}^{ab} are read from file 30. $W3$ is formed from the $(ai||bj)$ and the a_{ij}^{ab}

[Eq. (B.86)]. The $(ai||bj)$ are read from the transformed integral list on file 11, reorganized and stored on file 20, and combined with the a_{ij}^{ab} which are on file 30.

C. Integral Storage for Evaluation of MP4 Energy

In the final part of this section we consider in some detail the methods of storage of the various groups of integrals used in the MP4 calculation. The structure of the program allows reasonably efficient utilization of disk space, in particular there is considerable re-use of the largest file, file 20. The groups are $(ij||kl)$, a_{ij}^{ab}, $(ai||bj)$, $(ab||ci)$, $(ai||jk)$, and $(\mu\nu||\rho\sigma)$. We introduce two parameters, *LWIDTH* and *NSTRIP*. *LWIDTH* is the integer part of the square root of the number of virtual orbitals and *NSTRIP* is the smallest integer such that $NSTRIP \times LWIDTH \geq N$. For example, if the number of virtual orbitals is 31 then *LWIDTH* is 5 and *NSTRIP* is 7. The original purpose of these parameters was to divide up the various groups of integrals into *strips*, each strip being small enough to be held in core. A further benefit of this approach is that it appears to allow a straightforward migration of our code from a serial to a parallel computer system.

Storage of a_{ij}^{ab}: The a_{ij}^{ab} are divided up into $NSTRIP^2$ groups, each group containing $2n^2 \times LWIDTH^2$ elements. Each group can be thought of as a square on an $NSTRIP \times NSTRIP$ grid. Supposing the rows are labelled by *MASTRP* and the columns by *MBSTRP*, the square with the "coordinates" *(MASTRP, MBSTRP)* contains the set

$$\left\{\frac{(ai|bj)}{(\varepsilon_a+\varepsilon_b-\varepsilon_i-\varepsilon_j)}, -\frac{(aj|bi)}{(\varepsilon_a+\varepsilon_b-\varepsilon_i-\varepsilon_j)}\right\} \qquad (B.114)$$

with the index restrictions:

$$(MASTRP-1)\times LWIDTH+1 \leq (a-n-ncorb) \leq MASTRP \times LWIDTH$$

$$(MBSTRP-1)\times LWIDTH+1 \leq (b-n-ncorb) \leq MBSTRP \times LWIDTH$$

$$1 \leq i, j \leq n \qquad (B.115)$$

where *ncorb* is the number of frozen core orbitals. We note that both the "Coulomb" and "exchange" parts of the double-bar integrals are stored in the same group and that all indices have their full ranges. The storage requirement for the a_{ij}^{ab} is therefore $2n^2 \times LWIDTH^2 \times NSTRIP^2$, which is slightly larger than or equal to $2n^2N^2$. A *strip* is a row or column of squares. Each square is a logical record on file 30 and may span several physical records, each physical record containing 1024 double words.

Storage of $(ij||kl)$: The $(ij||kl)$ is the smallest group of integrals and is held in core. This is reasonable since this group requires only $2n^4$ words.

Storage of $(ai||bj)$: The $(ai||bj)$ are divided up into $NSTRIP^2$ groups, each group containing $2n^2 \times LWIDTH^2$ elements. Each group can be thought of as a square on an $NSTRIP$ by $NSTRIP$ grid. The square having for coordinates $(MASTRP, MBSTRP)$ contains the set

$$\{(ab|ij), -(aj|ib)\} \tag{B.116}$$

with the index restrictions:

$$(MASTRP - 1) \times LWIDTH + 1 \le (a - n - ncorb) \le MASTRP \times LWIDTH$$

$$(MBSTRP - 1) \times LWIDTH + 1 \le (b - n - ncorb) \le MBSTRP \times LWIDTH$$

$$1 \le i, j \le n \tag{B.117}$$

The storage requirement for these integrals is $2n^2 \times LWIDTH^2 \times NSTRIP^2$. Each square is a logical record on file 20 and may span several physical records, each physical record containing 1024 words.

Storage of $(ab||ci)$: The $(ab||ci)$ are also divided up into $NSTRIP^2$ groups, each group containing $2nN \times LWIDTH^2$ elements. Each group can be thought of as a square on an $NSTRIP$ by $NSTRIP$ grid. The square having for coordinates $(MASTRP, MBSTRP)$ contains the set

$$\{(ac|bi), -(ai|bc)\} \tag{B.118}$$

with the index restrictions :

$$(MASTRP - 1) \times LWIDTH + 1 \le (a - n - ncorb) \le MASTRP \times LWIDTH$$

$$(MBSTRP - 1) \times LWIDTH + 1 \le (b - n - ncorb) \le MBSTRP \times LWIDTH$$

$$1 \le i \le n$$

$$1 \le (c - n - ncorb) \le N \tag{B.119}$$

The storage requirement for these integrals is $2nN \times LWIDTH^2 \times NSTRIP^2$, slightly greater or equal to $2nN^3$. Each square is a logical record on file 20, in general spanning several physical records each physical record containing 1024 words.

Storage of $(ai||jk)$: The $(ai||jk)$ are divided into $NSTRIP$ groups, each group containing $2n^3 \times LWIDTH$ elements. The $MASTRP^{th}$ group contains the set

$$\{(aj|ik), -(ak|ij)\} \tag{B.120}$$

with the index restrictions:

$$(MASTRP - 1) \times LWIDTH + 1 \leq (a - n - ncorb) \leq MASTRP \times LWIDTH$$

$$1 \leq i, j, k \leq n \qquad (B.121)$$

The storage requirement for these integrals is $2n^3 \times LWIDTH \times NSTRIP$, slightly greater than or equal to $2n^3N$. Each square is a logical record on file 20, in general spanning several physical records.

Handling of $(\mu\nu||\rho\sigma)$: Finally, we consider the $(\mu\nu||\rho\sigma)$. This is the largest group of integrals and it requires special treatment. To illustrate this, if the same techniques as used for the other groups of integrals, i.e., no index restrictions and storage of both Coulomb and exchange components, were used for this group then $2m^4$ words of storage would be required, which is excessive. Our method of storing the $(\mu\nu||\rho\sigma)$ abandons the explicit storage of both the Coulomb and exchange components of a double-bar integral and also involves some index restrictions. We store the single-bar integrals $(\mu\rho|\nu\sigma)$ in $m(m+1)/2$ (logical) records of length $m(m+1)/2$, leading to a storage requirement of $m^2(m+1)^2/4$ words, i.e., the same as needed for the intermediate direct access file in a full two-electron integral transformation. For a given $(\mu\rho)(=\mu(\mu-1)/2+\rho$ if $\mu \geq \rho$ or $\rho(\rho-1)/2+\mu$ if $\rho > \mu)$, the $(\mu\rho)$ – th logical record contains all the $(\mu\rho|\nu\sigma)$ with $\nu \geq \sigma$. We note that the Coulomb component of $(\mu\nu||\rho\sigma)$ is on record number $(\mu\rho)$ and the exchange component is on record number $(\mu\sigma)$.

The most time-consuming steps in the MP4 calculation have been vectorized, namely the calculation of the contribution of triple and quadruple substitutions [Eqs. (B.93) to (B.105)] and third-order energy and fourth-order double substitutions contribution [Eqs. (B.83) to (B.86) and (B.92)]. The evaluation of the second-order energy and the contribution of single substitutions have not been vectorized since these steps are much less time-consuming than the others. We have done 'obvious' vectorization of all parts of the program, such as the initialization of large arrays, but the major effort was put into the time-consuming steps mentioned above.

D. Evaluation of $E_T^{(4)}$

Eqs. (B.93) and (B.94) define the contribution of triple substitutions to the fourth-order energy. The latter is the more time consuming, requiring $o(n^3N^4)$ operations, the former $o(n^3N^3)$. Since the triple substitution amplitudes W_{ijkabc} are antisymmetric with respect to index interchanges, it is necessary to calculate only those amplitudes with $c \geq b \geq a$ and $k \geq j \geq i$. (Since an amplitude is zero if any indices coincide, we can in fact make the further restriction $c > b > a$ and $k > j > i$). In place of Eq. (B.93), we may write:

$$E_T^{(4)} = - \sum_{c \geq b \geq a} \sum_{k \geq j \geq i} \frac{W_{ijkabc}^2}{(\varepsilon_a + \varepsilon_b + \varepsilon_c - \varepsilon_i - \varepsilon_j - \varepsilon_k)} \tag{B.122}$$

We note also that the six matrices:

$$S1_{ij,k} = \sum_{e=1}^{nvir} a_{ij}^{ae} (bc||ek) \tag{B.123}$$

$$S2_{ij,k} = \sum_{e=1}^{nvir} a_{ij}^{be} (ac||ek) \tag{B.124}$$

$$S3_{ij,k} = \sum_{e=1}^{nvir} a_{ij}^{ce} (ba||ek) \tag{B.125}$$

$$S4_{ij,k} = \sum_{m=1}^{nocc} (am||ij) a_{mk}^{bc} \tag{B.126}$$

$$S5_{ij,k} = \sum_{m=1}^{nocc} (bm||ij) a_{mk}^{ac} \tag{B.127}$$

$$S6_{ij,k} = \sum_{m=1}^{nocc} (cm||ij) a_{mk}^{ba} \tag{B.128}$$

where $1 \leq k \leq nocc$ and $1 \leq \text{ij} \leq nocc(nocc+1)/2$ (i.e. $1 \leq i \leq j \leq nocc$), are sufficient to calculate for a given a, b, and c the set of amplitudes

$$\{W_{ijkabc} \quad : \quad k \geq j \geq i\}.$$

Our algorithm involves looping over the unique sets of a, b, and c and for each set evaluating the matrices $S1$ to $S6$. From these we generate the unique amplitudes and therefore $E_T^{(4)}$.

When we make the transformation from spin to spatial orbitals, each of the six spin orbital Eqs. (B.123 to B.128) becomes eighteen spatial orbital equations. In all there are twenty spin cases to consider. For a RHF wavefunction there is a two-fold degeneracy among the spin cases and so only ten spin cases and only nine equations of the eighteen arising from each of Eqs. (B.123) to (B.128) need be considered.

Let us consider the evaluation of the spatial orbital components of **S1**. For a RHF wavefunction we may choose e to be an α orbital. We then have the following nine terms:

Spin for *ijkabc*	S1
$\alpha\alpha\alpha\alpha\alpha\alpha$	$([ai\|ej] - [aj\|ei])\ ((be\|ck) - (bk\|ce))$
$\alpha\alpha\beta\alpha\alpha\beta$	$([ai\|ej] - [aj\|ei])\ ((be\|ck)\qquad\qquad)$
$\alpha\alpha\beta\alpha\beta\alpha$	$([ai\|ej] - [aj\|ei])\ (\qquad\quad - (bk\|ce))$
$\beta\alpha\alpha\beta\alpha\alpha$	$([ai\|ej]\qquad\qquad)\ ((be\|ck) - (bk\|ce))$
$\beta\alpha\beta\beta\alpha\beta$	$([ai\|ej]\qquad\qquad)\ ((be\|ck)\qquad\qquad)$
$\beta\alpha\beta\beta\beta\alpha$	$([ai\|ej]\qquad\qquad)\ (\qquad\quad - (bk\|ce))$
$\alpha\beta\alpha\beta\alpha\alpha$	$(\qquad\quad - [aj\|ei])\ ((be\|ck) - (bk\|ce))$
$\alpha\beta\beta\beta\alpha\beta$	$(\qquad\quad - [aj\|ei])\ ((be\|ck)\qquad\qquad)$
$\alpha\beta\beta\beta\beta\alpha$	$(\qquad\quad - [aj\|ei])\ (\qquad\quad - (bk\|ce))$

Since α and β orbitals are identical, all of these terms are in fact linear combinations of the four terms:

$$[ai|ej](be|ck),\quad [ai|ej](bk|ce),\quad [aj|ei](be|ck),\ \text{and}\ [aj|ei](bk|ce).$$

So defining:

$$A1_{ij,e} = [ai|ej]$$

$$A2_{ij,e} = [aj|ei]$$

$$B1_{e,k} = (be|ck)$$

$$B2_{e,k} = (bk|ce)$$

$$C11_{ij,k} = \sum_e A1_{ij,e}\,B1_{e,k}$$

$$C12_{ij,k} = \sum_e A1_{ij,e}\,B2_{e,k}$$

$$C21_{ij,k} = \sum_e A2_{ij,e}\,B1_{e,k}$$

$$C22_{ij,k} = \sum_e A2_{ij,e} B2_{e,k}$$

where $ij = 1,2,..., n(n+1)/2$, all the components of $S1$ can be obtained by linear combinations of elements of $C11, C12, C21, C22$. A more preferable approach which slightly reduces the arithmetic and is a better vectorization strategy, follows from removing the $i \geq j$ index restriction. By doing so, it is possible to carry out only two matrix multiplications, but with larger matrices (and longer vectors). Explicitly for each a, b, and c triplet, we define:

$$A_{ij,e} = [ai \mid ej]$$

$$B1_{e,k} = (be \mid ck)$$

$$B2_{e,k} = (bk \mid ce)$$

$$C1_{ij,k} = \sum_e A_{ij,e} B1_{e,k}$$

$$C2_{ij,k} = \sum_e A_{ij,e} B2_{e,k}$$

A is n^2 by N and $B1$ and $B2$ are N by n. The outer loop variable is ij $(ij = (i-1) \times n + j)$ and so the vector length in the matrix multiplications is n^2.

The same strategy is used to generate the components of **S2** to **S6**. In evaluating components of **S4** to **S6** the **A** and **B** matrices are n^2 by n and n by n. The vector length in the matrix multiplications is still n^2 however.

The driver **MPW6** loops over the unique strips of virtual orbitals, reads the required integrals, and passes them to **MPW6EM** which evaluates a block of amplitudes and their contribution to $E_T^{(4)}$. The algorithm requires approximately $n^3N^4 + n^4N^3$ multiplications and the same number of additions.

E. Evaluation of $E_Q^{(4)}$

The contribution of quadruple substitutions is defined in Eqs. (B.95) to (B.109). The calculation proceeds by the evaluation of the X and Y matrices, which are contracted together to give the quadruples contribution to the energy, $E_Q^{(4)}$ [Eq. (B.105)]. The major steps are the generation of $X1$, $Y1$, $X4$, and $Y4$. The other matrices are generated easily from these [Eqs. (B.106) to (B.109)]. $X1$ and $Y1$ require $o(n^4N^2)$ operations while $X4$ and $Y4$ require $o(n^3N^3)$ operations.

$X1$ and $Y1$ have six spin cases in the general UHF case. For a RHF wavefunction the spin cases are doubly degenerate and only three must be considered. For $X1$ the equations are as follows:

$$X1_{ijkl}(\alpha\alpha\alpha\alpha) = \sum_{a,b=1}^{N_\alpha} [ab||ij][ab||kl](\varepsilon_a + \varepsilon_b - \varepsilon_i - \varepsilon_j)$$

$$X1_{ijkl}(\alpha\beta\alpha\beta) = \sum_{a=1}^{N_\alpha}\sum_{b=1}^{N_\beta}[ai|bj][ak|bl](\varepsilon_a + \varepsilon_b - \varepsilon_k - \varepsilon_l)$$

$$+ \sum_{a=1}^{N_\beta}\sum_{b=1}^{N_\alpha}[aj|bi][al|bk](\varepsilon_a + \varepsilon_b - \varepsilon_k - \varepsilon_l)$$

$$X1_{ijkl}(\alpha\beta\beta\alpha) = -\sum_{a=1}^{N_\alpha}\sum_{b=1}^{N_\beta}[ai|bj][al|bk](\varepsilon_a + \varepsilon_b - \varepsilon_k - \varepsilon_l)$$

$$-\sum_{a=1}^{N_\beta}\sum_{b=1}^{N_\alpha}[aj|bi][ak|bl](\varepsilon_a + \varepsilon_b - \varepsilon_k - \varepsilon_l)$$

The corresponding equations for $Y1$ are obtained by dropping the excitation energy $(\varepsilon_a + \varepsilon_b - \varepsilon_k - \varepsilon_l)$ from these equations. Since the α and β orbitals are identical, the $\alpha\alpha\alpha\alpha$ spin case is actually the sum of the $\alpha\beta\alpha\beta$ and $\alpha\beta\beta\alpha$ spin cases:

$$X1_{ijkl}(\alpha\alpha\alpha\alpha) = X1_{ijkl}(\alpha\beta\alpha\beta) + X1_{ijkl}(\alpha\beta\beta\alpha)$$

Also if we define the matrices:

$$A1_{ij,ab} = [ai|bj]$$

$$A2_{ij,ab} = -[aj|bi]$$

$$B1_{ab,ij} = [ai|bj](\varepsilon_a + \varepsilon_b - \varepsilon_i - \varepsilon_j)$$

$$B2_{ab,ij} = -[aj|bi](\varepsilon_a + \varepsilon_b - \varepsilon_i - \varepsilon_j)$$

$$C11_{ij,kl} = \sum_{a,b} A1_{ij,ab} B1_{ab,kl}$$

$$C12_{ij,kl} = \sum_{a,b} A1_{ij,ab} B2_{ab,kl}$$

$$C21_{ij,\,kl} = \sum_{a,\,b} A2_{ij,\,ab} B1_{ab,\,kl}$$

$$C22_{ij,\,kl} = \sum_{a,\,b} A2_{ij,\,ab} B2_{ab,\,kl}$$

It follows that:

$$X1_{ijkl}(\alpha\alpha\alpha\alpha) = C11_{ij,\,kl} + C12_{ij,\,kl} + C21_{ij,\,kl} + C22_{ij,\,kl}$$

$$X1_{ijkl}(\alpha\beta\alpha\beta) = C11_{ij,\,kl} + C22_{ij,\,kl}$$

$$X1_{ijkl}(\alpha\beta\beta\alpha) = C12_{ij,\,kl} + C21_{ij,\,kl}$$

The calculation of $Y1$ is similar. The definitions of $A1$ and $A2$ are the same, but $B1$ and $B2$ are given by:

$$B1_{ab,\,ij} = [ai\,|\,bj]$$

$$B2_{ab,\,ij} = -\,[aj\,|\,bi]$$

$X3$ and $Y3$ have only two spin case in the general UHF case. In the RHF case these are degenerate and so only one needs to be considered. It is obtained from $X1$ or $Y1$ as appropriate.

$$X3_{jk}(\alpha\alpha) = -\sum_{i} [X1_{ijki}(\alpha\alpha\alpha\alpha) + X1_{ijki}(\alpha\beta\beta\alpha)]$$

$$Y3_{jk}(\alpha\alpha) = -\sum_{i} [Y1_{ijki}(\alpha\alpha\alpha\alpha) + Y1_{ijki}(\alpha\beta\beta\alpha)]$$

The evaluation of $X4$ and $Y4$ proceeds similarly to the evaluation of $X1$ and $Y1$. Again there are six spin cases to be considered in the general UHF case, but only three need to be considered for a RHF wavefunction:

$$X4_{bcjk}(\alpha\alpha\alpha\alpha) = \sum_{a=1}^{N_\alpha}\sum_{i=1}^{n_\alpha} [ab\,||\,ij][ac\,||\,ik](\varepsilon_a + \varepsilon_c - \varepsilon_i - \varepsilon_k)$$

$$+ \sum_{a=1}^{N_\beta}\sum_{i=1}^{n_\beta} [ai\,|\,bj][ai\,|\,ck](\varepsilon_a + \varepsilon_c - \varepsilon_i - \varepsilon_k)$$

$$X4_{bcjk}(\alpha\beta\alpha\beta) = \sum_{a=1}^{N_\alpha}\sum_{i=1}^{n_\alpha}[ab||ij][ai|ck](\varepsilon_a + \varepsilon_c - \varepsilon_i - \varepsilon_k)$$

$$+ \sum_{a=1}^{N_\beta}\sum_{i=1}^{n_\beta}[ai|bj][ac||ik](\varepsilon_a + \varepsilon_c - \varepsilon_i - \varepsilon_k)$$

$$X4_{bcjk}(\alpha\alpha\beta\beta) = \sum_{a=1}^{N_\beta}\sum_{i=1}^{n_\alpha}[aj|bi][ak|ci](\varepsilon_a + \varepsilon_c - \varepsilon_i - \varepsilon_k)$$

The expressions for the spin cases of $Y4$ are obtained from these by removing the excitation energies $(\varepsilon_a + \varepsilon_b - \varepsilon_i - \varepsilon_k)$. As in the computation of $X1$ and $Y1$, the identity of α and β orbitals enables all three spin cases to be evaluated by four matrix multiplications. Specifically we define:

$$A1_{ck,\,ai} = [ai|ck](\varepsilon_a + \varepsilon_c - \varepsilon_i - \varepsilon_k)$$

$$A2_{ck,\,ai} = -\,[ak|ci](\varepsilon_a + \varepsilon_c - \varepsilon_i - \varepsilon_k)$$

$$B1_{ai,\,bj} = [ai|bj]$$

$$B2_{ai,\,bj} = -\,[aj|bi]$$

$$C11_{ck,\,bj} = \sum_{a,\,i} A1_{ck,\,ai} B1_{ai,\,bj}$$

$$C12_{ck,\,bj} = \sum_{a,\,i} A1_{ck,\,ai} B2_{ai,\,bj}$$

$$C21_{ck,\,bj} = \sum_{a,\,i} A2_{ck,\,ai} B1_{ai,\,bj}$$

$$C22_{ck,\,bj} = \sum_{a,\,i} A2_{ck,\,ai} B2_{ai,\,bj}$$

so that:

$$X4_{bcjk}(\alpha\alpha\alpha\alpha) = 2 \times C11_{ck,\,bj} + C12_{ck,\,bj} + C21_{ck,\,bj} + C22_{ck,\,bj}$$

$$X4_{bcjk}(\alpha\beta\alpha\beta) = 2 \times C11_{ck,\,bj} + C12_{ck,\,bj} + C21_{ck,\,bj}$$

$$X4_{bcjk}(\alpha\alpha\beta\beta) = C22_{ck,\,bj}$$

$Y4$ is evaluated in exactly the same way except $A1$ and $A2$ are now given by:

$$A1_{ck,\,ai} = [ai\,|\,ck]$$

$$A2_{ck,\,ai} = -\,[ak\,|\,ci]$$

Like **X3** and **Y3**, **X2** and **Y2** have only one essential spin case for a RHF wavefunction. These are given by:

$$X2_{bc}(\alpha\alpha) = \sum_j [X4_{bcjj}(\alpha\alpha\alpha\alpha) + X4_{bcjj}(\alpha\alpha\beta\beta)]$$

$$Y2_{bc}(\alpha\alpha) = \sum_j [Y4_{bcjj}(\alpha\alpha\alpha\alpha) + Y4_{bcjj}(\alpha\alpha\beta\beta)]$$

Each spin case of $X4$ and $Y4$ requires n^2N^2 words of memory and we do not presently hold the entire arrays in core, rather we evaluate them blockwise, $n^2 \times LWIDTH^2$ elements in each block. This is the reason for two outer loops over strips of virtual orbitals. The vector length in the matrix multiplications is $n \times LWIDTH$.

A couple of points should be made regarding the evaluation of the quadruple substitution contributions to the energy. First some time could be saved by storing both $(ab\,||\,ij)$ and $[ab\,||\,ij]$ integrals, thus avoiding the generation of the former from the latter "on the fly." Second, by assuming a RHF wavefunction, our algorithm does roughly 20% less arithmetic than a general UHF approach would require. A general UHF code can take advantage of the two-fold degeneracy among the six spin cases of $X1$, $Y1$, $X4$, and $Y4$, but would require five matrix multiplications to compute three essential spin cases, rather than four, as our RHF code does.

F. Evaluation of $E^{(3)}$

The third-order energy $E^{(3)}$ is evaluated using Eqs. (B.83) to (B.86). Presently $W1$ is evaluated not as in Eq. (B.84), however, but directly from AO integrals [Eqs. (B.111) to (B.113)]. The first step in the evaluation of $W1$ is the transformation of a_{ij}^{ab} to $\tilde{a}_{ij\rho\sigma}$ [Eq. (B.111)]. In matrix form this equation is:

$$\tilde{A} = C\,A\,C^T$$

This step requires $o(n^2N^3)$ operations. The next, and computationally most demanding step is the evaluation of $\tilde{W}$ [Eq. (B.112)]. This requires $o(n^2m^4)$ operations, where m is the number of basis functions. Formally $\tilde{W}$ has six spin cases. For a RHF wavefunction only three of these have to be computed since the other three make the same contribution to the energy. Introducing the notation:

$$\{\rho\sigma||ij\} = \{\rho i|\sigma j\} - \{\rho j|\sigma i\} = \tilde{a}_{ij\rho\sigma}$$

the expressions for the essential spin cases are:

$$\tilde{W}_{\mu\nu ij}(\alpha\alpha\alpha\alpha) = \sum_{\rho,\sigma}(\mu\nu||\rho\sigma)\{\rho\sigma||ij\}$$

$$\tilde{W}_{\mu\nu ij}(\alpha\beta\alpha\beta) = \sum_{\rho,\sigma}[(\mu\rho|\nu\sigma)\{\rho i|\sigma j\} + (\mu\sigma|\nu\rho)\{\rho j|\sigma i\}]$$

$$\tilde{W}_{\mu\nu ij}(\beta\alpha\alpha\beta) = -\sum_{\rho,\sigma}[(\mu\sigma|\nu\rho)\{\rho i|\sigma j\} + (\mu\rho|\nu\sigma)\{\rho j|\sigma i\}]$$

Since the AO integrals have no explicit dependence (we assign the α spin for bookkeeping purposes; μ and ν are tied to a and b which do have a spin dependence; the spin of a and b determine whether the MO coefficients in Eq. (B.113) are α or β coefficients; in a RHF formulation this point is not important of course), the $\beta\alpha\alpha\beta$ spin case can be generated from $\alpha\beta\alpha\beta$:

$$\tilde{W}_{\mu\nu ij}(\beta\alpha\alpha\beta) = -\tilde{W}_{\nu\mu ij}(\alpha\beta\alpha\beta)$$

This holds for UHF as well as RHF wavefunctions. For a RHF wavefunction further simplifications are possible. First:

$$\tilde{W}_{\mu\nu ij}(\alpha\alpha\alpha\alpha) = W_{\mu\nu ij}(\alpha\beta\alpha\beta) + W_{\mu\nu ij}(\beta\alpha\alpha\beta)$$

It follows that all three spin cases can be generated from the $\alpha\beta\alpha\beta$ spin case. Second, the spatial orbital label restriction $i \leq j$ can be applied. This is true for spin orbital labels in the general UHF case, but it is of no value unless i and j have the same spin.

Our algorithm for the evaluation of $\tilde{W}$ is as follows. We loop over the μ index, for each μ constructing the matrices:

$$A1_{\nu,\rho\sigma} = (\mu\rho|\nu\sigma)$$

$$A2_{\nu,\rho\sigma} = -(\mu\sigma|\nu\rho)$$

$$B1_{\rho\sigma,ij} = \{\rho i|\sigma j\}$$

$$B2_{\rho\sigma,ij} = \{\rho j|\sigma i\}$$

$\tilde{W}$ is obtained by two matrix multiplications:

$$\tilde{W}_{\mu\nu ij}(\alpha\beta\alpha\beta) = \sum_{\rho,\sigma}[A1_{\nu,\rho\sigma}B1_{\rho\sigma,ij} + A2_{\nu,\rho\sigma}B2_{\rho\sigma,ij}]$$

The vector length in the matrix multiplications is m, the number of basis functions.

As we have implemented it, this algorithm has a m^3 memory requirement (for $A1$ and $A2$). Although a few years ago this would have been a serious drawback, large memories are becoming more common, and we do not expect this requirement to be a major limitation, at least on modern vector processing mainframe computers. If memory is insufficient the algorithm could be adapted to reduce memory requirements by restricting the range of the ν index. With the present data organization, this would imply several reads of the AO integrals. Another approach, which would be particularly beneficial in cases where the integral list is sparse, would be to store both components of the double-bar AO integrals on a sequential file along with integral labels.

The final step in evaluating $W1$ is the transformation given by Eq. (B.113). This can be formulated as a similarity transformation

$$W1 = C^T \tilde{W} C$$

This is only an $o(n^2N^3)$ step and so much less costly than the evaluation of $\tilde{W}$. The calculation of $W2$ is an $o(n^4N^2)$ process and so is not as expensive as the evaluation of $W1$ or $W3$. As for $\tilde{W}$, analysis of spin cases shows that for a RHF wavefunction only one spin case must be calculated and the restriction $i \leq j$ applies. The relevant matrices are:

$$A1_{ij,kl} = (ik|jl)$$

$$A2_{ij,kl} = -(il|jk)$$

$$B1_{kl,ab} = [ak|bl]$$

$$B2_{kl,ab} = -[al|bk]$$

and $W2$ is obtained from:

$$W2_{ijab}(\alpha\beta\alpha\beta) = \sum_{k,l}[A1_{ij,kl}B1_{kl,ab} + A2_{ij,kl}B2_{kl,ab}]$$

In our present implementation, the ij loop, of length $n(n+1)/2$, is the outer loop. A better vectorization strategy would be to make the ab loop the outer loop, leading to a vector length of $LWIDTH^2$. However, since this step is in any case of little expense, only a very small performance improvement would be realized.

Finally we consider the evaluation of $W3$. This step has an $o(n^3N^3)$ dependence. The calculation is done in two steps, each giving two of the terms in Eq. (B.86). As for $W2$, for RHF wavefunctions only one spin case need be computed. The restriction $i \leq j$ also applies, but in this case it is not possible to take advantage of this in the matrix multiplications. The following matrices are used to calculate the first two terms of Eq. (B.86):

$$A1_{ai,\,ck} = [ai\,|\,ck]$$

$$A2_{ai,\,ck} = -\,[ak\,|\,ci]$$

$$B1_{ck,\,bj} = (kj\,|\,bc)$$

$$B2_{ck,\,bj} = -\,(kc\,|\,bj)$$

$$C11_{ai,\,bj} = \sum_{c,\,k} A1_{ai,\,ck} B1_{ck,\,bj}$$

$$C12_{ai,\,bj} = \sum_{c,\,k} A1_{ai,\,ck} B2_{ck,\,bj}$$

$$C21_{ai,\,bj} = \sum_{c,\,k} A2_{ai,\,ck} B1_{ck,\,bj}$$

$$C22_{ai,\,bj} = \sum_{c,\,k} A2_{ai,\,ck} B2_{ck,\,bj}$$

The second two terms are calculated similarly. The vector length in the matrix multiplications is $n \times LWIDTH$.

As it has become clear, a great deal of emphasis has been put onto designing a computer code well suited for vector processing. Comparisons with other codes performing similar calculations have shown that the present implementation is indeed very competitive with the others. Additionally it appears that the present scheme is amenable to efficient parallel processing.

C. Electronic Properties

Calculation of the electronic properties based on the electron density plays an increasingly important role in explaining the activity/structure relationship of molecules, and provides quantitative data concerning specific properties, thus forming the basis for a computer aided design of molecules and molecular assemblies. HONDO contains programs for the calculations of a series of properties. The list includes:

1. Dipole Moment
2. Quadrupole Moment
3. Octupole Moment
4. Mulliken Population Analysis
5. Valency and Bond Order Analysis
6. Spin Density at the Nuclei
7. Electrostatic Field at the Nuclei
8. Electrostatic Field Gradient at the Nuclei
9. Hyperfine Interactions (Fermi Contact Term and Anisotropic Interaction)
10. Orbital Localization
11. Electron and Spin Density Maps
12. Electrostatic Potential Map
13. Dipole Polarizability and Hyperpolarizabilities
14. Frequency Dependent (Hyper)polarizabilities

These properties can be calculated for SCF, MCSCF, and CI wavefunctions, in the latter case through the calculations of the natural orbitals. The program does not allow for calculation of properties from the perturbed wavefunctions arising in perturbation theory. Many other programs provide capabilities for the calculation of several of the properties in this list. Some also calculate other properties. The reader interested in the formalism associated with the various properties is referred to a few important papers.[31-33] The bond order and valency analysis,[34] to the best of our knowledge, is not provided by other popular programs. It is an extension of the population analysis, which yields bond order and valencies which indeed are in accord with classical concepts of chemistry, such as valence of atoms in molecules and order of bonding between pairs of atoms. In addition we have been interested recently in studying the response of molecules in the presence of an electric field.[33] Such response forms the basis of the non-linear optical properties of organic materials, an area of research which can greatly benefit from a computer aided approach to engineering. Our initial effort dealt with the calculation of static properties,[35] although frequency dependent polarizabilities are more closely related to the experimental situation. We summarize now the algorithms for response calculation.

1. Dipole Polarizability and Hyperpolarizabilities

(a) Theory: The energy of a molecule perturbed by a static, uniform field can be expanded[36] as:

$$E = E^{(0)} - \mu_i^{(0)} F_i - \frac{1}{2}\alpha_{ij} F_i F_j - \frac{1}{6}\beta_{ijk} F_i F_j F_k - \frac{1}{24}\gamma_{ijkl} F_i F_j F_k F_l - \ldots \qquad (C.1)$$

where the subscripts which identify the tensor components are summed over the cartesian axes x, y, and z. In the above equation $E^{(0)}$ is the unperturbed energy, F_i

is the component of the field in the i direction, $\mu^{(0)}$ is the permanent dipole moment of the molecule, α is the static dipole polarizability tensor, and β and γ are the first and second dipole hyperpolarizability tensors. These tensors describe the distortion of the molecule induced by the electric field. The elements of these tensors can thus be determined from derivatives of the energy with respect to electric field.

The molecular SCF energy for a closed shell system is given by:

$$E = V_{nuc} + \sum_{\mu\nu} D_{\mu\nu} (h_{\mu\nu} + F_{\mu\nu}) \tag{C.2}$$

where V_{nuc} is the nuclear repulsion energy, μ and ν run over basis functions, the density matrix elements $D_{\mu\nu}$, $h_{\mu\nu}$, and $F_{\mu\nu}$ are given by Eqs. (B.6) to (B.13).

For derivatives with respect to a static electric field (or field derivatives), and assuming field-independent basis functions, we note the following: the quantities V_{nuc} , $(\mu\nu|\rho\sigma)$ and $S_{\mu\nu}$ are constant with respect to electric field, and h does not have terms higher than first-order with respect to electric field. Using superscript to denote partial derivatives with respect to electric field, as in:

$$E^a \equiv \frac{\partial E}{\partial F_a} \tag{C.3}$$

the analytic derivative expressions to fourth-order are:

$$E^a = \sum_{\mu\nu} D_{\mu\nu} h^a_{\mu\nu} \tag{C.4}$$

$$E^{ab} = \sum_{\mu\nu} D^b_{\mu\nu} h^a_{\mu\nu} \tag{C.5}$$

$$E^{abc} = \sum_{\mu\nu} D^{bc}_{\mu\nu} h^a_{\mu\nu} \tag{C.6}$$

$$E^{abcd} = \sum_{\mu\nu} D^{bcd}_{\mu\nu} h^a_{\mu\nu} \tag{C.7}$$

After some algebraic manipulations, it can be shown that:

$$E^{abc} = 4\left[\left(\sum_{i}^{occ}\sum_{\mu\nu} C^{a}_{\mu i} F^{b}_{\mu\nu} C^{c}_{\nu i} - \sum_{ij}^{occ}\sum_{\mu\nu} C^{a}_{\mu i} S_{\mu\nu} C^{c}_{\nu j} \varepsilon^{b}_{ji}\right) + \{bca\} + \{cab\}\right] \tag{C.8}$$

$$\begin{aligned} E^{abcd} = 4\Bigg[\Bigg(&\sum_{i}^{occ}\sum_{\mu\nu}\left(C^{a}_{\mu i} F^{b}_{\mu\nu} C^{cd}_{\nu i} + C^{a}_{\mu i} F^{bd}_{\mu\nu} C^{c}_{\nu i} + C^{ad}_{\mu i} F^{b}_{\mu\nu} C^{c}_{\nu i}\right) \\ &- \sum_{ij}^{occ}\sum_{\mu\nu}\left(C^{a}_{\mu i} S_{\mu\nu} C^{c}_{\nu j} \varepsilon^{bd}_{ji} + C^{a}_{\mu i} S_{\mu\nu} C^{cd}_{\nu j} \varepsilon^{b}_{ji} + C^{ad}_{\mu i} S_{\mu\nu} C^{c}_{\nu j} \varepsilon^{b}_{ji}\right)\Bigg) \\ &+ \{bcad\} + \{cabd\}\Bigg] \end{aligned} \tag{C.9}$$

In the above equations the curly braces are used to abbreviate expressions that differ only by permutations of the superscripts. To evaluate the above expressions we need derivatives of D, F, C, and ε. They are given by:

$$D^{a}_{\mu\nu} = 2\sum_{i}^{occ}\left(C^{a}_{\mu i} C_{\nu i} + C_{\mu i} C^{a}_{\nu i}\right) \tag{C.10}$$

$$D^{ab}_{\mu\nu} = 2\sum_{i}^{occ}\left(C^{ab}_{\mu i} C_{\nu i} + C^{a}_{\mu i} C^{b}_{\nu i} + C^{b}_{\mu i} C^{a}_{\nu i} + C_{\mu i} C^{ab}_{\nu i}\right) \tag{C.11}$$

$$F^{a}_{\mu\nu} = h^{a}_{\mu\nu} + \sum_{\rho,\sigma} D^{a}_{\rho\sigma}\left[(\mu\nu|\rho\sigma) - \frac{1}{2}(\mu\rho|\nu\sigma)\right] \tag{C.12}$$

$$F^{ab}_{\mu\nu} = \sum_{\rho,\sigma} D^{ab}_{\rho\sigma}\left[(\mu\nu|\rho\sigma) - \frac{1}{2}(\mu\rho|\nu\sigma)\right] \tag{C.13}$$

Using the definitions:

$$C^{a} = C\, U^{a} \tag{C.14}$$

$$C^{ab} = C\, U^{ab} \tag{C.15}$$

the wavefunction and eigenvalue derivatives are obtained through the Coupled Perturbed Hartree-Fock (CPHF) equations:

$$U_{ij}^{a} = \frac{1}{\varepsilon_j - \varepsilon_i}\left[-\varepsilon_{ij}^{a} + \sum_{\mu\nu} C_{\mu i} F_{\mu\nu}^{a} C_{\nu j} \right] \tag{C.16}$$

$$U_{ij}^{ab} = \frac{1}{\varepsilon_j - \varepsilon_i}\left[-\varepsilon_{ij}^{ab} + \sum_{\mu\nu} C_{\mu i} F_{\mu\nu}^{ab} C_{\nu j} + \sum_{\mu\nu}\sum_{k} (C_{\mu i}F_{\mu\nu}^{a}C_{\nu k}U_{kj}^{b} + C_{\mu i}F_{\mu\nu}^{b}C_{\nu k}U_{kj}^{a}) \right] \tag{C.17}$$

resulting from differentiating Eq. (B.11). A noncanonical treatment is adopted, where only the off-diagonal occupied-virtual blocks of ε^a and ε^{ab} are zero valued. The above expression may thus be used for the occupied-virtual block of U^a and U^{ab}. The diagonal occupied-occupied and virtual-virtual blocks must satisfy derivatives of the orthonormality condition [Eq. (B.9)], so we chose for i and j belonging to the occupied space:

$$U_{ij}^{a} = 0 \tag{C.18}$$

$$U_{ij}^{ab} = -\frac{1}{2}\sum_{k} (U_{ki}^{a} U_{kj}^{b} + U_{ki}^{b} U_{kj}^{a}) \tag{C.19}$$

Substituting these into Eqs. (C.14) and (C.15) respectively, determines the diagonal blocks of ε^a and ε^{ab}.

(b) Implementation: The above equations have been implemented into HONDO. The program executes as follows:

■ E: the zeroth-order equations are solved as described in section B.

■ μ: the dipole moment operator integrals are calculated and the components of the permanent dipole moment are evaluated via Eq. (C.4) where $\mu_a^{(0)} = -E^a$.

■ α: the first derivative of the wavefunction is solved iteratively (with optional damping), using the first-order CPHF equations [Eq. (C.16)]. The elements of the dipole polarizability tensor are calculated via Eq. (C.5) where $\alpha_{ab} = -E^{ab}$.

■ β: the components of the dipole hyperpolarizability tensor are computed through Eq. (C.8), where $\beta_{abc} = -E^{abc}$, using the first-derivative wavefunction from above.

■ γ: the second-derivative wavefunction is solved iteratively (with optional damping), through the second-order CPHF equation [Eq. (C.17)]. The dipole second hyperpolarizability tensor is evaluated with Eq. (C.9), where $\gamma_{abcd} = - E^{abcd}$. Iterative solutions for the CPHF equations have been adopted so that large molecular systems, up to several hundred basis functions, could be considered.

(c) Frequency Dependent Hyperpolarizabilities: The formalism just presented can be modified to give dynamic or frequency dependent hyperpolarizabilities. As previously mentioned the starting point for the calculation of the static hyperpolarizabilities was the time-independent Hartree-Fock equation (B.11). We now consider a time-dependent perturbation

$$\mathcal{H}'(r_1, \ldots , r_n, t) = \mu{\cdot}\mathrm{F}(e^{+i\omega t} + e^{-i\omega t} + 1) \tag{C.20}$$

made up of the superposition of a static field with a monochromatic oscillating optical field of frequency ω and half amplitude of $\mu{\cdot}\mathrm{F}$ where $\mu = -\sum^{n} r_i$ is the usual dipole moment vector operator. Frenkel's time-dependent Hartree-Fock equation has the form

$$FC - i\frac{\partial}{\partial t}SC = SC\varepsilon \tag{C.21}$$

in terms of the external perturbations derived from Eq. (C.20).

Power series expansion of F, C, and ε gives rise to the frequency dependent CPHF equations which are very closely related to Eqs. (C.16 – C.17), and to energy derivatives which resemble Eqs. (C.4 – C.9). Thus we get the following polarizabilities:

static: $$\alpha_{ab}(0;0) = -\sum_{\mu\nu} D^{b}_{\mu\nu}(0)h^{a}_{\mu\nu} \tag{C.22}$$

dynamic: $$\alpha_{ab}(\mp\omega;\pm\omega) = -\sum_{\mu\nu} D^{b}_{\mu\nu}(\pm\omega)h^{a}_{\mu\nu} \tag{C.23}$$

We also get the following first-order hyperpolarizabilities:

Static:

$$\beta_{abc}(0;0, 0) = -\sum_{\mu\nu} D^{bc}_{\mu\nu}(0, 0)\, h^{a}_{\mu\nu} \tag{C.24}$$

Electrooptic Pockels effect (EOPE):

$$\beta_{abc}(\mp\omega;0, \pm\omega) = -\sum_{\mu\nu} D^{bc}_{\mu\nu}(0, \pm\omega)\, h^{a}_{\mu\nu} \tag{C.25}$$

Second harmonic generation (SHG):

$$\beta_{abc}(\mp 2\omega; \pm\omega, \pm\omega) = -\sum_{\mu\nu} D_{\mu\nu}^{bc}(\pm\omega, \pm\omega)\, h_{\mu\nu}^{a} \qquad (C.26)$$

Optical rectification (OR):

$$\beta_{abc}(0; \pm\omega, \mp\omega) = -\sum_{\mu\nu} D_{\mu\nu}^{bc}(\pm\omega, \mp\omega)\, h_{\mu\nu}^{a} \qquad (C.27)$$

Finally we have the following second-order hyperpolarizabilities:

Static:

$$\gamma_{abcd}(0;0,0,0) = -\sum_{\mu\nu} D_{\mu\nu}^{bcd}(0,0,0)\, h_{\mu\nu}^{a} \qquad (C.28)$$

Optical Kerr effect (OKE):

$$\gamma_{abcd}(\mp\omega;0,0,\pm\omega) = -\sum_{\mu\nu} D_{\mu\nu}^{bcd}(0,0,\pm\omega)\, h_{\mu\nu}^{a} \qquad (C.29)$$

DC-electric field induced second harmonic generation (EFISHG):

$$\gamma_{abcd}(\mp 2\omega;0,\pm\omega,\pm\omega) = -\sum_{\mu\nu} D_{\mu\nu}^{bcd}(0,\pm\omega,\pm\omega)\, h_{\mu\nu}^{a} \qquad (C.30)$$

Third harmonic generation (THG):

$$\gamma_{abcd}(\mp 3\omega;\pm\omega,\pm\omega,\pm\omega) = -\sum_{\mu\nu} D_{\mu\nu}^{bcd}(\pm\omega,\pm\omega,\pm\omega)\, h_{\mu\nu}^{a} \qquad (C.31)$$

DC-electric field induced optical rectification (EFIOR):

$$\gamma_{abcd}(0;0,\pm\omega,\mp\omega) = -\sum_{\mu\nu} D_{\mu\nu}^{bcd}(0,\pm\omega,\mp\omega)\, h_{\mu\nu}^{a} \qquad (C.32)$$

Intensity dependent refractive index (IDRI):

$$\gamma_{abcd}(\mp\omega;\pm\omega,\pm\omega,\mp\omega) = -\sum_{\mu\nu} D_{\mu\nu}^{bcd}(\pm\omega,\pm\omega,\mp\omega)\, h_{\mu\nu}^{a} \qquad (C.33)$$

Eqs. (C.22, C.24, C.28) are completely equivalent to Eqs. (C.5 – C.7). Algebraic manipulations similar to the ones that yielded Eqs. (C.8, C.9) can be carried out starting from Eqs. (C.24 – C.33). They lead to equations which formally appear more complicated than Eqs. (C.24 – C.33), yet are computationally less demanding. The computer implementation in HONDO of the calculation of all these dynamic effects follows closely the one described in the previous section. A complete account of the derivation and implementation of the above equations can be found in the works of Sekino and Bartlett[37] and Karna and Dupuis.[38]

D. Molecular Structure Options

With the advent of the analytical calculation of the derivatives of the energy with respect to the coordinates of the atoms (the forces acting on atoms) as well as of the higher derivatives, the interest in obtaining a more dynamical characterization of the motion of the nuclei on the potential energy surface has grown significantly over the past few years. We review here some of the options found in HONDO.

1. Equilibrium Structure Determination

Knowledge of the energy, energy derivatives, and (sometimes) energy hessian provides some very useful information to determine efficiently the preferred structure of the molecule, i.e. its most stable conformation. We are interested in minimizing E_{total} as written in Eq. (B.3). The energy is a function of $3Nuc$ coordinates, or $3Nuc - 6$ internal coordinates, for which we want to find the minimum. The geometry optimization algorithm is of the quasi-Newton type[39] in which the hessian matrix is updated during the course of the energy minimization. Presently this is the only algorithm implemented in HONDO, in contrast to algorithms which make use of an analytically computed hessian. The algorithm is described in the article of Spangler et al.[40] We note that options exist to carry out the optimization in the space of the internal coordinates as well as in the cartesian space, and that coordinates may be frozen during the optimization process (restricted optimization). This is a convenient feature if one is interested in generating specific 'cuts' of the potential surface. We note however that all $3Nuc$ gradients are computed, regardless of whether some coordinates are frozen or not. Thus the program is not presently designed to efficiently handle cases where a large fraction of the coordinates are frozen. In the current version of the program analytical gradients can be evaluated for all the SCF wavefunctions (RHF, UHF, ROHF, GVB, and MCSCF), and so geometry optimization using gradient information can be performed in all these cases. For other cases see the section on "non-gradient optimization" below.

2. Transition State Determination

In an analogous fashion, an automated procedure exists to determine the structure of transition states. The algorithm takes advantage of the energy derivatives, just as for the geometry optimization. The algorithm implemented in HONDO is the one developed by Cerjan and Miller,[41] modified to include updating of the hessian matrix during the search. This algorithm has been found to be very efficient in locating transition states, provided that we have a 'good' hessian matrix (a force constant calculated with a small basis set at the starting geometry is usually adequate). In some cases the user may have to resort to using the 'distinguished reaction coordinate' approach, whereby a series of optimizations are performed with an appropriately chosen coordinate being frozen at appropriately chosen values. From inspection of the potential energy curve, it is then possible to guess well the structure of the transition state. It is then best to return to the automated algorithm at that point. The method can be used in conjunction with all SCF wavefunctions, just as for geometry optimization.

3. Force Constant Calculation

Similarly, the Force method of Pulay[42] can be used in conjunction with all the SCF wavefunctions to determine the force constant matrix. The calculation is performed in the cartesian coordinate space, with the possibility of transforming the force matrix to the internal coordinate space at the end. Optionally the dipole and polarizability derivatives with respect to the cartesian coordinates can be calculated, yielding the infrared and Raman intensities. For closed shell SCF wavefunction, the program includes an analytical calculation of the second derivative matrix.[43] It is worth remarking that an option is provided for defining the symmetry adapted internal coordinates to be used in the vibrational analysis. By the same token, linear dependencies among internal coordinates can be removed as well.

4. Infrared and Raman Intensities Calculation

Dipole and polarizability derivatives with respect to the nuclear coordinates can be evaluated by themselves, in conjunction with any of the SCF wavefunctions. HONDO includes an implementation of the method proposed by Komornicki and McIver.[44] This approach is particularly useful when the force constant matrix is available from another source (force constant matrix calculated with a smaller basis set, or even experimentally determined force constant matrix). The derivatives thus calculated can be used to determine the infrared and Raman intensities.

5. Reaction Pathway Determination

HONDO contains an implementation of the algorithm of Schmid et al.[45] for determining the 'intrinsic reaction coordinate' (IRC) pathway, given the transition state structure and the vibrational mode corresponding to the imaginary frequency. The IRC path can be viewed as the steepest descent path (in an appropriate mass-weighted coordinate system) traced by a classical particle moving with infinitesimal velocity from the reactant state to the product state through the saddle point. In connection with the "reaction path hamiltonian" methodology[46] the IRC determination provides an attractive and practical method for obtaining dynamical information about chemical reactions. Currently HONDO does not include a link to a reaction path hamiltonian program.

Another benefit of the IRC lies in the definitive identification of reaction pathways. Often elementary reactions are characterized by the determination of the stationary point of the potential energy surface only. It is frequently difficult to establish with certainty (inferred from experiments or from calculations) which stationary structure belongs to which path. The IRC determination does just that.

Experience with the methodology indicates that tracing the IRC requires computation of resonances whose computing cost grows very fast with the size of the molecular system. Since we work in the context of ab initio (rather than semiempirical or other non ab initio) wavefunctions, we sought a technique which would accomplish the same thing as IRC but at a lower computing cost, albeit at a somewhat reduced accuracy. A dynamic "reaction coordinate" approach based on a time-dependent representation of the molecular motion (in contrast to the time-independent representation of the IRC) has been implemented in HONDO. The path connecting a saddle point and a local minimum energy structure can now be traced about five times faster, owing to a much reduced number of energy and gradient evaluations along the path.[47] Again this procedure may be applied in conjunction with any of the SCF wavefunctions.

6. Potential Surface Scan

When one is interested in obtaining molecular energies for a series of geometries, HONDO provides the option of stacking the geometries so that several energies can be computed in one job. This feature is especially useful for generating points on a potential energy surface for example.

7. Crossing Seam Minimum Energy Point Determination

The customary transition state discussed in Section 2 above is useful when the reaction is "adiabatic". In this case the splitting Δ at the avoided crossing (see Figure 1) is much larger than the thermal energy $k_B T$ such that the system remains an eigenfunction of the true hamiltonian and stays on the lower curve of Figure 1 throughout the reaction.

What do we do if the reaction of interest is "nonadiabatic"? Nonadiabatic processes are quite common in excited state chemistry and electron-transfer reactions.[49] In nonadiabatic processes the system traverses the avoided crossing while on the reactant surface (A in Figure 1) and then reverts back many times before making a transition to the product surface (B in Figure 1). If we assume that the transition satisfies the Franck-Condon principle[50] as well as the conservation of energy, then the transition must occur at the seam of the crossing surface between the two surfaces A and B. Hence, a more appropriate transition state for nonadiabatic reactions corresponds to the minimum energy point on the crossing seam surface. We refer to such a point as a "nonadiabatic transition state" to distinguish it from the traditional transition state.

HONDO has an implementation of the algorithm developed by Farazdel and Dupuis[51] for the determination of nonadiabatic transition states. If we denote the

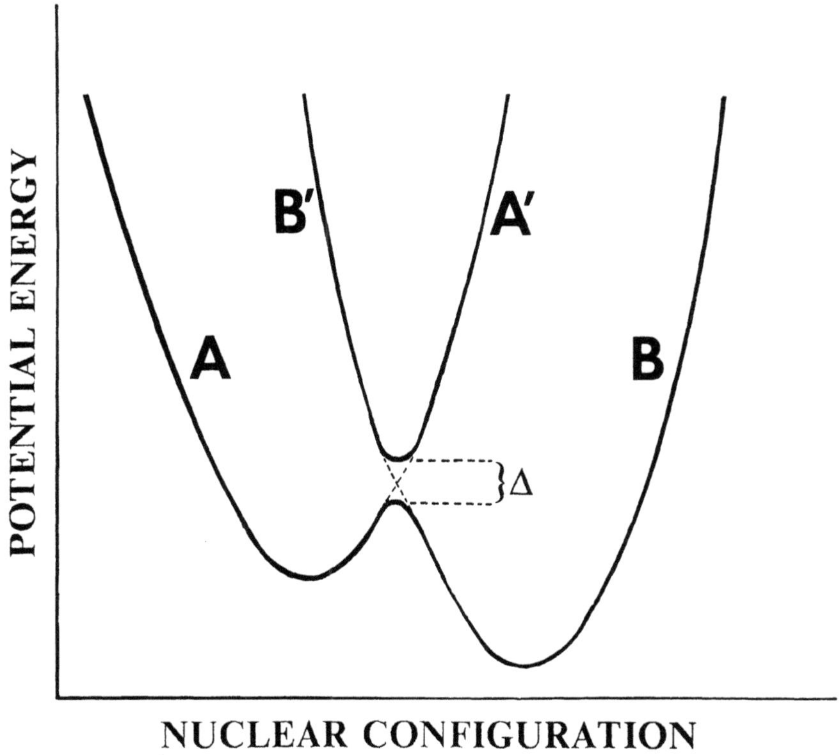

Figure 1. A one-dimensional schematic energy profile for the general reaction A → B. If initially the system is prepared in the state A and the abscissa (nuclear configuration) is increased in time, sweeping the energies through the avoided crossing, then according to the Landau-Zener[48] theory, the probability that the system will make a transition to state A′ is given by

$$P_{AA'} = 1 - P_{AB} \simeq \exp\left(-\frac{|\pi\Delta|^2}{h v (s_A - s_B)} \right).$$

Here Δ is the separation at the crossing, s_A and s_B are the slopes of the unperturbed surfaces A and B at the crossing, and v is the average (usually Boltzmann averaged) velocity with which the system moves through the crossing.

In the limit of large Δ we have $P_{AA'} \to 0$ or $P_{AB} \to 1$, and the reaction is said to be "adiabatic". In the opposite limit of $\Delta \to 0$, $P_{AA'} \to 1$ or $P_{AB} \to 0$, and the reaction is called "diabatic", i.e. there is no reaction! In the intermediate case (a "nonadiabatic" reaction) the reaction is neither purely adiabatic nor purely diabatic. Often the case of $\Delta \simeq 0$ is also called "quasi-diabatic".

---

energy of the quasi-diabatic (see the caption of Figure 1) states by $E_A(\underset{\sim}{R})$ and $E_B(\underset{\sim}{R})$, then the algorithm entails the minimization of a function of E_A and E_B

$$E = a\, E_A + b\, E_B \qquad (D.1)$$

with respect to the position of the nuclei $\underset{\sim}{R}$, subject to the constraint that $E_A(\underset{\sim}{R})$ be equal to $E_B(\underset{\sim}{R})$. In Eq. (D.1) a and b have been introduced to provide a unified formalism valid for searching both the seam surface itself or its symmetric subspace.[51] The energy equality constraint is treated by means of the method of Lagrange multipliers, and the optimization proceeds by a quasi-Newton algorithm.[39]

8. Non-Gradient Optimization

The non-gradient optimization program is designed to make it possible to determine the values of a few parameters which yield the lowest possible energy, and for which HONDO does not return an analytically calculated gradient. The program is set up to optimize one or a few gaussian exponents, although it can be tailored to any other type of optimization. In particular, since presently analytical gradients are not available for CI and MP2 wavefunctions, it is conceivable to modify the code to optimize geometrical parameters in conjunction with these two types of wavefunctions.

E. Other Options

1. Electron Transfer Reactions

(a) Introduction: Electron transfer (ET) processes have a central role in a variety of physical, chemical and biological systems, from semiconductors to key steps in photosynthesis and metabolism. ET is sometime termed "the simplest chemical reaction". In the so-called **intramolecular** (or unimolecular) ET, the electronic charge transfers from one site (the donor site D) of a molecule to another site (the acceptor site A) of the same molecule. For example

$$D^- - Sp - A \rightarrow D - Sp - A^- \tag{E.1}$$

$$D - Sp - A^+ \rightarrow D^+ - Sp - A \tag{E.2}$$

$$D^* - Sp - A \rightarrow D^+ - Sp - A^- \tag{E.3}$$

$$D - Sp - A^* \rightarrow D^+ - Sp - A^- \tag{E.4}$$

where Sp represents a rigid and usually neutral spacer. Eqs. (E.1) and (E.2) are simple charge-shift reactions and Eq. (E.3) is a charge-separation reaction from an excited donor while Eq. (E.4) is the same as Eq. (E.3) but from an excited acceptor.

In the case of **intermolecular** (or bimolecular) ET the transfer of electronic charge occurs between two molecular species. Studies have shown that often an intermolecular ET reaction, for example,

$$D + A \rightarrow D^+ + A^- \tag{E.5}$$

has the following three-step mechanism:

$$\mathrm{D} + \mathrm{A} \rightarrow \mathrm{D}|\mathrm{A} \qquad (E.6a)$$

$$\text{precursor} \qquad \mathrm{D}|\mathrm{A} \rightarrow \mathrm{D}^{+}|\mathrm{A}^{-} \qquad \text{successor} \qquad (E.6b)$$

$$\mathrm{D}^{+}|\mathrm{A}^{-} \rightarrow \mathrm{D}^{+} + \mathrm{A}^{-} \qquad (E.6c)$$

The first step is the diffusion of separated D and A toward each other to form the "precursor" complex molecule D|A . This is then followed by the actual ET within the precursor to form a "successor" complex molecule $D^{+}|A^{-}$. The last step is the separation of the successor to separated $D^{+} + A^{-}$. The experimental conditions usually can be set such that the first and the third steps are not rate determining, and the rate of the ET would be determined by the second step which is again an intramolecular ET reaction.

It is interesting to note that ET distances range from about 1 Å to more than 50 Å while ET times span some 20 orders of magnitude (from about a tenth of a femto seconds to a few hours). Understanding the origin of such a large variation in ET rates has prompted a tremendous amount of interest among researchers in different disciplines of physical sciences. At present time there are *classical,*[52] *semi-classical*[53] and *quantum mechanical*[54] theoretical models for ET processes. As an overview of different ET theories the reader is referred to Farazdel, et al[49] and references therein.

A key quantity in all different ET models is V_{AB} the ubiquitous many-electron electron transfer matrix element defined as

$$V_{AB} = (1 - S_{AB}^{2})^{-1} \, | H_{AB} - S_{AB} \, (H_{AA} + H_{BB})/2 \, |, \qquad (E.7)$$

where $S_{AB} = <\Psi_A|\Psi_B>$, $H_{AA} = <\Psi_A|H|\Psi_A>$, $H_{BB} = <\Psi_B|H|\Psi_B>$, $H_{AB} = <\Psi_A|H|\Psi_B>$ and H is the total electronic hamiltonian (i.e. excluding nuclear kinetic energy and nuclear repulsion terms from the total hamiltonian) of the system. All matrix elements in Eq. (E.7) are to be evaluated at a molecular geometry corresponding to either a nonadiabatic transition state (see Section D.7) or occasionally a saddle point (see Section D.2) depending upon the adiabaticity of the ET reaction (see Figure 1). Often in ET reactions, however, the so-called Condon approximation[55] holds, i.e. the value of V_{AB} is not very sensitive to the geometry as long as the distance between D and A in the molecule does not change. In such situations using an equilibrium geometry instead of the minimum point crossing seam would not be unreasonable. It is useful to know that $V_{AB} = \frac{1}{2}\Delta$ provided V_{AB} is evaluated at the crossing seam.

(b) Calculation of V_{AB}: Let Ψ_A and Ψ_B be the two symmetry-broken charge-localized (or simply "localized") quasi-diabatic UHF wavefunctions representing the reactant and the product states respectively. Being UHF wavefunctions, each constituent molecular spin-orbital of Ψ_A , $\mathbf{a} = (a_1, a_2, \ldots, a_N)$, has a nonzero overlap with each and every spin-orbital of Ψ_B , $\mathbf{b} = (b_1, b_2, \ldots, b_N)$, with the same spin, i.e., the overlap matrix **D**

$$\mathbf{D} = \int \mathbf{b}^{\dagger}\mathbf{a}d\tau \tag{E.8}$$

is not necessarily diagonal. As a result of this nonorthogonality, the expressions for direct calculation of matrix elements S_{AB} and H_{AB} which are needed for the computation of V_{AB} via Equation (E.7), becomes cumbersome. In order to simplify the evaluation, we use the method of the *corresponding orbital transformation* (56). In this procedure, each set of spin-orbitals **a** and **b** are linearly transformed

$$\hat{\mathbf{a}} = \mathbf{aV} \tag{E.9}$$

$$\hat{\mathbf{b}} = \mathbf{bU} \tag{E.10}$$

in such a way that the overlap matrix between the two new sets

$$\hat{\mathbf{d}} = \int \hat{\mathbf{b}}^{\dagger}\hat{\mathbf{a}}d\tau = \mathbf{U}^{\dagger}\mathbf{DV} \tag{E.11}$$

is diagonal. In our case **D** is a real matrix, hence Equation (E.11) is the singular value decomposition of **D** with diagonal elements of the diagonal matrix $\hat{\mathbf{d}}$ being the singular values. Matrices **U** and **V** are real and unitary. Here we develop expressions for the evaluation of S_{AB} and H_{AB} in terms of eigenvectors and overlap matrix of an LCAO-MO calculation.

Let us expand the molecular spin-orbitals **a** and **b** in terms of a common basis set $\chi = (\chi_1, \chi_2, \ldots, \chi_M)$, i.e.,

$$\mathbf{a} = \chi\mathbf{A} \tag{E.12}$$

and

$$\mathbf{b} = \chi\mathbf{B}. \tag{E.13}$$

Now using Eqs. (E.12), (E.13), (E.9) and (E.10), the one-electron contribution to H_{AB} [Eq. (22) of Ref. (56)] becomes

$$\begin{aligned} \Omega_{AB}{}^{(1)} &= C\sum_{i}^{N} < \hat{b}_i | \omega | \hat{a}_i > \mathrm{T}_{ii} \\ &= C\sum_{i}^{N}\sum_{\mu\nu}^{M} \hat{\mathrm{A}}_{\mu i}\,\mathrm{T}_{ii}\,\hat{\mathrm{B}}^{\dagger}_{i\nu}\,\omega_{\mu\nu} \\ &= C\sum_{\mu\nu}^{M} \mathrm{P}_{\mu\nu}\,\omega_{\mu\nu} \end{aligned} \tag{E.14}$$

where

$$\mathrm{T}_{ii} = \prod_{j \neq i}^{N} \hat{d}_{jj},$$

$$C = (\det \mathbf{U})(\det \mathbf{V}^{\dagger}),$$

$$\omega_{\mu\nu} = < \chi_{\mu} | \omega | \chi_{\nu} >$$

with ω denoting the one-electron operator and we have introduced three new matrices $\hat{\mathbf{A}} = \mathbf{AV}$, $\hat{\mathbf{B}} = \mathbf{BU}$ and $\mathbf{P} = \hat{\mathbf{A}}\mathbf{T}\hat{\mathbf{B}}^{\dagger}$. Note that $\mathbf{P}$ can be looked upon as a "generalized" density matrix in a sense that when there is no transformation (i.e., $\mathbf{U} = \mathbf{V} = \mathbf{I}$), $\mathbf{P}$ will reduce to the standard density matrix.

Similarly, the two-electron contribution [Eq. (23) of Ref. (56)] is

$$\begin{aligned} \Omega_{AB}{}^{(2)} &= C(\mathrm{prod})^{-1} \sum_{i<j}^{N} \mathrm{T}_{ii}\, \mathrm{T}_{jj} < \hat{b}_i \hat{b}_j | \omega(1,2)(1 - P_{12}) | \hat{a}_i \hat{a}_j > \\ &= \tfrac{1}{2} C(\mathrm{prod})^{-1} \sum_{\mu\nu\lambda\sigma}^{M} \mathrm{P}_{\lambda\mu}\, \mathrm{P}_{\sigma\nu} < \chi_{\mu}\chi_{\nu} | \omega(1,2)(1 - P_{12}) | \chi_{\lambda}\chi_{\sigma} > \end{aligned} \qquad (E.15)$$

where

$$\mathrm{prod} = \prod_{k=1}^{N} \hat{d}_{kk},$$

$\omega(1,2)$ denotes the Coulombic interaction between electrons 1 and 2; and P_{12} is the usual permutation operator. If we adopt the standard chemist's notation for two-electron integrals, the last equation may be written as

$$\Omega_{AB}{}^{(2)} = \tfrac{1}{2} C(\mathrm{prod})^{-1} \sum_{\mu\nu\lambda\sigma}^{M} \mathrm{P}_{\mu\nu}\, \mathrm{P}_{\lambda\sigma} \left\{ < \mu\nu | \lambda\sigma > - < \mu\sigma | \lambda\nu > \right\}. \qquad (E.16)$$

Here $< \mu\nu | \lambda\sigma > = < \chi_{\mu}\chi_{\lambda} | \omega(1,2) | \chi_{\nu}\chi_{\sigma} >$.

If we assume that each of the localized states Ψ_A and Ψ_B is represented by a single determinant spin unrestricted (UHF) wavefunction, then the overlap matrix $\mathbf{D}$ between orbitals with different spins vanishes, i.e.

$$\mathbf{D} = \begin{pmatrix} \mathbf{D}^{\alpha} & \mathbf{0} \\ \mathbf{0} & \mathbf{D}^{\beta} \end{pmatrix}.$$

Similar results are obtained for matrices $\mathbf{V}$, $\mathbf{U}$, $\hat{\mathbf{d}}$, $\mathbf{A}$, $\mathbf{B}$ and $\mathbf{P}$ with blocks $\mathbf{V}^{\alpha}$, $\mathbf{V}^{\beta}$, $\mathbf{U}^{\alpha}$, $\mathbf{U}^{\beta}$, and so on. As a consequence we have [from Eq. (E.14)]

$$\Omega_{AB}{}^{(1)} = C\left\{\sum_{\mu\nu}^{M} P^{\alpha}_{\mu\nu}\,\omega_{\mu\nu} + \sum_{\mu\nu}^{M} P^{\beta}_{\mu\nu}\,\omega_{\mu\nu}\right\}. \tag{E.17}$$

or

$$\Omega_{AB}{}^{(1)} = C\sum_{\mu\nu}^{M} P^{t}_{\mu\nu}\,\omega_{\mu\nu}, \tag{E.18}$$

where $\mathbf{P}^t = \mathbf{P}^\alpha + \mathbf{P}^\beta$. It is also assumed that the trivial spin integration in $\omega_{\mu\nu}$ is already performed.

Similarly, from Eq. (E.16) we get

$$\begin{aligned}\Omega_{AB}{}^{(2)} = \tfrac{1}{2}C(\text{prod})^{-1}\sum_{\mu\nu\lambda\sigma}^{M}\Big\{&P^{\alpha}_{\mu\nu}\,P^{\alpha}_{\lambda\sigma}\left[(\mu\nu\,|\,\lambda\sigma) - (\mu\sigma\,|\,\lambda\nu)\right]\\ &+P^{\alpha}_{\mu\nu}\,P^{\beta}_{\lambda\sigma}\,(\mu\nu\,|\,\lambda\sigma)\\ &+P^{\beta}_{\mu\nu}\,P^{\alpha}_{\lambda\sigma}\,(\mu\nu\,|\,\lambda\sigma)\\ &+P^{\beta}_{\mu\nu}\,P^{\beta}_{\lambda\sigma}\left[(\mu\nu\,|\,\lambda\sigma) - (\mu\sigma\,|\,\lambda\nu)\right]\Big\}\end{aligned} \tag{E.19}$$

or after some trivial manipulations we get

$$\Omega_{AB}{}^{(2)} = \tfrac{1}{2}C(\text{prod})^{-1}\sum_{\mu\nu\lambda\sigma}^{M}\left[P^{t}_{\mu\nu}\,P^{t}_{\lambda\sigma} - P^{\alpha}_{\mu\nu}\,P^{\alpha}_{\lambda\sigma} - P^{\beta}_{\mu\nu}\,P^{\beta}_{\lambda\sigma}\right](\mu\nu\,|\,\lambda\sigma). \tag{E.20}$$

Here $(\mu\nu\,|\,\lambda\sigma)$ is similar to $<\mu\nu\,|\,\lambda\sigma>$ except in the former the spin integration is already done. Note that Eqs. (E.18) and (E.20) bear a formal resemblance to that of standard one-electron and two-electron sums in LCAO-MO methods.

Our program module linked to HONDO consists of the following steps:

1. Calculate the overlap matrix $\mathbf{D}^\alpha$ from $\mathbf{D}^\alpha = \mathbf{B}^{\alpha\dagger}\mathbf{S}^\alpha\mathbf{A}^\alpha$, where $\mathbf{A}^\alpha$ and $\mathbf{B}^\alpha$ are the eigenvectors of the α-spin quasi-diabatic states and $\mathbf{S}^\alpha$ is the overlap matrix over atomic orbitals.
2. Perform the corresponding orbital transformation by computing the singular value decomposition of the real matrix $\mathbf{D}^\alpha$ as $\mathbf{D}^\alpha = \mathbf{U}^\alpha\hat{\mathbf{d}}^\alpha\mathbf{V}^{\alpha\dagger}$ and determine $\mathbf{U}^\alpha$, $\mathbf{V}^\alpha$ and the diagonal matrix $\hat{\mathbf{d}}^\alpha$
3. As an internal check, compute det$\mathbf{U}^\alpha$ and det$\mathbf{V}^\alpha$. The absolute value of both quantities should be unity since $\mathbf{U}^\alpha$ and $\mathbf{V}^\beta$ are unitary matrices.
4. Form matrices $\hat{\mathbf{A}}^\alpha$ and $\hat{\mathbf{B}}^\alpha$ as well as the diagonal matrix $\mathbf{T}$, using $\hat{\mathbf{A}}^\alpha = \mathbf{A}^\alpha\mathbf{V}^\alpha$, $\hat{\mathbf{B}}^\alpha = \mathbf{B}^\alpha\mathbf{U}^\alpha$ and

$$T_{ii} = \prod_{j\neq i}^{N} \hat{d}_{jj}$$

where N is the *total* number of α and β electrons, i.e. $N = n_\alpha + n_\beta$. Then compute the generalized density matrix $\mathbf{P}$:i1.generalizeddensitymatrixα by $\mathbf{P}^\alpha = \hat{\mathbf{A}}^\alpha \mathbf{T} \hat{\mathbf{B}}^{\alpha\dagger}$.

5. Compute S_{AB}^α from [Eq. (21) of Reference 56]

$$S_{AB}^\alpha = (\det \mathbf{U}^\alpha)(\det \mathbf{V}^{\alpha\dagger}) \prod_{j=1}^{n_\alpha} \hat{d}_{jj}^\alpha \tag{E.21}$$

where n_α is the number of α electrons.

6. Repeat steps 1 − 5 but this time for β-electrons.
7. Compute S_{AB} from

$$S_{AB} = S_{AB}^\alpha S_{AB}^\beta \tag{E.22}$$

8. Compute $\Omega_{AB}{}^{(1)}$ and $\Omega_{AB}{}^{(2)}$ using Eqs. (E.18) and (E.20) respectively. Then compute H_{AB} by $H_{AB} = \Omega_{AB}{}^{(1)} + \Omega_{AB}{}^{(2)}$ and finally V_{AB} using Eq. (E.7).

2. Effective Core Potentials

The ECP approach was developed by Khan et al.[57] to replace the inner shell electrons of atoms by an effective potential. This is in line with the accepted idea that in a molecule the electronic structure is dictated mostly by the valence electrons, and that the core orbitals resemble the atomic core orbitals. Potentials and associated valence basis sets have been developed for most of the periodic table, and are tabulated in the program just like the regular basis sets. In particular the potentials and basis set of Stevens et al.[58] can be invoked in HONDO by means of a key work for ease of use, although any other potentials and associated basis set can be used.

3. Representation of An External Field

It is sometimes necessary to try to represent by point charges the effect of a molecular environment. This amounts to including additional terms in the one-electron operator, and to solving the SCF equations with this modified operator. Such an option exits in HONDO. In other circumstances we may be interested in modeling the interaction of a molecule in the presence of an electric field. Prior to the development of analytical methods for calculating the (hyper)polarizability tensors, as outlined earlier, these tensors were calculated by finite difference of the energy or of the dipole moment, in accord with Eq. (C.1). The program offers the option to specify the magnitude of the uniform field applied onto the molecule, and the corresponding field operator is then added to the one-electron operator.

F. Miscellaneous Features

1. Integrals and Derivatives

(a) Electron Repulsion Integrals: HONDO's trademark and mainstay over the years has been the original numerical quadrature developed[1] specifically for the calculation of the ubiquitous electron repulsion integrals found in all ab initio electronic structure programs. The Rys quadrature algorithm outlined below, which combines efficiency and simplicity, has been credited as a key factor in the extraordinary development of the energy derivative methodology in the past ten years. Today the method is still very competitive with several different approaches proposed more recently.

It has been known since the early work of Boys that electron repulsion integrals over primitive cartesian basis functions $\eta = (x_1 - x_A)^{nx}$ $(y_1 - y_A)^{ny}$ $(z_1 - z_A)^{nz} \exp(-\alpha r_{1A}^2)$ can be expressed in the form:

$$(\eta_i\eta_j // \eta_k\eta_l) = \iint \eta_i(1)\eta_j(1)\frac{1}{r_{12}}\eta_k(2)\eta_l(2)d\tau_1 d\tau_2 = \sum_{m=o}^{L} C_m F_m(X) \qquad (F.1)$$

where $F_m(X)$ is related to the incomplete Gamma function:

$$F_m(X) = \int_0^1 t^{2m} \exp(-Xt^2)dt \qquad (F.2)$$

The value of X depends upon the exponential parameters and the position of the centers of the four gaussians. We rewrite Eq. (F.1) as

$$(\eta_i\eta_j // \eta_k\eta_l) = \int_0^1 P_L(t^2) \exp(-Xt^2)dt \qquad (F.3)$$

an integral which can be exactly evaluated by an N-point quadrature formula in accord with the theory of orthogonal polynomials:

$$(\eta_i\eta_j // \eta_k\eta_l) = \sum_{\alpha=1}^{N} P_L(t_\alpha^2)\, w_\alpha \qquad (F.4)$$

where t_α are the zeros of an appropriate Rys polynomial, and w_α are the corresponding weight factors. It can be shown that the polynomial $P_L(t^2)$ can be written as the product of 3 polynomials of t^2, each corresponding to a cartesian component x, y, or z of the electron coordinates. The numerical quadrature leads to the following final expression:

$$(\eta_i\eta_j // \eta_k\eta_l) = 2(\rho / \pi)^{1/2} \sum_{\alpha=1}^{N} I_x(u_\alpha)\, I_y(u_\alpha)\, I_z(u_\alpha)\, w_\alpha \qquad (F.5)$$

The three terms I_x, I_y, and I_z are themselves two-dimensional integrals for which one could work out algebraic formulas. An alternative approach has proven more desirable however which is based on recursion formulae. If we define $G(n, m)$ as:

$$G(n,m) = \iint_{-\infty}^{+\infty} dx_1 dx_2 (x_1 - x_i)^n (x_2 - x_k)^m \exp\left[-a(x_1 - x_A)^2 - b(x_2 - x_B)^2 - u^2(x_1 - x_2)^2 \right] \qquad (F.6)$$

we see that:

$$G(n,m)\exp(-G_x) = I_x(n_i + n_j,\ 0,\ n_k + n_l,\ 0,\ u) \qquad (F.7)$$

where G_x is a function of the x coordinates of the four centers. It can be shown that the G 's satisfy some 3-term recursion formulas:

$$G(n+1,m) = n\, B_{10}\, G(n-1,m) + m\, B_{00}\, G(n,m-1) + C_{00}\, G(n,m) \qquad (F.8)$$

$$G(n,m+1) = m\, B'_{01}\, G(n,m-1) + n\, B_{00}\, G(n-1,m) + C'_{00}\, G(n,m) \qquad (F.9)$$

To produce the list of desired I_x values for each t_α in the quadrature formula one first starts with the special cases $m = 0$, for $n = 1,2, \ldots$ and $n = 0$, for $m = 1,2, \ldots$. A 2-term relation is used for $G(n, 1)$ and $G(1, m)$. Finally the 3-term relation yields $G(n, m)$. The full list of $I(n_i, n_j, n_k, n_l)$ is generated by a 2-step process involving the transfer equations:

$$I_x(n_i, n_j, m, 0) = I_x(n_i + 1, n_j - 1, m, 0) + (x_i - x_j) I_x(n_i, n_j - 1, m, 0) \qquad (F.10)$$

$$I_x(n_i, n_j, n_k, n_l) = I_x(n_i, n_j, n_k + 1, n_l - 1) + (x_k - x_l) I_x(n_i, n_j, n_k, n_l - 1) \qquad (F.11)$$

(b) Integral Derivatives: We are concerned with the derivatives of the electron repulsion integrals with respect to the nuclear coordinates. From the form of a primitive cartesian gaussian function, it appears that the derivative with respect to its center is equal to a linear combination of two gaussian functions with angular momentum lowered and raised by one respectively:

$$\frac{\partial\, \eta(n_x, n_y, n_z)}{\partial x_A} = -\, n_x\, \eta(n_x - 1, n_y, n_z) + 2\,\alpha\, \eta(n_x + 1, n_y, n_z) \qquad (F.12)$$

The derivative of an electron repulsion integral involving η can be obtained as the linear combination of two intermediate integrals, each of them written as the sum of products of I_x, I_y , and I_z integrals. It is easy to see that when taking the derivative of the electron repulsion integral with respect to x_A the two intermediate integrals are built from the same I_y and I_z, but use different elements of the I_x array. The linear combination written above can be applied to the I_x components only, and therefore it is most convenient to create new integrals denoted I'_x , and similarly I'_y, and I'_z according to:

$$I'_x(n_x, ...) = -n_x I_x(n_x - 1, ...) + 2\alpha I_x(n_x + 1, ...) \qquad (F.13)$$

The final integral derivative has the form:

$$(\eta_i\eta_j // \eta_k\eta_l) = 2(\rho / \pi)^{1/2} \sum_{\alpha=1}^{N} I'_x(u_\alpha) I_y(u_\alpha) I_z(u_\alpha) w_\alpha \qquad (F.14)$$

Extension of this algorithm to higher derivatives is straightforward. Its implementation was first carried out in the GRADSCF ab initio program[59] which showed truly outstanding performance. We have recently implemented it in HONDO.

2. Point Group Symmetry

In this section we review the 'symmetrization' method which allows one to take advantage of molecular symmetry when calculating energy and energy derivatives.[7-9] The other well known approach to symmetry is to generate electron repulsion integrals over symmetry adapted basis functions. In this scheme many steps in the wavefunction calculation can be broken down into symmetry blocks. This approach is especially well suited to SCF calculations which are performed with an integral file significantly shorter. The overhead involved in calculating the symmetry adapted integrals is quickly recovered owing to the reduced length of the integral list. The transformation of the integrals over symmetry adapted atomic orbitals to integrals over molecular orbitals needed for the treatment of electron correlation also benefits greatly from symmetry blocking.

In contrast the 'symmetrization' method to be outlined below is well suited to cases where the overhead for symmetry blocking cannot be easily recovered, for example for very large molecules and in energy derivative calculations which are not iterative in nature. The 'symmetrization' method is quite general since it can handle symmetry point groups that have even degenerate irreducible representations. The method requires that the electron density obtained from the wavefunction have the symmetry of the molecule. In what follows we adhere to the following notations: the pair of indices A,a represents the a-th coordinate of center A; μ, ν, ρ, σ . . . represent atomic basis functions, i, j, k, l . . . represent molecular orbitals; q stands for any nuclear displacement. As an example we will deal with a closed shell energy expression. Extension to other wavefunctions is straightforward.

(a) Energy and Energy Derivatives: For a closed shell molecule, the energy and energy derivatives have the form:

$$E_{HF} = \sum_{\mu\nu} D_{\mu\nu} (\mu|h|\nu) + \frac{1}{2} \sum_{\mu\nu\rho\sigma} D_{\mu\nu} D_{\rho\sigma} [(\mu\nu|\rho\sigma) - \frac{1}{2}(\mu\rho|\nu\sigma)] + V_{Nuc} \qquad (F.15)$$

$$\frac{\partial E_{HF}}{\partial q} = \sum_{\mu\nu} D_{\mu\nu} \frac{\partial}{\partial q} (\mu|h|\nu) + \frac{1}{2} \sum_{\mu\nu\rho\sigma} D_{\mu\nu} D_{\rho\sigma} \frac{\partial}{\partial q} [(\mu\nu|\rho\sigma) - \frac{1}{2}(\mu\rho|\nu\sigma)]$$
$$+ \frac{\partial}{\partial q} V_{NUC} - \sum_{\mu\nu} W_{\mu\nu} \frac{\partial}{\partial q} (\mu|\nu). \qquad (F.16)$$

$$\frac{\partial^2 E_{HF}}{\partial p \partial q} = \sum_{\mu\nu} D_{\mu\nu} \frac{\partial^2}{\partial p \partial q} (\mu|h|\nu) + \frac{1}{2} \sum_{\mu\nu\rho\sigma} D_{\mu\nu} D_{\rho\sigma} \frac{\partial^2}{\partial p \partial q} [(\mu\nu|\rho\sigma) - \frac{1}{2}(\mu\rho|\nu\sigma)]$$
$$+ \frac{\partial^2}{\partial p \partial q} V_{NUC} - \sum_{\mu\nu} W_{\mu\nu} \frac{\partial^2}{\partial p \partial q} (\mu|\nu) - \sum_{\mu\nu} \frac{\partial W_{\mu\nu}}{\partial p} \frac{\partial}{\partial q} (\mu|\nu) \qquad (F.17)$$
$$+ \sum_{\mu\nu} \frac{\partial}{\partial p} D_{\mu\nu} \frac{\partial}{\partial q} (\mu|h|\nu) + \sum_{\mu\nu\rho\sigma} \frac{\partial}{\partial p} D_{\mu\nu} D_{\rho\sigma} \frac{\partial}{\partial q} [(\mu\nu|\rho\sigma) - \frac{1}{2}(\mu\rho|\nu\sigma)]$$

In Eqs. (F.15) to (F.17) V_{nuc} is the nuclear repulsion energy, $D_{\mu\nu}$ is a density matrix element expressed as a function of the molecular coefficients $C_{\mu i}$

$$D_{\mu\nu} = \sum_{i}^{occ} 2 C_{\mu i} C_{\nu i}, \qquad (F.18)$$

and $W_{\mu\nu}$ is the energy weighted density matrix element expressed in terms of orbital coefficients and orbital energies ε_i by:

$$W_{\mu\nu} = \sum_{i}^{occ} 2 \varepsilon_i C_{\mu i} C_{\nu i}. \qquad (F.19)$$

Let $F_{\mu\nu}$ denote the Fock matrix element:

$$F_{\mu\nu} = (\mu|h|\nu) + \sum_{\rho\sigma} D_{\rho\sigma} [(\mu\nu|\rho\sigma) - \frac{1}{2}(\mu\rho|\nu\sigma)] \qquad (F.20)$$

It is convenient to define a quantity $\tilde{F}_{\mu\nu}(Q, q)$ obtained by replacing in Eq. (F.20) the one- and two-electron integrals by their derivatives with respect to coordinate q. $\tilde{F}_{\mu\nu}$ is one of two terms which make up the derivative of $F_{\mu\nu}$. In practice $\tilde{F}_{\mu\nu}$ is used to set up the coupled-perturbed Hartree-Fock equations which in turn lead to the derivatives of the density matrix elements and of the Fock matrix elements. Point group symmetry can be used to reduce the computational task in calculating the Fock matrix elements $F_{\mu\nu}$ [Eq. (F.20)], the two-electron contribution to the gradient [Eq. (F.16)], the two-electron contribution to the hessian [Eq. (F.17)], and finally the two-electron contribution to the $\tilde{F}$ matrix.

(b) Basis Functions and Symmetry Transformations: We denote χ a real Cartesian basis function, χ^a and χ^{ab} the first and second derivatives of χ with respect to

A_x, A_y , and A_z when a equals 1, 2, and 3 respectively. Basis functions are grouped into shells. Functions in the same shell have the same A and total angular momentum λ, so an alternative notation is:

$$\chi\,(I,m) = \chi\,(A, n_x, n_y, n_z) \tag{F.21}$$

where I is the shell index. Let G be the point group of the molecule and $\hat{R}$ an element of G . Operator $\hat{R}$ maps point r into r', function $f(r)$ into function $f(r')$, and shell I into shell I'. Shells I and I' have the same angular momentum and their centers are related by:

$$A_{I'} = \hat{R}\,A_I \tag{F.22}$$

Basis function $\chi(I,m)$ maps into a linear combination of functions in shell I that is given by:

$$\hat{R}\,\chi\,(I,m) = \sum_{m'} \chi(I', m')\; R(\lambda;\, m', m) \tag{F.23}$$

since we insist that a shell be closed under rotation about its own center. The effect of a symmetry operation on a basis function derivatives χ^a and χ^{ab} are given by:

$$\hat{R}\,\chi^a(I,m) = \sum_{m'}\sum_{a'} \chi^{a'}(I', m')\;\; R(1;a',a)\;\; R(\lambda;m',m) \tag{F.24}$$

$$\hat{R}\,\chi^{ab}(I,m) = \sum_{m'}\sum_{a'b'} \chi^{a'b'}(I', m')\;\; R(1;a',a)R(1;b',b)R(\lambda;m',m) \tag{F.25}$$

(c) Symmetrization Operator: If χ denotes a row vector of basis functions and $\hat{R}\chi = \chi'$ the set of images of these functions under R', then χ provides a basis for an n by n matrix representation of the group

$$\chi' = \chi\;\, R \tag{F.26}$$

The density matrix defined by:

$$\rho(r) = 2\chi\, D\chi^+ \tag{F.27}$$

satisfies the equation:

$$D = RDR^+ \tag{F.28}$$

if the electron density $\rho(r)$ has the symmetry of the molecule. The Fock matrix satisfies the relation

$$F = R^+FR \tag{F.29}$$

Similarly we have shown that $\tilde{F}(A,a)$ satisfies the following property:

$$\tilde{F}(A,a) = \sum_{a'} R(1;a',a)R^{+}\tilde{F}(A',a')R \qquad (F.30)$$

where center A' is the image of center A under $\hat{R}$.

These two equations are the key equations for the definition of the 'symmetrization' projection operators. The Fock matrix, the energy first- and second-derivatives and the Fock matrix derivatives are obtained by symmetrization of skeleton Fock matrix, energy first- and second- derivatives, and Fock derivative matrix constructed from the list of unique electron repulsion integrals.

The symmetrization operator for F is given by:

$$F = (2g)^{-1} \sum_{R} R(F^{*} + F^{*+})\ R \qquad (F.31)$$

The operators for the energy first- and second-derivatives are:

$$E_A^a = g^{-1} \sum_{R} \sum_{a'} E_{A'}^{*a'}\ R(1;a',a) \qquad (F.32)$$

$$E_{AB}^{ab} = g^{-1} \sum_{R} \sum_{a'b'} E_{A'B'}^{*a'b'}\ R(1;a',a)R(1;b',b), \qquad (F.33)$$

and the operator for the $\tilde{F}$ has the form:

$$\tilde{F}(A,a) = (2g)^{-1} \sum_{R} \sum_{a'} R(1;a',a)R^{+} \left[\tilde{F}^{*}(A',a') + \tilde{F}^{*+}(A',a') \right] R \qquad (F.34)$$

The respective skeleton quantities labeled with an asterisk in the above equations are obtained from the list of symmetry unique integrals only, after these have been multiplied by their constituency factor, i.e. the number of integrals which they represent from the complete list of integrals. In practice, as the unique electron repulsion integrals and their derivatives are calculated and their constituency factors are determined, their contributions are added into the proper skeleton elements. Once all the integrals have been processed, and only then, the skeleton matrices are symmetrized. For all the quantities defined above we have found that computational times are roughly inversely proportional to the order of the point group.

Recently Carsky et al.[60] have shown that the list of unique integrals multiplied by their constituency factors can be used to perform the transformation of the electron repulsion integrals from the atomic orbital basis to the molecular orbital basis. This step is needed for the calculation of the energy of correlated wavefunctions. It is also needed for setting up the coupled perturbed equations which give the derivatives of the molecular orbital coefficients with respect to the nuclear coordinates. Thus it appears that all the computational steps involved in derivatives calculations can be accomplished from the petite list of integrals, i.e. the list of unique integrals. The implementation of the transformation of the petite list of integrals has been done in HONDO for abelian point groups only.

G. Illustrative Examples

From the description of the program given in the previous sections, it is apparent that HONDO offers a rather comprehensive collection of algorithms. Indeed we believe that not one single approach is applicable to all molecular systems, and that it is important to choose the best theoretical approach for each particular situation. With this in mind, we continue to incorporate new capabilities into HONDO as well as to improve the efficiency of the code by using algorithms that take best advantage of available computer architectures.

Specific physical situations require specific wavefunctions, and accordingly the program offers the user a variety of wavefunctions to choose from. Closed-shell systems which may be described by the Roothaan RHF formalism are numerous. Open-shell systems can often be described with the spin-unrestricted UHF or spin-restricted open-shell ROHF formalism. Diradicals, symmetry breaking structures, and excited states require MCSCF or CI wavefunctions. Numerous examples of the use of these several types of wavefunctions can be found in the literature.

In many cases it is necessary to use convergence aids in order to obtain convergence in the SCF process. Damping of the density matrix and extrapolation techniques are generally sufficient to bring about convergence. In more difficult cases, we mentioned the level shifting method of Saunders and Hillier.[13] As already indicated, level shifting is especially useful for calculations involving transition metal complexes, such as methylene(tricarbonyl)nickel $(CO)_3NiCH_2$. As a consequence, Spangler et al.[61] reported the first application of nonempirical molecular electronic structure theory to a realistic transition-metal carbene complex $(CO)_3NiCH_2$. This SCF study was carried out with a large and flexibly contracted basis set labeled Ni(15s11p6d/11s8p3d), C, O(9s5p/4s2p), H(5s/3s). The critical predicted equilibrium geometrical parameters were $R(Ni\text{-}CH_2) = 1.83$ Å, $\theta(HCH) = 108°$. The sixfold barrier of rotation about the Ni-CH_2 axis was found to be small (~ 0.2 kcal/mole).

GVB wavefunctions provide a convenient way to include a limited treatment of electron correlation effects, such as left-right, in-out, and angular correlation. The perfect pairing approximation was systematically evaluated[62] for the prediction of molecular structures, harmonic vibrational frequencies, and infrared intensities for saturated and unsaturated hydrocarbons. This theoretical model was found to slightly overemphasize left-right correlation effects resulting in bond lengths ~

0.01-0.02 Å longer than experiment, calculated harmonic frequencies within 4.5% of the observed fundamentals and 3.5% of the experimentally deduced harmonic frequencies. Similar conclusions hold for MCSCF wavefunctions such as the Fully Optimized Reaction Space (FORS) wavefunction of Ruedenberg et al.,[63] and the Complete Active Space (CAS) wavefunction of Roos et al.[24]

Although HONDO is not limited to MCSCF wavefunctions of the CAS type, these wavefunctions have the advantage that they can be conveniently defined by specifying only the orbitals which play an active role in the description of the physical system. In particular MCSCF are very efficacious for describing the electronic structure of radicals, especially resonance of valence structures which are often associated with symmetry breaking behaviors of wavefunctions, for describing bond breaking and bond formation in chemical reactions, and finally for improving electron densities, so yielding improved properties.

One example of the use of a MCSCF wavefunction to solve a broken symmetry dilemma can be found in the description of the allyl radical[64] CH_2CHCH_2. While the Hartree-Fock model incorrectly predicts a nonsymmetric structure, a (CAS) MCSCF wavefunction which includes all configurations obtained by distributing the three π electrons among the three π valence orbitals in all possible ways consistent with spin and spatial symmetry, yields a C_{2v} symmetry structure in accord with ESR evidence. The symmetry breaking vibrational mode has a frequency of about 1200 cm^{-1}.

A very similar approach was used in predicting the structure of the low lying states of the vinoxy radical[65] CH_2CHO. The ground state results from the resonance interaction of the ethenyloxy radical and of the (formyl)methyl radical, both with $^2A''$ symmetry. The valence bond picture of the former has a CC double bond with an out of plane unpaired electron localized on the oxygen atom while for the latter a CO double bond with an out of plane unpaired electron localized on the methylene carbon. The resonance interaction of these two structures yields the ground state and an excited state of same symmetry with T_e= 3.57 eV experimentally. The (3 electron + 3 π orbitals) wavefunctions described above gives a nearly quantitatively accurate description of the system. The ethenyloxy radical further has a $^2A'$ valence structure with no resonant counterpart. This structure is responsible for a near-infrared electronic state whose existence was predicted by the calculations prior to experimental observation.

MCSCF wavefunctions are a convenient way to obtain good zeroth-order descriptions of chemical reactions. For this purpose it is necessary to select the electrons which participate actively in the bond breaking and bond formation processes. In general a bonding orbital and an antibonding orbital can be associated with an electron pair and the MCSCF wavefunction is constructed by including all configurations obtained by distributing the active electrons among the active orbitals in all possible ways consistent with spin and spatial symmetry. Very often a molecular species is involved in several reaction channels, and different MCSCF wavefunctions may be needed for different reaction channels as the number of reactive electrons to describe several or all reaction channels at the same time becomes too large. This approach was used in the study of the H + O_3, OH + O_2, and O + HO_2 system[66] . In this case the same CAS wavefunction (7 electrons + 7 orbitals) could be used to describe all reaction channels considered.

Sometimes for better accuracy one may want to include additional electrons in the active space of the MCSCF wavefunction as was done for the study of the dissociation of formaldehyde[67] H_2CO to H_2 + CO. When one is able to include more electrons in the active space, the MCSCF wavefunction becomes large enough to yield very good accuracy. In most cases though, the MCSCF calculation needs to be followed by a large scale CI calculation for good accuracy. The most important configurations of the MCSCF wavefunction can be determined by performing a CI calculation using the same CAS configuration list, and the natural orbitals of the CAS wavefunction. The use of the natural orbitals removes the arbitrariness in the relative rotation of the active orbitals. The configurations with the largest coefficients of the CI expansion in the natural orbital space can be used as a reference set for a multireference CI calculation.

MCSCF wavefunctions may also yield improved charge distributions following a careful analysis of valence configurations which contribute to the description of the electronic state of interest. In a study of the ozone molecule[68] O_3, it is shown that a valence configuration which plays a lesser role in the structure determination possesses a high polarity. Even with a small coefficient this configuration plays a dramatic role in the description of the charge cloud, and is the key to obtaining accurate dipole moment and dipole moment derivative functions. Similar analyses are likely to be applicable to other systems.

Recently it has become important to extract from wavefunctions information which is not related to the energy only. A variety of property calculations are available with the present program. In particular, subroutines are included which perform a decomposition of the electron cloud and define bond orders and valencies. This is one partitioning scheme among several possible, which however yields quantities in qualitative accord with basic concepts of chemistry, as demonstrated for a series of prototypical systems.[34]

Another series of properties which can be calculated deal with the response of the electron cloud to an external electric field. The optical properties of organic materials is currently the subject of extensive research due to their potential technological applications in laser technology, integrated optics, and optical information processing. The present implementation is limited to closed-shell systems for which it is possible to calculate the polarizability, and the first and second hyperpolarizabilities. Until recently most theoretical work in this field was performed at a semi-empirical level of theory. The first systematic investigation of the accuracy of extended basis set SCF hyperpolarizabilities for large molecular systems has been carried out using HONDO. Systems considered included polyenes up to $C_{22}H_{24}$, the haloform series CHX_3, X = F, Cl, Br, I, and a series of nitroaniline derivatives[69-71] . Our preliminary findings are that α may qualitatively be predicted easily, that β is consistently too small by a factor of two and γ by a factor of six compared to experiment. Basis set and electron correlations are two sources of error in these studies. Correlation corrections appear to bring the agreement for β to 75% and for γ to about 50%[72] . For the haloform series and for the nitroaniline derivatives the calculated values reproduce the experimentally observed trends well.

The present version of HONDO has capabilities for studying electron transfer (ET) reactions including calculation of the omnipresent electron-transfer matrix element V_{AB} where A represents the state before and B the state after the transfer of electronic charge. Recent study of ET characteristics and their dependence on an external electric field for a typical rigid spiro $\pi - \sigma - \pi$ molecule — a "molecular electronic device" molecule — has revealed[49] that upon increasing the electric field strength F i) exothermicity increases nearly linearly in accord with a power series expansion of the energy of a molecule in the electric field (Eq. C.1) up to the first order in F; ii) intramolecular reorganization energy[73] (energy difference between A and B states but both at the B stable nuclear configuration) remains constant; iii) activation energy decreases; and iv) V_{AB} does not vary significantly. Additionally a simple model based on the traditional theory of a molecule in a weak electric field (Eq. C.1) is proposed which rationalizes the above behaviors.

In the studies highlighted above, in addition to the underlying research problems, one can find examples of other capabilities of the program, such as determination of vibrational spectra and related vibrational intensities, automated determination of transition state structure, and other features as well, all things which are secondary in the overall studies, but which however make the research possible in the first place.

References

1. M. Born and J.R. Oppenheimer, Ann.Physik **84**, 457 (1927).
2. M. Dupuis, J. Rys, H.F. King, J. Chem. Phys. **65**, 111 (1976).
3. W. Pauli, Z. Physik **31**, 765 (1925).
4. L. Pauling and E.B. Wilson, *Introduction to Quantum Mechanics*, McGraw-Hill, New York (1935).
5. C.C.J. Roothaan, Rev. Mod. Phys. **23**, 69 (1951) ; G.G. Hall, Proc. Roy. Soc. (London), **A205**, 541 (1951).
6. R.C. Raffinetti, Chem. Phys. Lett. **20**, 335 (1973).
7. M. Dupuis and H.F. King, Int. J. Quant. Chem., **11**, 613 (1976).
8. T. Takada, M. Dupuis, and H.F. King, J. Chem. Phys. **75**, 332 (1981).
9. T. Takada, M. Dupuis, and H. F. King, J. Comp. Chem. **4**, 234 (1983).
10. T.P. Hamilton and P. Pulay, J. Chem. Phys. **84**, 5728 (1986).

11. **MELDF**, L. McMurchie, S.T. Elbert, S. Langhoff, E.R. Davidson, and D. Feller, Indiana University, Bloomington, Indiana.
12. **GAUSSIAN-70**, W.J. Hehre, W.A. Lathan, M.D. Newton, R. Ditchfield, and J.A. Pople, Program no.236, Quantum Chemistry Program Exchange, Indiana University, Bloomington, Indiana.
13. V.R. Saunders and H.I. Hillier, Int. J. Quant. Chem. **7**, 699 (1973).
14. J.A. Pople and R.K. Nesbet, J. Chem. Phys. **22**, 571 (1954).
15. E.R. Davidson, Chem. Phys. Lett. **21**, 565 (1973).
16. F. Bobrowicz, and W.A. Goddard, III, in "Modern Theoretical Chemistry," Plenum, New York, 1977.
17. W.J. Hunt, P.J. Hay, and W.A. Goddard, III, J. Chem. Phys. **57**, 738 (1972).
18. I. Shavitt, Int. J. Quant. Chem. **S11**, 131 (1977); **S12**, 5 (1978).
19. B.R. Brooks, W.D. Laidig, P. Saxe, J.D. Goddard, Y. Yamaguchi, and H.F. Schaefer, III, J. Chem. Phys. **72**, 4652 (1980).
20. E.R. Davidson, J. Comp. Phys. **17**, 87 (1975).
21. B. Liu, in Proceedings of Workshop on "Numerical Algorithms in Chemistry: Algebraic Methods," LBL-8158, Lawrence Berkeley Laboratory, University of California, Berkeley, 1978.
22. D.R. Yarkony, Chem. Phys. Lett. **77**, 634 (1981).
23. B. Levy, Chem. Phys. Lett. **18**, 59 (1973).
24. B.O. Roos, P.R. Taylor, and P.E.M. Seigbahn, Chem. Phys. **48**, 157 (1980).
25. B.H. Lengsfield, III and B. Liu, J. Chem. Phys. **75**, 478 (1981).
26. J.A. Pople, J.S. Binkley, and R. Seeger, Int. J. Quant. Chem. **S10**, 1 (1976).
27. J.A. Pople, R. Seeger, and R. Krishnan, Int. J. Quant. Chem. **S11**, 149 (1977).
28. R. Krishnan and J.A. Pople, Int. J. Quant. Chem. **S14**, 91 (1978).
29. R. Krishnan, M.J. Frisch, and J.A. Pople, J. Chem. Phys. **72**, 4244 (1980).
30. C. Moller and M.S. Plesset, Phys. Rev. **46**, 618 (1934).
31. R.S. Mulliken, J. Chem. Phys. **23**, 1833 (1955)
32. D. Neumann, and J.W. Moskowitz, J. Chem. Phys. **49**, 2056 (1968).
33. C.E. Dykstra, and P. Jasien, Chem. Phys. Lett. **109**, 388 (1984).
34. H.O. Villar and M. Dupuis, Chem. Phys. Lett. **142**, 59 (1987).
35. G.J.B. Hurst, M. Dupuis, and E. Clementi, J. Chem. Phys., **89**, 385 (1988).
36. A.D. Buckingham, Adv. Chem. Phys. **12**, 107 (1967).
37. H. Sekino and R.J. Bartlett, J. Chem. Phys. **85**, 976 (1986).
38. S.P. Karna and M. Dupuis, J. Comput. Chem. **xx**, xxx (1990).
39. See for example, A. Szabo and N.S. Ostlund, *Modern Quantum Chemistry*, McGraw-Hill, New York (1989).
40. D. Spangler, I.H. Williams and G. Maggiora, J. Comput. Chem. **4**, 524 (1983).
41. A. Cerjan and W. Miller, J. Chem. Phys. **75**, 2800 (1981).
42. P. Pulay, Mol. Phys. **17**, 197 (1969); **18**, 473 (1970), **21**, 329 (1971).
43. J.A. Pople, R. Krishnan, H.B. Schlegel, and J.S. Binkley, Int. J. Quant. Chem., **S13**, 225 (1979).
44. A. Komornicki and J.W. McIver, Jr., J. Chem. Phys. **70**, 2014 (1979).
45. M. Schmid, M. Gordon and M. Dupuis, J. Chem. Phys. **66**, 2153 (1977); Int. J. Quant. Chem. **S13**, 225 (1979).

46. W.H. Miller, N.C. Handy, and J.E. Adams, J. Chem. Phys. **72**, 99 (1980).
47. S.A. Maluendes and M. Dupuis, (submitted to J. Chem. Phys.).
48. L. D. Landau, Phys. Z. Sowjetunion 1, 88 (1932); **2**, 46 (1932); C. Zener, Proc. R. Soc. London, Ser. A **137**, 696 (1932); **140**, 660 (1933).
49. A. Farazdel, M. Dupuis, E. Clementi and A. Aviram, *IBM Technical Report* KGN-200, September 21, 1989; J. Am. Chem. Soc. **112**, 4206 (1990).
50. J. Franck, Trans. Faraday Soc. **21**, 536 (1925); E. U. Condon, Phys. Rev. **32**, 858 (1928).
51. A. Farazdel and M. Dupuis, *IBM Technical Report* KGN-214, April 19, 1990; J. Comput. Chem. **xxx**, xxxx (1990).
52. R. A. Marcus, J. Chem. Phys. *24*, 966 (1956); Can. J. Chem. *37*, 155 (1959); Discuss. Faraday Soc. *29*, 21 (1960); N. S. Hush, Trans. Faraday Soc. *57*, 155 (1961); R. A. Marcus, J. Ann. Rev. Phys. Chem. *15*, 155 (1964); J. Chem. Phys. *43*, 679 (1965); Electrochim. Acta. *13*, 995 (1968); E. Waisman, G. Worry, and R. A. Marcus, J. Electroanal. Chem. *82*, 9 (1977).
53. B. S. Brunschwig, J. Logan, M. D. Newton, and N. Sutin, J. Am. Chem. Soc. *102*, 5798 (1980).
54. S. P. Dolin, E. D. German, and R. R. Dogonadze, J. Chem. Soc. Faraday Trans. 2 *73*, 648 (1977); N. R. Kestner, J. Logan, and J. Jortner, J. Phys. Chem. *78*, 2148 (1974); M. D. Newton, Int. J. Quantum Chem., Quantum Chem. Symp. *14*, 363 (1980).
55. J. Ulstrup, *Charge Transfer Processes in Condensed Media*; Springer-Verlag, New York (1979).
56. H. F. King, R. E. Stanton, H. Kim, R. E. Wyatt, and R. G. Parr, J. Chem. Phys. **47** 1936 (1967).
57. L.R Kahn, P. Baybutt, and D.G. Truhlar, J. Chem. Phys. **65**, 3826 (1976).
58. W.J Stevens, H. Bash, and M. Krauss, J. Chem. Phys. **81**, 6026 (1984).
59. **GRADSCF**, A. Komornicki, Polyatomics Research Institute, Mountain View, California.
60. P. Carsky, B.A. Hess, Jr., and L.S. Schaad, J. Comp. Chem. **5**, 280 (1984).
61. D. Spangler, J.J. Wendoloski, M. Dupuis, M.M.L. Chen, and H.F. Schaefer, III, J. Am. Chem. Soc. **103**, 3985 (1981).
62. M. Dupuis and J.J. Wendoloski, J. Chem. Phys. **80**, 5696 (1984).
63. L.M. Cheung, K.R. Sundberg, and K. Ruedenberg, Int. J. Quant. Chem. **16**, 1103 (1979).
64. T. Takada and M. Dupuis, J. Am. Chem. Soc. **105**, 1713 (1983).
65. M. Dupuis, J.J. Wendoloski, and W.A. Lester, Jr., J. Chem. Phys. **76**, 488 (1982).
66. M. Dupuis, G. Fitzgerald, B. Hammond, W.A. Lester, Jr., and H.F. Schaefer, III, J. Chem. Phys. **84**, 2691 (1986).
67. M. Dupuis, W.A. Lester, Jr., B.H. Lengsfield, III, and B. Liu, J. Chem. Phys. **79**, 6167 (1983).
68. M. Dupuis and W.A. Lester, Jr., Theor. Chim. Acta **71**, 255 (1971).
69. S.P. Karna, M. Dupuis, E. Perrin, and P.N. Prasad, (submitted to J. Chem. Phys.).

70. E. Perrin, P.N. Prasad, P. Mougenot, and M. Dupuis, J. Chem. Phys. **91**, 4728 (1989).
71. S.P. Karna and M. Dupuis, (submitted to Chem. Phys. Lett.).
72. C. Daniel and M. Dupuis, (submitted to Chem. Phys. Lett.).
73. Energy associated with changes in nuclear configuration of a molecule upon the transfer of the electronic charge from the "donor" site of the molecule to the "acceptor" site of the same molecule.

Chapter 7.

KGNMOL: A Program for Large Molecules and Molecular Interactions†

E. Clementi, G. Corongiu and S. Chakravorty

IBM Corporation
Scientific and Engineering Computations
Dept. 48B/428
Kingston, New York 12401, U.S.A.

1. Introduction

In this chapter we consider *ab initio* quantum chemistry for *large molecular systems*, in particular for repeated computations of two or more mutually interacting molecules in the computation of energy surfaces. These systems can often become "large" for *ab initio* quantum chemistry and in the range of 50 − 150 atoms, 200 − 400 electrons; notably *larger systems* are also discussed, mainly to indicate future trends in computational chemistry and to demonstrate our experimental lCAP-3090 supercomputer and the advantages of parallelism.

Most often large systems have no overall symmetry and the interactions can be weak, like in the case of hydrogen-bonded systems; in these cases *ab initio* quantum chemistry is probably at its worst. However, these are exactly the types of problems for which KGNMOL is written.[1-4] As mentioned in Chapter 1, KGNMOL is the present name (and version) of an older program named PPI,[5] developed at IBM-Poughkeepsie, New York, which, in turn, replaced a previous *ab initio* quantum chemistry program named IBMOL,[6-11] written in 1965 at the IBM Research Laboratory in San Jose, California and later rewritten at the Research Institute G. Donegani, Montedison s.p.a, Novara, Italy.

IBMOL was one of the pioneering *ab initio* programs in quantum chemistry; in its 1965 version[7] one finds the first use of contracted gaussian functions and the use of polarization functions; IBMOL was the first general geometry *ab initio* program to be essentially open-ended relative to the number of basis functions

---

† This chapter is a revised and extended edition of E. Clementi and G. Corongiu "KGNMOL: A Program for Molecular Interactions," Chapter 6 in MOTECC-89.

and, therefore, on the size of the molecule. Indeed, the 1965 version could handle up to 50 atoms, 800 primitives, 100 contracted of *s, p, d* and *f* type. However, it was possible to actually perform computations with about 50 atoms only few years later, in the early '70s. Because IBMOL aimed at the computation of energy surfaces of interacting molecules, like NH_3 interacting with HCl,[12-13] it had special *algorithms* to save integrals when: i) one molecule was kept at some fixed geometry while the second was placed at different distances and orientations; ii) one or a few atoms in a molecule were "moved" around to determine the best molecular geometry; and iii) the optimization of orbital exponents for a given basis set was requested.[14] During the last 25 years some of these features have been retained, some new ones have been added, and despite its age KGNMOL is still a useful program for large system computations and for special computational tasks like obtaining energy surfaces for *ab initio* interaction potentials (to be used, for example, in molecular dynamics and Monte Carlo simulations).

In the following we shall describe, in detail, some of the algorithms and report the expressions for the matrix elements which are given both for general gaussian functions or specialized for the geometrical basis set.[15-17] In writing this chapter we shall make liberal use of a number of papers, particularly those in Refs. 1−6. The basic theoretical formalism is presented only when needed since it is given in detail in the Hartree-Fock chapter (Chapter 2) and in the HONDO chapter (Chapter 6). There are, of course, many excellent volumes[18-22] for quantum chemistry which are assumed to be familiar to the reader.

2. Ab Initio Computation of Large Molecules

The computation of "large molecules" has come to have different meaning with the passage of time. In the mid-sixties, large molecules were of the size of pyridine[23] with minimal basis sets (7, 3/2, 1). In the early seventies a large molecule was of the size of a DNA base pair,[24] again with a minimal basis set. Today, in the nineties, a large molecule is of the size of 9 to 12 base pairs, again with a minimal basis set. Of course, the larger the basis set, the smaller the number of atoms and electrons of the "large" computation. Two observations can be made: each of the above computations required from 6 to about 12 hours of CPU time; secondly, one should not confuse quantum chemical "computation of a large molecule" with a "large computation." Among the latter we can mention those in molecular dynamics which are generally "trivially simple to execute," even if a week or more of CPU time and about three to five times that much for the elapsed time is often required. The easiness of the latter is due to the very simple *restart* nature of the computations and in the "simple" nature of the I/O requests. On the contrary, quantum mechanical computations of *large molecules* are instead characterized by big bursts of both computational time and I/O requirements. Another class of "large quantum chemical computations computations" is encountered with the effort required to derive intermolecular interaction potentials from fitting *many* ab-initio points on the interaction energy surface. Again, the quantitative meaning of "many" changes in time, from a few hundred in considering *amino acid and water*[25] in the mid-1970s, to about thirty thousand today in considering *amino acid and DNA fragments* in order to build up interaction potentials for nucleic acids interacting with proteins.[26]

With these *caveats*, let us consider a few examples of systems starting with a "medium" and moving toward a "large" molecule (by today's standards). We shall choose as "unit" of the systems considered the guanine-cytosine (G-C) base pair, stacked one above the other, n times, exactly as in the B-DNA geometry. We shall designate these molecular systems as $(G - C)_n$ and consider a few cases, namely $n = 3, 6, 9, 12$ with single-zeta (7, 3/2, 1), $n = 3$ with double-zeta (9, 5/4, 2) and $n = 1$ with a large triple-zeta basis set with polarization functions. In Table I we report the values of n for $(G - C)_n$, the number of atoms in the computed systems, of the electrons, how many primitive and contracted gaussians we have used, the thresholds for the integrals and SCFs energy (both in a.u.), the CPU time in minutes for the integrals (one- and two-electrons), the elapsed time for the entire SCF process, the number of integrals written on files, the number of integrals (over contracted functions) assuming the theoretical N^4 dependence, the "experimental" power x of N^x for the computation considered, and a specification on the hardware system (either IBM600E or IBM600S) in the lCAP-3090 configuration. Again, our software system was VM/XA.

We point out a few highlights: 1) a near to N^3 dependence, as expected,[27] but note that lowering the integral threshold to 1×10^{-7} would nearly double the number of integrals, 2) the large but *not forbidding* CPU computer time, and 3) the large but not forbidding storage requirements. The use of parallel I/O cuts down very much (about a factor of 12) the elapsed time relative to a sequential computation; a more accurate statement cannot be given since the work was performed with more than one job in the system and concomitant interference in the use of disks. On our lCAP-3090 with *four* 3090/600s the above CPU timing is expected to decrease by a factor of two; the share-memory, including the extended store of our system, makes it possible to keep a large fraction of the integrals in core, lessening the dependency on the use of disks. The $(G - C)_n$ with $n = 12$ was also computed using the ADD option (see below) on a $(G - C)_3$ onto a $(G - C)_9$ fragment: in this way the integral CPU time decreases from 969 minutes to 637 minutes.

For completeness, we report on a computation for the (G-C) base pair performed with a triple-zeta plus polarization basis from the geometrical basis set with (13, 8, 1/6, 3, 1) functions for the C, N, O atoms and (10, 1/3, 1) for the H atoms. These 947 primitives contracted to 459 functions, using 1×10^{-7} thresholds, yield a HF energy of -932.316363 a.u. Equivalently we obtain -539.543609 a.u. and -392.732893 a.u. for guanine and cytosine, respectively. Thus the HF interaction of G with C is ~1.09 eV, which is reduced to ~1.07 eV after BSSE correction. The (G-C) pair computation required 4.553×10^9 two-electron integrals on disks.

Table I. Toward "large molecules" by *ab initio* methods (CPU and elapsed times in minutes).

$(G-C)_n$	3	3	6	9	12
# atoms	87	87	174	261	348
# electrons	408	408	816	1224	1632
# primitives = N	1032	1548	2064	3096	4128
# contracted	315	630	630	945	1260
Int. Thresh.	1.D-6	1.D-6	1.D-6	1.D-6	1.D-6
SCF Thresh.	1.D-6	1.D-6	1.D-6	1.D-6	1.D-6
Int. (CPU t.)	14.0	49.5	101.2	332.2	969
SCF (Elaps. t.)	19.0	260.3	55.2	151.76	—
# written Int. $\times 10^7$	4.20	41.99	17.36	36.98	62.17
N / 10^{10}	0.1238	1.975	1.975	9.990	31.56
Power x of N_x	3.05	3.08	2.95	2.88	2.84
ICAP-3090	2(600E)	2(600E)	2(600E)	2(600S)	2(600S)

$N = (N^4 + 2N^3 + 3N^2 + 2N)/8$

To better appreciate Table I, we report in Table II data published by Clementi[27] in 1972. The molecular computations were presented as examples of computations on "large molecules." The computer used was an IBM-360/195; the basis set was minimal (7,3/2,1). The molecular systems of Table II are: one to six water molecules hydrating a Na^+ ion (with the Na-O distances of 5.25 a.u. and the H atoms pointing away from the sodium), cytosine, C, guanine, G, carbazole, CB, the guanine-cytosine pair, (G-C), 2:4:7 tri-nitro fluorenone, TNF, and (CB-TNF) complex (top to bottom of Table II). For the record, we recall that pyrrole, C_4NH_5, computed with IBMOL-1 in late 1966 on an IBM-7094, with a (7, 3/2, 1) basis required ~515 minutes.[7]

Comparing the $(G\text{-}C)_1$ with $(G\text{-}C)_9$ and $(G\text{-}C)_{12}$ (see Table I), we note that the latter two are about *ten* times larger than the first one; assuming a N^3 power for the integrals and noticing that the CPU timings of 1972 and 1989 are roughly comparable, we can conclude that there is about a 3×10^2 overall speed-up factor for the computations in 1989 relative to the one in 1972. Alternatively stated, it is as if *for KGNMOL-type codes* and for large molecular computations, each year we experienced an *average* speed-up factor of 15 to 20 for the last 17 years; this is the cumulative gain resulting from progress in CPU, I/O, system software, and application programming. *Relative to chemistry (number of atoms in the molecule), the cumulative speed-up is of about a factor of ten for the last 17 years, or less than*

Table II. Computation of "large molecules" on an IBM-360/195 in 1972.

Molecular System	# atoms	# electr.	# prim.	# contr.	CPU Time hr.	min.	sec.
$N_a^+ - H_2O$	4	20	38	12	–	–	13
$N_a^+ - (H_2O)_2$	7	30	60	19	–	–	40
$N_a^+ - (H_2O)_3$	10	40	82	26	–	1	51
$N_a^+ - (H_2O)_4$	13	50	104	33	–	3	48
$N_a^+ - (H_2O)_5$	16	60	126	40	–	6	40
$N_a^+ - (H_2O)_6$	19	70	148	47	–	10	45
$C_4\ N_3\ OH_5$; C	13	58	143	45	–	16	51
$C_5\ N_5\ OH_5$; G	16	78	191	90	–	35	09
$C_{12}\ N\ H_9$; CB	22	88	235	104	1	11	50
$C_9\ N_8\ O_2\ H_{10}$; $(G\text{-}C)_1$	29	136	334	105	2	3	–
$C_{13}\ N_3\ O_7\ H_5$; TNF	28	148	383	120	3	8	–
$C_{25}\ N_4\ O_7\ H_{14}$; (CB-TNF)	50	232	618	194	6	30	–

doubling the number of atoms year to year. This is sufficient, however, to move computational chemistry from the academic to the industrial environment. These very different viewpoints in "average speed-up" are intrinsic for N^3 or N^4 processes.

Let us now display graphically the computed molecular densities. For this task we do not have the possibility to compare "tables" of old with new data. We could note, however, that the variations in the electronic densities which take place during chemical reaction, for example $NH_3 + HCl \rightarrow NH_4\ Cl$,[12-13] were computed at many intermolecular distances and displayed on an IBM-2250 in 1968, resulting in the first animation of the chemical reaction obtained from *ab initio* computations.[28] Today we can do much more and much better, as described in Chapter 25 and shown immediately below.

In Figs. 1 to 4, we display $(G\text{-}C)_1$, $(G\text{-}C)_3$, $(G\text{-}C)_6$, and $(G\text{-}C)_9$ obtained by using ray-tracing models (see Chapter 25) and by computing the π and the total electronic density from triple-zeta plus polarization, double-zeta and minimal basis sets, respectively. The surface displayed corresponds to the electronic density value of 0.01 electrons.

The main result emerging from Tables I and II can be summarized as follows. We look *very confidently* at the next generation of computers where we expect, for example, even stronger parallelism, cluster to cluster and at the same time more and more flexible and powerful workstations. As a consequence we expect that the largest computation (by current standards) *of Table I will become "routine" within the next three to five years.* Later, in this chapter we shall proof conclusively that there seems to be no problem for obtaining Hartree-Fock functions even for "large" systems. The remaining but very pressing question now is: what is the outlook and forecast for obtaining *correlated energies?* This also will be discussed in this chapter.

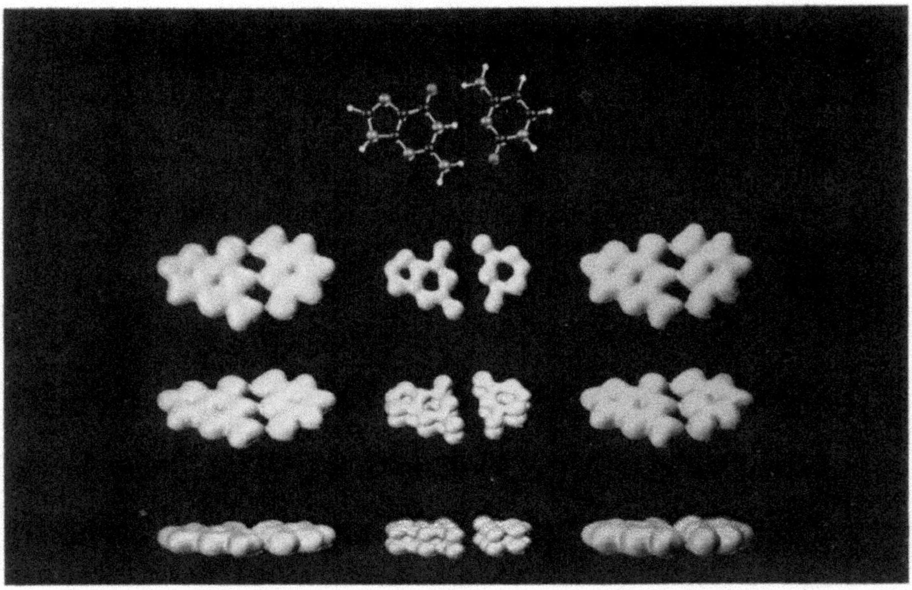

Figure 1. Ball and stick representation of the G-C complex at the top followed by σ, π and total electronic density projected onto the molecular, *xy* plane, rotated by 45° and by 80° (bottom three inserts). Basis set: triple-zeta plus polarization.

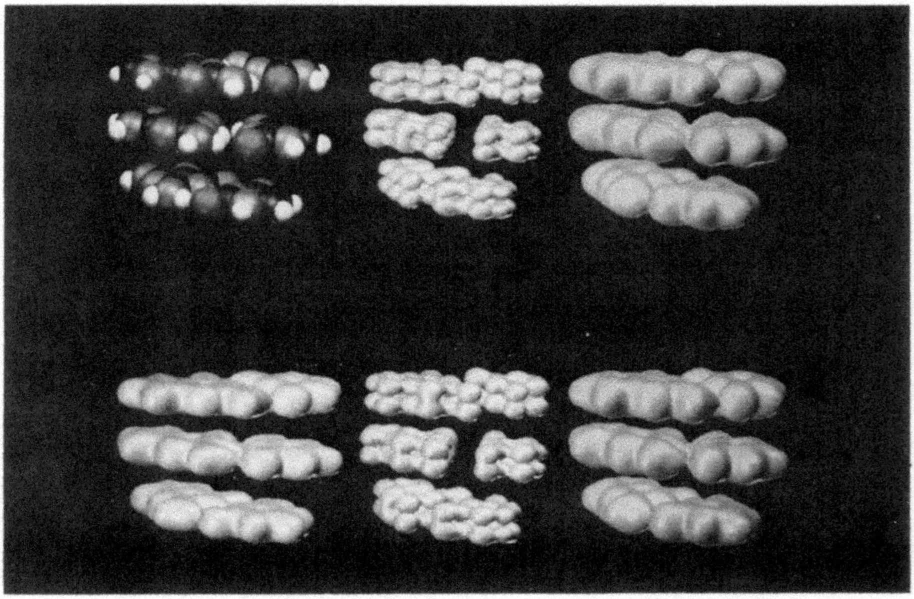

Figure 2. Space-filling, π and total electronic density for $(G\text{-}C)_3$, comparing minimal basis set (7,3/2,1) with double-zeta (9,5/4,2).

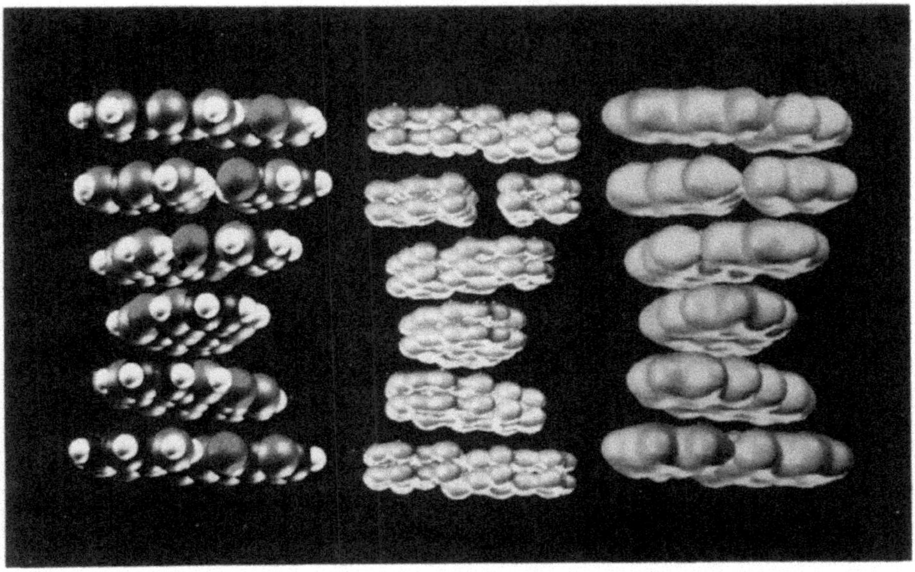

Figure 3. Space-filling, π and total electronic density for $(G\text{-}C)_6$ with minimal basis set (7,3/2,1).

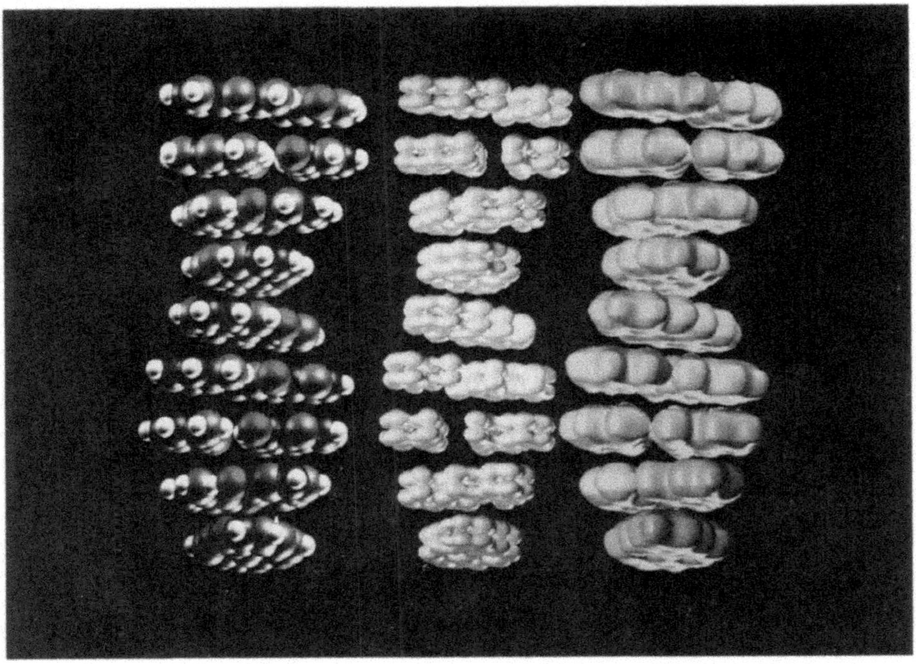

Figure 4. Space-filling, π and total electronic density for $(G\text{-}C)_9$ with minimal basis set (7,3/2,1).

3. Basis Sets: Gaussian Type Functions

Primitive Functions. The majority of current molecular quantum chemistry programs use Gaussian type functions (GTFs) as basis sets. These functions were first introduced in this context by Boys[29] and McWeeny.[30]

A generic unnormalized GTF in three dimensions is characterized by an exponent, α, a vector defining its center, $\vec{A}$, and an ordered triple of angular indices, $\vec{l} \equiv (l,m,n)$,

$$\begin{aligned} \eta(\alpha, \vec{A}, \vec{l}) &\equiv (x - x_A)^l (y - y_A)^m (z - z_A)^n \exp[-\alpha|\vec{r} - \vec{A}|^2], \\ &\equiv \eta(\alpha, \vec{A}, l, m, n) \\ &\equiv |A, \alpha, l, m, n|. \end{aligned} \tag{1}$$

The exponent is always taken to be real and positive. Typically, the centers of the GTFs are taken to be coincident with the nuclear positions so $\vec{A} = \vec{R}_A$.

If normalized functions are desired, one can simply scale using the normalization factor,

$$N(\alpha, \vec{l}) = \left[\left(\frac{2\alpha}{\pi} \right)^{3/2} \frac{(4\alpha)^{l+m+n}}{(2l-1)!!\,(2m-1)!!\,(2n-1)!!} \right]^{1/2}, \tag{2}$$

which is derived in Appendix A.

Despite the fact that the angular indices are not proper angular momentum labels, it is commonplace to assign another angular index $j = l + m + n$ to each primitive and refer to functions with $j = 0$ as s-type, $j = 1$ as p-type, $j = 2$ as d-type, etc., by analogy to atomic orbitals.

It is important to note that the function of Eq. (1) actually factorizes into three independent parts, each depending only on one of the three relative coordinates $x_A = x - A_x$, $y_A = y - A_y$, or $z_A = z - A_z$, but sharing the same exponent. In particular, the x_A dependent factor is,

$$X(\alpha, x_A, l) \equiv x_A^l \exp[-\alpha x_A^2]. \tag{3}$$

The other factors dependent on y_A and z_A are defined in the obvious way so that

$$\eta(\alpha, \vec{A}, \vec{l}) = X(\alpha, x_A, l)\, Y(\alpha, y_A, m)\, Z(\alpha, z_A, n) \tag{4}$$

By inspection of Eq. (2), the normalization factor for $X(\alpha, A_x, l)$ is just

$$N(\alpha, l) \equiv \left[\left(\frac{2\alpha}{\pi} \right)^{1/2} \frac{(4\alpha)^l}{(2l-1)!!} \right]^{1/2}. \tag{5}$$

As stated previously, the reason that GTFs are attractive choices for basis functions is that all of the matrix element calculations either reduce to evaluating simple algebraic formulae or well known special functions. This unique property follows from the fact that the product of two primitive GTFs, $\eta(\alpha_1, A, l_1)$ and $\eta(\alpha_2, \vec{B}, l_2)$, can be expressed as a linear combination of primitives with exponent

$$\gamma = \alpha_1 + \alpha_2$$

based on the center

$$\vec{P} = \frac{\alpha_1\vec{A} + \alpha_2\vec{B}}{\gamma}.$$

This constitutes the Gaussian Product Theorem given in Appendix B. Let

$$\vec{AB} \equiv \vec{A} - \vec{B}$$

denote the vector with components $(\overline{AB}_x, \overline{AB}_y, \overline{AB}_z)$ and length

$$\overline{AB}^2 \equiv \overline{AB}_x^2 + \overline{AB}_y^2 + \overline{AB}_z^2,$$

and similarly for $\vec{PA}$ and $\vec{PB}$. The final result of the Gaussian Product Theorem is the formula

$$\eta(\alpha_1, \vec{A}, l_1)\eta(\alpha_2, \vec{B}, l_2) = \exp[-\alpha_1\alpha_2\overline{AB}^2/\gamma] \sum_{i=0}^{l_1+l_2} \sum_{j=0}^{m_1+m_2} \sum_{k=0}^{n_1+n_2} x_P^i y_P^j z_P^k \exp[-\gamma r_P^2] \quad (6)$$
$$f(i, l_1, l_2, \overline{PA}_x, PB_x) f(j, m_1, m_2, \overline{PA}_y, \overline{PB}_y) f(k, n_1, n_2, \overline{PA}_z, \overline{PB}_z).$$

The coefficients $f(i, l_1, l_2, PA_x, PB_x)$ in Eq. (6) arise from the application of the binomial theorem to the polynomial part of the product,

$$(x + PA_x)^{l_1}(x + PB_x)^{l_2} = \sum_{i=0}^{l_1+l_2} f(i, l_1, l_2, PA_x, PB_x)x^i.$$

Contracted Functions. In general, gaussian functions are used in short linear combinations called "contracted gaussian functions" or "group" functions;[6-7] single gaussian functions are called "primitive" gaussian functions. IBMOL-Version 1 was the first molecular code to introduce the use of contracted functions.

The radial part of a contracted function depends on the choices of exponents and contraction coefficients for the associated primitives. Typically, these parameters are obtained by performing atomic SCF calculations on free atoms. Since the electron density in a molecule might be very different than that in a free atom, it is self-defeating to make the contracted functions resemble the atomic eigenfunctions too closely. The other extreme would be to avoid the contraction process completely by using the primitive functions as basis sets for the expansion of the molecular orbitals, thereby allowing maximum flexibility and accuracy. However, since the number of electron repulsion integrals which must be evaluated scales as the fourth power of the total number of basis functions, the cost of this approach is very high. Thus one uses a balanced approach, keeping in mind not only computational limitations but also that the basis set superposition error can become very large with improper contractions.

Often one imposes certain *restrictions* on the construction of contracted functions from a set of primitives in order to simplify the process of defining standard basis sets. In this chapter we will assume that a contraction is always constructed from a set of primitives with a common center and the same value of j. Furthermore, since the exponents and contraction coefficients for molecular calculations are most commonly determined by minimizing the total energy of the free atoms, it is natural to require that the contraction coefficients should depend only on j and α, but not on l, m, or n independently. The "limited" physical validity of these restrictions can be noted simply by recalling that molecules do hybridize "atomic" orbitals and that the $2p_\sigma$ are most likely different from the $2p_\pi$. Thus the above *restrictions* are useful to simplify the computation, but should be used with care.

Geometrical Basis Sets. The geometrical basis set is a basis of GTFs introduced by Clementi and Corongiu and was defined previously in Chapter 2. In this basis set the exponents for *all of the basis functions and for all of the atoms* within the molecule are taken as contiguous subsets of a geometric progression,

$$\alpha_k = \alpha_0 C^{k-1}, \tag{7}$$

for $k = 1$ to K where α_0 and C are real numbers to be determined variationally. Furthermore, there is a requirement for the exponents of the primitives associated with a contracted function with index j to be a contiguous subset of the exponents for primitives with index $j - 1$. This restriction is introduced only to simplify the programming task.

There are many advantages in using geometrical basis sets. Of primary interest is the fact that since there are several primitives with the same exponent contributing to different contracted functions, there is the opportunity to reduce the total number of arithmetical operations required for the calculation of the two-electron integrals by pretabulating various quantities. There is a second advantage, namely integrals with the same set of orbital exponents but different j values make use of some of the *same factors* (related to the angular part of the integral expression) which thus need to be computed only once. Third, the geometrical basis set appears to be ideal for Fock-Dirac computation in atoms, and there are evidences that this can be carried over to molecular Fock-Dirac computations. Finally, the geometrical basis set reintroduces in quantum chemistry the possibility to optimize the orbital exponent of a molecular basis set (see, for example, Chapter 2, Table V) in a relative simple way since all one has to optimize is reduced to two parameters α_0 and C. The geometrical basis sets are closely related to the even tempered basis sets[31] in turn based on Van Duijenveldt[32] work.

Basis Set Library for KGNMOL. There are many basis sets available in KGNMOL, some quite poor, but reported because of their popularity, some of intermediate quality and some notably good. In general, we can say that there are two criteria to judge the *goodness* of a basis set: first is its ability to yield a Harthree Fock function and second the ability to minimize the basis set superposition error. These are obvious and necessary but not sufficient criteria. A third criterion is that the basis set should be good for the specific physical property one is interested in. For example, if the property is molecular binding, then he must ensure a minimal (ideally zero) basis set superposition error (BSSE) and we return to the above two criteria. But one could be interested for example in hyperpolarizability and in this case he will need a much extended number of polarization functions.

In Table III we report the basis sets available in the KGNMOL Library for C, Si, and Ge atoms. We refer the reader to the I/O Documentation manuals for MOTECC-89 and MOTECC-90 (available to the MOTECC users) for the full library from H ($Z=1$) to Sr ($Z=38$). Note that it is some what arbitrary to prescribe basis sets which yield the Hartree-Fock limit as a criterion on the goodness of a molecular basis set; indeed much of this is due to experience and pragmatism. One point, however, should be stressed: if you are interested in computing molecular binding but you do not have the BSSE under control, then most likely you waste your time.

We recall that the ATOMSCF program, available from MOTECC, has the capability to optimize orbital exponents, and contraction coefficients, and it has a library of vector coupling coefficients for most atomic states. We would like, therefore, to conclude that orbital exponent optimization in molecules is a task which we should attend more and more.

4. The Hydrogen Molecule and the Water Dimer

The hydrogen molecule provides a simple but illustrative example of the importance of using proper basis functions. The water dimer provides an excellent example of the frustrations related to the search for a good basis set (without BSSE) needed to compute the correct binding. Starting with the hydrogen molecule, let us select a proper linear combination of basis functions to construct a set of spatial molecular orbitals with the required symmetry, Σ_g^+, for the ground state. In Fig. 5 we plot the Hartree-Fock function, keeping one of the two electrons fixed at $z=0$ whereas the second electron moves along the z axis. Notice that one of the hydrogen nuclei, H_A, is at -0.72 bohrs; the second, H_B, is at $+0.72$ bohrs,

Table III. Sample of basis sets from KGNMOL library.

Carbon		Silicon		Germanium	
Basis	Energy	Basis	Energy	Basis	Energy
STO 3G	-37.198393	STO 3G	-285.466187	GEOSMALL	-2070.263570
STO 4G	-37.475329	STO 4G	-287.185495	GEOSPTV	-2072.254980
3-21G	-37.480398	STO 5G	-287.614384	GEOSPTCV	2074.783510
STO 5G	-37.549193	STO 6G	-287.744608	GEOMEDIUM	-2074.933710
STO 6G	-37.572364	9,6/3,2	-288.376017	GEOLARGE	-2075.226730
7,3/2,1	-37.615997	GEOSMALL	-287.569760	GEOTRIPLEZ	-2075.317760
4-21G	-37.618601	GEOSPV	-288.578478	GEOEXTR	-2075.359672
GEOSMALL	-37.631449	GEOSPTCV	-288.741546	HF	-2075.3597
GEOSPTV	-37.632060	66-21G	-288.744663		
6-21G	-37.658318	66-31G	-288.749317		
GEOSPTCV	-37.663362	GEOMEDIUM	-288.785685		
9,5/4,2	-37.685193	13,10/6,5	-288.833193		
11,7/4,2	-37.686305	17,12/4,3	-288.841036		
13,8/5,3	-37.686722	GEOLARGE	-288.841980		
GEOLARGE	-37.687110	GEOTRPLEZ	-288.847989		
GEOTRPLEZ	-37.687156	GEOEXTR	-288.8543509		
GEOEXTR	-37.6886177	HF	-288.85436		
HF	-37.688619				

thus the midbond point is at the origin of the cartesian reference system. For additional discussion on H_2, see Chapter 5.

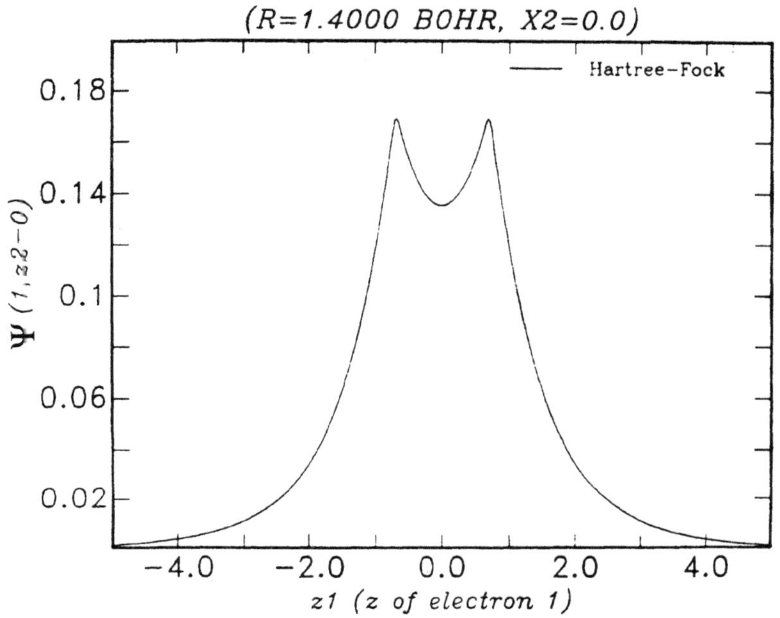

Figure 5. The hydrogen molecule at $|\vec{A} - \vec{B}| = R_e$. Electronic density from the Hartree-Fock function of Table I (18*s*,7*p*,6*d* geometrical basis set).

Assuming the molecule is in the orientation presented in Fig. 5, and that the centers used for defining the primitives coincide with the nuclear positions, it is easy to construct symmetry adapted linear combinations of atomic orbitals. It is clear that the linear combinations

$$\varphi_s(\vec{A}) + \varphi_s(\vec{B}) \qquad \varphi_{p_z}(\vec{A}) - \varphi_{p_z}(\vec{B}) \qquad \varphi_{d_{zz}}(\vec{A}) + \varphi_{d_{zz}}(\vec{B})$$

all have the proper spatial symmetry to mix into the ground state. It is also clear that we can use more than one φ_s or more than one φ_{p_z} or $\varphi_{d_{zz}}$, say *i*, *j* and *k* functions of the above three types. Notice that one could expand with *f*,*g*,*h*, ... functions. So, we could use a trial molecular orbital, MO, of the form

$$\phi_1 = \sum_i d_{s_i}[\varphi_{s_i}(\vec{A}) + \varphi_{s_i}(\vec{B})] + \sum_j d_{p_{zj}}[\varphi_{p_{zj}}(\vec{A}) - \varphi_{p_{zj}}(\vec{B})] + \sum_k d_{d_{zzk}}[\varphi_{d_{zzk}}(\vec{A}) + \varphi_{d_{zzk}}(\vec{B})] + \cdots$$

to define two molecular spin orbitals within the restricted Hartree-Fock approximation. The coefficients must then be determined by solving the Roothaan-Hartree-Fock equations. The MO has a (redundant) subscript, 1, indicating that it is the first (and only) occupied MO in the ground state wavefunction.

Table IV. Hydrogen, Water and its Dimer.

Computed Ground State Energies for H_2.

Calculation	s	p	d	f	R (bohr)	E_0 (hartree)
HF[a]	18				1.4000	-1.12853367
HF[a]	18	7			1.4000	-1.13356806
HF[a]	18	7	6		1.4000	-1.13362790
HF[a]	18	7	6	3	1.4000	-1.133629054
HF[b]	23	9	4	4	1.4000	-1.133629366
HF[c]	numerical				1.4000	-1.13362957
HF[d]	numerical				1.4000	-1.13362957
MP4[e]	6−3, 1, 1G (df, p)				1.4011	-1.16763
CI[f]	5	3	3	2	1.4011	-1.174142
CI[g]	13	7	2	1	1.4011	-1.174010
HCI[h]	15	7	2	1	1.4011	-1.17447467
HCI[i]	cofocal elliptic				1.4011	-1.17447588
Experimental data [j]					1.401	-1.174478

(a) Geometric basis set (by G. Corongiu, private communication).
(b) Spherical Gaussian calculation of Wells and Wilson.[33]
(c) Hartree-Fock limit from finite difference method of Laaksonen, et al.[34]
(d) Hartree-Fock limit from finite element method of Heinemann, et al.[35,36]
(e) J.A. Pople et al. (Ref. 37).
(f) CI with STO's basis set (Ref. 38).
(g) CI with geometrical basis set by Frye, Lie and Clementi[39] (see also Chapter 5).
(h) Hylleraas-CI by Frye, Lie and Clementi[39] (see also Chapter 5).
(i) Hylleraas-CI type calculation of Kolos et al.[40] employing a 249 term wavefunction with explicit correlation expressed in cofocal elliptic coordinates. A five term interpolation was used to obtain the energy at this separation to compare with the results of Frye, et al.
(j) Experimental data by G. Herzberg.[41]

(Continued)

Table IV *(continued)*. **Water and Water Dimer**

Hartree-Fock Energy for H_2O in a.u

a)	-76.06587	1972	E. Clementi and H. Popkie
			J. Chem Phys. **57**, 1077
	-76.067_{87}		Estimated limit at exp. geometry
b)	-76.064226	1975	B.J. Rosenberg and I. Shavitt
			J. Chem Phys. **63**, 2162
	-76.0675 ± 0.001		Estimated limit at exp. geometry
c)	-76.05869	1988	M.J. Frisch, J.E. Del Bene, J.S. Binkly, and H.F. Schaefer III, J. Chem Phys. **84**, 2279
d)	-76.067338	1987	D. Feller, C.M. Boyle, and E.R. Davidson
			J. Chem Phys. **86**, 3424
e)		1990	G. Corongiu and E. Clementi (unpublished)
			Geometrical b.s. (α_0 =0.04, C=1.19)
e1)	-76.026460		(O :28s, 18p/H: 18s)*
e2)	-76.065866		(O :28s, 18p, 8d/H: 18s, 8p)
e3)	-76.067417		(O :28s, 18p, 8d, 3f/H: 18s, 8p, 3d)
	-76.06755 ± 0.0001		Estimated limit at exp. geometry

Note: The Computations d) and e) are for the experimental water geometry.

Total Energy (non rel.) for H_2O in a.u.

b)	-76.339802		SD-CI
	-76.4375 ± 0.004		estimated limit
c)	-76.32428		MP2
	-76.33868		MP4
d)	-76.3963		MRSD-CI
e1)	-76.221957		MP2
e2)	-76.373318		MP2
e3)	-76.404040		MP2
f)	-76.2799	1976	H.J.Werner and W. Meyer, Mol. Phys. **31**, 885. CEPA but with neglect of is2 on O
g)		1990	E. Clementi and S. Chakravorty
			J. Chem. Phys. (in press)
	-76.406_5		CS′ Functional
	-76.413_6		B Functional
	-76.408_6		LYP Functional
	-76.420_9		CC Model

(continued)

Table IV *(continued)*. **Water and Water Dimer**

H.F. Binding for H_2O Dimer with (and without) BSSE in Kcal /mole (at different geometries)

c)	-3.43	(-3.79)		
h)	---	(-3.67)	1973	H. Popkie, H. Kistenmacher, and E. Clementi, J. Chem Phys. **59**, 1325
i)	-3.42	(-3.67)	1984	V. Carravetta, and E. Clementi, J. Chem Phys. **81**, 2686
j)	-3.731	(-3.779)	1988	K. Szalewicz, S.J. Cole, W. Kolos, and R.J. Barnett, J. Chem Phys. **89**, 3662
k)	-3.725	(-3.754)	1990	G. Corongiu (unpublished) (O: 17s, 12p, 4d, 2f/ H: 10s, 4p, 2d)

Total Binding for H_2O Dimer with and (without) BSSE in Kcal/mole (at different geometries)

c)	-4.57	(-5.30)		MP2
	-4.56	(-5.34)		MP4
i)	-5.76	(-6.01)		Density Functional (CS′)
j)	-4.73±0.2	(---)		MBPT
j)	-5.087	(-5.788)		MP2
k)	---	(-5.62)	1989	U. Niesar, G. Corongiu, M.J. Huang, M. Dupuis and E. Clementi, Int. J. Quantum Chem. Symp. **23**, 421 MP4

To exemplify this, the computed energies for the ground state of H_2 for various GTF basis sets are presented in Table IV. The first seven lines reports Hartree-Fock functions and energies (HF), followed by an MP4 energy, two configuration interaction functions and energies, and two Hylleraas-CI energies (HCI). Note the decrease in the energy as more basis functions of different types are added within the first group of HF results.

The second group of results[34-36] are exact numerical solutions to the Roothaan-Hartree-Fock equations, thus correspond to the solution with an infinitely large basis set (the so-called Hartree-Fock limit).

Let us now consider correlated functions. We recall the MP4 work by Pople et al.,[37] the CI work by Liu[38] and in our laboratory,[39] our Hylleraas-CI work[39] and, finally, Kolos[40] work. The experimental data are by Herzberg and Montfils,[41] revised later by Herzberg.[41] Future computations targeted to reach the experimental value will have to consider the non Born-Oppenheimer effect and relativistic corrections.

The data in the second part of Table IV collect a few significant computations on water and water dimer published in the last two decades. At the level of accuracy of these computations, the trimer of water and higher polymers would represent computations too demanding for today standards, exception made for those selecting density functional techniques. However, the latter is still somewhat deficient in its calibration.

Let us consider the Hartree-Fock for the water monomer. Its limit, in 1972, was predicted to be within 0.002 a.u. from the value of a), improved to ± 0.001 a.u. in 1975 and to ± 0.0001 a.u. in 1990, as shown in b) and e3). Computation d) is near to the HF limit, and computation e3) is essentially at the HF limit. Computation c) (on the other hand) is notably far from the HF limit.

The total energy of water is simply the sum of the ionization potentials of O and H plus the experimental binding energy of water; by so doing we introduce a minor error corresponding to the energy difference of the relativistic correction between H_2O and O. The best energy for which there is a wavefunction, is provided by computation d); the largest amount of correlation energy recovered by standard techniques (MP2) is from computation e3). Computations b) and f) are mainly of historical interest. Computations in g) have been limited to those density functionals which seems to have a statistically good reliability on a sample of about 50 molecules.

The dimer represents the beginning of the interesting chapter in chemical computations, which leads to liquid water, a solvent of enormous chemical interest. Here, generally one must compute an energy surface requiring hundreds of computations, as it was the case in h), i), and l), unless the interest is in testing a theoretical model, (j), or approaching a limit, (k).

5. Basis Set Superposition Error (BSSE)

As known, the BSSE is one of the true "challenges" in quantum chemistry which originates in the fact that basis sets are always *truncated*, i.e., never *complete*. In the early seventies (and sometimes even today) this problem was conveniently ignored by many quantum chemists and "suddenly" one could obtain "full agreement" with experimental energy data, even at the SCF level even better with very small basis sets! The "magic" was the BSSE which, properly "calibrated," could yield any targeted value (thus either the knowledge of experimental data or a good guess of it was most "useful"). To be sure, the problem has been known since the late 1950s when Mulliken and his group added a $2p_\sigma$ STO to polarize the hydrogen atom in a minimal basis set computation for the HF molecule in the ground state. The computed *binding energy improved substantially*; but it was then also realized one could *lower the energy at will*. The BSSE problem was unveiled and hence, as a first remedy, the *balanced* basis set concept was proposed and the HF molecule was used by Clementi[42] as a test case.

After a few years a major step forward was taken by Boys and Bernardi[43] who suggested using the *ghost orbitals* in computing the energy of the dissociation products — the *counter-poised,* CP, method.

In KGNMOL the computation of the BSSE is done automatically if the user so requires, saving human labor and computer time. The standard computation of the BSSE between two molecules A and B is performed as follows. Firstly one computes the supermolecule AB with the basis set $\{\alpha\}$ for the atoms on A and $\{\beta\}$ for B. The resulting energy is E(AB). Secondely, one computes the energy of A with both $\{\alpha\}$ and $\{\beta\}$ obtaining an energy E′(A). The operation is repeated for molecule B computed with $\{\beta\}$ and $\{\alpha\}$ yielding an energy E′(B). The interaction energy, I′(AB), of A with B is

$$I'(AB) = E(AB) - E'(A) - E'(B)$$

In this way, if $\{\beta\}$ contaminates $\{\alpha\}$ in computing the energy of A or conversely, if $\{\alpha\}$ contaminates $\{\beta\}$ in computing the energy of B, then this pollution of basis set can be quantitatively accounted for and it amounts to

$$BSSE(AB) = E(A) + E(B) - E'(A) - E'(B)$$

where E(A) and E(B) are simply the computations of A with $\{\alpha\}$ alone and of B with $\{\beta\}$ alone.

Notice that E(AB), E′(A) and E′(B) make all use of the composite basis set $[\{\alpha\} + \{\beta\}]$; therefore, one should not recompute the two electron integrals, which depend only on the basis set. However, one has to recompute the one electron integrals, since in computing E′(A) the $\{\beta\}$ basis are centered at the nuclear positions of B, which have *no nuclear charge* and equivalently for E′(B) relative to $\{\alpha\}$. Timing relative to this option are given in the next section.

6. Special Options for Adding Fragments or Molecules: ADD Option

As mentioned in the introduction, let us consider the following four very typical problems in quantum chemistry : a) construct the energy surface of molecule A interacting with molecule (or atom) B; b) expand the above problem to the case of *three* molecules, A, B and C; c) compare the energies of substituents for a given compound; and d) vary the geometry (e.g., bond angles, internuclear distances, torsions, internal rotations) of a fragment within a molecule, and e) add one or more basis to a set of a previous computation (for example the polarization functions), or add one at the time a give basis with different values of the orbital exponents, with the intent to optimize it.

In all these problems (except e)) we can partition the atoms of the entire system into two groups, A and B; within the first we place those atoms which maintain

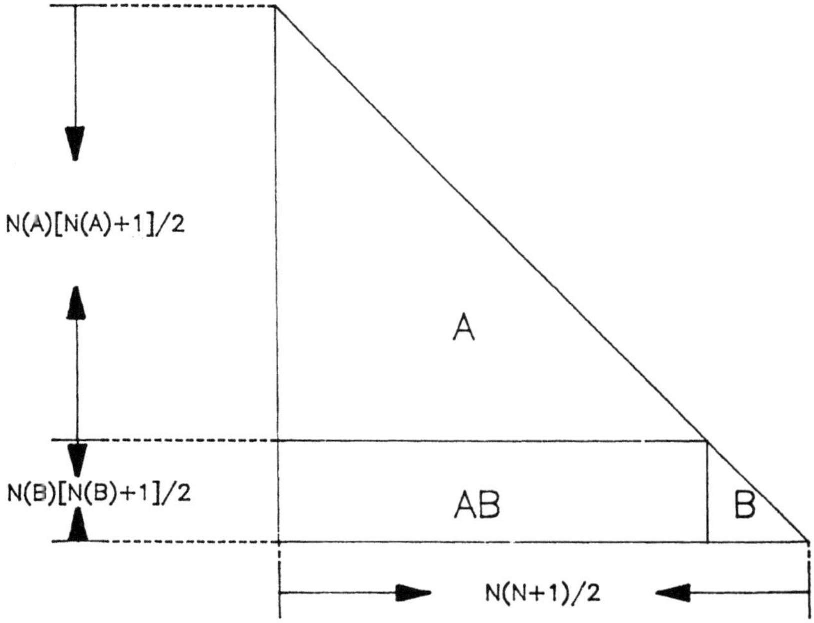

Figure 6. ℙ supermatrix. The top triangle refers to the $\mathbb{N}_A(\mathbb{N}_A + 1)/2$ matrix element which can be saved from computation to computation; the bottom trapezoid refers to the $\mathbb{N}_B(\mathbb{N}_B + 1)/2$ matrix elements needed to describe the *added* fragment and the $\mathbb{N}_A + \mathbb{N}_B$ matrix elements needed to compute the interaction between *A* and *B*.

the same position, designated $\{a_i\}$, and in the second group we place those atoms which "change" position, designed $\{b_i\}$, relative to a previous (or initial) computation. Equivalently in the case e) we call $\{\alpha\}$ the original basis set and $\{\beta\}$ the one containing the polarization function(s) or the function whose orbital exponent must be optimized. Let us consider the ℙ supermatrix (see, for example, Chapter 2) and designate $\{\alpha_i\}$ the basis sets describing the atoms $\{a_i\}$ of A and $\{\beta_i\}$ the basis sets describing the atoms $\{b_i\}$ of B; we see immediately that we can recover $[\mathbb{N}(A)(\mathbb{N}(A) + 1)/2]$ two-electron integrals out of a total of $[\mathbb{N}(\mathbb{N} + 1)/2]$ where $\mathbb{N}(A) = N(A)(N(A) + 1)/2$ and where $N(A)$ is the number of basis functions describing $\{a_i\}$ of A, and N is the number of basis functions describing the entire system, namely $N(A) + N(B)$ with $N(B)$ the number of basis functions for the $\{b_i\}$ of B. Figure 6 illustrates the saving in terms of the supermatrix ℙ. It follows immediately that, given two molecules, A and B, for which we wish to compute the interaction energy surface, it is most convenient to keep the larger molecules at a fixed position (thus its atoms correspond to the $\{a_i\}$ set) and change position and orientation of the smaller molecule (whose atoms correspond to the $\{b_i\}$ set). Equivalently, if one is interested in comparing a number of substituents, e.g., CH_3-CH_2-CH_2-CH_2X with X = H, F, Cl, etc., it is convenient to keep as "fixed" the large molecular fragment CH_3-CH_2-CH_2- and add to it first -CH_3, then -CH_2F and then -CH_2Cl. Depending on the specific nature of the problem, the molecular fragment in the above example could have been chosen as CH_3-CH_2-CH_2-C- or even CH_3-CH_2-CH_2-CH_2-. KGNMOL has an option such that the saving of integrals here indicated is done directly by the computer program.[14]

In the Input/Output Manual for MOTECC-89, we have given detailed examples on the use of the ADD option and on the automatic computation of the BSSE. Let us now report some CPU timing, which exemplifies the saving in computer time associated with these KGNMOL features.

In Tables V and VI we present CPU timing for computations using the "ADD" and the BSSE options available in KGNMOL.

The examples chosen are from computations needed to obtain the interaction potentials between proteins and DNA (presently about 32,000 *ab initio* computations of this type have been performed); this work could not have been considered feasible without the ADD and BSSE options. We shall consider a thymine-sugar fragment interacting with an iso-leucine residue (T-S and Ile), and as second example a phosphate-sugar-phosphate-sugar fragment of Z-DNA interacting with an arginine residue (P-S-P-S and Arg), an adenine-sugar fragment interacting with a serine residue (A-S and Ser), an adenine-sugar fragment interacting with one molecule of water (A-S and H_2O), and, finally, a phosphate-sugar-phosphate-sugar fragment from Z-DNA also interacting with a molecule of water (P-S-P-S and H_2O). The structure of these fragments is given in Fig. 7. Notice that the Arg, Ile and Ser residues are terminated with a $-NH_2$ group rather than the $-OH$ group (at the $-COOH$ end of the amino acid) and with a $-NH-CHO$ terminal (in place of the $-NH_2$ terminal group of the amino acid).

The timing reported in Table V were obtained using an IBM 3090/600E system working under VM/XA, with the *sequential* version of the KGNMOL program. The KGNMOL option of writing in double precision (and saving) the integrals on disk is here selected since, for the *same* fixed molecule, many different orientations and distances of the added molecule are considered over and over during a long time period of, for example, several weeks. In these circumstances, the KGNMOL version which keeps the integrals in the memory would be inappropriate; the latter program version is, however, very much faster in elapsed time, since it avoids writing the integrals on disk or reading the integrals from disk during the SCF and MP2 process. It is also noted that our department at IBM-Kingston seldom performs this type of computation with the sequential version of KGNMOL; the parallel version cuts down the I/O time substantially (up to 32 times faster) when up to 32 lCAP-3090s are used in parallel. In this way we can perform very "large" computations of the size of those discussed in the next section.

In Tables V and VI, for each computed molecular system, we report the two parts, namely the one kept fixed (FIXED, A), and the one added (ADDED, B), the number of atoms, primitive gaussians, contracted gaussians, the CPU time required to compute the integrals (both when we compute the BSSE and when we don't), the CPU time for the SCF (no BSSE) or for the SCF with CP correction (given by adding up the CPU times of three SCF computations, namely A with B, A with ghost of B, B with ghost of A) for the case of BSSE computations, and the number of two-electron integrals *written* in the integrals file, namely those larger than a preassigned threshold. Let us, however, add some comments before discussing the data of Tables V and VI.

As previously recalled, the number of integrals needed in a computation is proportional to the fourth power of the number, N, of basis sets, precisely $(\mathbb{N}(\mathbb{N}+1)/2)$ where $\mathbb{N} = N(N+1)/2$. However, some of these integrals are very small and, under proper conditions, they can be neglected. In general, if we neglect those two-electron integrals which are smaller than 10^{-4} a.u., the computed energy will be accurate to about 10^{-5} a.u., *at best*. Care should be taken for large systems where the number of very small integrals is large. For example, in a medium large molecule the guanine-cytosine complex, the computed energies using a (7,3/2,1) basis set are -928.624717 a.u., -928.625192 a.u., -928.625567 a.u., -928.625597 a.u., and -928.6225596 a.u. for integrals thresholds of 1×10^{-5}, 1×10^{-6}, 1×10^{-7}, 1×10^{-8}, and 1×10^{-9} respectively. The corresponding number of two-electron integrals over contracted functions are 1.27×10^{6}, $1.81 \times$

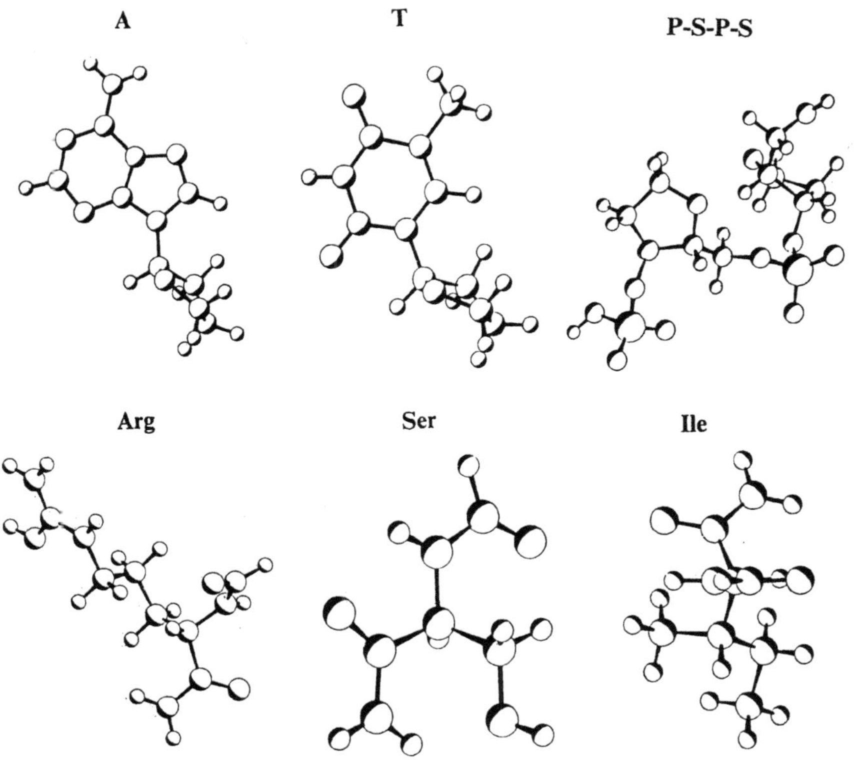

Figure 7. Molecular complexes used in the computations reported in Table V.

Table V. CPU timing in minutes for the ADD and BSSE options in KGNMOL.

Molecular System: T-S (FIXED, A) and Ile (ADDED, B)

				Integral Timing		SCF Timing		Written Int.
System	# Atoms	# Prim.	#Contr.	BSSE	no BSSE	BSSE	no BSSE	$\times 10^6$
A	26	272	82		7.57		1.96	2.02
A+B	51	504	151	28.62	26.88	24.99	9.63	9.32
A+B	51	504	151	26.77	25.04	24.39	9.33	8.35
A+B	51	504	151	25.23	23.47	22.90	8.51	7.64
A+B	51	504	151	24.30	22.54	21.59	8.08	7.12
A+B	51	504	151	24.08	22.33	21.81	8.38	6.76
A+B	51	504	151	24.13	22.37	20.86	8.13	6.53

Molecular System: P-S-P-S (FIXED, A) and Arg (ADDED, B)

				Integral Timing		SCF Timing		Written Int.
System	# Atoms	# Prim.	#Contr.	BSSE	no BSSE	BSSE	no BSSE	$\times 10^6$
A	41	462	141		22.51		5.87	7.02
A+B	70	746	226	56.71	51.59	56.04	21.68	18.50
A+B	70	746	226	54.81	49.69	53.64	20.77	17.59
A+B	70	746	226	53.37	48.26	51.88	20.10	16.89
A+B	70	746	226	52.28	47.14	50.59	19.56	16.35
A+B	70	746	226	51.60	46.50	49.54	19.21	15.94
A+B	70	746	226	51.16	46.07	48.99	18.97	15.62

10^6, 2.37×10^6, 2.93×10^6, and 3.46×10^6. In a larger example, the total energies computed for three guanine-cytosine base pairs, using a (9,5/4,2) basis, are -2794.544062 a.u. and -2794.369236 a.u. for integrals thresholds of 1×10^{-5} and 1×10^{-6} a.u., respectively. In KGNMOL the threshold value is set to 10^{-8} a.u. as the default value, but it is a "user" parameter. The value of the integral threshold has two functions; it ensures that no integral smaller than the threshold is *written in the integral file* or stored in memory and, in addition, it sets up inequalities criteria for the integral computation such that it is very unlikely that the program will *fully* compute an integral of value much smaller than indicated by the threshold; such criteria are used at the N^2, N^3 and N^4 level. For more details we refer to a work by Habitz and Clementi.[44] The saving obtained with the thresholds in computer time and storage for a large system can be very notable, since it

Table VI. CPU timing in minutes for the ADD and BSSE options in KGNMOL.

Molecular System: A-S (FIXED, A) and Ser (ADDED, B)

				Integral Timing		SCF Timing		Written Int.
System	# Atoms	# Prim.	#Contr.	BSSE	no BSSE	BSSE	no BSSE	$\times 10^6$
A	26	284	86		8.31		2.04	2.22
A+B	43	460	139	14.70	13.43	14.91	5.74	5.25
A+B	43	460	139	14.13	12.87	13.86	5.42	4.94
A+B	43	460	139	13.82	12.55	13.35	5.28	4.76
A+B	43	460	139	13.72	12.45	13.30	5.23	4.65
A+B	43	460	139	13.59	12.32	13.14	5.21	4.59

Molecular System: A-S (FIXED, A) and H_2O (ADDED, B)

				Integral Timing		SCF Timing		Written Int.
System	# Atoms	# Prim.	#Contr.	BSSE	no BSSE	BSSE	no BSSE	$\times 10^6$
A	26	284	84		8.31		2.04	2.22
A+B	29	308	93	1.87	1.44	4.87	2.08	2.28

Molecular System: P-S-P-S (FIXED, A) and H_2O (ADDED, B)

				Integral Timing		SCF Timing		Written Int.
System	# Atoms	# Prim.	#Contr.	BSSE	no BSSE	BSSE	no BSSE	$\times 10^6$
A	41	462	141		22.51		5.87	7.02
A+B	44	486	148	3.95	2.51	14.88	6.13	7.14

can reduce the number of computed integrals from about N^4 to about N^3. The N^3 dependency for the number of integrals in large systems was pointed out by Clementi[27] long ago.

The timing given in Table V needs additional specifications concerning the basis set. For the H atom, four primitive gaussians contracted to one have been used; for C, N, O the basis set is (7,3/2,1) and for P the basis set is (9,6/3,2) (see Ref. 32). The computations were performed as "stand-alone."

We take this occasion to note that KGNMOL has a rather extended library of basis sets, specifically the *small, medium,* and *large* geometrical basis sets for atoms from Z=1 to Z=54, and an *extra-large* geometrical basis set from Z=1 to Z=84. These basis sets are contracted to single-zeta, double-zeta, triple-zeta and split-valence. Also included are other basis sets for the Z=2 to Z=10 atoms, like

(7,3/2,1), (9,5/4,2), (11,7/4,2), (13,8/5,3) originally proposed by Van Duijenvelt,[32] and some of the bases proposed by Pople and his collaborators. The full input geometry and basis set for the example of Table V are available to the interested reader (all data needed to reproduce our computations). We take this occasion to stress that often we see in the literature timing for molecular computations which are "meaningless" because of lack of basic specifications (basis set, geometry, integral threshold, SCF convergency threshold, system environment, etc.).

Let us now return to Tables V and VI. Notice the relatively small saving when A and B are of the same size; for example, the full system A+B in the T-S and Ile case of Table V would use ~ (7.57 + 28.62) = 36.19 minutes for the integrals rather than 28.62 min., or about 20% more. However, consider P-S-P-S and H_2O in Table VI; here the saving is a decrease from ~ 22.51 + 3.95 = 26.46 to 3.95 minutes. It should be stressed that when one computes energies to map an energy surface, very often one computes *hundreds of points,* thus even a small saving, e.g. 10%, becomes most valuable. Notice also that Fig. 6 can be used for a very fast and rough estimate on the computer time-saving due to the ADD option. For example, A-S and Ser of Table VI are about equal in size; thus the area of B in the $\mathbb{P}$ supermatrix is about equal to the area of A, and about half of the area $A + B$. Therefore, the addition of B brings about a factor of three relative to the computation of A. This indeed can be verified by the data of Table VI.

7. Parallelization of KGNMOL and the MP2 Option

In the previous section we *"demonstrated"* that we can compute the Hartree-Fock functions of large molecular systems; but it is well known that the Hartree-Fock approximation is only a zero-order approximation. Thus, in this section we report on KGNMOL's CPU timings using a popular avenue to obtain correlation corrections, namely the MP2 approximation. The MP2 is widely used, even if a) it does not yield a correlated function, b) it can give, at best, only a fraction of the correlation energy, and c) it is heavily plagued by BSSE.

Let us start by considering a new way to exploit parallelism in quantum chemical programs. As known (see Chapter 1), parallel codes should be "load balanced" at each processor; otherwise, the overall "speed-up" might become very unattractive and even degenerate into a "slow-down." Indeed, improper load balancing, the presence of sequential tasks, and the communication and operating system's overheads are the main factors which bring about degradation in the parallelization efficiency. In the lCAP-3090 architecture we face another challenge: we must not only load-balance the CPUs of a given cluster, but we must also load-balance one cluster with all the other clusters in the lCAP system. Thus we talk of *intra-system* and *inter-system load balance*. Recently we have successfully implemented a dynamical inter-system as well as a dynamical intra-system load balance algorithm for the integrals, SCF and MP2 sections of KGNMOL.[45]

Dynamical intra-system load balancing. Let us consider the integral generation part of KGNMOL. *One* processor within a given system (or cluster) computes and stores in shared memory the T_1 and T_2 tables of Eqs. (68) and (69). The indices of a pair of gaussian functions (necessary to locate a specified quantity either in T_1 or in T_2) are allocated dynamically by *each* processor of the cluster, namely a processor, as soon as it becomes free (having performed its task), allocates a set of indices which are "locked out" to the other processors of the same cluster.

Dynamical inter-system load balancing. The *master* processor, executing on one of the clusters, assigns one *leader* processor to each cluster. All remaining processors are *slave* processors. Each *leader* processor computes and stores in the shared memory the T_1 and T_2 tables. Load balancing within the same cluster is achieved as above described for intra-system load balancing. Load balancing *across clusters* is achieved by requesting the *leader* processors to run over the indices of the outermost loop; however, the *master* processor reserves an input specified fraction of the indices (a "reservoir" of indices which corresponds to a specified percentage of all the indices) which are dynamically distributed to the first free *leader* processor. The parallelization efficiency thus achieved is *very high.*

Let us consider, as an example, $(G\text{-}C)_3$ with minimal basis set, with integrals and SCF thresholds of 1×10^{-6} a.u. and let us use an lCAP-3090 with only two clusters, an IBM-3090/600E and an IBM-3090/600S. Incidentally, the different speeds of the two clusters makes the inter-system load balancing even more interesting! Below we report *first* the results for dynamical inter-system load balancing, and *later* those for intra-system load balancing.

In Table VII we give the CPU and the elapsed time (in seconds) at each of the 12 processors for the computation and storage on disk files of the one- and two-electron integrals. In the last column we also provide the number of the integrals written in the integral file by each one of the 12 processors; processors #1 and #7 are the "*leader*" processors; processors #1 to #6 belong to the IBM-3090/600S and those from #7 to #12 belong to the IBM-3090/600E. The total elapsed time for all integrals is 917.02 seconds. The total elapsed time for 14 SCF iterations, needed to reach 1.0×10^{-6} a.u. SCF convergence, is 1082.37 seconds (of which 530.13 seconds are used in the sequential part of the code, presently non-parallelized). Notice that the CPU time of the two "leader processors" is ~15% higher than the CPU time of the "slave" processors due to the construction of the T_1 and T_2 tables; the very high balancing, especially of the 10 slave processors, is reflected both in the timing and in the number of integrals written in the integral file.

In Table VIII we report a simulation on the same molecule and with the same basis set as above, this time using only one cluster, namely an IBM-3090/600S. For this case the SCF process converged (in 14 iterations) after using 1560.5 seconds in elapsed time (of which 531.5 seconds are used in the sequential part of the code, presently non-parallelized). The total elapsed time for the integral section is 1560.52 seconds, thus clearly indicating (see sum of elapsed time for

Table VII. Dynamical Inter-System Load Balance for ICAP-3090 (time in seconds).

# CPU	1-Elect. Int. Time CPU	Elapsed	2-Elect. Int. Time CPU	Elapsed	No. 2-Elect. Integrals
1*	26.61	28.51	680.76	845.92	3678536
2	27.41	29.24	599.05	753.04	3683716
3	25.92	27.87	584.42	750.29	3617546
4	26.71	28.64	601.41	749.07	3787629
5	27.50	29.36	597.01	743.03	3742768
6	28.31	30.23	592.40	739.89	3771759
7*	34.57	35.82	737.41	875.12	3293967
8	35.53	36.97	628.57	760.50	3315505
9	28.30	29.58	631.77	759.73	3317781
10	28.96	30.15	627.56	757.36	3346292
11	29.74	30.77	618.61	756.76	3293504
12	30.50	31.57	613.54	756.01	3131346

*CPU #1 and CPU #7 are "leader" processors.

one- and two-electron integrals) that the code has a parallelization efficiency of about 100%. The number of integrals written by each CPU provides a complementary indication that nearly perfect load balancing was achieved.

Let us now consider the MP2 part of the code (for details on the MP2 theory see Chapter 6) and, as an example, let us consider the $(G\text{-}C)_1$ base pair using a double-zeta basis set. In Table IX we report the speed-up on an IBM-3090/600S using 5, 4, 2, 1 CPUs in parallel. Note that in the MP2 there is a large file of transformed integrals which needs to be written and read more than once from disk to expanded storage and from the latter to fast memory; this imposes a very notable burden on the system software, thus affecting the overall parallelization efficiency.

Table VIII. Dynamical Intra-System Load Balancing on an IBM-3090/600S (time in seconds).

# CPU	1-Elect. Int. Time CPU	Elapsed	2-Elect. Int. Time CPU	Elapsed	No. 2-Elect. Integrals
1	55.89	62.04	1192.92	1486.72	7077891
2	57.49	63.90	1092.95	1395.11	6912716
3	49.91	56.19	1121.23	1393.17	7073840
4	51.27	57.93	1116.22	1393.42	7136909
5	52.71	58.87	1093.23	1394.05	6900535
6	54.17	60.81	1092.73	1394.32	6878458

Table IX. Timing (minutes) and parallelization for MP2 in KGNMOL.

# Processors*	Integ. El. Time	Integ. Speed-up	Int. + SCF El. Time	Int. + SCF Speed-up	Int. + SCF + MP2 El. Time	Int. + SCF + MP2 Speed-up
1	44.55	1.00	66.75	1.00	494.18	1.00
2	22.54	1.98	35.60	1.88	274.75	1.80
4	11.18	3.98	19.42	3.44	182.23	2.71
5	10.06	4.43	18.73	3.56	179.54	2.75

*IBM-3090/600S

Before concluding we note 1) the speed-up obtained by parallelization (this makes MP2 computations more and more feasible), 2) that the MP2 takes about **eight** times longer (elapsed time) than the sum of the integrals and SCF processes, and 3) that only a fraction of the correlation energy can be recovered.

Therefore, for *large molecules* this very simple approach falls somewhat short of the required accuracy and, at the same time, it steps up considerably the computational cost relative to Hartree-Fock computations. For *small molecules*, the **MP2** is attractive at the computational level even if it has the above mentioned limitations.

8. Density Functionals for Molecules

The use of density functional models[46-50] for the correlation energy problem are becoming more and more popular with the chemistry community, as has been the case for some time with the solid state community. For the chemists, a main motivation stems from the realization that those methods which scale up with a power higher than four will not become usable for **large molecules**, notwithstanding past and future computer advances. In a previous section we prove that the Hartree-Fock, (HF) complexity corresponds to the *pragmatic upper limit for simulations of large molecules or for heavily repetitive simulations on medium size molecules,* for example those needed to scan the interaction potential hypersurface.

In this section we shall compare several density functionals.[52] We start with the one introduced by Lie and Clementi[53] (based on atomic correlation studies by Clementi and co-workers[54-58]) and designated as the *LC functional,* and the one introduced by Colle and Salvetti,[59] designated as *CS functional,* later reparametrized by Carravetta and Clementi[60]. To avoid confusion between different set of parameters, we refer to CS′ for the Colle-Salvetti with the Carravetta-Clementi paramters. Next we consider the functionals of Gunnarson and Lundqvist[61] and of Vosko Wilk and Nusair[62,63] with the self interaction corrections due to Stoll, Pavlidou and Preuss[64] and Savin, Stoll and Preuss[65], the *GL-SPP functional* and the *VWN-SPP functional.* Among the gradient corrected density functionals we consider the functional due to Perdew and Zunger[66] with the gradient term of Perdew[67](the *PZ functional*), the one due to Becke[68](the *B functional*), and the one due to Lee, Yang and Parr[69] (*LYP functional*). We note that Miehlich et al[70] have used these functionals to compute the correlation energies for atoms and atomization energies for a few molecules. In comparing about fifty molecules, we computed the HF densities with reasonably good quality basis sets and the functionals with consistent numerical integration techniques. Therefore our results should serve as useful platform for a comparison of the various density functional models.

Before concluding this introduction we wish to explain our overall strategy. We consider the used density functionals as a pragmatic solution to the determination of atomization energies and we consider "feasible" the computation of HF functions even for large molecules. In this spirit we can follow different strategies. **Firstly** we could concentrate our attention on functionals yielding both the exchange and correlation energy; **secondly** we could derive functionals yielding only the total correlation energy[53-69] (after all the HF exchange energy is available easily from HF computations) and **finally** we can derive functionals aimed at computing only the molecular extra correlation energy, ΔE_c , since from the Hartree-Fock model we can easily obtain the Hartree-Fock atomization energy. The three avenues aim at determining a progressively smaller and smaller amount of the many-body correction. It follows that a given **percent** error in the exchange plus correlation energy estimate (say 1%) might be chemically unacceptable but a similar percent error in only the correlation energy may be acceptable following the second or most likely the third avenue. This is the reason for our focus on the molecular extra correlation energy, ΔE_c, rather than on the the total "exchange-correlation" energy. For example, let us assume that the total correlation energy for one electron pair is about (1.5 ± 0.5) eV and that a given molecule has m electron pairs. If the molecule atomizes into K atoms each with P_k electron pairs, then the total molecular correlation energy is about $m(1.5 \pm 0.5)$ eV but ΔE_c is only $(m - \sum_k^K P_k)(1.5 \pm 0.5)$ eV.

One more comment could be added. Looking at the current density functional literature one might obtain the very wrong information that the field started with Thomas-Fermi model[71,72] and its elaborations and then had to wait for the Hohenberg and Kohn[73] work in 1964 and Kohn-Sham[74] paper of 1965. Nothing is more erroneous. First, it is obvious that the Coulomb energy can be written in terms of electronic density simply by inspection. In addition the idea that the exchange energy can be formulated as a density functional and that an electron gas is the appropriate system to investigate this possibility, goes back to W. Heisenberg[75] in 1927. Later, from 1934 to 1938, we have the famous papers by W. Wigner[51] where the correlation energy finds its first definition and its first formulation in term of density functionals. Following this, corrections for intermediate and high electron densities where proposed by Pines[76] and by Gell-Mann and Brueckner.[77] These two corrections and Wigner's original formulation where merged into a single expression by Gombas.[78] If one recalls that the correlation energy is mainly a kinetic energy effect (because of the virial theorem), then is clear that in the '30s it was fully appreciated that the kinetic energy and the potential energy for a many electron system could be written in term of a density functional. In the early '50s Slater[79] returned a number of times on this subject which eventually led to the X-α approximation.

Let us continue with our analyses of functionals. We start by recalling the algorithm for the functional of Lie and Clementi, viz. the LC functional[53] which uses as a starting point the Gombas functional, G functional.[78] We recall that the Gombas functional is a direct extension of the Wigner functional[51] (W functional), designed to include high density effects. The Wigner correlation energy[51] is

$$\varepsilon_c(\rho) = 0.0564\rho^{(1/3)}\ (0.0795 + \rho^{1/3})^{-1} \tag{8}$$

The G functional is given by the relation

$$\varepsilon_c(\rho) = a_1\ \rho^{1/3}(a_2 + \rho^{1/3})^{-1} + b_1 \ln(1 + b_2\rho^{1/3}) \tag{9a}$$

with $a_1 = 0.04096 \quad a_2 = 1.2 \quad b_1 = 0.02096 \quad b_2 = 2.29$

The LC functional is given below :

$$\varepsilon_c(\rho) = a'_1\ \rho_m^{1/3}(a'_2 + \rho_m^{1/3})^{-1} + b'_1 \ln(1 + b'_2\rho_m^{1/3}) \tag{9b}$$

where

$$\rho_m = \Sigma\bar{n}_i\rho_i \quad \text{and} \quad \bar{n}_i = n_i \exp\left[-0.5(2 - n_i)^2\right] \tag{10}$$

with $a'_1 = 0.02096 \quad a'_2 = 1.2 \quad b'_1 = 0.02096 \quad b'_2 = 2.29$

It must be noted that the LC functional is calibrated for special MCSCF functions, named as the Hartree-Fock Proper Dissociation, HFPD.[53] The use of ρ_m in place of the standard density ρ (the occupation numbers for the Hartree-Fock are $n_i = 0,1,2$ and for MCSCF are $0 \le n_i \le 2$) is to reduce the *interpair* correlation energy relative to the *intrapair* correlation; indeed, $\bar{n}_i$ has a value of two for a Hartree-Fock pair and a value smaller than one for a singly occupied orbital. We recall that recently Savin[80] emphasized this approach due to Lie and Clementi[53] and derived a new correlation energy density functional for CI wave functions. In this work we use the LC functional somewhat improperly, since we use HF rather than HFPD orbitals.

Next we consider the model due to Colle and Salvetti.[59] The expression for the correlation energy, E_c, was derived considering a wave function of N electrons that explicitly includes the electron-electron correlation[51]. With rough but reasonable approximations[59] for the one- and two-particle density matrices without spin, the E_c had been derived in terms of the SCF density matrices. In order to reproduce the correct value of E_c of the He atom, a single parameter fitting procedure was adopted, ending in the working equation

$$E_c = A\int \rho(\vec{R})\frac{1 + B\rho^{-8/3}(\vec{R})\left\{\nabla_r^2 P_2^{HF}[\vec{R} - \vec{r}/2; \vec{R} + \vec{r}/2]_{\vec{r}=0}\right\}}{1 + D\rho^{-1/3}(\vec{R})} e^{-C\rho^{-1/3}(\vec{R})} d\vec{R} \quad (11)$$

with

$A = -0.04981 \quad B = 0.066 \quad C = 0.253 \quad D = 0.349$

where ρ is the electron density, $\nabla_r^2 P_2^{HF}$ is the Laplacian of the SCF two-particle density matrix without spin with respect to the interelectronic distance $\vec{r} = \vec{r}_1 - \vec{r}_2$ evaluated at $\vec{R} = (\vec{r}_1 + \vec{r}_2)/2$, in the limit of vanishing interelectronic distance. The four constants in Eq. (11) are completely fixed by the value 2.29 of the single parameter q determined by the fitting procedure. It was observed[60] that the fitting procedure originally adopted[59] led to an underestimate of the the Coulomb hole. Carravetta and Clementi therefore refitted the CS functional to a different value for q and consequently a different set of constants[60] for the Eq. (11), namely

$q = 1.76 \quad A = -0.0578 \quad B = 0.062 \quad C = 0.253 \quad D = 0.543$

which has been used in this study. When we use the above constants we refer to the CS′ functional.

In the following we recall, the local spin density functionals proposed by Gunnarsson and Lundqvist[61] and of of Vosko, Wilk and Nusair[62] (VWN) who made use of Ceperley and Alder's[81] accurately determined electron liquid correlation energies. However these models tend to overestimate the correlation energy. Stoll, Pavlidou and Preuss[64] proposed a self interaction correction to such local density functional models and Savin *et al.*[65] applied it to atoms and molecules; we have followed their approach below. The expression for GL functional without the self interaction correction, in a.u., is:

$$\varepsilon_c(\rho_\uparrow, \rho_\downarrow) = \varepsilon_I(\rho_\uparrow, \rho_\downarrow)[1 - f(\zeta)] + \varepsilon_{II}(\rho_\uparrow, \rho_\downarrow)f(\zeta) \quad (12)$$

$$\varepsilon_i(\rho_\uparrow, \rho_\downarrow) = \frac{-C_i}{2}\left[(1 + x_i^3)\ln\left[1 + \frac{1}{x_i}\right] - x_i^2 + \frac{1}{2}x_i - \frac{1}{3}\right] \quad (i = I, II) \quad (13)$$

with

$$x_I = \frac{r_s}{11.4} \quad x_{II} = \frac{r_s}{15.9} \quad C_I = 0.0666 \quad C_{II} = 0.0406 \quad r_s = (4\pi\rho/3)^{-1/3}$$

and a switching function

$$f(\zeta) = \frac{\left[(1 + \zeta)^{4/3} + (1 - \zeta)^{4/3} - 2\right]}{(2^{4/3} - 2)} \quad (14a)$$

with

$$\zeta = (\rho_\uparrow - \rho_\downarrow)/(\rho_\uparrow + \rho_\downarrow). \quad (14b)$$

The VWN functional without the self interaction correction is as follows

$$\varepsilon_c(\rho_\uparrow, \rho_\downarrow) = \varepsilon_I(\rho_\uparrow,\rho_\downarrow) + \Delta\varepsilon_c(r_s, \zeta) \tag{15a}$$

$$\varepsilon_i(\rho_\uparrow, \rho_\downarrow) = A_i\left[\ln\frac{x^2}{X(x)} + \frac{2b}{Q}\tan^{-1}\frac{Q}{2x+b} - \frac{bx_o}{X(x_o)}\left(\ln\frac{(x-x_o)^2}{X(x)} + \frac{2(b+2x_o)}{Q}\tan^{-1}\frac{Q}{2x+b}\right)\right] \tag{15b}$$

where $x = (r_s)^{1/2}$,

$$Q = (4c_i - b_i^2)^{1/2} \quad X(x) = x^2 + b_i x + c_i \quad (i = I, II) \tag{15c}$$

and

$$\Delta\varepsilon_c(r_s, \zeta) = \varepsilon_{III}(\rho_\uparrow, \rho_\downarrow)\frac{f(\zeta)}{f''(0)}[1 + \beta_1(r_s)\zeta^4] \tag{15d}$$

$$\beta_1(r_s) = \frac{f''(0)}{\varepsilon_{III}(\rho_\uparrow, \rho_\downarrow)}\Delta\varepsilon_c(r_s, 1) - 1 \tag{15e}$$

$$\Delta\varepsilon_c(r_s, 1) = \varepsilon_I(\rho_\uparrow,\rho_\downarrow) - \varepsilon_{II}(\rho_\uparrow,\rho_\downarrow) \tag{15f}$$

and $f(\zeta)$ has been previously defined (Eqs. 14). The constants in Eqs. (15) are

$A_I = 0.0621841$	$A_{II} = 0.0310907$	$A_{III} = -0.033774$
$b_I = 3.72744$	$b_{II} = 7.06042$	$b_{III} = 1.131071$
$c_I = 12.9352$	$c_{II} = 18.0578$	$c_{III} = 13.0045$
$x_{o_I} = -0.10498$	$x_{oII} = -0.32500$	$x_{oIII} = -0.0047584$

The self interaction correction in the form prescribed by Stoll Pavlidou and Preuss (SPP) is defined as

$$E_c = \int \rho\varepsilon_c(\rho_\uparrow, \rho_\downarrow)d\tau - \int \rho_\uparrow\varepsilon_c(\rho_\uparrow, 0)d\tau - \int \rho_\downarrow\varepsilon_c(0, \rho_\downarrow)d\tau \tag{16}$$

Perdew and Zunger[66] have also shown the importance of the self interaction correction in Density Functional Theory.

We now summarize the expressions for the density functionals considered here containing gradients of electron densities. The fifth functional we consider is the one due to Perdew[67], namely the gradient correction grafted onto Perdew-Zunger's functional[66] (with the Ceperley-Alder parameters). The resulting functional is denoted as (PZ) and is given below (in a.u.).

$$E_c[\rho_\uparrow, \rho_\downarrow] = \int \varepsilon_c \cdot \rho d\tau + \int B(\rho_\uparrow, \rho_\downarrow) \mid \nabla\rho \mid^2 d\tau \tag{17a}$$

with

$$B(\rho_\uparrow, \rho_\downarrow) = B_{RG}(\rho) \exp(-b(\rho) \cdot f \cdot \mid \nabla\rho \mid \rho^{-7/6}) \cdot d^{-1}(\zeta) \tag{17b}$$

$$B_{RG}(\rho) = C(\rho)\rho^{-4/3} \tag{17c}$$

Here, $B_{RG}(\rho)$ is a coefficient obtained by Rasolt and Geldart[82]. And,

$$C(\rho) = c_1 + \frac{(c_2 + c_3 r_s + c_4 r_s^2)}{(1 + c_5 r_s + c_6 r_s^2 + c_7 r_s^3)} \tag{17c}$$

$$b(\rho) = (9\pi)^{1/6}(c_1 + c_2)/C(\rho) \tag{17d}$$

$$d(\zeta) = 2^{1/3}\left\{[(1+\zeta)/2]^{5/3} + [(1-\zeta)/2]^{5/3}\right\}^{1/2} \tag{17e}$$

For $r_s \geq 1$, $\varepsilon_i(\rho)$ is given by Ceperley's parametrization

$$\varepsilon_i(\rho) = \frac{\gamma_i}{(1 + \beta_i^1\sqrt{r_s} + \beta_i^2 r_s)} \qquad (i = I, II) \tag{17f}$$

and for $r_s < 1$, $\varepsilon_i(\rho)$ is given by

$$\varepsilon_i(\rho) = A_i \ln r_s + B_i + C_i r_s \ln r_s + D_i r_s. \qquad (i = I, II) \tag{17g}$$

The standard interpolation formula of von Barth and Hedin[83] in Eq. (12) is employed. All the constants in Eqs. (17) are listed below

$c_1 = 1.667 \times 10^{-3}$	$\gamma_I = -0.1423$	$\gamma_{II} = -0.0843$
$c_2 = 2.568 \times 10^{-3}$	$\beta_I^1 = 1.0529$	$\beta_{II}^1 = 1.3981$
$c_3 = 2.3266 \times 10^{-2}$	$\beta_I^2 = 0.3334$	$\beta_{II}^2 = 0.2611$
$c_4 = 7.389 \times 10^{-6}$	$A_I = 0.0311$	$A_{II} = 0.01555$
$c_5 = 8.723$	$B_I = -0.048$	$B_{II} = -0.0269$
$c_6 = 0.472$	$C_I = 0.0020$	$C_{II} = 0.0007$
$c_7 = 7.389 \times 10^{-2}$	$D_I = -0.0116$	$D_{II} = -0.0048$

and are taken from Perdew[67], Perdew and Zunger[66] and Savin et al[65].

The sixth functional considered is the one due to Becke[68] (B):

$$E_c = E_c^{\alpha\beta+\beta\alpha} + E_c^{\alpha\alpha} + E_c^{\beta\beta} \tag{18a}$$

$$E_c^{\alpha\beta+\beta\alpha} = -0.8\int \rho_\alpha \rho_\beta z_{\alpha\beta}^2[1 - \frac{\ln(1+z_{\alpha\beta})}{z_{\alpha\beta}}]d\vec{r} \tag{18b}$$

$$E_c^{\sigma\sigma} = -0.01\int \rho_\sigma D_\sigma z_{\sigma\sigma}^4[1 - \frac{2}{z_{\sigma\sigma}}\ln(1+\frac{z_{\sigma\sigma}}{2})]d\vec{r} \tag{18c}$$

where,

$$z_{\sigma\sigma} = c_{\sigma\sigma}2R_F^\sigma, \qquad z_{\alpha\beta} = c_{\alpha\beta}(R_F^\alpha + R_F^\beta), \qquad R_F^\sigma = \frac{1}{< s_\sigma^{-1} >} \tag{18d}$$

$$< s_\sigma^{-1} > = 3\left(\frac{3\rho_\sigma}{4\pi}\right)^{1/3} + 2a\frac{|\nabla\rho_\sigma|^2}{\rho_\sigma^{7/3}}\left(1 + b\frac{|\nabla\rho_\sigma|^2}{\rho_\sigma^{8/3}}\right)^{-4/5} \tag{18e}$$

$$D_\sigma = \tau_\sigma - \frac{1}{4}\frac{|\nabla\rho_\sigma|^2}{\rho_{\sigma,}} \qquad \tau_\sigma = \sum_{i_\sigma}|\nabla\psi_{i\sigma}|^2 \tag{18f}$$

The constants in Eqs. (18) are

$c_{\alpha\beta} = 0.63, \qquad c_{\sigma\sigma} = 0.96, \qquad a = 0.00375, \quad b = 0.007,$

Finally we consider the functional by Lee, Yang and Parr[69] (LYP) which has been simplified by Miehlich et al[70] so that all the $\nabla^2\rho$ terms were integrated out, since these terms contribute to the local form of the correlation energy density but cancel out after one partial integration. We have employed Hartree-Fock orbitals for the computation. Therefore the local behavior of the orbitals is not altered, which in turn justifies such an integration. The simplified form of the LYP functional is given below :

$$\begin{aligned}
E_c = &-a\int \frac{4}{1+d\rho^{-1/3}}\frac{\rho_\alpha\rho_\beta}{\rho} \\
&-ab\int \omega\Big\{\rho_\alpha\rho_\beta\Big[2^{11/3}C_F(\rho_\alpha^{8/3} + \rho_\beta^{8/3}) + \left(\frac{47}{18} - \frac{7}{18}\delta\right)|\nabla\rho|^2 \\
&-\left(\frac{5}{2} - \frac{1}{18}\delta\right)(|\nabla\rho_\alpha|^2 + |\nabla\rho_\beta|^2) \\
&-\frac{\delta-11}{9}\left(\frac{\rho_\alpha}{\rho}|\nabla\rho_\alpha|^2 + \frac{\rho_\beta}{\rho}|\nabla\rho_\beta|^2\right)\Big] \\
&-\frac{2}{3}\rho^2|\nabla\rho|^2 + \left(\frac{2}{3}\rho^2 - \rho_\alpha^2\right)|\nabla\rho_\beta|^2 + \left(\frac{2}{3}\rho^2 - \rho_\beta^2\right)|\nabla\rho_\alpha|^2\Big\}d\vec{r},
\end{aligned} \tag{19a}$$

where

$$\omega = \frac{\exp(-c\rho^{-1/3})}{1+d\rho^{-1/3}}\rho^{-11/3}, \quad \delta = c\rho^{-1/3} + \frac{d\rho^{-1/3}}{1+d\rho^{-1/3}}, \quad C_F = \frac{3}{10}(3\pi^2)^{2/3}, \qquad (19b)$$

and the constants in Eqs.(19) are

$$a = 0.04918, \quad b = 0.132 \quad c = 0.2533 \quad d = 0.349$$

We must note that the functionals reported do not represent the full list of functionals reported in literature, but only a reasonable, but arbitrary, sample. For a full list of functionals we refer the reader to the comprehensive volume by Kryachko and Ludena.[48]

Let us now compare how these functionals fare when applied to a good number of molecules (for atoms, see Chapter 2). The gaussian basis sets used in the SCF calculations with KGNMOL[84,85] are better than the Geotriplezeta quality basis of Urdaneta et al[16]. Polarization functions of d-type were added to yield near RHF quality wave functions and energies. For example in the case of linear molecules we have compared our results with the very accurate ones of McLean and Yoshimine[86] and those of Cade et al.[87] which have been obtained with optimized Slater type functions (STF). For non linear molecules our SCF energies seem to be superior or in close agreement with the ones available in the literature. We refer the reader to Ref. 88 for additional details on the molecular computations, like geometry, basis set, and experimental values.

We shall now compare the CS′, LC, GL-SPP, VWN-SPP, PZ, B and LYP **models** for the set of molecules reported in Table X. The computation of the functionals requires a rather short CPU time: the LC, GL-SPP and VWN-SPP functionals require, in general, less time than the CS′, PZ, B and LYP models since in the latter we also compute gradients of the density. In Table X we report the Hartree-Fock, the non-relativistic total energy, the atomization energy computed with Hartree-Fock, and with the CC, CS′, LC, GL-SPP, VWN-SPP, PZ, B and LYP functionals.

In Figures 8 and 9 we have plotted the ΔD curves for closed shell molecules ($\Delta D_{model} = D_{model} - D_e$) in addition to the values $D_{e,HF} - D_e = \Delta D_{HF}$. It may be noted that the molecules are ordered with increasing magnitude of ΔD_{HF}. The horizontal axis corresponds to the experimental values. From the figures it is evident that density functionals indeed estimate the correlation energy quite well barring the GL-SPP functional, which is a relatively old density functional model. It may be noted that for the GL-SPP model sometimes ΔE_c is negative indicating that in these cases the molecular correlation energy is less than the sum of atomic correlation energies. From the figures we can see that the gradient density functionals are better than the local density models. Considering the molecules in Figs. 8 and 9 and Table X, the average error for the HF, CS′, LC, GL-SPP, VWN-SPP, PZ, B and the LYP functionals are 4.85, 1.37, 2.04, 3.14, 1.2, 1.14, 1.09 and 1.32 eV, respectively. Note that the first functional reported in Figure 8 is the Clementi-

Chakravorty electron pair functional (CC), described in the following section but reported in the figure to facilitate comparison.

9. Clementi-Chakravorty Electron-Pair Functional

Recently a new algorithm has been proposed,[52,88] which we shall denote as the *CC functional,* where one uses atomic pair correlation data to compute the molecular correlation energy. The use of atomic pair correlation energies is based on the assumption that whereas there are no "atoms in molecules" at the Hartree-Fock level one could, as a crude approximation, use an "atoms in molecules" approach but only for the correlation energy problem. Alternatively stated, a dominant part of the correlation energy can be considered as a local effect and transferred from one iso-electronic system to a second iso-electronic system related to an equivalent atom in a different molecule.

In addition, we know: i) that the correlation energy is a functional of the electronic density,[48-51] ii) that there is no really large density fluctuation over the volume of a given molecule relative to simply summing the atomic densities, and iii) that any reasonable partitioning of the electronic density from a non-localized orbital description to a localized one will have only a relatively small effect on a density functional type computation of the correlation energy value.

A "brutally" simple and computationally most un-expensive localization of the electrons is the one proposed by R.S. Mulliken in his atomic gross population analysis[89]. Equivalently, we could consider any alternative algorithm which integrates the electronic density for volumes centered at the molecular nuclei and at the midpoint of bonds and extending over the entire molecular space; however, such integrations are a bit more computer time-consuming.[90]

In the new approach the total correlation energy is given as a sum of pair interactions; this will bring about shortcomings which might require the use of pair-pair corrections known to be of importance in describing multiple bonds.

To start with we partition the electronic density in each molecular orbital Φ_I by computing Mulliken's gross atomic population.[89] In order to put in evidence the correspondence between atomic populations and atomic pair correlation energies, we expand the Φ_I in terms of those atomic orbitals ϕ_a used to compute the atomic pair correlation energies.[91] Notice that the Φ_I are expressed — as usually done — by a linear combination of Gaussian-type functions, $\chi_a(A)$, centered at the nuclei of the molecule (e.g., the nucleus A). With obvious notation, we can define the molecular orbitals as

$$\Phi_I = \sum_A \sum_i c_{Ii}(A)\chi_i(A) \tag{20}$$

and the atomic orbital $\phi_a(A)$, also a linear combination of GTOs, as

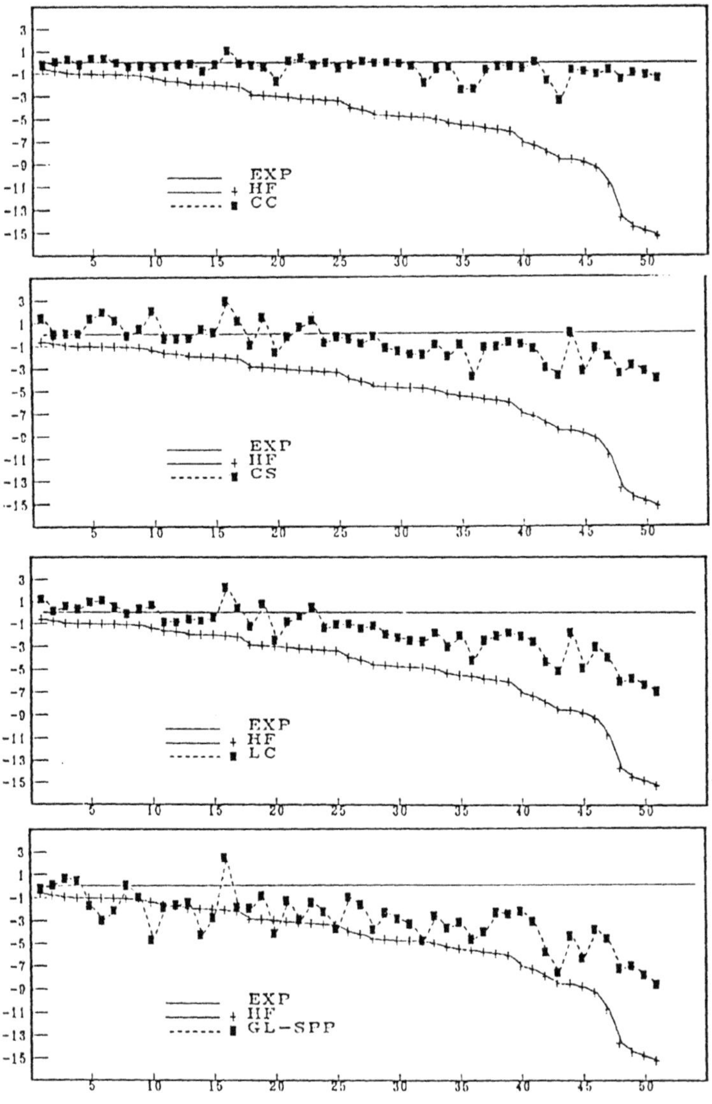

Figure 8. ΔD for the CC, CS′, LC and GL-SPP models (■ – – ■). The ΔD_{HF} curve (+--- +) is presented for reference in each plot. The horizontal axis corresponds to the experimental values. The molecules (closed shell only) are from Table IX and are listed.

1 → AlH	2 → BH	3 → Li_2	4 → LiH	5 → $LiCl$	6 → $NaCl$
7 → HCl	8 → H_2	9 → NaH	10 → Cl_2	11 → BF	12 → HF
13 → LiF	14 → ClF	15 → NaF	16 → B_2H_6	17 → H_2S	18 → BeO
19 → SiH_4	20 → F_2	21 → H_2O	22 → SiO	23 → PH_3	24 → CO
25 → CS	26 → CH_4	27 → NH_3	28 → PN	29 → C_2H_2	30 → HCN
31 → N_2	32 → MgO	33 → CH_2O	34 → C_3	35 → H_2O_2	36 → C_2
37 → CO_2	38 → C_2H_4	39 → CH_3OH	40 → C_2H_6	41 → C_3H_4	42 → N_2O
43 → O_3	44 → $(CH_3)_2S$	45 → C_2N_2	46 → $(CH_3)_2O$	47 → C_4H_6	48 → C_4H_5N
49 → C_6H_6	50 → C_5H_5N	51 → $C_4H_4N_2$			

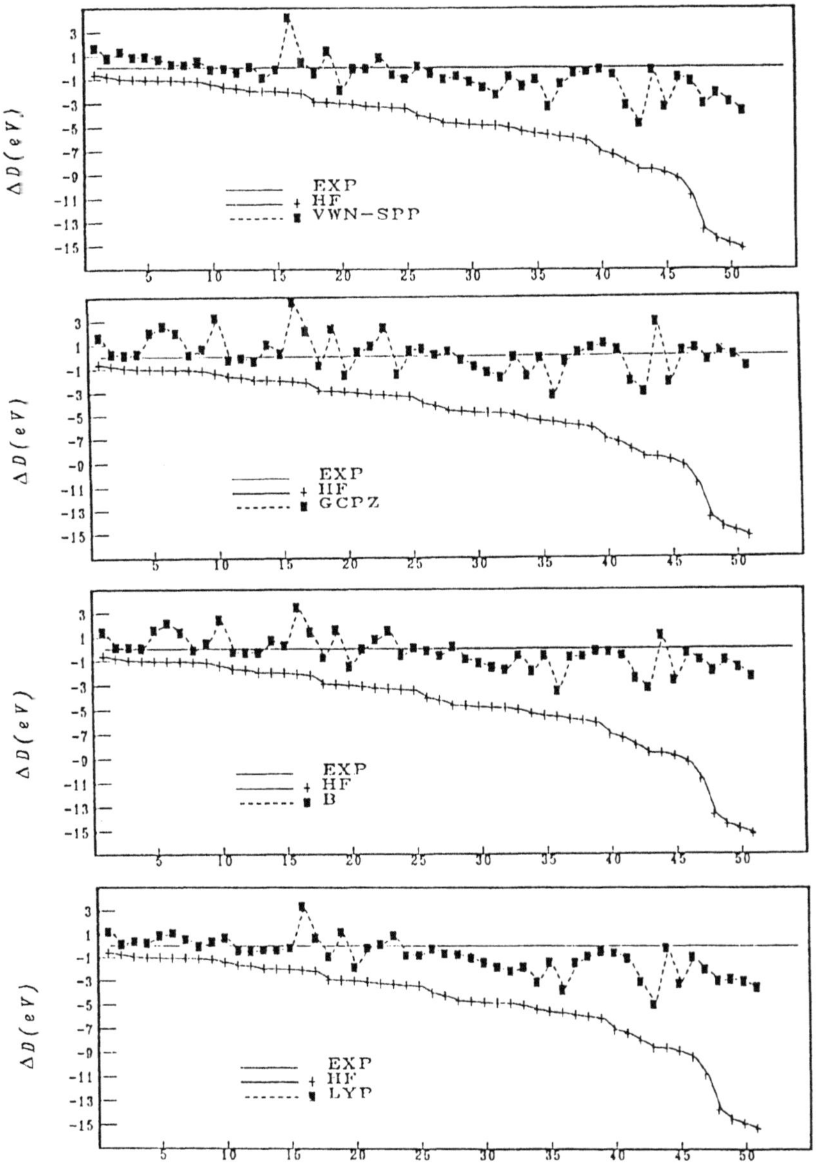

Figure 9. ΔD for the VWN-SPP, PZ, B and LYP models (■ – – ■). The ΔD_{HF} curve (+--- +) is presented for reference in each plot. The horizontal axis corresponds to the experimental values. The molecules are in the same order as in Fig. 8.

Table X. Molecular atomization energies (in eV) from density functional models and Hartree-Fock energies (in hartrees).

Formula[a]	AlH	BH	Li_2	LiH
- E(HF)	242.458204	25.130725	14.871089	7.987033
D_{HF}	2.389	2.789	0.170	1.486
D_e	3.010	3.580	1.140	2.520
$D_{CS'}$	4.479	3.605	1.225	2.583
D_{LC}	4.209	3.715	1.695	2.823
D_{GL-SPP}	2.829	3.685	1.805	2.973
$D_{VWN-SPP}$	4.649	4.355	2.445	3.373
D_{PZ}	4.619	3.825	1.275	2.763
D_B	4.389	3.705	1.265	2.563
D_{LYP}	4.179	3.705	1.495	2.763
Formula[b]	LiCl	NaCl	HCl	H_2
- E(HF)	467.047706	621.447328	460.104793	1.133491
D_{HF}	3.831	3.200	3.545	3.633
D_e	4.880	4.270	4.620	4.750
$D_{CS'}$	6.248	6.216	5.805	4.623
D_{LC}	5.808	5.356	5.135	4.633
D_{GL-SPP}	3.188	1.296	2.515	4.783
$D_{VWN-SPP}$	5.788	4.966	4.875	4.983
D_{PZ}	6.888	6.816	6.595	4.913
D_B	6.438	6.406	5.975	4.613
D_{LYP}	5.748	5.346	5.165	4.683
Formula[c]	NaH	Cl_2	BF	HF
- E(HF)	162.389517	918.988636	124.161391	100.066267
D_{HF}	0.934	1.078	6.201	4.381
D_e	2.120	2.510	7.890	6.120
$D_{CS'}$	2.620	4.548	7.468	5.712
D_{LC}	2.420	3.168	7.048	5.272
D_{GL-SPP}	1.140	-2.192	6.048	4.462
$D_{VWN-SPP}$	2.640	2.338	7.748	5.692
D_{PZ}	2.760	5.738	7.618	5.972
D_B	2.590	4.928	7.608	5.822
D_{LYP}	2.440	3.188	7.448	5.652
Formula[d]	LiF	ClF	NaF	B_2H_6
- E(HF)	106.986333	558.902393	261.369358	52.836507
D_{HF}	4.045	0.615	2.963	21.228
D_e	6.010	2.610	4.980	23.320
$D_{CS'}$	5.623	3.076	5.100	26.220
D_{LC}	5.413	1.896	4.531	25.619
D_{GL-SPP}	4.583	-1.624	2.260	25.770
$D_{VWN-SPP}$	6.063	1.756	4.811	27.540
D_{PZ}	5.663	3.606	5.211	27.869
D_B	5.713	3.296	5.231	26.759
D_{LYP}	5.673	2.256	4.791	26.669

a) Aluminium Hydride, Boron Hydride, Dilithium, Lithium Hydride
b) Lithium Chloride, Sodium Chloride, Hydrogen Chloride, Hydrogen
c) Sodium Hydride, Chlorine, Boron Fluoride, Hydrogen Fluoride
d) Lithium Fluoride, Chlorine Fluoride, Sodium Fluoride, Diborane

(continued)

Table X. *(continued)* Molecular atomization energies (in eV) from density functional models and Hartree-Fock energies (in hartrees).

Formula[a]	H_2S	BeO	SiH_4	F_2
- E(HF)	398.709255	89.446704	291.261300	198.761296
D_{HF}	5.826	1.834	11.248	-1.340
D_e	8.017	4.730	14.180	1.660
$D_{CS'}$	9.175	3.790	15.645	0.042
D_{LC}	8.445	3.490	14.976	-0.788
D_{GL-SPP}	6.165	2.790	13.316	-2.478
$D_{VWN-SPP}$	8.475	4.230	15.545	-0.228
D_{PZ}	10.115	3.990	16.505	0.102
D_B	9.405	4.000	15.746	0.192
D_{LYP}	8.695	3.830	15.376	-0.188
Formula[b]	H_2O	SiO	PH_3	CO
- E(HF)	76.066570	363.842981	342.489280	112.783429
D_{HF}	7.069	5.121	7.489	7.877
D_e	10.167	8.340	10.770	11.240
$D_{CS'}$	9.924	8.944	11.962	10.462
D_{LC}	9.364	8.024	11.302	9.842
D_{GL-SPP}	8.864	5.404	9.322	8.972
$D_{VWN-SPP}$	10.054	8.224	11.592	10.662
D_{PZ}	10.524	9.194	13.132	9.742
D_B	10.124	9.094	12.242	10.692
D_{LYP}	9.974	8.484	11.672	10.462
Formula[c]	CS	CH_4	NH_3	PN
- E(HF)	435.348384	40.215797	56.223203	395.179973
D_{HF}	4.519	14.386	8.802	1.799
D_e	7.930	18.398	13.050	6.440
$D_{CS'}$	7.668	17.925	12.238	6.218
D_{LC}	6.868	17.365	11.638	5.298
D_{GL-SPP}	4.188	17.405	11.408	2.658
$D_{VWN-SPP}$	7.008	18.525	12.568	5.478
D_{PZ}	8.428	19.025	13.148	6.828
D_B	7.988	18.235	12.498	6.588
D_{LYP}	7.148	18.135	12.388	5.728
Formula[d]	C_2H_2	HCN	N_2	MgO
- E(HF)	76.852036	92.910980	108.986760	274.369214
D_{HF}	13.000	8.818	5.095	-1.296
D_e	17.690	13.590	9.910	3.550
$D_{CS'}$	16.429	12.033	8.087	1.703
D_{LC}	15.749	11.353	7.417	0.993
D_{GL-SPP}	15.359	10.743	6.587	-1.217
$D_{VWN-SPP}$	16.989	12.393	8.267	1.313
D_{PZ}	17.379	12.733	8.557	1.763
D_B	16.829	12.373	8.367	1.813
D_{LYP}	16.619	12.133	8.097	1.373

a) Hydrogen Sulfide, Beryllium Oxide, Silane, Fluorine
b) Water, Silicon Oxide, Phosphine, Carbon Monoxide
c) Carbon Sulfide, Methane, Ammonia, Phosphorus Nitride
d) Acetylene, Hydrogen Cyanide, Nitrogen, Magnesium Oxide

(continued)

Table X. *(continued)* Molecular atomization energies (in eV) from density functional models and Hartree-Fock energies (in hartrees).

Formula[a]	CH_2O	C_3	H_2O_2	C_2
- E(HF)	113.912168	113.390816	150.843226	75.401045
D_{HF}	11.381	8.962	6.249	0.727
D_e	16.413	14.330	11.790	6.360
$D_{CS'}$	15.476	12.340	10.829	2.566
D_{LC}	14.596	11.310	9.739	2.166
D_{GL-SPP}	13.816	10.640	8.619	1.706
$D_{VWN-SPP}$	15.716	12.760	10.799	3.106
D_{PZ}	16.336	12.650	11.659	3.036
D_B	15.856	12.440	11.189	2.866
D_{LYP}	14.586	11.250	10.379	2.636
Formula[b]	CO_2	C_2H_4	CH_3OH	C_2H_6
- E(HF)	187.714195	78.061851	115.095363	79.262105
D_{HF}	11.250	18.710	16.366	24.160
D_e	17.080	24.650	22.470	31.220
$D_{CS'}$	15.900	23.499	21.701	30.299
D_{LC}	14.610	22.589	20.661	29.099
D_{GL-SPP}	13.060	22.329	20.021	28.999
$D_{VWN-SPP}$	15.740	24.169	22.141	31.039
D_{PZ}	16.600	24.949	23.171	32.219
D_B	16.390	24.039	22.281	30.939
D_{LYP}	15.640	23.789	21.981	30.629
Formula[c]	C_3H_4	N_2O	O_3	$(CH_3)_2S$
- E(HF)	115.907669	183.750802	224.342614	476.790668
D_{HF}	23.027	3.931	-2.117	25.068
D_e	30.380	11.867	6.451	33.650
$D_{CS'}$	29.065	8.838	2.749	33.676
D_{LC}	27.785	7.508	1.299	31.886
D_{GL-SPP}	27.255	6.048	-1.151	29.256
$D_{VWN-SPP}$	29.805	8.718	1.729	33.426
D_{PZ}	30.885	9.708	3.379	36.436
D_B	29.825	9.398	3.219	34.746
D_{LYP}	29.315	8.808	1.529	33.476
Formula[d]	C_2N_2	$(CH_3)_2O$	C_4H_6	C_4H_5N
- E(HF)	184.651390	154.122063	154.980209	208.876678
D_{HF}	12.995	25.606	33.515	33.425
D_e	21.840	34.953	44.280	47.090
$D_{CS'}$	18.506	33.650	42.203	43.529
D_{LC}	16.956	31.890	40.313	41.039
D_{GL-SPP}	15.486	31.120	39.633	39.819
$D_{VWN-SPP}$	18.556	34.150	43.093	44.069
D_{PZ}	19.596	35.330	44.873	46.729
D_B	19.196	34.540	43.323	45.219
D_{LYP}	18.646	33.990	42.283	44.139

a) Formaldehyde, Tricarbon, Hydrogen Peroxide, Dicarbon
b) Carbon Dioxide, Ethylene, Methanol, Ethane
c) Allene, Nitrous Oxide, Ozone, Dimethyl Sulfur
d) Cyanogen, Methyl Ether, trans 1,3 Butadiene, Pyrrole

(continued)

Table X. *(continued)* Molecular atomization energies (in eV) from density functional models and Hartree-Fock energies (in hartrees).

Formula[a]	C_6H_6	C_5H_5N	$C_4H_4N_2$
- E(HF)	230.783582	246.779711	262.767270
D_{HF}	45.190	39.299	33.175
D_e	59.668	54.129	48.430
$D_{CS'}$	56.817	50.792	44.415
D_{LC}	53.877	47.782	41.515
D_{GL-SPP}	52.617	46.312	39.815
$D_{VWN-SPP}$	57.507	51.242	44.775
D_{PZ}	60.057	54.142	47.515
D_B	58.677	52.532	46.075
D_{LYP}	56.847	51.112	44.925
Formula	HC	HN	OH
- E(HF)	38.278469	54.977445	75.418904
D_{HF}	2.485	2.114	3.051
D_e	3.650	3.400	4.630
D_{LC}	3.349	3.110	4.258
D_{GL-SPP}	3.315	2.890	4.056
$D_{VWN-SPP}$	4.115	3.830	5.146
D_{PZ}	2.845	3.860	5.266
D_B	3.925	3.630	5.136
D_{LYP}	3.665	3.530	4.966
Formula	O_2	NO	B_2
- E(HF)	149.656338	129.290241	49.088479
D_{HF}	1.163	2.277	0.871
D_e	5.210	6.610	3.000
D_{LC}	2.683	4.186	1.282
D_{GL-SPP}	1.943	3.338	1.163
$D_{VWN-SPP}$	3.913	5.148	2.272
D_{PZ}	4.443	5.688	1.762
D_B	4.393	5.528	1.782
D_{LYP}	2.903	5.208	1.483

a) Benzene, Pyridine, Pyrazine

$$\phi_a(A) = \sum_j d_{aj}\chi_j(A) \tag{21}$$

It follows that we can also write the molecular orbitals as linear combinations of the $\phi_a(A)$, the atomic orbitals namely

$$\Phi_I = \sum_A \sum_a b_{Ia}(A)\phi_a(A) \tag{22}$$

The $c_{Ii}(A)$ and d_{aj} are related to the $b_{Ia}(A)$ through the system of equations

$$< \Phi_I | \phi_b(B) > = \sum_A \sum_a b_{Ia} < \phi_a(A) | \phi_b(B) > \tag{23}$$

The Mulliken population for ϕ_a is obtained after solving for the b_{Ia}. In the following we express the gross atomic populations for the molecule as $p_a(A)$ giving explicitly an index for the atomic center A, and for the atomic orbital a or b (it designates the $1s$, $2s$, $2p_x$, $2p_y$, $2p_z$ atomic orbitals of Eq. 21).

The operational rule is to construct the configurations of the atoms in the molecule and compute the correlation energy of the molecule via the following expression

$$E_c(molecule) = \sum_A \sum_{a,b} (1/4)\ (p_a(A) + p_b(A))\eta_{ab} \tag{24a}$$

here, η_{ab} are " atoms in molecule " pair correlation energies[88] and the $p_a(A)$ are the Mulliken populations. This viewpoint follows closely the proposals by Clementi[92] and Hollister and Sinanoglu.[93] For the open shell systems we partition the gross population analyses for the closed and open shell separately, viz. $p_a^c(A)$ and $p_a^o(a)$ respectively. And we use the following equation instead

$$E_c(molecule) = \sum_A \sum_{a,b} (1/4)\ (p_a^c(A) + p_b^c(A))\eta_{ab} + \sum_A \sum_{a \neq b} (1/8)\ (p_a^o(A) + p_b^o(A))\eta_{ab} \tag{24b}$$

and finally the molecular extra correlation energy,

$$\Delta E_c = E_c(molecule) - \sum_A E_c(A) \tag{25}$$

where $E_c(A)$ is the atomic correlation energy for the dissociated atoms given in Ref. 88.

As obvious, the computation of $E_c(molecule)$ requires an insignificant amount of additional computer time over that required by the Hartree-Fock computation, essentially as much as is needed to obtain Mulliken's population analyses. As a consequence the computation is very fast so that it can be performed routinely. Note that the pair correlation energy computed by Clementi-Chakravorty corresponds to the ground state of the atoms, therefore to a fully dissociated molecule. Instead, for the molecule we need pair correlation energies corresponding to *atoms in the molecule*. For example, in CH_4, C_2H_2, C_2H_4, etc., the C atom experiences the presence of the electrons of the hydrogen atoms, as long ago pointed out by the "octet" rule in valence considerations where the carbon nucleus experiences 10 electrons and each hydrogen nucleus experiences 2 electrons. Thus, whereas at dissociation the carbon atom has configuration $1s^2(2s, 2p)^4$, within the CH_4 molecule it has a configuration which we could write as $1s^2(2s, 2p)^8$. Equivalently, the

Table XI. Molecular atomization energies (in eV) from Clementi-Chakravorty model and Hartree-Fock energies (in hartrees).

Formula	AlH	BH	Li_2	LiH
- E(HF)	242.458204	25.130725	14.871089	7.987033
D_{HF}	2.389	2.789	0.170	1.486
D_e	3.010	3.580	1.140	2.520
D_{CC}	2.800	3.637	1.405	2.334
Formula	LiCl	NaCl	HCl	H_2
- E(HF)	467.047706	621.447328	460.104793	1.133491
D_{HF}	3.831	3.200	3.545	3.633
D_e	4.880	4.270	4.620	4.750
D_{CC}	5.193	4.597	4.599	4.395
Formula	NaH	Cl_2	BF	HF
- E(HF)	162.389517	918.988636	124.161391	100.066267
D_{HF}	0.934	1.078	6.201	4.381
D_e	2.120	2.510	7.890	6.120
D_{CC}	1.823	2.031	7.514	5.918
Formula	LiF	ClF	NaF	B_2H_6
- E(HF)	106.986333	558.902393	261.369358	52.836507
D_{HF}	4.045	0.615	2.963	21.228
D_e	6.010	2.610	4.980	23.320
D_{CC}	5.880	1.865	4.808	24.415
Formula	H_2S	BeO	SiH_4	F_2
- E(HF)	398.709255	89.446704	291.261300	198.761296
D_{HF}	5.826	1.834	11.248	-1.340
D_e	8.017	4.730	14.180	1.660
D_{CC}	7.942	4.514	13.771	-0.025
Formula	H_2O	SiO	PH_3	CO
- E(HF)	76.066570	363.842981	342.489280	112.783429
D_{HF}	7.069	5.121	7.489	7.877
D_e	10.167	8.340	10.770	11.240
D_{CC}	10.346	8.782	10.565	11.232
Formula	CS	CH_4	NH_3	PN
- E(HF)	435.348384	40.215797	56.223203	395.179973
D_{HF}	4.519	14.386	8.802	1.799
D_e	7.930	18.398	13.050	6.440
D_{CC}	7.420	18.217	13.184	6.414
Formula	C_2H_2	HCN	N_2	MgO
- E(HF)	76.852036	92.910980	108.986760	274.369214
D_{HF}	13.000	8.818	5.095	-1.296
D_e	17.690	13.590	9.910	3.550
D_{CC}	17.712	13.512	9.645	1.761
Formula	CH_2O	C_3	H_2O_2	C_2
- E(HF)	113.912168	113.390816	150.843226	75.401045
D_{HF}	11.381	8.962	6.249	0.727
D_e	16.413	14.330	11.790	6.360
D_{CC}	15.867	13.949	9.389	4.032

(continued)

Table XI. *(continued)* Molecular atomization energies (in eV) from Clementi-Chakravorty model and Hartree-Fock energies (in hartrees).

Formula	CO_2	C_2H_4	CH_3OH	C_2H_6
- E(HF)	187.714195	78.061851	115.095363	79.262105
D_{HF}	11.250	18.710	16.366	24.160
D_e	17.080	24.650	22.470	31.220
D_{CC}	16.481	24.361	22.185	30.699
Formula	C_3H_4	N_2O	O_3	$(CH_3)_2S$
- E(HF)	115.907669	183.750802	224.342614	476.790668
D_{HF}	23.027	3.931	-2.117	25.068
D_e	30.380	11.867	6.451	33.650
D_{CC}	30.477	10.302	3.090	33.019
Formula	C_2N_2	$(CH_3)_2O$	C_4H_6	C_4H_5N
- E(HF)	184.651390	154.122063	154.980209	208.876678
D_{HF}	12.995	25.606	33.515	33.425
D_e	21.840	34.953	44.280	47.090
D_{CC}	21.085	33.948	43.678	45.624
Formula	C_6H_6	C_5H_5N	$C_4H_4N_2$	
- E(HF)	230.783582	246.779711	262.767270	
D_{HF}	45.190	39.299	33.175	
D_e	59.668	54.129	48.430	
D_{CC}	58.819	53.083	47.065	
Formula	HC	HN	OH	
- E(HF)	38.278469	54.977445	75.418904	
D_{HF}	2.485	2.114	3.051	
D_e	3.650	3.400	4.630	
D_{CC}	4.315	3.782	3.827	
Formula	O_2	NO	B_2	
- E(HF)	149.656338	129.290241	49.088479	
D_{HF}	1.163	2.277	0.871	
D_e	5.210	6.610	3.000	
D_{CC}	3.463	5.627	0.950	

hydrogen atoms in CH_4 experience a configuration which can be written as $1s^2$, different from the one at dissociation, namely $1s^1$. For this reason we need to choose appropriate pair correlation energies, Ne-like for carbon and He-like for hydrogen. With the atomic algorithm[91] we have tabulated such pair correlation energies, available to the interested reader.[88]

In Table XI we report the dissociation energies using the CC functional for the same molecules reported in Table X. One can notice that this very simple technique yields most reasonable predictions and is computationally very fast. The comparison of the density functionals of Section 8 with the electron pair functional here presented is provided by Fig. 8 where the first insert (top) are the results obtained with the CC functional; the average error of the CC functional is

0.62, namely for the limited sample of molecules analyzed is the best one. Note that we have not included pair-pair correlation corrections and this affects the molecules B_2H_6, F_2, MgO, H_2O_2, C_2, N_2O_4 and O_3 for which the Hartree-Fock configuration is a rather poor zero order function.

To conclude this and the preceding sections, we once more reiterate the well known fact that the Hartree-Fock technique is a N^4 to N^3 complexity algorithm (N being the number of basis functions), which increases at least to N^5 in conventional *ab initio* algorithms attempting to improve on it. On the other hand, the *density functionals generally have N^2 complexity.* Thus it follows that computations on large molecules *need to make use of some type of density functional.*

10. Program Organization and Routes

The overall structure of KGNMOL is an evolution from IBMOL-1.[6] The main sections are the computations of the one electron integrals (INT1), of the two electron integrals (INT2), of the SCF iterative program, of the MP2 program, of programs based on the SCF output, specifically population analysis, dipole, quadrupole and octopole moments, overlap population, potential energy, electric field, field gradient, diamagnetic susceptibility, magnetic shielding, line, plane and charge densities, Clementi-Chakravorty electron-pair functional, the density functionals of Becke, Colle-Salvetti, Gunnarson-Lundqvist, Vosko-Wilk-Nusair (the last two with the Stoll-Pavlidou-Preuss correction), Perdew-Zunger (with Perdew gradient), Lee-Yang-Parr, and Lie-Clementi.

To simplify its use, KGNMOL allows the user to specify its final goals, for example and SCF wavefunction, or an MP2 energy.

The available Routes are here listed, with their designation code.

INT1 Computes one electron integrals
INT2 Computes two electron integrals
SCF Computes SCF wavefunction
POP Population analysis
PRP One electron properties
DNST Electron densities and Laplacian
MP2 Computes MP2 energy
ECOR Clementi-Chakravorty electron pair functional
FUNCT Density functionals above listed
INT1&2 Computes the one and two electron integrals
MOLSCF Executes one, and two electron integrals and SCF programs
BSSE Automatic computation of the basis set superposition error.

The integral program can be executed with or without the ADD option; the SCF and MP2 programs must be notified accordingly. The Input/Output Documentation Manual for MOTECC-89, as well the one with special input, window oriented for the IBM/6000 workstations, provides over 30 input examples organized from simple and semi-automatic, to more complex. KGNMOL has been deposited with QCPE (number 538) and the latest version is available to the MOTECC-Club members. Note that the KGNMOL program is protected by IBM-Copyright. In the following sections we report the analytical expressions used in the electron integral programs of KGNMOL, details on the integral files and the main flow of the two electron integral package.

11. One-Electron Integral Formulae

We shall consider those matrix elements which are required during the SCF process, namely the overlap integrals (OIs), the kinetic energy integrals (KEIs), and the nuclear attraction integrals (NAIs). For completeness, the formulae for matrix elements over primitive and contracted functions are both given. This "re-typing" should help in finding misprints. The formulae for STOs are given in Ref. 44 for many-centers functions; for the one-center we refer to the chapter on Hartree-Fock for atoms.

Overlap Integrals. After expanding the many-electron wave function Ψ in MOs, ϕ, and those in turn in the atomic contracted functions, designated as φ, the following OIs of contracted functions must be evaluated:

$$S_{ij} = \int \varphi_i(1)\, \varphi_j(1)\, d\tau_1 \,, \tag{26}$$

where τ_1 is the integration variable over the three space coordinates of (dummy) electron 1. Using the Gaussian Product Theorem (see Appendix B and Fig. 10), and the property of GTFs which permits a separation in their cartesian components (Eq. 4), we can express Eq. (26) as:

$$\begin{aligned} S_{ij} &= \sum_{a,b} c_a\, c_b\, N(\alpha_1, \vec{l_1})\, N(\alpha_2, \vec{l_2}) \int d\tau_1\, \eta_a(\alpha_1, \vec{A}, \vec{l_1})\, \eta_b(\alpha_2, \vec{B}, \vec{l_2}) \\ &= \sum_{a,b} c_a\, c_b\, N(\alpha_1, \vec{l_1})\, N(\alpha_2, \vec{l_2})\, I_x\, I_y\, I_z\, \exp(-\alpha_1\, \alpha_2 \overline{AB}^2 / \gamma_1), \end{aligned} \tag{27}$$

where

$$I_x = \sum_{i=0}^{l_{x1}+l_{x2}} f(i, l_{x1}, l_{x2}, \overline{PA}_x, \overline{PB}_x) \int_{-\infty}^{\infty} dx_p\, x_p^i\, \exp(-\gamma_1\, x_p^2). \tag{28}$$

I_y and I_z have similar expressions.

The integrals in I_x can be calculated with the aid of Eq. (E7) in Appendix E. Note that the integrals with odd values of i vanish, so that:

$$I_x = \sqrt{\frac{\pi}{\gamma_1}} \sum_{i=0}^{[(l_{x1}+l_{x2})/2]} f(2i,\ l_{x1},\ l_{x2},\ \overline{PA}_x,\ \overline{PB}_x)\ \frac{(2i-1)!!}{(2\gamma_1)^i}. \tag{29}$$

Here and in the next sections $[n] \equiv$ largest integer $\leq n$.

Note that the summation index i ranges now from $0 \rightarrow [(l_{x1} + l_{x2})/2]$, and that the first variable of the function f is always even. Substitution of Eq. (29) and its y and z analogues into Eq. (27) yields the complete expression of S_{ij} in terms of GTFs:

$$\begin{aligned} S_{ij} =& \sum_{a,b} c_a\, c_b\, N(\alpha_1, \vec{l_1})\, N(\alpha_2, \vec{l_2}) \left(\sqrt{\frac{\pi}{\gamma_1}}\right)^3 \exp(-\alpha_1\alpha_2\overline{AB}^2/\gamma_1) \\ &\times \sum_{i=0}^{[(l_{x1}+l_{x2})/2]} f(2i,\ l_{x1},\ l_{x2},\ \overline{PA}_x,\ \overline{PB}_x)\ \frac{(2i-1)!!}{(2\gamma_1)^i} \\ &\times \sum_{j=0}^{[(l_{y1}+l_{y2})/2]} f(2j,\ l_{y1},\ l_{y2},\ \overline{PA}_y,\ \overline{PB}_y)\ \frac{(2j-1)!!}{(2\gamma_1)^j} \\ &\times \sum_{k=0}^{[(l_{z1}+l_{z2})/2]} f(2k,\ l_{z1},\ l_{z2},\ \overline{PA}_z,\ \overline{PB}_z)\ \frac{(2k-1)!!}{(2\gamma_1)^k}. \end{aligned} \tag{30}$$

The summation indices i, j, k have the following ranges:

$$i = 0 \rightarrow [(l_{x1} + l_{x2})/2],\quad j = 0 \rightarrow [(l_{y1} + l_{y2})/2],\quad k = 0 \rightarrow [(l_{z1} + l_{z2})/2].$$

The overlap integral in terms of primitive functions is

$$\begin{aligned} < A, \alpha_1, l_1, m_1, n_1 \mid B, \alpha_2, l_2, m_2, n_2 > = N_1 N_2 \left(\frac{\pi}{\alpha_1 + \alpha_2}\right)^{3/2} \\ \times \exp\left(-\frac{\alpha_1\alpha_2}{\alpha_1 + \alpha_2}\,\overline{AB}^2\right) \times \Sigma_x \Sigma_y \Sigma_z \end{aligned} \tag{31}$$

where

$$\Sigma_x = \sum_{i=0}^{[(l_1+l_2)/2]} f(2i,\ l_1,\ l_2,\ \overline{PA}_x,\ \overline{PB}_x)\ \frac{(2i-1)!!}{2^i(\alpha_1 + \alpha_2)^i} \tag{32}$$

and similar definitions for Σ_y and Σ_z.

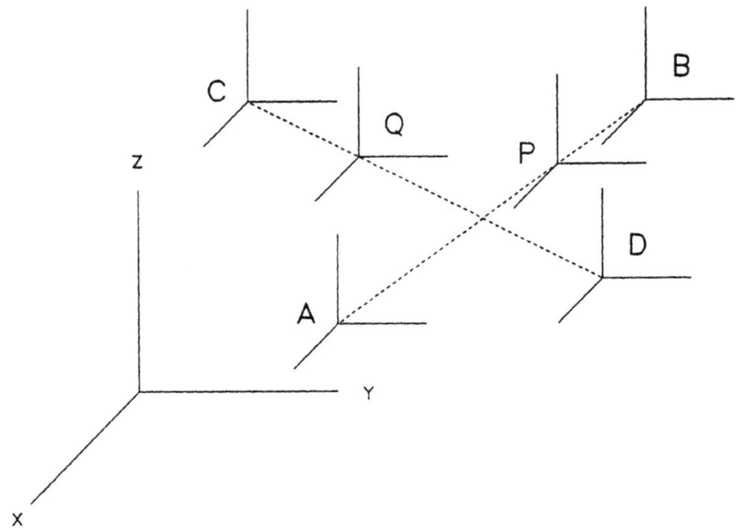

Figure 10. Illustration of the Gaussian Product Theorem. P is the new center of the GTFs originated from the product of GTFs centered one on atom A and another on atom B. Q is the new center of the GTFs originated from the product of GTFs centered one on atom C and another on atom D.

Kinetic Energy Integrals. The analytical form of the KEIs between the contracted functions φ_i and φ_j is:

$$T_{ij} = -\frac{1}{2}\int d\tau_1\, \varphi_i(1)\, \nabla_1^2\, \varphi_j(1), \tag{33}$$

where τ_1, φ_i and φ_j have the same meaning as in Eq. (26), and the Laplacian, ∇_1^2, operates on the space coordinates of (dummy) electron 1. Henceforth we will omit the subindex 1 in the Laplacian.

In terms of GTFs, T_{ij} becomes

$$T_{ij} = -\frac{1}{2}\sum_{a,b} c_a\, c_b\, N(\alpha_1, \vec{l}_1)\, N(\alpha_2, \vec{l}_2) \int d\tau_1\, \eta_a(\alpha_1, \vec{A}, \vec{l}_1)\, \nabla^2\, \eta_b(\alpha_2, \vec{B}, \vec{l}_2) \tag{34}$$

In order to evaluate the integral, $\nabla^2\, \eta(\alpha_2, \vec{B}, \vec{l}_2)$ has to be calculated:

$$\begin{aligned} \nabla^2\, \eta(\alpha_2, \vec{B}, \vec{l}_2) &= \left(\frac{\partial^2}{\partial x^2} + \frac{\partial^2}{\partial y^2} + \frac{\partial^2}{\partial z^2}\right) \eta(\alpha_2, \vec{B}, \vec{l}_2) \\ &= \frac{\partial^2}{\partial x^2}\eta(\alpha_2, \vec{B}, \vec{l}_2) + \frac{\partial^2}{\partial y^2}\eta(\alpha_2, \vec{B}, \vec{l}_2) + \frac{\partial^2}{\partial z^2}\eta(\alpha_2, \vec{B}, \vec{l}_2) \end{aligned} \tag{35}$$

In Eq. (35) we can focus our attention on just one of the space coordinates, for example x. Recall that the GTFs are separable in three similar space components (Eq. 4)

$$\frac{\partial^2}{\partial x^2}\eta(\alpha_2, \vec{B}, \vec{l}_2) = Y(\alpha_2, B_y, l_{y2})\, Z(\alpha_2, B_z, l_{z2})\frac{\partial^2}{\partial x^2}X(\alpha_2, B_x, l_{x2}), \tag{36}$$

$$\begin{aligned}\frac{\partial^2}{\partial x^2}X(\alpha_2, B_x, l_{x2}) &= \frac{\partial^2}{\partial x^2}x_B^{l_{x2}}\exp(-\alpha_2 x_B^2)\\ &= \frac{\partial}{\partial x}\left\{\frac{\partial}{\partial x}x_B^{l_{x2}}\exp(-\alpha_2 x_B^2)\right\}\end{aligned} \tag{37}$$

Since $\partial x_B = \partial x - \partial B_x = \partial x$, Eq. (37) becomes:

$$\begin{aligned}\frac{\partial^2}{\partial x^2}X(\alpha_2, B_x, l_{x2}) &= \frac{\partial}{\partial x_B}\left\{\frac{\partial}{\partial x_B}x_B^{l_{x2}}\exp(-\alpha_2 x_B^2)\right\}\\ &= \frac{\partial}{\partial x_B}\left\{l_{x2}\, x_B^{l_{x2}-1}\exp(-\alpha_2 x_B^2) - 2\alpha_2\, x_B^{l_{x2}+1}\exp(-\alpha_2 x_B^2)\right\}\\ &= \left\{l_{x2}\,(l_{x2}-1)\, x_B^{l_{x2}-2} - 2\alpha_2\,(2l_{x2}+1)x_B^{l_{x2}} + 4\alpha_2^2\, x_B^{l_{x2}+2}\right\}\exp(-\alpha_2 x_B^2).\end{aligned} \tag{38}$$

Substitution of Eq. (38) into Eq. (36) yields:

$$\frac{\partial^2}{\partial x^2}\eta(\alpha_2, \vec{B}, \vec{l}_2) = \left\{l_{x2}(l_{x2}-1)\, x_B^{-2} - 2\alpha_2(2l_{x2}+1) + 4\alpha_2^2\, x_B^2\right\}\eta(\alpha_2, \vec{B}, \vec{l}_2). \tag{39}$$

Subsequently this expression, and its y and z analogues, can be written into Eq. (35):

$$\begin{aligned}\nabla^2\eta(\alpha_2, \vec{B}, \vec{l}_2) = \Big\{&l_{x2}\,(l_{x2}-1)\, x_B^{-2} + l_{y2}\,(l_{y2}-1)\, y_B^{-2} + l_{z2}(l_{z2}-1)z_B^{-2}\\ &- 2\alpha_2\,(2l_{x2} + 2l_{y2} + 2l_{z2} + 3) + 4\alpha_2^2(x_B^2 + y_B^2 + z_B^2)\Big\}\,\eta(\alpha_2, \vec{B}, \vec{l}_2).\end{aligned} \tag{40}$$

Note that the outcome of the operation of the Laplacian on a GTF is a linear combination of GTFs with different angular momenta. Let us define the following vectors:

$$\vec{l}_2^{\,x+} = (l_{x2}+2,\ l_{y2},\ l_{z2}),\quad \vec{l}_2^{\,y+} = (l_{x2},\ l_{y2}+2,\ l_{z2}),\quad \vec{l}_2^{\,z+} = (l_{x2},\ l_{y2},\ l_{z2}+2),$$

$$\vec{l}_2^{\,x-} = (l_{x2}-2,\ l_{y2},\ l_{z2}),\quad \vec{l}_2^{\,y-} = (l_{x2},\ l_{y2}-2,\ l_{z2}),\quad \vec{l}_2^{\,z-} = (l_{x2},\ l_{y2},\ l_{z2}-2).$$

We can now rewrite Eq. (40):

$$
\begin{aligned}
&\nabla^2 \eta(\alpha_2, \vec{B}, \vec{l_2}) \\
&\quad = l_{x2}(l_{x2}-1)\, \eta(\alpha_2, \vec{B}, \vec{l_2^{x-}}) + l_{y2}(l_{y2}-1)\, \eta(\alpha_2, \vec{B}, \vec{l_2^{y-}}) \\
&\quad + l_{z2}(l_{z2}-1)\, \eta(\alpha_2, \vec{B}, \vec{l_2^{z-}}) + 4\alpha_2^2 \{\eta(\alpha_2, \vec{B}, \vec{l_2^{x+}}) + \eta(\alpha_2, \vec{B}, \vec{l_2^{y+}}) \\
&\quad + \eta(\alpha_2, \vec{B}, \vec{l_2^{z+}})\} - 2\alpha_2(2l_{x2} + 2l_{y2} + 2l_{z2} + 3)\, \eta(\alpha_2, \vec{B}, \vec{l_2}).
\end{aligned}
\tag{41}
$$

Substitution of Eq. (41) into Eq. (34) yields T_{ij}.

The kinetic energy integral in terms of primitive GTF is given by

$$
\begin{aligned}
&< A, \alpha_1, l_1, m_1, n_1 \mid -\frac{1}{2}\nabla^2 \mid B, \alpha_2, l_2, m_2, n_2 > \\
&\quad = N_1 N_2 \Big[\alpha_2 \{2(l_2 + m_2 + n_2) + 3\} < A, \alpha_1, l_1, m_1, n_1 \mid B, \alpha_2, l_2, m_2, n_2 > \\
&\quad - 2\alpha_2^2 \Big\{ < A, \alpha_1, l_1, m_1, n_1 \mid B, \alpha_2, l_2 + 2, m_2, n_2 > \\
&\qquad + < A, \alpha_1, l_1, m_1, n_1 \mid B, \alpha_2, l_2, m_2 + 2, n_2 > \\
&\qquad + < A, \alpha_1, l_1, m_1, n_1 \mid B, \alpha_2, l_2, m_2, n_2 + 2 > \Big\} \\
&\quad - \frac{1}{2}\Big\{ l_2(l_2 - 1) < A, \alpha_1, l_1, m_1, n_1 \mid B, \alpha_2, l_2 - 2, m_2, n_2 > \\
&\qquad + m_2(m_2 - 1) < A, \alpha_1, l_1, m_1, n_1 \mid B, \alpha_2, l_2, m_2 - 2, n_2 > \\
&\qquad + n_2(n_2 - 1) < A, \alpha_1, l_1, m_1, n_1 \mid B, \alpha_2, l_2, m_2, n_2 - 2 > \Big\} \Big].
\end{aligned}
\tag{42}
$$

Nuclear Attraction Integrals. The NAIs over the contracted functions can be written as:

$$
V_{ij}^C = Z_C \int d\tau_1\, \varphi_i(1)\, r_{1C}^{-1}\, \varphi_j(1), \tag{43}
$$

where τ_1, φ_1 and φ_j have the same meaning as in Eq. (26), and r_{1C} and Z_C are the nuclear-electron distance and the nuclear charge, respectively.

Expressed in terms of GTF primitives, V_{ij}^C becomes:

$$
V_{ij}^C = Z_C \sum_{a,b} c_a\, c_b\, N(\alpha_1, \vec{l_1})\, N(\alpha_2, \vec{l_2}) \int d\tau_1\, \eta_a(\alpha_1, \vec{A}, \vec{l_1})\, r_{1C}^{-1}\, \eta_b(\alpha_2, \vec{B}, \vec{l_2}) \tag{44}
$$

Application of the Gaussian Product Theorem transforms Eq. (44) into:

$$V_{ij}^{C}= Z_C \sum_{a,b} c_a\, c_b\, N(\alpha_1, \vec{l_1})\, N(\alpha_2, \vec{l_2})\ \exp(-\alpha_1\,\alpha_2\,\overline{AB}^2/\gamma_1)$$
$$\times \sum_{i=0}^{l_{x1}+l_{x2}} \sum_{j=0}^{l_{y1}+l_{y2}} \sum_{k=0}^{l_{z1}+l_{z2}} f(i, l_{x1}, l_{x2}, \overline{PA}_x, \overline{PB}_x) \tag{45}$$
$$f(j, l_{y1}, l_{y2}, \overline{PA}_y, \overline{PB}_y)\, f(k, l_{z1}, l_{z2}, \overline{PA}_z, \overline{PB}_z)$$
$$\times \int d\tau_1\, x_P^i\, y_P^j\, z_P^k\, r_{1C}^{-1} \exp(-\gamma_1\, r_P^2)$$

All indices and symbols have already been defined above.

The use of the Fourier (or the Laplace) transform of r_{1C}^{-1}, as well as of Eq. (C4) (Appendix C), is required for the evaluation of the integral in Eq. (45). The calculation of the two-electron-repulsion-integrals (ERIs) proceeds in a way very much like that of the NAIs, and a detailed treatment of the procedure will be given in the section dedicated to the ERIs. The final expression for V_{ij}^{C} is:

$$V_{ij}^{C} = Z_C \sum_{a,b} c_a\, c_b\, N(\alpha_1, \vec{l_1})\, N(\alpha_2, \vec{l_2}) \frac{2\pi}{\gamma_1} \exp(-\alpha_1\,\alpha_2\,\overline{AB}^2/\gamma_1)$$
$$\times \sum_{i,r,u} A_{i,r,u}(l_{x1}, l_{x2}, A_x, B_x, C_x, \gamma_1) \sum_{j,s,v} A_{j,s,v}(l_{y1}, l_{y2}, A_y, B_y, C_y, \gamma_1) \tag{46}$$
$$\sum_{k,t,w} A_{k,t,w}(l_{z1}, l_{z2}, A_z, B_z, C_z, \gamma_1)\, F_\kappa(\gamma_1\, \overline{CP}^2),$$

where $F_\kappa(t)$ is the Incomplete Gamma Function described in Appendix D,

$$\kappa = i + j + k - 2(r + s + t) - u - v - w,$$

and

$$A_{i,r,u}(l_{x1}, l_{x2}, A_x, B_x, C_x, \gamma_1)$$
$$= (-1)^i f(i, l_{x1}, l_{x2}, \overline{PA}_x, \overline{PB}_x) \frac{(-1)^u\, i!\, (\overline{CP}_x)^{i-2(r+u)}}{r!\, u!\, (i-2r-2u)!\, (4\gamma_1)^{r+u}}.$$

$A_{j,s,v}$ and $A_{k,t,w}$ are similarly defined in terms of the y and z components. The summations in Eq. (46) are over the following ranges:

$$\begin{array}{lll} i = 0 \rightarrow l_{x1} + l_{x2} & j = 0 \rightarrow l_{y1} + l_{y2} & k = 0 \rightarrow l_{z1} + l_{z2} \\ r = 0 \rightarrow [i/2] & s = 0 \rightarrow [j/2] & t = 0 \rightarrow [k/2] \\ u = 0 \rightarrow [(i-2r)/2] & v = 0 \rightarrow [(j-2s)/2] & w = 0 \rightarrow [(k-2t)/2]. \end{array} \tag{47}$$

The nuclear attraction integrals, written in terms of primitive GTFs, are:

$$
\begin{aligned}
&< A,\ \alpha_1, l_1, m_1, n_1 \mid \frac{1}{r_{1c}} \mid B, \alpha_2, l_2, m_2, n_2 > \\
&= N_1 N_2 \left(\frac{2\pi}{\alpha_1+\alpha_2}\right) \exp\left(-\frac{\alpha_1\alpha_2}{\alpha_1+\alpha_2}\overline{AB}^2\right) \\
&\times \sum_{i,r,u} A_{i,r,u}(l_1, l_2, A_x, B_x, C_x, \gamma) \sum_{j,s,v} A_{j,s,v}(m_1, m_2, A_y, B_y, C_y, \gamma) \\
&\sum_{k,t,w} A_{k,t,w}(n_1, n_2, A_z, B_z, C_z, \gamma)\, F_\kappa(\gamma \cdot \overline{CP}^2),
\end{aligned}
\tag{48}
$$

where

$$\gamma = \alpha_1 + \alpha_2; \quad \epsilon = \frac{1}{4\gamma}; \quad \kappa = i + j + k - 2(r+s+t) - u - v - w;$$

and

$$A_{i,r,u} = (-)^i f_i(l_1, l_2, \overline{PA}_x, \overline{PB}_x) \frac{(-)^u\, i!\, (\overline{CP}_x)^{i-2(r+u)}\, \epsilon^{r+u}}{r!\, u!\, (i-2r-2u)!}. \tag{49}$$

$A_{j,s,v}$, $A_{k,t,w}$ are similarly defined in terms of the y- and z-components. The summations in Eq. (48) are over the following ranges:

$i = 0 \rightarrow l_1 + l_2$	$j = 0 \rightarrow m_1 + m_2$	$k = 0 \rightarrow n_1 + n_2$
$r = 0 \rightarrow [i/2]$	$s = 0 \rightarrow [j/2]$	$t = 0 \rightarrow [k/2]$
$u = 0 \rightarrow [(i-2r)/2]$	$v = 0 \rightarrow [(j-2s)/2]$	$w = 0 \rightarrow [(k-2t)/2]$

12. Two-Electron Integral Formulae

General Analytical Formulae

Several papers and reviews of classical methods of calculating electron repulsion integrals are available.[2,5,6,94-99] In addition, there has recently been considerable interest in using recursion schemes;[100] finally, we refer to Chapter 6 for yet another way to compute the two-electron integrals[101].

The ERIs of contracted functions are defined as

$$ERI = (ij \mid kl) = \int\int d\tau_1\, d\tau_2\, \varphi_i(1)\, \varphi_j(1)\, (r_{12})^{-1}\, \varphi_k(2)\, \varphi_l(2), \tag{50}$$

where $(r_{12})^{-1}$ denotes the distance between two electrons with dummy indices 1 and 2, $\varphi_i(1)$ is the contracted function φ_i for electron with dummy index 1; $\varphi_k(2)$ has an analogous definition; τ_1 and τ_2 are the integration variables over the space coordinates of electron 1 and 2, respectively.

Expansion of the AOs (φ's) in primitives renders:

$$ERI = \sum_{a,b,c,d} c_a\, c_b\, c_c\, c_d\, N(\alpha_1, \vec{l}_1)\, N(\alpha_2, \vec{l}_2)\, N(\alpha_3, \vec{l}_3)\, N(\alpha_4, \vec{l}_4)$$
$$\times \int\int d\tau_1\, d\tau_2\, \eta_a(\alpha_1, \vec{A}, \vec{l}_1)\, \eta_b(\alpha_2, \vec{B}, \vec{l}_2)\, (r_{12})^{-1}\, \eta_c(\alpha_3, \vec{C}, \vec{l}_3)\, \eta_d(\alpha_4, \vec{D}, \vec{l}_4). \tag{51}$$

As in the one-electron-integrals, we use GTFs as primitive functions; consequently it is possible to apply the Gaussian Product Theorem (Appendix B). This is done for each pair of GTFs describing the same electron, i.e., for the pair $\eta_a(\alpha_1, \vec{A}, \vec{l}_1)\, \eta_b(\alpha_2, \vec{B}, \vec{l}_2)$ and the pair $\eta_c(\alpha_3, \vec{C}, \vec{l}_3)\, \eta_d(\alpha_4, \vec{D}, \vec{l}_4)$, respectively.

From now on it will be assumed that the first pair yields a linear combination of GTFs, all centered on P with exponent γ_1 (see Appendix B):

$$\gamma_1 = \alpha_1 + \alpha_2 \tag{52}$$

$$\vec{P} = \frac{\alpha_1 \vec{A} + \alpha_2 \vec{B}}{\gamma_1}, \tag{53}$$

while the second pair of GTF are centered on Q, and γ_2 is their exponent:

$$\gamma_2 = \alpha_3 + \alpha_4, \tag{54}$$

$$\vec{Q} = \frac{\alpha_3 \vec{C} + \alpha_4 \vec{D}}{\gamma_2}. \tag{55}$$

Eq. (51) becomes, in terms of GTFs:

$$
\begin{aligned}
ERI = & \sum_{a,b,c,d} c_a\, c_b\, c_c\, c_d\, N(\alpha_1, \vec{l}_1)\, N(\alpha_2, \vec{l}_2)\, N(\alpha_3, \vec{l}_3)\, N(\alpha_4, \vec{l}_4) \\
& \times \exp(-\alpha_1\, \alpha_2\, \overline{AB}^2/\gamma_1)\, \exp(-\alpha_3\, \alpha_4\, \overline{CD}^2/\gamma_2) \\
& \times \sum_{i=0}^{l_{x1}+l_{x2}} f(i, l_{x1}, l_{x2}, \overline{PA}_x, \overline{PB}_x) \sum_{j=0}^{l_{y1}+l_{y2}} f(j, l_{y1}, l_{y2}, \overline{PA}_y, \overline{PB}_y) \\
& \times \sum_{k=0}^{l_{z1}+l_{z2}} f(k, l_{z1}, l_{z2}, \overline{PA}_z, \overline{PB}_z) \sum_{i'=0}^{l_{x3}+l_{x4}} f(i', l_{x3}, l_{x4}, \overline{QC}_x, \overline{QD}_x) \\
& \times \sum_{j'=0}^{l_{y3}+l_{y4}} f(j', l_{y3}, l_{y4}, \overline{QC}_y, \overline{QD}_y) \sum_{k'=0}^{l_{z3}+l_{z4}} f(k', l_{z3}, l_{z4}, \overline{QC}_z, \overline{QD}_z) \\
& \times \int\!\!\int d\tau_1\, d\tau_2\, x_{P_1}^{i}\, y_{P_1}^{j}\, z_{P_1}^{k}\, x_{Q_2}^{i'}\, y_{Q_2}^{j'}\, z_{Q_2}^{k'}\, (r_{12})^{-1} \exp(-\gamma_1\, r_{P_1}^2 - \gamma_2\, r_{Q_2}^2).
\end{aligned}
\tag{56}
$$

We recall that $\overline{PA}_x$, $\overline{PA}_y$, $\overline{PA}_z$ are the components of the vector $\vec{PA} \equiv \vec{P} - \vec{A}$, and that the f function comes from the binomial expansion as explained in Appendix B. Analogous definition hold for the vectors $\vec{PB}$, $\vec{QC}$, and $\vec{QD}$.

The subindex P_1 of the space coordinates and r^2 in Eq. (56) refers to the position of (dummy) electron 1 with respect to center P. Subindex Q_2 is defined in a similar way.

Eq. (56) allows the interpretation of the ERI as a Coulomb type interaction between two charge distributions: one centered on P (between A and B), and the other centered on Q (between C and D) as shown in Fig. 10.

The integral in Eq. (56) (which we will denote by I), can be computed either by using the Laplace transform or the Fourier transform of $(r_{12})^{-1}$. The latter is more general and is discussed in Appendix C.

The Fourier transform of $(r_{12})^{-1}$ is

$$
(r_{12})^{-1} = \left(\frac{1}{2\pi^2}\right) \int d\vec{k}\; k^{-2} \exp(\mathrm{i}\, \vec{r}_{P_1} \cdot \vec{k}) \exp(-\mathrm{i}\, \vec{r}_{Q_2} \cdot \vec{k}) \exp(\mathrm{i}\, \vec{PQ} \cdot \vec{k}),
$$

where

$$
\vec{r}_{12} = \vec{r}_1 - \vec{r}_2 = (\vec{r}_1 - \vec{P}) - (\vec{r}_2 - \vec{Q}) + (\vec{P} - \vec{Q}) = \vec{r}_{P_1} - \vec{r}_{Q_2} + \vec{PQ},
$$

and $\mathrm{i} = \sqrt{-1}$. (Beware of the difference between the complex number i and the index variable i.)

Substitution of this expression in the integral part (I) of Eq. (56) yields

$$
\begin{aligned}
\mathrm{I} &= \int\int d\tau_1\, d\tau_2\, x_{P_1}^{i}\, y_{P_1}^{j}\, z_{P_1}^{k}\, x_{Q_2}^{i'}\, y_{Q_2}^{j'}\, z_{Q_2}^{k'}\, (r_{12})^{-1} \exp(-\gamma_1\, r_{P_1}^2 - \gamma_2\, r_{Q_2}^2) \\
&= \left(\frac{1}{2\pi^2}\right) \int d\vec{k}\; k^{-2}\, I_x\, I_{x'}\, I_y\, I_{y'}\, I_z\, I_{z'} \exp(\mathrm{i}\, \vec{PQ} \cdot \vec{k}),
\end{aligned}
\tag{57}
$$

where

$$I_x = \int_{-\infty}^{\infty} dx_{P_1}\, x_{P_1}^{i} \exp(-\gamma_1\, x_{P_1}^2 + \mathrm{i}\, k_x\, x_{P_1}),$$

$$I_{x'} = \int_{-\infty}^{\infty} dx_{Q_2}\, x_{Q_2}^{i'} \exp(-\gamma_2\, x_{Q_2}^2 + \mathrm{i}\, k_x\, x_{Q_2}),$$

and similarly for I_y, $I_{y'}$, I_z and $I_{z'}$. We recall from Eq. (56) that the indices have the following ranges:

$$i = 0 \rightarrow l_{x1} + l_{x2}, \qquad i' = 0 \rightarrow l_{x3} + l_{x4}.$$

Using Eq. C3 and the definition of the Hermite polynomial of Appendix C, we can calculate the integrals of I_x and $I_{x'}$:

$$I_x = \mathrm{i}^{i} \sqrt{\pi}\; i!\, \exp(-k_x^2/4\gamma_1)\, \gamma_1^{-1/2} (4\gamma_1)^{-i/2} \sum_{r=0}^{\left[\frac{i}{2}\right]} \frac{(-1)^r \{2(4\gamma_1)^{-1/2}\}^{i-2r}\, k_x^{i-2r}}{r!\,(i-2r)!}, \tag{58a}$$

$$I_{x'} = \mathrm{i}^{i'} \sqrt{\pi}\; i'!\, \exp(-k_x^2/4\gamma_2)\, \gamma_2^{-1/2} (4\gamma_2)^{-i'/2} \sum_{r'=0}^{\left[\frac{i'}{2}\right]} \frac{(-1)^{r'} \{2(4\gamma_2)^{-1/2}\}^{i'-2r'}\, k_x^{i'-2r'}}{r'!\,(i'-2r')!}. \tag{58b}$$

The ranges of the indices are:

$$i = 0 \rightarrow l_{x1} + l_{x2}, \qquad i' = 0 \rightarrow l_{x3} + l_{x4},$$

$$r = 0 \rightarrow \left[\frac{i}{2}\right], \qquad r' = 0 \rightarrow \left[\frac{i'}{2}\right].$$

Recall that $[n]$ is the largest integer $\leq n$. Similar expressions can be found for I_y, $I_{y'}$, I_z and $I_{z'}$.

Substituting these results in I (Eq. 57) and using Eq. C4 of Appendix C, one obtains

$$\mathrm{I} = \delta \frac{1}{\pi^2} \int_0^1 ds\, s^{-3}\, J_x\, J_y\, J_z, \tag{59}$$

where

$$\delta = \frac{\gamma_1 + \gamma_2}{4\gamma_1\,\gamma_2},$$

and

$$\begin{aligned} J_x = {} & \pi\, \mathrm{i}^{i} \,(-\mathrm{i})^{i'} (\gamma_1\,\gamma_2)^{-1/2} (4\gamma_1)^{-i/2} (4\gamma_2)^{-i'/2}\, i!\, i'! \\ & \times \sum_{r=0}^{\left[\frac{i}{2}\right]} \sum_{r'=0}^{\left[\frac{i'}{2}\right]} \frac{(-1)^{r+r'} \{2\,(4\gamma_1)^{-1/2}\}^{i-2r} \{2\,(4\gamma_2)^{-1/2}\}^{i'-2r'}}{r!\, r'!\, (i-2r)!\, (i'-2r')!} \\ & \times \int_{-\infty}^{\infty} dk_x\, k_x^{i+i'-2r-2r'} \exp(-\delta\, s^{-2}\, k_x^2 + \mathrm{i}\, \overline{PQ}_x\, k_x), \end{aligned} \tag{60}$$

and similarly for J_y and J_z. $\overline{PQ}_x$ is the x component of $\vec{PQ} \equiv \vec{P} - \vec{Q}$.

Note that s is a dummy integration variable, introduced when an exponential form was converted into an integral (see Eq. C4).

The integration over the k-space can be accomplished using Eq. C3, so, for example, J_x becomes

$$\begin{aligned} J_x = {} & 2\pi^{3/2}\,(-1)^{i}\, i!\, i'!\, (\gamma_1 + \gamma_2)^{-1/2}\, s \sum_{r=0}^{\left[\frac{i}{2}\right]} \sum_{r'=0}^{\left[\frac{i'}{2}\right]} \frac{(4\gamma_1)^{r-i} (4\gamma_2)^{r'-i'} (i+i'-2r-2r')!}{r!\, r'!\, (i-2r)!\, (i'-2r')!} \\ & \times \sum_{u=0}^{[(i+i'-2r-2r')/2]} \frac{(-1)^{u}\, \overline{PQ}_x^{\,i+i'-2r-2r'-2u}}{u!\, (i+i'-2r-2r'-2u)!\, \delta^{i+i'-2r-2r'-u}} \\ & s^{2(i+i'-2r-2r'-u)} \exp\left(-\frac{\overline{PQ}^2}{4\delta}\, s^2\right). \end{aligned} \tag{61}$$

The indices i, i', r and r' have the same ranges as defined above. The new index u ranges from $0 \to \left[\frac{i+i'}{2} - r - r'\right]$. Similar expressions can be derived for J_y and J_z.

Substitution of Eq. (61) and its analogues into I (Eq. 59) and thereupon into Eq. (56) renders

$$ERI = \sum_{a,b,c,d} c_a\, c_b\, c_c\, c_d\, N(\alpha_1, \vec{l_1})\, N(\alpha_2, \vec{l_2})\, N(\alpha_3, \vec{l_3})\, N(\alpha_4, \vec{l_4})$$

$$\times \frac{2\pi^{5/2}}{\gamma_1\, \gamma_2\, (\gamma_1 + \gamma_2)^{1/2}} \exp(-\alpha_1\, \alpha_2\, \overline{AB}^2/\gamma_1) \exp(-\alpha_3\, \alpha_4\, \overline{CD}^2/\gamma_2) \tag{62}$$

$$\times \sum_{i,i',r,r',u} B_{i,i',r,r',u} \sum_{j,j',s,s',v} B_{j,j',s,s',v} \sum_{k,k',t,t',w} B_{k,k',t,t',w}\, F_\kappa\left(\frac{\overline{PQ}^2}{4\delta}\right),$$

where

$$\kappa = i + i' + j + j' + k + k' - 2(r + r' + s + s' + t + t') - u - v - w,$$

and $F_\kappa(t)$ is the incomplete Gamma function defined in Appendix D.

For the description of the B functions it is useful to introduce the following variables:

$$\Theta_x = f(i, l_{x1}, l_{x2}, \overline{PA}_x, \overline{PB}_x) \frac{i!\, \gamma_1^{r-i}}{r!\, (i - 2r)!}, \tag{63a}$$

and

$$\Theta_{x'} = f(i', l_{x3}, l_{x4}, \overline{QC}_x, \overline{QD}_x) \frac{i'!\, \gamma_2^{r'-i'}}{r'!\, (i' - 2r')!}, \tag{63b}$$

where f is the same function as used before (Eq. B4). The y, y', z and z' components are likewise defined.

So, for example, $B_{i,i',r,r',u}$ is given by Eq. (64) given immediately below,

$$B_{i,i',r,r',u} = (-1)^{i'} \Theta_x \Theta_{x'} \frac{(2\,\delta)^{2(r+r')} (i + i' - 2r - 2r')!\, (-1)^u\, \delta^u\, \overline{PQ}_x^{\,i+i'-2r-2r'-2u}}{(4\delta)^{i+i'} u!\, (i + i' - 2r - 2r' - 2u)!}.$$

$B_{j,j',s,s',v}$ and $B_{k,k',t,t',w}$ are obtained by substituting the corresponding indices in Eq. (64).

The summations in Eq. (62) are over the following ranges

$$i = 0 \rightarrow l_{x1} + l_{x2},$$

$$i' = 0 \rightarrow l_{x3} + l_{x4},$$

$$r = 0 \rightarrow \left[\frac{i}{2}\right],$$

$$r' = 0 \rightarrow \left[\frac{i'}{2}\right],$$

$$u = 0 \rightarrow \left[\frac{i+i'}{2} - r - r'\right].$$

The summation indices over the other coordinates have similar ranges.

As in the previous integral equations, we repeat the two-electron integral expressions, this time in terms of primitive GTFs:

$$\begin{aligned}&< (A,\alpha_1, l_1, m_1, n_1),\ (B, \alpha_2, l_2, m_2, n_2) \mid \frac{1}{r_{12}} \mid (C, \alpha_3, l_3, m_3, n_3)\ (D, \alpha_4, l_4, m_4, n_4) > \\ &= N_1 N_2 N_3 N_4 \frac{2\pi^2}{\gamma_1\gamma_2} \left(\frac{\pi}{\gamma_1+\gamma_2}\right)^{1/2} \times \exp\left(- \frac{\alpha_1\alpha_2}{\alpha_1+\alpha_2} \overline{AB}^2 - \frac{\alpha_3\alpha_4}{\alpha_3+\alpha_4} \overline{CD}^2 \right) \\ &\sum_{i_1, i_2, r_1, r_2, u} B_{i_1, i_2, r_1, r_2, u} \sum_{j_1, j_2, s_1, s_2, v} B_{j_1, j_2, s_1, s_2, v} \\ &\sum_{k_1, k_2, t_1, t_2, w} B_{k_1, k_2, t_1, t_2, w}\ F_\kappa\left(\frac{\overline{PQ}^2}{4\delta}\right)\end{aligned} \tag{65}$$

where

$$\begin{aligned}&\gamma_1 = \alpha_1 + \alpha_2;\ \ \gamma_2 = \alpha_3 + \alpha_4;\ \ \delta = \frac{1}{4}\left(\frac{1}{\gamma_1} + \frac{1}{\gamma_2}\right); \\ &P_i = \frac{\alpha_1 A_i + \alpha_2 B_i}{\alpha_3 + \alpha_4};\ \ Q_i = \frac{\alpha_3 C_i + \alpha_4 D_i}{\alpha_3 + \alpha_4};\ i = x, y, z; \\ &\kappa = i_1 + i_2 + j_1 + j_2 + k_1 + k_2 - 2(r_1 + r_2 + s_1 + s_2 + t_1 + t_2) - u - v - w;\end{aligned} \tag{66}$$

and

$$\begin{aligned}B_{i_1, i_2, r_1, r_2, u} &= B_{i_1, i_2, r_1, r_2, u} \times (l_1, l_2, A_x, B_x, P_x, \gamma_1 \mid l_3, l_4, C_x, D_x, Q_x, \gamma_2) \\ &= (-1)^{i_2} f(i_1, l_1, l_2, \overline{PA}_x, \overline{PB}_x) f(i_2, l_3, l_4, \overline{QC}_x, \overline{QD}_x) \\ &\times \frac{i_1!\, i_2!\, \gamma_1^{r_1}\, \gamma_2^{r_2}\, (2\delta)^{2(r_1+r_2)}}{\gamma_1^{i_1}\, \gamma_2^{i_2}\, (4\delta)^{i_1+i_2}\, r_1!\, r_2!\, (i_1 - 2r_1)!\, (i_2 - 2r_2)!} \\ &\times \frac{[i_1 + i_2 - 2r_1 - 2r_2]!\, (-1)^u\, \overline{PQ}_x^{\,i_1+i_2-2(r_1+r_2+u)}\, \delta^u}{u!\, [i_1 + i_2 - 2(r_1 + r_2 + u)]!}\end{aligned} \tag{67}$$

$B_{j_1, j_2, s_1, s_2, t}$ and $B_{k_1, k_2, t_1, t_2, w}$ are similarly defined in terms of the y- and z-components respectively.

The summations in Eq. (65) are over the following ranges:

$$i_1 = 0 \rightarrow l_1 + l_2$$

$$i_2 = 0 \rightarrow l_3 + l_4$$

$$r_1 = 0 \rightarrow [i_1/2]$$

$$r_2 = 0 \rightarrow [i_2/2]$$

$$u = 0 \rightarrow \left[\frac{i_1 + i_2}{2} - r_1 - r_2\right]$$

and similarly for the indices of the summations over the y- and z-components.

13. Special Formulas for Integrals Involving s- and p-Type Functions

In the formulas of this section, the subscripts i, j, k, and l define the axis of a p-type orbital function, and each is to be replaced with the value x or y or z as appropriate. These expressions were used in the first version of IBMOL (1965) to speed up the integral evaluation. It is rather obvious that the "special formulae" code is much simpler than the general one. It should also be obvious that the special formulae, if extended to d and f functions, would become a truly cumbersome task of algebra first and coding later. For large molecules the next speed-up is obtained by separating each of the expressions given below into one-, two-, three-, and four-center expressions, which was done in IBMOL-6.

The Overlap Integral.

$$S^{oo}(a, b) = < S_a \mid S_b > = N_a N_b \left(\frac{\pi}{a+b}\right)^{3/2} \exp\left(- \frac{ab}{a+b} \overline{AB}^2\right)$$

$$S^{io}(a, b) = < P_{ia} \mid S_b > = - \frac{b}{a+b} (A_i - B_i) S^{oo}(a, b)$$

$$S^{oj}(a, b) = < S_a \mid P_{jb} > = \frac{a}{a+b} (A_j - B_j) S^{oo}(a, b).$$

$$S^{ij}(a, b) = < P_{ia} \mid P_{jb} > = \left\{ \frac{1}{2}(a+b)\, \delta_{ij} - \frac{ab}{(a+b)^2} (A_i - B_i)\,(A_j - B_j) \right\} S^{oo}(a, b)$$

Here, as throughout this section, δ_{ij} is the Kronecker delta.

The Kinetic Integral. We first define a set of functions:

$$K^{oo}(a, b) = \frac{3ab}{a+b} - \frac{2a^2b^2}{(a+b)^2} \overline{AB}^2$$

$$K^{io}(a, b) = -\frac{2ab^2}{(a+b)^2}(A_i - B_i)$$

$$K^{oj}(a, b) = \frac{2a^2b}{(a+b)^2}(A_j - B_j)$$

$$K^{ij}(a, b) = \frac{ab}{(a+b)^2}\delta_{ij}.$$

In terms of these K-functions and the S-functions given above, the kinetic integrals are defined as follows:

$$T^{oo}(a, b) = < S_a \mid -\frac{1}{2}\Delta \mid S_b > = K^{oo}(a, b)S^{oo}(a, b)$$

$$T^{io}(a, b) = < P_{ia} \mid -\frac{1}{2}\Delta \mid S_b > = K^{io}(a, b)S^{oo}(a, b) + K^{oo}(a, b)S^{io}(a, b)$$

$$T^{oj}(a, b) = < S_a \mid -\frac{1}{2}\Delta \mid P_{jb} > = K^{oj}(a, b)S^{oo}(a, b) + K^{oo}(a, b)S^{oj}(a, b)$$

$$T^{ij}(a, b) = < P_{ia} \mid -\frac{1}{2}\Delta \mid P_{jb} > = K^{ij}(a, b)S^{oo}(a, b) + K^{io}(a, b)S^{oj}(a, b) + K^{oj}(a, b)S^{io}(a, b) + K^{oo}(a, b)S^{ij}(a, b).$$

The Nuclear Attraction Integral. Again we define a set of intermediate functions. These use the Incomplete Gamma Function defined in Appendix D. In each case the argument, $t = (a + b)\,\overline{PC}^2$.

$$L^{oo}(c;a,b) = F_o(t)$$

$$L^{io}(c;a,b) = L^{oi}(c;a,b) = (C_i - P_i)F_i(t)$$

$$L^{ij}(c;a,b) = (P_i - C_i)(P_j - C_j)F_2(t) - \frac{1}{2(a+b)}\delta_{ij}\,F_1(t).$$

The nuclear attraction integrals may now be defined:

$$V^{oo}(a, b) = < S_a \mid \frac{1}{r_c} \mid S_b > = \Theta \sum_c S^{oo}(a, b)L^{oo}(c;a,b)$$

$$V^{io}(a, b) = < P_{ia} \mid \frac{1}{r_c} \mid S_b > = \Theta \sum_c \{S^{io}(a,b)L^{oo}(c;a,b) + S^{oo}(a,b)L^{io}(c;a,b)\}$$

$$V^{oj}(a, b) = < S_a \mid \frac{1}{r_c} \mid P_{jb} > = \Theta \sum_c \{S^{oj}(a,b)L^{oo}(c;a,b) + S^{oo}(a,b)L^{oj}(c;a,b)\}$$

$$V^{ij}(a,b) = < P_{ia} \mid \frac{1}{r_c} \mid P_{jb} > = \Theta \sum_c \{S^{ij}(a,b)L^{oo}(c;a,b) + S^{io}(a,b)L^{oj}(c;a,b)$$
$$+ S^{oj}(a,b)L^{io}(c;a,b) + S^{oo}(a,b)L^{ij}(c;a,b)\},$$

where $\Theta = \frac{2}{\pi^{1/2}} (a+b)^{1/2}$.

Note that every additive term of these formulas contains either explicitly or implicitly the term $S^{00}(a,b)$. Thus all normalization factors are included by the definition of this term. This applies also to the formulas for the kinetic integrals above.

The Electron Repulsion Integral. In this case also the special formulas are simplified by first defining a set of intermediate functions. In these formulas the centers P and Q are previously defined. The argument of the Incomplete Gamma Function in these formulas is

$$t = \frac{(a+b)(c+d)}{(a+b+c+d)} \overline{PQ}^2.$$

To abbreviate these formulas, the following conventions are used throughout:

$S_1 = a + b;\quad S_2 = c + d;\quad S_4 = S_1 + S_2 = a + b + c + d.$

The intermediate functions are defined as:

$$G^{oooo}(t) = F_o(t)$$

$$G^{iooo}(t) = G^{oioo}(t) = \frac{S_2}{S_4} (P_i - Q_i)\, F_1(t)$$

$$G^{ooio}(t) = G^{oooi}(t) = \frac{S_1}{S_4} (P_i - Q_i)\, F_1(t)$$

$$G^{ijoo}(t) = \frac{S_2}{S_4} \left\{ \frac{S_2}{S_4} (P_i - Q_i)(P_j - Q_j)F_2(t) - \delta_{ij} \frac{1}{2S_1} F_1(t) \right\}$$

$$G^{ooij}(t) = \frac{S_1}{S_4} \left\{ \frac{S_1}{S_4} (P_i - Q_i)(P_j - Q_j)F_2(t) - \delta_{ij} \frac{1}{2S_2} F_1(t) \right\}$$

$$G^{iojo}(t) = G^{oijo}(t) = G^{oioj}(t) = G^{iooj}(t)$$
$$= -\frac{1}{S_4} \left\{ \frac{S_1 S_2}{S_4} (P_i - Q_i)(P_j - Q_j)F_2(t) - \frac{1}{2}\delta_{ij}F_1(t) \right\}$$

$$G^{ijko}(t) = G^{ijok}(t) = \frac{S_2}{S_4^2}\left\{\frac{S_1S_2}{S_4}(P_i - Q_i)(P_j - Q_j)(P_k - Q_k)F_3(t)\right.$$

$$\left. - \frac{1}{2}[\delta_{ij}(P_k - Q_k) + \delta_{ik}(P_j - Q_j) + \delta_{jk}(P_i - Q_i)]\, F_2(t)\right\}$$

$$G^{oijk}(t) = G^{iojk}(t) = \frac{-S_1}{S_4^2}\{(P_i - Q_i)(P_j - Q_j)(P_k - Q_k)F_3(t)$$

$$ - \frac{1}{2}[\delta_{ij}(P_k - Q_k) + \delta_{ik}(P_j - Q_j) + \delta_{jk}(P_i - Q_i)]F_2(t)\}$$

$$G^{ijkl}(t) = \frac{1}{S_4^2}\left\{\frac{S_1^2S_2^2}{S_4^2}(P_i - Q_i)(P_j - Q_j)(P_k - Q_k)(P_l - Q_l)F_4(t)\right.$$

$$ - \frac{1}{2}\frac{S_1S_2}{S_4}\left[\delta_{ij}(P_k - Q_k)(P_l - Q_l) + \delta_{ik}(P_j - Q_j)(P_l - Q_l)\right.$$

$$ + \delta_{il}(P_j - Q_j)(P_k - Q_k) + \delta_{ik}(P_i - Q_i)(P_l - Q_l)$$

$$\left. + \delta_{jl}(P_i - Q_i)(P_k - Q_k) + \delta_{kl}(P_i - Q_i)(P_j - Q_j)\right] F_3(t)$$

$$\left. + \frac{1}{4}[\delta_{ij}\delta_{kl} + \delta_{ik}\delta_{jl} + \delta_{il}\delta_{jk}]\, F_2(t)\right\}.$$

The electron repulsion integrals can now be defined in terms of the G-functions and the S-functions, normalization factors again being included by the explicit or implicit use of the product $S^{00}(a,b)\ S^{00}(c,d)$. An abbreviated notation will be introduced whose meaning will be apparent to the reader. The multiplicative term Λ appearing in each formula is defined as:

$$\Lambda = 2\left(\frac{S_1S_2}{\pi S_4}\right)^{1/2}$$

$$< S_aS_b \mid S_cS_d > = < S_aS_b \mid \frac{1}{r_{12}} \mid S_cS_d > = \Lambda S^{oo}(a,b)S^{oo}(c,d)G^{oooo}(t)$$

$$= \Lambda S_{ab}^{oo}S_{cd}^{oo}G^{oooo}$$

$$< P_{ia}S_b \mid S_cS_d > = \Lambda\left\{S_{ab}^{io}S_{cd}^{oo}G^{oooo} + S_{ab}^{oo}S_{cd}^{oo}G^{iooo}\right\}$$

$$< P_{ia}S_b \mid P_{kc}S_d > = \Lambda\left\{S_{ab}^{io}[S_{cd}^{ko}G^{oooo} + S_{cd}^{oo}G^{ooko}] + S_{ab}^{oo}[S_{cd}^{oo}G^{ioko} + S_{cd}^{ko}G^{iooo}]\right\}$$

$$< P_{ia}P_{jb} \mid S_cS_d > = \Lambda\left\{S_{ab}^{ij}S_{cd}^{oo}G^{oooo} + S_{ab}^{io}S_{cd}^{oo}G^{ojoo}\right.$$

$$\left. + S_{ab}^{oj}S_{cd}^{oo}G^{iooo} + S_{ab}^{oo}S_{cd}^{oo}G^{ijoo}\right\}$$

$$< P_{ia}F_{jb} \mid P_{kc}S_d > \; = \Lambda \{ S_{ab}^{ij} [S_{cd}^{ko} G^{oooo} + S_{cd}^{oo} G^{ooko}]$$
$$+ S_{ab}^{io} [S_{cd}^{ko} G^{ojoo} + S_{cd}^{oo} G^{ojko}] + S_{ab}^{oj} [S_{cd}^{ko} G^{iooo}$$
$$+ S_{cd}^{oo} G^{ioko}] + S_{ab}^{oo} [S_{cd}^{ko} G^{ijko}] \}$$

$$< P_{ia}P_{jb} \mid P_{kc}P_{ld} > \; = \Lambda \{ S_{ab}^{ij} [S_{cd}^{kl} G^{oooo} + S_{cd}^{ko} G^{oool}$$
$$+ S_{cd}^{ol} G^{ooko} + S_{cd}^{oo} G^{ookl}] + S_{ab}^{io} [S_{cd}^{kl} G^{ojoo} + S_{cd}^{ko} G^{ojol}$$
$$+ S_{cd}^{ol} G^{ojko} + S_{cd}^{oo} G^{ojkl}] + S_{ab}^{oj} [S_{cd}^{kl} G^{iooo} + S_{cd}^{ko} G^{iool}$$
$$+ S_{cd}^{ol} G^{ioko} + S_{cd}^{oo} G^{iokl}] + S_{ab}^{oo} [S_{cd}^{kl} G^{ijoo} + S_{cd}^{ko} G^{ijol}$$
$$+ S_{cd}^{ol} G^{ijko} + S_{cd}^{oo} G^{ijkl}] \}$$

It should be obvious from the above equations that precomputation of the S and G terms will result in a drastic saving of computer time, especially for geometrical basis sets.

14. Two-Electron Integrals with Explicit Factorization of the One-Electron Dependent Terms

Equation (62) is a general formula for the ERI over GTF primitives. It has been used in this form in the early IBMOL programs. Posterior version of IBMOL[10] rearranged the formula in order to take advantage of the fact that some factors are common to several integrals, so their computation need not be repeated.

A major modification was introduced by Habitz and Clementi[5] in which the idea of considering the ERI as the interaction of two charge distributions[44] was exploited by precomputing two files which are dependent on either electron 1 or 2 only and which contain most of the terms necessary to complete the integration over the radial part of the integral. In this way the only step with a fourth power dependency on the size of the basis set is the merging of the files, together with the multiplication with the appropriate elements of the Incomplete Gamma Function. This program structure allows, at an early stage, checks on the size of the integrals so that very small contributions can be neglected by not storing them. Also, the densities can be used as an estimate for the upper limit of the value of the integral so that it can be decided in advance, whether or not an integral is to be computed, using as criterion a preassigned threshold value.

The formulae are modified in order to keep, as long as possible, the terms concerning electron 1 separated from those concerning electron 2 in such a way that two files (tables $T_1(\vec{n})$ and $T_2(\vec{n}')$) containing the information about each electron can be built up:

$$T_1(\vec{n}) = X_1(n_x)Y_1(n_y)Z_1(n_z)\, c_a\, c_b\, N(\alpha_1, \vec{l_1})\, N(\alpha_2, \vec{l_2})\, \gamma_1^{-1}(4\pi^5)^{1/4} \times \exp(-\alpha_1\alpha_2\overline{AB}^2/\gamma_1), \tag{68}$$

$$T_2(\vec{n}') = X_2(n_x')Y_2(n_y')Z_2(n_z')\, c_c\, c_d\, N(\alpha_3, \vec{l_3})\, N(\alpha_4, \vec{l_4})\, \gamma_2^{-1}(4\pi^5)^{1/4} \times \exp(-\alpha_3\alpha_4\overline{CD}^2/\gamma_2), \tag{69}$$

where $\vec{n} = (n_x, n_y, n_z)$, $\vec{n}' = (n_x', n_y', n_z')$, and

$$X_1(n_x) = \sum_{i=0}^{l_{x1}+l_{x2}} (-1)^i i!\, f(i, l_{x1}, l_{x2}, \overline{PA}_x, \overline{PB}_x) \sum_{r=0}^{\left[\frac{i}{2}\right]} \frac{1}{r!\,(i-2r)!\,(4\gamma_1)^{i-r}}, \tag{70}$$

$$X_2(n_x') = \sum_{i'=0}^{l_{x3}+l_{x4}} (+1)^{i'} i'!\, f(i', l_{x3}, l_{x4}, \overline{QC}_x, \overline{QD}_x) \sum_{r'=0}^{\left[\frac{i'}{2}\right]} \frac{1}{r'!\,(i'-2r')!\,(4\gamma_2)^{i'-r'}} \tag{71}$$

The summations in Eqs. (70) and (71) are limited to all i and r (i' and r') combinations which lead to the same value of n_x (n_x') under the condition $n_x = i - 2r$ ($n_x' = i' - 2r'$). The Y and Z parts are expressed in a similar way.

The merging of the two tables is the actual integral calculation:

$$ERI = \sum_{a,b,c,d} \sum_{\vec{n}} T_1(\vec{n}) \sum_{\vec{n}'} T_2(\vec{n}')\, H(\vec{n} + \vec{n}'), \tag{72}$$

where summations have the following ranges

$$n_x = 0 \rightarrow l_{x1} + l_{x2}, \quad n'_x = 0 \rightarrow l_{x3} + l_{x4},$$

and similarly for the y and z indices. Indices a, b, c and d are defined as in Eq. (51).

The H function has the fourth power dependency because it contains elements of both electrons 1 and 2.

$$H(\vec{m}) = \sum_{u=0}^{\left[\frac{m_x}{2}\right]} h_x(u, m_x) \sum_{v=0}^{\left[\frac{m_y}{2}\right]} h_y(v, m_y) \sum_{w=0}^{\left[\frac{m_z}{2}\right]} h_z(w, m_z)\, G(\kappa), \tag{73}$$

with

$$\kappa = m_x + m_y + m_z - u - v - w, \quad \vec{m} \equiv (m_x, m_y, m_z),$$

where $m_x = n_x + n_x'$, $m_y = n_y + n_y'$ and $m_z = n_z + n_z'$,

$$G(\kappa) = \delta^{-\kappa} \sqrt{\frac{1}{\gamma_1 + \gamma_2}}\; F_\kappa\!\left(\frac{\overline{PQ}^2}{4\delta}\right), \tag{74}$$

$$h_x(u, m_x) = (-1)^u\, m_x!\, [u!\, (m_x - 2u)!]^{-1}\, \overline{PQ}_x^{\,m_x - 2u} \tag{75}$$

and similarly for h_y and h_z.

Note that the indices u, v, w have the same meaning (and ranges) as the indices with the same name in Eq. (62). This implies that $m_x = i + i' - 2r - 2r'$ and similarly for the other two components.

15. Geometrical Basis Sets and Their Use

Geometrical basis sets were introduced in order to make feasible molecular computations for large chemical systems (see Chapter 2, section 5).

As we will explain below, all the exponents for a geometrical basis set are functions of two parameters, α_0 and $\mathbf{C}$. Moreover, the exponents in a geometrical basis set have a first order dependency in the $\mathbf{C}$ constant and a second order dependency in the α_0 parameter. These facts offer a realistic possibility to perform orbital exponent optimization for molecular systems. In the KGNMOL program we have the possibility to perform automatic orbital exponent optimization.

Let us consider an *atomic* function with orbital symmetries denoted as 1, ..., λ, ..., μ. We define a reference set

$$\{\alpha\}_{\mathbf{R}} = \alpha(1), \;\ldots,\; \alpha(i), \;\ldots,\; \alpha(f), \;\ldots,\; \alpha(N)$$

using the following relation:

$$\alpha_j = \alpha(j) = \alpha_0 \mathbf{C}^{j-1}, \quad j = 1 \rightarrow N, \tag{76}$$

where α_0 and $\mathbf{C}$ are real numbers.

From this definition, it follows that any pair of α's are related in such a way that we can write each one of them in terms of the other. Defining $d_{i,j} = i - j$ and using Eq. (76) we obtain

$$\begin{aligned} \alpha_i &= \alpha_0 \mathbf{C}^{i-1} = \alpha_0 \mathbf{C}^{d_{i,j} + j - 1} = \alpha_0 \mathbf{C}^{j-1} \mathbf{C}^{d_{i,j}}, \\ \alpha_j &= \alpha_0 \mathbf{C}^{j-1} = \alpha_0 \mathbf{C}^{d_{j,i} + i - 1} = \alpha_0 \mathbf{C}^{i-1} \mathbf{C}^{d_{j,i}}, \end{aligned} \tag{77}$$

or

$$\alpha_i = \alpha_j \mathbf{C}^{d_{i,j}} = \alpha_j \mathbf{C}^{d_{j,i}},$$
$$\alpha_j = \alpha_i \mathbf{C}^{d_{j,i}} = \alpha_i \mathbf{C}^{d_{i,j}}. \tag{78}$$

A geometrical basis set allows us to have extensive pre-calculations during the computation of the integrals because of two important facts:

- All the orbital exponents of primitives in different atoms belong to the same reference set of orbitals $\{\boldsymbol{\alpha}\}_{\mathbf{R}}$.
- Given an atom, all the different symmetry sets of orbital exponents must satisfy

$$\{\boldsymbol{\alpha}\}_{\mathbf{R}} \supset \{\boldsymbol{\alpha}\}_1 \supset \{\boldsymbol{\alpha}\}_2 \supset \ldots \supset \{\boldsymbol{\alpha}\}_\lambda \supset \ldots \supset \{\boldsymbol{\alpha}\}_\mu, \tag{79}$$

where each of these sets is an increasing ordered set referenced by a set of successive integers $\{\boldsymbol{\iota}\}_\lambda$. Among these sets the same contention scheme occurs:

$$\{\boldsymbol{\iota}\}_{\mathbf{R}} \supset \{\boldsymbol{\iota}\}_1 \supset \{\boldsymbol{\iota}\}_2 \supset \ldots \supset \{\boldsymbol{\iota}\}_\lambda \supset \ldots \supset \{\boldsymbol{\iota}\}_\mu . \tag{80}$$

This constraint is graphically shown in Fig. 11.

In the computation of the two-electron-integrals the Gaussian Product Theorem is applied (Appendix B). It is very simple to show that for a geometrical basis set of size N and with parameters α_0 and $\mathbf{C}$ we can write

$$\gamma_1 = \alpha_1 + \alpha_2 = \alpha_0 \mathbf{C}^{i_1 - 1}(1 + \mathbf{C}^{d_{i_2, i_1}}) = \alpha_1(1 + \mathbf{C}^{d_{i_2, i_1}}) \tag{81}$$

$$\vec{P} = \frac{\vec{A} + \mathbf{C}^{d_{i_2, i_1}} \vec{B}}{1 + \mathbf{C}^{d_{i_2, i_1}}} \tag{82}$$

$$\vec{PA} = - \frac{\mathbf{C}^{d_{i_2, i_1}} \vec{AB}}{1 + \mathbf{C}^{d_{i_2, i_1}}} \tag{83a}$$

and

$$\vec{PB} = \frac{\vec{AB}}{1 + \mathbf{C}^{d_{i_2, i_1}}} \tag{83b}$$

From the above expressions we can easily obtain a very useful relation between the two vectors $\vec{PA}$ and $\vec{PB}$:

$$\vec{PA} = - \mathbf{C}^{d_{i_2, i_1}} \vec{PB}, \tag{84}$$

thus,

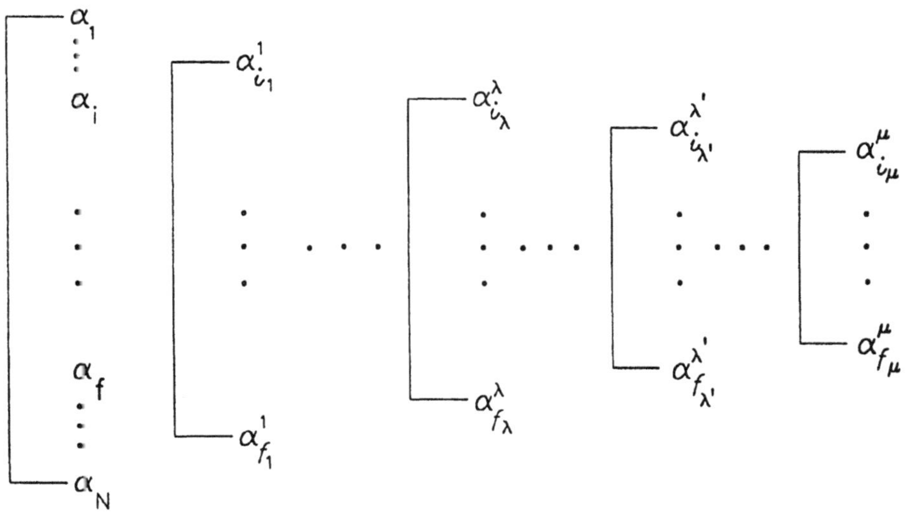

Figure 11. Contention Scheme among the $\{\alpha\}_\lambda$, with λ from 1 to μ. $\alpha^\lambda_{i_\lambda} \equiv$ the lowest orbital exponent that belongs to $\{\alpha\}_\lambda$, and $\alpha^\lambda_{f_\lambda} \equiv$ the highest orbital exponent that belongs to $\{\alpha\}_\lambda$; i_λ and f_λ stand for the initial and final indices of the set $\{\alpha\}_\lambda$, respectively.

$$\frac{\overline{PA_x}}{\overline{PB_x}} = \frac{\overline{PA_y}}{\overline{PB_y}} = \frac{\overline{PA_z}}{\overline{PB_z}} = -\mathbf{C}^{d_{i_2,i_1}} \tag{85}$$

Placing the cartesian origin of axis in P, i.e., for $\vec{P} = (0,\ 0,\ 0)$, the ratios

$$\frac{A_x}{B_x} = \frac{A_y}{B_y} = \frac{A_z}{B_z} = -\mathbf{C}^{d_{i_2,i_1}} \tag{86}$$

hold. Note that these ratios have the same value for all pairs of exponents having the same size of the index interval.

Let us assume that i and j are indices of exponents of functions centered on A and B, respectively. Eq. (82) can be generalized for an arbitrary pair of i and j:

$$\vec{P}_{\alpha_i,\alpha_j} = \frac{\vec{A} + \mathbf{C}^{d_{j,i}}\vec{B}}{1 + \mathbf{C}^{d_{j,i}}} \quad \text{if } j \geq i \text{ and} \tag{87a}$$

$$\vec{P}_{\alpha_i,\alpha_j} = \frac{\vec{B} + \mathbf{C}^{d_{i,j}}\vec{A}}{1 + \mathbf{C}^{d_{i,j}}} \quad \text{if } j < i. \tag{87b}$$

Let us consider the case in which the function centered on B has the largest exponent, i.e., $j \geq i$. The following relations can be set:

$$\begin{aligned}
&\vec{P}_{\alpha_1,\alpha_1} = \vec{P}_{\alpha_2,\alpha_2} = \quad \ldots \quad \ldots \quad \ldots \quad = \vec{P}_{\alpha_N,\alpha_N} = \frac{\vec{A}+\vec{B}}{2}\\
&\vec{P}_{\alpha_1,\alpha_2} = \vec{P}_{\alpha_2,\alpha_3} = \quad \ldots \quad \ldots \quad = \vec{P}_{\alpha_{N-1},\alpha_N} = \frac{\vec{A}+\mathbf{C}\vec{B}}{1+\mathbf{C}}\\
&\quad \vdots \\
&\vec{P}_{\alpha_i,\alpha_j} = \vec{P}_{\alpha_{i+1},\alpha_{j+1}} = \quad \ldots \quad = \vec{P}_{\alpha_{N-(j-i)},\alpha_N} = \frac{\vec{A}+\mathbf{C}^{d_{j,i}}\vec{B}}{1+\mathbf{C}^{d_{j,i}}}\\
&\quad \vdots \\
&\vec{P}_{\alpha_1,\alpha_N} = \frac{\vec{A}+\mathbf{C}^{d_{N,1}}\vec{B}}{1+\mathbf{C}^{d_{N,1}}}
\end{aligned} \tag{88}$$

A similar scheme can be found when the largest exponent function is located on A ($j < i$). In this case, Eq. (87b) should be used.

We remark that all elements of the diagonal (top line) have the same value of $\vec{P}$. The same holds for the elements of a given co-diagonal. Consequently, in order to have all possible values of $\vec{P}$ it is only necessary to calculated the elements of the first row and the first column of the diagram. Hence, instead of calculating $N_A \times N_B$ elements, only $N_A + N_B - 1$ are required.

The function f, which is part of the linear coefficients in Eq. B4 can now be expressed as:

$$f(i,l_{x1},l_{x2},\overline{PA}_x,\overline{PB}_x) = \sum_{j=\max(0,i-l_{x2})}^{\min(i,l_{x1})} \frac{l_{x1}!\,l_{x2}!(-\mathbf{C}^{d_{i_2,i_1}}\overline{PB}_x)^{l_{x1}-j}\,\overline{PB}_x^{l_{x2}-i+j}}{j!\,(l_{x1}-j)!\,(i-j)!\,(l_{x2}-i+j)!}$$

$$f(i,l_{x1},l_{x2},\overline{PA}_x,\overline{PB}_x) = \overline{PB}_x^{l_{x1}+l_{x2}-i} \sum_{j=\max(0,i-l_{x2})}^{\min(i,l_{x1})} \frac{l_{x1}!\,l_{x2}!(-\mathbf{C}^{d_{i_2,i_1}})^{l_{x1}-j}}{j!\,(l_{x1}-j)!\,(i-j)!\,(l_{x2}-i+j)!}$$

and using Eqs. (83)

$$\begin{aligned}
f(i,l_{x1},l_{x2},\overline{PA}_x,\overline{PB}_x) = &\left(\frac{\overline{AB}_x}{1+\mathbf{C}^{d_{i_2,i_1}}}\right)^{l_{x1}+l_{x2}-i}\\
&\sum_{j=\max(0,i-l_{x2})}^{\min(i,l_{x1})} \frac{l_{x1}!\,l_{x2}!\,(-\mathbf{C}^{d_{i_2,i_1}})^{l_{x1}-j}}{j!\,(l_{x1}-j)!\,(i-j)!\,(l_{x2}-i+j)!}
\end{aligned} \tag{89}$$

Therefore, for fixed centers A and B the value of the function f depends upon the vectors $\vec{l_1}$ and $\vec{l_2}$ and d_{i_2, i_1}. Hence, for a given combination of vectors $\vec{l_1}$ and $\vec{l_2}$, the same set of relations as described in Eq. (88) can be deduced for f.

Let us point out some of the computational savings resulting from the use of the geometrical basis set. Let us start with Eq. (51) and four primitive functions with exponents α_1, α_2, α_3 and α_4, centered in atoms A, B, C and D, respectively.

Using a geometrical basis set with parameters α_0 and **C**, we can set

$$\alpha_1 \in \{\boldsymbol{\alpha}\}_{l_A}, \quad \alpha_2 \in \{\boldsymbol{\alpha}\}_{l_B}, \quad \alpha_3 \in \{\boldsymbol{\alpha}\}_{l_C}, \quad \alpha_4 \in \{\boldsymbol{\alpha}\}_{l_D},$$

where l_I ($I = A, B, C$ or D) stands for the orbital symmetry l in center I.

Using Eq. (86), the exponents α_1, α_2, α_3, α_4 can be written as

$$\alpha_1 = \alpha_0 \mathbf{C}^{i_1 - 1}, \quad \alpha_2 = \alpha_0 \mathbf{C}^{i_2 - 1}, \quad \alpha_3 = \alpha_0 \mathbf{C}^{i_3 - 1}, \quad \alpha_4 = \alpha_0 \mathbf{C}^{i_4 - 1}, \tag{90}$$

where i_1, i_2, i_3, i_4 $\in \{\boldsymbol{\iota}\}_{\mathbf{R}}$. These four indices can be ordered. Let us assume that they follow $i_1 \leq i_2$ and $i_3 \leq i_4$.

According to definitions (81) and (82), γ_1, γ_2, $\vec{P}$ and $\vec{Q}$ can be written as:

$$\gamma_1 = \alpha_1 + \alpha_2 = \alpha_0 \mathbf{C}^{i_1 - 1}(1 + \mathbf{C}^{d_{i_2, i_1}}) = \alpha_1(1 + \mathbf{C}^{d_{i_2, i_1}}) \tag{91a}$$

$$\gamma_2 = \alpha_3 + \alpha_4 = \alpha_0 \mathbf{C}^{i_3 - 1}(1 + \mathbf{C}^{d_{i_4, i_3}}) = \alpha_3(1 + \mathbf{C}^{d_{i_4, i_3}}) \tag{91b}$$

$$\vec{P} = \frac{\vec{A} + \mathbf{C}^{d_{i_2, i_1}} \vec{B}}{1 + \mathbf{C}^{d_{i_2, i_1}}} \tag{92a}$$

$$\vec{Q} = \frac{\vec{C} + \mathbf{C}^{d_{i_4, i_3}} \vec{D}}{1 + \mathbf{C}^{d_{i_4, i_3}}}. \tag{92b}$$

The formulae for ERI presented in the last section seem especially well suited to be expressed in terms of geometrical basis set functions. For given pairs of exponents (α_1, α_2 and α_3, α_4), there are two major levels in which extensive pre-computations can be performed during the evaluation of Eq. (72) with the use of geometrical basis sets. The first level occurs while generating the density dependent files, Eq. (68) and (69), and the second one is during the merge of these two files, Eq. (72).

First Level of Pre-tabulations. With the use of the Eqs. (91a) to (92b), $T_1(\vec{n})$ and $T_2(\vec{n}')$, Eqs. (68) and (69) can be expressed as:

$$T_1(\vec{n}) = X_1(n_x) Y_1(n_y) Z_1(n_z)\, c_a\, c_b\, N(\alpha_1, \vec{l_1})\, N(\alpha_2, \vec{l_2}) \times \frac{(4\pi^5)^{1/4}}{\alpha_0 C^{i_1-1}(1 + C^{d_{i_2, i_1}})} \exp\left(-\frac{\alpha_0 C^{i_1-1} C^{d_{i_2, i_1}} \overline{AB}^2}{1 + C^{d_{i_2, i_1}}}\right) \quad (93a)$$

and

$$T_2(\vec{n}') = X_2(n_x') Y_2(n_y') Z_2(n_z')\, c_c\, c_d\, N(\alpha_3, \vec{l_3})\, N(\alpha_4, \vec{l_4}) \times \frac{(4\pi^5)^{1/4}}{\alpha_0 C^{i_3-1}(1 + C^{d_{i_4, i_3}})} \exp\left(-\frac{\alpha_0 C^{i_3-1} C^{d_{i_4, i_3}} \overline{CD}^2}{1 + C^{d_{i_4, i_3}}}\right) \quad (93b)$$

with $\vec{n} = (n_x, n_y, n_z)$, $\vec{n}' = (n_x', n_y', n_z')$, and

$$X_1(n_x) = \sum_{i=0}^{l_{x1}+l_{x2}} (-1)^i\, i! \left(\frac{\overline{AB}_x}{1 + C^{d_{i_2, i_1}}}\right)^{l_{x1}+l_{x2}-i} \times \sum_{j=\max(0, i-l_{x2})}^{\min(i, l_{x1})} \frac{l_{x1}!\, l_{x2}!\, (-C^{d_{i_2, i_1}})^{l_{x1}-j}}{j!\,(l_{x1}-j)!\,(i-j)!\,(l_{x2}-i+j)!} \times \sum_{r=0}^{\left[\frac{i}{2}\right]} \frac{1}{r!\,(i-2r)!\,(4\alpha_0 C^{i_1-1}(1 + C^{d_{i_2, i_1}}))^{i-r}}, \quad (94a)$$

and

$$X_2(n_x') = \sum_{i'=0}^{l_{x3}+l_{x4}} (+1)^{i'}\, i'! \left(\frac{\overline{CD}_x}{1 + C^{d_{i_4, i_3}}}\right)^{l_{x3}+l_{x4}-i'} \times \sum_{j'=\max(0, i'-l_{x4})}^{\min(i', l_{x3})} \frac{l_{x3}!\, l_{x4}!\, (-C^{d_{i_4, i_3}})^{l_{x3}-j'}}{j'!\,(l_{x3}-j')!\,(i'-j')!\,(l_{x4}-i'+j')!} \times \sum_{r'=0}^{\left[\frac{i'}{2}\right]} \frac{1}{r'!\,(i'-2r')!\,(4\alpha_0 C^{i_3-1}(1 + C^{d_{i_4, i_3}}))^{i'-r'}}. \quad (94b)$$

The summations in Eqs. (94a) and (94b) are limited to all i and r (i' and r') combinations which lead to the same value of n_x (n_x') under the conditions: $n_x = i - 2r$ and $n_x' = i' - 2r'$. The Y and Z parts are expressed in a similar way.

In these equations we can see that T_1 becomes a function of the two centers A and B, of $\bar{l}$ and l_2, of the initial index i_1 (lowest exponent of the pair), and of the difference d_{i_2,i_1}. Analogous considerations can be made for T_2.

Second level of pre-tabulations. The second level of pre-tabulation lies where the final integral is calculated by merging the two tables $T_1(\vec{n})$ and $T_2(\vec{n}')$ and combining them in a proper way with the $H(\vec{m})$ function (Eq. 73). The critical point of this level is the combination of the four exponents α_1, α_2, α_3, α_4. This combination appears in the h functions (Eq. 75) and in the $G(\kappa)$ function (Eq. 74). In the h functions, the four exponent dependency is found in the components of the vector $\underline{PQ}$, which connects centers P and Q of the new combined primitives. The four exponent dependency in the $G(\kappa)$ function occurs in $\overline{PQ}^2$ and δ.

Using Eqs. (92), we find that

$$\vec{PQ} = \frac{\vec{AC} + \mathbf{C}^{d_{i_2,i_1}}\vec{BC} + \mathbf{C}^{d_{i_4,i_3}}\vec{AD} + \mathbf{C}^{d_{i_4+i_2,i_3+i_1}}\vec{BD}}{\mathbf{C}_{1234}}, \tag{95}$$

where

$$\mathbf{C}_{1234} = 1 + \mathbf{C}^{d_{i_2,i_1}} + \mathbf{C}^{d_{i_4,i_3}} + \mathbf{C}^{d_{i_4+i_2,i_3+i_1}}.$$

Thus, $\vec{PQ}$ depends upon d_{i_2,i_1} and d_{i_4,i_3}. The dependency of $\overline{PQ}^2$ on the four exponents is straightforward to derive from the dependency of the components of $\underline{PQ}$ upon these four exponents.

Concerning the variable δ, we can assume without losing generality that the following order exists: $i_1 \leq i_3$, i.e., the lowest exponent of the quartet has been arbitrarily assigned to center A.

Under this consideration the variable

$$\delta = \frac{\gamma_1 + \gamma_2}{4\gamma_1\gamma_2}$$

can be written as:

$$\delta = \frac{\mathbf{C}_\Sigma}{4\alpha_0\mathbf{C}^{i_1-1}\mathbf{C}^{d_{i_3,i_1}}\mathbf{C}_{1234}}, \tag{96}$$

where $\mathbf{C}_{1234}$ is the denominator in Eq. (95) and

$$\mathbf{C}_\Sigma = 1 + \mathbf{C}^{d_{i_2,i_1}} + \mathbf{C}^{d_{i_3,i_1}} + \mathbf{C}^{d_{i_4,i_1}}$$

therefore, δ is a function of the differences d_{i_2, i_1}, d_{i_3, i_1} and d_{i_4, i_1}, as well as of the first index i_1, i.e., the index of the lowest exponent for the given quartet of exponents.

Using the relations (96) and (95), we obtain that the argument t of the Incomplete Gamma Function can be calculated by

$$t = \frac{\overline{PQ}^2}{4\delta} = \frac{\alpha_0 \mathbf{C}^{i_1 - 1} \mathbf{C}^{d_{i_3, i_1}}}{\mathbf{C}_{1234} \mathbf{C}_\Sigma} (\overline{PQ}'^2_x + \overline{PQ}'^2_{y'} + \overline{PQ}'^2_z), \tag{97}$$

where

$$\vec{PQ'} = (\overline{PQ}_{x'}, \overline{PQ}_{y'}, \overline{PQ}_{z'}) = \vec{PQ}\, \mathbf{C}_{1234}.$$

Thus, t depends on the centers A, B, C and D; on d_{i_2, i_1}, d_{i_3, i_1} and d_{i_4, i_1}; and on the lowest index i_1.

Summarizing,

$$\begin{aligned} \vec{PQ} &= \vec{PQ}(\alpha_0, \mathbf{C}, d_{i_2, i_1}, d_{i_4, i_3}, \vec{A}, \vec{B}, \vec{C}, \vec{D}), \\ \delta &= \delta(\alpha_0, \mathbf{C}, i_1, d_{i_2, i_1}, d_{i_3, i_1}, d_{i_4, i_1}), \\ t &= t(\alpha_0, \mathbf{C}, i_1, d_{i_2, i_1}, d_{i_3, i_1}, d_{i_4, i_1}, \vec{A}, \vec{B}, \vec{C}, \vec{D}). \end{aligned}$$

By substituting relations (95) – (97) into the equation (74) for $G(\kappa)$ we obtain that

$$G = G(\alpha_0, \mathbf{C}, i_1, d_{i_2, i_1}, d_{i_3, i_1}, d_{i_4, i_1}, \vec{A}, \vec{B}, \vec{C}, \vec{D}, l_{x1}, l_{x2}, l_{x3}, l_{x4}).$$

While merging the two tables $T_1(\vec{n})$ and $T_2(\vec{n}')$ and combining them with the appropriate $H(\vec{m})$ we are going to have large amounts of repeated computations, since for each quartet of atoms, identical combinations of exponents are found in different contractions. Hence, when we are calculating the integrals centered on a given quartet of atoms (see next section at Fig. 13), a great deal of repetition can be avoided in the determination of the values of $G(\kappa)$, one of the most time-consuming sections of the ERI computation. Presently, however, most of the simplifications associated with the use of geometrical basis sets has not been implemented. A first step, however, has been considered as described in the following section.

16. A New Loop Structure for Uncontracted Geometrical Basis Sets

From section 7 we know that the only step with fourth power dependency is the computation of the H functions and the merging of the tables T_1 and T_2 (see Eqs. 68, 69) with the appropriate H value (see Eq. 73). From Eq. (73), one can see that the H functions are dependent on both the coordinates and exponents of the four centers So, for a given four centers, the H functions are only dependent on the quartet of exponent. For uncontracted geometrical basis sets GBSs, there are many integrals which share the same combination of exponents, therefore, many integrals share the same H function. We note that the repeated use of the same functions (related to the angular integration) was realized and exploited in the "special *s* and *p*" integral expressions used by Clementi and Davis in the first version of the IBMOL molecular program.[6,7]

As an illustration, Figure 12 shows how a H function of $(pp|ss)$ integrals is shared with those of $(ps|ss)$ and $(ss|ss)$ integrals. Hence, one computes only one set of H functions, those corresponding to the highest total angular momentum, and uses it for the entire ERI block (a shell quartet). In this way, a great deal of repetition can be avoided in the evaluation of H functions. For instance, if each atom of a given quartet has an *spdf* shell, there are altogether 160,000 (20^4) ERIs and the evaluation of only 455 values of the H function is required, whereas the old scheme requires 26,628 values for the H functions. Consequently, there is a large saving in CPU-time at this incomplete level of the computation. With this data organization, one has also the advantage of long loops in both T_1 and T_2 which is ideal for vector computers. For example, for the case of a $(sp,sp|sp,sp)$ integral block, the length of the T_1 and T_2 loops is about 20 to 45.

In order to take advantage of the time saving in the H function evaluations, all the basis functions of an uncontracted GBS are grouped according to shells. Thus, each exponent of the basis set belongs to an *s*, *sp*, *spd* or *spdf* shell. In addition, table values and related information are grouped according to exponent (or shell) pairs in building the density tables.

The new loop structure, that will be used as an option in the forthcoming version of the KGNMOL program, is shown in Figure 13. For a given four centers (A,B,C,D), the program loops over all possible exponent pairs of the left (AB) and right (CD) center pairs. For a given choice of the four exponents, the program determines the highest angular momentum needed for the shell quartet and computes the required H functions. Then the table merging for the entire integral block is performed. The innermost part of the routine is the merging of T_1 and T_2 with the appropriate H function; this most simple loop requires the longest time fraction in computing the ERIs.

Let us give numerical examples and the corresponding computer time. In the following all calculations were carried out on an IBM 3090/400 base machine using 1 processor. In Table XII, we summarize the CPU time required for *the evaluation*

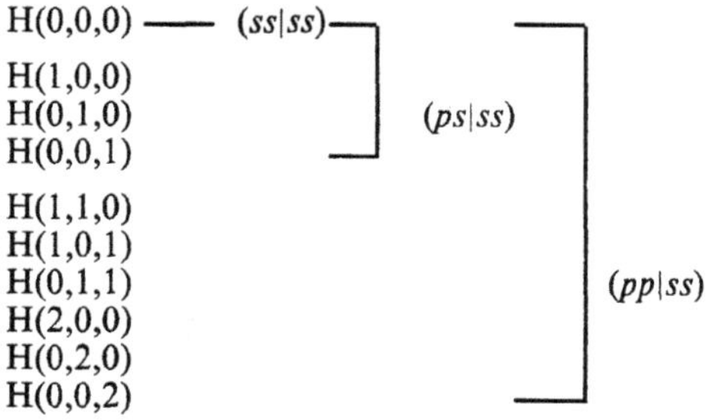

Figure 12. The sharing of a set of H functions of (*pp*|*ss*) integrals with those of (*ps*|*ss*) and (*ss*|*ss*) integrals.

of H functions for both the standard code and the present option. The examples chosen have different numbers of atoms and different basis set sizes. The table shows that in all cases there is large improvement in the new program, ranging from 6 to 30 times speedup over the old program. As one might have expected, greater savings may be achieved for basis sets which contain higher angular momentum (see Table XII). The improvement in cases with *spdf* type basis sets (20-29 times) are greater than those with *spd* basis sets (12-20 times) which in turn are greater than those with *sp* basis sets (6-11 times).

```
for a given four centers (A,B,C,D)
loop over all exponent pairs (αA(i) αB(j)) of left center pair AB
    loop over all exponent pairs (αC(k) αD(l)) of right center pair CD
        determine the highest angular momentum needed
        calculate the required H functions
        evaluate all integrals of the shell quartet using the following two loops
        loop over all table values (T1) of αAαB
            loop over all table values (T2) of αCαD
                ERI = ERI + T1 × T2 × H
            endloop
        endloop
    endloop
endloop
```

Figure 13. Loop structure of INT2 routine for the new option.

In Table XIII, the total CPU times for the same set of examples are provided for the evaluation of the ERIs. A comparison of the two tables indicates that the trend of improvement in the total run time is opposite to that of the H function evaluation. From the results in Table XIII, the greatest saving, about a factor of two, appears in ERIs involving only *s* and *p* functions, and the smallest for ERIs involving *s, p, d* and *f* functions. The difference in trends can be readily understood in terms of the percentage of total run time spent in the H function computation. For basis sets involving only *s* and *p* functions, about 50% (from 42 to 54 %) of the total time is spent in the H function evaluation, therefore, it leads to an overall speedup of a factor of two. On the other hand, for *spdf* type basis sets, most of the computer time is in the table merging rather than in the computation

Table XII. CPU time (in seconds) for H function evaluation[a]

molecule	basis set[b]	KGNMOL timing		
		standard	option	ratio[c]
CO	sp(1)	8.09	1.07	7.6
HCN	sp(1)	17.67	2.20	8.0
$(H_2O)_2$	sp(2)	13.43	2.28	5.9
C_3H_4	sp(2)	59.14	7.49	7.9
C_2H_5NO	sp(2)	178.90	16.79	10.7
CO	spd(1)	26.00	1.39	18.7
HCN	spd(1)	57.90	3.09	18.7
$(H_2O)_2$	spd(2)	36.06	3.04	11.9
C_3H_4	spd(2)	142.79	10.02	14.3
C_2H_5NO	spd(2)	591.72	26.27	19.8
CO	spdf(1)	45.78	2.00	22.9
HCN	spdf(1)	107.48	4.37	24.6
$(H_2O)_2$	spdf(2)	81.01	4.03	20.1
C_3H_4	spdf(2)	305.94	14.39	21.3
C_2H_5NO	spdf(2)	2491.89	87.43	28.5

[a] integral cutoff = 10^{-8} a.u.
[b] sp(1): α_0 = 0.016, **C** = 2.6, C, N and O: (13s,8p), and H: (10s);
sp(2): α_0 = 0.012, **C** = 3.0, C, N and O: (9s,5p), and H: (6s);
spd(1): α_0 = 0.016, **C** = 2.6, C, N and O: (13s,8p,3d), and H: (10s,3p);
spd(2): α_0 = 0.012, **C** = 3.0, C, N and O: (9s,5p,2d), and H: (6s,1p);
spdf(1): α_0 = 0.016, **C** = 2.6, C, N and O: (13s,8p,3d,1f), and H: (10s,3p,1d); and
spdf(2): α_0 = 0.012, **C** = 3.0, C, N and O: (9s,5p,2d,1f), and H: (6s,2p);
[c] ratio = previous / present.

Table XIII. Total CPU time (in seconds) for INT2 evaluation.[a]

molecule	basis set[b]	KGNMOL timing		
		standard	option	ratio[c]
CO	sp(1)	19.47	10.07	1.93
HCN	sp(1)	36.26	18.11	2.00
$(H_2O)_2$	sp(2)	23.20	11.36	2.04
C_3H_4	sp(2)	111.33	55.21	2.02
C_2H_5NO	sp(2)	404.79	211.28	1.92
CO	spd(1)	98.64	56.40	1.75
HCN	spd(1)	194.56	108.28	1.80
$(H_2O)_2$	spd(2)	106.43	61.13	1.74
C_3H_4	spd(2)	463.35	268.90	1.72
C_2H_5NO	spd(2)	2984.18	1863.79	1.60
CO	spdf(1)	228.22	147.05	1.55
HCN	spdf(1)	475.64	310.14	1.53
$(H_2O)_2$	spdf(2)	361.91	246.61	1.47
C_3H_4	spdf(2)	1475.64	1019.77	1.45
C_2H_5NO	spdf(2)	15689.02	10558.37	1.49

[a] integral cutoff = 10^{-8}.
[b] see footnote of Table I.
[c] ratio = previous / present.

of the H functions (from 16 to 20 %). This accounts for the fact that, even if a speedup of 20 to 30 times is achieved in the H function evaluation, the overall improvement for *spdf* cases is less than that for *sp* and *spd* basis sets. We note that a slight improvement of the total run time also arises from the longer T_1 and T_2 vectors in the table merging step. In summary, the new algorithm represents a significant advance over the previous method.

Finally, in Table XIV, we compare timings of the present algorithm against the previous scheme and other integral packages. No point group symmetry is taken into account in the computations. The three example chosen employed *sp, spd* and *spdf* type basis sets for HF, HCN and $(H_2O)_2$, respectively. In both GAUSSIAN 86 and HONDO-8 packages, the option for the axis-switch method[102] is used for ERIs involving only s and p functions, and the Rys-polynomial method[101] for ERIs involving higher symmetry (d or f) functions.

Table XIV. CPU time (in seconds) for different integral programs.[a,b]

program	molecule					
	HF[c]		HCN[d]		$(H_2O)_2$[e]	
KGNMOL option	22.7	(1.0)	31.3	(1.0)	12.5	(1.0)
KGNMOL standard	34.9	(1.5)	57.6	(1.8)	26.0	(2.1)
HONDO-8	52.2	(2.3)	81.0	(2.6)	41.5	(3.3)
GAUSSIAN 86	74.9	(3.3)	126.3	(4.0)	69.2	(5.6)

[a] integral cutoff = 10^{-10} a.u.
[b] ratio compared to the present program (in parentheses).
[c] α_0 = 0.016, C = 2.6, F: (13s,8p,2d,1f), and H: (10s,2p,1d).
[d] α_0 = 0.012, C = 3.0, C and N: (11s,6p,2d), and H: (8s,1p).
[e] α_0 = 0.012, C = 3.0, O: (9s,5p), and H: (6s).

We have also considered the implementation of the present algorithm for ERI calculation using *contracted* geometrical basis sets.[103] Because of the unconstrained contraction scheme, the contracted GBSs do not allow an *sp, spd* or *spdf* shell formalism. A program organization which requires pre-tabulation of all H functions (once a given four centers is chosen) has been implemented. The storage and retrieval of the H functions reduce the efficiency; preliminary results suggests a speedup of only 15 to 35% for contracted GBSs is possible. Further details of the new program (organization, input and output), for both contracted and uncontracted geometrical basis sets, can be found in Ref. 103.

Acknowledgments

The data on dynamic load balancing and MP2 performance are due to Dr. P.L. Cristinziano, and the data on the new loop structure of the integral evaluation to Dr. M.W. Wong, whose help is here gratefully acknowledged. Since KGNMOL is the present level of the PPI and of IBMOL molecular packages, one of us (E.C.) would like to take the occasion of this chapter and thank the many colleagues and collaborators with whom he shared the hope of new ways to broaden chemistry on the base of mathematical models and computer simulations.

Appendix 7A.
Normalization of Gaussian Type Functions

A normalized GTF is defined as:

$$\eta(\alpha, \vec{A}, \vec{l}\,) = N(\alpha, \vec{l}\,) x_A^{l_x} y_A^{l_y} z_A^{l_z} \exp(-\alpha r_A^2).$$

$N(\alpha, \vec{l}\,)$ is the normalization constant and it is defined such that

$$\int d\tau\, \eta^2(\alpha, \vec{A}, \vec{l}\,) = 1, \tag{A1}$$

so

$$N^{-2}(\alpha, \vec{l}\,) = \int_{-\infty}^{\infty}\int_{-\infty}^{\infty}\int_{-\infty}^{\infty} dx_A\, dy_A\, dz_A\, x_A^{2l_x} y_A^{2l_y} z_A^{2l_z} \exp(-2\alpha(x_A^2 + y_A^2 + z_A^2)) \tag{A2}$$
$$= I_x I_y I_z,$$

where

$$I_x = \int_{-\infty}^{\infty} dx_A\, x_A^{2l_x} \exp(-2\alpha\, x_A^2)$$

and similarly for I_y and I_z.

Using Eq. E7 one obtains

$$I_x = (2\alpha)^{-(2l_x+1)/2}\, \Gamma\left(\frac{2l_x + 1}{2}\right). \tag{A3}$$

Hence

$$N^2(\alpha, \vec{l}\,) = \frac{\left(\frac{2\alpha}{\pi}\right)^{3/2} (4\alpha)^{l_x + l_y + l_z}}{(2l_x - 1)!!\,(2l_y - 1)!!\,(2l_z - 1)!!}. \tag{A4}$$

Appendix 7B.
The Gaussian Product Theorem

The Gaussian product theorem states that the product of two GTFs having different centers, A and B, is a linear combination of GTFs, all with the same exponent γ and the same center P on the line AB.

Suppose $\eta(\alpha_1, \vec{A}, \vec{l}_1)$ and $\eta(\alpha_2, \vec{B}, \vec{l}_2)$ are two unnormalized GTFs centered on A and B, respectively. Their product is:

$$\eta(\alpha_1, \vec{A}, \vec{l}_1)\, \eta(\alpha_2, \vec{B}, \vec{l}_2) = x_A^{l_{x1}}\, x_B^{l_{x2}}\, y_A^{l_{y1}}\, y_B^{l_{y2}}\, z_A^{l_{z1}}\, z_B^{l_{z2}} \exp(-\alpha_1 r_A^2 - \alpha_2 r_B^2). \qquad (B1)$$

Consider, to start with, that $\vec{l}_1 = \vec{l}_2 = \vec{0} \equiv (0,0,0)$ (i.e., the GTFs are 1s functions) so that only the exponential part has to be taken into account. Noting that

$$r_A^2 = (\vec{r} - \vec{A}) \cdot (\vec{r} - \vec{A}), \qquad (B2)$$

and defining $\gamma = \alpha_1 + \alpha_2$, and

$$\vec{P} = \frac{\alpha_1 \vec{A} + \alpha_2 \vec{B}}{\gamma},$$

it is possible to show that

$$\begin{aligned} \exp(-\alpha_1 r_A^2 - \alpha_2 r_B^2) &= \exp(-\alpha_1\alpha_2 \overline{AB}^2/\gamma)\, \exp(-\gamma r_P^2) \\ &= \exp(-\alpha_1\alpha_2 \overline{AB}^2/\gamma)\, \eta(\gamma, \vec{P}, \vec{0}). \end{aligned} \qquad (B3)$$

$\eta(\gamma, \vec{P}, \vec{0})$ is an 1s GTF centered on $\vec{P}$ with exponent γ.

If $\vec{l}_1 \neq \vec{0}$ and/or $\vec{l}_2 \neq \vec{0}$ then extra factors of the type $x_A^{l_{x1}}\, x_B^{l_{x2}}$ appear. It is convenient to express the coordinates relative to the point P so that

$$\vec{r}_A = (\vec{r} - \vec{A}) = (\vec{r} - \vec{P}) + (\vec{P} - \vec{A}) = \vec{r}_P + (\vec{P} - \vec{A}),$$

and

$$x_A = x_P + PA_x.$$

The other coordinates can be expressed in a similar way.

Using the binomial theorem, which states that

$$(x + A)^l\,(x + B)^m = \sum_{i=0}^{l+m} f(i,l,m,A,B)x^i,$$

where

$$f(i,l,m,A,B) = \sum_{j=\max(0,i-m)}^{\min(i,l)} \frac{l!\;m!\;A^{l-j}B^{m-i+j}}{j!\;(l-j)!\;(i-j)!\;(m-i+j)!},$$

one finds that

$$x_A^{l_{x1}}\,x_B^{l_{x2}} = \sum_{i=0}^{l_{x1}+l_{x2}} f(i,l_{x1},l_{x2},\overline{PA}_x,\overline{PB}_x)x_P^i, \tag{B4}$$

and equivalent equations hold for the y and z coordinates.

Substitution of Eqs. B3 and B4 and its analogues into Eq. B1 renders the general formula:

$$\begin{aligned}\eta(\alpha_1,\vec{A},\vec{l}_1)\,\eta(\alpha_2,\vec{B},\vec{l}_2) &= \exp(-\alpha_1\alpha_2\overline{AB}^2/\gamma)\\ &\times \sum_{i=0}^{l_{x1}+l_{x2}} f(i,l_{x1},l_{x2},\overline{PA}_x,\overline{PB}_x)\,x_P^i\,\exp(-\gamma\,x_P^2)\\ &\times \sum_{j=0}^{l_{y1}+l_{y2}} f(j,l_{y1},l_{y2},\overline{PA}_y,\overline{PB}_y)\,y_P^j\,\exp(-\gamma\,y_P^2)\\ &\times \sum_{k=0}^{l_{z1}+l_{z2}} f(k,l_{z1},l_{z2},\overline{PA}_z,\overline{PB}_z)\,z_P^k\,\exp(-\gamma z_P^2).\end{aligned} \tag{B5}$$

B5 can be rearranged in the following way

$$\eta(\alpha_1, \vec{A}, \vec{l_1})\, \eta(\alpha_2, \vec{B}, \vec{l_2}) = \exp(-\alpha_1\alpha_2\overline{AB}^2/\gamma)$$
$$\times \sum_{i=0}^{l_{x1}+l_{x2}} \sum_{j=0}^{l_{y1}+l_{y2}} \sum_{k=0}^{l_{z1}+l_{z2}} f(i, l_{x1}, l_{x2}, \overline{PA}_x, \overline{PB}_x)$$
$$\times f(j, l_{y1}, l_{y2}, PA_y, PB_y)\, f(k, l_{z1}, l_{z2}, PA_z, PB_z)\, x_P^i\, y_P^j\, z_P^k \exp(-\gamma r_P^2) \tag{B6}$$
$$= \sum_{\vec{l}} d_{\vec{l}}\; \eta(\gamma, \vec{P}, \vec{l}),$$

where $\vec{l} = (i, j, k)$.

Appendix 7C. Integrals Related to the Gamma Function

In the deduction of integrals over GTFs the following definite integrals are needed:

$$\int_{-\infty}^{\infty} dx \ \exp(-ax^2 + bx) = \sqrt{\frac{\pi}{a}} \ \exp\left(\frac{b^2}{4a}\right), \qquad (C1)$$

with $a > 0$.

This equality can readily be proved by substituting $z = x - \frac{b}{2a}$ and using Eq. E7.

$$\int_{-\infty}^{\infty} dx \ \exp(-ax^2 + \mathrm{i}bx) = \sqrt{\frac{\pi}{a}} \ \exp\left(-\frac{b^2}{4a}\right), \qquad (C2)$$

where $\mathrm{i} = \sqrt{-1}$.

Eq. C2 is obtained by expanding exp (i bx) in power series.

A generalization of Eq. C2 is

$$\int_{-\infty}^{\infty} dx \, x^n \ \exp(-ax^2 + \mathrm{i}\, bx) = \mathrm{i}^n \left(\frac{1}{\sqrt{4a}}\right)^n H_n\left(\frac{b}{\sqrt{4a}}\right) \sqrt{\frac{\pi}{a}} \ \exp\left(-\frac{b^2}{4a}\right), \qquad (C3)$$

where $H_n(z)$ is the Hermite polynomial, defined as

$$H_n(z) = n! \sum_{k=0}^{\left[\frac{n}{2}\right]} \frac{(-1)^k \, (2z)^{n-2k}}{k! \, (n-2k)!},$$

where [m] is the largest integer $\leq$ m. For a proof of Eq. C3 see Ref. 9 and references therein.

Another integral property used in the derivation of ERI is

$$\exp(-ax^2) = 2ax^2 \int_0^1 ds\, s^{-3} \exp(-ax^2 s^{-2}), \qquad (C4)$$

where $a > 0$.

Although the integrals involved in the Fourier transform are not related to the Gamma function, the Fourier transform of r^{-1}, where

$$r = \sqrt{\vec{r} \cdot \vec{r}}\ , \text{ and } \vec{r} = (x_1, x_2, \ldots, x_n),$$

will be discussed next.

In one dimension, the Fourier transform $F(x)$ of $f(k)$ is defined as

$$F(x) = \frac{1}{\sqrt{2\pi}} \int_{-\infty}^{\infty} dk\, f(k) \exp(\mathrm{i}xk).$$

The generalization for n dimensions is:

$$F(x_1, \ldots, x_n) = \frac{1}{\sqrt{2\pi}} \int d\vec{k}\ f(k_1, \ldots, k_n) \exp(\mathrm{i}\vec{r} \cdot \vec{k}),$$

where $d\vec{k} = dk_1, \ldots, dk_n$.

Particularly for the 3-dimensional case it holds that

$$\sqrt{\frac{\pi}{2r^2}} = \sqrt{\left(\frac{1}{2\pi}\right)^3} \int d\vec{k}\ k^{-2} \exp(\mathrm{i}\vec{r} \cdot \vec{k}), \qquad (C5a)$$

or

$$\frac{1}{r} = \left(\frac{1}{2\pi^2}\right) \int d\vec{k}\ k^{-2} \exp(\mathrm{i}\vec{r} \cdot \vec{k}). \qquad (C5b)$$

Appendix 7D.
Incomplete Gamma Function

The Incomplete Gamma Function is a definite integral that cannot be evaluated in closed form. To compute it, asymptotic formulas, recursion formulas and continued fractions together with tabular interpolation are commonly used.

Definition of The Incomplete Gamma Function

The F function is defined by the expression

$$F_v(t) = \int_0^1 dx\, x^{2v} \exp(-tx^2), \qquad (D1)$$

where t, $v \geq 0$ and are real and integer variables, respectively. According to the last formula we have that the greater the value of t, the smaller the value of the function $F_v(t)$ for a given value of v. When t is zero we can express the integral in a closed form such as

$$F_v(0) = (2v+1)^{-1}. \qquad (D2)$$

Use of Taylor Series Expansions

A Taylor expansion is very useful to calculate $F_v(t)$ because the derivative of the function changes the order of the function. This fact is readily seen when we apply Leibniz's rules to differentiate Eq. D1 with respect to t. Doing that differentiation we find

$$\frac{\partial F_v(t)}{\partial t} = \int_0^1 dx\, x^{2v} \frac{\partial \exp(-tx^2)}{\partial t} = -F_{v+1}(t). \qquad (D3)$$

Therefore, using a Taylor Series expansion for the function $F_v(t)$ about ω_0, we obtain

$$F_v(t+\omega_0) = \sum_{i=0}^{\infty} \frac{(-t)^i F_{v+i}(\omega_0)}{i!}, \qquad (D4)$$

if we have $\omega_0 = 0$ and we substitute Eq. D2, we get

$$F_\nu(t) = \sum_{i=0}^{\infty} \frac{(-t)^i}{i!\,(2\nu + 2i + 1)}.$$

This power series is unconditionally convergent for all values of t due to the fact that the greater the value of ν the smaller the value of $F_\nu(t)$.

Asymptotic Series For the Incomplete Gamma Function

Let us define $\Theta_\nu(t)$ by

$$\Theta_\nu(t) = \int_1^{\infty} dx\, x^{2\nu} \exp(-tx^2). \qquad (D5)$$

With the aid of $\Theta_\nu(t)$, the $F_\nu(t)$ can be written as

$$F_\nu(t) = \frac{\Gamma(\nu + 1/2)}{2t^{\nu+1/2}} - \Theta_\nu(t). \qquad (D6)$$

Integrating Eq. D5 by parts we obtain

$$\Theta_\nu(t) = \frac{\exp(-t) + (2\nu - 1)\,\Theta_{\nu-1}(t)}{2t}.$$

Given the preceding relation and that $\Theta_\nu(t)$ is a positive quantity, the inequality

$$\Theta_\nu(t) < \frac{\exp(-t)}{2t}$$

is satisfied whenever $\nu < 0$. If we apply the recurrence relation D7 we obtain a series

$$\Theta_n(t) = \exp(-t)\left[\frac{1}{2t} + \frac{(2\nu - 1)}{(2t)^2} + \frac{(2\nu - 1)(2\nu - 3)}{(2t)^3} + \cdots + \frac{(2\nu - 1)\ldots(2\nu - 2i + 3)}{(2t)^i}\right] + \mathbf{R},$$

where

$$\mathbf{R} = \Theta_{\nu-i}\frac{(2\nu - 1)\ldots(2\nu - 2i + 1)}{(2t)^i}.$$

The term $\mathbf{R}$ can be bounded according to the expression

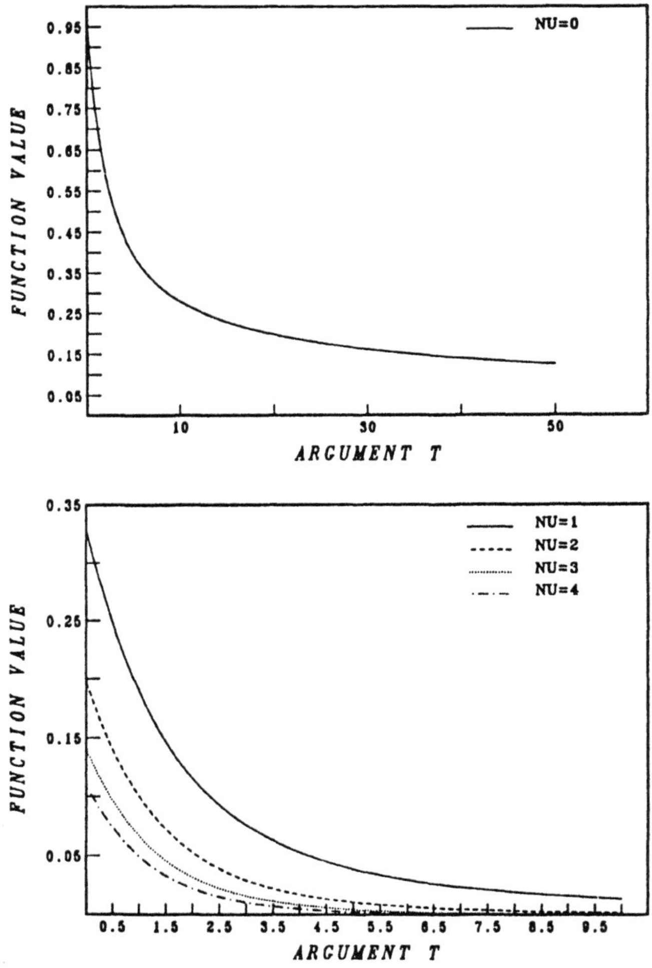

Figure 7. (TOP) Incomplete Gamma Function for NU = 0. $F_{NU}(T)$ for NU = 0 and $T \in [0, 50]$. (BOTTOM) Incomplete Gamma Function for NU = 1 to 4. $F_{NU}(T)$ for NU = 1 → 4 and $T \in [0, 10]$.

$$|\mathbf{R}| < \left| \frac{(2\nu - 1) \ldots (2\nu - 2i + 1)\ \exp(-t)}{(2t)^{i+1}} \right| \quad \text{if } i \geq \nu.$$

Therefore, the absolute value of the remainder **R** in the previous equation is smaller than the absolute value of the first term omitted from the series (the i + 1-th term), given that $i \geq m$, and we have an asymptotic series.

Use of Tabular Interpolation

By using one of the above methods, one can construct a table of $F_\nu(t_i)$ for a fixed set of argument values t_i. For a given argument t_i, this is done by determining the value of $F_\nu(t_i)$ for the maximum integer ν_{max} and recurring down using equations D3 or D6 to obtain all the values of $F_\nu(t_i)$ for ν varying from ν_{max} to 0.

Once we have constructed a table of $F_\nu(t_i)$ the value of $F_\nu(t)$ may be obtained from it by interpolation. However, because for small values of t, $F_\nu(t)$ is a rapidly varying function, with comparatively large high order derivatives, a high order interpolation scheme and/or a closely spaced table will be required. Since for large argument the function becomes slowly varying, the table must have a grid of increasing interval. For t-values greater than 44, a special formula for the asymptotic expression

$$F_\nu(t) = \frac{\Gamma(\nu + 1/2)}{2t^{(\nu+1/2)}}$$

is used.

We report graphically $F_0(0) \rightarrow F_0(50)$ (Fig. 7, top) and $F_1(0) \rightarrow F_1(10)$ up to $F_4(0) \rightarrow F_4(10)$ (Fig. 1, bottom). From these graphs one can appreciate the reason for using asymptotic expressions and gain some insight in how to define thresholds for eliminating integrals smaller than a pre-defined value.

Appendix 7E.
The Complete Gamma Function

The Complete Gamma Function is defined through the definite integral

$$\Gamma(x) = \int_0^\infty du\, u^{x-1} \exp(-x). \qquad (E1)$$

The following convenient relationships are easily proved:

$$\Gamma(x) = (x-1)\,\Gamma(x-1) \quad x > 1, \qquad (E2)$$

$$\Gamma(1) = 1, \qquad (E3)$$

$$\Gamma\left(\frac{1}{2}\right) = \sqrt{\pi}\,. \qquad (E4)$$

If n is an integer, then

$$\Gamma(n) = (n-1)! \quad n > 1,\ 0! \equiv 1, \qquad (E5)$$

$$\Gamma\left(n + \frac{1}{2}\right) = \frac{(2n-1)!!}{2^n}\sqrt{\pi} \quad n > 0,\ \ -1!! \equiv 1 \qquad (E6)$$

and if k is an odd integer, then $k!! \equiv 1 \times 3 \times 5 \times \cdots \times k$.

If n is an integer and $a > 0$, then

$$\int_{-\infty}^{\infty} dx\, x^n \exp(-ax^2) = \begin{cases} 0 \ \text{ if } n \text{ is odd}, \\ a^{-(l+1)/2}\,\Gamma\left(\dfrac{l+1}{2}\right) \ \text{ if } n \text{ is even}. \end{cases} \qquad (E7)$$

References

1. R. Gomperts and E. Clementi, IBM Research Report, **KGN-118** (June, 1988).
2. E. Clementi, R. Gomperts and C. Urdaneta, IBM Research Report, **KGN-53** (November, 1986).
3. E. Clementi, S. Chakravorty, G. Corongiu, D. Schneider and D. Wong (in preparation).
4. R. Gomperts and E. Clementi, KGNMOL in Quant. Chem. Program Exch., deposited as Program #538 (1988).
5. P. Habitz and E. Clementi, Comp. Phys. Comm., **29**, 301 (1983).
6. E. Clementi and D.R. Davies, J. Comp. Physics, **1**, 223 (1966).
7. D.R. Davies and E. Clementi, "IBMOL: Computation of Wave Functions of General Geometry," Special IBM-Research Report, San Jose, Calif. (1965).
8. A. Veillard and E. Clementi, "IBMOL: Version 2," Special IBM-Research Report, San Jose, Calif. (1966).
9. (a) E. Clementi and J.W. Mehl, "IBM System/360, IBMOL-5 Program, General Information Manual," IBM-Research Report, **RJ853** (May, 1971); (b) E. Clementi and J.W. Mehl, "IBMOL-5 Program, System Manual," IBM-Research Report, **RJ1244** (July, 1973); c) E. Clementi and J.W. Mehl, "IBMOL-5 Program, User's Guide," IBM-Research Report, **RJ889** (July, 1971).
10. E. Ortoleva, G. Castiglione and E. Clementi, IBMOL-7, Computer Phys. Comm., **12**, 337 (1980).
11. D.R. Davies and E. Clementi, IBMOL in Q.C.P.E. as Program #92 (1966).
12. E. Clementi, J. Chem. Phys., **46**, 3851 (1967); ibid., **47**, 2323 (1967).

13. E. Clementi and J.N. Gayles, J. Chem. Phys., **47**, 3837 (1967).
14. E. Clementi, Int. J. Quantum Chem., **1S**, 307 (1967).
15. E. Clementi and G. Corongiu, Chem. Phys. Letters, **90**, 359 (1982).
16. C. Urdaneta, G. Corongiu and E. Clementi, IBM-Research Report, **KGN-133** (August, 1987).
17. See this volume, Chapter 2, Tables V to XI.
18. M. Weissbluth, "Atoms and Molecules," Academic Press, New York (1978).
19. M. Tinkham, "Group Theory and Quantum Chemistry," McGraw-Hill, New York (1964).
20. G.H.F. Diercksen, B.T. Sutcliffe and A. Veillard, Eds., "Computational Techniques in Quantum Chemistry and Molecular Physics," Reidel, Dordrecht (1975).
21. H.F. Schaefer, III, Ed., "Methods of Electronic Structure Theory," Vol. 3, Plenum Press, New York (1977).
22. J.C. Slater, "Quantum Theory of Matter," McGraw-Hill Book Co., New York (1968).
23. E. Clementi, J. Chem. Phys., **47**, 4493 (1967).
24. E. Clementi, J. Mehl and W. von Niessen, J. Chem. Phys., **54**, 508 (1971).
25. E. Clementi, F. Cavallone and R. Scordamaglia, J. Am. Chem. Soc., **99**, 5531 (1977).
26. M. Aida and G. Corongiu, private communications.
27. E. Clementi, Proc. Nat. Acad. Sci., U.S.A., **69**, 2942 (1972).
28. See, for example, *The New York Times,* October 10, 1968, and *Time Magazine,* November 8, 1968.
29. S.F. Boys, Proc. Roy. Soc., **A200**, 542 (1950).
30. R.F. McWeeny, Proc. Roy. Soc., **A196**, 215 (1946).
31. M.W. Schmidt and K. Ruedenberg, J. Chem. Phys., **71**, 3951 (1979).
32. F. Van Duijenveldt, IBM Tech. Report, **RJ-945** (1972).
33. B.H. Wells and S. Wilson, J. Phys. B., **22**, 1285 (1989).
34. L. Laaksonen, P. Pyykkö and D. Sundholm, Int. J. Quant. Chem., **23**, 319 (1983).
35. H.D. Heinemann, D. Kolb and B. Fricke, Chem. Phys. Lett., **137**, 180 (1987).
36. H.D. Heinemann, D. Kolb and B. Fricke, Phys. Rev. A, **38**, 4994 (1988).
37. J.A. Pople, B.T. Luke, N.J. Frisch and J.S. Binkley, J. Phys. Chem., **89**, 2198 (1985).
38. B. Liu, J. Chem. Phys., **80**, 581 (1984).
39. D. Frye, G.C. Lie and E. Clementi, J. Chem. Phys. **91**, 2369 (1989).
40. W. Kolos, K. Szalewicz and H. Monkhorst, J. Chem. Phys., **84**, 3259 (1986).
41. G. Herzberg and A. Montfils, J. Mol. Spec., **5**, 482 (1960). Revised work in: G. Herzberg, Phys. Rev. Lett., **23**, 1081 (1970).
42. E. Clementi, J. Chem. Phys., **36**, 750 (1962)
43. S.F. Boys and F. Bernardi, J. Mol. Phys., **19**, 553 (1970).
44. P. Habitz and E. Clementi in "ETO Multicenter Molecular Integrals," C.A. Weatherford and H.W. Jones, Eds., Raidel, Boston (1982).
45. G. Corongiu and P.L. Cristinziano, to be published.

46. For a recent review dealing with electron correlation see *Methods in Computational Physics*, edited by S. Wilson (Plenum, New York, 1987), Vol. I
47. A.C. Hurley, *Electron Correlation in Small Molecules* (Academic, New York, 1976) also see H.F. Schaefer III, Editor, *Methods of Electronic Structure Theory*, (Plenum, New York, 1977).
48. For a recent and comprehensive review refer to E.S. Kryachko and E.V. Ludena, *Energy Density Functional Theory of many-electron systems*, (Kluwer Academic, Dordrecht, 1990).
49. "Density Matrices and Density Functionals," Proc. of the Coleman Symp., R. Erdahl and V.H. Smith, R., Eds., Reidel, Dordrecht (1987).
50. E. Clementi, Proc. Nat. Acad. Sci., U.S.A., **69**, 2942 (1972).
51. E.P. Wigner, Phys. Rev., **46**, 1002 (1934); E.P. Wigner and F. Seitz, Phys. Rev., **43**, 804 (1933); Trans. Faraday Soc., **34**, 678 (1938).
52. E. Clementi and S. Chakravorty, J. Chem. Phys. (in press, 1990)
53. G.C. Lie and E. Clementi, J. Chem. Phys., **60**, 1275 (1974) and J. Chem. Phys., **60**, 1288 (1974).
54. E. Clementi in "Computers and their Role in the Physical Sciences," S. Fernbach and A. Taub, Eds., Gordon and Breach Science Publishers, New York, pp. 503 – 542 (1970).
55. E. Clementi, IBM J. Res. and Dev., **9**, 2 (1965).
56. E. Clementi in "Chemistry of Cyano Group," Z. Rappoport, Ed., John Wiley & Sons, New York (1970).
57. E. Clementi and C. Salez, in "V Coloquio Internacional de Quimicos Teoricos de Expresion Latina," IBM Res. Lab., San Jose, Calif., Table III, p. 115 (1974).
58. E. Clementi and H. Popkie, J. Chem. Phys., **57**, 4870 (1972). H. Kistenmacher, H. Popkie and E. Clementi, J. Chem. Phys., **59**, 5842 (1973); J. Chem. Phys., **60**, 4455 (1974). 3050 (1966).
59. R. Colle and O. Salvetti, Theor. Chim. Acta (Ber.), **37**, 329 (1975); ibid. **53**, 55 (1979); J. Chem. Phys., **79**, 1404 (1983).
60. V. Carravetta and E. Clementi, J. Chem. Phys., **81**, 2646 (1984).
61. O. Gunnarson and B.I. Lundqvist, Phys. Rev. B., **13**, 4274 (1976).
62. S.H. Vosko, L. Wilk and M. Nusair, Can. J. Phys., **58**, 1200 (1980).
63. J.B. Lagowski and S.H. Vosko, J. Phys. B., **21**, 203 (1988).
64. H. Stoll, C.M.E. Pavlidou and H. Preuss, Theor. Chem. Acta, **49**, 143 (1978).
65. A. Savin, H. Stoll and H. Preuss, Theor. Chim. Acta, **70**, 407 (1986).
66. J.P. Perdew and A. Zunger, Phys. Rev. B., **23**, 5048 (1981).
67. J.P. Perdew, Phys. Rev. B., **33**, 8822 (1986).
68. A.D. Becke, J. Chem. Phys., **88**, 1053 (1988).
69. C. Lee, W. Yang and R.G. Parr, Phys. Rev. B. **37**, 785 (1988).
70. B. Miehlich, A. Savin, H. Stoll and H. Preuss, Chem. Phys. Lett., **157**, 200 (1989).
71. L.H. Thomas, Proc. Camb. Phil. Soc., **23**, 542 (1927).
72. E. Fermi, Rend. Accad. Lincei, **6**, 602 (1927).
73. P. Hohenberg and W. Kohn, Phys. Rev., **136B**, 864 (1964).

74. W. Kohn and L.J. Sham, Phys. Rev., **140A**, 1133 (1965).
75. W. Heisenberg, Z. Physik, **38**, 39 (1926); *ibid.* **39**, 499 (1926); *ibid.* **41**, 239 (1927).
76. D. Pines, Solid State Physics, pp.374, Academic Press, New York (1955).
77. D. Gell-Mann and K.A. Brueckner, Phys. Rev. **106**, 364 (1957).
78. P. Gombas, *Pseudopotentiale*, Springer Verlag, Wien (1967).
79. J.C. Slater, Phys. Rev., **81**, 3853 (1965); Adv. Quantum Chem. **6**, 1 , (1972); *Electronic Structure of Atoms and Molecules*, Tech. Rep. no.**3**, Solid State and Molecular Theory Group, M.I.T. (Feb. 15, 1953).
80. A. Savin, Int. J. Quant. Chem. Symp., **22**, 59 (1988), J. de. Chim. Phys., **84**, 757 (1989).
81. D.M. Ceperley and B.J. Alder, Phys. Rev. Lett., **45**, 566 (1980); also see D.M. Ceperley, Phys. Rev. B., **18**, 3126 (1978).
82. M. Rasolt and D.J.W. Geldart, Phys. Rev. B, **13**, 1477 (1976).
83. U. von Barth and L. Hedin, J. Phys. C., **5**, 1629 (1972).
84. E. Clementi and G. Corongiu in Chapter 6 of "MOdern TEchniques for Computational Chemistry - MOTECC-89", E. Clementi Editor, ESCOM, Leiden 1989.
85. G. Corongiu, E. Clementi, S. Chakravorty and P.L. Cristinziano, KGNMOL- Input/Output Documentation from MOTECC Input/ Output Documentation, S. Chin, E. Clementi, G. Corongiu, M. Dupuis, D. Frye, D. Logan, A. Mohanty and V. Sonnad (1990).
86. A.D. McLean and M. Yoshimine, Supp. to IBM J. Res. Dev. (1967).
87. P.E. Cade and W. Huo, At. Data and Nucl. Data Tables **12**, 415 (1973).
88. E. Clementi and S. Chakravorty, IBM Tech. Rep., KGN-215, (April 19, 1990).
89. R.S. Mulliken, J. Chem. Phys., **23**, 1833, 1841, 2338, 2343 (1955).
90. S. Antoci and L. Mihich, Gazz. Chim. Ital. **108**, 383 (1978).
91. S. Chakravorty and E. Clementi, Phys. Rev. A, **39**, 2290 (1989).
92. E.Clementi, J. Chem. Phys., **39**, 1 (1963).
93. C. Hollister and O. Sinanoglu, J. Am. Chem. Soc., **88**, 13 (1966).
94. See V.R. Saunders in "Computational Techniques in Quantum Chemistry and Molecular Physics," G.H.F. Diercksen, B.T. Sutcliffe and A. Veillard, Eds., Reidel, Dordrecht (1975).
95. See V.R. Saunders in "Methods in Computational Physics," G.H.F. Diercksen and S. Wilson, Eds., Reidel, Dordrecht (1983).
96. D. Hegarty and G. van der Velde, Int. J. Quant. Chem., **23**, 1135 (1983).
97. D. Hegarty in "Advanced Theory and Computational Approaches to the Electronic Structure of Molecules," Reidel, Dordrecht (1984).
98. S. Obara and A. Saika, J. Chem. Phys., **84**, 3963 (1986).
99. I. Shavitt, Methods in Computational Physics, **1**, 1 (1963).
100. S. Obara and A. Saika, J. Chem. Phys., **89**, 1540 (1988).
101. (a) M. Dupuis, J. Rys and H.F. King, J. Chem. Phys., **65**, 111 (1976); (b) H.F. King and M. Dupuis, J. Comput. Phys. **21**, 44 (1976); (b) J. Rys, M. Dupuis and H.F. King, J. Comput. Chem., **4**, 154 (1983).
102. J.A. Pople and W.J. Hehre, J. Comput. Phys., **27**, 161 (1978).

103. M.W. Wong, G. Corongiu and E. Clementi, IBM Research Report (April 1990).

Chapter 8.

SIRIUS: A General Purpose Direct Second Order MCSCF Program

Hans Jϕrgen Aa. Jensen
Department of Chemistry
University of Aarhus
DK-8000 Aarhus C, Denmark

Hans Ågren
Institute of Quantum Chemistry
University of Uppsala
Box 518, S-75120 Uppsala, Sweden

Jeppe Olsen
Theoretical Chemistry, Chemistry Center
University of Lund
Box 124, S-22100 Lund, Sweden

A. Introduction

Many recent advances in the multiconfiguration self-consistent field (MCSCF) description of molecules have made MCSCF a useful tool in practical applications. Routine optimization of MCSCF wave functions is now a realistic proposition even for configuration spaces with more than one million configurations. These advances have opened new possibilities for a theoretical description of electronic structures and properties as well as for the understanding of various phenomena associated to molecules. This has already been manifested by many recent MCSCF studies. The general purpose MCSCF program SIRIUS,[1-6] described in the present chapter, has accomplished these possibilities by using a direct, fully second order and step-restricted optimization algorithm.

The MCSCF method dates back to the early days of quantum chemistry.[7] Like other *ab initio* electronic structure methods for solving the many-body problem of molecules it has experienced a substantial development both concerning its theoretical basis and its implementation. Until a decade ago the overall utility of MCSCF was limited due to lack of reliable and fast convergence algorithms, in particular for applications requiring large wave functions. Accordingly, only a few studies used MCSCF as a stand-alone method. Instead, MCSCF was mostly used for pre-optimization of orbitals in larger scale correlated calculations.

The prospects for MCSCF have changed drastically during the last decade because of theoretical progress along several lines. We summarize this progress in a four-fold way: (1) The introduction of a unitary parametrization of orbital rotations, (2) the introduction of direct CI algorithms, (3) the development of direct second order MCSCF methods, and (4) the implementation of the restricted step optimization method. In addition to these achievements one should also mention other aspects that have served as motifs for the development of MCSCF, in particular the emphasis that lately has been put on analytical response theory and on calculations of molecular properties in general. We refer to recent reviews[8-12] for an overview of the historic developments.

The novel feature with the present MCSCF method, as implemented in the SIRIUS program, was the unification of the four important developments outlined above. It is thus unitarily parametrized and density matrix driven with CSF or determinant based formulations for handling of long configurational expansions. This makes it possible to handle larger wave functions in a fully coupled second order optimization, and therefore provides a possibility for more accurate wave functions. The second order optimization gives the possibility to obtain sharp and controllable convergence in a relatively few number of iterations on the energy hypersurface. This requires full account of the Hessian energy matrix, which for large CI expansions can only be obtained by using iterative techniques with direct evaluation of the Hessian matrix times a vector without explicit construction of the Hessian. The unitary parametrization removed previous problems with orthogonality and constrained optimization, and made it easy to formulate a second order algorithm. The use of second order expansions forms the basis for reliable optimization with the restricted step algorithm.[13] The direct CI methods for full CI's introduced with the graphical unitary group approach (GUGA)[14-17] made it possible to handle long configuration expansions in the MCSCF model. Recently even better performance has been obtained with new determinant CI's,[18,19,6] and this is used in SIRIUS. The one-index transformations allow direct large scale MCSCF with full second order optimization, in particular, with full account of the configuration-orbital coupling block.

The direct CI algorithms are particularly efficient for full CI, and the complete active space model (CASSCF), where a full CI is performed in the subset of active orbitals, has therefore become the prominent MCSCF model, *c.f.* review by Roos.[10] (Full CI in the valence orbitals has also been introduced by Ruedenberg and coworkers[20,21] under the name of fully optimized reaction space: FORS.) The CASSCF model has the conceptual advantage that the calculation is defined by a selection of the active orbitals and the number of active electrons, which at least partly can be carried out on physical grounds. With this progress in MCSCF methodology it is possible to include important static correlation in the reference wave function, even for quite complicated cases, and thus to provide an adequate zeroth order or better description. However, the CASSCF model has a major drawback: the rapid growth in number of configurations as a function of active orbitals and electrons. Many of the most accurate *ab initio* calculations of today are therefore carried out in a two-fold fashion: an initial MCSCF calculation to optimize the most important orbitals and to select reference configurations is followed by, for example, a singles and doubles multiconfiguration-reference CI (MR-SD CI) calculation to take account of the dynamic correlation energy. However, with the recent extension of efficient direct CI and MCSCF techniques to an occupation restricted active configuration space (RASSCF),[6] which is included in SIRIUS, it may often be possible to use MCSCF as the final wave function.

In addition to MCSCF (and the special cases Hartree-Fock and configuration interaction, HF and CI), the SIRIUS program can also evaluate Møller-Plesset second order[22] (MP2) energies. However, the MP2 module is primarily considered an auxiliary algorithm to aid in the design of the MCSCF wave function and to provide a good initial guess of the MCSCF orbitals. This is accomplished by evaluation of the MP2 natural orbitals through second order. It is also possible to freeze a selected number of core orbitals, giving the second order correlation energy and natural orbitals based on the valence electrons only. This may often be more relevant than the full MP2 result because standard basis sets do not contain orbitals which will give an adequate description of the correlation of the core electrons.

1. Why MCSCF?

The primary goal of the MCSCF model is to provide adequate accuracy with a more compact wave function, and thus computationally more feasible, than the configuration interaction model (CI) where the orbitals are frozen according to some criterium, *e.g.* canonical Hartree-Fock orbitals or MCSCF orbitals from a small MCSCF. The simultaneous variational optimization of the occupied orbitals and the configuration coefficients allows the orbitals to change shape in the given configurations instead of having to add new configurations with new orbitals describing the change in shape. This flexibility in orbital shape makes MCSCF much more powerful than CI in describing *e.g.* polarization of the electronic cloud caused by an external field (*i.e.* in describing response properties). The MCSCF model is *general* in the sense that it may be used for excited states, core hole states, transition states and so on as well as regular ground states at equilibrium geometries.

An important feature of the MCSCF model is that MCSCF wave functions can describe dissociation correctly. This property stems from the inherent *self-adjustable* character of the MCSCF wave function that can handle *a priori* unknown electronic structures away from the ground state equilibrium, that is, in situations when neither experimental nor theoretical data are available, nor is intuition at hand to design the wave function. For ground state species at equilibrium geometry the aufbau principle and the molecular orbital picture usually holds. Here the wave function is in general strongly dominated by a single Slater determinant, the Hartree-Fock determinant, implying that perturbational oriented approaches are well suited to catch the missing correlation contribution to the property of interest. For excited states, or for reactive and transition states or other points on the potential energy surface, a break-down of the molecular orbital picture is more a rule than an exception. Any theoretical treatment should, therefore, ideally handle equilibrium, reactive or transition states on the same footing. In particular the making or breaking of molecular bonds and the description of electronic structures at transition states constitute problematic aspects in perturbational approaches which can be handled correctly by MCSCF. In such cases there is often a sharp variation of density and wave function, the latter containing mixture of covalent and ionic terms and near-degeneracies of electronic configurations, *etc*. MCSCF thus starts out with a correct treatment of the static part of the electronic correlation and can gradually be enlarged to encompass the dynamical correlation; it is thus possible with RAS-MCSCF to obtain a larger fraction of the dynamical correlation than previously has been the case for MCSCF.

Because of its fully variational character an optimized MCSCF wave function lends itself excellently for use as a reference wave function in static or dynamic property calculations. For any quantity, possibly apart from the total energy of a given geometrical point, the variational MCSCF approach offers many analytical as well as computational advantages. New formulations of response function theory, in conjunction with a unitary parametrization of the wave function including also its time dependence, has led to a number of theoretical simplifications and efficient approaches that require a fully variational, *viz*. MCSCF, character of the reference wave functions.[23] Analytical response theory, as opposed to *e.g.* finite differences, expresses the response property in terms of the property gradient vector, the MCSCF gradient (which is zero), and the MCSCF Hessian. As a result, a number of properties, like molecular gradients and Hessians, polarizabilities, dipole derivatives, *etc*., can be computed with techniques similar to those used in wave function optimizations. The equations to be solved when evaluating second order molecular properties are similar to those that show up in Newton-Raphson MCSCF optimization, in some cases the operations are identical. Thus with these approaches the inherent accuracy of MCSCF wave functions is directly "transferable" to a number of molecular properties; the MCSCF contains *analytical transferability* between wave functions and properties. The analytical transferability also implies that for a given level of approximation for the wave function, determined by the particular choice of 1-particle and N-particle basis sets, a number of properties may be obtained exactly (within the model) and simultaneously. Such analytical approaches provide various rigorous criteria to check the calculations, and it also gives the possibility to obtain a large body of data for different properties in one and the same batch of calculations. This helps in identifying a confidence level for predictions.

The attractive features of MCSCF can thus be coined as "variational" and "general" (including transition and excited states), "self-adjustable" and "analytically transferable". As shown in the next paragraphs, the inherent potential of MCSCF is fully utilized in a second order formulation. On the negative side one can mention lack of size-consistency, complex programming and calculations, and the fact that the choice of an adequate configuration space makes attempts at a black-box implementation dubious. Concerning the size-consistency problem one finds that the MCSCF wave function behaves better than the corresponding CI due to the orbital reoptimization, and one should also qualify its relative merits in this respect by the fact that at this time no computational feasible method is available that treats correctly both dissociation and size-consistency.

2. Summary of program features

The underlying MCSCF theory, the program implementation, and the use of auxiliary algorithms for efficient optimization of large scale MCSCF wave functions is described in the present chapter. Various aspects of the program and its auxiliary algorithms are demonstrated by a series of sample applications. In this subsection we first summarize the main program features.

SIRIUS is a general purpose MCSCF program based on a direct, fully second order and step-restricted optimization algorithm. The direct character refers to the iterative solution of an MCSCF eigenvalue equation by means of successive linear transformations where the norm-extended Hessian matrix is multiplied onto a trial vector without explicitly constructing the Hessian. This allows the use of very large electronic wave functions. A norm-extended optimization algorithm is utilized in which the number of negative eigenvalues of the Hessian is monitored and in which optimization control is based on the trust region concept. The iterative solution of the eigenvalue equation is accomplished with a generalization of the Davidson-Liu simultaneous expansion method for iterative calculation of an eigenvector. The NEO algorithm guarantees convergence to the lowest state of a symmetry, and, to our experience, it also shows reliable convergence for excited states. The MCSCF wave function may be specified as complete active space or restricted active space, the latter including as a special case the so-called second order wave functions with single and double excitation spaces.

Main attributes. The program can be attributed three main labels

i) *second order:* It is second order because it is based on the second-order Taylor expansion of the electronic energy in all variational parameters, *viz.* configuration coefficients and orbital rotations. This implies quadratic convergence and the possibility for sharp convergence.

ii) *direct:* It is direct because it gives the solution of an MCSCF eigenvalue equation by means of successive linear transformations where the full norm-extended Hessian is multiplied by a trial vector without explicitly constructing the Hessian. This permits optimization of large wave functions with many configurations for which an explicit construction of the Hessian would be prohibitively large, both with respect to CPU time and memory requirements.

iii) *step restricted:* The program is step-restricted because it performs a controlled "walk" on the energy hypersurface in which each step is restricted to a closed region, a hypersphere. The size of the hypersphere is dynamically updated through a feed-back procedure leading to guaranteed convergence for the lowest state of a given symmetry, and most often to a well-behaved convergence for the lowest excited states.

Parametrization. The SIRIUS program is parametrized for a spin-independent electronic Hamiltonian and complete active space (CAS) and restricted active space (RAS) wave functions. It uses by default configuration state functions (CSF's) in the many-electron basis set, with the option to work in a Slater determinant (SD) basis. The CAS wave functions are characterized by an orbital division into inactive, active, and secondary spaces that are doubly occupied, fractionally occupied, and empty, respectively. The inactive orbitals are optimized but not correlated; the active orbitals constitute the space for all possible electronic distributions that fulfill space and spin requirements and occupation restrictions (for RAS), and are thus both correlated and optimized. For RAS wave functions the active space is subdivided into three spaces, which will be denoted RAS I, RAS II, and RAS III. In SIRIUS, a minimum as well as a maximum number of electrons are specified both RAS I and RAS III , the distribution of electrons in the RAS II space is then determined by the conservation of the total number of active (correlated) electrons in the system. Optionally, the user can be guided in the choice of orbital division by precalculations with MP2 or iterative natural orbitals CI (INO-CI) routines (which also will produce a better initial guess for the MCSCF orbitals). For a pictorial idea of typical orbital selections for CAS and RAS calculations see Fig.1.

Optimization. SIRIUS is equipped with norm-extended optimization (NEO) as well as Newton-Raphson optimization (NR) algorithms. The NEO algorithm implemented for trust region optimization is used in the global region, while the NR algorithm has a relatively small radius of convergence and is therefore only used in the local region. The NR algorithm serves as a CPU saving option in the local region when optimizing excited states. NEO is the most efficient way known to implement the direct, restricted-step second order optimization. The advantage with the NEO algorithm may be summarized as:

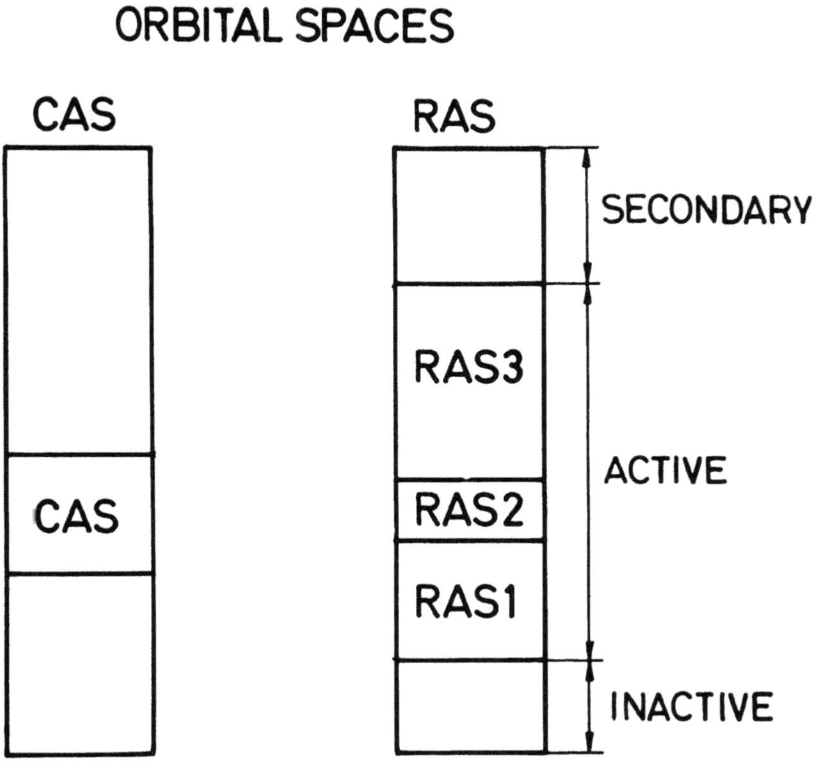

Figure 1. Schematic division of orbital spaces for complete active space (CAS) and restricted active space (RAS) wave functions.

a) Dynamical update of the damping/level-shift so that the step l is equal to the trust radius.

b) The level-shift is automatically in the correct interval.

c) It is quadratically convergent.

d) It is equal to CI if there are no orbital rotations, *i.e.* convergence in one iteration.

e) It gives the second order energy predictions without any significant extra calculations.

Convergence characteristics. The following convergence characteristics are associated with the program.

a) Guaranteed ground state convergence.

b) Reliable convergence for the lowest excited states.

c) Guaranteed correct state convergence (correct Hessian index).

d) Sharp convergence (for properties).

e) Root flipping problem avoided.

f) Core hole state convergence without variational collapse.

Auxiliary algorithms. The program is equipped with auxiliary algorithms for efficient optimization. These can be summarized as:

a) Improved start guesses for orbitals and wave functions given by HF, MP2, CI or INO-CI initial calculations.

b) Orbital absorption steps for fixed CI vectors.

c) A split trial vector algorithm.

d) Optimal orbital trial vectors.

e) Transform orbitals to Fock-type orbitals or to natural (CAS) or pseudo-natural orbitals (RAS) between macro iterations, with counterrotation of the CI vector.

f) Explicit construction of important parts of the CI Hamiltonian matrix to generate better trial vectors than the Davidson-Liu algorithm.

Other features.

a) Configuration coefficients are treated directly in the configuration state function (CSF) or Slater determinant (SD) basis such that a standard direct CI handling can be used.

b) One-index integral transformations account for the effects of the orbital Hessian and the configuration-orbital coupling block.

c) Fock matrices are used to treat the inactive, doubly occupied, orbitals, and inactive orbitals can be handled with two-electron integrals in atomic basis.

d) For closed shell and one open shell restricted Hartree-Fock a step-restricted quadratically convergent self-consistent field, QC-SCF, which only uses two-electron integrals in a.o. basis is obtained as a special case.

e) The program is implemented with D_{2h} symmetry, but can also handle simple degeneracies.

Restarts and interfaces. SIRIUS provides the possibility for restart from the beginning of the current macro iteration. It interfaces to several separate post-programs for calculations of spectra and properties, examples of these are given in Sec. F. Converged MCSCF wave functions are saved on disk for further processing by any other property programs of interest. SIRIUS interfaces at present to any program for calculation of one- and two-electron integrals that uses the MOLECULE[24] format. We have used MOLECULE,[24] HERMIT[25] and STOCOS,[26] the two former ones using multicenter Gaussian functions, the latter using one-center oscillating type (Slater trigonometric) type functions. However, other formats for the one- and two-electron integrals over basis functions can easily be adapted.

B. Theory

The non-relativistic, time independent electronic Schrödinger equation,

$$\hat{\mathscr{H}}\Psi = E\Psi, \tag{1}$$

for a fixed nuclear geometry forms the basis for a large part of quantum chemical modelling of chemical phenomena (justified through the Born-Oppenheimer approximation[27]). We will specifically consider orbital-based quantum chemistry, where one selects a basis set of orbitals ("atomic orbitals," AO's) for the description of the electronic motion, and where the wave function is built from products of these orbitals, so-called Slater determinants.

Within this framework, a quantum chemical calculation that is initiated in order to throw light on a specific phenomenon related to a specific molecule will involve the following steps. First, the geometry of the molecule must be specified. Second, a set of basis orbitals, the 1-particle basis set, must be selected, and third, the wave function approximation, the N-particle basis set, must be determined. The last two are separate choices, but not independent. (For instance, if the Hartree-Fock approximation will give sufficient information, no correlating orbitals are needed in the basis set.) If a multiconfigurational self-consistent field (MCSCF) wave function, including Hartree-Fock and configuration interaction as special cases, is an appropriate choice, the SIRIUS program will variationally optimize the configuration coefficients as well as the occupied orbitals. In the MO-LCAO approximation, the molecular orbitals are built as linear combinations of the atomic orbitals.

The basic assumption is that of a spin-independent electronic Hamiltonian which is projected onto the Fock space defined by the chosen set of orbital basis functions through an outer projection. This will be elaborated in the following subsection, which contains some general considerations about the Hamiltonian in orbital based quantum chemistry. Next, the MCSCF wave function will be defined and parametrized, and in subsection 3 we discuss the design of an MCSCF calculation, that is, the selection of configurations. In the final two subsections the restricted-step second order MCSCF optimization is introduced followed by an overview of how this may be implemented in the direct iterative NEO formulation.

The basic implementation is described in Sec. C and the configuration interaction theory used in SIRIUS is presented in Sec. D. In Sec. E auxiliary algorithms for faster convergence are elucidated and Sec. F reviews some applications.

1. Orbital based quantum chemistry: the Hamiltonian

The formalism of second quantization is ideally suited for description of orbital based electronic structure models. In second quantization the theoretical expression for the Hamiltonian has the same form as it has in conventional quantum chemistry program codes, namely containing terms of nuclear repulsion, the usual one- and two-electron integrals, and, indirectly, the overlap integrals:

$$\hat{\mathscr{H}} = V^{nn} + \sum_{pq} h_{pq}\hat{E}_{pq} + \frac{1}{2}\sum_{pqrs} (pq \mid rs)\hat{e}_{pq,rs} \tag{2}$$

where $\hat{E}_{pq}$ and $\hat{d}_{pq,rs}$ are the spinless one- and two-electron operators:

$$\hat{E}_{pq} = \sum_{\sigma} \hat{a}^{\dagger}_{p\sigma}\hat{a}_{q\sigma} \tag{3}$$

and

$$\begin{aligned}\hat{e}_{pq,rs} &= \sum_{\sigma\sigma'} \hat{a}^{\dagger}_{p\sigma}\hat{a}^{\dagger}_{r\sigma'}\hat{a}_{s\sigma'}\hat{a}_{q\sigma} \\ &= \hat{E}_{pq}\hat{E}_{rs} - \delta_{qr}\hat{E}_{ps}\end{aligned} \tag{4}$$

The operators $\hat{a}^{\dagger}_{p\sigma}$ and $\hat{a}_{p\sigma}$ are the creation and annihilation operators associated with the spin-orbital $\varphi_{p\sigma}$, and they obey the usual anti-commutation rules, see *e.g.* Ref. (28). The summations run over an orthonormal set of so-called molecular orbitals (MO's) which span the space of orbital basis functions (MO-LCAO), and the Hamiltonian is thus an outer projection on the orbital basis set selected for the calculation. In the limit of a *complete* set of orbital basis functions the second quantized Hamiltonian is equivalent to the conventional Hamiltonian in the "first quantization" Schrödinger picture in terms of differential operators and multiplicative operators.

The Hamiltonian and other operators are in second quantization represented by the projection of the full "first quantized" operators onto the space spanned by the molecular orbitals, that is, the space in which the calculations are actually carried out. This means that derived formulas are directly programmable; the formulas are always expressed in terms of integrals, so it is possible to make manipulations at the operator level which translate into new computational expressions. The domain of the operators is called the Fock space, which is spanned by all possible states which can be built from the set of basis orbitals. Slaters rules for the Hamiltonian matrix elements between two Slater determinants are built into the operator algebra through the anticommutation rules, and they are therefore automatically always used to reduce the number of terms in the formulas. Furthermore, note that the number of electrons does not enter the Hamiltonian in second quantization, the state vectors carry all the occupation information (therefore, second quantization is also called the occupation number representation). The same Hamiltonian is used for calculations on, for example, a neutral molecule, the cation after ionization, or the anion after electron transfer. Of course we must ensure that the selected orbital basis set has sufficient flexibility to describe both the initial and final states of interest.

Finally, a useful flexibility in second quantization is the possibility to move different aspects as time dependence or geometry dependence to the Hamiltonian or to the wave function, whatever is computationally most convenient. (This may be compared to the choice of Schrödinger or Heisenberg picture.) An example of the flexibility is that the nuclear dependence of the basis orbitals can be moved to the Hamiltonian, and the physical relaxation of the wave function is then separated from the geometry dependence of the basis set. This modified Hamiltonian fulfills the Hellmann-Feynman theorem, and a geometrical distortion may thus be treated in the same way as, for example, electric field perturbations, and standard static response function theory can be applied to evaluate molecular gradients and Hessians as well as higher derivatives for a number of different wave function types (reviewed by Helgaker and Jørgensen [29]). Another example is the transfer of the orbital optimization to the Hamiltonian in MCSCF optimization, thus regarding the configurations as formal configurations (see Panin[30] for a mathematical definition).

We finish by noting that we now see that one way of looking at MO-LCAO models is to regard the choice of basis orbitals as the choice of a model Hamiltonian, as a projection of the full Hamiltonian on the finite orbital basis. The exact solutions to this model operator are the full configuration interaction (FCI) solutions, and the object of the MCSCF model is to obtain information about the important parts of the FCI solutions with much less computational effort. Of course this will only be of interest if the FCI solution would give an adequate description of the problem at hand, and the selection of orbital basis functions is therefore, in short, an important step which can make the rest of the calculation meaningless. It is for instance essential that the AO's can describe the extent of the electronic cloud: the basis must contain sufficiently compact as well as sufficiently diffuse orbitals. The linear combinations of AO's must also be able to form nodal planes for description of polarizing and correlation effects, that is, the basis must contain polarization functions. It is clear that the issues in selection of basis functions are the same in all orbital-based correlated calculations, variational or perturbational, and this will not be further discussed here.

Another mathematical realization of the same model is the state vector representation ("the CI Hamiltonian"):

$$\hat{\mathscr{H}}^{CI} = \sum_{\mu\nu} \mid \mu > < \mu \mid \hat{\mathscr{H}} \mid \nu > < \nu \mid \tag{5}$$

where the sums run over the full CI space. Although equivalent at this level, this formulation does not give the possibility to develop direct MCSCF, in fact, expressions will be in terms of the CI Hamiltonian which we do not want to construct explicitly.

2. Parametrization and Orbital Classes

The present program addresses the MCSCF optimization, that is, the simultaneous optimization of the configuration coefficients and orbitals:

$$\mid MC > = \exp(-\hat{\kappa})\left(\sum_{\mu} c_{\mu} \mid \phi_{\mu} >\right) \tag{6}$$

where $\{\mid \phi_{\mu} > \}$ is the selected set of configuration basis functions and $\{c_{\mu}\}$ are the expansion coefficients to be optimized. The operator $\exp(-\hat{\kappa})$, where

$$\hat{\kappa} = \sum_{s<r} \kappa_{sr} \hat{E}^{-}_{sr} \tag{7}$$

and

$$\hat{E}^{-}_{sr} = \hat{E}_{sr} - \hat{E}_{rs} \tag{8}$$

is the now well-established exponential parameterization of the optimization.[8] The variation of the parameters $\{\kappa_{sr}\}$ corresponds to all possible orthogonal orbital rotations as $\hat{E}^{-}_{sr}$ generators for proper orthogonal rotations (special orthogonal group). In this parametrization the orbitals are always orthonormal, which makes it is possible to use unconstrained optimization techniques.

The configuration basis functions are either configurational state functions (CSF's), which are spin-adapted combinations of Slater determinants (SD's), or Slater determinants. The use of CSF's instead of SD's will usually reduce the length of the configuration vectors by one third or more, thus requiring less disk storage, and will furthermore remove the risk of converging to an unwanted spin state. However, the kernel of the program always uses SD's, which is better suited for matrix operations in the computation intensive parts, and the program will switch back and forth between the two representations as required. This aspect will be elaborated in Sec. D.2.

For applications of the MC wave function in, for instance, response property calculations, we believe an exponential parametrization of the orthogonal rotations of the configuration space is to be preferred. How to use an orthogonal exponential parametrization of the configuration coefficients, still based on matrix elements between the CSF's, has been described by Olsen *et al.*[8] and by Lengsfield and Liu.[31,32] However, for the MCSCF optimization itself, this is unnecessary and we will use the configuration amplitudes c_μ directly as parameters.[1] Since this parameter space of configuration amplitudes includes a redundancy related to scaling (normalization) an orthogonal projection $\hat{P}$ which removes any component in the redundant direction must be applied at appropriate places. Applying $\hat{P}$ to the gradient and to the Hessian leads to a second-order convergent Newton-Raphson optimization similar to the one by Lengsfield and Liu.[31,32] An alternative to this procedure is to keep the norm-related variable combination in the Hessian (hence the notion of norm-extended optimization (NEO)). The *n*th eigenpair of this slightly modified Hessian can then be used to define another second-order convergent optimization scheme, which is intrinsically connected to our preferred approach of using a restricted-step algorithm.[1] A more explicit description is given in the next section. The augmented Hessian approach (AH)[33,34] is a special case of NEO. We have shown how the extra parameter can be identified and how to implement the restricted-step algorithm. The reason for the necessity of a "slightly modified" Hessian is that only real configuration amplitudes with real variations are considered instead of complex amplitudes with complex variations, *i.e.*, not $\left(\frac{\partial}{\partial c_i^*}, \frac{\partial}{\partial c_i}\right)$ but $\left(\frac{\partial}{\partial c_i^*} + \frac{\partial}{\partial c_i}\right)$.

We want to optimize the energy functional

$$E(\mathrm{x}) = \frac{< MC \mid \hat{\mathscr{H}} \mid MC >}{< MC \mid MC >}, \tag{9}$$

where $\hat{\mathscr{H}}$ is the non-relativistic Hamiltonian given in Eq. (2) and x is the combined set of configuration coefficients c and orbital rotation parameters κ from Eq. (6)B6a. This is equivalent to finding the x for which $E(\mathrm{x})$ is stationary, that is, for which the gradient is a zero vector (the generalized Brillouin theorem, GBT), and for which the Hessian has the correct structure: positive definite for the lowest state of a symmetry, one negative eigenvalue for the next state in the same symmetry, *etc.*

(a) Orbital Classes: For the specification of the configuration space, the molecular orbitals are divided into three primary classes: inactive orbitals, active orbitals, and secondary orbitals. The inactive orbitals are doubly occupied (and thus not correlated), the active orbitals have occupation between zero and two, and the secondary orbitals are empty. All occupied orbitals, that is both the inactive and active orbitals, are optimized in the MCSCF calculation. Occasionally it may be advantageous to introduce two additional classes: frozen and deleted orbitals. The latter term is used about orbitals which are deleted from the basis functions, examples are the 3*s* component of Cartesian *d*-type orbitals and numerically linear dependent orbitals. Frozen orbitals designate doubly occupied orbitals which are *not* allowed to relax in the MCSCF optimization. They can be used as a kind of extended core potential for transition elements or it may be desirable to freeze the core orbitals to reduce basis set superposition error.

The SIRIUS program system is designed to handle two types of MCSCF efficiently: complete active spaces (CAS) and restricted active spaces (RAS). The MCSCF calculation is specified through user specification of active orbitals and number of "active" electrons, and for RAS in addition restrictions in occupation of these orbitals. The CAS wave function is a full CI in the active orbital subset, where the active electrons are distributed in all possible ways among the active orbitals. The CAS wave function is size consistent, provided the active orbitals have been selected in a symmetric way among the constituent atoms, and has the possibility for correct dissociation. Unfortunately, this choice will not in general give a balanced treatment of a potential energy surface because the amount of correlation which is included can vary a lot for different geometries.[10,11] The severest problem with the CAS model is the very fast increase in the number of CSF's as a function of the number of active electrons and the number of active orbitals.

(b) Restricted active space wave functions. A modern MCSF program, like SIRIUS, can employ CI expansions with up to perhaps ten million CI parameters, and it is therefore important to have very efficient routines for handling CI expansions of this size. Another prerequisite is that the CI codes must allow CI expansions more general than full CI in a set of active orbitals, in order to allow, at least partly, inclusion of dynamic correlation effects in MCSCF calculations. As previously mentioned, the SIRIUS program supports the use of the so-called RAS expansions, where the active space is divided into three spaces, RAS I, RAS II, and RAS III and a lower (upper) limit is placed on the number of electrons in RAS I (RAS III). The SIRIUS codes contain a slight generalization of the RAS concept, by allowing the specification of both an upper and a lower limit on the allowed number of electrons in RAS I as well as RAS III. In addition to the multireference excitation schemes, as multireference singles and doubles CI, this allows the use of CI spaces that employ $\sigma - \pi$ separation and CI expansions relevant for core-hole spectroscopy.

CI calculations can be done in a configuration state function (CSF) basis, where each parameter corresponds to an eigenfunction of $\hat{S}^2$, or in a Slater determinant (SD) basis, where each parameter corresponds to an eigenfunction of the spin projection on a given axis, $\hat{S}_z$. The main advantages of the CSF approach are that a well-defined spin multiplicity is ensured and that the number of CSF's is considerable smaller than the correponding number of SD's. The main advantage of the SD approach is that the it allows the construction of very fast codes; the SD based codes employed in SIRIUS are faster than any other published codes for RAS expansions.

An obvious way of combining the two approaches is to store all CI vectors in the CSF basis, thereby using minimal disk space and ensuring a well-defined spin multiplicity, and transform to the SD basis and carrying out the direct CI calculation in the SD basis, thereby allowing the use of the fast SD codes, and finally transforming the result of the direct CI iteration back to the CSF basis. This is the procedure followed in SIRIUS.

3. Design of the MCSCF wave function

From the parametrization introduced in the previous sections it follows that after the choice of basis orbitals, the next issue is the selection of the orbital classes and through this selection, the configuration space. As a background for this, the problems with multiple stationary points are discussed. This leads also to considerations about which configurations are needed for an adequate description (*i.e.*, one which returns the desired information about the chemical problem to be studied) Next we focus the attention on the use of Møller-Plesset second order perturbation theory (MP2) as both orbital and configuration generator.[5] This method is only applicable to closed-shell electronic states, and the possibility of using iterative natural orbital CI (INO-CI) in the general case is briefly discussed. Finally, some remarks on the description of excited states are given.

(a) Multiple Stationary Points: When an MCSCF calculation is carried out on the N'th state of a given symmetry, the Hessian index of that state (stationary point) must be N-1.[8] As the MCSCF energy hypersurface is expressed in terms of linear (configuration) and non-linear (orbital) parameters, many stationary points may exist on the MCSCF hypersurface with index N-1. Not all of these are good representations of the N'th state, some of them may be spurious (unphysical), but often different stationary points describe different correlation aspects of the molecule. Limitations in resources and economic considerations require a selection of which aspects should be described in detail. Several authors have discussed criteria to decide which of the multiple stationary points that can be accepted on general grounds. Golab *et al.*[35] suggest the solution should be stable towards imaginary perturbations, as is true for the full CI result. (This cannot be checked with the current version of SIRIUS.) Wavefunctions not passing this test are said to contain a triplet or singlet spin instability (instability in triplet or singlet space). As an example, for Hartree-Fock a triplet instability means that unrestricted Hartree-Fock would lead to lower energy than the restricted Hartree-Fock state.

Energy criteria should only be used with caution, even for ground state optimizations. In a calculation on HCl[5] the desired solution was not the lowest minimum found. The local minimum for the CASSCF wavefunction where the valence electrons in 3*s* and 3*p* were correlated had about 0.04 a.u. higher energy than another minimum where 2*s* and 2*p* were correlated. In many cases the use of MP2 or INO-CI as orbital and configuration generator significantly reduces the risk of converging to an unwanted minimum, as discussed below.

For excited states, a commonly used criterion is to require all N-1 negative eigenvalues of the Hessian to be in the configuration variable subspace.[11] This ensures the variational upper bound theorem. However, this requirement can be too restrictive, for example, if one wants to investigate a larger region of an excited state potential surface. Root-flipping is defined as the occasion where at least one of the negative eigenvalues is in the space of orbital rotations. Root-flipping is not desirable although it does not by itself imply that the wavefunction is a poor description of the excited state. For instance, one may obtain an excellent description of core hole states with SIRIUS, even though these MCSCF wave functions will have many negative eigenvalues in the space of orbital rotations. Root flipping may also be acceptable in low-lying excited states which have a different correlation structure than the ground state, and avoidance of root flipping would require many extra active orbitals that only would give insignificant improvements in the excited state wave function.

(b) Selection of Configuration Space: It is important to describe well those (natural) orbitals, which are active in the experiment, especially those which in the limit of full CI has a big change in occupation. When the change is primarily in shape, it can usually be described on the Hartree-Fock level (kept doubly occupied), if high accuracy is not needed. For a chemical reaction the minimal requirement is that both reactant and product states are covered. Usually the valence orbitals are of primary interest, but not always. Core photoionization provides an example where correlation of the core orbitals can be significant (especially when symmetry necessitates more than one configuration to locate the core hole on one atom, for example the $O_2(1\sigma_g^{-1}$ state). Another example is calculation of spin polarization at the nucleus, which might be important for accurate analysis of, for example, an ESR experiment.

This viewpoint makes it very appealing to choose the configuration space based on a choice of *active* orbitals, and this is the philosophy behind the CASSCF model. In the CASSCF model a subspace full CI is performed with the active electrons in the space of active orbitals, and the advent of direct CI techniques have made such models attractive. As advantages count that the "arbitrary" choice of configurations is limited to a choice of active orbitals rather than individual configurations and the flexibility to obtain dissociation into the correct fragment states with size consistency because of the subspace full CI. The major drawback is the growth in size as a function of number of electrons and orbitals. Another problem is that it can be difficult *a priori* to select a set of active orbitals, which gives an unbiased description of a whole potential energy surface different orbitals may be important in different sectors.[10,11]

An intuitive guide to selection of active orbitals can be obtained from the electron pair picture of the molecular electronic distribution, and a standard way to choose a configuration space has been to include in the active space one correlating orbital for each strongly occupied active orbital.[9,10] Care should be exercised, and Roos[10] illustrates through examples some of the more complicated bonding situations which may occur. An adequate description then requires more correlating orbitals in the active space. An analysis of the converged MCSCF wave function in terms of an electron pair picture (and its breakdown) can often give qualitative insight in the nature of the converged wave function and is very useful as such. Shepard[11] also reviews some examples on occasions where one correlating orbital for each strongly occupied is insufficient. He lists three cases, discontinuity caused by shift of correlation (*e.g.*, from bond to lone pair), when $(A \cdot \; B \cdot)$ competes with $(A^+ \; B^-)$, and when two qualitatively different correlation effects compete. His solutions are (1) include lone pair orbitals in active space with a correlating orbital, (2) orbital doubling (two correlating orbitals for each strongly occupied), and (3) include extra active orbitals (*e.g.*, π-orbitals for Li_2.

Many effects cannot be satisfactorily described with CASSCF because of the rapid growth in number of configurations when higher accuracy is sought. A multi-reference configuration interaction calculation on top of the CASSCF calculation seems to be able to give the higher accuracy, and this led to the introduction of a new MCSCF model in SIRIUS: the restricted active space "RASSCF"[6] which encompasses the typical MR-CI configuration spaces and, for example, $\sigma - \pi$ separation and core hole restrictions. The aim is to obtain high accuracy in the MCSCF calculation and, thus, eliminate the need to follow the MCSCF calculation by a larger CI calculation. The cost is the loss of size consistency.

(c) Second Order Møller -Plesset Perturbation Theory as Orbital and Configuration Generator: For a closed-shell molecule, the natural orbitals from second order Møller-Plesset perturbation theory, MP2, are a useful source for guidance in the choice of active orbitals. This has been discussed and numerically tested in Ref.(5). The MP2 orbitals also provide a good and cost-effective starting point for the MCSCF optimization, in addition the MP2 orbitals significantly reduce the risk of converging to an unwanted minimum.

For ground state calculations on molecules where Hartree-Fock is a reasonably good approximation, the magnitude of the natural orbital (NO) occupation numbers in an MP2 calculation gives a good estimate for choosing balanced and physically reasonable configuration spaces. The occupation number analysis should be supplemented with an analysis of the structure of the MP2 NO's before the active space is chosen. It is also possible to freeze a selected number of core orbitals, giving the second order correlation energy and natural orbitals based on the valence electrons only. This will give an even better starting point, and furthermore reduce the risk of including undesired core-correlating orbitals in the active space.

The investigation in Ref. (5) leads to the following conclusions. For many molecular systems, for example for H_2O, the simplest balanced choice based on the magnitude of the MP2 NO occupation numbers coincides with the choice of one correlating orbital for each strongly occupied. However, in some cases (for example HCl) the occupation numbers of the MP2 NO's show that for the simplest balanced choice of a CAS configuration space it is necessary to include two correlating orbitals for some of the strongly occupied orbitals. The MP2 NO occupancies can also guide the choice of more accurate CAS configuration spaces, but for larger molecular systems it can be difficult to select reasonable configuration spaces using a CAS wave function, and the more general RAS model is recommended for such cases. For example, a RAS model is necessary to get an accurate MCSCF description of benzene, which goes beyond a π-electron active space, without going to very large, expensive configuration spaces.

The choice of configuration space based on MP2 NO orbitals is also vindicated by the result found in Ref. (5) that the MCSCF NO occupation numbers are close to the MP2 numbers and that the MCSCF NO's have large overlap with the MP2 NO's. The MP2 NO's therefore provide a very good initial guess for the orbitals, the MCSCF calculation can be expected to be in the local region from the first iteration and fast second order convergence will be observed for the whole calculation. Tests also indicate that for large scale CI calculations on closed shell systems with a dominant single configuration the MP2 NO's often constitute a better choice than MCSCF orbitals to reduce the size of unaccounted orbital relaxation.

When MP2 is a good model, it can almost be used to define a "black box" for generating configurations and orbital for MCSCF optimization, the program would then choose the configuration space automatically based on MP2 NO occupations, within specified or default constraints. However, this is not to be recommended, because it is likely to give bad choices some times. We recommend that the user analyzes the MP2 results and makes a qualified choice based on that analysis.

(d) Iterative natural orbital configuration interaction as orbital and configuration generator: For molecular systems which are open shell or with dominant multiconfiguration character the MP2 model cannot be used. This includes almost all excited states. For these cases SIRIUS provides the option to use the NO's from a configuration interaction calculation, that is, to perform one iteration in an iterative natural orbital CI scheme (INO-CI). The simplest useful choice would typically be a SD-CI calculation. The disadvantage of the CI's compared to perturbation theory is that a CI requires a full integral transformation while MP2 NO's only need integrals with at most three unoccupied indices (unoccupied with respect to the Hartree-Fock).

4. Restricted-step, second-order MCSCF optimization

We want to optimize the energy functional $E(\mathbf{x})$ given in Eq. (9). For this purpose the energy functional is expanded to second order in the non-redundant variables around the current expansion point (CEP), $\mathbf{x}^{(k)}$:

$$\Delta E^{(2)}(\mathbf{y}) = E^{(2)}(\mathbf{x}) - E(\mathbf{x}^{(k)}) = \mathbf{g}^{\prime t}\mathbf{y} + \frac{1}{2}\mathbf{y}^{t}\mathbf{H}'\mathbf{y} \tag{10}$$

where $\mathbf{y} = \mathbf{x} - \mathbf{x}^{(k)}$ is the step, $\mathbf{g} = \mathbf{Pg}$ is the gradient at the CEP, and

$$\mathbf{H}' = \mathbf{PHP} \tag{11}$$

is the Hessian at the CEP. The projector $\mathbf{P}$

$$\mathbf{P} = \mathbf{I} - \mathbf{O} = \mathbf{I} - \mathbf{x}^{(k)}\mathbf{x}^{(k)t} \tag{12}$$

removes the redundant CEP component ($\mathbf{I}$ is the identity matrix).

Restricted step optimization: Second order optimization schemes are based on using $\Delta E^{(2)}$ as an approximation to ΔE:

$$\Delta E(\mathbf{y}) = E(\mathbf{x}) - E(\mathbf{x}^{(k)}) = \Delta E^{(2)}(\mathbf{y}) + R^{(2)}(\mathbf{y}) \tag{13}$$

Here $R^{(2)}$ is the remainder from the second order Taylor expansion, the effect of all higher-order terms. The applicability of such an approximation is clearly dependent on the ratio between the remainder $R^{(2)}$ and the second order prediction $\Delta E^{(2)}$; only when $R^{(2)}$ is less important than $\Delta E^{(2)}$ can the second order expansion be trusted to correctly describe the structure of the parameter hypersurface. First we describe how the optimal step $\mathbf{y}^{R}$ based on knowledge of $\Delta E^{(2)}$ may be determined, and then the expressions for the gradient and Hessian needed to evaluate $\Delta E^{(2)}$ are presented. The Newton step $\mathbf{y}^{N}$ is defined through optimizing the quadratic function $\Delta E^{(2)}$, which has exactly one stationary point (provided the Hessian $\mathbf{H}'$ is non-singular)

$$\mathbf{y}^{N} = -(\mathbf{H}')^{-1}\mathbf{g} \tag{14}$$

If $\mathbf{H}'$ has the correct structure and $\Delta E^{(2)}(\mathbf{y}^{N})$ is an acceptable approximation to $\Delta E^{(2)}(\mathbf{y})$, then this is the optimal optimization step we can take based on the information in $\Delta E^{(2)}(\mathbf{y}^{N})$. If these conditions are not fulfilled the region of space must be restricted to a neighborhood of the CEP; $\| \mathbf{y} \| \leq r_{\text{trust}}$, where $\Delta E^{(2)}(\mathbf{y})$ is an acceptable approximation to $\Delta E(\mathbf{y})$. This leads to a restricted step algorithm.[13] We choose to use the standard Euclidean 2-norm such that this region of space, which we will call the trust region, becomes a hyper-sphere in the parameter space. The task is to find the optimal point in this closed region of space, and for minimization we shall minimize $\Delta E^{(2)}(\mathbf{y})$ for $\mathbf{y}$ belonging to the trust region. This always has a solution, because the trust region is a closed region. The solution is either a stationary point in the interior (the Newton step $\mathbf{y}^{N}$) or a stationary point on the boundary. Using a Lagrange multiplier one can show[36] that the optimal point $\mathbf{y}^R$ always can be written as:

$$\mathbf{y}^R = -(\mathbf{H}' - \nu\mathbf{I})^{-1}\mathbf{g} \tag{15}$$

with a specific, unique value of ν. If the optimal step is the Newton step then $\nu = 0$, otherwise ν assumes *the one* value between $-\infty$ and the lowest eigenvalue of $\mathbf{H}'$ which makes $\| \mathbf{y}(\nu) \| = r_{\text{trust}}$.

The remaining ingredient in the restricted-step algorithm is a procedure for updating the size of the trust region without a lot of calculations of the exact energy change. We start by guessing a trust radius, perform one iteration and compare the actual energy change to the predicted energy change:

$$r(\mathbf{y}) = \frac{\Delta E(\mathbf{y}^R)}{\Delta E^{(2)}(\mathbf{y}^R)} = 1 + \frac{R^{(2)}(\mathbf{y}^R)}{\Delta E^{(2)}(\mathbf{y}^R)} \tag{16}$$

If $r(\mathbf{y}^R)$ is close to unity then $R^{(2)}$ is much smaller than $\Delta E^{(2)}$ and the second order expansion is a good approximation to the exact surface. We then increase the trust radius and take a longer step in next iteration. If the ratio $r(\mathbf{y}^R)$ is negative, then the remainder $R^{(2)}$ is both bigger than $\Delta E^{(2)}$ and has opposite sign, and the second order approximation is clearly inadequate. In this case we must reject the step and calculate a new step from the previous expansion point based on a smaller trust region.

With a restricted-step algorithm along these lines, based on a trust region and a dynamical update of this trust region, the MCSCF optimization has guaranteed convergence to a minimum in a finite number of steps (to a given convergence tolerance greater than zero). Sec. C.2 specifies the implementation in SIRIUS. For more details on restricted-step algorithms and proof of convergence theorems we refer to chapter 5 in the monograph by Fletcher.[36]

The gradient and the Hessian: We have two types of variables to optimize, the configuration amplitudes and the molecular orbitals, parametrized with the amplitudes themselves and with the non-redundant orbital rotations. In the rest of the chapter we will use ^{c} or ^{o} as superscripts for vectors when the formula only deals with the configuration or orbital rotation parameters, with respectively. The vectors ^{c} superscripts are extended with zeroes for the orbital rotation parameters and the vectors with ^{o} superscripts are extended with zeroes for the configuration amplitudes when they enter in formulas where all parameters are included. A vector $\mathbf{v}$ of all the variables will thus be the sum of $\mathbf{v}^c$ and $\mathbf{v}^o$

$$\mathbf{v} = \mathbf{v}^C + \mathbf{v}^O \tag{17}$$

We define a macro iteration as each time we change the current expansion point (CEP). In the kth macro iteration the CEP equals

$$\mathbf{x}^{c(k)} = \mathbf{x}^{c(k)} + \mathbf{x}^{o}(k) = \mathbf{x}^{c(k)} \tag{18}$$

where $\mathbf{x}^{c}(k)$ is the configuration amplitudes and $\mathbf{x}^{o}(k)$ is a zero vector. The latter reflects the fact that an integral transformation is performed in the beginning of each macro iteration such that the orbitals in the MO integrals and the orbitals of the CEP are the same. (Incidentally, this makes the optimization non-linear, which should be born in mind when comparing to books on numerical analysis. They almost exclusively deals with linear optimization. This is *e.g.* important if use of Hessian update methods is investigated.)

The derivation of the expressions for the gradient and Hessian in NEO optimization was described in Ref. (1). The gradient evaluated at the CEP, $\mathbf{x}^{c}(k)$ is

$$g^{c}_{\mu} = \frac{\partial E}{\partial c_{\mu}} = 2(<0 \mid \hat{\mathscr{H}} \mid \phi_{\mu}> - E^{CEP} c_{\mu}^{(k)}) \tag{19}$$

$$g^{o}_{sr} = \frac{\partial E}{\partial \kappa_{sr}} = <0 \mid [\hat{E}^{-}_{sr}, \hat{\mathscr{H}}] \mid 0> = 2 <0 \mid [\hat{E}_{sr}, \hat{\mathscr{H}}] \mid 0> \tag{20}$$

and the Hessian evaluated at the CEP, $\mathbf{x}^{c}(k)$ is:

$$\mathbf{H} = \mathbf{K} - 2(\mathbf{g}\mathbf{x}^{t} + \mathbf{x}\mathbf{g}^{t}) = \mathbf{K} - 2(\mathbf{g}\mathbf{x}^{ct} + \mathbf{x}^{c}\mathbf{g}^{t}) \tag{21}$$

where

$$K^{cc}_{\mu,\nu} = 2\left(<\phi_{\mu} \mid \hat{\mathscr{H}} \mid \phi_{\nu}> - E^{CEP}\delta_{\mu\nu}\right) \tag{22}$$

$$K^{co}_{\mu,sr} = 2 <0 \mid [\hat{E}^{-}_{sr}, \hat{\mathscr{H}}] \mid \phi_{\mu}> \tag{23}$$

$$K^{oo}_{sr,ut} = <0 \mid [\hat{E}^{-}_{sr}, \hat{E}^{-}_{ut}, \hat{\mathscr{H}}] \mid \phi_{\mu}> \tag{24}$$

We have used the short-hand notations

$$c^{(k)}_{\mu} = x^{c}_{\mu}(k) \tag{25}$$

$$\mid 0> = \sum_{\mu} c^{(k)}_{\mu} \mid \phi_{\mu}> \tag{26}$$

where $\mid 0>$ is the CEP, and we have defined

$$E^{CEP} = E(\mathbf{x}^{(k)}) = <0 \mid \hat{\mathscr{H}} \mid 0> \tag{27}$$

as the energy expectation value at the CEP. The CEP will by convention always be normalized, that is $<0 \mid 0> = 1$. Note that the $\mathbf{K}^{cc}$ block is basically two times the CI Hamiltonian matrix.

5. The direct iterative NEO algorithm

To simplify notation we drop the $^{(k)}$ superscript designating macro iteration no. k in this section. The NEO algorithm utilizes the auxiliary matrix $^{\beta}\mathbf{L}$ defined as:

$$\begin{aligned}^{\beta}\mathbf{L} &= \mathbf{K} + (\beta - 1)(\mathbf{g}^{c}\mathbf{x}^{t} + \mathbf{x}\mathbf{g}^{ct}) + (\beta - 2)(\mathbf{g}^{o}\mathbf{x}^{t} + \mathbf{x}\mathbf{g}^{ot}) \\ &= \mathbf{H}' + \beta(\mathbf{g}\mathbf{c}^{t} + \mathbf{c}\mathbf{g}^{t})\end{aligned} \tag{28}$$

where $\mathbf{H}' = \mathbf{PHP} = \mathbf{PKP}$, $\mathbf{c}$ is the CEP, and was defined in Eq. (12). This matrix has the following properties[1]:

$$\mathbf{O}^{\beta}\mathbf{L} = \beta\mathbf{x}\mathbf{g}^{t} \tag{29}$$

$$\mathbf{O}^{\beta}\mathbf{L}\mathbf{O} = 0 \tag{30}$$

$$\mathbf{P}^{\beta}\mathbf{L}\mathbf{P} = \mathbf{PHP} = \mathbf{H}' \tag{31}$$

We have used $\mathbf{g}^{t}\mathbf{c} = 0$. Following Ref. (1) we rewrite the eigenvalue equation:

$$^{\beta}\mathbf{L}^{\beta}\mathbf{z}_i = {}^{\beta}\lambda_i{}^{\beta}\mathbf{z}_i \tag{32}$$

as

$$\mathbf{O}^{\beta}\mathbf{L}^{\beta}\mathbf{z}_i = {}^{\beta}\lambda\mathbf{O}^{\beta}\mathbf{z}_i \tag{33}$$

$$\mathbf{P}^{\beta}\mathbf{L}(\mathbf{O} + \mathbf{P})^{\beta}\mathbf{z}_i = {}^{\beta}\lambda_i\mathbf{P}^{\beta}\mathbf{z}_i \tag{34}$$

where $\mathbf{O} + \mathbf{P} = \mathbf{I}$, the identity. Employing Eqs. (29-31) and performing some rearrangement we obtain:

$$^{\beta}\lambda_i = \beta a_1^{-1}(\mathbf{g}^{t\beta}\mathbf{z}_i) \tag{35}$$

$$(\beta a_1)^{-1}\mathbf{P}^{\beta}\mathbf{z}_i = (\mathbf{H}' - {}^{\beta}\lambda_i\mathbf{P})^{-1}\mathbf{g} \tag{36}$$

Here

$$a_1 = \mathbf{x}^{(k)t\beta}\mathbf{z}_i \tag{37}$$

is the overlap between the NEO eigenvector and the CEP vector. Thus step

$$^{\beta}\mathbf{y}_i^{NEO} = (\beta a_1)^{-1}\mathbf{P}^{\beta}\mathbf{z}_i \tag{38}$$

based on the eigenvector $^{\beta}\mathbf{z}_i$ is equivalent to the restricted step

$$\mathbf{y}^{R} = -(\mathbf{H}' - {}^{\beta}\lambda_i\mathbf{P})^{-1}\mathbf{g} \tag{39}$$

which might also have been found solving the set of linear equations:

$$(\mathbf{H}' - {}^{\beta}\lambda_i \mathbf{P})\mathbf{y}^R = -\mathbf{g} \tag{40}$$

The advantage of solving the eigenvalue Eq. (32) instead of this set of linear equations is that for NEO the level shift automatically lies in the correct interval. If one solves the set of linear equations it is necessary first to calculate the lowest eigenvalue(s) of the Hessian to find the valid range for the level shift. For large wave function calculations the update of the level shift would therefore involve extra costly calculations, comparable to one microiteration cycle. In the NEO eigenvalue method, on the other hand, the number of negative eigenvalues of the projected Hessian $\mathbf{H}'$ is monitored without ever explicitly determining any eigenvalues of $\mathbf{H}'$. In NEO the implicit level shift ${}^{\beta}\lambda$ and the step length norm $\|{}^{\beta}\mathbf{y}^{NEO}\|$ are functions of β and the optimal step is obtained by choosing β so that the step length is equal to the trust radius. This can be done very efficiently through a dynamical adjustment of β in the micro iterations as shown in section C.3. In the local region β will be kept at a fixed value and this leads to quadratic convergence behavior.[1]

The program can be thought of in two levels: i) The macro iteration "walk" on the parameter surface towards the desired stationary point, and ii) The micro iteration solution of a specific eigenvector-eigenvalue problem which gives the optimal step for each macro iteration. In Sec. C we describe one cornerstone in this program, namely the efficient use of a restricted-step second order optimization algorithm with the NEO implementation for the macro iterations. Another cornerstone is the use of direct techniques to solve the NEO eigenvector-eigenvalue problem, that is, the iterative solution of the eigenvalue equation by means of the linear transformations:

$$\sigma_j = {}^{\beta}\mathbf{L}\mathbf{b}_j \tag{41}$$

instead of setting the matrix ${}^{\beta}\mathbf{L}$ up explicitly. This saves both CPU time and memory, and makes large MCSCF calculations possible. We use a generalization of the algorithm which was developed by Davidson[37] or the block extension for several simultaneous expansion vectors that has been described by Liu,[38] see Appendix D. Each calculation of one linear transformation or a block of linear transformations defines one micro iteration, that is one iteration towards the lowest eigenvector(s) of ${}^{\beta}\mathbf{L}$. We block Eq. (41) after configuration and orbital rotation variables:

$$\begin{pmatrix} \sigma^c \\ \sigma^o \end{pmatrix} = \begin{pmatrix} {}^{\beta}\mathbf{L}^{cc} & {}^{\beta}\mathbf{L}^{co} \\ {}^{\beta}\mathbf{L}^{oc} & {}^{\beta}\mathbf{L}^{oo} \end{pmatrix} \begin{pmatrix} \mathbf{b}^c \\ \mathbf{b}^o \end{pmatrix} \tag{42}$$

The direct linear transformations are expressed through a set of four formulas, one for each block of ${}^{\beta}\mathbf{L}$. They describe the σ_j as a function of $\mathbf{b}_j$ without using the matrix elements of ${}^{\beta}\mathbf{L}$ explicitly (reordering of summations). The explicit formulas for the four blocks are derived from the gradient and Hessian formulas given in Eqs. (19-27), the result is

$$\sum_{\nu} {}^{\beta}L^{cc}_{\mu\nu}b^{c}_{\nu} = \sum_{\nu} [2(<\phi_{\mu} \mid \hat{\mathscr{H}} \mid \phi_{\nu}> - E^{CEP}\delta_{\nu\mu}) + (\beta - 1)(c_{\mu}g_{\nu} + g_{\mu}c_{\nu})]b_{\nu}$$
$$= 2(<\phi_{\mu} \mid \hat{\mathscr{H}} \mid B> - E^{CEP}b_{\mu}) + (\beta - 1)[c_{\mu}(\sum_{\nu} g_{\nu}b_{\nu}) + g_{\mu}(\sum_{\nu} c_{\nu}b_{\nu})] \tag{43}$$

$$\sum_{r>s} {}^{\beta}L^{co}_{\mu,\, sr}b^{o}_{sr} = \sum_{r>s}[2<0 \mid [\hat{E}^{-}_{sr}, \hat{\mathscr{H}}] \mid \phi_{\mu}> + (\beta - 2)c_{\mu}g_{sr}]b_{sr}$$
$$= 2<0 \mid [\sum_{r>s} b_{sr}\hat{E}^{-}_{sr}, \hat{\mathscr{H}}] \mid \phi_{\mu}> + (\beta - 2)c_{\mu}(\sum_{r>s} g_{sr}b_{sr}) \tag{44}$$
$$= 2<0 \mid \tilde{\mathscr{H}} \mid \phi_{\mu}> + (\beta - 2)c_{\mu}(\sum_{r>s} g_{sr}b_{sr})$$

$$\sum_{\nu} {}^{\beta}L^{oc}_{sr,\, \nu}b^{c}_{\nu} = \sum_{\nu}[2<0 \mid [\hat{E}^{-}_{sr}, \hat{\mathscr{H}}] \mid \phi_{\nu}> + (\beta - 2)g_{sr}c_{\nu}]b_{\nu}$$
$$= 2<0 \mid [E^{-}_{sr}, \hat{\mathscr{H}}] \mid B> + (\beta - 2)g_{sr}(\sum_{\nu} c_{\nu}b_{\nu}) \tag{45}$$
$$= 2\check{g}_{sr} + (\beta - 2)g_{sr}(\sum_{\nu} c_{\nu}b_{\nu})$$

$$\sum_{t>u} {}^{\beta}L^{oo}_{sr,\, ut}b^{o}_{ut} = \sum_{t>u} <0 \mid [\hat{E}^{-}_{sr}, \hat{E}^{-}_{ut}, \hat{\mathscr{H}}] \mid 0> b_{\perp}$$
$$= <0 \mid \left[E^{-}_{sr}, [\sum_{t>u} b_{ut}\hat{E}^{-}_{ut}, \hat{\mathscr{H}}] \right] \mid 0>$$
$$+ \frac{1}{2}\sum_{t>u} <0 \mid \left[[\hat{E}^{-}_{sr}, \hat{E}^{-}_{ut}], \hat{\mathscr{H}} \right] \mid 0> b_{ut} \tag{46}$$
$$= <0 \mid [\hat{E}^{-}_{sr}, \tilde{\mathscr{H}}] \mid 0> + \frac{1}{2}\sum_{t}(g_{ts}b_{rt} - g_{tr}b_{st})$$
$$= \tilde{g}_{sr} + \frac{1}{2}\sum_{t}(g_{ts}b_{rt} - g_{tr}b_{st})$$

We have here introduced the one-index transformed gradient $\tilde{\mathbf{g}}$ and the transition gradient $\check{\mathbf{g}}$. The latter is evaluated over a symmetric transition density matrix:

$$<0 \mid [\hat{E}^{-}_{sr}, \hat{\mathscr{H}}] \mid B> = <0 \mid [\hat{E}_{sr}, \hat{\mathscr{H}}] \mid B> + <B \mid [\hat{E}_{sr}, \hat{\mathscr{H}}] \mid 0>. \tag{47}$$

Eq. (43) is a standard direct CI and Eq. (44) is a direct CI with a modified hamiltonian $\tilde{\mathscr{H}}$ (described in the next paragraph). The direct CI steps as well as the density matrix constructions for $\mathbf{g}$, $\tilde{\mathbf{g}}$, and $\check{\mathbf{g}}$ are described in Sec. D. In Appendix A the Eqs. (45) and (46) are reduced to the program implemented expressions.

The one-index transformed Hamiltonian $\tilde{\mathscr{H}}$ is implicitly defined in the direct sigma vector equations given above. Explicitly we derive $\tilde{\mathscr{H}}$ as:

$$\tilde{\mathscr{H}} = \left[\sum_{r>s} b_{sr} \hat{E}^{-}_{sr}, \hat{\mathscr{H}} \right] = \sum_{pq} \tilde{h}_{pq} \hat{E}_{pq} + \frac{1}{2} \sum_{pqrs} (\widetilde{pq \mid rs}) \hat{e}_{pq,rs} \tag{48}$$

where

$$\tilde{h}_{pq} = \sum_{t} (h_{tq} b_{pt} + h_{pt} b_{qt}) \tag{49}$$

and

$$(\widetilde{pq \mid rs}) = \sum_{t} \left[(tq \mid rs) b_{pt} + (pt \mid rs) b_{qt} + (pq \mid ts) b_{rt} + (pq \mid rt) b_{st} \right] \tag{50}$$

where t runs over all orbitals. These equations are derived using the commutation relations between the operators $\hat{E}^{-}_{pq}$, $\hat{E}_{pq}$ and $\hat{e}_{pq,rs}$.

The recognition of a one-index transformed Hamiltonian in the treatment of the coupling block[8] is a key point in the formulation of a second order MCSCF, in fact it is indispensable for a direct handling of the coupling block. An explicit calculation of the coupling block is prohibitive for accurate MC functions with many configurations, both because of storage requirements and calculation time. In the formulation given here the one-index transformed and the normal Hamiltonians, $\tilde{\mathscr{H}}$ and $\hat{\mathscr{H}}$, can be processed in a completely parallel fashion since they use the same direct CI coupling elements, $A^{\mu\nu}_{pq}$ and $A^{\mu\nu}_{pq,rs}$ (given below in Eqs. (53-54)).

The element $< \phi_\mu \mid \hat{\mathscr{H}} \mid B >$ in Eq. (45) represents a direct CI iteration with the $\mathbf{b}^c$ trial vector:

$$\begin{aligned} \phi_\mu \mid \hat{\mathscr{H}} \mid B > &= \sum_{\nu} < \phi_\mu \mid \hat{\mathscr{H}} \mid \phi_\nu > b_\nu \\ &= \sum_{\nu} \sum_{pq} A^{\mu\nu}_{pq} h_{pq} b_\nu + \frac{1}{2} \sum_{pqrs} A^{\mu\nu}_{pq,rs} (pq \mid rs) b_\nu \\ &= \sum_{\nu} \left[E^C \delta_{\mu\nu} + \sum_{uv} A^{\mu\nu}_{uv} F^C_{uv} + \frac{1}{2} \sum_{uvxy} A^{\mu\nu}_{uv,xy} (uv \mid xy) \right] b_\nu \end{aligned} \tag{51}$$

where E^C is the inactive energy

$$E^C = \sum_{i} (h_{ii} + F^C_{ii}) \tag{52}$$

and

$$A_{pq}^{\mu\nu} = < \mu \mid \hat{E}_{pq} \mid \nu > \tag{53}$$

$$A_{pq,rs}^{\mu\nu} = < \mu \mid \hat{e}_{pq,rs} \mid \nu > \tag{54}$$

are the direct CI coupling coefficients. They are further reduced in Sec. D where the CI theory used in SIRIUS is described. The $\mathbf{F}^C$ matrix, defined in Eq. (64) below, is a standard Fock matrix and $< \phi_\mu \mid \tilde{\mathscr{H}} \mid 0 >$ is a direct CI iteration with the one-index transformed Hamiltonian and the CI coefficients of the CEP.

The direct evaluation of the configuration block $^{\beta}\mathbf{L}^{cc}\mathbf{b}^c$ is a standard direct CI. In the present program we treat the whole problem, the orbital block and the orbital-configuration coupling block as well as the above-mentioned configuration block with direct methods.[8,1] In this way it is possible to perform a second-order, controlled optimization even for large configuration expansions where the explicit calculation of the orbital-configuration coupling is prohibitive. This algorithm thus "solves" the orbital part of the MCSCF problem in the sense that it will be possible to do an MCSCF optimization if it is possible to do the corresponding CI. The computer time needed for performing the MCSCF will nearly be a constant times the effort of performing the CI, and, as we wrote in connection with our first larger calculation,[2] the impact of future improvements for the CI method, that is the handling of symbolic density matrix elements, can directly be conveyed to MCSCF. This has clearly been demonstrated by the development from graphical unitary group approach (GUGA) based inital SIRIUS implementation[17,2,3] to the current determinant based implementation.[6]

C. Implementation

In this section we go into more details with the implementation of the general ideas presented in section B. We begin with an overview of the basic steps in the NEO algorithm for MCSCF optimization:

0 Select basis set, configuration space, convergence threshold, and so on. Initial guess. Set macro iteration count $k := 0$.

1 Increment macro iteration count $k := k + 1$. Integral transformation.

2 Energy and gradient evaluation. Check for convergence, if $\|\mathbf{g}\| < thrgrd$ then exit macro iteration loop.

3 Step control (when $k > 1$). If the step is not acceptable then decrement macro iteration counter by 1; solve the reduced NEO matrix from the previous iteration for a reduced trust radius; save this modified step; and go back to step 1. Otherwise continue.

4 Solve NEO eigenproblem iteratively by projecting the NEO matrix onto a basis of trial vectors. The NEO matrix over the trial vectors constitutes the reduced NEO matrix. These iterations, "the micro iterations," proceed as follows:

4.0 First trial vector is defined as the CEP, the linear transformation of which is β times the gradient. Set up initial guess(es) for eigenvector(s). Set micro iteration count $j := 0$.

4.1 Increment micro iteration count $j := j + 1$. Calculate the linear transformation by the NEO matrix of the trial vector(s).

4.2 Add these trial vectors to the reduced NEO matrix. Solve the reduced NEO eigenproblem for current β value.

4.3 When at least one trial vector of configuration type as well as one of orbital type have been selected, compare the step size of the step defined by the eigenvector of the reduced NEO matrix with the trust radius. If they don't agree update β. However, β will not be reduced below a preset minimum, β equal to the minimum corresponds to the case when the step restriction becomes inactive, that is, when the Newton step is shorter than the trust radius.

4.4 Calculate the residual vector and check for convergence of micro iterations. We use a linear threshold $\min(0.2\,\|\mathbf{g}\|, 0.1)$ in the global region and a quadratic threshold $\max(8\,\|\mathbf{g}\|^2, 0.9 \times \mathit{thrgrd})$ in the local region. For excited state optimization, the NEO eigenvectors corresponding to lower excited states will be converged to the global threshold. In the global iterations the step does not need to be accurately determined as long as the general direction and step length is reasonable, while an accurate step vector is essential for second-order convergence behavior in the local region. If the NEO eigenvector is converged to the current threshold, exit the micro iterations and go to step 5.

4.5 Use the generalized Davidson-Liu algorithm to find the next trial vectors from the residual vectors. Go back to step 4.1.

5 Rotate orbitals and transform to Fock type and natural orbitals. Counterrotate CI vector and save the new orbitals and CI vector(s). Save the reduced NEO matrix, calculated in step 4, for step 3. Go back to step 1.

In subsection 1 we will discuss the specification of the MC wave function and we derive the computational expressions for the evaluation of the gradient. In particular we will discuss the savings obtained through the treatment of the inactive orbitals by the use of generalized Fock matrices as intermediate entities.

In subsection 2 the evaluation of the step determined by the reduced NEO matrix as well as the calculation of the second-order prediction of the energy change is described together with the step control. One special feature is the saving of the reduced NEO matrix from the previous macro iteration, such that a revised step can be calculated without further integral transformations if the step turned out to be too long.

In subsection 3 the micro iterations are described in more detail. In particular we discuss the direct MCSCF step, the linear transformations by the NEO matrix, and the dynamical update of the β damping factor (step 4.3 above). In subsection 4 we briefly analyze the integral transformations needed for our direct restricted-step second order MCSCF optimization with the reductions achieved by using Fock matrices. Finally, in subsection 5 the step-control algorithm for ground and excited state optimization is described in some detail.

1. Orbital classes, gradient, and NEO transformation

To summarize: in Section B.2 the orbitals were divided into three groups: fully occupied (inactive) orbitals, partly occupied (active) orbitals, and empty (secondary) orbitals. In addition we can specify frozen or deleted orbitals. All occupied orbitals are also called primary orbitals. We will use the following notation for the orbitals:

p,q,r,s for general orbitals
i,j,k,l for inactive orbitals
u,v,x,y for active orbitals
a,b,c,d for secondary orbitals,

conforming to the notation used by Roos, Siegbahn and coworkers.[39,40] The handling of Fock matrices for the gradient evaluation, including the treatment of the inactive orbitals, is similar to the one used in Ref. (40).

We define a generalized Fock matrix

$$F_{pq} = \sum_{\sigma} < 0 \mid \hat{a}^{\dagger}_{p\sigma}[\hat{a}_{q\sigma}, \hat{\mathscr{H}}] \mid 0 > . \tag{55}$$

By inserting the Hamiltonian defined in Eq. (2) we arrive at:

$$F_{pq} = \sum_{r} D_{pr} h_{qr} + \sum_{rst} P_{pr,st}\,(qr \mid st) \tag{56}$$

where D_{pr} and $P_{pr,st}$ are the spinless one- and two-electron reduced density

$$D_{pq} = <0 \mid \hat{E}_{pq} \mid 0> \tag{57}$$

$$P_{pq,rs} = <0 \mid \hat{e}_{pq,rs} \mid 0> \tag{58}$$

This generalized Fock matrix describes the effective, attractive potential for an electron to be ionized and was introduced by Hinze[41] in another context. The eigenvalues of this Fock matrix in the metric given by the one-electron density matrix D are approximations to ionization potentials, the extended Koopmans' theorem (EKT).[42]

Elements of both density matrices are obviously zero if any index is secondary, and the Fock element F_{pq} is consequently zero if the first index p is a secondary orbital. The only non-zero elements in D and P involving inactive indices are[40]:

$$D_{ip} = 2\delta_{ip} \tag{59}$$

$$P_{ij,pq} = 2\delta_{ij}D_{pq} - 2\delta_{iq}\delta_{jp} \tag{60}$$

$$P_{pj,jq} = 4\delta_{ip}\delta_{jq} - 2\delta_{ij}D_{pq} \tag{61}$$

Hence only **D** and **P** elements with all indices active need to be explicitly stored. Using these facts the non-zero Fock elements can be expressed as:

$$F_{iq} = 2(F^C_{iq} + F^V_{iq}) = 2F^D_{iq} \tag{62}$$

$$F_{vq} = \sum_u D_{vu}F^C_{uq} + F^Q_{vq} \tag{63}$$

Here $\mathbf{F}^C$ and $\mathbf{F}^V$ are the inactive (core) and active (valence) Fock matrices,† respectively:

$$F^C_{pq} = h_{pq} + 2\sum_i \mathbb{P}_{pq,ii} \tag{64}$$

$$F^V_{pq} = \sum_{vu} D_{vu}\mathbb{P}_{pq,vu} \tag{65}$$

$$F^D_{pq} = F^C_{pq} + F^V_{pq} \tag{66}$$

where $\mathbb{P}$ is the P super matrix[43]:

---

† The matrices are named core and valence because that is a typical division; however, a specific calculation might very well have, *e.g.*, core orbitals among the active orbitals.

$$\mathbb{P}_{pq,rs} = (pq \mid rs) - \frac{1}{4}[(pr \mid qs) + (ps \mid qr)], \tag{67}$$

and

$$F^{Q}_{yq} = \sum_{uvx} P_{yu,vx}(qu \mid vx). \tag{68}$$

The calculation of the $\mathbf{F}^C$, $\mathbf{F}^V$, $\mathbf{F}^Q$ matrices are described in Appendix B. Using the definition in Eq. (55) and comparing with the definition of the orbital gradient in Eq. (20) one finds that

$$\begin{aligned} g_{sr} &= <0 \mid [\hat{E}^{-}_{sr}, \hat{\mathscr{H}}] \mid 0> = 2<0 \mid [\hat{E}_{sr}, \hat{\mathscr{H}}] \mid 0> \\ &= 2(F_{sr} - F_{rs}). \end{aligned} \tag{69}$$

These definitions lead to four different computational expressions for the orbital gradient; one for each of the inactive-active, the inactive-secondary, the active-secondary and the active-active rotations (*c.f.* Appendix A). If the wave function is complete in the active space (CAS) the active-active rotations are redundant optimization parameters, however, the gradient elements for active-active rotations are needed for the correction terms in Eq. (46) (the **B** term defined in Eq. (A4)). No extra integral transformation is necessary as the integrals needed for the active-active rotations also are needed for the CI part. The actual evaluation of the gradient is carried out with the following five steps:

1. Calculate D_{uv} and $P_{uv,xy}$.
2. Calculate F^C_{pq} and F^V_{pq} using atomic super matrix integrals.
3. Read the m.o. integral file, calculate F^Q_{yq} and store $(uv \mid xy)$ (the latter for the configuration part of the gradient).
4. Calculate the orbital gradient from Eq. (69) as described in Appendix A.
5. Calculate configuration gradient from Eq.(19), a standard direct CI iteration with the configuration amplitudes of the CEP.

All the entities mentioned in these five steps are saved since they are needed for the linear NEO transformations in the micro iterations. Because the linear transformation of a direct MCSCF step, Eqs. (43-46), is expressed in terms of gradient type elements, the calculation very similar to the gradient evaluation.

2. *The macro iterations, step calculation, and step control*

The step, that is the change in the configuration coefficients and the rotation of the orbitals, is determined by the appropriate eigenvector of the NEO matrix ${}^{\beta}\mathbf{L}$ (*c.f.* Eqs. (28-40)) and, as will be shown below, the predicted energy change can be determined from the corresponding eigenvalue of the NEO matrix.

Because the eigenvector is solved for iteratively, it is only determined to a specified accuracy; this accuracy need not be so stringent when far from convergence as in the last macro-iterations. The eigenvector of the reduced NEO matrix that determines the step corresponds to the lowest eigenvalue if converging to the lowest state of the specified spin and space symmetry, and to the second lowest eigenvalue if converging to the second state, etc. Note that even if this vector is not converged to the eigenvector of the full NEO matrix, all the predictions based on the reduced NEO matrix, including the predicted energy change, are still exact, and we thus keep the correct trust radius control even with "poor" convergence of the micro iterations. On the other hand, for excited state calculations we lose control if the micro iteration convergence is too poor, and the total convergence could then become unpredictable. In this latter case the prediction might be a correct prediction for a wrong direction.

The dimension m of the reduced NEO matrix L^r is usually much less than the total number of variables. We select the first trial vector $\mathbf{b}_1$, that is the first basis vector of the reduced NEO matrix, as the current expansion point $\mathbf{x}^{(k)}$ (CEP). The trial vectors are always chosen to be orthonormal. The approximate eigenvector we get from the reduced NEO matrix is

$$^{\beta}\mathbf{L}^r\mathbf{z} = {}^{\beta}\lambda\mathbf{z} \tag{70}$$

$$\mathbf{z} = \sum_{i=1}^{m} a_i\mathbf{b}_i \tag{71}$$

where $\mathbf{z}$ is normalized: $\Sigma_i a_i^2 = 1$. The vector $\mathbf{a}$ is the eigenvector of the reduced NEO matrix and can be obtained from a standard program for the eigenvalues and eigenvectors of small matrices. Using Eq. (38) the step becomes

$$\mathbf{y} = \frac{\gamma}{\beta a_1}\mathbf{Pz} = \frac{\gamma}{\beta a_1}\sum_{i=2}^{m} a_i\mathbf{b}_i \tag{72}$$

Here γ is an additional step reduction factor used in calculations towards excited states, for which a stationary point on the trust sphere does not always exist. For calculations on the lowest state, γ will always be equal to one. We insert the obtained step in the second-order Taylor expansion for the energy, Eq. (10), and use Eqs. (28-38) to relate this prediction for the energy change to a_1, β, γ and the eigenvalue $^{\beta}\lambda$:

$$\begin{aligned}\Delta E^{(2)} &= \mathbf{g}^t\mathbf{y} + \frac{1}{2}\mathbf{y}^t\mathbf{H}'\mathbf{y} \\ &= \gamma(\beta^{-1})^{\beta}\lambda + \frac{1}{2}\gamma^2(\beta a_1)^{-2}\mathbf{z}^t\mathbf{H}'\mathbf{z} \\ &= \left[(\gamma - \gamma^2)\beta^{-2} + \frac{1}{2}\gamma^2(\beta a_1)^{-2}\right]^{\beta}\lambda,\end{aligned} \tag{73}$$

where we have used:

$$\begin{aligned} \mathbf{z}^t\mathbf{H}'\mathbf{z} &= \mathbf{z}^t\mathbf{PLPz} \\ &= \mathbf{z}^t\mathbf{Lz} - \mathbf{z}^t(\mathbf{OLO} + \mathbf{OLP} + \mathbf{PLO})\mathbf{z} \\ &= {}^{\beta}\lambda - 2a_1\beta(\mathbf{g}^t\mathbf{z}) = {}^{\beta}\lambda(1 - 2a_1^2) \end{aligned} \tag{74}$$

The step length, which is to be compared against the trust radius, can also very easily be evaluated from the reduced problem. The step length is the square root of:

$$\begin{aligned} \mathbf{y}^t\mathbf{y} &= \gamma^2(\beta a_1)^{-2}\sum_{i=2}^{m} a_i^2 \\ &= \gamma^2(\beta a_1)^{-2}(1 - a_1^2) = \gamma^2\beta^{-2}(a_1^{-2} - 1) \end{aligned} \tag{75}$$

By the choice of $\mathbf{b}_1$ as the CEP we have confined the β dependence solely to the first column of the $\mathbf{L}^r$ matrix. The $\mathbf{L}^r$ matrix is saved with $\beta = 1$, and the first column of $\mathbf{L}^r$ is simply multiplied by β before calling the eigenvalue routine.

At the end of the macro iteration the β damping factor is varied such that the squared step length according to Eq. (75) lies within an acceptable tolerance (*e.g.* 10%) of the squared trust radius. Through the specific choice of $\mathbf{b}_1$ we have accomplished that ${}^{\beta}\mathbf{L}^r$ is always a valid projection of ${}^{\beta}\mathbf{L}$ for any β and that ${}^{\beta}\mathbf{L}^r$ has a simple dependence of β. Eq. (75) thus always gives the correct squared step length based on the current approximate eigenvector of ${}^{\beta}\mathbf{L}$ from ${}^{\beta}\mathbf{L}^r$. Furthermore, if the next macro iteration shows that the trust radius was too long, we can reuse the reduced NEO matrix to find a new step corresponding to the new trust radius by varying β. In this way, by saving ${}^{\beta}\mathbf{L}^r$ and all the basis vectors $\mathbf{b}$ at the end of a macro iteration, we can take a "back step" without transforming integrals back to the previous mo's. This simple and cheap back step together with the dynamical update of β described in the next subsection is one of the advantages of the NEO algorithm.

3. The micro iterations, the dynamical update of the damping factor, and the direct MCSCF step

As mentioned above, the specific choice of $\mathbf{b}_1$ implies that the reduced NEO matrix ${}^{\beta}\mathbf{L}^r$ is always a correct projection of the full ${}^{\beta}\mathbf{L}$ matrix, and this fact enables us to minimize the number of micro iterations used to converge towards a step of the wrong length. As soon as the $\mathbf{L}^r$ matrix is fairly stabilized, the β value can be adjusted in each micro iterations so that the step length for the current approximate eigenvector is within the specified tolerance of the trust radius. Practical experiments indicate that it is reasonable to begin adjusting β when both a configuration trial vector and an orbital trial vector have been added to the set of trial vectors. A simple secant procedure is used to find β for ground state optimization.

We use a generalization of the Davidson-Liu algorithm[37,38] to find the next trial vector(s) in each micro iteration, and for this we need the diagonal of the ${}^{\beta}\mathbf{L}$ matrix explicitly. From Eqs. (43) and (46) we get:

$$^{\beta}L^{cc}_{\mu\mu} = 2[< \phi_\mu \mid \hat{\mathscr{H}} \mid \phi_\mu > - E^{CEP} + (\beta - 1)g_\mu c_\mu] \tag{76}$$

$$^{\beta}L^{oo}_{sr,sr} = < 0 \mid [\hat{E}^{-}_{sr}, \hat{E}^{-}_{sr}, \hat{\mathscr{H}}] \mid 0 > \tag{77}$$

One notes that the configuration section of the diagonal is β dependent. SIRIUS may use both exact and approximate orbital diagonal elements, where the latter gives savings in the integral transformations (*c.f.* Sec. C.4 below). Explicit expressions for the exact as well as approximate orbital diagonal elements of Eq. (72) can be found in Appendix C.

The direct MCSCF step, that is the linear transformation of Eq. (41)

$$\sigma_j = {}^{\beta}\mathbf{L}\mathbf{b}_j \tag{78}$$

was expressed in terms of gradient type contributions in Eqs. (43,46) and Appendix A. This means that we can calculate the σ vectors using the gradient routines by supplying them with the usual Hamiltonian and a transition density matrix for Eqs. (43) and (45) and with the one-index transformed Hamiltonian and the usual density matrix for Eqs. (44) and (46). This calculation corresponds to the first term in each of the four equations, the remaining terms are simple scalar products and vector additions with known vectors. In practice we have coded special routines to calculate the linear transformations for obtaining the σ vectors, where we exploit the possibility to calculate the different gradient type terms in parallel.

4. *Integral Transformations*

One crucial point in the implementation is the calculation of the one-index transformed inactive and active Fock matrices for the σ vector evaluation with orbital trial vectors, that is $\mathbf{F}^{C}$ and $\mathbf{F}^{V}$ of Eqs. (64) and (65) with one-index transformed integrals (designated $\tilde{\mathbf{F}}^{C}$ and $\tilde{\mathbf{F}}^{V}$). This can be done completely with atomic super matrix integrals, that is, no m.o. integrals are needed for this operation. We discuss how this can be achieved in Appendix B. In fact, no m.o. integrals at all are needed for a closed shell or a one open shell restricted Hartree-Fock optimization, and a quadratically-convergent step-restricted Hartree-Fock method is obtained as a special case of our general MCSCF implementation This is equivalent to the QC-SCF method (quadratically convergent SCF) of Bacskay[44] but with the additional benefit of restricted-step convergence control.

We can now tabulate which m.o. integrals are needed at different levels ("order" here denotes the order in the Taylor expansion of the energy):

zeroth order (energy, CI):
(uv | xy)

first order (gradient):
(uv | py)

second order, direct:
a) (uv | pq), (au | pv), (iu | pv)
b) for the exact diagonal of orbital Hessian we also need (ai | ai) and (ii | aa)

second order with explicit orbital Hessian:
(st | pq), (sp | tq) where s,t are any occupied orbital.

Frozen orbitals, that is inactive orbitals which are kept frozen during the optimization, can always be completely excluded from the integral transformation. Frozen orbitals have typically been used to counteract basis set superposition errors.

5. *Step-Control Algorithm*

We have implemented the following step-control algorithm for ground state optimization (the ratio r was defined in Eq. (16)):

1. if $r < r_{reject}$ decrease trust radius, reject step and find new step with reduced trust radius;
2. $r \geq r_{reject}$ accept step with the following conditions;
   a. if $r \leq r_{\min}$ decrease trust radius;
   b. if $r_{\min} < r \leq r_{good}$ no change in trust radius;
   c. if $r_{good} < r$ increase trust radius.

For excited state optimization it is necessary to keep stricter control and we have used:

1. if $r \leq r_{\min}$ or $r \geq 2 - r_{\min}$ decrease trust radius and reject step;
2. $r_{good} \leq r \leq 2 - r_{good}$ increase trust radius and accept step;
3. else no change in trust radius and step accepted.

For ground state optimization we have used $r_{reject} = 0.25$, $r_{\min} = 0.40$ and $r_{good} = 0.80$, and for excited states $r_{reject} = 0.40$, $r_{\min} = 0.60$ and $r_{good} = 0.80$.

The parameters include also the initial value and the minimum value for the damping factor β (1.0 and 0.2); the initial value for the trust radius (0.7) and the increment and decrement factors for the update of the trust radius (1.2 and 0.67) and the convergence threshold (10^{-5}). It is emphasized that the overall convergence features (and the convergence guarantee) are independent of the exact choice of these initial values, since the important ones anyway are dynamically updated. In most cases the default values are quite appropriate. In specific cases one can, however, lose one or two iterations by a poor initial choice; *e.g.* a too pessimistic initial value of the trust region may cost an iteration for the adjustment to a larger value that corresponds to a smoother character of the energy hypersurface than initially expected.

The micro iteration convergence thresholds are dynamically dependent on the energy gradient, which means that far from (macro) convergence the iterative solution of the eigenvalue equation is only solved approximately, while close to the variational optimum the threshold for micro iteration convergence is set to a small value. Also the number of specified roots to be simultaneously expanded has an effect on the convergence rate. Ideally one should expand simultaneously all those roots that correspond to near degenerate roots (in practice near degenerate diagonal elements) of the initial Hessian L-matrix. Alternatively (and computationally cheaper) the Hessian diagonal used to get new trial vectors may be expanded with an explicit subblock of the Hessian, see Appendix D. For the nth excited state of a symmetry a minimum of n simultaneously expanded roots are evidently needed in the global region in order to control convergence to the desired state. However, all but the nth root need only to be very approximately converged, in SIRIUS we use to a residual norm of 10^{-1}.

D. Direct configuration interaction theory

Three types of calculations involving information from the CI expansions occur in SIRIUS:

1. Direct CI
   $\sigma_I = \Sigma_K G_{I,K} C_K$
   where G is either the Hamiltonian or a one-index transformed Hamiltonian.

2. Density matrix calculation
   $\rho^1_{ij} = < L | E_{ij} | R >$ (one-body spin reduced density matrix)
   $\rho^2_{ij,kl} = < L | e_{ij,kl} | R >$ (two-body spin reduced density matrix)
   $D^S_{ij} = < L | \hat{a}^\dagger_{i\alpha}\hat{a}_{j\alpha} - \hat{a}^\dagger_{i\beta}\hat{a}_{j\beta} | R >$ (one body spin density matrix)
   The density matrices used in SIRIUS are symmetrized combinations of the above matrices in order to assure maximal computational symmetry.

3. Counter rotation of CI coefficients.
   Rotations of orbitals which only mix orbitals belonging to the same subblock can be done in order to assure, that the diagonal subblocks of the Fock matrix or one body density matrix are diagonal. In order to retain the same

total wave function the CI coefficients must also be modified. Efficient ways of performing the CI rotations have recently been published.[45,46]

In Section 1 we outline our use of CSF's and show how the CSF-SD transformation can be carried out in an efficient manner. In Section 2 we describe how the SD's of a RAS CI expansion can be written in terms of products of α- and β-strings, allowing a very compact way of storing the information needed to evaluate single or multiple excitations out from a CI expansion. In Section 3, we present our SD based CI algorithms, and show how the splitting of the determinants in strings naturally gives a highly vectorized algorithm. In Section 4, we discuss the construction of density matrices, and Section 5 discusses how the counterrotations of the orbitals can be carried out.

1. RAS-CI expansions in a CSF basis

The construction of the CSF's corresponding to a RAS-CI expansion with a given spin multiplicity can be divided into two steps: the construction of all orbital configurations that fulfill the RAS constraints, and the subsequent expansion of each orbital configuration in terms of CSF's. A configuration is defined by the occupation of each orbital, while a CSF in addition requires the specification of the spincoupling of the occupied orbitals. In this section we will discuss the expansion of configurations in terms of CSF's and SD's and show how the transformation between the CSF representation and the SD representation can be carried out.

The expansion of orbital configurations in terms of CSF's or SD's is conveniently discussed in terms of prototypes. A prototype configuration is defined by the number of doubly occupied orbitals and the number of singly occupied orbitals, without specifying the actual occupied orbitals. For a system with 2n + 1 electrons, prototypes with 1,3,...,2n + 1 unpaired electrons occur, *in toto* n prototypes, and for a system with 2n electrons there are prototypes with 0,2,...,2n unpaired electrons, *in toto* n prototypes. The number of prototypes occuring in a given RAS expansion may be smaller due to the specified spin multiplicity (*e.g.*, the spin multiplicity 3 cannot be obtained in configurations with all orbitals doubly occupied).

Each prototype configuration can be expanded in terms of a formal set of spin adapted functions, the prototype CSF's. We will use the convention that all doubly occupied orbitals first are coupled to singlet spin functions, so we only need to consider the explicit spin coupling of the singly occupied orbitals. For a configuration with N unpaired electrons, the number of CSF's with spin multiplicity $2S + 1$ is given by Weyl's formula

$$N_c = F(N, S) = \binom{N}{\frac{1}{2}N - S} - \binom{N}{\frac{1}{2}N - S - 1}. \tag{74}$$

Numerous ways of expressing the CSF's have been invented and reinvented. Referring to Hinze[47] and Pauncz[48] for general references, we will here restrict us to the so-called genealogical basis of CSF's . In the genealogical approach, each CSF is built up as a sequence of intermediate spin couplings, starting from a single electron which is an $S = \frac{1}{2}$ spin state, the coupling of the next electron can give an $S = 0$ or $S = 1$ spin state, and so on. In the genealogical representation each prototype CSF $|t>^c$ can be represented as a sequence of 'spin-up' or 'spin-down' couplings, which can be represented by a vector $\mathbf{t}^c$, with $t_i^c = +1(-1)$ if the coupling of the i'th electron to the $(i-1)$'th spin function increases (decreases) the spin. The CSF's obtained in this way are orthonormal. The number of times spin-up couplings occur minus the number of times spin-down couplings occur is related to the total spin as

$$n_+ - n_- = 2S. \tag{75}$$

Considering a prototype with 3 unpaired electrons, there are two doublet CSF's

$$\begin{aligned} &|++->^c \\ &|+-+>^c \end{aligned} \tag{76}$$

and one quartet

$$|+++>^c. \tag{77}$$

The genealogical CSF's can be expanded in SD's. To do this we introduce a set of prototype determinants $|t>^d$ with N unpaired electrons and a spin projection M_s. The number of determinants with N unpaired electrons and spinprojection M_s is given by

$$N_d = \binom{N}{\frac{1}{2}(N+2M_s)}\binom{N}{\frac{1}{2}(N-2M_s)}. \tag{78}$$

A given SD can be represented by a vector $\mathbf{t}^d$, with $t_i^d = +1(-1)$ if spin orbital i is an α (β) spin orbital. The number of occurrences of $t_i^d = +1$ minus the number of occurrences of $t_i^d = -1$, equals 2 times the spin projection M_s. Corresponding to the previous example with $M_s = \frac{1}{2}$, three SD's can be constructed

$$\begin{aligned} &|++->^d \\ &|+-+>^d \\ &|-++>^d \end{aligned} \tag{79}$$

The expansion of a given CSF in terms of the corresponding set of SD's can be written

$$|{}^{i}\mathbf{t}^{\mathbf{c}}>^{c} = \sum_{j} d_{ji} |{}^{j}\mathbf{t}^{\mathbf{d}}>^{d} \tag{80}$$

where the index i labels the different spin couplings, and j labels the different spin orbital combinations. A very elegant algorithm for expanding the M_s component of a given genealogical CSF were given by Graebenstetter *et al.*[49] In order to obtain the expansion coefficient d_{ji} one first constructs the accumulated arrays

$$\begin{aligned} {}^{j}T_k^d &= \sum_{l=1}^{k} {}^{j}t_l^d \\ {}^{i}T_k^c &= \sum_{l=1}^{k} {}^{i}t_l^c \end{aligned} \tag{81}$$

Since the M_s component must be lower than the spin in each intermediate coupling one has the inequality

$${}^{j}T_k^d \leq {}^{i}T_k^c \tag{82}$$

If this equation is fulfilled, the coefficient d_{ji} is obtained as

$$d_{ji} = \prod_{k=1}^{N} a_k({}^{i}\mathbf{t}^{\mathbf{c}}, {}^{j}\mathbf{t}^{\mathbf{d}}) \tag{83}$$

with the a_k coefficients defined as

$$\begin{aligned}
{}^{i}t_k^c = +1,\ {}^{j}t_k^d = +1:\ a_k({}^{i}\mathbf{t}^{\mathbf{c}}, {}^{j}\mathbf{t}^{\mathbf{d}}) &= \sqrt{\frac{{}^{i}T_k^c + {}^{j}T_k^d}{2\,{}^{i}T_k^c}} \\
{}^{i}t_k^c = +1,\ {}^{j}t_k^d = -1:\ a_k({}^{i}\mathbf{t}^{\mathbf{c}}, {}^{j}\mathbf{t}^{\mathbf{d}}) &= \sqrt{\frac{{}^{i}T_k^c - {}^{j}T_k^d}{2\,{}^{i}T_k^c}} \\
{}^{i}t_k^c = -1,\ {}^{j}t_k^d = +1:\ a_k({}^{i}\mathbf{t}^{\mathbf{c}}, {}^{j}\mathbf{t}^{\mathbf{d}}) &= -\sqrt{\frac{{}^{i}T_k^c - {}^{j}T_k^d + 1}{2\,{}^{i}T_k^c + 2}} \\
{}^{i}t_k^c = -1,\ {}^{j}t_k^d = -1:\ a_k({}^{i}\mathbf{t}^{\mathbf{c}}, {}^{j}\mathbf{t}^{\mathbf{d}}) &= \sqrt{\frac{{}^{i}T_k^c + {}^{j}T_k^d + 1}{2\,{}^{i}T_k^c + 2}}
\end{aligned} \tag{84}$$

In SIRIUS, all configurations corresponds to a given prototype are consecutively stored. Each configuration can then be expanded in terms of CSF's or SD's by using the information stored for the given prototype. In general several actual configurations correspond to a given prototype, so the actual CSF's can be indexed as $|i, j, k>^{c}$ with k being the prototype, j being the configuration number for the given prototype and i being a specific spin coupling for the given configuration. A given SD can be represented as $|i', j, k>^d$, with j and k again labelling prototype and configuration, respectively, and i' representing a given spin orbital product. A given CI expansion can be represented in terms of CSF's or SD's

$$
\begin{aligned}
|0> &= \sum_{i,j,k} {}^kC^c(i,j)\,|i,j,k>^c \\
|0> &= \sum_{i',j,k} {}^kC^d(i',j)\,|i',j,k>^d
\end{aligned}
\tag{85}
$$

Using Eq. (80) one can transform between the CI vectors in the CSF- and SD-representations

$$
\begin{aligned}
{}^kC^d(i',j) &= \sum_{i} {}^kC^c(i,j)\,{}^kd(i',i) \\
{}^kC^c(i,j) &= \sum_{i'} {}^kC^d(i',j)\,{}^kd(i',i)
\end{aligned}
\tag{86}
$$

where ${}^k\mathbf{d}$ is the transformation matrix for prototype k. The transformation between the CSF and SD respresentations can thus be accomplished through a matrix multiplication. Large vector lengths are in general assured by processing all configurations corresponding to a given prototype simultaneously.

The number of operations needed to transform the CI coefficients of a given configuration between the two representations is $N_c \times N_d$, with N_c and N_d given by Eqs. (74) and (78), respectively. The number of multiplications needed to transform a complete CI expansion between the two representations can be estimated as the total number of CSF's times an average number of SD's per configuration. This operation count will normally be significantly smaller than the number of multiplications used in a direct CI iteration, and since the transformation furthermore is performed with matrix multiplications, it will only constitute a small fraction of the time needed for a direct CI iteration.

2. *Slater determinants and strings*

Consider a RAS expansion with n electrons and spinprojection M_s. All determinants in the expansion then contain a number of α spin orbitals given by

$$n_\alpha = \frac{1}{2}(n + 2M_s) \tag{87}$$

and a number of β spin orbitals given by

$$n_\beta = \frac{1}{2}(n = 2M_s) \tag{88}$$

By writing all α spin orbitals before all β spin orbitals, each determinant can be written as

$$|I_\alpha I_\beta> = \alpha(I_\alpha)\beta(I_\beta)\,|\,Vac> \tag{89}$$

where $\alpha(I_\alpha)$ is an α string, *i.e.*, an ordered (ascending order chosen) product of creation operators for α spin orbitals and $\beta(I_\beta)$ is a β string, *i.e.*, an ordered product (ascending order chosen) of creation operators for β spin orbitals. The effective use of considering SD's as products of α- and β-strings was pioneered by Handy.[18,19] For a full CI expansion with n electrons in N orbitals and $M_s = 0$ the number of α- and β-strings is

$$\binom{N}{n/2},$$

while the number of SD's (neglecting symmetry) is

$$\binom{N}{n/2}^2$$

so that the number of strings equals the square root of the number of SD's. The splitting of SD's into products of α- and β-strings allows us to write the CI vector as a matrix $C(I_\alpha, I_\beta)$ with $I_\alpha(I_\beta)$ labeling the α- (β-) strings.

It is convienient to obtain a graphical representation and an ordering of the α- and β-strings, corresponding to a given set of RAS constraints. Let us study either α- or β-strings with n electrons and N orbitals. Each string can be represented as N ordered pairs $((k, m_k), k = 1, N)$, with k being an orbital index and m_k being the number of orbitals in the string in the range 1 to k. This set of ordered pairs can be pictured as a path in a diagram by drawing lines (arcs) between the points (vertexes) $((k, m_k), k = 1, N)$. All paths start at (0,0) and end at (N,n). Consider the case with n=3 and N=5, with RAS I consisting of the first 2 orbitals, RAS II containing orbital 3, and RAS III containing orbital 4 and 5. The minimum number of electrons in RAS I is set to 1, and the largest allowed number of electrons in RAS III is set to 2. In Fig. 2 we have drawn a diagram for this case. Two parths are shown in Fig. 2, corresponding to the strings containing orbitals (2 4 5) and (1 2 3), respectively. The path of all other allowed strings are contained in the polygon spanned by these two strings.

The paths of the diagram can be ordered. To do this we first define the vertex weight $W(k,m)$ as the number of allowed paths that starts from (0,0) and ends at (k,m), without considering the paths from (k,m) to (N,n). The weight of (0,0) is set to 1. Since all paths that pass through (k,m) must pass through vertex (k-1,m-1) or (k-1,m), the vertex weights are obtained as

$$W(k, m) = W(k-1, m-1) + W(k-1, m) \tag{90}$$

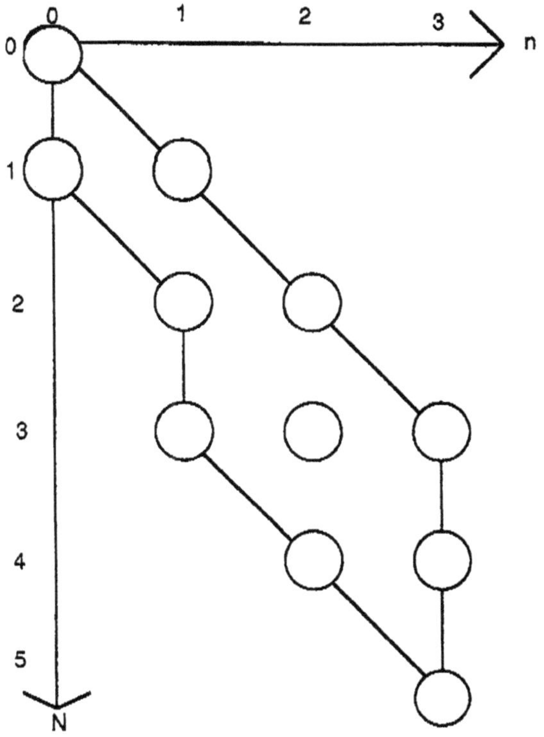

Figure 2. The shape of the string graph for a case with 3 electrons distributed in 5 orbitals. For RAS constraints, see text.

with the weights of all vertexes outside the allowed polygon set to 0. From the vertex weight one can define the arcweights of the arc joining (k-1,m-1) and (k,m) as

$$Y_1(k, m) = W(k, m - 1) \tag{91}$$

while the arcweight of the arc joining vertexes (k,m) and (k-1,m), having equal numbers of electrons, is set to zero

$$Y_0(k, m) = 0 \tag{92}$$

The index of a path (string) $((k, m_k), k = 1, N)$ is given as

$$I(\mathbf{m}) = \sum_{k=1}^{N} Y_{m_h - m_{h-1}}(k, m_k) \tag{93}$$

The way of ordering discussed here is the so-called reverse lexical ordering and it is only one of several possible ways of ordering the strings. We refer to Duch[50] for a more general discussion. The vertex weights and the arc weights corresponding to the string space of Fig. (2) are given in Fig. (3). The string (1 2 3) has

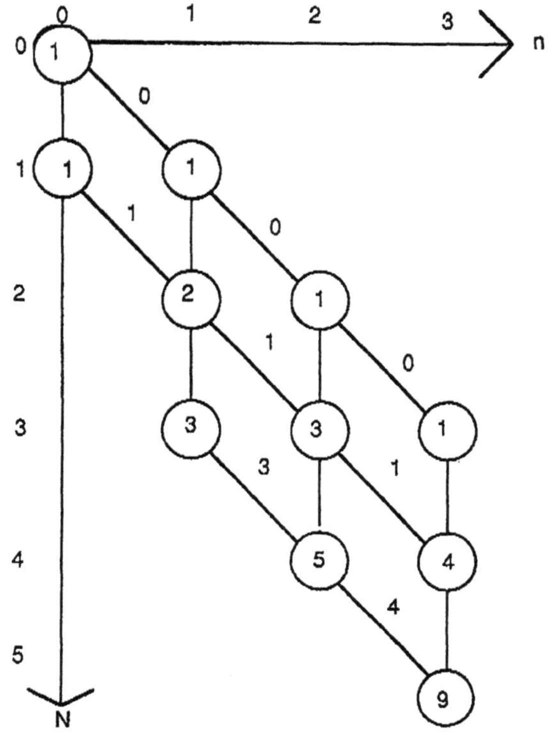

Figure 3. Arc- and vertex- weights for a string graph with 3 electrons distributed in 5 orbitals. For RAS constraints, see text.

the lowest index, 1, and the string (2 4 5) has the highest index, 9. The paths inside the polygon of Fig. (CI.1) have indices 2 to 8.

It is often necessary to deviate from the above ordering, thus making the above ordering only a first step in finding the index of a given string. It is convenient to have all strings with the same symmetry stored consecutively. It is then very simple to impose spatial symmetry on the CI vector. In order to describe RAS restrictions it is useful to reorder the strings of a given symmetry, so that strings with the identical numbers of electrons in RAS I and RAS III are stored consecutively. The set of α- and β-strings are thus divided into subsets, with each subset having a specific symmetry and a specific number of electrons in RAS I and RAS III. The restrictions of the CI expansion to be of given symmetry and fulfilling the RAS constraints are then realized by restricting which subblocks of α- and β-strings, that can be combined. This ordering allow us to process CI vectors with matrix- or vector-oriented tools.

In order to calculate density matrices and direct CI transformation, efficient ways of calculating the effect of one- and two-body excitations times a SD must be provided. Since all the operators used in SIRIUS are built up as sum of the elementary operators $\hat{a}^\dagger_{i\alpha}\hat{a}_{j\alpha}$ and $\hat{a}^\dagger_{i\beta}\hat{a}_{j\beta}$, one can tabulate excitations of the α- and β-strings separately, instead of tabulating information about excitations from SD's. This is one of the great advantages of the string approach. In the codes of SIRIUS, lists of the nonvanishing internal single excitations are constructed and stored;

$$\begin{aligned} |\alpha(J_\alpha)> &= sgn(ij)\hat{a}^\dagger_{i\alpha}\hat{a}_{j\alpha}|\alpha(I_\alpha)> \\ |\beta(J_\beta)> &= sgn(ij)\hat{a}^\dagger_{i\beta}\hat{a}_{j\beta}|\beta(I_\beta)> \end{aligned} \tag{94}$$

Double and higher excitations can then be obtained from successive table lookups. A straightforward implementation of single and double CI requires triple excited intermediate strings. One of the appealing features of our implementation of RAS, is that these extra intermediate strings are avoided by a suitable ordering of the multiple excitations. We refer to our original work[6] for further details.

3. *Direct CI for RAS expansions*

In a direct CI iteration, one calculates

$$\sigma(I_\alpha, I_\beta) = \sum_{J_\alpha J_\beta} < \beta(J_\beta)\alpha(J_\alpha)|\hat{\mathscr{H}}|\alpha(I_\alpha)\beta(I_\beta) > C(J_\alpha, J_\beta) \tag{95}$$

without the explicit construction and storage of the CI Hamiltonian matrix $< \beta(J_\beta)\alpha(J_\alpha)|\hat{\mathscr{H}}|\alpha(I_\alpha)\beta(I_\beta) >$. Introducing the expansion of the generators E_{ij} as

$$E_{ij} = E^\alpha_{ij} + E^\beta_{ij} \tag{96}$$

and using the definition of $\hat{\mathscr{H}}$ of Eq. (48), the σ-vector of Eq. (41) is written as a sum of three vectors

$$\sigma(I_\alpha, I_\beta) = \sigma^1(I_\alpha, I_\beta) + \sigma^2(I_\alpha, I_\beta) + \sigma^3(I_\alpha, I_\beta) \tag{97}$$

where

$$\begin{aligned} \sigma^1(I_\alpha, I_\beta) &= \sum_{J_\beta}\sum_{kl} < \beta(J_\beta)|E^\beta_{kl}|\beta(I_\beta) > g_{kl}C(I_\alpha, J_\beta) \\ &+ \frac{1}{2}\sum_{J_\beta}\sum_{(ij)\geq(kl)} < \beta(J_\beta)|E^\beta_{ij}E^\beta_{kl}|\beta(I_\beta) > (ij|kl)C(I_\alpha, J_\beta) \end{aligned} \tag{98}$$

$$\begin{aligned} \sigma^2(I_\alpha, I_\beta) &= \sum_{J_\alpha}\sum_{kl} < \alpha(J_\alpha)|E^\alpha_{kl}|\alpha(I_\alpha) > g_{kl}C(J_\alpha, I_\beta) \\ &+ \frac{1}{2}\sum_{J_\alpha}\sum_{(ij)\geq(kl)} < \alpha(J_\alpha)|E^\alpha_{ij}E^\alpha_{kl}|\alpha(I_\alpha) > (ij|kl)C(J_\alpha, I_\beta) \end{aligned} \tag{99}$$

$$\sigma^3(I_\alpha, I_\beta) = \sum_{J_\alpha J_\beta} \sum_{ijkl} < \alpha(J_\alpha) | E_{kl}^\alpha | \alpha(I_\alpha) > \cdot < \beta(J_\beta) | E_{ij}^\beta | \beta(I_\beta) > (ij|kl) C(J_\alpha, J_\beta) \tag{100}$$

The one electron integrals g_{kl} occurring in Eqs. (98) and (99) are obtained from the regular one electron integrals as

$$\begin{aligned} g_{kl} &= h_{kl} - \sum_{j<k} (kj|jl) - \frac{1}{1+\delta_{kl}} (kk|kl) \quad \text{for } k \geq l \\ g_{kl} &= h_{kl} - \sum_{j<k} (kj|jl) \quad \text{for } k < l \end{aligned} \tag{101}$$

The occurrence of the non-hermitian integrals g_{kl} is caused by the asymmetry introduced by restricting the summation over two electron terms in σ^1 and σ^2.

Let us consider the construction of σ^1 in the case where no RAS restrictions and no spatial symmetry restrictions are imposed, so all α-strings and all β-strings can be combined. W note that the matrix elements in σ^1 is independent of I_α. This can be exploited by structuring the loops as follows

Loop over I_β
  Initialize an array $F(J_\beta)$ to zero
  Loop over excitations E_{kl}^β from $|\beta(I_\beta)>$
    $|\beta(K_\beta)> = sgn(kl) E_{kl}^\beta | \beta(I_\beta)>$
    $F(K_\beta) = F(K_\beta) + sgn(kl) g_{kl}$
    Loop over excitations E_{ij}^β from $|\beta(K_\beta)>$ with $(ij) \geq (kl)$
      $|\beta(J_\beta)> = sgn(ij) E_{ij}^\beta | \beta(K_\beta)>$
      $F(J_\beta) = F(J_\beta) + \frac{1}{2} sgn(kl) sgn(ij) (ij|kl)$
    End of loop over E_{ij}^β
  End of loop over E_{kl}^β
  $\sigma^1(I_\alpha, I_\beta) = \Sigma_{J_\beta} F(J_\beta) C(I_\alpha, J_\beta)$; vectorized over I_β
End of loop over I_α

The construction of the intermediate vector **F** allows summation of all terms combining two strings. The time consuming step is the multiplication of the sparse vector **F** with the matrix of CI coefficients. The number of multiplications involved in this step is to leading order

$$N_1 = N_d \frac{n_\beta^2 (N - n_\beta)^2}{4} \tag{102}$$

which is obtained by realizing that the number of nonvanishing terms in **F** equals the number of strings that differ from $\beta(I_\beta)$ by 0, 1, or 2 spin orbitals. In Eq. (102), N_d is the number of Slater determinants and N is the number of active orbitals The σ^2 vector can be constructed with a very similar algorithm. In order to ensure vectorization with stride 1, the CI matrix, **C**, is first transposed. The multiplication count for the construction of σ^2 becomes

$$N_2 = N_d \frac{n_\alpha^2 (N - n_\alpha)^2}{4} \tag{103}$$

The third part of σ, the σ^3-vector is less trivial to vectorize. A gather and scatter operation can be inserted in order to avoid indirect addressing in the innermost loop. The corresponding algorithm becomes:

Loop over kl
    set up L(I), R(I), and sgn(I) defined by
    $|\alpha[L(I)]> = E_{kl}^{\alpha}|\alpha[R(I)]> sgn(I)$
    $C'(I, J_\beta) = C(L(I), J_\beta)sgn(I)$; vectorized gathering
    Loop over I_β
        Loop over excitations E_{ij}^{β} from $|\beta(I_\beta)>$
            $|\beta(J_\beta)> = sgn(ij)E_{ij}^{\beta}|\beta(I_\beta)>$
            $F(J_\beta) = F(J_\beta) + sgn(ij)(ij|kl)$
        End of loop over E_{ij}^{β}
        $V(I) = \Sigma_{J_\beta} F(J_\beta)C'(I, J_\beta)$; vectorized over I
        $\sigma^3(R(I), I_\beta) = \sigma^3(R(I)I_\beta) + V(I)$; vectorized scattering
    End of loop over I_β
End of loop over kl

The construction of **V** from **F** and **C′** is usually the most operation intensive part of the construction of σ^3. The multiplication count for this part is through leading order

$$N_3 = N_d n_\alpha n_\beta (N - n_\alpha)(N - n_\beta) \tag{104}$$

The gathering and the scattering operations each involve $N_d n_\alpha (N - n_\alpha)$ operations. A comparison of the operation count of this algorithm with previous algorithms is given in Ref. 6.

The previous discussion was for FCI expansions. For general RAS expansions only certain combinations of subblocks are allowed. The loops for generating σ^1, σ^2, and σ^3 can easily be modified to allow the generation of only the relevant subblocks. Abelian symmetry is implemented in a sim fashion. We refer to Olsen et al.[6] for further details.

For systems with $M_s = 0$ the coefficient matrix obeys

$$C(I_\alpha, I_\beta) = (-1)^S C(I_\beta, I_\alpha) \tag{105}$$

which implies that

$$\begin{aligned} \sigma^2(I_\alpha, I_\beta) &= (-1)^S \sigma^1(I_\beta, I_\alpha) \\ \sigma^3(I_\alpha, I_\beta) &= (-1)^S \sigma^3(I_\beta, I_\alpha) \end{aligned} \tag{106}$$

These relations eliminate the calculation of σ^2 and halves the part of σ^3 that needs to be constructed. As a result, the operation count for $M_s = 0$ states is reduced by a factor of two.

The most operation intensive part of the discussed direct CI algorithms is the matrix times sparse vector operations occurring in the innermost loop. The capability of the IBM VS2 compilers to vectorize over outer loops can be used to implement this operation as follows:

Task: Obtain $V(I) = \Sigma_J C(I,J)F(J)$, F being sparse

```
Loop over segments of I
   Zero a local array t
   Loop over J for which F(J) ≠ 0
       Loop over I in segment
           t(I) = t(I) + C(I,J)*F(J)
       End of loop over I in segment
   End of loop over J
   V(I) = t(I)
End of loop over segments of I
```

The local array **t** ensures the use of the compound instruction on the IBM 3090/VF, performing one addition and one multiplication in a single cycle. The use of vectorization over the outer loops minimizes the traffic between the vector registers of the CPU and cache or main memory, since multiple loadings of the vector **V** are avoided. This implementation of the matrix times sparse vector routine runs with 30-60 MFLOPS, although the routine is coded in normal FORTRAN.

4. Construction of density matrices

The one- and two-body density matrices are needed for the calculation of the orbital gradient, the orbital-orbital Hessian block times an orbital vector, and the orbital-CI Hessian block times a CI vector. Two different CI expansions can be involved

$$
\begin{aligned}
|L> &= \sum_{I_\alpha I_\beta} C_l(I_\alpha, I_\beta)\,|I_\alpha, I_\beta> \\
|R> &= \sum_{I_\alpha I_\beta} C_r(I_\alpha, I_\beta)\,|I_\alpha, I_\beta>
\end{aligned}
\tag{107}
$$

The corresponding one-body density matrix

$$D_{ij} = < L|E_{ij}|R > \tag{108}$$

is obtained as a sum of tv erms

$$D_{ij} = D^1_{ij} + D^2_{ij} \tag{109}$$

with

$$
\begin{aligned}
D^1_{ij} &= < L|E^\alpha_{ij}|R > \\
D^2_{ij} &= < L|E^\beta_{ij}|R >
\end{aligned}
\tag{110}
$$

A loop structure for the e ient construction of $\mathbf{D}^2$ is

Zero $\mathbf{D}^2$
loop over $\beta(I_\beta)$
  loop over excitations . out from $|\beta(I_\beta)>$
    $|\beta(J_\beta)> = sgn(ij)$ $\beta(I_\beta)>$
    $D^2_{ij} = D^2_{ij} + sgn(ij)\Sigma$ $(I_\alpha, J_\beta)C_r(I_\alpha, I_\beta)$ vectorized over I_α
  end of loop over excit ons E^β_{ij}
end of loop over $\beta(I_\beta)$

The algorithm can be in mented so all excitations E^β_{ij} and the corresponding strings $\beta(J_\beta)$ are obtained d saved. The loop with the inner product of $C_l(*, J_\beta)$ with $C_r(*, I_\beta)$ can then be realized so that segments of $C_r(*, I_\beta)$ only needs to be loaded to vector registers a single time. This allows efficient vectorization on the IBM 3090/VF. A very similar loop for constructing $\mathbf{D}^1$ can be constructed. In order to obtain vectorization with stride one, it is necessary to transpose C_l and C_r. In order to generalize the above loop structure to include RAS constraints, it is just necessary to restrict the loop over I_α to be over subsets where $C_l(I_\alpha, J_\beta)$ and $C_l(I_\alpha, I_\beta)$ are included in the RAS expansion. An attractive feature of the above is that the spin-density matrix is trivially obtained as $\mathbf{D}^s = \mathbf{D}^1 - \mathbf{D}^2$.

All the two electron density matrices used in SIRIUS can be derived from the general form

$$d_{ij,kl} = < L \mid E_{ij}E_{kl} - \delta_{jk}E_{il} \mid R > \tag{111}$$

The two body density matrix has the permutational symmetry

$$d_{ij,kl} = d_{kl,ij} \tag{112}$$

which can be used to restrict the calculation of $\mathbf{d}$ to index combinations with $(ij) \geq (kl)$, where (ij) is some ordering index for orbital pair (ij). The two body density matrix can be separated into spin components

$$d_{ij,kl} = d^1_{ij,kl} + d^2_{ij,kl} + d^3_{ij,kl} + d^3_{kl,ij} - \delta_{jk}D_{il} \tag{113}$$

with

$$\begin{aligned} d^1_{ij,kl} &= < L \mid E^\alpha_{ij}E^\alpha_{kl} \mid R > \\ d^2_{ij,kl} &= < L \mid E^\beta_{ij}E^\beta_{kl} \mid R > \\ d^3_{ij,kl} &= < L \mid E^\beta_{ij}E^\alpha_{kl} \mid R > \end{aligned} \tag{114}$$

In addition to the general permutation symmetry of Eq. (112), $\mathbf{d}^1$ and $\mathbf{d}^2$ satisfy

$$\begin{aligned} d^1_{ij,kl} - \delta_{jk}D^1_{il} &= d^1_{il,kj} - \delta_{lk}D^1_{ij} \\ d^2_{ij,kl} - \delta_{jk}D^2_{il} &= d^2_{il,kj} - \delta_{lk}D^2_{ij} \end{aligned} \tag{115}$$

which can be used to further reduce the explicit evaluation of $d^1_{ij,kl}$ and $d^2_{ij,kl}$ to index combinations with $j \geq 1$.

The construction of $\mathbf{d}^2$ can be accomplished as follows

Zero $\mathbf{d}^2$
Loop over betastrings $\beta(I_\beta)$
  Loop over excitations E^β_{kl} out from $\mid \beta(I_\beta) >$
    $\mid \beta(K_\beta) > = sgn(kl)E^\beta_{kl} \mid \beta(I_\beta) >$

Loop over excitations E^{β}_{ij} out from $|\beta(K_\beta)>$ $((ij) \geq (kl), j \geq 1)$

$|\beta(J_\beta)> = sgn(ij)E^{\beta}_{ij}|\beta(K_\beta)>$

$d^2_{ij,kl} = d^2_{ij,kl} + sgn(ij)sgn(kl)\Sigma_{I_\alpha} C_l(I_\alpha, J_\beta)C_r(I_\alpha, I_\beta)$

end of loop over excitations E^{β}_{ij} out from $|\beta(K_\beta)>$

end of loop over excitations E^{β}_{kl} out from $|\beta(I_\beta)>$

end of loop over betastrings $\beta(I_\beta)$

The above loop is constructed so the same column $C_r(*, I_\beta)$ is multiplied with several columns of C_l. It is again important that the code is written in such a way that only a single load of $C_r(*, I_\beta)$ to vector registers is performed. We have attained this by first constructing lists of ij,kl,J_β, and signs, and then calculating the inner products over I_α for all strings J_β. The construction of $\mathbf{d}^1$ follows a very similar algorithm. In order to vectorize with stride one, it is again necessary to transpose $\mathbf{C}_l$ and $\mathbf{C}_r$. The construction of $\mathbf{d}^3$ is more complicated since all indices of strings differ. By performing outer gatherings, it is possible to obtain a loop structure without any indirect addressing in the innermost loop

zero $\mathbf{d}^3$

loop over excitations E^{α}_{kl}

set up $L(I)$, $R(I)$, $sgn(I)$ so $|\alpha(L(I))> = E^{\alpha}_{kl}|\alpha(R(I))> sgn(I)$

$C'(I, I_\beta) = C_l(L(I), I_\beta)sgn(I)$

loop over betastrings $\beta(I_\beta)$

$V(I) = C_r(R(I), I_\beta)$

loop over excitations E^{β}_{ij} out from $|\beta(I_\beta)>$

$|\beta(J_\beta)> = sgn(ij)E^{\beta}_{ij}|\beta(I_\beta)>$

$d^3_{ij,kl} = d^3_{ij,kl} + sgn(ij)\Sigma_I C'(I, J_\beta)V(I)$

end of loop over excitations E^{β}_{ij} out from $|\beta(I_\beta)>$

end of loop over betastrings $\beta(I_\beta)$

end of loop over excitations E^{α}_{kl}

In order to have efficient vectorization we have again separated the parts so that all strings $\beta(J_\beta)$ are obtained first and then the actual inner products are obtained.

The operation count (number of multiplications) n^1, n^2, n^3 for constructing $\mathbf{d}^1, \mathbf{d}^2, \mathbf{d}^3$, respectively, are through leading order

$$
\begin{aligned}
n^1 &= \frac{1}{4} n_\alpha^2 (N - n_\alpha)^2 N_{dct} \\
n^2 &= \frac{1}{4} n_\beta^2 (N - n_\beta)^2 N_{dct} \\
n^3 &= n_\beta (N - n_\beta) n_\alpha (N - n_\alpha) N_{dct}
\end{aligned}
\tag{116}
$$

The operation counts are identical to the counts obtained for the similar direct CI loops on Section 3.

In cases where the $M_s = 0$ component of a system is used, additional simplifications can be obtained:

$$
\begin{aligned}
d^1_{ij,kl} &= d^2_{ij,kl} \\
d^3_{ij,kl} &= d^3_{kl,ij}
\end{aligned}
\tag{117}
$$

The use of the above relations eliminated the need for constructing $\mathbf{d}^1$ and halves the part of $\mathbf{d}^3$ for states with $M_s = 0$.

5. *Counter rotations of CI coefficients*

The active orbitals of a CAS state can be transformed into orbitals, that diagonalize the active part of the $\mathbf{F}^C + \mathbf{F}^V$ or the density matrix. For a RAS expansion, the diagonal subblocks of the one body density or the $\mathbf{F}^C + \mathbf{F}^V$ matrix can be made diagonal. The CI expansion corresponding to the new set of orbitals can be obtained by a new CI calculation, but this is rather expensive for a large CI expansion. In a very elegant work by Malmquist[45,46] a formalism was obtained that allows the direct transformation of the CI coefficients correponding to a given orbital transformation. The method also allows the calculation of transition moments between RAS expansions with different orbital sets. Since the amount of work involved in the algorithm is comparable to a direct CI iteration with only a one body operator, the explicit transformation requires only a small fraction of the time needed for a full direct CI iteration.

The basic manipulation of the CI coefficients in the algorithm is an ordered product of excitations

$$
|0'\rangle = \prod_{k=1,N} \left(\sum_i t_{ik} E_{ik}\right) |0\rangle \tag{118}
$$

Since the α- and β-spin orbitals commute, Eq. (118) can be rewritten as

$$
|0'\rangle = \prod_{k=1,N} \left(\sum_i t_{ik} E^\alpha_{ik}\right) \prod_{k=1,N} \left(\sum_i t_{ik} E^\beta_{ik}\right) |0\rangle \tag{119}
$$

The excitations of Eq. (119) can be carried out very efficiently in the Slater determinant basis. Consider the part with excitations of spin orbitals with β spin. An algorithm for this part is

Zero an array C′ with length of CI expansion
Loop over i
  loop over I_β
    zero an array F
    Loop over excitations E^β_{ki} out from $|\beta(I_\beta)>$
      $|\beta(J_\beta)> = sgn(ki)E^\beta_{ki}|\beta(I_\beta)>$
      $F(J_\beta) = F(J_\beta) + sgn(ki)*t_{ik}$
    end of loop over excitations E^β_{ki} out from $|\beta(I_\beta)>$
    $C'(I_\alpha, I_\beta) = C'(I_\alpha, I_\beta) + \Sigma_{J_\beta} C(I_\alpha, J_\beta)$; vectorized over I_α
  end of loop over I_β
  $C(I_\alpha, I_\beta) = C(I_\alpha, I_\beta) + C'(I_\alpha, I_\beta)$
End of loop over i

The central part of this algorithm is again a matrix times a sparse vector. It was shown in Sec. 3 how this operation can be carried out very efficiently on the IBM 3090/VF.

The α-part of the transformation can be carried out in a similar fashion. It is again convenient to use transposed arrays ($C(I_\beta, I_\alpha)$ instead of $C(I_\alpha, I_\beta)$) in order to vectorize with stride one.

E. Auxiliary Optimization Algorithms

In this section several auxiliary algorithms are considered.[4,5] The sample calculations in Ref. (4) illustrate some of the improvements which may be obtained.

1. Split configuration and orbital trial vectors

Explicit solutions to the Norm Extended (NEO) or the Newton-Raphson (NR) equations for finding optimal steps can only be obtained for small configuration spaces, for large configuration spaces iterative techniques must be used as outlined in Sec. B. In both the direct NEO approach and the direct NR approach, the solution vector is obtained with an algorithm that requires linear transformations on trial vectors $\mathbf{b}_j$ with ${}^\beta\mathbf{L}$ according to Eq. (41): $\sigma_j = {}^\beta\mathbf{L}\mathbf{b}_j$. (In NR optimization $\beta = 0$.) Using superscripts ^{c} and ^{o} to denote the configuration and orbital parts, respectively, of trial vectors $\mathbf{b}_j$ and the NEO Hessian matrix ${}^{\beta}\mathbf{L}$ the linear transformation can be split in configuration and orbital parts

$$\begin{pmatrix} {}^{\beta}\mathbf{L}^{cc} & {}^{\beta}\mathbf{L}^{co} \\ {}^{\beta}\mathbf{L}^{oc} & {}^{\beta}\mathbf{L}^{oo} \end{pmatrix} \begin{pmatrix} \mathbf{b}^{c} \\ \mathbf{b}^{o} \end{pmatrix} = \begin{pmatrix} {}^{\beta}\mathbf{L}^{cc}\,\mathbf{b}^{c} \\ {}^{\beta}\mathbf{L}^{oc}\,\mathbf{b}^{c} \end{pmatrix} + \begin{pmatrix} {}^{\beta}\mathbf{L}^{oc}\,\mathbf{b}^{o} \\ {}^{\beta}\mathbf{L}^{oo}\,\mathbf{b}^{o} \end{pmatrix} \tag{120}$$

It was shown in Sec. B.5 that the individual terms on the right-hand side can be expressed in terms of gradients with one-index integrals or with transition densities, with no mixing of the contributions from the configuration and orbital parts of the trial vector. It follows that the linear transformations can be implemented separately on the configuration $\mathbf{b}_j^c$ and the orbital $\mathbf{b}_j^o$ part of trial vector $\mathbf{b}_j$ with essentially no loss of efficiency. This fact can be used to split the generalized Davidson algorithm to solve the NEO eigenvalue equation or to split a modified conjugate gradient algorithm for solving an NR-like equation. In the split trial vector algorithms the trial vectors are divided into configuration and orbital trial vectors ($\mathbf{b}_j = \mathbf{b}_j^c$ or $\mathbf{b}_j = \mathbf{b}_j^o$), and the solution vector is determined in the reduced space generated by the respective algorithms. Split trial vector algorithms have the advantage of allowing trial vectors to be added to the part of the solution space where convergence is most difficult. For example, when slow convergence appears in the orbital part of the solution space more orbital vectors may be added than configuration vectors. This flexibility is important as it may save linear transformations on configuration trial vectors. The splitting of the trial vectors also allows us to develop a very efficient algorithm for finding orbital trial vectors as will be discussed below in Sec. E.2.

In the NEO algorithm, let us assume that we have performed n split Davidson iterations and thus generated a set of n orthonormal trial vectors, for example;

$$\{\mathbf{b}_j : j = 1, \ldots, n\} = \left\{ \begin{pmatrix} \mathbf{b}^{c} \\ 0 \end{pmatrix}_1 \begin{pmatrix} 0 \\ \mathbf{b}^{o} \end{pmatrix}_2 \begin{pmatrix} 0 \\ \mathbf{b}^{o} \end{pmatrix}_3 \begin{pmatrix} \mathbf{b}^{c} \\ 0 \end{pmatrix}_4 \cdots \begin{pmatrix} 0 \\ \mathbf{b}^{o} \end{pmatrix}_n \right\} \tag{121}$$

and the corresponding linear transformed vectors

$$\{\sigma_j : j = 1, \ldots, n\} = \left\{ \begin{pmatrix} \sigma^{c} \\ \sigma^{o} \end{pmatrix}_1 \begin{pmatrix} \sigma^{c} \\ \sigma^{o} \end{pmatrix}_2 \begin{pmatrix} \sigma^{c} \\ \sigma^{o} \end{pmatrix}_3 \begin{pmatrix} \sigma^{c} \\ \sigma^{o} \end{pmatrix}_4 \cdots \begin{pmatrix} \sigma^{c} \\ \sigma^{o} \end{pmatrix}_n \right\} \tag{122}$$

By choosing $\mathbf{x}^{(k)}$ as the first trial vector, the corresponding linear transformed vector is $\beta\mathbf{g}$ (*c.f.* $\mathbf{g} = \hat{\mathbf{P}}\mathbf{g}$ and Eq. (28)), and all the β dependence is confined to the first trial vector. Equations (121) and (122) may be used to set up a reduced eigenvalue equation

$$\begin{pmatrix} \mathbf{L}^{cc\beta R} & {}^{\beta}\mathbf{L}^{coR} \\ \mathbf{L}^{oc\beta R} & {}^{\beta}\mathbf{L}^{ooR} \end{pmatrix} \begin{pmatrix} \mathbf{z}^{cR} \\ \mathbf{z}^{oR} \end{pmatrix} = {}^{\beta}\lambda^{R} \begin{pmatrix} \mathbf{z}^{cR} \\ \mathbf{z}^{oR} \end{pmatrix} \tag{123}$$

where

$${}^{\beta}L_{ij}^{R} = \mathbf{b}_i^{t}\sigma_j \tag{124}$$

The reduced eigenvalue equation determines the optimal solution $(z)^R$ in the basis of trial vectors in Eq. (121). In the split Davidson algorithm the trial vector generated by the Davidson algorithm is divided into an orbital and a configuration trial vector, and one or both of these trial vectors are orthonormalized and added to the basis in Eq. (121). The iterative procedure is continued until the norm of the residual vector is less than the requested threshold. Note that if the value of β is updated during the sequence of iterations the corresponding update of the reduced eigenvalue equation is obtained by multiplying the first row and column with a constant β'/β, where β' is the updated value of the bet parameter.

If the NR algorithm is used, let us now assume that in the nth step of a split modified conjugate algorithm we have generated a set of n orthonormal trial vectors (Eq. (121)) which are orthogonal to $\mathbf{x}^{(k)}$ and we have generated the linear transformed vectors (Eq. (122)) according to Eq. (120). The reduced linear equation

$$\mathbf{g}^R + (\mathbf{H}'^R - \nu\mathbf{I})\delta^R = 0 \tag{125}$$

where

$$g_i^R = \mathbf{b}_i^t\mathbf{g} \tag{126}$$

$$H'^R_{ij} = \mathbf{b}_i^t\sigma_j \tag{127}$$

determines the optimal solution vector

$$\delta^R = \begin{pmatrix} \delta^{cR} \\ \delta^{oR} \end{pmatrix} \tag{128}$$

in the basis in Eq. (121). The residual vector

$$\mathbf{r}^R = \mathbf{g} + (\mathbf{H}' - \nu\hat{\mathbf{P}})\delta^R \tag{129}$$

divided by the diagonal element of $(\mathbf{H}' - \nu\hat{\mathbf{P}})$ generates an improved configuration or orbital trial vector, and the iterative procedure is continued until the norm of the residual vector is smaller than a specified tolerance.

2. *Optimal orbital trial vectors*

When solving the NR equation or the NEO eigenvalue equation iteratively using the split trial vector algorithm outlined in the previous subsection, it is desirable to use an efficient algorithm for finding orbital trial vectors in order to save expensive direct CI steps. Below we describe such an algorithm that allows orbital vectors to be determined that are optimal for an approximation to the configuration coefficients in the NR or NEO equation.

Consider initially the solution of the NR equation and assume that the set of configuration coefficients represents the current approximation to the solution vector, for example δ^{cR} of Eq. (129). Optimal orbitals for this set of configuration coefficients may be determined by substituting

$$\begin{pmatrix} \delta^{cR} \\ \mathbf{b}^{o} \end{pmatrix}$$

as solution vector to the level shifted NR equations (Eq. 125) and then solve the second component

$$\mathbf{g}^{o} + (\mathbf{H}'^{oo} - \nu\mathbf{I})\mathbf{b}^{o} + \mathbf{H}'^{oc}\,\delta^{cR} = 0 \tag{130}$$

for the optimal orbital vector $\mathbf{b}^{o}$. This equation is a set of linear equations

$$\mathbf{f}^{o} + (\mathbf{H}'^{oo} - \nu\mathbf{I})\mathbf{b}^{o} = 0 \tag{131}$$

with a modified gradient

$$\mathbf{f}^{o} = \mathbf{g}^{o} + \mathbf{H}'^{oc}\,\delta^{cR} \tag{132}$$

This set of linear equations has the dimension of the orbital space and may therefore easily be solved. The vector $\mathbf{H}'^{oc}\,\delta^{cR}$ may be constructed from the saved linear transformed vectors of Eq. (122) without carrying out direct CI steps.

Optimal orbital vectors may be determined in the NEO algorithm in a similar way. From Eqs. (36-40) and (123) it is seen that the level shift in direct NEO is ${}^{\beta}\lambda^{R}$. The current approximation to the configuration coefficients may be determine from the same equations to be $\alpha\hat{\mathbf{P}}\mathbf{z}_{i}^{cR}$. From Eq. (31) it is clear that

$${}^{\beta}\mathbf{L}^{oc}(\alpha\hat{\mathbf{P}}\mathbf{z}_{i}^{cR}) = \mathbf{H}'^{oc}(\alpha\hat{\mathbf{P}}\mathbf{z}_{i}^{cR}) \tag{133}$$

When $\mathbf{x}^{(k)}$ is chosen as the first trial vector the projection on the solution vector $\mathbf{z}^{R}$ may simply be implemented by excluding the first component of $\mathbf{z}^{cR}$ when constructing $\mathbf{H}'^{oc}(\alpha\hat{\mathbf{P}}\mathbf{z}_{i}^{cR})$ from the linear transformed vectors in Eq. (122). Optimal orbital vectors can thus also be determined in NEO without carrying out direct CI steps.

3. *Convergence of solution vectors in direct NEO and NR algorithms*

When direct algorithms are used to find the solution vector significant savings in the computational cost can be obtained if we know *a priori* how accurate the solution vectors need to be determined. The optimal accuracy clearly differs whether the calculation is in the local or the global region. In the global region a step on the trust sphere in the right direction is adequate, whereas in the local region the solution vector must be sufficiently accurate to maintain the quadratic convergence. Assume initially that local convergence is obtained with a NR algorithm and that the solution vector δ^R in the kth macro iteration is converged to a residual accuracy $\| \mathbf{r}^R \|$ where

$$\mathbf{r}^R = \mathbf{g} + \mathbf{H}' \,\delta^R \tag{134}$$

The level shift is zero because we are in the local region. The parameter set for the $k + 1$'th iteration point becomes

$$\mathbf{x}^{(k+1)} = \mathbf{x}^{(k)} + \delta^R \tag{135}$$

Substituting δ^R of Eq. (134) in Eq. (135) gives

$$\mathbf{x}^{(k+1)} = \mathbf{x}^{(k)} - \mathbf{H}'^{-1}\mathbf{g} + \mathbf{H}'^{-1}\mathbf{r}^R \tag{136}$$

The first two terms represent the exact NR solution which gives quadratic convergence in the error vector[36]

$$\xi^{(k)} = \mathbf{x}^{(k)} - \eta \tag{137}$$

where η denotes the stationary point. Quadratic convergence will thus be obtained if the last term gives errors which are quadratic in the error vector. Since the gradient $\mathbf{g}$ has errors which are of first order in ξ we must converge to an accuracy of

$$\| \mathbf{r}^R \| \leq \mathbf{k} \,\| \mathbf{g} \|^2 \tag{138}$$

where k is an error constant, to maintain quadratic convergence.

More insight into the convergence behavior can be obtained by considering a Taylor expansion of the gradient with $\mathbf{x}^{(k)}$ as expansion point

$$\mathbf{g}(\mathbf{x}^{(k)} + \delta) = \mathbf{g} + \mathbf{H}'\delta + O(\delta^2) = \mathbf{r}(\delta) + O(\delta^2) \tag{139}$$

A comparison of Eq. (134) and Eq. (139) shows that the difference between the residual and the gradient at the new expansion point $\mathbf{x}^{(k+1)} = \mathbf{x}^{(k)} + \delta$ is caused by second order terms. An estimate of the error constant k in Eq. (138) can be judged from the agreement between the residual and the gradient at the new expansion point. If they differ we have exceeded the region where the first order terms dominate. Our numerical experience has shown that it is reasonable to converge to a residual accuracy below $10 \,\| \mathbf{g} \|^2$ (to use an error constant $\mathrm{k} = 10$). A smaller error constant has in general very little, if any, effect on the number of Newton-Raphson iterations that is used in a second order calculation. If a larger error constant is used, extra Newton-Raphson iterations are generally required as the quadratic convergence of the Newton-Raphson algorithm is not fully utilized.

Eq. (139) gives a very accurate criterion for terminating a sequence of micro iterations near wave function convergence. For example, if we want to converge a calculation to a gradient tolerance 10^{-5}, we may replace in the last iteration the quadratic convergence threshold (which is below 10^{-5}) with 10^{-5} since the linear estimate which the residual gives of the gradient at the new expansion point is extremely accurate. In the global region where the NEO algorithm is applied, we have used as criterion for convergence that the residual is smaller than $0.2 \| \mathbf{g} \|$. The reason is that further accuracy in the solution vector is unimportant compared to deviations resulting from using a second order Taylor expansion to represent the energy hypersurface.

4. Transformation to natural and Fock type orbitals

The total energy is invariant with respect to a rotation within the inactive and within the secondary space and for a CAS wave function also with respect to a rotation within the active space. For a RAS wave function the energy is only invariant with respect to rotations within each of the three active subspaces: RAS I, RAS II, and RAS III. This freedom in the choice of orbitals can be used to transform the orbitals to improve diagonal dominance in the Hessian matrix in each MCSCF iteration. Diagonal dominance is important as the convergence of the direct iterative algorithms strongly depends on diagonal dominance. The set of orbitals that gives optimal diagonal dominance is not well defined. A choice which may be justified based on algebraic considerations and numerical experience is to rotate the inactive and the secondary orbitals such that the sum of the inactive and active Fock matrices is diagonal in the inactive and in the secondary space (Fock type orbitals) and to rotate the active orbitals to get natural orbitals. In the rest of this section we examine in more detail this choice of orbitals. In Ref. (8), a review, the orbital part of the Hessian matrix is expressed in terms of inactive and active Fock matrices and additional terms. Inspection of these formulas shows that the above choice of orbitals makes many off-diagonal contributions zero. For example the inactive- secondary block of the orbital Hessian becomes zero except for some two-electron terms. These two-electron terms are known to be small in the limit of a single configuration calculation. The transformation to natural orbitals is also motivated by the analysis of Löwdin[51] who showed that natural orbitals give the most compact configuration expansion representation of the state in a certain sense. The analysis of Löwdin is not directly concerned with diagonal dominance in the configuration part of the Hessian (*i.e.*, the CI Hamiltonian), however, it indicates that natural orbitals constitute a well defined set to use which is very likely to be better than the "arbitrary" set of orbitals that otherwise is obtained in the second-order calculation. The numerical test calculations we have carried out support the above choice of orbitals.[4]

The tests show that it is worth the effort to transform the orbitals in each iteration to obtain natural and Fock type orbitals as a significant reduction thereby can be obtained in the number of trial vectors. We should also mention that the transformation to natural orbitals introduces changes in the configuration coefficients and that these changes may easily be calculated as described in Refs. 45 and 46 and in Sec. D.5 above. This calculation of the changes in configuration coefficients is of the same order as the construction of the one-electron reduced density matrix.

5. *Intermediate optimization of orbitals for fixed configuration coefficients*

When a second order MCSCF calculation is initiated it is usually observed that very large orbital rotations are carried out compared to the change in the configuration coefficients. Most of the the large orbital rotations can be absorbed in the global region through intermediate orbital optimizations for fixed configuration coefficients. Intermediate orbital optimizations are very important when the the direct CI steps constitute the most time consuming part in a MCSCF calculation. We describe below an algorithm for intermediate optimization of the orbitals for fixed configuration coefficients.[52,4]

In the MCSCF calculations where the direct CI steps are the time consuming part of the calculation it is advantageous to carry out intermediate optimizations of the orbitals for fixed configuration coefficients. Intermediate optimization of the orbitals requires no direct CI steps and may be implemented as a restricted step optimization in the orbital subspace. The second order Taylor expansion for fixed configuration coefficients is the orbital part of the Taylor expansion in Eq. (10)

$$\Delta E^{(2o)}(\kappa) = \mathbf{g}^{ot}\,\kappa + \frac{1}{2}\kappa^t\,\mathbf{H}'^{oo}\kappa. \tag{140}$$

In ground state calculations the full Hessian $\mathbf{H}'$ is positive definite at convergence and the subblock $\mathbf{H}'^{oo}$ must also be positive definite. For ground states Eq. (140) can thus straightforwardly be used to carry out intermediate optimizations of the orbitals. In calculations on the nth excited state the Hessian matrix $\mathbf{H}'$ must have index n at convergence. If the configuration subblock $\mathbf{H}'^{cc}$ has index n then $(\mathbf{H}'^{oo} - \mathbf{H}'^{oc}(\mathbf{H}'^{cc})^{-1}\mathbf{H}'^{co})$ must be positive definite.[34] In calculations where the coupling between the orbital and configuration spaces is small and where $\mathbf{H}'^{cc}$ has no small eigenvalues we may expect that $\mathbf{H}'^{oc}(\mathbf{H}'^{cc})^{-1}\mathbf{H}'^{co}$ compared to $\mathbf{H}'^{oo}$ and that $\mathbf{H}'^{oo}$ is positive definite by itself. If $\mathbf{H}'^{oo}$ is positive definite then intermediate orbital optimizations can also straightforwardly be carried out for excited states using restricted step optimization. If $\mathbf{H}'^{oo}$ is not positive definite then intermediate orbital optimization may produce worse orbitals. In excited state optimization SIRIUS will only carry out an intermediate optimization of the orbitals if no "root flipping" has occurred, that is, the reference vector is the proper root in the CI space.

In ground state calculations we have included the active-active orbital rotation parameters in the intermediate orbital optimizations. The active-active orbital rotation parameters are only redundant when the coupling between the orbital and configuration space is considered. In excited state calculations the active-active orbital rotation parameters are not included as these parameters may result in a variational collapse to the ground state.

When $\mathbf{H}'^{oo}$ is positive definite restricted step orbital optimization may be implemented as a NEO type optimization where the orbital parameters are determined by solving the eigenvalue problem

$$\begin{pmatrix} 0 & \gamma\tilde{\mathbf{g}}^{ot} \\ \gamma\mathbf{g}^{o} & \mathbf{H}'^{oo} \end{pmatrix} \begin{pmatrix} (1/\gamma) \\ \delta^{o} \end{pmatrix} = \omega \begin{pmatrix} (1/\gamma) \\ \delta^{o} \end{pmatrix} \tag{141}$$

for its lowest root ω. The second component of this equation

$$\mathbf{g}^{o} + (\mathbf{H}'^{oo} - \omega\mathbf{I})\delta^{o} = 0 \tag{142}$$

is a level shifted NR equation where the eigenvalue ω is the level shift and the eigenvector with its first component normalized to $1/\gamma$ determines the orbital rotation parameters δ. The lowest root is calculated in Eq. (140) because

$$\begin{pmatrix} 0 & \gamma\tilde{\mathbf{g}}^{ot} \\ \gamma\mathbf{g}^{o} & \mathbf{H}'^{oo} \end{pmatrix} \tag{143}$$

is of a dimension one larger than $\mathbf{H}'^{oo}$ and therefore separates the roots of $\mathbf{H}'^{oo}$. The desired length of the orbital step is obtained by adjusting the parameter γ. The second order change in the electronic energy is then[4]

$$\Delta E^{o(2)}(\kappa) = \frac{1}{2}\omega(1/\gamma^2 + \kappa^t \kappa) \tag{144}$$

If an orbital step is rejected we have for simplifying reasons chosen to reduce the step by scaling: $\kappa \rightarrow k\kappa$ where k is the factor chosen to decrease the trust radius for a rejected step. The second order energy change corresponding to the scaled step is

$$\Delta E^{o}(2)(k\kappa) = \frac{1}{2}\omega k^2 (\frac{2-k}{k\gamma^2} + \kappa^t\kappa) \tag{145}$$

The determination of the solution to Eq. (141) requires either an explicit construction of $\mathbf{H}'^{oo}$ or linear transformations $\mathbf{H}'^{oo}\,\mathbf{b}^{o}$. We have implemented the latter approach in our program. A linear transformation can be expressed in terms of an orbital gradient with one-index transformed integrals. An efficient evaluation of orbital gradients which requires the smallest possible integral transformation is obtained by expressing the orbital gradient in terms of generalized Fock matrices as described by Siegbahn *et al.*[40] Evaluation of Fock matrices with one-index transformed integrals is described in Appendix B.

The integral transformation to the molecular orbital basis for $\mathbf{H}'^{oo}\mathbf{b}^{o}$ depends on the orbital operators which are included in the optimization. In the most general case where inactive-active, inactive-secondary, active-active and active-secondary orbital operators (operator level 1) are considered MO integrals with two general indices and two indices confined to the active space are needed (integral transformation A). If we exclude the active-secondary orbital operators we obtain operator level 2 which requires MO integrals with all four indices confined to the active space (integral transformation B). Finally, operator level 3 is obtained if the inactive-secondary orbital operators are excluded from the operators in operator level 2. Operator level 3 will internally optimize the occupied orbitals, however, no reduction in the integral transformation is obtained when the inactive-secondary orbital operators are excluded from operator level 2 as integral transformation B is required in both cases. By default SIRIUS will employ intermediate orbital optimization at operator level 1.

6. *Initial guess of orbitals and configuration coefficients*

Initial guess of orbitals: In an MCSCF calculation where no good initial guess of orbitals is available from *e.g.* a previous calculation at a close-lying geometry, or from an MP2 or INO-CI calculation as suggested in Sec. B.3, a set of canonical Hartree-Fock orbitals may seem to be a reasonable choice for initial orbitals. However, when the basis set contains diffuse functions, diffuse virtual canonical Hartree-Fock (HF) orbitals will be determined, which have orbital energies close to zero. If these diffuse virtual orbitals are included in the initial active space unnecessary slow convergence will be encountered in the MCSCF macro iteration as the diffuse canonical HF orbitals have little resemblance with the converged correlating MCSCF orbitals. A better set of initial Hartree-Fock orbitals is obtained by requiring the virtual HF orbitals to diagonalize the one-electron Hamiltonian. This is equivalent to neglecting the two-electron repulsion terms when determining the virtual HF orbitals. The neglect of the two-electron repulsion terms lower the orbital energies of all the low-lying virtual orbitals except the diffuse ones which basically will remain unchanged. The virtual Hartree-Fock orbitals that diagonalize the one-electron Hamiltonian is for that reason a better choice than virtual canonical Hartree-Fock orbitals to get an initial set of active orbitals. The calculations performed in Ref. (4) clearly illustrate how important it is to avoid the undesired diffuse orbitals in the initial set of active orbitals.

Initial CI iterations for frozen orbitals: The simplest choice is to use the single configuration corresponding to the lowest diagonal element in the configuration block of the Hessian. However, the CPU time spent to perform a few CI iterations before initiating the MCSCF macro iterations is often saved several times in the following MCSCF optimization. It should be clear if we consider the extreme example of starting an MCSCF calculation with converged MCSCF orbitals. If the MCSCF calculation is started with the single configuration of lowest energy, the calculation will probably use 3-4 macro iterations before it is converged, while it would be converged immediately if the CI was converged beforehand. Test calculations in Ref. (4) show that it is reasonable to start the MCSCF calculations with 3 direct CI iterations. Similar experience has been described by Werner and Knowles.[52]

An improved initial guess can also be obtained by using the lowest eigenvalue(s) and corresponding eigenvector(s) of the explicit subspace CI Hamiltonian described in Appendix D.

As discussed in Sec. B.3 good initial guess for the orbitals can often be obtained with MP2 or INO-CI. In this case much CPU time is saved by doing, *e.g.*, 3 initial CI iterations for frozen orbitals.

7. Optimization of core hole states

In this section we describe the optimization of core hole states in molecules. Such states are encountered in various spectroscopies associated to inner shell ionization at high excitation energies. Examples of spectroscopies involving core hole states are photoelectron, electron energy loss, Auger emission, X-ray emission and related spectroscopies. A pictorial representation of the physical procesess underlying these spectroscopies is given in Fig. 4. Application of SIRIUS for calculation of core hole derived spectra is briefly reviewed in Sec. F.2. Due to the localized nature of core holes core electron spectroscopies, being either of emission or absorption ty pe exhibit some particular features, such as the mapping of local densities. A characterization of the electronic structure of these core hole states calls for some special considerations. First, a substantial orbital relaxation follows the creation of a core hole and a state specific method such as MCSCF is therefore best suited, while perturbational oriented methods have difficulties in describing the relaxation and correlation simultaneously. Second, the states are embedded in electronic continua. The continuum interaction gives rise to an energy contribution, the calculation of which requires the explicit calculation of continuum orbitals. This is a difficult task to undertake for molecular (non-local) potentials, however, in most cases this energy contribution is small enough to be neglected. Third, core hole states are subject to variational collapse. For calculations of discrete excited states in the low energy regime the number of negative Hessian eigenvalues is known *a priori.* This is, however, usually not the case for core hole states or other higher excited states which are embedded in the continuum. When a finite basis is used to describe these states the continuum has a discrete representation and the states will have a finite, but in general unpredictable, Hessian index. We therefore need other criteria than those based on explicit knowledge of the hessian index to converge to states embedded in the continuum.

With the tools developed in direct second order MCSCF it has been possible to design an efficient procedure to optimize core hole states without variational collapse. For each core hole containing state, whether the main core hole state or a shake-up state, an intermediate calculation is carried out with the opened core

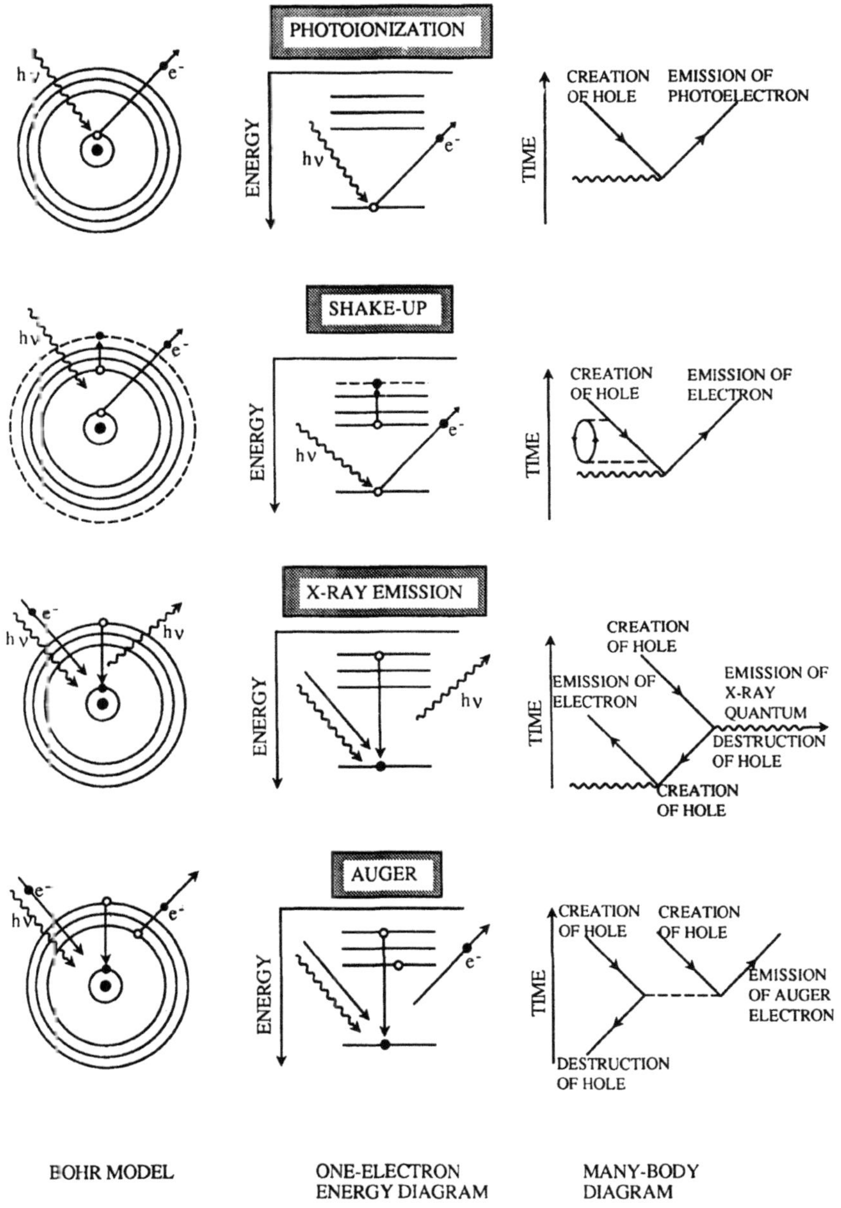

Figure 4. Pictorial representations of core electron spectra.

orbital kept frozen. The orbitals of the ground state calculation transformed to Fock type or MP2 type orbitals can be used as initial guesses. In this limited variational space the main core hole and the first, second, ... shake-up states have the same Hessian index as an ordinary ground state, respectively first, second,...

excited states, and can be optimized using the same methods as for the calculations on ground and excited states described in the previous sections (see example in Sec. F.2). The intermediate optimization takes the MCSCF wave function to a local region of the full core hole state calculation, where also the opened core orbital is relaxed. The large energy separation between core and valence hole states is responsible for that the frozen core orbital solution lies in the local energy region for the full orbital optimization. The Newton-Raphson algorithm with full coupling between configuration and orbital variables (with zero level shift) does straightforwardly converge to the relaxed core hole state as this algorithm does not require knowledge of the correct Hessian index in the local region. The use of the NEO algorithm or other algorithms that follow a particular root number in the configuration space to obtain local second order convergence cannot be recommended because these algorithms require explicit knowledge about the Hessian structure (furthermore it would be very CPU demanding to converge to, say, the 640'th excited state).

The relaxation among valence orbitals following the removal of a core electron leads to a valence electronic structure resonant with that of the equivalent core species (*e.g.* nitrogen for carbon core hole and fluorine for oxygen core hole). The corresponding relaxation energy is large, roughly 25 eV for first row species, and originates essentially in orbital relaxation. The application of the orbital absorption algorithm, described above, is therefore particularly beneficial for core-hole state calculations since it quickly picks up the major part of the relaxation energy before full MCSCF iterations involving all configuration or determinantal parameters are initiated.

It is not possible to both fully relax and correlate the opened core orbital without variational collapse. Therefore a single occupancy restriction for the core orbital is imposed throughout. We thus do not account for the semi-internal excitations involving this orbital. The interaction with the discrete core hole configuration and the semi-internal configurations is strong enough to generate an Auger effect, but is generally not strong enough to noticeably perturb the core hole state energy. Using Feshbach[53] or many-body[54,55] theory the single occupancy restriction is predicted to raise the core hole energy by only few tenths of an eV for first row species. The large energy separation between half-filled and filled core hole states is the reason to the smallness of this interaction. Only in extreme cases, for example when super Coster-Kronig transitions are energetically allowed,[56] we can anticipate a noticeable displacement of the core hole state by the semi-internal excitations.[53]

Important applications of the procedure described above are found in ΔSCF calculations of core ionization potentials and their chemical shifts. As briefly commented in Sec. F.2, the ΔSCF approximation is very useful for component and structural analysis of larger organic compounds and can be used in conjunction with photoelectron spectra as a "finger-print" tool. Like ground states core hole states have traditionally been optimized within the Hartree-Fock (SCF) approximation by a series of diagonalizations of the (open-shell) Fock operator,[57,58] a procedure which is linear with respect to the residual error vector. A maximum overlap criterion has often been used to identify the singly occupied core orbital from iteration to iteration.[58] Since the core orbital has negligible overlap with any other orbital (the core hole state is also almost completely orthogonal to any other self-consistent state) this procedure does not collapse variationally. However, for larger molecules the density of orbital eigenvalues grow progressively larger, and as for ground state optimizations such straight Fock diagonalizations may lead to oscillations or even divergences. Since Fock matrix diagonalizations have become the bottleneck for traditional SCF calculations of at least small to medium-sized molecules, techniques of various kinds have been employed to improve convergence. Examples thereof are eigenvalue shifts,[59] dynamical damping procedures,[60] or extrapolation and interpolation of densities.[58] However, such procedures are not infallible. As shown in the present chapter quadratic SCF can guarantee convergence for the states lowest of its symmetry. The procedure outlined above constitutes two direct SCF calculations, direct in terms of the solutions of the eigenvalue equations given by a series of linear transformations involving the exact orbital Hessian. It is step-restricted for global convergence and quadratically and sharply convergent locally. In the first intermediate calculation where the opened core orbital is kept frozen the norm-extended optimization takes the SCF expansion point to the local energy region for full orbital optimization obtained with the Newton-Raphson technique. Like for general MCSCF, this procedure gives a safe and efficient way to obtain the correct SCF solution for core hole states and therefore ΔSCF solution for core electron ionization potentials. An example is given in Sec. F.2 where results for core electron binding energy shifts for model molecules of polymethylmethacrylate (PMMA)[61,62] are briefly discussed.

F. Survey of Applications

The present MCSCF program has been used in a broad variety of contexts for calculation of electronic structures, spectra and reactions for systems of sizes ranging from Li^- over cyclobutadiene and cytosine to molecular clusters in solution. In these applications the SIRIUS program have been used either as a stand-alone program or as a main program providing wave functions for different property post-programs. The sharp and stable convergence, the analytical transferability, and the fact that direct techniques allow applications to large wave functions have served as incitements for use of the SIRIUS program in these contexts. Below, in the first sub-section, we give some examples of applications. We limit ourselves by presenting these examples only in the form of a list, and we refer to the original articles for any computational or formal aspects. After this list of applications we briefly review in the second sub-section the use of SIRIUS as a stand-alone program in one particluar field, namely the interpretation of core hole derived spectra of molecules. Such spectra provide some unusual features for studies of electronic and conformational structures.

1. List of general applications

(a) Gradients, Hessians and second order properties: SIRIUS wave functions have been used for the calculation of molecular gradients, Hessians and other time-independent first and second order properties such as IR intensities and polarizabilities. These calculations are examplified by Refs. (63,64,29). SIRIUS has been used in connection with potential energy surface walks and searches for transition and equilibrium structures, see Refs. (65-67) for examples. The main motif behind the use of direct second order MCSCF as a main program in this context is given by the "analytic transferability" between wave functions and properties. This feature follows from the fact that the relaxation contribution to the second order properties is obtained by solving an equation system that is identical to the one used for the wave function optimization. The same electronic Hessian is used as generating matrix in the linear transformation, the only difference being that the MCSCF gradient is replaced by the gradient expressions for the perturbation of interest. The use of large-scale MCSCF wave functions for calculations of molecular Hessians is given in Ref. (63), calculations of MCSCF dipole-moment derivatives in Ref. (64).

(b) Potential energy searches: Investigations of full potential energy surfaces, including conformations, isomerization, transition state searches and and reaction paths, is examplified by Ref. (65) where the $N_2 + H_2 \rightarrow N_2H_2$ concerted and isomerization reactions were studied. Walking on potential energy surfaces using gradient extremal and trust radius algorithms were demonstrated for methylene-amine.[66] Applications paths for protonization and proton affinities have been given in Ref. (68).

The MCSCF method is not only general in the sense that it can treat equilibrium and transition states on the same footing and give correct dissociation, but also because it handles general open-shell and excited states. This makes it possible to carry out conformational studies and surface walks also for excited states. A demonstration of this for various excited states of H_2CO^+ and its deuterated species was given in Ref. (69). In that work it was also shown how locally expanded MCSCF gradients and Hessians can be used for solving the quantum equations of motions and to derive autocorrelation functions. These functions provide a time-dependent description of the molecular motions on a potential energy hypersurface, and also the vibronic spectra. An example of this is given by calculations of multidimensional vibronic spectra on formaldehyde presented in Ref. (69).

(c) Discrete and continuum spectra, dynamic properties: SIRIUS has been used as provider of wave functions for calculation of time-dependent response properties. This has been achieved with multi-configurational linear response techniques derived for large MCSCF wave functions.[70] In these applications the solution of two-component generalized egenvalue equations were achieved by direct techniques. Polarization propagator calculations have provided optical and UV excitation spectra with high accuracy. Such spectra have been obtained for systems of different sizes,[70-74,53] viz. Li^-, CH^+, N_2, ozone, ethylene, benzene, urea and cytosine. Different types of wave functions have been employed, HF, CAS and RAS wave functions giving the RPA and MCLR approaches. Also other dynamic properties such as dynamic polarizabilities and C_6 dispersion coefficients have been obtained with MCLR. As example we mention the calculation of the static and dynamic polarizabilities of Li^-, reaching for very high accuracy.[75] Despite its apparent simplicity the electronic structure of Li^- has posed a very tough challenge for modern quantum methods. Widely disparate results were obtained for the static polarizability, a quantity which is unknown experimentally. Various scattering, variational and perturbational approaches were employed, in the latter case using perturbation or coupled cluster theory methods of various orders. In Ref. (75) it was shown that with MCSCF reference wave functions the MCLR method[70] gave close to exact basis set results already for a very small size of the MC wave function. This points at the strength of variational methods used in combination with response theory for calculation of properties of species with complicated and *a priori* unknown electronic structures.

MCSCF reference wave functions are not only useful as reference for linear response methods when these aim at discrete spectra, but also when cross sections in the continuum and in particular continuum shape resonances are studied. With MCSCF coupled to linear response it is possible transcend the static exchange and the strong orthogonality approximations commonly employed and to describe the continuum with full correlation between the escaping electron and the residual ion. This generalization is important especially at low excitation energies and for threshold phenomena. MCLR can be used in different ways, either to straightforwardly produce primitive eigenspectra, or to provide the spectral moments of various orders, or for producing pseudospectra fulfilling quadrature requirements. From these quantities the correct energy normalized continuum can be obtain by Stieltjes techniques. Demonstrations of this for small molecules have been given in Refs. (53,76).

(d) Solvent effects: SIRIUS wave functions have been used in conjunction with a new approach for studying solvent effects, namely the multiconfigurational self-consistent reaction-field method, MCSCRF.[77] In this approach the atom, molecule or supermolecule is assumed to be surrounded by a homogeneous continuous medium described by its macroscopic dielectric constant. The MCSCF wave function is quadratically optimized with respect to all variational parameters in the presence of the polarizable medium. The generalization to multi-configuration cavity wave functions opened the possibilities to study solvation effects for open-shell states, excited states and for various electronic spectra. The first applications thus concerned solvation shifts of ionization energies of various solvated species, water, methanol, formamide and benzene.[78] Other applications have involved the construction of Born-Haber cycles,[79] and solvation red shifts in the radiative decay of singlet molecular oxygen.[80] With the Born-Haber cycles the reorganization contribution to the solvation energy is unravelled. Thermochemical analysis was employed to correlate intra- or intermolecular ionization shifts of different but related quantities. Ref. (68) used MCSCF energies for a thermochemical analysis of gas phase proton affinities and core electron binding energy shifts. The thermochemical models were applied for prediction of proton affinities of carbon, nitrogen and oxygen containing compounds, and for the site of protonation in some cases. Model calculations of infinite screening and ionization potentials of metallic clusters were given in Ref. (81).

(e) Ozone: Finally we mention an example where SIRIUS wave functions have been used for a whole range of properties of one and the same species, namely the ozone molecule. In Ref. (74) investigations were carried out for potential energy surfaces, properties and spectra of singlet (X^1A_1) and triplet (a^3B_2) ozone by means of MCSCF/MCLR analytical response theory calculations. MCSCF analytical gradients and Hessians were used to locate equilibrium and transition state structures, and to obtain associated vibrational and rotational constants, infrared intensities and dipole moments. By means of MC linear response functions electronic excitation energies and oscillator strengths, static and dynamic polarizabilities as well as dispersion C_6 coefficients are obtained. In general good agreement is achieved for CAS reference wave fucntions, in some cases within experimental error margins, for properties where experimental data are known. For potential energy searches where experimental data mostly are lacking, on the other hand, it is clear that the level of theory has to be increased to RAS in order to gain a high confidence level at the difficult parts of the surfaces. A very low IR intensity for triplet ozone was predicted, something which may have a bearing on the ozone depletion problem.

2. Applications on core electron spectra

In this section we briefly review the use of SIRIUS for analyzing electronic spectra associated with the creation or decay of core hole states in molecules. Core electron spectroscopy poses some special features that serve as interesting problems for computational analysis. The localized nature of core holes provide the possibilities to map both conformational and electronic structures using suitable energy and intensity models in conjunction with the spectral analysis. Furthermore, since core hole states preserve the valence electron structure, one finds that quantities such as ionization potentials and chemical shifts do not show the strong computational size dependency as do absolute state energies or low energy valence spectra. This widens the range of applications for structural analysis to larger molecular systems. Core hole states are interesting not only for analysis of electronic and conformational structures, *i.e.* properties of the discrete states, but also for studies of the dynamics of the photoionization or the decay processes themselves. The electronic structure studies have focussed on interpretations in terms of MO theory and many-electron theory. There has also been large efforts to explore eletron-nuclear couplings, vibronic fine structure and molecular conformations that can be derived form these quantities. Direct applications on the dynamics of core ionization and decay processes, *e.g.*, by means of scattering theory calculations are, however, still rather scarce.

There are some particular difficulties associated with the calculation of core hole states. The creation of a core hole state is accompanied by a substantial orbital relaxation, core hole states are short-lived and they are embedded in one or more electronic continua. These features present complications compared to calculations of ordinary low-lying states or of low-energy optical or UV spectra. The qualitative interpretation of processes involving core hole states is not always straightforward. By virtue of the equivalent core notion, the transition takes place between two entirely different chemical species, which implies that the optimal parameters of the wave functions, *viz.* molecular orbitals and configurations, for the initial and final wave functions will be quite different. It is therefore not clear if one in principle should interpret a core hole derived spectrum in terms of the neutral ground-, the initial-, or the final states or in terms of a mixture of these states.

Due to the large orbital relaxation, sets of mutually non-orthogonal orbitals must be used. This leads to a non-orthogonality problem in the evaluation of transition matrix elements between determinantal wave functions. In a Hartree-Fock scheme this is manageable and has traditionally been solved by means of co-factor calculations. However, for correlated wave functions of even moderate accuracy the co-factor calculations to get transition matrix elements require a prohibitively large computational effort. Most of the previous studies of core electron spectra of molecules have been carried out by means of self-consistent field and configuration interaction methods. These studies have shown that the orbital relaxation effect is crucial and that a correlated treatment of the transition matrix elements always must be performed with mutually non-orthogonal orbitals in order to keep a compact and manageable size of the configuration expansions. The demand for state-specific MO optimizations puts an effective limit to the size of the wave functions that can be used in such calculations. Although some attempts have been made to attack these spectra with propagator oriented approaches, see Ref. (55), only state specific variational methods have been applied to any extent. The reason is the difficulty to treat the large orbital relaxation and at the same time apply an operator manifold that contains core-valence transfer operators. Furthermore, some core electron processes are delicately dependent on the precise nature of the electron correlation, and would require higher order propagator methods if the unrelaxed ground state is used as reference state. On the other hand the state specific approaches for these kinds of spectra have suffered from very costly evaluations of either the transition density matrix (soft X-ray emission) or wave function overlap (shake-up) using sets of non-orthogonal orbitals and at the same time containing large configurational expansions. Even with the application of some efficient schemes for co-factor calculations,[82,83] a "brute force" calculation of the matrix elements soon becomes prohibitive. The use of different orbital bases has thus presented a problem concerning the interpretation and computation. The interpretation is basically because the final state simulates another chemical species (equivalent core approximation) and standard MO arguments may not apply. The computation because of the non-orthogonality problem for transition density matrices or the (generalized) overlap amplitudes. The latter problem has recently been solved for a large class of wave functions,[45,79,84,85] implying that the time bottle-neck now lies in the optimization of the wave functions and not in the matrix elements.

Various techniques have been developed within the scope of the SIRIUS program package for the calculation and interpretation of core hole derived spectra. The spectra are divided as absorption type spectra, such as photoionization, photodetachment or electron energy loss spectra, and emission type spectra, such as soft X-ray emission, Auger emission and autoionization spectra. Applications to these types of spectra have been carried out. Below, we shortly mention some of the more important applications of the SIRIUS program in this context. A pictorial representation of these spectroscopic processes in terms of the Born model, one-electron diagrams and many-body diagrams is given in Fig. 4. The applications reviewed are divided into the following subsections:

(a) ΔSCF calculations of chemical shifts.
(b) The molecular shake-up phenomenon.
(c) Soft X-ray emission of free molecules.
(d) Auger emission of free molecules.
(e) Vibronic coupling effects.

(a) ΔSCF calculations of chemical shifts: In this section we shortly demonstrate how ΔSCF and ΔMCSCF solutions for core ionization potentials and chemical shifts may be obtained with SIRIUS. We exemplify thereby also the usefulness of the ΔSCF approximation for component and structural analysis of larger organic compounds, here model molecules of polymethylmethacrylate (PMMA). As shown in Ref. (61) these molecules serve as examples where the surface sensitivity of XPS can be used to investigate bio-compatible materials, in this case the cleanliness of intra-ocular eye-lenses. As empirically demonstrated by the recorded spectra[61] the core ionization potentials follow trends associated to functional groups, something which make them potentially interesting as "finger print" tools for component and structural analysis. As is demonstrated in Ref. (61) high-resolution XPS spectra of organic compounds generally exhibit chemically shifted bands superposing each other and containing *a priori* unknown band asymmetries and stoichiometric intensities. Reliable theoretical predictions of ionization potentials are therefore desirable either for verifying assignments of strongly shifted bands or for "resolving" compound structures containing several overlapping bands with small shifts. Calculations are also necessary to establish links between measured values for different quantities and the structural properties.

Fig. 5 shows carbon core electron shifts for the PMMA related molecules taken from Ref. (61,62). They correlate quite well with experimental values for different compounds. This correlation indicates that final state relaxation is the main cause

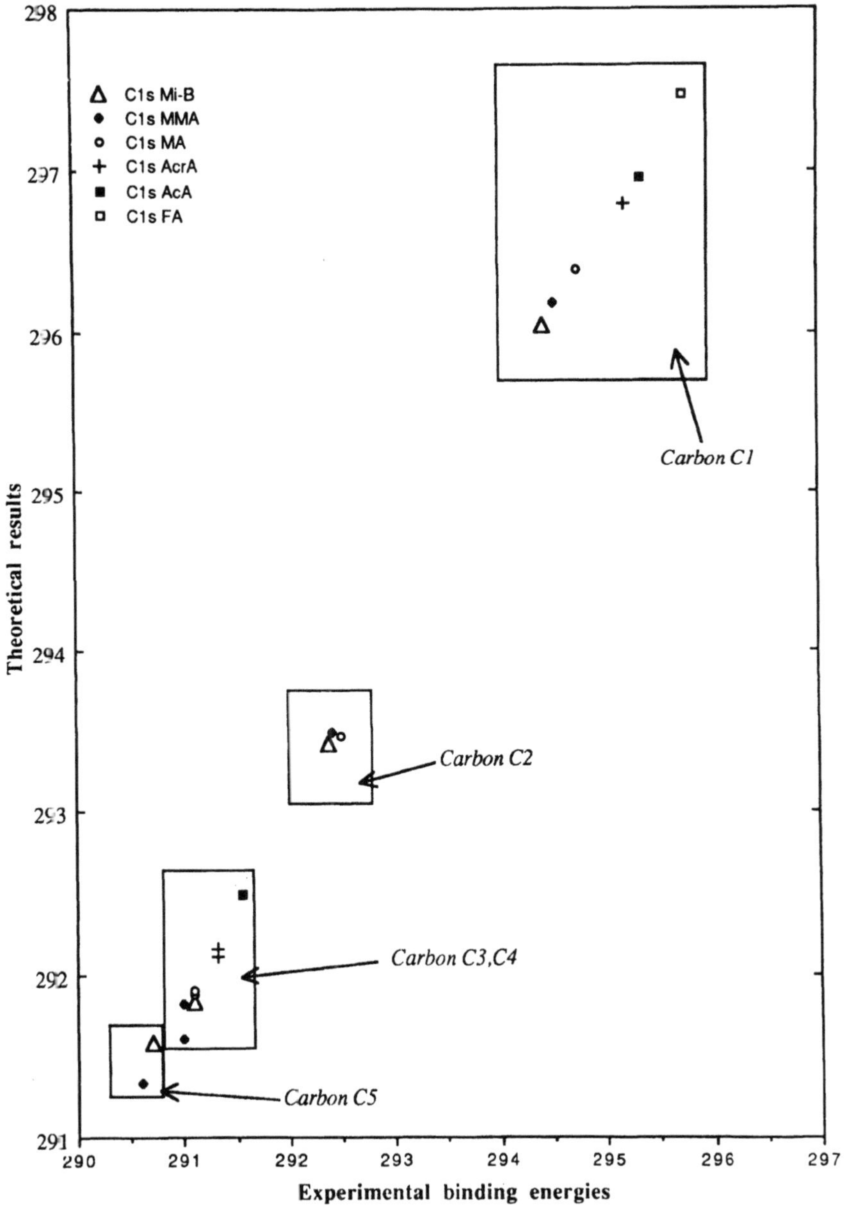

Figure 5. Correlation between ΔSCF and experimental chemical shifts of carbon core electron ionization potentials of some model molecules for PMMA.[61]

for the shifts, and that the size dependency of this relaxation is very similar for the two cases. For larger molecules in the series C1*s* and O1*s* bands are shifted to successively smaller binding energies because the final core state becomes better stabilized by the extraatomic polarization. In general one finds linear relationships for Koopmans energies, ΔSCF energies (at core frozen and at core optimized levels) with respect to the experimental energies. Linear relations of this kind for Koopmans and relaxation energies have been observed and discussed in connection with XPS measurements on several compounds.[86] The main difference between one-particle, Koopmans, and experimental ionization potentials are known to be due to relaxation, correlation, and relativistic effects. With a ΔSCF representation for ionization potentials, we distinguish relaxation energy from the correlation energy if the latter is defined as expectation value differences between Hartree-Fock (one-determinantal) and exact wave functions, taken differentially between ground and ionized states. In contrast to valence ionization where this definition leads to relaxation and correlation energies that are roughly equal in magnitude, albeit mostly of opposite sign, relaxation is known to dominate core ionization potentials. In contrast to valence ionization no correlating valence pairs are broken upon core ionization which leads to large cancellations in the correlation energy. The core-valence pair contribution and the relaxation contribution to the correlation energy error depends mostly on the radial contraction around the ionized core, and does not increase unwieldy for larger compounds. It is thus clear that the relative success of the ΔSCF results is based on cancellation of errors, namely that the valence electron correlation basically is "untouched" upon core ionization. This fact that makes the ΔSCF method a potentially interesting tool for component analysis of larger organic compounds in conjunction with analysis of high resolution XPS spectra. The relativistic error is dominated by the core contribution, and therefore fairly constant and determined only by the type of atom ionized.

The remaining errors in core ionization potentials refer to vibronic and discrete-continuum interaction effects. Only the former is significant and is actually the dominant correction to the electronic absolute ΔSCF values which are consistently 1 to 1.5 eV too large for PMMA related and other organic conmpounds. Finally, the effect of the finite basis set should be taken into account. The compactness of both ground and core hole states, without any need of describing "correlating" orbitals of diffuse character make the prospect for ΔSCF calculations rather pleasant in this respect. Tests show that given the flexibility of an extended polarization basis sets on the ionized species, only standard double zeta basis sets are needed for the remaing parts of the molecule in order to obtain reliable results. Particularities, such as the specific contraction schemes are not very crucial.[87]

Ref. (62) reports results and performance of second order SCF used to optimize core hole states in PMMA related molecules. The two-step procedure described in section D is used for some 30 core ionization potentials. Using a gradient of 10^{-6} the calculations need only 5 frozen core orbital iterations and 4 iterations for fully relaxing all orbitals including the core orbital. These numbers are slightly altered only for a couple of cases, thus manifesting that the opmization procedure decribed in Sec. E.7 is very safe for states imbedded in a continuum. It is interesting to note that the core optimized values not only improves the absolute ionization energies but also the chemical shift predictions. In fact the deviation seems to closely follow those when the equivalent core approximation is applied.[68] The present findings thus suggest that full orbital relaxation of all orbitals is called for when overlapping features such as "backbone" structure should be resolved and assigned.

The performance of an MCSCF calculation for core hole states and shake-up states was described in Ref. (4) and further studied in Ref. (62). Provided standard triple zeta plus polarization basis set and valence CAS wave function were used accurate core ionization potentials are indeed given: the $C1s^{-1}$ and $O1s^{-1}$ potentials of CO, deviations from experimental results of only few tenths of an eV are obtained.[4] In such calculations the core hole wave function contains the same active space as the ground state, but with the core hole orbital added to the active space and kept singly occupied. The change in energy is obtained by relaxing the core hole orbital varies between 0.02 to 0.05 au for different shake-up states. As for the ΔSCF results quoted above core relaxed values show an improved agreement with experiment. Again agreement should be qualified from the fact that no relativistic effects are included but also that there is an uncertainty of a few tenths of an eV in the precise setting of the vertical ionization potential from the experimental spectra. Summarizing this subsection we conclude that the procedures implemented in SIRIUS provide safe optimization of states embedded in continua. It was also demonstrated that ΔSCF and ΔMCSCF approximations are capable of obtaining correct chemical shifts to a few tenths of an eV and that they can be used for component analysis for a large class of molecules.

(b) The molecular shake-up phenomenon: Since the introduction of monochromatized X-ray excitation[88] core electron shake-up phenomenon has been extensively studied with photoelectron spectroscopy.[89] The shake-up spectrum consists of a complex satellite structure following the main core hole peak, and contains a wealth of resolved structures hardly matched by electronic spectroscopy in the low-energy regime. The analysis of this phenomenon is substantially more complex than ordinary UV or optical spectra, due to several contributing dynamic and intrinsic effects. The presence of a core hole introduces an extended region of excitations with a high density of states characterized by multiple hole-particle character of the wave functions. The role of multiple excitations calls for correlation schemes with excitation patterns of high orders. Like other core electron spectroscopies the orbital description of the initial and final states differs due to the substantial relaxation. As mentioned in the previous section the relaxation effect introduces extra considerations both concerning the interpretation (c.f. equivalent core notion) and concerning calculation of matrix elements.

The shake-up phenomenon is commonly analyzed in the framework of the sudden approximation.[90] It is obtained from Fermi's golden rule with a single channel and strong orthogonality condition imposed on the continuous photoelectron and neglect of energy variations of the photoelectron transition elements. Within this approximation the basic matrix element to evaluate is given by overlap amplitude:

$$X_p^{(n)} = < \Psi_f^{(n)}(N-1) \mid \hat{a}_p \mid \Psi_i^{(0)}(N) > \quad (146)$$

For a state-by-state analysis of core electron shake-up, initial state correlation is important.[91] An explanation is given by a simple hole-particle expansion of the wave functions:

$$\mid \Psi^x > = c_0^x \mid \Phi_0^x > + \sum_{i,j} c_i^{j,x} \mid \Phi_i^{j,x} > + \text{ higher order excitations} \quad (147)$$

Here $x = gs$ (ground state), $x = 1s$ (main hole state), or $x = ss$ (shake-up state). The following obvious coefficient relations apply: $c_0^{gs} >> c_i^{j,gs}$, $c_0^{1s} >> c_i^{j,1s}$, and $c_0^{ss} << c_i^{j,ss}$. Using frozen orbitals and the Slater-Condon rules the sudden approximation intensity

$$I = \mid X_p^{(n)} \mid^2 = \mid < \Psi^{x=1s,ss} \mid a_{1s} \mid \Psi^{gs} > \mid^2 \quad (148)$$

simplifies to

$$I(gs - x) = \mid c_0^{gs} * c_0^x + \sum_{i,j} c_i^{j,gs} * c_i^{j,x} \mid^2 \quad (149)$$

Considering the relations of the coefficients given above, one finds that for $x = 1s$, main core electron ionization, the intensity is well guided by $\mid c_0^{gs} * c_0^x \mid^2$. However, for the shake-up states ($x = ss$) *at least* one coefficient product in the second sum is of the same order as $c_0^{gs} * c_0^x$. Initial state correlation is thus decisive for state-specific intensities, but also for total shake cross sections. For unsaturated species, in particular π electron systems, the initial state correlation effect is, however, even more accentuated. This is because shake-up states of unsaturated species are to a large extent governed by the energetically possible double excitations, which are siginificant in the corresponding ground state wave functions (c.f. Brillouin theorem).

In the SIRIUS applications to shake-up spectra the sudden approximation overlap amplitudes have been evaluated according to a scheme developed in Ref. (84). It assumes configuration state functions and uses a GUGA (graphical unitary group) approach to evaluate the generalized overlap amplitudes (GOA's): $X_p^{(n)} = < \Psi_f^{(n)}(N-1) \mid \hat{a}_p \mid \Psi_i^0(N) >$ where $\Psi_i^0(N)$ and $\Psi_f^{(n)}(N-1)$ denote complete active space wave functions for ground and shake-up states, respectively. It has recently also been updated for determinants and restricted active space (RAS) wave functions.[85] Basically it performs: i) A biorthogonalization of the orbitals following Malmquist[45,46] bringing the overlap matrix into diagonal form; ii) An expansion of the initial ground and final shake-up states to a larger configuration vector which includes both states; iii) A transformation and a scaling of the sorted CI vectors to match the biorthogonal MO basis; iv) The final evaluation of the GOA is then obtained as a simple scalar product between two CI vectors. In general one chooses the (GUGA or determinant) graph for the overlap calculation as the graph for the (N-1) core hole calculation with the single occupancy restriction removed. The CSF's (determinants) of $\hat{a}_p \mid \Psi_i^0(N) >$ is then assigned to this graph followed by an assignment of $\mid \Psi_f^{(n)}(N-1) >$ to the same graph. In this scheme the overlap element corresponding to the "overlap graph" is considered as a special transition density element. After biorthogonalization the new CI vectors $\mid aC^{\alpha}(ji) >$ and $\mid C^{\beta}(j) >$ pertaining to the CSF's $\phi^{\alpha}(i)$ and $\phi^{\beta}(j)$ are obtained by scaling: $C^{\beta}(j, \mu)d_t^{n_t(\mu)}$ where $n_t^{(\mu)}$ is the occupation number of orbital t in CSF μ, and d_t denotes the corresponding diagonal density matrix element. After this scaling the sudden overlap GOA's are obtained as a simple dot product between two CI vectors: $X_p^{(n)} = \Sigma_{\mu} C^{\alpha}(i, \mu) C^{\beta}(j, \mu)$. All operations are quick, and the computational effort to obtain the GOA's is negligible compared to the optimization of the wave functions. Corrections for state overlap were achieved by a symmetric orthogonalization procedure, however, no Hamiltonian corrections are carried out.

SIRIUS applications on the shake-up phenomenon have been carried out for some first row molecules, diatomic and hydrides.[4,84,91-92] Some larger compounds have been addressed such as carbonyl containing molecules for which charge transfer shake-up peaks were studied and characterized.[91] The appearance of such peaks was associated to donor and acceptor character of molecular orbitals and to the character of the groups participating in the shake-up processes. Easily polarizable systems with high mobility of charge, *e.g.* those with aromatic pendant groups show particularly intensive charge transfer shake-up features. Models based on simple electrostatics and on charge screening of the core hole was investigated and compared with large scale MCSCF calculations. Features that fingerprint spectra of functional groups were established. For shake-up spectra in general we find that different systems, such as atoms, hydrides, organic molecules with or without saturated bonds, require different levels of sophistication in the theoretical analysis, and that different parts of one and the same spectrum are manifested differently with respect to effects such as orbital relaxation, many-electron contributions and molecular bonding. It is clear that theory lags experiment for the shake-up phenomenon. CASSCF has a restricted range of applicability, and it is still an open question how far we can reach with the type of RASSCF and RAS CI schemes descirbed in this chapter.

(c) Soft X-ray emission of free molecules: Soft x-ray emission spectra are governed by dipole emission between initial, well localized, core hole states and final states with holes among the valence orbitals. Soft X-ray transitions thus take place from the penultimate main shell, *i.e.* the outermost core shell, and the valence shell. Since the lifetime broadening of hole states in these shells is comparatively small, the molecular soft X-ray spectra not only resolves structure corresponding to separate molecular orbitals, but also corresponding to vibrational fine structure.[94]

The soft X-ray transitions take place between states that, at least in a first approximation, can be considered bound. This means that the full Fermi golden rule expression can be computed without further approximations. The calculations are not so demanding from the point of view of selecting 1-particle or N-particle basis sets. For the purpose of interpretation there are several levels of approximations that we impose on the theoretical analysis: i) one-center approximation using the MO model; ii) Hartree-Fock frozen orbital one- and two center contributions; iii) separate state optimized (open-shell) Hartree-Fock wave function including relaxation; iv) correlated wave functions with separately optimized (open-shell) Hartree-Fock orbitals; and v) separate state calculations with simultaneous optimization of all (orbital and configurational) parameters. Of particular interest is here the limitation of the first and crudest of the approximation levels, namely the frozen orbital one-center model, since the validity of this model is the basis for the "interpretability" of the X-ray emission experiment in terms of electronic structure theory of the ground state molecule.

The soft X-ray spectral intensities are governed by the Einstein coefficients for spontaneous emission, which in Hartree units are given as

$$W_{ij} = \frac{4\alpha E_{ij}^3}{3c^2} T^2 \tag{150}$$

where c is the speed of light, α refers to the fine structure constant, E_{ij} denotes the transition energy, and T is the transition moment matrix element. The one-electron transition moment operator $\hat{T}$ is expressed in second quantization as

$$\hat{T} = \sum_{p,q} < \varphi_p^i \mid \hat{t} \mid \varphi_q^j > \hat{E}_{pq} \tag{151}$$

and the corresponding matrix elements as

$$T_{ij} = \sum_{p,q} < \varphi_p^i \mid \hat{g} \mid \varphi_q^j > < \psi^i \mid \hat{E}_{pq} \mid \psi^j > = \sum_{p,q} P_{pq} D_{pq} \tag{152}$$

where ψ^i and ψ^j denote the total electronic wave functions for the initial and final state of the transition. The transition moment is thus given as the trace of a property matrix multiplied by the first order density matrix. We evaluate the electric dipole transition moment both in the length and velocity formulations, which for exact wave functions are equivalent since

$$< \psi^i \mid \nabla_r \mid \psi^j > = (E_j - E_i) < \psi^i \mid \hat{r} \mid \psi^j > \tag{153}$$

Comparison of these two values provides an indication of the quality of the utilized wave functions and the reliability of the theoretical prediction.

An efficient way to calculate the property matrices was devised in Ref. (95). The calculation of the transition density matrix follows much the same way as the generalized overlap amplitude in shake-up. For X-ray spectra we consider the calculation of three types of GUGA graphs, which, except for the inner shell orbitals, contain an identical set of active orbitals: 1) symmetry-constrained graph for 1- and 2-electron density matrices with the inner orbital singly occupied, 2) symmetry-constrained graph with the inner orbital doubly occupied, and 3) graph with no symmetry for 1-electron transition denisty matrix (TDM) and all occupancies allowed for the inner orbital. The wave functions are optimized for the initial and final states using the first and second graph, respectively, and the so obtained optimal CSF-vectors are expanded in no-symmetry form matching the third graph. A similar scheme has been implemented also for determinants and for restricted active space (RAS) wave functions. The use of determinants simplifies the operations considerably, and it is always advantageous to transform to determinant basis before evaluating the TDM, even if the converged wave functions are expressed in CSF basis. As for shake-up the orbitals are subject to a biorthogonal transformation, and the CI vectors subsequently to a scaling. For soft X-ray spectra we also have to consider the inactive part of the TDM, which after scaling becomes:

$$D_{kk}^{ij} = < \psi^i \mid \hat{E}_{kk} \mid \psi^j > = 2S^{ij} \tag{154}$$

where S^{ij} is the full overlap between states i and j, which is obtained as an ordinary dot product between the scaled CI vectors,

$$S^{ij} = \sum_{\mu} c_i^{\alpha}(\mu) c_j^{\beta''}(\mu) \tag{155}$$

The active part of the TDM is evaluated from

$$D_{uv}^{ij} = < \psi^i \mid \hat{E}_{uv} \mid \psi^j > = \sum_{\mu v} c_i^{\alpha}(\mu) A_{uv}^{\mu v} c_j^{\beta''}(v) \tag{156}$$

where

$$A_{uv}^{\mu v} = < \mu \mid \hat{E}_{uv} \mid v > \tag{157}$$

is the one-electron coupling element. Since one has to repeat this procedure for each component of all three dipole forms in the evaluation of the transition moment while using one single density matrix, it is more efficient to perform a contravariant backtransformation of the density matrix from MO to AO basis rather than transforming each different property matrix in the opposite way.

$$\mathbf{Tr}(\mathbf{D}_{MO}\mathbf{P}_{MO}) = \mathbf{Tr}(\mathbf{D}_{AO}\mathbf{P}_{AO}) \tag{158}$$

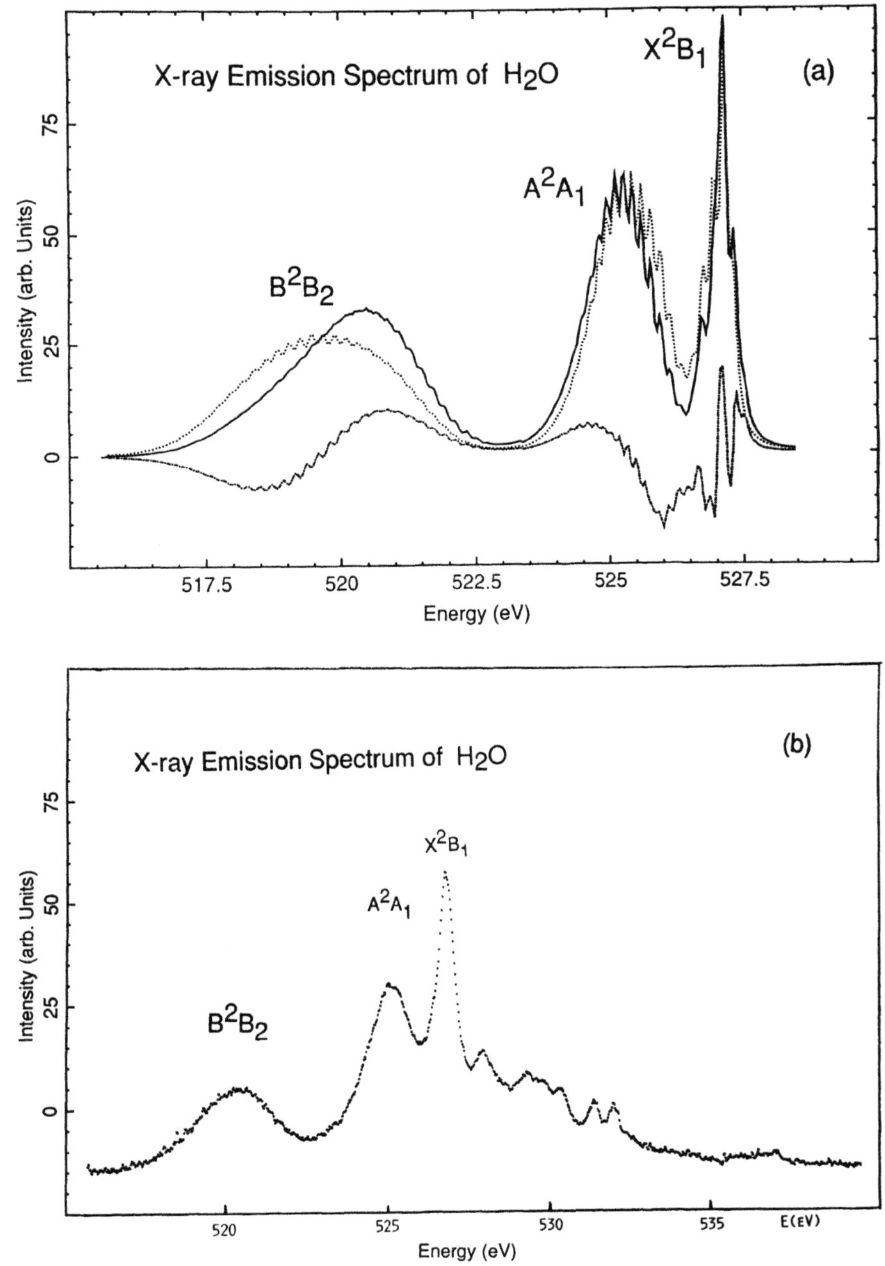

Figure 6. Calculation of the spectral function for X-ray emission of water.[96]

Since separate MCSCF optimization of each state leads to non-orthogonality of the total wave functions and the magnitude of the transition moments may be of the same order of magnitude as the overlap between the MCSCF states, we have employed an orthogonalization procedure that takes care of this problem. We have chosen a symmetric orthogonalization

$$\mathbf{T}' = \mathbf{S}^{-1/2}\,\mathbf{T} \tag{159}$$

where $\mathbf{S}$ is the state overlap matrix. Symmetric orthogonalization is not biased towards any of the states, and we therefore prefer it to a Gram-Schmidt orthogonalization.

Applications including the algorithms sketched above have been carried out for several first row molecules.[79,95-99] These applications do not only treat intensities and energies of soft X-ray spectra but also satellite energetics,[99] vibronic fine structure[96,98] and angular distributions for fluorescence X-ray emission.[79] An illustration of a computed X-ray spectrum, including vibronic fine structure and so-called lifetime-vibrational interference effects is given in Fig. 6. By and large, the one-center model works well. This is an important conclusion when trying to deduce information on electronic structure from X-ray emission spectra. Taking CO as illustration one finds that transitions from the carbon "lone-pair" 5σ orbital are very intense in the carbon spectrum whereas transitions from the basically oxygen-like 4σ orbital are strong in the oxygen spectrum. The delocalized 1π orbital is represented in both spectra. These rules of thumb for calculation of X-ray intensities can thus be considered adequate for spectra of a number of first and second row molecules. Improvements in terms of two-center contributions and use of separately optimized orbitals give numerical corrections, which mostly are small enough for the simpler interpretation to remain valid. The most important correction model is given by the relaxation effect which seemingly is larger than the correlation effect. With the one-center model we obtain roughly a 25% agreement with spectral intensities, with separate optimized CASSCF about 10% agreement, which also is the typical gauge invariance obtained in that case. However, one should point out that even though the one-center model previously has been used in conjunction with Hartree-Fock canonical MO's of the ground state, it applies conceptually just as well to correlated wave functions. In this case it refers to the local character of the 1-particle density matrix pertaining to a certain symmetry, *e.g.* to local p-symmetry for $K\alpha$ spectra, and to the molecular orbitals that diagonalize these matrices. Probably, the core-valence transition density matrices and the "bi-natural" orbitals diagonalizing these are the quantities that carry "maximum interpretability" for soft X-ray spectra. For somewhat larger compounds it seems desirable to include Hamiltonian interactions, something which is not trivial for reduced spaces. Perhaps the most efficient way is to apply a common set of orbitals for the final states (but non-orthogonal to the initial set of orbitals) and to expand spectroscopically pure states which are dynamically correlated by means of RAS CI wave functions. Implementations of these features have now been carried out.

(d) Auger emission of free molecules: For a metastable core hole state there are decay channels alternative to X-ray emission. These channels emerge as an effect of the Hamiltonian interactions with electronic continua of ionic states with at least one additional charge unit. This interaction gives rise to the Auger effect, *i.e.* the expel of electrons with characteristic energies dependent only on the type of system, but independent of the ionizing agent. If the initial state is neutral this effect is coined as autionization. The Auger spectra have mostly been used as a probe for element analysis. In theoretical research it has also been used for analysis of the electronic structures and of the many-body character of ionic states, but also as a probe for the dynamics of electron-molecule scattering processes. For molecules it has almost exclusively been used in the first of these two respects, i.e. for the study of electronic structures of doubly charged molecular ions. As for the spectroscopies alluded to above one can derive the working equations for the Auger effect starting out from Fermi's golden rule (here often called Wentzel's ansatz[100]), and impose a single channel and strong orthogonality condition for the outgoing electron. After some simplifications[101] the Auger transition amplitude reads

$$A_{fi} = \frac{1}{2}\sum_{qrs} < uq \| rs > < LL \mid \hat{a}_q^\dagger \hat{a}_s \hat{a}_r \mid K > = \frac{1}{2}\sum_{qrs} \tau^u_{qrs}\chi_{qrs} \tag{160}$$

where u denotes the continuum Auger orbital, τ^u_{qrs} the Auger transition moment (orbital transition element) and χ_{qrs} the generalized overlap amplitude. Like photoelectron spectra one can qualitatively analyze Auger spectra in terms of the orbital transition elements and the overlap amplitudes. The entity corresponding to the orbital transition element in the shake-up phenomenon (and which is neglected sudden approximation), is evidently needed to compute for Auger cross sections. Some simplifications, *e.g.* approximation of this matrix element in terms of local densities referring to the site of core ionization, have been used in model calculations.[102] However, in general its calculation requires an explicit determination of the continuum orbitals for the Auger electron.[103] The second factor in Eq. (160), the overlap element, is on the other hand only dependent of the electronic structure, in particular the many-body character, of the residual states. The energetics of the spectrum is dependent only on the separation of the final valence double hole states. Therefore, the analysis of Auger spectra gives valuable probes of such states in molecules.

Several investigations have been carried for Auger or problems associated to the Auger effect with the SIRIUS program.[96,103-106] The applications include examples, where the break-down and hole-mixing character of the Auger final state wave functions have been explored. Also vibronic fine structure and lifetime-vibrational interference effects[96,104] analogous to those obtained in soft X-ray emission have been explored. Investigations have also been conducted for resonance and dissociative Auger effects, which lead to spectra with some remarkable features.[106] From the calculations it is found that Auger spectra of free molecules show considerable structure over the wide range of energy that covers all possible double hole states among valence levels. Due to lack of strict selection rules the density of two-hole states reached by the Auger transitions rises sharply with the number of electrons, especially for systems with a low point-group symmetry. Both from the experimental and theoretical points of view the consequence of this is that a state by state analysis is aggravated. For smaller species the outermost low kinetic energy part may exhibit fine vibronic structure, while the spectra grow progressively more complicated for higher kinetic energies with overlapping features that are hard to discern even at high resolution.

In contrast to ground states and single hole states, where the dynamical electron correlation dominates and a quasi-particle picture holds, it is found the two-hole states are subject to strong static correlation. The static correlation is due to internal and, in particular, semi-internal excitations generated by simultaneous excitations in the occupied and unoccupied manifolds of orbitals. Thus these semi-internal excitations couple "down" (hole-hole, or h-h) and "up" (hole-particle, h-p) excitations producing 3h-2p configurations that are near-degenerate with the main 2h hole "Koopmans" configurations. It can easily be seen from an orbital energy diagram that there are many such transitions that produce configurations which are nearly degenerate with the main 2-hole configurations, in particluar for π electron systems with small excitation gaps. Actually, only for the outermost two-hole states a quasi-particle picture holds with a strong leading Hartree-Fock configuration and with a dynamic rather than a static correlation correction. For most of the spectrum of two-hole states, however, there is a break-down of the quasi-particle picture. The near-degeneracy of states also leads to non-adiabatic coupling effects. Thus for doubly charged electron-rich molecules there is often a simultaneous breakdown of both the molecular orbital (aufbau) and the Born-Oppenheimer pictures.

We divide the interaction into those where there is a complete break-down of the molecular orbital picture and those where we have hole-mixing effects. The latter case implies *electronic interference effects* for the Auger states. For Auger we thus have both electronic and vibronic interference effects. The breakdown operates in the major part of the kinetic energy interval covering the inner-inner and inner-outer valence states. The quasi-particle picture is valid only for hydrides for which innermost, inner-inner $2s^{-2}$ states, still are subject to the breakdown effect. This static correlation in the low-kinetic energy end is, however, still much stronger than for the inner valence single hole states in the corresponding photoelectron spectra.

(e) Non-adiabatic and vibronic coupling: In this subsection we briefly review some examples of SIRIUS calculations of non-adiabatic or vibronic coupling. Actually, core hole states are subject to vibronic coupling for all molecules containing an element of symmetry. This is is so because core hole states are quasi-degenerate in the high symmetry point group and couples therefore over one (or more) vibrational modes leading to a lowering of the point group. Thus core hole states of polyatomics are symmetry broken and localized due to such "pseudo Jahn-Teller" effects. An example of this is given by the $O1s^{-1}\Sigma_u$ and Σ_g states of CO_2^+ that couple over the antisymmetric stretch, thereby lowering the symmetry from

$D_{\infty h}$ to $C_{\infty v}$.[107,108] For homonuclear diatomics one has to apply a time-dependent picture of the ionization process, to obtain a localized description of the ionization event. Using MCSCF wave functions of high symmetry limited configurational expansion can be used to compensate for lack of relaxation at the Hartree-Fock level.[109] However, computationally it is always more efficient at any level to use symmetry-broken solutions. Convergence with respect to the size of the (correlated) wave function is always better for the symmetry broken solution, not only for state energies, but also for energy gradients and transition moments.[110] The calculations of core hole states quoted above are thus symmetry broken, see *e.g.* Refs. (111). For calculation of the Renner-Teller effect with SIRIUS, see Ref. (112). In this subsection we deal with vibronic effects associated to static and dynamic non-adiabatic coupling of the wave functions *i.e.* physical effects for which a beyond Born-Oppenheimer picture is implied. Such vibronic coupling effects are not particularly associated with core hole states, but rather to high energy valence or core valence states or other energy regions with a comparatively high density of states. The applications contain, however, computational algorithms related to those used in sections F.2(a) to F.2(d).

The break-down of the BO approximation is manifested in a number collisional and spectroscopic processes, and forms the underlying mechanism for charge transfer and for vibronic coupling and symmetry breaking. The algorithm used by SIRIUS is based on numerical differentiation. The calculation of the first and second order non-adiabatic coupling elements

$$N_{ij}^{(1)}(Q_0) = < \Psi_i(q;Q) \mid \frac{\partial \Psi_j}{\partial Q}(q;Q) >_{Q=Q_0} \tag{161}$$

$$N_{ij}^{(2)}(Q_0) = < \Psi_i(q;Q) \mid \frac{\partial^2 \Psi_j}{\partial Q^2}(q;Q) >_{Q=Q_0} \tag{162}$$

is achieved by expanding the basic overlap element

$$K_{ij}(Q_0, x) = < \psi_i(q; Q_0 - x) \mid \psi_j(q; Q_0 + x) > \tag{163}$$

symmetrically over the coupling point. This leads to one order higher accuracy than with a straight numeric differentiation, both for first and second derivatives. The coupling constants are then evaluated as

$$N_{ij}^{(1)}(Q_0) = \frac{1}{4x}[K_{ij}(Q_0, x) - K_{ji}(Q_0, x)] + O(x^2) \tag{164}$$

$$N_{ij}^{(2S)}(Q_0) = \frac{1}{4x}[K_{ij}(Q_0, x) + K_{ji}(Q_0, x) - 2\delta_{ij}] + O(x^2) \tag{165}$$

$$N_{ij}^{(2A)}(Q_0) = \frac{dN_{ij}^{(1)}}{dQ}(Q_0) = \frac{1}{x}[N_{ij}^{(1)}(Q_0 + \frac{x}{2}) - N_{ij}^{(1)}(Q_0 - \frac{x}{2})] + O(x^2) \tag{166}$$

for the first, the symmetric and antisymmetric parts of the second order coupling constants respectively. Symmetric orthogonalization is imposed on $\Psi_i(q, Q)$ and $\Psi_j(q, Q)$ which modulates the coupling constants by a simple scheme.[113] The techniques used to evaluate the overlap elements are similar to those described in section F.2(b). For the diagonal Born-Oppenheimer (BO) correction to the electronic energy of state i we obtain similarly

$$B_{ii}(Q_0) = (\frac{1}{2M}) < \Psi_i(q; Q) \mid \frac{\partial^2 \Psi_i}{\partial Q^2}(q; Q) >_{Q=Q_0} \tag{167}$$

There have been three kinds of applications of these formulas associated to the SIRIUS program:

i) Calculation of non-adiabatic coupling constants for ion-atom collisions (*e.g.* Na + $Li^+ \rightarrow Na^+$ + Li in Ref. 113). Role of correlation for the calculation of the basic non-adiabatic coupling elements was investigated. The investigation also included a close examination of the stability in the numerical procedure.

ii) Formulation of a diabatic model for photoionization with application to the much debated 27 eV structure in the photoelectron spectrum of acetylene.[114] The first order non-adiabatic coupling constants were used to define the diabatic electronic states which diagonalize the nuclear kinetic operator instead of the electronic Hamiltonian. It was found that the diabatic representation was more appropriate than the adiabatic one for interpreting this spectrum.

iii) Calculations of dynamic vibronic coupling and the solution of the coupled Scrödinger equation. Application has been made on the dynamic vibronic coupling between the $B^2\Sigma_u^+$ and $C^2\Sigma_g^+$ states of CO_2^+.[115] Here spurious, probably symmetry forbidden, transitions are observed in the photoelectron spectrum. A conjecture was made that this is due to a vibronic coupling mechanism between the B and C states over the antisymmetric stretch motion. In order to prove this the vibronic spectra pertinent to the dynamic non-Born-Oppenheimer potentials were calculated by solving the multi-channel Schrödinger equation for the $B^2\Sigma_u^+$ and $C^2\Sigma_g^+$ ionic states, where the in-going first and second order non-adiabatic coupling constants as well as the Born-Oppenheimer potential surfaces were obtained from large CAS MCSCF calculations. The wave function is expanded as

$$\psi(Q, q) = \sum_i \chi_i(Q)\phi_i(Q; q) \tag{168}$$

where the sum in principle is evaluated over all solutions ϕ_i of the electronic Schrödinger equation. The natural way to go *beyond* the BO Approximation, *i.e.* to take into account the movement of the nuclei, is to solve the resulting equations for χ_i, obtained when the wave function proposed in Eq. (168) is inserted in the Schrödinger equation. This procedures assumes differentiability in the Q space. In the case of only one nuclear degree of freedom, the resulting equations become:

$$\left((\frac{-1}{2M})\frac{\partial^2 \chi_j(Q)}{\partial Q^2} + (E_j(Q) - E)\chi_j(R) \right)\delta_{ij} + \\ (\frac{-1}{2M})\sum_i \left(\chi_i(Q) < \phi_j | \frac{\partial^2}{\partial Q^2} | \phi_i > + 2\frac{\partial \chi_i(Q)}{\partial Q} < \phi_j | \frac{\partial}{\partial Q} | \phi_i > \right) = 0 \tag{169}$$

where the brackets indicate integration over the q coordinates. The i and j run over all terms in Eq. (E18). This number is for practical reasons finite in numerical calculations. This set of coupled differential equations is thus called the multi-channel Schrödinger Equation.[116] If all χ_i's and ϕ_i's are separately normalized to unity, a weight factor c_i appears in Eq. (169) describing the relative mixing of the different channels. From these weight factors it is straightforward to derive a modified expression for the vibronic intensity spectrum, see Ref. (115). It is seen clearly from the equations given above that the key entities beyond the BO approximation are the first- and second-order non-adiabatic coupling constants, i.e. the matrix elements of the first and second derivatives with respect to Q in Eq. (E18).

The results confirmed the conjecture about a vibronic coupling mechanism in CO_2^+. An analogous similar investigation solving coupled channel equations has also been applied to a spectroscopic problem for HBr^+, see Ref (117). In this case the spin-orbit interaction was used as a (model) coupling parameter. In both these examples it became evident that although qualitative confirmation on the origin of the observed features could be established, the results were very sensitive to the computational procedure. For solving the multichannel Schrödinger equation we envisage large scale RAS calculations as one requisite for obtaining converged quantitative results. This goes for the examples given above and presumably for great many of the standing problems associated to interpretation of fine structure in electronic spectra of species such as first row hydrides or diatomics. How accurately we can simulate these effects through the computational techniques we have today is still an open question.

Acknowledgments

We want to thank our collaborators over the years for assistance and inspiration, in particular Prof. P. Jørgensen and Prof. T.U. Helgaker. We thank Prof. Jørgensen for valuable contributions to the SIRIUS program. We also want to thank Prof. B.O. Roos for letting us use his matrix-oriented second order integral transformation routines in SIRIUS. In 1983, we had the opportunity to use the GUGA program of Prof. P.E.M. Siegbahn and the CASSCF programs of B.O. Roos, P.E.M. Siegbahn, P.R. Taylor, A. Heiberg, and J. Almlöf as the starting point for the development of SIRIUS. We are obliged to these authors for making their programs available to us and for many inspiring discussions over the years.

Appendix 8A.
Calculation of the Orbital Part of Gradient Vectors and Sigma Vectors

In this appendix we give the explicit expressions for the construction of the orbital part of the gradient and σ vectors in terms of Fock and density matrices. The notation used here was explained in Sec. C. The formula for the orbital part of the gradient was (Eq. (69))

$$g_{sr} = 2(F_{sr} - F_{rs}), \tag{A1}$$

and the orbital part of the σ vectors were expressed in terms of gradient-like elements (Eqs. (43)-(46)):

$$\sigma_{sr}^{(j^c)} = (\check{F}_{sr} - \check{F}_{rs}) \tag{A2}$$

for a configuration trial vector (assumed to be orthogonal to the reference vector c) and

$$\sigma_{sr}^{(j^o)} = 4(\widetilde{F}_{sr} - \widetilde{F}_{rs}) + B_{sr} \tag{A3}$$

for an orbital trial vector. We have defined the following matrix element:

$$B_{sr} = \frac{1}{2}\sum_t (b_{ts}^{(j)} g_{tr} - b_{tr}^{(j)} g_{ts}) \tag{A4}$$

where $\mathbf{b}^{(j)}$ is the j'th trial vector and $\mathbf{g}$ the orbital part of the gradient vector. In this appendix this is expanded to the final, implemented computational expressions using the expressions for the generalized Fock matrix given in Eqs. (55-69).

The orbital part of gradient and sigma vectors is divided into four types, one for each of inactive-active (iu), inactive-secondary (ia), active-active (uv) and active-secondary rotations (ua). Defining

$$\mathbf{F}^D = \mathbf{F}^C + \mathbf{F}^V \tag{A5}$$

we arrive at the following expressions for the orbital gradient:

$$g_{iu} = 4F^D_{iu} - 2\sum_x (D_{ux}F^C_{xi}) - 4F^Q_{ui} \tag{A6a}$$

$$g_{ia} = 4F^D_{ia} \tag{A6b}$$

$$g_{uv} = 2\sum_x (D_{ux}F^C_{xv} - D_{vx}F^C_{xu}) - 4(F^Q_{uv} - F^Q_{vu}) \tag{A6c}$$

$$g_{ua} = 2\sum_x (D_{ux}F^C_{xa}) + 4F^Q_{ua} \tag{A6d}$$

where x runs over all active orbitals.

By replacing the Fock type elements with the corresponding elements $\check{F}_{pq}$ calculated with the transition density matrices from Eq. (47) we obtain the final expressions for the orbital part of the sigma vector from a configuration trial vector (which is orthonormal to **c**):

$$\sigma^{(j)}_{iu} = 8\check{F}^V_{iu} - 4(\sum_x \check{D}_{ux}F^C_{xi}) - 4\check{F}^Q_{ui} \tag{A7a}$$

$$\sigma^{(j)}_{ia} = 8\check{F}^V_{ia} \tag{A7b}$$

$$\sigma^{(j)}_{uv} = 4\sum_x (\check{D}_{ux}F^C_{xv} - \check{D}_{vx}F^C_{xu}) + 4(\check{F}^Q_{uv} - \check{F}^Q_{vu}) \tag{A7c}$$

$$\sigma^{(j)}_{ua} = 4(\sum_x \check{D}_{ux}F^C_{xa}) + 4\check{F}^Q_{ua} \tag{A7d}$$

We have used $\check{F}^D_{rs} = F^C_{rs} < 0 \mid B^{(j)} > + \check{F}^V_{sr} = \check{F}^V_{rs}$ because the overlap $< o \mid B^{(j)} >$ between the MC CEP coefficients and the configurational trial vector is assumed to be zero. Analogously, by replacing the Fock type elements with the corresponding elements F_{pq} calculated with one-index transformed integrals according to Eqs. (49-50) we obtain the final expressions for the orbital part of the sigma vector from an orbital trial vector:

$$\sigma^{(j)}_{iu} = 4\tilde{F}^D_{iu} - 2(\sum_x D_{ux}\tilde{F}^C_{\xi}) - 2\tilde{F}^Q_{iu} + B_{iu} \tag{A8a}$$

$$\sigma^{(j)}_{ia} = 4\tilde{F}^D_{ia} + B_{ia} \tag{A8b}$$

$$\sigma^{(j)}_{uv} = 2\sum_x (D_{ux}\tilde{F}^C_{xv} - D_{vx}\tilde{F}^C_{xu}) + 2(\tilde{F}^Q_{uv} - \tilde{F}^Q_{vu}) + B_{uv} \tag{A8c}$$

$$\sigma_{ua}^{(j)} = 2(\sum_{x} D_{ux}\tilde{F}_{xa}^{C}) + 2\tilde{F}_{ua}^{Q} + B_{ua} \tag{A8d}$$

where the tilde [~] marks that the quantity (integral, Fock- or $\mathbf{F}^Q$-matrix) has undergone a one-index transformation and the "check" accent [v] denotes that it has been evaluated over transition density matrices. The $\mathbf{F}^Q$-matrix elements were defined by Eq. (68). The $\tilde{\mathbf{F}}^Q$-matrix elements are similarly defined with one-index transformed two-electron integrals. $\mathbf{F}^C$ and $\mathbf{F}^V$ are the inactive and active Fock matrices defined by Eqs. (64) and (65) respectively.

Appendix 8B.
Calculation of Generalized Fock Matrices from Regular and One-Index Transformed Integrals

In this appendix we describe the calculation of the auxiliary matrices using for calculation of energy gradients and Hessian transformations. These are the inactive and active Fock matrices $\mathbf{F}^C$ and $\mathbf{F}^V$ defined by Eqs. (64) and (65) and the $\mathbf{F}^Q$ matrix, defined by Eq. (68), evaluated either with standard integrals or with one-index transformed integrals.

We divide the one-electron density matrix $\mathbf{D}$ into two parts $\mathbf{D}^C$ and $\mathbf{D}^V$, ($\mathbf{D} = \mathbf{D}^C + \mathbf{D}^V$), where $\mathbf{D}^C$ is the density from the inactive orbitals (with 2 in the diagonal for the inactive orbitals and 0 elsewhere) and $\mathbf{D}^V$ is the density from the active orbitals. The inactive and active Fock matrices are then written as

$$F^C_{pq} = h_{pq} + \sum_i D^C_{ii} \mathbb{P}_{pq,ii} = h_{pq} + 2\sum_i \mathbb{P}_{pq,ii} \tag{B1}$$

$$F^V_{pq} = \sum_{uv} D^V_{uv} \mathbb{P}_{pq,uv} \tag{B2}$$

and can be backtransformed to atomic orbital (AO) basis as

$$F^{C,AO}_{\alpha\beta} = h_{\alpha\beta} + \sum_{\gamma\delta} D^{C,AO}_{\gamma\delta} \mathbb{P}_{\alpha\beta,\,\gamma\delta} \tag{B3}$$

$$F^{V,AO}_{\alpha\beta} = \sum_{\gamma\delta} D^{C,AO}_{\gamma\delta} \mathbb{P}_{\alpha\beta,\,\gamma\delta} \tag{B4}$$

here $\mathbb{P}_{pq,rs}$ and $\mathbb{P}_{\alpha\beta,\,\gamma\delta}$ denote P super matrix integrals in MO and AO basis, respectively (defined by Eq. (67)). Eq. (B3) expresses the Fock matrix commonly used in standard restricted Hartree-Fock programs. In order to avoid tedious four-index integral transformations, we evaluate these Fock matrices by means of two two-index transformations: first by transforming the density matrices to AO basis; second, by reading the atomic integral file using Eqs. (B3) and (B4); and third, by transforming the AO fock matrices to MO basis. Using the one-index transformed integrals as defined by Eqs. (49-50) the one-index Fock matrix $\tilde{\mathbf{F}}^V$ is constructed as:

$$
\begin{aligned}
\tilde{F}^V_{pq} &= \sum_{uv} D^V_{uv}\tilde{\mathbb{P}}_{pq,uv} \\
&= \sum_{uv} D^V_{uv}\Big(\sum_t \mathbb{P}_{tq,uv}b_{pt} + \mathbb{P}_{pt,uv}b_{qt} + \mathbb{P}_{pq,tv}b_{ut} + \mathbb{P}_{pq,ut}b_{vt}\Big) \\
&= \sum_t (F^V_{tq}b_{pt} + F^V_{pt}b_{qt}) + \sum_{uvt} D^V_{uv}(\mathbb{P}_{pq,tv}b_{ut} + \mathbb{P}_{pq,ut}b_{vt}) \\
&= \sum_t (F^V_{tq}b_{pt} + F^V_{pt}b_{qt}) + 2\sum_{\perp}\Big(\sum_v D^V_{uv}b_{vt}\Big)\mathbb{P}_{pq,ut}
\end{aligned}
\tag{B5}
$$

where we have made use of the relations $D^V_{uv} = D^V_{vu}$ and $\mathbb{P}_{pq,tv} = \mathbb{P}_{pq,vt}$. In the same way one finds

$$
\begin{aligned}
\tilde{F}^C_{pq} &= \sum_t (F^C_{tq}b_{pt} + F^C_{pt}b_{qt}) + 2\sum_{it} (D^C_{ii}b_{it})\mathbb{P}_{pq,it} \\
&= \sum_t (F^C_{tq}b_{pt} + F^C_{pt}b_{qt}) + 4\sum_{it} b_{it}\mathbb{P}_{pq,it}
\end{aligned}
\tag{B6}
$$

where t runs over all orbitals in Eqs. (B5) and (B6). One thus finds that the inactive and active Fock matrices with one-index transformed integrals can be evaluated by means of a one-index transformation of the Fock matrix with standard integrals plus another Fock-matrix type contribution with a modified density matrix. Note that the second term in the first summation in Eq. (B5) as well as in Eq. (B6) is the transpose of the first term. This term is the last term in Eqs. (B5) and (B6), and it can also be transformed to AO basis. The $\tilde{\mathbf{F}}^C$ and $\tilde{\mathbf{F}}^V$ matrices can thus be computed completely from the atomic supermatrix integrals.

One also needs $\check{\mathbf{F}}^V$ with a transition density matrix $\check{\mathbf{D}}^V$ (*c.f.* Eq. (A7a)). $\check{\mathbf{D}}^V$ can also be transformed to AO basis and $\check{\mathbf{F}}^V$ can then be constructed directly from the atomic supermatrix integrals, and we do not need any four-index transformed integrals to compute these either. Finally, one also needs one-index transformed $\mathbf{F}^Q$ matrices for the orbital part of the σ vector. These are obtained from Eq. (68) with the ordinary two-electron integrals in MO basis replaced by their one-index transformed analogues. This term is the only one that requires integrals with two unoccupied indices.

Appendix 8C.
The Diagonal of the Orbital Hessian Matrix

In this appendix we list the expressions for the diagonal of the orbital part of the Hessian matrix. In the direct MCSCF scheme the full Hessian is never explicitly calculated, however, the Hessian diagonal is needed to get new trial vectors in each macro iteration. The orbital diagonal is subdivided into elements describing inactive-active, inactive-secondary, active-active and active-secondary rotations. The notation used here is the same as in sections B and C.

$$\begin{aligned} L^{oo}_{iu,iu} &= <0 \mid [\hat{E}^{-}_{iu}, \hat{E}^{-}_{iu}, \hat{\mathscr{H}}] \mid 0> \\ &= 2\sum_{v,x}[P'_{uu,vx}(vx \mid ii) + 2P'_{ux,uv}(vi \mid \xi)] \\ &\quad + 4\sum_{v}(\delta_{uv} - D_{uv})[3(vi \mid ui) - (uv \mid ii)] \\ &\quad + 2D_{uu}F^{C}_{ii} + 4(F^{D}_{uu} - F^{D}_{ii}) - 2F_{uu} \end{aligned} \tag{C1}$$

$$\begin{aligned} L^{oo}_{ia,ia} &= <0 \mid [\hat{E}^{-}_{ia}, \hat{E}^{-}_{ia}, \hat{\mathscr{H}}] \mid 0> \\ &= 4[3(ai \mid ai) - (aa \mid ii)] + 4(F^{D}_{aa} - F^{D}_{ii}) \end{aligned} \tag{C2}$$

$$\begin{aligned} L^{oo}_{uv,uv} &= <0 \mid [\hat{E}^{-}_{uv}, \hat{E}^{-}_{uv}, \hat{\mathscr{H}}] \mid 0> \\ &= \sum_{x,y}[4P'_{vx,vy}(ux \mid uy) + 4P'_{ux,uy}(vx \mid vy) - 8P'_{ux,vy}(ux \mid uy) \\ &\quad + 2P'_{vv,xy}(uu \mid xy) + 2P'_{uu,xy}(vv \mid xy) - 4P'_{uv,xy}(uv \mid xy)] \\ &\quad + 2(D_{vv}F^{C}_{uu} + D_{uu}F^{C}_{vv} - 2D_{uv}F^{C}_{uv} - F_{uu} - F_{vv}) \end{aligned} \tag{C3}$$

$$\begin{aligned} L^{oo}_{ua,ua} &= <0 \mid [\hat{E}^{-}_{ua}, \hat{E}^{-}_{ua}, \hat{\mathscr{H}}] \mid 0> \\ &= 2\sum_{v,x}[P'_{uu,vx}(aa \mid vx) + 2P'_{ux,uv}(ax \mid av)] + 2D_{uu}F^{C}_{aa} - 2F_{uu} \end{aligned} \tag{C4}$$

where

$$\mathbf{F}^{D} = \mathbf{F}^{C} + \mathbf{F}^{V} \tag{C5}$$

and

$$P'_{pq,rs} = \frac{1}{2}(P_{pq,rs} + P_{pq,sr}) \tag{C6}$$

is the symmetrized second order density matrix.

As default an approximate orbital diagonal is used in SIRIUS, where all non-Fock matrix elements are neglected, except $\mathbf{F}^C$ is replaced by $\mathbf{F}^D$ in Eqs. (C1-C4) (this corresponds to approximating $P_{uv,xy}$ with $D_{uv}D_{xy} - \frac{1}{2}D_{uy}D_{xv}$).

Appendix 8D.
Iterative Solution of the CI Eigenvalue Problem

The optimal CI coefficients are eigenvectors of the eigenvalue equation

$$\mathbf{HC} = E\mathbf{C} \tag{D1}$$

where $\mathbf{H}$ is the CI Hamiltonian matrix, $\mathbf{C}$ contains the optimal CI coefficients and E is the corresponding energy. It is only necessary to obtain a few eigensolutions of Eq. (D1); in order to converge to the n'th eigensolution n eigensolutions must normally be obtained. The use of matrix dimensions of up to several millions makes the use of iterative techniques to solve Eq. (D1) mandatory. The use of iterative schemes is facilitated by the fact that only a few of the extreme eigenvectors are sought.

The iterative techniques currently used in quantum chemistry build up a subspace and solve the eigenvalue equation exactly in this subspace. In this section, we will describe the construction of the basis vectors of the subspace. Methods for obtaining basis vectors can be obtained as approximate solutions to the Newton equations for the energy function

$$E(\mathbf{C}) = \frac{\mathbf{C}^t\mathbf{HC}}{\mathbf{C}^t\mathbf{C}} \tag{D2}$$

which has the eigenvectors of Eq. (D1) as the stationary points. Assume that $\mathbf{C}$ is a normalized approximation to the eigenvector with E being the corresponding energy. The correction needed to obtain an improved normalized eigenvector approximation can be restricted to be orthogonal to $\mathbf{C}$ so one can write all possible variations as

$$\mathbf{d} = (1 - \mathbf{C}\mathbf{C}^t)\mathbf{e} \tag{D3}$$

with $\mathbf{e}$ being an arbitrary vector. The gradient of $E(\mathbf{C})$ with respect to $\mathbf{d}$ becomes

$$\begin{aligned}\mathbf{f} &= 2(1 - \mathbf{C}\mathbf{C}^t)(\mathbf{H} - E)\mathbf{C} \\ &= 2(\mathbf{H} - E)\mathbf{C}\end{aligned} \tag{D4}$$

and the Hessian matrix with respect to $\mathbf{d}$ becomes

$$\begin{aligned}\mathbf{G} &= 2(1 - \mathbf{C}\mathbf{C}^t)(\mathbf{H} - E)(1 - \mathbf{C}\mathbf{C}^t) \\ &= 2(\mathbf{H} - E) + 2(\mathbf{C}(\mathbf{H} - E)\mathbf{C})\begin{pmatrix} 0 & -1 \\ -1 & 0 \end{pmatrix}\begin{pmatrix} \mathbf{C}^t \\ \mathbf{C}^t(\mathbf{H} - E) \end{pmatrix}\end{aligned} \tag{D5}$$

The Hessian matrix of Eq. (D5) has $\mathbf{C}$ as an eigenvector with eigenvalue zero.

In actual calculations, the exact Hessian is approximated by a matrix, $\mathbf{G}_{\mathbf{apr}}$, that is obtained by replacing the exact Hamilton matrix $\mathbf{H}$ with an approximate Hamilton matrix, $\mathbf{H}_0$

$$\mathbf{G}_{\mathbf{apr}} = 2(\mathbf{H}_0 - E) + 2(\mathbf{C}(\mathbf{H}_0 - E)\mathbf{C})\begin{pmatrix} 0 & -1 \\ -1 & 0 \end{pmatrix}\begin{pmatrix} \mathbf{C}^t \\ \mathbf{C}^t(\mathbf{H}_0 - E) \end{pmatrix} \tag{D6}$$

It is noted that also $\mathbf{G}_{\mathbf{apr}}$ has $\mathbf{C}$ as an eigenvector with eigenvalue zero. The approximate Newton step becomes

$$\mathbf{d} = -(\mathbf{H}_0 - E)^{-1}(\mathbf{H} - E - E_1)\mathbf{C} \tag{D7}$$

with

$$E_1 = \frac{\mathbf{C}^t(\mathbf{H}_0 - E)^{-1}(\mathbf{H} - E)\mathbf{C}}{\mathbf{C}^t(\mathbf{H}_0 - E)^{-1}\mathbf{C}} \tag{D8}$$

In order to obtain Eqs. (D7) and (D8) we used the Sherman-Morrison formula giving the inverse of a matrix plus a low-rank matrix. The approximate Newton equation is well-defined although $\mathbf{G}_{\mathbf{apr}}$ is singular since the $\mathbf{f}$ is orthogonal to $\mathbf{C}$ and since we require $\mathbf{d}$ to be orthogonal to $\mathbf{C}$. It is common use to set E_1 to zero so the correction becomes

$$\mathbf{d} = -(\mathbf{H}_0 - E)^{-1}(\mathbf{H} - E)\mathbf{C} \tag{D9}$$

which is the so-called Davidson scheme.[37] In several respects, the full correction vector Eq. (D7) is to be preferred. Consider the limit where the matrix $\mathbf{H}_0$ becomes $\mathbf{H}$. The Davidson correction becomes

$$\mathbf{d} = -\mathbf{C} \tag{D10}$$

while the complete correction vector becomes

$$\mathbf{d} = -\mathbf{C} - E_1(\mathbf{H} - E)^{-1}\mathbf{C} \tag{D11}$$

While the Davidson correction fails to introduce any new information in this limit, the complete correction introduces an inverse iteration direction, giving cubic convergence (for information about inverse iteration see, *e.g.*, Parlett.[118] In order to assure than an improved approximation to $\mathbf{H}$ yields improved convergence, the full correction vector of Eq. (D7) must be used.

Until now, the exact nature of $\mathbf{H}_0$ has not been discussed. It is well known that the major part of the Hamilton matrix is diagonally dominated if canonical orbitals are used. It is therefore a natural choice to use the diagonal of $\mathbf{H}$ as $\mathbf{H}_0$. This diagonal approximation can be augmented with the exact Hamiltonian matrix in a subspace of the most important configurations. In SIRIUS, one can use a more general choice of $\mathbf{H}_0$. A subspace P is selected in which a more elaborate approximation to $\mathbf{H}_0$ than a diagonal will be constructed. The P-subspace can be further divided into three subspaces, P1, P2, P3. The complete Hamiltonian is set up in the combined P1, P2 subspace. The P1-P3 subblock is calculated exactly, while the P2-P3 subblock is set to zero, and the P3-P3 subblock is approximated with a diagonal. This scheme allows the use of P3 subspaces containing several thousand parameters, while the P1, P2 subspaces normally are restricted to contain at most a few hundred parameters. It is our experience that this choice of $\mathbf{H}_0$ is superior to a pure diagonal $\mathbf{H}_0$, especially in cases where only a few trial vectors can be retained.

References

1. H.J.Aa. Jensen and P. Jørgensen, J. Chem. Phys., **80**, 1204 (1984).
2. H.J.Aa. Jensen and H. Ågren, Chem. Phys. Letters, **110**, 140 (1984).
3. H.J.Aa. Jensen and H. Ågren, Chem. Phys., **104**, 229 (1986).
4. H.J.Aa. Jensen, P. Jørgensen, and H. Ågren, J. Chem. Phys., **87**, 451 (1987).
5. H.J.Aa. Jensen, P. Jørgensen, H. Ågren, and J. Olsen, J. Chem. Phys., **88**, 3834 (1988).
6. J. Olsen, B.O. Roos, P. Jørgensen, and H.J.Aa. Jensen, J. Chem. Phys., **89**, 2185 (1988).
7. D.R. Hartree, W. Hartree, and B. Swirles, Phil. Trans. Roy. Soc., **A238**, 229 (1939).
8. J. Olsen, D.L. Yeager, and P. Jørgensen, Adv. Chem. Phys., **54**, 1 (1983).
9. P.E.M. Siegbahn, Faraday Symposia of the Chemical Society, **19**, 97 (1984).
10. B.O. Roos, Adv. Chem. Phys., **69**, 399 (1987).
11. R. Shepard, Adv. Chem. Phys., **69**, 63 (1987).
12. H.-J. Werner, Adv. Chem. Phys., **69**, 1 (1987).
13. P. Jørgensen, P. Swanström, and D.L. Yeager, J. Chem. Phys., **78**, 347 (1983).
14. J. Paldus, Phys. Rev. A., **14**, 1620 (1976).
15. I. Shavitt, Int. J. Quant. Chem., **S12**, 5 (1978).
16. P.E.M. Siegbahn, J. Chem. Phys., **72**, 1647 (1980).
17. P.E.M. Siegbahn, Chem. Phys. Letters, **109**, 417 (1984).
18. N.C. Handy, Chem. Phys. Letters, **74**, 280 (1980).
19. P.J. Knowles and N.C. Handy, Chem. Phys. Letters, **111**, 315 (1984).
20. K. Ruedenberg and K.R. Sundberg, in "Quantum Science," J.-L. Calais, O. Goscinski, J. Linderberg and Y. Öhrn, eds., Plenum, New York (1976).
21. K. Ruedenberg, M.W. Schmidt, M.M. Gilbert, and S.T. Elbert, Chem. Phys., **71**, 41,51,65 (1982).
22. C. Møller and M.S. Plesset, Phys. Rev., **46**, 618 (1934).
23. J. Olsen and P. Jørgensen, J. Chem. Phys., **82**, 3235 (1985).
24. J. Almlöf, USIP nr. **74-29**, University of Stockholm (1974).
25. T.U. Helgaker, "HERMIT" program, unpublished (1986).
26. V. Carravetta and I. Cacelli, the PHOTO12 program is described in Chapter 12 of this volume (1990).
27. M. Born and R. Oppenheimer, Ann. d. Phys., **84**, 457 (1927).
28. P. Jørgensen and J. Simons, "Second Quantization-Based Methods in Quantum Chemistry," Academic, Newe York (1981).
29. T.U. Helgaker and P. Jørgensen, Adv. Quant. Chem., **19**, 183 (1988).
30. A.L Panin, Yu.G. Khait, and N.N. Gorinchoy, Int. J. Quant. Chem., **30**, 509 (1986).
31. B.H. Lengsfield III and B. Liu, J. Chem. Phys., **75**, 478 (1981).
32. B.H. Lengsfield III, J. Chem. Phys., **77**, 4073 (1982).
33. D. Yarkony, Chem. Phys. Letters, **77**, 624 (1981).

34. R. Sheppard, I. Shavitt, and J. Simons, J. Chem. Phys., **76**, 4543 (1982).
35. J.T. Golab, D.L. Yeager, and P. Jϕrgensen, Chem. Phys., **78**, 175 (1983); J.T. Golab, D.L. Yeager, and P. Jϕrgensen , Chem. Phys., **93**, 83 (1985).
36. R. Fletcher, "Practical Methods of Optimization," Vol. 1, Wiley, New York (1980).
37. E. Davidson, J. Comput. Phys., **17**, 87 (1975).
38. B. Liu, "The Simultaneous Expansion Method," in Proceedings of Workshop on *Numerical Algorithms in Chemistry: Algebraic Methods*, p. 49 (National Resource for Computations in Chemistry, Berkely, 1978).
39. B.O. Roos, P.R. Taylor, and P.E.M. Siegbahn, Chem. Phys., **48**, 157 (1980).
40. P.E.M. Siegbahn, A. Heiberg, J. Almlöf, and B.O. Roos, J. Chem. Phys., **74**, 2384 (1981).
41. J. Hinze, J. Chem. Phys., **50**, 6424 (1973).
42. D. Day, Int. J. Quant. Chem., **8**, 501 (1974).
43. C.C.J. Roothaan and P.S. Bagus, Methods Comput. Phys., **2**, 47 (1963).
44. G.B. Bacskay, Chem. Phys., **61** 385 (1980).
45. P.-Å. Malmquist, Int. J. Quant. Chem., **30**, 479 (1986).
46. P.-Å. Malmquist and B.O. Roos, Chem. Phys. Letters, **155**, 189 (1989).
47. "The Unitary Group for the Evaluation of Electronic Energy Matrix Elements," J. Hinze, ed., Springer, Berlin (1981).
48. R. Pauncz, "Spin Eigenfunctions Construction and Use," Plenun, New York (1979).
49. J.E. Graebenstetter, T.J. Tseng, and F. Grein, Int. J. Quant. Chem., **10**, 143 (1976).
50. W. Duch, "GRMS or Graphical Representation of Model Spaces," Springer, Berlin (1986).
51. P.-O. Löwdin, Phys. Rev., **97**, 1474 (1955).
52. H.-J. Werner and P. Knowles, J. Chem. Phys., **82**, 5053 (1985).
53. V. Carravetta, H. Ågren, H.J.Aa. Jensen, P. Jϕrgensen , and J. Olsen, J. Phys. B., **22**, 2133 (1989).
54. L.S. Cederbaum, W. Domcke, and J. Schirmer, Phys. Rev. A., **22**, 206 (1980).
55. J. Schirmer, G. Angona, S. Svensson, D. Nordfors, and U. Gelius, J. Phys. B., **20**, 6031 (1987).
56. S. Svensson, N. Mårtensson, E. Basdilier, P.-Å. Malmquist, U. Gelius, and K. Siegbahn, Physica Scripta, **14**, 141 (1976).
57. P.S. Bagus, Phys. Rev., **139**, A619 (1965).
58. MOLECULE-ALCHEMY program package, J. Almlöf, P.S. Bagus, B. Liu, D. MacLean, U.I. Wahlgren, and M. Yoshimine, IBM Research Laboratory; see also IBM Research Report **RJ-1077** (1972).
59. M.C. Zerner and M. Hehenberger, Chem. Phys. Letters, **61**, 550 (1979).
60. G. Karlström, Chem. Phys. Letters, **67**, 348 (1979).
61. A. Naves de Brito, M.P. Keane, N. Correia, S. Svensson, U. Gelius, and B.J. Lindberg, J. Electron. Spectrosc., to be published.
62. A. Naves de Brito, N. Correia and H. Ågren, to be published.

63. T.U. Helgaker, H.J.Aa. Jensen, and P. Jørgensen, J. Chem. Phys., **84**, 6280 (1986).
64. T.U. Helgaker, J. Almlöf, H.J.Aa. Jensen, and P. Jørgensen , J. Chem. Phys., **84**, 6266 (1986).
65. H.J.Aa. Jensen, P. Jørgensen, and T.U. Helgaker, J. Chem. Phys., **85**, 3917 (1986).
66. H.J.Aa. Jensen, P. Jørgensen, and T.U. Helgaker, J. Am. Chem. Soc., **109**, 2895 (1987).
67. P. Jørgensen, H.J.Aa. Jensen, and T.U. Helgaker, Theor. Chim. Acta, **73**, 55 (1988).
68. D. Nordfors, N. Mårtensson, and H. Ågren, J. Electron. Spectrosc., submitted for publication, 1990.
69. A. Cesar, H. Ågren, T.U. Helgaker, H.J.Aa. Jensen, and P. Jørgensen, in preparation.
70. P. Jørgensen, H.J.Aa. Jensen, and J. Olsen, J. Chem. Phys., **89**, 3654 (1988).
71. H.J.Aa. Jensen, H. Koch, P. Jørgensen, and J. Olsen, Chem. Phys., **119**, 297 (1988).
72. J. Olsen, A.M. Sanchez de Meras, H.J.Aa. Jensen, and P. Jørgensen , Chem. Phys. Letters, **154**, 380 (1989).
73. H. Ågren, C. Medina-Llanos, K. Mikkelsen, and H.J.Aa. Jensen, Chem. Phys. Letters, **153**, 322 (1989).
74. D. Nordfors, Thesis, University of Uppsala (1990).
75. H. Ågren, J. Olsen, H.J.Aa. Jensen, and P. Jørgensen, Phys. Rev. A., **40**, 2265 (1989).
76. H. Ågren, V. Carravetta, H.J.Aa. Jensen, and P. Jørgensen , in preparation.
77. K. Mikkelsen, H. Ågren, H.J.Aa. Jensen, and T.U. Helgaker, J. Chem. Phys., **89**, 3086 (1988).
78. C. Medina-Llanos, H. Ågren, K. Mikkelsen, and H.J.Aa. Jensen, J. Chem. Phys., **90**, 6422 (1989).
79. H Ågren, A. Flores-Riveros, and H.J.Aa. Jensen, Physica Scripta, **40**, 745 (1989).
80. K. Mikkelsen and H. Ågren, J. Phys. Chem., to be published (1990).
81. D. Nordfors, N. Mårtensson, and H. Ågren, Phys. Rev. B., **39**, 12922 (1988).
82. F. Prosser and S. Hagstrom, Int. J. Quant. Chem., **2**, 89 (1968).
83. H. Ågren, R. Arneberg, J. Müller, and R. Manne, Chem. Phys., **83**, 53 (1984).
84. H. Ågren and H.J.Aa. Jensen, Chem. Phys. Letters, **137**, 431 (1987).
85. J. Olsen, unpublished (1990).
86. J.J Pirreaux, S. Svensson, E. Basilier, P.-Å. Malmquist, U. Gelius, R. Caudano, and K. Siegbahn, Phys. Rev. A., **14**, 2133 (1976).
87. H. Ågren, S. Svensson, and U.I. Wahlgren, Int. J. Quant. Chem., **11**, 317 (1977).

88. U. Gelius, E. Basilier, S. Svensson, T. Bergmark, and K. Siegbahn, J. Electron. Spectrosc., **2**, 405 (1974).
89. U. Gelius, J. Electron. Spectrosc., **5**, 985 (1974).
90. T. Åberg, Ann. Acad. Sci. Fenn. Ser. A., **308**, 7 (1969).
91. D. Nordfors, A. Nilsson, N. Mårtensson, S. Svensson, U. Gelius, and H. Ågren, J. Electron. Spectrosc., to be published (1990).
92. V. Carravetta, H. Ågren, and A. Cesar, Chem. Phys. Letters, **148**, 210 (1988).
93. V. Carravetta, H. Ågren, D. Nordfors, and S. Svensson, Chem. Phys. Letters, **152**, 190 (1988).
94. J. Nordgren and H. Ågren, Comments on Atomic and Molecular Physics, **14**, 203 (1984).
95. H. Ågren and A. Cesar, to be published.
96. A. Cesar, H. Ågren, and V. Carravetta, Phys. Rev. A., **40**, 2187 (1989).
97. H. Ågren, A. Flores-Riveros, P. Glanz, and J. Nordgren, to be published.
98. A. Cesar, H. Ågren, P. Glanz, and J. Nordgren, J. Phys. B., to be published (1990).
99. R. Brammer, J. Chem. Phys., **87**, 1153 (1987).
100. G. Wentzel, Z. Physik, **43**, 521 (1927).
101. R. Manne and H. Ågren, Chem. Phys., **93**, 201 (1985).
102. H. Siegbahn, L. Asplund, and P. Kelfve, Chem. Phys. Letters, **35**, 330 (1975).
103. V. Carravetta and H. Ågren, Phys. Rev. A., **35**, 1022 (1987).
104. A. Cesar, H. Ågren, A. Naves de Brito, M.P. Keane, P. Baltzer, S. Svensson, L. Karlsson, P.G. Fournier and J. Fournier, J. Mol. Struct., to be published (1990).
105. A. Cesar, H. Ågren, C.M. Liegener, A. Naves de Brito, M.P. Keane, N. Correia, S. Svensson, L. Karlsson, in preparation.
106. A. Naves de Brito, M.P. Keane, S. Svensson, L. Karlsson, H. Aksela, and S. Aksela, in preparation.
107. W. Domcke and L.S. Cederbaum, Chem. Phys., **25**, 189 (1977).
108. J. Nordgren, L. Selander, L. Petersson, C. Nordling, K. Siegbahn, and H. Ågren, J. Chem. Phys., **76**, 3928 (1982).
109. S. Bagus, and B.O. Roos, Chem. Phys. Letters, **82**, 505 (1981).
110. J. Müller, H. Ågren, and O. Goscinski, Chem. Phys., **38**, 349 (1979).
111. N. Wassdahl, G. Bray, P. Glanz, P. Johansson, J. Nordgren, A. Flores-Riveros, and H. Ågren, in preparation.
112. A. Cesar, M.B. Huang, and H. Ågren, Chem. Phys. Letters, **162**, 84 (1989).
113. H. Ågren, A. Flores-Riveros, and H.J.Aa. Jensen, Phys. Rev. A., **34**, 4606 (1986).
114. A. Flores-Riveros, H. Ågren, R. Brammer, and H.J.Aa. Jensen, J. Chem. Phys., **85**, 6270 (1986).
115. M.A. Natiello, O. Vahtras, A. Engelmann, and H. Ågren, to be published.
116. M.A. Natiello and A.R. Engelmann, in "Lecture Notes in Physics," **325**, Erkki Brändas and Nils Elander, eds., Springer, Berlin, 329 (1989).

117. N. Correia, M. Natiello, H. Ågren, M. Keane, S. Svensson, and L. Karlsson, to be published.
118. B.N. Parlett, "The Symmetric Eigenvalue Problem," Prentice-Hall, New York (1980).

Chapter 9.

MOLCAS: A General Purpose Quantum Chemistry Program System for Correlated Wavefunctions

B.O. Roos, G. Karlström,
P.-Å. Malmqvist, A.J. Sadlej

Department of Theoretical Chemistry
Chemical Centre
P.O.B. 124
S-221 00 LUND, Sweden

and

P.-O. Widmark
S&TC group, ACIS
IBM Sweden

A. Introduction

In this section a brief outline will be given of the program system MOLCAS, a general purpose program system for electronic structure calculations with special emphasis on correlated wavefunctions. MOLCAS is the result of many years of work involving a number of scientists. The different programs contributing to MOLCAS stem from different sources[1] and have been assembled into a common software by the quantum chemistry group at Lund in cooperation with IBM Sweden under a joint study contract.[2]

MOLCAS includes two approaches to the electron correlation problem. The first approach applies in cases where the wavefunction is well represented by a Hartree-Fock determinant. Electron correlation can in this case be successfully treated using either Many-Body Perturbation Theory (MBPT)[3-4] or the Coupled Pair Functional (CPF) method.[5-6] Alternatively singles and doubles configuration interaction (SDCI) from a single reference configuration can be performed, but this procedure is likely to lead to less accurate results. A very efficient and fast method, for larger systems, where the above methods cannot be applied, is to use second order perturbation theory, and this option is also provided by MOLCAS.

The computational schemes mentioned above are valid approaches to the electronic structure problem in cases, where the wavefunction is dominated by a single electronic configuration. The most common area of application of these techniques are therefore studies of molecules in their ground electronic state, at geometries close to equilibrium, and to weak interactions between such systems. Only a limited part of the chemical problems of interest for theoretical studies are then covered by these methods. The electronic wave function is in many cases more complex and includes several strongly interacting electronic configurations. Such cases are very frequent for excited states, and occur in most cases where chemical bonds are broken or formed. These *near degeneracies* can even occur at ground state equilibrium geometries for some first row molecules (C_2 and ozone are good examples. The near degeneracies explain the bi-radical character of these systems). Even more common is this situation for systems involving transition metal atoms.

The second approach to the electronic structure problem in MOLCAS deals with the latter situation. The basic feature here is the multi-configuration SCF (MCSCF) technique. The so-called Complete Active Space (CAS) SCF method was developed around ten years ago[7-10] and forms the cornerstone of the multi-configurational approach. An extension of the CASSCF method, called the Restricted Active Space (RAS) SCF method,[11] is implemented in MOLCAS. The RASSCF method allows a more flexible wavefunction by introducing two new orbital subspaces with restrictions on the number of holes and particles, respectively. The RASSCF program is to be used to find the zeroth order approximation to the wavefunction. Normally such a wavefunction is restricted to the orbital space that contains the near degeneracies. Dynamic correlation effects are added in a second step. MOLCAS provides two possibilities for such calculations: Multi-Reference (MR) CI[12] and the Average Coupled Pair Function (ACPF) method.[13] Both these types of wavefunctions are built from configurations, which are singly and doubly excited with respect to a set of reference configurations. This set is normally taken as a subset of the configurations generated by the preceding RASSCF calculation. The CASCF-MRCI approach has in a number of applications shown to give very accurate molecular properties and seems to be the best approximation to the full CI wavefunction available today, except maybe for simple closed shell molecular systems.[14]

MOLCAS also provides a program for the calculation of transition properties between different electronic states at the RASSCF level of approximation. A special method (the RASSCF State Interaction (RASSI) method[15-16]) has been developed which makes it practical to compute the Hamiltonian matrix in a basis consisting of RASSCF wavefunctions obtained for different electronic states. In general such wavefunctions are non-orthogonal and use different (non-orthogonal) one-electron bases. The major use of the RASSI program is for calculation of transition moments, but it has also been used for studies of weak intermolecular interactions and charge transfer in transition metal complexes.

Included in MOLCAS is also a program for the calculation of first order electric and magnetic properties for wavefunctions obtained by any of the methods included in the software. The property program can also be used to compute first order relativistic corrections.

MOLCAS communicates with the user via a user interface program. In the VM/XA SP version of the program presently released, this interface is written in REXX. The separate programs contained in MOLCAS are essentially separate units, which communicates with each other through files. The task of keeping track of these files is taken care of by the user interface shell. This shell includes a set of commands used to set up and run a specific job. The user further has access to an on-line help menu, which includes descriptions of all programs and inputs as well as help in setting up a job stream.

MOLCAS exists today in a version adapted to run under the VM/XA SP operating system, but can easily be modified to run under other versions of VM. It will soon be implemented also for MVS and AIX.

B. Program Descriptions

This section contains brief descriptions of all programs included in the MOLCAS system. The theoretical background for the methods used in these programs will not be given here. Instead the reader is referred to the original articles and to the other sections of this book.

1. Integral Evaluation and Handling

The use of any *ab initio* quantum chemical program must start with the generation of a set of one- and two-electron integrals over the atomic orbital (AO) basis set used in the calculation. As most other program systems MOLCAS builds the AO basis set from a set of cartesian gaussian functions centered (in principle) on the individual atoms of the molecular system. These, so-called, Contracted Gaussian Type Functions (CGTF) can be of two types. Either a segmented contraction scheme is used, where the primitive gaussian functions are partitioned into subsets and contracted functions are formed within each set, or a general contraction scheme is used, where each contracted function is formed as a linear combination over all primitive functions, for each angular quantum number.[17] These generally contracted basis functions have been obtained from calculations on atoms, normally including correlation effects, and in some cases also polarization.[18] They are therefore in a sense true atomic orbitals and the basis sets obtained in this way have been named ANO (atomic natural orbital) basis sets. They are superior to most segmented basis sets in describing correlation effects in molecular system.[17-19] A library of ANO type basis functions for first and second row atoms is included in the MOLCAS software.[18] It will in the near future be extended to cover all atoms up Kr.

The integrals are evaluated using the program MOLECULE[20-21] in a version vectorized to run efficiently on the IBM 3090 vector facility. MOLECULE has been specially designed to run effectively for generally contracted basis sets. One of the unique features of MOLECULE is symmetry adaption. The program computes integrals over symmetry adapted basis functions for all subgroups of D_{2h}. The symmetry adaption of the basis set is done automatically by the program and the user only has to specify, in the input, the appropriate symmetry operations needed to construct the point group. This symmetry adaption has several advantages. First it creates a compact list of two-electron integrals. Moreover, it becomes trivial to carry the symmetry adaption over to subsequent steps in the calculation. This is especially important for the integral transformation programs, in which one symmetry block of integrals is handled at a time. However, also in the RASSCF and MRCI programs can symmetry be fully exploited (up to D_{2h}), resulting in large efficiency gains. This symmetry adaption is carried through to all programs included in MOLCAS.

The integrals generated by MOLECULE are stored in (almost) random order with indices. The packing convention used for the indices, in the present version of the program, limits the number of contracted basis functions to 255. This may seem to be a small number, considering the large basis set calculations which today can be performed at the SCF level of theory. However, MOLCAS is designed to perform calculations at higher levels of approximations, where the restrictions on the basis set are normally more severe (CASSCF calculations can, however, be performed using basis sets with more than 200 CGTF's).

The MOLCAS system uses the integrals generated by MOLECULE in two different ways. First they are used to generate Fock matrices. This is, of course, done in the SCF program, but also in the RASSCF, MRCI, CPF, and MBPT programs. In RASSCF Fock matrices are used to describe the interaction within the inactive orbital space and between the active and inactive orbitals. In the same spirit, the MRCI, CPF, and MBPT programs use Fock matrices for the orbitals frozen in the calculation. Actually, in the latter case the Fock matrix is generated already at the transformation step, and replaces the bare nucleus Hamiltonian. Thus the frozen orbitals do not have to be specified at all when the transformation of the basic integrals to an MO basis has been performed. MOLCAS computes Fock matrices from one-electron integrals and a supermatrix generated from the AO two-electron integrals. The MOLCAS program SUPCON generates this supermatrix from the MOLECULE integrals, either as an ordered list or with indices. The ordered list is sometimes longer, but the subsequent construction of the Fock matrix can be effectively vectorized. The generation of the supermatrix is performed in place without the use of any extra sorting space. This is an important feature, since disk space can sometimes be a limiting factor in calculations using large basis sets.

The transformation of the two-electron integrals from AO to MO basis is performed from an ordered set of integrals. Thus an ordering step is needed. As for the supermatrix generation, this ordering is performed directly in place without the use of any intermediate bin sort file. The expense is slightly increased in CPU and I/O times (some buffers have to be transferred to and from disk more than once) but this loss is more than balanced by the gain in disk storage. The total time used for the construction of the supermatrix and the integral ordering represents, anyway, only a minor part of the total computation time. Sometimes not all integrals have to be ordered. When, for example, the ordered list is to be used in a subsequent RASSCF calculation, only those symmetry blocks of integrals are needed for which there are active orbitals. To be more specific, in a π-electron calculation, where all the sigma orbitals are inactive, ordered integrals are only needed for symmetry blocks with all four indices in π symmetry. The same saving is possible in the CI or MBPT type of programs, where only symmetry blocks corresponding to orbital actually used in the calculation have to be constructed.

All these features have been introduced to save disk space. Experience shows that the disk space available on many installations is often the limiting factor in large scale applications, not the CPU time. With the procedure outlined about it is necessary to store at least two two-electron files on disk: the supermatrix file and the list of ordered integrals. Considering the limitation in disk space, especially on workstations, an approach that works with only one two-electron integral list may be advantageous. In such an approach an ordered list of *packed* two-electron integrals is constructed directly by the integral generating program. This list is used both for Fock matrix construction and in AO to MO transformation step. Such a procedure is presently being tested for the next release of the MOLCAS software and shows a great promise. The somewhat increased CPU time is more than balanced by the decrease in I/O time and the large reduction in disk storage. The packed integral list is for moderately large calculations (around 100-150 basis functions) only one third of the original MOLECULE file in length. This is then the only AO integral list used in calculation.

2. The Two-Electron Transformation Program MOTRA

The two-electron transformation program (MOTRA) in MOLCAS is a straightforward application of the n^5 procedure. The actual transformation is carried through as a sequence of matrix multiplications performed by means of the ESSL subroutine library. MOTRA thus runs at the top speed of most vector computers (around 80 mflops on the IBM 3090VF). The efficiency of the program is further enhanced by the use of symmetry. Since MOLECULE produces integrals over a symmetry adapted basis set, the transformation can be performed symmetry block by symmetry block. The total computer time needed is therefore approximately proportional to the fifth power of the largest number of basis functions in any of the irreducible representations. This results in a considerable savings in CPU time even in the case of a two dimensional point group. As a result the transformation step will very seldom constitute the bottleneck in a calculation. On the contrary, it will in most applications only represent a minor part of the total computer time.

The transformation is performed in two steps. Two of the indices are first transformed. A resorting of these half transformed integrals is then performed before the remaining two indices are transformed. The sorting is done in core, if possible. Otherwise an intermediate disk file is generated. The advantages of the symmetry blocking is apparent also in this step, since the latter option normally is only necessary for the largest symmetry blocks of integrals, if at all. The final integrals in the molecular orbital basis are stored as an ordered set, one symmetry block at a time. The ordering of the integrals is a prerequisite for an effective use of them in subsequent programs, where they are normally used as matrices in vector operations.

3. The Self Consistent Field (SCF) Program

The MOLCAS system is not intended for large scale SCF applications. Such calculations are today preferably to be done using direct methods, where the basic integrals are computed when needed and never stored on disk. The major use of the SCF program in MOLCAS is to generate molecular orbitals for electron correlation calculations and, of course, for calibration of basis sets and estimation of correlation effects.

The program is used to generate closed shell wavefunctions. Open shell calculations (for the high spin case) can be performed using the RASSCF program with high efficiency. Low spin open shell SCF calculations are of little value and should be avoided. Instead, the different possible spin-couplings, which appear in such cases, should be allowed for, resulting in a true MCSCF wavefunction.

The SCF program has stable and rapid convergence properties and does not need any starting orbitals. The initial guess is obtained by diagonalizing the bare nucleus Hamiltonian (a standard procedure in most SCF programs). The convergence of the standard relaxation scheme, that constitutes the basic SCF cycle quite often exhibits poor convergence and even divergence. The program has two types of convergence accelerating schemes, dynamic damping[22] and an energy gradient minimization algorithm, very similar to the DIIS method by Pulay.[23] The dynamic damping yields very rapid convergence in the first few iterations, but the effect decreases closer to the converged result. The energy gradient minimization technique on the other hand, assumes a quadratic energy hypersurface and is therefore not very efficient far from convergence. On the other hand, it exhibits excellence convergence properties close to the converged solution, where the energy hypersurface is nearly quadratic. This technique is switched on by the program when the maximum change of a density matrix element is below a certain threshold. In this way the strength of both techniques are effectively used. Using these convergence accelerator methods a calculation normally converges even in pathological cases like multiply charged negative ions and dissociated system.

The program has a feature of specifying non-integer "occupation numbers" without any restrictions. The wavefunction will, of course, lose its physical meaning, when these occupation numbers differ from two and zero, but the technique can be used for generating starting orbitals for open-shell and multi-configurational calculations.

The virtual space obtained by diagonalizing the Fock matrix normally yields poor starting orbitals for MCSCF calculations. Improved starting orbitals can often be obtained by instead diagonalizing the bare nucleus Hamiltonian in the virtual orbital space. More compact orbitals are usually obtained in this way, which are more suited as starting orbitals.

4. The Restricted Active Space (RAS) SCF Program

The RASSCF program in MOLCAS performs multi-configurational SCF calculations using the restricted active space SCF method.[11,34] RASSCF is an extension of the Complete Active Space (CAS) approach, where the wavefunction is obtained as a full CI expansion in an active orbital space.[7-10] The RASSCF method is based on a partitioning of the occupied molecular orbitals into different groups:

1. *Inactive orbitals*

   These orbitals are doubly occupied in all configurations.

2. *Active orbitals*

   These orbitals are subdivided into three separate groups:

2a. *RAS1 orbitals*

   With n_1 orbitals the total occupation is allowed to vary between $2n_1$ and $2n_1 - x_1$ where x_1 is a small number. A maximum number of holes is thus allowed in this subspace.

2b. *RAS2 orbitals*

   n_2 orbitals. All possible occupations are allowed in this subspace (the old CAS space).

2c. *RAS3 orbitals*

   n_3 orbitals. A total occupation varying between 0 and x_3 is allowed, where x_3 is a small number.

RASSCF is thus an extension of the CASSCF method, such that a certain excitation level is allowed in orbitals "below" the CAS orbitals and in orbitals "above". Typical RAS wavefunctions are: singles, doubles, ... CI with a single reference state (x_1 and x_2 = 1, 2, ..., $n_2 \neq 0$); etc. The RASSCF program thus allows multi-configurational SCF calculations to be performed on a large variety of wavefunctions.

The major drawback of the CASSCF approach was the steep increase in the number of configuration state functions (CSF's) with the number of active orbitals. For example, distributing twelve electrons among twelve active orbitals yields 226512 singlet CSF's, when no reduction due to spatial symmetry is assumed. Adding one more active orbital increases the number to 736164. This strong dependence of the size of the MC expansion on the number of active orbitals makes it difficult to perform CASSCF calculations with more than around twelve active orbitals, at least in cases where the number of active electrons is about the same as the number of active orbitals. In most applications this limit does not pose a problem, but there are situations where it would be desirable to be able to increase the number of correlating orbitals in the MCSCF wavefunction.

Normally dynamic correlation effects are included in the calculation in a subsequent step, where a multi-reference (MR) CI calculation is performed using selected CFS's from the CASSCF wavefunction as reference configurations. The CASSCF calculation is then restricted to include only static correlation effects, and is used to determine the orbitals.

However, cases exist where the shape of the strongly occupied orbitals depend on dynamical correlation effects. A typical example of this phenomenon is found for negative ions, which in many cases are only bound when a large fraction of the dynamical correlation is taken into account. Orbital optimization at a correlated level is then the only consistent approach. The polarity in a chemical bond is often sensitive to dynamic correlation effects. The calculation of e.g. a dipole moment may in such cases pose difficulties if the orbitals have been obtained at a too low level of approximation. A third example relates to the calculation of wavefunctions for exited states of molecules. The shape of orbitals occupied in the excited state may in some cases depend strongly on dynamic correlation effects. A drastic example is given by the V state in the ethene molecule, which has the dominant electronic configuration $(\pi_u\pi_g)_S$. The π_g orbital is at the HF level of approximation very diffuse. It contracts slowly when more and more correlation is included.[25]

The RAS CI method has previously been implemented using a determinant based CI expansion.[24] The MOLCAS implementation is in terms of spin adapted CSF's. The CI module has been designed to be able to handle large Full CI (FCI) problems efficiently. While it has been shown that determinant expansions offer the most readily vectorizable implementation, the increase in the expansion length, and the fact that the wavefunction is not inherently a spin eigenstate, remains a disadvantage. The alternative is to look for improvements in the Graphical Unitary Group Approach (GUGA),[26] which allow the CI expansion to increase without increasing the list of coupling coefficients beyond feasible limits. A method has been found that achieves this goal. The basic idea is to split the graph into an upper and a lower part and compute explicit coupling coefficients for the two halves separately (split-GUGA).[11]

A major problem with a spin-adapted basis, in the context of FCI, is the generation, storage and use of a large number of coupling coefficients. In particular, since the number of two-electron coupling coefficients increases very rapidly, it soon became obvious that the explicit storage of them would seriously limit the size of such calculations. To circumvent this limitation the factorization of the two-electron coupling coefficients into products of one-electron coupling coefficients is a first step. This can be very efficiently implemented on a vector computer, but still places a realistic limit of around 50,000 CSF's to the size of the FCI due to storage requirements. A further reduction in the size of the external storage required can be achieved by a factorization of the one-electron coupling coefficients. One way to do this is a straight forward dissection of the segment product form of the coupling coefficients, which is already at the heart of the GUGA method. The idea is to keep sorted, not the entire set of finished products, but a much smaller set of half-finished products to be combined when needed. This approach has already been used for several years in our transition density programs.[15] Combined with a new storage scheme for the CI coefficients, these half-finished products, or new set of coupling coefficients, makes it possible to implement a directly vectorized sigma vector formation in terms of row or column DAXPY operations on a matrix.

The orbital optimization in the RASSCF program has been implemented using an improved version of the approximate super-CI method, which had been used earlier for CASSCF calculations.[7,9] This method has in many applications shown good convergence properties, although the asymptotic convergence is only first-order. The virtue of the method lies in its computational efficiency, and the fact that the computational effort is almost independent of the number of inactive orbitals. Only a limited set of transformed two-electron integrals is needed, namely those with three of the four indices corresponding to active orbitals. This method is therefore especially suited to treat large systems with many electrons and large basis sets. The first order convergence can be improved by the use of quasi-Newton update methods.[27] Such a scheme has been implemented here and results in greatly improved convergence properties in most applications. The update is made on the inverted approximate Hessian. The most important effect of the update is to reintroduce the CI coupling into the orbital optimization step.

Convergence of the orbital optimization procedure is normally good for CASSCF type wavefunctions, but problems can occur in calculations on excited states, especially when several states are close in energy. Further, convergence can be slow in some true RAS calculations due to strong CI coupling effects. It should also be noted that the program has not been optimized for SDCI calculations with a large number of orbitals in RAS1 and RAS3. Such calculations should therefore be avoided with the present version of the program. The next release of MOLCAS will include a RASSCF program, which is much improved in this respect.

The largest CASSCF calculation hitherto performed with the program contained more than 700000 CSF's. Calculations of this size are, however, time consuming.

The table presented below gives CI iteration times for some typical CASSCF calculations run on an IBM 3090-170S/VF. Typically 2-5 CI iterations are used in a RASSCF macroiteration.

Number of CSF's	CPU time per CI iteration (sec)
28848	15
92406	60
268800	227
716523	878

The CAS(RAS)SCF method represents in many applications an intermediate step, which is followed by a multi-reference CI (or ACPF) calculation, where the most important of the configurations constituting the RASSCF wavefunction, are used as reference configurations. In this two level treatment of electron correlation the static correlation effects are treated at the RASSCF level, while dynamic correlation (including cluster corrections) are handled in the subsequent MRCI step. Typical cases, where such a partitioning of the electron correlation is natural , is in spectroscopic applications (excited states) and in studies of chemical reactions (transition states). It has been shown in s number of applications that such an approach, if carefully performed, gives results very close to full CI within the given basis set.[14] An analysis of a number of different types of CASSCF calculations can be found in Ref. 10 where the virtues as well as the problems inherent to the MCSCF approach are illustrated.

5. The RAS State Interaction (RASSI) Program

The RASSI (RAS State Interaction) program forms matrix elements of the Hamiltonian and other operators in a wavefunction basis, which consists of individually optimized CI expansions from the RASSCF program. Also it solves the Schrödinger equation within the space of these wavefunctions. If only matrix elements of some one-electron operator(s), such as the dipole transition moments, are required, the calculation of Hamiltonian matrix elements and the transformation to the eigenbasis of this matrix can be skipped.

An important usage of the RASSI program is for the calculation of electronic transition moments. When the states involved have the same symmetry, it is imperative that the transformation to the eigenstate basis is performed. The reason is that the individually optimized RASSCF states are interacting and non-orthogonal, resulting in an error in the computed transition matrix element, which is the difference in electric dipole moment times the overlap of any two states involved. For excited states, this overlap is often in the order of 10%.

Association between individually optimized states and the exact electronic eigenstates is often not clear, when the calculation involves several excited states. The reason is that the different sates each uses a different set of orbitals. The state interaction calculation gives an unambiguous set of non-interacting and orthonormal eigenstates to the projected Schrödinger equation, and also the overlaps between the original RASSCF wavefunctions and the eigenstates. The latter is a very efficient diagnostic, since it describes the RASSCF states in terms of one single wavefunction basis set.

To make the last point clear, assume the following situation. Three RASSCF calculations have been performed, one where the lowest state is optimized, one for the first excited state, and one for the second excited state, all in the same symmetry. The active orbitals are fairly much mixed around, so a simple inspection of the CI coefficients is insufficient for comparing the states. Assume also that the three lowest CI roots have been computed for each set of orbitals. It may now happen that the second root of each calculation is a fair approximation to the exact 2nd eigenstate and the same with the 3rd root, or possibly that the order gets interchanged in one or two of the calculations. In such a case, a RASSI calculation, using the nine roots as the wavefunction basis, will give three improved solutions close to the original ones, and of course, six more, that can be considered to be the 'removed garbage'. The overlap integrals will confirm that each of the input states consist mainly of one particular out of the three lowest eigenstates. This situation is the one normally assumed, if no further information is available.

However, experience shows that another type of situation may also occur. The RASSI calculation may well result in 4 lowest roots that overlap strongly with the three original RASSCF states. This may well change the assignments and may also give valuable information about the importance of some state. The natural orbitals of the eigenstates may show that the active space used in the RASSCF calculation was in some way inappropriate. The latter situation occurs when the active orbitals of the different calculations do not span approximately the same space.

Another troublesome situation is also solved by the RASSI method. The analysis of the original states in terms of RASSI eigenstates may reveal, that the three optimized RASSCF states consist of mainly two low energy RASSI eigenstates. This shows immediately the presence of spurious local minima in the original calculation. Also in this case, the natural orbitals will probably offer a clue to the solution of the problem.

A further application is the preparation of a suitable orbital basis for subsequent MRCI calculations. Note that such an application allows also the use of badly converged RASSCF wave functions, or of RASSCF wavefunctions containing multiple minima solutions close to a common exact eigenstate. In effect, the RASSI program cleans up the situation by decreasing the errors due to bad convergence (pushing the error to the 'garbage part' of the spectrum). This requires, of course, that the set of input states (nine in the above example) provides flexibility enough to remove at least a major part of the error. As one would expect, this is usually the case: the erratic non-convergent, or the too slowly convergent, error mode is to a large extent spanned by the few lowest RASSCF wavefunctions.

Finally, there are situations where there is no problem to obtain adiabatic RASSCF solutions, but where it is still imperative to use RASSI natural orbitals in a subsequent MRCI. Consider the case of a chemical bond involving a transition metal atom. It is not unusual that such a bond involves two or more electronic states of the atom. These states may interact strongly, at least within some range of interatomic distances. Here, an MCSCF solution, such as RASSCF, will have at least two very different solutions, one associated with each configuration of the transition metal atom. Using one set of orbitals, one electronic state has a reasonably described potential energy surface, while other states gets pushed far up in energy. With a second set of orbitals another electronic state gets correctly described. In no calculation with a single set of orbitals do we obtain the avoided crossings, describing the switch from one diabatic state to the other. The only way to accomplish this is via a RASSI calculation. In such calculations it is also possible to introduce an empirical correction, by shifting the diagonal elements of the RASSI Hamiltonian such that the atomic spectrum in the separated atom limit is correctly described.

A matrix element of the RASSI Hamiltonian is constructed from a non-unitary rotation of the involved orbital sets to a bi-orthonormal basis. The new CI coefficients are obtained from a series of one-electron transformations. The first and second order transition density matrices can then easily be obtained. The implementation of this step of the calculation is very effective. Actually the major part of the computer time is spent in the transformation of the two-electrom integrals to the bi-orthonormal basis, and the formation of the Fock matrix for the inactive orbital pairs. The reader is referred to the original articles for more details on the method.[15-16] In practice calculations can be performed with up to around 20 terms in the RASSCF wavefunction basis. Also very long RAS-CI expansions can without problems be used as the wavefunction basis. Practical applications have been performed using wavefunctions comprising up to 50,000 CSF's and this is certainly not an upper limit.

6. The Configuration Interaction Programs

MOLCAS contains options for Coupled Pair Functional (CPF[5] or MCPF[6]), multi-reference SDCI,[12] and Average Coupled Functional (ACPF) calculations.[13] These programs have originally been written by members of the quantum chemistry group in Stockhom[28] and have been adapted to run on the IBM 3090 system by the present authors. They are all based on the direct CI method[29] with the vector coupling coefficients generated using the unitary group approach.[26] The

CPF and MCPF wavefunctions are generated from a single, open or closed shell reference functions, while MR-SDCI and ACPF wavefunctions use several reference configurations. The wavefunction is in all cases obtained as a linear combination of all CSF's that can be generated by single, or single and double replacements in the reference configurations. For a more detailed discussion of these methods, the reader is referred to the literature.

CPF and MCPF calculations are started by generating a set of molecular orbitals by using the SCF program (closed shell case) or by the RASSCF program (open shell case). The CPF method is probably the most straightforward method available today for electron correlation calculations in cases where the wavefunction is dominated by a single configuration. A number of recent applications has shown the strength of the method. It has, for example, recently been used to compute the ring-ring distance in the ferrocene molecule with high accuracy. Several earlier attempts to solve this problem with other methods have failed.[30]

The first step in an MRCI calculation is the generation of molecular orbitals and reference configurations by means of the RASSCF programs. In most cases the near degeneracy effects can be incorporated through CAS type wavefunctions, but in some cases it might be advantageous to include dynamic correlation (through the RAS concept) already in this step of the calculation. The selection of reference configurations has to be done with great care, in order to achieve a balanced treatment of dynamical correlation effects. Normally one is interested in a relative energy (the dissociation energy, an excitation energy, etc.). Consider, for example, MRCI calculations, for two different electronic states. One interesting quantity is then the excitation energy. In order to obtain an accurate value, one must balance the calculation such that the same fraction of the correlation energy is obtained for both states. To do this is not trivial. However, one method that seems to work well, is to choose the reference configurations, such that the total weight of the reference configurations is the same in the two final wavefunctions. A detailed analysis of such a procedure has been given in a recent MRCI study of the electron affinity of the methylene radical.[31] Also, the larger the reference weight is, the closer is the MRCI calculation to the basis set limit, the full CI. It is not advisable to do calculations where the reference weight is smaller than around 90%.

If the resulting MRCI wavefunction yields non-refernce configurations with large weight ($< 1\%$), this is an indication that the reference set is badly chosen, or that the active orbital set in the proceeding RAS(CAS)SCF calculation is not appropriate. All these considerations put a limit to which systems one can treat with the present program and hope to obtain accurate results for the correlation energy. The maximum size for practical applications depends, of course, on the computer used, especially on available central storage and disk capacity. A realistic estimate for the present version of the code is 2-3 million CSF's. A new MRCI program is presently under implementation and will be included in MOLCAS-2. This program will be able to handle about ten times as many CSF's.

7. Many Body Perturbation Theory (MBPT) Programs

Included in MOLCAS is a set of MBPT programs devised to perform complete 4th order MBPT calculations of the correlation energy including single, double, triple and quadruple excitations. The program can also perform linear coupled cluster calculations with double excitations only (LCCD). The MBPT code has been generated by a modification of the MBPT programs developed by the Bratislav Group of Quantum Chemistry.[32] For references concerning the development of those programs see Ref. 32-34. A general description of the MBPT programming philosophy and some more references can be found in Ref. 3-4.

Limited order many body perturbation theory remains a valid approach to the electron correlation problem only in such cases, where the wavefunction is dominated by a single (Hartree-Fock) configuration. The present implementation is further restricted to the closed shell case. The calculation of the correlation energy is for such cases not of much interest in itself. The major contribution of the MBPT method therefore appears to be in the area of molecular properties, i.e. molecular equilibrium geometries, force fields, electric moments and polarizabilities, and intermolecular forces. Some molecular properties can conveniently be computed using the finite field approach and MOLCAS provides the possibility for such calculations.

When limiting the MBPT calculations to low order, as is always done, a very careful consideration of the appropriateness of the single-reference state assumption is advisable. The violation of this assumption may considerably affect the behavior of the whole series and can make higher-order terms quite important. Hence, it is advisable that in all those cases where the experience is not good enough, the single determinant description should be checked against some small MCSCF or CI calculation. If the weight of the leading RHF determinant in the resulting CI wavefunction is less than about 90%, the low-order MBPT calculations are not recommended. It means that one should not draw any subtle conclusions from the low-order MBPT data for such systems.

One should not be very optimistic with regard to the use of the finite-order single-reference MBPT methods in very accurate calculations of atomic and molecular correlation energies. The convergence of the MBPT series is too slow for that purpose. In order to achieve the final accuracy better than 10^{-3} au even the fifth-order expansion may not be sufficient. One can hope to expect, however, that the differences in the correlation energy arising from a small perturbation, e.g., small geometry distortions or weak external fields, should either converge much faster or should become negligible in all but a few lowest-order terms. Hence the major field of application of the (low) finite-order MBPT procedures should be the study of the electron correlation contributions to atomic and molecular properties resulting from the perturbation of the given system. Such applications can be expected to be more successful and more reliable than the attempts to calculate the best correlation energy for many-electron systems. Closely related and conceptually similar are the foundations of the MBPT studies of the electron correlation contributions to the energy of weak intermolecular interactions.

For basis sets with more than 100 functions the complete fourth-order MBPT calculations become time consuming and require a lot of disk space. Much less costly and sufficiently useful might then be some incomplete 4th order results, e.g. those limited to single, double, and quadruple excitations. In extreme cases one can limit the MBPT calculation to second order only. A number of applications found in the recent literature shows that even such a simple approach may in many cases give valuable information. Such calculations are much simpler to perform than higher order MBPT. MOLCAS includes a special unit, MBPT2, for second order perturbation theory calculations. It is devised to perform second order MBPT calculations of the correlation energy for closed shell SCF reference functions. In comparison with the complete 4th order MBPT program, the present limited version is based on a full use of molecular symmetry and uses a built in incomplete 4-index transformation: only the (vv/oo) and (vo/vo) integrals are calculated. This code is very fast and can be used with the largest basis sets that MOLCAS can handle.

8. The Molecular Properties Program

The natural orbitals and occupation numbers produced by all wavefunction generators available in MOLCAS can be used as input to the property program, which computes a number of first order one-electron properties. This program also produces integrals needed for finite-field calculations and for the calculations of transition moment integrals in RASSI. The following molecular properties can be obtained:

1. Electric dipole moments
2. Electric quadruple moments
3. Electric fields at the position of the nuclei
4. Electric field gradients at the position of the nuclei
5. Diamagnetic shielding
6. Relativistic corrections (Darwin and mass-velocity terms).

One-electron matrices can be produced for finite field calculations using the following (external) perturbation operators:

1. The dipole moment operator
2. The electric field operator
3. The quadrupole moment operator
4. The electric gradient operator.

Preprocessing of the one-electron integral file for transition property calculations with the RASSI program can be performed for the following operators:

1. The dipole moment operator
2. The velocity (gradient) operator
3. The angular momentum operator.

C. A Timing Example, The Pyrimidine Molecule

In this section, a CASSCF and MRCI calculation of several excited states of the pyrimidine molecule is used to give some timing examples for the MOLCAS system. All calculations have been performed with the IBM 3090 170S/VF system. To be able to make, at least approximate comparisons to other computers, it might be enough to mention that the maximum speed obtained for matrix multiplication using the ESSL library is around 80 Mflops. Matrix multiplication is one of the key operations in many of the MOLCAS programs and should therefore serve well for comparing different computers. It should also be mentioned that MOLCAS has not yet been implemented to run in parallel mode.

The pyrimidine calculations were made at one geometry only (the experimental) and used a basis set of the ANO type: for carbon, and nitrogen a primitive 12s8p3d basis set was contracted to 3s3p1d and for hydrogen 6s was contracted to 2s. This gives in total 348 primitive gaussian functions contracted to 116. All calculations were performed in C_{2v} symmetry and the basis set splits into the four irreducible representations a_1, b_2, b_1, and a_2 as 50,36,18,12, where the two last representations correspond to π-orbitals.

The MOLECULE program generates 6,040,235 two-electron integrals over the contracted functions. The total CPU time used in this step was 40.2 minutes. These integrals were ordered by the program INTORD resulting in a list of 12,073,632 integrals. The CPU time used was 1.2 minutes. The reason for the much larger number of ordered integrals is that the list is squared in order to be able to store the integrals as a set of matrices, which is the natural input for the AO to MO transformation program. The disk space required is, however, not much larger, since the ordered integrals are stored without indices. The program SUPMAT was used to produce 2,376,007 supermatrix elements. The computer time used in this step was 1.1 minutes. The transformation of the AO two-electron integrals to MO basis requires 1.5 minutes of CPU time. This is a full transformation. In the CASSCF iterations (see below) a limited transformation is performed, involving only the π-orbitals. The CPU time used for such a transformation is of course negligible.

Next a series of CASSCF calculations were performed. All σ-orbitals were in these calculations considered to be inactive. Twelve active orbitals of π symmetry was used to correlate six π electrons. They split in C_{2v} symmetry as 8,4 in the b_1, and a_2 representations, respectively. The total number of CSF's of $^1A_1$ symmetry was 7954. The corresponding number of $^1B_2$ was 7776. The π-π^* excited singlet states belong to these two symmetries. CASSCF calculations were performed for several of these states. Here we shall only discuss the timings for one of these calculations, an optimization with respect to the average of the four lowest $^1A_1$ states. This calculation converged in 18 macroiterations, starting with orbitals obtained in a previous SCF calculation on the ground state. The energy was then converged to 10^{-8} au and the energy gradient to 10^{-5} au. In each macroiteration between 3 and 8 CI iterations were needed to converge the four lowest roots of the secular problem. The CPU time used is completely determined by the CI step and is therefore proportional to the number of required CI roots and the number of iterations used to converge them. The CPU time used in a macroiteration therefore varies, between 2 and 5 minutes, and the total time used was 72 minutes. Obviously a calculation on the ground state would require approximately one fourth of this time, a calculation on the second root half the time, etc. An average calculation for the third and fourth roots only would, however, require about the same time as the above calculation, since also in this case has four roots to be extracted in the CI step.

The orbitals obtained in the CASSCF calculations were used to perform MRCI calculations for the ground state and a number of excited states. Here we shall report some of the data from a series of calculations on the first excited state of $^1B_2$ symmetry. The calculations were made to check the convergence of the CI energy with the number of reference states. The orbitals were taken from a CASSCF calculation (using 6 active π-orbitals) on the same state. All single and double excitations out of a given set of reference configurations were included in the MRCI wavefunction, however, with one restriction: only one hole was allowed among the inactive σ-orbitals. Thus the calculation accounts for electron correlation effects between the π electrons and also between π and σ electrons, but not between the σ electrons. This limitation was used in order to keep the size of the MRCI expansion within tractable limits, and since it was believed that the last type of correlation effects would be rather constant for different π excited states. Some data from the calculations on the 1^1B_2 state are presented in the table below.

Calculation	1	2	3	4	5
Number of ref. CSF's	6	7	8	9	10
Weight of the ref. space	.865	.892	.915	.936	.955
Number of CSF's	641698	742829	806561	892318	940276
Total energies +262 au					
MRCI	-.854271	-.859245	-.861110	-.864782	-.865730
MRCI+Q	-.915127	-.917084	-.913802	-.910939	-.912445
Excitation energies (eV)					
MRCI	5.46	5.32	5.48	5.46	5.51
MRCI+Q	4.87	4.90	5.12	5.20	5.16
CPU time (min)	224	256	295	301	320

The weight of the reference space gives the fraction of the CASSCF wavefunction used as the reference CSF's. Notice that the total number of CSF's increases only slowly with the number of reference states and also that the CPU time is almost proportional to this number. The table shows the convergence of the total energy and the energy relative to the ground state as a function of the weight of the reference space. The excitation energies have been computed by comparing with ground state energies obtained from MRCI wavefunctions with (almost) the same weight of the reference space. The cluster corrected (MRCI+Q) energies are obtained by means of a multi-configurational analogue to the Davidson correction. The energies shown in the table converge slowly with the number of reference configurations. The convergence is, however, considerably faster for the excitation energy, which has converged to within 0.1 eV already in the third calculation, where the weight of the reference space, compared to the full CAS reference space is 91.5%. This is true both for the MRCI energy and the cluster corrected energy. The experimental value for the 1^1B_2 excitation energy is 5.12 eV.

Similar calculations as the one illustrated above has been performed for all the singly excited valence states of pyrimidine and also for the lowest n-π^* excited states. For some of the higher excited states the size of the MRCI wavefunction was about 2×10^6 . These calculations also include calculations of the transition moments (using the RASSI method), and electric properties of all the excited states. Detailed results will be presented elsewhere.[36]

D. Concluding Remarks

The MOLCAS quantum chemistry software provides a large variety of programs for electronic structure calculations. It has a basis set library for first and second row atoms, which allows accurate work to be done in an effective way. This library will soon be extended up to Kr. MOLCAS provides a balanced treatment of systems where near-degeneracy effects are of importance. The spirit of MOLCAS is to include the most important terms in the wavefunction already at the zeroth level of approximation. Thus Hartree-Fock wavefunctions are only used as the starting point for electron correlation calculations in such cases, where a single configuration dominates. In other cases a multi-configurational approach is recommended. The area of application is therefore large and includes spectroscopy, chemical reactions, transition metal chemistry, etc.

However, the flexibility of the MOLCAS software also prevents it from being a black box. To use MOLCAS requires a certain knowledge about quantum chemical calculations in general and an insight into the chemistry of the problem being investigated. In our opinion these are rather natural requirements for anyone involved in any type of quantum chemical research.

A number of important programs are still lacking in this first release of the MOLCAS software. Most important of them is probably programs for energy gradient calculations at different levels of theory. Such programs will be implemented into the next release, which will be based on a new integral code with largely improved efficiency.[37] This release will also include the possibility to perform full CI calculations with wavefunctions comprising several hundred million determinants[38] and improved codes for RASSCF and MRCI calculations. In general a factor of more than two is expected for the computational efficiency of the next generation of programs in the MOLCAS software.

References

1. Appropriate references to the authors of the different codes in MOLCAS will be given in connection with the description of the different programs.
2. The following persons have been involved in the construction of the MOLCAS software: Gunnar Karlström, Per-Åke Malmqvist, Björn Roos, Andrzej Sadlej (University of Lund), and Per-Olof Widmark (IBM, Sweden).
3. P. Cársky and M. Urban, *Ab Initio Calculations. Methods and Applications in Chemistry*. Lecture Notes in Chemistry, Vol 16, Springer, Berlin (1980).
4. M. Urban, I. Cernusak, V. Kellö, and J. Noga, in: *Methods in Computational Chemistry,* (ed.S. Wilson) Vol. 1, p.117, Plenum Press, New York (1987).
5. R. Ahlrichs, P. Scharf, and C. Erhardt, J. Chem. Phys. **82**, 890 (1985).
6. D. P. Chong and S. R. Langhoff, J. Chem. Phys. **84**, 5606 (1986).
7. B. O. Roos, P. R. Taylor, and P. E. M. Siegbahn, Chem. Phys. **48**, 157 (1980).
8. P. E. M. Siegbahn, J. Almlöf, J. Heiberg, and B. O. Roos, J. Chem. Phys. **74**, 2384 (1981).
9. B. O. Roos, Int. J. Quantum Chem. Symp. **14**, 175, (1980).
10. For a recent review of the CASSCF method see: B. O. Roos in *Ab Initio Methods in Quantum Chemistry II* (ed. K. P. Lawley), John Wiley and sons, Ltd., New York (1987).
11. P.-Å. Malmqvist, A. Rendell, and B. O. Roos, J. Phys. Chem. **00**, 0000 (1990).
12. P. E. M. Siegbahn, J. Chem. Phys. **72**, 1647, (1980).
13. R. J. Gdanitz and R. Ahlrichs, Chem. Phys. Letters, **143**, 413 (1988).
14. C. W. Bauschlicher, Jr., S. R. Langhoff, and P. R. Taylor, Adv. Chem. Phys. (in press).
15. P.-Å. Malmqvist, Int. J. Quantum Chem. **30**, 479 (1986).
16. P.-Å. Malmqvist and B. O. Roos, Chem. Phys. Letters, **155**, 189 (1989).
17. J. Almlöf and P. R. Taylor, J. Chem. Phys. **86**, 4070, (1986).
18. P.-O.Widmark and B. O. Roos, Theoret. Chim. Acta **00** 0000 (1990).
19. J. Almlöf, T. Helgaker, and P. R. Taylor, J. Phys. Chem. **92**, 3029 (1988).
20. J. Almlöf, USIP Rep 72-09 (September 1972); USIP Rep 74-29 (December 1974), Institute of Physics, University of Stockholm, Sweden.
21. J. Almlöf, *A vectorized MOLECULE code.* Final report on the project supported by the NASA Ames grant No. NCC2-158 (Dec. 1984).
22. G. Karlström. Chem. Phys. Letters **67**, 348 (1979).

23. P. Pulay, Chem. Phys. Letters, **73**, 393 (1980); J. Comp. Chem. **3**, 556 (1982).
24. J. Olsen, B. O. Roos, P. Jϕrgensen, and H. J. Aa. Jensen, J. Chem. Phys. **89**, 2185 (1988).
25. R. Lindh and B. O. Roos, Int. J. Quantum Chem. **35**, 813 (1989).
26. I. Shavitt. Int. J. Quantum Chem. Symp. **11**, 133 (1977); ibid. **12**, 5 (1978).
27. J. Olsen, D. L. Yeager and P. Jϕrgensen, Adv. Chem. Phys. **54**, 1 (1983).
28. The main authors are P.E.M. Siegbahn and M.R. Blomberg (Department of Theoretical Physics, University of Stockholm, Vanadisvägen 9, S-113 46 Stockholm, Sweden).
29. B. O. Roos, Chem. Phys. Letters **15**, 153 (1972).
30. J. Almlöf, (private communication).
31. T. Noro and M. Yoshimine, J. Chem. Phys. **91**, 3012 (1989).
32. The main authors are: M. Urban, V. Kellö, J. Noga, and I. Cernusak (Department of Physical Chemistry, Comenius University, 842 15 Bratislava, Czechoslovakia).
33. M. Urban, I. Hubac, V. Kellö, and J. Noga, J. Chem. Phys. **72**, 3378 (1980).
34. V. Kellö and M. Urban, Int. J. Quantum Chemistry, **18**, 1431 (1980).
35. J. Noga, Comput. Phys. Commun. **29**, 117 (1983).
36. P.-Å. Malmqvist, A. Rendell, and B. O. Roos, (to be published).
37. R. Lindh, SEWARD, a program for one- and two-electron integral calculations using generally contracted spherical guaussian type basis functions.
38. J. Olsen, (to be published).

Chapter 10.

MELD: A *M*any *El*ectron *D*escription

E.R. Davidson

Department of Chemistry
Indiana University
Bloomington, IN 47405, U.S.A.

1. Introduction

MELD is a molecular configuration interaction program developed primarily for the study of electronic excitation spectra and molecular properties. It is a "classical" brute-force program which makes no use of unitary groups, direct methods, or many-body perturbation theory. The advantage retained, over more modern methods, is the flexibility to describe a wide variety of open-shell states.

MELD is truly a melding of many different program segments put together originally by L. McMurchie, S. Elbert, S. Langhoff, and E. Davidson for a CDC computer. The original program was written in FORTRAN IV with many machine language subroutines. This was converted to run on a VAX by D. Rawlings. At the same time the program was changed to FORTRAN 77 and all machine language coding was removed along with most machine dependency. Finally, David Feller worked extensively on all parts of the program. He documented the code and wrote a user's guide. He also worked on data compression methods to reduce the size of the integral and CI matrix files.

Rather than develop a heavily overlayed program, MELD was developed as a sequence of programs which communicate with each other through disk files. Flow through the program steps is controlled by JCL. This allows great flexibility in the type of calculation done and in breaking a long job into several shorter runs. The principal program segments are associated with formation of integrals over Gaussian basis functions, choosing molecular orbital coefficients by a spin-restricted SCF calculation, formation and diagonalization of a configuration interaction matrix, and calculation of a density matrix and properties. All input is free-format key-word driven and is read and checked before any calculations are done. The program makes explicit use of symmetry blocking to the irreducible representations of subgroups of D_{2h}. The theoretical aspects of MELD have been reported and documented in various scientific journals.[1-14]

2. Integrals[1,2]

An unnormalized Gaussian function on center A is written as

$$\phi(n, l, m, \alpha_A, A) = x_A^n y_A^l z_A^m \exp(-\alpha_A r_A^2) \tag{1}$$

where (x_A, y_A, z_A) are the components of the vector $\mathbf{r}_A = \mathbf{r} - \mathbf{A}$ with norm r_A. The normalization constant for this function is

$$N_{nlm}(\alpha_A) = N_n(\alpha_A) N_l(\alpha_A) N_m(\alpha_A) \tag{2}$$

where

$$N_k(\alpha) = (2\alpha/\pi)^{1/4}(4\alpha)^{k/2}[(2k-1)!!]^{-1/2} \tag{3}$$

Such functions are referred to as $s, p, d, \ldots$ when $L = n + l + m$ is 0,1,2,...., respectively. Contracted basis functions are formed from the ϕ's as

$$f_{nlmA} = \sum_{\alpha_A} C^L(\alpha_A) N_{nlm}(\alpha) \phi(n, l, m, \alpha_A, A) \tag{4}$$

where $C^L(\alpha_A)$ is independent of n,l,m (for fixed L).

Efficient computation of integrals requires that f_{nlmA} which involve the same α's and the same nucleus be treated as sets. For maximum efficiency these sets should be large enough that calculation of the auxiliary functions common to a block of integrals becomes insignificant.

For the sake of simplicity, the formulas in this chapter will be given only for integrals over ϕ's rather than f's, in order to avoid writing explicitly the sums over α. To this end it is convenient to define the charge distribution Ω_{IJ} of two functions $\phi_I(n,l,m, \alpha_A, A)$ and $\phi_J(n', l', m', \alpha_B, B)$ as

$$\Omega = \phi_I \phi_J = x_A^n x_B^{n'} y_A^l y_B^{l'} z_A^m z_B^{m'} \exp[-(\alpha_A r_A^2 + \alpha_B r_B^2)] \tag{5}$$

The key to the utility of Gaussians is the well-known identity which transforms the above exponential to a single exponential about a center **P** on the line segment AB:

$$\exp[-(\alpha_A r_A^2 + \alpha_B r_B^2)] = E_{IJ} \exp(-\alpha_p r_p^2) \tag{6}$$

where

$$\mathbf{P} = (\alpha_A \mathbf{A} + \alpha_B \mathbf{B})/(\alpha_A + \alpha_B) \tag{7}$$

$$\alpha_p = \alpha_A + \alpha_B \tag{8}$$

and

$$E_{IJ} = \exp\{-\alpha_A\alpha_B(\alpha_A + \alpha_B)^{-1} \mid \mathbf{A} - \mathbf{B} \mid^2\} \tag{9}$$

Since molecular calculations usually require absolute rather than relative accuracy, all integrals involving Ω_{IJ} can be neglected if the constant E_{IJ} is sufficiently small.

The products $x_A^n x_B^{n'}$, $y_A^l y_B^{l'}$, and $z_A^m z_B^{m'}$ could be converted into polynomials in x_p, y_p, and z_p using relations like

$$x_A = x_p + PA_x \tag{10}$$

where

$$x_A = x - A_x \tag{11}$$

$$x_p = x - P_x \tag{12}$$

and

$$PA_x = P_x - A_x. \tag{13}$$

It is more convenient in what follows, however, to define

$$\Lambda_j(x_p; \alpha_p) \exp(-\alpha_p x_p^2) = (\partial/\partial P_x)^j \exp(-\alpha_p x_p^2), \tag{14}$$

from which it follows that Λ_j is related to the Hermite polynomial H_j by

$$\Lambda_j(x_p; \alpha_p) = \alpha_p^{j/2} H_j(\alpha_p^{1/2} x_p). \tag{15}$$

The utility of the Λ's is obvious—they will allow the charge distribution to be written as a sum of derivatives with respect to the coordinates of **P** and the derivatives can be taken outside of any integral over electronic coordinates.

Now let us find the coefficients for expanding $x_A^n x_B^{n'}$ in the Λ's:

$$x_A^n x_B^{n'} = \sum_{n=0}^{n+n'} d_N^{nn'} \Lambda_N(x_p; \alpha_p). \tag{16}$$

The recursion relation for the Hermite polynomials is

$$\xi H_N(\xi) = NH_{N-1}(\xi) + 1/2H_{N+1}(\xi). \tag{17}$$

Consequently

$$x_A\Lambda_N(x_p; \alpha_p) = N\Lambda_{N-1} + PA_x\Lambda_N + 1/2\Lambda_{N+1}/\alpha_p. \tag{18}$$

The recursion relations on the $d_N^{nn'}$ are then easily seen to be

$$d_N^{n+1,n'} = (2\alpha_p)^{-1} d_{N-1}^{nn'} + PA_x d_N^{nn'} + (N+1) d_{N+1'}^{nn'} \tag{19}$$

$$d_N^{n,n'+1} = (2\alpha_p)^{-1} d_{N-1}^{nn'} + PB_x d_N^{nn'} + (N+1) d_{n+1'}^{nn'} \tag{20}$$

where

$$d_0^{00} = 1. \tag{21}$$

Similarly we can write

$$y_A^l y_B^{l'} = \sum_{L=0}^{l+l'} e_L^{ll'} \Lambda_L(y_p; \alpha_p) \tag{22}$$

and

$$z_A^m z_B^{m'} = \sum_{M=0}^{m+m'} f_M^{mm'} \Lambda_M(z_p; \alpha_p) \tag{23}$$

so that

$$\Omega_{IJ} = E_{IJ} \sum_{NLM} d_N^{nn'} e_L^{ll'} f_M^{mm'} \Lambda_N(x_p) \Lambda_L(y_p) \Lambda_M(z_p) \exp(-\alpha_p r_p^2). \tag{24}$$

It is often convenient to write Ω_{IJ} as

$$\Omega_{IJ} = \sum_k D_k \Lambda_{N_k}(x_p) \Lambda_{L_k}(y_p) \Lambda_{M_k}(z_p) \exp(-\alpha_p r_p^2), \tag{25}$$

where

$$D_k = E_{IJ} d_{N_k}^{nn'} e_{L_k}^{ll'} f_{M_k}^{mm'}. \tag{26}$$

In these equations k indexes all (N,L,M) combinations for which D is nonzero and the charge distribution is specified by a list of the nonzero D_k and the corresponding (N_k, L_k, M_k). In practice the $C^L(\alpha_A)C^L(\alpha_B)$ and the normalization are also incorporated into the D_k.

Some integrals which must be evaluated involve derivatives of the basis functions. Consequently, it is convenient to define the additional charge distributions

$$G_{IJ} = \phi_I \nabla \phi_J \tag{27}$$

$$T_{IJ} = \nabla \phi_I \cdot \nabla \phi_J \tag{28}$$

and

$$Q_{IJ} = (\nabla \phi_I) \times (\nabla \phi_J). \tag{29}$$

The x component of G_{IJ} is easily obtained from

$$\partial\phi(n', l', m', \alpha_B, B)/\partial x = n'\phi(n' - 1, l', m', \alpha_B, B) - 2\alpha_B\phi(n' + 1, l', m', \alpha_B, B) \tag{30}$$

so that

$$G_{IJ}^{(x)} = \sum_k g_k^{(x)} \Lambda_{N_k} \Lambda_{L_k} \Lambda_{M_k} \exp(-\alpha_p r_p^2) \tag{31}$$

with

$$g_k^{(x)} = E_{IJ}(n' d_{N_k}^{n,n'-1} - 2\alpha_B d_{N_k}^{n,n'+1}) e_{L_k}^{ll'} f_{M_k}^{mm'}. \tag{32}$$

The distribution T_{IJ} may be expanded in an analogous manner to give

$$T_{IJ} = \sum_k t_k \Lambda_{N_k} \Lambda_{L_k} \Lambda_{M_k} \exp(-\alpha_p r_p^2) \tag{33}$$

with

$$t_k = (t_k^{xx} + t_k^{yy} + t_k^{zz}) E_{IJ} \tag{34}$$

$$t_k^{xx} = [nn' d_{N_k}^{n-1,n'-1} - 2n\alpha_B d_{N_k}^{n-1,n'+1} - 2n'\alpha_A d_{N_k}^{n+1,n'-1} + 4\alpha_A \alpha_B d_{N_k}^{n+1,n'+1}] e_{L_k}^{ll'} f_{M_k}^{mm'}, \tag{35}$$

and similar expressions for t_k^{yy} and t_k^{zz}. The z component of Q_{IJ} may be similarly shown to be of the form

$$Q_{IJ}^{(z)} = \sum_k q_k^{(z)} \Lambda_{N_k} \Lambda_{L_k} \Lambda_{M_k} \exp(-\alpha_p r_p^2), \tag{36}$$

where

$$q_k^{(z)} = f_{M_k}^{mm'} E_{IJ} q_k^{xy}. \tag{37}$$

The q_k^{xy} obtained from $\partial\phi_I/\partial x \; \partial\phi_J/\partial y - \partial\phi_I/\partial y \; \partial\phi_J/\partial x$ involve new ϕ's of total powers of x, y, and z higher by two than the starting ones. These highest powers cancel since

$$\begin{aligned}\phi(n+1, l, m, \alpha_A, A)\phi(n', l'+1, m', \alpha_B, B) - \phi(n, l+1, m, \alpha_A, A)\\ \phi(n'+1, l, m', \alpha_B, B)\\ = (x_A y_B - y_A x_B)\phi(n, l, m, \alpha_A, A)\phi(n, l, m, \alpha_B, B)\\ = [AB_y\phi(n+1, l, m, \alpha_A, A) - AB_x\phi(n, l+1, m, \alpha_A, A)]\\ \phi(n, l, m, \alpha_B, B).\end{aligned} \tag{38}$$

Consequently,

$$\begin{aligned}q_k^{xy} &= l'(nd_{N_k}^{n-1,n'} - 2\alpha_A d_{N_k}^{n-1,n'})e_{L_k}^{l,l'-1} - 2n\alpha_B d_{N_k}^{n-1,n} e_{L_k}^{l,l'+1}\\ &- l(n'd_{N_k}^{n,n'-1} - 2\alpha_B d_{N_k}^{n,n'+1})e_{L_k}^{l-1,l'} - 2n'\alpha_A d_{N_k}^{n,n'-1} e_{L_k}^{l+1,l'}\\ &+ 4\alpha_A\alpha_B[AB'_y d_{N_k}^{n-1,n'} e_{L_k}^{l,l'} - AB'_x d_{N_k}^{n,n} e_{L_k}^{l+1,l'}].\end{aligned} \tag{39}$$

It is important to note that different sets of (N_k, L_k, M_k) appear in Eqs. 26, 27, 31, 33 and 36. Also, larger values of $N + L + M$ appear in the derivatives than in the charge distribution.

It is clear that one-electron integrals to be evaluated all reduce to the form

$$\begin{aligned}[NLM\,|\,\theta] = \int \theta(\mathbf{r}_1)\Lambda_N(x_{1p}; \alpha_p)\Lambda_L(y_{1p}; \alpha_p)\Lambda_M(z_{1p}; \alpha_p)\\ \times\ \exp(-\alpha_p r_{1p}^2)d\tau_1.\end{aligned} \tag{40}$$

Likewise the basic two-electron integrals all take the form

$$\begin{aligned}[NLM\,|\,\theta\,|\,N'L'M'] = \int \theta(\mathbf{r}_1\mathbf{r}_2)\Lambda_N(x_{1p}; \alpha_p)\Lambda_L(y_{1p}; \alpha_p)\Lambda_M(z_{1p}; \alpha_p)\\ \times\ \exp(-\alpha_p r_{1p}^2)\Lambda_{N'}(x_{2Q}; \alpha_Q)\Lambda_{L'}(y_{2Q}; \alpha_Q)\Lambda_{M'}(z_{2Q}; \alpha_Q)\\ \times\ \exp(-\alpha_Q r_{2Q}^2)d\tau_1 d\tau_2.\end{aligned} \tag{41}$$

The one-electron integrals can be further classified as (a) those that can be done in closed form, and (b) those that require the same numerically approximated auxiliary functions as the two-electron integrals.

A. One-Electron Integrals, Closed Form

The basic integral of this type is the one-dimensional integral

$$\int dx \Lambda_N(x_p; \alpha)\exp(-\alpha x_p^2) = \delta_{N,0}(\pi/\alpha)^{1/2} \tag{42}$$

(the $\delta_{N,0}$ arises from the orthogonality of the Hermite polynomials). Thus the overlap integral is simply

$$[NLM\,|\,1] = \delta_{N,0}\delta_{L,0}\delta_{M,0}(\pi/\alpha_p)^{3/2}. \tag{43}$$

The relation

$$x_c\Lambda_N(x_p;\alpha_p) = N\Lambda_{N-1} + \tfrac{1}{2}\Lambda_{N+1}/\alpha_p + PC_x\Lambda_N \tag{44}$$

then gives

$$[NLM\,|\,x_c] = (\delta_{n,1} + PC_x\delta_{N,0})\delta_{L,0}\delta_{M,0}(\pi/\alpha_p)^{3/2} \tag{45}$$

with similar results for y_c and z_c. The second moments are just

$$[NLM\,|\,x_c^2] = [2\delta_{N,2} + 2PC_x\delta_{N,1} + (PC_x^2 + \tfrac{1}{2}\alpha_p^{-1})\delta_{N,0}] \\ \delta_{L,0}\delta_{M,0}(\pi/\alpha_p)^{3/2} \tag{46}$$

and

$$[NLM\,|\,x_cy_c] = (\delta_{N,1} + PC_x\delta_{N,0})(\delta_{L,1} + PC_y\delta_{L,0})\delta_{M,0}(\pi/\alpha_p)^{3/2} \tag{47}$$

with similar results for y_c^2, z_c^2, x_cz_c, or y_cz_c.

The kinetic energy and gradient matrix elements are given directly by the formula for $[NLM\,|\,1]$. Notice that only the t_k (and g_k) coefficients for $(N_k, L_k, M_k) = (0,0,0)$ are required to evaluate the kinetic energy and gradient matrix elements.

B. Other One-Electron Integrals

The basic integral in this category is the nuclear attraction integral $[NLM\,|\,r_c^{-1}]$. From the definition of the Λ's, this may be written as

$$[NLM\,|\,r_c^{-1}] = (\partial/\partial P_x)^N(\partial/\partial P_y)^L(\partial/\partial P_z)^M[000\,|\,r_c^{-1}]. \tag{48}$$

The integral $[000\,|\,r_c^{-1}]$ was shown by Boys[15] to be given by

$$[000\,|\,r_c^{-1}] = (2\pi/\alpha_p)F_O(T), \tag{49}$$

where

$$T = \alpha_p CP^2, \tag{50}$$

and

$$F_O(T) = \int_O^1 \exp(-Tu^2)du. \tag{51}$$

If we define the auxiliary function R_{NLM} by

$$R_{NLM} = (\partial/\partial P_x)^N(\partial/\partial P_y)^L(\partial/\partial P_z)^M F_O(T) \tag{52}$$

then

$$[NLM\,|\,r_c^{-1}] = (2\pi/\alpha_p)R_{NLM}. \tag{53}$$

The computation of R_{NLM} will be described in a later section.

Matrix elements of the components of the electric field, such as $x_c r_c^{-3}$, can be evaluated in two ways which illustrate the tricks needed for more complicated integrals. First

$$x_c r_c^{-3} = \partial r_c^{-1}/\partial C_x \tag{54}$$

where it should be noted that differentiation is with respect to the nuclear position. Therefore

$$[NLM\,|\,x_c r_c^{-3}] = (\partial/\partial C_x)[NLM\,|\,r_c^{-1}]. \tag{55}$$

But since T depends on $C_x - P_x$,

$$(\partial/\partial C_x)g(T) = -\,(\partial/\partial P_x)g(T) \tag{56}$$

for any function $g(T)$. Hence,

$$[NLM\,|\,x_c r_c^{-3}] = -\,(2\pi/\alpha_p)R_{N+1,L,M}. \tag{57}$$

Alternatively this integral can be evaluated by noting that

$$x_c r_c^{-3} = -\,\partial r_c^{-1}/\partial x. \tag{58}$$

Hence, by integration by parts,

$$[NLM\,|\,x_c r_c^{-3}] = [\partial(NLM)/\partial x\,|\,r_c^{-1}]. \tag{59}$$

But from the definition of Λ_N,

$$\partial(N,L,M)/\partial x = -\,(N+1,L,M),$$

so

$$[NLM\,|\,x_c r_c^{-3}] = -\,(2\pi/\alpha_p)R_{N+1,L,M}. \tag{60}$$

The components of the electric field gradient are similarly obtained using

$$(\partial^2/\partial C_x^2 - \partial^2/\partial C_y^2)r_c^{-1} = 3(x_c^2 - y_c^2)r_c^{-5}, \tag{61}$$

$$1/3(2\partial^2/\partial C_z^2 - \partial^2/\partial C_x^2 - \partial^2/\partial C_y^2)r_c^{-1} = (3z_c^2 - r_c^2)r_c^{-5}, \tag{62}$$

and

$$(\partial^2/\partial C_x \partial C_y)r_c^{-1} = 3x_c y_c r_c^{-5}. \tag{63}$$

Notice these formulas are written so as to avoid problems with the delta function which arises in $\nabla^2 r_c^{-1}$. Hence the electric field gradient integrals are given by

$$[NLM|3(x_c^2 - y_c^2)r_c^{-5}] = (2\pi/\alpha_p)(R_{N+2,L,M} - R_{N,L+2,M}), \tag{64}$$

$$[NLM|(3z_c^2 - r_c^2)r_c^{-5}] = (2\pi/\alpha_p)(2R_{N,L,M+2} - R_{N+2,L,M} - R_{N,L+2,M})/3, \tag{65}$$

$$[NLM|3x_c y_c r_c^{-5}] = (2\pi/\alpha_p)R_{N+1,L+1,M}. \tag{66}$$

Matrix elements over the one-electron spin-orbit operator may be evaluated by two different methods. If the space part of the basic spin-orbit operator is considered to be $r_c^{-3}\mathbf{r}_c \times \nabla$ then the z component, for example, is $r_c^{-3}(x_c\partial/\partial y - y_c\partial/\partial x)$. Matrix elements of this operator can be evaluated by combining the results given above for the electric field with the g_k expansion coefficients given previously for the gradient. Alternatively integration by parts gives

$$< \phi_I | r_c^{-3}\left(x_c\frac{\partial}{\partial y} - y_c\frac{\partial}{\partial x}\right) | \phi_J > = \left[\frac{\partial\phi_I}{\partial x}\frac{\partial\phi_J}{\partial y} - \frac{\partial\phi_I}{\partial y}\frac{\partial\phi_J}{\partial x} | r_c^{-1}\right]. \tag{67}$$

The spin-orbit integral then reduces to using the q_k coefficients in summing $[NLM|r_c^{-1}]$ matrix elements.

C. Two-Electron Integrals

The simplest integral in this category is the electron repulsion

$$[NLM|r_{12}^{-1}|N'L'M'] = (\partial/\partial P_x)^N(\partial/\partial P_y)^L(\partial/\partial P_z)^M(\partial/\partial Q_x)^{N'}(\partial/\partial Q_y)^{L'} (\partial/\partial Q_z)^{M'}[000|r_{12}^{-1}|000] \tag{68}$$

Boys evaluated the basic integral as

$$[000|r_{12}^{-1}|000] = \lambda F_O(T) \tag{69}$$

where

$$\lambda = 2\pi^{5/2}\alpha_p^{-1}\alpha_Q^{-1}(\alpha_p + \alpha_Q)^{-1/2} \tag{70}$$

and

$$T = \alpha_p \alpha_Q (\alpha_p + \alpha_Q)^{-1} PQ^2. \tag{71}$$

Because T involves only the combination $\mathbf{P} - \mathbf{Q}$,

$$(\partial/\partial Q_x)^N g(T) = (-\partial/\partial P_x)^N g(T) \tag{72}$$

for any function $g(T)$. Hence

$$[NLM|r_{12}^{-1}|N'L'M'] = \lambda(-1)^{N'+L'+M'} R_{N+N',\,L+L',\,M+M'}. \tag{73}$$

Just as for the electric field and field gradient, integrals over $x_{12}r_{12}^{-3}$, $x_{12}y_{12}r_{12}^{-5}$, etc., are easily evaluated. For example

$$[NLM|x_{12}r_{12}^{-3}|N'L'M'] \\ = -\lambda(-1)^{N'+L'+M'} R_{N+N'+1,\,L+L'+1,\,M+M'}, \tag{74}$$

$$[NLM|3x_{12}y_{12}r_{12}^{-5}|N'L'M'] \\ = \lambda(-1)^{N'+L'+M'} R_{N+N'+1,\,L+L'+1,\,M+M'}, \tag{75}$$

$$[NLM|3(x_{12}^2 - y_{12}^2)r_{12}^{-5}|N'L'M'] \\ = \lambda(-1)^{N'+L'+M'}(R_{N+N'+2,L+L',\,M+M'} - R_{N+N',\,L+L'+2,\,M+M'}), \tag{76}$$

$$[NLM|(3z_{12}^2 - r_{12}^2)r_{12}^{-5}|N'L'M'] \\ = \lambda(-1)^{N'+L'+M'}(2R_{N+N',\,L+L',M+M'+2} \\ - R_{N+N'+2,L+L',M+M'} - R_{N+N',\,L+L'+2,\,M+M'})/3. \tag{77}$$

Integrals over these r_{12}^{-5} operators appear in calculation of spin-spin interaction matrix elements.

The space part of the two-electron spin-orbit operator has the form $r_{12}^{-3}\mathbf{r}_{12}x\nabla_1$. The z component of this operator is then

$$r_{12}^{-3}(x_{12}\frac{\partial}{\partial y_1} - y_{12}\frac{\partial}{\partial x_1}).$$

Integration by parts yields a result similar to that obtained for the one-electron integral,

$$< \phi_I(1)\phi_K(2)|r_{12}^{-3}(x_{12}\frac{\partial}{\partial y_1} - y_{12}\frac{\partial}{\partial x_1})\phi_J(1)\phi_L(2) > \\ = [\frac{\partial \phi_I}{\partial x_1}\frac{\partial \phi_J}{\partial y_1} - \frac{\partial \phi_I}{\partial y_1}\frac{\partial \phi_J}{\partial x_1}|r_{12}^{-1}|\phi_K\phi_L]. \tag{78}$$

Consequently, this integral is given simply as a sum over $[NLM|r_{12}^{-1}|N'L'M']$ using coefficients q_k for the IJ orbitals and D_k for the KL orbitals. Alternatively, of course, the integral could be written as a sum over integrals like $[NLM|x_{12}r_{12}^{-3}|N'L'M']$ with g_k coefficients for the IJ orbitals.

D. Auxiliary Functions

The function R_{NLM} is defined as

$$R_{NLM} = (\partial/\partial a)^N(\partial/\partial b)^L(\partial/\partial c)^M \int_0^1 e^{-Tu^2} du \tag{79}$$

where

$$T = \alpha(a^2 + b^2 + c^2). \tag{80}$$

By direct chain-rule differentiation an explicit formula for R can be found:

$$R_{NLM} = \sum_{n=0}^{[N/2]} \sum_{l=0}^{[L/2]} \sum_{m=0}^{[M/2]} a^{N-2n} b^{L-2l} c^{M-2M} \times \frac{N!}{(2n)!!(N-2n)!} \frac{L!}{(2l)!!(L-2l)!} \frac{M!}{(2m)!!(M-2m)!} F_{N+L+M-n-l-m}. \tag{81}$$

For generating a table of all R_{NLM} up to some maximum N+L+M, as is needed in doing blocks of integrals, recursion relations are more useful. These can be found from introduction of the more general integral:

$$R_{NLMj} = (-\alpha^{1/2})^{N+L+M}(-2\alpha)^j \int_0^1 u^{N+L+M+2j} H_N(\alpha^{1/2}au) H_L(\alpha^{1/2}bu) H_M(\alpha^{1/2}cu) e^{-Tu^2} du.$$

Let us first note that

$$R_{000j} = (-2\alpha)^j F_j(T), \tag{82}$$

where

$$F_j(T) = \int_0^1 u^{2j} \exp(-Tu^2) du. \tag{83}$$

From the recursion relations for the Hermite polynomials, it follows that

$$R_{0,0,M+1,j} = cR_{0,0,M,j+1} + MR_{0,0,M-1,j+1} \tag{84}$$

$$R_{0,L+1,M,j} = bR_{0,L,M,j+1} + LR_{0,L-1,M,j+1} \tag{85}$$

$$R_{N+1,L,M,j} = aR_{N,L,M,j+1} + NR_{N-1,L,M,j+1}. \tag{86}$$

Thus the desired R_{NLM} (given as R_{NLM0}) can be generated from a table of $F_j(T)$ for all j between 0 and the maximum N+L+M.

Shavitt[16] has given several formulas useful for evaluating $F_j(T)$. Rapid and accurate evaluation of this function for a wide range of j and T requires some care, however. Our best program at present evaluates $F_j(T)$ by different formulas depending on T.

For $0 < T < 12$ and $0 \leq j \leq J$, $F_J(T)$ is first evaluated using the seven term Taylor expansion

$$F_J(T) = \sum_{k=0}^{6} F_{J+k}(T^*)(T^* - T)^k/k! \tag{87}$$

where $F_{J+k}(T^*)$ has been pretabulated for T* at intervals of 0.1. With $|T^* - T| \leq 0.05$ this formula has a relative accuracy of 10^{-13} for all J $\leq$ 16. The downward recursion relation

$$F_j(T) = [2TF_{j+1}(T) + \exp(-T)]/(2j+1) \tag{88}$$

can then be used to obtain all $F_j(T)$.

For the range $12 < T < 30$ we note that

$$F_0(T) = 1/\pi^{1/2}/T^{1/2} - \int_1^{\infty} \exp(-Tu^2)du. \tag{89}$$

If the integral $\int_1^{\infty} \exp(-Tu^2)du$ is now expressed as $\exp(-T)g/T$, then $F_0(T)$ can be found with a relative accuracy of 10^{-14} from the modified asymptotic series for g:

$12 < T < 15$

$$g = 0.4999489092 - 0.2473631686\,T^{-1} + 0.321180909\,T^{-2} - 0.3811559346\,T^{-3}; \tag{90}$$

$15 < T < 18$

$$g = 0.4998436875 - 0.24249438\,T^{-1} + 0.24642845\,T^{-2}; \tag{91}$$

18 < T < 24

$$g = 0.499093162 - 0.2152832\, T^{-1}; \tag{92}$$

24 < T 30

$$g = 0.490. \tag{93}$$

Upward recursion, which is unstable for small T or large j, can be used to obtain $F_j(T)$ with a relative accuracy of 10^{-12} for $j \leq 16$ and $T \leq 12$.

For T > 30, $F_0(T) = 1/2\pi^{1/2}/T^{1/2}$ combined with upward recursion is accurate to 14 significant figures. Finally, for T > 2J + 36 the exact upward recursion

$$F_{j+1}(T) = (2T)^{-1}[(2j+1)F_j(T) - \exp(-T)] \tag{94}$$

can be replaced by

$$F_{j+1}(T) = (2T)^{-1}(2j+1)F_j(T) \tag{95}$$

without loss of accuracy.

The series for g was derived from the definition

$$g = Te^{-T}\int_1^{\infty} du \exp(-Tu^2)$$

by the change of variable $v = 2T(u-1)$ which gives

$$2g = \int_0^{\infty} dv \exp[-v(1+\alpha v)] \tag{96}$$

with $\alpha = (4T)^{-1}$. Since decreasing accuracy in g is required as T increases, a Taylor's series expansion of g,

$$g = \Sigma(n!)^{-1}\partial^n g/\partial\alpha^n|_{\alpha^*}(\alpha - \alpha^*)^n, \tag{97}$$

was found with α^* corresponding to the smallest T in the interval. Truncation and rearrangement of this series gave Eqs. 90-93. As is well known in the special case $\alpha^* = 0$, this is really an asymptotic series for g; but, for $|\alpha - \alpha^*|$ sufficiently small, the truncated series gives g to the accuracy required for finding $F_0(T)$ quite precisely.

E. Charge Distributions

The integral computation is driven by a table of charge distributions containing all of the expansion coefficients D_k, the orbital exponents α_p, and charge distribution centers **P**. This list is parsed as it is generated to eliminate small charge distributions. By eliminating both one-electron as well as two-electron integrals involving small charge distributions, a "charge preserving" balance is maintained which allows fairly large thresholds to be set for large molecules. The threshold is normally kept fairly small (10^{-12}) for extended basis sets on small molecules. For double zeta plus polarization basis sets on large molecules, a threshold as large as 10^{-6} is often used. In addition, the program has a second threshold for deleting small two-electron integrals after they are computed. This does not preserve charge balance and does not save time in the integral computation. This threshold cannot be set much larger than 10^{-8} for large molecule calculations.

In addition, the integral calculation is adapted to the point group symmetry for sub-groups of D_{2h}. This is accomplished in several steps. The first step occurs in the input where basis functions are assigned only to symmetry unique atoms. Concepts from the theory of double cosets are then used to generate all unique symmetry related charge distributions in the form $R(f_iSf_j)$ where R and S are point group operators.[3] Finally a minimum set of symmetry related blocks of two electron integrals $[f_iTf_j|R(f_kSf_l)]$ are computed and written contiguously to the integral file. Thus, not only are all f's belonging to the same $n + l + m$ treated as a block, but also all f's related to these by symmetry.

F. Pseudo Potentials[2]

The integral program also allows addition of a pseudo potential to the nuclear-electron potential energy for each nucleus. This pseudo potential has the form[2]

$$U(r) = U_{L+1}(r) + \sum_{l=0}^{L} \sum_{m=-1}^{l} |lm> [U_l(r) - U_{L+1}(r)] < lm| \tag{98}$$

where $U_l(r)$ is expanded in a series of Gaussians as

$$U_l = N_c/r + r^{-2}\Sigma d_{jl}r^{n_j} \exp(-\xi_j r^2) \tag{99}$$

and N_c is the number of core electrons and $n_j \geq 0$.

This pseudo potential results in two distinct types of integrals (which we also refer to as type 1 and type 2).

$$\chi_{AB} = \int d\tau \phi_A r_C^{n'-2} \exp(-\xi r_C^2) \phi_B \tag{100}$$

and

$$\gamma_{AB} = \sum_{m=-l}^{l} \int_0^\infty dr_C [\int d\Omega_C \phi_A Y_{lm}(\Omega_C)] r_C^{n'} \exp(-\xi r_C^2) \times [\int d\Omega_C Y_{lm}(\Omega_C)\phi_B], \tag{101}$$

where the Y_{lm} are real, orthonormal spherical polynomials; χ_{AB} refers to the U_{L+1} term in the potential and γ_{AB} to the $U_l - U_{L+1}$ terms.

The reduction of χ_{AB} proceeds by transforming the exponential parts of ϕ_A and ϕ_B to center C in the following manner:

$$\exp(-\alpha_A r_A^2) = \exp(-\alpha_A r_C^2 - 2\alpha_A \mathbf{CA}\cdot\mathbf{r}_C - \alpha_A |\mathbf{CA}|^2), \tag{102}$$

where

$$\mathbf{CA} = \mathbf{C} - \mathbf{A}. \tag{103}$$

If we now define

$$D_{ABC} = 4\pi N(n_A, l_A, m_A, \alpha_A) N(n_B, l_B, m_B, \alpha_B) \exp(-\alpha_A|\mathbf{CA}|^2 - \alpha_B|\mathbf{CB}|^2) \tag{104}$$

$$\mathbf{k} = -2(\alpha_A \mathbf{CA} + \alpha_B \mathbf{CB}) \tag{105}$$

and

$$\alpha = \alpha_A + \alpha_B + \xi \tag{106}$$

χ_{AB} is simplified to

$$\chi_{AB} = \frac{D_{ABC}}{4\pi} \int d\tau r_C^{n'-2} \exp(\alpha r_C^2) \exp(\mathbf{k}\cdot\mathbf{r}_C) x_A^{n_A} y_A^{1_A} z_A^{m_A} x_B^{n_B} y_B^{1_B} z_B^{m_B} \tag{107}$$

The next step is to expand $\exp(\mathbf{k}\cdot\mathbf{r}_C)$ in spherical doordinates:

$$\exp(\mathbf{k}\cdot\mathbf{r}_C) = 4\pi \sum_{\lambda=0}^{\infty} \sum_{\mu=-\lambda}^{\lambda} M_\lambda(kr_C) Y_{\lambda\mu}(\theta_k, \phi_k) Y_{\lambda\mu}(\theta_C, \phi_C) \tag{108}$$

where M_λ is a modified spherical Bessel function of the first kind:

$$M_\lambda(x) = x^\lambda \left(\frac{1}{x}\frac{d}{dx}\right)^\lambda \frac{\sinh x}{x} \tag{109}$$

$$= i^\lambda j_\lambda(-ix). \tag{110}$$

Transforming $x_A, y_A, z_A, x_B, y_B, z_B$ to point C and separating variables of integration we obtain

$$\chi_{AB} = D_{ABC} \sum_{a=0}^{n_A} \sum_{b=0}^{1_A} \sum_{c=0}^{m_A} \sum_{d=0}^{n_B} \sum_{3=0}^{1_B} \sum_{f=0}^{m_B} \binom{n}{a}A \binom{1}{b}A \binom{m}{c}A \binom{n}{d}B \binom{1}{e}B \binom{m}{f}B$$
$$\times\ CA_x^{n_A-a} CA_y^{1_A-b} CA_z^{m_A-c} CB_x^{n_B-d} CB_y^{1_B-e} CB_z^{m_B-f} \tag{111}$$
$$\times \sum_{\lambda=0}^{\infty} \Omega_\lambda^{a+d,b+e,c+f} Q_\lambda^{a+b+c+d+e+f+n'}(k, \alpha)$$

where

$$\hat{x} = x_C/r_C, \text{ etc.} \tag{112}$$

and where the angular integral is defined as

$$\Omega_\lambda^{IJK} = \sum_{\mu=-\lambda}^{\lambda} Y_{\lambda\mu}(\Omega_k) \int d\Omega \hat{x}^I \hat{y}^J \hat{z}^K Y_{\lambda\mu}(\Omega) \tag{113}$$

and the radial integral as

$$\Omega_\lambda^N(k,a) = \int_0^\infty dr\, r^n \exp(-\alpha r^2) M_\lambda(kr) \tag{114}$$

The product of powers of $\hat{x}_C$, $\hat{y}_C$, and $\hat{z}_C$ in the angular integral may be expanded in a sum of spherical polynomials of orders up to I+J+K and differing from I+J+K by a multiple of 2. By orthogonality, then, the sum over λ may be truncated at a+b+c+d+e+f, and (a+b+c+d+e+f) $-\ \lambda$ must be even.

The reduction of the type 2 integral, γ_{AB}, proceeds in a manner similar to that of χ_{AB}. When the exponential parts of ϕ_A and ϕ_B are transformed to point C we obtain

$$\gamma_{AB} = \frac{D_{ABC}}{4\pi} \sum_{m=-l}^{l} \int_0^\infty dr_C \left[\int d\Omega_C x_A^{n_A} y_A^{l_A} z_A^{m_A} \exp(\mathbf{k}_A \cdot \mathbf{r}_C) Y_{lm}(\Omega_C)\right]$$
$$\times\ r_C^{n'} \exp(-\alpha r_C^2) \left[\int d\Omega_C x_B^{n_B} y_B^{l_B} z_B^{m_B} \exp(\mathbf{k}_B \cdot \mathbf{r}_C) Y_{lm}(\Omega_C)\right] \tag{115}$$

where D_{ABC} and α are defined as before, and

$$\mathbf{k}_A = -2\alpha \mathbf{CA} \tag{116}$$

$$\mathbf{k}_B = -2\alpha \mathbf{CB} \tag{117}$$

Transforming $x_A, y_A, z_A, x_B, y_B, z_B$ to center C, and reexpressing $\exp(\mathbf{k}_A \cdot \mathbf{r}_C)$ and $\exp(\mathbf{k}_B \cdot \mathbf{r}_C)$, γ_{AB} becomes

$$\gamma_{AB} = 4\pi D_{ABC} \sum_{a=0}^{n_A} \sum_{b=0}^{l_A} \sum_{c=0}^{m_A} \sum_{d=0}^{n_B} \sum_{e=0}^{l_B} \sum_{f=0}^{m_B} \binom{n}{a}A \binom{l}{b}A \binom{m}{c}A \binom{n}{d}B \binom{l}{e}B \binom{m}{f}B$$
$$\times CA_x^{n_A-a} CA_y^{l_A-b} CA_z^{m_A-c} CB_x^{n_B-d} CB_y^{l_B-e} CB_z^{m_B-f} \tag{118}$$
$$\times \sum_{\lambda=0}^{\infty} \sum_{\lambda'=0}^{\infty} Q_{\lambda\lambda'}^{a+b+c+d+e+f+n'}(k_A, k_B, \alpha) \sum_{m=-l}^{l} \Omega_{\lambda lm}^{abc} \Omega_{\lambda' lm}^{def}$$

where the angular integral $\Omega_{\lambda lm}^{abc}$ is given by

$$\Omega_{\lambda lm}^{abc} = \sum_{\mu=-\lambda}^{\lambda} Y_{\lambda\mu}(\Omega_k) \int \frac{d\Omega}{4\pi} \hat{x}^a \hat{y}^b \hat{z}^c Y_{\lambda\mu}(\Omega) Y_{lm}(\Omega) \tag{119}$$

and the radial integral $Q_{\lambda\lambda'}^{N}$, is given by

$$Q_{\lambda\lambda'}^{N}(k_A, k_B, \alpha) = \int_0^{\infty} dr\, r^N \exp(-\alpha r^2) M_\lambda(k_A r) M_{\lambda'}(k_B r) \tag{120}$$

As with the type 1 angular integral, $\Omega_{\lambda lm}^{abc}$ may be reexpressed by expanding $x^a y^b z^c$ as a sum of spherical polynomials of order up to a+b+c and differing from a+b+c by a multiple of 2. Therefore, using the vector sum rule for spherical polynomials, the only nonzero terms in the sum over λ are

$$\max(l - a - b - c, 0) \le \lambda \le l + a + b + c \tag{121}$$

and likewise for λ'. Also consistent with the first type of angular integral, $l + a + b + c - \lambda$ must be even.

Evaluation of the Angular Integrals. To evaluate the angular integrals we first expand the real orthonormal spherical polynomials Y_{lm} in terms of $\hat{x}$, $\hat{y}$, and $\hat{z}$:

$$Y_{\lambda\mu} = \sum_{r,s,t}^{\lambda} y_{rst}^{\lambda\mu} \hat{x}^r \hat{y}^s \hat{z}^t, \qquad r + s + t = \lambda. \tag{122}$$

The complete angular integrals are then

$$\Omega_{\lambda}^{IJK} = \sum_{\mu=-\lambda}^{\lambda} \left[\sum_{r,s,t}^{\lambda} y_{rst}^{\lambda\mu} \hat{k}_x^r \hat{k}_y^s \hat{k}_z^t \right] \times \sum_{r,s,t}^{\lambda} y_{rst}^{\lambda\mu} \int d\Omega\, \hat{x}^{I+r} \hat{y}^{J+s} \hat{z}^{K+t} \tag{123}$$

$$\Omega_{\lambda l m}^{abc} = \sum_{\mu=-\lambda}^{\lambda} \left[\sum_{r,s,t}^{\lambda} y_{rst}^{\lambda\mu} \hat{k}_x^r \hat{k}_y^s \hat{k}_z^t \right] \times \sum_{r,s,t}^{\lambda} \sum_{u,v,w}^{l} y_{rst}^{\lambda\mu} y_{uvw}^{lm} \int d\Omega \hat{x}^{a+r+u} \hat{y}^{b+s+v} \hat{z}^{c+t+w} \tag{124}$$

The evaluation of the integral is straightforward:

$$\begin{aligned} (4\pi)^{-1} \int d\Omega \hat{x}^i \hat{y}^j \hat{z}^k &= 0, && \text{i, j, or k odd} \\ &= \frac{(i-1)!!(j-1)!!(k-1)!!}{(i+j+k+1)!!} && \text{i, j, and k even} \end{aligned} \tag{125}$$

Type 1 Radial Integral, $Q_l^n(k, \alpha)$. Gradshteyn and Ryzhik reexpress the type 1 radial integral as[17]

$$Q_l^n(k, \alpha) = \sqrt{\pi}\, k^l 2^{-l-2} \alpha^{-(l+n+1)/2} R \phi((l+n+1)/2; l+3/2; k^2/4\alpha) \tag{126}$$

where R is the ratio of gamma functions,

$$\begin{aligned} R = \Gamma((l+n+1)/2)/\Gamma(l+3/2) &= \frac{\sqrt{\pi}\,(l+n-1)!!}{2(2l+1)!!}, && n+l \text{ even,} \\ &= \frac{(l+n-1)!!}{(2l+1)!!}, && n+l \text{ odd} \end{aligned} \tag{127}$$

and ϕ is the degenerate hypergeometric function. The confluent hypergeometric series for ϕ is

$$\phi(a,b,z) = 1 + \frac{a}{b}\frac{z}{1!} + \frac{a(a+1)}{b(b+1)}\frac{z^2}{2!} + \cdots \tag{128}$$

The resulting expression for $Q_l^n(k, \alpha)$ is equivalently obtained by substitution of a power series for the modified spherical Bessel function $M_l(kr)$ in Eq. (114). An asymptotic series for ϕ is given by

$$\phi(a;b;z) = R^{-1} z^{a-b} \exp(z) \left[\frac{(b-a)(1-a)}{1!} z^{-1} + \frac{(b-a)(b-a+1)(1-a)(2-a)}{2!} z^{-2} + \cdots \right] \tag{129}$$

Although this series diverges, the magnitudes of the terms decrease until a minimum is reached, at which point the partial sum represents a best approximation to ϕ. Summing to this minimum gives 12-figure accuracy for the following n and z:

n	0	1	2	3	4	5	6	7	>7
z	≥ 31.	28.	25.	23.	22.	20.	19.	18.	15

Note that the asymptotic form truncates for $n + l$ even and $n \geq l + 2$ providing an exact analytical expression for ϕ. An exactly equivalent form for these n and l is found by noting[7]

$$\phi(a;b;z) = \exp(z)\phi(b - a;b; -z) \tag{130}$$

Substitution of the confluent hypergeometric series yields

$$\phi(a:b:z) = \exp(z)\left[1 - \frac{(b-a)z}{b1!} + \frac{(b-a)(b-a+1)}{b(b+1)2!}z^2 + \cdots\right] \tag{131}$$

In contrast to the asymptotic series, evaluation of this form for $n + l$ even and $n \geq l + 2$ present no problems for small z. For other n and l, however, when it does not terminate, Eq. (131) is not useful, owing to differencing.

Recursion relations were derived and implemented to allow most Q_l^n's to be calculated from just a few starting values. Referring back to Eq. (113) and discussion, we first note that only $Q_\lambda^{a+b+c+d+e+f+n'}$ for which $a + b + c + d + e + f - \lambda$ is even are required, as all others are paired with vanishing angular integrals. Secondly, we note a recursion relation on $M_l(x)$:

$$M_l(x) = M_{l-2}(x) - \frac{(2l-1)}{x} M_{l-1}(x) \tag{132}$$

Using this relation and integration by parts, a number of recursion relations on the Q_l^n may be derived:

$$Q_l^{l+2} = \frac{k}{2\alpha} Q_{l-1}^{l+1}, \tag{133A}$$

$$Q_l^n = \frac{1}{k}[2\alpha Q_{l-1}^{n+1} - (n + l - 1)Q_{l-1}^{n-1}] \tag{133B}$$

$$Q_l^n = \frac{1}{(n+l+1)}[2\alpha Q_l^{n+2} - kQ_{l+1}^{n+1}] \tag{133C}$$

$$Q_l^n = \frac{1}{2\alpha}[(n + l - 1)Q_l^{n-2} + kQ_{l-1}^{n-1}] \tag{133D}$$

$$Q_l^n = \frac{1}{2\alpha}\{(n + l - 3)Q_{l-2}^{n-2} + [k - (2l-1)\frac{2\alpha}{k}]Q_{l-1}^{n-1}\} \tag{133E}$$

$$Q_l^n = \frac{1}{(n+l+1)}\{2\alpha Q_{l+2}^{n+2} - [k - (2l+3)\frac{2\alpha}{k}]Q_{l+1}^{n+1}\} \tag{133F}$$

Upon examination of the asymptotic form of Q_l^n, Eqs. (133C) and (133F) are found to give serious differencing errors for $k^2/4\alpha$ large. Likewise, the alternating

A. Large $K^2/4\alpha$, n even

n	1 0	1	2	3	4
0	S				
1		B			
2	S		B		
3		A		B	
4	D		A		B
5		D		A	
6	D		D		A

C. Small $k^2/4\alpha$, n even

n	1 0	1	2	3	4
0	C				
1		C			
2	S		C		
3		A		C	
4	D		A		C
5		D		A	
6	D		D		A

B. Large $K^2/4\alpha$, n odd

n	1 0	1	2	3	4
0					
1	S				
2		S			
3	D		E		
4		D		E	
5	D		D		E
6		D		D	

D. Small $k^2/4\alpha$, n odd

n	1 0	1	2	3	4
0					
1	F				
2		F			
3	D		F		
4		D		S	
5	D		S		S
6		D		D	

Figure 1. Recurrence algorithm for the type 1 radial integral Q_l^n. S indicates the appropriate series given in Section IV. A,B,C,D,E,F refer to recursion relations, Eqs. (133A) – (133F).

series reveals that Eqs. (133B) and (133E) have a differencing problem for $k^2/4\alpha$ small. Fig. 1 gives separate stable recurrence schemes for small and large $k^2/4\alpha$. Switching from one to the other at $k^2/4\alpha = 3.0$ yields a relative accuracy of 10^{-13} in the Q_l^n.

Double Power Series. A double power series for the type 2 radial integral, $Q^N_{\lambda\lambda'}(k_A, k_B, \alpha)$, is suggested by substitution of power series for both modified spherical Bessel functions appearing in Eq. (120). From Ambramowitz and Stegun[18]

$$M_\lambda(z) = z^\lambda \sum_{j=0}^{\infty} \frac{(z/\sqrt{2}\,)^{2j}}{j\,!\,(2\lambda+1+2j)\,!\,!} \tag{134}$$

Therefore

$$Q^N_{\lambda\lambda'}(k_A, k_B, \alpha) = k_A^\lambda k_B^{\lambda'} \sum_{j=0}^{\infty} \frac{(k_A/\sqrt{2}\,)^{2j}}{j\,!\,(2\lambda+1+2j)\,!\,!} \sum_{i=0}^{\infty} \frac{(k_B/\sqrt{2}\,)^{2i}}{i\,!\,(2\lambda'+1+2i)\,!\,!} \int_0^\infty dr\, r^{N+\lambda+\lambda'+2j+2i} \exp(-\alpha r^2) \tag{135}$$

The integral is evaluated as

$$\int_0^{\infty} dr\, r^M \exp(-\alpha r^2) = \frac{(M-1)!!}{(2\alpha)^{(M+1)/2}} \sqrt{\frac{\pi}{2}} \quad \text{M even}$$
$$= \frac{(M-1)!!}{(2\alpha)^{(M+1)/2}} \quad \text{M odd} \tag{136}$$

After some rearrangement $Q^N_{\lambda\lambda'}$ becomes

$$Q^N_{\lambda\lambda'}(k,k,\alpha) = \frac{k_A^{\lambda} k_B^{\lambda'}}{(2\alpha)^{(n+\lambda+\lambda'+1)/2}} \sum_{I=0}^{\infty} \left(\frac{k_A^2}{4\alpha}\right)^I (n+\lambda+\lambda'+2I-1)!!\, T_I \tag{137}$$

where

$$T_I = \sum_{i=0}^{I} \frac{(k_B^2/k_A^2)^i}{(I-i)!\, i!\, (2\lambda+1+2I-2i)!!\, (2\lambda+1+2i)!!} \tag{138}$$

T_I is now simply related to the hypergeometric function $F(a,b{:}c;z)$:

$$T_I = [(2\lambda+1+2I)!!\, I!\, (2\lambda'+1)!!]^{-1} \times F(-I, -\lambda-1/2-I; \lambda'+3/2; k_B^2/k_A^2) \tag{139}$$

where we have used

$$F(a,b;c;z) = \sum_{i=0}^{\infty} \frac{a!!\, b!!}{c!!} \frac{z^i}{i!} \tag{140}$$

Recursion relations on the F's (Ref. 18, p. 558) allow a recursion relation to be derived for T_I:

$$T_{I+1} = (\beta + \gamma_z) T_I + \delta(1-z^2) T_{I-1} \tag{141}$$

where

$$\delta = -\frac{(\lambda+\lambda'+2I+3)}{(I+1)(2\lambda+3+2I)(2\lambda'+3+2I)(\lambda+\lambda'+2I+1)(\lambda+\lambda'+I+2)} \tag{142}$$

$$\beta = \frac{1}{(I+1)(2\lambda+3+2I)} - I(2\lambda+1+2I)\delta \tag{143}$$

$$\gamma = \frac{1}{(I+1)(2\lambda'+3+2I)} - I(2\lambda'+1+2I)\delta \tag{144}$$

$$z = k_B^2/k_A^2 \tag{145}$$

Single Power Series. A single power series for the radial integral is found by substituting a power series for just one of the modified spherical Bessel functions:

$$Q^N_{\lambda\lambda'}(k_A, k_B, \alpha) = \frac{k_A^\lambda}{\alpha^{(N+\lambda+1)/2}} \sum_{j=0}^{\infty} \frac{(k_A^2/2\alpha)^j}{j!\,(2\lambda+1+2j)!!} Q_\lambda^{N+\lambda+2j}(k_B/\sqrt{\alpha}, 1) \quad (146)$$

Evaluation of the type 1 radial integral Q_l^n for arbitrary n has already been discussed; however, the scheme is only practical when these quantities are obtained with a minimum of effort. An upwards recursion relation on n is found in a manner similar to the other recursion relations presented in Section IV:

$$Q_l^n(k, \gamma) = \frac{1}{\gamma}\left[\left(\frac{k^2}{4\alpha} + \frac{2n-5}{2}\right)Q_l^{n-2} + \frac{(l-n+4)(l+n-3)}{4}Q_l^{n-4}\right] \quad (147)$$

Thus, only $Q_\lambda^{N+\lambda}$ and $Q_\lambda^{N+\lambda+2}$ are required initially to compute an arbitrary number of terms in the series.

Owing to the power series expansion in $k_A^2/2\alpha$, one would expect the method to be inefficient when this quantity is large. When it is small, however, one might expect the method to be rapidly convergent, regardless of the size of $k_B^2/2\alpha$. Such is not the case for the following reasons. We may extract $\exp(k_B^2/4\alpha)$ from the $Q_\lambda^{N+\lambda+2j}$, hopefully leaving quantities that cannot become too large. We compare this with $\exp[(k_A + k_B)^2/4\alpha]$ that is extracted from the points and weights expression. It is apparent that the possibly large crossterm $\exp(k_A k_B/2\alpha)$ is still hidden in Eq. (146). Not only can this result in overflows, the number of terms in the series may be prohibitive. For these reasons, an effective upper limit to the utility of this method was found to be $(k_A + k_B)^2/2\alpha = 100$ when approximately 70 terms are required to give $Q^N_{\lambda\lambda'}$ (arbitrary N, λ, λ') to an accuracy of 10^{-13}.

Gaussian Points and Weights Method. We can write the modified spherical Bessel function $M_l(z)$ in exponential form as

$$M_l(z) = \frac{1}{2z}\, R_l(-z)\exp(z) - (-1)^l R_l(z)\exp(-z) \quad (148)$$

where

$$R_l(z) = \sum_{k=0}^{l} \frac{(l+k)!}{k!\,(l-k)!}(2z)^{-k} \quad (149)$$

For large z, $M_l(z)$ becomes simply the first term in Eq. (148). Thus, when $k_A/\sqrt{\alpha}$ and $k_B/\sqrt{\alpha}$ are large, the type 2 radial integral is approximated by (after a change of variable $r \to r/\sqrt{\alpha}$

$$Q_{\lambda\lambda'}^{N}(k_A, k_B, \alpha) \simeq \frac{1}{4k_A k_B}\int_0^{\infty} dr\left(\frac{r}{\sqrt{\alpha}}\right)^{N-2} \exp\left(-r^2 + \frac{A^k}{\sqrt{\alpha}}r + \frac{B^k}{\sqrt{\alpha}}r\right) \tag{150}$$

This form immediately suggests the use of a Gaussian points and weights scheme. We proceed by differentiating the integrand to find a maximum at

$$r_C = \frac{1}{4}(k_A + k_B)/\sqrt{\alpha} \pm \frac{1}{2}\left[\frac{1}{4}(k_A + k_B)^2/\alpha + 2(N-2)\right]^{1/2} \tag{151}$$

For the range of $\frac{1}{2}(k_A + k_B)/\sqrt{\alpha}$ for which the method ultimately proved practical, the effect of the $2(N-2)$ term was very small. Therefore, in the interest of keeping r_C independent of N, we approximate the maximum as

$$r = \frac{1}{2}(k_A + k_B)/\sqrt{\alpha} \tag{152}$$

A change of variables $t = r - r_C$ should minimize the number of points in the numerical integration:

$$Q_{\lambda\lambda'}^{N} = \int_{-r_C}^{\infty} dt\, f(t, r_C, k_A, k_B, \alpha)\exp(-t^2) \tag{153}$$

where

$$f(t,r,k,k,\alpha) = \left(\frac{t+r_C}{\sqrt{\alpha}}\right)^N M_\lambda\left[\frac{k_A}{\sqrt{\alpha}}(t+r_C)\right] M_\lambda\left[\frac{k_B}{\sqrt{\alpha}}(t+r_C)\right] \times \exp\left[-2r_C t - r_C^2\right]$$

We now extract

$$\exp\left[\frac{k_A}{\sqrt{\alpha}}(t+r_C)\right] \text{ and } \exp\left[\frac{k_B}{\sqrt{\alpha}}(t+r_C)\right]$$

from M_λ and $M_{\lambda'}$, respectively (to give M'_λ and $M'_{\lambda'}$). Then f reduces to

$$f(t, r_C, k_A, k_B, \alpha) = \left(\frac{t+r_C}{\sqrt{\alpha}}\right)^N M'_\lambda\left[\frac{k_A}{\sqrt{\alpha}}(t+r_C)\right] M'_\lambda\left[\frac{k_B}{\sqrt{\alpha}}(t+r_C)\right] \exp(r_C^2) \tag{154}$$

This maneuver forces M'_λ and $M'_{\lambda'}$ to be of reasonable magnitude and allows $\exp(r_C^2)$ to be extracted and combined directly with the exponential in D_{ABC}. $M'_l(z)$ is calculated using Eqs. (148) and (149) for $z > 5.0$. For $z > 16.1$, only the first term is required. When $z \leq 5.0$, the power series in Eq. (134) eliminates differencing problems.

Equation (153) suggests calculating zeros of polynomials orthogonal with weight function $\exp(-r^2)$ over the integration range $[-r_C, \infty]$. It is inconvenient, however, to recalculate these zeros for each r_C. For sufficiently large r_C, $f(-r_C)$ is negligible compared with $f(0)$ and we may employ the integration range $[-\infty, \infty]$. Thus, within this approximation the orthogonal polynomials are simply the Hermite polynomials. A table of the zeros and weights for up to 20 degree polynomials is found in Abramowitz and Stegun.[18]

The number of integration points required for a given accuracy decreases with increasing $(k_A + k_B)^2/2\alpha$, showing considerable dependence on λ, λ' and N, also. The following conservative scheme produced $Q^N_{\lambda\lambda'}$ for all λ, λ' and N to a relative accuracy of 10^{-13}.

Range of $(k_A + k_B)/2\alpha$	Number of points
$[10^2, 10^3]$	20
$[10^3, 10^5]$	10
$> 10^5$	5

Equations (153) and (154) may be used to calculate a crude approximation to $Q^N_{\lambda\lambda'}$. Using only the first term in $R_l(z)$ to calculate $M_l(z)$,

$$Q^N_{\lambda\lambda'} \simeq \frac{\exp(r_C^2)}{4\alpha^{(1/2)(N-2)} k_A k_B} \int_{-r_C}^{\infty} dt (t + r_C)^{N-2} \exp(-t^2) \tag{155}$$

Approximating $t + r_c$ as r_C and the integration limits as $[-\infty, \infty]$, we arrive at

$$Q^N_{\lambda\lambda'} \simeq \exp(r_C^2)\left(\frac{r_C}{\sqrt{\alpha}}\right)^{N-2} \sqrt{\pi}/(4k_A k_B) \tag{156}$$

This expression may then be used to determine whether a particular term in Eq. (99) is negligible, before any effort is spent calculating the possibly large number of radial integrals.

G. Symmetrized Integrals

The program segment SYMTRN forms one and two electron integrals over symmetry orbitals. For subgroups of D_{2h}, the group orbitals are defined as[3]

$$g_{iaA} = \sum_R \Gamma_i(R) R F_{aA} \tag{157}$$

where the f_{aA} are a symmetry unique set of basis functions. Here $\Gamma_i(R)$ is the character in irreducible representation "i" for the operator R. The sum runs over the left coset of the subgroup which sends nucleus A into itself. Thus, each R maps A onto a different, unique nucleus $R(A)$. For D_{2h} subgroups there are no degenerate irreducible representations and all characters are ± 1. The program numbers the operators from $1-8$ so that the binary representation of the operator index minus 1 has three bits (cba), which give the prescription for forming the operator as a product of reflections $\sigma_{yz}^a\sigma_{xz}^b\sigma_{xy}^c$ in the cartesian axes. It is assumed that the molecular symmetry elements have been properly aligned with the cartesian frame. The irreducible representations are similarly numbered from 1 to 8 so that the irrep index minus 1 has three bits (gfe), with the $\Gamma_i(R)$ for operator (abc) given by $\Gamma = (-1)^{ae+bf+cg}$.

The integrals $[g_{iaA}g_{jbB} | g_{ncC}g_{mdD}]$ are evaluated using equations derived using double cosets and projector operators. *In the following discussion, these equations are simplified and specialized to* D_{2h} *and its subgroups.*

The ordinary projector operator

$$P_i = n_G^{-1}\sum_G \Gamma_i(G)G, \tag{158}$$

where the sum runs over the entire group $\{G\}$ (of dimension n_G). Now consider the subgroup $\{U\}$ of operators which sends nucleus A into itself. Then, for cartesian Gaussian orbitals,

$$Uf_{aA} = C_a(U)f_{aA} \tag{159}$$

where $C_a(U)$ is ± 1. If $\{Q\}$ is the set of distinct operators in the left coset of $\{U\}$ then

$$P_i = n_G^{-1}\sum_Q\sum_U \Gamma_i(Q)\Gamma_i(U)QU \tag{160}$$

and

$$P_i f_{aA} = n_G^{-1}\sum_U \Gamma_i(U)C_a(U)\sum_Q \Gamma_i(Q)Qf_{aA}. \tag{161}$$

The sum of U then becomes

$$\sum_U \Gamma_i(U)C_a(U) = n_U\Delta_{ia} \tag{162}$$

where Δ_{ia} is zero if symmetry i cannot be made from orbital f_a, and unity if it can be. Also $n_Q n_U = n_G$ so

$$\Delta_{ia} g_{iaA} = n_Q P_i f_{aA} \tag{163}$$

In the following discussion, we will only consider the case where $\Delta_{ia} = 1$ so that f_{aA} can actually yield a group orbital of symmetry Γ_i. For D_{2h}, $\{Q\}$ will itself be a subgroup so

$$g_{iaA} = n_Q P_{i\{Q\}} f_{aA}. \tag{164}$$

One electron integrals with an operator symmetric under $\{G\}$ can be evaluated using double cosets. Consider an operator h with $[G,h] = 0$, then

$$I_1 = < g_{iaA} | h | g_{jbB} > = n_R n_S < P_i f_{aA} | h | P_j f_{bB} > \tag{165}$$

Here A and B are assumed to have associated subgroups $\{U\}$ and $\{V\}$ which leave them invariant, and associated distinct left cosets $\{R\}$ and $\{S\}$. From $P_i^2 = P_i$, $P_i h = h P_i$, and $P_i P_j = \delta_{ij} P_i$,

$$I_1 = \delta_{ij} n_Q n_R \; < f_{aA} | h P_i | f_{bB} > \tag{166}$$

Now define the double coset of $\{U\}$ and $\{V\}$ as a minimum set of distinct point group operators $\{L\}$ such that $\{U\}\{L\}\{V\}$ generates $\{G\}$. For the Abelian subgroups of D_{2h} considered here, each element of $\{G\}$ will be generated $\lambda_L = \dim\,[\{U\} \cap \{V\}]$ times. Hence

$$P_i = n_G \lambda_L^{-1} \sum_U \sum_L \sum_V \Gamma_i(ULV)ULV \tag{167}$$

Using $\Gamma_i(XY) = \Gamma_i(X)\Gamma_i(Y)$ gives

$$I_1 = \delta_{ij} n_G \lambda_L^{-1} \sum_L \Gamma_i(L) C_b(L) \; < f_{aA} | h | f_{bL}(B) > \tag{168}$$

Notice that λ_L is the dimension of the subgroup which leaves both A and B invariant. For D_{2h} and its subgroups $\{L\}$ is also a subgroup of dimension $n_G \lambda_L / n_U n_V$ and

$$P_i\{L\} = n_L^{-1} \sum_L \Gamma_L(L)L \tag{169}$$

is the projector in that subgroup, Hence, for this case

$$I_1 = \delta_{ij} n_R n_S \; < f_{aA} | h | \Gamma_{i,\{L\}} f_{bB} > \tag{170}$$

The charge distribution $(P_i f_{aA})(P_j f_{bB})$ can be simplified somewhat. First note that

$$(P_if_{aA})(P_jf_{bB}) = P_{ij}(f_aP_jf_b) \tag{171}$$

where P_{ij} is the projection onto the direct product irreducible representation $i \otimes j$. Also note that

$$P_k(f_aXf_b) = \Gamma_k(X)P_k(f_bXf_a) \tag{172}$$

for these Abelian groups with $X^{-1} = X$. Hence

$$P_{ij}(f_{aA}P_jf_{bB}) = P_{ij}(f_{aA}P_{j,\,\{L\}}f_{bB}) \tag{173}$$

The two electron integral is

$$\begin{aligned} I_2 &= [g_{iaA}g_{jbB} \,|\, g_{ncC}g_{mdD}] \\ &= n_Qn_Rn_Sn_T[P_if_{aA}P_jf_{bB} \,|\, P_nf_{cC}P_mf_{dD}] \end{aligned} \tag{174}$$

where centers C and D have associated subgroups $\{W\}$ and $\{X\}$, cosets $\{S\}$ and $\{T\}$, and double coset $\{M\}$. Introducing the double cosets $\{L\}$ of $\{U\}$, $\{V\}$ and $\{M\}$ of $\{W\}$, $\{X\}$, gives

$$I_2 = n_Qn_Rn_Sn_T[P_{ij}(f_{aA}P_{j\{L\}}f_{bB}) \,|\, P_{nm}(f_{cC}P_{m\{M\}}f_{dD})] \tag{175}$$

This can be further reduced by considering the subgroup $\{U\} \cap \{V\}$ which leaves A and $L(B)$ invariant and the subgroup $\{W\} \cap \{X\}$ which leaves C and M(D) invariant. If $\{N\}$ is the double coset of these two subgroups, then

$$I_2 = \delta_{ij,\,nm}n_Qn_Rn_Sn_T[f_{aA}P_{j\{L\}}f_{bB} \,|\, P_{nm,\,\{N\}}(f_{cC}P_{m\{M\}}f_{dD})] \tag{176}$$

In practice the factor n_Sn_Tn/n_M is incorporated into the definition of the charge distribution so that the sums implied in Eq. (176) involve only ± 1 factors. After the integrals over basis functions are converted into integrals over symmetry orbitals, they are sorted by the module USORT into standard order blocked on the irreducible representations i,j,n,m. At this point, labels identifying the integrals are removed and the integrals are packed into standard order. Optionally, zero integrals may be suppressed and additional words whose bit patterns give the locations of the zeros are added to the file.

3. Self Consistent Field[4]

This segment of the program is designed to find self-consistent-field wavefunctions for closed shells and spin and equivalence restricted open shells. It also has a number of options regarding the choice of the virtual orbitals, and methods of calculation. It also can do coupled Hartree-Fock calculations with a variety of operators and can restrict itself to using spherical harmonic combinations of the cartesian Gaussians.

The usual closed shell Hartree-Fock equations may be written, for the molecular orbital

$$\phi_m^a = \sum_i g_{ia} c_{im}^a \tag{177}$$

as

$$\mathbf{F}^a \mathbf{c}_m^a = \varepsilon_m^a \mathbf{S}^a c_m^a \tag{178}$$

where

$$\mathbf{F}^a = \mathbf{h}^a + \mathbf{J}^a - 1/2\mathbf{K}^a \tag{179}$$

and

$$J_{ij}^a = \sum_b \sum_k \sum_l [iaja \mid kblb] P_{kl}^b \tag{180}$$

$$K_{ij}^a = \sum_b \sum_k \sum_l [iaka \mid jblb] P_{kl}^b \tag{181}$$

$$P_{kl}^a = \sum_m c_{km}^a c_{lm}^a n_m^a \tag{182}$$

Here "a" and "b" refer to irreducible representations and i, j, k, l to group orbitals, g_{ia}, within those representations and n_m^a are occupation numbers. The generalized eigenvalue problem is solved by making a transformation

$$\mathbf{G} = \mathbf{W}^T \mathbf{F} \mathbf{W} \tag{183}$$

where

$$\mathbf{W}^T \mathbf{S} \mathbf{W} = 1 \tag{184}$$

and **W** is the matrix of vectors from the previous iteration.

The initial **W** is determined in a variety of ways. An initial guess to **C** may be specified in the input, taken from an output file of a previous Hartree-Fock calculation, or estimated by a Hückel calculation. It may be projected onto a specific subspace of the basis set specified by giving the linear combinations of basis functions which are to be used as the vector space for the calculation. This allows restriction of the calculation to spherical harmonics as well as other options.

The closed shell equations are solved iteratively. The program provides the capability of damping by averaging the Fock operators from successive iterations with an internally defined damping factor. Normally this internal damping factor becomes small near convergence. In exceptional cases it is necessary to restrict the damping to a large non-zero value in order to achieve convergence. The program also provides the capability of extrapolation of the Fock matrix based on the elements of three successive Fock operators. This accelerates convergence when the convergence is slow, but monotonic. The program also provides for some orbitals to be "frozen" at their input coefficients (after Schmidt orthogonalization).

The program can produce normal canonical virtual orbitals. It can also provide IVO's following the definition of Hunt and Goddard. These are generally better for describing excited states. It can also produce "**K**" orbitals by diagonalizing the operator $\lambda\mathbf{F} - \mathbf{K}$ (for $\lambda \simeq 0.04$) within the virtual space. For cases where some of the eigenvalues of this operator are positive, a second diagonalization of $-\mathbf{K}$ is usually done within this subspace of positive eigenvalues of $\lambda\mathbf{F} - \mathbf{K}$. These "**K**" orbitals are a greatly improved starting point for CI calculations.

If coupled Hartree-Fock is desired, the program will add to the usual h operator the operator $\Sigma e_i\theta_i$ where e_i are field strengths and θ_i are chosen from the list of property operators (dipole, quadrupole, field, field gradient). This can be useful for obtaining properties at the SCF level since the choice $e_i = 0$ simply gives $< \theta_i >$ for the SCF wavefunction.

Open shell spin-restricted calculations can be done by either the "one-Hamiltonian" or "two-Hamiltonian" method. In the two-Hamiltonian method, there is actually one extra Fock operator for each fractionally occupied orbital. The Fock operator for the doubly occupied orbitals remains as defined above with some n_m^a less than two (and possibly fractional if there is spatial degeneracy). For the open shells an extra operator is defined as

$$\Delta_q^a = \sum_b \sum_m (\gamma_{qm}\mathbf{J}_m^b + \tfrac{1}{2}\beta_{qm}\mathbf{K}_m^b) \tag{185}$$

where $\mathbf{J}_m$ and $\mathbf{K}_m$ are Coulomb and exchange operators generated from the charge density of orbital m, and the sum runs over all open shells. The Fock operator for orbital q is then $\mathbf{F}^a + \Delta_q^a$. The coefficients γ_{qm} and β_{qm} for a spin-eigenfunction (linear combination of determinants) is obtained from $(n_q^a)^{-1}\delta E/\delta\phi_q^a - \mathbf{F}^a$. For a spatially degenerate Ψ, E is replaced by the formal expression for E, the average energy of the degenerate wavefunctions where all degenerate orbitals appear in an equivalent way. If Lagrange multipliers θ_{ji} are introduced to preserve orthogonality, the equation of variation of E gives

$$\mathbf{F}^a\mathbf{c}_m^a = \varepsilon_m^a\mathbf{S}^a\mathbf{c}_m^a + \sum_j \theta_{jm}\mathbf{S}^a\mathbf{c}_j^a \tag{186}$$

for the closed shells, and

$$\mathbf{F}^a_m \mathbf{c}^a_m = \varepsilon^a_m \mathbf{S}^a \mathbf{c}^a_m + \sum_j \theta_{jm} \mathbf{S}^a \mathbf{c}^a_j \tag{187}$$

for open shells.

These can be solved by noting that

$$\theta_{jm} = \theta^*_{mj} = (c^a_j)^T \mathbf{F}^a_j \mathbf{c}^a_m \qquad j > m \tag{188}$$

If additional zero terms are added to preserve Hermiticity,

$$\{\mathbf{F}^a - \sum_j [\mathbf{S}^a \mathbf{c}^a_j (\mathbf{c}^a_j)^T \mathbf{F}^a_j + \mathbf{F}^a_j \mathbf{c}^a_j (\mathbf{c}^a_j)^T \mathbf{S}^a]\} \mathbf{c}^a_m = \varepsilon^a_m \mathbf{S}^a \mathbf{c}^a_m \tag{189}$$

gives an effective Fock operator for the doubly occupied orbital. For the open shells,

$$\begin{aligned}(\mathbf{Q}^a_m)^T \left\{ \mathbf{F}^a_m - \sum_{j>m} [\, \mathbf{S}^a \mathbf{c}^a_j (\mathbf{c}^a_j)^T \mathbf{F}^a_j + \mathbf{F}^a_j \mathbf{c}^a_j (\mathbf{c}^a_j)^T \mathbf{S}^a] \right\} \mathbf{Q}^a_m \mathbf{c}^a_m \\ = \varepsilon_m (\mathbf{Q}^a_m)^T \mathbf{S}^a \mathbf{Q}^a_m \mathbf{c}^a_m\end{aligned} \tag{190}$$

where

$$\mathbf{Q}^a_m = \mathbf{1} - \sum_{j<m} \mathbf{c}^a_j (\mathbf{c}^a_j)^T \mathbf{S}^a \tag{191}$$

This means that each open shell is diagonalized in the virtual space left over from the previously determined orbitals.

There are some cases where this formalism does not apply. For open shell singlets with two half-filled orbitals in the same symmetry the correct treatment requires that the orbitals be allowed to be non-orthogonal. In that case the program solves for a wavefunction of the form (closed shell) $(uv_i + v_i u)$. This is done in a variationally stable way so that the energy is a true minimum for the ith state without danger of "variational collapse" to a lower state of the same form. The program then transforms either to the orthogonal orbitals $s = u + v$ and $t = u - v$ in terms of which the wavefunction has the form (closed) $[a(st + ts) + b(s^2 - t^2)]$ or to natural orbitals for which the wavefunction is (closed) $[as^2 - bt^2]$.

The program can also find ground state wavefunctions of the form (closed shell) $(as^2 - bt^2)$ where "s" and "t" are orbitals of different symmetry. Notice that when

s and t are of the same symmetry, this would be equivalent to the "split shell" wavefunction. Also, for ground state wavefunctions, this is equivalent to one GVB pair.

Certain hole and/or particle states cannot be obtained by the two-Hamiltonian method. For some of these, a one-Hamiltonian method is preferred. For this method it is best to assume that all **F** and **S** matrices have already been transformed to **G** and **1** matrices using the best current (orthogonal) guess to the orbitals. Then, for each symmetry, a single new operator can be constructed which will be diagonal at convergence. For example, consider a state with a "hole" in orbital μ below the Fermi surface. If the orbitals are partitioned into blocks D, S, V (double, single, or virtual occupancy),

$$\mathbf{G}' = \begin{array}{lll} G & i \in D, & j \in V \\ G_\mu & i \in S, & j \in V \\ G^* & i \in S, & j \in D \end{array}$$

Here $G^* = 2\mathbf{G} - \mathbf{G}_\mu$ is the Fock operator which would have been written for the fully closed shell system. Its appearance here is a reflection of Koopmann's theorem. The remaining undefined elements have no effect on the wavefunction at convergence, but are chosen to optimize convergence by making the eigenvectors of $\mathbf{G}'$ give the approximate transformation matrix between the current orbitals and the converged orbitals. A suitable choice for hole states is

$$\mathbf{G}' = \begin{array}{rl} G^* & i, j \in D \\ G^* & i, j \in S \\ G_\mu + \delta_{ij} J_{\mu\mu} & i, j \in V \end{array}$$

The primary advantage of this one-Hamiltonian method is variational stability. One can solve for core hole states without danger of collapse to a hole in a higher energy orbital. This method can be generalized to treat particle states with one electron excited into a high energy orbital, and to particle-hole states. For particle states, the off-diagonal blocks are given by the same equations as for hole states. For the diagonal blocks with the particle in orbital ϕ_m, the choice

$$\mathbf{G}' = \begin{array}{rl} G^* - \delta_{ij} J_{\mu\mu} & i, j \in D \\ G_\mu & i, j \in S \\ G_\mu & i, j \in V \end{array}$$

produces a variationally stable method.

4. Integral Transformation

In preparation for a configuration interaction (CI) calculation, the integrals over group orbitals (symmetry adapted basis) must be transformed into integrals over molecular orbitals. For large molecules, one frequently relegates certain molecular orbitals to the core. That is, the CI is done with a "frozen core" of molecular orbitals from which no excitations are allowed. The transformation program takes advantage of this by moving the effect of these orbitals into a modified nuclear-nuclear repulsion

$$V'_{NN} = V_{NN} + \sum_{\text{core}} < \phi_i | h + J_{\text{core}} - 1/2K_{\text{core}} | \phi_i > \tag{192}$$

and a modified nuclear-electron attraction

$$(V'_{Ne})_{ij} = < \phi_i | V_{Ne} + J_{\text{core}} - 1/2K_{\text{core}} | \phi_j > \tag{193}$$

After this is done (with similar changes in property integrals), no further reference is needed to the core molecular orbitals and the active orbitals are renumbered with orbital "1" being the first active orbital. The transformation program also assumes that some of the virtual orbitals may not be used in the CI so they can be omitted from the transformation.

The actual transformation is done as matrix multiplication. First a symmetry block of integrals over group orbitals is repartitioned,

$$R_{iajb}^{(mcnd)} = [iajb \,|\, mcnd] \tag{194}$$

The R matrix is then transformed;

$$R' = (\mathbf{W}^a)^T [\mathbf{R}(\mathbf{W}^b)]. \tag{195}$$

The resulting R' are written to a direct access file in a chain sort and brought back into computer memory as

$$U_{mcnd}^{(i'aj'b)} = [i'aj'b \,|\, mcnd] \tag{196}$$

where the $'$ denotes molecular orbital indices. This is then transformed

$$U' = (\mathbf{W}^c)^T [\mathbf{U}(\mathbf{W}^d)] \tag{197}$$

This again is written to a direct access file and brought back into memory so it can be put out to disk as a sequential file.

At present, the program has an inefficiency at this point. The integral file produced by TRNX is not the order desired for the CI program so the file must pass through SORTIN which puts it into the desired order. Following comments by Bowen Liu, the final order no longer has integrals blocked by symmetry. Rather, the file consists of

(a) integrals needed to form diagonal CI matrix elements — h_{ii}, $[ij\,|\,ij]$, $[ii\,|\,jj]$.

(b) integrals needed to form CI matrix elements when the configurations differ by a single excitation. For each i, j pair, this is h_{ij}, $[ij\,|\,kk]$, $[ik\,|\,jk]$ (all k).

(c) integrals needed to form CI matrix elements when the configurations differ by a double excitation. For each $ijkl$, a triad of integrals are stores — $[ij\,|\,kl]$, $[ik\,|\,jl]$, $[sub\ <i\ sub\ lambda>\,|\,jk]$.

5. Configuration Interaction

The configuration interaction program employs a brute-force procedure. Based on the input, a list of configurations is generated. From these, a sub-list may be selected by second-order perturbation theory. Then the CI matrix elements are generated and written to a file.

The configuration input possibilities are varied. The configurations may simply be read in the input or from previous output files. Alternatively one or more configurations can be specified and all single and double excitations generated. Finally, a list of sets of orbitals with ranges of occupation of electrons for each set can be specified and all configurations generated. From the list of generated configurations some sublist can be selected based on rules about excitation levels from a list of configurations. For every configuration generated, all spin-couplings are normally used. Optionally, only the one "interacting subspace" spin coupling may be used [5] These programs are presently limited to spin S of no more than 6, and no more than 8 open shells.

The program will form a zero'th wavefunction by diagonalizing the CI hamiltonian in some subspace. This may produce coefficients for several excited states. The program will then select doubly-excited configurations from outside the reference space based on either the estimated second-order Epstein-Nesbet energy[6]

$$E_I^{(2)} = \frac{|<\Phi_I|H|\Psi_j^0>|^2}{(H_{II} - E_j^0)} \tag{198}$$

or on first order wavefunction coefficients. Singly excited configurations are normally always retained, although they can be selected. Finally, the program will generate a CI matrix using the reference space plus selected configurations.

For up to 6 open shells the CI matrix elements are generated by explicit formulas. That is, the configurations $\Phi_{\nu I}$ associated with a list of singly and doubly occupied orbitals listed in increasing order of the MO index within each symmetry $\{\phi_{1,1}\ \phi_{2,1} \dots \phi_{1,2}\ \phi_{2,2} \dots\}$ is defined as

$$\Phi_{\nu,\, I} = A[\phi_{i_1}(1)\phi_{i_2}(2) \dots \phi_{i_N}(N)\Xi_\nu] \tag{199}$$

where Ξ_ν is a spin eigenfunction formed as a linear combination of spin products. Clearly, in $\Phi_{\nu,\, I}$ all doubly occupied orbitals can be moved to the beginning of the list without a change of sign. If this is done,

$$\Phi_{\nu I} = A[\phi_{i_1}(1)\alpha,\ \phi_{i_1}(2)\beta, \dots \phi_{i_k}(2k-1)\alpha\phi_{i_k}(2k)\beta\phi_{j_1}(2k+1) \\ \dots \phi_{j_n}(N)\Xi_\nu(2k+1 \dots N)$$

where Ξ is a standard spin eigenfunction for $N-2k$ electrons. The standard orthonormal spin eigenfunctions were generated by successive vector coupling of the last electron to a spin eigenfunction of the previous electrons according to the usual branching diagram. These Ξ have the form

$$\Xi_\nu(1 \dots p) = \sum_Q \chi_{q_1}(1)\chi_{q_2}(2) \dots \chi_{q_r}(p)C_{Q,\, \nu} \tag{200}$$

where the χ_i are α or β.

A block of the Hamiltonian associated with orbital products $\{I\}$ and $\{J\}$ contains the matrix elements

$$H_{\nu I,\, \mu J} = < \phi_{\nu I} | H | \phi_{\mu J} > \tag{201}$$

These may always be expressed as

$$H_{\nu I,\, \mu J} = \sum_\lambda U_\lambda \Gamma^\lambda_{\nu\mu} \tag{202}$$

where λ indexes a few one and two electron integrals, $\Gamma^\lambda_{\nu\mu}$ are fixed coefficients which are independent of the integrals and of the orbital indices. They depend only on the pattern of the excitation and intervening singly occupied orbitals in I and J. That is $a^2bcd \rightarrow b^2cd^2$ has a definite set of Γ coefficients. These are built into tables the program accesses for up to 6 unpaired electrons.

For 7 or 8 unpaired electrons, the Γ tables became too extensive. The program then relapses to an older procedure. The matrix elements are first written as

$$H_{\nu I,\, \mu J} = \sum_P \sum_Q C_{Q\nu} C'_{P\mu} < D_{QI} | H | D_{PJ} > \tag{203}$$

where the D_{QI} are Slater determinants. Then the turn-over rule is used to reduce the number of terms in the double sum,[7]

$$H_{\nu I,\mu J} = \sum_{\xi} \sum_{P} B_{\xi\nu} C'_{P\mu} < D_{\xi I} | H | D_{P\mu} > \tag{204}$$

Here ξ runs over a sublist of the D_{QI} Slater determinants of dimension equal to the number of spin-couplings possible (the same as the number of ν).

In order to limit the number of times the integral file must be read, the program actually forms a "formula" file. This contains only a list of the (I,J) pairs which will give rise to non-zero H_{IJ} and the location of the needed integrals on the integral file. This formula file is chain sorted to a random access file where each chain contains a list of all the H_{IJ} which can be formed from a given memory load of integrals. The CI matrix element file then contains the H_{IJ} in a more-or-less random order.

The CI program can also find the B_K estimates of the energy.[8] For a given reference space, the effective hamiltonian

$$H_{IJ}^{eff} = H_{IJ} + \sum_{k} < \Phi_I | H | \Phi_K > \; < \Phi_K | H | \Phi_J > / (E^* - H_{KK}) \tag{205}$$

is formed and diagonalized to give a quasi-degenerate second order estimate of the energy.

6. Eigenvalue Program

The segment RTSM finds the lowest few eigenvalues and eigenvectors of the CI matrix.[9] The algorithm is based on an expansion of the true eigenvectors in a set of orthonormal vectors $\{b_i\}_{i=1}^{K}$ arranged in the columns of a matrix B as

$$\mathbf{x}_j = \mathbf{B}\mathbf{c}_j \tag{206}$$

with

$$\mathbf{B}^T\mathbf{H}\mathbf{B}\mathbf{c}_j = E_j\mathbf{c}_j \tag{207}$$

The E_j are related to the true eigenvalues λ_j of $\mathbf{H}$ by

$$E_j \geq \lambda_j \tag{208}$$

provided both the E_j and λ_j are numbered in increasing order. Further, if an additional column is added to $\mathbf{B}$, to increase its dimension to $K + 1$,

$$E_j^K \geq E_j^{K+1} \geq \lambda_j \tag{209}$$

Hence, convergence of the energies are monotomic non-increasing as the set **B** is augmented.

The vectors **b** are chosen by first order perturbation theory. Suppose the residual $\mathbf{r}_i$ is defined by

$$\mathbf{r}_i = (\mathbf{H} - E_i)\mathbf{x}_i \tag{210}$$

then (for $E_i \neq H_{ii}$)

$$x_{ki}^{(1)} = r_{ki}/(E_i - H_{ii}) \tag{211}$$

The next b_i is then chosen to be the component of $x_i^{(1)}$ outside the current vector space

$$\mathbf{y} = (\mathbf{1} - \mathbf{B}\mathbf{B}^T)\mathbf{x}^{(1)} \tag{212}$$

$$\mathbf{b}_{k+1} = \mathbf{y}/||y|| \tag{213}$$

In the present implementation of the program, only one **b** vector is added each iteration. This requires that one of the $\mathbf{x}_i$ be chosen as the target vector for improvement on that iteration. This vector is chosen as the one with the maximum $|c_{Ki}|$ from the previous iterations.

The program is organized for matrices which are large compared to the memory of the computer. For that reason, only two vectors of the dimension of the CI are allowed in memory at once. In order to accomplish this both the $\mathbf{b}_i$ and the $\mathbf{HB}_i$ vectors are stored on disk. The matrix $\mathbf{B}^T\mathbf{HB}$ is formed incrementally by bordering the existing $\mathbf{B}^T\mathbf{HB}$ with $\mathbf{B}^T(\mathbf{Hb}_{k+1})$.

Occasionally the matrix **B** may become inconveniently large. When this happens, the program replaces the current $\{\mathbf{b}_i\}$ with $\{\mathbf{Bc}_i\}$ and the current $\{\mathbf{HB}_i\}$ with $\{(\mathbf{HB})\mathbf{c}_i\}$. Care is taken at this step to preserve orthogonality. Also $(\mathbf{Bc}_j)^T(\mathbf{HBc}_i)$ is computed to reinitialize $\mathbf{B}^T\mathbf{HB}$ rather than setting it to $E_i\delta_{ij}$.

At present, the program tests for convergence on $|c_{Ki}| \leq T$ for the target vector from which $\mathbf{b}_K$ was constructed. The correlation energy is then usually converged to a relative error T^2. The diagonalization program also prepares an analysis of the results. An extrapolated energy is estimated from

$$E_{\text{est}} = (E_{CI} - E_{REF})(E^{(2)}/E^{(2)}_{\text{kept}})[C_o^2/(2C_o^2 - 1] + E_{REF} \tag{214}$$

Here E_{REF} is the zero'th-order energy found in CISTAR for the reference space wavefunction, E_{CI} is the RTSIM eigenvalue, $E^{(2)}$ is the second order perturbation theory result, $E^{(2)}_{\text{kept}}$ is the portion of the second order perturbation theory corresponding to the configurations kept in the CI, and C_o^2 is the sum of the squares of the coefficients in the reference space. The factor $E^{(2)}/E^{(2)}_{\text{kept}}$ extrapolates the CI to account for the doubly excited configurations not included. The factor $C_o^2/(2C_o^2 - 1)$ extrapolates the CI to account for multiple unlinked higher excitations (quadrupole, hexapole, etc.).[10]

The orbital products may be ordered on either their contribution to the wavefunction norm or to the energy. The contribution to the norm is defined as $[\Sigma x_{\nu I}^2]^{1/2}$ where the sum is over the spin coupling associated with the same orbital product. Similarly, the energy contribution is estimated from a second order formula

$$\delta E_I = \sum_{\nu\mu}(H_{\nu I\mu I} - E\delta_{\nu\mu})x_{\nu I}x_{\mu I} \tag{215}$$

which approximates the energy change if these configurations are omitted.

The major problem with perturbation selected CI is that it necessarily is a subset of SDCI which is not size consistent nor size extensive. While it is possible to estimate the correction for this[10] in some cases, a more reliable method would be desirable. A step in that direction is obtained by partitioning the CI matrix as

$$\mathbf{H} = \begin{pmatrix} \mathbf{H}_o & \mathbf{h}^T \\ \mathbf{h} & \mathbf{G} \end{pmatrix} \tag{216}$$

where $\mathbf{H}_o$ is the reference space matrix , $\mathbf{G}$ is the matrix in the selected space, and $\mathbf{h}$ is the connecting matrix. Then the eigenvector can also be partitioned as

$$\mathbf{x} = \begin{pmatrix} \mathbf{a} \\ \mathbf{b} \end{pmatrix} \tag{217}$$

with intermediate normalization $a^T a = 1$. The partitioned eigenvalue equation is then

$$\begin{aligned} \mathbf{H}_o\mathbf{a} + \mathbf{h}^T\mathbf{b} &= \mathrm{E}\mathbf{a} \\ \mathbf{ha} + \mathbf{Gb} &= \lambda\mathbf{b} \end{aligned} \tag{218}$$

For $\lambda = E$, this is the usual matrix eigenvalue problem. But for $\lambda = E_o$, this gives

$$\mathbf{b} = (\lambda\,\mathbf{1} - \mathbf{G})^{-1}\mathbf{ha} \tag{219}$$

which is the Hylleraas variation perturbation theory result. In linearized coupled cluster, LCC, theory, "a" is found from

$$\mathbf{H}_o\mathbf{a} = E_o\mathbf{a} \tag{220}$$

and E is given by

$$E = \mathbf{a}^T\mathbf{H}_o\mathbf{a} + \mathbf{a}^T\mathbf{h}^T\mathbf{b} \tag{221}$$

Alternatively, this can be modified to resemble B_K theory (and other quasi-degenerate models) by determining "a" from

$$[\mathbf{H}_o + \mathbf{H}^T(\lambda\,\mathbf{1} - \mathbf{G})^{-1}\mathbf{h}]\mathbf{a} = \mathrm{Ea} \tag{222}$$

The value of λ can either be chosen as the eigenvalue of $\mathbf{H}_o$ or as the expectation value $\mathbf{a}^T\mathbf{H}_o\mathbf{a}$. The LCC method is size-consistent, but not accurate when "**a**" differs from the eigenvector of $\mathbf{H}_o$. The choice of "**a**" as the eigenvector of $\mathbf{H}_{eff}$ is not size-consistent, but is more accurate in nearly degenerate cases.

7. Molecular Properties

The program computes a variety of one-electron properties as expectation values of the respective operator. The operators now considered are molecular dipole and quadrupole moments, p^4, nuclear force and field gradient, and nuclear delta function.

The properties program, MOLINT, first forms the density matrix in the MO basis. This may then be diagonalized to give the natural orbitals in the MO and AO basis for iterative natural orbital calculations.[12] The properties are found from the $Tr\rho O$ for each operator O with both the total density $\rho_\alpha + \rho_\beta$ and the unpaired spin density $(\rho_\alpha - \rho_\beta)/2S$ (where S is the total spin). The field gradient and anisotropic hyperfine tensors are diagonalized to give the irreducible components. For properties found from the total charge density, the electron contribution is subtracted from the nuclear contribution.

There are also program segments for computing transition properties. TMOM uses two wavefunctions and any of the property operators to form $< \Psi_1 | O | \Psi_2 >$. This is restricted to the case that both Ψ_1 and Ψ_2 are expanded in the same MO basis. Another program, SPNORB, evaluates the transition matrix elements of the effective spin orbit operator, $\Sigma\zeta_A \mathbf{l}_{ia} \cdot \mathbf{s}_i r_{iA}^{-3}$. The segment PES evaluates[13] the Dyson orbital $< \Psi_{N-1} | \Psi_N > = \phi$ and MOMAVGLT plots its spherically averaged momentum distribution $\int | \phi(\mathbf{p})^2 | d\Omega_p/4\pi$.

The program OCCUP evaluates Mulliken and Löwdin populations and occupation numbers from a SCF or CI wavefunction.[14] It will find populations of user defined atomic orbitals or fragment orbitals defined as linear combinations, spanning one or more atomic centers, of the original atom-centered basis functions. This assumes that fragment calculations have been carried out in the same basis to define meaningful orbitals onto which the CI density matrix can be projected. If the density matrix in the original basis is P, then the projected density matrix is $P' = S^{-1}U^TPUS^{-1}$, where S is the overlap matrix of the fragment orbitals, and U is the overlap between the fragment orbitals and the original basis functions. The trace of $P'S$ will be N if the fragment basis spans the full space of the original functions. Otherwise, $Tr(P'S) < N$ and the defect represents electrons outside the fragment orbital space. The matrix $P'S$ is partitioned in the Mulliken scheme into $\Sigma P'_{ij}S_j$, which represents the population in fragment orbital i.

References

1. L.E. McMurchie and E.R. Davidson, J. Comput. Phys., **26**, 218 (1978).
2. L.E. McMurchie and E.R. Davidson, J. Comput. Phys., **44**, 289 (1981).
3. E.R. Davidson, J. Chem. Phys., **62**, 400 (1975).
4. E.R. Davidson, Chem. Phys. Letters, **21**, 565 (1973); E.R. Davidson, J. Chem. Phys., **57**, 1999 (1972); L.Z. Stenkamp and E.R. Davidson, Theo. Chim. Acta, **30**, 283 (1973); C.F. Jackels and E.R. Davidson, Int. J. Quantum Chem., **8**, 707 (1974); H.L. Hsu, E.R. Davidson and R.M. Pitzer, J. Chem. Phys., **65**, 609 (1976); E.R. Davidson and L.Z. Stenkamp, Int. J. Quantum Chem. Symp., **10**, 21 (1976); C.F. Bender, V. McKoy and E.R. Davidson, J. Chem. Phys., **67**, 2191 (1977); D. Feller and E.R. Davidson, J. Chem. Phys., **74**, 3977 (1981).
5. K. Iberle and E.R. Davidson, J. Chem. Phys., **76**, 5385 (1982).
6. E.R. Davidson, J. Chem. Phys., **48**, 3169 (1968); E.R. Davidson and C.F. Bender, Chem. Phys. Letters, **59**, 369 (1978); E.R. Davidson, L.E. Nitzsche, and L.E. McMurchie, Chem. Phys. Letters, **62**, 467 (1979).
7. E.R. Davidson, Int. J. Quantum Chem., **8**, 83 (1974).
8. L.E. Nitzsche and E.R. Davidson, J. Chem. Phys., **68**, 3103 (1978); L.E. Nitzsche and E.R. Davidson, J. Am. Chem. Soc., **100**, 7201 (1978); E.R. Davidson, L.E. McMurchie, and S.J. Day, J. Chem. Phys., **74**, 5491 (1981); D.C. Rawlings and E.R. Davidson, Chem. Phys. Letters, **98**, 424 (1983).
9. E.R. Davidson, J. Comput. Phys., **17**, 87 (1975); E.R. Davidson in *Proceedings of the NATO Advanced Study Institute on Methods in Computational Physics*, G.H.F. Diercksen and S. Wilson, Eds., Reidel, Dordrecht, p. 95 (1983); E.R. Davidson, Comput. Phys. Comm., **53**, 49 (1989).
10. S. Langhoff and E.R. Davidson, Int. J. Quantum Chem., **8**, 61 (1974); E.R. Davidson, in *The World of Quantum Chemistry*, R. Daudel and B. Pullman, Eds , Reidel Publ. Co., pp. 17-30 (1974); E.R. Davidson and D.W. Silver, Chem. Phys. Letters, **52**, 403 (1977).
11. R.J. Cave and E.R. Davidson, J. Chem. Phys., **88** (9), 5770 (1988); R.J. Cave and E.R. Davidson, J. Chem. Phys., **89**, 6798 (1988).
12. C.F. Bender and E.R. Davidson, J. Phys. Chem., **70**, 2675 (1966); *Reduced Density Matrices in Quantum Chemistry*, E.R. Davidson, Academic Press (1976).
13. A.O. Bawagan, C.E. Brion, E.R. Davidson and D. Feller, Chem. Phys., **113**, 19 (1987).
14. E.R. Davidson, J. Chem. Phys., **46**, 3320 (1967).
15. S.F. Boys, Proc. Roy. Soc., **A200**, 542 (1950).
16. I. Shavitt, Mtehods of Computational Physics, **2**, 1 (1963).

17. I.S. Gradshteyn and I.M. Ryzhik, *Tables of Integrals Series and Products*, A. Jefferey, Ed., 4th Edition, Acadamic Press (1980).
18. M. Abramowitz and I.A. Stegun, *Handbook of Mathematical Functions*, Dover.
19. W.J. Hunt and W.A. Goddard, Chemy. Phys. Lett., **3**, 414 (1969).

Chapter 11.

ALCHEMY II: A Research Tool for Molecular Electronic Structure and Interactions

A.D. McLean, M.Yoshimine, B.H. Lengsfield†,
P.S. Bagus, and B. Liu

IBM Research Division
Almaden Research Center
650 Harry Road
San Jose, CA 95120-6099, U.S.A.

Introduction

ALCHEMY II[1] is a recent version of the collection of programs, written primarily at IBM's Almaden Research Center, which have been the dominant research tool of the computational chemistry group in that laboratory for over twenty years. Early versions of ALCHEMY[2] pioneered the use of sorting procedures, introduced into computational chemistry by Yoshimine,[2,3] and now universally used, for transformations that were once seen as a crippling bottleneck. Later versions brought in superior MCSCF algorithms[4] and introduced symbolic and prototype direct CI programs[4,5] similar in efficiency to, but more flexible than, the widely heralded GUGA[6] based programs.

ALCHEMY II complements the HONDO, KGNMOL, and SIRIUS *ab initio* quantum chemistry codes also available in MOTECC-90. While it overlaps the functions of these codes in several respects, it adds integral generation for Slater-type basis functions for linear molecules, flexibility in the choice of configurational bases for MCSCF and CI calculations, and transition moments between electronic states, of the same or of different symmetry, whose configuration spaces are generated from common orbitals.

This chapter will not be comprehensive. Background and general development is amply covered in other chapters, particularly those on HONDO and KGNMOL. We restrict our presentation to a listing of the program modules contained in MOTECC-90 together with a discussion of an application, weak molecular inter-

† Present address: Lawrence Livermore Laboratory, Livermore CA.

actions, which has been a long term research theme at Almaden and which exemplifies the need for the level of flexibility in configuration space generation that we make available in the program. As an Appendix, we give the previously unpublished MCSCF algorithm of Liu and Lengsfield, still at the forefront of MCSCF development even though several years old. Another Appendix describes the only program module not written by our research group, the Gaussian integral generator, MOLECULE, kindly made available by authors Jan Almlöf and Peter Taylor.

ALCHEMY II is under active development; program modules incorporating analytical derivative codes for finding stationary points and reaction pathways on potential energy surfaces and for electronic properties, as well as determinant based MCSCF and CI, will be available in the next version. While discussing future development, we note that similar capabilities will be added to, or already exist in, other major quantum chemistry codes. Clearly, no one code will be the best in all aspects of a quantum chemistry calculation. Indeed, it would cripple the research efforts of any group to keep their code at near perfection in all aspects; yet at present there is no convenient way to use the best features of the several major codes in doing a calculation.

We are addressing this problem by building an architecture, ARCHEM, which establishes a framework for communication between programs which are either loosely coupled (each major program is made available as a black box) or tightly coupled (modules inside a program are accessible so that a computational path through modules of different major codes can be chosen). Under ARCHEM a scientist will generate a non-redundant input that describes a calculation, specify the computational path for the calculation, and receive output both informative and usable for future calculation. With varying degrees of success this aim has been achieved within a major program (GAUSSIAN, CADPAC, HONDO, or ALCHEMY, as examples); ARCHEM will take one more step toward an optimum user interface and will make conveniently available the resources of several major programs. This long overdue attempt to bring order into the house of *ab initio* computational chemistry is being done in collaboration with the research group of Bjorn Roos at Lund University.

ALCHEMY II Program Modules

In the program modules listed, calculations using gaussian basis sets can make use of up to D_{2h} symmetry. Calculations with linear Slater-type basis sets make full use of $C_{\infty v}$ symmetry and partial use of $D_{\infty h}$ symmetry.

VMOL3 or MOL3: This is the Gaussian integral generator of Almlöf and Taylor described in Appendix 11A. It handles generally contracted (needed for Atomic Natural Orbital (ANO) basis sets) with up to i-type basis functions. Segmented contracted basis sets are treated as a special case of generally contracted. Spherical or cartesian basis sets can be requested. VMOL3 is the vectorized version, considerably faster than the non-vectorized MOL3.

PINT3: Program for 1-electron property integrals over generally contracted gaussians. Complements MOL3.

LINT50: Integral generator for Slater-type basis functions for linear molecules only. Handles angular quantum numbers up to $l=8$. An outline of the computational method is available in reference 2. Two electron $1/r_{12}$ integrals, 1-electron energy integrals, and a wide variety of 1-electron property integrals are available.

VOT3 or OT3: Vectorized or non-vectorized versions of a program that performs a wide variety of order and transform functions, accepting input from the integral generators listed above. This module also contains the code for generating population matrices, including those needed for distributed multipole analyses, and performing localization transformations needed in the study of weak molecular interactions.

SCFM50 or SCFL50: Single configuration SCF program, notable for the wide variety of open shells that can be handled[7] The same function is provided as a special case of the MCSCF module described below, but sometimes it is more convenient to use this module. SCFM50 is the version for gaussian basis sets; SCFL50 the version for linear Slater.

SYMBOLM: This module generates configuration spaces, and prepares symbolic energy expressions (for CI calculations[5] or Prototype CI energy expressions (for MCSCF calculations, see Appendix 11B). This module is at the heart of ALCHEMY II; it provides the flexibility needed for exploratory electronic structure calculations and does it using efficient algorithms.

VMCSCF or MCSCF: Vectorized or non-vectorized MCSCF program that uses rapidly convergent second order procedures; routinely used for MCSCF configuration spaces of up to 10000 configuration state functions (CSF's). The computing procedures are described at length in Appendix 11B.

INTSRT and CI2D: Module INTSRT sorts canonically ordered integrals over the orbital basis for a CI calculation into the order required to match the symbolic energy expressions (SYMBOLM output) for efficient processing in module CI2D which does a direct CI calculation, where the Hamiltonian matrix is not explicitly formed. Several lowest eigenstates can be requested. Configuration spaces of several hundred thousand are routine; some calculations with up to two million CSF's have been done.

TMSYMB: Generates the 1-particle density expressions for matrix elements connecting electronic states of different symmetries.

TMDSSO: Transforms integrals and computes transition dipoles connecting electronic states of different symmetry. The code for transition dipoles connecting states of the same symmetry is contained in module CI2D.

COPYV: This is a utility program facilitating vector preparation for input to the various ALCHEMY II modules.

Weak Molecular Interactions

This section outlines a hierarchy of variational CI fragment interaction energy calculations, shows where the interacting correlated fragments (ICF) model of Liu and McLean[8] fits into the hierarchy, and demonstrates how investigation of that model requires the flexibility in configuration space generation that we have built into ALCHEMY. At the same time, it argues for wider use of the model.

We consider the interaction of two molecular fragments, A and B. Each fragment, and we specifically discuss A, has a wave function expanded in a configuration space generated from orbitals which we classify as C_A (core), I_A (internal, or active), and X_A (external). Core orbitals are fully occupied in each generated configuration; internal orbitals are strongly occupied orbitals possibly augmented by leading correlating orbitals, normally determined by SCF or MCSCF calculations which optimize the dominant part of the fragment wave function; external orbitals complete the remainder of the space for the chosen basis set. A's configuration space is generated by assigning its n_A active electrons (total number minus the number of core electrons) to active and external orbitals:

$$\begin{array}{cc} I_A & X_A \\ & \\ n_A & 0 \\ n_A - 1 & 1 \\ n_A - 2 & 2 \\ n_A - 3 & 3 \\ \cdot & \cdot \\ \cdot & \cdot \end{array} \qquad (1)$$

Assignment n_A 0 generates zeroth order configurations, $n_A - 1$ 1 generates first order configurations, etc.; the number of external electrons specifies the order of a configuration. An n-th order fragment wave function is expanded in configuration spaces through order n.

When molecular fragments A and B interact to form a supermolecule, the configuration space in which the supermolecule wave function is expanded should contain, at least, the direct product of the configuration spaces of the two fragments. And, to increasing levels of accuracy, interacting zeroth, first, and second order supermolecule wave functions can be expanded in configuration spaces derived from direct products of zeroth, first, and second order fragment configuration spaces.

The **interacting zeroth order fragments** supermolecule configuration space must be

$$\begin{array}{ccc} I_A & I_B & X_{AB} \\ \leftarrow n_A + n_B \rightarrow & & 0 \end{array} \tag{2}$$

The interacting zeroth order wave function, therefore, comes from a complete CI in the $I_A \oplus I_B$ basis. Added to the 'neutral' configurations derived from electron assignments $n_A\, n_B\, 0$ to $I_A\, I_B\, X_{AB}$, are all 'ionic' electron assignments such as $n_A - 1\;\; n_B + 1\;\; 0$, for example. The orbital space X_{AB}, unused in the interacting zeroth order configuration space, is $X_A \oplus X_B$.

Interacting zeroth order supermolecule wave functions will have substantial orbital sensitivity. Suppose, for example, that I_A and I_B are optimized in fragment MCSCF calculations; then the computed interaction energy, as a function of fragment separation, will be quite different when these orbitals are used for all fragment separations to that obtained when I_A and I_B are reoptimized in supermolecule MCSCF calculations at each separation.

Note that because the zeroth order fragment and interacting zeroth order supermolecule configuration spaces are complete, orbital orthogonality constraints, imposed in carrying out the calculations, have no effect on the variationally determined total energies, and therefore no effect on the interaction energy.

The **interacting first order fragments** supermolecule configuration space must be

$$\begin{array}{ccccc} I_A & & I_B & & X_{AB} \\ \leftarrow & n_A + n_B & & \rightarrow & 0 \\ \leftarrow & n_A + n_B - 1 & & \rightarrow & 1 \\ n_A - 1 & & n_B - 1 & & 2 \end{array} \tag{3}$$

complete, including all ionic terms, through zeroth and first order, but restricted in second order to simultaneous electron excitations from I_A and I_B. This restriction makes an interacting first order wave function size consistent; at dissociation, the supermolecule energy is the sum of first order fragment energies. Orbital sensitivity in interacting first order wave functions is much less than in interacting zeroth order because the complete first order space in the former largely takes care of orbital relaxation effects; these, at least for weak interactions, will be taken care of just as well by first order mixing in a calculation where I_A and I_B are not reoptimized as in one where they are reoptimized in a supermolecule MCSCF calculation.

The importance of the second order terms in the interacting first order wave function must be emphasized. In weak interactions they describe the dispersion force; in strong interactions they contain the bulk of the molecular extra correlation energy due to inter-fragment electron correlations, present in the molecule but not in the fragments, associated with bond formation.

Equally important, the second order terms in the interacting first order configuration space must be constructed from **localized orbitals**. Because this second order space is not complete, its contribution to the interacting first order supermolecule wave function depends on both the localization algorithm and the imposition of orbital orthogonality constraints.

From a conceptual point of view, the interacting first order wave function restricts intra-fragment correlation effects to those contained in the internal orbitals. No such restriction is made on inter-fragment correlation effects, which come from the full orbital space; in fact, the desired orbital space is one whose extension can yield no further inter-fragment correlation.

The **interacting second order fragments** supermolecule configuration space must be

$$\begin{array}{cccc} I_A & & I_B & X_{AB} \\ \leftarrow & n_A + n_B & \rightarrow & 0 \\ \leftarrow & n_A + n_B - 1 & \rightarrow & 1 \\ \leftarrow & n_A + n_B - 2 & \rightarrow & 2 \\ n_A - 1 & & n_B - 2 & 3 \\ n_A - 2 & & n_B - 1 & 3 \\ n_A - 2 & & n_B - 2 & 4 \end{array} \qquad (4)$$

complete through second order, and restricted in third and fourth order to products of localized single with localized double excitations, and localized double with localized double excitations. In actual calculation, with the exception of systems with very few electrons (four or less) this model cannot be pursued computationally without approximation; normally, it is approximated by neglecting third and fourth order contributions. That is, an interacting second order supermolecule wave function is approximated by a second order supermolecule wave function. Neglect of the third and fourth order terms results in a loss of size consistency; the dissociated energy computed from a second order wave function is higher than the sum of second order fragment energies.

Interacting second order and second order supermolecule interaction energies will agree only when the effect of the third and fourth order terms on the former is the same at all fragment separations. The difference between the two will decrease rapidly as leading intra-fragment correlating orbitals are added to the internal orbital sets I_A and I_B; in fact, as this is done, the interaction energy converges very rapidly to the complete CI interaction energy, as long as basis set superposition error (BSSE) adjustments are made. (See our discussion below).

The ICF model of Liu and McLean is the interacting first order fragments model. Liu and McLean proposed this model as one which balanced the addition of intra-fragment and inter-fragment correlation effects as fragment orbitals were added to the internal sets, defining a series of calculations which they claimed would be rapidly convergent to the BSSE adjusted complete CI interaction energy; the 'true' interaction contained in the given basis for orbital expansion. They named the model 'Interacting Correlated Fragments' since this series converged as a function of the amount of electron correlation included in the fragment descriptions.

Of course, as intra-fragment correlating orbitals are moved from the external set into I_A and I_B, interacting zeroth order, interacting first order, and interacting second order will all generate interaction energies that converge to the complete CI result; the question is which offers the most computationally efficient path to get there. For van der Waals and hydrogen bonded interactions we believe that the ICF model defines that most efficient path. We point to our results so far on He_2[8], Be_2[9], Mg_2[10], Me-H_2O[11] and Ar_2[12] as evidence, realizing that a much wider range of systems (hydrogen bonded, open shell, ionic, for example) must be studied before our belief can be forcefully defended.

So far, there has been no systematic investigation of interacting first order fragment interaction energies for strong bonding; a tantalizing area for research where the most interesting effect to investigate will be that of the much larger overlap of localized I_A and I_B orbitals in the second order configuration space.

The virtue of the ICF model is clear: it keeps to a minimum the amount of intra-fragment electron correlation that is brought into a calculation for some desired level of accuracy in the interaction energy. Only essential terms are introduced into the supermolecule wave function; these can then be systematically dissected to identify the qualitative origins of, and quantitative contributions to, the interaction energy. Unfortunately, even though ICF configuration spaces are orders of magnitude smaller than for interacting second order, ICF calculations can generally only be done without approximation when there is no intra-fragment correlation contained in the internal orbitals. Later in this section we show how the flexibility built into ALCHEMY II's configuration space generator is necessary to implement ICF approximations when the full ICF configuration space leads to calculations that exceed available computing limits. But first we need to review the treatment of BSSE.

Basis set superposition error adjustments are necessary to remove the effects of basis set deficiencies which are not uniform as a function of inter-fragment separation. These deficiencies exist in both the orbital and configurational bases as we now indicate.

Internal orbitals I_A in the supermolecule, if optimized, have fragment B's basis set available for improved orbital expansions. The improvement coming from this availability is obviously a function of inter-fragment distance and orientation; it can only disappear when A's basis is complete, at least for I_A expansions. This completeness requirement becomes more severe when fragment correlating orbitals are moved into I_A as the level of a particular order of interacting fragment calculation is raised. We have referred[8] to this non-uniformity in the quality of the orbital basis as **orbital basis inconsistency** (OBI). More important, as measured by the size of the effect, is the non-uniformity in the quality of the configurational space as a function of inter-fragment separation. The n_A electron A component of the supermolecule wave function is improved in the supermolecule over its description in the dissociated fragment by use of n_A-particle functions built partially, or completely, from B orbitals. This effect will exist in interacting zeroth order, ICF, and interacting second order wavefunctions, even when the orbital basis is complete. It does not disappear until calculations reach the level of complete CI in a complete basis; in practice, never. We have referred[8] to this effect as **configurational basis inconsistency** (CBI).

Equally, OBI and CBI exists for the B component of supermolecule wave functions. OBI and CBI together make up the widely discussed BSSE. The OBI and CBI components of BSSE are not independent; two calculations which yield the same interaction energy via the same BSSE adjustment can have different OBI and CBI components of that BSSE adjustment. This would be the case, for example, in ICF calculations based first on the use of the same fragment orbitals for all supermolecule fragment separations (no OBI by construction), and second on orbitals reoptimized for each separation (where there must be an OBI unless the orbital basis is complete for internal orbital expansions).

The key to correct thought in making BSSE adjustments in calculated interaction energies is the recognition that, if complete CI calculations were possible, the complete CI interaction energy for a given inter-fragment separation would be the difference between the complete CI supermolecule energy and the sum of complete CI fragment energies **each computed in the same supermolecule basis**. The treatment of BSSE in actual calculations must smoothly approach this complete CI limit as the sequence of calculations (an ICF sequence, or an interacting second order sequence, for example) approaches complete CI.

This means that fragment A calculation in an interacting zeroth order calculation must be done in configuration space

$$\begin{array}{ccc} I_A & I_B & X_{AB} \\ \leftarrow & n_A \rightarrow & 0 \end{array} \tag{5}$$

in an interacting first order fragment interaction calculation in configuration space

$$\begin{array}{ccc} I_A & I_B & X_{AB} \\ \leftarrow & n_A \rightarrow & 0 \\ \leftarrow & n_A - 1 \rightarrow & 1 \end{array} \tag{6}$$

and in an interacting second order calculation in configuration space

$$\begin{array}{ccc} I_A & I_B & X_{AB} \\ \leftarrow & n_A \rightarrow & 0 \\ \leftarrow & n_A - 1 \rightarrow & 1 \\ \leftarrow & n_A - 2 \rightarrow & 2 \end{array} \tag{7}$$

These calculations must be done, **at each supermolecule conformation,** with all B's nuclear charges set to zero. Clearly, an equivalent treatment of B is mandated.

There are two important points to be made here, especially for ICF interaction energies, and we illustrate them in a specific example; the He – H_2O interaction at the lowest ICF1 level, where there are no fragment correlating orbitals in the internal space. Without BSSE adjustment, the fragment energies would be those of single configurations; the supermolecule calculation would have in its second order configuration space the simultaneous single excitations from the single zeroth order supermolecule configuration, whose contribution goes to zero at dissociation. The interaction is properly considered, in a large enough orbital basis, as the interaction of Hartree-Fock fragments. Yet our prescription for the BSSE adjusted interaction energy calls not for Hartree-Fock fragment calculations, but calculations in which one fragment's energy is lowered by correlation effects contained in configurations built partially from B orbitals. The effect is large. Since, at first acquaintance, such fragment calculations are hard to accept, explicit illustration of the effect and discussion of its validity is one of the points we wish to make. The other is that BSSE adjustments can change dramatically with basis set; yet after the adjustments the agreement between interaction potentials calculated in quite differently constituted basis sets is remarkably improved. This, in itself, is a compelling argument for the adjustment.

The He – H_2O Interaction

A recent study of pressure broadening of H_2O microwave lines due to collisions with helium atoms [11] looked at at the sensitivity of the pressure broadening cross sections to variation of the potential energy surface. Surfaces were generated with two quite different basis sets (designated GISS and IBM, reflecting the laboratories of origin), two different wave function models (ICF1 and MP4), both with and without BSSE adjustments. These eight surfaces are reproduced here, in Table I, to illustrate the effect of BSSE adjustments and to provide a concrete example for discussing them. We introduce the Table with a minimal description; for further details the reader is referred to the original paper.

Table I gives surfaces for the interaction of a helium atom positioned at spherical polar coordinates (r, θ, ϕ,) in a coordinate frame whose origin is at the center of mass of a rigid water molecule. The water molecule is in the yz plane, with the oxygen atom on the positive z-axis which bisects the HOH angle. Thus, a helium approach with $\theta = 0$ is an in-plane approach to the lone pair water electrons, $\theta = 180$ is along the bisector towards the hydrogens; approaches with $\phi = 0$ are 'broadside', and with $\phi = 90$ are 'edge on', to the water molecule.

Both orbital bases are contracted gaussians; GISS has a O: 9s,6p,2d; H.He: 5s,2p, composition chosen to minimize the OBI component of BSSE, while IBM is a more traditional basis with composition O: 4s,3p,2d,1f; H,He: 3s,2p,1d, heavily contracted in the functions describing atomic character and with a more extensive polarization component. The IBM basis is expected to, and does, have larger BSSE.

The ICF1 supermolecule and fragment calculations for each supermolecule conformation are expanded in configuration spaces

$$\begin{array}{llcccccc}
 & & 1s_{He} & & (OH)_1(OH_2)lp_1lp_2 & & & X \\
 & & \leftarrow & & 10 & & \rightarrow & 0 \\
 & & \leftarrow & & 9 & & \rightarrow & 1 \\
 & & 1 & \leftarrow & & 7 & \rightarrow & 2 \\
\text{supermolecule} & H_2O & \leftarrow & & 2 & & \rightarrow & 0 \\
 & He & \leftarrow & & 1 & & \rightarrow & 1 \\
 & & \leftarrow & & 8 & & \rightarrow & 0 \\
 & & \leftarrow & & 7 & & \rightarrow & 1
\end{array} \qquad (8)$$

where in the He calculation the H_2O nuclear charges are set to zero, and in the H_2O calculation the He nuclear charge is set to zero. These calculations give the BSSE adjusted interaction energy; without adjustment, the separated fragment energy is simply the sum of two single configuration energies, each in its own basis. The internal orbitals are given a localized description: $1s_{He}$ for helium; two OH bond and two lone pair orbitals, labeled $(OH)_1$,$(OH)_2$, lp_1, and lp_2, for water. The localization, needed to produce orthonormal orbitals for the second order ICF space, is based on population matrix diagonalization.[8]

The MP4 calculations can be regarded as approximations to interacting second order; in this case with the same internal orbital space as ICF1. They are size consistent Moller-Plesset through fourth order perturbation theory calculations based on a single zeroth order configuration. Comparison of ICF1, with essen-

Table I. Water-helium potentials. ICF1 and MP4 uncorrected potentials are compared in two basis sets (GISS and IBM) in the first four columns. Parallel BSSE adjusted potentials are shown in the remaining four columns. Helium is positioned in spherical polar coordinates from the center of mass of a rigid water molecule, as defined in the text. Energies in cm^{-1}, distances in bohr, and angles in degrees.

			Unadjusted				BSSE Adjusted			
			GISS Basis		IBM Basis		GISS Basis		IBM Basis	
r	θ	ϕ	ICF1	MP4	ICF1	MP4	ICF1	MP4	ICF1	MP4
4.0	0	0	1669.8	1866.2	1577.4	1742.5	2034.4	1913.4	1952.1	1811.0
4.0	45	0	1753.8	2003.9	1652.5	1851.3	2169.5	2057.9	2086.8	1947.5
4.0	45	90	1240.3	1419.0	1161.3	1305.8	1524.0	1450.5	1456.0	1368.7
4.0	90	0	1686.9	2028.4	1540.3	1791.4	2131.4	2070.4	2045.9	1954.4
4.0	90	90	2133.7	2369.0	2051.8	2270.6	2523.3	2426.6	2451.9	2337.8
4.0	135	0	1683.6	2010.9	1553.9	1815.9	2120.4	2090.8	2036.6	1979.3
4.0	135	90	3727.7	4341.5	3639.9	4181.9	4503.2	4461.2	4454.5	4372.8
4.0	180	0	1981.5	2325.5	1848.5	2137.8	2437.6	2393.1	2357.0	2287.5
4.0	120	0			1487.5	1746.6			1978.2	1917.3
4.0	120	45			2251.5	2575.0			2812.5	2715.9
4.0	120	90			3734.0	4246.3			4506.6	4410.5
4.5	0	0	503.4	603.1	457.1	539.7	663.3	633.5	625.1	583.5
4.5	45	0	552.2	680.9	498.0	594.8	742.5	719.4	702.4	662.8
4.5	45	90	338.5	423.8	300.6	368.0	458.4	445.8	428.8	408.9
4.5	90	0	561.7	739.8	465.3	576.6	763.2	769.0	723.0	707.2
4.5	90	90	606.2	709.0	564.8	662.8	769.4	748.2	734.7	706.4
4.5	135	0	544.9	709.9	466.1	577.9	747.0	768.2	706.8	710.7
4.5	135	90	1182.7	1480.8	1108.4	1340.8	1510.4	1528.2	1477.4	1473.7
4.5	180	0	631.5	803.8	548.9	679.0	836.5	851.0	800.1	800.7
5.0	0	0	120.6	167.5	97.8	136.5	188.9	187.4	172.6	165.2
5.0	45	0	144.4	205.1	114.5	156.7	230.2	232.9	211.8	205.9
5.0	45	90	60.2	98.3	41.7	70.8	109.7	113.6	98.4	99.0
5.0	90	0	164.0	250.4	99.3	137.9	253.3	270.8	234.7	240.4
5.0	90	90	125.1	164.8	106.0	146.5	191.2	192.2	177.1	175.6
5.0	135	0	146.2	222.0	96.1	130.1	239.2	264.8	221.1	237.2
5.0	135	90	317.9	449.3	269.7	357.7	458.5	483.3	437.0	451.7
5.0	180	0	165.5	244.1	108.9	155.6	255.1	277.5	238.8	254.2
5.0	120	0			91.6	124.9			222.6	222.6
5.0	120	45			160.7	217.0			286.5	296.2
5.0	120	90			269.2	348.5			417.8	426.8

(Continued)

tially no intra-fragment correlation, and MP4, with a high percentage of intra-fragment correlation, both containing essentially the same inter-fragment correlation, makes an interesting illustration of the effect of intra-fragment correlation on the energy surface. In the $He - H_2O$ interaction, as is typically observed, it makes the interaction more repulsive. BSSE adjustments to MP4 are simply made by doing MP4 fragment calculations in the supermolecule basis at each inter-fragment separation. There is nothing else that we can do; a pre-

Table I (cont.). Water-helium potentials. ICF1 and MP4 uncorrected potentials are compared in two basis sets (GISS and IBM) in the first four columns. Parallel BSSE adjusted potentials are shown in the remaining four columns. Helium is positioned in spherical polar coordinates from the center of mass of a rigid water molecule, as defined in the text. Energies in cm^{-1}, distances in bohr, and angles in degrees.

			Unadjusted				BSSE Adjusted			
			GISS Basis		IBM Basis		GISS Basis		IBM Basis	
r	θ	ϕ	ICF1	MP4	ICF1	MP4	ICF1	MP4	ICF1	MP4
5.5	0	0	6.9	27.3	-5.1	11.4	35.6	40.4	29.2	31.5
5.5	45	0	17.2	43.5	-1.1	14.2	55.7	62.9	47.7	51.0
5.5	45	90	-13.6	2.5	-24.4	-13.4	6.6	12.8	2.6	7.7
5.5	90	0	32.1	71.2	-12.7	-7.4	70.7	85.1	62.7	71.3
5.5	90	90	-6.6	6.3	-15.0	0.8	19.6	24.9	14.9	19.6
5.5	135	0	17.6	49.2	-17.8	-18.8	61.1	79.5	53.3	67.0
5.5	135	90	47.6	100.0	17.2	45.5	106.9	125.3	97.6	111.5
5.5	180	0	18.8	51.3	-23.6	-15.9	57.6	74.5	50.1	63.7
6.0	0	0	-19.7	-11.4	-27.6	-21.8	-7.7	-2.8	-9.8	-5.8
6.0	45	0	-15.4	-5.0	-29.0	-26.1	1.7	7.7	-1.3	3.1
6.0	45	90	-25.8	-19.3	-33.2	-30.0	-17.4	-12.4	-18.7	-13.9
6.0	90	0	-6.5	9.8	-38.5	-45.3	9.9	19.4	7.0	13.8
6.0	90	90	-31.4	-28.6	-35.0	-28.8	-21.0	-16.1	-22.4	-17.6
6.0	135	0	-21.5	-5.3	-44.1	-56.0	-0.4	15.3	1.0	10.0
6.0	135	90	-21.6	-2.6	-43.2	-36.5	2.9	14.3	-0.5	9.1
6.0	180	0	-19.6	-7.3	-51.8	-58.4	-2.2	8.3	-5.7	3.4
6.5	0	0	-20.8	-17.7	-27.4	-26.1	-15.9	-12.3	-16.2	-12.8
6.5	45	0	-18.9	-15.2	-30.3	-31.6	-11.7	-7.5	-12.3	-8.7
6.5	45	90	-22.2	-19.9	-27.6	-27.1	-18.7	-15.2	-18.9	-15.3
6.5	90	0	-14.4	-8.2	-36.8	-44.8	-7.4	-1.8	-8.0	-3.5
6.5	90	90	-28.4	-28.5	-29.8	-27.2	-24.1	-20.5	-24.5	-20.9
6.5	135	0	-21.7	-18.1	-41.1	-52.1	-10.9	-4.4	-11.9	-6.5
6.5	135	90	-30.1	-24.2	-46.2	-46.3	-20.3	-13.5	-21.6	-15.6
6.5	180	0	-24.4	-20.2	-46.6	-55.0	-15.9	-9.8	-17.5	-11.9
7.0	0	0	-17.0	-15.2	-21.6	-21.5	-15.1	-11.9	-14.0	-11.5
7.0	45	0	-15.2	-14.1	-23.9	-25.6	-12.4	-9.7	-12.2	-9.7
7.0	45	90	-16.2	-15.8	-20.1	-19.9	-14.8	-12.4	-14.5	-12.1
7.0	90	0	-13.3	-11.4	-27.3	-32.6	-10.2	-7.0	-10.0	-7.2
7.0	90	90	-21.1	-21.6	-21.4	-20.1	-19.0	-16.5	-19.0	-16.5
7.0	135	0	-18.2	-17.6	-29.9	-36.4	-12.3	-8.7	-12.5	-9.3
7.0	135	90	-24.3	-23.0	-35.0	-36.5	-20.4	-16.5	-21.0	-17.3
7.0	180	0	-20.5	-19.4	-33.4	-38.8	-15.8	-12.2	-16.3	-13.0

(Continued)

scription that pools the internal B orbitals into the internal space for the fragment, as called for in configuration space (7), cannot be implemented, for multi-reference MP4 codes do not exist. However, in interacting second order,

Table I (cont.). Water-helium potentials. ICF1 and MP4 uncorrected potentials are compared in two basis sets (GISS and IBM) in the first four columns. Parallel BSSE adjusted potentials are shown in the remaining four columns. Helium is positioned in spherical polar coordinates from the center of mass of a rigid water molecule, as defined in the text. Energies in cm^{-1}, distances in bohr, and angles in degrees.

			Unadjusted				BSSE Adjusted			
			GISS Basis		IBM Basis		GISS Basis		IBM Basis	
r	θ	ϕ	ICF1	MP4	ICF1	MP4	ICF1	MP4	ICF1	MP4
7.5	0	0	-11.7	-11.3	-15.3	-15.1	-11.0	-9.3	-10.5	-8.8
7.5	45	0	-11.1	-10.8	-16.5	-17.3	-10.1	-8.4	-9.6	-7.9
7.5	45	90	-11.4	-11.7	-13.6	-13.3	-10.8	-9.2	-10.4	-8.8
7.5	90	0	-10.4	-9.9	-17.6	-20.0	-8.9	-7.0	-8.4	-6.8
7.5	90	90	-15.0	-15.3	-14.6	-13.8	-13.6	-11.9	-13.4	-11.7
7.5	135	0	-13.5	-13.9	-19.0	-21.8	-10.2	-8.1	-10.0	-8.2
7.5	135	90	-17.6	-17.7	-23.3	-24.4	-15.9	-13.5	-16.1	-13.8
7.5	180	0	-15.3	-15.4	-21.3	-23.6	-12.6	-10.4	-12.5	-10.5
8.0	0	0	-8.2	-8.2	-10.1	-9.7	-8.0	-6.8	-7.6	-6.4
8.0	45	0	-7.9	-7.9	-10.5	-10.6	-7.5	-6.4	-7.1	-6.0
8.0	45	90	-7.9	-8.5	-9.0	-8.4	-7.7	-6.6	-7.3	-6.2
8.0	90	0	-7.6	-7.6	-10.6	-11.3	-6.9	-5.8	-6.4	-5.4
8.0	90	90	-10.5	-10.8	-10.0	-9.4	-9.4	-8.3	-9.2	-8.1
8.0	135	0	-9.5	-10.1	-11.4	-12.1	-7.7	-6.5	-7.4	-6.3
8.0	135	90	-12.5	-13.0	-14.7	-15.1	-11.4	-9.9	-11.4	-9.9
8.0	180	0	-10.9	-11.3	-13.0	-13.5	-9.3	-8.0	-9.0	-7.7
10.	45	0	-2.2	-2.0	-2.0	-1.8	-2.1	-1.9	-2.0	-1.7
10.	45	90	-2.1	-2.1	-2.0	-1.7	-2.1	-1.8	-2.0	-1.7
10.	135	0	-2.3	-2.2	-2.1	-1.8	-2.2	-1.9	-2.0	-1.8
10.	135	90	-3.1	-3.4	-2.9	-2.6	-2.9	-2.6	-2.7	-2.4
12.	0	0	-0.8	-0.6	-0.7	-0.6	-0.8	-0.6	-0.6	-0.6
12.	90	0	-0.7	-0.6	-0.6	-0.5	-0.7	-0.6	-0.6	-0.5
12.	90	90	-0.8	-0.7	-0.7	-0.6	-0.8	-0.7	-0.7	-0.6
12.	180	0	-0.9	-0.7	-0.8	-0.7	-0.8	-0.7	-0.8	-0.7

and MP4, we show in the next section that expansion of the fragment internal space is not crucial.

We can now look critically at Table I and make the following observations:

1. The blocks of data for r = 6.5 and 7.0 bohr describe the interaction in the region of maximum attraction. The maximum difference between the ICF1 BSSE adjusted entries in the two basis sets (columns 5 and 7) is 1.6 cm^{-1} at 6.5 bohr and 0.5 cm^{-1} at 7.0 bohr; mean differences for the eight data points in each block are 0.8 cm^{-1} and 0.1 cm^{-1}, respectively. Without BSSE adjustment (columns 1 and 3) these differences would be 22.2 cm^{-1} and 14.0 cm^{-1} maximum, and 13.1 cm^{-1} and 8.6 cm^{-1} mean. Conclusion: the BSSE adjust-

ment has made the ICF1 potentials in the two basis sets superimposable to within approximately $1 cm^{-1}$ at these separations. Before adjustment, the differences between the two potentials were as large as the adjusted interaction energies! Because the composition of the two basis sets is so different we can also conclude that, in this region of the energy surface, the adjusted energies themselves are most likely converged; that is, larger basis sets will not give BSSE adjusted energies differing from those of Table I to an accuracy of order $1 cm^{-1}$.

2. Comparison of columns 3 an 7, showing the size of the BSSE adjustment for ICF1 calculations in the IBM basis, is just as devastating; adjustments of up to $30 cm^{-1}$ at 6.5 bohr and $18 cm^{-1}$ at 7.0 bohr, bigger than the adjusted interaction energies themselves, are seen.

3. In the attractive 6.5 to 7.0 bohr region the MP4 calculations show strikingly similar results. In the two basis sets, after BSSE adjustment, the MP4 interaction potentials are superimposable within a range about twice the ICF1 range; before adjustment the differences were about 50% larger than for ICF1. The magnitude of the adjustment in the IBM basis is also about 50% larger than in the ICF1 calculations.

4. We see, then, that in the attractive region the BSSE adjusted potentials are near the basis set limits for ICF1 (within $\sim 1 cm^{-1}$) and MP4 (within $\sim 2 cm^{-1}$) model calculations. The IBM basis BSSE adjusted potentials (columns 7 and 8) are our best; we see a significant difference between the ICF1 and MP4 models, certainly greater than any uncertainty in each of the potentials. The MP4 interaction is less attractive, as anticipated; MP4 differs from ICF1 primarily in its inclusion of intra-fragment correlation, which we have observed to increase the repulsive forces in all interactions so far studied. The two potentials probably bracket the true one, although this would have to be proved by actual calculation. The MP4 model does have the potential advantage of working with a better H_2O charge distribution (it gives a H_2O dipole moment closer to the experimental value) so that contributions to the interaction energy from terms involving permanent electric moments will be superior. This is true, too, of the interacting second order model which MP4 is approximating, and there can be no dispute that, with a given internal space, interacting second order (done without approximation) is superior to interacting first order (ICF). However, we are at an end as far as MP4 calculations can go. We can only increase the basis set in order to more closely approach the MP4 basis set limit, and this will not tell us much. ICF calculations, on the other hand, can be upgraded by moving important intra-fragment correlating orbitals into the internal space, ultimately to converge to the true interaction energy. At least, in principle. Actual calculations will have to be approximated, as we will outline in the next section, and will need to show that the approximations do not do more damage than the improvement gained by upgrading the ICF level.

5. On the repulsive wall the situation is obscured by the fact that in this region of the interaction the IBM basis is superior to the GISS; the GISS potential is further away from the basis set limit than the IBM in both ICF1 and MP4 calculations. But even so, the adjusted ICF1 surfaces in the two basis sets are closer to each other than the unadjusted. The magnitude of the adjustments is large. The same is true of the MP4 surfaces. Qualitatively, Table I shows that the BSSE adjusted ICF1 surface has a steeper repulsive wall and a deeper attractive region than the BSSE adjusted MP4, implying a crossover region which the Table shows occurs at around 5.0 bohr.

We now proceed from this numerical demonstration of the folly of calculating weak interactions without BSSE adjustment to a closer look at why, in this ICF1 He - H_2O interaction, designed as the interaction of two uncorrelated Hartree-Fock fragments, the correct BSSE adjustment brings in intra-fragment correlation.

First, consider as an extreme example, the interaction of a Hartree-Fock helium atom with a complete CI water molecule. The needed ICF calculations would be done in the configuration space

		$1s_{He}$	B_{H_2O}	X_{He}
		←	10 →	0
		←	9 →	1
		1	7	2
supermolecule	H_2O	←	2 →	0
	He	←	1 →	1
		←	8 →	0
		←	7 →	1

(9)

In configuration space (9), the orbital set B_{H_2O} contains all of the water basis functions, treated as internal orbitals; the external orbitals X_{He} span the He orbital basis complementary to $1s_{He}$. The calculation is treated as a 10 electron correlation problem with the oxygen core electrons factored out. It is apparent in this extreme case that there is a great deal of He atomic correlation contained in the supermolecule calculation. Most of it is contained in the first order electron assignment 0 9 1 to orbital sets $1s_{He}$, B_{H_2O}, and X_{He}, arising from 0 1 1 ⊗ 0 8 0 where the full CI space of H_2O is multiplied by He double electron excitations from its dominant configuration, one into B_{H_2O} and one into X_{He}. (The assignment 0 10 0 arising from 0 2 0 ⊗ 0 8 0 also contains He atomic correlation, but until B_{H_2O} becomes very large this will be be less important). Equally apparent is the fact that this intra-fragment correlation in the supermolecule is being compensated for in the BSSE adjusted He fragment calculation specified in configuration space (9'). We see there that precisely these same electron assignments, 0 1 1 and 0 2 0, which multiply the complete H_2O configuration space in the supermolecule, are present in the fragment calculation. Without the BSSE adjustment the fragment calculation would have been in configuration space (10),

$$\begin{array}{cc} 1s_{He} & X_{He} \\ \\ 2 & 0 \\ 1 & 1 \end{array} \tag{10}$$

which simply gives the He contribution to the supermolecule energy at infinite separation, essentially the Hartree-Fock energy. An alternative BSSE adjustment to that shown in configuration space (9) is that of configuration space (11),

$$\begin{array}{ccc} 1s_{He} & B_{H_2O} & X_{He} \\ \\ 2 & \leftarrow & 0 \rightarrow \\ 1 & \leftarrow & 1 \rightarrow \end{array} \tag{11}$$

where the full supermolecule orbital basis is used but, by augmenting the He external space rather than the internal, the key intra-fragment correlating assignments 0 1 1 and 0 2 0 are missed, and the computed He energy is still essentially the Hartree-Fock energy. We must accept configuration space (9) as the optimum one for calculating the He BSSE adjustment. Parallel considerations apply, of course, to the H_2O BSSE adjustment.

This extreme case was presented to show that intra-fragment correlation can most certainly be brought into a fragment in a supermolecule, in the case that the fragment at dissociation is described at the Hartree-Fock level. We chose this case to temporarily avoid the complaint that usually surfaces in discussing interacting Hartree-Fock fragment calculations such as our ICF1 He – H_2O, namely, that since the internal H_2O orbitals in configuration space (9) are strongly occupied they should not be available to the He and should, in some manner, be removed from the He fragment calculation. By removing them, the most important of the intra-fragment He correlating terms in our BSSE adjusted configuration space (9) are taken out. Let us dispel this unfounded notion once and for all.

Because the BSSE adjusted fragment calculation involves complete zeroth and first order spaces, the 'forbidden' space made inaccessible to the He by removal of H_2O strongly occupied orbitals is limited to that part of the H_2O strongly occupied orbital space orthogonal to the He internal orbitals. Therefore, by bringing He correlating orbitals into the internal space as the ICF level is upgraded, and the overlap of the H_2O strongly occupied orbitals with the He internal orbital space gets larger, the 'forbidden' space gets smaller and its being unavailable to He has less effect. Ultimately, in the limit of a large enough He internal space, the 'forbidden' space is reduced to zero. To say that the 'forbidden' space can be used when it derives from He orbitals but not when it derives from H_2O strongly occupied orbitals is absurd.

Generalizing to the interaction of fragments A and B, **the only thing that makes any part of B's space unavailable to A is the presence of B itself.** In the BSSE adjusted fragment calculation, with B absent, the full supermolecule orbital space is available to B, without restriction.

We have just seen how important it is to make the supermolecule internal orbitals the internal orbitals of the BSSE adjusted fragment calculation in ICF calculations. How then do we justify the MP4 BSSE adjustments of Table I which, based on a single configuration zeroth order space, do not use a zeroth order space built from supermolecule internal orbitals? To answer this we recall that MP4 is approximating interacting second order; BSSE adjusted MP4 fragment calculations are the equivalent of, taking the He fragment as example, doing the BSSE adjusted interacting second order in configuration space

$$\begin{array}{ccccc} 1s_{He} & & (OH)_1(OH_2)lp_1lp_2 & & X \\ 2 & \leftarrow & 2 & \rightarrow & 0 \\ 1 & \leftarrow & 1 & \rightarrow & 1 \\ 0 & \leftarrow & 0 & \rightarrow & 2 \end{array} \tag{12}$$

with the water nuclear charges set to zero, rather than in the better configuration space prescribed in (7),

$$\begin{array}{ccccc} 1s_{He} & & (OH)_1(OH_2)lp_1lp_2 & & X \\ \leftarrow & & 2 & \rightarrow & 0 \\ \leftarrow & & 1 & \rightarrow & 1 \\ \leftarrow & & 0 & \rightarrow & 2 \end{array} \tag{13}$$

We see that the second order space in (12) contains the key electron assignment 0 1 1 which in the preferred (13) appears in first order. The less important 0 2 0, a zeroth order assignment in (13), also appears in the second order space of (12). Thus (12) does contain the most important intra-fragment correlations and is a reasonable approximation to (13). In general, moving B's internal orbitals into the external space for the BSSE adjusted fragment A calculation is justified in interacting second order; it is completely unjustified in interacting first order.

Approximate ICF Configuration Spaces

A continuing research program in our laboratory is to investigate how big a role the ICF model can play in determining intermolecular interactions. While we have given an ICF overview in the previous sections, we have not paid attention to the many details that are faced in carrying out such studies. These have to do primarily with orbital choice, localization, and the approximation of ICF configuration spaces, both for the supermolecule and the BSSE adjusted fragment. In this section we discuss configuration space approximation, since this is where the flexibility we have built into the SYMBOLM module of ALCHEMY II comes into play.

As soon as intra-fragment correlating orbitals are brought into the internal space of ICF calculations, the configuration spaces will most probably have to be approximated. We found this even for four electron correlation problems such as He_2.[8] A more demanding case, Ar_2 treated as a 16-electron correlation problem, was taken up by McLean, Liu, and Barker.[12] That study was primarily concerned with an improved description of the repulsive wall; while it is still the most accurate work available, it is clear that the approach used there has some problems that we feel can be solved. First, the wrong BSSE adjustments were made, and second, approximations to the ICF2 configuration space (in which the 3d intra-fragment correlating orbitals on each atom were brought into the internal space) were approached from the wrong point of view. The approximations used there introduce problems of size consistency, and unclear prescriptions for BSSE adjusted fragment calculations. These authors looked directly at approximating the ICF supermolecule configuration space, and systematically included classes of configurations with the aim of converging to the unapproximated ICF limit. We now believe that this approach becomes computationally intractable before a desirable level of convergence is reached. McLean[13] has changed this approach to one that starts from the fragments, just as our earlier discussion of unapproximated interacting fragments of a given order started with the fragments (configuration space (1)). It is this preferred way of looking at the problem that we outline here.

The first order approximation to fragment A, used in generating an approximate ICF supermolecule configuration space with fragment B, is made in configuration space (14),

$$
\begin{array}{lccc}
 & S_A & C_A & X_A \\
I_A: & n_A & 0 & 0 \\
 & n_A-1 & 1 & 0 \\
 & n_A-1 & 0 & 1 \\
 & & & \\
II_A: & n_A-2 & 2 & 0 \\
 & n_A-2 & 1 & 1 \\
 & n_A-3 & 3 & 0 \\
 & n_A-3 & 2 & 1
\end{array}
\qquad (14)
$$

where the internal orbitals have been divided into two subsets, the strongly occupied S_A, and the weakly occupied C_A (containing intra-fragment correlating orbitals). The approximate first order fragment configuration space (14) is divided into two, as shown. I_A contains the dominant electron assignment n_A 0 0 and single excitations generated from that assignment; II_A contains the electron assignment $n_A -2\ 2\ 0$ (containing the bulk of the intra-fragment electron correlation) and single excitations from that assignment. We note that configuration space I_A in (14) is the first order space, without approximation, at a lower ICF level where there are no internal weakly occupied orbitals. Just as the presence of electron assignment $n_A\ 0\ 0$ requires the single excitations $n_A -1\ 1\ 0$ and $n_A -1\ 0\ 1$ to generate second order dispersion terms in the supermolecule configuration space, so the presence of $n_A -2\ 2\ 0$ requires single excitations from it to generate necessary additional dispersion terms in the supermolecule when intra-fragment correlation is added. In other words, none of the II_A electron assignments in (14) can be ignored.

The supermolecule configuration space generated from (14) and its B counterpart consists of four subspaces: $I_A \otimes I_B$, $II_A \otimes I_B$, $I_A \otimes II_B$, and $II_A \otimes II_B$. These subspaces are

	S_A	S_B	C_A	C_B	X_{AB}	
$I_A \otimes I_B$:	n_A	n_B	$\leftarrow 0$	$\rightarrow$	0	
	n_A	$n_B -1$	$\leftarrow 1$	$\rightarrow$	0	
	n_A	$n_B -1$	$\leftarrow 0$	$\rightarrow$	1	
	$n_A -1$	n_B	$\leftarrow 1$	$\rightarrow$	0	(15)
	$n_A -1$	$n_B -1$	$\leftarrow 2$	$\rightarrow$	0	
	$n_A -1$	$n_B -1$	$\leftarrow 1$	$\rightarrow$	1	
	$n_A -1$	n_B	$\leftarrow 0$	$\rightarrow$	1	
	$n_A -1$	$n_B -1$	0	0	2	

	S_A	S_B	C_A	C_B	X_{AB}	
$II_A \otimes I_B$:	$n_A -2$	n_B	$\leftarrow 2$	$\rightarrow$	0	
	$n_A -2$	$n_B -1$	$\leftarrow 3$	$\rightarrow$	0	
	$n_A -2$	$n_B -1$	$\leftarrow 2$	$\rightarrow$	1	
	$n_A -2$	n_B	$\leftarrow 1$	$\rightarrow$	1	
	$n_A -2$	$n_B -1$	1	0	2	(16)
	$n_A -3$	n_B	$\leftarrow 3$	$\rightarrow$	0	
	$n_A -3$	$n_B -1$	$\leftarrow 4$	$\rightarrow$	0	
	$n_A -3$	$n_B -1$	$\leftarrow 3$	$\rightarrow$	1	
	$n_A -3$	n_B	$\leftarrow 2$	$\rightarrow$	1	
	$n_A -3$	$n_B -1$	2	0	2	

$$
\begin{array}{lccccc}
 & S_A & S_B & C_A & C_B & X_{AB} \\
I_A \otimes II_B: & n_A & n_B-2 & \leftarrow 2 & \rightarrow & 0 \\
 & n_A-1 & n_B-2 & \leftarrow 3 & \rightarrow & 0 \\
 & n_A-1 & n_B-2 & \leftarrow 2 & \rightarrow & 1 \\
 & n_A & n_B-2 & \leftarrow 1 & \rightarrow & 1 \\
 & n_A-1 & n_B-2 & 0 & 1 & 2 \\
 & n_A & n_B-3 & \leftarrow 3 & \rightarrow & 0 \\
 & n_A-1 & n_B-3 & \leftarrow 4 & \rightarrow & 0 \\
 & n_A-1 & n_B-3 & \leftarrow 3 & \rightarrow & 1 \\
 & n_A & n_B-3 & \leftarrow 2 & \rightarrow & 1 \\
 & n_A-1 & n_B-3 & 0 & 2 & 2
\end{array}
\tag{17}
$$

and

$$
\begin{array}{lccccc}
 & S_A & S_B & C_A & C_B & X_{AB} \\
II_A \otimes II_B: & n_A-2 & n_B-2 & \leftarrow 4 & \rightarrow & 0 \\
 & n_A-2 & n_B-2 & \leftarrow 3 & \rightarrow & 1 \\
 & n_A-2 & n_B-3 & \leftarrow 5 & \rightarrow & 0 \\
 & n_A-2 & n_B-3 & \leftarrow 4 & \rightarrow & 1 \\
 & n_A-2 & n_B-2 & 1 & 1 & 2 \\
 & n_A-2 & n_B-3 & 1 & 2 & 2 \\
 & n_A-3 & n_B-2 & \leftarrow 5 & \rightarrow & 0 \\
 & n_A-3 & n_B-2 & \leftarrow 4 & \rightarrow & 1 \\
 & n_A-3 & n_B-3 & \leftarrow 6 & \rightarrow & 0 \\
 & n_A-3 & n_B-3 & \leftarrow 5 & \rightarrow & 1 \\
 & n_A-3 & n_B-2 & 2 & 1 & 2 \\
 & n_A-3 & n_B-3 & 2 & 2 & 2
\end{array}
\tag{18}
$$

Notice that the second order classes in (15) - (18) are restricted to simultaneous excitations of an A and a B electron. This preserves size consistency in the approximated ICF configuration space. In zeroth and first order classes all possible assignments of electrons in the combined set of $C_A \oplus C_B$ orbitals is made.

How are these configuration spaces to be used? As mentioned earlier, $I_A \otimes I_B$ in (15) is simply a rewrite of the unapproximated ICF space in a calculation with no weakly occupied orbitals in the internal space. Call this an ICF1 calculation as we did in the He - H_2O example. We will always be able to handle this. The trick is to find the effect of intra-fragment correlation on the ICF1 interaction potential in a stepwise manner, assuming that the effects of different intra-fragment correlations are additive. Spot checks on additivity can be made by including several different correlation effects in a single calculation, when that is computationally feasible.

Consider Ar_2, and suppose we want to find the effect of 3d atomic correlation on the ICF1 interaction energy. To minimize the amount of computation this can be done one atom at a time; thus if we correlate atom A we set $S_A = 3s_A, 3p_A$, $S_B = 3s_B, 3p_B$, $C_A = 3d_A$, C_B is empty, and external space X_{AB} is the remainder of the supermolecule orbital space. The calculation is done in configuration space $(I_A \oplus II_A) \otimes I_B$ of (15)-(18). In this calculation, we have a 3d correlated A atom interacting with a Hartree-Fock B atom. Not surprisingly, the appropriate BSSE adjusted atom calculations are in configuration space (19),

$$
\begin{array}{lccccc}
 & S_A & S_B & C_A & C_B & X_{AB} \\
I_A: & \leftarrow & n_A \rightarrow & \leftarrow & 0 \rightarrow & 0 \\
 & \leftarrow & n_A - 1 \rightarrow & \leftarrow & 1 \rightarrow & 0 \\
 & \leftarrow & n_A - 1 \rightarrow & \leftarrow & 0 \rightarrow & 1 \\
II_A: & \leftarrow & n_A - 2 \rightarrow & \leftarrow & 2 \rightarrow & 0 \\
 & \leftarrow & n_A - 2 \rightarrow & \leftarrow & 1 \rightarrow & 1 \\
 & \leftarrow & n_A - 3 \rightarrow & \leftarrow & 3 \rightarrow & 0 \\
 & \leftarrow & n_A - 3 \rightarrow & \leftarrow & 2 \rightarrow & 1
\end{array}
\tag{19}
$$

where the BSSE adjusted Hartree-Fock atom will be calculated in I_A of (19), and the BSSE correlated atom in $I_A \oplus II_A$ of (19). The BSSE adjusted Hartree-Fock atom calculation is just what it would be in an ICF1 calculation. When there are many electrons these BSSE adjusted configuration spaces can become astronomical in size, orders of magnitude larger than the supermolecule calculation. Fortunately, this is easily handled. In configuration space (19) the electron assignments into $S_A \oplus S_B$ are restricted to those with 0, 1, and 2, electrons in S_B, making sure that the S_A orbitals are optimum A orbitals, not the localized supermolecule orbitals.

The effect of $3d_B$ correlation on the ICF1 interaction will be the same as that of $3d_A$, and the combined effect should, to good approximation , be the sum. This would be exactly true if, in a calculation with $C_A = 3d_A$ and $C_B = 3d_B$ in configuration space $(I_A \oplus II_A) \otimes (I_B \oplus II_B)$ the inclusion of $II_A \otimes II_B$ and the simultaneous presence of $I_A \otimes II_B$ and $II_A \otimes I_B$ has no effect. Exploratory calculations on Ar_2[13] show every indication that this is the case. It is especially pleasing that the second order terms in $II_A \otimes II_B$, which involve simultaneous excitations from the weakly correlating orbitals, are small, for these have the largest CSF counts. We could proceed further to look at the effect of 4s and 4p correlation independently of 3d, and then check the effect of intershell intra-fragment correlations by putting them all into a single calculation.

Our aim has been to take this discussion far enough to establish that we have a framework in which we can systematically trace the source of corrections to the ICF1 interaction energy and to show that such studies require the ability to divide the orbital space into numerous subsets with large numbers of different electron assignments to them. Further approximations in ICF calculations may call for the division of the external orbital space into subsets, as well as for placing restrictions on the angular momentum coupling of electrons within and between orbital subsets, but such matters are best left to a more specialized presentation than this. Other calculations, not normally of weak interactions, may specify reference configurations and ask to retain only those configurations up to a given level of electron excitation from them. All of these possibilities exist in ALCHEMY II.

Appendix 11A.
The MOLECULE Integral Generator

Jan Almlöf
Department of Chemistry
University of Minnesota
and
Peter R. Taylor
ELORET Institute
NASA Ames Research Center

MOLECULE is a vectorized program for the computation of symmetry-adapted one- and two-electron integrals over generally contracted spherical harmonic Gaussian basis functions. It has been interfaced with the ALCHEMY II program modules by Yoshimine.

MOLECULE generates one- and two-electron integrals in a basis of symmetry-adapted functions for the point group D_{2h} and its subgroups. These symmetry-adapted functions are comprised of generally contracted Gaussian orbitals, with either spherical harmonic or Cartesian angular parts. The program can handle up to $l=6$ (i-type functions), but this is easily extended if required.

The two-electron integral evaluation in MOLECULE is highly vectorized. The general scheme involves computation of integrals over shells of primitive Cartesian Gaussians, followed by a contraction transformation, a transformation to spherical harmonics (if desired), and then the formation of appropriate linear combinations of these integrals to give symmetry integrals. The transformation steps are implemented as conventional n^5 transformations in terms of matrix multiplications. The values of n encountered are discussed below. The symmetry-adaptation "transformation" has a lower dependence of m^3, where m is the number of non-trivial symmetry operations.[14,15] The most complicated manipulations arise in the computation of primitive integrals, but the overall dependence is no worse than k^4, where k is the number of primitive Gaussian shells. We now consider these various steps in somewhat more detail.

The computation of integrals over primitive Gaussians is carried out using a recursive scheme with features in common with that of Obara and Saika.[16] A given integral is written as a sum of products of three terms that depend on only one of the Cartesian axes each and a term involving the standard auxiliary function $F_m(t)$. The integrals are computed over shell groups, where a shell group comprises not only all functions that differ only in their azimuthal quantum numbers, but all exponent values that contribute to a given set of contracted functions. Thus, for instance, the auxiliary function calculations are carried out using vectors of length k^4, for k primitive exponents: "extrinsic vectorization" as it is termed by Saunders. In practice, such long vectors may lead to excessive storage requirements, and the program includes a multipass feature that reduces the vector lengths by powers of k until the calculation can fit in the available memory. This multipassing does not increase the number of operations in any consequential way, but it may eventually reduce program efficiency if the vector lengths become short enough. In addition, since calculation of the auxiliary function is time-consuming, compression/expansion techniques are used to eliminate negligibly small argument values and to facilitate the use of different algorithms for different argument subranges.

Once primitive integrals have been evaluated, they are contracted. If the full k^4 vector length was used in the primitive integral evaluation, this contraction is simply a conventional "n^5" transformation. If the number of contracted functions in a group is n, the actual requirements are dominantly those of the nk^4 step. Note that the matrix multiplications in the contraction are organized with inner loops of length up to k^3 , rather than merely k.[17] This is desirable since the relatively small number of primitives used in higher angular functions otherwise leads to rather inefficient matrix multiplications. Where the code is to multipass over primitives, the partial contractions are performed as primitive integrals become available.

After contraction, it is possible to transform from Cartesians to unnormalized spherical harmonics if desired. In this case it is possible to retain the lower angular "contaminants" (e.g. the 3s component from the d Cartesians) or to delete them. Indeed, it is possible to delete any component the user chooses. Once again, this transformation is carried out as a series of matrix multiplications, and again the innermost loops are n^3, rather than n, in length.

Finally, the contracted integrals in either the Cartesian or spherical harmonic basis are combined linearly to yield integrals over symmetry-adapted basis functions. This is not performed as a transformation, since it involves only the addition or subtraction of terms, and it involves only the third power of the number of symmetry operations. The symmetry-adapted integrals are written out with labels.

Appendix 11B.

Algorithm for Large Scale Second Order MCSCF Calculations

B. Liu and B. H. Lengsfield

We present a matrix formulation of the second order MCSCF method, developed for efficient calculation with large one-particle basis sets and large CI expansions. A dynamic convergence threshold has been found for the iterative solution of the second order MCSCF linear equations which minimizes the number of iterations required in each MCSCF iteration. The computational effort for the Hessian matrix trial vector product has been reduced by one third by the inclusion of one third order term which does not affect the convergence behavior of the linear equations.

I. Introduction

The Multi-Configuration Self-Consistent-Field (MCSCF) method has long been recognized as an effective way of obtaining zeroth order wave function in the calculation of potential energy surfaces and molecular properties. MCSCF wave functions also have special advantages in the computation of first and second derivatives of the energy with respect to nuclear displacements. However, the utility of the MCSCF calculations has been limited because they have been plagued by convergence problems. Development of second order MCSCF procedures based on the exponential unitary transformation method has eliminated this bottleneck making it possible now to obtain MCSCF wave functions with several thousand configurations for both ground and excited electronic states.[18].

Because of the wide range of problems that a MCSCF method can address, it is desirable to formulate an algorithm in the most general manner possible without sacrificing computational efficiency. In particular it is desirable to have the capability to handle both large one-particle basis sets and large configuration interaction (CI) expansions. Flexibility in the choice of configurations to be included is also important. The most time consuming parts of a large MCSCF calculation are: 1) integral transformation; 2) CI calculation; 3) Hessian construction; and 4) iterative solution of the MCSCF equations. Parts 2) thru 4) also require considerable data retrieval from peripheral memory devices which makes the calculation inefficient on supercomputers. In this Appendix we report 1) a new procedure for the construction of the orbital part of the Hessian, which significantly reduces the integral transformation time, total Hessian construction time and IO requirements, 2) an efficient, new method for calculating the CI coupling contributions to the MCSCF equations, and 3) an improved method for solving the system of

MCSCF linear equations which minimizes the number of iterations required. We also report a prototype-CI method which drastically reduces the amount of information that must be stored on IO devices for the CI calculation and is comparable in efficiency to modern direct CI methods, but more suitable for typical MCSCF calculations. Together, these developments lead to an efficient MCSCF algorithm for large basis set and large CI expansion calculations. The MCSCF program in ALCHEMY II has been constructed using this algorithm.

II. Overview of the MCSCF Method

Our basic approach to the MCSCF problem has been described in reference 18. In this section, we give a brief review of this direct MCSCF algorithm. The new developments are presented in sections III through V.

Consider a general wave function of the form

$$\Psi^{\mu} = \sum_{I}^{N} \omega_I^{\mu} \Phi_I \tag{II.1}$$

whose energy is given in terms of the unique elements of the one- and two-particle density matrices,

$$E^{\mu} = \sum_{i \geq j} D_{ij}^{\mu\mu} < \phi_i \mid \tilde{h} \mid \phi_j > + \sum_{\substack{i \geq j \\ (ij) \geq (kl)}} \sum_{k \geq l} D_{ijkl}^{\mu\mu} < \phi_i \phi_j \mid \tilde{g} \mid \phi_k \phi_l > , \tag{II.2a}$$

or alternatively

$$E^{\mu} = < \omega^{\mu} \mid H \mid \omega^{\mu} > , \tag{II.2b}$$

where $\{\Phi_I\}$ represents a basis of configuration state functions (CSF's); $\{\underset{\sim}{\omega}^{\mu}\}$ the set of CI coefficients for the state μ $\{\phi_i\}$ a set of molecular orbitals; $\tilde{h}$ and $\tilde{g}$ the usual one and two electron operators comprising the nonrelativistic Hamiltonian, $\tilde{H}$ and $\underset{\sim}{H}$ is the projection of this Hamiltonian into our basis of CSF's, i.e. $H_{IJ} = < \Phi_I \mid \tilde{H} \mid \Phi_J >$.

1. Second Order MCSCF Equation

By varying the molecular orbitals and CI vector in Eq. (II.2), we seek a stationary point on the energy hypersurface. The variational parameters are the unique elements of the generators of the exponential unitary transformations that rotate the orbitals and CI vectors,

$$\tilde{U}_{ORB} = e^{\Delta} \cong \tilde{1} + \tilde{\Delta} + \frac{1}{2}\tilde{\Delta}^2 \tag{II.3a}$$

and

$$\tilde{U}_{CI} = e^{\gamma} \cong \tilde{1} + \tilde{\gamma} + \frac{1}{2}\tilde{\gamma}^2 \tag{II.3b}$$

The generators $\tilde{\Delta}$ and $\tilde{\gamma}$ are antisymmetric, i.e. $\tilde{\Delta}^+ = -\tilde{\Delta}$ and $\tilde{\gamma}^+ = -\tilde{\gamma}$. They contain only nonredundant terms, namely only those rotations that alter the energy of our wave function.

If we subject the CI vector and orbitals in Eq. (II.2) to the unitary transformations, retain only terms up to second order in γ and Δ, and drop the state label μ we obtain

$$E' = E_0 + < g_{ORB} \,|\, \Delta > + < g_{CI} \,|\, \gamma > + \frac{1}{2} < \Delta \,|\, b \,|\, \Delta > + \frac{1}{2} < \gamma \,|\, a \,|\, \gamma >$$
$$+ \frac{1}{2}[< \Delta \,|\, c^+ \,|\, \gamma > + < \gamma \,|\, c \,|\, \Delta >], \tag{II.4}$$

where $|\,\gamma >$ and $|\,\Delta >$ are vectors containing the nonredundant elements of $\tilde{\gamma}$ and $\tilde{\Delta}$ $|\, g_{ORB} >$ and $|\, g_{CI} >$ are the orbital and CI gradient vectors; and b, a, and c are the orbital Hessian, the CI Hessian, and the CI-orbital coupling portion of the Hessian, respectively.

If we differentiate this expression with respect to the unique, nonredundant elements of Δ and γ and set the resulting expression to zero, we obtain

$$\begin{pmatrix} B & C^+ \\ C & A \end{pmatrix} \begin{pmatrix} |\,\Delta > \\ |\,\gamma > \end{pmatrix} = \begin{pmatrix} -\,|\,G_{ORB} > \\ -\,|\,G_{CI} > \end{pmatrix}, \tag{II.5}$$

where we have divided the gradient and Hessian terms by two for convenience;

i.e. $B = \frac{1}{2}b$, $C = \frac{1}{2}c$, etc.

2. Orbital gradient

The orbital gradient vector is given by a sum over all unique density matrix elements,

$$| G_{ORB} > = \frac{1}{2} \sum_{\alpha} D_{\alpha}^{\mu\mu} | I^{\alpha} > , \tag{II.6}$$

where α is a composite index labeling one and two electron integrals and $D_{\alpha}^{\mu\mu}$ denotes unique elements of the one and two particle density matrices. The components of the vector $| I^{ij} >$, labeled by nonredundant orbital rotations ma, are given by

$$I_{ma}^{ij} = \delta_{im} < \phi_a | \tilde{h} | \phi_j > \rho_{ia} + \delta_{jm} < \phi_i | \tilde{h} | \phi_a > \rho_{ja}, \tag{II.7}$$

with

$$\rho_{ia} = \left\{ \begin{array}{lll} 1 & & i > a \\ -1 & \text{for} & i < a \\ 0 & & i = a \end{array} \right\} \tag{II.8}$$

For two electron terms, the components of $| I^{ijkl} >$ are given by

$$I_{ma}^{ijkl} = \delta_{im} < \phi_a \phi_j | \tilde{g} | \phi_k \phi_l > \rho_{ia} + \delta_{jm} < \phi_i \phi_a | \tilde{g} | \phi_k \phi_l > \rho_{ja}$$

$$+ \delta_{km} < \phi_i \phi_j | \tilde{g} | \phi_a \phi_l > \rho_{ka} + \delta_{lm} < \phi_i \phi_j | \tilde{g} | \phi_k \phi_a > \rho_{la}. \tag{II.9}$$

In Eqs. (II.7 and II.9) and in what follows, the indices i, j, k, l, m, and n refer to orbitals occupied in the CI wave function, while a and b refer to all orbitals in the calculation.

3. CI gradient

The CI gradient vector is given by

$$| G_{CI} > = \sum_{\alpha} < \alpha > | T^{\mu\alpha} > , \tag{II.10}$$

where $< \alpha >$ refers to a one or two electron integral and $| T^{\mu\alpha} >$ is a vector of transition density matrix elements with components

$$T_{\nu}^{\mu\alpha} = D_{\alpha}^{\mu\nu} \rho_{\mu\nu}. \tag{II.11}$$

4. Orbital Hessian

The orbital Hessian is a square matrix whose elements, labeled by two orbital rotations, are given by

$$B_{(ia)(jb)} = \sum_{\alpha} \{ b^{\alpha}_{(ia)(jb)} + \xi_{iajb}[f^{\alpha}_{ij}\delta_{ab} + f^{\alpha}_{ib}\delta_{aj} + f^{\alpha}_{ab}\delta_{ij}]\} \tag{II.12}$$

for $i < a, j < b$, and $(ia) \geq (jb)$, where

$$\xi_{ijij} = \xi_{ikjk} = \xi_{ikil} = -1, \tag{II.13a}$$

$$\xi_{ijjl} = \xi_{ijki} = -1, \tag{II.13b}$$

and

$$\xi_{ijkl} = 0. \tag{II.13c}$$

Here again, α denotes one and two electron integral labels, and the matrices in Eq. (II.12) are given by

$$b^{ij}_{(ma)(nb)} = \frac{1}{2} D^{\mu\mu}_{ij} < \phi_a \mid \tilde{h} \mid \phi_b > (\rho_{ia}\rho_{jb}\delta_{im}\delta_{jn} + \rho_{ib}\rho_{ja}\delta_{in}\delta_{jm}), \tag{II.14}$$

$$\begin{aligned} b^{ijkl}_{(ma)(nb)} = \frac{1}{2} D^{\mu\mu}_{ijkl} \\ [< \phi_a\phi_b \mid \tilde{g} \mid \phi_k\phi_l > (\rho_{ia}\rho_{jb}\delta_{im}\delta_{jn} + \rho_{ib}\rho_{ja}\delta_{in}\delta_{jm}) \\ + < \phi_i\phi_j \mid \tilde{g} \mid \phi_a\phi_b > (\rho_{ka}\rho_{lb}\delta_{km}\delta_{ln} + \rho_{kb}\rho_{la}\delta_{kn}\delta_{lm}) \\ + < \phi_a\phi_j \mid \tilde{g} \mid \phi_b\phi_l > \rho_{ia}\rho_{kb}\delta_{im}\delta_{kn} + < \phi_b\phi_j \mid \tilde{g} \mid \phi_a\phi_l > \rho_{ib}\rho_{ka}\delta_{in}\delta_{km} \\ + < \phi_a\phi_j \mid \tilde{g} \mid \phi_k\phi_b > \rho_{ia}\rho_{lb}\delta_{im}\delta_{ln} + < \phi_b\phi_j \mid \tilde{g} \mid \phi_k\phi_a > \rho_{ib}\rho_{la}\delta_{in}\delta_{lm} \\ + < \phi_i\phi_a \mid \tilde{g} \mid \phi_b\phi_l > \rho_{ja}\rho_{kb}\delta_{jm}\delta_{kn} + < \phi_i\phi_b \mid \tilde{g} \mid \phi_a\phi_l > \rho_{jb}\rho_{ka}\delta_{jn}\delta_{km} \\ + < \phi_i\phi_a \mid \tilde{g} \mid \phi_k\phi_b > \rho_{ja}\rho_{lb}\delta_{jm}\delta_{ln} + < \phi_i\phi_b \mid \tilde{g} \mid \phi_k\phi_a > \rho_{jb}\rho_{la}\delta_{jn}\delta_{lm}] \end{aligned} \tag{II.15}$$

$$f^{ij}_{ma} = \frac{1}{4} D^{\mu\mu}_{ij} [< \phi_a \mid \tilde{h} \mid \phi_j > (1 + \delta_{ia})\delta_{im} + < \phi_i \mid \tilde{h} \mid \phi_a > (1 + \delta_{ja})\delta_{jm}], \tag{II.16}$$

and

$$\begin{aligned} f^{ijkl}_{ma} = \frac{1}{4} D^{\mu\mu}_{ijkl} \\ [< \phi_a\phi_j \mid \tilde{g} \mid \phi_k\phi_l > (1 + \delta_{ia})\delta_{im} + < \phi_i\phi_a \mid \tilde{g} \mid \phi_k\phi_l > (1 + \delta_{ja})\delta_{jm} \\ + < \phi_i\phi_j \mid \tilde{g} \mid \phi_a\phi_l > (1 + \delta_{ka})\delta_{km} + < \phi_i\phi_j \mid \tilde{g} \mid \phi_k\phi_a > (1 + \delta_{la})\delta_{lm}] \end{aligned} \tag{II.17}$$

Here ma runs over all symmetry allowed orbital rotations including the redundant rotations, except rotations between virtual (unoccupied) orbitals.

5. CI-Orbital Coupling

The CI-orbital coupling portion of the Hessian, C, is given by the direct product between a vector of transition density matrix elements, $T^{\mu\alpha}$, Eq. (II.11), and the I vector defined by Eqs. (II.7) and (II.9):

$$C = \sum_{\alpha} | T^{\mu\alpha} > < I^{\alpha} | , \tag{II.18}$$

where again α runs over all unique elements of the one and two particle density matrices.

6. CI Hessian

The CI Hessian, A, is given by

$$A_{(\mu v)(\mu o)} = < \omega^{v} | H | \omega^{o} > \rho_{\mu v} \rho_{\mu o}, \quad v \neq o \tag{II.19}$$

and

$$\begin{aligned} A_{(\mu v)(\mu v)} &= < \omega^{v} | H | \omega^{v} > - < \omega^{\mu} | H | \omega^{\mu} > \\ &\equiv E^{v} - E^{\mu}, \end{aligned} \tag{II.20}$$

where the index μ, labels the CI state being optimized, v and o label the remainder of the CI basis.

7. Large CI Expansions

The formalism presented above requires that the orbital-CI Hessian C, the CI Hessian A and the CI gradient $| G_{CI} >$ be constructed by first transforming the Hamiltonian H from the CSF basis into a basis of CI vectors. For large CI expansion the one and two index transformation of C and A, respectively, are too costly. This problem can be avoided[19] by constructing these matrices in the CSF basis, and then solving, instead of (II.5), the following equation:

$$\left[\begin{pmatrix} 1 & 0 \\ 0 & M \end{pmatrix} \begin{pmatrix} B & \overline{C}^{+} \\ \overline{C} & H - E^{\mu} \end{pmatrix} \begin{pmatrix} 1 & 0 \\ 0 & M \end{pmatrix} + z \begin{pmatrix} 0 & 0 \\ 0 & L \end{pmatrix} \right] \begin{pmatrix} | \Delta > \\ | \delta\omega > \end{pmatrix} = \begin{pmatrix} - | G_{ORB} > \\ MH | \omega^{\mu} > \end{pmatrix}, \tag{II.21}$$

where z is an arbitrary, but nonzero constant,

$$\overline{C} = \sum_{\alpha} | t^{\mu\alpha} > < I^{\alpha} | , \tag{II.22}$$

and

$$t_I^{\mu\alpha} = \sum_J K_{IJ}^{\alpha} \omega_J^{\mu}, \tag{II.23}$$

with K_{IJ}^{α} the coupling constants for the Hamiltonian matrix element H_{IJ}. The new vector $| t^{\mu\alpha} >$ is simply related to the vector $| T^{\mu\alpha} >$ defined in Eq. (II.11) by

$$T_v^{\mu\alpha} = \sum_I \omega_I^v t_I^{\mu\alpha}. \tag{II.24}$$

We have also introduced the projection operators,

$$L = | \omega^{\mu} > < \omega^{\mu} | \tag{II.25}$$

and

$$M = 1 - L. \tag{II.26}$$

In this formulation, the improved CI vector is given by

$$| \omega^{\mu} >' = | \omega^{\mu} > - | \delta\omega > - \frac{1}{2} < \delta\omega | \delta\omega > | \omega^{\mu} > . \tag{II.27}$$

When the dimension of the Newton-Raphson matrix equation, (II.21), is large it becomes advantageous to use an iterative method for its solution. In these methods, the most time consuming step is the multiplication of a trial vector by the matrix. For large CI expansions, it is efficient not to explicitly construct the CI-orbital coupling part of the Hessian, but to use the following expressions to directly compute its contributions to the matrix-vector product:

$$\bar{C} | b_{\Delta} > = \sum_{\alpha} | t^{\mu\alpha} > < I^{\alpha} | b_{\Delta} > \tag{II.28}$$

and

$$\bar{C}^{+} | b_{\gamma} > = \sum_{\alpha} | I^{\alpha} > < t^{\mu\alpha} | b_{\gamma} > , \tag{II.29}$$

where $| b_{\Delta} >$ and $| b_{\gamma} >$ are, respectively, the orbital and CI parts of the trial vector multiplied by the projection operators. Equation (II.29) is equivalent to Eq. (II.6) for the construction of the orbital gradient; the only difference being the density matrix element $D_{\alpha}^{\mu\mu}$ has been replaced by a transition density matrix element $< t^{\mu\alpha} | b_{\gamma} >$. In other words, the product $\bar{C}^{+} | b_{\gamma} >$ can be constructed as the orbital gradient by using the transition density matrix between $| \omega^{\mu} >$ and $| b_{\gamma} >$. Similarly, Eq. (II.28) can be viewed as the construction of the CI gradient, Eq. (II.10), using updated integrals $< I^{\alpha} | b_{\Delta} >$. Furthermore, making use of Eq. (II.23), we can rewrite Eq. (II.28) as

$$\bar{C} | b_{\Delta} > = \delta H | \omega^{\mu} > , \tag{II.30}$$

where δH is the Hamiltonian matrix constructed from the updated integrals $< I^{\alpha} | b_{\Delta} >$. This equation is particularly useful when an efficient CI program, such as our Prototype CI program, is available.

Finally, we note that, in using Eq. (II.21), it is important not to explicitly construct M(H-E)M, but to perform a series of matrix-vector multiplications. The multiplication of M onto a trial vector can be done very rapidly as it only involves a scalar product of two vectors followed by the multiplication of a vector by a scalar.

III. Matrix Formulation of Orbital Hessian

The equations given in the preceding section for orbital gradient and Hessian are not ideal for modern computers. In this section we cast the calculation of these quantities in a matrix formulation which is particularly suited for fast vector and array processors, and present a new computational scheme which significantly reduces the two-electron integral transformation time.

1. Orbital Hessian

Let us define a matrix $\underline{b}$ such that

$$\underline{b}_{ijab}\rho_{ia}\rho_{jb} = \sum_{\alpha} b^{\alpha}_{(ia)(jb)} = b_{(ia)(jb)}, \tag{III.1}$$

where the b^{α} matrix is defined by Eqs. (II.14) and (II.15). This expression constitutes the first of two terms contributing to the orbital Hessian, Eq. (II.12). The $\underline{b}$ matrix can be expressed as a sum of matrix products as follows:

$$\underline{b} = \underline{b}^{1} + \underline{D}^{J}\underline{J} + \underline{D}^{K}\underline{K} + \underline{D}^{K^{t}}\underline{K}^{t}, \tag{III.2}$$

where

$$\underline{b}^{1}_{ijab} = \frac{1}{2}D_{ij}(1 + \delta_{ij}) < \phi_a | \tilde{h} | \phi_b >, \tag{III.3}$$

$$\underline{D}^{J}_{ijkl} = \frac{1}{2}D_{ijkl}[1 + \delta_{ij,kl}](1 + \delta_{ij}), \tag{III.4}$$

$$\underline{J}_{klab} = < \phi_a\phi_b | \tilde{g} | \phi_k\phi_l >, \tag{III.5}$$

$$\underline{D}^{K}_{ijkl} = \frac{1}{2}D_{ikjl}[1 + \delta_{ij,kl}](1 + \delta_{ik})(1 + \delta_{jl}), \tag{III.6}$$

$$\underline{K}_{klab} = < \phi_a\phi_k | \tilde{g} | \phi_b\phi_l >, \tag{III.7}$$

$$\underline{D}^{K^t}_{ijkl} = \frac{1}{2} D_{iljk}(1 - \delta_{kl})(1 + \delta_{sub\ <i\ sub\ lambda>})(1 + \delta_{jk}), \tag{III.8}$$

and

$$\underline{K}^t_{klab} = < \phi_a \phi_l \mid \tilde{g} \mid \phi_b \phi_k > . \tag{III.9}$$

The summation implied in the matrix equation, (III.2), is over $k \geq l$.

Let us define another matrix $\underline{P}$ whose elements are given by

$$\underline{P}_{ia} = \sum_j \sum_b [\underline{b}^1 + \underline{D}^J \underline{J}]_{ijab} \delta_{jb}. \tag{III.10}$$

The Fock matrix appearing in the second term that contributes to the orbital Hessian in Eq. (II.12) is given by

$$F_{ia} = \sum_\alpha f^\alpha_{ia} = \frac{1}{4}(1 + \delta_{ia}) \underline{P}_{ia}, \tag{III.11}$$

and the orbital gradient elements of Eq. (II.6) by

$$G_{ia} = \frac{1}{2} \rho_{ia} \underline{P}_{ia}. \tag{III.12}$$

The matrix expression given above for the orbital Hessian, Eq. (III.2), applies in a case of no symmetry. Additional expressions needed to treat molecules with non-degenerate point group symmetry are given below. A four-symmetry block of the density matrix, where i, j, k and l all belong to different symmetries, makes contributions to six off-diagonal blocks of the orbital Hessian. Here, by diagonal block we mean a block of the orbital Hessian where all $\mathbf{B}_{(ma)(nb)}$ has m, a, n and b in the same symmetry, and by off-diagonal we mean a block where m and a belong to one symmetry while n and b belong to another symmetry. We note that m and a must belong to the same symmetry, and so do n and b, because no rotation between orbitals belonging to different symmetries are allowed. There are two Coulomb type contributions, for example,

$$b_{(ia)(jb)} = \frac{1}{2} \rho_{ia} \rho_{jb} \sum_k \sum_l D_{ijkl} < \phi_a \phi_b \mid \tilde{g} \mid \phi_k \phi_l > , \tag{III.13}$$

and four exchange type contributions, for example,

$$b_{(ia)(kb)} = \frac{1}{2} \rho_{ia} \rho_{kb} \sum_j \sum_l D_{ijkl} < \phi_a \phi_j \mid \tilde{g} \mid \phi_b \phi_l > . \tag{III.14}$$

A two symmetry block of the density matrix, where i and j belong to one symmetry and k and l belong to another, makes contributions to two diagonal blocks and one off-diagonal block of the orbital Hessian. There are two Coulomb type contributions, for example,

$$b_{(ia)(jb)} = \frac{1}{2}\rho_{ia}\rho_{jb}\sum_{k}\sum_{l} D_{ijkl}(1+\delta_{ij}) < \phi_a\phi_b \mid \tilde{g} \mid \phi_k\phi_l > , \qquad \text{(III.15)}$$

and one exchange type contribution

$$b_{(ia)(kb)} = \frac{1}{2}\rho_{ia}\rho_{kb}\sum_{j}\sum_{l} D_{ijkl}(1+\delta_{ij})(a+\delta_{kl}) < \phi_a\phi_j \mid \tilde{g} \mid \phi_b\phi_l > . \qquad \text{(III.16)}$$

A two symmetry block of the density matrix, where i and k belong to one symmetry and j and l belong to another, makes contributions to two diagonal blocks and one off-diagonal block of the orbital Hessian. There are two Coulomb type contributions, for example,

$$b_{(ia)(jb)} = \frac{1}{2}\rho_{ia}\rho_{jb}\sum_{k}\sum_{l} D_{ijkl}(1+\delta_{ij,kl}) < \phi_a\phi_b \mid \tilde{g} \mid \phi_k\phi_l > , \qquad \text{(III.17)}$$

two exchange type contributions, for example,

$$b_{(ia)(kb)} = \frac{1}{2}\rho_{ia}\rho_{kb}\sum_{j}\sum_{l} D_{ijkl}(1+\delta_{ij,kl}) < \phi_a\phi_j \mid \tilde{g} \mid \phi_b\phi_l > , \qquad \text{(III.18)}$$

and one exchange type contribution

$$b_{(ia)(lb)} = \frac{1}{2}\rho_{ia}\rho_{lb}\sum_{j}\sum_{k} D_{ijkl}(1+\delta_{ik,jl}) < \phi_a\phi_j \mid \tilde{g} \mid \phi_b\phi_k > . \qquad \text{(III.19)}$$

The equations (III.13-19) can easily be cast into the matrix form of Eq. (III.2) .

2. Integral transformation

The expressions given in section III.1 requires electron repulsion integrals between molecular orbitals. More precisely, all two electron integrals involving two or more molecular orbitals occupied in the MCSCF wave function are needed. As the molecular orbitals are expanded in terms of a set of basis functions and we usually have available only electron repulsion integrals between these basis functions, it is necessary, at the beginning of each MCSCF iteration, to perform a four-index transformation to obtain the integrals between molecular orbitals. For small to medium sized CI expansion, the time needed for an MCSCF calculation is dominated by this integral transformation. In cases of high molecular symmetry, a significant portion of this transformation time can be reduced if we transform only two indices to the occupied molecular orbitals leaving two indices over basis functions, construct an orbital Hessian for rotations between occupied molecular orbitals and basis functions, and then transform this Hessian to one for rotations between molecular orbitals.

Let the molecular orbitals be expanded in a set of basis functions $\{\chi\}$

$$\phi_i = \sum_p c_p^i \chi_p \, . \tag{III.20}$$

The orbital gradient and Hessian for rotation between occupied orbitals and basis functions are given by the equations in Section III.1 with indices a and b replaced everywhere by p and q, respectively, ϕ_a and ϕ_b replaced by χ_p and χ_q, respectively, and all ρ_{ip} factors replaced by 1. The orbital gradient between molecular orbitals is then given by

$$G_{ia} = \rho_{ia} \sum_p G_{ip} c_p^a, \tag{III.21}$$

and the orbital Hessian given by

$$B_{(ia)(jb)} = \rho_{ia}\rho_{jb} \sum_{p,q} B_{(ip)(jq)} c_p^a c_q^b \tag{III.22}$$

The advantage of this approach is quite easily understood. We are simply trading off integral transformation for gradient and Hessian transformation. In cases of high molecular symmetry, there are many more integral blocks to be transformed than there are Hessian blocks, because nonzero integrals may arise from orbitals in up to four symmetries, while the orbitals labeling the Hessian elements belong to at most two symmetries. For a calculation with no symmetry, the two approaches are about equal; although with the basis function approach it is much easier to take advantage of the fact that we only need orbital Hessian matrix elements between nonredundant orbital rotations.

IV. Iterative solution of MCSCF equations

The second order MCSCF system of linear equations, dimensioned the number of nonredundant orbital rotations plus the number of CSF's, must be solved by an iterative method in cases of large CI expansions. In this section we review such a method, describe a new method for determining just how accurately the linear equations must be solved in each MCSCF iteration, and a method for reducing the computational effort in forming the matrix vector product required in the solution of the linear equations.

1. Iterative solution of a system of linear equations

Consider a system of N linear equations

$$VX = W, \tag{IV.1}$$

where V is an NxN matrix, X and W are column vectors of dimension N. We seek a solution vector X which satisfies this equation.

Construct a set of n vectors $\{X^i\}$ such that

$$X^1 = D^{-1}W, \tag{IV.2a}$$

$$X^i = Y^i - \sum_{j=1}^{i-1} \frac{< Y^i | X^j >}{< X^j | X^j >} \tag{IV.2b}$$

where

$$Y^i = D^{-1}VX^{i-1} \tag{IV.2c}$$

and D is the diagonal part of the matrix V. Projection of Eq. (IV.1) into the space spanned by $\{X^i\}$ results in a system of n linear equations

$$vx = w, \tag{IV.3}$$

where

$$v_{ij} = < X^i | V | X^j > \tag{IV.4}$$

and

$$w_i = < X^i | W > . \tag{IV.5}$$

An approximate solution to Eq. (IV.1) can be obtained from the solution vector of Eq. (IV.3):

$$\tilde{X} = \sum_{i=1}^{n} x_i X^i \tag{IV.6}$$

Thus, one increases n until a sufficiently accurate approximate solution is obtained. The most time consuming part of each iteration is in forming the product VX^{j-1} in Eq. (IV.2c).

2. Convergence threshold

There are two tests of convergence of the iterative process described in the preceding section. The first involves the calculation of

$$T1_n = < V\tilde{X} - W \mid V\tilde{X} - W > . \tag{IV.7}$$

This quantity gives an absolute measure of the accuracy of the approximate solution $\tilde{X}$. But it requires the matrix-vector product $V\tilde{X}$ which is costly to construct. While this product can be obtained inexpensively if all products VX^i constructed in preceding iterations are stored on some peripheral memory device, the real time needed to retrieve these vectors can become undesirably long.

The second method involves calculating the quantity

$$T2_n = \frac{x_n^2 < X^n \mid X^n >}{S} \tag{IV.8}$$

where

$$S = \sum_{i=1}^{n} x_i^2 < X^i \mid X^i > . \tag{IV.9}$$

This quantity is a measure of the contribution of X^n to the approximate solution $\tilde{X}_n$. Thus, a small $T2_n$ does not guarantee a smaller $T2_{n+1}$. However, practical experience shows that T2 usually turns out to be within a factor of 2 of T1. Since the cost of calculating T2 is essentially negligible, it is the preferred practical measure of convergence. To be absolutely safe, one can switch over to T1 when T2 is below the desired convergence threshold for the linear equations.

In a given MCSCF iteration the accuracy needed in the solution to the Newton-Raphson equations depends on the accuracy of the MCSCF wave function at that iteration. One measure of the convergence of an MCSCF wave function is provided by S, Eq.(IV.9), the sum of squares of the elements of the matrices $\tilde{\Delta}$ and $\tilde{\gamma}$, obtained from the solution of the linear equations. In each MCSCF iteration, S is a measure of the convergence of the input MCSCF wave function. In a second order MCSCF calculation, we expect S to decrease quadratically in successive iterations. In other words, the output wave function of an MCSCF iteration should be accurate to S^2, if the input wave function is accurate to S and if the linear equations have been solved sufficiently accurately. However, it is unnecessary to solve the linear equations beyond an accuracy of

$$x_n^2 < X^n \mid X^n > \leq Th = S^2 \tag{IV.10}$$

because the second order MCSCF iterative process cannot give better than quadratic convergence in any event. Equation (IV.10) provides a dynamic threshold for convergence of the iterative solution of the linear equations and ensures that no extra iterations are performed. Also, one can safely require Th to be no smaller than the convergence threshold of the overall MCSCF process. However, to be safe, it is desirable to require that at least two significant figures are obtained in the solution vector, i.e. $\mathrm{Th} \leq 0.01\mathrm{S}$.

3. The CI portion of VX

The CI portion of the VX^i product is, according to Eqs. (II.21) and (II.30), given by

$$| Z > = \delta \,|\, \omega > + (H - E) \,|\, \delta\omega > , \qquad \text{(IV.11)}$$

where $| \delta\omega >$ is the CI portion of $| X^i >$. If we combine the Hamiltonian and the updated Hamiltonian, the CI vector and the CI portion of the X vector, we have

$$Z + \delta H \delta\omega = (H + \delta H)(\,| \omega > + | \delta\omega >\,) - | G_{CI} > - E \,|\, \delta\omega > - \delta E \,|\, \omega > , \qquad \text{(IV.12)}$$

where

$$\delta E = < \omega \,|\, \delta H \,|\, \omega > . \qquad \text{(IV.13)}$$

The second term in the left hand side of Eq. (IV.12) is a third order term scaled by $| \delta\omega >$. Since the magnitudes of the vectors X^i have no effect on the approximate solution $\tilde{X}$, as is obvious from Eqs. (IV.3-6), they can be made so small that

$$Z + \delta H \delta\omega \cong Z. \qquad \text{(IV.14)}$$

The Eq. (IV.12) is to be preferred over Eq. (IV.11) because it involves only one matrix vector multiplication instead of the two required in the latter. Test calculations have shown that normalizing the X^i vectors to 1×10^{-4} makes the two equations equivalent in accuracy while (IV.12) requires only half the computational effort.

V. The Prototype CI Method

1. Overview

Most large MCSCF calculations involve large numbers configurations generated by distributing electrons in a relatively small set of active orbitals. The configurations selected are usually not easily described as single and double excitations from a small set of reference configurations. Modern direct CI methods,[5,20] designed mainly for calculations involving single and double excitations from a set of reference configurations, are therefore not efficient for MCSCF calculations. Conventional CI methods, requiring explicit construction and storage of all Hamiltonian matrix elements, are impractical for large (20,000 configuration) CI calculations.

In the prototype CI method, expressions for the nonzero elements of the Hamiltonian matrix are stored by prototype on a peripheral memory device. The one and two-electron integrals between the active orbitals are stored in memory. In each iteration in the solution of the matrix eigenvalue problem, the prototype matrix expressions are read in and combined with the appropriate integrals to form all the matrix elements belonging to the same prototype. These matrix elements, once formed, are used to make their contributions to the matrix-vector product and then discarded. Thus, the only IO involved in each CI iteration is a single pass over the prototype matrix expressions. The requirement that all active integrals reside in main storage is not a severe one since the active orbital space is small. Furthermore, a simple sort of the integrals, before the CI calculation, will allow them to be stored and retrieved with the prototype expressions thus eliminating the main storage requirement at a very small cost of increased IO. The prototype-CI method has the additional advantages that it is completely general in the choice of configurations to be included in a calculation and it can take full advantage of molecular symmetry, i.e. it is not limited to Abelian groups.

2. Prototype configurations

The description of the prototype CI method begins with the definition of prototype configurations. A prototype configuration is specified by the number of occupied orbitals, m, and m quartets (n_a, l_a, j_a, k_a, $a = 1,m$) giving the occupation, orbital symmetry label, and the range of actual orbital serial numbers each orbital can take on, respectively. A prototype configuration represents all actual configurations given by

$$\prod_{a=1}^{m}(i_a\, l_a)^n$$

with

$$j_a \leq i_a \leq k_a$$

To avoid redundant actual configurations, the following restrictions must be applied:

$$l_a \leq l_{a+1},$$

$$n_a \geq n_{a+1} \text{ if } l_a = l_{a+1},$$

and

$$i_a < i_{a+1}.$$

The prototype configurations form a compact representation of the actual configurations. For example, the 1317 configurations that can arise from distributing 8 electrons in 8 nondegenerate active orbitals of the same symmetry can be represented by the following 5 prototype configurations:

$$\begin{array}{ll} a & 1\,2\,3\,4 \\ n_a & 2\,2\,2\,2 \\ l_a & 1\,1\,1\,1 \\ j_a & 1\,1\,1\,1 \\ k_a & 8\,8\,8\,8 \end{array} \tag{V.1}$$

$$\begin{array}{ll} a & 1\,2\,3\,4\,5 \\ n_a & 2\,2\,2\,1\,1 \\ l_a & 1\,1\,1\,1\,1 \\ j_a & 1\,1\,1\,1\,1 \\ k_a & 8\,8\,8\,8\,8 \end{array} \tag{V.2}$$

$$\begin{array}{ll} a & 1\,2\,3\,4\,5\,6 \\ n_a & 2\,2\,1\,1\,1\,1 \\ l_a & 1\,1\,1\,1\,1\,1 \\ j_a & 1\,1\,1\,1\,1\,1 \\ k_a & 8\,8\,8\,8\,8\,8 \end{array} \tag{V.3}$$

$$\begin{array}{ll} a & 1\,2\,3\,4\,5\,6\,7 \\ n_a & 2\,1\,1\,1\,1\,1\,1 \\ l_a & 1\,1\,1\,1\,1\,1\,1 \\ j_a & 1\,1\,1\,1\,1\,1\,1 \\ k_a & 8\,8\,8\,8\,8\,8\,8 \end{array} \tag{V.4}$$

and

$$\begin{array}{ll} a & 1\,2\,3\,4\,5\,6\,7\,8 \\ n_a & 1\,1\,1\,1\,1\,1\,1\,1 \\ l_a & 1\,1\,1\,1\,1\,1\,1\,1 \\ j_a & 1\,1\,1\,1\,1\,1\,1\,1 \\ k_a & 8\,8\,8\,8\,8\,8\,8\,8 \end{array} \tag{V.5}$$

With each prototype configuration is associated an orthonormal set of prototype configuration state functions (CSF's) given as linear combinations of Slater determinants. Specification of the 1, 1, 2, 5 and 14 CSF's ,respectively, for the prototype configurations in the example completes a compact representation of the 2535 possible singlet CSF's. It turns out to be convenient for the CI calculation to follow the description of each prototype configuration by a list of actual configurations arising from it, each given by m actual orbital indices. This provides a convenient method for configuration selection; one simply list the actual configurations to be included in the CI instead of all the configurations generated from the prototype.

3. Prototype matrices

Consider the Hamiltonian matrix elements arising from the interaction between two prototype configurations with N′ and N″ CSF's, respectively. The interaction between actual configurations I and J belonging to the first and second prototype configurations, respectively, form an N′XN″ submatrix. Of course, if the actual configurations I and J differ by more than two orbital occupations, the submatrix is null. How the occupied orbitals of configuration I are matched to the occupied orbitals of configuration J can be described by a match vector of dimension m′, the number of occupied orbitals in I. The b-th component of this vector gives the position of the occupied orbital in configuration J that has the same symmetry and orbital index as the b-th orbital in configuration I. A zero entry means that the index of the b-th orbital in I is not found in J. For example, the match vector between the actual configurations $1^2 2^2 3^2 4^2$ and $3^2 4^2 1^1 2^1 5^1 6^1$ is (3,4,1,2). Two actual configuration pairs that have the same match vector give rise to submatrices that are identical in form; they differ only in the orbital labels of the integrals contributing to the matrix elements. The matrix expressions, integrals and coupling coefficients, for one actual configuration pair can be obtained from that of the other simply by replacing the integral labels by the indices of the orbitals occupying the same positions in the other actual configuration pair. Therefore, all actual configuration pairs arising from a prototype configuration pair can be grouped by their match vectors. For each of these groups, it is sufficient to evaluate and store the matrix expression for one actual configuration pair. If we label the integrals by the position the orbitals occupy in the actual configurations, instead of the actual orbital indices, keep a list of all the actual configuration pairs (serial numbers) belonging to the group, and a list of all the actual configurations, the expressions for all submatrices in that group can be easily generated. The representative matrix expression described above is referred to as a prototype matrix.

There are actually two match vectors for each pair of actual configurat ions depending on which of the two configurations is taken as reference. Thus, in the example used above, the other match vector is (3,4,1,2,0,0,). However, it makes no difference which way we choose to define the match vector as long as the same definition is used for each prototype configuration pair.

In a large CI calculation where the matrix eigenvalue problem must be solved by an iterative method, the key step in each iteration is the multiplication of a trial eigenvector by the Hamiltonian matrix. If we assume the central memory is large enough to hold all the active orbital integrals and two CI vectors, then we can read the prototype matrices into central memory one at a time, construct all the actual matrix elements from the prototype matrix, and make their contributions to the matrix-vector product. However, this is not the most efficient way of proceeding. We describe below a scheme in which the prototype matrices are sorted according to whether the configuration pair differ by 0, 1 or 2 electron occupations, which leads to more efficient processing of each kind of prototype matrices in each CI iteration.

4. Prototype diagonal matrices

The prototype diagonal matrices, occupation difference 0, are further divided into diagonal and off diagonal elements. The diagonal elements have nonzero contributions from only one-electron integrals $h_i = (i|i)$, Coulomb integrals $J_{ij} = (ii|jj)$ and exchange integrals $K_{ij} = (ij|ij)$ where i and j run over all occupied orbitals of the configuration of interest. There are altogether m one-electron integral contributions and m^2 two-electron contributions to each diagonal element, where m is the number of occupied orbitals. All one-electrons h_i, with i a closed shell orbital, make the same contribution, i.e. has the same coupling coefficient, to all the diagonal elements of the prototype matrix. Furthermore the coupling constant is the same for all closed shell orbitals that have the same degeneracy factor. The same is true for closed-closed and closed-open Coulomb integrals, and all the closed-closed and closed-open exchange integrals. Some of the open-open Coulomb and exchange integrals may also behave the same way. However, this can only be determined by examining their coupling coefficients. Therefore the information about the construction of this common contribution to the diagonal elements only need to be stored once on peripheral memory and its calculation need not be duplicated for each element.

The off-diagonal elements of a prototype diagonal matrix have nonzero contributions from the open-open J and K integrals only. It follows that the off-diagonal elements of actual matrices arising from actual configurations that differ only in closed-shell orbitals must be identical. Therefore it is desirable to presort the actual configuration list associated with each prototype matrix to collect together all pairs that have the same off-diagonal elements. It will then be possible to calculate the unique matrices one at a time and make their contributions to the matrix-vector product for all the actual configuration pairs necessary.

Each off-diagonal element of a prototype diagonal matrix has contributions from at most one J and one K integrals for Abelian point groups. Each integral may have the same coupling coefficient for several matrix elements. It is therefore advantageous to form a nonredundant integral-coefficient product list to minimize the number of multiplications. Also, the expression for each element then contains only a list of pointers into this nonredundant product list, instead of pointers to both integrals and coupling coefficients.

It is not necessary to have all the actual configurations in central memory when processing the prototype diagonal matrices. They can be retrieved from peripheral memory as needed because only a single scan per CI iteration is needed. However, it is necessary to have all the one-electron, J and K integrals in the central memory.

5. Prototype singles matrices

The elements of a prototype singles matrix, occupation difference 1, have nonzero contribution from one one-electron integral $h_{ij} = (i|j)$ where i and j are orbitals containing the unmatched electron between the configuration pair, and the two-electron integrals (nn|ij) and (ni|nj) where n runs over all the occupied orbitals of the configurations of interest. It follows that all actual singles matrices are different, unlike the situation found in the off-diagonal elements of the diagonal matrices.

All prototype singles matrix elements can be decomposed into a true singles part which has contributions from all the integrals given above and a true doubles part which has contribution from two two-electron integrals only for Abelian groups. The true singles part are, within a scale factor, the same in all the matrix elements, the scale factor being the coupling coefficient for the one-electron integral h_{ij} Thus it is sufficient to calculate the true single part for one matrix element only. The true singles contribution to a different matrix element is simply the first contribution multiplied by the ratio of the one-electron integral coupling coefficient for the second matrix element over that of the first one. Therefore it is only necessary to store information about the true singles part for one matrix element and the one-electron integral coupling coefficient for each element. Taking advantage of this simplification, we can reduce the amount the information stored on peripheral memory for prototype singles matrices and avoid duplicate computation in processing them.

As in the diagonal matrix case, it is advantageous to construct a nonredundant integral-coupling coefficient list for the two-electron part of the singles matrices. In this context, it is desirable to sort the actual configuration pairs belonging to a prototype matrix by the unique one-electron integral that characterizes their interaction. This way, we reduce the memory requirement of the integral-coupling coefficient list to a minimum. But unlike the diagonal case, all the actual configurations should be stored in the central memory when processing the singles matrices. It is also necessary to keep all the one-electron integrals and all (nn|ij) and (ni|nj) integrals in central memory. The IO required to retrieve these quantities when needed is costly.

6. Prototype doubles matrices

The elements of a prototype doubles matrix, orbital occupation difference 2, have nonzero contributions from three two-electron integrals for Abelian point groups: (ij|kl), (ik|jl) and (il|jk) where i, j, k and l are the unmatched orbitals between the configuration pair of interest. Furthermore, each matrix element has only two nonzero integral contributions. Since these matrix elements do not depend on the other occupied orbitals of the configuration pair, all actual matrices belonging to the same prototype doubles matrix which have the same unmatched actual orbitals must be the same. Therefore it is again desirable to presort the actual configuration list associated with each prototype matrix to collect together all pairs that have the same elements, so that it is possible to calculate the unique matrices one at a time and make their contributions to the matrix-vector product for all the actual configuration pairs necessary. This sort is simply done by collecting together all actual configuration pairs that differ by the same actual orbitals occupations. We can either store the actual indices of the actual integrals with the unique matrices or the actual integrals themselves. In the latter case it would not be necessary to keep the active two-electron integrals in the central memory. In either case, the processing of the prototype doubles matrices, unlike the diagonal and singles matrices, does not need the actual configurations.

7. Discussion

An actual prototype CI calculation proceeds in three steps. First, the configurations and CSF's to be included in the CI are generated in terms of prototype configurations, and the expressions for prototype matrices generated by comparison of Slater determinants. Second, prototype matrices are sorted in the manner described into diagonal elements, off-diagonal elements of the diagonal matrices, singles matrices and doubles matrices Third, the actual CI calculation using the sorted prototype matrix expressions and transformed active orbital integrals is performed. The first two steps only need to be done once for each choice of configuration list. The third step of course must be performed once for each choice of nuclear configurations.

The prototype CI method described above give rise to 1) a compact expression for the elements of the Hamiltonian matrix and 2) an algorithm for the explicit construction of these matrix elements that is so efficient that it is practical to repeat the work in each CI iteration, thus eliminating the need to store the lengthy Hamiltonian matrix in peripheral memory and suffer the large IO cost for its retrieval. The price we pay to achieve these advantages is that, during the CI calculation, we must keep a list of all the actual configurations in central memory. This requirement, however, is quite insignificant compared to the storage requirement of the matrix eigenvalue problem of two CI vectors in central memory, if there are many CSF's associated with each configuration. Therefore this method is the most effective for calculations with many open shells as in the case of most large MCSCF calculations.

In a MCSCF calculation where the Newton-Raphson equations are solved by an iterative method, it is necessary to from the product between the Hessian and a trial vector. The product between the orbital part of the Hessian with the orbital part of the trial vector, is straightforward once the orbital Hessian has been constructed. The product between the CI parts of the Hessian and the trial vector involves essentially the product between the CI Hamiltonian and the CI part of the trial vector.[19] It has been shown that the product between the orbital-CI coupling part of the Hessian and the CI part of the trial vector, which contributes to the orbital part of the product vector, can be expressed as orbital gradients constructed from a transition density matrix constructed from the CI wave function and the CI part of the trial vector.[1] Also, the product between the CI-orbital coupling part of the Hessian and the orbital part of the trial vector, which contributes to the CI part of the product vector, can be expressed as a product between the Hamiltonian matrix constructed from a updated list of active integrals and the CI wave function.[1] The prototype CI method is easily extended to provide for efficient calculation of density and transition density matrices. Thus the prototype CI method not only provides for efficient large CI calculations, it also provides a efficient way of calculating the Hessian-trial vector product needed in the iterative solution of the Newton-Raphson equations in a second order MCSCF calculation.

References

1. ALCHEMY II authors are M. Yoshimine, B.H. Lengsfield III, P.S. Bagus, A.D. McLean, and B. Liu.
2. A.D. McLean in "Proceedings of the Conference on Potential Energy Surfaces in Chemistry", Ed. William A. Lester, Jr., IBM Research report RA-18 (1971).
3. M. Yoshimine, IBM Research Report RJ555 (1969); J. Comp. Phys., **11**, 333 (1973).
4. Appendix 11B of this chapter.
5. B. Liu and M. Yoshimine, J. Chem. Phys., **74**, 612 (1981).
6. See, for example, I. Shavitt, in "Modern Theoretical Chemistry, Vol III, Methods of Electronic Structure Theory", ed. H.F. Schaefer III (Plenum, New York 1977).
7. P.S. Bagus, IBM Research Report RJ1077 (1972).
8. B. Liu and A.D. McLean, J. Chem. Phys., **91**, 2348 (1989).
9. B. Liu and A.D. McLean, J. Chem. Phys. **72**, 3418 (1980); B.H. Lengsfield, A.D. McLean, M. Yoshimine, and B. Liu, *ibid* **79**, 1891 (1983).
10. H. Partridge, C.W.Bauschlicher, L.G.M. Pettersson, A.D. McLean, B.Liu, M.Yoshimine, and A. Komornicki, J. Chem. Phys., **92**, 5377 (1990).
11. S. Green, D.J. DeFrees, and A.D. McLean, J. Chem. Phys., submitted.
12. A.D. McLean, B. Liu,and J.A. Barker, J. Chem. Phys., **89**, 6339 (1988).
13. A.D. McLean, unpublished.
14. J. Almlöf, MOLECULE integral program, University of Stockholm Institute of Physics Report 74-29 (1974).
15. E. R. Davidson, J. Chem. Phys., **62**, 400 (1975).
16. S. Obara and A. Saika, J. Chem. Phys., **84**, 3963 (1986).
17. J. Almlöf and P. R. Taylor, in Advanced Theories and Computational Approaches to the Electronic Structure of Molecules, ed. C. E. Dykstra (Reidel, Dordrecht, 1984).
18. B. H. Lengsfield III, J. Chem. Phys., **77**, 4073 (1982).
19. B. H. Lengsfield III and B. Liu, J. Chem. Phys., **75**, 478 (1981).

20. P. E. M. Siegbahn, J. Chem. Phys. **72**, 1647 (1980); B. R. Brooks and H. F. Schaefer III, J. Chem. Phys., **70**, 5092 (1979).

Chapter 12.

Continuum by L^2 Methods: Molecular Photoionization Cross Section

I. Cacelli
Scuola Normale Superiore
Piazza dei Cavalieri, 56126 Pisa, Italy

V. Carravetta and A. Rizzo
I. C. Q. E. M. del CNR
Via Risorgimento 35, 56126 Pisa, Italy

R. Moccia
Dipartimento di Chimica e Chimica Industriale, Università di Pisa,
Via Risorgimento 35, 56126 Pisa, Italy

1. Introduction

Molecular electronic continuum states play an important role in several relevant processes studied in different investigation fields: photochemistry, astrophysics, plasma and laser physics. Electron emission is one of the most efficient decay mechanisms for molecules excited above the ionization limit. Based on this effect, different electron spectroscopy techniques have been rapidly developed in the last 30 years, offering a powerful tool to study molecular electronic structure. Low energy electron scattering by gaseous molecules is fundamental in understanding a wide variety of phenomena connected for instance to the development of molecular gaseous laser and of plasma systems.

Most of the several quantum chemical *ab initio* methods presently available have been applied to the calculation of bound molecular states. They are generally based on the projection of the Schrödinger equation on a finite set of square integrable (L^2) functions. The unquestionable success of such techniques in describing the structure of rather complicated systems containing even tens of electrons, has recently drawn the attention of theoreticians who generally used, for the electronic continuum, "scattering" techniques which present several difficulties in the application to molecules because of the multi-center nature of the problem.

Methods based on the description of the molecular electronic continuum by L^2 functions have some definitive advantages over the traditional scattering methods. They can easily allow different levels of approximation in the calculations that besides may be carried out by the well established computer codes already developed to treat bound-state problems. Even at the lowest approximation level, realistic multi-center non-local potentials like the Hartree-Fock potential can be easily adopted for the description of the electron-molecule(ion) interaction. Moreover, also the standard scattering methods may have some benefit by projecting the equations on a finite set of L^2 functions, thus reducing the algorithm to a sequence of matrix operations that can be efficiently implemented on modern supercomputers.

The present chapter will describe some L^2 methods and computer codes that have been developed for the description of the electronic continuum in small molecules[1]. Because a large part of our investigations has been concerned with the calculation of molecular *differential and integral* photoionization cross sections for absorption of one or two photons, these application will be particularly stressed. It should be noticed anyway that essentially the same methods have been successfully applied to *ab initio* quantitative studes, both for atoms and for molecules, of the Auger effect[2], of shake-up/off processes in core ionization[3,4], of resonances in atomic photoionization[5].

The calculation of the ionization cross section requires the knowledge of the appropriate transition matrix elements between a bound state, usually the ground state, and the continuous electronic states. This may be accomplished either by performing separate calculations for the states involved or by a direct evaluation of the transition density matrix via, for instance, the Equation Of Motion (*EOM*) method[6-10]. It should be added that while the differential cross section requires the detailed representation of the matrix elements between a bound state and all the degenerate states in the continuum, the integral cross section requires the less demanding oscillator strength density.

As previously said, the lack of spherical symmetry in molecules curtails the possibility of applying straightforwardly all the methods developed and extensively used to study the photoionization of atoms, like the Close Coupling method[11] or the many-body methods for the continuum[12]. Among the methods which resort to a separate particularly adapted computational technique for the molecular states in the continuum exploiting variational or other techniques[13,14] the most extensively applied are: the Schwinger Variational (*SV*) method[15,16], the Continuum Multiple Scattering (*CMS*) method[17] and the Linear Algebraic (*LA*) method[18,19] while the most used techniques which evaluate directly the transition density matrix are based upon the several versions of the *EOM*, the simplest being the Random Phase Approximation (*RPA*)[6].

The Stieltjes Moment Theory (*SMT*) method[20] instead is not directly concerned with the computation of the states in the continuum since its aim is to extract from an L^2 basis set calculation of the transition probabilities the oscillator strength density in the energy region properly belonging to the continuum. Thus it may be applied either to the results yielded by separate calculations of the initial and final states or to those arising from a direct evaluation of the transition density matrix.

Since the emphasis here is upon the possibility to make the most out of purely L^2 basis set results, no details are given of the *SV*, *CMS* and *LA* methods which actually imply either the use or the evaluation of continuous states. The interested reader may find elsewhere excellent descriptions of these techniques[15-18] and several valuable applications[16,19,21,22]. Here we restrict the presentation to methods based upon the reactance **K**-matrix theory[23-25] and the *SMT*; both methods, although with different degree of accuracy, offer the means to elicit from an L^2 basis set calculation the properties and the details of the continuum.

In the following we shall describe the main characteristics of these two methods and of the related algorithms that have been implemented in a package of computer codes for the *ab initio* calculation of molecular photoionization cross sections. More extensive informations about the applications can be found in a series of studies of the ionization process in hydrides of the second and third row[26-31], while a very detailed presentation of the formulae can be found in a recently published review paper[1].

2. Molecular states in the electronic continuum

In the framework of the B&O approximation and of a purely electrostatic hamiltonian the molecular continuum states which concern the present report may be labeled according to their asymptotic behaviour as

$$\langle \mathbf{x}, \mathbf{Q}, \mathbf{\Omega} | \{E_I, \Gamma_{\mu_P}, S_I, \vec{k}, S, M_S\}, \{v\}, \{J\} \rangle = \Phi^{(-)}_{E_I, \Gamma_{\mu_P} S_I, \vec{k}_e, S, M_S}(\mathbf{x}, \mathbf{Q}) \chi^I_{\{v\}}(\mathbf{Q}) D^I_{\{J\}}(\mathbf{\Omega}) \tag{2.1}$$

where $\Phi^{(-)}$ is the global electronic w.f. satisfying the boundary conditions appropriate to photoionization, χ^I the vibrational and D^I the rotational w.f. of the parent ion; the other symbols have the following meaning:

- $\mathbf{x}$ - set of the space-spin electronic coordinates $(x_i \equiv \vec{r}_i, \sigma_i)$;
- $\mathbf{Q}$ - set of the internal nuclear coordinates;
- $\mathbf{\Omega}$ - set of the Euler angles specifing the orientation of the parent ion fixed body frame with respect to the laboratory frame;
- $\{v\}$ - set of the vibrational quantum numbers;

- $\{J\}$ - set of the rotational quantum numbers;

- E_I - electronic energy level of the parent ion;

- $\Gamma_{\mu I}$ - irreducible representation and component of the parent ion electronic state referred to a nuclear geometry $\mathbf{Q}_0$;

- $\vec{k}_e$ - wave vector of the photoelectron;

- S, M_S - labels of the total electronic spin state.

In what follows the set of the parent ion indices E_I, $\Gamma_{\mu I}$, S_I will be concisely indicated by a letter j (e.g. $E_{Ia}, \Gamma_{\mu Ia}, S_{Ia} \equiv j_a$). Since we are interested in transitions essentially electronic in character we shall adopt from the start the approximation of considering the rotational degrees of freedom as classical. Thus, in order to make comparisons with gas-phase experiments the calculated observable quantities will be averaged upon all the (equiprobable) orientations (see Appendix). In order to have an analytical expression of the dependence of the transition matrix elements upon the orientation $\hat{k}_e$ of the wave vector with respect to the body-fixed frame, it is highly convenient to adopt a partial wave expansion of the continuum electronic w.f.

$$\Phi^{(-)}_{j,\varepsilon,\hat{k}} > = \sum_{l,m} \Phi^{(-)}_{j,\varepsilon,l,m} > Y^{*}_{l,m}(\hat{k}_e) \tag{2.2}$$

where the photoelectron direction, specified by the unit vector $\hat{k}_e$, is referred to the molecular frame.

In this report the attention will be devoted primarily to the electronic continuum and most of the formulae are written for the Vertical Transition Approximation (*VTA*) or fixed nuclei approximation which, for photon energies sufficiently higher than the ionization thresholds, represents an acceptable simplification. In the *VTA* all vibrational indices disappear and it will be sufficient to specify the frozen equilibrium nuclear geometry $\mathbf{Q}_0$ of the initial state; the symmetry labels of the initial and of the parent ion electronic states will then refer to the same molecular symmetry point group. Since in the *VTA* the only particles involved in the processes are the electrons, it is possible to express the one-photon and two-photon ionization differential cross sections, with a_0 and t_0 denoting atomic units (a.u.) of length and time and α the fine structure constant, as (see Appendix and Fig. 1)

$$\frac{\partial \sigma^{(1)}_{j,\varepsilon,\hat{k}_e \leftarrow \{0\}}(\omega_1, \mu_1)}{\partial \Omega_{\hat{k}_e}} = (2\pi)^2 a_0^2 \alpha \omega_1^g \times \frac{1}{12\pi} \sum_J (-1)^J (2J+1) C(1, J, 1; -\mu_1, 0) P_J(\cos\theta_{\hat{k}_e}) F_{j,J} \tag{2.3a}$$

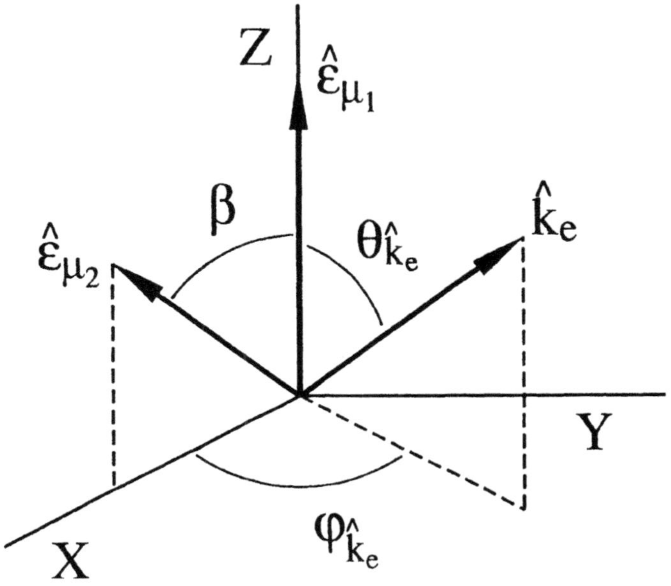

Figure 1. The coordinate system employed to describe the angular distribution of photoelectrons. The two polarization vectors $\hat{\varepsilon}_{\mu_1}$, $\hat{\varepsilon}_{\mu_2}$ lie in the xz plane.

$$\frac{\partial\sigma^{(2)}_{j,\,\varepsilon,\,\hat{k}_e \leftarrow \{0\}}(\omega_1, \mu_1, \omega_2, \mu_2; \beta)}{\partial\Omega_{\hat{k}_e}} = (2\pi)^3 a_0^4 t_0 \alpha^2 \omega_1^g \omega_2^g$$
$$\times \sum_{J,M}\sum_{L}\sum_{j_1,j_2} A_{\mu_1,\mu_2}(L, J, M; j_1, j_2) F_{j,j_1,j_2,J}(\omega_1, \omega_2) Y_{L,0}(\beta, 0) Y_{J,M}(\hat{k}_e) \tag{2.3b}$$

where all the molecular quantities $F_{j,J}$, $F_{j,j_1,j_2,J}$ are defined in terms of the matrix elements

$$\mathbf{M}^{(1)}_{j,\,\varepsilon,\,l,\,m,\,\leftarrow\{0\}}(\mu) = \langle \Psi_{\{0\}} | \hat{O}_\mu | \Phi^{(-)}_{j,\,\varepsilon,\,l,\,m} \rangle_{\mathbf{Q}_0} \tag{2.4a}$$

$$\mathbf{M}^{(2)}_{j,\,\varepsilon,\,l,\,m,\,\leftarrow\{0\}}(\omega_1, \mu_1, \omega_2, \mu_2) = \langle \Psi_{\{0\}} | \hat{O}^{(2)}_{\mu_1,\mu_2}(\omega_1, \omega_2) | \Phi^{(-)}_{j,\,\varepsilon,\,l,\,m} \rangle_{\mathbf{Q}_0} \tag{2.4b}$$

Assuming the validity of the Dipole Approximation and within the limits of the Lowest Order Perturbation Theory (*LOPT*)[32-34]

$$\hat{O}^{(2)}_{\mu_1,\mu_2}(\omega_1, \omega_2) = (1 + \hat{P}_{1,2})\hat{O}_{\mu_1}\hat{G}^{+}_{M}(E_0 + \omega_2)\hat{O}_{\mu_2} \tag{2.5}$$

where $\hat{G}^{+}_{M}(E_0 + \omega_2)$ is the appropriate resolvent of the isolated molecule, E_0 is the vibrational - plus - electronic energy level of the initial state and $\hat{O}_\mu$ stands for one of the possible standard expressions of the radiation-matter interaction operator.

The main problem consists therefore in the evaluation of the one- and two-photon transition matrix elements.

The computation of the matrix elements may be achieved by various procedures requiring the determination of the states in the continuum. This problem, which is thoroughly examined in several treatises on scattering theory, may be solved by various methods which generally start from the definition of "unperturbed" photoelectron waves (derived, for instance, from some model hamiltonian) and/or the use of one of the well known variational methods suited to the continuum[13,14,23,35].

3. The K-matrix technique

Here we present a procedure based upon the energy variational approach, originally pionered by Fano[36], which may be adapted to the usual L^2 basis set approach of quantum chemistry. In addition no orthogonality constraints will be invoked the aim being to realize the fastest convergence with the basis size.

The method, similarly to the Close-Coupling employed for atoms, is essentially a Configuration Interaction (*CI*) carried out with a basis set formally including the continuum.

3.1 The partial wave channels

The asymptotic states $\Phi^{(-)}_{j,\varepsilon,l,m}$ of eq. (2.2) describing a singly ionized parent ion plus a photoelectron may be labeled, as said, by specifying the ion state and the photoelectron partial wave. Since we assume that the electrostatic hamiltonian is satisfactory to our purposes, only total states of definite spin multiplicity, equal to that of the initial state, will be considered and it should be understood that the spin of the photoelectron is coupled to the total spin S_j of the parent ion.

Thus if $\Phi^{+}_{j,M'S}$ is the state of the parent ion considered

$$\phi_j(x_1, x_2, \dots, x_{N-1}, \sigma_N) = \sum_{m_S} C(S_j, 1/2, S; M_S - m_S, m_S)\Phi^{+}_{j,M_S - m_S}(x_1, x_2, \dots, x_{N-1}, \sigma_N)\,|\,1/2, m_S> \qquad (3.1.1)$$

is the spin coupled ion state, and the global *asymptotic* state corresponding to a specific "partial wave channel" (*pwc*), identified by *j,l,m*, will be given by

$$\varphi_{E,j,\varepsilon,l,m}(x_i, \dots, x_N) = C\hat{A}\phi_j(x_1, x_2, \dots, x_{N-1}, \sigma_N)u_{j,\varepsilon,l,m}(r_N)Y_{l,m}(\hat{r}_N) \quad (3.1.2)$$

with

$$E = E_j^+ + \varepsilon$$

C a suitable normalization constant and $\hat{\mathbf{A}}$ the antisymmetrizer. The one electron radial states $u_{j,\varepsilon,l,m}(r)$ may be obtained by several techniques; for instance by solving, either directly or variationally in an L^2 basis, the eigenvalue equation of an appropriate one-electron effective hamiltonian projected upon the space spanned by the $Y_{l,m}$ spherical harmonics. In this case the effective hamiltonian must be defined in order to yield for large values of r (r larger than the radius of the region where practically all of the ion charge is contained) a potential $V_{j,eff}(\vec{r})$ given by

$$V_{j,eff}(\vec{r}) = \int d\vec{r}\,'\rho_{jj}(\vec{r}\,')\frac{1}{|\vec{r} - \vec{r}\,'|} + \text{pot. due to nuclei} \qquad (3.1.3)$$

where ρ_{jj} is the electronic charge density due to the parent ion state. Alternatively they may be obtained by diagonalizing the molecular electrostatic hamiltonian in a basis (3.1.2) corresponding to an assigned *pwc* where, as before, the $u_{j,\varepsilon,l,m}(r)$ are approximated by linear combination of aptly chosen one-electron basis functions formally including continuous indices. Each *pwc* calculation is expected to yield a bound orbital series going into a Rydberg series plus a continuum starting at a threshold E_{t_j} independent from l,m.

In the following, when no ambiguity may arise, the *pwc* state labels j,ε,l,m in (3.1.2) will be concisely designated by a lower case roman letter a,b,c,d for open channels ($\varepsilon > 0$) and by capital roman letters for both open and closed channels ($\varepsilon < 0$).

The normalization of the $u_{j,\varepsilon,l,m}$ will be to unity for closed channels and to $\delta(\varepsilon - \varepsilon')$ for open channels. If the u are expressed as linear combinations of L^2 basis functions the normalization for unit energy interval will be obtained by a procedure described in Section 3.2 .

The required states will be finally obtained by diagonalizing the electrostatic hamiltonian in a basis including all available *pwc* functions (3.1.2) plus a localized basis, "localized channel", *lc* , which is necessary in order to

i) describe the short range correlation;

ii) ensure that the continuum states are orthogonal to the lower lying states; thus, if symmetry allowed, the localized basis must include the ground state as well;

iii) afford the description of resonances corresponding, for instance, to localized doubly excited states of the neutral system.

It should be emphasized that since no orthogonality constraints are invoked the *lc* and the *pwc*'s will not be orthogonal. In the following , although not strictly nec-

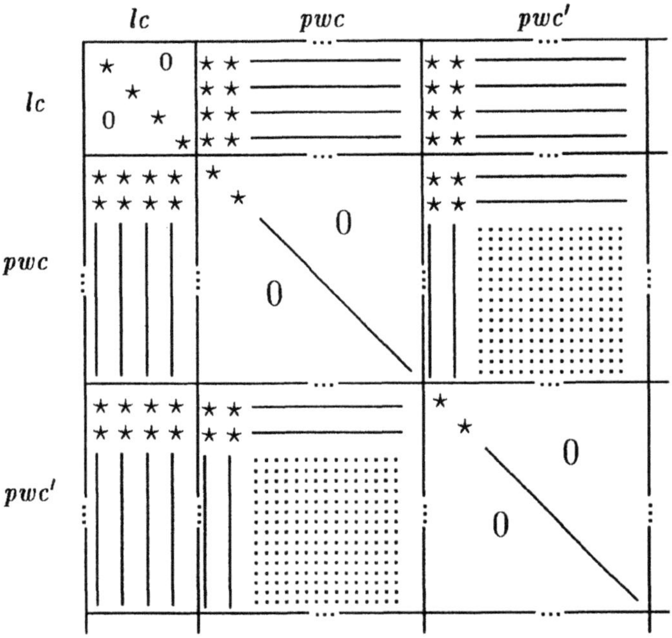

Figure 2. Pattern of the Hamiltonian **H** and the overlap **S** matrices in a basis formed by localized states (*lc*) and partial wave channel (*pwc*) states; see section 3.1 for explanation.

essary, it will be assumed that the channel functions of the *pwc* as well as those of the *lc* are such as to give a diagonal representation of the electronic hamiltonian.

The localized basis functions, "target states", are chosen such as to become totally negligible in the region of space where it may be safely assumed that the interaction between the photoelectron and the parent ion, given by a potential (3.1.3), reduces practically to a coulombic one.

In this region and beyond ($r \to \infty$) the one-electron radial orbitals $u_a(r)$ corresponding to positive energies ε should be expressible as shifted Coulomb waves, i.e.

$$u_a(r) \underset{r\to\infty}{\Longrightarrow} \cos\,[\Delta_a(k_a)]\{F_{l_a,\,\varepsilon}(k_a r) + tg[\Delta_a(k_a)]G_{l_a,\,\varepsilon}(k_a r)\} \tag{3.1.4}$$

where

$$\frac{k_a^2}{2} = \varepsilon = E - E_a > 0.$$

Thus

$$u_a(r) \underset{r\to\infty}{\Longrightarrow} \sqrt{\frac{2}{\pi}}\,\frac{1}{\sqrt{k_a}}\,\frac{1}{r}\sin\left[\theta_a(r) + \Delta_a(k_a)\right]$$

with

$$\theta_a(r) = k_a r + \frac{1}{k_a}\ln 2k_a r + \sigma_{l_a}(k_a) - l_a\frac{\pi}{2}$$

where $\sigma_{l_a}(k_a)$ is the Coulomb wave phase shift and the $\Delta_{j,\,l_a,\,m_a}(k_a)$ is the additional phase shift of the actual wave due to the short-range interaction.

The Hamiltonian **H** and the overlap **S** matrices in a basis formed by a *lc*-basis plus a number of *pwc*-basis sets (discrete plus continuous indices) will have a structure of the form shown in Fig. 2 , where it has been assumed that each channel basis diagonalizes the hamiltonian. The variational coefficients $C_{\varepsilon A,\,Eb}$ of the eigenvector corresponding to the degenerate component labeled by the *pwc* index b of the energy level E in the continuum (E at least higher than the lowest among the thresholds E_{t_j}) must satisfy

$$\sum_A \sum\!\!\!\!\!\!\int d\varepsilon(\mathbf{H}_{\varepsilon' D,\,\varepsilon A} - E\,\mathbf{S}_{\varepsilon' D,\,\varepsilon A})C_{\varepsilon A,\,Eb} = 0 \tag{3.1.5}$$

where the $\sum\!\!\!\!\!\int d\varepsilon$ means summation upon the discrete values plus integration upon the continuous values. According to the **K**-matrix theory[37] these coefficients may be expressed as

$$C_{\varepsilon A,\,Eb} = \delta_{Ab}\delta(\varepsilon - E) + \frac{P}{E-\varepsilon}\mathbf{K}_{\varepsilon A,\,Eb} \tag{3.1.6}$$

where P indicates the Cauchy principal value, the $\mathbf{K}_{\varepsilon A,\,Eb}$ are the elements of the scattering off-shell reaction **K**-matrix and the degeneracy of the eigenvalue E will be given by the number of the available *pwc*'s b open at the energy E considered ($E > E_{t_b}$). The substitution of (3.1.6) into (3.1.5) leads to

$$\begin{aligned}\mathbf{K}_{\varepsilon' D,\,Eb} = &\sum_{A(\neq D)} \sum\!\!\!\!\!\!\int d\varepsilon(\mathbf{H}_{\varepsilon' D,\,\varepsilon A} - E\mathbf{S}_{\varepsilon' D,\,\varepsilon A})\frac{P}{E-\varepsilon}\mathbf{K}_{\varepsilon A,\,Eb}\\ &+(1-\delta_{Db})(\mathbf{H}_{\varepsilon' D,\,Eb} - E\mathbf{S}_{\varepsilon' D,\,Eb})\end{aligned} \tag{3.1.7}$$

as the condition which must be satisfied by the **K** matrix elements.

From eq. (3.1.7) it may be inferred that:

i) the square matrix $\mathbf{K}(E)$ (**K** matrix on the energy shell) formed with the elements $\mathbf{K}_{Ea,\,Eb}$ of order equal to the number of open available channels is hermitian[38];

ii) if the **H** and the **S** matrix elements are smooth functions of the energy indices in the continuous ranges, then the $\mathbf{K}_{\varepsilon' D, Eb}$ elements are also expected to be smooth functions of the left energy index ε'. It should be noted that the **H** and the **S** matrix elements may be made smooth functions of the energy indices by either a careful choice of the basis functions or of the effective potential $V_{j, eff}(\vec{r})$ (3.1.3) (see also sec. 3.2);

iii) rapid variations of $\mathbf{K}_{\varepsilon' D, Eb}$ are expected when the second energy index E is close to the discrete values ε belonging to some closed channel (different from b), i.e. close to a resonance.

The normalization of the eigenvectors (3.1.6) may be checked and it is found that

$$\sum_A \!\!\!\!\!\!\int d\varepsilon'' \sum_D \!\!\!\!\!\!\int d\varepsilon' C^*_{\varepsilon'' A, E'b} \mathbf{S}_{\varepsilon'' A, \varepsilon' D} C_{\varepsilon' D, Ef} = \delta(E' - E)\left[\delta_{b,f} + \pi^2 \sum_c \mathbf{K}^*_{Ec, Eb} \mathbf{K}_{Ec, Ef}\right]. \tag{3.1.8}$$

Thus the set of the degenerate eigenvectors $\{\mathbf{U}_{..,Ea}\mathbf{U}_{..,Eb}\}$ corresponding to the open available *pwc* properly normalized to $\delta(E' - E)$ may be expressed as[1]

$$\{\mathbf{U}(E)\} = \{\mathbf{C}(E)\}\left\{[\mathbf{I} + \pi^2 \mathbf{K}^\dagger(E)\mathbf{K}(E)]\right\}^{-\frac{1}{2}}.$$

In order to have final states $|\psi^{(-)}_{Ea}>$ satisfying the appropriate boundary conditions for the photoionization process[39], i.e. ingoing waves in all channels outgoing wave only in channel a, we observe that our approximate states should be expressed as[1]

$$|\psi^{(-)}_{Ea}> \simeq \sum_D \!\!\!\!\!\!\int d\varepsilon\, |\varphi_{\varepsilon D}> \mathbf{U}^{(-)}_{\varepsilon D, Ea} \tag{3.1.9}$$

where the vectors $\{\mathbf{U}^{(-)}(E)\}$ are given by

$$\{\mathbf{U}^{(-)}(E)\} = \{\mathbf{C}(E)\}[\mathbf{I} - i\pi\mathbf{K}(E)]^{-1} \tag{3.1.10}$$

To evaluate the differential cross section and the branching ratios we need a proper linear combination of the (3.1.9) which will behave asymptotically like the product of a specific ion state j_a times a coulombic plane wave in the $\hat{k}_a$ direction plus ingoing spherical waves in all channels[39]. The required states $|\psi_{E, j_a, \vec{k}}>$ which have this behaviour may be expressed as[1]

$$|\psi^{(-)}_{E, j_a, \vec{k}_a}> = \sum_{l,m} i^l e^{-i[\sigma_l(k_a) + \Delta_{ja, l, m}(k_a)]} Y^*_{l, m}(\hat{k}) |\psi^{(-)}_{E, j_a, l, m}> \tag{3.1.11}$$

and the matrix elements with a generic bra $< B \mid$ is

$$< B | \psi^{(-)}_{E, j_a, \vec{k}_a} > = \sum_d \!\!\!\!\!\!\int d\varepsilon < B | \varphi_{\varepsilon D} > \sum_{l,m} \mathbf{U}^{(-)}_{\varepsilon, D; E, j_a, l, m} \, \imath^l e^{-\imath[\sigma_l(k_a) + \Delta_{j_a, l, m}(k_a)]} Y^*_{l,m}(\hat{k}). \qquad (3.1.12)$$

It will be found advantageous in the following to define the states

$$\Phi^{(-)}_{j_a, \varepsilon_a, l, m} > = \sum_D \!\!\!\!\!\!\int d\varepsilon | \varphi_{\varepsilon D} > \mathbf{U}^{(-)}_{\varepsilon, D, E, j_a, l, m} \, \imath^l e^{-\imath[\sigma_l(k_a) + \Delta_{j_a, l, m}(k_a)]} \qquad (3.1.13)$$

by which the (3.1.12) may be recast as

$$< B | \psi^{(-)}_{E, j_a, \vec{k}_a} > = \sum_{l,m} < B | \Phi^{(-)}_{j_a, \varepsilon_a, l, m} > Y^*_{l,m}(\hat{k}). \qquad (3.1.14)$$

The $| \phi^{(-)}_{j_a, \varepsilon_a, l, m} >$ correspond to the electronic part of the states of eq. (2.2) and enter in the expressions of the one- and two-photon transition matrix elements.

3.2 The L^2 basis set approximation

The key requirement for the applicability of an L^2 technique is a regular behaviour of all quantities which are represented upon a finite basis set. The presence of the $P/(E - \varepsilon)$ in the expression of the variational coefficients (3.1.6) relieves the finite basis from the burden of representing the fastest variations with the energy E which is henceforth described analytically. Thus the **K**-matrix elements may be safely interpolated on the first energy index which runs over the basis eigenvalues. Moreover, the properties depending upon the continuum spectrum of each *pwc* must have a smooth energy dependence. In this way it is possible to find the solutions of eq. (3.1.7) by using an L^2 basis set to represent the u orbitals, which will yield a completely discrete representation (that may be taken as the grid values) of the hamiltonian and overlap matrices. To accomplish this procedure we are confronted with the problem stated below :

i) obtain the proper continuum normalization factor and phase shift $\Delta_{j_a m}(k_a)$ for the variationally determined orbitals;

ii) calculate the matrix elements $\mathbf{H}_{\varepsilon' D, Eb}$ and $\mathbf{S}_{\varepsilon' D, Eb}$ for arbitray values of E and perform in an appropriate way, using the discretized set of values available, the energy integration of eq. (3.1.7) that involves a Cauchy principal value.

All these problems may be dealt with satisfactorily if the representation of the orbitals u is achieved by the use of an L^2 basis set capable of producing many closely separated states in the continuum with the appropriate asymptotic behaviour (see eq. 3.1.4) in a region of space where the effective potential due to the parent ion is satisfactorily approximated by a coulombic one. Assuming for the time being that this is so, the fitting of the orbital u in this region, with a shifted Coulomb wave eq. (3.1.4), will provide both the normalizing factor and the phase shift. Some numeric examples given in sec. 6 will show that this procedure affords a satisfactory accuracy.

Once the problems of the normalization factor and of the phase shift $\Delta_{j,l,m}(k)$ have been dealt with, the problem ii) may be solved by noting that the matrix elements corresponding to the closely spaced wavepackets yielded by the energy variational calculations should provide enough grid points for an accurate polynomial interpolation to evaluate both the $(\mathbf{H}_{\varepsilon' D,\varepsilon A} - E\,\mathbf{S}_{\varepsilon' D,\varepsilon A})$ and the $\mathbf{K}_{\varepsilon A,Eb}$ for arbitrary values of ε. Thus if we define

$$(\mathbf{H}_{\varepsilon' D,\varepsilon A} - E\mathbf{S}_{\varepsilon' D,\varepsilon A})(1-\delta_{DA}) = \mathbf{V}(E)_{\varepsilon' D,\varepsilon A}$$

eq. (3.1.7) takes the form

$$\mathbf{K}_{\varepsilon' D,Eb} - \sum_{A}\sum\!\!\!\!\!\!\int d\varepsilon \mathbf{V}(E)_{\varepsilon' D,\varepsilon A}\frac{P}{E-\varepsilon}\mathbf{K}_{\varepsilon A,Eb} = \mathbf{V}_{\varepsilon' D,Eb} \tag{3.2.1}$$

and the integration for the grid points $\varepsilon' D = \varepsilon'_j D$ may be conveniently carried out, for each channel,

$$\sum\!\!\!\!\!\!\int d\varepsilon \mathbf{V}(E)_{\varepsilon_j D,\varepsilon A}\frac{P}{E-\varepsilon}\mathbf{K}_{\varepsilon A,Eb} = \sum_{J(\varepsilon_J<E_{t_A})}\frac{\mathbf{V}_{\varepsilon_j D,\varepsilon_J A}\mathbf{K}_{\varepsilon_J A,Eb}}{E-\varepsilon_J} + \sum_{i(\varepsilon_i>E_{t_A})}\int_{\varepsilon_i}^{\varepsilon_{i+1}}\mathbf{V}(E)_{\varepsilon_j D,\varepsilon A}\frac{P}{E-\varepsilon}\mathbf{K}_{\varepsilon A,Eb}$$

where the ε_J are the bound states energies of the *pwc* A and the $\varepsilon_j, \varepsilon_{j+1}...$ are the grid points in the continuum supplied by the variational calculation. The values of $\mathbf{V}_{\varepsilon_j D,\varepsilon A}$ and $\mathbf{K}_{\varepsilon A,Eb}$ in each interval may be approximated by an n-dependent polynomial expansion around the midpoint $\bar{\varepsilon}_i = (\varepsilon_i + \varepsilon_{i+1})/2$ obtained for instance by a Lagrange interpolation of the $(n+1)$ closest grid points. Thus if we define the variable

$$x = \varepsilon - \bar{\varepsilon}_i$$

the interpolated value $F(\varepsilon)$ $(\varepsilon_i \le \varepsilon < \varepsilon_{i+1})$ of $f(\varepsilon)$ ($\mathbf{V}_{\varepsilon_j D,\varepsilon A}$ or $\mathbf{K}_{\varepsilon A,Eb}$) may certainly be expressed as

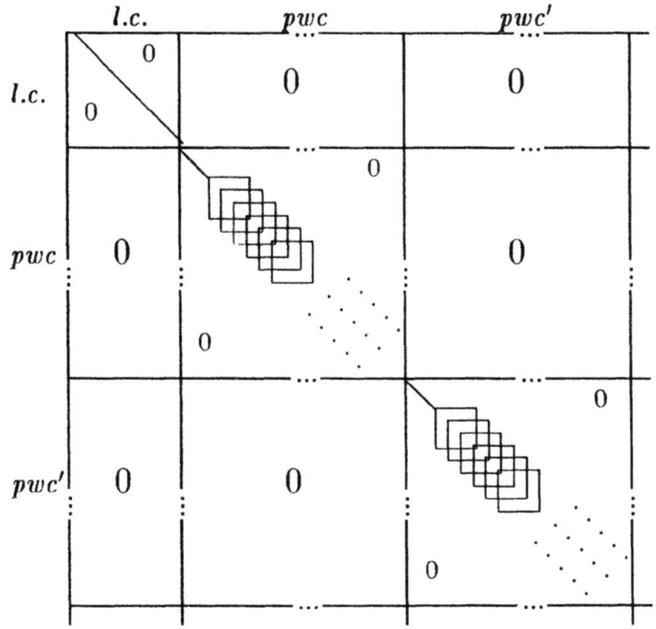

Figure 3. Pattern of the matrix **P** in a basis formed by localized states (*lc*) and partial wave channel (*pwc*) states; see section 3.2 for explanation.

$$F(\varepsilon) = \sum_{\nu=0}^{n} \sum_{\mu=0}^{n} \alpha_{\nu,\,\mu}(i) f(\varepsilon_\nu) x^{\mu}$$

The energy integral for each interval ($\varepsilon_i \div \varepsilon_{i+1}$) then becomes

$$\int_{\varepsilon_i}^{\varepsilon_{i+1}} \mathbf{V}(E)_{\varepsilon_j D,\, \varepsilon A} \frac{P}{E-\varepsilon} \mathbf{K}_{\varepsilon A,\, Eb} = \sum_{\nu} \sum_{\nu'} \mathbf{V}(E)_{\varepsilon_j D,\, \varepsilon_\nu A} \mathbf{P}^{(i)}_{\nu,\, \nu'}(E) \mathbf{K}_{\varepsilon_{\nu'} A,\, Eb}$$

where

$$\mathbf{P}^{(i)}_{\nu,\, \nu'}(E) = \sum_{\mu} \sum_{\mu'} \alpha_{\nu\mu}(i) \alpha_{\nu'\mu'}(i) \int_{\varepsilon_i - \bar{\varepsilon}_i}^{\varepsilon_{i+1} - \bar{\varepsilon}_i} \frac{P}{E + \bar{\varepsilon}_i - x} x^{\mu+\mu'} dx$$

may be considered as a block of a block diagonal matrix structured as shown in Fig. 3.

By exploiting the foregoing approximations eq. (3.2.1) becomes a finite dimensional simultaneous linear equation system

$$[\mathbf{I} - \mathbf{V}(E)\mathbf{P}(E)]\mathbf{K}_{..,\, Eb} = \mathbf{V}(E)_{..,\, Eb} \qquad (3.2.2)$$

where the (non-symmetrical) coefficient matrix $[\mathbf{I} - \mathbf{V}(E)\mathbf{P}(E)]$ is the same for all the open *pwc*'s. The determinant $|\mathbf{I} - \mathbf{V}(E)\mathbf{P}(E)|$ of the coefficient matrix in eq. (3.2.2) at values of E higher than the first ionization threshold is expected to have real roots near the eigenvalues of the closed channel subspaces corresponding to the localized part of an autoionising state. It may also be noted that the $\mathbf{K}_{\beta, Ea}$ yielded by (3.2.2) will vanish at $E = \varepsilon_\beta$ and the ratio $\mathbf{K}_{\beta, Ea}/(E - \varepsilon_\beta)$ remains finite. At $E = E^{(\beta)}$ the elements of the **K**-matrix will diverge but the matrix elements (3.1.12) will be finite since the normalization factors (see eq. (3.1.6) and (3.1.10)) will eliminate the divergence. It can be added that an accurate description of the electronic resonances is obtained by the determination of the poles of the scattering matrix that may be expressed in terms of the $\mathbf{K}(E)$[1].

3.3 RPA matrix elements in the continuum

In sec. 1 we remarked the possibility of computing directly the one electron transition matrix elements through the implementation of an *EOM* for linear response[6]. In this section we present a numerical procedure which allows to obtain all quantities needed to compute the differential cross section from the results of an *EOM* calculation carried in an L^2 basis set of orbitals. Although there are in the literature some examples of *RPA* equations solved in the continuum[40] they are restricted to atoms where the spherical symmetry affords considerable simplification of the problem. Here the presentation is limited to the *RPA* for singlet closed shell cases and for Abelian molecular point symmetry groups; the extension to other cases may be realized by applying the same general procedures outlined below.

Let a, b, c ... indicate the set of *pwc* indices $j\ l\ m$ (i.e. $a \equiv j_a l_a m_a$) where j stands for an occupied *SCF* orbital and l , m specify the assigned partial wave of an orbital projected upon the virtual orbital set of the *SCF* ground state problem. The appropriate excitation *RPA* operators are thus defined as

$$\hat{\mathbf{O}}^\dagger_{Eb} = \sum_D \int d\varepsilon (\mathbf{X}_{\varepsilon D,\, Eb} \hat{a}^\dagger_{\varepsilon D} \hat{a}_{jD} + \mathbf{Y}_{\varepsilon D,\, Eb} \hat{a}^\dagger_{jD} \hat{a}_{\varepsilon D}) \tag{3.3.1}$$

where the index b labels the degenerate components of the final level corresponding to the *excitation energy E* and the singlet spin coupling scheme is understood. The coefficients **X** and **Y** are determined by the *RPA* equation, with discrete plus continuous indices,

$$\begin{pmatrix} \mathbf{M} & \mathbf{Q} \\ \mathbf{Q}^* & \mathbf{M}^* \end{pmatrix} \begin{pmatrix} \mathbf{X}_{\cdot,\, Eb} \\ \mathbf{Y}_{\cdot,\, Eb} \end{pmatrix} = E \begin{pmatrix} \mathbf{I} & \mathbf{0} \\ \mathbf{0} & -\mathbf{I} \end{pmatrix} \begin{pmatrix} \mathbf{X}_{\cdot,\, Eb} \\ \mathbf{Y}_{\cdot,\, Eb} \end{pmatrix}. \tag{3.3.2}$$

The orbital basis for each *pwc* is determined by the solution of the static-exchange approximation (*SEA*)[41] equation

$$\hat{P}_{l_a, m_a}\hat{P}_{virt}(\hat{f} - \hat{J}_{j_a} + 2\hat{K}_{j_a})\hat{P}_{virt}\hat{P}_{l_a, m_a}\psi_{\varepsilon_{orb.}, j_a, l_a, m_a} = \varepsilon_{orb.}\psi_{\varepsilon_{orb.}, j_a, l_a, m_a} \qquad (3.3.3)$$

projected upon the l, m spherical harmonic subspace of the virtual orbitals. In the above equation $\hat{f}$ is the ground state Fock operator and it should be emphasized that in spite of the lack of orthogonality between the excited orbitals related to different occupied orbital indices j the *RPA* eq. (3.3.2) will retain the same structure. With this choice the matrix elements of the *RPA* eq. (3.3.2) may be written as

$$\mathbf{M}_{\varepsilon'd', \varepsilon d} = \delta_{d', d}\delta_{\varepsilon', \varepsilon}\varepsilon + \mathbf{V}_{\varepsilon'd', \varepsilon d} \qquad (3.3.4)$$

where now ε represents the "unperturbed" excitation energy given as the difference between the orbital energy $\varepsilon_{orb.}$ and the ionization threshold, which is now the energy ε_a of the occupied orbital of the corresponding *pwc*, and

$$\begin{aligned}\mathbf{V}_{\varepsilon'd', \varepsilon d} =& \delta_{j_{d'}, j_d}(1 - \delta_{l_d m_d, l_{d'}, m_{d'}}) < \varepsilon'd' | \hat{f} | \varepsilon d > \\ &+ (1 - \delta_{d, d'})\{2[\varepsilon'd' \, j_d | j_{d'} \, \varepsilon d] - [\varepsilon'd' \, \varepsilon d | j_{d'} \, j_d]\}\end{aligned} \qquad (3.3.5a)$$

$$\mathbf{Q}_{\varepsilon'd', \varepsilon d} = 2[j_{d'} \, \varepsilon'd' | j_d \, \varepsilon d] - [j_{d'} \, \varepsilon d | j_d \, \varepsilon'd'] \qquad (3.3.5b)$$

For energies corresponding to excitation in the continuum eq. (3.3.2) may be solved by a procedure similar to the **K**-matrix technique described in sec. 3.2 by adopting the following expressions for the **X** and **Y** *RPA* coefficients

$$\mathbf{X}_{\varepsilon d, Eb} = \delta_{d, b}\delta(\varepsilon - E) + \frac{P}{E - \varepsilon}\mathbf{K}_{\varepsilon d, Eb} \qquad (3.3.6a)$$

$$\mathbf{Y}_{\varepsilon d, Eb} = \mathbf{L}_{\varepsilon d, Eb}. \qquad (3.3.6b)$$

It is easy to verify that, using the definitions (3.3.4), (3.3.5a) and (3.3.5b) these last expressions, once substituted in eq. (3.3.2), yield the following pair of coupled equations

$$\mathbf{K}_{\varepsilon'd', Eb} - \sum_d \int d\varepsilon \left[\mathbf{V}_{\varepsilon'd', \varepsilon d}\frac{P}{E - \varepsilon}\mathbf{K}_{\varepsilon d, Eb} + \mathbf{Q}_{\varepsilon'd', \varepsilon d}\mathbf{L}_{\varepsilon d, Eb}\right] = \mathbf{V}_{\varepsilon'd', Eb} \qquad (3.3.7a)$$

$$- (\varepsilon' + E)\mathbf{L}_{\varepsilon'd', Eb} - \sum_d \int d\varepsilon \left[\mathbf{Q}^*_{\varepsilon'd', \varepsilon d}\frac{P}{E - \varepsilon}\mathbf{K}_{\varepsilon d, Eb} + \mathbf{V}^*_{\varepsilon'd', \varepsilon d}\mathbf{L}_{\varepsilon'd', Eb}\right] = \mathbf{Q}_{\varepsilon'd', Eb}. \qquad (3.3.7b)$$

Applying the same procedures employed in sect. 3.2 to eqs. (3.3.7a), (3.3.7b), the determination of the required **K** and **L** is reduced to the solution of the finite dimensional linear equation system

$$\begin{pmatrix} 1 - \mathbf{VP}(E) & -\mathbf{QR} \\ -\mathbf{Q}^*\mathbf{P}(E) & -(1E+\boldsymbol{\varepsilon}) - \mathbf{V}^*\mathbf{R} \end{pmatrix} \begin{pmatrix} \mathbf{K}_{.,Eb} \\ \mathbf{L}_{.,Eb} \end{pmatrix} = \begin{pmatrix} \mathbf{V}_{.,Eb} \\ \mathbf{Q}_{.,Eb} \end{pmatrix} \tag{3.3.8}$$

where $\mathbf{P}(E)$ is the "integration" matrix previously defined, $\mathbf{R}$ is a similar matrix but for the absence of the $P/(E-\varepsilon)$ and $\boldsymbol{\varepsilon}$ is the diagonal matrix of the "unperturbed" excitation energies.

The discretized vectors $\mathbf{K}_{.,Eb}$ and $\mathbf{L}_{.,Eb}$, solutions of eq. (3.3.8), may then be used to obtain the transition matrix elements to the continuum

$$\begin{aligned} &< \Psi_{\{0\}} | \hat{O}^{(1)} | \varphi_{Ea} > \simeq \\ &< j_a | \hat{O}^{(1)} | Ea > + \sum_d \sum_{\mu,\nu} < j_d | \hat{O}^{(1)} | \varepsilon_\mu d > \mathbf{P}^d_{\mu,\nu}(E) \mathbf{K}_{\varepsilon_\nu d, Ea} + \\ &< j_d | \hat{O}^{(1)} | \varepsilon_\mu d > {}^*\mathbf{R}^d_{\mu,\nu} \mathbf{L}_{\varepsilon_\nu d, Ea} \end{aligned} \tag{3.3.9}$$

It may be shown that the vectors (3.3.6a) and (3.3.6b) yielded by eq. (3.3.8) are not normalized to unit energy range and that the correct *RPA* normalization to unit energy range

$$\sum_d \int d\varepsilon \left(\mathbf{X'}^*_{\varepsilon d Ea} \mathbf{X'}_{\varepsilon d, bE'} - \mathbf{Y'}^*_{\varepsilon d, Ea} \mathbf{Y'}_{\varepsilon d, E'b} \right) = \delta(E - E')\delta_{a,b}, \tag{3.3.10}$$

is obtained by the following transformation[1]

$$< \Psi_{\{0\}} | \hat{O}^{(1)}_\mu | \Phi^{(-)}_{Ea} > = \sum_d < \Psi_{\{0\}} | \hat{O}^{(1)}_\mu | \varphi_{Ed} > [1 - i\pi \mathbf{K}(E)]^{-1}_{d,a}. \tag{3.3.11}$$

The introduction of the phase shift factor according to eq. (3.1.13), yields the *RPA* version of the matrix elements $\mathbf{M}^{(1)}_{j,\varepsilon,lm,\leftarrow\{0\}}(\mu)$, of eq. (2.4a), required to compute (see Appendix) the differential and the integral cross sections for a one photon absorption process.

4. One- and two-photons transition matrix elements

4.1 One-photon transitions

In the dipole approximation the minimal coupling prescription[32,34,42] for the radiation-matter interaction gives the following form of the operator $\hat{O}_\mu$ to be used in the calculation of the transition matrix elements

$$\hat{O}^{p}_{\mu} = - \varepsilon_{\mu} \cdot \sum_{i} \frac{q_i}{m_i} \hat{\vec{p}}_i \qquad (4.1.1)$$

where the index i runs over all particles with charge q_i mass m_i and momentum $\hat{\vec{p}}_i$; ε_μ stands for the photon polarization; the operator $\hat{O}^p_\mu$ is said to be the "velocity gauge" (VG) version of the transition operator. It is well known that by the Göeppert-Mayer transformation[43] , which is a gauge transformation[32], it is possible to use the electric-dipole operator

$$\hat{O}^{l}_{\mu} = - \varepsilon_{\mu} \cdot \sum_{i} q_i \hat{\vec{r}}_i \qquad (4.1.2)$$

which represents the "length gauge" (LG) form of the transition operator. The fact that the kinetic energy in the Göeppert-Mayer hamiltonian is expressible in terms of the canonical momenta only[32] while the gauge invariant momenta are needed in the minimal coupling version has raised considerable debate about what should be, on physical grounds, the preferable form of the transition operator[44]. The question appears to be settled by now[45] in the sense that on physical grounds there is no preferred gauge; thus, provided exact states are employed, the various possible gauges should give equal results. The fact that the observable quantities cannot depend upon the gauge transformations is clearly obvious but, as it has been said[32] " *... l'equivalence entre les diverses descriptions peut paraître à première vue bien élémentaire et même superflue. En fait il n'en est rien. Leur traduction dans un cas concret est loin d'être toujours facile, et donne lieu quelquefois à des interprétations incorrectes ou à des erreurs subtiles.* "

Thus the problem scales down, from a numerical viewpoint, to that of choosing the gauge most adequate to the approximate states employed in the calculation. It has been argued, originally by Chandrasekhar[46] and subsequently by Bethe and others[47,48], that the velocity formula gives results generally less influenced by the changes in the wave functions particularly if it is energy-variational determined. From the calculations carried out on atomic systems, by far the most numerous and most accurate available, the above contention is supported by the results when the transition involves changes in the principal quantum number while for transitions not involving changes in the principal quantum number the length formula might be preferable[49,50]. This simple criterion, transferred to the molecular photoionization cross section, should suggest the use of VG as the preferred gauge.

It must be added that some forms of the *EOM* method, like for instance the simple version of the *RPA*, will guarantee, in the limit of a complete set of one-electron orbitals, and for the fixed nuclei approximation, the equality of the results.

It remains however to consider the effect of the B&O approximation. For the one-photon process these effects may be practically accounted for by calculating the Franck Condon factors (*FC*) for the transition[1,51], while for the two-photon process some additional comments are required due to the presence of the resolvent.

4.2 Two-photon transitions

While the evaluation of the one-photon transition amplitudes entails only a simple one-body operator, the two-photon transition operator (2.5a)[32,34,52] is a more complicated object which presents serious difficulties both from a theoretical and from a computational viewpoint. Thus this section will be devoted to the study of the approximations introduced in the evaluation of the two-photon transition matrix elements

$$\mathbf{M}^{(2)}_{j,\,\varepsilon,\,l,\,m,\,\{v\}\leftarrow\{0\}}(\omega_1,\mu_1,\omega_2,\mu_2) = <\Psi_{\{0\}}\,|\,\hat{O}^{(2)}_{\mu_1,\mu_2}(\omega_1,\omega_2)\,|\,\Phi^{(-)}_{j,\,\varepsilon,\,l,\,m}\chi^{j}_{\{v\}}> . \tag{4.2.1}$$

where the vibrational motion has been considered in the framework of the **B&O** approximation.

If the non-adiabatic terms of the molecular hamiltonian[51] may be neglected and the the dependence of the electronic transition matrix elements upon the nuclear coordinates is disregarded a "vibrational" resolvent in eq. (4.2.1) may be singled out, i.e.

$$\mathbf{M}^{(2)}_{j,\,\varepsilon,\,l,\,m,\,\{v\}\leftarrow\{0\}}(\omega_1,\mu_1,\omega_2,\mu_2)\simeq$$
$$(1+\hat{P}_{1,2})(\sum_M <\Phi_{\{\lambda_0\}}\hat{O}_{\mu_1}\Phi_{\{\lambda_M\}}>\hat{Q}_{M,\,\{v\}\leftarrow\{0\}}(\omega_1)<\Phi_{\{\lambda_M\}}\hat{O}_{\mu_2}\Phi^{(-)}_{j,\,\varepsilon,\,l,\,m}>) \tag{4.2.2}$$

where $\Phi_{\{\lambda_0\}}>$ is the ground electronic state, $\Phi_{\{\lambda_M\}}>$ is an excited electronic state and λ_0 , λ_M indicate the sets of electronic quantum numbers. In eq. (4.2.2)

$$\hat{Q}_{M,\,\{v\}\leftarrow\{0\}}(\omega) = \lim_{\eta\to 0^+}\sum_{\{v_M\}}\frac{<\chi^{0}_{\{v_0\}}\chi^{M}_{\{v_M\}}><\chi^{M}_{\{v_M\}}\chi^{j}_{\{v\}}>}{\omega-(E_{M,\,\{v_M\}}-E_{0,\,\{v_0\}})+i\eta} \tag{4.2.3}$$

involves a summation over the complete set of *FC* factors for each electronic excited state. The $E_{M,\{v_M\}}$ are the (B&O) energies of the vibrational levels of the M-th electronic state, $E_{0,\{v_0\}}$ being the ground state vibrational level energy. The evaluation of the transition amplitude (4.2.1) would thus involve the knowledge of the vibrational resolvent (4.2.3) or eventually of the complete set of *FC* factors coming out from its spectral representation. Computing (4.2.3) by approximating directly the FC factors even for a restricted subset of electronic states is not a trivial problem. Therefore other approximations are usually introduced, i.e.

- the *FC* overlap integrals are assumed to have a narrow distribution around the vertical transition value;

- the photon energies are mantained sufficiently far from any resonant condition with the intermediate states;

- if the photons energies are sufficiently far from any of the excitation energies, an average value, corresponding to the vertical excitation may be substituted in the denominator of (4.2.3) ; with the same approximations the electronic transition moments are evaluated at the initial state equilibrium geometry;

- a summation over the vibrational levels of the final state is supposed to be performed on the two-photon squared amplitude, what allows to drop from the amplitude the *FC* overlap between the ground electronic state vibrational wavefunction and the final electronic state vibrational wavefunction.

The two-photon transition amplitude (4.2.2) is thus reduced to

$$\begin{aligned}
&\mathbf{M}^{(2)}_{j,\,\varepsilon,\,l,\,m,\,\{v\}\leftarrow\{0\}}(\omega_1,\mu_1,\omega_2,\mu_2) \simeq \\
&(1+\hat{P}_{1,2})\sum\!\!\!\!\int_M \frac{<\Phi_{\{\lambda_0\}}\hat{O}_{\mu_1}\Phi_{\{\lambda_M\}}><\Phi_{\{\lambda_M\}}\hat{O}_{\mu_2}\Phi^{(-)}_{j,\,\varepsilon,\,l,\,m}>}{\omega_2 - E^{v}_M + E_{0,\,\{v_0\}}} \\
&= (1+\hat{P}_{1,2})<\Phi_{\{\lambda_0\}}\hat{O}_{\mu_1}\hat{G}^{+}_{elec}(\omega_2+E_{0,\,\{v_0\}})\hat{O}_{\mu_2}\Phi^{(-)}_{j,\,\varepsilon,\,l,\,m}>_{\mathbf{Q}_0} \\
&= \mathbf{M}^{(2)}_{j,\,\varepsilon,\,l,\,m\leftarrow\{0\}}(\omega_1,\mu_1,\omega_2,\mu_2)
\end{aligned} \tag{4.2.4}$$

where $E^{v}_M - E_{0,\{v_0\}}$ coincides with the vertical excitation energy and the transition matrix elements are computed at a set of nuclear fixed coordinates $\mathbf{Q_0}$, usually the ground electronic state equilibrium coordinates.

The evaluation of the transition amplitude (4.2.4) for molecules through techniques exploiting approximate though analytic expressions of the resolvent $G^{+}_{elect}(\omega + E_{0,\{v_0\}})$ or through the solution of Dalgarno & Lewis type equations[53] is in practice made prohibitively complicated by the lower symmetry of the molecule with respect to the atoms. The easiest way to compute the matrix element (4.2.4) is thus through a projection of the resolvent on a finite set of normalized states approximating the correct eigenstates.

Let's thus write

$$\mathbf{M}^{(2)}_{j,\varepsilon,l,m\leftarrow\{0\}}(\omega_1,\mu_1,\omega_2,\mu_2) = (1+\hat{P}_{1,2})\int dx'_a dx_a dx'_b dx_b \hat{O}_{\mu_1}(x_a,x'_a)\hat{O}_{\mu_2}(x_b,x'_b)$$
$$\times \sum_M \frac{<\Phi_{\{\lambda_0\}}\hat{\psi}^\dagger(x'_a)\hat{\psi}(x_a)\Phi_{\{\lambda_M\}}><\Phi_{\{\lambda_M\}}\hat{\psi}^\dagger(x'_b)\hat{\psi}(x_b)\Phi^{(-)}_{j,\varepsilon,l,m}>}{\omega_2 - E^{\upsilon}_M + E_{0,\{\upsilon_0\}}} \tag{4.2.5}$$

$\hat{\psi}^\dagger(x)$ being the field operator generating a particle in the spin-space point x. This expression, which involves the transition density matrices

$$X^0_M(x_a,x'_a) = <\Phi_{\{\lambda_0\}}\hat{\psi}^\dagger(x'_a)\hat{\psi}(x_a)\Phi_{\{\lambda_M\}}> \tag{4.2.6}$$

$$X^M_{j,\varepsilon,l,m}(x_b,x'_b) = <\Phi_{\{\lambda_M\}}\hat{\psi}^\dagger(x'_b)\hat{\psi}(x_b)\Phi^{(-)}_{j,\varepsilon,l,m}> \tag{4.2.7}$$

is suitable for the introduction of the now widely diffuse Response techniques, i.e. of those techniques involving the study of the response of a reference state to a general time-dependent external perturbation.

The matrix elements of (4.2.6) may be obtained by the Linear Response function (often indicated as Retarded Polarization Propagator or Two-time Green's Function) through its residues while the excitation energies $\omega_M = E^{\upsilon}_M - E_{0,\{\upsilon_0\}}$ are obtained through its poles. When the reference exact state is approximated by its Single-Configuration *SCF* description, the well known *RPA* (or *TDHF*) is obtained. By taking

$$\hat{O}^\dagger_M = \sum_{p,q}\{\mathbf{X}_{pq,M}\hat{a}^\dagger_p\hat{a}_q + \mathbf{Y}_{pq,M}\hat{a}^\dagger_q\hat{a}_p\} \tag{4.2.8}$$

as the excitation operator the transition moments between the reference state $\Phi_{\{\lambda_0\}}>$ and an excited state $\Phi_{\{\lambda_M\}}>$ are then given by the following expression

$$\begin{aligned} <\Phi_{\{\lambda_0\}}|\hat{O}_\mu|\Phi_{\{\lambda_M\}}> &= \int dx_a dx'_a \hat{O}_\mu(x'_a,x_a)X^0_M(x_a,x'_a) \\ &= <\Phi_{\{\lambda_0\}}|[\hat{O}_\mu,\hat{O}^\dagger_M]_-|\Phi_{\{\lambda_0\}}>. \end{aligned} \tag{4.2.9}$$

The electronic excited state - excited state transition moments can be obtained from the residues at two poles of the Quadratic Response function[54-56]. On the other hand by exploiting the formal properties of the excitation operators $\hat{O}_M$ and the form of the solution of the *EOM* equation, it is possible to write[56]

$$\begin{aligned} < \Phi_{\{\lambda_M\}} | \hat{O}_\mu | \Phi_{\{\lambda'_M\}} > &= \int dx_b dx'_b \hat{O}_\mu(x'_b, x_b) X^M_{M'}(x_b, x'_b) \\ &= < \Phi_{\{\lambda_0\}} | [\hat{O}_M, \hat{O}_\mu, \hat{O}^\dagger_{M'}]_- | \Phi_{\{\lambda_0\}} > \\ &+ \frac{1}{2} < \Phi_{\{\lambda_0\}} | [[\hat{O}_M, \hat{O}^\dagger_{M'}]_-, \hat{O}_\mu]_+ | \Phi_{\{\lambda_0\}} > \end{aligned} \tag{4.2.10}$$

where

$$[A, B, C]_- = \frac{1}{2}[A, [B, C]_-]_- + \frac{1}{2}[[A, B]_-, C]_- \tag{4.2.11}$$

Upon substitution of the expression (4.2.8) for the *RPA* excitation operator, eq. (4.2.10) is easily written in terms of the transition density matrix elements $X_{pq,M}$, $Y_{pq,M}$. The reader will find elsewhere[54,55] the explicit formulas in the case of a *MCSCF* approximation for the reference state and of a Multiconfigurational form of the excitation operator. Here we report the simplified expression taken by eq. (4.2.10) when both the reference state and the excitation operators are approximated at the *SCF/RPA* level:

$$\begin{aligned} &< \Phi_{\{\lambda_m\}} | \hat{O}_\mu | \Phi_{\{\lambda'_M} > = \\ &\sum_{i,\nu,\nu'} \left\{ X^*_{\nu i,M} X_{\nu' i,M'} < \nu | \hat{O}_\mu | \nu' > + Y^*_{\nu i,M} Y_{\nu' i,M'} < \nu' | \hat{O}_\mu | \nu > \right\} \\ &- \sum_{i,i',\nu} \left\{ X^*_{\nu i,M} X_{\nu' i',M'} < i' | \hat{O}_\mu | i > + Y^*_{\nu i,M} Y_{\nu i',M'} < i | \hat{O}_\mu | i' > \right\} \\ &+ \delta_{M,M'} < \Phi_{\{\lambda_0\}} | \hat{O}_\mu | \Phi_{\{\lambda_0\}} > \end{aligned} \tag{4.2.12}$$

with the greek indexes ν,ν' running over the particle states and i,i' over the hole states. In our computations the virtual orbitals may be generally non orthogonal, while the occupied orbitals, being taken from the *SCF* representation of the neutral molecular system, are orthonormal. Equation (4.2.12) is slightly modified by the insertion of the appropriate overlap integrals, i.e.

$$\begin{aligned} &< \Phi_{\{\lambda_M\}} | \hat{O}_\mu | \Phi_{\{\lambda'_M\}} > \simeq \\ &\sum_{i,\nu,\nu'} \{ X^*_{\nu i,M} X_{\nu' i,M'} < \nu | \hat{O}_\mu | \nu' > + Y^*_{\nu i,M} Y_{\nu' i,M'} < \nu' | \hat{O}_\mu | \nu > \} \\ &- \sum_{i,i',\nu,\nu'} \{ X^*_{\nu i,M} X_{\nu' i',M'} < i' | \hat{O}_\mu | i > < \nu | \nu' > + \\ &Y^*_{\nu i,M} Y_{\nu' i',M'} < i | \hat{O}_\mu | i' > < \nu' | \nu > \} \\ &+ \delta_{M,M'} < \Phi_{\{\lambda_0\}} | \hat{O}_\mu | \Phi_{\{\lambda_0\}} > \end{aligned} \tag{4.2.13}$$

The expressions are further simplified if the final state is approximated as an antisymmetrized product of the ion wavefunction and of the ejected electron channel-wave $|j, \varepsilon, l, m>$ i.e. $\Phi_{j,\varepsilon,l,m}> \simeq \hat{a}^{\dagger}_{j,\varepsilon,l,m}\,\hat{a}_j\,\Phi_{\{\lambda_0\}}>$ Eq. (4.2.13) thus becomes:

$$\begin{aligned}
&<\Phi_{\{\lambda_M\}}|\hat{O}_\mu|\Phi_{j,\varepsilon,l,m}> \simeq \\
&\sum_{\nu} X^{*}_{\nu j, M}\Big\{<\nu|\hat{O}_\mu|j,\varepsilon,l,m> + \\
&<\nu|j,\varepsilon,l,m><\Phi_{\{\lambda_0\}}|\hat{O}_\mu|\Phi_{\{\lambda_0\}}>\Big\} \\
&-\sum_{i,\nu} X^{*}_{\nu i, M}<\nu|j,\varepsilon,l,m><j|\hat{O}_\mu|i>.
\end{aligned} \tag{4.2.14}$$

The resolvent contained in eq. (2.5a) will display poles for ω_1 (ω_2) equal to one of the discrete transition energies of the molecule. These intermediate resonances yield divergent two-photon cross sections due to the limits of the *LOPT* and only by including higher terms in the perturbation expansion, which will take into account the finite widths of the discrete levels, it is possible to recover a more physical behaviour[33,34,57,58]

It may be verified that *exploiting the resonance condition* the *LOPT* form of the transition operator (2.5a) guarantees gauge invariant results for the transition amplitudes

$$\mathbf{M}^{(2)}_{f\leftarrow a} = <\Phi_f|(1+\hat{P}_{1,2})\hat{O}_{\mu_1}\hat{G}^{+}_0(E_{0,\{\upsilon_0\}}+\omega_2)\hat{O}_{\mu_2}|\Phi_i> \tag{4.2.15}$$

if exact initial $\Phi_i>$ and final $\Phi_f>$ eigenstates are used. In the dipolar approximation, the $\mathbf{M}^{(2)}_{f\leftarrow i}$ evaluated employing the *VG* form of the transition operator is equal to the $\mathbf{M}^{(2)}_{f\leftarrow i}$ evaluated with the *LG* form of the transition operator times the product of the photon energies[1] i.e.

$$\left[\mathbf{M}^{(2)}_{f\leftarrow i}\right]_{VG} = \omega_1\omega_2\left[\mathbf{M}^{(2)}_{f\leftarrow i}\right]_{LG}.$$

This result is quite general and *in the dipolar approximation* it holds for any *n*-photon process evaluated in the *LOPT*[34]. Limiting the following comments to the *VTA* we observe that eq. (4.2.4) may be written as

$$\begin{aligned}
\left[\mathbf{M}^{(2)}_{f\leftarrow i}\right]_{VG,(LG)} &= (1+\hat{P}_{1,2})\sum{}'\left[\frac{<\Phi_f|\hat{O}^{p,(l)}_{\mu_1}|\Phi_c><\Phi_c|\hat{O}^{p,(l)}_{\mu_2}|\Phi_i>}{\omega_2-(E_c-E_i)}\right]_{Q_0} - \\
&(1+\hat{P}_{1,2})\left[\frac{<\Phi_f|\hat{O}^{p,(l)}_{\mu_2}|\Phi_i>}{\omega_1}\Big(<\Phi_i|\hat{O}^{p,(l)}_{\mu_1}|\Phi_i> - <\Phi_f|\hat{O}^{p,(l)}_{\mu_1}|\Phi_f>\Big)\right]_{Q_0}
\end{aligned} \tag{4.2.16}$$

where the $|\Phi_c>$ are the eigenstates of the electronic hamiltonian $\mathscr{H}_{elec}(\mathbf{Q}_0)$ and the prime on the summation indicates that the contributions due to the initial and final states have been singled out. These quantities which for parity conservation should be zero, owing to the VTA do not vanish if evaluated in the *LG*. As a matter of fact they are proportional to the the difference between the initial and the final state dipole electric moments at $\mathbf{Q}_0$ which, considering that the final state lies in the continuum, may become huge. For photon energies far from any intermediate resonance the transition amplitudes are expected to be finite; this means that, with the *LG*, the summation upon all the other states in (4.2.16) is required to almost cancel the electric dipole of the final state in the continuum. It is understandable therefore that the *VG*, which does not require this cancellation, is less likely to give numerical inaccuracy. The conclusion is that, in this respect, the *VG* should be the preferred choice for the molecular cases.

5. Integral cross section by L^2 methods

We discussed, in sect.s 2, 3.1 and 3, the form of the continuum final state in a photoionization process by adopting a partial wave expansion of the photoelectron orbital and by introducing a **K**-matrix in order to account for the coupling of the different partial channels. On the other hand, the **K**-matrix method, taking into account the infinite degeneracy of the continuum final level, brings information that is not strictly necessary if we want to calculate a physical quantity not depending on the angular distribution of the emitted electron, for example the integrated cross section of the process. In this case we can adopt an L^2 method that allows to obtain simply the oscillator strength of the excitation, or better the oscillator strength density as a function of the excitation energy in the electronic continuum. The oscillator strength here defined is obtained by integrating over the photoelectron angular parameters i.e. it corresponds to the summation of all the degenerate partial wave contributions l,m of eq. (3.1.13). The L^2 variational methods of quantum chemistry that are widely applied to the calculation of bound states can be easily used to get such quantities. As a matter of fact the finite set of eigenstates with eigenvalues larger then the ion energy, that are obtained for instance in a *CI* calculation, will offer a discretized description of the electronic continuum. This approach is essentially equivalent to directly diagonalizing the projected hamiltonian of eq. (3.1.5) without adopting the **K**-matrix to formally take into account the continuous indices. The discretized states in the continuum so obtained will not have of course the correct continuum degeneracy, but can be used to calculate approximate excitation energies from the ground state and the related oscillator strengths that will form a "discretized variational spectrum". It may be observed that such discrete spectrum cannot, of course, describe even qualitatively the structure of a continuum; moreover, because the final states are variationally determined, the distribution of the discretized oscillator strengths on the energy axis is strongly dependent on the L^2 basis set employed.

5.1 Stieltjes imaging

Among the possible techniques for extracting the continous oscillator strength density from a discretized spectrum, the moment method named Stieltjes Imaging (*SI*) has certainly been the most applied, following the pioneering work of Langhoff and coworkers[20,59-61], for the calculation of integrated one-photon ionization cross sections and it has been satisfactorily extended to the study of other discrete-continuum transition processes [2,3,28]. The *SI* approach is essentially based on the hypothesis that the variational discretized spectrum, even if clearly inadequate to describe the structure of a continuum, can anyway give converging approximations at least of the lowest spectral moments. Only a short description of the *SI* technique is briefly sketched here, several extensive presentations being available in the literature[20,59]

Starting from the Kramers - Heisenberg expression of the polarizability as a function of a complex frequency z,

$$\alpha(z) = \int_0^{\infty} \frac{df(\varepsilon)}{\varepsilon^2 - z^2} \qquad (5.1.1)$$

where

$$\frac{df(\varepsilon)}{d\varepsilon} = \sum_i^{discrete} f_i \delta(\varepsilon - \varepsilon_i) + g(\varepsilon) \qquad (5.1.2)$$

is the oscillator strength density containing contributions both from the discrete and the continuous electronic spectrum and ε is the excitation energy from the ground state, the cross section $\sigma(\omega)$ is proportional to the imaginary part of the polarizability for real values ω of the frequency or, alternatively, to the oscillator strength density in the continuum

$$\sigma(\omega) = \frac{4\pi\omega}{c} Im[\alpha(\omega)] = \frac{2\pi^2}{c} g(\omega) \qquad (5.1.3)$$

Eq. (5.1.1) is particularly convenient for showing that the polarizability is a Stieltjes integral as far as $df(\varepsilon)/d\varepsilon \geq 0$ in the integration range, which is obviously true for excitations from the ground state. According to the Stieltjes approach, this kind of integral can be convergently approximated, also on the real axis, as

$$\alpha(z) = \int_0^{\infty} \frac{dF^{(n)}(\varepsilon)}{\varepsilon^2 - z^2} + R_n(z^2) \qquad (5.1.4)$$

where $F^{(n)}(\varepsilon)$ is an approximate cumulative oscillator strength multistep function

$$F^{(n)}(\varepsilon) = \begin{cases} 0 & 0 < \varepsilon < \varepsilon_1^{(n)} \\ \sum_{j=1}^{l} f_j^{(n)} & \varepsilon_l^{(n)} < \varepsilon < \varepsilon_{l+1}^{(n)} \\ \sum_{j=1}^{n} f_j^{(n)} & \varepsilon_n^{(n)} < \varepsilon \end{cases} \tag{5.1.5}$$

given in terms of a pseudospectrum $\{\varepsilon_j^{(n)}, f_j^{(n)}\}$ that is defined by the following $2n$ equations

$$S(-2k) = \sum_{j=1}^{n} \frac{f_j^{(n)}}{\left[\varepsilon_j^{(n)}\right]^{2k}} \qquad k = 1, 2, \dots 2n. \tag{5.1.6}$$

Here the $S(-2k)$ are spectral moments of the true oscillator strength function $f(\varepsilon)$:

$$S(-j) = \int_0^{\infty} \frac{df(\varepsilon)}{\varepsilon^j} \tag{5.1.7}$$

The convergence properties of expression (5.1.5) can be more exactly expressed as[61]

$$F^{(n)}(\varepsilon - 0) < F^{(n+1)}(\varepsilon - 0) \leq f(\varepsilon) \leq F^{(n+1)}(\varepsilon + 0) < F^{(n)}(\varepsilon + 0) \tag{5.1.8}$$

so that, by differentiating the approximate cumulative function $F^{(n)}(\varepsilon)$, an approximate oscillator strength density $dF^{(n)}(\varepsilon)/d\varepsilon$, that will converge to the correct one on increasing the order n of the pseudospectrum, can be obtained. The derivative of $F^{(n)}(\varepsilon)$ in the energy continuum $g^{(n)}(\varepsilon)$ will be expressed, according to the Stieltjes approach, as

$$g^{(n)}(\varepsilon) = \begin{cases} 0 & 0 < \varepsilon < \varepsilon_1^{(n)} \\ \dfrac{1}{2}\dfrac{f_{i+1}^{(n)} + f_i^{(n)}}{\varepsilon_{i+1}^{(n)} - \varepsilon_i^{(n)}} & \varepsilon_i^{(n)} < \varepsilon < \varepsilon_{i+1}^{(n)} \\ 0 & \varepsilon_n^{(n)} < \varepsilon \end{cases} \tag{5.1.9}$$

which defines, through eq. (5.1.3), a convergent approximation for the cross section. The calculation of $\sigma(\omega)$ is then directly connected to the knowledge of the lowest spectral moments, which, as mentioned before, are supposed to be obtained, with enough accuracy, by L^2 calculations. These calculations will in general produce a "variational" spectrum $\{\varepsilon_i, f_i\}$ that, as observed before, is strongly basis set dependent in the continuum portion of the spectrum. On the contrary, the pseudospectrum $\{\varepsilon_j^{(n)}, f_j^{(n)}\}$, *as far as the approximated spectral moments can be considered to converge to the correct ones*, is virtually independent of the basis set. It may be recalled that the pseudo excitation energies and oscillator strengths correspond to abscissas and weights of a generalized Gaussian quadrature of order n for the integral in (5.1.1), taking $df(\varepsilon)/d\varepsilon$ as a weight function. The moment problem expressed by eq. (5.1.6) can be linearized by a Paddè approximant to give the polarizability integral

$$\alpha(z) = \sum_{k=1}^{n} S(-2k) z^{2(k-1)} + R_n(z^2) = \sum_{j}^{n} \frac{f_j^{(n)}}{(\varepsilon_j^{(n)})^2 - z^2} + R_n(z^2) \\ = \frac{P_{n-1}(z^2)}{Q_n(z^2)} + R_n(z^2) \tag{5.1.10}$$

with

$$P_{n-1}(z^2) = \sum_{i=0}^{n-1} a_i^{(n)} z^{2i} \qquad Q_n(z^2) = 1 + \sum_{i=1}^{n} b_i^{(n)} z^{2i}. \tag{5.1.11}$$

The pseudospectrum $\{\varepsilon_j^{(n)}, f_j^{(n)}\}$ is then obtained from the roots and residues of the Paddè approximant. The coefficients $a_i^{(n)}$ and $b_i^{(n)}$ can be obtained by solving a set of linear equations

$$\sum_{L=1}^{n} S(-2M + 2L - 2) b_L^{(n)} = -S(-2M - 2) \quad n \leq M \leq (2n - 1) \tag{5.1.12}$$

$$a_M^{(n)} = \sum_{L=0}^{M} S(-2M + 2L - 2) b_L^{(n)} \quad 0 \leq M \leq (n - 1) \tag{5.1.13}$$

easily derived from eq. (5.1.10). A non trivial solution of eq. (5.1.12) can always be found in so far as the $S(-2k)$'s are moments of a non decreasing distribution, but the numerical implementation of eqs. (5.1.12) (5.1.13) is not convenient if a very large number of moments is employed. In fact the spectral moments are highly redundant and in order to avoid numerical problem they must be computed with high accuracy. Several suggestions can be found in the literature for overcoming this problem if the input of the *SI* algorithm is a "variational" spectrum[60,62,63]. All the proposed algorithms are essentially based on the use of more stable quantities, instead of the spectral moments, to compute the Paddè approximants; nevertheless it can be said that none of these techniques is completely trouble free if pseudospectra of high order are computed and the use of double precision in the computer implementation of the algorithms is higly recommended. We have adopted the approach suggested by Langhoff[60] that is based on a continued fraction approximation of a Stieltjes integral different from the polarizability, so that not only the even, but also the odd negative moments are involved in the Taylor expansion. The main equations of the algorithm starting from an L^2 spectrum $\{\varepsilon_i, f_i\}$ are here presented for completeness

$$
\begin{aligned}
&Q_n(z) = (1 - \alpha_n z)Q_{n-1}(z) - z^2\beta_{n-1}Q_{n-2}(z) \\
&P_{n-1}(z) = (1 - \alpha_n z)P_{n-2}(z) - z^2\beta_{n-1}P_{n-3}(z) \\
&Q_{-1}(z) = 0 \quad Q_0(z) = 1 \quad P_{-1}(z) = 0 \quad P_0(z) = \beta_0 \\
&\alpha_n = \frac{1}{\beta_0\beta_1 \dots \beta_{n-1}} \sum_{i=1}^{N} \varepsilon_i^{1-2n} Q_{n-1}^2(\varepsilon_i) f_i \quad \alpha_0 = 0 \\
&\beta_n = \frac{1}{\beta_0\beta_1 \dots \beta_{n-1}} \sum_{i=1}^{N} \varepsilon_i^{-2n} Q_n^2(\varepsilon_i) f_i \quad \beta_0 = S(0) \\
&Q_n(\varepsilon_j^{(n)}) = 0 \qquad f_j^{(n)} = -\frac{P_{n-1}(\varepsilon_j^{(n)})}{\varepsilon_j^{(n)} Q'_n(\varepsilon_j^{(n)})}
\end{aligned}
\tag{5.1.14}
$$

5.2 Generalization of the Stieltjes imaging

The *SI* technique here reviewed for the calculation of photoionization cross sections can be in principle generalized to any discretized "spectrum" formed by "energy" values and "intensities" as far as these are positive quantities. It has been effectively applied to the calculation of Auger intensities,[2] shake-off intensities[3] as well as two-photon ionization cross sections.[28]

As shown in Appendix, the two-photon cross section may be written[1] in terms of positive definite quantities $G_{j,i}^{(2)}(\omega_1, \omega_2)$, Eq. (A.13), that can be regarded as proportional to partial two-photon oscillator strengths, explicitely depending in this case, not only on the final state energy but also on the energies and polarization of the absorbed photons. By an L^2 calculation we can compute these partial oscillator strengths, for fixed values of the two-photon parameters, for a discretized set of final states and form with them and the corresponding final state energy values a "variational spectrum" to which the previously described *SI* procedure can easily be applied. The *SI* will actually generate a set of derivatives, i.e. of two-photon oscillator strength densities values at energies generally different from $E_a + \omega_1 + \omega_2$ and a simple interpolation will be employed in order to match the resonance condition.

Essentially the same procedure can be followed to extimate by the Golden Rule

$$
\Gamma_\varepsilon = 2\pi \mid < \Phi_d | H - E_d | \Phi_\varepsilon > |^2 \tag{5.2.1}
$$

the linewidth Γ_{Ed} of a resonance structure in the photoionization spectrum due for instance, to the presence of a discrete state $|\Phi_d >$ belonging to a closed channel, embedded and interacting with the continuum states $\Phi_\varepsilon >$ of open channels. The positive quantity Γ_ε in Eq. (5.2.1) can be computed for the discrete set of continuum energies ε_j obtained by an L^2 calculation. The *SI* procedure applied to the "spectrum" $\{\varepsilon_j, \Gamma_{\varepsilon j}\}$ will produce Stielties derivative values that, interpolated at the energy E_d, will offer a reasonable approximation to the resonance linewidth.

This method has been effectively applied to the calculation, within Wentzel's ansatz approximation, of molecular Auger spectra.[2] From this point of view the *SI* technique can be seen as a practical and mathematically well defined procedure to obtain the correct continuum normalization starting from a quite general discretized spectrum formed by "energies" and "strengths" obtained from an L^2 calculation.

The constraint, mentioned before, of using positive definite quantities for the "strengths" can in effect be easily circumvented by simple modifications of the numerical procedure. A more serious limit of the *SI* technique can be seen in its "energy resolution capability." Any discretized L^2 representation of the electronic continuum will be, in general, characterized by a certain "resolution" that is of course proportional to the number of L^2 states per unit energy interval. As already observed, this density can be quite different in the different energy regions and it is certainly basis set dependent; in principle, not necessarily in practice, it can be increased by increasing the dimension of the basis set in order to have a better energy resolution. The pseudospectrum, optimized in the sense of being a "quadrature spectrum," will offer, as previously discussed, a more uniform and basis set independent representation of the continuum, but it cannot improve the energy resolution because, in general, its dimension will be much less than the dimension of the input variational spectrum. This is strictly connected to the fact that only a limited number of the computed low order spectral moments will satisfactorily approximate the correct values and will consequentely be employed in the *SI* calculation. As a matter of fact the *SI* procedure allows to obtain only a discrete set of Stieltjes derivative values i.e. of strength density values, as shown by Eq. (5.1.9), that associating the value of each step of the hystogram to the midpoint of the corrisponding interval, may be generally fitted very well by simple functions in order to get the correct continuous dependence of the strength density upon the energy. This unfortunately will in general smooth out any narrow resonance structure eventually present in the spectrum; a drawback that can avoided only by making the pseudospectrum in the energy region around the resonance structure denser, if possible. This can sometimes be achieved by increasing the order of the pseudospectrum i.e. the number of employed spectral moments. Unfortunately this number cannot be increased at will if consistency of the final results with respect to the dimension of the pseudospectrum is desired.

The so called Stieltjes-Tchebycheff procedure,[60,61] that differs from the plain *SI* technique in the sense that one of the pseudospectrum energies may be varied continuosly, has been considered a natural extension of *SI* in order to compute continuos values of the Stieltjes derivative. Unfortunately, as verified by us as well as by other authors, the Stieltjes-Tchebycheff algorithm is affected by numerical instabilities that easily introduce artificial "modulations" of the computed strength density which make this technique, at least in the present numerical implementations, pratically useless for obtaining clear information about the presence of resonance structure in the spectrum. In other words the presence of a narrow resonance will make almost impossible extracting a reasonable continuum normalization from a L^2 discretized spectrum in the neighborhood of the resonance position.

6. Computational aspects

Both the **K**-matrix and the Stieltjes Imaging technique outlined in the previous sections, can be applied at different levels of accuracy in the description of initial and final states. Nevertheless the independent particle approximation should be considered not only the most largely applied approximation but also the preferred starting point for any higher level calculation.

In adopting such approximation the main difficulty in the numerical implementation of both methods can be traced to the description, by L^2 functions, of the continuum orbital. The Static-Exchange Approximation[41] is probably the most appropriate way of calculating the emitted electron orbital, in the sense that the continuum orbital is variationally optimized for the Coulomb plus exchange potential of the ionic target. Approximating the final state as $|\Phi_{j,\varepsilon}> \simeq \hat{a}^{\dagger}_{j,\varepsilon}\,\hat{a}_j\,|\Phi_{\{\lambda_0\}\,SCF}>$ corresponds to a description of the scattering final state by its asymptotic form, as already discussed in section 2. The effect of the electronic correlation may be introduced by taking into account the interaction of *SEA* states corresponding to different parent ion states plus multi-excited localized states. This is easily realized, as discussed in sect. 3, by the **K**-matrix approach that is essentially a *CI* in the electronic continuum. The Stieltjes Imaging technique, on the other hand, having as an input the discretized excitation spectrum obtained by a previous independent calculation, can of course be applied at any level of approximation of initial and excited states or, by an *EOM* approach, of the directly determined transition density matrices.

6.1 The basis set

The requirements of the two methods with respect to the description of the continuum orbital are not exactly the same. As discussed in previous sections, the practical implementation of both techniques is based on a discretization of the continuum energy that should consist in energy values as close as possible to each other because of the interpolation procedure that is employed in the **K**-matrix method and because of the low energy resolution capability of the Stieltjes Imaging technique. The matrix elements to interpolate, as well as the oscillator strengths, being expressed by integrals involving both bound and continuum orbitals, are generally well approximated if the L^2 "continuum" orbital can reproduce in a reasonable way the exact continuum orbital in the molecular space where the largest contribution to such integrals is collected. Differing from the Stieltjes Imaging, the **K**-matrix technique has the additional requirement of the calculation of the phase shift Δ due to the short range potential and of the appropriate normalization constant. Whatever numerical procedure is applied to obtain such phase shift from an L^2 calculation, it certainly implies to have an L^2

continuum orbital that can be considered a good approximation to the correct one for distances from the ionization center well above the molecular dimensions.

The basis functions commonly employed in quantum chemistry calculations, i.e. Gaussian (*GTO*) and Slater (*STO*) type orbitals, can hardly satisfy the first requirement of giving a discretized spectrum dense enough on the energy axis. This is directly connected to the overlap among the basis functions that rapidly leads to redundancy when the basis set dimension is increased. For the same reason they are even less adequate for the second requirement of describing the asymptotic behaviour of the continuum orbital. We use a completely L^2 approach employing standard *STO*'s, hydrogenic functions

$$R_{nl\zeta}(r) = A_{nl\zeta}\left(\frac{2\zeta r}{n}\right)^{l} L_{n+l}^{2l+1}\left(\frac{2\zeta r}{n}\right)e^{-\frac{\zeta r}{n}} \tag{6.1.1}$$

and a new type of basis functions (*STOCOS*) formed by *STO* times an oscillating factor,

$$\chi_{n\lambda k}(r) = B_{n\lambda k} r^{n-1} e^{-\lambda r} \cos(kr) \tag{6.1.2}$$

that can easily mimic, using small values for λ, the long range behaviour of a scattering orbital. It should be noted that, even for very small values of parameter λ, the basis set redundancy may be avoided because of the trigonometric factor in the *STOCOS* functions and a dense spectrum, with energies roughly scaled as the squares of the values of parameter k, may be obtained. The one- and two-electron integrals that are involved in the numerical implementation of the *SEA* as well as of any standard quantum chemistry calculation can be easily computed with the *STOCOS*'s provided that all functions have the same origin[64]. This naturally suggests to use a one-center expansion (*OCE*) basis set both for bound and continuum molecular orbitals.

The *OCE* approximation applied to the study of molecules can be considered satisfactory for systems having only one heavy atom; for this reason we restricted the field of our applications to hydrides of the second and third row. Table I collects some ground state properties calculated at the *SCF* level for NH_3 and H_2O using an *OCE-STO* basis set and a multicenter Gaussian basis set. Figs. 4,5 present a comparison of maps in the molecular plane, for the total density and the $3a_1$ orbital of the H_2O molecule respectively, obtained by employing a standard double zeta Gaussian basis set and a basis set of 66 *STO*'s, having l values up to 7 in both cases. The two calculations show only small discrepancies; the *OCE* approximation can be considered quite adequate to our purposes in the sense that gives a reasonably good description of the bound orbitals and consequently of the Static-Exchange potential of the ion target. This is confirmed also by the results reported in Table I.

However the main advantage of using a *STOCOS* basis set for molecular calculations is, of course, the improved description of the continuum orbital. In order

Table I. Total SCF energies and orbital energies by *OCE STOs* and multi center *GTOs* basis sets.

	H_2O			NH_3	
	OCE[a]	GTO[b]		OCE[c]	GTO [b]
Energy	−76.0138	−76.0005	Energy	−56.1167	−56.1597
$1b_1$	13.63	13.73	$3a_1$	11.36	11.25
$3a_1$	15.62	15.15	$1e$	16.62	16.79
$1b_2$	19.15	19.10			
$2a_1$	36.52	36.71	$2a_1$	30.66	31.02
$1a_1$	559.16	558.94	$1a_1$	421.92	422.34

a): Basis set of 60 *STOs* with l up to 7.
b): Basis set from S. Huzinaga and Y. Sakai, J. Chem. Phys., **50**, 1371, (1969). (13,7/4,2) for *O* and *N* plus (4/2) for *H*.
c): Basis set of 66 *STOs* with l up to 7

to esemplify the numerical accuracy that can be reached by employing such functions we present in Fig. 6 a comparison between results obtained by diagonalizing the hydrogenic hamiltonian projected on our L^2 basis set and the exact hydrogen continuum. The calculation was performed by using 8 hydrogenic functions for the lowest bound states plus 26 *STOCOS* functions with the λ parameter equal to 0.1 and the k parameter equally spaced in the range $0.2 \div 1.2$. The L^2 continuum orbital are, except for a normalization constant, practically undistinguishable from the correct Coulomb waves on a quite large interval of distances from the nucleus; Fig. 6 shows, for different energies, only the "tail" region where the L^2 orbital is damped by the exponential factor contained in the basis function. It should be noted that, because of the linear combination of several basis functions forming each variational orbital, such damping is much less effective than it could be expected considering the value adopted for the parameter λ.

In Table II and Fig. 7 we present some results of a typical partial wave *SEA* calculation for the $3a_1$ hole of the water molecule. A basis set of 36 *STO*'s has been used for the *SCF* calculation of the molecule ground state; the bound orbital set so obtained was employed to define the *SEA* hamiltonian that in this case has the following expression

$$\hat{h}_{3a_1} = \hat{f} - \hat{J}_{3a_1} + 2\hat{K}_{3a_1} \qquad (6.1.3)$$

where $\hat{f}$, $\hat{J}$ and $\hat{k}$ are the ground state Fock operator and the Coulomb and exchange operators respectively. The *SEA* one-electron hamiltonian was then diagonalized on a larger basis set obtained by adding to the one for the bound orbitals 4 hydrogenic functions and 54 *STOCOS* functions with $l = 2$, $m = -1$,

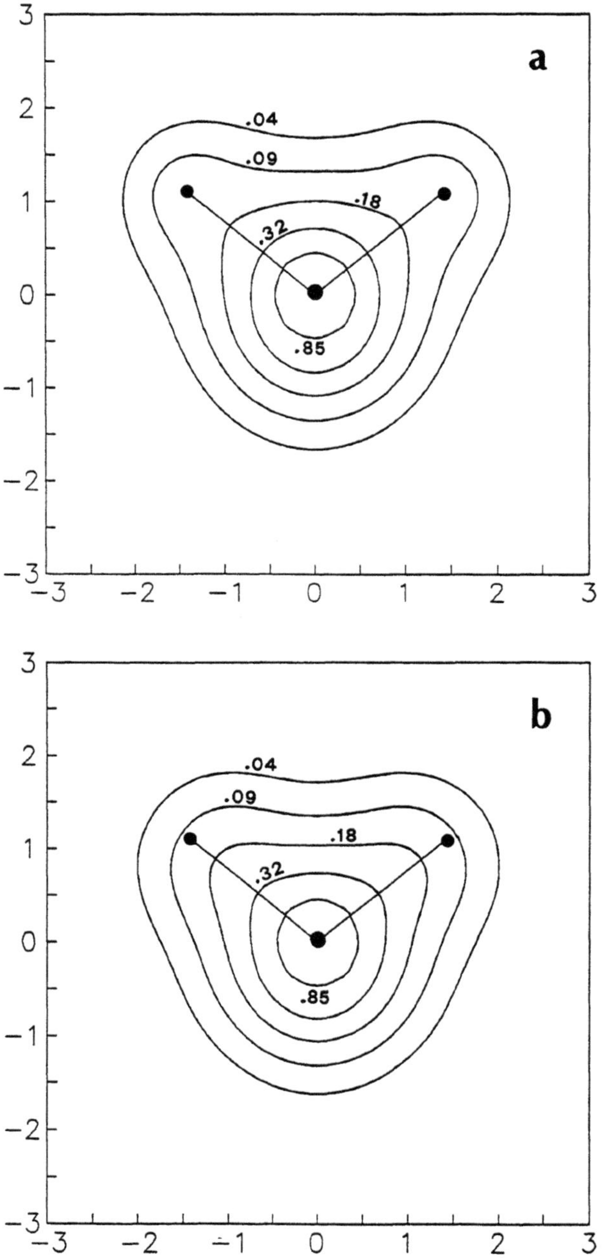

Figure 4. Map of the electron density of H_2O in the molecular plane: *SCF* calculations performed by using a *GTO* multicenter basis set (*a*) and an *OCE-STO* basis set (*b*); see section 6 for details.

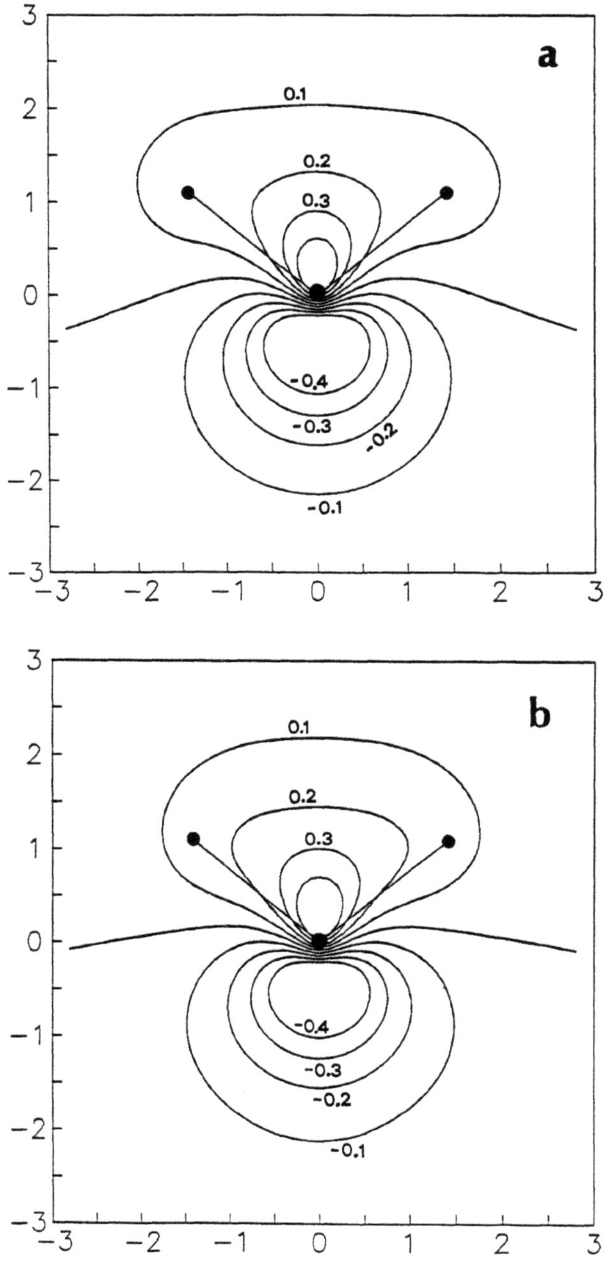

Figure 5. Map of the $3a_1$ orbital of H_2O in the molecular plane: *SCF* calculations performed by using a *GTO* multicenter basis set (*a*) and an *OCE-STO* basis set (*b*); see section 6 for details.

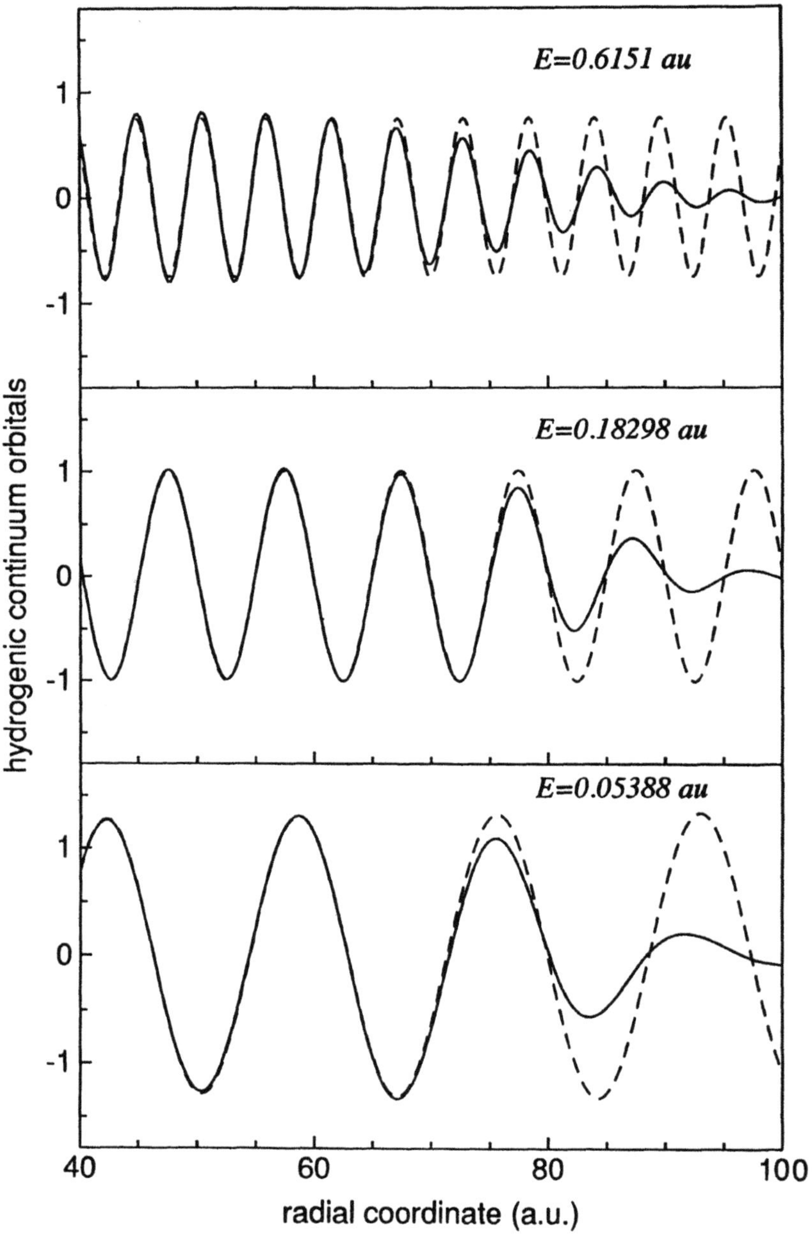

Figure 6. Comparison of exact Coulomb waves (full line) and L^2 Coulomb waves projected on a basis of *STOCOS* functions (dotted line) for different energies; see section 6 for details.

Table II. Calculated normalization constants for unit energy range and phase shifts (radians) for the *d* waves of the $3a_1 \rightarrow b_2$ (B_2) partial wave channel of H_2O.

Basis set 1[a]				Basis set 2[b]			
ε(a.u.)	$1/\sqrt{\Delta\varepsilon}$	C^c_{fitt}	δ(rad)	ε(a.u.)	$1/\sqrt{\Delta\varepsilon}$	C^d_{fitt}	δ(rad)
0.1403	5.773	5.778	0.489	0.1490	3.996	4.003	0.493
0.2053	5.342	5.341	0.541	0.2171	3.690	3.691	0.548
0.2805	4.993	4.996	0.588	0.2959	3.449	3.450	0.596
0.3658	4.704	4.705	0.628	0.3852	3.251	3.252	0.636
0.4612	4.460	4.461	0.659	0.4851	3.085	3.087	0.667
0.5669	4.250	4.249	0.682	0.5954	2.942	2.945	0.688
0.7445	3.983	3.984	0.700	0.7162	2.817	2.820	0.700
0.8756	3.831	3.832	0.704	0.8474	2.707	2.709	0.706
1.0170	3.696	3.696	0.702	0.9891	2.609	2.612	0.705
1.1685	3.574	3.574	0.695	1.1412	2.522	2.525	0.700
1.3302	3.463	3.463	0.687	1.3036	2.443	2.447	0.690
1.5020	3.363	3.363	0.676	1.4762	2.372	2.376	0.680
1.6839	3.271	3.271	0.666	1.6591	2.307	2.311	0.668
1.8758	3.187	3.187	0.657	1.8521	2.247	2.252	0.659
2.0777	3.109	3.109	0.649	2.0553	2.192	2.200	0.649
2.2896	3.037	3.037	0.645	2.2684	2.140	2.149	0.640
2.5115	2.970	2.969	0.643	2.4918	2.092	2.088	0.631
2.7433	2.907	2.910	0.644	2.7255	2.045	2.013	0.632
2.9850	2.847	2.838	0.648	2.9703	1.997	1.932	0.654
3.2369	2.790	2.810	0.656	3.2269	1.947	1.878	0.703
3.4995	2.731	2.683	0.656	3.4977	1.888	1.898	0.772
3.7748	2.660	2.762	0.682	3.7880	1.813	2.049	0.826

a): Basis with 7 *STOs* plus 54 oscillating functions with $n=3$, $\zeta=0.177$, $k_{min}=0.26$, $k_{max}=3.00$ and k increment 0.05.
b): Basis with 7 *STOs* plus 27 oscillating functions with $n=3$, $\zeta=0.355$, $k_{min}=0.26$, $k_{max}=3.00$ and k increment 0.10.
c): Obtained by a fitting with a Coulombic wave in the range 30-35 a.u. (see text).
d): Obtained by a fitting with a Coulombic wave in the range 22-27 a.u. (see text).

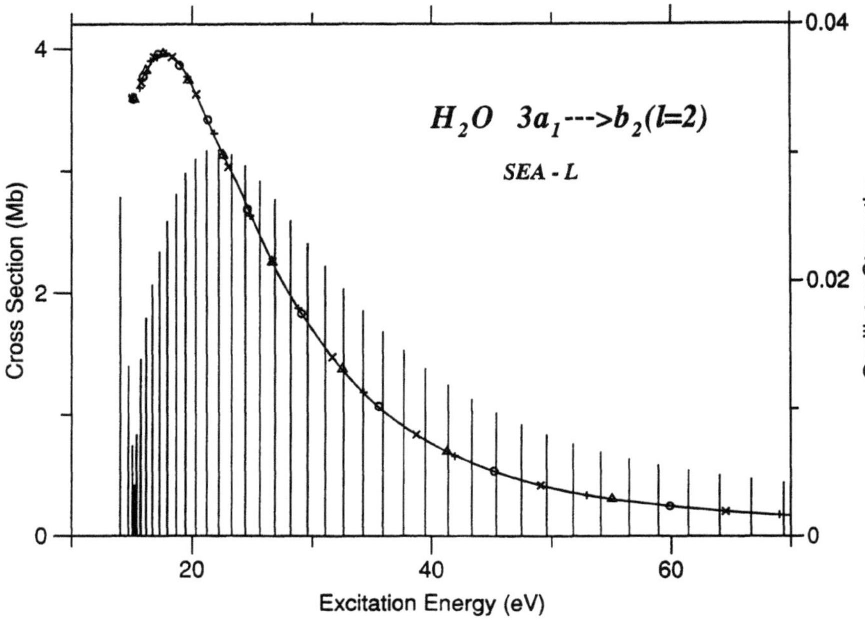

Figure 7. Bar diagram of the discretized oscillator strength distribution obtained by a *SEA* calculation for the $3a_1 \rightarrow b_2$ ($l = 2, m = -1$) partial wave channel of H_2O, different symbols indicating Stieltjes imaged values of the cross section that are fitted by the continuous line; see section 6 for details.

$\lambda = 0.177$ and k values equally spaced in the range $0.26 \div 3.00$. The bar diagram in Fig. 7 shows the discretized oscillator strength distribution obtained from the described calculation for the partial wave channel $3a_1 \rightarrow b_2$ (partial channel $l = 2$, $m = -1$). It should be stressed that both the closely spaced values of energy and the very smooth behaviour of the calculated distribution are a positive consequence of the peculiar basis set employed.

As shown in Fig. 6 the main advantage of using a *STOCOS* basis set is however in the description of the asymptotic behaviour of the continuum orbitals. The matching in the asymptotic region between the L^2 orbital and a shifted Coulomb function, allows to obtain at the same time both the factor $C_{jlm}(\varepsilon)$ to normalize the variational orbital for unit energy interval and the short range phase shift $\Delta_{jlm}(\varepsilon)$ that is necessary in the differential cross section calculation. Table II shows an example of results obtained in this way by fitting each L^2 continuum orbital $\phi_{3a_1 lm\varepsilon_v}(r)$ (partial wave $l = 2$, $m = -1$) obtained by the described $3a_1$ hole calculation, to a linear combination of the regular and irregular Coulomb waves corresponding to the same energy value

$$\phi_{3a_1lm\varepsilon_v}(r) = \frac{1}{C_{3a_1lm}(\varepsilon_v)}\{F_{l,\,m,\,\varepsilon_v}(r)\cos\left[\Delta_{jlm}(\varepsilon_v)\right] + G_{l,\,m,\,\varepsilon_v}(r)\sin\left[\Delta_{jlm}(\varepsilon_v)\right]\} \tag{6.1.4}$$

The second and sixth columns in Table II contain the normalization factors derived by estimating the continuum density of states around the energy value ε_j as

$$\rho(\varepsilon_j) = \frac{2}{\varepsilon_{j+1} - \varepsilon_{j-1}} \tag{6.1.5}$$

The normalization factors obtained in this way compare quite favourably with the ones in the third and seventh columns that have been calculated by the, in principle more accurate, fitting procedure in the asymptotic region.

6.2 One photon ionization

As already observed in section 5, in order to calculate only the integrated cross section a direct diagonalization of the *SEA* hamiltonian projected on the basis set including the different partial waves followed by the Stieltjes Imaging of the obtained discretized spectrum is practically equivalent to the more involved **K**-matrix calculation.

An example of an oscillator strength distribution obtained by the direct diagonalization of the *SEA* hamiltonian matrix, including partial waves with l up to 3, is reported in Fig. 8 together with the integrated cross section for the photoionization of the $3a_1$ orbital of H_2O calculated by the Stieltjes Imaging procedure. The cross section, not reported for clarity, calculated by the **K**-matrix technique for the diagonalization of the *SEA* hamiltonian matrix on the same basis set is practically identical to the one in Fig. 8.

The continuos line, superimposed to the $3a_1 \rightarrow b_2$ ($l = 2$) oscillator strength distribution in Fig. 7, representing the contribution to the $3a_1$ photoionization cross section due to the single $l = 2$, $m = -1$ partial wave, was obtained by the *SI* procedure. It is quite clear, looking at the smoothness of the calculated distribution, that the *SI* calculation could be, in principle, avoided and that the normalization for unit energy range that is necessary in the calculation of the cross section, could be obtained for instance by the approximated expression for the continuum density of states reported in eq. (6.1.5). This is certainly impossible if the oscillator strength distribution has been obtained by a *SEA* calculation coupling different partial waves as it is the one reported in Fig. 8. Several numerical procedure could be suggested in order to extimate the cross section shape from such a distribution. Among them the *SI* procedure, giving, as already discussed in section 5.1, principal representations of the oscillator strength distribution with remarkable properties, can be considered the most convenient one to obtain the contin-

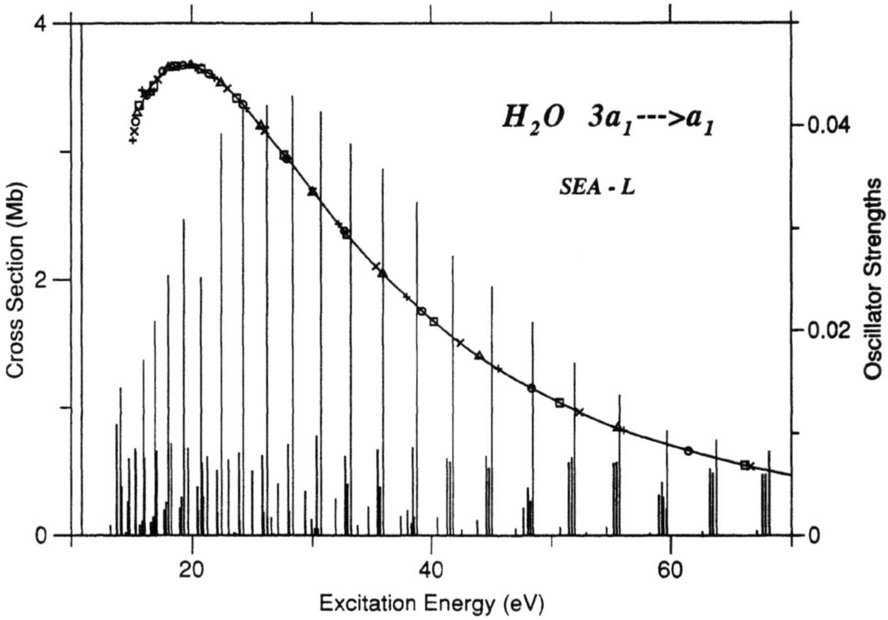

Figure 8. Bar diagram of the discretized oscillator strength distribution obtained by a *SEA* calculation for the $3a_1 \rightarrow a_1$ channel of H_2O including partial waves with l up to 3. Different symbols indicate Stieltjes imaged values of the cross section that are fitted by the continuous line; see section 6 for details.

uous oscillator strength density $df/d\varepsilon$ in a convergent way. This is clearly shown in Fig. 8 where the values of the cross section obtained by pseudospectra of different order are indicated by different markers; the full set of values can be perfectly fitted by a single smooth function of the energy representing the final result. The shape resonance just above the photoionization threshold is quite well outlined by the *SI* calculation. As discussed in sec. 5.1 this is possible, despite the intrisic low resolution capability of the *SI* technique, because of the high density of oscillator strengths obtained by the L^2 calculation in that energy region and because of the typically large width of a shape resonance.

It must be emphasized that, differently from the L^2 **K**-matrix approach, which may guarantee accurate results sweeping through a resonance no matter how narrow, the *SI* would require an increasingly unmanageable basis size with the decreasing of the resonance width.

Besides some care is required in the *SI* of spectra obtained by multichannel calculations, in the usually employed vertical approximation, when "discrete" excited states having strong oscillator strengths are weakly coupled to the electronic

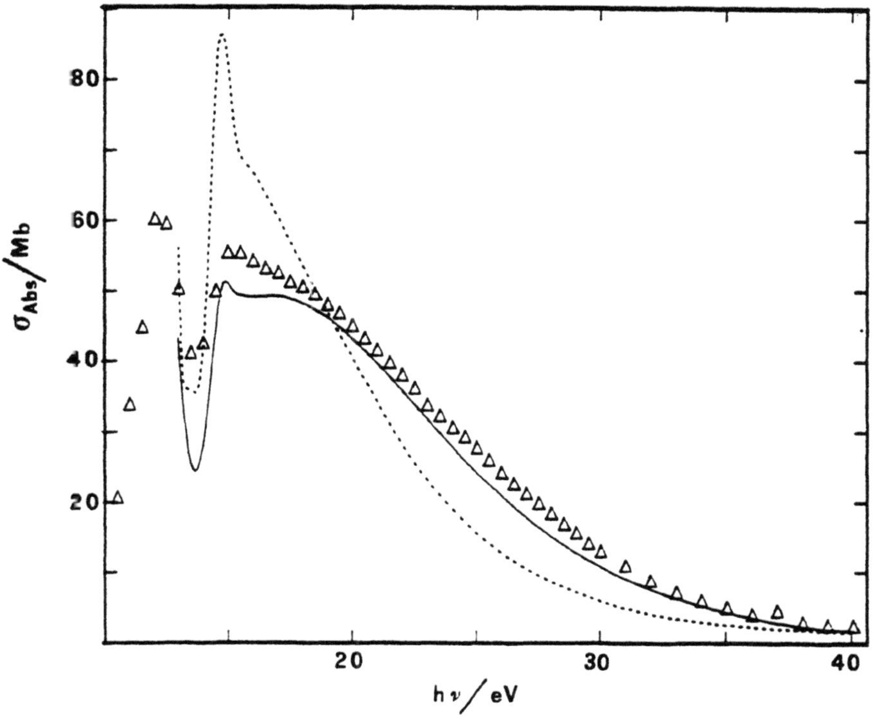

Figure 9. Photoabsorption cross section of *HCl* by Multi-Channel *RPA* (full line) and Single Channel *RPA* (dotted line) calculations and experimental energy loss data (triangles) (S.Daniel, Y.Iida, F.Carnovale and C.E. Brion, Chem. Phys. **83**, 319 (1984)).

continua of open channels. In this case the behaviour of the cross section close to the resonance may be strongly dependent on the order of the pseudospectrum considered or, in other terms, because of the weak interaction, the resonance linewidth may be too small for the energy resolution capability of the calculation. The *SI* procedure may eventually yield a fictiously large linewidth that will be clearly dependent on the number of spectral moments employed. Such structures may be in deceiving agreement with the experimental data just because of the low energy resolution of the experiment or because of the presence of other decay mechanisms, due to coupling between electronic and nuclear motions, that, of course, are not included in a vertical approximation calculation. Computational problems of this kind have been encounterd in the *SI* of multichannel spectra of H_2O[27] NH_3[27] and H_2S[29], where the resonance linewidths have been estimated by the simple *SI* procedure presented in section 5.2. We suggested[29] that in order to give a more reasonable account of the effect of such "quasi-stationary" states on the photoabsorption cross section, it is more convenient to cancel their contribution to the discretized oscillator strength distribution and then add it back, in some appropriate way, to the final Stieltjes cross section profile.

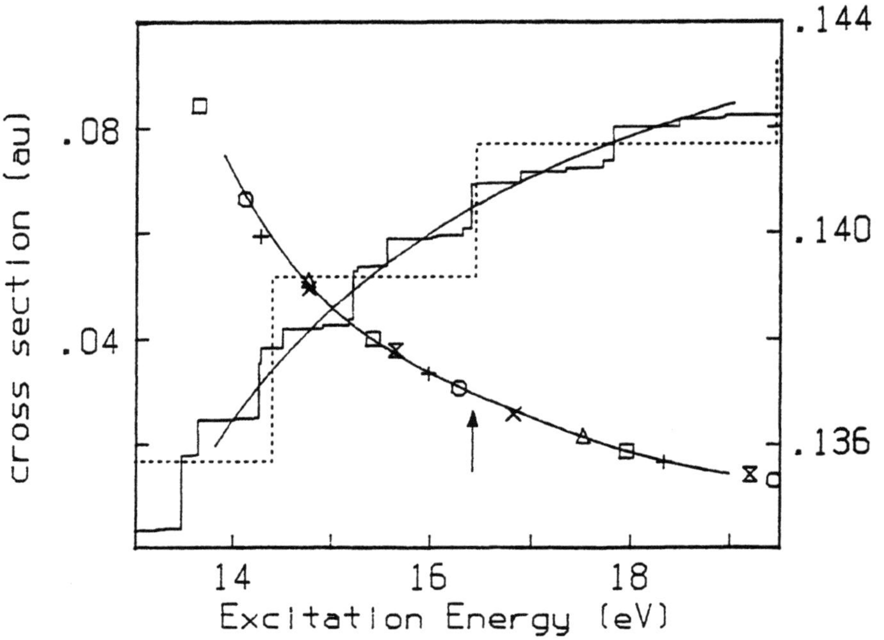

Figure 10. Stieltjes calculation for the $1b_1 \rightarrow a_1$ channel of H_2O in the *LG* gauge and photon energy equal to 8.2 eV. Symbols mark Stieltjes imaged derivative values corresponding to pseudospectra of order 12 to 18, while the smooth decreasing curve is obtained by a least square fitting of such values. Both refer to the left ordinate scale. The full-line histogram is built with the discretized spectrum obtained by Multi-Channel *RPA* for the intermediate states, while the broken line histogram derives from the pseudospectrum of order 16. Both histograms refer to the right ordinate scale. The increasing smooth curve is the integral of the fitted cross section scaled in order to facilitate the comparison with the full-line histogram. The vertical arrow indicates the Stieltjes derivative value corresponding to the resonance condition.

Most of the published *SI* calculations of photoionization cross sections for small molecules have been performed in the independent channel approximation. Our studies of the photoionization process in small hydrides have clearly shown that this approximation, although satisfactory for the second row hydrides[27] for instance, may be sometimes quite inadequate. The comparison of the multichannel and single-channel photoabsorption cross section of *HCl* reported in Fig. 9 is a clear example of the importance of taking into account the coupling of different excitation channels. A similar strong channel coupling effect was also observed in the study of the photoabsorption process in H_2S[29].

While there is little hope to extract, by the *SI* technique, theoretical asymmetry parameters from an L^2 calculations coupling directly different partial waves, it is worth to devise some approximate technique to distinguish partial channel contributions to the integrated cross section from a multi-channel calculation. We

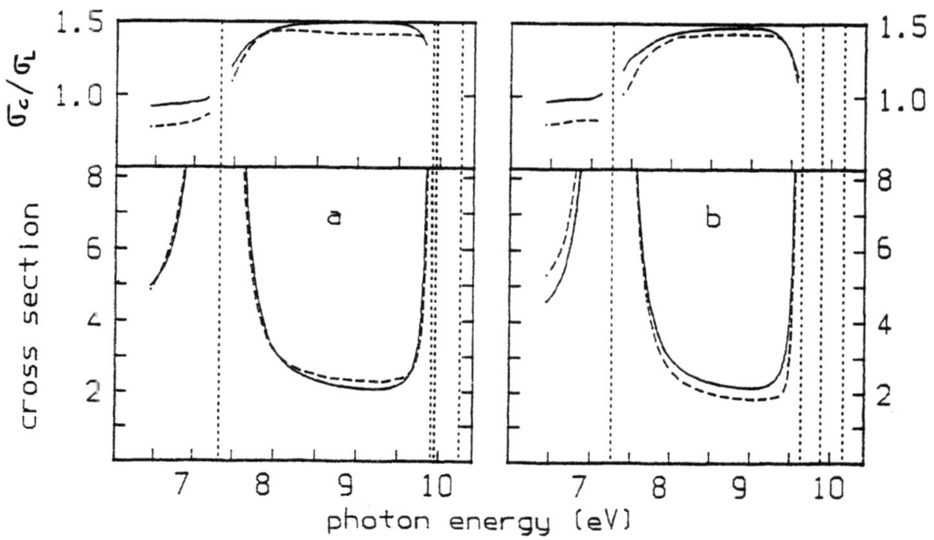

Figure 11. Total two-photon ionization cross section for linearly polarized photons σ_L and ratio σ_C/σ_L with the circularly polarized two-photon cross section; intermediate states by Single Channel *SEA* (*a*) and Multi-Channel *RPA* (*b*), broken line for the *LG* results and full line for the *VG* results.

suggested to perform multi-channel calculations projected on the basis of the single channel excited states[26,27,29] and to regard the states which in the single channel approximation lay above the first ionization threshold as the only ones that may contribute to the photoionization process. Thus, in the case of an *RPA* calculation, the fraction f_{Al} of the multi-channelionization oscillator strength pertaining to each channel A has been estimated through the following formulae

$$f_{Al} = f_l w_{Al} \tag{6.2.1}$$

$$w_{Al} = \sum_k (|\mathbf{X}_{Ak,l}|^2 - |\mathbf{Y}_{Ak,l}|^2) \tag{6.2.2}$$

where the summation runs over all single channel states of channel A having excitation energies above the first ionization threshold with the exception of the forementioned resonant states. It may be observed that, differently from other proposed methods of partitioning a multi-channel spectrum to obtain partial photoionization cross sections[65,66], the present one does not neglect the interference term contributions so that the spectral moments $S(-k)$ employed in the total photoionization cross section calculation are not altered.

As already discussed in sec. 3 the **K**-matrix technique, differently from the *SI* technique, describes correctly the degeneracy of excited states in the continuum

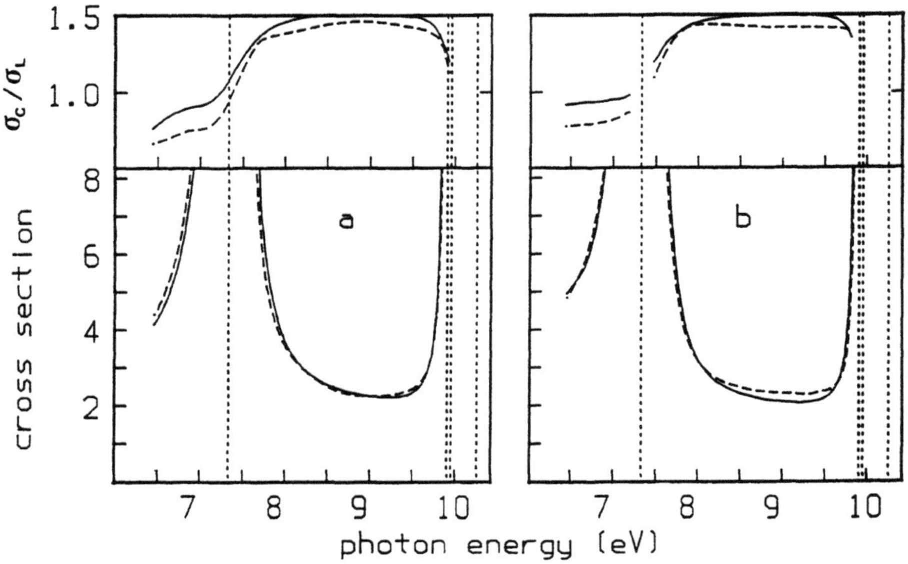

Figure 12. Total two-photon ionization cross section for linearly polarized light $\sigma_L^{(2)}$ and ratio $\sigma_C^{(2)}/\sigma_L^{(2)}$ between the circularly and linearly polarized two-photon cross sections by **K**-matrix (*a*) or Stieltjes Imaging (*b*) techniques.

even when the practical implementation of the method is based on the use of an L^2 basis set. This allows not only to calculate the differential cross section but also, when channel coupling is included, to select in a rigorous way the contribution pertaining to each photoionization channel just evaluating all the matrix elements in eq. (3.1.4) that correspond to the considered channel. This provides, of course, a clear way of obtaining theoretical photoionization branching ratios.

6.3 Two photon ionization

The serious theoretical and computational difficulties connected to the approximation of the resolvent have already been discussed in sec. 4.2; in the present section it may be worth pointing out some computational details about the application of the *SI* procedure to the calculation of the two-photon integrated cross section[28]. This last property, as was already observed in Appendix, may be in fact written in terms of molecular quantities that are positive definite[1]. By this last feature they may be considered a sort of two-photon oscillator strength that, by an L^2 calculation at fixed values of the two photons parameters, can be obtained for a discretized set of final states. In this way a generalized primitive spectrum is formed, to which the *SI* technique is applied according to the procedure described in sect. 5.2. A clear example of the numerical stability of this procedure is shown in Fig. 10 where some numerical details of an *SI* calculation[28] of the $1b_1$ two-photon ionization cross section of H_2O are reported. The full-line histogram is

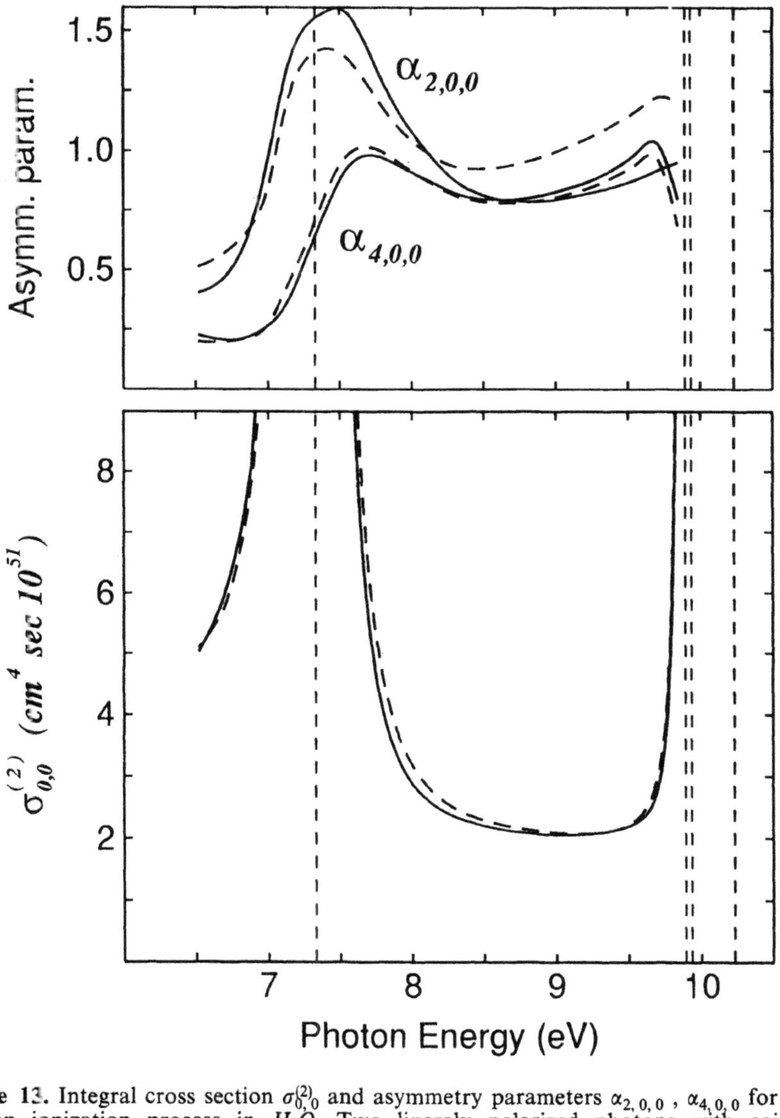

Figure 13. Integral cross section $\sigma^{(2)}_{0,0}$ and asymmetry parameters $\alpha_{2,0,0}$, $\alpha_{4,0,0}$ for a two-photon ionization process in H_2O. Two lineraly polarized photons with coincident polarization vectors ($\beta = 0$). Broken line: *VG*. Full line: *LG*.

given by the sum of the two-photon oscillator strengths in *LG* gauge, corresponding to discretized *SEA* final states, while the broken line histogram corresponds to the 16th order principal representation. Different symbols indicate Stieltjes derivative values obtained from spectral principal representations of order from 13 to 18 and the smooth decreasing function represents the least square fitting of such values. The fitting accuracy is confirmed by the shape of the integral (full line) curve that adequately describes the behaviour of the full line

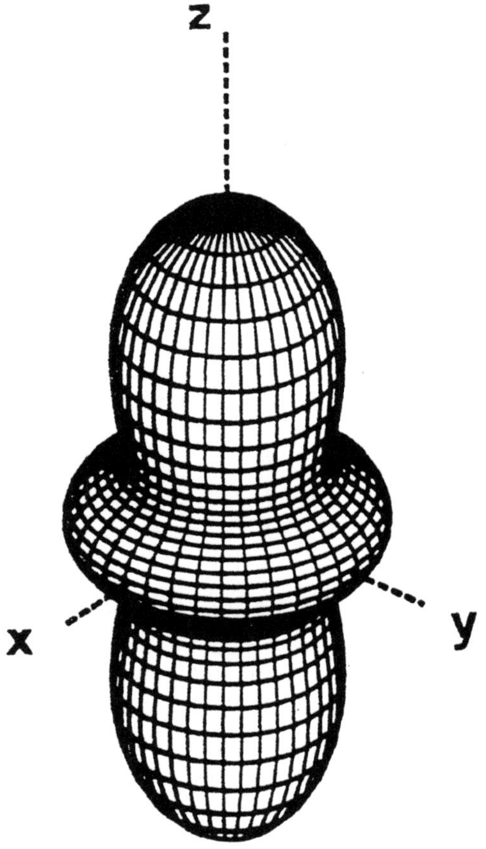

Figure 14. Photoelectron angular distribution for a two-photon ionization process in H_2O. Two linearly polarized photons with an energy of 9.02 eV, at an angle $\beta = 0$. *VG* results.

histogram. The two-photon ionization cross section is given by the value of the fitting function at the resonance energy ($\omega = 8.2$ eV in the present case) indicated by the vertical arrow.

The total integrated cross section for the absorption of two linearly polarized photons is reported in Fig. 11. The divergent behaviour of the calculated $\sigma^{(2)}$ at any discrete electronic excitation energy of the molecule is, of course, due to the adopted vertical transition *LOPT* approximation as discussed in sec. 4.2.

The same result can be obtained, of course, by the **K**-matrix approach to the final state description[30], as it is clearly shown in Fig. 12 where the integrated ionization cross section for the absorption of two linearly polarized photons in H_2O is reported. The more complex calculations that are necessary in the **K**-matrix

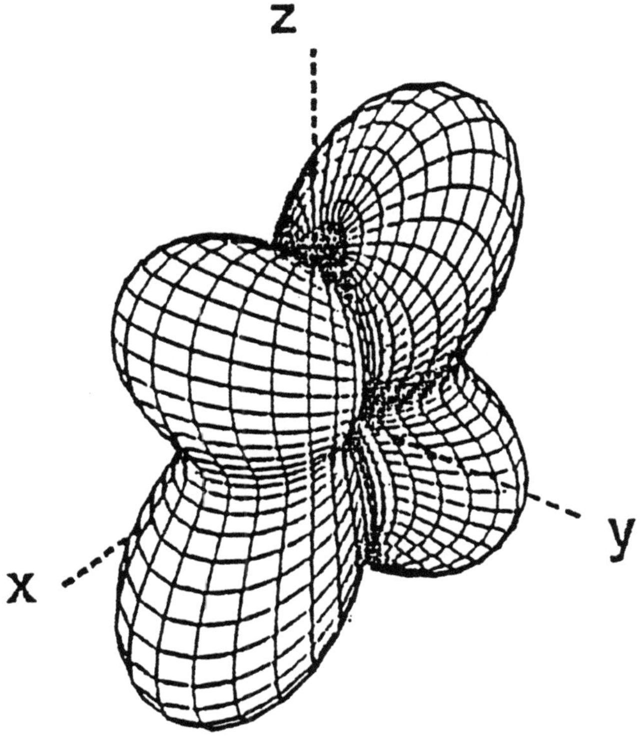

Figure 15. Photoelectron angular distribution for a two-photon ionization process in H_2O. Two linearly polarized photons with an energy of 9.02 eV, at an angle $\beta = 90$ *VG* results.

approach, essentially of the same order of difficulty that is encounterd in the one-photon calculations, are justified by the possibility of describing the angular distribution of the photoelectrons. This last quantity shows, as a function of the photon energy, polarization and geometrical setup, a rather structured behaviour which may be very useful to characterize the molecular electronic distribution.

As mentioned in Appendix, for an angle $\beta = 0$ (two photons with coincident polarization) the angular distribution of photoelectron may be described in terms of only two parameters: $\alpha_{j,2,\mu_1,\mu_2}(\varepsilon)$ and $\alpha_{j,4,\mu_1,\mu_2}(\varepsilon)$ where $\mu_1 = \mu_2 = 0$ for linearly polarized photons and $\mu_1 = \mu_2 = \pm 1$ for circularly polarized photons. The physically essential requisite of positivity for the Two-Photon Ionization Differential Cross Section (*TPIDCS*) furnishes definite relationships between the two asymmetry parameters α_2, α_4 that are restricted to a simlpy connected region in the (α_4, α_2) plane[1]. A useful description of the behaviour of the *TPIDCS* of a system at varying photoelectron energies may be obtained drawing the line connecting all (α_4, α_2) pairs at increasing photoelectron energies in such plane. The resulting curve, showing more or less sharp discontinuities in correspondence to intermediate states resonances, may provide an immediate overview of the complexity of the *TPIDCS* distributions for that system.

The two asymmetry parameters for the $1b_1$ ionization of H_2O[31] obtained by the theoretical and computational approach described in the previous sections, are plotted in Fig. 13 for the case of two lineraly polarized photons, together with the corresponding integrated cross sections. The experimental evidence of a large band in the UV absorption spectrum corresponding to the dissociative $1b_1 \rightarrow 4a_1$ excited state (the first excited state responsible for the resonance at a vertical excitation energy of 7.33 eV in the figure) and the large number of intermediate bound states whose excitation energy is greater than 9.8 eV, limit the range of validity of our results. On the other hand, the introduction as a first correction of the vibrational contribution, essentially through the insertion of the Franck-Condon factors in the calculation of the transition matrix elements, apart from resolving the vibrational structure of the resonances, is not expected to modify noticeably the overall behaviour of the curves of Fig. 13 even at close distance from the intermediate state resonances.

For the case of two linearly polarized photons of Fig. 13 where both $\alpha_{j,2,\mu_1,\mu_2}(\varepsilon)$ and $\alpha_{j,4,\mu_1,\mu_2}(\varepsilon)$ are almost always positive, the ratio $\alpha_{j,2,\mu_1,\mu_2}(\varepsilon) / \alpha_{j,4,\mu_1,\mu_2}(\varepsilon)$ goes below the value of 15/6 at some photon energy after the first resonance. In this case[1] the angular distribution has the typical pattern shown in Fig. 14, where the ratio

$$\frac{4\pi}{\sigma^{(2)}_{0,0}(\varepsilon)} \frac{\partial\sigma^{(2)}_{0,0}}{\partial\Omega_{\hat{k}_e}}$$

is displayed for the absorption of two linearly polarized photons of energy 9.02 eV (photoelectron energy of .20 a.u.).

Finally Fig. 15 displays the angular distribution of photoelectrons for a choice of angle β different from zero, after the absorption of two linearly polarized photons. The photon energy is again 9.02 eV. As β increases from 0° to 90°, the angular distribution loses its cylindrical symmetry and shows an increasing tendency to distortion. When $\beta \neq 0$ and in the case of at least one non-linearly polarized photon the symmetry with respect to the xz plane (i.e. the plane containing the $\hat{\mu}_1$ and $\hat{\mu}_2$ versors) is broken, this being related to the existence of a non zero imaginary part of the "molecular" parameter $F_{j,0,2,2}(\varepsilon)$ [1]. This asymmetry was, in the case we investigated, very small and quite probably hardly detectable, its contribution being, on the average, two orders of magnitude smaller than the symmetric one.

Appendix 12A.

Rotationally averaged differential cross sections

The rotational averaging of the photoionization cross section has been extensively discussed elsewhere[24,31,67,68], both for one- and two-photon ionization. Here only a brief summary of the main formulae will be given.

We consider photoionization due to absorption of photons whose individual polarization my be either linear or circular. In the dipole approximation we may assume without loss of generality that the first photon has an energy ω_1 and polarization μ_1, being directed along the z axis. The second photon - which is involved in the two-photon process - has an energy ω_2 and polarization μ_2 lying in the xz plane. The geometrical arrangement is shown in Fig. 1, where the angle between the two polarization directions is indicated by β. It is important to remember that the polarization directions of Fig. 1 are the actual directions of the polarization vector for linearly polarized photons ($\mu_1, \mu_2 = 0$) and of the propagation vector for circularly polarized photons ($\mu_1, \mu_2 = \pm 1$).

In the fixed nuclei approximation, making use of the straightforward algebra of the coupling of the angular momenta, the following formulas are obtained[1] ($g = 1$ in the Length Gauge, *LG*, and $g = -1$ in the Velocity Gauge, *VG*) for the rotationally averaged One-Photon Ionization Differential Cross Section (*OPIDCS*)

$$\frac{\partial\sigma^{(1)}_{j,\mu_1}(\omega_1)}{\partial\Omega_{\hat{k}_e}} = \frac{\omega_1^g \pi}{3c}\sum_J(-1)^J(2J+1)C(1, J, 1; -\mu_1, 0)P_J(\cos\theta_{\hat{k}_e})F_{j,J}(\varepsilon) \tag{A.1}$$

and the Two-Photon Ionization Differential Cross Section (*TPIDCS*)

$$\frac{d\sigma^{(2)}_{j,\mu_1,\mu_2,\beta}(\omega_1,\omega_2)}{d\Omega_{\hat{k}_e}} = \frac{(2\pi)^3(\omega_1\omega_2)^g}{c^2}\sum_J\sum_L\sum_M\sum_{j_1,j_2}A_{\mu_1,\mu_2}(L, J, M; j_1, j_2)F_{j,j_1,j_2,J}(\varepsilon)Y_{L,M}(\beta,0)Y_{J,M}(\hat{k}_e). \tag{A.2}$$

It should be noted that in these last two equations we have used a notation slightly simpler than that employed in eq.s (2.3a) and (2.3b). $\theta_{\hat{k}_e}$ is the angle between the polarization direction and the direction of photoejection of the electron having energy $\varepsilon = \omega_1 - I_{pj}$, I_{pj} being the ionization potential. The "molecular" quantities $F_{j,J}(\varepsilon)$ are given by

$$F_{j,J}(\varepsilon) = \sum_{l_a, l_b} \sum_{m_a, m_b} \sum_{\mu_a, \mu_b} f_J^{(1)}(l_a, m_a, \mu_a, l_b, m_b, \mu_b) \mathbf{M}^{(1)*}_{j,\varepsilon, l_a, m_a, \leftarrow\{0\}}(\mu_a) \mathbf{M}^{(1)}_{j,\varepsilon, l_b, m_b, \leftarrow\{0\}}(\mu_b) \quad (A.3)$$

where the definition (2.4a) of the transition matrix elements has been used and

$$f_J^{(1)}(l_a, m_a, \mu_a, l_b, m_b, \mu_b) = \delta_{m_a + \mu_a, m_b + \mu_b} C(l_a, J, l_b; 0, 0) C(l_a, J, l_b; m_a, m_b - m_a) C(1, J, 1; \mu_b, \mu_a - \mu_b). \quad (A.4)$$

The "molecular" quantities $F_{j,j_1,j_2,J}(\varepsilon)$ are defined as follows

$$F_{j,j_1,j_2,J}(\varepsilon) = \sum_{L_a, L_b} \sum_{\mu_a, \mu_b} \sum_{\nu_a, \nu_b} \sum_{m} f_J^{(2)}(L_a, \mu_a, \nu_a, L_b, \mu_b, \nu_b, m; j_1 j_2)$$
$$\times \mathbf{M}^{(2)*}_{j,\varepsilon, L_a, m-\mu_a-\nu_a, \leftarrow\{0\}}(\omega_1, \mu_a, \omega_2, \nu_a) \mathbf{M}^{(2)}_{j,\varepsilon, L_b, m-\mu_b-\nu_b, \leftarrow\{0\}}(\omega_1, \mu_b, \omega_2, \nu_b) \quad (A.5)$$

where

$$f_J^{(2)}(L_a, \mu_a, \nu_a, L_b, \mu_b, \nu_b, m; j_1, j_2) = (-1)^{\mu_a + \nu_a + \mu_b + \nu_b + m}[(2L_a + 1)(2L_b + 1)]^{\frac{1}{2}}$$
$$\times C(1, 1, j_1; \mu_a, \nu_a) C(1, 1, j_2; \mu_b, \nu_b) C(L_a, L_b, J; m - \mu_a - \nu_a, \mu_b + \nu_b - m) \quad (A.6)$$
$$\times C(L_a, L_b, J; 0, 0) C(j_1, j_2, J; -\mu_a - \nu_a, \mu_b + \nu_b).$$

and the $A_{\mu_1,\mu_2}(L, J, M; j_1 j_2)$ coefficients have the following expression

$$A_{\mu_1,\mu_2}(L, J, M; j_1, j_2) = (-1)^{\mu_1 - \mu_2} C(1, 1, L; -\mu_2, \mu_2)[(2J+1)^3(2L+1)]^{-\frac{1}{2}}$$
$$\sum_{\mu} \{C(1, 1, L; -\mu, M + \mu) C(1, 1, j_1; \mu_1, \mu) C(1, 1, j_2; \mu_1, M + \mu) \quad (A.7)$$
$$C(j_1, j_2, J; -\mu - \mu_1, M + \mu + \mu_1)\}.$$

By integrating eq. (A.1) upon the solid angle $\Omega_{\hat{k}_e}$ the integral photoionization cross section is obtained

$$\sigma_j^{(1)}(\omega_1) = \frac{4\pi^2}{c} \omega_1^g \frac{1}{3} F_{j,0}(\varepsilon) = \frac{4\pi^2}{c} \omega_1^g \frac{1}{3} G_j^{(1)} \quad (A.8)$$

which does not depend upon μ_1 and with

$$G_j^{(1)} = \sum_{l,m,\mu} | < \Phi_{\{0\}} | \hat{O}_\mu | \Phi_{j,\varepsilon,l,m} > |^2. \quad (A.9)$$

The integration upon $\Omega_{\hat{k}_e}$ of eq. (A.2) provides the quantity $B^{(2)}_{j,\mu_1,\mu_2,\beta}(\omega_1, \omega_2)$

$$B^{(2)}_{j,\mu_1,\mu_2,\beta}(\omega_1,\omega_2) = \sum_{l,m}(-1)^{m+\mu_2}\frac{\sqrt{4\pi}}{2l+1}C(1,1,l;\mu_1,m)^2$$
$$\times\, G^{(2)}_{j,l}(\omega_1,\omega_2)\sum_{\lambda}\frac{1}{\sqrt{2\lambda+1}}C(1,1,\lambda;-\mu_2,\mu_2)C(1,1,\lambda;-m,m)Y_{\lambda,0}(\beta,0) \quad (A.10)$$

with

$$G^{(2)}_{j,l}(\omega_1,\omega_2) = \sum_{L,M,m'}\left|\sum_{\mu'}C(1,1,l;\mu',m'-\mu')\mathbf{M}^{(2)}_{j,\varepsilon,L,M,\leftarrow\{0\}}(\omega_1,\mu',\omega_2,m'-\mu')\right|^2 \quad (A.11)$$

required to obtain, from eq. (A.2), the integral cross section

$$\sigma^{(2)}_{j,\mu_1,\mu_2,\beta}(\omega_1,\omega_2) = (2\pi)^3\frac{\omega_1^g\omega_2^g}{c^2}B^{(2)}_{j,\mu_1,\mu_2,\beta}(\omega_1,\omega_2). \quad (A.12)$$

For $\beta = 0$ the previous expression reduces to

$$\sigma^{(2)}_{j,\mu_1,\mu_2,\beta=0}(\omega_1,\omega_2) = (2\pi)^3\frac{\omega_1^g\omega_2^g}{c^2}\sum_{l}\frac{C(1,1,l;\mu_1,\mu_2)^2}{2l+1}G^{(2)}_{j,l}(\omega_1,\omega_2). \quad (A.13)$$

It should be remarked that according to (A.9) and (A.11) the quantities $G_j^{(1)}$ and $G^{(2)}_{j,l}(\omega_1,\omega_2)$ are positive definite and do not depend upon the final *pwc*'s phase shifts $\Delta_{j,L,M}(k)$ see eq. (3.1.13). These properties make the $\sigma_j^{(1)}(\omega)$ and, with a slightly more involved procedure, $\sigma^{(2)}_{j,\mu_1,\mu_2,\beta}(\omega_1,\omega_2)$ amenable to a Stieltjes Imaging approach as discussed in sec. 5.

The rotationally averaged *OPIDCS* for linearly or circularly polarized photons in molecules that do not exhibit optical activity takes thus the well known form[1]

$$\frac{d\sigma^{(1)}_{j,\mu_1}(\omega_1)}{d\Omega_{\hat{k}_e}} = \frac{\sigma_j^{(1)}(\omega_1)}{4\pi}[1+\alpha_{j,2,\mu_1}(\varepsilon)P_2(\cos\theta_{\hat{k}_e})] \quad (A.14)$$

where $\sigma_j^{(1)}(\omega)$ is given by eq. (A.8). Using both eq. (A.1) and eq. (A.14) the asymmetry parameter $\alpha_{j,2,\mu_1}(\varepsilon)$ is expressed as

$$\alpha_{j,2,\mu_1}(\varepsilon) = 5C(1,2,1;-\mu_1,0)\frac{F_{j,2}(\varepsilon)}{F_{j,0}(\varepsilon)} \quad (A.15)$$

and for non-polarized light

$$\alpha_{j,\,2,\,n.p.}(\varepsilon) = \sqrt{5/2}\,\frac{F_{j,\,2}(\varepsilon)}{F_{j,\,0}(\varepsilon)} \tag{A.16}$$

The *TPIDCS* shows equally interesting properties[1]; for $\beta = 0$ it reduces to the well known expression

$$\frac{\partial\sigma^{(2)}_{j,\,\mu_1,\,\mu_2,\,\beta=0}(\varepsilon)}{\partial\Omega_{\hat{k}_e}} = \frac{\sigma^{(2)}_{j,\,\mu_1,\,\mu_2,\,\beta=0}(\varepsilon)}{4\pi}\Big\{1 + \alpha_{j,\,2,\,\mu_1,\,\mu_2,\,\beta=0}(\varepsilon)P_2(\cos\theta_{\hat{k}_e}) + \alpha_{j,\,4,\,\mu_1,\,\mu_2,\,\beta=0}(\varepsilon)P_4(\cos\theta_{\hat{k}_e})\Big\} \tag{A.17}$$

where the two asymmetry parameters $\alpha_{j,\,2,\,\mu_1,\,\mu_2,\,\beta=0}(\varepsilon)$ and $\alpha_{j,\,4,\,\mu_1,\,\mu_2,\,\beta=0}(\varepsilon)$ have been introduced. In detail

$$\sigma^{(2)}_{j,\,\mu_1,\,\mu_2,\,\beta=0}(\varepsilon) = -\frac{(2\pi)^3(\omega_1\omega_2)^g}{c^2}\frac{1}{\sqrt{3}}\Big\{(-1)^{\mu_1}\delta_{\mu_1,\,-\mu_2}F_{j,\,0,0,0}(\varepsilon) + C(1,1,1;\mu_1,\mu_2)^2F_{j,\,1,1,\,0}(\varepsilon) - \sqrt{\frac{3}{5}}\,C(1,1,2;\mu_1,\mu_2)^2F_{j,\,2,2,\,0}(\varepsilon)\Big\} \tag{A.18a}$$

$$\begin{aligned}\alpha_{j,\,2,\,\mu_1,\,\mu_2,\,\beta=0}(\varepsilon) &= \frac{(-1)^{\mu_1-\mu_2}}{5\sigma^{(2)}_{j,\,\mu_1,\,\mu_2,\,\beta=0}(\varepsilon)}\\ &\times\{C(1,1,2;\mu_1,\mu_2)^2C(2,2,2;-\mu_1-\mu_2,\mu_1+\mu_2)F_{j,\,2,\,2,\,2}(\varepsilon)\\ &-\frac{2}{3}(-1)^{\mu_1}\delta_{\mu_1,\,-\mu_2}\,C(1,1,2;\mu_1,\mu_2)\mathscr{R}[F_{j,\,2,\,0,\,2}(\varepsilon)]\\ &+C(1,1,1;\mu_1,\mu_2)^2C(1,1,2;-\mu_1-\mu_2,\mu_1+\mu_2)F_{j,\,1,\,1,\,2}(\varepsilon)+\\ &+C(1,1,1;\mu_1,\mu_2)C(1,1,2;\mu_1,\mu_2)C(1,2,2;-\mu_1-\mu_2,\mu_1+\mu_2)\mathscr{R}[F_{j,\,1,\,2,\,2}(\varepsilon)]\}\end{aligned} \tag{A.18b}$$

$$\begin{aligned}\alpha_{j,\,4,\,\mu_1,\,\mu_2,\,\beta=0}(\varepsilon) &= \frac{(-1)^{\mu_1-\mu_2}}{9\sigma^{(2)}_{j,\,\mu_1,\,\mu_2,\,\beta=0}(\varepsilon)}\\ &\times F_{j,\,2,\,2,\,4}(\varepsilon)C(1,1,2;\mu_1,\mu_2)^2C(2,2,4;-\mu_1-\mu_2,\mu_1+\mu_2).\end{aligned} \tag{A.18c}$$

References

1. I. Cacelli, V. Carravetta, A. Rizzo, and R. Moccia, Computer Phys. Rep. (in press).
2. V. Carravetta and H. Ågren, Phys. Rev. A, **35**, 1022 (1987).
3. H.Ågren and V.Carravetta, J. Chem. Phys., **87**, 370 (1987).
4. V.Carravetta, J. Phys. B : At. Mol. Opt. Phys., **21**, 1777 (1988).
5. R. Moccia and P. Spizzo, in "Numerical Determination of the Electronic Structure of Atoms, Diatomic and Polyatomic Molecules," M. Defranceschi and J. Delhalle, Eds., NATO ASI Series C271 (Kluwer Acad. Publishers Dordrecht 1989).
6. D.J. Rowe, Rev. Mod. Phys., **40**, 153 (1968).
7. D.L. Yeager and P. Jφrgensen, Chem. Phys. Lett., **65**, 77 (1979).
8. R. McWeeny, Int. J. Quantum Chem., **23**, 405 (1983).
9. J. Oddershede, P. Jφrgensen, and D. L. Yeager, Comp. Phys. Rep., **2**, 33 (1984).
10. E. Dalgaard, J. Chem. Phys., **72**, 826 (1980).
11. R.J.W. Henry and L. Lipsky, Phys. Rev., **153**, 51 (1967).
12. H.P. Kelly, Chem. Phys. Lett., **20**, 547 (1973).
13. B.H. Brandsen, "Atomic Collision Theory", (W.A.Benjamin Inc., New York 1970).
14. R.K. Nesbet, "Variational Methods in Electron-Atom Scattering Theory", (Plenum Press, New York 1980).
15. R.R. Lucchese and V. McKoy, Phys. Rev. A, **21**, 112 (1980).
16. R.R. Lucchese, G. Raseev, and V. McKoy, Phys. Rev. A, **25**, 2572 (1982)
17. D. Dill and J.L. Dehmer, J. Chem. Phys., **61**, 692 (1974); P.M. DiHerman, D. Dill, and J.L. Dehmer, *ibid.*, **76**, 5703 (1982).
18. L.A. Collins and B.I. Schneider, Phys. Rev. A, **24**, 2387 (1981); *ibid.* **24**, 1264 (1981).
19. L.A. Collins and B.I. Schneider, Phys. Rev. A, 29, 1695 (1984).
20. P.W. Langhoff, in "Theory and Application of Moment Methods in Many-Fermions Systems," B.J. Dalton, S.M. Grimes, J.P. Vary, and S.A. Williams, Eds., (Plenum, New York, 1980), pp. 191-212 and references therein.
21. M. Braunstein and V. McKoy, J. Chem. Phys., **90**, 1535 (1989).
22. J. Schirmer, M. Braunstein and V. McKoy, Phys. Rev. A, **41**, 283 (1990).
23. M.L. Goldberger and K.M. Watson, "Collision Theory", (John Wiley and Sons, Inc., New York 1964).
24. I. Cacelli, V. Carravetta and R. Moccia, J. Chem. Phys., **85**, 7038 (1986).
25. R. Moccia and P. Spizzo, Phys. Rev. (in press).
26. I. Cacelli, V. Carravetta, and R. Moccia, Mol. Phys., **59**, 385 (1986).
27. I. Cacelli, V. Carravetta, R. Moccia, and A. Rizzo, J. Phys. Chem., **92**, 979 (1987).
28. I. Cacelli, V. Carravetta, R. Moccia, and A. Rizzo, J. Chem. Phys., **89**, 7301 (1983).
29. I. Cacelli, V. Carravetta, and R. Moccia, Chem. Phys., **120**, 51 (1988).

30. I. Cacelli, V. Carravetta, A. Rizzo, and R. Moccia, Chem. Phys. Lett., **155**, 210 (1989).
31. I. Cacelli, V. Carravetta, A. Rizzo, and R. Moccia, J. Chem. Phys., **92**, 2883 (1990).
32. C. Cohen-Tannoudji, J. Dupont-Roc, and G. Grynberg, "Introduction à l′ électrodinamique quantique," (InterEditions/Edition du CNRS, 1987).
33. C. Cohen-Tannoudji, J. Dupont-Roc, and G. Grynberg, "Processus d′interaction entre photons et atomes," (InterEditions/Edition du CNRS, 1988).
34. F.H.M. Faisal, "Theory of Multiphoton Processes," (Plenum Press, New York, 1987).
35. J.R. Taylor, "Scattering Theory," Cap. 13, 20, (John Wiley and Sons, Inc., New York, 1972).
36. U. Fano, Phys. Rev., **124**, 1866 (1961).
37. U. Fano and J.W. Cooper, Rev. Mod. Phys., **40**, 441 (1968).
38. C.J. Joachain, "Quantum Collision Theory," Chapter 16, (North-Holland Publishing Co., New York, 1975).
39. G. Breit and H. A. Bethe, Phys. Rev., **93**, 888 (1954).
40. M. Ya Amusia and N. A. Cherepkov, "Case Studies in Atomic Physics," **5**, 47 (1975).
41. W.J. Hunt and W.A. Goddard III, Chem. Phys. Lett., **3**, 414 (1969).
42. A. Messiah "Mécanique Quantique," Cap.XVII, (Dunod Paris, 1960).
43. M. Göeppert-Mayer, Ann. Phys. (Leipzig), **9**, 273 (1931).
44. See for example D. H. Kobe and Kuo-Ho Yang, Phys. Rev. A, **32**, 952 (1985) and references therein.
45. Y. Aharonov and C.K. Au, Phys. Rev. A, **20**, 1553 (1979); Y. Aharonov and C.K. Au, Phys. Letters, **86A**, 269 (1981); W.P. Healy, Phys. Rev. A, **22**, 2891 (1980); J. Bergou, J. Phys.B: At. Mol. Phys., **16**, L647 (1983); J.R. Ackeralt and P.W. Milonni, J. Opt. Soc. Am., **B1**, 116 (1984).
46. S. Chandrasekhar, Astrophys. J., **102**, 223 (1945).
47. H.A. Bethe and E.E. Salpeter, "Quantum mechanics of one- and two-electron atoms", (Springer, Berlin, 1957).
48. H.A. Bethe and R.W. Jackiv, "Intermediate Quantum Mechanics", (Benjamin, New York, 1968).
49. R.J.S. Crossley, Adv. At. Molec. Phys., **5**, 237 (1969).
50. R.J.S. Crossley, Physica Scripta, **T8**, 117 (1984).
51. G. Fischer, "Vibronic Coupling", (Academic Press, London, 1984).
52. S.H. Lin, Y. Fujimura, H.J. Neusser and E.W. Schlag, "Multiphoton Spectroscopy of Molecules," (Ac. Press, Orlando, Cal., 1984).
53. A. Dalgarno and J.T. Lewis, Proc. Roy. Soc. Ser. A, **233**, 70 (1956).
54. J. Olsen and P. Jϕrgensen, J. Chem. Phys., **82**, 3235 (1985).
55. D. Lynch, M.F. Herman, and D.L. Yeager, Chem. Phys., **64**, 69 (1982).
56. P. Jϕrgensen, P. Swanstrϕm, D.L. Yeager, and J. Olsen, Int. J. Quantum Chem., **23**, 959 (1983) .

57. S.N. Dixit, P. Lambropoulos, and P. Zoller, Phys. Rev. A, **24**, 318 (1981); S.N. Dixit and P. Lambropoulos, *ibid.*, A, **27**, 861 (1983); S.N. Dixit and V. McKoy, J. Chem. Phys., **82**, 3546 (1985).
58. H. Rudolph, D.L. Lynch, S.N. Dixit, and V. McKoy, J. Chem. Phys., **84**, 6657 (1986); H. Rudolph and V. McKoy, J. Chem. Phys., **91**, 2235 (1989); H. Rudolph and V. McKoy, *ibid.*, **91**, 7995 (1989).
59. P.W. Langhoff and C.T. Corcoran, J. Chem. Phys., **61**, 146 (1974).
60. P.W. Langhoff, C.T. Corcoran, J.S. Sims, F. Weinhold, and R.M. Glover, Phys. Rev. A, **14**, 1042 (1976).
61. P.W. Langhoff, in "Electron-Molecule and Photon-Molecule Collisions," T.N Rescigno, B.V. McKoy, and B. Schneider, Eds., (Plenum, New York, 1979), pp. 183-224 and references therein.
62. R.K. Nesbet, Phys. Rev. A, **14**, 1065 (1976).
63. F. Müller-Plathe and G. Dierksen, Phys. Rev. A, **40**, 696 (1989).
64. V. Carravetta and I. Cacelli, IBM Research Report, KGN-196 (1989).
65. G.R.J. Williams and P.W. Langhoff, Chem. Phys. Lett., **60**, 201 (1979).
66. G.R.J. Williams and P.W. Langhoff, Chem. Phys. Lett., **78**, 21 (1981).
67. J.C. Tully, R.S. Berry, and B.J. Dalton, Phys. Rev., **176**, 95 (1968).
68. J.C. Hansen and R.S. Berry, J. Chem. Phys., **80**, 4078 (1984).

Chapter 13.

Dirac-Fock Self-Consistent Field Calculations for Closed-Shell Molecules with Kinetic Balance and Finite Nuclear Size

A. Mohanty and E. Clementi

IBM Corporation
Scientific and Engineering Computations
Dept. 48B/428
Kingston, New York 12401, U.S.A.

1. Introduction

The method of basis set expansion was extended by Kim[1] to the relativistic study of many-electron atoms over two decades ago. Recently there has been an increasing interest[2-20] in the development of theoretical and computational techniques for solving the Dirac-Fock equation for atoms and molecules. One of the fundamental problem associated with the Dirac-Fock calculations for the many-electron system is the so called "variational collapse" or "the finite basis set disease"[21] caused by the unbounded nature of the kinetic energy operator in the Dirac Hamiltonian. However, for the case of atoms, it is now established[2-20] that with the "kinetic balance" condition[2-7] imposed on the basis functions, one can avoid the variational collapse, thus the relativistic calculations for atoms can be routinely performed using standard nonrelativistic techniques[22]. Grant et al.[7-11] have used kinetically balanced Slater basis functions to obtain results for atomic structure that are comparable in accuracy to the numerical DF limit[23]. Recently we have reported [18-20] Dirac-Hartree-Fock atomic structure calculations using geometric-type Gaussian basis functions. As has been shown by us[18-20] (see chapter 3) the kinetic balance constraint, along with finite nucleus size approximation, provides rapid convergence in atomic structure calculations, and the results for neutral atoms (Z=2-86) are in excellent agreement with the numerical DF limit[23,24], even when Gaussian basis functions are used.

Dirac-Fock SCF theory for closed- shell molecular systems was first developed by Malli and Oreg[25] where each molecular spinor is expanded as a linear combination of basis atomic spinors. Matsuoka et al.[26] proposed a Dirac-Fock multi-configuration SCF theory for closed-shell molecules and Ayoma et al.[27] carried out DF SCF calculations for some diatomic and polyatomic molecules using spherical-type Gaussian functions.

The idea of implementing the kinetic balance condition in DF calculations for molecular structure is particularly important if one has to test the variational stability of such calculations. However, to date only a few molecular Dirac-Fock calculations using basis set expansion method, which includes the kinetic balance condition, have been reported. Lee and McLean[28] made the first attempts on heavy atom containing diatomic molecules, such as AuH and AgH, with kinetic balance condition as the criterion for basis set selection. Malli et al.[29,30] have carried out, in a frozen core approximation, DF SCF calculations for diatomics involving atoms of sixth row elements gold(Z=79) to bismuth(Z=83), e.g. AuH, TlH etc. In the formulation of Malli and Pyper,[29] kinetic balance condition is automatically satisfied as their atomic basis spinors are generated by the numerical program of Grant et al.[23] Aerts et al. have reported[31] DF molecular structure calculations for SiH_4, SnH_4, and GeH_4 using an extended kinetic balance condition[13]. Recently Laksonnen et al.[32] have published matrix Dirac-Hartree-Fock calculations for the HCl molecule with kinetically balanced Gaussian basis functions where they have used the small-component basis functions in the uncontracted form. It should be noted that all these calculations have been performed in the point nucleus model.

For analytic SCF molecular structure calculations, except for simple diatomic cases, Gaussian basis functions are universally used mainly because of the ease with which the multi-center integrals can be evaluated in analytic forms. However, for DF SCF calculations, spherical Gaussian basis functions are indeed the natural choices[33-35] for describing the correct boundary conditions near the nucleus, where relativistic contributions are significant. In addition the use of geometric-type Gaussian basis functions in relativistic SCF calculations in a finite nuclear size model provide a special advantage; that is, one can systematically extend the basis to completeness while studying the convergence properties[18-20] of DHF energies. Geometric Gaussian basis functions have been proposed by Clementi et al.[36] in order to make molecular computations for large chemical systems feasible. Moreover, a new algorithm for efficient evaluation of two-electron integrals using uncontracted geometric-type basis functions has been developed recently by Wong et al.[37], with potential applications in molecular DF computations.

In this chapter we demonstrate the use of kinetically balanced geometric Gaussian basis functions in DF molecular structure calculations. The molecular spinors are expanded as a linear combination of atomic basis spinors of Gaussian-type functions, with the small component related to the large component by the kinetic balance condition. It should be stressed that in our approach, the small-component basis spinors are matched with the corresponding the large component basis spinors in contracted form, thus assuring a clean separation[7] of the positive and the negative energy spectrum - a condition essential, if spurious solutions are to be carefully avoided in molecular DF calculations. This type of one-to-one matching is also necessary to enforce variational safety and to obtain the proper nonrelativistic limit[7-11] of the DF solution. Furthermore, because of the geometric nature of the basis spinors, which are kinetically balanced, we show that not only all the multi-center integrals involving the small-components can be related to corresponding integrals involving the large components, but also they are directly obtainable from the standard nonrelativistic cartesian-type integrals for which an efficient computational method exist[37]. We have derived a general transformation between solid spherical harmonic Gaussian-type functions and the cartesian Gaussian functions, which allows us to express all the matrix elements in simple algebraic form for computational purposes. Finally, the effect of finite nucleus size is readily incorporated following the approach of Matsuoka[38].

The organization of the rest of the paper is as follows. In section 2 we review the Dirac-Coulomb Hamiltonian of a molecular system and present the Fock equation for closed-shell molecular structure calculations. In section 3 we describe our choice of molecular basis spinors and show how these can be simply constructed from the atomic basis spinors. In section 4 we give explicit expressions for the Fock-matrix elements in terms of the primitive one- and two-electron integrals involving atomic basis spinors of Gaussian-type functions. In section 5 we evaluate the one- and two-electron matrix elements involving the small component and show how these small-component matrix elements are obtained from previously computed large-component matrix elements. The transformations of the spherical Gaussian functions to the Cartesian form is derived in section 6 and here we point out how, using such a transformation, one can easily recover all the relativistic one- and two-electron matrix elements from a set of nonrelativistic matrix elements obtained using Cartesian-type Gaussian basis functions. Finally in section 7 we give our conclusions. In Appendix A we report Matsuoka's[38] corrections to the nuclear attraction integrals by taking the finite nuclear size into account. In appendix B we describe the rotation matrix elements as applied to evaluate the general nuclear attraction integrals described by Matsuoka[38].

2. Preliminary

For a molecular system the electronic energy E is given by the expectation value of the electronic Hamiltonian $\mathscr{H}_e$:

$$E = \frac{<\Phi|\mathscr{H}_e|\Phi>}{<\Phi|\Phi>}. \tag{2.1}$$

Φ in Eq. (2.1) represents a many-electron wave function which, for a closed-shell molecule of N electrons, is an antisymmetrised product of molecular spinors ψ_i, namely,

$$\Phi = (N!)^{-1/2}\sum_{\Pi}(-1)^{\pi}\Pi[\psi_1(\vec{r}_1)...\psi_N(\vec{r}_N)], \tag{2.2}$$

where Π represents a permutation of the subscript labelling the electrons and π its parity.

In the "relativistic"or Dirac-Fock formalism,[25-27] the electronic Hamiltonian in Born-Oppenheimer approximation (neglecting the constant repulsive energy between the nuclei of the system in a fixed geometry) is taken to be the Dirac-Coulomb Hamiltonian defined (in atomic units (a.u.) $e = m = \hbar = 1$, $c \simeq 137.036$) by

$$\mathscr{H}_{DC} \equiv \mathscr{H}_e = \sum_i \mathscr{H}_D(i) + \sum_{i<j} \frac{1}{r_{ij}}, \tag{2.3}$$

where

$$\mathscr{H}_D(i) = c\vec{\alpha}_i \cdot \vec{p}_i + (\beta_i - 1)c^2 + V_{nuc}(r_i). \tag{2.4a}$$

In Eq. (2.4a) Dirac operators $\vec{\alpha}$ and β are expressed by the matrices[39]

$$\vec{\alpha} = \begin{pmatrix} 0 & \vec{\sigma} \\ \vec{\sigma} & 0 \end{pmatrix} \text{ and } \beta = \begin{pmatrix} I & 0 \\ 0 & -I \end{pmatrix}, \tag{2.4b}$$

where $\vec{\sigma}$ stands for the three Pauli matrices, I is the 2×2 unit matrix, and we have subtracted the rest mass energy to get the binding energy.

The potential due to the nuclei of the system is given by

$$V_{nuc}(r_i) \equiv \sum_C V_C(i), \tag{2.5a}$$

where $V_C(i)$ denotes the nuclear attraction potential associated with the nucleus C and the summation index C ranges over all nuclei. Assuming the nucleus C of the molecule to be a uniform charge distribution of radius R_C, the nuclear attraction potential due to nucleus C can be written[33] as

$$\begin{aligned} V_C(i) &= -\left(\frac{Z_C}{2R_C}\right)\left[3 - \frac{r_{iC}^2}{R_C^2}\right], \quad r_{iC} < R_C, \\ &= -\frac{Z_C}{r_{iC}}, \quad r_{iC} \geq R_C, \end{aligned} \tag{2.5b}$$

where the radius of the nucleus R_C is given by[24] $R_C = 2.2677 \times 10^{-5} A_C^{1/3}$, A_C being the atomic mass of nucleus C.

The molecular spinors $\{\psi_i\}$ are expanded in terms of four-component basis spinors χ_p as

$$\psi_i = \sum_{p=1}^{n} C_{ip}\chi_p, \tag{2.6}$$

where C_{ip} are the expansion coefficients. The spinors $\{\psi_i\}$ obey the orthonormality constraint:

$$<\psi_i|\psi_j> = \delta_{ij}. \tag{2.7}$$

For a closed-shell molecular system the Dirac-Fock self consistent field equations are written in the form:[25-27]

$$FC_i = \varepsilon_i SC_i, \tag{2.8}$$

where ε_i is the orbital energy of the ith molecular spinor, F is the relativistic Fock matrix and S is the overlap matrix, the elements of which are defined in the subsequent sections.

If we define the molecular spinors in terms of the large and the small components as[27]

$$\psi_i = \begin{bmatrix} \psi_i^L \\ i\psi_i^S \end{bmatrix}, \tag{2.9}$$

the expansion given by Eq. (2.6) can be recast in the form

$$\psi_i^L = \sum_{p=1}^{n} C_{ip}^L \chi_p^L \tag{2.10a}$$

and

$$\psi_i^S = \sum_{p=1}^{n} C_{ip}^S \chi_p^S. \tag{2.10b}$$

Note that the choice of phase for the small component as shown in Eq.(2.9) is same as in reference 27.

The small component basis spinors χ_p^S are related to the large component basis spinors χ_p^L according to the kinetic balance condition[2-7];

$$\chi_p^S = (\vec{\sigma} \cdot \vec{p})\ \chi_p^L. \tag{2.11}$$

The specific choice of these basis spinors will be presented in the next section.

The elements of the Fock matrix are given by[25,26]

$$F_{pq} = \begin{bmatrix} V_{pq}^{LL} + J_{pq}^{LL} - K_{pq}^{LL} & c\Pi_{pq}^{LS} - K_{pq}^{LS} \\ c\Pi_{pq}^{SL} - K_{pq}^{SL} & V_{pq}^{SS} + J_{pq}^{SS} - K_{pq}^{SS} - 2c^2 S_{pq}^{SS} \end{bmatrix} \tag{2.12a}$$

where

$$S_{pq} = \begin{bmatrix} S_{pq}^{LL} & 0 \\ 0 & S_{pq}^{SS} \end{bmatrix}. \tag{2.12b}$$

The matrix elements shown in Eq. (2.12) can be expressed in terms of the basis spinors $\{\chi_p\}$ as follows:

(1) The overlap matrices:

$$S_{ApBq}^{TT} = < \chi_{Ap}^{T} | \chi_{Bq}^{T} >, \tag{2.13a}$$

$$S_{ApBq}^{\bar{T}\bar{T}} = < \chi_{Ap}^{\bar{T}} | \chi_{Bq}^{\bar{T}} >. \tag{2.13b}$$

(2) The nuclear attraction matrices:

$$V_{ApBq}^{TT} = \sum_{C} < \chi_{Ap}^{T} | V_C(i) | \chi_{Bq}^{T} >, \tag{2.14a}$$

$$V_{ApBq}^{\bar{T}\bar{T}} = \sum_{C} < \chi_{Ap}^{\bar{T}} | V_C(i) | \chi_{Bq}^{\bar{T}} >. \tag{2.14b}$$

(3) The kinetic energy matrices:

$$\Pi_{ApBq}^{T\bar{T}} = < \chi_{Ap}^{T} | \vec{\sigma} \cdot \vec{p} | \chi_{Bq}^{\bar{T}} >, \tag{2.14c}$$

$$\Pi_{ApBq}^{\bar{T}T} = < \chi_{Ap}^{\bar{T}} | \vec{\sigma} \cdot \vec{p} | \chi_{Bq}^{T} >. \tag{2.14d}$$

(4) The two-electron Coulomb matrices:

$$J_{pq}^{TT} = \sum_{rs} [D_{rs}^{\bar{T}\bar{T}} J_{pqrs}^{TT\bar{T}\bar{T}} + D_{rs}^{TT} J_{pqrs}^{TTTT}] \tag{2.15a}$$

and

$$K_{pq}^{TT'} = \sum_{rs} [D_{rs}^{TT'} K_{pqrs}^{TT'TT'}], \tag{2.15b}$$

where the density matrices are defined by[26]

$$D_{pq}^{TT'} = \sum_{i} C_{ip}^{T*} C_{iq}^{T'}. \tag{2.16}$$

The general expressions for the direct Coulomb and exchange supermatrix elements are given by

$$J_{pqrs}^{TTT'T'} = \iint \chi_p^{T\dagger}(\vec{r}_1)\chi_q^{T}(\vec{r}_1)\frac{1}{r_{12}}\chi_r^{T'\dagger}(\vec{r}_2)\chi_s^{T'}(\vec{r}_2)\, d\vec{r}_1\, d\vec{r}_2 \tag{2.17a}$$

and

$$K_{pqrs}^{TT'TT'} = \iint \chi_p^{T\dagger}(\vec{r}_1)\chi_r^{T'}(\vec{r}_1)\frac{1}{r_{12}}\chi_q^{T\dagger}(\vec{r}_2)\chi_s^{T'}(\vec{r}_2)\, d\vec{r}_1\, d\vec{r}_2. \tag{2.17b}$$

In describing the Fock matrix elements above, as in chapter 3, we have adopted the notation of Grant et al.[7-10] where the superscripts $T\overline{T}$ represents a pair of LS or SL and the superscripts T and T' are either L or S; i.e., TT' may be LL, LS, SL, SS and TT may be either LL or SS.

The total energy of the system is represented by

$$E = \sum_{TT'}\sum_{pq} h_{pq}^{TT'} D_{pq}^{TT'} + \frac{1}{2}\sum_{TT'}\sum_{pq}\sum_{rs}\left[D_{pq}^{TT} J_{pqrs}^{TT\overline{T}\overline{T}} D_{rs}^{\overline{T}\overline{T}} - D_{pq}^{TT'} K_{pqrs}^{TT'\,TT'} D_{rs}^{TT'}\right], \tag{2.18}$$

where

$$h_{pq}^{LL} = V_{pq}^{LL}, \tag{2.19}$$

$$h_{pq}^{SL} = c\Pi_{pq}^{SL} = h_{qp}^{LS*} \tag{2.20}$$

and

$$h_{pq}^{SS} = V_{pq}^{SS} - 2c^2 S_{pq}^{SS}. \tag{2.20}$$

The nonrelativistic limit of the SCF equation Eq. (2.8) can be obtained by setting the speed of light, $c = \infty$, and by using the kinetic balance condition[2-7] Eq. (2.11). Following Matsuoka et al.,[27] the SCF equations for both large and the small components can be written from Eqs. (2.8) and (2.12) in a matrix notation as

$$(V^{LL} + J^{LL} - K^{LL})C_i^L - (c\Pi^{SL} + K^{SL})C_i^S = \varepsilon_i S^{LL} C_i^L \tag{2.21a}$$

and

$$(\Pi^{LS} - \frac{1}{c}K^{LS})C_i^L - \left[\frac{1}{c}(V^{SS} + J^{SS} - K^{SS}) - 2cS^{SS}\right]C_i^S = \frac{1}{c}\varepsilon_i S^{SS} C_i^S. \tag{2.21b}$$

In the limit $c = \infty$, Eq. (2.21b) can be approximated as

$$C_i^S \simeq \frac{1}{2c}(S^{SS})^{-1}\Pi^{SL}C^L. \tag{2.22}$$

In deriving Eq. (2.22) the positive definiteness of the overlap matrix has been used. Substituting Eq. (2.22) in (2.21a) we have

$$\left[V^{LL} + J^{LL} - K^{LL} - \Pi^{LS}(S^{SS})^{-1}\Pi^{SL}/2\right]C_i^L = \varepsilon_i S^{LL} C_i^L . \tag{2.23}$$

This is same as Kim's[1] reduction of relativistic SCF equation for atoms to the nonrelativistic form. However, using the identities:

$$(\sigma \cdot \nabla)^2 = \nabla^2 I \tag{2.24}$$

and

$$\sum_{pq} | \chi_p^S > (S^{SS})_{pq}^{-1} < \chi_q^S | = 1, \tag{2.25}$$

Eq. (2.23) is reduced to the familiar nonrelativistic HF equation. Hence by having a complete set of basis spinors we will have the correct nonrelativistic limit. As has been discussed by Grant[7] Eqs. (2.25) and (2.24) are precisely what one obtains by imposing the kinetic balance condition, Eq. (2.11), on the basis sets, which satisfy the standard boundary conditions.[7]

3. Choice of Basis Spinors

The basis spinors are constructed from atomic spinors in the central field approximation as proposed by Grant et al.[22] viz,

$$\Psi_{n\kappa m}(\vec{r}) = \frac{1}{r}\begin{pmatrix} P_{n\kappa}(r)\chi_{\kappa m}(\theta, \phi) \\ iQ_{n\kappa}(r)\chi_{-\kappa m}(\theta, \phi) \end{pmatrix}. \tag{3.1}$$

The quantum number κ, which defines the symmetry of the relativistic spinors, is given by

$$\kappa = -2(j - l)(j + 1/2), \tag{3.2}$$

where l is the orbital quantum number and j = l + 1/2 or l - 1/2 is the total angular momentum quantum number. In Eq. (3.1) the two-component spinors, $\chi_{\kappa m}$ and $\chi_{-\kappa m}$ are used to characterize the angular states. For $\kappa < 0$

$$\chi_{\kappa m} = \begin{bmatrix} C_1^-(jm) & Y_{j-1/2}^{m_j-1/2} \\ C_2^-(jm) & Y_{j-1/2}^{m_j+1/2} \end{bmatrix} \tag{3.3a}$$

and for $\kappa > 0$

$$\chi_{\kappa m} = \begin{bmatrix} C_1^+(jm) & Y_{j+1/2}^{m_j-1/2} \\ C_2^+(jm) & Y_{j+1/2}^{m_j+1/2} \end{bmatrix} \tag{3.3b}$$

where the Clebsch-Gordan (C.G.) coefficients are defined by

$$C_1^-(jm_j) \equiv \sqrt{\frac{j+m_j}{2j}}, \quad C_2^-(jm_j) \equiv \sqrt{\frac{j-m_j}{2j}}, \tag{3.4a}$$

and

$$C_1^+(jm_j) \equiv \sqrt{\frac{j-m_j+1}{2j+2}}, \; C_2^+(jm_j) \equiv -\sqrt{\frac{j+m_j+1}{2j+2}}. \tag{3.4b}$$

Note that the phase choices for the C.G. coefficients defined in Eq. (3.4) are same as in reference 39. For the case of atoms, the radial functions are expanded in terms of the basis functions as

$$P_{n\kappa}(r) = \sum_i d^L_{\kappa i} g^L_{\kappa i}(r), \tag{3.5a}$$

and

$$Q_{n\kappa}(r) = \sum_i d^S_{\kappa i} g^S_{\kappa i}(r), \tag{3.5b}$$

where the summation index i runs over the number of basis functions N. $g^L_{\kappa i}(r)$ and $g^S_{\kappa i}(r)$ are the Gaussian basis functions belonging to the large and the small component respectively, and $d^L_{\kappa i}$ and $d^S_{\kappa i}$ are the corresponding expansion coefficients.

For the large component, the Gaussian-type basis functions are of the form:[18,19]

$$g^L_{\kappa p}(r) = N^L_{\alpha_p} r^{n_\kappa} e^{-\alpha_i r^2} \tag{3.6}$$

where $n_\kappa = (l+1)$ and $N^L_{\alpha_p}$ is the normalisation factor for the large component.

For the small component, because of spherical symmetry, application of Eq. (2.11) to Eq. (3.6) leads to the expression

$$g^S_{\kappa p}(r) = N^S_{\alpha_p} \left(\frac{d}{dr} + \frac{\kappa}{r}\right) g^L_{\kappa p}(r), \tag{3.7}$$

where $N^S_{\alpha_p}$ is the corresponding normalisation factor for the small component.

However, for molecules , since the system is no longer spherically symmetric, we expand the molecular spinors in terms of the basis spinors having explicit angular dependence defined with respect to coordinate systems centered on each atom. The large component basis spinors are obtained by multiplying the angular part $\chi_{\kappa m}$ (Eq. (3.3a) or (3.3b)) with Eq. (3.6). For the small component we use the kinetic balance condition (2.11), the right-hand side of which can be put in the form

$$(\vec{\sigma} \cdot \vec{p})\chi^L_p = (\vec{\sigma} \cdot \vec{p})(\vec{\sigma} \cdot \hat{r})(\vec{\sigma} \cdot \hat{r})\chi^L_p, \tag{3.8}$$

where we have used the identity $(\vec{\sigma}\cdot\hat{r})(\vec{\sigma}\cdot\hat{r})=1$. Combining Eqs. (3.1), (3.5) and (3.8) with the identity:[39]

$$(\vec{\sigma}\cdot\hat{r})\ \chi_{\kappa m}=\chi_{-\kappa m},\tag{3.9}$$

we can express the normalized large- and small-component basis spinors centered at A with angular momentum quantum numbers $j_A m_{jA}$ as follows: (a) $\kappa<0,\ l_A=j_A-1/2,$

$$\chi^{L^-}_{Ap}=N^{L^-}_{\alpha_p}\ r_A^{j_A-1/2}\begin{bmatrix}C_1^-(j_A m_{jA})\ Y^{m_{jA}-1/2}_{j_A-1/2}\\ C_2^-(j_A m_{jA})\ Y^{m_{jA}+1/2}_{j_A-1/2}\end{bmatrix}e^{-\alpha_p r_A^2},\tag{3.10a}$$

$$\chi^{S^-}_{Ap}=N^{S^-}_{\alpha_p}\ N^{L^+}_{\alpha_p}\ r_A^{j_A+1/2}\begin{bmatrix}C_1^+(j_A m_{jA})\ Y^{m_{jA}-1/2}_{j_A+1/2}\\ C_2^+(j_A m_{jA})\ Y^{m_{jA}+1/2}_{j_A+1/2}\end{bmatrix}e^{-\alpha_p r_A^2},\tag{3.10b}$$

and (b) $\kappa>0,\ l_A=j_A+1/2$,

$$\chi^{L^+}_{Ap}=N^{L^+}_{\alpha_p}\ r_A^{j_A+1/2}\begin{bmatrix}C_1^+(j_A m_{jA})\ Y^{m_j-1/2}_{j_A+1/2}\\ C_2^+(j_A m_{jA})\ Y^{m_{jA}+1/2}_{j_A+1/2}\end{bmatrix}e^{-\alpha_p r_A^2},\tag{3.11a}$$

$$\chi^{S^+}_{Ap}=N^{S^+}_{\alpha_p}N^{L^-}_{\alpha_p}\ [(j_A+3/2+\kappa_A)-2\alpha_p r_A^2)]\ r_A^{j_A-1/2}\begin{bmatrix}C_1^-(j_A m_{jA})\ Y^{m_{jA}-1/2}_{j_A-1/2}\\ C_2^-(j_A m_{jA})\ Y^{m_{jA}+1/2}_{j_A-1/2}\end{bmatrix}e^{-\alpha_p r_A^2},\tag{3.11b}$$

where the Clebsch-Gordan coefficients, C_i^{mp}, for $i=1,2$ have been defined in Eq. (3.2) and the radial coordinates are defined with respect to each center, viz, $\vec{r}_A=\vec{r}-\vec{A}$. The normalization constants for the large and the small components are given by[19]

$$N^{L^\mp}_{\alpha_p}=\left[\frac{\sqrt{\pi}}{2^{2(l_A+1)+3/2}}\ \frac{(2l_A+1)!!}{\alpha_p^{(2l_A+3)/2}}\right]^{-1/2},\tag{3.12a}$$

$$N^{S^+}_{\alpha_p}=(N^{L^+}_{\alpha_p}/N^{L^-}_{\alpha_p})\left[\frac{\alpha_p}{(2l_A+1)}\left\{(4\kappa_A^2-1)+4(l_A+1+\kappa_A)\right\}\right]^{-1/2},\tag{3.12b}$$

and

$$N^{S^-}_{\alpha_p}=-1,\tag{3.12c}$$

where $l_A=j_A\mp 1/2$ for $\kappa<0$ and $\kappa>0$. We should note that the choice of the phase of the normalization constant given in Eq. (3.12c) is a result of identity defined by Eq. (3.9).

It is clear from Eqs. ((3.10b) and (3.11b)) that the same set of exponents have been used for the large as well as the small component. Furthermore, the form of the small component as shown in Eqs. ((3.10b) and (3.11b)) should indicate that these are in the contracted form; i.e. in particular, the terms on the right hand side of Eq.(3.11b) represent one basis function instead of two. As has been pointed by Grant[7] this type of one-to-one matching between the large and the small component is essential for a clean separation of positive and negative energy solutions, thus avoiding spurious solutions in the DF spectrum. Such a separation of the positive and negative energy spectrum is also crucial[7] for correct nonrelativistic limit of the DF solution as well as for avoiding the variational collapse. Our recent results[18-20] for DF solution of neutral atoms(Z=2-86) using such matched basis sets have corroborated such behavior of DF solutions numerically.

The exponents α_p's are generated according to

$$\alpha_p = \alpha_0 \, \beta^{p-1}, \; p = 1,2, \dots N, \tag{3.13}$$

where α_0 and β are parameters determined by the basis set size N and the particular many-electron system being considered.

The geometric basis sets defined by Eq. (3.13) were introduced by Clementi et al.[36] for nonrelativistic atomic and molecular computations. The employment of geometric-type Gaussian basis functions in DF atomic structure calculations gives excellent results as has been shown by us.[18,19] Also one has the added advantage of of being able to systematically extend the basis to near completeness while studying the convergence properties[18-20] of DF calculations, particularly in the finite-size nucleus model.

A few comments are in order with regard to the use of Gaussian basis functions in DF calculations for many-electron systems, particularly in a finite-size nucleus model. As noted earlier, Gaussian functions are indeed the natural solutions[33] for describing relativistic wavefunctions for atoms near the nucleus. The use of Gaussian basis sets have an added advantage, particularly in the case of molecules, since kinetically balanced calculations on various heavy atoms have shown greatly accelerated convergence[18-20], particularly in a finite-size nucleus model. Indeed the relaxation of the cusp conditions near the nucleus leads to a faster convergence with fewer number of basis functions, as has been demonstrated by us for many-electron atoms[18-20]. The result is not at all surprising since it has been proven in general by Rose[34] that for a homogenous charged sphere model of the nucleus, the exact solution of the large and small components can be expressed as an infinite series of quadratures. Therefore, in the limit of a complete basis set of Gaussian type functions that can essentially represent this series exactly, the DF calculations can be made to yield results of any desired degree of accuracy; the only approximation is in terminating the series with a finite number of terms. Furthermore, it has been shown by Ishikawa et al.[12] that the behaviour of the $P_{1/2}$

an $S_{1/2}$ solution near the nucleus for the finite-size nucleus is precisely what one obtains by a kinetically balanced Gaussian basis function expansion.; consequently, one needs optimal exponents much smaller in number for a correct description of $P_{1/2}$ and $S_{1/2}$ solutions. In the same token, by extending the finite-size nucleus calculations to many-electron atoms, we have shown that a similar convergence pattern is observed when geometric-type Gaussian basis functions are used. Obviously, this property of accelerated convergence makes geometric-type Gaussian basis functions all the more suitable for molecular DF computations where the economy of basis sets is of paramount importance.

4. Evaluation of One- Electron and Two-Electron Matrix Elements

Expressions for one- and two-electron matrix elements are obtained by substituting the basis functions given in Eqs. (3.10a) -(3.11b) in the expressions for Fock matrix element (Eqs. (2.13a)-(2.14d) and (2.17a)-(2.17b). In order to make the notation tractable we introduce indices i,j,k, and l which take values $\mp$ depending on the signs of the κ values, and label the basis spinors according to their kappa values; i.e., χ_p^{L+} and χ_p^{L-} are the basis functions for the large components with $\kappa > 0$ and $\kappa < 0$, with χ_p^{S+} , and χ_p^{S-} as the corresponding small components. Using this notation the Fock matrix elements are expressed as follows:

(1) The overlap integrals (large component):

$$S_{ApBq}^{L^iL^j} = <\chi_{Ap}^{Li}|\chi_{Bq}^{Lj}> = N_{\alpha_p}^{L^i}N_{\alpha_q}^{L^j}$$
$$\Big[C_1^i(j_Am_{jA})C_1^j(j_Bm_{jB}) < Al_Am_{lA}\alpha_p|Bl_Bm_{lB}\alpha_q> \tag{4.1}$$
$$+ C_2^i(j_Am_{jA})C_2^j(j_Bm_{jB}) < Al_Am_{lA}+1\alpha_p|Bl_Bm_{lB}+1\alpha_q>\Big],$$

where

$$l_A \equiv j_A \mp 1/2,\ l_B \equiv j_B \mp 1/2, \tag{4.2a}$$

and

$$m_{lA} \equiv m_{jA} - 1/2,\ m_{lB} \equiv m_{jB} - 1/2. \tag{4.2b}$$

In Eq. (4.1) the primitive one-electron integrals are defined by

$$< Al_Am_{lA}\alpha_p|Bl_Bm_{lB}\alpha_q> =$$
$$\int r_{1A}^{l_A}\, r_{1B}^{l_B}\, e^{-\alpha_p r_{1A}^2}\, e^{-\alpha_p r_{1B}^2}\, Y_{l_A}^{m_{lA}^*}(\theta_{1A}\phi_{1A})\ Y_{l_B}^{m_{lB}}(\theta_{1B}\phi_{1B})\ d\vec{r}_1\,, \tag{4.3}$$

where $\vec{r}_{1A} = \vec{r}_1 - \vec{A}$ and $\vec{r}_{1B} = \vec{r}_1 - \vec{B}$.

In order to facilitate the notation for the small component matrix elements we define the following operators:

$$\Lambda_{Ap}^{+} = N_{\alpha_p}^{S^+} \; [(j_A + 3/2 + \kappa_A) - 2\alpha_p(\frac{-\partial}{\partial\alpha_p})], \tag{4.4a}$$

$$\Lambda_{Bq}^{+} = N_{\alpha_q}^{S^+} \; [(j_B + 3/2 + \kappa_B) - 2\alpha_q(\frac{-\partial}{\partial\alpha_q})], \tag{4.4b}$$

$$\Lambda_{Cr}^{+} = N_{\alpha_r}^{S^+} \; [(j_C + 3/2 + \kappa_C) - 2\alpha_r(\frac{-\partial}{\partial\alpha_r})], \tag{4.4c}$$

$$\Lambda_{Ds}^{+} = N_{\alpha_s}^{S^+} \; [(j_D + 3/2 + \kappa_D) - 2\alpha_s(\frac{-\partial}{\partial\alpha_s})] \tag{4.4d}$$

and

$$\Lambda_{Ap}^{-} = N_{\alpha_p}^{S^-}, \;\; \Lambda_{Bq}^{-} = N_{\alpha_q}^{S^-}, \;\; \Lambda_{Cr}^{-} = N_{\alpha_r}^{S^-}, \;\; \Lambda_{Ds}^{-} = N_{\alpha_s}^{S^-}. \tag{4.4e}$$

(2) The overlap Integrals (small-component):

$$S_{ApBq}^{S^iS^j} = < \chi_{Ap}^{S^i} \mid \chi_{Bq}^{S^j} > = \Lambda_{Ap}^{i} \;\; \Lambda_{Bq}^{j} \; [S_{ApBq}^{L^{\bar{i}}L^{\bar{j}}}]. \tag{4.5a}$$

(3) The nuclear attraction integrals (large component):

$$V_{ApBq}^{L^iL^j} = \sum_C < \chi_{Ap}^{L^i} \mid V_C(i) \mid \chi_{Bq}^{L^j} > \; = N_{\alpha p}^{L^i} N_{\alpha q}^{L^j}$$

$$\sum_C \Big[C_1^i(j_A m_{jA}) C_1^j(j_B m_{jB}) < A l_A m_{lA} \alpha_p \mid V_C \;\; (i) \mid B l_B m_{lB} \alpha_q > \tag{4.6}$$

$$+ \; C_2^i(j_A m_{jA}) C_2^j(j_B m_{jB}) < A l_A m_{lA} + 1 \alpha_p \mid V_C \;\; (i) \mid B l_B m_{lB} + 1 \alpha_q > \Big].$$

(4) The nuclear attraction integrals (small component):

$$V_{ApBq}^{S^iS^j} = \sum_C < \chi_{Ap}^{S^i} \mid V_C(i) \mid \chi_{Bq}^{S^j} > \; = \Lambda_{Ap}^{i} \;\; \Lambda_{Bq}^{j} \; [V_{ApBq}^{L^iL^j}]. \tag{4.7}$$

(5) The kinetic energy integrals The kinetic energy integrals are related to the overlap integrals involving the components as follows, viz,

$$\Pi_{ApBq}^{S^iL^j} = S_{ApBq}^{S^iS^j} / N_{\alpha_q}^{S^j}, \;\;\; \Pi_{ApBq}^{L^iS^j} = S_{ApBq}^{S^iS^j} / N_{\alpha_p}^{S^i}. \tag{4.8}$$

We should note that in all these expressions the small component basis spinors are utilized in the unnormalized form when differentiation with respect to the exponents is taken. This is evident from the form of the operators given in Eqs. (4.4a)-(4.4d). The expressions for all the one-electron matrix elements involving small-components are given in section 5.

(b)Two-electron direct Coulomb and Exchange Matrix Elements

The most basic of all the two-electron integrals is the four-center direct Coulomb integral involving the large components. This integral is defined as

$$J^{L^iL^jL^kL^l}_{ApBqCrDs} = \iint \chi^{L^{i\dagger}}_{Ap}(\vec{r}_{1A})\, \chi^{L^j}_{Bq}(\vec{r}_{1B})\, \frac{1}{r_{12}}\, \chi^{L^{k\dagger}}_{Cr}(\vec{r}_{2C})\, \chi^{L^l}_{Ds}(\vec{r}_{2D})\, d\vec{r}_1\, d\vec{r}_2 \tag{4.9}$$

where $\vec{r}_{1A} = \vec{r}_1 - \vec{A}$, $\vec{r}_{1B} = \vec{r}_1 - \vec{B}$, $\vec{r}_{2C} = \vec{r}_2 - \vec{C}$, and $\vec{r}_{2D} = \vec{r}_2 - \vec{D}$.

Here, as before, $\chi^{L^i}_{Ap}$, $\chi^{L^j}_{Bq}$, $\chi^{L^k}_{Cr}$, and $\chi^{L^l}_{Ds}$ denote the large component spinors centered at A,B,C and D respectively , and the indices i,j,k, and l take values $\mp$ depending on the signs of the κ values. There are 16 types of direct Coulomb matrix elements each with 4 terms. The most general form of the direct term involving large-components combining Eqs. (4.9), (3.10a) and (3.12a)) can be written as

$$\begin{aligned} J^{L^iL^jL^kL^l}_{ApBqCrDs} = {} & N^{L^i}_{\alpha p} N^{L^j}_{\alpha q} N^{L^k}_{\alpha r} N^{L^l}_{\alpha s} \\ & \times \Big[C^i_1(j_A m_{jA}) C^j_1(j_B m_{jB}) C^k_1(j_C m_{jC}) C^l_1(j_D m_{jD}) \\ & < A l_A m_{lA} \alpha_p B l_B m_{lB} \alpha_q \,||\, C l_C m_{lC} \alpha_r D l_D m_{lD} \alpha_s > \\ & + C^i_1(j_A m_{jA}) C^j_1(j_B m_{jB}) C^k_2(j_C m_{jC}) C^l_2(j_D m_{jD}) \\ & < A l_A m_{lA} \alpha_p B l_B m_{lB} \alpha_q \,||\, C l_C m_{lC} + 1 \alpha_r D l_D m_{lD} + 1 \alpha_s > \\ & + C^i_2(j_A m_{jA}) C^j_2(j_B m_{jB}) C^k_1(j_C m_{jC}) C^l_1(j_D m_{jD}) \\ & < A l_A m_{lA} + 1 \alpha_p B l_B m_{lB} + 1 \alpha_q \,||\, C l_C m_{lC} \alpha_r D l_D m_{lD} \alpha_s > \\ & + C^i_2(j_A m_{jA}) C^j_2(j_B m_{jB}) C^k_2(j_C m_{jC}) C^l_2(j_D m_{jD}) \\ & < A l_A m_{lA} + 1 \alpha_p B l_B m_{lB} + 1 \alpha_q \,||\, C l_C m_{lC} + 1 \alpha_r D l_D m_{lD} + 1 \alpha_s > \Big], \end{aligned} \tag{4.10}$$

where the primitive two-electron integral is defined by

$$\begin{aligned} & < A l_A m_{lA} \alpha_p B l_B m_{lB} \alpha_q \,||\, C l_C m_{lC} \alpha_r D l_D m_{lD} \alpha_s > = \\ & \iint r^{l_A}_{1A}\, r^{l_B}_{1B}\, r^{l_C}_{2C}\, r^{l_D}_{2D}\, e^{-\alpha_p r^2_{1A}}\, e^{-\alpha_q r^2_{1B}}\, e^{-\alpha_r r^2_{2C}}\, e^{-\alpha_s r^2_{2D}} \\ & \overset{*}{Y}{}^{m_{lA}}_{l_A}(\theta_{1A},\phi_{1A})\, Y^{m_{lB}}_{l_B}(\theta_{1B},\phi_{1B})\, \overset{*}{Y}{}^{m_{lC}}_{l_C}(\theta_{2C},\phi_{2C})\, Y^{m_{lD}}_{l_D}(\theta_{2D},\phi_{2D})\, d\vec{r}_1\, d\vec{r}_2 \end{aligned} \tag{4.11}$$

and

$$l_A \equiv j_A \mp 1/2,\ l_B \equiv j_B \mp 1/2,\ l_C \equiv j_C \mp 1/2,\ l_D \equiv j_D \mp 1/2 \tag{4.12}$$

define the values of the total angular momentum of the spinors for positive and negative values of κ at the corresponding centers. By direct verification one can express all other matrix elements corresponding to the two-electron integrals in terms of the basic integrals of LLLL type given in Eq. (4.10). To do so we define indices $\bar{i}, \bar{j}, \bar{k}, \bar{l}$, which take values $\pm$ when the corresponding indices i,j,k,l take the conjugate values $\mp$(for relevant signs of κ). We can then write all two-electron matrix elements in a unified form as follows:

$$J^{L^iL^jS^kS^l}_{ApBqCrDs} = \Lambda^k_{Cr}\, \Lambda^l_{Ds}\, J^{L^iL^jL^{\bar{k}}L^{\bar{l}}}_{ApBqCrDs}, \tag{4.13a}$$

$$J^{S^iS^jL^kL^l}_{ApBqCrDs} = \Lambda^i_{Ap}\ \Lambda^j_{Bq} \left[J^{S^iS^jL^kL^l}_{ApBqCrDs} \right], \tag{4.13b}$$

and

$$J^{S^iS^jS^kS^l}_{ApBqCrDs} = \Lambda^i_{Ap}\ \Lambda^j_{Bq}\ \Lambda^k_{Cr}\ \Lambda^l_{Ds} \left[J^{L^iL^jL^kL^l}_{ApBqCrDs} \right]. \tag{4.14}$$

The exchange integrals are easily obtained from the direct integrals by exchange of indices which in the most general form can be written as

$$K^{T^iT^jT^kT^l}_{ApBqCrDs} = J^{T^iT^kT^jT^l}_{ApCrBqDs}, \tag{4.15}$$

where T and T' can assume index L or S . Explcit forms for integrals $J^{L^iL^jS^kS^l}_{ApBqCrDs}$ and $J^{S^iS^jS^kS^l}_{ApBqCrDs}$ in terms of the primitive integrals involving spherical Gaussians will be discussed in section 5.

5. Evaluation of Small-Component Matrix Elements

In this section we give explicit expressions for the Fock-matrix elements involving the small-component basis spinors in terms of the primitive integrals given in section 4.

(1) One-Electron Integrals

For evaluation of the one-electron integrals we define the following general matrix elements:

$$\begin{aligned} &< \delta_A A l_A m_{lA} \alpha_p | \delta_B B l_B m_{lB} \alpha_q > \equiv \\ &\int r_{1A}^{l_A+2\delta_A}\ r_{1B}^{l_B+2\delta_B}\ e^{-\alpha_p r_{1A}^2}\ e^{-\alpha_q r_{1B}^2}\ Y_{l_A}^{m_{lA}*}(\theta_{1A},\phi_{1A})\ Y_{l_B}^{m_{lB}}(\theta_{1B},\phi_{1B})\ d\vec{r}_1 \\ &= \left[(\frac{-\partial}{\partial\alpha_p})^{\delta_A}\ (\frac{-\partial}{\partial\alpha_q})^{\delta_B} \right] < A l_A m_{lA} \alpha_p | B l_B m_{lB} \alpha_q > . \end{aligned} \tag{5.1}$$

From Eq. (5.1) one-electron Fock-matrix elements of operator $\hat{h}$ can be written as

$$\begin{aligned} &< \delta_A A l_A m_{lA} \alpha_p | \hat{h}(\vec{r}_1) | \delta_B B l_B m_{lB} \alpha_q > = \\ &\int r_{1A}^{l_A+2\delta_A}\ r_{1B}^{l_B+2\delta_B}\ \hat{h}(\vec{r}_1) e^{-\alpha_p r_{1A}^2}\ e^{-\alpha_q r_{1B}^2}\ Y_{l_A}^{m_{lA}*}(\theta_{1A},\phi_{1A})\ Y_{l_B}^{m_{lB}}(\theta_{1B},\phi_{1B})\ d\vec{r}_1 \\ &= \left[(\frac{-\partial}{\partial\alpha_p})^{\delta_A}\ (\frac{-\partial}{\partial\alpha_q})^{\delta_B} \right] < A l_A m_{lA} \alpha_p | \hat{h} | B l_B m_{lB} \alpha_q > . \end{aligned} \tag{5.2}$$

Combining Eq. (5.1) with Eq. (4.1) we can define a general matrix element related to the overlap integral as

$$\left[(\frac{-\partial}{\partial\alpha_p})^{\delta_A}(\frac{-\partial}{\partial\alpha_q})^{\delta_B}\right]S^{L^iL^j}_{ApBq}=$$
$$= C^i_1(j_Am_{jA})C^j_1(j_Bm_{jB}) < \delta_A A l_A m_{lA}\alpha_p | \delta_B B l_B m_{lB}\alpha_q > \qquad (5.3a)$$
$$+ C^i_2(j_Am_{jA})C^j_2(j_Bm_{jB}) < \delta_A A l_A m_{lA}+1\alpha_p | \delta_B B l_B m_{lB}+1\alpha_q >$$
$$\equiv S^{L^iL^j}_{\delta Ap\delta Bq}\,.$$

Similarly for the nuclear-attraction integrals given in Eq. (4.6a) we obtain

$$\left[(\frac{-\partial}{\partial\alpha_p})^{\delta_A}(\frac{-\partial}{\partial\alpha_q})^{\delta_B}\right]V^{L^iL^j}_{ApBq}$$
$$= \sum_C \Big[C^i_1(j_Am_{jA})C^j_1(j_Bm_{jB}) < \delta_A A l_A m_{lA}\alpha_p | V_C\ (i) | \delta_B B l_B m_{lB}\alpha_q > \qquad (5.3b)$$
$$+ C^i_2(j_Am_{jA})C^j_2(j_Bm_{jB}) < \delta_A A l_A m_{lA}+1\alpha_p | V_C\ (i) | \delta_B l_B m_{lB}+1\alpha_q > \Big]$$
$$\equiv V^{L^iL^j}_{\delta Ap\delta Bq}\,.$$

With the definition Eq. (5.3a) we can write all the one-electron overlap matrix elements involving the small-components from the general expression Eq. (4.5) as follows:

$$S^{S^-S^-}_{ApBq} = \Lambda^-_{Ap}\ \Lambda^-_{Bq}\ [S^{L^+L^+}_{ApBq}] = N^{S^-}_{\alpha_p}N^{S^-}_{\alpha_q}\ S^{L^+L^+}_{\delta Ap\delta Bq}\ [\delta_A=0,\delta_B=0], \qquad (5.4a)$$

$$S^{S^+S^+}_{ApBq} = \Lambda^+_{Ap}\ \Lambda^+_{Bq}\ [S^{L^-L^-}_{ApBq}] = N^{S^+}_{\alpha_p}N^{S^+}_{\alpha_q}\ \times$$
$$\Big\{(j_A+3/2+\kappa_A)(j_B+3/2+\kappa_B)\ S^{L^-L^-}_{\delta Ap\delta Bq}\ [\delta_A=0,\delta_B=0]$$
$$-2\alpha_q(j_A+3/2+\kappa_A)\ S^{L^-L^-}_{\delta Ap\delta Bq}\ [\delta_A=0,\delta_B=1] \qquad (5.4b)$$
$$-2\alpha_p(j_B+3/2+\kappa_B)S^{L^-L^-}_{\delta Ap\delta Bq}\ [\delta_A=1,\delta_B=0]+4\alpha_p\alpha_q S^{L^-L^-}_{\delta Ap\delta Bq}\ [\delta_A=1,\delta_B=1]\Big\},$$

$$S^{S^-S^+}_{ApBq} = \Lambda^-_{Ap}\ \Lambda^+_{Bq}\ [S^{L^+L^-}_{ApBq}] = N^{S^-}_{\alpha_p}N^{S^+}_{\alpha_q}$$
$$\times\Big\{(j_B+3/2+\kappa_B)S^{L^+L^-}_{\delta Ap\delta Bq}\ [\delta_A=0,\delta_B=0]-2\alpha_q S^{L^+L^-}_{\delta Ap\delta Bq}\ [\delta_A=0,\delta_B=1]\Big\} \qquad (5.4c)$$

and

$$S^{S^+S^-}_{ApBq} = \Lambda^+_{Ap}\ \Lambda^-_{Bq}\ [S^{L^-L^+}_{ApBq}] = N^{S^+}_{\alpha_p}N^{S^-}_{\alpha_q}$$
$$\times\Big\{(j_A+3/2+\kappa_A)S^{L^-L^+}_{\delta Ap\delta Bq}\ [\delta_A=0,\delta_B=0]-2\alpha_p S^{L^-L^+}_{\delta Ap\delta Bq}\ [\delta_A=1,\delta_B=0]\Big\}. \qquad (5.4d)$$

Similarly from Eqs. (4.6) and (5.3b) nuclear attraction matrix elements we have

$$V_{ApBq}^{S^-S^-} = \Lambda_{Ap}^- \; \Lambda_{Bq}^- \; [V_{ApBq}^{L^+L^+}] = N_{\alpha_p}^{S^-} N_{\alpha_q}^{S^-} \; V_{\delta Ap\delta Bq}^{L^+L^+} \; [\delta_A = 0, \delta_B = 0], \tag{5.5a}$$

$$\begin{aligned} V_{ApBq}^{S^+S^+} &= \Lambda_{Ap}^+ \; \Lambda_{Bq}^+ \; [V_{ApBq}^{L^-L^-}] = N_{\alpha_p}^{S^+} N_{\alpha_q}^{S^+} \; \times \\ &\Big\{(j_A + 3/2 + \kappa_A)(j_B + 3/2 + \kappa_B) \; V_{\delta Ap\delta Bq}^{L^-L^-} \; [\delta_A = 0, \delta_B = 0] \\ &-2\alpha_q(j_A + 3/2 + \kappa_A) \; V_{\delta Ap\delta Bq}^{L^-L^-} \; [\delta_A = 0, \delta_B = 1] \\ &- 2\alpha_p(j_B + 3/2 + \kappa_B) \; V_{\delta Ap\delta Bq}^{L^-L^-} \; [\delta_A = 1, \delta_B = 0] + 4\alpha_p\alpha_q V_{\delta Ap\delta Bq}^{L^-L^-} \; [\delta_A = 1, \delta_B = 1]\Big\}, \end{aligned} \tag{5.5b}$$

$$\begin{aligned} V_{ApBq}^{S^-S^+} &= \Lambda_{Ap}^- \; \Lambda_{Bq}^+ \; [V_{ApBq}^{L^+L^-}] = N_{\alpha_p}^{S^-} N_{\alpha_q}^{S^+} \; \times \\ &\Big\{(j_B + 3/2 + \kappa_B) V_{\delta Ap\delta Bq}^{L^+L^-} \; [\delta_A = 0, \delta_B = 0] - 2\alpha_q V_{\delta Ap\delta Bq}^{L^+L^-} \; [\delta_A = 0, \delta_B = 1]\Big\}, \end{aligned} \tag{5.5c}$$

and

$$\begin{aligned} V_{ApBq}^{S^+S^-} &= \Lambda_{Ap}^+ \; \Lambda_{Bq}^- \; [V_{ApBq}^{L^-L^+}] = N_{\alpha_p}^{S^+} N_{\alpha_q}^{S^-} \; \times \\ &\Big\{(j_A + 3/2 + \kappa_A) V_{\delta Ap\delta Bq}^{L^-L^+} \; [\delta_A = 0, \delta_B = 0] - 2\alpha_p V_{\delta Ap\delta Bq}^{L^-L^+} \; [\delta_A = 1, \delta_B = 0]\Big\}. \end{aligned} \tag{5.5d}$$

For the evaluation of the matrix elements of the kinetic energy operator we can use the general relation (4.8) with Eqs. (5.4a)-(5.4d).

(2) Two-electron Matrix elements (small component)

To evaluate these for the small component we define the general primitive two-electron integral

$$\begin{aligned} &\left[\left(\frac{-\partial}{\partial \alpha_p}\right)^{\delta_A} \left(\frac{-\partial}{\partial \alpha_q}\right)^{\delta_B} \left(\frac{-\partial}{\partial \alpha_r}\right)^{\delta_C} \left(\frac{-\partial}{\partial \alpha_s}\right)^{\delta_D} \right] < A l_A m_{lA} \alpha_p B l_B m_{lB} \alpha_q | \, | C l_C m_{lC} \alpha_r D l_D m_{lD} \alpha_s > = \\ &\iint r_{1A}^{l_A + 2\delta_A} \; r_{1B}^{l_B + 2\delta_B} \; r_{2C}^{l_C + 2\delta_C} \; r_{2D}^{l_D + 2\delta_D} \; e^{-\alpha_p r_{1A}^2} \; e^{-\alpha_q r_{1B}^2} \; e^{-\alpha_r r_{2C}^2} \; e^{-\alpha_s r_{2D}^2} \\ &\times Y_{l_A}^{m_{lA}*}(\theta_{1A}\phi_{1A}) \; Y_{l_B}^{m_{lB}}(\theta_{1B}\phi_{1B}) \; Y_{l_C}^{m_{lC}*}(\theta_{2C}\phi_{2C}) \; Y_{l_D}^{m_{lD}}(\theta_{2D}\phi_{2D}) \; d\vec{r}_1 \; d\vec{r}_2 \\ &\equiv < \delta_A A l_A m_{lA} \alpha_p \delta_B B l_B m_{lB} \alpha_q | \, | \delta_C C l_C m_{lC} \alpha_r \delta_D D l_D m_{lD} \alpha_s >. \end{aligned} \tag{5.6}$$

By using (5.6) the basic two-electron integral defined in section 3 can be generalised to

$$\left[(\frac{-\partial}{\partial\alpha_p})^{\delta_A}(\frac{-\partial}{\partial\alpha_q})^{\delta_B}(\frac{-\partial}{\partial\alpha_r})^{\delta_C}(\frac{-\partial}{\partial\alpha_s})^{\delta_D}\right] J^{L^iL^jL^kL^l}_{ApBqCrDs} = N^{S^i}_{\alpha_p}N^{S^j}_{\alpha_q}N^{S^k}_{\alpha_r}N^{S^l}_{\alpha_s}$$
$$\times\Big[C^i_1(j_Am_{jA})C^j_1(j_Bm_{jB})C^k_1(j_Cm_{jC})C^l_1(j_Dm_{jD})$$
$$<\delta_AAl_Am_{lA}\alpha_p\delta_BBl_Bm_{lB}\alpha_q||\delta_CCl_Cm_{lC}\alpha_r\delta_DDl_Dm_{lD}\alpha_s>$$
$$+\,C^i_1(j_Am_{jA})C^j_1(j_Bm_{jB})C^k_2(j_Cm_{jC})C^l_2(j_Dm_{jD})$$
$$<\delta_AAl_Am_{lA}\alpha_p\delta_BBl_Bm_{lB}\alpha_q||\delta_CCl_Cm_{lC}+1\alpha_r\delta_DDl_Dm_{lD}+1\alpha_s> \tag{5.7}$$
$$+\,C^i_2(j_Am_{jA})C^j_2(j_Bm_{jB})C^k_1(j_Cm_{jC})C^l_1(j_Dm_{jD})$$
$$<\delta_AAl_Am_{lA}+1\alpha_p\delta_BBl_Bm_{lB}+1\alpha_q||\delta_CCl_Cm_{lC}\alpha_r\delta_DDl_Dm_{lD}\alpha_s>$$
$$+\,C^i_2(j_Am_{jA})C^j_2(j_Bm_{jB})C^k_2(j_Cm_{jC})C^l_2(j_Dm_{jD})$$
$$<\delta_AAl_Am_{lA}+1\alpha_p\delta_BBl_Bm_{lB}+1\alpha_q||\delta_CCl_Cm_{lC}+1\alpha_r\delta_DDl_Dm_{lD}+1\alpha_s>\Big]$$
$$\equiv J^{L^iL^jL^kL^l}_{\delta Ap\delta Bq\delta Cr\delta Ds}\,.$$

The primitive two-electron integral in Eq. (4.10) thus can be identified as

$$J^{L^iL^jL^kL^l}_{\delta Ap\delta Bq\delta Cr\delta Ds}[\delta_A=0,\,\delta_B=0,\,\delta_C=0,\,\delta_D=0] = J^{L^iL^jL^kL^l}_{ApBqCrDs}\,. \tag{5.8}$$

By direct application of Eq. (5.6) one can express all the small component matrix elements corresponding to the two-electron integrals in terms of the general integrals of the LLLL type given by Eq. (5.6). A few examples are given below as a proto type way of calculating these matrix elements in algebraic form, which is suitable for computation.

(i) Evaluation of matrix elements of the type $J^{L^iL^jS^kS^l}_{ApBqCrDs}$

Defining

$$(j_A+3/2+\kappa_A) = (n_{\kappa_A}+1), \tag{5.9a}$$

$$(j_B+3/2+\kappa_B) = (n_{\kappa_B}+1), \tag{5.9b}$$

$$(j_C+3/2+\kappa_C) = (n_{\kappa_C}+1), \tag{5.9c}$$

$$(j_D+3/2+\kappa_D) = (n_{\kappa_D}+1), \tag{5.9d}$$

and expanding Eq. (4. 10a), the following expressions are obtained for all four matrix elements:

$$J^{L^iL^jS^-S^-}_{ApBqCrDs} = \Lambda^-_{Cr}\,\Lambda^-_{Ds}\left[J^{L^iL^jL^+L^+}_{ApBqCrDs}\right]$$
$$= N^{S^-}_{\alpha_r}N^{S^-}_{\alpha_s}J^{L^iL^jL^+L^+}_{\delta Ap\delta Bq\delta Cr\delta Ds}[\delta_A=0,\,\delta_B=0,\,\delta_C=0,\,\delta_D=0], \tag{5.10a}$$

$$J^{L^iL^jS^-S^+}_{ApBqCrDs} = \Lambda^-_{Cr}\ \Lambda^+_{Ds}\left[J^{L^iL^jL^+L^-}_{ApBqCrDs}\right] = N^{S^-}_{\alpha_r}\ N^{S^+}_{\alpha_s}\ \times$$
$$\{(n_{\kappa_D}+1)\ J^{L^iL^jL^+L^-}_{\delta Ap\delta Bq\delta Cr\delta Ds}[\delta_A=0, \delta_B=0, \delta_C=0, \delta_D=0] \tag{5.10b}$$
$$-\ 2\alpha_s\ J^{L^iL^jL^+L^-}_{\delta Ap\delta Bq\delta Cr\delta Ds}[\delta_A=0, \delta_B=0, \delta_C=0, \delta_D=1]\},$$

$$J^{L^iL^jS^+S^-}_{ApBqCrDs} = \Lambda^+_{Cr}\ \Lambda^-_{Ds}\left[J^{L^iL^jL^-L^+}_{ApBqCrDs}\right] = N^{S^+}_{\alpha_r}\ N^{S^-}_{\alpha_s}\ \times$$
$$\{(n_{\kappa_C}+1)\ J^{L^iL^jL^-L^+}_{\delta Ap\delta Bq\delta Cr\delta Ds}[\delta_A=0, \delta_B=0, \delta_C=0, \delta_D=0] \tag{5.10c}$$
$$-\ 2\alpha_r\ J^{L^iL^jL^-L^+}_{\delta Ap\delta Bq\delta Cr\delta Ds}[\delta_A=0, \delta_B=0, \delta_C=1, \delta_D=0]\}$$

and

$$J^{L^iL^jS^+S^+}_{ApBqCrDs} = \Lambda^+_{Cr}\ \Lambda^+_{Ds}\left[J^{L^iL^jL^-L^-}_{ApBqCrDs}\right] = N^{S^+}_{\alpha_r}\ N^{S^-}_{\alpha_s}\ \times$$
$$\{(n_{\kappa_C}+1)\ (n_{\kappa_D}+1)\ J^{L^iL^jL^-L^-}_{\delta Ap\delta Bq\delta Cr\delta Ds}[\delta_A=0, \delta_B=0, \delta_C=0, \delta_D=0]$$
$$-\ 2\alpha_r\ (n_{\kappa_D}+1)\ J^{L^iL^jL^-L^-}_{\delta Ap\delta Bq\delta Cr\delta Ds}[\delta_A=0, \delta_B=0, \delta_C=1, \delta_D=0] \tag{5.10d}$$
$$-\ 2\alpha_s\ (n_{\kappa_C}+1)\ J^{L^iL^jL^-L^-}_{\delta Ap\delta Bq\delta Cr\delta Ds}[\delta_A=0, \delta_B=0, \delta_C=1, \delta_D=0]$$
$$+\ 4\alpha_r\ \alpha_s\ J^{L^iL^jL^-L^-}_{\delta Ap\delta Bq\delta Cr\delta Ds}[\delta_A=0, \delta_B=0, \delta_C=1, \delta_D=1]\}.$$

Evaluation of the two-electron integrals of the type $J^{S^iS^jL^kL^l}_{ApBqCrDs}$ follows from Eqs. (5.10a)–(5.10d) by the symmetry operations:

$$\begin{array}{l} A \to C,\ C \to A,\ B \to D,\ D \to B, \\ p \to r,\ r \to p,\ q \to s,\ s \to q \end{array} \tag{5.11a}$$

and

$$i \to k,\ k \to i,\ j \to l,\ l \to j. \tag{5.11b}$$

(ii) The matrix elements of the type $J^{S^iS^jS^kS^l}_{ApBqCrDs}$

These can be evaluated in a straightforward manner by using Eq. (4.14) as follows:

$$J^{S^-S^-S^-S^-}_{ApBqCrDs} = [\Lambda^-_{Ap}\ \Lambda^-_{Bq}\ \Lambda^-_{Cr}\ \Lambda^-_{Ds}\ ]J^{L^+L^+L^+L^+}_{ApBqCrDs} = N^{S^-}_{\alpha_p}\ N^{S^-}_{\alpha_q}\ N^{S^-}_{\alpha_r}\ N^{S^-}_{\alpha_s}$$
$$\times\ J^{L^+L^+L^+L^+}_{\delta Ap\delta Bq\delta Cr\delta Ds}[\delta_A=0, \delta_B=0, \delta_C=0, \delta_D=0], \tag{5.12}$$

$$J^{S^-S^-S^+S^-}_{ApBqCrDs} = \Lambda^-_{Ap}\ \Lambda^-_{Bq}\ \Lambda^+_{Cr}\ \Lambda^-_{Ds}\ [J^{L^+L^+L^-L^+}_{ApBqCrDs}] = N^{S^-}_{\alpha_p}\ N^{S^-}_{\alpha_q}\ N^{S^+}_{\alpha_r}\ N^{S^-}_{\alpha_s}\ \times$$
$$\{(n_{\kappa_C}+1)\ J^{L^+L^+L^-L^+}_{\delta Ap\delta Bq\delta Cr\delta Ds}[\delta_A=0, \delta_B=0, \delta_C=0, \delta_D=0] \tag{5.13a}$$
$$-\ 2\alpha_r\ J^{L^+L^+L^-L^+}_{\delta Ap\delta Bq\delta Cr\delta Ds}[\delta_A=0, \delta_B=0, \delta_C=1, \delta_D=0]\},$$

$$J_{ApBqCrDs}^{S^-S^-S^-S^+} = \Lambda_{Ap}^- \Lambda_{Bq}^- \Lambda_{Cr}^- \Lambda_{Ds}^+ [J_{ApBqCrDs}^{L^+L^+L^+L^-}] = N_{\alpha_p}^{S^-} N_{\alpha_q}^{S^-} N_{\alpha_r}^{S^-} N_{\alpha_s}^{S^+} \times$$
$$\{(n_{\kappa_D}+1)\ J_{\delta Ap\delta Bq\delta Cr\delta Ds}^{L^+L^+L^+L^-}[\delta_A=0, \delta_B=0, \delta_C=0, \delta_D=0] \tag{5.13b}$$
$$-2\alpha_s\ J_{\delta Ap\delta Bq\delta Cr\delta Ds}^{L^+L^+L^+L^-}[\delta_A=0, \delta_B=0, \delta_C=0, \delta_D=1]\},$$

$$J_{ApBqCrDs}^{S^+S^-S^-S^-} = \Lambda_{Ap}^+ \Lambda_{Bq}^- \Lambda_{Cr}^- \Lambda_{Ds}^- [J_{ApBqCrDs}^{L^-L^+L^+L^+}] = N_{\alpha_p}^{S^+} N_{\alpha_q}^{S^-} N_{\alpha_r}^{S^-} N_{\alpha_s}^{S^-} \times$$
$$\{(n_{\kappa_A}+1)\ J_{\delta Ap\delta Bq\delta Cr\delta Ds}^{L^-L^+L^+L^+}[\delta_A=0, \delta_B=0, \delta_C=0, \delta_D=0] \tag{5.13c}$$
$$-2\alpha_p\ J_{\delta Ap\delta Bq\delta Cr\delta Ds}^{L^-L^+L^+L^+}[\delta_A=1, \delta_B=0, \delta_C=0, \delta_D=0]\},$$

$$J_{ApBqCrDs}^{S^-S^+S^-S^-} = \Lambda_{Ap}^- \Lambda_{Bq}^+ \Lambda_{Cr}^- \Lambda_{Ds}^- [J_{ApBqCrDs}^{L^+L^-L^+L^+}] = N_{\alpha_p}^{S^-} N_{\alpha_q}^{S^+} N_{\alpha_r}^{S^-} N_{\alpha_s}^{S^-} \times$$
$$\{(n_{\kappa_B}+1)\ J_{\delta Ap\delta Bq\delta Cr\delta Ds}^{L^-L^+L^+L^+}[\delta_A=0, \delta_B=0, \delta_C=0, \delta_D=0] \tag{5.13d}$$
$$-2\alpha_q\ J_{\delta Ap\delta Bq\delta Cr\delta Ds}^{L^+L^-L^+L^+}[\delta_A=0, \delta_B=1, \delta_C=0, \delta_D=0]\},$$

$$J_{ApBqCrDs}^{S^-S^-S^+S^+} = \Lambda_{Ap}^- \Lambda_{Bq}^- \Lambda_{Cr}^+ \Lambda_{Ds}^+ [J_{ApBqCrDs}^{L^+L^+L^-L^-}] = N_{\alpha_p}^{S^-} N_{\alpha_q}^{S^-} N_{\alpha_r}^{S^+} N_{\alpha_s}^{S^+} \times$$
$$\{(n_{\kappa_C}+1)\ (n_{\kappa_D}+1)\ J_{\delta Ap\delta Bq\delta Cr\delta Ds}^{L^+L^+L^-L^-}[\delta_A=0, \delta_B=0, \delta_C=0, \delta_D=0]$$
$$-2\alpha_r\ (n_{\kappa_D}+1)\ J_{\delta Ap\delta Bq\delta Cr\delta Ds}^{L^+L^+L^-L^-}[\delta_A=0, \delta_B=0, \delta_C=1, \delta_D=0] \tag{5.14a}$$
$$-2\alpha_s\ (n_{\kappa_C}+1)\ J_{\delta Ap\delta Bq\delta Cr\delta Ds}^{L^+L^+L^-L^-}[\delta_A=0, \delta_B=0, \delta_C=0, \delta_D=1]$$
$$+4\alpha_r\ \alpha_s\ J_{\delta Ap\delta Bq\delta Cr\delta Ds}^{L^+L^+L^-L^-}[\delta_A=0, \delta_B=0, \delta_C=1, \delta_D=1]\},$$

$$J_{ApBqCrDs}^{S^+S^+S^-S^-} = \Lambda_{Ap}^+ \Lambda_{Bq}^+ \Lambda_{Cr}^- \Lambda_{Ds}^- [J_{ApBqCrDs}^{L^-L^-L^+L^+}] = N_{\alpha_p}^{S^+} N_{\alpha_q}^{S^+} N_{\alpha_r}^{S^-} N_{\alpha_s}^{S^-} \times$$
$$\{(n_{\kappa_A}+1)\ (n_{\kappa_B}+1)\ J_{\delta Ap\delta Bq\delta Cr\delta Ds}^{L^-L^-L^+L^+}[\delta_A=0, \delta_B=0, \delta_C=0, \delta_D=0]$$
$$-2\alpha_p\ (n_{\kappa_B}+1)\ J_{\delta Ap\delta Bq\delta Cr\delta Ds}^{L^-L^-L^+L^+}[\delta_A=1, \delta_B=0, \delta_C=0, \delta_D=0] \tag{5.14b}$$
$$-2\alpha_q\ (n_{\kappa_B}+1)\ J_{\delta Ap\delta Bq\delta Cr\delta Ds}^{L^-L^-L^+L^+}[\delta_A=0, \delta_B=1, \delta_C=0, \delta_D=0]$$
$$+4\alpha_p\ \alpha_q\ J_{\delta Ap\delta Bq\delta Cr\delta Ds}^{L^-L^-L^+L^+}[\delta_A=1, \delta_B=1, \delta_C=0, \delta_D=0]\},$$

$$J^{S^+S^-S^+S^-}_{ApBqCrDs} = \Lambda^+_{Ap}\ \Lambda^-_{Bq}\ \Lambda^+_{Cr}\ \Lambda^-_{Ds}\ [J^{L^-L^+L^-L^+}_{ApBqCrDs}] = N^{S^+}_{\alpha_p} N^{S^-}_{\alpha_q} N^{S^+}_{\alpha_r} N^{S^-}_{\alpha_s} \times$$
$$\{(n_{\kappa_A}+1)\ (n_{\kappa_C}+1)\ J^{L^-L^+L^-L^+}_{\delta Ap\delta Bq\delta Cr\delta Ds}[\delta_A=0, \delta_B=0, \delta_C=0, \delta_D=0]$$
$$-2\alpha_p\ (n_{\kappa_C}+1)\ J^{L^-L^+L^-L^+}_{\delta Ap\delta Bq\delta Cr\delta Ds}[\delta_A=1, \delta_B=0, \delta_C=0, \delta_D=0] \qquad (5.14c)$$
$$-2\alpha_r\ (n_{\kappa_A}+1)\ J^{L^-L^+L^-L^+}_{\delta Ap\delta Bq\delta Cr\delta Ds}[\delta_A=0, \delta_B=0, \delta_C=1, \delta_D=0]$$
$$+4\alpha_p\ \alpha_r\ J^{L^-L^+L^-L^+}_{\delta Ap\delta Bq\delta Cr\delta Ds}[\delta_A=1, \delta_B=0, \delta_C=1, \delta_D=0]\},$$

$$J^{S^+S^-S^-S^+}_{ApBqCrDs} = \Lambda^+_{Ap}\ \Lambda^-_{Bq}\ \Lambda^-_{Cr}\ \Lambda^+_{Ds}\ [J^{L^-L^+L^+L^-}_{ApBqCrDs}] = N^{S^+}_{\alpha_p} N^{S^-}_{\alpha_q} N^{S^-}_{\alpha_r} N^{S^+}_{\alpha_s} \times$$
$$\{(n_{\kappa_A}+1)\ (n_{\kappa_D}+1)\ J^{L^-L^+L^+L^-}_{\delta Ap\delta Bq\delta Cr\delta Ds}[\delta_A=0, \delta_B=0, \delta_C=0, \delta_D=0]$$
$$-2\alpha_p\ (n_{\kappa_D}+1)\ J^{L^-L^+L^+L^-}_{\delta Ap\delta Bq\delta Cr\delta Ds}[\delta_A=1, \delta_B=0, \delta_C=0, \delta_D=0] \qquad (5.14d)$$
$$-2\alpha_s\ (n_{\kappa_A}+1)\ J^{L^-L^+L^+L^-}_{\delta Ap\delta Bq\delta Cr\delta Ds}[\delta_A=0, \delta_B=0, \delta_C=0, \delta_D=1]$$
$$+4\alpha_p\ \alpha_s\ J^{L^-L^+L^+L^-}_{\delta Ap\delta Bq\delta Cr\delta Ds}[\delta_A=1, \delta_B=0, \delta_C=0, \delta_D=1]\},$$

$$J^{S^-S^+S^-S^+}_{ApBqCrDs} = \Lambda^-_{Ap}\ \Lambda^+_{Bq}\ \Lambda^-_{Cr}\ \Lambda^+_{Ds}\ [J^{L^+L^-L^+L^-}_{ApBqCrDs} =] = N^{S^-}_{\alpha_p} N^{S^+}_{\alpha_q} N^{S^-}_{\alpha_r} N^{S^+}_{\alpha_s} \times$$
$$\{(n_{\kappa_B}+1)\ (n_{\kappa_D}+1)\ J^{L^+L^-L^+L^-}_{\delta Ap\delta Bq\delta Cr\delta Ds}[\delta_A=0, \delta_B=0, \delta_C=0, \delta_D=0]$$
$$-2\alpha_q\ (n_{\kappa_D}+1)\ J^{L^+L^-L^+L^-}_{\delta Ap\delta Bq\delta Cr\delta Ds}[\delta_A=0, \delta_B=1, \delta_C=0, \delta_D=0] \qquad (5.14e)$$
$$-2\alpha_s\ (n_{\kappa_B}+1)\ J^{L^+L^-L^+L^-}_{\delta Ap\delta Bq\delta Cr\delta Ds}[\delta_A=0, \delta_B=0, \delta_C=0, \delta_D=1]$$
$$+4\alpha_q\ \alpha_s\ J^{L^+L^-L^+L^-}_{\delta Ap\delta Bq\delta Cr\delta Ds}[\delta_A=1, \delta_B=0, \delta_C=0, \delta_D=1]\},$$

$$J^{S^-S^+S^+S^-}_{ApBqCrDs} = \Lambda^-_{Ap}\ \Lambda^+_{Bq}\ \Lambda^+_{Cr}\ \Lambda^-_{Ds}\ [J^{L^+L^-L^-L^+}_{ApBqCrDs} =] N^{S^-}_{\alpha_p} N^{S^+}_{\alpha_q} N^{S^+}_{\alpha_r} N^{S^-}_{\alpha_s} \times$$
$$\{(n_{\kappa_B}+1)\ (n_{\kappa_C}+1)\ J^{L^+L^-L^-L^+}_{\delta Ap\delta Bq\delta Cr\delta Ds}[\delta_A=0, \delta_B=0, \delta_C=0, \delta_D=0]$$
$$-2\alpha_q\ (n_{\kappa_C}+1)\ J^{L^+L^-L^-L^+}_{\delta Ap\delta Bq\delta Cr\delta Ds}[\delta_A=0, \delta_B=1, \delta_C=0, \delta_D=0] \qquad (5.14f)$$
$$-2\alpha_r\ (n_{\kappa_B}+1)\ J^{L^+L^-L^-L^+}_{\delta Ap\delta Bq\delta Cr\delta Ds}[\delta_A=0, \delta_B=0, \delta_C=1, \delta_D=0]$$
$$+4\alpha_q\ \alpha_r\ J^{L^+L^-L^-L^+}_{\delta Ap\delta Bq\delta Cr\delta Ds}[\delta_A=0, \delta_B=1, \delta_C=1, \delta_D=0]\},$$

$$
\begin{aligned}
J_{ApBqCrDs}^{S^+S^+S^+S^-} &= \Lambda_{Ap}^{+}\ \Lambda_{Bq}^{+}\ \Lambda_{Cr}^{+}\ \Lambda_{Ds}^{-}\ [J_{ApBqCrDs}^{L^-L^-L^-L^+}] = N_{\alpha_p}^{S^+} N_{\alpha_q}^{S^+} N_{\alpha_r}^{S^+} N_{\alpha_s}^{S^-} \times \\
&\{(n_{\kappa_A}+1)\ (n_{\kappa_B}+1)\ (n_{\kappa_C}+1)\ J_{\delta Ap\delta Bq\delta Cr\delta Ds}^{L^-L^-L^-L^+}[\delta_A=0, \delta_B=0, \delta_C=0, \delta_D=0] \\
&- 2\alpha_p\ (n_{\kappa_B}+1)\ (n_{\kappa_C}+1)\ J_{\delta Ap\delta Bq\delta Cr\delta Ds}^{L^-L^-L^-L^+}[\delta_A=1, \delta_B=0, \delta_C=0, \delta_D=0] \\
&- 2\alpha_q\ (n_{\kappa_A}+1)\ (n_{\kappa_C}+1)\ J_{\delta Ap\delta Bq\delta Cr\delta Ds}^{L^-L^-L^-L^+}[\delta_A=0, \delta_B=1, \delta_C=0, \delta_D=0] \\
&- 2\alpha_r\ (n_{\kappa_A}+1)\ (n_{\kappa_B}+1)\ J_{\delta Ap\delta Bq\delta Cr\delta Ds}^{L^-L^-L^-L^+}[\delta_A=0, \delta_B=0, \delta_C=1, \delta_D=0] \qquad (5.15a) \\
&+ 4\alpha_p\ \alpha_q\ (n_{\kappa_C}+1)\ J_{\delta Ap\delta Bq\delta Cr\delta Ds}^{L^-L^-L^-L^+}[\delta_A=1, \delta_B=1, \delta_C=0, \delta_D=0] \\
&+ 4\alpha_p\ \alpha_r\ (n_{\kappa_B}+1)\ J_{\delta Ap\delta Bq\delta Cr\delta Ds}^{L^-L^-L^-L^+}[\delta_A=1, \delta_B=0, \delta_C=1, \delta_D=0] \\
&+ 4\alpha_q\ \alpha_r\ (n_{\kappa_A}+1)\ J_{\delta Ap\delta Bq\delta Cr\delta Ds}^{L^-L^-L^-L^+}[\delta_A=0, \delta_B=1, \delta_C=1, \delta_D=0] \\
&- 8\alpha_p\ \alpha_q\ \alpha_r\ J_{\delta Ap\delta Bq\delta Cr\delta Ds}^{L^-L^-L^-L^+}[\delta_A=1, \delta_B=1, \delta_C=1, \delta_D=0]\},
\end{aligned}
$$

$$
\begin{aligned}
J_{ApBqCrDs}^{S^+S^+S^-S^+} &= \Lambda_{Ap}^{+}\ \Lambda_{Bq}^{+}\ \Lambda_{Cr}^{-}\ \Lambda_{Ds}^{+}\ [J_{ApBqCrDs}^{L^-L^-L^+L^-}] = N_{\alpha_p}^{S^+} N_{\alpha_q}^{S^+} N_{\alpha_r}^{S^-} N_{\alpha_s}^{S^+} \times \\
&\{(n_{\kappa_A}+1)\ (n_{\kappa_B}+1)\ (n_{\kappa_D}+1)\ J_{\delta Ap\delta Bq\delta Cr\delta Ds}^{L^-L^-L^+L^-}[\delta_A=0, \delta_B=0, \delta_C=0, \delta_D=0] \\
&- 2\alpha_p\ (n_{\kappa_B}+1)\ (n_{\kappa_D}+1)\ J_{\delta Ap\delta Bq\delta Cr\delta Ds}^{L^-L^-L^+L^-}[\delta_A=1, \delta_B=0, \delta_C=0, \delta_D=0] \\
&- 2\alpha_q\ (n_{\kappa_A}+1)\ (n_{\kappa_D}+1)\ J_{\delta Ap\delta Bq\delta Cr\delta Ds}^{L^-L^-L^+L^-}[\delta_A=0, \delta_B=1, \delta_C=0, \delta_D=0] \\
&- 2\alpha_s\ (n_{\kappa_A}+1)\ (n_{\kappa_B}+1)\ J_{\delta Ap\delta Bq\delta Cr\delta Ds}^{L^-L^-L^+L^-}[\delta_A=0, \delta_B=0, \delta_C=0, \delta_D=1] \qquad (5.15b) \\
&+ 4\alpha_p\ \alpha_q\ (n_{\kappa_D}+1)\ J_{\delta Ap\delta Bq\delta Cr\delta Ds}^{L^-L^-L^+L^-}[\delta_A=1, \delta_B=1, \delta_C=0, \delta_D=0] \\
&+ 4\alpha_p\ \alpha_s\ (n_{\kappa_B}+1)\ J_{\delta Ap\delta Bq\delta Cr\delta Ds}^{L^-L^-L^+L^-}[\delta_A=1, \delta_B=0, \delta_C=0, \delta_D=1] \\
&+ 4\alpha_q\ \alpha_s\ (n_{\kappa_A}+1)\ J_{\delta Ap\delta Bq\delta Cr\delta Ds}^{L^-L^-L^+L^-}[\delta_A=0, \delta_B=1, \delta_C=0, \delta_D=1] \\
&- 8\alpha_p\ \alpha_q\ \alpha_s\ J_{\delta Ap\delta Bq\delta Cr\delta Ds}^{L^-L^-L^+L^-}[\delta_A=1, \delta_B=1, \delta_C=0, \delta_D=1]\},
\end{aligned}
$$

$$J^{S^-S^+S^+S^+}_{ApBqCrDs} = \Lambda^-_{Ap}\ \Lambda^+_{Bq}\ \Lambda^+_{Cr}\ \Lambda^+_{Ds}\ [J^{L^-L^-L^-L^+}_{ApBqCrDs}] = N^{S^-}_{\alpha_p} N^{S^+}_{\alpha_q} N^{S^+}_{\alpha_r} N^{S^+}_{\alpha_s} \times$$
$$\{(n_{\kappa_B}+1)\ (n_{\kappa_C}+1)\ (n_{\kappa_D}+1)\ J^{L^+L^-L^-L^-}_{\delta Ap\delta Bq\delta Cr\delta Ds}[\delta_A=0, \delta_B=0, \delta_C=0, \delta_D=0]$$
$$-2\alpha_s\ (n_{\kappa_B}+1)\ (n_{\kappa_C}+1)\ J^{L^+L^-L^-L^-}_{\delta Ap\delta Bq\delta Cr\delta Ds}[\delta_A=0, \delta_B=0, \delta_C=0, \delta_D=1]$$
$$-2\alpha_q\ (n_{\kappa_C}+1)\ (n_{\kappa_D}+1)\ J^{L^+L^-L^-L^-}_{\delta Ap\delta Bq\delta Cr\delta Ds}[\delta_A=0, \delta_B=1, \delta_C=0, \delta_D=0]$$
$$-2\alpha_r\ (n_{\kappa_B}+1)\ (n_{\kappa_D}+1)\ J^{L^+L^-L^-L^-}_{\delta Ap\delta Bq\delta Cr\delta Ds}[\delta_A=0, \delta_B=0, \delta_C=1, \delta_D=0] \qquad (5.15c)$$
$$+4\alpha_s\ \alpha_q\ (n_{\kappa_C}+1)\ J^{L^+L^-L^-L^-}_{\delta Ap\delta Bq\delta Cr\delta Ds}[\delta_A=0, \delta_B=1, \delta_C=0, \delta_D=1]$$
$$+4\alpha_s\ \alpha_r\ (n_{\kappa_B}+1)\ J^{L^+L^-L^-L^-}_{\delta Ap\delta Bq\delta Cr\delta Ds}[\delta_A=0, \delta_B=0, \delta_C=1, \delta_D=1]$$
$$+4\alpha_q\ \alpha_r\ (n_{\kappa_D}+1)\ J^{L^+L^-L^-L^-}_{\delta Ap\delta Bq\delta Cr\delta Ds}[\delta_A=0, \delta_B=1, \delta_C=1, \delta_D=0]$$
$$-8\alpha_s\ \alpha_q\ \alpha_r\ J^{L^+L^-L^-L^-}_{\delta Ap\delta Bq\delta Cr\delta Ds}[\delta_A=0, \delta_B=1, \delta_C=1, \delta_D=1]\}$$

$$J^{S^+S^-S^+S^+}_{ApBqCrDs} = \Lambda^+_{Ap}\ \Lambda^-_{Bq}\ \Lambda^+_{Cr}\ \Lambda^+_{Ds}\ [J^{L^-L^+L^-L^-}_{ApBqCrDs} =] N^{S^+}_{\alpha_p} N^{S^-}_{\alpha_q} N^{S^+}_{\alpha_r} N^{S^+}_{\alpha_s} \times$$
$$\{(n_{\kappa_A}+1)\ (n_{\kappa_C}+1)\ (n_{\kappa_D}+1)\ J^{L^-L^+L^-L^-}_{\delta Ap\delta Bq\delta Cr\delta Ds}[\delta_A=0, \delta_B=0, \delta_C=0, \delta_D=0]$$
$$-2\alpha_p\ (n_{\kappa_C}+1)\ (n_{\kappa_D}+1)\ J^{L^-L^+L^-L^-}_{\delta Ap\delta Bq\delta Cr\delta Ds}[\delta_A=1, \delta_B=0, \delta_C=0, \delta_D=0]$$
$$-2\alpha_s\ (n_{\kappa_A}+1)\ (n_{\kappa_C}+1)\ J^{L^-L^+L^-L^-}_{\delta Ap\delta Bq\delta Cr\delta Ds}[\delta_A=0, \delta_B=0, \delta_C=0, \delta_D=1]$$
$$-2\alpha_r\ (n_{\kappa_A}+1)\ (n_{\kappa_D}+1)\ J^{L^-L^+L^-L^-}_{\delta Ap\delta Bq\delta Cr\delta Ds}[\delta_A=0, \delta_B=0, \delta_C=1, \delta_D=0] \qquad (5.15d)$$
$$+4\alpha_p\ \alpha_s\ (n_{\kappa_C}+1)\ J^{L^-L^+L^-L^-}_{\delta Ap\delta Bq\delta Cr\delta Ds}[\delta_A=1, \delta_B=0, \delta_C=0, \delta_D=1]$$
$$+4\alpha_p\ \alpha_r\ (n_{\kappa_B}+1)\ J^{L^-L^+L^-L^-}_{\delta Ap\delta Bq\delta Cr\delta Ds}[\delta_A=1, \delta_B=0, \delta_C=1, \delta_D=0]$$
$$+4\alpha_s\ \alpha_r\ (n_{\kappa_A}+1)\ J^{L^-L^+L^-L^-}_{\delta Ap\delta Bq\delta Cr\delta Ds}[\delta_A=0, \delta_B=0, \delta_C=1, \delta_D=1]$$
$$-8\alpha_p\ \alpha_s\ \alpha_r\ J^{L^-L^+L^-L^-}_{\delta Ap\delta Bq\delta Cr\delta Ds}[\delta_A=1, \delta_B=0, \delta_C=1, \delta_D=1]\}$$

and

$$
\begin{aligned}
J^{S^+S^+S^+S^+}_{ApBqCrDs} &= \Lambda^+_{Ap}\,\Lambda^+_{Bq}\,\Lambda^+_{Cr}\,\Lambda^+_{Ds}\,[J^{L^-L^-L^-L^-}_{ApBqCrDs}] = N^{S^+}_{\alpha_p}\,N^{S^+}_{\alpha_q}\,N^{S^+}_{\alpha_r}\,N^{S^+}_{\alpha_s} \times \\
&\{(n_{\kappa_A}+1)\,(n_{\kappa_B}+1)\,(n_{\kappa_C}+1)\,(n_{\kappa_D}+1)\,J^{L^-L^-L^-L^-}_{\delta Ap\delta Bq\delta Cr\delta Ds}[\delta_A=0,\delta_B=0,\delta_C=0,\delta_D=0] \\
&-2\alpha_p\,(n_{\kappa_B}+1)\,(n_{\kappa_C}+1)\,(n_{\kappa_D}+1)\,J^{L^-L^-L^-L^-}_{\delta Ap\delta Bq\delta Cr\delta Ds}[\delta_A=1,\delta_B=0,\delta_C=0,\delta_D=0] \\
&-2\alpha_q\,(n_{\kappa_A}+1)\,(n_{\kappa_C}+1)\,(n_{\kappa_D}+1)\,J^{L^-L^-L^-L^-}_{\delta Ap\delta Bq\delta Cr\delta Ds}[\delta_A=0,\delta_B=1,\delta_C=0,\delta_D=0] \\
&-2\alpha_r\,(n_{\kappa_A}+1)\,(n_{\kappa_B}+1)\,(n_{\kappa_D}+1)\,J^{L^-L^-L^-L^-}_{\delta Ap\delta Bq\delta Cr\delta Ds}[\delta_A=0,\delta_B=0,\delta_C=1,\delta_D=0] \\
&-2\alpha_s\,(n_{\kappa_A}+1)\,(n_{\kappa_B}+1)\,(n_{\kappa_C}+1)\,J^{L^-L^-L^-L^-}_{\delta Ap\delta Bq\delta Cr\delta Ds}[\delta_A=0,\delta_B=0,\delta_C=0,\delta_D=1] \\
&+4\alpha_p\,\alpha_q\,(n_{\kappa_C}+1)\,(n_{\kappa_D}+1)\,J^{L^-L^-L^-L^-}_{\delta Ap\delta Bq\delta Cr\delta Ds}[\delta_A=1,\delta_B=1,\delta_C=0,\delta_D=0] \\
&+4\alpha_p\,\alpha_r\,(n_{\kappa_B}+1)\,(n_{\kappa_D}+1)\,J^{L^-L^-L^-L^-}_{\delta Ap\delta Bq\delta Cr\delta Ds}[\delta_A=1,\delta_B=0,\delta_C=1,\delta_D=0] \\
&+4\alpha_q\,\alpha_r\,(n_{\kappa_A}+1)\,(n_{\kappa_D}+1)\,J^{L^-L^-L^-L^-}_{\delta Ap\delta Bq\delta Cr\delta Ds}[\delta_A=0,\delta_B=1,\delta_C=1,\delta_D=0] \qquad (5.16) \\
&+4\alpha_r\,\alpha_s\,(n_{\kappa_A}+1)\,(n_{\kappa_B}+1)\,J^{L^-L^-L^-L^-}_{\delta Ap\delta Bq\delta Cr\delta Ds}[\delta_A=0,\delta_B=0,\delta_C=1,\delta_D=1] \\
&+4\alpha_p\,\alpha_s\,(n_{\kappa_B}+1)\,(n_{\kappa_C}+1)\,J^{L^-L^-L^-L^-}_{\delta Ap\delta Bq\delta Cr\delta Ds}[\delta_A=1,\delta_B=0,\delta_C=0,\delta_D=1] \\
&+4\alpha_q\,\alpha_s\,(n_{\kappa_A}+1)\,(n_{\kappa_C}+1)\,J^{L^-L^-L^-L^-}_{\delta Ap\delta Bq\delta Cr\delta Ds}[\delta_A=0,\delta_B=1,\delta_C=0,\delta_D=1] \\
&-8\alpha_p\,\alpha_q\,\alpha_r\,(n_{\kappa_D}+1)\,J^{L^-L^-L^-L^-}_{\delta Ap\delta Bq\delta Cr\delta Ds}[\delta_A=1,\delta_B=1,\delta_C=1,\delta_D=0] \\
&-8\alpha_p\,\alpha_r\,\alpha_s\,(n_{\kappa_B}+1)\,J^{L^-L^-L^-L^-}_{\delta Ap\delta Bq\delta Cr\delta Ds}[\delta_A=1,\delta_B=0,\delta_C=1,\delta_D=1] \\
&-8\alpha_q\,\alpha_r\,\alpha_s\,(n_{\kappa_A}+1)\,J^{L^-L^-L^-L^-}_{\delta Ap\delta Bq\delta Cr\delta Ds}[\delta_A=0,\delta_B=1,\delta_C=1,\delta_D=1] \\
&-8\alpha_p\,\alpha_q\,\alpha_s\,(n_{\kappa_C}+1)\,J^{L^-L^-L^-L^-}_{\delta Ap\delta Bq\delta Cr\delta Ds}[\delta_A=1,\delta_B=1,\delta_C=0,\delta_D=1] \\
&+16\alpha_p\,\alpha_q\,\alpha_r\,\alpha_s\,J^{L^-L^-L^-L^-}_{\delta Ap\delta Bq\delta Cr\delta Ds}[\delta_A=1,\delta_B=1,\delta_C=1,\delta_D=1]\}.
\end{aligned}
$$

As pointed out earlier in section 4 the exchange integrals are easily obtained from the direct integrals by an exchange of indices (see Eq.(4.15)). Also for computational purposes, it should be noted that the exchange integrals of the LL and SS type defined in Eqs. (2.17b) and (4.15) can be represented[26,27] in a symmetrized form:

$$K^{T^iT^jT^kT^l}_{ApBqCrDs} = \frac{1}{2}(K^{T^iT^jT^kT^l}_{ApBqDsCr} + K^{T^iT^jT^kT^l}_{ApBqCrDs}). \qquad (5.17)$$

6. Evaluation of Primitive Integrals in Cartesian Form

The one-and two-electron primitive integrals (Eqs. (5.2) and (5.6)) involving solid spherical harmonic Gaussian-type Functions can be evaluated using the expressions given by Harris[40] and also by the formalism developed by Fieck[41,42]. Recently Matsuoka[43] has generalised Fieck's formulation[42] for use in relativistic molecular integral calculations and has examined the numerical aspects of such computations. However, the computational efficacy of the formulations by Harris[40] has not yet been examined in practice, even in nonrelativistic molecular calculations. Moreover, in molecular DF calculations, while computing matrix elements involving the small components, the derivatives of the primitive integrals with respect to exponents (Eqs. (5.1) and (5.7)) will inevitably appear in a DF formalism which includes the kinetic balance condition. And it is these derivatives which make the computation of the integrals more cumbersome within the approach described by Harris[40]. Alternatively, for specific l values, one could use the well-known[44] expansions of solid spherical harmonics in terms of cartesian variables x, y, z, and then evaluate the integrals by standard non-relativistic methods, using these expansion coefficients. Here we present a general derivation of the solid spherical harmonics in terms of cartesian cordinates x,y and z, in complex combination, and prove that all the matrix elements involving the small and the large component are directly obtainable from a set of standard nonrelativistic matrix elements of cartesian-type Gaussian basis functions.

Following Edmonds[44] we write the spherical harmonics $Y_l^m(\theta, \phi)$ in terms of associated Legendre polynomials and express the solid spherical harmonics as

$$r^l Y_l^m(\theta, \phi) = (-1)^m \sqrt{\frac{(2l+1)}{4\pi}} \sqrt{\frac{(l-m)!}{(l+m)!}} \times (x+iy)^m r^{l-m} \left[(1-t^2)^{-m/2} P_l^m(t)\right], \tag{6.1}$$

where

$$t = \cos\theta. \tag{6.2}$$

In Eq. (6.1) the terms in the bracket can be related to the orthogonal Jacobi polynomial[44]. $P_{l-m}^{(m,m)}(t)$ by the identity:

$$(1-t^2)^{-m/2} P_l^m(t) = 2^{-m} \frac{(2m+n)!}{(n+m)!} P_{l-m}^{(m,m)}(t), \tag{6.3}$$

where $n = (l-m)$. The Jacobi Polynomial $P_{l-m}^{(m,m)}(t)$ has various representations in hypergeometric functions[45]. Using these representations one can show that[45] these polynomial are related to hyperspherical polynomial $P_n^{(m,m)}(t)$ in the following way

$$\begin{aligned} P_n^\lambda(t) &= \frac{\Gamma(m+1)\Gamma(n+2m+1)}{\Gamma(2m+1)\Gamma(n+m+1)} P_n^{(m,m)}(t) \\ &= \frac{\Gamma(\lambda+1/2)\Gamma(n+2\lambda)}{\Gamma(2\lambda)\Gamma(n+\lambda+1/2)} P_n^{(\lambda-1/2,\lambda-1/2)}(t), \end{aligned} \tag{6.4}$$

where $\lambda = m + 1/2$. Moreover, using the relations of Jacobi polynomial to the hypergeometric functions one has the explicit representation due to Szego[45]

$$P_n^\lambda(t) = \sum_{k=0}^{[n/2]} (-1)^k \frac{\Gamma(n-k+\lambda)}{\Gamma(\lambda)\Gamma(k+1)\Gamma(n-2k+1)} (2t)^{n-2k}. \tag{6.5}$$

Combining Eqs. (6.1), (6.4) and (6.5), we write the solid spherical harmonic Gaussian basis functions in the following form:

$$r^l Y_l^m(\theta, \phi) e^{-\alpha r^2} = b(lm) \sum_{k=0}^{[n/2]} C_{mk}^n (x+iy)^m \ (z)^{n-2k} (r)^{2k} e^{-\alpha r^2}, \tag{6.6}$$

where

$$C_{mk}^n = (-1)^k (2)^{n-2k} \frac{\Gamma(n-k+m+1/2)}{\Gamma(m+1/2)\Gamma(k+1)\Gamma(n-2k+1)}, \tag{6.7}$$

and

$$b(lm) = (-1)^m \sqrt{\frac{(2l+1)}{4\pi}} \sqrt{\frac{(l-m)!}{(l+m)!}} \frac{\Gamma(2m+1)}{\Gamma(m+1)} 2^{(-m)}. \tag{6.8}$$

The expression Eq. (6.6) can be generalized for the solid spherical harmonics as

$$\begin{aligned}(\frac{-\partial}{\partial\alpha})^\delta [r^l Y_l^m(\theta, \phi) e^{-\alpha r^2}] &= r^{l+2\delta} Y_l^m(\theta, \phi) e^{-\alpha r^2} \\ &= b(lm) \sum_{k=0}^{[n/2]} C_{mk}^n (x+iy)^m \ (z)^{n-2k} (r)^{2k+2\delta} e^{-\alpha r^2},\end{aligned} \tag{6.9}$$

where δ takes the value 1 or 0 depending on whether the derivatives are involved or not.

By using the binomial expansion of the terms such as $(x+iy)^m$ and $r^{2k+2\delta}$, the right hand side of Eq.(6.9) can be expanded in cartesian form. For example, the solid spherical harmonic centered around atom A can be expanded in cartesian form as

$$\begin{aligned}& r_A^{l_A+2\delta_A} Y_{l_A}^{m_{lA}}(\theta_A, \phi_A) e^{-\alpha_p r_A^2} = \\ & b(lm) \sum_{k_A=0}^{[n_A/2]} C_{m_{lA} k_A}^{n_A} (x_A+iy_A)^{m_{lA}} z_A^{n_A-2k_A} (r_A^2)^{k_A+\delta_A} e^{-\alpha_p r_A^2} \\ & = b(lm) \sum_{\alpha_A=0}^{m_A} \sum_{\beta_A=0}^{k_A+\delta_A} \sum_{\gamma_A=0}^{k_A+\delta_A-\beta_A} \sum_{k_A=0}^{[n_A/2]} (i)^{\alpha_A} C_{m_A k_A}^{n_A} \binom{m_A}{\alpha_A} \binom{k_A+\delta_A}{\beta_A} \binom{k_A+\delta_A-\beta_A}{\gamma_A} \\ & \left[x_A^{2k_A+2\delta_A-2\beta_A-2\gamma_A+m_A-\alpha_A} y_A^{2\gamma_A+\alpha_A} z_A^{n_A-2k_A+2\beta_A}\right] e^{-\alpha_p r_A^2}.\end{aligned} \tag{6.10}$$

In order to compactify the notation we define the summation over the variables associated with each center A as follows:

$$\begin{aligned}\sum_A W_A \equiv b(l_A m_{lA}) & \sum_{\alpha_A=0}^{m_A} \sum_{\beta_A=0}^{k_A+\delta_A} \sum_{\gamma_A=0}^{k_A+\delta_A-\beta_A} \sum_{k_A=0}^{[n_A/2]} \\ & (i)^{\alpha_A} C_{m_A k_A}^{n_A} \binom{m_A}{\alpha_A} \binom{k_A+\delta_A}{\beta_A} \binom{k_A+\delta_A-\beta_A}{\gamma_A}\end{aligned} \tag{6.11a}$$

and

$$\sum_A W_A^* \equiv b(l_A m_{lA}) \sum_{\alpha_A=0}^{m_A} \sum_{\beta_A=0}^{k_A+\delta_A} \sum_{\gamma_A=0}^{k_A+\delta_A-\beta_A} \sum_{k_A=0}^{[n_A/2]} (-i)^{\alpha_A} C_{m_A k_A}^{n_A} \binom{m_A}{\alpha_A} \binom{k_A+\delta_A}{\beta_A} \binom{k_A+\delta_A-\beta_A}{\gamma_A}, \tag{6.11b}$$

with similar notations for expansions around centers B , C and D . Using the above short hand notation, the matrix elements of one-electron operators $\hat{h}$ (which could be either the identity matrix for the overlap integrals or $V_C(i)$ for the nuclear attraction integrals) can be written as

$$< A l_A m_{lA} \alpha_p \delta_A | \hat{h} | B l_B m_{lB} \alpha_q \delta_B > = \sum_A \sum_B W_A^* \; W_B < \eta(A\vec{l}_A \alpha_p) | h | \eta(B\vec{l}_B \alpha_q) > ' \tag{6.12}$$

where the summations over A and B are defined in Eqs. (6.11a) and (6.11b) and the indices $l_{xA,}l_{yA}$ and l_{zA} of the the unnormalised primitive Gaussians defined by

$$\eta(A\vec{l}_A \alpha_p) = x_A^{l_{xA}} \; y_A^{l_{yA}} \; z_A^{l_{zA}} \; e^{-\alpha_p r_A^2}, \tag{6.13}$$

are given as follows:

$$l_{xA} = 2k_A + 2\delta_A - 2\beta_A - 2\gamma_A + m_A - \alpha_A, \tag{6.14a}$$

$$l_{yA} = 2\gamma_A + \alpha_A \tag{6.14b}$$

and

$$l_{zA} = n_A - 2k_A + 2\beta_A, \tag{6.14c}$$

with similar expressions for indices l_{xB}, l_{yB} and l_{zB}. etc.

Here the integrals over the primitive cartesian Gaussians are the standard nonrelativistic-type for which well known techniques[36] already exist.

In order to evaluate the two-electron matrix elements we write the standard four-center primitive integral (Eq. (5.6)) in the form

$$< A l_A m_{lA} \alpha_p \delta_A \; B l_B m_{lB} \alpha_q \delta_B | \, | C l_C m_{lC} \alpha_r \delta_C \; D l_D m_{lD} \alpha_q \delta_D > = \sum_A \sum_B \sum_C \sum_D W_A^* \; W_B W_C^* \; W_D < \eta(A\vec{l}_A \alpha_p) \; \eta(B\vec{l}_B \alpha_q) | \frac{1}{r_{12}} | \eta(C\vec{l}_C \alpha_r) \; \eta(D\vec{l}_D \alpha_s) >, \tag{6.15}$$

where the the appropriate summation indices are explained in Eq. (6.11) and the indices l_A, l_B, l_C, and l_D are noted in Eq. (6.14). It is clear that these reduce to the nonrelativistic-type[36] four-center two-electron integrals. Furthermore Eqs. (6.12) and (6.15) complete the proof that all the relativistic matrix elements involving the large and the small components are directly obtainable from the standard nonrelativistic-type matrix elements which uses cartesian Gaussian functions as the basis.

A few comments are now in order. The expressions given in Eq. (6.12) and (6.15) are for $m_A > 0, m_B > 0, m_C > 0, m_D > 0$. However, for negative values of m one can use the relation[44]

$$Y_l^{-m} = (-1)^m Y_l^{m*}, \tag{6.16}$$

to generate similar expressions to evaluate one- electron and two-electron integrals involving Y_l^{-m} .

7. Concluding Remarks

In this paper we have presented a method to incorporate the kinetic balance condition in DHF calculations of molecular structure. The large component of the molecular spinor is expanded as a linear combination of atomic basis spinors of Gaussian-type functions. The small component of the molecular basis spinor is related to the large component by the kinetic balance condition, and is matched with the large component basis in the contracted form. The employment of these matched basis sets of Gaussian basis spinors should provide the correct nonrelativistic limit, and the variational safety needed in relativistic molecular structure calculation. As we have recently demonstrated in case of atoms, this type of one-to-one matching between the large and the small-component is essential to avoid variational collapse in basis set calculations in the DHF approximation. We have also emphasized the advantage of employing the geometric basis functions of Gaussian type in molecular structure calculations. As a consequence of the geometric nature of the basis functions, one can not only express the Fock matrix elements involving the small- component in terms of the previously evaluated matrix elements involving the large-components, but also one can extend the basis towards completeness. This last point is crucial in studying the convergence properties of the DHF calculations in finite-size nucleus models, where the relaxation of cusp condition near the origin due to finite-size results in a fewer basis functions with a faster convergence rate. In our approach the finite nucleus-size effects can readily be incorporated by adding the corrections to the nuclear attraction integrals following Matsuoka[38]. Work is in progress to include open-shell and therefore molecular symmetry[46] within the DF SCF formalism described above; the numerical results will also be published shortly[47].

Acknowledgement

The authors would like to express their sincere thanks to Prof. G. Malli for carefully checking many of the formulas presented in the original manuscript. Any remaining errors are, of course, authors' responsibility. One of the authors (A. M.) would like to thank Prof. G. Malli for stimulating discussions on the subject. Thanks are also due to Dr. Daniel Frye and Dr. Abbas Farazdel for their useful comments and suggestions during the course of the preparation of the manuscript.

Appendix 13A.

Finite Nuclear Size Corrections

In this appendix we follow Matsuoka[38] to summarize additional matrix elements that arise when finite nucleur size is taken into account. For the point nucleus case we have $V_C(i) = -Z_C/r_{iC}$, whereas for a nucleus of uniform spherical charge distribution of radius R_C the potential is modified to a form given by Eq. (2.5b). We follow Matsuoka[38] to compute the matrix elements of the modified potential given by Eq. (2.5b).

Let us recall that the nuclear attraction matrix elements are given by Eqs.((4.6a)-(4.6b) and (5.5a)-(5.5d)). We define an unnormalised primitive Gaussian basis function in the form

$$\xi_A(\vec{r}_A) \equiv r_A^{l_A+2\delta_A} Y_{l_A}^{m_{lA}}(\theta_A, \phi_A) e^{-\alpha_p r_A^2} \equiv <r|\delta_A A l_A m_{lA} \alpha_p>. \qquad (A.1)$$

Then the correction to the general nuclear attraction matrix elements (Eqs.(4.6a)-(4.6b) and (5.5a-5.5d)) can be easily computed by evaluating the matrix element of the operator $V_C(i)$ given by (2.5b) between two primitive Gaussians centered at A and B, viz,

$$<\delta_A A l_A m_{lA} \alpha_p | V_C(i) | \delta_B B l_B m_{lB} \alpha_q> = \int \xi_A^* (\vec{r}_{1A}) \xi_B (\vec{r}_{1B}) V_C(\vec{r}_{1C}) d\vec{r}_1. \qquad (A.2)$$

Following Matsuoka[38] we redefine the potential in Eq. (2.5b) in terms of two other potentials:

$$V'_C(i) = -(Z_C)\left[\frac{3}{2R_C} - \frac{r_{iC}^2}{2R_C^3} - \frac{1}{r_{iC}}\right], \quad r_{iC} < R_C \qquad (A.3a)$$
$$= 0 \qquad\qquad r_{iC} \geq R_C$$

and

$$V''_C(i) = -\frac{(Z_C)}{r_{iC}}. \qquad r_{iC} > 0 \qquad (A.3b)$$

The potential in Eq. (A.3a) gives the correction due to the finite size of the nucleus and the potential in Eq.(A.3b) is the usual Coulomb potential for the point nucleus. Hence, the finite nucleus correction to the primitive integral (A.2) can be readily incorporated by making the substitutions:

$$\begin{aligned} <\delta_A A l_A m_{lA} \alpha_p | V_C(i) | \delta_B B l_B m_{lB} \alpha_q> &= <\delta_A A l_A m_{lA} \alpha_p | V''_C(r_{iC}) | \delta_B B l_B m_{lB} \alpha_q> \\ &+ <\delta_A A l_A m_{lA} \alpha_p | V'_C(i) | \delta_B B l_B m_{lB} \alpha_q>, \end{aligned} \qquad (A.4)$$

in Eqs.(4.6a)-(4.6b) and Eqs. (5.5a)-(5.5d). Note that $\delta_A = \delta_B = 0$ for the matrix elements given by Eqs. (4.6a)-(4.6b). From Eq.(A.4) it is clear that the first term in Eq. (A.4) involving the point nucleus can be evaluated by the standard nonrelativistic methods[36] as discussed in section 4. However, for the second term, there are three types of additional matrix elements which have been calculated by Matsuoka[38] and for the sake of completeness are reported below.

(1) One-center integral

$$\begin{aligned} \int \xi^*_C(\vec{r}_{1C})\xi_{C'}(\vec{r}_{1C})V'_C(\vec{r}_{1C})d\vec{r}_1 &= \delta_{l_C l_{C'}}\delta_{m_{lC} m_{lC'}}(-Z_C/2) \\ \times \quad &\{R_C^{2\delta+2}[3F_{\delta+1}(\alpha R_C^2) - F_{\delta+2}(\alpha R_C^2) - 2V_\delta(\alpha R_C^2)]\}, \end{aligned} \qquad (A.5)$$

where

$$\delta = \delta_C + \delta_{C'} + l_C, \quad \alpha = \alpha_p + \alpha_q \qquad (A.6)$$

and the auxilliary functions are defined by

$$F_m(t) = \int_0^1 u^{2m} \exp(-tu^2)\, du \qquad (A.7)$$

and

$$\begin{aligned} V_m(t) &= \int_0^1 u^{2m+1} \exp(-tu^2)\, du \\ &= \frac{m!}{2t^{m+1}} (1 - e^{-t}\sum_{i=0}^{m} \frac{t^i}{i!}). \end{aligned} \qquad (A.8)$$

The expression (A.5) is a generalization of the expressions given by us in chapter 3, section 8, where we treated specifically the finite-size ncleus effects for atoms.

(2) Two-center hybrid integral

The basis function ξ_B centered on B are expanded in terms of spherical harmonics around center C as

$$\xi_B(\vec{r}_{1B}) = \sum_{lm} \alpha_{lm}(r_{1C}) Y_l^m(\theta_{1C}, \phi_{1C}), \tag{A.9}$$

with

$$\alpha_{lm}(r_{1C}) = \sum_{n=0}^{\infty} a_{2n}^{(lm)} \, r_{1C}^{2n+l}. \tag{A.10}$$

The constant coefficients $a_{2n}^{(lm)}$ have been obtained by Suzuki[48,49], which is given below.

Using Eqs.(A.9)and (A.10), the two-center hybrid integral are evaluated in a form: similar to Eq.(A.5) as

$$\int \xi_C^* (\vec{r}_{1C}) \xi_B (\vec{r}_{1B}) V'_C(\vec{r}_{1C}) d\vec{r}_1 =$$
$$(-Z_C/2) \sum_{n=0}^{\infty} \left\{ R_C^{2\delta'+2} a_{2n}^{(l_C m_{lC})} [3F_{\delta'+1}(\alpha R_C^2) - F_{\delta'+2}(\alpha R_C^2) - 2V_{\delta'}(\alpha R_C^2)] \right\}, \tag{A.11}$$

where

$$\delta' = n + \delta_C + l_C, \tag{A.12}$$

and the auxilliary function $V_m(t)$ is defined by Eq. (A.8)

(3) Two-center Coulomb integral and the general three-center integral

By expanding the product of two Gaussians in the form of another Gaussian Matsuoka [38] obtained

$$\xi_A^* (\vec{r}_{1A}) \xi_B (\vec{r}_{1B}) = \sum_{l_C m_{lC}} \beta_{l_C m_{lC}}(r_{1C}) Y_{l_C}^{m_{lC}}(\theta_{1C}, \phi_{1C}), \tag{A.13}$$

where

$$\beta_{l_C m_{lC}}(r_{1C}) = \sum_{n=0}^{\infty} b_{2n}^{(l_C m_{lC})} r_{1C}^{l_C+2n}. \tag{A.14}$$

Hence we can write the general three-center integral as

$$\int \xi_A^* (\vec{r}_{1A}) \xi_B (\vec{r}_{1B}) V'_C(\vec{r}_{1C}) d\vec{r}_1 =$$
$$(-Z_C) \, (-\sqrt{\pi}) \sum_{n=0}^{\infty} \left\{ R_C^{2n+2} [3b_{2n}^{(00)}/(2n+3)(2n+5)(n+1)] \right\}, \tag{A.15}$$

where,

$$b_0^{(00)} = = \sqrt{4\pi}\,\Omega_{AB}(r_C), \tag{A.16}$$

with

$$\Omega_{AB}(r_C) = \lim_{r_{1C}\to 0} \xi_A^* \xi_B. \tag{A.17}$$

As pointed out by Matsuoka[38], the coefficients $b_{2n}^{0\,0}$, for $n \geq 1$ contain higher derivatives of Ω_{AB}; however, the magnitude of the second plus higher order terms in the series is $< 10^{-16}$. a.u. Therefore, to the precision of $< 10^{-16}$ a.u., the integral (A.15) can be approximated by n=0 term to give[38]

$$\int \xi_A^* (\vec{r}_{1A})\xi_B(\vec{r}_{1B})V'_C(\vec{r}_{1C})d\vec{r}_1 = (-Z_C)(\frac{-2\pi}{5})\ \Omega_{AB}(r_C)R_C^2, \tag{A.18}$$

which can represent accurately the needed finite size nucleus correction to the above integrals. A similar analysis can be applied to the two-center Coulomb integral defined by

$$\int \xi_A^* (\vec{r}_{1A})\xi_A(\vec{r}_{1A})V'_C(\vec{r}_{1C})d\vec{r}_1. \tag{A.19}$$

The calculation of the coefficients a_{2n}^{lm} is straightforward due to the formula already given by Matsuoka[38] (using eqs. (6.1) and (6.4) of Suzuki[48]), viz,

$$\begin{aligned} a_{2n}^{l_A m_{lA}} = {} & 2\gamma(l_B m_{1B} \mid l_A) \\ & \times \sum_s h_{l_A+n,\,s}(l_B m_{1B} \mid l_A) R_{AB}^{s+l_B-l_A-2n-1} \frac{g^s(R_{AB})}{s!}, \quad m_{lA} = m_{1B} \geq 0. \end{aligned} \tag{A.20}$$

In eq. (A.20), R_{AB} is the distance between the two centers A and B,

$$g^s(R_{AB}) = [(\frac{\partial}{\partial R})^s g(R)]_{R=R_{AB}}, \tag{A.21}$$

where

$$g(R) = \frac{R^{l_A+2\delta_A}}{R^{l_A-1}} \exp^{-\alpha R^2}, \tag{A.22}$$

and $\gamma(l_B m_{1B} \mid l_A)$ defined in Eq. (2.9) of Suzuki[48] has the form, viz,

$$\begin{aligned} \gamma(l_B m_{1B} \mid l_A) = {} & (-1)^{l_B} \frac{(2l_B-1)!!\,(2l_A-1)!!}{2^{l_B+l_A+1}} \\ & \times \left[\frac{(2l_B+1)\,(2l_A+1)}{(l_B-m_{1B})!(l_B+m_{1B})!(l_B-m_{1B})!(l_B+m_{1B})!} \right]^{1/2}. \end{aligned} \tag{A.23}$$

The h-coefficients have been computed by using Suzuki's formula[48] (a misprint in Eq. (5.6) of reference 48 has been corrected)

$$h_{n,s}(LM|l) = \sum_{k=0}^{[(2n-s)/2]} \sum_{K=[(2n-s+1)/2]-k}^{L+l-k} \tag{A.24}$$
$$\{(2K)![(2n-2k+1)(2n-2k-s)!(2K-2n+2k-s)!]^{-1}\} b_{Kk}(LM|l),$$

where the coefficients $b_{Kk}(LM|l)$ can be expressed by a single sum due to Rashid[50] (see Eq. 16 of Suzuki[49]), viz,

$$b_{Kk}(LM|l) = \frac{(L-M)!\,(L+M)!\,(l-M)!\,(l+M)!(L-K-1/2)!(l-k-1/2)!}{(L+l-K-k)!(L-1/2)!K!\,(l-1/2)!\,k!}$$
$$\times \sum_{s=0}^{L-M} \frac{(-L+K+k+s-1/2)!}{s!\,(L-M-s)!(l-L+s)!(-L+s-1/2)!(L+M-s)!}. \tag{A.25}$$

The h-coefficients are given in a simplified form by Eq. 27 of Suzuki'[49], viz,

$$h_{n,2n-i}(LM|l) = \frac{(L-M)!\,(L+M)!\,(l-M)!\,(l+M)!}{2(L-1/2)!\,(l-1/2)!}$$
$$\sum_{k=0}^{[i/2]} \sum_{K=[(i+1)/2]-k}^{L+l-k} \sum_{s=0}^{L-M} \frac{(n-k-1/2)!\,(K-1/2)!}{(n-k+1/2)!}$$
$$\times \frac{1}{([i/2]-k)!([(i+1)/2]-k-1/2)!(K+k-[(i+1)/2])!(K+k-[i/2]-1/2)!}$$
$$\times \frac{(L-K-1/2)!(l-k-1/2)!(-L+K+k+s-1/2)!}{(L+l-K-k)!k!s!(L-M-s)!(l-L+s)!(-L+s-1/2)!(L+M-s)!},$$

where n runs over all non-negative integers, i ranges from 0 to min $\{2n, 2(L+l)\}$, and $[i/2]$ denotes the quotient of i/2. As pointed out by Matsuka[38] and Suzuki[49] the summation index s runs actually from max $\{0, L-l\}$ to L-M; hence the h-coefficients can be evaluated for any value of L and l. However, for practical calculations the summation over s only extends to the limited range since the dominant corrections to the nuclear attraction integral for DF calculations are expected to come from l_A or $l_B = 0,1$.

As has been stressed by Matsuoka[38] the formula (A.20) is only valid if center **B** is on the + Z-axis with A as the origin of the coordinate system; when center **B** is on the - Z-axis, these coefficients are to be multiplied by a factor $(-1)^{(l_A+l_B)}$. For a general case one has to evaluate the rotation matrix elemnts (see Eqs. (6.1) and (6.2) of Suzuki [48]) which we describe in Appendix B.

Appendix 13B.
Eulerian Angles and Rotation Matrix Elements

In this appendix we follow Harris[40] to summarize the rotation matrix elements and apply these to evaluate the general two-electron nuclear attraction integrals[38,48] described in appendix A.

We follow the convention by Edmondson[44] to describe a right-handed coordinate system such that a counterclockwise rotation by an angle $\pi/2$ about the z axis carries the positive x axis into the position originally occupied by the positive y axis. Then the Eulerian angles (α, β, γ) describe the following ordered operations on a set of coordinate axes:

(1) Rotation by an angle α counterclockwise about the z axis.

(2) Rotation by an angle β, $0 \leq \beta \leq \pi$ counterclockwise about the new y axis.

(3) Rotation by an angle γ counterclockwise about the new z axis.

Following Harris[40] we define a coordinate system centered at a point P, with polar axis $\theta = 0$ along the directed line $(\vec{B} - \vec{A})$. Such an axial system, which we shall characterize by Eulerian angles $(\alpha_P, \beta_P, \gamma_P)$, defines the coordinates (r_P, θ_P, ϕ_P) and we shall denote this as the quantized axes. In terms of $\vec{A}$ and $\vec{B}$, the Eulerian angels are given by[40]

$$\cos\alpha_P = \frac{B_x - A_x}{R_{AB}\sin\beta_P}, \quad \sin\alpha_P = \frac{B_y - A_y}{R_{AB}\sin\beta_P}, \tag{B.1}$$

$$\cos\beta_P = \frac{B_z - A_z}{R_{AB}} \quad and \ \gamma_P = 0, \tag{B.2}$$

where $R_{AB} = |\vec{B} - \vec{A}|$.

Furthermore, we adopt Harris'[40] notation and define $(\alpha_{PA}, \beta_{PA}, \gamma_{PA})$ as the rotation which carries axes denoted by $(\alpha_A, \beta_A, \gamma_A)$ to the quantized system of axes denoted by $(\alpha_P, \beta_P, \gamma_P)$.

Even if the spherical harmonics $Y_{l_A}^{m_{lA}}(\theta_A, \phi_A)$ are not quantized in the direction $(\vec{A} - \vec{B})$, it still can be expressed in terms of spherical harmonics $Y_{l_A}^{m'_{lP}}(\theta_P, \phi_P)$ and the rotation matrix $D^l_{m_{lA}m'_{lP}}(\alpha_{PA}, \beta_{PA}, \gamma_{PA})$ as follows:

$$Y_{l_A}^{m_{lA}}(\theta_A, \phi_A) = \sum_{m'_{lP}=-m_{lA}}^{|m_{lA}|} Y_{l_A}^{m'_{lP}}(\theta_P, \phi_P) D^{l}_{m_{lA}m'_{lP}}(\alpha_{PA}, \beta_{PA}, \gamma_{PA}). \qquad (B.3)$$

We shall follow Harris[40] to define elements of the rotation matrix in (B.3) as

$$D^{l}_{mm'}(\alpha, \beta, \gamma) = e^{im\alpha+im'\gamma} d^{l}_{mm'}(\cos\beta), \qquad (B.4)$$

where

$$d^{l}_{mm'}(t) = (l+m)!(l-m')! \sum_{s} \frac{(-1)^s(\frac{1-t}{2})^{s-m/2+m'/2}(\frac{1+t}{2})^{l+m/2-m'/2-s}}{s!(l-m'-s)!(s+m'-m)!(l+m-s)!}. \qquad (B.5)$$

Eq. (B.5) is valid for both m *and* $m' \geq 0$. For negative values of m and m' one can use

$$d^{l}_{mm'}(t) = d^{l}_{-m-m'}(t) = (-1)^{l+m+m'} d^{l}_{-mm'}(-t) = (-1)^{l+m+m'} d^{l}_{m-m'}(-t). \quad (B.6)$$

The relation between the Eulerian angles $(\alpha_A, \beta_A, \gamma_A)$, $(\alpha_P, \beta_P, \gamma_P)$ and $(\alpha_{PA}, \beta_{PA}, \gamma_{PA})$, have been obtained by Harris[40,51] using straightforward trigonmetric manipulations. These are:

$$\cos\beta_{PA} = \cos\beta_A \cos\beta_P + \sin\beta_A \sin\beta_P \cos(\alpha_P - \alpha_A) \qquad (B.7)$$

and

$$\alpha_{PA} = -\gamma_A + \theta, \qquad (B.8)$$

where

$$\sin\theta = \frac{\sin\beta_P \sin(\alpha_P - \alpha_A)}{\sin\beta_{PA}} \qquad (B.9)$$

and

$$\cos\theta = \frac{-\cos\beta_P + \cos\beta_A \cos(\beta_{PA})}{\sin\beta_A \sin\beta_{PA}}. \qquad (B.10)$$

Similarly

$$\gamma_{PA} = \gamma_P - \chi, \qquad (B.11)$$

where

$$\sin\chi = \frac{\sin\beta_A \sin(\alpha_P - \alpha_A)}{\sin\beta_{PA}} \qquad (B.12)$$

and

$$\cos\chi = \frac{\cos\beta_A - \cos\beta_P \cos(\beta_{PA})}{\sin\beta_P \sin\beta_{PA}}. \qquad (B.13)$$

Specification of the range $0 \leq \beta_{PA} \leq \pi$ insures that Eq. (B.4) is single valued.

If $\beta_{PA} = 0$ or π, one may use instead of Eqs. (B.8) and (B.11),

$$\alpha_{PA} = (\alpha_P - \alpha_A)\{sign(\cos\beta_A)\} \tag{B.14}$$

and

$$\gamma_{PA} = \gamma_P - \gamma_A \cos\beta_{PA}. \tag{B.15}$$

If $\beta_{PA} \neq 0$ or π, but, $\sin\beta_A$ or $\sin\beta_P = 0$, then Eq. (B.7) can be modified to give

$$\cos\beta_{PA} = \cos\beta_A \cos\beta_P. \tag{B.16}$$

The results for other angles are given by

$$\cos\theta = \frac{-\sin\beta_A \cos\beta_P + \cos\beta_A \sin\beta_P \cos(\alpha_P - \alpha_A)}{\sin\beta_{PA}} \tag{B.17}$$

and

$$\cos\chi = \frac{\cos\beta_A \sin\beta_P - \sin\beta_A \cos\beta_P \cos(\alpha_P - \alpha_A)}{\sin\beta_{PA}}. \tag{B.18}$$

It should be noted that in general $(\alpha_{PA}, \beta_{PA}, \gamma_{PA}) \neq (\alpha_{AP}, \beta_{AP}, \gamma_{AP})$; rather,

$$\alpha_{AP} = \pi - \gamma_{PA}, \tag{B.19}$$

$$\gamma_{AP} = \pi - \alpha_{PA} \tag{B.20}$$

and

$$\beta_{AP} = \beta_{PA}. \tag{B.21}$$

References

1. Y.K.Kim, Phys. Rev., **154**, 17 (1967).
2. F.Solliec, C.Mijoule and J.M.Leclereq, Chem. Phys.Letters, **85**, 190 (1982).
3. R.E. Stanton and S. Havriliak, J. Chem. Phys., **81**, 1910 (1984).
4. Y. Ishikawa, R.C. Binning and K.M. Sando, Chem. Phys. Letters, **101**, 111 (1983).
5. Y. Ishikawa, R.C. Binning and K. M. Sando, Chem. Phys. Letters, **105** 189 (1984); **117** 444 (1985).
6. K.G. Dyall, I.P. Grant and S. Wilson, J. Phys., **B17**, 493 (1984); ibid **L45**; ibid **B17**, 1201.
7. I.P. Grant, J. Phys., **B19**, 3187 (1986).
8. I.P. Grant and H.M. Quiney, Adv. in Atom. and Mol. Phys., **23**, 37 (1988).
9. H.M. Quiney, I.P. Grant and S. Wilson, J. Phys., **B20**, 1413 (1987).

10. H.M. Quiney, I.P. Grant and S. Wilson, "On the Relativistic Many-Body Perturbation Theory of Atomic and Molecular Electronic Structure," to be published in *Lecture Notes in Chemistry*, U. Kaldor, Ed., Springer Verlag, Berlin.
11. H.M. Quiney, I.P. Grant and S. Wilson, J. Phys., **B22**, L-15 (1989).
12. Y. Ishikawa, R. Baretty and R.C. Binning, Jr., Chem. Phys. Letters, **121**, 130 (1985); Y Ishikawa and H.M. Quieny, Int. J. Quant. Chem. Sympo. **21**, 523 (1987).
13. P.J.C. Aerts and W.C. Nieuwpoort, Chem. Phys. Letters, **113**, 165 (1985).
14. O. Visser, P.J.C. Aerts, D. Hegarty and W.C. Nieuwpoort, Chem. Phys. Letters, **134**, 34 (1987).
15. O. Matsuoka, M. Klobukowski and S. Huzinaga, Chem. Phys. Letters, **113**, 395 (1985).
16. O. Matsuoka and S. Huzinaga, Chem. Phys. Letters, **140**, 567 (1987).
17. S. Okada and O. Matsuoka, "Relativistic well-tempered Gaussian basis sets for Helium through Mercury", J. Chem. Phys. **91(7)**, 4193 (1989).
18. A.K. Mohanty and E. Clementi, Chem. Phys. Letters, **157**, 348 (1989).
19. A.K. Mohanty and E. Clementi, (see this volume Chapter 3).
20. A.K. Mohanty and E. Clementi, "Kinetically Balanced Geometric Gaussian Basis Sets for Relativistic Atoms", J. Chem. Phys. (in press).
21. W.Kutzelnigg, Int. J. Quan. Chem. **25** 107 (1984).
22. C.C.J.Roothaan, Rev. Mod. Phys. **23**, 69 (1951); **32**179,(1960).
23. I.P. Grant, B.J. McKenzie, P.H. Norrington, D.F. Mayers and N.C. Pyper, Comput. Phys. Communication, **21**, 207 (1980).
24. J.P. Desclaux, Comput. Phys. Communication, **9**, 31 (1975); J.P. Desclaux, Atomic and Nuclear Data Tables, **12**, 311 (1973).
25. G. Malli and J. Oreg, J. Chem. Phys. **68**, (1979) 529.
26. O. Matsuoka, N. Suzuki,T. Ayoma,G. Malli, J. Chem. Phys. **73**, 1320 (1980).
27. T. Ayoma, H. Yamakawa, and O. Matsuoka, J. Chem. Phys. **73** 1328 (1980).
28. Y.S. Lee and A.D. McLean, J. Chem. Phys. **36**, 735 (1982)
29. G. L. Malli and N.C. Pyper, Proc. R. Soc. London, Ser. **A407**, 377 (1986).
30. A.F. Ramos, N.C. Pyper, G. L. Malli, Phys. Rev. A, **38**, 2729 (1986).
31. O. Visser, P.J.C. Aerts, L. Visscher and W.C. Nieuwpoort, Progress in Fock-Dirac Methodology; " Open-Shell States of Polyatomic Molecules", Presented at the 30th Sanibel Symposia, held at St. Augustine, Florida, March 1990. Also see P.J.C. Aerts and W.C. Nieuwpoort, Physica Scripta, **36** 433 (1986): P.J.C. Aerts and W.C. Nieuwpoort, Int. J, Quant. Chem. Symposium, **19** 19, 267 (1986).
32. L.Laaksonen, I.P. Grant and S. Wilson, J. Phys., **B21**, 1969 (1988).
33. A.Bohr and V. F. Weisskopf Phys. Rev. **77**, 77 (1950); F.G. Werner and J.A. Wheeler, Phys. Rev. **109**, 126 (1958);
34. M.E. Rose Phys. Rev. **82**,389 (1951); for solution of Dirac equation in a central potential, particularly in a homogeneously charged sphere model of the nucleus, see for example, "Relativistic Electron Theory" by M.E. Rose, John Wiley & sons, Inc., 1961 (page 240).
35. I.P.Grant, Adv. Phys. **19** 747 (1970).

36. G. Corongiu, R. Wong, S. Chakravorty and E. Clementi, (see this volume chapter 7).
37. M.W. Wong, G. Corongiu and E. Clementi , "Two-electron Integral Evaluation for Uncontracted Geometrical-Type Gaussian Functions", submitted to Journal of Computational Chemistry.
38. O.Matsuoka, Chem. Phys. Letters, **140**, 362 (1987).
39. "Relativistic Quantum Mechanics" by J.D. Bjorken and S. D. Drell, McGraw-Hill Book Company (1964).
40. F. E. Harris, Rev. Modern Physics, **33** , 558 (1963).
41. G. Fieck, J. of Phys. **B, 12** , 1063 (1979).
42. G. Fieck, Theoret. Chim. Acta **54** , 223 (1980).
43. O.Matsuoka, J.Chem. Phys., **92**, 4364 (1990).
44. For Properties of spherical harmonics , see for example, "Angular Momentum in Quantum Mechanics" by A. R. Edmonds, Princeton University press, 1974. The expressions for the solid spherical harmonics (for specific l and m values) in terms of cartesian coordinates can be found in page 124.
45. "Orthogonal Polynomials" by G. Szego, Published by American Mathematical Society, Fourth Edition, (1975).
46. G. Malli, A.Mohanty and E. Clementi (in preparation)
47. A.Mohanty, G. Malli and E. Clementi (in preparation)
48. N. Suzuki, J. Math. Phys. **25**, 1133 (1984).
49. N. Suzuki, J. Math. Phys. **26**, 3193 (1985).
50. M.A.Rashid, J. Math. Phys. **22**, 271 (1985).
51. For errata in the calculation of rotation angles, see F.E. Harris and H.H. Michels, J. Chem. Phys. **45** , 116 (1966).

Chapter 14.

First-Principles Molecular Dynamics

Michele Parrinello
IBM Research Division
Zurich Research Laboratory
CH-8803 Rüschlikon, Switzerland

Recent progress has made possible the practical formulation of a first-principles Molecular Dynamics scheme in which the interatomic potential is explicitly derived from the electronic ground-state within Density Functional theory in the local density approximation. This approach is not only useful in simulating atomic dynamics but also in solving the ground-state problem for a generic atomic aggregate. In such a context, it allows a general minimization scheme to be formulated based on the concept of simulated annealing. In this note we review briefly the formulation of this new approach for molecular dynamics and density functional calculations.

1. Introduction

Computer simulations based on Molecular Dynamics (MD) is a very powerful tool for the study of complex matter systems.[1,2] Such techniques allow us to calculate the atomic trajectories and thus simulate the microscopic behavior of condensed systems. Equilibrium and non-equilibrium properties can then be extracted. These simulations are a very valuable supplement to both theory and experiment. In particular, microscopic details of the atomic motion, which are not directly available in real experiments, are accessible to computer simulations. This makes MD very attractive for material science since, in principle, realistic models of the structure and of the properties of materials may be studied with MD techniques in a wide area of current interest.

MD assumes that atomic dynamics is governed by the laws of classical mechanics. This is a valid approximation if quantum effects on the nuclear motion can be neglected and if a strict adiabatic separation between nuclear and electronic degrees of freedom holds, i.e., the electrons can be considered as being always in the ground-state corresponding to the instantaneous nuclear configuration. In most cases of interest these conditions are well satisfied. However, explicit knowledge of the electronic ground-state is required for a correct description of the interatomic potential. This is a difficult many-body quantum problem which is made even more complicated by the Fermi statistics of the electrons. A great simplification in this respect is provided by Density Functional (DF) theory[3,4,5] which allows us to recast the many-body problem into a set of self-consistent single-particle equations. These may be solved, in practice, by adopting the Local Density Approximation (LDA) for exchange and correlation.[4-7] This is the most commonly used approach to date for electronic structure calculations in condensed matter physics. For not too strongly correlated systems, LDA works surprisingly well as demonstrated by the great success enjoyed by first-principles quantum mechanical LDA calculations in predicting the properties of simple structures.[5,6] However, until very recently the computational complexity of quantum mechanical LDA energies and forces was considered prohibitive for direct applications to MD simulations. For this reason effective potentials which are fitted either to empirical data or to a limited set of quantum mechanical calculation have been widely used to model the interatomic interaction in MD simulations. This approach is appropriate for chemically inert systems like rare gases, but it is clear that one must go beyond effective potentials in order to simulate systems of interest to material science where non-trivial chemical rebonding effects occur. It seems particularly difficult to find sufficiently general effective potentials that work for different states of matter aggregation and for a wide class of materials. In fact, the effective potential approach suffers from an important limitation: the strong correlation existing in nature between local electronic structure, i.e., bonding properties and atomic dynamics, is missed as well as the effect of atomic dynamics on electronic properties. A conceptually more satisfactory approach has been proposed recently and here we call it combined Molecular Dynamics Density Functional (MD-DF) scheme.[8] In such a scheme one considers a fictitious dynamical system whose potential energy surface is an appropriate functional of both electronic and ionic degrees of freedom. The electronic degrees of freedom are the single-particle orbitals of DF theory. When the above energy functional is minimized with respect to electronic degrees of freedom, one recovers the Born-Oppenheimer (BO) potential energy surface to be used in MD simulations of the physical trajectories of the nuclei. Various kinds of dynamical evolutions are possible in the fictitious parameter space, depending on the problem of interest. If only electronic dynamics is used, it provides an alternative framework for ground-state electronic structure calculations, which is particularly useful in large scale problems. If coupled nuclear and electronic dynamics is considered, the scheme provides a convenient formulation for the simultaneous relaxation of both nuclear and electronic degrees of freedom. Global minimization in the fictitious parameter space may be performed quite efficiently by using the concept of simulated annealing.[9] Since minima of the fictitious potential energy surface correspond to minima of the physical BO surface, one can generate in this way the relevant ground-state configurations of a given atomic aggregate. Furthermore, under appropriate conditions the difference in the time scale of the dynamics of nuclear and electronic parameters can be kept so large that nuclear trajectories, initially lying on the BO surface, will deviate appreciably from the BO surface only after times much longer than the MD time step and, in some cases, even longer than the MD observation time. Therefore, the combined MD-DF approach is an efficient scheme to generate from first-principles the correct BO dynamics of the nuclei. Here *correct* means the same dynamics that one would generate by computing first-principles quantum mechanical energies and forces for each one of the many nuclear configurations occurring in a MD run.

The chapter is organized as follows. In Sec. 2 we outline the construction of the BO potential energy surface within DF theory. In Sec. 3 we consider simple dynamical approaches to minimize the energy functional and we stress the convenience of the dynamical point of view for large scale electronic problems. The general formulation of the MD-DF scheme is presented in Sec. 4, in which we discuss both global optimization of the BO surface via simulated annealing and *ab initio* MD with BO trajectories generated via the MD-DF approach. Some present limitations of the method and some possible future directions are discussed in Sec. 5.

2. The interatomic potential within DF theory

According to DF theory, the total ground-state energy $\Phi(\{\mathbf{R}_I\})$ of a system of interacting electrons and nuclei, corresponding to the nuclear configuration $\{\mathbf{R}_I\}$, is a unique functional of the electronic density $n(\mathbf{r})$[3]. By varying the nuclear configuration $\{\mathbf{R}_I\}$, Φ defines the BO potential energy surface for the nuclei.

We express $n(\mathbf{r})$ in terms of N doubly occupied single-particle orbitals, i.e.

$$n(\mathbf{r}) = 2\sum_i^{cc} |\psi_i(\mathbf{r})|^2, \tag{1}$$

where the factor 2 stands for double occupancy. We are assuming here that all the occupation numbers can be taken as equal. This is the case when we do not consider incompletely filled degenerate states. The ground-state energy can be found by minimizing the functional $E[\{\phi_i\}, \{\mathbf{R}_I\}]$ with respect to the *electronic degrees of freedom* $\{\phi_i\}$, i.e.

$$\Phi(\{\mathbf{R}_I\}) = \min_{\{\psi_i\}} E[\{\psi_i\}, \{\mathbf{R}_I\}] \ . \tag{2}$$

The functional $E[\{\psi_i\}, \{\mathbf{R}_I\}]$ is given by

$$\begin{aligned} E[\{\psi_i\}, \{\mathbf{R}_I\}] =& 2\sum_i^{occ}\int d\mathbf{r}\psi_i^*(\mathbf{r})\left(-\frac{1}{2}\nabla^2\right)\psi_i(\mathbf{r}) + \int d\mathbf{r}V^{ext}(\mathbf{r})n(\mathbf{r}) \\ & \frac{1}{2}\int d\mathbf{r}d\mathbf{r}'\frac{n(\mathbf{r})n(\mathbf{r}')}{|\mathbf{r}-\mathbf{r}'|} + E^{xc}[n] + \frac{1}{2}\sum_{I\neq J}\frac{Z_I Z_J}{|\mathbf{R}-\mathbf{R}'|} \end{aligned} \tag{3}$$

Atomic units $e = \hbar = m_e = 1$ are used throughout the chapter. $E^{xc}[n]$ is the exchange-correlation energy and $V^{ext}(\mathbf{r})$ is the total external potential felt by the electrons, i.e. the total electrostatic potential of the nuclei in an all-electron formulation. Z_I are nuclear charges. The single-particle orbitals ψ_i are subject to the orthonormality constraints:

$$\int d\mathbf{r}\psi_i^*(\mathbf{r})\psi_j(\mathbf{r}) = \delta_{ij} \ . \tag{4}$$

Core electrons do not contribute to chemical bonding, but constitute a considerable complication because they introduce length scales much shorter than the interatomic distance. We use here the standard way of integrating out core states, namely we replace nuclei plus core electrons with appropriate pseudo-ions.[10] Then, in Eq. (3), $V^{ext}(\mathbf{r})$ becomes a sum of ionic pseudopotentials and Z_I become ionic pseudocharges. Experience has demonstrated that a particular class of pseudopotentials, which are called norm-conserving,[11,12] are able to describe with great accuracy the properties of atoms in many different chemical environments. Calculations performed on molecules, surfaces and bulk solids have shown that norm-conserving pseudopotentials give structural properties, such as cohesive energies, bond distances and vibrational frequencies, which are practically identical to those obtained in all-electron calculations of similar numerical accuracy. Norm-conserving pseudopotentials are non-local since they act differently on the different angular momentum components of the wavefunction. For the sake of simplicity we shall neglect in what follows this technical complication and we shall limit ourselves to local pseudopotentials. A discussion on how to efficiently include non-local pseudopotentials in the MD-DF scheme can be found in Refs. 13 and 14.

The standard way of solving Eq. (2), subject to the constraints given in Eq. (4), consists in solving the associated Euler-Lagrange equations, i.e.

$$H\psi_i(\mathbf{r}) = \varepsilon_i\psi_i(\mathbf{r}) \ , \tag{5}$$

where:

$$H = -\frac{1}{2}\nabla^2 + V^{ext}(\mathbf{r}) + V^H(\mathbf{r}) + \mu^{xc}(\mathbf{r}) \ . \tag{6}$$

Here

$$V^H(\mathbf{r}) = \int d\mathbf{r}' \frac{n(\mathbf{r}')}{|\mathbf{r} - \mathbf{r}'|}$$

is the Hartree potential and

$$\mu^{xc}(\mathbf{r}) = \frac{\delta E^{xc}[n]}{\delta n(\mathbf{r})}$$

is the exchange-correlation potential. The Schrödinger-type Eqs. (5) are called Kohn-Sham (KS) equations and provide the theoretical framework of most self-consistent electronic structure calculations. By expanding the ψ_i in terms of a basis set the KS equations are converted into a matrix eigenvalue problem. In the plane-wave pseudopotential formalism[10,15,6,7] we have:

$$\psi_i(\mathbf{r}) = \sum_{\mathbf{G}} c^i_{\mathbf{G}} \exp(i\mathbf{G}\cdot\mathbf{r}) \ . \tag{7}$$

The sum over **G** in Eq. (7) is usually truncated to include only M plane waves up to a cutoff G_{max}, which determines the accuracy of the calculation. The use of plane waves assumes periodic boundary conditions, as is customary in MD simulations of bulk systems. Notice that in Eq. (7) we are assuming that the unit cell is so large that we can neglect band dispersion in the Billlouin zone. Using Eq. (7), Eqs. (5) become:

$$\sum_{\mathbf{G}'} H_{\mathbf{G}\mathbf{G}'} c^i_{\mathbf{G}'} = \varepsilon^i_{\mathbf{G}} c^i_{\mathbf{G}} \ , \tag{8}$$

with the Hamiltonian matrix given by:

$$H_{\mathbf{G}\mathbf{G}'} = \frac{1}{2}\mathbf{G}^2 \delta_{\mathbf{G},\,\mathbf{G}'} + \mu^{xc}_{\mathbf{G}-\mathbf{G}'} + V^{ext}_{\mathbf{G}-\mathbf{G}'} + V^{H}_{\mathbf{G}-\mathbf{G}'} \ . \tag{9}$$

In the standard approach, Eqs. (5) are solved by repeated diagonalization of the matrix in Eq. (9). Each diagonalization requires $O(M^3)$ operations. In plane-wave calculations M is usually much greater than N and the numerical cost of the diagonalization becomes rapidly intolerable as the size of the system is increased.

3. A dynamical approach to energy functional minimization

For large scale problems a more convenient formulation is provided by a different approach which does not make explicit use of the KS equations. In this approach we regard the DF minimization as a non-linear minimization problem. Minimization is then achieved by introducing a fictitious dynamics in the space of electronic degrees of freedom, i.e. the ψ_i's.

The simplest dynamics that one can conceive in this respect is provided by steepest descents (SD), which in the present case can be formulated in terms of first order equations in the fictitious time variable:

$$\dot{\psi}_i(\mathbf{r}, t) = -\frac{1}{2}\frac{\delta E}{\delta\psi^*_i(\mathbf{r}, t)} \ . \tag{10}$$

Here the dot indicates time derivative and the functional derivative

$$\frac{\delta E}{\delta \psi_i^*(\mathbf{r}, t)} = 2H\psi_i(\mathbf{r}, t) \ .$$

Here H depends non-linearly on the ψ_i, (Eq. 6). The orthonormality constraints (Eq. 4) must be added to Eq. (10), which define a trajectory in the space of the electronic degrees of freedom where, at each step, one moves a little bit down the *gradient*

$$\frac{\delta E}{\delta \psi_i^*(\mathbf{r}, t)} \ .$$

In this way the system is driven towards a minimum. Experience has shown that at fixed nuclear configuration E has only one minimum and therefore the SD procedure leads to the absolute minimum. Note that one achieves in this dynamical way the objective of self-consistently minimizing E, without ever having to perform explicit diagonalizations.

In the numerical implementation a SD step consists of the following:

$$\psi_i(\mathbf{r}, t + \Delta t) = \psi_i(\mathbf{r}, t) = \Delta t H\psi_i(\mathbf{r}, t) + constraints \ . \tag{11}$$

The elementary time step Δt in Eq. (11) is a parameter that fixes the time scale and therefore governs the convergence rate of the SD scheme. The $\psi_i(\mathbf{r}, t = 0)$ constitute the initial trial state. This must be non-orthogonal to the ground-state. Since E is invariant under unitary transformations in the space of the occupied states, the orthogonalization scheme to be used in Eq. (11) is fixed by convenience. A simple Gram-Schmidt procedure is clearly one possibility.

The basic SD step consists of acting on the ψ_i with the Hamiltonian H followed by an orthonormalization of the new ψ_i so obtained. In a plane-wave formulation the action of the Hamiltonian on the wavefunctions requires $O(NM^2)$ floating point operations, whereas the orthonormalization requires $O(N^2M)$. Such a counting can be improved by partitioning the Hamiltonian into its kinetic and potential energy components. The kinetic energy term is diagonal in reciprocal space so that its action on the wavefunctions requires only $O(NM)$ operations. The action of a local potential on the Fourier coefficients of the wavefunctions is a convolution that can be conveniently calculated in real space by means of Fast Fourier Transform (FFT) techniques,[8] requiring $O(NM \log M)$ operations. In conclusion, the big dimension M enters only linearly or via $M \log M$ in the count of the operations. This gives a significant improvement over standard diagonalization techniques where M enters as M^3. For sufficiently large systems the terms growing as N^2M dominate and the numerical cost of the algorithm still grows as the $(size)^3$ of the system. An additional advantage of the present formulation is that the Hamiltonian H is never explicitly required, permitting a significant saving in memory occupation.

The efficiency of the SD scheme is controlled by the number of steps that are necessary to achieve convergence. This can be quite large, especially in low symmetry situations, where the straightforward SD scheme appears inferior to other iterative approaches for large scale electronic structure calculation. Various attempts to improve the efficiency of the SD approach have been made, using either higher order integration formulae[16,17] or modified SD procedures.[14,18] Recently, a very efficient scheme has been obtained by adapting the well known method of conjugate gradient (CG) minimization[19] to the problem of minimization of the energy functional.[14] Also preconditioning methods that further speed up the rate of convergency have been proposed.[20]

An important point that we want to stress here is that straightforward SD, although at times quite inefficient, constitutes a very simple dynamical approach to minimization. One of its virtues is that it can be easily generalized to more complex systems, having more degrees of freedom. For instance, one may consider the parameter space consisting of both electronic and nuclear degrees of freedom. In this case the SD equations read:

$$\begin{aligned} \bar{\mu}_i \dot{\psi}_i(\mathbf{r}) &= -\frac{1}{2}\frac{\delta E}{\delta \psi_i^*(\mathbf{r}, t)} + constraints \\ \overline{M}_I \dot{\mathbf{R}}_I &= -\frac{\partial E}{\partial \mathbf{R}_I(t)}\,, \end{aligned} \tag{12}$$

where the 'masses' $\bar{\mu}_i$ and $\overline{M}_I$ have been introduced in order to account for the difference in time scales associated with the different parameters. Eqs. (12) define a mass-weighted SD trajectory on the E surface that moves downhill from the starting point to a nearby local minimum. By varying the ψ_i at fixed ionic positions there is a single minimum on the E surface which coincides with a point on the BO potential energy surface Φ defined in Eq. (2). As a consequence, the local minima of the E surface are also local minima of the Φ surface. Eqs. (12) can be used to optimize complex atomic structures by relaxing, simultaneously, electronic and nuclear degrees of freedom. In conventional gradient methods this is accomplished by following a trajectory on the Φ surface, which means that perfect electronic self-consistency is required for any nuclear configuration. By using Eqs. (12) one follows instead a trajectory on the fictitious E surface, and the optimal relative relaxation rate for electrons and ions can be achieved by adjusting the respective mass parameters. The scheme may be easily generalized to include additional degrees of freedom like the volume or the shape of the unit cell. Simultaneous relaxation of electrons and ions substantially equivalent to that defined in Eqs. (12) has been used successfully in several problems involving local optimization of complex structures.

The dynamical minimization approach that we have just described has two important advantages: (i) it allows us to treat efficiently large systems within the plane-wave pseudopotential method, and (ii) it allows us to relax simultaneously nuclear and electronic degrees of freedom. The limitation of the scheme is that it is a method for local rather than global optimization. It is well known that the potential energy surface Φ has many minima. To deal with this problem one needs a global optimization method. We show in the next section that a satisfactory solution to the global optimization problem is based on statistical mechanics and it is intimately connected with the simulation of finite temperature properties.

4. Molecular dynamics in the coupled electron-ion parameter space

In Sec. 3 we have seen that the introduction of a dynamics based on first-order differential equations has led us to the introduction of very useful computational schemes. Here we show that the introduction of a Newtonian dynamics (second order differential equations) in the DF parameter space further enhances our computational capabilities and allows us to implement a DF based MD scheme. To this purpose we consider the parameters Ψ_i and $\mathbf{R}_I$ in the E functional to be dependent on time and write the Lagrangian:[8]

$$L = 2\sum_i^{occ}\int d\mathbf{R}\mu_i\left|\dot{\psi}_i(\mathbf{r})\right|^2 + \frac{1}{2}\sum_I M_I\dot{\mathbf{R}}_I^2 - E[\{\psi_i\}, \mathbf{R}_I] + 2\sum_{ij}\Lambda_{ij}\left(\int d\mathbf{r}\psi_i^*(\mathbf{r})\psi_j(\mathbf{r}) - \delta_{ij}\right), \tag{13}$$

where M_I are the physical ionic masses and μ_i are arbitrary parameters of appropriate units. In what follows we will consider a unique parameter μ for the electronic degrees of freedom, independent from the electronic state, even though this is by no means necessary. In Eq. (13) we have both a classical kinetic energy term $K_e = 2\sum_i^{occ} \int d\mathbf{r}\mu|\dot{\psi}_i(\mathbf{r})|^2$ associated with the electronic parameters ψ_i and a classical kinetic energy term $K_I = \frac{1}{2}\sum M_I\dot{\mathbf{R}}_I^2$ associated with the nuclear coordinates. K_e and K_I measure the rate of variation of the respective degrees of freedom in the coupled electron-ion parameter space. The Lagrange multipliers Λ_{ij} are used to impose the orthonormality constraints (Eq. 4) that, in the language of classical mechanics, are just simple holonomic constraints.

The Lagrangian in Eq. (13) generates a dynamics for the parameters ψ_i and $\mathbf{R}_I$ through the equations of motion:

$$\mu\ddot{\psi}_i(\mathbf{r}, t) = -\frac{1}{2}\frac{\delta E}{\delta\psi_i^*(\mathbf{r}, t)} + \sum_j \Lambda_{ij}\psi_j(\mathbf{r}, t) \tag{14a}$$

$$M_I\ddot{\mathbf{R}}_I = -\frac{\partial E}{\partial\mathbf{R}_I(t)} \ . \tag{14b}$$

The Lagrange multipliers in Eq. (14a) are defined by the constraint conditions (Eq. 4), yielding:[17]

$$\Lambda_{ij} = H_{ji} - \mu \int d\mathbf{r} \dot{\psi}_j^*(\mathbf{r}, t)\dot{\psi}_i(\mathbf{r}, t), \tag{15}$$

where $H_{ji} = \int d\mathbf{R}\psi_j^*(\mathbf{r}, t)H\psi_i(\mathbf{r}, t)$. The matrix $\mathbf{\Lambda} = [\Lambda_{ij}]$ in Eq. (15) is Hermitian and corresponds to a well defined choice for the orthonormalization procedure. This should be contrasted with the case of SD dynamics where the choice of the orthonormalization procedure is irrelevant (see Sec. 2). Here it is not since, while the potential energy E is invariant under unitary transformations in the space of the occupied electron states, the kinetic energy K_e and consequently the Lagrangian L are not.

Equation (15) ensures that the *analytical* solution of Eq. (14a) satisfies the constraints at all $t > t_0$ if it satisfied them at $t = t_0$. In practice, however, Eqs. (14) are *numerically* integrated by standard MD techniques.[2] In the present case, it is convenient to use a simple low order formula such as the Verlet algorithm, in order to save on memory occupation. We have then:

$$\psi_i(\mathbf{r}, t + \Delta t) = -\psi_i(\mathbf{r}, t - \Delta t) + 2\psi_i(\mathbf{r}, t) + \frac{(\Delta t)^2}{\mu}\left(-\frac{1}{2}\frac{\delta E}{\delta\psi_i^*(\mathbf{r}, t)} + \sum_j \Lambda_{ij}\psi_j(\mathbf{r}, t)\right) \tag{16a}$$

$$\mathbf{R}_I(t + \Delta t) = -\mathbf{R}_I(t - \Delta t) + 2\mathbf{R}_I(t) - \frac{(\Delta t)^2}{M_I}\frac{\partial E}{\partial \mathbf{R}_I(t)} . \tag{16b}$$

The analytical expression in Eq. (15) for the matrix of constraint $\mathbf{\Lambda} = [\Lambda_{ij}]$ cannot be used here. In fact, numerical integration errors lead to a rapid deterioration of the constraints. This difficulty can be avoided by adopting the scheme proposed in Ref. 21, where the general problem of incorporating holonomic constraints in MD simulations is addressed. According to Ref. 21, the Lagrangian multipliers should be treated as parameters determined by the requirement that the trajectory satisfies to numerical accuracy the constraints at each time step. The Λ_{ij} so obtained depend upon the specific algorithm used to integrate the trajectory. In our case the $\psi_i(t + \Delta t)$ are given by Eqs. (16a), where the $\psi_i(t)$ and the $\psi_i(t - \Delta t)$ are supposed to be orthonormal. It is convenient to define also $\overline{\psi}_i(t + \Delta t)$, i.e., the wavefunctions at the time $t + \Delta t$ as given by Eqs. (16a) without considering the action of the forces of constraint:

$$\overline{\psi}_i(\mathbf{r}, t + \Delta t) = -\psi_i(\mathbf{r}, t - \Delta t) + 2\psi_i(\mathbf{r}, t) + \frac{(\Delta t)^2}{\mu}(-H\psi_i(\mathbf{r}, t)) . \tag{17}$$

By imposing the constraint conditions on the $\psi_i(t + \Delta t)$ we obtain, for the hermitian matrix

$$\mathbf{X} = \frac{(\Delta t)^2}{\mu}\Lambda^*,$$

the following equation:

$$\mathbf{X}\mathbf{X}^\dagger + \mathbf{X}\mathbf{B} + \mathbf{B}^\dagger\mathbf{X}^\dagger = \mathbf{1} - \mathbf{A}, \tag{18}$$

where dagger indicates hermitian conjugate and **1** indicates the identity matrix. The matrices **A** and **B** are given by:

$$\begin{aligned} A_{ij} &= \int d\mathbf{r}\overline{\psi}_i^*(\mathbf{r}, t+\Delta t)\overline{\psi}_j(\mathbf{r}, t+\Delta t) \\ B_{ij} &= \int d\mathbf{r}\psi_i^*(\mathbf{r}, t)\overline{\psi}_j(\mathbf{r}, t+\Delta t) \ . \end{aligned} \tag{19}$$

Equation (18) is a quadratic equation for **X** or equivalently for **Λ**. A convenient way of solving it is by iteration, taking into account that Δt is small. It is straightforward to show that:

$$\begin{aligned} \mathbf{A} &= 1 + O(\Delta t^2) \\ \mathbf{B} &= 1 + O(\Delta t) \end{aligned} \tag{20}$$

Then to lowest order in Δt:

$$\mathbf{X}^{(0)} = \frac{1}{2}(1 - \mathbf{A}) \tag{21}$$

and an accurate solution can be obtained by iterating:

$$\mathbf{X}^{(n)} = \frac{1}{2}\left[1 - \mathbf{A} + (1-\mathbf{B})*(1-\mathbf{B}^\dagger) - (1 - \mathbf{B} - \mathbf{X}^{(n-1)})*(1 - \mathbf{B}^\dagger - \mathbf{X}^{(n-1)})\right] \tag{22}$$

Provided that Δt is sufficiently small only a very few iterations are required to satisfy the constraints to machine accuracy.

Equations (14) allow us to sample the complex parameter space of the ψ_i and $\mathbf{R}_I$ with the MD techniques used in statistical mechanics simulations.[1,2] In particular, the equilibrium value $< K >$ of the classical kinetic energy $K = K_e + K_I$ can be calculated as the temporal average over the trajectories generated by the equations of motion (Eqs. 14) and related to the temperature of the system by suitable normalization. By variations of the velocities, i.e., the $\dot{\psi}_i$ and the $\dot{\mathbf{R}}_I$, we can make the temperature of the system vary. In this way we can make our fictitious dynamical system undergo various thermal treatments such as annealing and quenching. During such processes all relevant degrees of freedom, both nuclear and electronic, are relaxed simultaneously. In particular, we can set up a dynamical process that brings the system to its equilibrium state at $T = 0$ by slowly reducing the temperature. We call such a process dynamical simulated annealing. It is just a specific application of the concept of simulated annealing, introduced by Kirkpatrick, Gelatt and Vecchi,[9] to solve complex optimization problems. Here we differ from the original simulated annealing methodology in that we use MD rather than Monte Carlo (MC) to explore configuration space.

As discussed in Ref. 9, simulated annealing is a method for global optimization. At variance with steepest descents, which is a local method, it offers a means of exploring globally the energy surface and, with some luck, locating the absolute minimum. This is important if the ions are allowed to move; in such a case E can have several minima and local optimization methods such as SD can easily miss some of the relevant ones. Often it is not realized how complicated the energy surface can be, even in systems with a very small number of ionic degrees of freedom. In such circumstances simulated annealing is possibly the most efficient tool.

We now indicate how the Newtonian dynamics in Eq. (14) can be very usefully employed also as a means of simulating BO trajectories for the nuclei via our fictitious dynamics and thus perform what we call *ab initio* MD simulations.

The correct classical dynamical equations for the nuclei are:

$$M_I \ddot{\mathbf{R}}_I = -\frac{\partial \Phi}{\partial \mathbf{R}_I} \tag{23}$$

In general, the nuclear trajectories generated by Eqs. (14) and those generated by Eqs. (23) do not coincide. However, one can play with the parameter μ in order to have a much shorter time scale for the electronic degrees of freedom than for the nuclei. In this case nuclear trajectories, initially lying on the BO surface, will deviate from the BO surface only after times that are significantly longer than the MD time step and, in some cases, even longer than the MD observation time. This dynamics may be called *classical adiabatic dynamics*. This dynamic decoupling between ionic and electronic degrees of freedom can be maintained for times much longer than simulation times if the system is an insulator or a semiconductor. When the system is metallic we have instead observed a systematic tendency to thermal equilibration between electron and nuclear subsystems. This is related to the existence of empty electorn states which are degenerate with the occupied states.[22,23] Due to the difference in time scales between electrons and nuclei, the thermal equilibration proceeds at a slow rate and our fictitious dynamics is useful even for metals. In this case, however, the electronic degrees of freedom must be reminimized several times during an MD run.

5. Conclusions

The scheme described here has been applied in a wide variety of problems and has proven to be able to: (i) compute ground-state electronic properties of large and/or disordered systems at the level of state-of-the-art electronic structure calculations; (ii) perform *first-principle* MD simulations where the only assumptions are the validity of classical mechanics to describe ionic motion, the Born-Oppenheimer approximation to separate nuclear and electronic coordinates and the LDF approximation.

We have also acquired a more precise understanding of the limitations of the new approach. MD-DF runs are much heavier than conventional MD runs both in terms of CPU time and of memory occupation. With supercomputers, feasible simulations involve cells containing up to about 150 bulk atoms for systems like silicon. In this size range the computational cost is dominated by operations whose cost grows as the third power of the number of atoms N_A. This is a consequence of the plane-wave representation used for the electronic orbitals ψ_i. Due to the delocalized character of the plane waves, operations like those required to reorthogonalize the orbitals involve $O(N_A^3)$ operations. The effectiveness[24] with which the scheme can be parallelized somewhat alleviates this problem. A possible way around the severe $O(N_A^3)$ dependence is to resort to a localized basis representation. This, however, entails additional problems related to accuracy and efficiency which have to be solved.

Another limitation of plane waves refers to the difficulty in dealing simultaneously with different length scales, as is the case with transition atoms. In conventional electronic structure calculations one solves this problem by using either localized or augmented basis sets. In neither case have the practical problems arising in relation with the MD-DF approach been solved yet. It may well be that the solution relies on the use of novel approaches which are quite unconventional for electronic structure calculations.

A third and more fundamental limitation of the MD-DF approach concerns the conditions under which classical adiabatic dynamics occurs. We mentioned in Sec. 4 that adiabaticity is not perfect for metals, where we always observed a well defined tendency to thermal equilibration among electrons and nuclei in the fictitious classical system. This means that, at the same time, the electrons get hotter and the nuclei colder. During such a process the classical kinetic energy of the electrons increases continuously and, correspondingly, the nuclear trajectories deviate from the BO surface. This can be opposed by systematically minimizing the electronic degrees of freedom before intolerable deviations from the BO surface can occur. In this way, however, one generates nuclear trajectories that do not obey Eqs. (23). In particular, the average nuclear temperature associated with such trajectories is not constant as it should be for a system in thermodynamic equilibrium. In order to correct for such an unwanted behavior, use can be made of one of the possible MD techniques to control the temperature.[2]

Acknowledgments

It is a pleasure to thank R. Car for the fun that we have had together in the many years of fruitful collaboration.

References

1. A. Rahman in *Correlation Functions and Quasi-Particle Interactions in Condensed Matter*, edited by J.W. Halley, Plenum, New York (1978).
2. For a recent overview see also *Molecular-Dynamics Simulation of Statistical-Mechanical Systems*, edited by G. Ciccotti and W.G. Hoover, North-Holland, Amsterdam (1986).
3. P. Hohenberg and W. Kohn, Phys. Rev., **136**, B864 (1964).
4. W. Kohn and L.J. Sham, Phys. Rev., **140**, A1133 (1965).
5. See, for instance, the review articles in *Theory of the Inhomogeneous Electron Gas*, edited by S. Lundqvist and N.H. March, Plenum, New York (1983).
6. G.P. Srivastava and D. Weaire, Adv. Phys., **36**, 463 (1987).
7. R.O. Jones and O. Gunnarsson, Rev. of Mod. Phys., **61**, 689 (1989).
8. R. Car and M. Parrinello, Phys. Rev. Lett., **55**, 2471 (1985).
9. S. Kirkpatrick, G.D. Gelatt, Jr. and M.P. Vecchi, Science, **220**, 671 (1983).
10. M.L. Cohen and V. Heine, Solid State Phys., **24**, 37 (1970).
11. D.R. Hamann, M. Schlüter and C. Chiang, Phys. Rev. Lett., **43**, 1494 (1979).
12. G.B. Bachelet, D.R. Hamann and M. Schlüter, Phys. Rev., **B26**, 4199 (1982).
13. D.C. Allan and M.P. Teter, Phys. Rev. Lett., **59**, 1136 (1987).
14. I. Štich, R. Car, M. Parrinello and S. Baroni, submitted for publication.
15. M.T. Yin and M.L. Cohen, Phys. Rev., **B26**, 5668 (1982).
16. R. Car, M. Parrinello and W. Andreoni, in Ref. 37, as well as A.R. Williams and J. Soler, Bull. Am. Phys. Soc., **32**, 562 (1987), have applied to SD equations the integration formula originally derived for Newton's equations by M.C. Payne, J.D. Joannopoulos, D.C. Allen, M.P. Teter and D.M. Vanderbilt, Phys. Rev. Lett., **56**, 2656 (1986).
17. R. Benedek, B.I. Min and J. Garner, Phys. Rev. B., **39**, 4853 (1989).
18. M.P. Teter and D.C. Allan, preprint.
19. W.M. Press, B.P. Flannery, S.A. Teukolsky and W.T. Vetterling, *Numerical Recipes*, Cambridge Univ. Press, p. 301 (1986); L.A. Hageman and D.M. Young, *Applied Iterative Methods, Computer Science and Applied Mathematics*, Academic Press, New York, p. 138 (1981).
20. M.P. Teter, M.C. Payne and D.C. Allan, Phys. Rev. B., **40**, 12255 (1989).
21. J.P. Ryckaert, G. Ciccotti and H.J.C. Berendsen, J. Comput. Phys., **23**, 327 (1977).
22. I Štich, R. Car, and M. Parrinello, Phys. Rev. Lett., **63**, 2240 (1989).
23. G. Galli, R.M. Martin, R. Car and M. Parrinello, Phys. Rev. Lett., **63**, 988 (1989).
24. B.P. Feuston and E. Clementi, IBM preprint **KGN-222** (1990).

Chapter 15.

LCAO Ab Initio Band Structure Calculations for Polymers

J.M. André, J.L. Brédas†, J. Delhalle, D.J. Vanderveken*, D.P. Vercauteren* and J.G. Fripiat

Laboratoire de Chimie Théorique Appliquée
*Laboratoire de Chimie Moléculaire Structurale
Facultés Universitaires Notre-Dame de la Paix
61 rue de Bruxelles
B-5000 Namur, Belgium

1. Introduction

Since the first theoretical works of the sixties on LCAO techniques in polymer quantum chemistry,[1-6] the field has known a rapid development. Periodic systems are computationally easier to treat than aperiodic ones since the translational symmetry can be fully exploited in order to reduce to manageable dimensions the formidable task of computing electronic states of an extended system.

Ab initio programs for polymers are available and are currently applied in several groups like Erlangen,[7] Vienna,[8] Budapest,[9] Torino,[10] Kingston,[11,12] and Namur.[13,14] However, due to different computational strategies, those programs do not produce rigorously the same numbers even for polymers studied with the same basis sets and within exactly the same geometrical conformations. This is a result of different cut-off procedures of the integrals and originates mainly from an internal structure which is, in all cases, a logical extension of molecular programs. It is illuminating to note that all polymer packages use standard "molecular" strategies taken from the IBMOL, KGNMOL, GAUSSIAN or other series.

Furthermore, it has to be realized that sophisticated *ab initio* methods, already time consuming for middle- and large-size molecules, become very onerous when

† Now permanently at: Service de Chimie des Matériaux Nouveaux, Université de Mons-Hainaut, Avenue Maistriau 21, B-7000 MONS (Belgium).

applied to polymers of chemical or biological interest. In this sense, the use of double-zeta or polarization LCAO basis is still exceptional.

The conclusion is that additional works should be considered in order to develop efficient *ab initio* programs for polymers which would request much smaller computing times than the existing ones. In our code, this has been achieved by implementing into a general system the new fast techniques for evaluating integrals over Gaussian basis, the explicit use of helical symmetry, and efficient methods of computing long-range electrostatic effects. This requested a general methodology, more "polymer-minded" than "molecule-minded".

This chapter is organized as follows:

1. For the sake of completeness, we define a periodic model of a polymeric chain and recall the principles of calculation of LCAO band structures.

2. We discuss the important question of cut-off procedures in existing programs in relation with the electrostatic balance between the nucleus- and electron-electron interactions. We present the multipole expansion used, in practice, for evaluating long-range Coulomb interactions. We also discuss the exchange behavior and, in particular, pay attention to the short- or long-rangeness character of the exchange contributions.

3. We investigate the ways of reducing computing times by use of symmetry elements other than translational (e.g., screw axes).

4. We introduce a simulated *ab initio* algorithm called the VEH (Valence Effective Hamiltonian) technique that, by its very simple numerical implementation, allows us to scan easily the band structures of stereoregular polymers.

5. We present an interactive graphics interface for the ordering of the band structures and representation of the density of states with a particular emphasis on reaching high quality output for publication.

The discussed topics are implemented into an original computer program.

2. The Periodic Model of a Polymer Chain

The standard quantum mechanical treatment of polymers is based on a periodic model of the polymeric chain. To introduce this concept, three model chains have been consistently used throughout the literature. They are represented in Fig. 1. The first model is a linear chain of hydrogen atoms. It has appeared as the extension to infinite systems of the first quantum mechanical calculations on an hydrogen atom and of the first calculation on an hydrogen molecule. Historically, a second model is that of an infinite polyene and results from the standard

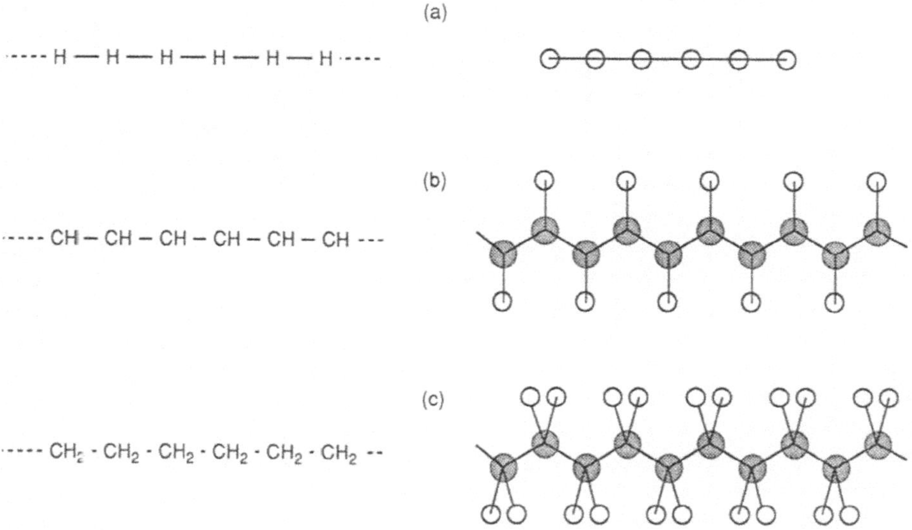

Figure 1. Model of (a) a chain of hydrogen atoms, (b) polyacetylene, and (c) polyethylene.

π-electron approximation popular since the work of Erich Hückel in the early 1930s.[15] The third model is that of a polyethylene chain issued from the full machinery of quantum mechanical treatment of all-electron or all-valence electron systems. Pioneering quantum mechanical calculations on a polyethylene chain were carried out around the end of the 1960s.[16,17]

The linear chain of hydrogen atoms turns out to be a somewhat academic case since its existence is highly speculative. The infinite polyene chain is practically more attractive. On one side, it is the limiting term in the series of linear conjugated polyene molecules: ethylene, butadiene, hexatriene .... a classical set studied in all elementary textbooks on quantum chemistry. On the other side, a renewed important interest stems from the discovery of highly electrically conducting polyacetylene at the end of the 1970s. Finally, polyethylene now plays in polymer quantum chemistry a role similar to those of the helium atom in atomic physics and the hydrogen molecule in molecular physics.

Some limitations are usually introduced in standard quantum polymeric calculations. They are sketched in Fig. 2 in the case of polyethylene modeled as a zigzag chain built from methylene groups. A first limitation is that we usually consider an isolated chain; in practice, the chain is never isolated but does exist in a liquid or solid-state environment. A second limitation is that the chain is taken to be infinite and perfectly stereoregular. The fact of considering the chain as infinite is not a severe limitation since polymers can have very large molecular weights; in the case of conventional polymers like polyethylene or polypropylene,

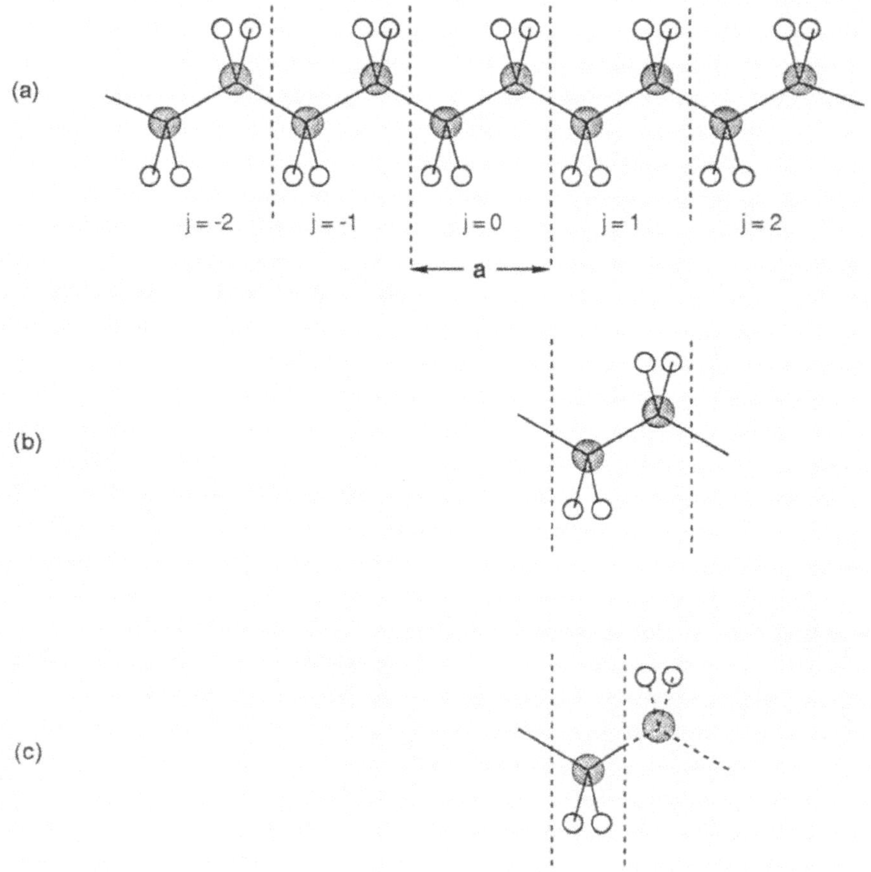

Figure 2. Molecular model of polyethylene. (a) chain and numerotation, (b) translational unit cell, and (c) helical unit cell.

molecular weights in the range of $10^5 - 10^6$ dalton are achievable. The model also neglects chain end effects; the importance of such effects has not yet been studied in detail. The crudest assumption of all is that of perfect stereoregularity of the linear chain. However, this allows one to take into account the translational symmetry and to apply concepts of condensed matter physics in order to get a rather complete description of the electronic structure of polymers.

3. Principles of LCAO Band Structure Calculations on Polymers

The condensed matter physicist makes full use of the possible translational symmetry of the lattice and uses the language of Brillouin zones introduced in the classical work by Bloch in 1928.[18] In this context, the so-called Bloch functions (molecular orbitals for an infinite one-dimensional (1D) chain) are eigenfunctions of a translation operator and as such are expressed in terms of a wave number, k, defined in the reciprocal space. For systems having one-dimensional periodicity and a periodic electronic density:

$$\rho(\mathbf{r} + \mathrm{ja}) = \rho(\mathbf{r})$$

where a is the length of the polymer unit cell in direct space, Bloch theorem states the phase relation of the orbitals at points related by periodicity:

$$\phi_n(k, \mathbf{r} + ja) = e^{ikja}\phi_n(k, \mathbf{r})$$

Thus, the associated "orbital" energies are functions of k:

$$\varepsilon_n = \varepsilon_n(k)$$

and can be plotted with respect to k. The representation of the corresponding dispersion curves as a function of k is called an energy band. Standard theorems of solid state physics demonstrate that those energy bands are periodic in the 1D-reciprocal space:

$$\varepsilon_n(k) = \varepsilon_n(k + 1g)$$

where g is the reciprocal translation unit length ($=2\pi/a$). Thus, a given orbital $\phi_n(k, \mathbf{r})$ is identical at a k point and at its translation $k + 1g$. The calculation of the energy bands needs therefore to be performed only for a range of k values equivalent to a single unit cell of the reciprocal lattice. The length spanned by the maximum interval between all possible vectors is called the first *Brillouin zone* of the polymer. Another corollary of Bloch theorem states that the energy bands are symmetric with respect to $k=0$:

$$\varepsilon_n(k) = \varepsilon_n(-k)$$

The calculation is thus simplified since, using a symmetrized part of the reciprocal space (the first Brillouin zone ranging from $-\pi/a$ to $+\pi/a$), it needs to be explored from $k=0$ to $k=+\pi/a$ (i.e., half the first Brillouin zone).

An alternative form of Bloch theorem:

$$\phi_n(k, \mathbf{r} + ja) = e^{ikja}\phi_n(k, \mathbf{r})$$

is obtained when we factorize the Bloch orbitals $\phi_n(k, \mathbf{r})$ into their periodic contributions $[u_n(k, \mathbf{r}) = u_n(k, \mathbf{r} + ja)]$ and nonperiodic free-electron-like contributions $[e^{ikz}]$:

$$\phi_n(k, \mathbf{r}) = e^{ikz}\, u_n(k, \mathbf{r})$$

Conventionally, the periodicity axis is chosen to be along the z axis. The LCAO Bloch form $\phi(\mathrm{k}, \mathbf{r})$ of a single atomic orbital $\chi(\mathbf{r}-\mathbf{P})$ is thus written as:

$$\phi(\mathrm{k}, \mathbf{r}) = \Sigma_j \chi(\mathbf{r} - \mathbf{P} - \mathrm{ja})$$

$$= N_C^{-1/2} \Sigma_j e^{ikja} c_j \chi(\mathbf{r} - \mathbf{P} - \mathrm{ja})$$

where $N_C^{-1/2}$ is the normalization factor for a polymer containing N_C cells.

The LCAO Bloch form which describes the polymer orbital extending over the whole polymer as a Bloch combination of functions centered at the atomic nuclei of the polymer chain is then:

$$\phi_n(\mathrm{k}, \mathbf{r}) = \Sigma_j \Sigma_p c_{npj} \chi_p(\mathbf{r} - \mathbf{P} - \mathrm{ja})$$

$$= N_C^{-1/2} \Sigma_j e^{ikja} \Sigma_p c_{np}(k) \chi_p(\mathbf{r} - \mathbf{P} - \mathrm{ja})$$

where n is the band index and k the position vector (wave number) in the first Brillouin zone. The polymer orbitals have imaginary components except in two high symmetry points of the first Brillouin zone (point Γ at $k=0$ and point H at $k=\pi/a$) where the Bloch exponential e^{ikja} reduces to real values:

$$\phi_n(\Gamma, \mathbf{r}) = N_C^{-1/2} \Sigma_j \{\Sigma_p c_{np}(0) \chi_p(\mathbf{r} - \mathbf{P} - \mathrm{ja})\}$$

$$\phi_n(\mathrm{H}, \mathbf{r}) = N_C^{-1/2} \Sigma_j (-1)^j \{\Sigma_p c_{np}(\pi/a) \chi_p(\mathbf{r} - \mathbf{P} - \mathrm{ja})\}$$

The last two relations are useful for getting quick sketches of the band structures of simple polymeric structures.

From a conceptual point of view, it appears that polymer quantum chemistry is an ideal field for cooperation between condensed matter physicists and molecular quantum chemists. There exists a common interpretation in the discussions concerning orbital energies, orbital symmetry, and gross charges by chemists and solid-state physicists who use terms less familiar to the chemist like first Brillouin zone, dependence of "wave function" (in fact the monoelectronic wave function is called an orbital by the chemist) with respect to wave vector k, Fermi surfaces, Fermi contours, and density of states.

However, polymer quantum chemistry is not, strictly speaking, one-dimensional physics. It must be emphasized that even if we deal with 1D periodic systems, the orbitals are truly 3D, so that polymer quantum chemistry is not a reduction of solid-state physics to a single dimension space. In the strictly 1D physics, the systems are periodic in one dimension and have 1D wave functions. The three-dimensionality of the basic Bloch orbitals renders invalid important theorems applicable to purely 1D systems as elegantly shown by McCubbin and Teemull.[19] A typical example is that of the presence of extrema in energy bands which should only occur at the center and the edges of the Brillouin zone in a strictly 1D system. For polymers, even in simple cases like the linear zigzag polyethylene chain, some extrema of the energy bands are encountered at arbitrary positions in the first Brillouin zone which are not high symmetry points.

Turning to applications, it is clear from the previous discussion that the numerical procedure must combine equations from molecular quantum chemistry and solid-state physics. In molecular quantum chemistry, a molecular orbital is expanded in terms of basis functions; secular systems of equations and determinants are solved; their eigenvalues are the orbital energies. From the LCAO coefficients, charges and bond orders (projection of the density matrices onto the limited basis used) are calculated.

In polymer quantum chemistry, we take into account the lattice periodicity; the orbitals, the systems of equations, and the determinants are no longer real but have imaginary components. This introduces lattice sums which have to be evaluated by adequate procedures. Indeed, from Bloch theorem, we know that a polymeric orbital has the LCAO form:

$$\phi_n(k, \mathbf{r}) = N_C^{-1/2} \Sigma_j e^{ikja} \Sigma_p c_{np}(k) \chi_p(\mathbf{r} - \mathbf{P} - ja)$$

By forming the expectation value of a given one-electron operator and by using the variational procedure for getting the best LCAO coefficients $c_{np}(k)$, we obtain for the whole polymer the secular system of equations:

$$\Sigma_p c_{np}(k)\{\Sigma_j e^{ikja} [h^j_{pq} - \varepsilon_n(k) S^j_{pq}]\} = 0$$

where the matrix elements h^j_{pq} and the overlap integrals S^j_{pq} are respectively defined by:

$$h^j_{pq} = \int \chi_p(\mathbf{r} - \mathbf{P})h(\mathbf{r})\chi_q(\mathbf{r} - \mathbf{Q} - ja)\, dv = \int \chi_p(\mathbf{r})h(\mathbf{r})\chi^j_q(\mathbf{r})dv$$

$$S^j_{pq} = \int \chi_p(\mathbf{r} - \mathbf{P})\chi_q(\mathbf{r} - \mathbf{Q} - ja)\, dv = \int \chi_p(\mathbf{r})\chi^j_q(\mathbf{r})dv$$

The compatibility condition of the previous system of equations is the vanishing of the secular determinant:

$$|\{\Sigma_j e^{ikja} [h^j_{pq} - \varepsilon(k) S^j_{pq}]\}| = 0$$

Solving this equation to get the unknown band energies $\varepsilon_n(k)$ produces in the reduced scheme the band structure $\varepsilon(k)$ as a multivalued function of k. We note also that h^j_{pq} is a matrix element of the one-electron operator h between the atomic orbital χ^j_p centered in the origin unit cell and the atomic orbital χ_q centered in cell j. S^j_{pq} has the same meaning for the unit overlap. The key problem is to obtain the exact numerical value or an approximate value of the various matrix elements over the basis functions. Fortunately, it is known from the form of the AOs that these matrix elements should decrease exponentially with the distance between their orbital centers and thus give rise to a natural convergency of the

Chart 1. A comparison of some basic formulas of molecular and polymer quantum chemistry.

Molecular Quantum Chemistry	Polymer Quantum Chemistry
Orbital:	
$\phi_j(\mathbf{r}) = \Sigma_p c_{jp}\,\chi_p(\mathbf{r}-\mathbf{P})$	$\phi_n(k,\mathbf{r}) = N^{-1/2}\Sigma_j e^{ikja}\,\Sigma_p c_{np}(k)\times$ $\chi_p(\mathbf{r}-\mathbf{P}-ja)$
Secular system:	
$\Sigma_p c_{jp}\,(h_{pq}-\varepsilon_j S_{pq}) = 0$	$\Sigma_p c_{np}(k)\{\Sigma_j e^{ikja}\,[h^j_{pq}-\varepsilon_n(k)\,S^j_{pq}]\}$ $\equiv\Sigma_p c_{np}(k)\{h_{pq}(k)-\varepsilon_n(k)\,S_{pq}(k)\} = 0$
Secular determinant:	
$\mid (h_{pq}-\varepsilon\,S_{pq}) \mid = 0$	$\mid\{\Sigma_j e^{ikja}[h^j_{pq}-\varepsilon(k)\,S^j_{pq}]\}\mid$ $\equiv\mid\{h_{pq}(k)-\varepsilon_n(k)\,S_{pq}(k)\}\mid = 0$
One-electron Hartree-Fock operator:	
$h(\mathbf{r}_i) = -(1/2)\nabla^2$ $-\Sigma_A Z_A\,\mid\mathbf{r}-\mathbf{R}_A\mid^{-1}$ $+\Sigma_j^{occ}\{2J_j(\mathbf{r})-K_j(\mathbf{R})\}$	$h(\mathbf{r}_i) = -(1/2)\nabla^2$ $-\Sigma_j\Sigma_A Z_A\,\mid\mathbf{r}-\mathbf{R}_A-ja\mid^{-1}$ $+\Sigma_j^{occ}\int dk\{2J_j(k,\mathbf{r})-K_j(k,\mathbf{r})\}$
Coulomb repulsion operator:	
$J_j(\mathbf{r})\,\phi_i(\mathbf{r}) =$ $\{\int dr'\,\phi_j(\mathbf{r}')\,\phi_j(\mathbf{r}')\mid\mathbf{r}-\mathbf{r}'\mid^{-1}\}\phi_i(\mathbf{r})$	$J_j(k,\mathbf{r})\,\phi_i(k',\mathbf{r}) =$ $\{\int dr'\,\phi_j^*(k,\mathbf{r}')\,\phi_j(k,\mathbf{r}')\mid\mathbf{r}-\mathbf{r}'\mid^{-1}\}\phi_i(k',\mathbf{r})$
Exchange interaction operator:	
$K_j(\mathbf{r})\,\phi_i(\mathbf{r}) =$ $\{\int dr'\,\phi_j(\mathbf{r}')\,\phi_i(\mathbf{r}')\mid\mathbf{r}-\mathbf{r}'\mid^{-1}\}\phi_j(\mathbf{r})$	$K_j(k,\mathbf{r})\,\phi_i(k',\mathbf{r}) =$ $\{\int dr'\,\phi_j^*(k,\mathbf{r}')\,\phi_i(k',\mathbf{r}')\,\mid\mathbf{r}-\mathbf{r}'\mid^{-1}\}\phi_j(k,\mathbf{r})$
Integrals between molecular or polymeric orbitals:	
$J_{ij} = \int dr\,\phi_i(\mathbf{r})\,J_j(\mathbf{r})\,\phi_i(\mathbf{r})$ $= \int dr\,\phi_i(\mathbf{r})\int dr'\,\phi_j(\mathbf{r}')\,\phi_j(\mathbf{r}')\mid\mathbf{r}-\mathbf{r}'\mid^{-1}\,\phi_i(\mathbf{r})$ $= \int dr\int dr'\phi_i(\mathbf{r})\,\phi_i(\mathbf{r})\mid\mathbf{r}-\mathbf{r}'\mid^{-1}\,\phi_j(\mathbf{r}')\,\phi_j(\mathbf{r}')$	$J_{ij}(k,k') = \int dr\,\phi_i^*(k,\mathbf{r})\,J_j(k',\mathbf{r})\,\phi_i(k,\mathbf{r})$ $= \int dr\,\phi_i^*(k,\mathbf{r})\int dr'\,\phi_j^*(k',\mathbf{r}')\,\phi_j(k',\mathbf{r}')\times$ $\mid\mathbf{r}-\mathbf{r}'\mid^{-1}\phi_i(k,\mathbf{r})$ $= \int dr\int dr'\phi_i^*(k,\mathbf{r})\,\phi_i(k,\mathbf{r})\mid\mathbf{r}-\mathbf{r}'\mid^{-1}\times$ $\phi_j^*(k',\mathbf{r}')\,\phi_j(k',\mathbf{r}')$
$K_{ij} = \int dr\,\phi_i(\mathbf{r})\,K_j(\mathbf{r})\,\phi_i(\mathbf{r})$ $= \int dr\,\phi_i(\mathbf{r})\int dr'\,\phi_j(\mathbf{r}')\,\phi_i(\mathbf{r}')\mid\mathbf{r}-\mathbf{r}'\mid^{-1}\phi_j(\mathbf{r})$ $= \int dr\int dr'\phi_i(\mathbf{r})\,\phi_j(\mathbf{r})\mid\mathbf{r}-\mathbf{r}'\mid^{-1}\,\phi_j(\mathbf{r}')\,\phi_i(\mathbf{r}')$	$K_{ij}(k,k') = \int dr\,\phi_i^*(k,\mathbf{r})\,K_j(k',\mathbf{r})\,\phi_i(k,\mathbf{r})$ $= \int dr\,\phi_i^*(k,\mathbf{r})\int dr'\,\phi_j^*(k',\mathbf{r}')\,\phi_i(k,\mathbf{r}')\times$ $\mid\mathbf{r}-\mathbf{r}'\mid^{-1}\,\phi_j(k',\mathbf{r})$ $= \int dr\int dr'\phi_i^*(k,\mathbf{r})\,\phi_j(k',\mathbf{r})\mid\mathbf{r}-\mathbf{r}'\mid^{-1}\times$ $\phi_j^*(k',\mathbf{r}')\,\phi_i(k,\mathbf{r}')$

Chart 1 (continued). A comparison of some basic formulas of molecular and polymer quantum chemistry.

Molecular Quantum Chemistry	Polymer Quantum Chemistry
Integrals between atomic orbitals:	
$S_{pq} = \int \chi_p(\mathbf{r})\chi_q(\mathbf{r})\, dv$	$S^j_{pq} = \int \chi_p(\mathbf{r} - \mathbf{P} - 0a)\, \chi_q(\mathbf{r} - \mathbf{Q} - ja) dv$ $= \int \chi_p(\mathbf{r})\chi^j_q(\mathbf{r}) dv$
$h_{pq} = \int \chi_p(\mathbf{r})h(\mathbf{r})\chi_q(\mathbf{r})\, dv$	$h^j_{pq} = \int \chi_p(\mathbf{r} - \mathbf{P} - 0a)\, h(\mathbf{r}) \times$ $\chi_q(\mathbf{r} - \mathbf{Q} - ja) dv$ $= \int \chi_p(\mathbf{r})h(\mathbf{r})\chi^j_q(\mathbf{r})\, dv$
$= T_{pq} - \Sigma_A Z_A V_{pq\|A}$	$= T^j_{pq} - \Sigma_h \Sigma_A Z_A V^j_{pq}\|^h_A$
$+ \Sigma_r \Sigma_s D_{rs}\{2(pq\|rs) - (pr\|qs)\}$	$+ \Sigma_h \Sigma_l \Sigma_r \Sigma_s D^{hl}_{rs}\{2(pq^j\|r^h s^l) - (pr^h\|q^j s^l)\}$
$T_{pq} = \int \chi_p(\mathbf{r})\{-1/2\Delta\}\chi_q(\mathbf{r})\, dv$	$T^j_{pq} = \int \chi_p(\mathbf{r} - \mathbf{P} - 0a)\{-1/2\Delta\} \times$ $\chi_q(\mathbf{r} - \mathbf{Q} - ja)\, dv$ $= \int \chi_p(\mathbf{r})\{-1/2\Delta\}\chi^j_q(\mathbf{r})\, dv$
$V^j_{pq}\|^h_A = \int \chi_p(\mathbf{r})\|\mathbf{r} - \mathbf{R}_A\|^{-1}\chi_q(\mathbf{r})\, dv$	$V_{pq\|A} = \int \chi_p(\mathbf{r} - \mathbf{P} - 0a)\|\mathbf{r} - \mathbf{R}_A - ha\|^{-1} \times$ $\chi_q(\mathbf{r} - \mathbf{Q} - ja)\, dv$ $= \int \chi_p(\mathbf{r})\|\mathbf{r} - \mathbf{R}_A - ha\|^{-1}\chi^j_q(\mathbf{r})\, dv$
$(pq\|rs) = \int\int \chi_p(\mathbf{r})\chi_q(\mathbf{r})\|\mathbf{r} - \mathbf{r}'\|^{-1} \times$ $\chi_r(\mathbf{r}')\chi_s(\mathbf{r}')\, dv\, dv'$	$(pq^j\|r^h s^l) = \int\int \chi_p(\mathbf{r} - \mathbf{P} - 0a) \times$ $\chi_q(\mathbf{r} - \mathbf{Q} - ja)\|\mathbf{r} - \mathbf{r}'\|^{-1} \times$ $\chi_r(\mathbf{r}' - \mathbf{R} - ha)\chi_s(\mathbf{r}' - \mathbf{S} - la)\, dv\, dv'$ $= \int\int \chi_p(\mathbf{r})\chi^j_q(\mathbf{r})\|\mathbf{r} - \mathbf{r}'\|^{-1}\chi^h_r(\mathbf{r}')\chi^l_s(\mathbf{r}')\, dv\, dv'$
Results:	
$\rightarrow \quad \varepsilon_1, \varepsilon_2, \varepsilon_3, \ldots$	$\rightarrow \quad \varepsilon_1(k), \varepsilon_2(k), \varepsilon_3(k), \ldots$
$\rightarrow \quad D_{pq} = \Sigma_j\, c_{jp}\, c_{jq}$	$\rightarrow \quad D^h_{pq} = a/2\pi\, \Sigma_n$ $\int dk\, c^*_{np}(k)c_{nq}(k)e^{ikha}$

lattice sums appearing in the secular systems and in the secular determinant. It is to be noted that the dimensions of the matrix equations to be solved are equal to the number of atomic orbitals per unit cell, the effect of the infinite lattice being included in the naturally convergent but formally infinite sums. A comparison of the basic formulas used in molecular and polymer quantum chemistry is given in Chart 1.

The previous equations show that in a general way a polymer can be considered as a large molecule. Using this property, it is likely that the usual methods of quantum chemistry could be used to investigate the electronic structure of polymers. As a consequence, it is correct to introduce the well-known approximations of molecular quantum chemistry into the formalism of polymeric orbitals (e.g., the use of contracted Gaussian orbitals). LCAO-based *ab initio* calculations require the definition of a basis set of size N and imply the calculations of integrals or matrix elements between the N basis functions of the basis set. For atoms, all the basis functions are centered on the same location and the integrals are said to be one-centered. For a molecule or a polymeric cluster of M atoms, the N^2 (or more exactly the $N(N+1)/2$ when taking into account the hermitian redundancy) kinetic integrals:

$$T_{pq} = \int \chi_p(\mathbf{r})\{-1/2\nabla^2\}\chi_q(\mathbf{r})\,dv$$

can be either one- or two-centered while the $M{*}N(N+1)/2$ nuclear attraction integrals:

$$V_{pq|A} = \int \chi_p(\mathbf{r})\,|\mathbf{r}-\mathbf{R}_A|^{-1}\chi_q(\mathbf{r})\,dv$$

can be one-, two-, or three-centered. Finally, one finds $N(N+1)/2{*}[N(N+1)/2+1]/2$, i.e., approximately $N^4/8$, either one-, two-, three- or four-centered electron repulsion integrals:

$$(pq\,|\,rs) = \int\int \chi_p(\mathbf{r})\chi_q(\mathbf{r})\,|\mathbf{r}-\mathbf{r}'|^{-1}\chi_r(\mathbf{r}')\chi_s(\mathbf{r}')dvdv'$$

For large molecules or large basis sets, a huge number of integrals has to be calculated, which constitutes the bottleneck of first principles *ab initio* SCF-MO-LCAO calculations on large oligomeric systems.

In the infinite regular polymeric chain as in the molecular or oligomeric cases, nonempirical methods are dependent on the fourth power of the number of orbitals considered. If N_C is the number of cells for which all the overlap and Fock elements are greater than a given threshold (e.g., 10^{-6} au) and N is the size of the basis set in a unit cell, we have to calculate $N_C \cdot N(N+1)/2$ Fock integrals and overlap terms. The repulsion operator due to its two-electron character gives rise to much more complex matrices. In usual schemes, those matrices (of number N_C^3) contain $N^2(N+1)^2/8$ integrals as a first approximation. As a consequence, the time consuming part of an *ab initio* method has a general $N_C^3N^4/8$ dependence.

In the *ab initio* methods, all the necessary integrals are correctly evaluated up to infinity in terms of lattice sums as soon as the geometry and the basis functions are defined. A correct computation is still a formidable task at the present time. Indeed, it is proven that long range effects are very important as soon as the unit cell contains permanent dipoles. Multipole expansions must be used in order to get a satisfactory balance of the electrostatic interactions.

As indicated previously, the LCAO polymer equations are based on the fact that Bloch theorem implies equivalent relations between coefficients (csubnp^0, c^h_{nq}) of atomic orbitals χ^0_p, χ^h_q respectively centered in the origin unit cell (0) and in another unit cell (h) separated by a (ha) translation:

$$c^h_{np} = e^{ikha}\, c^0_{np}$$

Hereafter, lower indices (p, q, r, or s) refer to the labeling of a given orbital (χ_p, χ_q, χ_r, or χ_s) within a unit cell while upper indices (j, h, l) refer to the position of a given unit cell. When applying Hartree-Fock or Self-Consistent-Field (SCF) concepts, we consider the motion of a single electron in the field of fixed nuclei and in the averaged Coulomb and Pauli (exchange) fields of all the electrons. In Chart 1 we drop the 0 upper indices and do not detail the extent of the summations.

4. The Electrostatic Balance between the Nucleus- and Electron-Electron Interactions

As already mentioned, the calculation of the matrix elements of the Fock operator in the selected basis is in itself a difficult problem due to the presence of long-range interactions in the electron-nucleus attraction and in the electron-electron repulsion and exchange interactions. To clearly state this crucial point, we rewrite the detailed form of a single matrix element:

$$h^j_{pq} = \int \chi_p(\mathbf{r})h(\mathbf{r})\chi^j_q(\mathbf{r})dv$$

$$= T^j_{pq} \leftarrow \text{ kinetic}$$

$$- \Sigma_h\Sigma_A Z_A V^j_{pq} |^h_A \leftarrow \text{ nuclear attraction}$$

$$+ \Sigma_h\Sigma_l\Sigma_r\Sigma_s\, 2\, D^{hl}_{rs}\, (pq^j \,|\, r^h s^l) \leftarrow \text{ electron repulsion}$$

$$- \Sigma_h\Sigma_l\Sigma_r\Sigma_s\, D^{hl}_{rs}\, (pr^h \,|\, q^j s^l) \leftarrow \text{ exchange}$$

The elements of the density matrices are defined by the summation over all the doubly occupied electronic bands and their integration over the first Brillouin zone of the polymer:

$$D^{h}_{pq} = a/2\pi\, \Sigma^{occ}_{n} \int dk\, c_{np}{}^{*}(k)c_{nq}(k)e^{ikha}$$

In order to discuss the long-range behavior of the electrostatic interactions, we rearrange the previous equation into:

$$h^{j}_{pq} = T^{j}_{pq} \leftarrow \text{ kinetic}$$

$$+\Sigma_h\Sigma_l\Sigma_r\Sigma_s\, 2\, D^{hl}_{rs}\{(pq^{j}\,|\,r^{h}s^{l}) \leftarrow \text{ Coulomb electron-electron}$$

$$-\, n_0^{-1}S^{hl}_{rs}V^{j}_{pq}\,|\,^{h'}_{A}\Sigma_A Z_A V pq^{j}\,|\,A^{h}\} \leftarrow \text{ Coulomb electron-nuclei}$$

$$-\, \Sigma_h\Sigma_l\Sigma_r\Sigma_s\, D^{hl}_{rs}\, (pr^{h}\,|\,q^{j}s^{l}) \leftarrow \text{ exchange}$$

where n_0 is the number of electrons per unit cell. In the last equation, we have rescaled the electron-nuclei attraction term to introduce the same double summation over h and l as in electron-electron Coulomb terms by making use of the normalization-like condition:

$$\Sigma_h\Sigma_r\Sigma_s\, 2D^{hl}_{rs}S^{hl}_{rs} = \Sigma_l\Sigma_r\Sigma_s\, 2D^{0l}_{rs}S^{0l}_{rs} = \Sigma_l\Sigma_r\Sigma_s\, 2D^{l}_{rs}S^{l}_{rs} = n_0$$

In the programs developed early on, not enough attention had been paid to the convergency properties of the contribution which is controlled by the balance between the electron density and the sum of the nuclear charges:

$$\Sigma_h\Sigma_l\Sigma_r\Sigma_s\, 2D^{hl}_{rs}\{(pq^{j}\,|\,r^{h}s^{l}) - n_0^{-1}S^{hl}_{rs}V^{j}_{pq}\,|\,^{h}_{A}\Sigma_A Z_A V^{j}_{pq}\,|\,A^{h}\} =$$

$$\int dv\, \chi_p(\mathbf{r})\{\int dv'\, \rho(\mathbf{r}')/|\mathbf{r}-\mathbf{r}'|^{-1} - \Sigma_h\Sigma_A Z_A\,|\mathbf{r}-\mathbf{R}_A|^{-1}\}\chi^{j}_{q}(\mathbf{r})$$

The electron density is defined for polymers by:

$$\rho(\mathbf{r}) = \Sigma_h\Sigma_l\Sigma_r\Sigma_s\, 2D^{hl}_{rs}\, \chi^{h}_{r}(\mathbf{r})\chi^{l}_{s}(\mathbf{r})$$

$$= \Sigma_h\Sigma_l\Sigma_r\Sigma_s\, 2D^{0l-h}_{rs}\chi^{h}_{r}(\mathbf{r})\chi^{l}_{s}(\mathbf{r})$$

where D^{0l-h}_{rs} is the density matrix element between atomic orbital χ_r in the origin cell 0 and atomic orbital χ_s in cell l-h. The summations over h and l extend over all the unit cells of the polymer chain, those over r and s over the atomic orbitals present in one cell.

No serious difficulties were anticipated other than due to the nature and the size of the atomic basis set such as encountered in molecular quantum chemistry. A close analysis shows that the formula for the electron density already points to interesting translational symmetry relations. It must be realized that in this formula, the translational symmetry of the periodic polymer is only implicitly introduced by the identities between translationally equivalent elements of the density matrix:

$$D_{rs}^{hl} = D_{rs}^{0l-h}$$

Two lattice summations acting on all terms are, in principle, to be performed. However, an explicit use of the translational periodicity can be made by a straightforward manipulation[20] when using a new index m = l − h:

$$\begin{aligned}\rho(\mathbf{r}) &= \Sigma_h \Sigma_l \Sigma_r \Sigma_s \, 2D_{rs}^{0l-h} \, \chi_r^h(\mathbf{r}) \chi_s^l(\mathbf{r}) \\ &= \Sigma_h \Sigma_m \Sigma_r \Sigma_s \, 2D_{rs}^{0m} \, \chi_r^h(\mathbf{r}) \chi_s^{h+m}(\mathbf{r}) \\ &= \Sigma_m \Sigma_r \Sigma_s \, 2D_{rs}^{0m} \, \Sigma_h \, \chi_r^h(\mathbf{r}) \chi_s^{h+m}(\mathbf{r})\end{aligned}$$

This leads to an important simplification of the formula by letting the h index act only on the atomic orbitals. The use of the same idea for the two-electron integrals also results in a significant simplification of the integral formulas and a decrease in the number of integrals to be stored.[20] This speeds up considerably both the integral and SCF parts of a polymer program with respect to the traditional approach, thereby making feasible routine calculations on polymers of "chemically" interesting size. It is striking that little attention has been paid to formulations such as this one although it was already implicit in a paper by Pisani and Dovesi in 1980[21] but without direct reference to the electronic density.

From straight chemical intuition, it was usually believed that the range of effectively interacting cells which should be explicitly considered in the actual calculations would be very small though no consistent rule for the truncation of j, h, and l summations could actually be supplied.

The electrostatic attraction and repulsion terms:

$$+ \Sigma_h \Sigma_l \Sigma_r \Sigma_s \, 2D_{rs}^{hl} \, (pq^j \,|\, r^h s^l) - \Sigma_h \Sigma_A Z_A V_{pq}^j \,|\, _A^h$$

are simply the Coulomb energy of a negative charge distribution:

$$P_{pq}^j(\mathbf{r}) = \chi_p(\mathbf{r}) \chi_q^j(\mathbf{r})$$

in the field of the positive framework of all the nuclei and the negative electron distribution:

$$P_{rs}^{l-h}(\mathbf{r}) = \chi_r(\mathbf{r}) \chi_s^{l-h}(\mathbf{r})$$

Due to cell neutrality, the total charge per cell is zero so that the electrostatic contributions of cells far away from the center of the charge distribution $P_{pq}^j(\mathbf{r})$ goes to zero. In this sense, even if individual nuclear attraction and electron repulsion integrals decrease very slowly as a function of the distance (r^{-1} dependence), there is a strict cancellation at medium and large distances between the diverging nuclear attraction and electron repulsion terms since positive nuclear charges exactly compensate the negative electron density of an equal number of electrons.

This is easily illustrated by investigating the contributions of electrostatic interactions from cell h with the density $P^{j}_{pq}(\mathbf{r})$:

$$\int P^{j}_{pq}(\mathbf{r}_1)[-\Sigma_A|\mathbf{r}_1-\mathbf{R}_A-ja|^{-1}+\Sigma_l\Sigma_r\Sigma_s 2D^{hl}_{rs}\int P^{l-h}_{rs}(\mathbf{r}_2)|\mathbf{r}_1-\mathbf{r}_2|^{-1}d\mathbf{r}_2]d\mathbf{r}_1$$

If the center of the charge density $P^{j}_{pq}(\mathbf{r}_1)$ is $\mathbf{G}^{j}_{pq}$ and the one of $P^{l-h}_{rs}(\mathbf{r}_2)$ is $\mathbf{G}^{l-h}_{rs}$ and if $\mathbf{G}^{j}_{pq}$ and $\mathbf{G}^{l-h}_{rs}$ are sufficiently far apart then $\mathbf{r}_{12}\approx\mathbf{r}_{1A}\approx|\mathbf{R}_A-\mathbf{G}pq^{j}|$. As a consequence, we can reasonably approximate the former expression by:

$$\approx S^{j}_{pq}|\mathbf{R}_A-\mathbf{G}^{j}_{pq}|^{-1}[\Sigma_A Z_A+\Sigma_l\Sigma_r\Sigma_s 2D^{0l}_{rs}S^{0l}_{rs}]$$

This leads to an absolute cancellation since the two terms within the braces are, from left to right, the total nuclear charge and the total number of electrons (normalization condition) per unit cell.

This type of argumentation which is in principle correct[22] has much delayed a formal but necessary mathematical analysis on the exact convergence properties of these series, to define better the numerical applicability of the LCAO equations for polymer chains.

However, it must be pointed out that as early as 1956, Löwdin[23] had already stressed the formal difficulties which would arise in the numerical solutions of this type of equation. Furthermore, starting with the work of O'Shea and Santry[24] in 1974 and Ukrainski[25] in 1975, it became gradually apparent that model chains embody also size-related difficulties, mathematically expressed as conditionally convergent lattice sums. This is due to the fact that the centers of positive (nuclear) charges and negative (electron) distributions do not always coincide and that, in almost every case, small but non-negligible local dipoles do exist. In such a case, the mathematical analysis shows that the convergency of nucleus and electron-electron interactions goes as:

$$\Sigma^{N}_{n=1}\{1/(n+\delta)-1/n\}$$

which can be very slow if the separation of charges δ is significant.

Several solutions to this problem have been investigated by the Namur group. A first technique[26] makes use of the approximation of the F_0 function which is involved in the calculations of polycenter integrals over Gaussian functions.[27] It is thus only valid for the formalism where those functions are found, i.e., mainly for Gaussian-type integrals. If this approximation is used in the infinite summations of the LCAO formalism, it is possible to isolate quantities which are immediately recognized as traditional Madelung expressions. Those Madelung summations, characteristic of the system and of the basis set used, can be evaluated by a direct application of the Laplace transformation[28] or of the Fourier representation method.[29-31]

This has led to another elegant way, using the Fourier transformation technique for evaluating the exact cancellation[32] of the direct space matrix elements. This procedure has, however, not been extended since it intrinsically induces some incoherence between the direct space representation of the matrix elements and the reciprocal working space of the Fourier transform.

A more easily implemented solution is based on the multipole expansion which also handles correctly those long-range interactions. A complete formulation has been given for stereoregular polymers[33] and for periodic helices[34] and is summarized below. It only implies the precise evaluation of the multipole moments of the charge distribution in the unit cell (or in the reduced unit cell for helical polymers). The technique rests on the multipole expansion of the interaction operator r_{ij}^{-1}, within the Fock matrix element. This procedure is valid as soon as there is no significant overlap between the charge distributions whose interactions are to be calculated. This technique has the advantage of taking into account the long-range effects on the Fock matrix elements directly in the framework of the exact LCAO formalism. It is thus possible to perform the iterations to relax the charge distribution of the system in its own field. In this way, the induced multipole interactions are automatically adjusted. Exact summations up to infinity are carried out in such a way as to minimize the round-off errors. Selective control on the convergence and, more specifically, on the quantities one can afford to compute (e.g., electric moments) is possible and suitable. The method is straightforward to implement into computer programs since the necessary quantities are one-electron integrals which are easily calculated. The number of such integrals grows as the second power of the basis set size and is proportional to the number of interacting cells explicitly considered in the calculation, i.e., N^2N_C. The computing time is thus negligible with respect to conventional techniques where, as previously cited, a number of $N^4N_C^3$ terms is to be computed. The method has the additional advantage of being physically appealing through the multipole moments which constitute a traditional framework for interpreting electrostatic interactions.

5. Multipole expansion for Long-Range Coulomb Interactions

As noted in the previous section, in actual polymer calculations it is obviously impossible to deal with very large values of the number of unit cells since the two-electron part of an LCAO polymer calculation involves an enormous number of integrals. There is a need for limiting that number of unit cells to some amenable value N (empirically ranging from 1 in the nearest neighbor approximation to about 5). Long-range interactions behave like conditionally and slowly convergent series and are actually significantly contributing far beyond this number N. An analysis to this problem has solved this question by the use of a multipole expansion. The form of the matrix elements is cast into the form:

$$h^j_{pq} = T^j_{pq} + \sum_{l=-N}^{N} \sum_{r,s} 2D^l_{rs} \sum_{h=-N}^{N} [(pq^j | r^h s^{h+l}) - \frac{1}{n_0} S^l_{rs} \sum_A Z_A V^j_{pq} |^h_A]$$
$$+ L^j_{pq}(N) - \sum_h \sum_l \sum_r \sum_s D^{l-h}_{rs} (pr^h | q^j s^l)$$

with:

$$L^j_{pq}(N) = \{ \sum_{h=-\infty}^{\infty} - \sum_{h=-N}^{N} \} [- \sum_A Z_A V^j_{pq} |^h_A + \sum_{l=h-N}^{h+N} \sum_{r,s} D^{hl}_{rs}(pq^j | r^h s^l)]$$

The detailed deduction of the multipole expansion gives:

$$L^j_{pq}(N) = \sum_{k=0}^{\infty} \sum_{l=0}^{\infty} U^{j(k,\ell)}_{pq} a^{-(k+\ell+1)} \Delta_N^{(k+\ell+1)}$$

where:

$$\Delta_N^{(n)} = \sum_{h=1}^{\infty} h^{-n} - \sum_{h=1}^{N} h^{-n} = \zeta(n) - \sum_{h=1}^{N} h^{-n}$$

$\zeta(n)$ is the Riemann zeta function, $U^{j(k,\ell)}_{pq}$ is the interaction of the 2^k-pole and 2^l-pole moments:

$$U^{j(k,\ell)}_{pq} = \sum_{m=-s(k,\ell)}^{s(k,\ell)} (k+\ell)!(-1)^m[(-1)^k + (-1)^\ell].$$
$$[(k+|m|)!(\ell+|m|)!]^{-1} M^{j(k,m)}_{pq} M^{(\ell,m)*}$$

In the above expressions, s is equal to the smallest numbers between k and ℓ. Capital letter $M^{(\ell,m)}$ refers to the mth component of the 2^ℓth electric moment expressed in spherical coordinates and related to the charge distributions associated with the orbital product $P^j_{pq}(\mathbf{r})$:

$$M^{j(k,m)}_{pq} = < \chi_p(\mathbf{r}) | r^k P_k^{|m|} \cos(\theta) e^{im\phi} | \chi^j_q(\mathbf{r}) >$$

or to the total charge (electrons + nuclei) associated with each translational unit:

$$M^{(k,m)} = -\sum_{A} Z_A r_A^k P_k^{|m|}(\cos\theta_A)e^{im\phi_A} + \sum_{j=-N}^{N}\sum_{p,q} D_{pq}^j < \chi_p(\mathbf{r})\,|\,r^k P_k^{|m|}(\cos\theta)e^{im\phi}\,|\,\chi_q^j(\mathbf{r}) >$$

Because of electroneutrality constraints:

$$U_{pq}^{j(k,0)} = 0$$

and to avoid coordinate dependence of the results, truncation is performed to sum only $U_{pc}^{j(t,\ell)}$ terms of identical $k+\ell+1$ values. We obtain the working formula:

$$L_{pq}^j(N) = \sum_{k=3,5,\ldots;\,k=\mathrm{odd}} \Delta_N^{(k)}\, a^{-k} \sum_{\ell=1}^{k-1} U_{pq}^{j(k-\ell-1,\ell)}$$

where a selective control can be made by the N and k parameters.

6. The Short- or Long-Rangeness Character of the Exchange Contribution

Important and recent analysis of the exchange term has been given in the literature starting with important papers by Ukrainski[24] and by Piela, André, Fripiat and Delhalle,[35] completed by more detailed studies by Calais and Delhalle,[36-39] Monkhorst,[40] and Monkhorst and Kertesz.[41] We summarize here a simple approach of the problem.

The exchange contribution:

$$-\Sigma_h\Sigma_l\Sigma_r\Sigma_s\, D_{rs}^{hl}(pr^h\,|\,q^js^l) = -\Sigma_h\Sigma_l\Sigma_r\Sigma_s\, D_{rs}^{0l-h}(pr^h\,|\,q^js^l)$$

corrects for the self-repulsion included in the Coulomb part of the SCF one-electron operator as magistrally demonstrated in solid state calculations by J.C. Slater.[42] It also includes the effect of the Pauli principle on the independent electron model. Clearly, the magnitude of the integrals $(pr^h\,|\,q^js^l)$ is determined, in the same sense as indicated above, by the distributions:

$$P_{pr}^h = \chi_p(\mathbf{r})\chi_r^h(\mathbf{r}) \text{ and } P_{qs}^{jl} = \chi_q^j(\mathbf{r})\chi_s^l(\mathbf{r})$$

and by the distance between their center of charges. The terms in h and l summations are significant near the origin cell 0 for the summation over h and near the cell j for the summation over l:

$$
\begin{aligned}
&-\Sigma_h\Sigma_l\Sigma_r\Sigma_s\, D^{hl}_{rs}(pr^h\,|\,q^js^l)\\
&\approx -\Sigma_r\Sigma_s\ \{\ D^{0j}_{rs}(pr^0\,|\,q^js^j)\\
&\quad + D^{0j-1}_{rs}\,(pr^1\,|\,q^js^j) + D^{0j+1}_{rs}\,(pr^{-1}\,|\,q^js^j) + \cdots\\
&\quad + D^{0j+1}_{rs}\,(pr^0\,|\,q^js^{j+1}) + D^{0j-1}_{rs}\,(pr^0\,|\,q^js^{j-1}) + \cdots\ \}\\
&\approx D^{0j}_{pq}\,(pp^0\,|\,q^jq^j)
\end{aligned}
$$

This points out that exchange contributions can be important for Fock matrix elements between widely separated orbitals since the decrease of both two-center repulsion integrals $(pp^0\,|\,q^jq^j)$ and elements of density matrices D^{0j}_{pq} is very slow with distance. In non-metallic cases, D^{j}_{pq} decays exponentially and in metallic cases it decays like $|j|^{-1}$. Up to rather recently, little had been undertaken to characterize the convergency of D^{0j}_{pq} with respect to j. However, troublesome numerical results have triggered such an analysis.

From the few existing studies, mainly based on simplified method such as the Hückel approach, it was known that D^{0j}_{pq} can decay as slowly as $(-1)^j|j|^{-1}$ leading to a $(-1)^j|j|^{-2}$ asymptotic decay of the exchange terms as first reported by Ukrainski[25] in a semiempirical work on a metallic polyacetylene model. In the previously cited analysis of *ab initio* results on an hydrogen chain model,[35] we have gained insight into the exchange behavior and reached the conclusions that poor convergency of the exchange energy is connected to the HOMO-LUMO separation (energy gap). In cases where the Hartree-Fock scheme produces good quality results (saturated polymers, insulators, ...), it is observed that the exchange potential is of a short-range nature. By contrast, in metallic situations, essentially characterized by a degeneracy or a near-degeneracy of the HOMO and LUMO levels, the exchange potential is of a very long-range nature. However, this situation precisely corresponds to the case where a single-determinant approximation to the ground state wave functions does not even provide a qualitatively correct description of the system.

The understanding of the long-range behavior of the exchange term was further developed by a study due to Monkhorst and Kertesz.[41] Using a two-band model system within the Hückel methodology, they recovered the $(-1)^j|j|^{-2}$ asymptotic decay of exchange in the half-filled band (metallic) situation and established a simple relationship between band gap, band width and an (exponential) large-j behavior of the density matrix elements in the filled band (insulating) situation.

More recently, Delhalle, Fripiat and Harris[43-45] have provided a general analysis of the convergency of D^{0j}_{pq} in terms of basic theorems on the convergency of Fourier series coefficients. They have shown that the convergency of D^{0j}_{pq} is essentially determined by the analytic properties of $D_{pq}(k)$:

$$D_{pq}(k) = \Sigma_j D^j_{pq} e^{ikja}$$

throughout the first Brillouin zone. In general, it can be stated that fastly decaying matrix elements D^j_{pq} correspond to insulators with large band gaps and, at the other extreme, metallic systems will lead to poorly converging D^j_{pq}.

A typical example has been published recently.[46] Minimal basis set calculations on the metallic infinite chain of hydrogen atoms have been performed within the computational framework we just described to illustrate the dependence of the Hartree-Fock energy bands on the summation of exchange contributions. The numerical results show the gradual decay of the density of states at the Fermi level as the number of terms in the exchange lattice sums is increased. This last point is a typical deficiency of the Hartree-Fock techniques as applied to infinite metallic systems, independent of their dimensionality.

7. Use of Screw Symmetry in Polymer Calculations

Many polymers exist in helical conformations, which leads to large translational units. For instance, the stable conformation of polypropylene is an helix which includes three asymmetric units, $-[CH_2 - CH(CH_3)]-$, per turn; the corresponding translational cell includes 27 atoms (18 hydrogens and 9 carbons). Even in the nearest neighbor approximation, a significant *ab initio* calculation would be a formidable task. A way to reduce this numerical effort is to consider explicitly the helical symmetry as it was proposed by Imamura and Fujita[47] and by Blumen and Merkel.[48]

The common helix designation[49] L*M/t specifies the number L of skeletal atoms in the asymmetric unit of the chain and the number of such asymmetric units M per t turns of the helix in the translational repeat. The translation period is denoted by a, and $\tau = a/M$ is the length of the reduced (helical) period which we hereafter call reduced unit cell or simply cell. The whole polymer is generated from this unit by applying screw symmetry operations $S^j(\mathbf{r}) = D(j\alpha)(\mathbf{r}) + j\,\boldsymbol{\tau}$ where $\alpha = 2\pi t/M$ (or $-2\pi t/M$) for a right- (or left-) handed helix. $D(\alpha)$ is a matrix representing a clockwise rotation around the z axis by the angle α; $\boldsymbol{\tau}$ is a vector of component $(0,0,\tau)$ directed along z. Chart 2 compares the usual equations of polymer quantum chemistry with and without helical translations. Since the size of the unit cell in the direct space is reduced by considering the helical operations, k can be regarded as a wave vector in the extended Brillouin zone $(-\pi/\tau \leq k \leq +\pi/\tau)$.

A point worth mentioning is that some electron integrals have to be rotated to take into account the helical symmetry as seen from the expression of the Bloch orbitals in Chart 2. The s- and p_z-type orbitals (the latter one being oriented along the polymer chain axis z) are invariant under the helical rotation. The p_x- and

Chart 2. A comparison of some basic formulas of polymer quantum chemistry with and without helical rotation.

Without helix	With helix
Orbital:	
$\phi_n(k, \mathbf{r}) = N^{-1/2}\Sigma_j\, e^{ikja}\Sigma_p\, c_{np}(k)\chi_p(\mathbf{r} - \mathbf{P} - ja)$	$\phi_n(k, \mathbf{r}) = N^{-1/2}\Sigma_j\Sigma_p\, e^{ikj\tau}c_{np}(k)\chi_p \times \{D(-ja)[\mathbf{r} - S^j(\mathbf{P})]\}$
Secular system:	
$\Sigma_p\, c_{np}(k)\{\Sigma_j\, e^{ikja}[h^j_{pq} - \varepsilon_n(k)\, S^j_{pq}]\}$	
$\Sigma_p\, c_{np}(k)\{\Sigma_j\, e^{ikj\tau}[h^j_{pq} - \varepsilon_n(k)\, S^j_{pq}]\}$	
$\equiv\Sigma_p\, c_{np}(k)\{h_{pq}(k) - \varepsilon_n(k)\, S_{pq}(k)\} = 0$	$\equiv\Sigma_p\, c_{np}(k)\{h_{pq}(k) - \varepsilon_n(k)\, S_{pq}(k)\} = 0$
Secular determinant:	
$\mid\{\Sigma_j\, e^{ikja}[h^j_{pq} - \varepsilon(k)\, S^j_{pq}]\}\mid$	$\mid\{\Sigma_j\, e^{ikj\tau}[h^j_{pq} - \varepsilon(k)\, S^j_{pq}]\}\mid$
	$\equiv\mid\{h_{pq}(k) - \varepsilon_n(k)\, S_{pq}(k)\}\mid = 0$

p_y-rotated orbitals are obtained from cartesian p_x and p_y ones by the following relations:

$$p_x^j = p_x^j \cos(j\alpha) + p_y^j \sin(j\alpha)$$

$$p_y^j = -\, p_x^j \sin(j\alpha) + p_y^j \cos(j\alpha)$$

The form of the matrix elements is cast into the same form as in section 5:

$$h^j_{pq} = T^j_{pq} + \sum_{l=-N}^{N} \sum_{r,s} 2D^l_{rs} \sum_{h=-N}^{N} [(pq^j \mid r^h s^{h+l}) - \frac{1}{n_0} S^l_{rs} \sum_A Z_A V^j_{pq} \mid ^h_A]$$

$$+ L^j_{pq}(N) - \sum_h \sum_l \sum_r \sum_s D^{l-h}_{rs}\,(pr^h \mid q^j s^l)$$

with

$$L^j_{pq}(N) = \{ \sum_{h=-\infty}^{\infty} - \sum_{h=-N}^{N} \} [- \sum_A Z_A V^j_{pq} \mid ^h_A + \sum_{l=h-N}^{h+N} \sum_{r,s} D^{hl}_{rs}(pq^j \mid r^h s^l)]$$

The detailed deduction of the multipole expansion gives now:

$$L^{j}_{pq}(N) = \sum_{k=0}^{\infty}\sum_{\ell=0}^{\infty} Upq^{j(k,\ell)}(N)\,\tau^{-(k+\ell+1)}$$

where:

$$U^{j(k,\ell)}_{pq}(N) = \sum_{m} 2(-1)^{\ell+m}(k+\ell)![(k+|m|)!(\ell+|m|)!]^{-1}.$$

$$w(k+\ell+1)\Delta_{k+\ell+1}(N,m,\alpha)M^{*j(k,m)}_{pq}(0)M^{(\ell,m)}(0)$$

In the above expressions, $w(k+\ell+1)$ takes the value 1 for $k+\ell+1$ odd; $+i$ for $k+\ell+1$ even and right-handed helix; $-i$ for $k+\ell+1$ even and a left-handed helix. The functions $\Delta_{k+\ell+1}(N,m,\alpha)$ are defined by:

$$\Delta_{k+\ell+1}(N,m,\alpha) = \zeta(k+\ell+1,m\alpha) - \sum_{h=1}^{N} \cos(hm\alpha)h^{-(k+\ell+1)}$$

$$\zeta(k+\ell+1,m\alpha) = \sum_{h=1}^{\infty} \cos(hm\alpha)h^{-(k+\ell+1)}$$

if $k+\ell+1$ is odd, or:

$$\Delta_{k+\ell+1}(N,m,\alpha) = \zeta(k+\ell+1,m\alpha) - \sum_{h=1}^{N} \sin(hm\alpha)h^{-(k+\ell+1)}$$

$$\zeta(k+\ell+1,m\alpha) = \sum_{h=1}^{\infty} \sin(hm\alpha)h^{-(k+\ell+1)}$$

if $k+\ell+1$ is even.

The functions $\Delta_{k+\ell+1}(N,m,\alpha)$ have to be evaluated once since the arguments are only related to the type of helix and to the conditions of the calculations (i.e., the input number of cells N). As in section 5, capital letter $M^{(\ell,m)}$ refers to the mth component of the 2^{ℓ}th electric moment expressed in spherical coordinates and related to the charge distributions associated with the orbital product $P^{j}_{pq}(\mathbf{r})$ or to the total charge (electrons and nuclei) associated with each translational unit. Explicit working formulas are given in Ref. 50.

8. A Simulated Ab Initio Technique: The Valence Effective Hamiltonian (VEH)

Even minimal basis sets calculations are very time consuming when applied to large systems of practical interest owing to the huge number of two-electron integrals to be calculated and stored. Thus, there is also an imperative need for simplified methods that could produce results of *ab initio* quality more economically. The Valence Effective Hamiltonian, the so-called VEH method, originally developed for molecules by Nicolas and Durand[51] and extended to polymers by André et al.,[52] has turned out to be a versatile tool for describing valence electronic structures of hydrocarbons with the same type of accuracy as double-zeta *ab initio* SCF calculations.

Generally speaking, an *effective* or *model* Hamiltonian is intended to simulate a given "exact" Hamiltonian H in some subspace of the Hilbert space spanned by the eigenvectors of H. Given a scalar product in the operator space, one can set up a variational principle to determine the "best" effective Hamiltonian of a given functional form. In the particular case of the VEH method, one simulates the Fock Hamiltonian of an *ab initio* valence pseudopotential calculation. The trial one-electron Hamiltonian is taken as the sum of the kinetic energy and a sum of effective atomic potentials of the atoms within their specific chemical environment and is written in the polymer case as:

$$h(\mathbf{r}) = -1/2\Delta + \Sigma_h \Sigma_A V_A^h(\mathbf{r})$$

The summations over h and A are, respectively, extended over cells and atoms within cells. For computational reasons, the atomic effective potentials are chosen as Gaussian projectors:

$$V_A^h(\mathbf{r}) = \Sigma_l \Sigma_m \Sigma_{tu} C_{lmtu}^A \, |{}_{lmt}^{Ah}> <{}_{lmu}^{Ah}|$$

The summations over l and m define the angular dependence of the projector (l=0 for s-type; l=1 for p-type). For computational facility, the projectors are normalized Gaussians:

$$|{}_{lmt}^{Ah}> = \kappa_{lmt}^{Ah}(\mathbf{r})$$
$$= N(x - x_A)^n (y - y_A)^{n'} (z - z_A - ha)^{n''} \exp[-\alpha_{lmt}^A (\mathbf{r} - \mathbf{R}_A - ha)^2]$$

In such a relation, the direction of periodicity is the z direction, ha is a translation in the direct space, $\mathbf{R}_A$ (with components x_A, y_A, z_A) is the center of the Gaussian. Evidently, $l=m=0$ means an s-projector ($n=n'=n''=0$) while $l=1$ represents a p projector (for example, p_x is obtained by putting $n=1$, $n'=n''=0$). The parameters of the method are therefore the Gaussian coefficients C_{lmtu}^A and exponents α_{lmt}^A. They are determined by a least-square fitting on a series of well-chosen pattern molecules (ethane, butadiene, ...). The potentials can be of single-zeta quality (Σ_{tu} limited to a single term) or of double-zeta quality (Σ_{tu} extended to four terms, two diagonal and two equal cross terms). Furthermore, they can be spherical (isotropic) if the coefficients C^{A}_{lmtu} for a given atom and a given l value are independent of m. Alternatively, they are nonspherical (anisotropic) if the various angular dependencies on the same atom (p_x, p_y, p_z) have different weights C^{A}_{lmtu} and different exponents α^{A}_{lmt}. In those conditions, the matrix elements of the VEH methodology become:

$$h^{j}_{pq} = T^{j}_{pq} - \Sigma_h \Sigma_A \Sigma_l \Sigma_m \Sigma_{tu}\, C^{A}_{lmtu} < p^0 \mid ^{Ah}_{lmt} > < ^{Ah}_{lmu} \mid q^j >$$

where the integrals $< p^0 \mid ^{Ah}_{lmt} >$ and $< ^{Ah}_{lmu} \mid q^j >$ are mixed integrals between LCAO atomic basis orbitals and the effective potential projectors:

$$< p^0 \mid ^{Ah}_{lmt} > = \int \chi_p(\mathbf{r} - \mathbf{P} - 0a)\kappa^{Ah}_{lmt}(\mathbf{r} - \mathbf{R}_A - ha)\, dv$$

$$= \int \chi_p(\mathbf{r})\kappa^{Ah}_{lmt}(\mathbf{r})\, dv$$

$$< ^{Ah}_{lmu} \mid q^j > = \int \kappa^{Ah}_{lmu}(\mathbf{r} - \mathbf{R}_A - ha)\chi_q(\mathbf{r} - \mathbf{Q} - ja)\, dv$$

$$= \int \kappa^{Ah}_{lmu}(\mathbf{r})\chi^{j}_{q}(\mathbf{r})\, dv$$

The VEH technique does reduce the integral evaluation to the calculation of kinetic and overlap integrals between Gaussian functions only (as long as GTOs are chosen as atomic basis). As soon as the integral part over the atomic functions is completed, the integrals between Bloch functions are evaluated. The complex eigenvalue problem producing the band structure is solved for selected k points. Since it further avoids the iterative cycles inherent to all SCF calculations, the method is very fast and economic in computing time. After obtaining the band structures, the standard programs described in section 9 are used to evaluate such electronic properties as the density of states. The method is free from any empirical parameter and, by its very principle, generally gives valence orbital energies comparable to *ab initio* ones depending on the quality of the effective potentials. A list of usual VEH effective atomic potentials in several molecular environments is given in the appendix.

In polymers, the VEH method has turned out to produce energy gaps of excellent quality. This is a clear indication that part of the electron-hole correlation is included into the virtual one-electron levels by that methodology. The Koopmans transition energies are thus reduced and, accordingly, the calculated values of the polarizabilities are larger (in agreement with the experimental trends). The same behavior has also been observed in semiempirical polarizability CNDO-type calculations (where the experimental transition energies is a basic ingredient of the parametrization through the correlation decrease of empirically evaluated one-electron integrals).

9. Particular Aspects of Computer Implementation

9.1 Band Structure Calculations

Several *ab initio* molecular SCF-MO-LCAO programs have been written to date and are largely distributed. Among those, let us cite POLYATOM,[53] and the IBMOL or KGNMOL,[54] GAUSSIAN,[55] HONDO,[56] and TURBOMOLE[57] series.

The practice of polymer computations is not so standardized. The Fock and overlap matrices are computed at a given level of approximation (semiempirical or *ab initio*). The calculation is made in direct space. The k-dependent matrices are diagonalized in the reciprocal space and, if necessary, a self-consistent procedure is turned on. Diagonalizations are achieved with the very efficient and self-contained routing CBORIS[58] which uses successively the Cholesky decomposition, the Householder similarity transformation, and the QL algorithm. If difficulties of linear dependence origin are encountered, it is preferable to use the direct transformation into orthogonal Löwdin's orbitals as explained in section 9.3. In all cases, the output consists of the energy bands and of the form of the molecular orbitals of the polymer. An interactive graphical communication can be initiated for:

1. ordering the energy bands,

2. plotting the standard electronic properties, such as band structures, band widths, density of states, simulations of electron spectra, electron densities,

3. calculating electron indices as charges and bond orders, and

4. determining conformations or other properties.

Automatic programs taking into account the effects of long-range interactions are fully implemented as previously described in sections 5 and 7.

9.2 Summation over Polymeric States and Integrations over First Brillouin Zones

In molecules, the MO energy levels are discrete and constitute a finite set. If we gradually go to the infinite system, the number of elementary cells goes up to ∞. The wave number k becomes a continuous variable within the interval $[-\pi/a, +\pi/a]$. The summation over the discrete set of MOs is replaced by a continuous integration between $-\pi/a$ and $+\pi/a$ over the first Brillouin zone.

Since the states due to the N cells of the polymer are unequivocally described by a length equal to $2\pi/a$ of reciprocal space, the relation between the number of states dN and a length dk of the reciprocal space is:

$$dN = (Na/2\pi)\, dk$$

If we take into account the fact that each "orbital" state can be occupied by two electrons of opposite spins, the relation between the number of electrons dn and a length dk of the reciprocal space becomes:

$$dn = (Na/\pi)\, dk$$

It is then easy to state the relationship:

$$\frac{1}{N}\sum_{k} = \frac{1}{2\pi/a}\int_{-\pi/a}^{+\pi/a} dk$$

$$\frac{1}{N}\sum_{k} f_k = \frac{1}{2\pi/a}\int_{-\pi/a}^{+\pi/a} f(k)\, dk$$

In practice, the integrals are evaluated by numerical integration over the first Brillouin zone. The complex eigenvalue problem is solved at 12 non-equidistant k-points in half the Brillouin zone. These 12 k-points corresponds to the abscissas of a 12-point Gauss-Legendre quadrature necessary to numerically compute the density matrix D:

$$D^{j}_{pq} = \frac{a}{2\pi}\int_{BZ} dk\ \exp(ikja)\{\sum_{n=1}^{n_0} C^{*}_{np}(k)\, C_{nq}(k)\}$$

$$= \sum_{i=1}^{12} W_i\ \exp(ik_i ja)\{\sum_{n=1}^{n_0} C^{*}_{np}\,(k_i)\, C_{nq}\,(k_i)\}$$

n_0 being the number of occupied bands.

In the last equation, the k_i and W_i are respectively the abscissas and the weights of the Gauss-Legendre quadrature. They are listed in Table I. The use of this quadrature scheme is only valid if the cell-index j is relatively small because of the presence of an oscillatory exponential function in the integrand.

9.3 Basis Set Linear Dependence

In band structure calculations, a near linear dependence error often occurs when the lowest eigenvalues of the overlap matrix S(k) are of the order of magnitude of 10^{-2} or smaller, while linear dependence problems start in the case of molecules only if the corresponding eigenvalue is as small as 10^{-6}.

The reason of this different behavior between 1D band structure and molecular calculations lies in the truncation of the infinite lattice sums which we have to

Table I. Weights and roots of the Legendre polynomial of degree 12 as needed for numerical integrations in half the Brillouin zone.

i	k_i (radians)	A_i
1	0.0289 6448 7993	0.0235 8766 8193
2	0.1506 1226 1496	0.0534 6966 2997
3	0.3614 3603 4181	0.0800 3916 4272
4	0.6482 3944 1541	0.1015 8371 3361
5	0.9930 0795 9289	0.1167 4626 8269
6	1.3740 8014 8713	0.1245 7352 2906
7	1.7675 1250 4877	0.1245 7352 2906
8	2.1485 8469 4301	0.1167 4626 8269
9	2.4933 5321 2049	0.1015 8371 3361
10	2.7801 5661 9408	0.0800 3916 4272
11	2.9909 8039 2093	0.0534 6966 2997
12	3.1126 2816 5596	0.0235 8766 8193

evaluate. Indeed, the problem can usually be removed if the lattice sums are properly truncated.[59]

Inherent linear dependencies can also be removed by the canonical orthogonalization proposed by Löwdin.[60] In an orthogonalization procedure, it exists several ways to choose the transformation matrix $\mathbf{T}(k)$ which is used to transform the generalized eigenvalue problem:

$$\mathbf{h}(k)\mathbf{C}(k) = \mathbf{S}(k)\mathbf{C}(k)\mathbf{E}(k)$$

into the classical eigenvalue equations:

$$\mathbf{h}'(k)\mathbf{U}(k) = \mathbf{U}(k)\mathbf{E}(k)$$

where:

$$\mathbf{h}'(k) = \mathbf{T}^{\dagger}(k)\mathbf{h}(k)\mathbf{T}(k)$$

and:

$$U(k) = T^{-1}(k)\, C(k)$$

The matrix T(k) must obey the relation:

$$T^{\dagger}(k)S(k)T(k) = 1$$

In Löwdin's canonical procedure, the matrix T(k) is defined as:

$$T(k) = W(k)s^{-1/2}(k)$$

where W(k) and s(k) are the eigenvectors and the (diagonal) eigenvalues matrices of S(k), respectively. If a given eigenvalue $s_i(k)$ approaches zero, i.e., if a linear dependence exists in the Bloch basis set, the matrix T(k) will have a column with very large values and will not be well behaved. A way to circumvent this problem is to eliminate the columns in the matrix W(k) corresponding to small eigenvalues and, consequently, to work within a reduced transformed orthonormal Bloch basis set. The truncated new functions will span exactly the same region of space as the original one if the eliminated eigenvalues were exactly zero. In eliminating the columns corresponding to small eigenvalues ($< 10^{-4}$), only a small part of the basis set is lost.[61]

9.4 Band Indexing Difficulty

Another interesting point to mention in actual calculations of band structures is the band indexing difficulty. In the reduced zone scheme, the band structure $\varepsilon_n(k)$ is a multivalued function of k and the successive values of n for any given k correspond to increasing levels: first to the inner core electrons, then to valence bands, and finally to unoccupied bands. Due to the prohibitive computing time associated with the complex diagonalization $h(k)\, C_n(k) = S(k)\, C_n(k)\, \varepsilon_n(k)$, the energy bands are calculated for discrete wave number values only, $\varepsilon_n(k_i)$. Bands are usually given an index n, increasing in order of increasing energy, but difficulties arise because the rule breaks down at degeneracy points. Two possibilities exist: either band crossing or avoided crossing. To these two possibilities correspond entirely different physical realities; in the first case, there is no band gap, whereas an energy gap does exist in the second case. If point symmetry exists within the unit cell, the problem can be solved by group theory, but when there is a low symmetry and a large number of bands, this becomes practically untractable. Currently, the band indexing is generally carried out by visual inspection or, in a more effective way, by using an interactive graphics interface with the knowledge of the local derivatives $\varepsilon'_n(k_i)$ of the energy eigenvalues. A general technique, easily applicable, for determining locally the k-derivative of all bands in each k-point is obtained by considering the scalar product[62] :

$$\varepsilon_n(k) = C_n^{\dagger}(k)\, h(k)\, C_n(k)$$

with the normalization condition:

$$1 = C_n^\dagger(k)\, S(k)\, C_n(k)$$

Hence,

$$\frac{d\varepsilon_n(k)}{dk} = \frac{dC_n^\dagger(k)}{dk}\,\mathbf{h}(k)\, C_n(k) + C_n^\dagger(k)\,\frac{d\mathbf{h}(k)}{dk}\, C_n(k) + C_n^\dagger(k)\,\mathbf{h}(k)\,\frac{dC_n(k)}{dk}$$

After straightforward matrix manipulation, we get the working formula:

$$\frac{d\varepsilon_n(k)}{dk} = C_n^\dagger(k)\,\frac{d\mathbf{h}(k)}{dk}\, C_n(k) - \varepsilon_n(k)\, C_n^\dagger(k)\,\frac{dS(k)}{dk}\, C_n(k)$$

which is simplified in an orthogonal atomic or Bloch basis to:

$$\frac{d\varepsilon_n(k)}{dk} = C_n^\dagger(k)\,\frac{d\mathbf{h}(k)}{dk}\, C_n(k)$$

Since the k-dependence of $\mathbf{h}(k)$ and $S(k)$ matrices is very simple:

$$\frac{dh_{pq}(k)}{dk} = \frac{d}{dk}\sum_j e^{ikja}\, h_{pq}^j = ia\sum_j je^{ikja}\, h_{pq}^j$$

$$\frac{dS_{pq}(k)}{dk} = \frac{d}{dk}\sum_j e^{ikja}\, S_{pq}^j = ia\sum_j je^{ikja}\, S_{pq}^j$$

The procedure has been implemented into an algorithm of general use for polymer consideration. In a practical way, a band structure calculation program is linked to a graphical device and the variation of $\varepsilon_n(k)$ are graphically represented by a rotating fixed length vector centered at $\{\varepsilon_n(k), k\}$ points. Interactive capabilities for zooming various energy zones have been added when the separation between two or more bands is not large enough. Experience has shown that nearly all kinds of bands can be labeled by this procedure and the time and the difficulties of data manipulations are very much reduced.

The following technique is used for the ordering of the band structures. For the first k-point, k_1, the bands are ordered in increasing energy, i.e., band index m is equal to the energy index n. For the next k-point, k_i, a search is performed for each band to find the energy index n corresponding to that band m. Knowing the energy values ε_n and their derivatives ε'_n at each k_i-point, it is possible to extrapolate the energies at k_{i-1}:

$$E_n(k_{i-1}) = \varepsilon_n(k_i) - \varepsilon'_n(k_i)*\Delta k$$

where Δk represents the step between two successive k-points. Similarly, we estimate the energy of the band m at the k_i-point:

$$E_m(k_i) = \varepsilon_m(k_{i-1}) + \varepsilon'_m(k_{i-1})^*\Delta k$$

For each band, we search the best agreement between the extrapolated energies (E) and the calculated energies (ε) at the k_{i-1} and k_i points on the basis of the least-square deviation:

$$\Delta_{i-1} = \varepsilon_m(k_{i-1}) - E_n(k_{i-1})$$

$$\Delta_i = E_m(k_i) - \varepsilon_n(k_i)$$

$$R_n = [\Delta_{i-1}]^2 + [\Delta_i]^2$$

The smallest R_n gives us the energy index n for the band m at the k_i-point. This procedure is repeated for all following k-points.

9.5 Density of States Calculations

As soon as the band structure $\varepsilon_n(k)$ has been obtained according to one of the various methods described above, the problem of getting relevant information to be compared to experimental data has to be considered. Indeed, making guesses or interpretations on properties is the most fruitful and rewarding aspect of theoretical studies. Numerous experimental quantities can be correlated with terms deduced from electronic charge distributions and from band structure calculations. However, a band structure is not measured directly and one has to apply some transformations to bring the calculated data to a form readily comparable to experiment. This is typical of the electronic density of states (DOS) distributions. The DOS distribution has its own meaning since in polymeric materials, the number of atoms is very large and the eigenvalues are very dense and bounded; it is therefore more convenient to deal with energy level distribution functions than with individual values. The DOS is defined as the number of allowed energy levels per energy unit:

$$D(E) = \frac{a}{\pi} \sum_{n=1}^{n_0} \left| \frac{dk}{d\varepsilon_n(k)} \right|_{\varepsilon_n(k)=E}$$

Such a function has the advantage of providing a synthetic description of one-electron levels and of being closely related to experiment. Though conceptually simple, electronic DOS distributions are not straightforward to evaluate numerically. General computer algorithms for polymers have been proposed.[63,64] Whenever the first derivative of $\varepsilon_n(k)$, $\varepsilon'_n(k)$, goes to zero, there is a large contribution to the DOS, giving rise to singularities in the DOS spectrum. We know in advance there are at least three values of the wave number k for which $\varepsilon'_n(k)$ vanishes identically, i.e., $k=0$ and $k = \pm\pi/a$. However, other extrema can occur but their number and location cannot be predicted *a priori*. When implementation by computer has to be considered, one has to turn to a new function $D(E_i)$, the DOS histogram. It exhibits the same behavior as D(E) but gives rise to finite peak height only. It is an average of D(E) over some energy interval ΔE:

$$D(E_i) = (\Delta E)^{-1} \int_{E_i - \frac{\Delta E}{2}}^{E_i + \frac{\Delta E}{2}} D(E)\, dE$$

$$E_i = E_0 + i\,\Delta E, \quad i = 0, 1, 2, \ldots, \quad \Delta E > 0$$

The value of ΔE is chosen according to a particular experiment to interpret. $D(E_i)$ is obviously the limited function of $D(E_i)$ when ΔE goes to zero. $D(E_i)$ is the height of the ith rectangular box centered at E_i; it represents an element of the histogram of the DOS. Various approaches to compute these $D(E_i)$ have been proposed. The histogram method involves generating the eigenvalues $\varepsilon_n(k)$ of the system at a large number of points in the Brillouin zone and then counting the results into boxes that subdivide the energy range. Another approach, more effective in the sense of saving computing time and numerical stability, is currently preferred. Basically, it consists of an analytical or continuous integration over small intervals $[k_i - \Delta k/2, \quad k_i + \Delta k/2]$ located at wave number values for which a diagonalization has been requested.

DOS distributions can be used to test results on energy band structure obtained from various methods. It is known that derivatives enhance variations of a function; therefore, it is no wonder a DOS distribution is a more sensitive test on energy bands since it involves the inverse of the first derivative $\varepsilon'_n(k)$.

9.6 Graphics Interface: BandDos

BandDos is an in-house highly interactive program for band structure ordering and display of the density of states as well as its convolution. The program is written in FORTRAN and uses the IBM graPHIGS™ software[65-67] based on the PHIGS standard. It optimally runs on an IBM 5080 workstation equipped with a 5085 or 5086 processor. All dialogues between the user and the graphics application are designed around the mouse, two menus, and LPFKeys. A panel describes the functionalities associated to the mouse and the LPFKeys.

The graphic screen is designed to be divided in three views, the left part for the band structure, the middle one for the density of states, and the right part for the convolution of the density of states (Fig. 3). The band structures are separated into three classes: core, valence, and unoccupied. All energy values for each k-point are represented by a marker and a small rotating fixed length vector showing its local derivative. In the initial state, the bands are ordered by increasing energy values. Two options allow to show or hide these energy markers and derivatives lines. The automatic ordering procedure described in section 9.4 is called just by selecting the *Ordering* option in the main menu (Fig. 4). A manual ordering of the band structure is also foreseen. By picking two energies values at

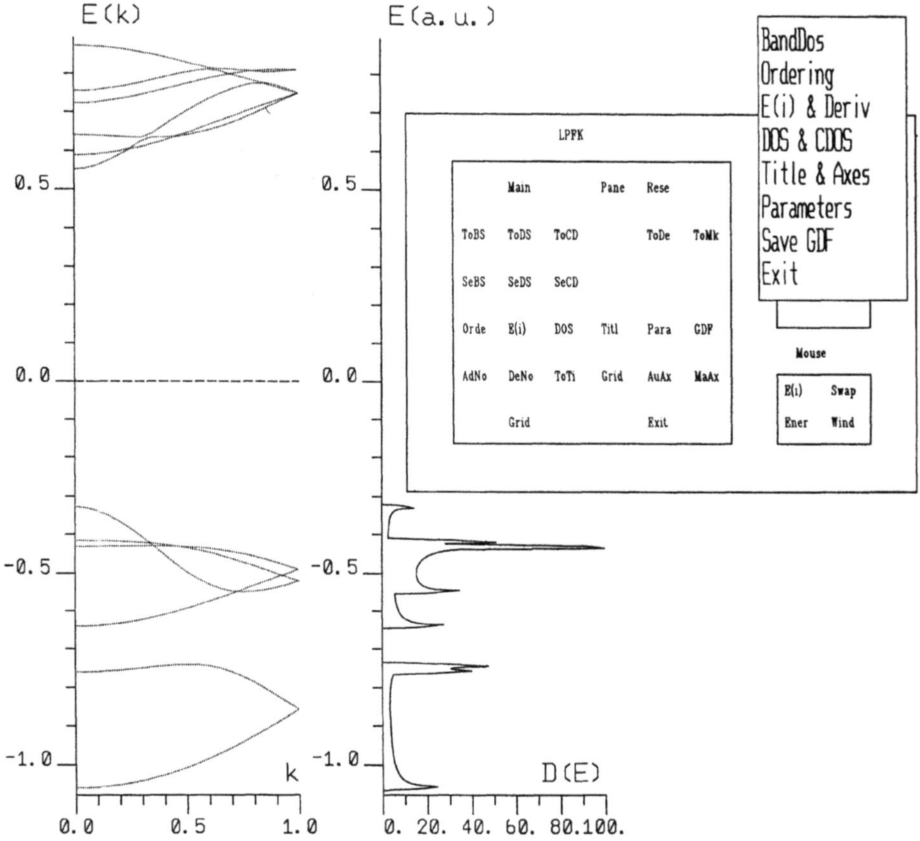

Figure 3. Band structure, density of states, and convolution of density of states for a polyethylene chain computed at the *ab initio* level. The main menu and the panel indicating the actions of the various function keys are displayed on the right side.

the same k-point, the corresponding bands are swapped from this k-point to the last (π/a). The *E(i)* option displays a table with the energies and derivatives around the last picked energy marker on the upper right side of the screen.

At any time, the computation of the density of states and its convolution corresponding to the current band structure may be performed. By default, they are displayed in two other views with the same energy scale than the band structure. Windowing facilities allow the user to zoom, rescale, and redefine the view disposition just by defining another window with the mouse. Each of the three views, band structure, density of states, convolution, may be viewed or hidden individually.

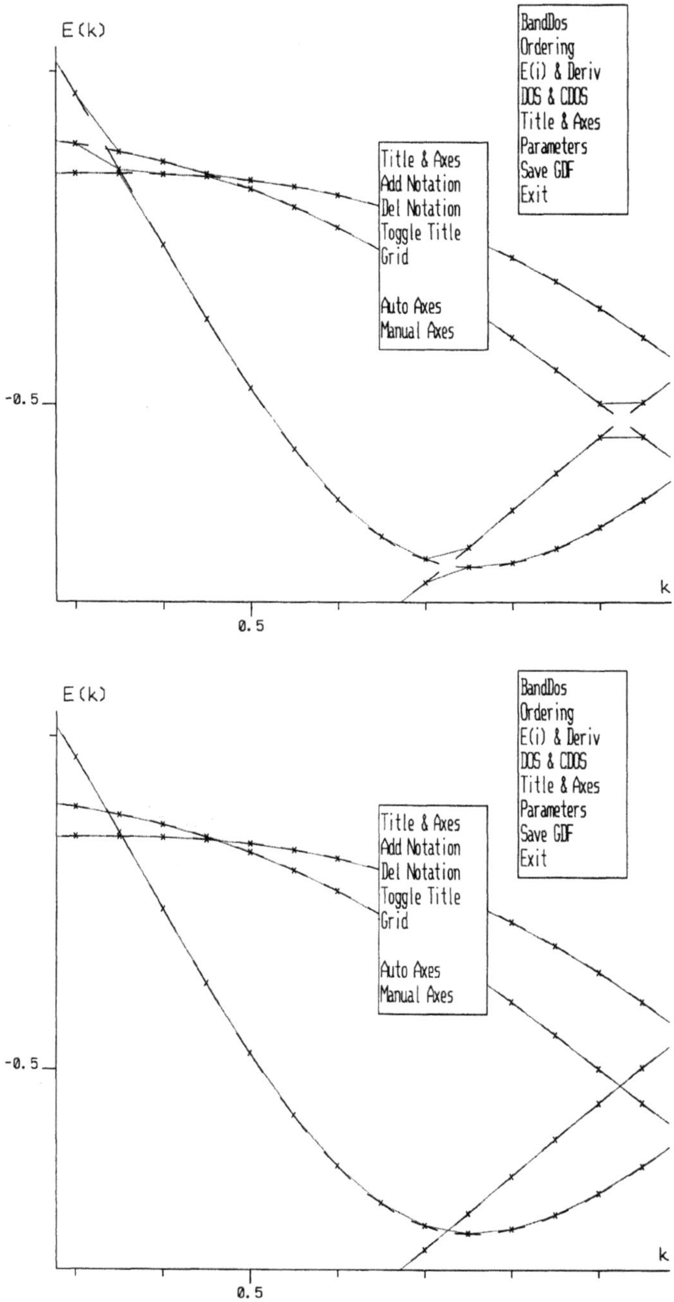

Figure 4. Screen display before and after automatic ordering or the band structure for polyethylene computed at the *ab initio* level.

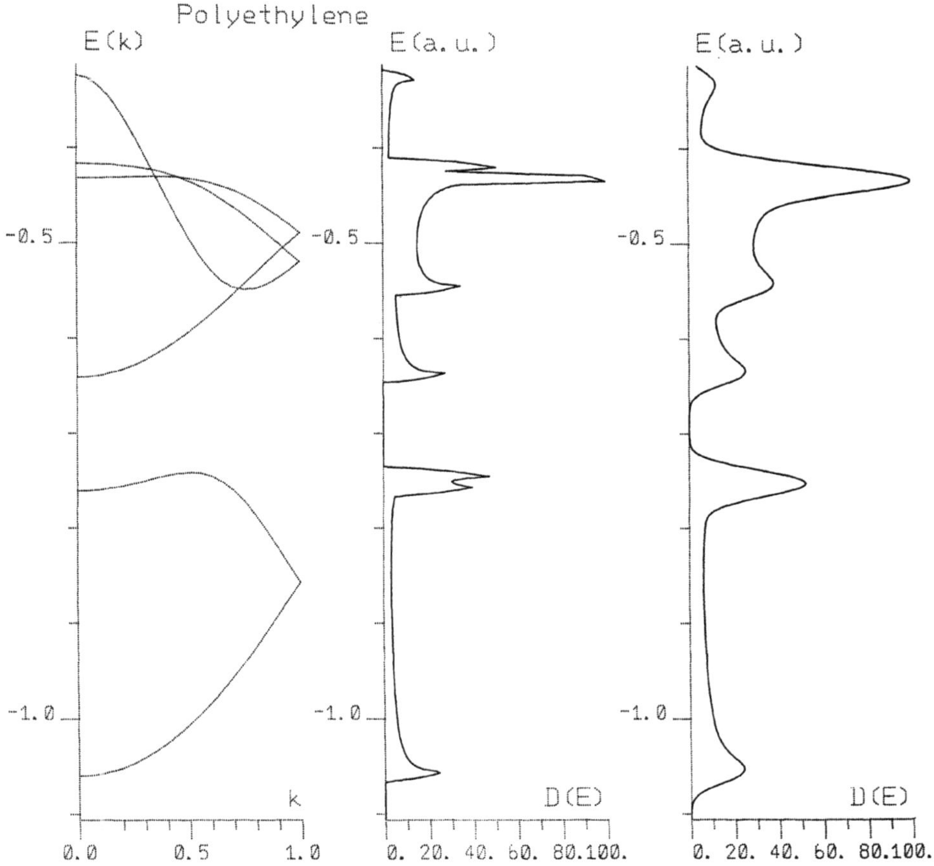

Figure 5. Band structure, density of states, and convolution of density of states for the polyethylene chain computed at the ab initio level.

A special menu, *Title & Axes*, is dedicated for annotations on the screen. Any notations (input via the keyboard) may be placed by dragging with the mouse anywhere on the screen with or without the help of a grid. Options allow to show or hide them totally or individually at a time. A default title given at the beginning of the application is added on the top of the screen. The axes labelling is automatic, although options allow to define new marking rules (first, last, number of tick marks, format of labels, ...). A special option loads a parameter file containing graphic information such as a color table, the color indices, linetypes, ... used for bands, density of states, ... Finally, a particular option creates a metafile in the GDDM format (filetype = ADMGDF)[65] for high quality drawings on pen-plotter (Fig. 5).

10. Applications

Many applications of the quantum theory applied to the electronic structure of polymers can be found in the literature. For the sake of simplicity, we refer to some review papers which describe the present state-of-the-art of quantum chemical calculations on stereoregular polymers.[68-74]

We conclude by referring to the now "classical" example of the band structure of polyethylene calculated by a full *ab initio* LCAO methodology and represented in Fig. 5. The tracing and the plotting of the energy bands and of its corresponding density of states has been obtained using the algorithm available in MOTECC-90.

Appendix 15A.

Tables of VEH Parameters

Single-zeta potentials

Isotropic[52]

Atom A	l,m	α^A_{lm}	C^A_{lm}
C	s	0.60	−1.21
	p	1.12	−1.99
H	s	1.40	−1.30

Double-zeta potentials

1° *Anisotropic*[52]

Atom A	l,m	α^A_{lm1}	α^A_{lm2}	C^A_{lm11}	C^A_{lm22}	$C^A_{lm12} = C^A_{lm21}$
C	s	1.39	0.35	−1.79	−0.47	+0.20
	p_σ	3.25	0.64	−4.66	−0.54	+0.18
	p_π	1.00	0.30	−3.49	−0.84	+1.14
H	s	7.00	0.60	−4.41	−0.43	+0.22

2° *Isotropic*

Carbon and hydrogen[75]

Atom A	l,m	α^A_{lm1}	α^A_{lm2}	C^A_{lm11}	C^A_{lm22}	$C^A_{lm12} = C^A_{lm21}$
C	s	3.63	0.33	−10.075	−1.178	+2.077
	p	6.50	0.52	−11.468	−0.443	−0.185
H	s	5.00	0.49	−4.696	−0.500	+0.646

Nitrogen, sulfur, and oxygen and carbon linked to nitrogen, sulfur, and oxygen.
Nitrogen[76]
Sulfur[77]
Oxygen (B. Thémans, unpublished results (1986)).

Atom A	l,m	α^{A}_{lm1}	α^{A}_{lm2}	C^{A}_{lm11}	C^{A}_{lm22}	$C^{A}_{lm12} = C^{A}_{lm21}$
S	s	1.00	0.30	+3.683	+1.521	−3.454
	p	0.90	0.20	−1.899	−0.047	−0.017
C(-S)	s	3.63	0.33	−6.506	−0.514	+0.354
	p	6.50	0.52	−12.380	−0.474	−0.067
N	s	1.50	0.50	+1.520	+0.790	−2.149
	p	3.00	0.35	−4.789	+0.046	−0.353
C(-N)	s	3.63	0.33	−0.486	+0.222	−1.710
	p	6.50	0.52	−1.637	+0.244	−2.834
O	s	5.00	0.50	−8.559	−1.387	+1.107
	p	2.89	0.07	−5.364	−0.066	−0.045
C(-O)	s	3.63	0.33	−5.347	−0.224	−0.212
	p	6.50	0.52	−4.315	+0.063	−2.192

3° *Isotropic*

Carbon and hydrogen
B. Thémans, unpublished results (1986).

Atom A	l,m	α^{A}_{lm1}	α^{A}_{lm2}	C^{A}_{lm11}	C^{A}_{lm22}	$C^{A}_{lm12} = C^{A}_{lm21}$
C	s	0.60	0.20	−1.721	−0.479	+0.515
	p	6.50	0.52	−9.441	−0.188	−0.927
H	s	5.00	0.49	−3.433	−0.347	+0.178

References

1. André, J.M., L. Gouverneur and G. Leroy, Int. J. Quant. Chem., **1**, 427 (1967).
2. André, J.M., L. Gouverneur and G. Leroy, Int. J. Quant. Chem., **1**, 451 (1967).
3. André, J.M., J. Chem. Phys., **50**, 1536 (1969).
4. Del Re, G., J. Ladik and G. Biczo, Phys. Rev., **155**, 997 (1967).
5. Ladik, J., Acta Phys. Hung., **18**, 173 (1965).
6. Ladik, J., Acta Phys. Hung., **18**, 185 (1965).
7. Suhai, S., and J. Ladik, Solid State Commun., **22**, 227 (1977).
8. Karpfen, A., Int. J. Quant. Chem., **19**, 1297 (1981).
9. Kertesz, M., Acta Phys. Acad. Sci. Hung., **41**, 127 (1976).

10. Pisani, C., R. Dovesi and C. Roetti, *Hartree-Fock Treatment of Crystalline Systems*, Springer Verlag, Berlin (1988).
11. André, J.M., D.P. Vercauteren, V.P. Bodart, J.L. Brédas, J. Delhalle and J.G. Fripiat, *Documentation for an ab initio Polymer Program (PLH)*, IBM IS&TG, **POK-28** (1983).
12. Otto, P., E. Clementi and J. Ladik, IBM Technical Report, **POK-13** (1982).
13. André, J.M., Comput. Phys. Comm., **1**, 391 (1970).
14. André, J.M., V.P. Bodart, J.L. Brédas, J. Delhalle and J.G. Fripiat, in *Quantum Theory of Polymers: Solid State Aspects*, J. Ladik and J.M. André (Eds.), NATO-ASI Series C123, Reidel Publishing Co., 1-21 (1984).
15. Hückel, E., Z. Physik., **60**, 423 (1930); **70**, 204 (1931); **76**, 628 (1932).
16. André, J.M., and G. Leroy, Chem. Phys. Lett., **5**, 71 (1970).
17. McCubbin, W.L., and R. Manne, Chem. Phys. Lett., **2**, 230 (1968).
18. Bloch, F., Z. Physik., **52**, 555 (1928).
19. McCubbin, W.L., and F.A. Teemull, Phys. Rev., **A6**, 2478 (1972).
20. André, J.M., D.P. Vercauteren and J.G. Fripiat, J. Comput. Chem., **5**, 349 (1984).
21. Pisani, C., and R. Dovesi, Int. J. Quant. Chem., **17**, 501 (1980).
22. André, J.M., and J. Delhalle, in *Quantum Theory of Polymers*, J.M. André, J. Delhalle and J. Ladik (Eds.), NATO-ASI Series C39, D. Reidel Publishing Co., 1 et sq. (1978).
23. Löwdin, P.O., Advan. Phys., **5**, 1 (1956).
24. O'Shea, F., and D.P. Santry, Chem. Phys. Lett., **25**, 164 (1974).
25. Ukrainski, I.I., Theoret. Chim. Acta, **38**, 139 (1975).
26. Delhalle, J., J.M. André, C. Demanet and J.L. Brédas, Chem. Phys. Lett., **54**, 186 (1978).
27. Suhai, S., and J. Ladik, Solid State Commun., **22**, 227 (1977).
28. Delhalle, J., J.G. Fripiat and L. Piela, Int. J. Quant. Chem., **S14**, 431 (1980).
29. Harris, F.E., and H. Monkhorst, Phys. Rev., **B2**, 4400 (1970).
30. Harris, F.E., J. Chem. Phys., **56**, 4422 (1972).
31. Harris, F.E., in *Theoretical Chemistry: Advances and Perspectives*, H. Eyring and D. Henderson (Eds.), Vol. 1, Academic Press, New York, 147 et sq (1975).
32. André, J.M., J.G. Fripiat, C. Demanet, J.L. Brédas and J. Delhalle, Int. J. Quant. Chem., **S12**, 233 (1978).
33. Delhalle, J., L. Piela, J.L. Brédas and J.M. André, Phys. Rev., **B22**, 6254 (1980).
34. Piela, L., J.M. André, J.L. Brédas and J. Delhalle, Int. J. Quant. Chem., **S14**, 405 (1980).
35. Piela, L., J.M. André, J.G. Fripiat and J. Delhalle, Chem. Phys. Lett., **77**, 143 (1981).
36. Calais, J.L., and J. Delhalle, in *Understanding Molecular Properties*, J. Avery et al. (Eds.), D. Reidel Publishing Co., 511 et sq. (1987).
37. Delhalle, J., and J.L. Calais, J. Chem. Phys., **85**, 5286 (1986).
38. Delhalle, J., and J.L. Calais, Phys. Rev., **B35**, 9460 (1987).
39. Delhalle, J., and J.L. Calais, Int. J. Quant. Chem., **S21**, 115 (1987).

40. Monkhorst, H.J., Phys. Rev., **B20**, 1504 (1979).
41. Monkhorst, H.J., and M. Kertesz, Phys. Rev., **B24**, 3015 (1981).
42. Slater, J.C., Phys. Rev., **81**, 385 (1951).
43. Delhalle, J., J.G. Fripiat and F.E. Harris, Int. J. Quant. Chem., **S18**, 141 (1984).
44. Delhalle, J., and F.E. Harris, Int. J. Quant. Chem., **22**, 219 (1985).
45. Delhalle, J., and F.E. Harris, Phys. Rev., **B31**, 6755 (1985).
46. Delhalle, J., M.H. Delvaux, J.G. Fripiat, J.M. André and J.L. Calais, J. Chem. Phys., **88**, 3141 (1988).
47. Imamura, A., and H. Fujita, J. Chem. Phys., **61**, 115 (1974).
48. Blumen, A., and C. Merkel, Phys. Status Solidi, **B83**, 425 (1977).
49. Wunderlich, B., *Macromolecular Physics*, Vol. 1, Academic Press, New York (1973).
50. André, J.M., D.P. Vercauteren, V.P. Bodart and J.G. Fripiat, J. Comput. Chem., **5**, 535 (1984).
51. Nicolas, G., and Ph. Durand, J. Chem. Phys., **70**, 2020 (1979); **72**, 453 (1980).
52. André, J.M., L.A. Burke, J. Delhalle, G. Nicolas and Ph. Durand, Int. J. Quant. Chem., **S13**, 283 (1979).
53. Csizmadia, I.G., M.C. Harrison, J.W. Moscowitz, S. Seung, B.T. Sutcliffe and M.P. Barnett, POLYATOM, Quantum Chemistry Program Exchange, **11**, 47 (1964); Newmann, D.B., H. Basch, R.L. Korregay, L.C. Snyder, J. Moskowitz, C. Hornback and P. Liebman, POLYATOM 2, Quantum Chemistry Program Exchange, **11**, 199 (1973).
54. Veillard, A., IBMOL 4, Computer Program, IBM Research Laboratory, San Jose, Calif. (1968); E. Ortoleva, G. Castiglione and E. Clementi, IBMOL 6, Comput. Phys. Comm., **19**, 337 (1980); R. Gomperts and E. Clementi, KGNMOL, Quantum Chemistry Program Exchange, Bulletin **8**, 538 (1988).
55. Hehre, W.J., W.A. Lathan, M.D. Newton, R. Ditchfield and J.A. Pople, GAUSSIAN 70, Quantum Chemistry Program Exchange, **11**, 236 (1973); J.S. Binkley, R. Whiteside, P.C. Hariharan, R. Seeger, W.J. Hehre, M.D. Newton and J.A. Pople, GAUSSIAN 76, Quantum Chemistry Program Exchange, **11**, 368 (1978); J.S. Binkley, R.A. Whiteside, R. Krishnan, R. Seeger, D.J. DeFrees, H.B. Schlegel, S. Topiol, L.R. Kahn and J.A. Pople, GAUSSIAN 80, Quantum Chemistry Program Exchange, **13**, 406 (1981); J.S. Binkley, M.J. Frisch, K. Rahgavachari, D.J. DeFrees, H.B. Schlegel, R.A. Whiteside, E.M. Fluder, R. Seeger and J.A. Pople, GAUSSIAN 82, Carnegie Mellon Univ., Pittsburgh, PA; M.J. Frisch, J.S. Binkley, H.B. Schlegel, K. Rahgavachari, R.L. Martin, J.J.P. Stewart, F.W. Bobrowicz, D.J. DeFrees, R. Seeger, R.A. Whiteside, D.J. Fox, E.M. Fluder and J.A. Pople, GAUSSIAN 86, Release C, Carnegie-Mellon University, Pittsburgh, PA.
56. Dupuis, M., J. Rys and H.F. King, HONDO, Quantum Chemistry Program Exchange, **11**, 336, 338 (1977); **13**, 401 (1981); M. Dupuis, J.D. Watts and H.O. Villar, HONDO7, Quantum Chemistry Program Exchange, Bulletin **8**, 544 (1988).
57. Aldrichs, R., M. Bär, M. Häser, H. Horn and C. Kölmel, Chem. Phys. Lett., **162**, 165 (1989).

58. Zupan, J., Ann. Soc. Sci. Bruxelles, Ser. 1, **89**, 337 (1975).
59. Suhai, S., P.S. Bagus and J. Ladik, Chem. Phys., **68**, 467 (1982).
60. Löwdin, P.O., Rev. Mod. Phys., **39**, 259 (1967).
61. Szabo, A., and N.S. Ostlund, *Modern Quantum Chemistry*, McGraw-Hill (1989).
62. André, J.M., J. Delhalle, G.S. Kapsomenos and G. Leroy, Chem. Phys. Lett., **14**, 485 (1972).
63. Delhalle, J., and S. Delhalle, Int. J. Quant. Chem., **11**, 349 (1977).
64. Delhalle, J., D. Thelen and J.M. André, Computers and Chemistry, **3**, 1 (1979).
65. IBM graPHIGS Library Manual, *Programmer's Reference for graPHIGS*, **SC33-8104**, IBM Corporation, Neighborhood Road, Kingston, New York 12401.
66. IBM graPHIGS Library Manual, *Understanding graPHIGS*, **SC33-8102**, IBM Corporation, Neighborhood Road, Kingston, New York 12401.
67. IBM graPHIGS Library Manual, *Writing Applications with graPHIGS*, **SC33-8103**, IBM Corporation, Neighborhood Road, Kingston, New York 12401.
68. André, J.M., in *Electronic Structure of Polymers and Molecular Crystals*, J.M. André and J. Ladik (Eds.), NATO-ASI Series B9, Plenum Press, 1-21 (1975).
69. André, J.M., Adv. Quant. Chem., **12**, 65 (1980).
70. André, J.M., in *Large Finite Systems*, J. Jortner, A. Pullman and B. Pullman (Eds.), D. Reidel Publishing Co., Dordrecht, 277-288 (1987).
71. André, J.M., J.L. Brédas and J. Delhalle, in *Biological and Artificial Intelligence Systems*, E. Clementi and S. Chin (Eds.), ESCOM, Leiden, 109-219 (1988).
72. André, J.M., in *Strategies for Computer Chemistry*, C. Tosi (Ed.), Kluwer Academic Publishers, 45-77 (1989).
73. André, J.M., and J. Delhalle, in *Self-Consistent Field: Theory and Applications*, R. Carbo and P. Klobukowski (Eds.), Elsevier Science Publishers (1990).
74. André, J.M., in *Molecular Level Calculations of the Structures and Properties of Non-Crystalline Polymers*, J. Bicerano (Ed.), M. Dekker, Inc., New York (1990).
75. Brédas, J.L., R.R. Chance, R. Silbey, G. Nicolas and Ph. Durand, J. Chem. Phys., **75**, 255 (1981).
76. Brédas, J.L., B. Thémans and J.M. André, J. Chem. Phys., **78**, 6137 (1983).
77. Brédas, J.L., R.R. Chance, R. Silbey, G. Nicolas and Ph. Durand, J. Chem. Phys., **77**, 371 (1982).

Chapter 16.

BNDPKG2: A Linear Combination of Gaussian Orbitals (LCGO) Band Structure Program for Cubic Crystals With One Atom Per Unit Cell

N.E. Brener, J. Callaway
Department of Physics and Astronomy

and

J.M. Tyler
Department of Computer Science

Louisiana State University
Baton Rouge, LA 70803, U.S.A.

1. Introduction

The calculation of the energy levels (bands) of electrons is the foundation for our knowledge of the electronic structure of solids. It is possible to determine experimentally, at least approximately, the energy distribution of single particle levels (usually called the density of states) and thus test the quality of the theoretical results. Band computations often yield the electron (spatial) density distribution directly and may serve as the basis for subsequent calculations of the total energy, lattice constant, electron momentum distribution, optical conductivity, and so on. These quantities may also be compared with experiment.

A quantum chemist who peruses the published literature concerning energy band calculations is likely to perceive two significant differences with respect to studies of molecular properties with which he may be familiar: (1) a greater emphasis on single particle states and properties and consequently lesser quantitative concern with electron correlation, and (2) a change in the conceptual basis of the work from Hartree-Fock to density functional theory.[1] The former is primarily the consequence of the greater size and complexity of the systems considered. Ordinary configuration interaction methods are not practical for solids. The latter results from some unpleasant pathologies which occur in the application of the Hartree-Fock equations to solids, particularly metals.[2]

The linear combination of Gaussian orbitals (LCGO) method[3] expands the wave function of a single electron state in Gaussian orbitals. The expansion is constructed so that the periodicity requirements (Bloch's theorem) are automatically satisfied. Let $u_i(\vec{r})$ be either a single Gaussian orbital, centered at the origin, or a linear combination of such functions (a contraction). The phased linear combination

$$\varphi_i(\vec{k},\vec{r}) = \frac{1}{\sqrt{N}} \sum_{\mu} \exp(i\vec{k} \cdot \vec{R}_\mu)\, u_i(\vec{r} - \vec{R}_\mu), \tag{1a}$$

obeys Bloch's theorem and so is characterized by wave vector $\vec{k}$. The sum runs over N (a large number) unit cells centered at lattice vectors $\vec{R}_\mu$. If there is only one atom in each unit cell, which is a limitation in the current program (BNDPKG 2) i may designate an orbital (or a component of an orbital) on that atom; but if there is more than one atom (M say) in the cell located at positions $\vec{d}_n (n = 1, \dots, M)$ with respect to the local origin, we write

$$\varphi_{in}(\vec{k},\vec{r}) = \frac{1}{\sqrt{N}} \sum_{\mu} \exp(i\vec{k} \cdot \vec{R}_\mu)\, u_i(\vec{r} - \vec{R}_\mu - \vec{d}_n), \tag{1b}$$

where u_i is the i'th orbital or component on the atom located at d_n.

Because φ obeys Bloch theorem, it can be used as a basis function for the crystal wavefunction $\psi_l(\vec{k},\vec{r})$ (l is a band index), which, in the case of one atom per unit cell. is written as

$$\psi_l(\vec{k},\vec{r}) = \sum_i c_{il}(\vec{k})\, \varphi_i(\vec{k},\vec{r}). \tag{2}$$

The expansion coefficients c_{il} are the eigenvectors of the Hamiltonian matrix. The number of bands that can be obtained in this way is the dimension of the Hamiltonian matrix which is $\sum n_j$, where n_j is the number of independent orbitals or components included on atom j (functions with the same radial behavior but different angular dependences are counted as independent).

A brief comparison of the LCGO method with some other methods in current use may be appropriate. The most common methods in use at present are the Green's function method[4] (also called KKR for Korringa-Kohn-Rostoker); the augmented plane wave methods (APW),[4] and, particularly for crystals with several atoms per unit cell, linearized versions of these[5]: LMTO (linear combination of muffin tin orbitals), LAPW (linearized augmented plane wave), and ASW (augmented spherical wave) methods. The basic KKR and APW methods expand the wave function in sets of functions which depend on energy: the solutions of the radial Schrödinger equation for different angular momenta inside a sphere inscribed in a unit cell. Combinations of these solutions (multiplied by spherical harmonics) are jointed to plane waves in interstitial portions of the cell. The result is that the matrix elements of the Hamiltonian depend on energy as well as wave vector. Then it is necessary to find the physical energy eigenvalues one at a time, by looking for the roots of a determinantal equation. The linearized methods which are now quite popular avoid this problem by expanding the matrix elements about some reference energy through first order. Then it becomes possible to obtain all relevant eigenvalues for a given $\vec{k}$ with one diagonalization. Approximations are thereby introduced and the resulting bands are accurate only within some range of energy near the selected reference energy. The linearized methods are, however, quite fast and reasonably accurate for simple materials (compact structures, small number of atoms per unit cell). For example, there have been numerous highly successful applications to a wide variety of elemental metals[5] but results of different groups who have considered more complicated materials such as $YBa_2Cu_3O_7$ show significant differences.[6]

The LCGO is, in contrast, a linear method. The matrix elements of the Hamiltonian depend on the wave vector $\vec{k}$ and not on energy. The greatest advantages of LCGO are apparent when one wishes to go beyond the band structure itself to the calculation of quantities like charge and spin form factors, the Compton profile, and the optical conductivity, as mentioned above. These require determination of matrix elements of operators (in the case of the optical conductivity for example, one needs elements of the momentum operator $\vec{p}$). These quantities may be computed quite straightforwardly and accurately using LCGO because the wave functions are combinations of simple functions which permit analytic evaluation of the necessary integrals. In the example above, analytic expressions for the matrix elements of the momentum operator between Gaussians can easily be obtained.

2. The LCGO Method for Energy Band Calculations

This section contains a description of the theory underlying the BNDPKG 2 computer program. However, discussion of specific computational methods is deferred to section 3. Atomic units with energies in Rydbergs are used throughout this chapter.

It is evident from Eqs. (1) and (2) that we will face a generalized eigenvalue problem of the form

$$\sum_j H_{ij}(\vec{k})c_{jl}(k) = E_l(\vec{k}) \sum_j S_{ij}(\vec{k})c_{jl}(\vec{k}), \tag{3}$$

where H and S are the Hamiltonian and overlap matrices respectively (and a single index has been used to enumerate the elements of the orbital basis). The matrix elements of the Hamiltonian are

$$H_{ij}(\vec{k}) = \int \varphi_i^*(\vec{k}, \vec{r}) H \varphi_j(\vec{k}, \vec{r}) d^3r$$
$$= \sum_\rho \exp(i\vec{k} \cdot \vec{R}_\rho) E_{ij}(\vec{R}_\rho) \tag{4a}$$

where

$$E_{ij}(\vec{R}_\rho) = \int u_i^*(\vec{r}) H u_j(\vec{r} - \vec{R}_\rho) d^3r. \tag{4b}$$

The periodicity of the Hamiltonian,

$$H(\vec{r}) = H(\vec{r} + \vec{R}_\rho), \tag{5}$$

($\vec{R}_\rho$ is any lattice vector) was used in obtaining the second line of Eq. (4a). The elements of the overlap matrix are given by similar expressions

$$S_{ij}(\vec{k}) = \sum_\rho \exp(\vec{k} \cdot \vec{R}_\rho) S_{ij}(\vec{R}_\rho), \tag{6a}$$

$$S_{ij}(\vec{R}_\rho) = \int u_i(\vec{r}) u_j(\vec{r} - \vec{R}_\rho) d^3r. \tag{6b}$$

Although Eq. (6b) obviously involves only two center integrals, a traditional decomposition of Eq. (4b) involves expressing the crystal potential, which is incorporated in the Hamiltonian, as a sum of contributions from different cells, thereby leading to three center integrals. To avoid these it is convenient to expand the potential in a Fourier series,

$$V(\vec{r}) = \sum_s V(\vec{K}_s) \exp(i\vec{K}_s \cdot \vec{r}) \tag{7}$$

in which the $\vec{K}_s$ are reciprocal lattice vectors, and $V(\vec{K}_s)$ is referred to as a Fourier coefficient of potential. We may write (4b) as:

$$E_{ij}(\vec{R}_\rho) = T_{ij}(\vec{R}_\rho) + V_{ij}(\vec{R}_\rho) \tag{8a}$$

where

$$T_{ij}(\vec{R}_\rho) = \int u_i(\vec{r})(-\nabla^2) u_j(\vec{r} - \vec{R}_\rho) d^3r, \tag{8b}$$

and

$$V_{ij}(\vec{R}_\rho) = \sum_s V(\vec{K}_s) S_{ij}(\vec{K}_s, \vec{R}_\rho). \tag{8c}$$

The $S_{ij}(\vec{K}_s, \vec{R}_\rho)$ are called generalized overlap integrals, and are given by equations similar to (6b):

$$S_{ij}(\vec{K}_s, \vec{R}_\rho) = \int u_i(\vec{r}) \exp(i\vec{K}_s \cdot \vec{r}) u_j(\vec{r} - \vec{R}_\rho) d^3r. \tag{8d}$$

All the quantities involved in Eqs. (8) are two center integrals. It is an important fact that the S_{ij} can be computed analytically when Gaussian orbitals are used for the u's in the equations.

A difficulty arises in connection with the use of Eq. (7). The contributions to the potential $V(\vec{r})$ from the nucleus and tightly bound inner shell electrons converges too slowly when an expansion in plane waves is employed. If the contribution from large reciprocal lattice vectors makes a significant contribution to V_{ij} (this is determined by the exponents in the Gaussian orbitals), it is desirable to evaluate the matrix element through a real space integral rather than a Fourier expansion. Details are given in Ref. 3.

Since the crystal potential is determined in part by the electron distribution, it is apparent that reliable band calculations must be made self consistently. This requires an iterative process in which the charge distribution calculated in one step (iteration) of the calculation is used to determine the potential for the next step. Here the LCGO method also seems to have an advantage. Methods such as KKR and APW, mentioned previously, are real space methods and require a precise representation of the potential in the unit cell. This, in turn, requires an accurate solution of Poisson's equation which, in view of the often complex geometry, may be complicated to obtain. These problems have often been bypassed by introducing the ansatz that the crystal potential is spherically symmetric within a sphere surrounding each atom, and constant within the interstitial region between spheres. This is the "muffin tin" approximation. The errors in this approximation are not large for close packed structures, (example, face centered cubic elemental metals), but become more important when one considers open structures (the diamond lattice, for example), or atoms of different sizes. In contrast, the LCGO does not use any approximation of this type. Poisson's equation (in atomic units)

$$\nabla^2 V = 8\pi\rho \tag{9}$$

is transformed to yield an algebraic relation between the Fourier coefficients of potential and those of the charge density, $\rho(K_s)$:

$$V(\vec{K}_s) = -8\pi\, \rho(\vec{K}_s)/K_s^2 \tag{10}$$

A limiting process, which we will not describe in detail here, leads to a finite and readily calculable value for $K_s = 0$. Of course it has to be noted that Eq. (9) applies only to the electrostatic contribution to the total potential: the exchange-correlation contribution is determined by the local charge density, and has to be considered separately.

The calculation of the Fourier coefficients of the charge density is quite straightforward within the LCGO method.[3,7] We find

$$\begin{aligned}\rho(\vec{K}_s) &= (2\pi)^{-3} \sum_l \int d^3k \left[\int |\psi_l(\vec{k}, \vec{r})|^2 \exp(-i\vec{K}_s \cdot \vec{r}) d^3r \right] \\ &= (2\pi)^{-3} \sum_l \int d^3r \; c^{*}_{sub\, <i\ sub\ lambda>}(\vec{k}) S_{ij}(\vec{k}, \vec{K}_s) c_{jl}(\vec{k}),\end{aligned} \tag{11}$$

and

$$S_{ij}(\vec{K}_s, \vec{k}) = \sum_\sigma \exp(i\vec{k} \cdot \vec{R}_\sigma) S_{ij}(\vec{K}_s, \vec{R}_\sigma). \tag{12}$$

In Eq. (11), the band sum and Brillouin zone integral include occupied states only. These equations imply that it is quite straightforward to compute the change in the electrostatic potential from iteration to iteration. Note that the quantities $S_{ij}(\vec{k}, \vec{K}_s)$ need be computed only once (at the beginning of the iterative cycle).

Unfortunately, the change in the exchange-correlation potential is not so simple to compute. One encounteres potentials $V_{xc}(\rho)$ which are given by complicated expressions obtained by analytic fits to numerical calculations. The simplest case is the exchange only (Kohn-Sham) potential

$$V_x = -4\left(\frac{3\rho}{8\pi}\right)^{1/3}. \tag{13}$$

The problem here is the dependence on $\rho(\vec{r})^{1/3}$. The best way to handle V_{xc} is, in principle, just to reconstitute the charge density at each iteration, calculate $V_{xc}(\vec{r})$, and recalculate its Fourier transform. V_{xc} is much smoother than the Coulomb potential, as is already apparent from Eq. (13), so no numerical problems are encountered. The unavoidable difficulty is that the calculation of $\rho(\vec{r})$ is time consuming. Since the exchange correlation potential is much smaller than the Coulomb potential in most of the cell, it has proved to be satisfactory in many cases simply to linearize V_{xc},

$$V_{xc}(\vec{r}) = V_{xc}[\rho^{(0)}(\vec{r})] + \left(\frac{\partial V_{xc}}{\partial \rho}\right)_0 [\rho(\vec{r}) - \rho^{(0)}(\vec{r})]. \tag{14}$$

Here, $\rho^{(0)}$ is the initially assumed charge density. The change in the Fourier coefficient of the exchange correlation potential produced by a change $\delta\rho$ in the charge density is then

$$\delta V_{xc}(\vec{K}_s) = \sum_t U(\vec{K}_s - \vec{K}_t)\delta\rho(\vec{K}_t), \tag{15a}$$

where

$$U(\vec{K}_s - \vec{K}_t) = \frac{1}{\Omega}\int \exp(-i(\vec{K}_s - \vec{K}_t)\cdot\vec{r})\left(\frac{\partial V_{xc}}{\partial\rho}\right)_0 d^3r, \tag{15b}$$

and Ω is the volume of the unit cell. The derivative of the exchange correlation potential is evaluated at the start of the iterations using the initial charge density, and its Fourier coefficients are also worked out. It is apparent that Eqs. (15) are useful only if the change in charge density is not large.

Detailed expressions for the integrals involved in the above formalism are given in Ref. 3. The exchange correlation potential that is currently built into the programs is the von Barth-Hedin potential[8] as parametrized by Rajagopal et al.[9] Relativistic contributions are not included in the current program.

3. BNDPKG 2

BNDPKG 2 is a program package that calculates self-consistent energy bands, wave functions, the density of states, and Fermi surface cross sections for metal crystals with *cubic symmetry and one atom per unit cell.* The cubic symmetry is used to reduce the calculation time in different parts of the program by factors of up to 48. Ferromagnetic (but not anti-ferromagnetic) order is allowed for these metals.

The original serial version of BNDPKG was written for an IBM OS/MVT operating system in 1977 and was later modified for an IBM MVS/XA operating system. This serial version of BNDPKG was used as the starting point for BNDPKG 2. Band calculations have been done for many different metals with cubic symmetry using this system[10-18] and sample results are presented in section 4.

BNDPKG 2 is divided into two major sections. One is used for ferromagnetic metals and the other for paramagnetic metals. Each of these sections has three parts. The first part (SCFINIT) of each has a modest amount of input data and computes files used in both the second part (SCF) and third part (BAND). SCF creates additional data which, together with data from SCFINIT, is used by BAND to generate the energy bands, density of states, and Fermi surface cross sections. SCFINIT requires less than 20 minutes and is vectorized. The decision

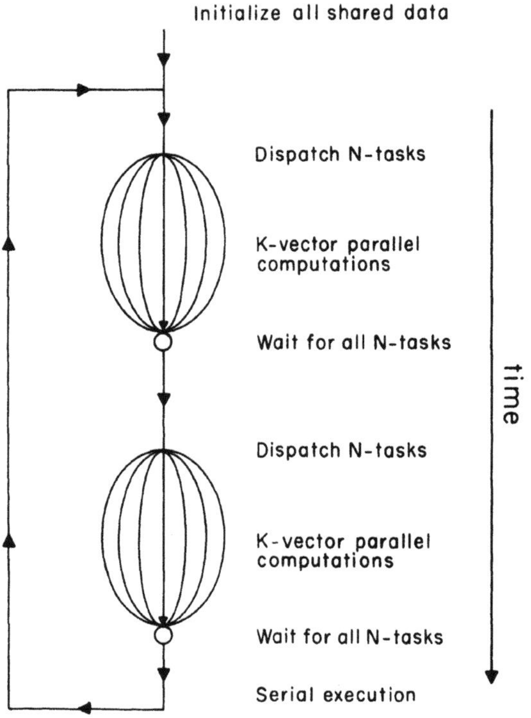

Figure 1. Schematic diagram of SCF processing.

was made to run this serially because of the short amount of run-time required. BAND also requires a relatively small amount of time and hence is also run serially.

SCFINIT is executed once per band calculation. It explicitly determines: the Fourier coefficients for the crystalline potential from an assumed charged density; the energy (E_{ij}) and overlap (S_{ij}) integrals; the $\vec{k}$-dependent Hamiltonian ($H_{ij}(\vec{k})$) and overlap ($S_{ij}(\vec{k})$) matrices; and the generalized overlap matrix ($S_{ij}(\vec{k}, \vec{K})$).

After SCFINIT has run to completion, SCF can be run. SCF is the interactive procedure that leads to self-consistency. This part requires several hours of computer time to obtain to a self-consistent result; in some calculations, e.g. chromium, this required 200 iterations for the desired convergence.

The original BNDPKG was analyzed to determine its vector potential. Included in this analysis was the consideration of the physical problem being solved, the solution method used, and the algorithms. The vectorization of the serial programs was done first. At the same time the vector processing potential was being determined, changes were being made in SCF that would allow parallel execution, i.e., any inherent data independence in this problem was preserved. Where effective the design, logic, and data were changed to make efficient use of the vector

Table I. Execution times for FCC.

3090-400 Times for FCC Chromium			
CPU's	Processing	CPU-time	Elapsed Time
1	Scalar	495.9 sec	504.1 sec
1	Vectorized	171.9 sec	174.6 sec
2	Vector-Parallel	174.7 sec	91.6 sec
3	Vector-Parallel	174.9 sec	62.6 sec
4	Vector-Parallel	175.1 sec	48.3 sec

Table II. Execution times for BCC.

3090-400 Times for BCC Chromium			
CPU's	Processing	CPU-time	Elapsed Time
1	Scalar	323.5 sec	328.6 sec
1	Vectorized	107.6 sec	116.4 sec
2	Vector-Parallel	108.4 sec	57.0 sec
3	Vector-Parallel	108.4 sec	39.1 sec
4	Vector-Parallel	108.5 sec	29.8 sec

facilities. All input/output statements were evaluated and removed if not essential to the presentation of the final results. Any input/output statement that would be in the parallel code was moved by rewriting major portions of the system.

Because this would be run on a virtual memory operating system, additional changes were made to exploit this architecture. Each algorithm was studied and changed, when possible, to localize its computation to a small number of virtual memory pages. Input/output statements were again reviewed to determine whether the use of virtual I/O for these would improve performance. For example, one direct access data set that contained the reciprocal lattice vectors was changed to be entirely read into an array and subsequently used throughout execution. This data is managed by VIO which resulted in improvement in processing times.

The original BNDPKG version has approximately 8,000 FORTRAN statements. Different sections of the code were actually timed in execution to accurately determine where the computer time was being consumed during execution. It was determined that almost all the time was being consumed in the SCF calculation (SCF). So, the maximum tuning effort was devoted to this code and this focus was on the overall logic, the data organization, and especially the DO-loop structures. The number of times each section of code was entered and the amount of

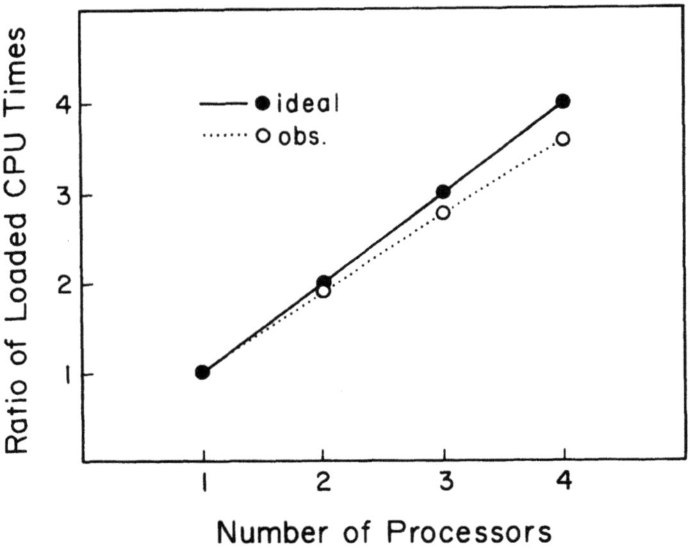

Figure 2. Speed-up for FCC.

computer time consumed in each were recorded. This information was used to determine the code to rewrite for the best improvement in performance.

After the code was selected to be rewritten, the standard FORTRAN compiler options were used to determine the "goodness" of the vectorization after each revision. In the most compute intense parts, the number of compound vector

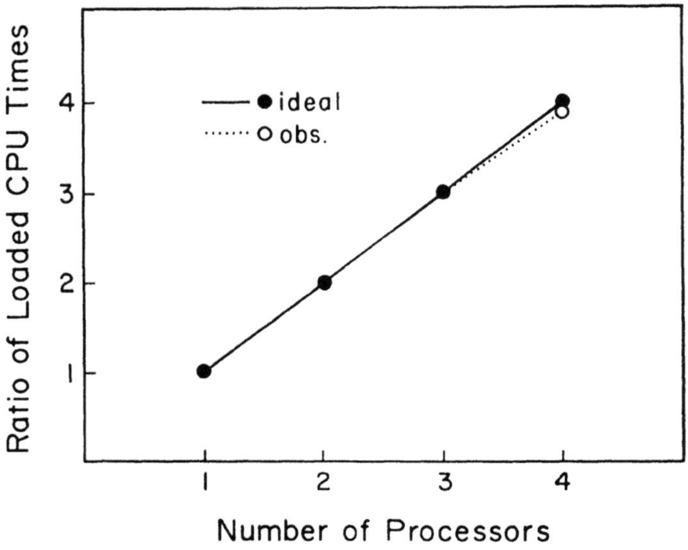

Figure 3. Speed-up for BCC.

Table III. Orbital exponents of the Gaussian basis set for molybdenum.

s	p	d	f
564811.58	4706.3308	172.7583	0.8
84340.172	1104.2933	51.005487	
19223.887	361.17122	19.267661	
5640.427	139.14298	7.9229414	
1935.0108	58.356511	3.3785846	
736.03701	25.821389	1.3293987	
304.54661	10.718475	0.49531799	
134.49152	4.7732916	0.15895673	
50.36175	2.0262457		
22.461279	0.88764641		
7.711217	0.34208279		
3.9263438	0.088		
1.8287531			
0.89138977			
0.39496469			
0.085897748			

instructions and loads of the vector registers were determined. If these could be improved, the code was changed to promote fewer vector register loads and/or more compound vector instructions. These compilations were repeated until an excellent level of vectorization was obtained. Loop reversals, simplified index expressions, and rearranged orders for some of the calculations were required and done.

After excellent vector performance was obtained in FORTRAN, it was determined that the IBM Engineering and Scientific Library (ESSL) should be used because it provides additional performance gains. ESSL has especially good performance for matrix multiplication (DGEMUL), matrix-vector multiplications (DAXPY, DYAX, ...), and eigenvalue-eigenvector determination (DSLEV). All of these ESSL routines were used.

SCF was divided into four separate major calculations, as shown in Fig. 1. The two most compute intensive of these four depend on the $\vec{k}$'s in one-forty-eighth of the Brillouin zone (BZ). These two sections of the code can be characterized as lengthy (with respect to computer time) DO-loops with $\vec{k}$ as the index. The FCC and BCC lattices have 89 and 55 $\vec{k}$'s, respectively. The underpinning physical problem provided the very useful fact that each $\vec{k}$ is independent in that part of the Brillouin zone being considered. Computationally, this means that they are independent and can be calculated in parallel with the sharing of some data.

The parallel implementation was done for the FCC metals by sectioning the 89 $\vec{k}$-vectors into groups that each have equal numbers of vectors. This is done auto-

Table IV. Orbital exponents of the Gaussian basis set for rhodium.

s	p	d	f
666938.460	5411.7578	224.17203	0.8
101663.950	1296.6564	66.353851	
23507.471	420.90869	25.11933	
6770.9733	160.23018	10.373975	
2274.1640	67.515001	4.4478314	
860.12987	30.400995	1.8266017	
356.22922	13.622169	0.71420590	
157.23074	6.253729	0.24613681	
60.823076	2.7461248		
27.200755	1.1965637		
9.9950277	0.46435074		
5.0587150	0.11		
2.0451786			
1.0271646			
0.4537890			
0.10732168			

matically with a subroutine. If four processors are available for parallel computing, this subroutine will section the vectors into 23, 22, 22, and 22 vectors. The section of 23 vectors will always be started first to improve load balancing. The same subroutine was used for the BCC metals. This subroutine provides good load balancing to the number of processors available which can change from run to run (e.g. IBM 3090-400 to IBM 3090-600).

For this application a large grain size is desirable and the sectioning with the $\vec{k}$-vectors provided this. Each $\vec{k}$ was done in parallel with a minimal amount of data sharing and with almost no interference (see results). SCF was re-structured to have two major subroutines, each of which can be run in parallel. These subroutines were started in a DO-loop with the DISPATCH statement and the shared data was passed in COMMON. Each subroutine obtained its sectioning with a starting $\vec{k}$ and an ending $\vec{k}$. The WAIT FOR statement was used to synchronize the calculation after each occurrence of a DISPATCH.

The vectorization and parallel speedup characteristics of the SCF part of BNDPKG2 were measured by executing on an IBM 3090-400E computer in scalar and in vector. The actual physical problem that provided the input was to determine the lattice constant at the ferromagnetic transition for both FCC and BCC chromium. These results show that the BCC calculation requires less time than the FCC calculation (55 $\vec{k}$-vectors vs. 89 $\vec{k}$-vectors).

Table I (FCC) and Table II (BCC) summarize the results of these measurements by presenting CPU-time (in seconds) for one iteration in the SCF part of

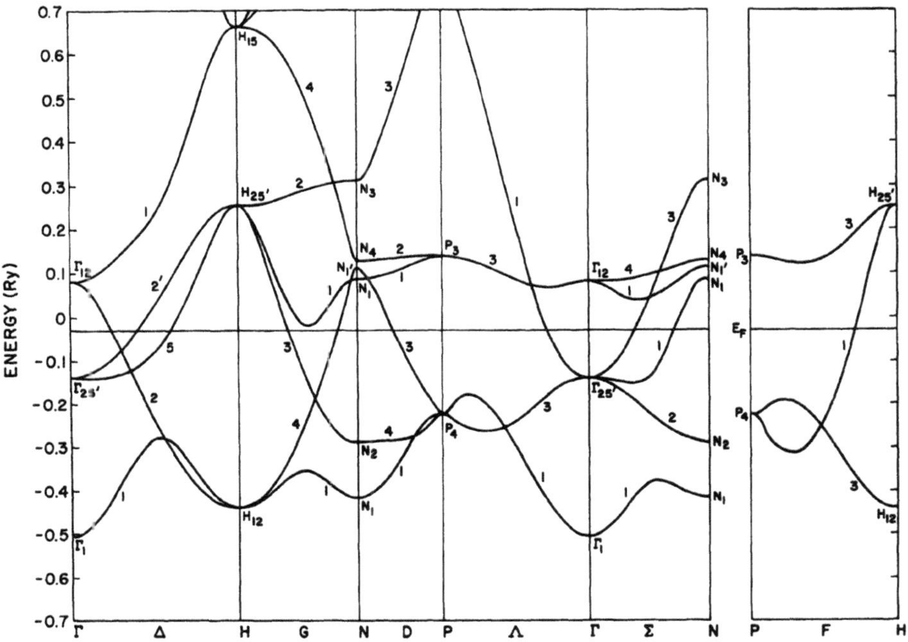

Figure 4. Energy bands of BCC molybdenum.

BNDPKG 2. An effective parallel speed-up of approximately 3.9 (for four processors) was obtained when compared to the version executing on the same processor in uniprocessor mode for BCC, and a slightly smaller speed-up was obtained for the FCC case. The graphic presentations of Tables I and II are given in Figs. 2 and 3.

4. Sample BNDPKG 2 Calculations

In this section, results are presented for BNDPKG 2 calculations on two 4d transition metals, body-centered-cubic (BCC) molybdenum (Mo)[16] and face-centered-cubic (FCC) rhodium (Rh)[17], henceforth referred to as Mo and Rh, respectively. The main input for a BNDPKG 2 calculation consists of the following: 1) the crystal structure (either BCC or FCC); 2) the magnetic state of the crystal (either ferromagnetic or paramagnetic); 3) the lattice constant; 4) the starting electronic configuration; 5) the basis set used to generate the starting potential, which is called the overlapping atomic densities (OAD) potential; and 6) the main basis set used to generate the self-consistent potential and the energy bands. The crystal structure and magnetic state of the elemental metals, which are the materials that can be studied by BNDPKG 2, are well known and can be obtained from many references. The magnetic state of both Mo and Rh is paramagnetic, which means

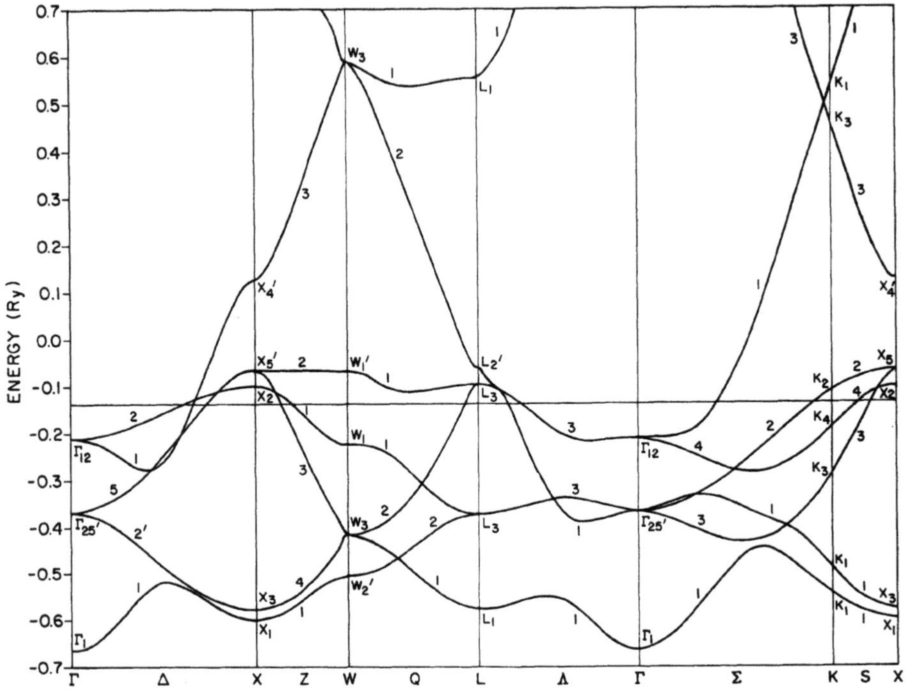

Figure 5. Energy bands of FCC rhodium.

that the paramagnetic versions of the SCFINIT, SCF, and BAND modules were used for the band structure calculations. The lattice constants of the elemental metals can be obtained by extrapolating the room temperature lattice constants, obtained from Wyckoff,[19] to T = 0 K using measured thermal expansion coefficients. This procedure yields lattice constants of 5.943 a.u. and 7.174 a.u. for Mo and Rh, respectively.

The starting electronic configuration is determined by the user, based on the expected final or self-consistent configuration. While BNDPKG 2 should converge to the same self-consistent electronic configuration for many different starting configurations, an appropriate choice of starting configuration can significantly speed up the convergence process. An electronic structure of the free atom is a good choice for the starting configuration of the crystalline calculation. The starting configurations used for the Mo and Rh calculations are $4d^5$ $5s^1$ and $4d^8$ $5s^1$, respectively. The starting potential is that obtained by constructing a spatial superposition of spherically averaged atomic charge densities. The Fourier coefficients of the Coulomb potential of this arrangement can easily be found from the atomic wave functions. However, the determination of the Fourier coefficients of the exchange correlation potential requires numerical integration. The method for this is discussed in Ref. 10.

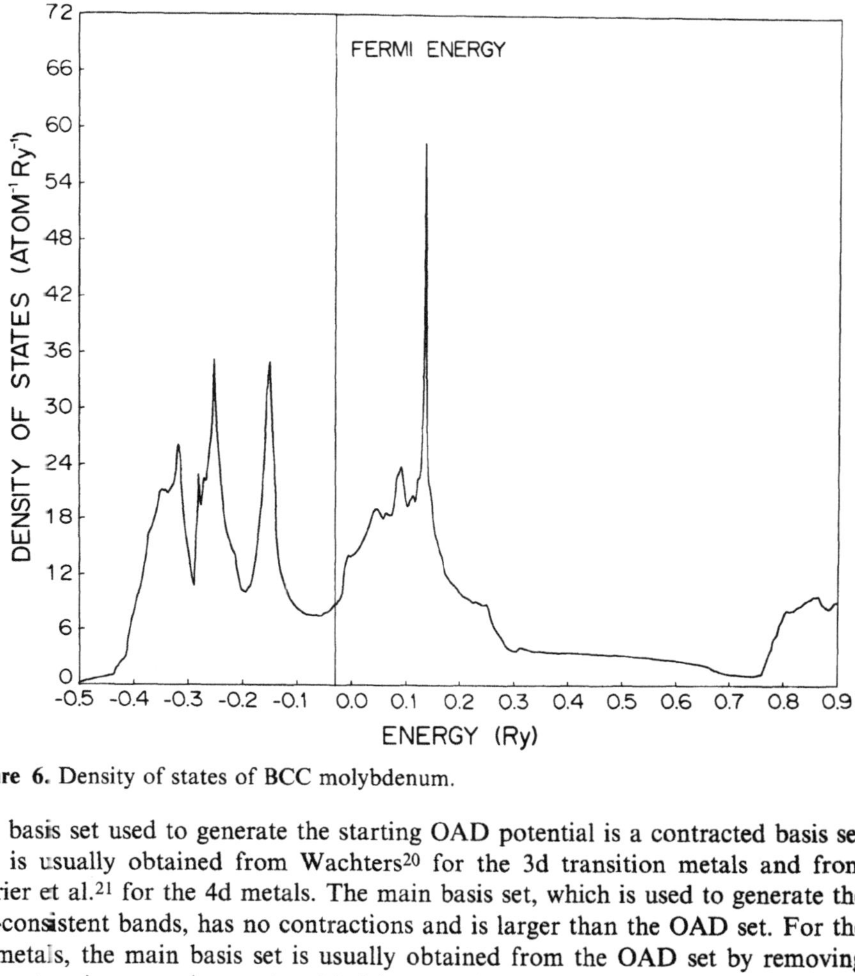

Figure 6. Density of states of BCC molybdenum.

The basis set used to generate the starting OAD potential is a contracted basis set and is usually obtained from Wachters[20] for the 3d transition metals and from Poirier et al.[21] for the 4d metals. The main basis set, which is used to generate the self-consistent bands, has no contractions and is larger than the OAD set. For the 3d metals, the main basis set is usually obtained from the OAD set by removing all contractions, so that each orbital is treated as a separate basis function, and then deleting the most diffuse *s* orbital and adding a diffuse *p* orbital (which is also given in Wachters) and an *f* orbital. In the case of the 4d metals, the main basis set is usually obtained from the uncontracted basis set given in Poirier et al. by deleting the most diffuse *s* orbital and adding a diffuse *p* orbital and an *f* orbital. The main basis sets used for the Mo and Rh calculations are given in Tables III and IV.

As mentioned previously, all of the above input is read by the SCFINIT module of BNDPKG 2, which in turn generates several data sets that are read by the SCF module. SCF then generates the self-consistent energy bands at a certain number of $\vec{k}$-vectors (55 for BCC, 89 for FCC) in the irreducible part of the Brillouin zone (BZ). The module BAND then uses the output of SCF, together with some of the data sets created by SCFINIT, to generate the self-consistent energy bands at a larger number of $\vec{k}$-vectors (506 for BCC, 505 for FCC) in the irreducible section of the BZ. This larger number of $\vec{k}$-vectors enables one to obtain

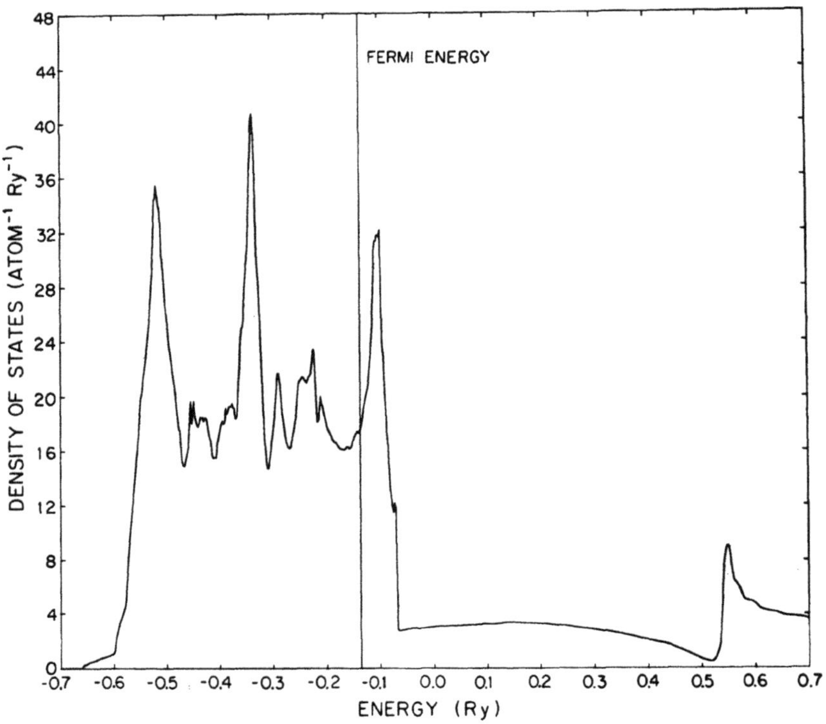

Figure 7. Density of states of FCC rhodium.

smoother graphs and more accurate results for the energy bands and related properties, all of which are generated by BAND. In particular, BAND produces graphs of the energy bands along symmetry axes, graphs of the density of states, and cross section diagrams of the Fermi surface, as well as numerical data for all of these quantities.

Figures 4 and 5 show the energy bands of Mo and Rh, respectively, along symmetry axes in the BZ. In a general way, the band structures of these and other (3d, 4d, 5d) transition metals can be described in terms of hybridization of a broad s-p band (width roughly 1 Ry) with a narrower d band complex. The Fermi level runs through the d bands, rising toward the top as the atomic number increases, and leading to a complicated Fermi surface of several sheets.

Other important quantities that can be obtained from the energy bands are the widths of the s-d bands, the d bands, the occupied portion of the s-d bands, and the occupied portion of the d bands. In the case of BCC Mo, these four quantities are given respectively by the energy differences $H_{25'} - \Gamma_1$, $H_{25'} - H_{12}$, $E_F - \Gamma_1$, and $E_F - H_{12}$ in Fig. 4, and have respective values of .761 Ry, .693 Ry, .647 Ry, and .555 Ry. In the case of FCC Rh, the corresponding four quantities are given respectively by the energy differences $X_5 - \Gamma_1$, $X_5 - X_1$, $E_F - \Gamma_1$, and $E_F - X_1$ in Fig. 5, and have respective values of .597 Ry, .533 Ry, .526 Ry, and .462 Ry.

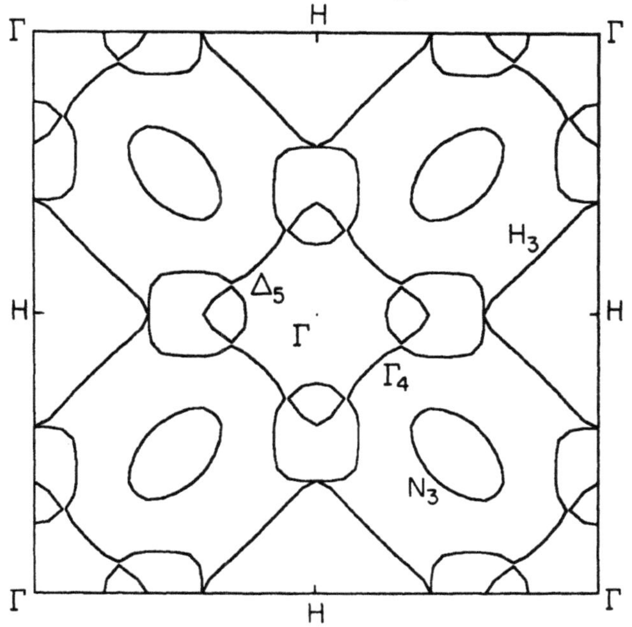

Figure 8. Cross sections of the BCC molybdenum Fermi surface in a (100) plane.

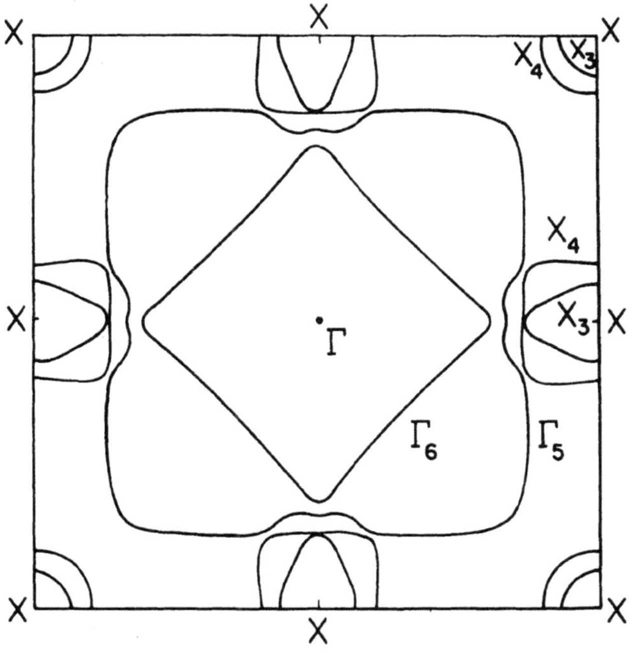

Figure 9. Cross sections of the FCC rhodium Fermi surface in a (100) plane.

Note the tendency of the bands to become narrower as the atomic number increases. The above values are in good agreement with experimental results where the latter exist.

One of the most important properties related to the band structure is the density of states (DOS), which is given for Mo and Rh in Figs. 6 and 7, respectively. In the case of Mo, the sharp peak above the Fermi energy is due to the flat portions of the energy bands around the state P_3, while for Rh, the peak that lies just above the Fermi energy is due to the flat portions of the band structure around the state L_3. The presence of DOS peaks above the Fermi level suggests that a ferromagnetic state will develop with an increase in the lattice constant. Computationally, this is indeed found to be the case for both materials,[16,17] with the onset of ferromagnetism in Mo and Rh occurring at lattice constant expansions of 36% and 14%, respectively. Note that the onset of ferromagnetism occurs at a smaller percentage expansion for Rh than for Mo, corresponding to the fact that the DOS peak is closer to the Fermi level in Rh than in Mo.

Another important property related to the energy bands is the Fermi surface. The cross sections of the Mo and Rh Fermi surfaces in the (100) plane are given in Figs. 8 and 9, respectively. The main features of the Mo Fermi surface are the electron jack (Γ_4) centered at Γ, the hole octahedra (H_3) at H, and the ellipsoidal hole surfaces (N_3) at N, while the main structures of the Rh Fermi surface include two large electron surfaces (Γ_5 and Γ_6) centered at Γ and two hold surfaces (X_3 and X_4) at X. All of the above Fermi surface structures are related to the intersections of the energy bands with the Fermi level, which are shown in Figs. 4 and 5. Calculations of the extremal areas of the Mo and Rh Fermi surface cross sections are in good agreement with experimental measurements of these quantities.

References

1. For a review of density functional theory, see J. Callaway and N.H. March, Sol. St. Phys., **38**, 135 (1984).
2. Probably the most important is the vanishing of the density of states at the Fermi energy, which leads to an incorrect temperature dependence for the electron specific heat. This was noticed more than 50 years ago: J. Bardeen, Phys. Rev., **50**, 1098 (1936).
3. C.S. Wang and J. Callaway, Comp. Phys. Commun., **14**, 327 (1978).
4. For an introduction, see J. Callaway, "Quantum Theory of the Solid State," Academic Press, New York, Chapter 4 (1974).
5. O.K. Andersen in "The Electronic Structure of Complex Systems," P. Phariseau and W. Temmerman, Eds., Plenum Press, New York (1984).
6. W.E. Pickett, Rev. Mod. Phys., **61**, 433 (1989). Contrast F. Herman, R.V. Kasowski and W.Y. Hsu, Phys. Rev., **B36**, 6904 (1987).

7. J. Callaway and J.L. Fry in "Computational Methods in Band Theory," P.M. Marcus, J.F. Janak and A.R. Williams, Eds., Plenum Press, New York, p.512 (1971).
8. U. von Barth and L. Hedin, J. Phys., **C5**, 1629 (1972).
9. A.K. Rajagopal, S.P. Singhal and J. Kimball (unpublished), as quoted by A.K. Rajagopal in "Advances in Chemical Physics," Vol. 41, G.I. Prigogine and S.A. Rice, Eds., Wiley, New York, p. 59 (1979).
10. P. Blaha and J. Callaway, Phys. Rev., **B32**, 7664 (1985).
11. D.G. Laurent, C.S. Wang and J. Callaway, Phys. Rev., **B17**, 455 (1978).
12. D.G. Laurent, J. Callaway, J.L. Fry and N.E. Brener, Phys. Rev., **B23**, 4977 (1981).
13. J. Callaway and C.S. Wang, Phys. Rev., **B16**, 2095 (1977).
14. C.S. Wang and J. Callaway, Phys. Rev., **B15**, 298 (1977).
15. A.R. Jani, N.E. Brener and J. Callaway, Phys. Rev., **B38**, 9425 (1988).
16. A.R Jani, G.S. Tripathi, N.E. Brener and J. Callaway, Phys. Rev., **B40**, 1593 (1989).
17. G.S. Tripathi, N.E. Brener and J. Callaway, Phys. Rev., **B38**, 10454 (1988).
18. H. Chen, N.E. Brener and J. Callaway, Phys. Rev., **B40**, 1443 (1989).
19. R.W.G. Wyckoff, Crystal Structures, Vol. 1, Wiley, New York (1965).
20. A.J.H. Wachters, J. Chem. Phys., **52**, 1033 (1970).
21. R. Poirier, R. Kari, and I.G. Csizmadia, Physical Science Data, Vol. 34: Handbook of the Gaussian Basis Set, Elsevier, New York (1985).

Chapter 17.

Monte Carlo and Molecular Dynamics Simulations†

E. Clementi, G. Corongiu, M. Aida, U. Niesar, and G. Kneller

IBM Corporation
Scientific and Engineering Computations
Dept. 48B/428
Kingston, New York 12401, U.S.A.

"Then, from these forces, by other propositions, which are also mathematical, we deduce the motions of the planets, the comets, the moon, and the sea. I wish we could derive the rest of the phenomena of nature by the same kind of reasoning from mechanical principles; for I am induced by many reasons to suspect that they may all depend upon forces."

— Isaac Newton (1686)

1. Introduction

Just before the turn of the century, it was generally believed that all the fundamental problems of physics had been solved. Given the interaction forces between matter, physicists believed that we could know everything in the universe by solving the set of Newtonian equations for all the particles. The fate of our universe was predetermined by the initial conditions.

As it turns out, however, physics seems to be far richer than humanity can comprehend. The twentieth century not only witnesses the discoveries of quantum mechanics and relativity, but also finds the emergence of chaotic behaviors from non-linear differential equations which points out the possibility that the future may not be completely predictable after all. Bearing these in mind, we now turn to the more practical sides of physical problems. What Newton said more than 200 years ago is basically still correct now, and forms the basis for a new branch of computational physics called *molecular dynamics* simulations. In this approach, we assume that all the properties of a substance, for example water, are com-

† This chapter is a revision and extension of the chapter by E. Clementi, G. Corongiu, G.C. Lie, U. Nieser and P. Procacci, "Monte Carlo and Molecular Dynamics Simulations", Chapter 8 in MOTECC-89.

pletely determined by the underlining interaction forces between the water molecules. The "other propositions" that are necessary for this approach are precisely those equations of motion that Newton put down in *Principia Mathematica* where the above quotation appeared in the preface. It should be noted, however, that there are phenomena, such as the solvation of electrons, whose simulations require the use of quantum mechanics principles.[1] In most applications where the thermal de Broglie wave lengths are much smaller than particle sizes and inter-particle spacings, classical Newtonian simulations should suffice. Thus as it stands, the *molecular dynamics* (MD) simulations are generally referred to classical systems.

While Newton was able to figure out the forces between heavenly bodies, it was with the discovery of quantum mechanics in the 1920's that we finally understand the interaction forces between atoms and molecules. In our global approach to science, we try to obtain these forces from the first principles of quantum mechanics by carrying out *ab initio* calculations. The methods of these calculations have been described extensively in detail elsewhere in this book. It should be pointed out, however, that a majority of the researchers in the field of molecular dynamics simulations uses what are generally referred to as *empirical* or *semi-empirical* potentials. These potentials are obtained by adjusting parameters in the potential expressions so that some known experimental or theoretical results are reproduced.

Even with the forces known, Newton's wish could still not be realized in practice before the advent of supercomputers. The problem is associated with the sheer number of particles that one has to deal with. Even with the periodic boundary conditions to be discussed later in the chapter, the number of equations of motion one has to solve to obtain meaningful physical results remains at hundreds or thousands. These equations can now be routinely integrated over tens of thousands of steps in a modern fast computer, thus making the study of many static and dynamic properties of substances possible. Although study of the tide formations in the sea from MD simulation is still far beyond the reach of present day computers, nor need a MD description because of the different time scale, fluid flows through pipes or past obstacles have already been investigated from the forces between the constituent atoms as discussed in Chapter 18.

In Chapter 19 we present dynamic effects in fluids at a different time scale, using Brownian dynamics. Thus, the time scale *first*, the computer power *as distant second* are clearly the basic discriminants for different methods, *i.e.* molecular dynamics, Brownian dynamics, cellular automata and fluid dynamics.

In this chapter, we will discuss the general techniques, such as integration schemes, periodic boundary conditions, properties calculations, *etc.*, that are commonly employed in molecular dynamics simulations. For the sake of completeness, a less powerful but also widely used method for the study of static macroscopic properties of a many particle system, *Monte Carlo*, will also be briefly described. This will then be followed with specific research examples carried out in our laboratory over the past few years. All of these examples include water: the first two investigate the properties of liquid water from new *ab initio* water-water interaction potentials, the third one compare a MD simulation carried out with and without solvent and the last example the behavior of water in a crystal of myoglobin. However, since potential functions form the starting point of any MD simulation, we will first address how they are generated from the results of *ab initio* calculations.

2. Interaction Potentials

Given a system of N particles, the total interaction energy can be written as:

$$\Delta E = \sum_{i>j} V(i,j) + \sum_{i>j>k} V(i,j,k) + \ldots.. + V(1,2,3,\ldots.,n) \tag{1}$$

where the first term is the two-body term, the second the three-body, and so on till the last term which contains all the bodies of the system. The assumption that ΔE is equal to $\Sigma V(i,j)$ is known as the *pair-wise additive approximation;* the remaining terms of the series are often referred to as *nonadditive corrections.* The basic idea in Eq. (1) is that, in a many particle system, there are entities recognizable as molecules. As often underlined in this book, it is the quantum mechanical approach that defines an atom or a molecule. In principle, all methods outlined in Chapters 6 to 11 can be used to determine interaction potentials between different molecules.

In the following, we will summarize our method for deriving interaction potentials[2] and we will give, as working example, the *ab initio* interaction potentials for biological systems, recently developed in our department.

The general procedure for obtaining *ab initio* potentials consists of five stages: 1) definition of the system; 2) selection of conformations; 3) *ab initio* computations; 4) selection of a functional form; 5) fitting procedure. The stages 2, 3, and 5 are executed more than once.

The first stage, definition of the sub-system, is a crucial one for a *large* system such as water-protein, since even with today's supercomputers we still can not perform routinely quantum chemistry calculations of a macromolecule without simplifications. The first step consists, therefore, in partitioning the system in sub-units of tractable size. However, the sub-units have to be chosen such as to preserve as close as possible the chemical and physical integrity of the whole. For example, in the water-protein system, the sub-unit may consist of water and amino acids. However, if we represent the alanine residue with only an isolated $-CH_3$ group to reduce the computations, it would give a completely different chemical system than the one we are interested in.

The second step involves the selection of the conformations and/or configurations to perform the *ab initio* computations. In this task, both chemical intuition and graphics play a very important role. A point to be kept in mind, during the selection, is that both attractive and repulsive interaction energy configurations are equally important.

In the third stage a method for calculating the conformation energies is chosen. In principle, it can be any of those underlined in previous chapters. However, the system size and the number of computations to be performed dictate the method and the basis set choice, being this step the most CPU intensive part of the process.

In the fourth step, a suitable mathematical function is selected. Among the many possible analytical forms, we comment here on the one we have adopted very often, but not always, for intermolecular pairwise potentials in biological systems. Given two molecules M and N, the interaction potential $V(M, N)$ might contain explicitly both an angular and a radial dependency. Mostly for simplicity reasons we have selected to use only radial functions depending on the interatomic distance $r(i, j)$, with i and j atoms belonging to molecules M and N, respectively. Regarding the form of the radial function, one has to compromise between two conflicting requirements. For accuracy and flexibility, one would like to use a fairly long series of terms. On the other hand, the longer the series, the larger the number of associated fitting parameters, and the larger the number of machine cycles required to compute interaction energies in a later simulation. We have often settled for a three term expansion of the form

$$V(M, N) = \sum_{ij} \left[-\frac{A_{ij}^{ab}}{r_{ij}^{6}} + \frac{B_{ij}^{ab}}{r_{ij}^{12}} + \frac{C_{ij}^{ab} q(i) q(j)}{r_{ij}} \right] \tag{2}$$

where A, B, and C are fitting parameters for atom i of class a and charge $q(i)$ in molecule M and for atom j of class b and charge $q(j)$ in molecule N. The parameter C (often equal to 1) should not be interpreted as related to the dielectric constant; it represents an average correction to the charges $q(i)$ and $q(j)$ which are obtained from computation of molecular wave functions of molecules M and N at infinite separation. It is obvious that the value of the net charge is a function of the internuclear separation and relative orientation of M and N. The parameters A and B are not to be equated with the formally similar Lennard-Jones parameters. The quantities A, B, and C are mutually dependent. The class identifiers are assigned to atoms according to their chemical environment. We refer elsewhere[2,3] and to the next section for a detailed description of the class definitions.

Once the analytical form of the potential has been chosen, a fitting to the interaction energies previously computed is performed in the fifth step. The form of the selected analytical function and the algorithm used in the fitting procedure are both important. It should be noted that obtaining "the best fit," *i.e.* the smallest value of the standard deviation, is always necessary but not sufficient, since the computed potential may give unreasonable binding energies in regions of the potential energy hypersurface which were poorly sampled. The quality of the fit and the completeness of the configurations sampled in the third step can be assessed by computing local minima in the potential hypersurface. This is usually achieved by starting with unbiased configurations, typically by random generation of coordinates, followed by a minimum search algorithm, for example a quasi-Newton method. A good potential must be able to predict all important configurations with interaction energies close to those computed in the third step, and other reasonable configurations not included in it. If new minima (or large deviations) are found, *ab initio* calculations for these configurations are performed, and steps 2, 3, and 5 are repeated as many times as needed, until convergence is reached.

By proceeding in the way described above, it is reasonable to believe that all important configurations are sampled, and the potential can be used with some confidence in further applications. Notice, however, that a potential "good" for energy prediction does not necessarily imply equal "goodness" in different properties.

In Chapter 1 we present a summary of all available potentials computed in this laboratory. The interested reader can refer to the given references for details.

2.1. Ab initio Pair Potentials for DNA-Protein Interactions

The specificity of the interactions of protein with nucleic acid, of protein with protein and of nucleic acid with nucleic acid play important roles in biological systems.[4] In order to understand these biological processes, a systematic investigation is necessary, and, in this context, theoretical studies can be of importance to obtain quantitative information at the molecular level. Molecules and atoms are the components of biological systems: quantum mechanical methods, so successful to describe electrons and nuclei in molecular and atomic chemical reactions, are also useful to understand the mechanism of biological processes at the molecular level. A systematic and rigorous theoretical treatment, however, is still impracticable, because of the size and complexity of the biological systems. To perform global simulations in biological system, therefore, we are compelled to look to other methods, which on one hand are less rigorous and on the other, should be much less time-consuming than the direct application of the *ab initio* molecular orbital method. Eventually one might attempt to combine these two techniques as, for example, pioneered in a study by Bolis *et al.* in the reaction of papain.[5]

In this section, we discuss *ab initio* pair potentials, modeled as to represent the interaction energies obtained by *ab initio* SCF calculations. In contrast to empirical energy functions,[6-8] whose parameters are fitted by referring largely to experimental values, the *ab initio* pair potentials are fully based on *ab initio* SCF interaction energies. In this way we can obtain the needed reference data (computed interaction energies) for any desired configuration simply by performing *ab initio* SCF calculations. Indeed scanning the entire potential surfaces is possible and the larger the number of the reference data, the more reliable the potential. Once a set of potential parameters is available, it can be applied in several problems concerning biological systems, using Monte Carlo and/or molecular dynamics simulations.

Here, we discuss *ab initio* pair potential for the interaction between DNA and proteins. Both DNA and proteins are large macro-molecules, however, they can be decomposed into several sub-units. In addition, each atom of DNA and proteins can be classified according to the chemical environment of the surrounding which makes up a sub-unit. Therefore, we can approach the interactions of DNA and proteins by a systematic study. An analytical pair potential, designed to be able to predict the interaction energies between any unit of DNA and proteins, is expected to correctly predict the interaction between DNA and proteins. Our long range goal is to perform global simulations for any biological system composed of DNA and protein. Therefore, any sub-unit of DNA and protein must be taken into account in constructing the *ab initio* pair potential, and this requires the computation of *ab initio* SCF computations systematically for all those sub-units, ensuring a full scanning of each potential energy surface. This study represents the most complex and systematic effort of *large-scale computer simulations* aimed at obtaining the interaction between biological molecules by means of the *ab initio* molecular orbital method. The study to obtain, by quantum mechanical computations, the interaction of water with proteins and nuclic acids - performed about 15 years ago - was about equivalent as human effort, but over one order of magnitude fewer *ab initio* energies (and all on smaller system) were computed: this is a result of progress both in computer hardware and application software.

Model Molecules

The SCF interaction energies between model molecules in various conformations are used as reference data for a non-linear least-square fitting yielding the set of parameters for the *ab initio* pair potentials.

As model molecules, we selected the fundamental and repeating units of DNA and proteins, as described below. The models have been chosen keeping in mind that all the classes of the atoms in proteins and nucleic acids (see below in Classification of Atoms) must be included in the model molecule sets, to ensure that the pair potential can describe the interactions between any protein and any DNA.

To represent DNA, we used two models: one is for the base-sugar portion, the other for the polydeoxyribonucleotide chain portion. For the former portion, we used, as models, cytosine-sugar (Fig. 1a), guanine-sugar (Fig. 1b), thymine-sugar

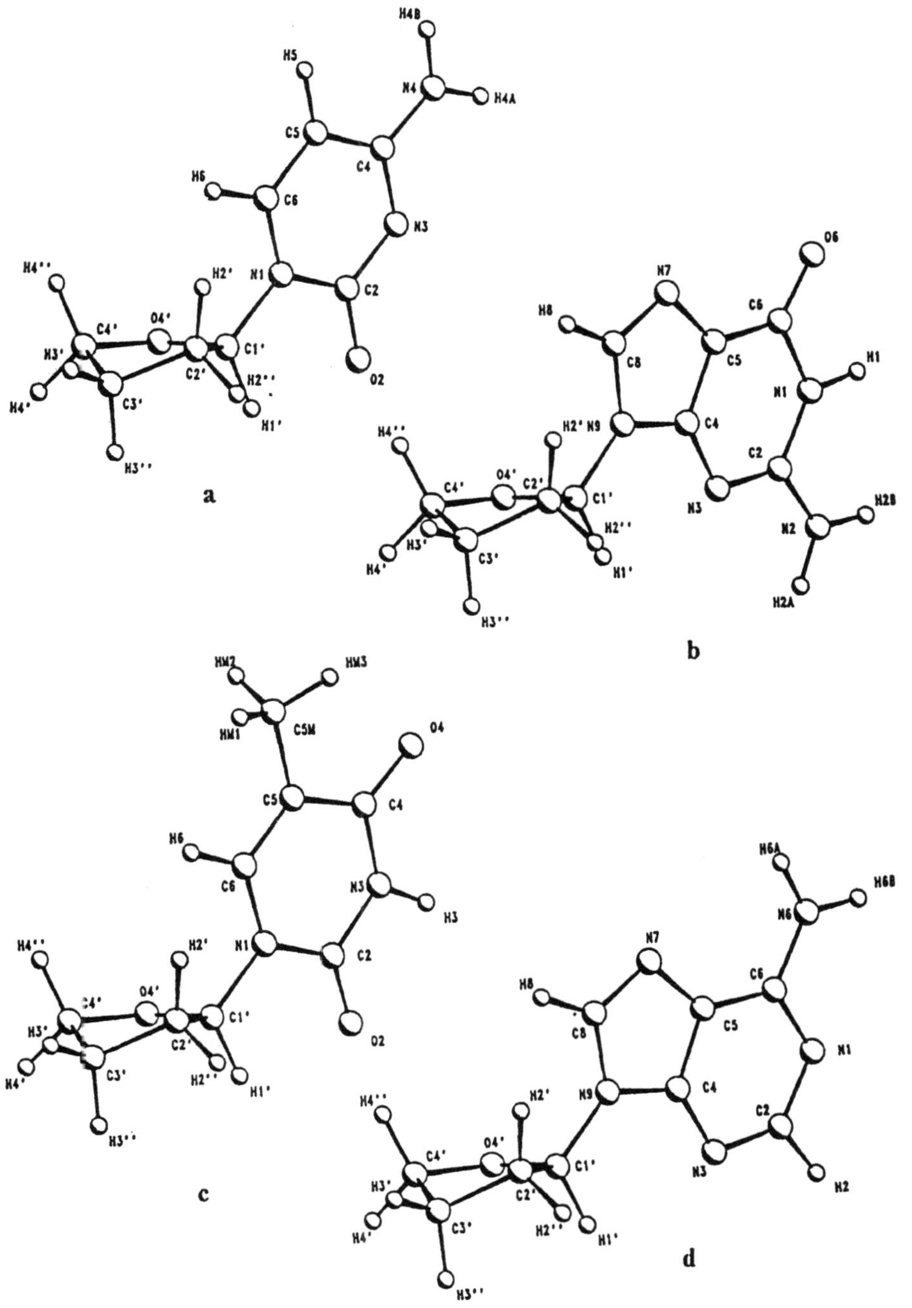

Figure 1. Structures of the base-sugar model molecules: a) Cytosine-sugar; b) Guanine-sugar; c) Thymine-sugar; d) Adenine-sugar.

(Fig. 1c), and adenine-sugar (Fig. 1d). Their geometries were taken from the X-ray data of B-DNA;[9] the orientation about the glucosidic bond is *anti* and the sugar puckering mode is $C_{2'}$-*endo*. For the guanine-sugar model molecule, we adopted an additional conformation: The geometry was taken from Z-DNA;[10] the orientation about the glucosidic bond is *syn* and the sugar puckering mode is $C_{3'}$-*endo*. For the polydeoxyribonucleotide chain portion, we used two conformations: one was taken from B-DNA, and the other was taken from the d(CpGpC) portion of Z-DNA.

The protein backbone is simply a repetition of the basic amino acid unit, which is as follows;

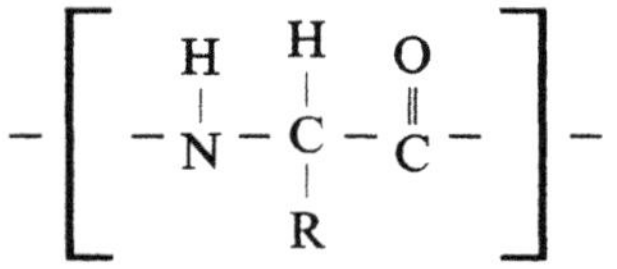

where, R describes each side chain of the 20 naturally occurring amino acids. Each amino acid residue is influenced by the neighboring residues. In order to take into account the effect of the neighbors, we add to each model molecule the two ends HCO- and -NH_2, as follows;

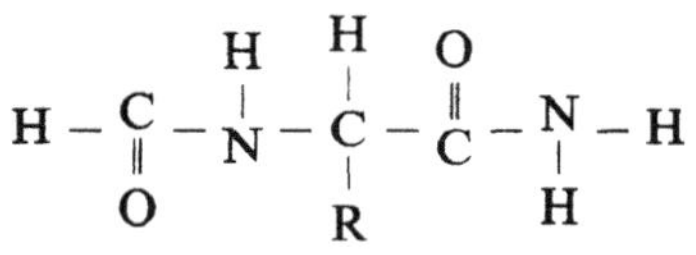

Among the 20 different amino acid residues, 3 of them have cationic forms (lysine, arginine and histidine), and 2 of them have anionic forms (glutamic acid and aspartic acid); these charged amino acids are used as additional model molecules. As model molecule for the residue with the disulfide bond, we used cystine with two sets of the HCO- and -NH_2 ends. For proline, additional model molecules are adopted:

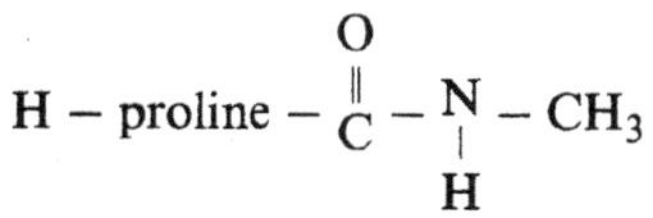

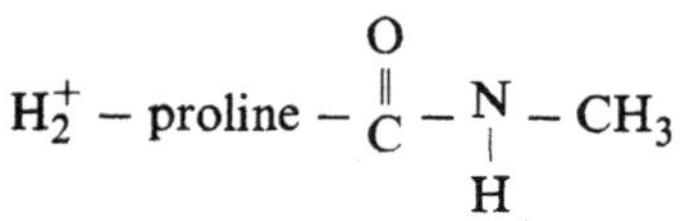

In total, we have used 28 different model molecules in order to represent all the repeating units of proteins. The geometries of the model molecules were taken from the experimental X-ray data,[8] and the torsion angles ϕ and ψ have been chosen in such a way as to have different conformations, for example, α-helix or β-sheets; the torsion angle ω has been always imposed to be 180°.

Ab Initio Calculations

The interaction energy between two model molecules A and B is defined as

$$\Delta E(AB) = E(AB) - E(A) - E(B)$$

where the terms on the right hand side are the total energies of the supermolecule, AB, and individual molecules, A and B, respectively. Since we are concerned with an order of 10^4 computations on large systems, the calculations have been done with a standard minimal basis set (7,3/2,1). The counterpoise method has been applied in order to reduce the basis set superposition errors. The calculations have been done with the IBM-3090/600 computer, using the KGNMOL program[11] mainly in its parallel version, and with heavy reliance on the ADD option (see MOTECC-90, Chapter 7).

For each combination of two model molecules, many conformations were generated with the following operational rules:
1) One of the two molecules is kept fixed at the origin of a cartesian reference system.
2) A rotation, defined by three Eulerian angles, is performed on the second molecule. The three Eulerian angles are systematically varied in order to generate a number of different relative orientations of the two molecules.
3) The center of mass of the second molecule is placed along 26 different directions radially from the origin, using a spherical polar coordinate system. For each direction, the interaction energies are calculated for conformations whose distance varies by the increment of 1 a.u..

The SCF interaction energies with the same fixed molecule were calculated using the ADD option of KGNMOL. The total number of combinations among model molecules is 123; for each of these combinations, the SCF interaction energies of about 180 conformations, were calculated. After the above procedure, additional calculations were included as described below in Fitting Procedure. *In total we have computed 31768 ab initio interaction energies between the model molecules,* likely the most extensive computation in quantum chemistry.

Classification of Atoms

In the determination of *ab initio* pair potential, atoms with the same atomic number (Z value), but different environments are differentiated by being assigned to different classes. Once the classification is done for all the atoms in the system, there is no more need to identify further the molecules or fragments which contain these atoms.

The atoms in DNA and proteins are differentiated according to the following classification criteria:
1) Atomic number (Z value) which provides the first basic distinction among different atoms : for example, H, C, N, O, P, S, *etc.* (Z=1,6,7,8,15,16, *etc*, respectively).
2) Hybridization of the atom : sp^3, sp^2, *etc.*
3) Environment of the atom in the model molecule : Two atoms with the same atomic number but bound to two different sets of atoms are classified into different categories. In the case of the hydrogen atom, the environment of the second neighbor atoms is also taken into consideration, thus we can distinguish between different types of hydrogen bonds.

In Table I, the classes are summarized, with the respective average values of electron density, obtained with Mulliken population analysis,[12] using the minimal basis set (7,3/2,1). We recall that the electron density is one of the indices which reflect the electronic structure of the molecule. After the classification described above, we had the confirmation that those atoms which correspond to the same class have close values of electron density. The carbon atoms have been classified into 32 categories, the nitrogen atoms into 13, the phosphorus atom into 1, the oxygen atoms into 8, the sulfur atoms into 3, and the hydrogen atoms into 19. Therefore, the total number of the classes is 76. Previous classification of the atoms in proteins and nucleic acids interacting with a single molecule of water have been given,[2,3] our present classification overlaps extensively with the previous one.

Table I. Classification of atoms. First column: Class name. Second column: Environment of the atom (marked with a *) belonging to the given class. Third column: Typical value of the electron density.

Class	Type	ρ	Class	Type	ρ
Carbon atom					
CS01	H H – C* – S H	6.77	CHP2	H H H – C* – N+ – H C H	6.30
CMET	H H – C* – C H	6.61	CHH3	H H – C* – C O	6.21
CMEN	H H – C* – N H	6.51	COPM[2]	H H – C* – O – P	6.23
CS02	H H – C* – C S	6.55	CTER	C H – C* – C C	6.20
CHH1	H H – C* – C C	6.42	CALP	C / H – C* – N C \	6.11
CHP1[1]	H O H – C* – C (−1) C O	6.45	CANP	C H H – C* – N+ – H C C	6.13
CHH2	H H – C* – N C	6.30	CHYD	C H – C* – O C	6.01
CHHP	H H H – C* – N+ – H C C	6.33	CGLC	N H – C* – O C	5.86

(Continued)

Table I. *(cont.)* Classification of atoms. First column: Class name. Second column: Environment of the atom (marked with a *) belonging to the given class. Third column: Typical value of the electron density.

Class	Type	ρ	Class	Type	ρ
CFRM	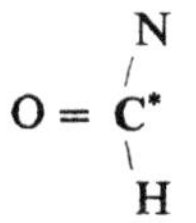	5.75	CCD4	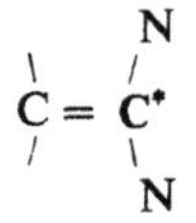	5.74
CKE1	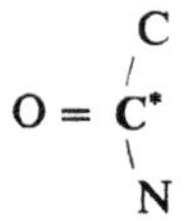	5.55	CCH1	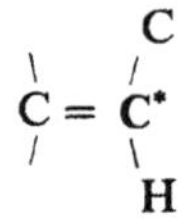	6.2
CKE2	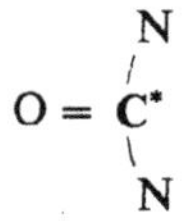	5.46	CCH2	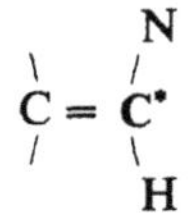	6.0
CCBX	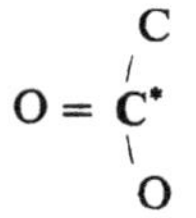	5.46	CNNH	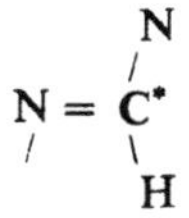	5.95
CCBM[1]	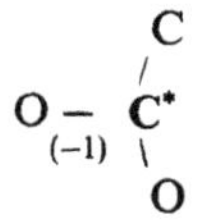	5.56	CNNP[3]	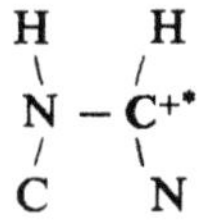	5.83
CCD1	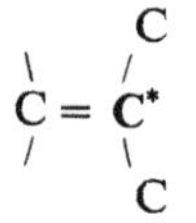	6.04	CAM1	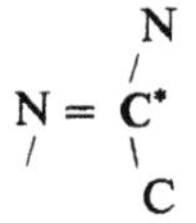	5.69
CCD2	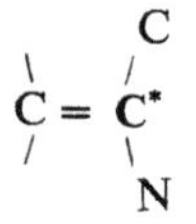	6.0	CAM2	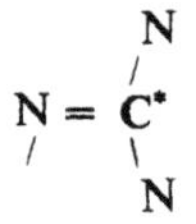	5.52
CCD3	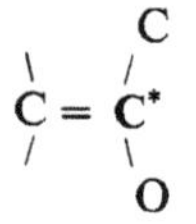	5.81	CAMP[4]	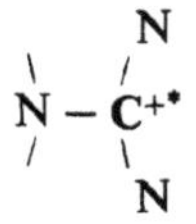	5.44

(Continued)

Table I. *(cont.)* Classification of atoms. First column: Class name. Second column: Environment of the atom (marked with a *) belonging to the given class. Third column: Typical value of the electron density.

Class	Type	ρ	Class	Type	ρ
Nitrogen atom					
NAM1[5]	H \ N* − C / H	7.68	NCC6[7]	C \ N* = C	7.41
NAM2[5]	C / H − N* \ C	7.53	NCC5[8]	C \ N* = C	7.30
NAM3[5]	C / C − N* \ C	7.36	NAP1	H | H − N+* − C | H	7.78
NAM4[6]	H \ N* − C / H	7.63	NAP2	H | H − N+* − C | C	7.58
NAM5[6]	C / H − N* \ C	7.46	NAP3[3]	H H \ / N* − C+ / \ C N	7.48
NIMI	H \ N* = C	7.58	NAP4[4]	H N \ / N* − C+ / \ H N	7.67
			NAP5[4]	H N \ / N* − C+ / \ C N	7.50
Phosphorus atom					
POM1	−O | O − P* − O | (−1) O	13.09			

(Continued)

Table I. *(cont.)* Classification of atoms. First column: Class name. Second column: Environment of the atom (marked with a *) belonging to the given class. Third column: Typical value of the electron density.

Class	Type	ρ	Class	Type	ρ
Oxygen atom					
OKET	/ O* = C \	8.44	OETR	C / O* \ C	8.42
OKEF	| O* = C – H	8.42	OHYD	C / O* \ H	8.52
OPM1	–O | O – P – O* | | O*(–1)	8.90	OPC1	P / O* \ C	8.67
OCBM[1]	O* | / – C – C (–1) | \ O*	8.62	OPH1	P / O* \ H	8.78
Sulfur atom					
SC01	C / S* \ H	16.00	SSCC	C \ S* – S* \ C	15.89
			SC02	C / S* \ C	15.79
Hydrogen atom					
HALI[9]	| H* – C – C |	0.8	HNP1	H* | H* – N+ – C | H*	0.53

(Continued)

Table I. *(cont.)* Classification of atoms. First column: Class name. Second column: Environment of the atom (marked with a *) belonging to the given class. Third column: Typical value of the electron density.

<table>
<tr><th>Class</th><th>Type</th><th>ρ</th><th>Class</th><th>Type</th><th>ρ</th></tr>
<tr><td>HCOP[10]</td><td>H*
|
H* – C –
|
O – P</td><td>0.8</td><td>HNP2</td><td>H*
|
H* – N+ – C
|
C</td><td>0.54</td></tr>
<tr><td>HGLC[11]</td><td>N
|
H* – C – O
|
C</td><td>0.74</td><td>HAP3[3]</td><td>H* H
\ /
N – C+
/ \
C N</td><td>0.52</td></tr>
<tr><td>HFRM</td><td>|
H* – C = O</td><td>0.81</td><td>HAP4[4]</td><td>H* N
\ /
N – C+
/ \
H* N</td><td>0.59</td></tr>
<tr><td>HCAR[12]</td><td>|
H* – C =</td><td>0.78</td><td>HAP5[4]</td><td>H* N
\ /
N* – C+
/ \
C N</td><td>0.60</td></tr>
<tr><td>HCNP[3]</td><td>H H*
\ /
N – C+
/ \
C N</td><td>0.66</td><td>HS01</td><td>H*
\
S – C</td><td>0.9</td></tr>
<tr><td>HAM1[13]</td><td>H*
\
N – C
/
H*</td><td>0.64</td><td>HIMI</td><td>H*
\
N = C</td><td>0.75</td></tr>
<tr><td>HAM2[13]</td><td>C
/
H* – N
\
C</td><td>0.62</td><td>HHYD</td><td>H*
\
O – C</td><td>0.60</td></tr>
<tr><td>HAM4[14]</td><td>H
\
N* – C
/
H</td><td>0.73</td><td>HOP1</td><td>H*
\
O – P</td><td>0.73</td></tr>
</table>

(Continued)

Table I. *(cont.)* Classification of atoms. First column: Class name. Second column: Environment of the atom (marked with a *) belonging to the given class. Third column: Typical value of the electron density.

Class	Type	ρ	Class	Type	ρ
HAM5[14)]	H* – N(–C)(–C)	0.69			

1) Atoms of anionic carboxyl group.
2) $C_{5'}$ atom in DNA.
3) Atoms in protonated histidine 5-membered ring, for example.
4) Atoms in protonated arginine residue, for example.
5) Nitrogen atom in plane conformation.
6) Nitrogen atom in pyramidal conformation.
7) Nitrogen atom of this type in 6-membered ring.
8) Nitrogen atom of this type in 5-membered ring.
9) Aliphatic hydrogen atom.
10) Hydrogen atom attached to $C_{5'}$ atom near phosphate group.
11) Hydrogen atom attached to the glucosidic bond.
12) Hydrogen atom in aromatic ring.
13) Hydrogen atom attached to the nitrogen atom in plane conformation.
14) Hydrogen atom attached to the nitrogen atom in pyramidal conformation.

Analytical Form of the Pair Potentials

A simple, easy to compute analytical form in the pair potential is a necessary requirement for Monte Carlo or molecular dynamics simulations. At the same time, the form of the potential must be optimal for a good fit of the SCF energies. After trials with different forms, including those standards presented in literature (12,6,1 type, for example), we adopted an analytical pair potential of the form:

$$\mathbf{V}_{MN} = \sum_i \sum_j \left(\frac{\mathbf{B}_{ij}^{ab}}{r_{ij}^{10}} + \frac{\mathbf{S}_{ij}^{ab}}{r_{ij}^{4}} + \frac{\mathbf{T}_{ij}^{ab}}{r_{ij}^{2}} + \frac{\mathbf{Q}_{ij}^{ab}}{r_{ij}} \right) \tag{3}$$

where

$$\mathbf{B}_{ij}^{ab} = \mathbf{b}_{a,i} \times \mathbf{b}_{b,j}$$

$$\mathbf{S}_{ij}^{ab} = -\,(\mathbf{q}'_i \times \mathbf{s}_{a,i})\;(\mathbf{q}'_j \times \mathbf{s}_{b,j})$$

$$\mathbf{T}_{ij}^{ab} = (\mathbf{q}'_i \times \mathbf{t}_{a,i})\;(\mathbf{q}'_j \times \mathbf{t}_{b,j})$$

$$\mathbf{Q}_{ij}^{ab} = 332.0 \times \mathbf{q}'_{a,i} \times \mathbf{q}'_{b,j}$$

$$\mathbf{q}'_{a,i} = \mathrm{q}_i + \boldsymbol{\delta}_{a,i}$$

Here, M and N are fragments (model molecules) of DNA and proteins. V_{MN} is the total interaction energy (kcal/mole) between molecules M and N, i is the i-th atom in molecule M belonging to class a, j is the j-th atom in molecule N belonging to class b, and r_{ij} is the distance in Å between the two atoms i and j, and $\mathbf{b}_{a,i}$, $\mathbf{s}_{a,i}$, $\mathbf{t}_{a,i}$, and $\boldsymbol{\delta}_{a,i}$ are the fitting parameters of the class a. Finally, q_i represents the charges of atoms i , obtained with Mulliken population analysis.[12]

It is noteworthy that the index of the parameters corresponds to only one class: the form of the parameters is $\mathbf{b}_{a,i}$ not $\mathbf{b}_{ab,ij}$, for example. This most important constraint brings about a very large decrease of the number of the parameters needed to reproduce the SCF interaction energies between pairs of model molecules related to DNA and protein. In the above equations, the parameters $\mathbf{b}_{a,i}$, $\mathbf{s}_{a,i}$ and $\mathbf{t}_{a,i}$ are constrained to be positive whereas the parameter $\boldsymbol{\delta}_{a,i}$ can be positive or negative. In Eq. (3), $\mathbf{B}_{ij}^{ab}$ has a positive sign, whereas the signs of S_{ij}^{ab} and T_{ij}^{ab} are related to the sign of the charges of the atoms i and j.

The fourth term in the right side of Eq. (3) corresponds to the electrostatic energy, which is one of the major contributions to the total interaction energy. The parameter $\boldsymbol{\delta}$ is determined to ensure that the effective charge, $\mathbf{q}'_{a,i}$, represents reasonably well the electrostatic energy in Eq. (3) (see Fitting Procedure). The effective charges, the original Mulliken charges and the classes for the atoms of the model molecules of Fig. 1 are shown in Table II. Analogous tables for all 33 model molecules can be found in Ref. 13.

Fitting Procedure

The SCF interaction energies corresponding to the various conformations between the model molecules, each of which describes a fragment either of DNA or of a protein, are the "reference data" for a non-linear least-square fitting computer program.

As previously mentioned, the electrostatic energy represents the largest contribution to the interaction energy and it is the dominant term for conformations with large distance between the two molecules. To determine the parameters $\boldsymbol{\delta}$, we have chosen those *ab initio* interaction energies corresponding to conformations in which the shortest intermolecular distance, r_{ij}, between the nearest atoms i and j for the two molecules is larger than 7.0 a.u.. Keeping fixed the values of the $\boldsymbol{\delta}$ parameter, we then determined the **b**, **s** and **t** parameters, using the interaction energies of all the conformations.

This procedure is repeated many times. Once a set of parameters is available, the following steps are performed:
1) Iso-energy contour maps are computed using the analytical potentials.

Table II. Classes and charges in the model molecules. First column: Name of atom as shown in Fig.1. Second column: Charge of atom, obtained from Mulliken population analysis, with the (7,3/2,1) basis set. Third column: Name of class as shown in Table I. Fourth column: Effective charge of atom (see text).

a) Cytosine-sugar

Atom	q	Class	q′
H4″	0.2038	HALI	0.0738
C4′	-0.2216	CHH3	-0.0282
H4′	0.2207	HALI	0.0907
O4′	-0.4102	OETR	-0.2411
C3′	-0.4251	CHH1	-0.1538
H3′	0.2221	HALI	0.0921
H3″	0.2189	HALI	0.0889
C2′	-0.4233	CHH1	-0.1520
H2′	0.1998	HALI	0.0698
H2″	0.2300	HALI	0.1000
C1′	0.1332	CGLC	0.1178
H1′	0.2680	HGLC	0.0896
N1	-0.3760	NAM3	-0.1260
C2	0.5394	CKE2	0.2204
O2	-0.4456	OKET	-0.3814
N3	-0.4114	NCC6	-0.2247
C4	0.3096	CAM1	0.0200
N4	-0.6696	NAM1	-0.1958
H4A	0.3685	HAM1	0.1912
H4B	0.3435	HAM1	0.1662
C5	-0.3154	CCH1	-0.0965
H5	0.2070	HCAR	0.0044
C6	-0.0030	CCH2	0.2442
H6	0.2368	HCAR	0.0342

b) Guanine-sugar

Atom	q	Class	q′
H4″	0.2107	HALI	0.0807
C4′	-0.2184	CHH3	-0.0250
H4′	0.2172	HALI	0.0872
O4′	-0.4095	OETR	-0.2404
C3′	-0.4353	CHH1	-0.1640
H3′	0.2267	HALI	0.0967
H3″	0.2200	HALI	0.0900
C2′	-0.4262	CHH1	-0.1549
H2′	0.2219	HALI	0.0919
H2″	0.2281	HALI	0.0981
C1′	0.1386	CGLC	0.1232
H1′	0.2516	HGLC	0.0732
N9	-0.3492	NAM3	-0.0992
C8	0.0547	CNNH	0.0422
H8	0.2156	HCAR	0.0130
C4	0.2630	CCD4	0.2529
N3	-0.4222	NCC6	-0.2355
C2	0.4751	CAM2	0.0964
N2	-0.6811	NAM1	-0.2073
H2A	0.3731	HAM1	0.1958
H2B	0.3476	HAM1	0.1703
N1	-0.5416	NAM2	-0.3310
H1	0.3763	HAM2	0.3418
C6	0.4265	CKE1	0.2423
O6	-0.4217	OKET	-0.3575
C5	-0.0396	CCD2	-0.1456
N7	-0.3021	NCC5	-0.1556

c) Thymine-sugar

Atom	q	Class	q′
H4″	0.2062	HALI	0.0762
C4′	-0.2208	CHH3	-0.0274
H4′	0.2207	HALI	0.0907
O4′	-0.4123	OETR	-0.2432
C3′	-0.4259	CHH1	-0.1546
H3′	0.2242	HALI	0.0942
H3″	0.2184	HALI	0.0884
C2′	-0.4236	CHH1	-0.1523
H2′	0.2077	HALI	0.0777
H2″	0.2282	HALI	0.0982
C1′	0.1352	CGLC	0.1198
H1′	0.2584	HGLC	0.0800
N1	-0.3862	NAM3	-0.1362
C6	-0.0317	CCH2	0.2155
H6	0.2390	HCAR	0.0364
C5	-0.0962	CCD1	-0.1644
C5M	-0.6387	CMET	-0.2149
HM1	0.2074	HALI	0.0774
HM2	0.2266	HALI	0.0966
HM3	0.2443	HALI	0.1143
C4	0.4312	CKE1	0.2470
O4	-0.4021	OKET	-0.3379
N3	-0.5567	NAM2	-0.3461
H3	0.3907	HAM2	0.3562
C2	0.6021	CKE2	0.2831
O2	-0.4462	OKET	-0.3820

d) Adenine-sugar

Atom	q	Class	q′
H4″	0.2094	HALI	0.0794
C4′	-0.2211	CHH3	-0.0277
H4′	0.2211	HALI	0.0911
O4′	-0.4090	OETR	-0.2399
C3′	-0.4250	CHH1	-0.1537
H3′	0.2251	HALI	0.0951
H3″	0.2174	HALI	0.0874
C2′	-0.4268	CHH1	-0.1555
H2′	0.2151	HALI	0.0851
H2″	0.2304	HALI	0.1004
C1′	0.1376	CGLC	0.1222
H1′	0.2583	HGLC	0.0799
N9	-0.3520	NAM3	-0.1020
C8	0.0791	CNNH	0.0666
H8	0.2259	HCAR	0.0233
N7	-0.3363	NCC5	-0.1898
C5	-0.0081	CCD2	-0.1141
C6	0.3005	CAM1	0.0109
N6	-0.6697	NAM1	-0.1959
H6A	0.3610	HAM1	0.1837
H6B	0.3581	HAM1	0.1808
C4	0.2569	CCD4	0.2468
N3	-0.3683	NCC6	-0.1816
C2	0.0756	CNNH	0.0631
H2	0.2278	HCAR	0.0252
N1	-0.3828	NCC6	-0.1961

2)*Ab initio* calculations are done for some of the new local minima given by the iso-energy contour maps.
3) The new points are included in the fitting procedure.
The above procedure was repeated until we obtained an overall good agreement between the *ab initio* and fitted energies.

For the final set of parameters, the standard deviation is 1.5 kcal/mole, for interaction energies covering the range from -250 kcal/mole to +20 kcal/mole. It is worth noting that, since we have computed 31768 interaction energies and we have a total of 304 (= 76 × 4) independent parameters to be determined, we used more than 100 reference data per parameter.

Discussion

As mentioned before, many potential energy functions are available for biological systems[6-8,14] even commercially. In parameterizing those potential energy functions, experimental data from crystallography and spectroscopy were used with various combinations of *ab initio* molecular orbital calculations. These potentials are used widely in research of molecular biology, and recently, an assessment on the quality of those energy functions has been published.[15,16] In general, these potentials are not sensitive enough to reproduce the very specific interaction between atoms in different hybridizations and environments, bearing the parameterization for the "non-bonded" interactions too limited. In order to represent the complexity of the interactions between biological molecules, an energy function with higher specificity is required, and this constitutes one of the main motivation for this study.

Our main objective is to present a new analytical pair potential yielding a superior description of interactions between DNA and proteins. The simplicity of the analytical form is necessary (for computational efficiency in Monte Carlo or MD simulations), but at the same time, the analytical form should represent the interaction energy with reasonable resolution. We should not emphasize the individual terms in Eq. (3) since without a very direct and unique physical meaning. In other words, the variables in Eq. (3) do not have much theoretical ground, exception made for the charges obtained by the Mulliken population analysis in the *ab initio* molecular orbital method.

As previously noticed, the electrostatic energy is one of the major contributions to the interaction energy. As it is known, the Mulliken population analysis is the simplest technique to describe the electron distribution in a molecule. The Mulliken charges, however, are known to be basis-set dependent, and are not as reliable as we might wish. To improve on the use of the Mulliken charges, attempts have been presented to define net atomic charges designed to reproduce the electrostatic potential of a molecule.[17,18] We selected a different approach. As a convention, we shall call the fourth term in Eq. (3) "the electrostatic energy." Strictly speaking, this assumes that a molecule can be described by a distribution of point charges. Obviously, this assumption is not true, and other terms in Eq. (3) are introduced to correct the assumption. However, when the distance between two molecules is very large, the interaction energy is dominated by the electrostatic energy. The parameter values in the fourth term in Eq. (3) were determined by considering in the fitting procedure only the configurations with far-away molecules. Then, by keeping these values constant, the other parameters have been determined as described in the previous section. The final fitting is excellent for the conformations in which the distance r_{ij} between two molecules is larger than 7.0 a.u.. This indicates that the fourth term in Eq. (3) can be identified with the true electrostatic energy.

It is interesting to compare our "effective charges" with atomic partial charges derived from a high-resolution X-ray data set.[19] There are a few differences between the theoretical charges (obtained with several molecular orbital methods) and those derived from experiments. Small differences are not too important; however, notable is the difference for the phosphorus atom in the phosphate group of the nucleic acid. Almost all theoretical methods show high positive charges on the phosphorus atom, while the experimental charge is only slightly positive.[19] The value of our "effective charge" for the phosphorus atom is quite similar to the one experimentally derived.

Before concluding, we want to briefly mention that molecular dynamics simulations are presently in progress using the *ab initio* pair potentials described here for the non-bonded interaction, along with the potentials for bending and torsional barrier.[20] These simulations are performed to obtain a preliminary proof of the over-all reliability of our potentials and include formation of α-helix and β-sheet structures and also interactions of simple molecules with an oligonucleotide in water solution.

3. The Monte Carlo Method

The Monte Carlo (MC) techniques are generally used to evaluate numerically, by means of random sampling (hence the term "Monte Carlo"), multidimensional integrals which are too difficult to solve with analytical methods. These methods can be applied also to study properties of condensed matters, where we are often interested in thermodynamical averages obtained over probability distributions.

For a N particle system, the thermodynamic average of a function $F(\mathbf{q}^N)$, which depends only on the configurational variables $\mathbf{q}^N = (\mathbf{q}_1, \mathbf{q}_2, ..., \mathbf{q}_N)$, is given by

$$< F > = \int F(\mathbf{q}^N) p(\mathbf{q}^N) d\mathbf{q} \tag{4}$$

where p is the probability density function. It is generally impossible to evaluate this integral directly for a many particle system because there are too many degrees of freedom, even if we know the probability function. Therefore, some sort of importance sampling is required. The algorithm first proposed by Metropolis and co-workers[21] allows a series of random walks to be generated in configuration space according to the probability density distribution function p. The book edited by Berne is an excellent review of the methodology and applications.[22]

Assuming N molecules in a given configuration and in a given volume at temperature T, the Metropolis algorithm consists essentially in iterating the following steps: 1) a molecule is chosen randomly; 2) a trial move is attempted, *i.e.*, the molecule is displaced and rotated; 3) the change in energy ΔE of the system is evaluated; 4) if the energy is decreased the trial move is accepted with unit probability, and if the energy is increased the move is accepted with probability $\exp(-\Delta E/k_BT)$ and rejected with probability $1-\exp(-\Delta E/k_BT)$. Experience in Monte Carlo simulation shows that the maximum allowed particle displacement and rotation should be adjusted to give an acceptance probability of about 50%.

The full process is repeated many times (order of millions) to form a sequence of configurational states over which properties of the system can be averaged. The Metropolis algorithm can be used to sample any density function regardless of complexity and in any number of dimensions. The main shortcoming, relative to molecular dynamics (see next section), is that it cannot deal with time dependent properties. A rigorous description of the statistics of Monte Carlo Metropolis algorithms has been provided by Kalos and Whitlock.[23]

The equilibrium properties of a thermodynamic ensemble can be obtained from Monte Carlo simulation by

$$G=\frac{\int g(X)f(X)dX}{\int f(X)dX}=\sum_{n=1}^{N}\frac{g(X_n)}{N} \tag{5}$$

where G could, for example, be the energy or radial distribution function, and N is the number of Monte Carlo steps. In practice, for each quantity of interest, a different number of steps in the random walk may have to be generated because convergence is reached at varying rates. The successive X's in the random walk are not independent and in most circumstances there is positive correlation, giving an artificially high variance. The number of steps to be taken is found by investigation, for example by averaging some property of the system every, say, m steps, and observing the behavior of G with increasing walk length. A decision is then made about where in the random walk G has converged.[23] Often, the convergence is monitored using mean values of some properties computed over independent sections or blocks of the Markov chain. The sections have to be long enough such that possible correlations between them can be neglected. Once the simulation is completed, the average property values should be ideally constant and their statistical error is computed in terms of the standard deviation:

$$\sigma^2 = \frac{1}{k(k-1)} \sum_{j=1}^{k} [<G(X_N)^2>_j - <G(X_N)>_j^2] \tag{6}$$

where k is the number of partial averages.

A more rigorous way to check the system convergence is to compare results gathered from a set of simulations, carried out by adopting completely different starting configurations. Because of the ergodic hypothesis, the different simulations have to converge to the same average values.

One more point has to be mentioned: the choice of the ensemble. The canonical ensemble, with fixed N,V,T variables, is the most commonly used; however, simulations have been carried out also for the grand canonical[24] and the isothermal isobaric[25] ensembles. In the grand canonical ensemble the variables μ,V,T (μ is the chemical potential) are kept fixed and N is a quantity sampled in the MC chain. In the isothermal isobaric ensemble, the quantities N,P,T are fixed and the variable V is sampled; a variation of V implies a variation in the configuration variables $\mathbf{q}^N$ which have to be scaled according to the new value of V.

Simulations in the canonical ensemble have been performed in our laboratory for liquid water,[26-28] solvation of the monovalent cations and anions,[29] small molecules,[30] and biological systems.[31]

4. The Molecular Dynamics Method

There are many texts available[32-36] on the fundamentals of classical molecular dynamics computer simulation. Its scope is rapidly being extended to genetics[14,36-39] and hydrodynamics.[40-43] This section is meant to provide an overview of these fundamentals.

The computer simulation starts on the assumption that the motion of particles can be treated with classical dynamics according to Newton's equations of motion. For an isolated system composed of N particles, the translational motions are given by

$$m_i \frac{d^2 \mathbf{r}_i}{dt^2} = \mathbf{F}_i(\mathbf{r}_1, \mathbf{r}_2, ..., \mathbf{r}_N) \tag{7}$$

where

$$\mathbf{F}_i = -\nabla_i V(\mathbf{r}_1, \mathbf{r}_2, ..., \mathbf{r}_N) \ , \tag{8}$$

m_i is the mass and $\mathbf{r}_i$ the position of the particle i. $\mathbf{F}$ is the net force acting on the particle arising from the potential V at the instant t. Assuming, for simplicity, a pairwise additive potential, and the absence of external forces, then

$$V(\mathbf{r}_1, \mathbf{r}_2, ..., \mathbf{r}_N) = \sum_{ij} V_{ij}(\mathbf{r}_{ij}) \; ; \qquad \mathbf{r}_{ij} = \mathbf{r}_i - \mathbf{r}_j \tag{9}$$

and the force acting on particle i is given by

$$\mathbf{F}_i = \sum_j \mathbf{F}_{ij} \tag{10}$$

where

$$\mathbf{F}_{ij} = -\mathbf{F}_{ji} = -\frac{\partial V_{ij}(\mathbf{r}_{ij})}{\partial \mathbf{r}_{ij}} \tag{11}$$

Sometimes the basic equation of motion is taken to be the Langevin equation.[44] This uses the theory of Brownian motion to reduce the complexity of the problem, at the cost of classical rigor and trajectory detail. In the simple rotational or translational Langevin equations the intermolecular potential energy is simplified into a combination of viscous retardation and stochastic force (or torque). For example, in the translational case the Newton equation is replaced by the Langevin equation

$$m\,\dot{\mathbf{v}}_i = \mathbf{F}_i = -\,m\,\beta\,\mathbf{v}_i + \mathbf{W}_i \tag{12}$$

where β is the friction coefficient, and $\mathbf{W}_i$ the random force. The dynamical problem comes down to approximating the potential energy surface with viscous retardation and random fluctuation. We refer to Chapter 19 for a detailed discussion on Brownian motion and application examples.

Molecular dynamic simulations usually consist of 3 steps. The first one is the initialization. In this step, each moving particle is put into a regular lattice point with fixed orientation and given a random velocity consistent with the input temperature. The velocities are immediately renormalized so that the simulation system will have a zero net linear momentum. The simulation is then carried out with frequent renormalization of the velocities to the given input temperature. The purpose of this stage is to relax the system by removing the excessive energy through renormalization to equilibrium.

The second stage of the simulation is the continuation of the first but without renormalization of the velocities to ensure that a true equilibrium state is reached. Note that unlike the Monte Carlo method, the input temperature in the molecular dynamics simulation is not a control parameter: in the micro-canonical ensemble the total energy, the volume, and the number of particles are constant quantities. Therefore the final simulation temperature will in general not turn out to be the input temperature. But because of the renormalization done during the first stage, the final temperature will be close to it. During this stage of the simulation the particle's history is collected for later analysis. Only positions and velocities are usually stored, since most of the important and interesting properties can be obtained from them.

The last phase of the simulation is the analysis of the stored history to produce static as well as dynamic properties of the system.

4.1. Numerical Integration for the Equations of Motion

The essence of the method of computer simulation is to take advantage of the speed and accuracy of the modern digital computer to integrate numerically the equations of motion for hundreds or even thousands of molecules. There are many different ways of achieving this objective, with different efficiency and conveniency. The problem consists essentially in integrating ordinary second order differential equations by a finite difference approach. From the particle positions and velocities at time t, one attempts to obtain, with reasonable accuracy, the same quantities at a later time $t + \Delta t$. Because of the finite step size, Δt, the solution will be only approximate; therefore the particle trajectories will not be precise and will eventually diverge after some time. It is evident that the choice of Δt is important and depends on the solution method adopted and on the simulated system. In all the proposed algorithms, one compromises between the desired accuracy of the trajectories and the size of the time step.

The most commonly used methods are briefly described below.

In the *Gear predictor-corrector method* [45] the numerical integration of Newton differential equations is achieved in two successive stages. First, the molecular center of mass position, $\mathbf{r}$, and its derivatives are "predicted" for the time $t + \Delta t$ from the position and derivatives at the previous time step t, according to a Taylor expansion truncated at the order m

$$\left(\frac{d^k\mathbf{r}}{dt^k}\right)_{predicted} = \sum_{J=k}^{m} \frac{J!\Delta t^J}{k!(J-k)!} \frac{d^J\mathbf{r}(t)}{dt^J} \tag{13}$$

In the "correction" step, the forces computed at the predicted positions are used to correct the derivatives of all orders according to

$$\left(\frac{d^k\mathbf{r}}{dt^k}\right)_{corrected} = \left(\frac{d^k\mathbf{r}}{dt^k}\right)_{predicted} + C_k^m\left[\frac{\Delta t^2}{2}\frac{\mathbf{F}(\mathbf{r}^{pred})}{m} - \left(\frac{d^2\mathbf{r}}{dt^2}\right)_{predicted}\right] \tag{14}$$

Values of the order-dependent correction constants C_k^m can be found in the specialist literature.[46]

The *Verlet algorithm*,[47] introduced in the late sixties, is one of the most stable and popular numerical algorithms for the integration of the center of mass motion of a diffusing molecule. The method uses the positions and accelerations at the time t and the positions at the time $t - \Delta t$ to predict the positions at the time $t + \Delta t$. The integration scheme, based on a Taylor expansion of the 3-*rd* order, is described by the following equation:

$$\mathbf{r}_i(t + \Delta t) = 2\mathbf{r}_i(t) - \mathbf{r}_i(t - \Delta t) + \Delta t^2 \ddot{\mathbf{r}}_i(t) \tag{15}$$

From this equation it is apparent that the velocities are not needed to compute the trajectories; however, when needed, for example for computing the kinetic energy of the system, they can be obtained according to the basic rule of differentiation

$$\dot{\mathbf{r}}(t) = \frac{\mathbf{r}(t + \Delta t) - \mathbf{r}(t - \Delta t)}{2\Delta t} \tag{16}$$

It should be noted that, whereas the positions obtained in Eq. (15) are affected by an error of the order Δt^4, the velocities in Eq. (16) are subject to an error of the order Δt^2.

An alternative method for deriving more accurate velocities is essentially a modification of the Verlet algorithm, called *leapfrog*.[48] This scheme involves velocities at half time step, that is

$$\dot{\mathbf{r}}_i(t + \frac{1}{2}\Delta t) = \dot{\mathbf{r}}_i(t - \frac{1}{2}\Delta t) + \ddot{\mathbf{r}}_i(t)\Delta t \tag{17}$$

$$\mathbf{r}_i(t + \Delta t) = \mathbf{r}_i(t) + \dot{\mathbf{r}}_i(t + \frac{1}{2}\Delta t)\Delta t \tag{18}$$

The accelerations at time t are calculated from the coordinates at the same time t. The velocities of the half forward step in Eq. (17) are then computed and substituted into Eq. (18) to obtain the new positions, at the time $t + \Delta t$. If the velocities $\dot{\mathbf{r}}_i$ are needed at the time t, they can be computed from the mean of the velocities at $t - (1/2)\Delta t$ and $t + (1/2)\Delta t$.

The Gear predictor-corrector and Verlet algorithms have been compared in Ref. 49, where the root mean square fluctuations and drift of the total energy of a model system have been monitored for different step sizes. While at short time steps the Gear algorithm gives smaller total energy fluctuations than the Verlet algorithm, they have comparable accuracy for intermediate time steps. Finally at larger time steps the Verlet algorithm becomes more stable. The lower accuracy of the Verlet algorithm at short step sizes is due to the low order integration scheme. At larger time steps the higher order derivatives of the Gear algorithm induce destabilization effects, probably because of excess fluctuations in the force field. In the case of the Verlet algorithm this shortcoming is less severe because the memory of previous time steps is lost earlier, the method provides more flexibility in adjusting the particles' trajectories.

The algorithms previously described can be used to integrate the translational motion of diffusing molecules. A different scheme has to be applied for rotational motion. If the rigid body approximation is adopted, the rotational motion in the principal molecular frame is described by Euler's equations:

$$\begin{aligned} I_{xx}\dot{\omega}_x - (I_{yy} - I_{zz})\omega_y\omega_z &= T_x \\ I_{yy}\dot{\omega}_y - (I_{zz} - I_{xx})\omega_z\omega_x &= T_y \\ I_{zz}\dot{\omega}_z - (I_{xx} - I_{yy})\omega_x\omega_y &= T_z \end{aligned} \tag{19}$$

where I_{xx}, I_{yy}, I_{zz} represent the principal moments of inertia, $\boldsymbol{\omega}$ the angular velocity and $\mathbf{T}$ the torque acting on the molecule.

Equations free of singularities can be obtained using Hamilton's quaternions which describe the orientation of a molecule with respect to a laboratory frame. Similar techniques as used for the integration of translational motion can be applied. The Gear predictor-corrector routine has been implemented by Evans[50] and Sonnenschein[51] while a leapfrog-like algorithm has been proposed by Fincham.[52]

While the algorithms described above integrate only the free degrees of motion of a molecule, the *SHAKE* algorithm[53] follows a different line. The idea in this technique is to keep some geometrical constraints in a system of principally free moving atoms. This is especially advantageous for large molecules where some arbitrary selected degrees of freedom (for example bond lengths) are meant to be fixed while all other movements are allowed.

Assuming that all geometrical constraints can be represented in terms of fixed atomic distances, the method uses undetermined Lagrangian multipliers to represent forces which are required to keep these distances constant. The SHAKE algorithm then applies an iterative scheme to correct the particle's motions until the constraints are satisfied within a given tolerance.

Although this method might not be as rigorous as the direct integration of the free degrees of motion, it appears that the loss in accuracy is essentially not increased in comparison to the unconstrained integrator. Its strength is evident in large molecules where the free degrees of motion can impossibly be isolated and integrated alone.

4.2. Calculation of Properties

One of the basic premises of the classical mechanics is that once we know the phase space motion of a physical system, all of its macroscopic properties can be deduced. Thus we can, in principle, obtain all the properties from the phase space trajectory of a molecular dynamics simulation. In practice, however, due to the limited phase space actually traveled in a given simulation, some properties may not be easily obtainable to the accuracy one wants. In the following, we will review methods to calculate some of the properties which are commonly studied in the molecular dynamics simulations.

These properties are usually classified as either *static* or *dynamic*. Only the dynamic properties are related to time-dependent trajectories. An example of a static property is the heat capacity, which depends on the fluctuations of the kinetic energy from instant to instant. A well-known example of a dynamic property is the self-diffusion coefficient, which is determined by the correlation of motions at different times. Static properties can also be obtained from Monte Carlo simulations.

Another way to classify the properties is according to whether they are manifestations of *single molecular motions* or results of *collective phenomena*. A good example of the former is the self-diffusion coefficient, while a good example of the latter is the dielectric constant. For a given run of a MD simulation, single particle properties are invariably much easier to evaluate to a given accuracy, because they can be averaged over all the molecules in the system, in addition to averages over different time steps.

With the above understanding, let us now review the methods for the calculations of the properties from a given MD phase trajectory. We will first discuss the static properties, followed by dynamic properties. Note that the micro-canonical ensemble will be assumed in the following discussion.

Translational kinetic and potential energies of the system can be easily calculated, respectively, by:

$$K_{tr} = < \frac{1}{2}\sum_{i}^{N} m_i \dot{\mathbf{r}}_i^2 > \tag{20}$$

and

$$V = < \sum_{i}^{N} \phi_i > \tag{21}$$

where $< ... >$ means averaging over the time steps, N is the number of molecules in the system, ϕ_i the potential energy of a molecule i, and m_i the mass.

The *temperature* of the simulation system can be obtained from the equipartition principle. For example, the translational temperature can be calculated from

$$T_{tr} = \frac{K_{tr}}{(3/2)Nk_B} \tag{22}$$

Rotational kinetic energy and temperature are similarly defined. The total kinetic energy of the system becomes

$$U = V + K_{tr} + K_{rot} \tag{23}$$

and the temperature of the system is the mean of translational and rotational temperature. In the case where molecular vibrations are also included in the simulation, their contribution should also be taken into account.

The internal *pressure* of the system is calculated from the virial theorem:

$$P = \frac{1}{V}(Nk_B T + W) \tag{24}$$

where W is the virial of the system,

$$W = -\frac{1}{3} < \sum_i^N \mathbf{r}_i \frac{\partial}{\partial \mathbf{r}_i} \phi_i > = -\frac{1}{3} < \sum_i^N \sum_{j>i}^N \mathbf{r}_{ij} \frac{\partial}{\partial \mathbf{r}_{ij}} \phi_{ij} > \tag{25}$$

The constant volume *heat capacity* can be obtained from potential or kinetic energy fluctuations. For a system with n degrees of freedom per particle it can be calculated from:[54]

$$C_v = R\frac{n}{2}\left[1 - \frac{< K^2 > - < K >^2}{(n/2)Nk_B^2 T^2}\right]^{-1} \tag{26}$$

where R is the gas constant anf N the number of molecules in the system. One of the most important static properties that can be obtained from MD simulations is the pair *radial distribution function*, since it is widely available from experimental measurements. It is a function usually written as $g_{\alpha\beta}$, which gives the probability of finding a particle α at a distance r away from a particle β. The function is usually normalized to 1 at large r. It is the deviation of this function from uniformity that gives rise to the scattering of X-ray or neutrons. Therefore, the theoretical radial distribution functions can be compared to the X-ray and/or neutron scattering data to check the reliability of the potential model. The radial distribution function can be calculated by dividing the interval within $r = 0$ and the cutoff radius into segments of size δr and counting the number of atomic pairs (α, β) with distance falling in a given range. The results are to be averaged over different time steps. Let $n(r)$ be the number of pairs which are r_i to $r_i + \delta r$ apart, then

$$g_{\alpha\beta}(r) = \frac{V}{N} < \frac{n(r)}{4\pi r^2 \delta r} > \tag{27}$$

where V is the simulation volume.

Radial distribution functions are related to the elastic X-ray or neutron *(coherent) scattering intensities* by[26,55]

$$I(q) = f^2 + \sum_i \sum_j f_i f_j \int_0^\infty 4\pi r^2 \frac{N}{V} [g_{ij}(r) - 1] \frac{\sin(qr_{ij})}{qr_{ij}} dr \tag{28}$$

where f^2 is the coherent self part of the scattering intensity. The variable q is related to the scattering angle 2θ and the wavelength λ of the incident beam by $q = 4\pi \sin\theta/\lambda$. f_i and f_j are the scattering amplitudes of atoms on different molecules. These values are independent of q in the case of neutron scattering and a function of q for X-ray scattering. If one is interested in the inter-molecular contributions only, then it is more convenient to work with the *structural* function $H(q)$ which is defined by

$$H(q) = I(q) - f^2 \tag{29}$$

The *dielectric constant* ε is related to the fluctuations of the dipole moments of the system

$$\mathbf{M} = \sum_{i=1}^{N} \boldsymbol{\mu}_i \ , \tag{30}$$

where $\boldsymbol{\mu}$ is the dipole moment of a single molecule in the liquid environment. The fluctuation of $\mathbf{M}$ is defined by the Kirkwood factor g_K :

$$g_K = \frac{<\mathbf{M}^2> - <\mathbf{M}>^2}{N\mu^2} \tag{31}$$

for a field-free system.

In a computer simulation the fluid, whose dielectric permittivity is to be calculated, can be thought of as a large sphere made up of replicas of the simulation cell (see section 5 for details). This system is enclosed in a medium of dielectric permittivity ε_o, which is polarized by the dipole moments of the unit cell comprising the fluid. This interaction appears in the hamiltonian of the fluid being simulated and causes the Kirkwood factor to depend on ε_o. The relation between the dielectric permittivity and the mean square net dipole is dependent on the internal field correction, one version being[56]

$$\frac{(\varepsilon - 1)(2\varepsilon_o + 1)}{3(\varepsilon + 2\varepsilon_o)} = \frac{4\pi\rho\mu^2}{9k_BT} \; g_K(\varepsilon_o) \tag{32}$$

where ρ is the number density.

Theoretically one would like to use a surrounding medium with the same dielectric permittivity as the simulated system. Of course, this dielectric constant is not known in advance so that one usually chooses ε_o to be similar to the expected value. However, it can be shown that the resulting dielectric constant is independent of ε_o, but other properties may be affected.

Computation of the dielectric constant using the above formalism often runs into considerable problems associated with very large fluctuations of the Kirkwood factors, making it necessary to perform long simulations in order to get reliable results.

The dynamical properties are most conveniently studied via the time correlation function which relates a property A to a property B at a later time t. It is calculated at discrete time steps according to:

$$< A(0)\, B(t) > \equiv < A(t_0)\, B(t_0 + t) >_{t_0} \tag{33}$$

where $A(t)$ and $B(t)$ are dynamical variables of the system, such as position, velocity, density, *etc.* The time average in this case denotes an averaging over different time origins t_0 within a simulation run.

The *velocity autocorrelation function*

$$C_{vv}(t) = < \mathbf{v}_i(0) \cdot \mathbf{v}_i(t) > \tag{34}$$

is often of great interest because its time integral is related to the self diffusion coefficient, whereas its temporal Fourier transform is a measure of the "density of states."

The *self-diffusion coefficient D* can be obtained from the velocity autocorrelation function by

$$D = \frac{1}{3} \int_0^\infty C_{vv}(t)\, dt \tag{35}$$

or from the so-called Einstein relation

$$D = \lim_{t \to \infty} \frac{< |r_i(0) - r_i(t)|^2 >}{6t} \tag{36}$$

While the self-diffusion coefficient provides a good measure for translational mobility, it is often of interest to find an equivalent property for rotational diffusion. It is available through the *orientational autocorrelation functions*

$$C_l\,(t) = < P_l\,(\mathbf{n}_i(0) \cdot \mathbf{n}_i(t)) > \tag{37}$$

where P_l is the l-th Legendre polynomial and $\mathbf{n}_i$ is a fixed unit vector in the molecule. These correlation functions usually decay exponentially after a few molecular collisions. Their rates of decay are called relaxation times, and can often be measured experimentally. For example, the relaxation times for $C_2(t)$ may be compared with the half-width of Raman scattering bands, while $C_1(t)$ corresponds to infrared band shapes.[57] Correlation times obtained by integrating the orientational correlation functions can be compared with experimental data, for example the integral of $C_2(t)$ with NMR relaxation times.[58]

Infrared and Raman spectra can be obtained for systems in which the time evolution of the dipole moment and polarizability can be described classically.

The absorption cross section of infrared spectroscopy $\alpha(\omega)$ is proportional to the absorption line shape $I(\omega)$:[34,58,59]

$$\alpha(\omega) \sim \omega(1 - e^{\hbar\omega/kT})\, I(\omega) \tag{38}$$

$$I(\omega) \sim \int_{-\infty}^{+\infty} < \mathbf{M}(0)\cdot\mathbf{M}(t) > \; e^{-i\omega t} dt \tag{39}$$

where $\mathbf{M}$ is the total dipole moment of the system under investigation. This simplifies to averages over single molecule dipole moment autocorrelation functions for localized vibrations [using $C_1(t)$ from Eq. (37)].

The Raman spectra can be treated in a similar way. The polarized and depolarized components of the differential scattering cross section for scattering into a frequency range $d\omega$ and a solid angle $d\,\Omega$ are available through[34,59,60]

$$\frac{d^2\sigma_{pol}}{d\omega\, d\,\Omega}\lambda^4 \sim \int_{-\infty}^{+\infty} < \bar{\alpha}(0)\bar{\alpha}(t) > \; e^{-i\omega t} dt \tag{40a}$$

$$\frac{d^2\sigma_{depol}}{d\omega\, d\,\Omega}\lambda^4 \sim \int_{-\infty}^{+\infty} < \mathrm{trace}(\tilde{\boldsymbol{\beta}}(0)\tilde{\boldsymbol{\beta}}(t)) > \; e^{-i\omega t} dt \tag{40b}$$

where

$$\bar{\alpha} = \frac{1}{3}\mathrm{trace}(\tilde{\boldsymbol{\alpha}}) \quad ; \quad \tilde{\boldsymbol{\alpha}} = \bar{\alpha}\,\tilde{\mathbf{1}} + \tilde{\boldsymbol{\beta}}$$

$\tilde{\boldsymbol{\alpha}}$ is the total polarizability tensor of the system which is decomposed into a spherical part $\bar{\alpha}\tilde{\mathbf{1}}$ and a traceless anisotropic part $\tilde{\boldsymbol{\beta}}$, λ is the wavelength of the scattered radiation and $\tilde{\mathbf{1}}$ is the unit matrix.

In some special cases the depolarized component simplifies to an orientational autocorrelation function [$C_2(t)$ from Eq. (37)] but this is not valid in general.[34]

The *density* and the particle *current correlation functions* are two of the most important functions in the study of dynamical properties of fluids. The microscopic *density* and its Fourier component are defined as:

$$\rho(\mathbf{r}, t) = \sum_{\alpha=1}^{N} \delta\,[\mathbf{r} - \mathbf{r}_\alpha(t)\,] \tag{41a}$$

$$\rho(\mathbf{q}, t) = \frac{1}{\sqrt{N}} \sum_{\alpha=1}^{N} e^{i\mathbf{q}\cdot\mathbf{r}_\alpha(t)} \tag{41b}$$

The *particle current* and its Fourier components are defined as

$$\mathbf{j}(\mathbf{r}, t) = \sum_{\alpha=1}^{N} v_\alpha(t)\delta[\mathbf{r} - \mathbf{r}_\alpha(t)] \tag{42a}$$

$$\mathbf{j}(\mathbf{q}, t) = \frac{1}{\sqrt{N}} \sum_{\alpha=1}^{N} v_\alpha\,(t) e^{i\mathbf{q}\cdot\mathbf{r}_\alpha(t)} \tag{42b}$$

The Fourier components can be separated into two parts: Longitudinal (parallel to the $\mathbf{q}$ vector) and transverse (perpendicular to the $\mathbf{q}$ vector).

Information concerning the frequency dependent shear viscosity can be obtained from the transverse *current correlation function*

$$J_t(q, t) = \frac{1}{N} \sum_\alpha \sum_\beta < v_{\zeta\alpha}(0) v_{\zeta\beta}(t)\; e^{-i\mathbf{q}\cdot\,(\mathbf{r}_\alpha(0) - \mathbf{r}_\beta(t))} > \tag{43}$$

where $v_{\zeta\alpha}$ is the velocity component of molecule α along an axis ζ perpendicular to the $\mathbf{q}$ vector. The initial value of this correlation function should be the mean square velocity, $k_B T/m$. The temporal Fourier transform is a frequency dependent function which, in the hydrodynamic limit, has a Lorentzian shape centered at zero frequency, implying an exponential dissipation of the transverse current fluctuations. However, for some fluids the damping may be slowed sufficiently at higher q to sustain the propagation of shear waves, which may then be seen in the simulation as peaks in the frequency spectrum. Our study indicates that water is a liquid with such characteristics. It should be noted that it is extremely difficult to measure the propagation of shear waves experimentally.

4.3. Properties Related to Neutron Scattering Experiments

Neutron scattering experiments are a well established technique to study condensed matter systems. Since neutrons have the relation between energy $E = \hbar\omega$ and momentum $\vec{p} = \hbar k$, given by

$$E = \frac{\hbar^2 k^2}{2m} \tag{44}$$

where m is the neutron mass of 1.008μ, they can be used to study spatial correlations on the Å-scale with diffraction type experiments while time correlations on the *ps*-scale can be studied through inelastic scattering experiments.

In the field of biochemistry neutrons are widely used to study the structure and dynamics of biomolecules. A review on diffraction methods is given by Wlodlawer.[61] To elucidate dynamical effects in biomolecules and the solvent around them, more and more inelastic neutron scattering experiments are done.[62-67] As far as one considers phenomena on the *ps*-timescale, these experiments can be combined with MD simulations to give a better understanding of the complicated dynamics of biomolecules in solution. In fact, quite recently inelastic neutron scattering experiments are combined with MD simulations of proteins, to see if their dynamical behavior on the *ps*-timescale seen in the neutron scattering experiment can be reproduced from the MD simulation.

The basic quantity measured in inelastic neutron scattering experiments is the differential scattering cross section per solid angle and unit energy interval:

$$\frac{d^2\sigma}{d\,\Omega d\omega} = N \cdot b^2 \frac{k}{k_0} S(\mathbf{q}, \omega) \tag{45}$$

With k and k_0 we denote the momentum of the scattered and unscattered neutrons (divided by $\hbar$), N is the number of scatterers, b is the scattering length of the nuclei, and $S(\mathbf{q}, \omega)$ is the *dynamic structure factor.* Due to the dispersion relation (Eq. 44) we have the following relation between the modulus of the momentum transfer $q = |\mathbf{k}_0 - \mathbf{k}|$ and the corresponding energy transfer $\hbar\omega$

$$q = k_0 \sqrt{2 - \frac{\hbar\omega}{E_0} - 2\sqrt{1 - \frac{\hbar\omega}{E_0}} \cos(\Theta)} \tag{46}$$

where Θ is the scattering angle and E_0 is the energy of the unscattered neutrons. In the case of scattering from isotropic samples the dynamic structure factor is a function of q .

The inelastic cross section can be expressed as the sum of the coherent and incoherent parts:

$$\frac{d^2\sigma}{d\,\Omega d\omega} = \left(\frac{d^2\sigma}{d\,\Omega d\omega}\right)_{coh} + \left(\frac{d^2\sigma}{d\,\Omega d\omega}\right)_{inc} \tag{47a}$$

$$\left(\frac{d^2\sigma}{d\,\Omega d\omega}\right)_{coh} = N \cdot b_{coh}^2 \frac{k}{k_0} S_{coh}(\mathbf{q}, \omega) \tag{47b}$$

$$\left(\frac{d^2\sigma}{d\,\Omega d\omega}\right)_{inc} = N \cdot b_{inc}^2 \frac{k}{k_0} S_{inc}(\mathbf{q}, \omega) \tag{47c}$$

Experimentally it is possible to measure separately the coherent and the incoherent sections.

The dynamic structure factors can be written as the Fourier-transform of corresponding time correlation functions, usually called the coherent and incoherent *intermediate scattering functions*:

$$S_{coh}(\mathbf{q}, \omega) = \frac{1}{2\pi}\int_{-\infty}^{+\infty} dt\; e^{-i\omega t} F_{coh}(\mathbf{q}, t) \tag{48}$$

$$F_{coh}(\mathbf{q}, t) = \frac{1}{N}\sum_{\alpha, \beta} b_{\alpha, coh}\; b_{\beta, coh} < e^{-i\mathbf{q} \cdot \mathbf{r}_\alpha(0)} e^{i\mathbf{q} \cdot \mathbf{r}_\beta(t)} > \tag{49}$$

$$S_{inc}(\mathbf{q}, \omega) = \frac{1}{2\pi}\int_{-\infty}^{+\infty} dt\; e^{-i\omega t} F_{inc}(\mathbf{q}, t) \tag{50}$$

$$F_{inc}(\mathbf{q}, t) = \frac{1}{N}\sum_{\alpha} b_{\alpha, inc}^2 < e^{-i\mathbf{q} \cdot \mathbf{r}_\alpha(0)} e^{i\mathbf{q} \cdot \mathbf{r}_\alpha(t)} > \tag{51}$$

The indices α and β are running over all atoms in the system and $\mathbf{r}_\alpha$ and $\mathbf{r}_\beta$ are the corresponding positions. The parameters $b_{\alpha, coh}$ and $b_{\alpha, inc}$ are, respectively, the coherent and incoherent scattering length of atom α. From Eq. (41b), defining the spatial Fourier transform of the (weighted) *total* and *single particle density* by

$$\mathbf{B}(\mathbf{q}, t) = \frac{1}{\sqrt{N}}\sum_{\alpha} b_{\alpha, coh}\; e^{i\mathbf{q} \cdot \mathbf{r}_\alpha(t)} \tag{52}$$

$$\mathbf{b}_\alpha(\mathbf{q}, t) = b_{\alpha, inc}\; e^{i\mathbf{q} \cdot \mathbf{r}_\alpha(t)} \tag{53}$$

The intermediate scattering functions can be written in a more compact manner in terms of autocorrelation functions of these quantities:

$$F_{coh}(\mathbf{q}, t) = < \mathbf{B}(-\mathbf{q}, 0)\mathbf{B}(\mathbf{q}, t) > \tag{54}$$

$$F_{inc}(\mathbf{q}, t) = \frac{1}{N}\sum_{\alpha} < \mathbf{b}_\alpha(-\mathbf{q}, 0)\mathbf{b}_\alpha(\mathbf{q}, t) > \tag{55}$$

It is obvious that the coherent dynamic structure factor or -- equivalent -- the *coherent* intermediate scattering function provide information over the *collective* behavior of the system under investigation, whereas the corresponding *incoherent* quantities reflect the *single particle* motion.

From the general quantum mechanical definition of time correlation functions of two observables **A** and **B**,

$$c_{AB} \doteq <\mathbf{A}(0)\mathbf{B}(t)> = \frac{1}{Z}\mathrm{tr}\left\{e^{-\beta\mathbf{H}}\mathbf{A}(0)\; e^{\frac{i}{\hbar}\mathbf{H}t}\mathbf{B}(0)e^{\frac{-i}{\hbar}\mathbf{H}t}\right\} \tag{56}$$

$$Z = \mathrm{tr}\{e^{-\beta\mathbf{H}}\} \tag{57}$$

where $\beta = 1/k_BT$, and **H** is the Hamiltonian of the system, one obtains the following relations:

$$c^*{}_{AB}(t) = c_{B^+A^+}(-t) \tag{58}$$

$$c_{AB}(t) = c_{BA}(-t + i\beta\hbar) \tag{59}$$

where $\mathbf{A}^+$, $\mathbf{B}^+$ are the adjoint operators of **A** and **B** . Applied to the intermediate scattering functions this means

$$F^*(\mathbf{q}, t) = F(\mathbf{q}, -t) \tag{60}$$

$$F(\mathbf{q}, t) = F(-\mathbf{q}, -t + i\beta\hbar) \tag{61}$$

From Eq. (60) it follows that the dynamic structure factor is real and Eq. (61) is reflected in the relation of *detailed balance*:

$$S(\mathbf{q}, \omega) = e^{\beta\hbar\omega}S(-\mathbf{q}, -\omega) \tag{62}$$

In the classical limit ($\hbar \to 0$) we see from Eqs. (60), (61) and (62) that $F(\mathbf{q}, t)$ and $S(\mathbf{q}, \omega)$ are real and symmetric functions with respect to t and ω, if they are symmetric with respect to $\mathbf{q}$. This means that the system under consideration is invariant under spatial reflection. This is always true for isotropic systems where the intermediate scattering function and the dynamic structure factor depend on $|\mathbf{q}|$, or if we look at intermediate scattering functions and dynamic structure factors which are averaged over all $\mathbf{q}$ with the same modulus.

If we replace the second operator in a correlation function $c_{AB}(t)$, defined as in Eq. (56), by the time derivative of **B** we obtain according to Eq. (59), and noting that $\partial_t c_{AB}(t) = c_{A\dot{B}}(t)$:

$$\partial_t c_{AB}(t) = c_{A\dot{B}}(t) = c_{\dot{B}A}(-t + i\beta\hbar) \tag{63}$$

From Eqs. (63) and (59) we get for the second time derivative of $c_{AB}(t)$

$$\partial_t^2 c_{AB}(t) = -c_{\dot{A}\dot{B}}(t) \tag{64}$$

If this relation is applied to the coherent intermediate scattering function, one obtains:

$$-\partial_t^2 F_{coh}(\mathbf{q}, t) = q^2 C_l(\mathbf{q}, t) \tag{65}$$

$$C_l(\mathbf{q}, t) = < j_l(-\mathbf{q}, 0)\; j_l(\mathbf{q}, t) > \tag{66}$$

$$j_l(\mathbf{q}, t) = \mathbf{n}_q \cdot \mathbf{j}(\mathbf{q}, t) = \frac{1}{\sqrt{N}} \sum_\alpha b_{\alpha, coh}(\mathbf{n}_q \cdot \mathbf{v}_\alpha(t))\; e^{i\mathbf{q} \cdot \mathbf{r}_\alpha(t)} \tag{67}$$

We see that the second time derivative of the coherent intermediate scattering function is proportional to the autocorrelation function of the *longitudinal current density*, which is the projection of the total microscopic current density onto the unit vector $\mathbf{n}_q$ parallel to $\mathbf{q}$. The correlation function $C_l(\mathbf{q}, t)$ and its Fourier transform $\tilde{C}_l(\mathbf{q}, \omega)$ are very important for the study of sound modes in liquids, since they show up more pronounced in $C_l(\mathbf{q}, \omega)$ than in the dynamic structure factor itself.[68] In the incoherent case we have corresponding to Eqs. (65) to (67)

$$-\partial_t^2 F_{inc}(\mathbf{q}, t) = \frac{1}{N} \sum_\alpha q^2 C_{l,\alpha}(\mathbf{q}, t) \tag{68}$$

$$C_{l,\alpha}(\mathbf{q}, t) = <j_{l,\alpha}(-\mathbf{q}, 0)\; j_{l,\alpha}(\mathbf{q}, t)> \tag{69}$$

$$j_{l,\alpha}(\mathbf{q}, t) = \mathbf{n}_a \cdot \mathbf{j}_\alpha(\mathbf{q}, t) = b_{\alpha, inc}(\mathbf{n}_q \cdot \mathbf{v}_\alpha(t)) e^{i\mathbf{q} \cdot \mathbf{r}_\alpha(t)} \tag{70}$$

Since the intermediate scattering functions and the corresponding dynamic structure factors are connected via a Fourier transform with respect to time, we have from Eqs. (65) and (68):

$$\omega^2 S_{coh}(\mathbf{q}, \omega) = q^2 \tilde{C}_l(\mathbf{q}, \omega) \tag{71}$$

$$\omega^2 S_{inc}(\mathbf{q}, \omega) = \frac{1}{N} \sum_\alpha q^2\, \tilde{C}_{l,\alpha}(\mathbf{q}, \omega) \tag{72}$$

In isotropic systems, where the density and current correlation functions depend only on $q = |\mathbf{q}|$, $C_l(q,t)$ and $C_{l,\alpha}(q,t)$ can be written as

$$C_l(q,t) = \frac{1}{3} <\mathbf{j}_l(-\mathbf{q}, 0) \cdot \mathbf{j}_l(\mathbf{q}, t)> \tag{73}$$

$$C_{l,\alpha}(q,t) = \frac{1}{3} <\mathbf{j}_{l,\alpha}(-\mathbf{q}, 0) \cdot \mathbf{j}_{l,\alpha}(\mathbf{q}, t)> \tag{74}$$

In this case the following relation between the incoherent dynamic structure factor and the Fourier transform of the single particle velocity autocorrelation functions (VACF) holds:

$$\lim_{q \to 0} \frac{\omega^2}{q^2} S_{inc}(q, \omega) = \frac{1}{3N} \sum_\alpha b_{\alpha, inc}^2 \cdot \frac{1}{2\pi} \int_{-\infty}^{+\infty} dt\; e^{-i\omega t} < \mathbf{v}_\alpha(0) \cdot \mathbf{v}_\alpha(t) > \tag{75}$$

We see that $C_{l,\alpha}(q,t)$ is a natural generalization of the VACF of particle α.

To study the *static* properties of a given system, another set of quantities connected to the dynamic structure factor is important: its *moments*, given by

$$<\omega^n>_{coh}(\mathbf{q}) \doteq \int_{-\infty}^{+\infty} d\omega\ \omega^n S_{coh}(\mathbf{q}, \omega) = (-i)^n\, \partial_t^n\, F_{coh}(\mathbf{q}, t)\,|_{t=0} \tag{76}$$

$$<\omega^n>_{inc}(\mathbf{q}) \doteq \int_{-\infty}^{+\infty} d\omega\ \omega^n\, S_{inc}(\mathbf{q}, \omega) = (-i)^n\, \partial_t^n\, F_{inc}(\mathbf{q}, t)\,|_{t=0} \tag{77}$$

Due to the fact that according to Eq. (62) the dynamic structure factor, calculated from classical ensemble or time averages is an *even* function in ω, all odd classical moments vanish for systems with inversion symmetry. The most important moment is the zero-th moment $<\omega^0>_{coh}(\mathbf{q})$, which is called *static structure factor*:

$$S(\mathbf{q}) = <\omega^0>_{coh}(\mathbf{q}) = F(\mathbf{q}, 0) \tag{78}$$

The static structure factor can be obtained from neutron diffraction experiments,[69] see in addition Eq (28). A discussion of higher moments can be found in Ref. 69.

The dynamic structure factor $S(\mathbf{q}, \omega)$ reduces to the static structure factor at zero frequency. In the hydrodynamic limit ($q \to 0$), the dynamic structure factor often consists of three peaks at $\omega = 0$ and $\pm\, c_s\, q$, where c_s is the adiabatic speed of sound. The central peak is the Rayleigh peak and its width gives a measure of the thermal diffusiveness. The side peaks are called Brillouin peaks and represent the propagation and damping of sound in the fluids. As q increases, these peaks usually broaden and eventually disappear.

To obtain the coherent/incoherent dynamic structure factor of a liquid or a macromolecule from a MD simulation one has first to calculate the corresponding intermediate scattering functions from the trajectories of the atoms involved. This means one has to replace the quantum mechanical averages in Eqs. (54) and (55) by *time averages*. Abbreviating $t = m \cdot \Delta t$ as m, Δt being the time difference between the MD configurations, one obtains the following expressions:

$$F_{coh}(\mathbf{q}, m) \cong <B(-\mathbf{q}, 0)B(\mathbf{q}, m)>_{cl} = \frac{1}{N_t - m} \sum_{k=0}^{N_t - m - 1} B(-\mathbf{q}, k)B(\mathbf{q}, k + m) \tag{79}$$

$$\begin{aligned} F_{inc}(\mathbf{q}, m) &\cong \frac{1}{N}\sum_{\alpha} <b_\alpha(-\mathbf{q}, 0)b_\alpha(\mathbf{q}, t)>_{cl} \\ &= \frac{1}{N}\sum_{\alpha}\left\{\frac{1}{N_t - m} \sum_{k=0}^{N_t - m - 1} b_\alpha(-\mathbf{q}, k)b_\alpha(\mathbf{q}, k + m)\right\} \end{aligned} \tag{80}$$

With N_t we denote the number of configurations in the analysis, and the indices k and m are running from 0 to $N_t - 1$. Consistent with Eq. (61) we have $F(\mathbf{q}, m) = F(-\mathbf{q}, |m|)$ for $m \in [-N_t + 1, -1]$. Note that in Eqs. (79) and (80) we have introduced the notation $<\ldots>_{Cl}$ to emphasize the fact that we are using a classical representation of the system.

To calculate the classical time correlation functions in Eqs. (79) and (80) in a very efficient way we used the *Fast Correlation Algorithm* which works as follows, for an arbitrary representative of a stationary process $u(t)$, given at $t = k \cdot \Delta t, \quad k \in [0, N_t - 1]$.[70,71] The estimate of its autocorrelation function, which is usually written as

$$\begin{aligned} C_{uu}(m) &= \frac{1}{N_t - m} \cdot \sum_{k=0}^{N_t - m - 1} u^*(k)u(k+m), && \text{for m } \in [0, N_t - 1] \\ C_{uu}(m) &= C^*_{uu}(|m|), && \text{for m } \in [-N_t + 1, -1] \end{aligned} \tag{81}$$

can also be expressed in the following form ($m \in [-N_t + 1, N_t - 1]$) (see Ref. 71):

$$C_{uu}(m) = \frac{1}{N_t - |m|} \cdot \frac{1}{2N_t} \sum_{n=0}^{2N_t - 1} e^{i \frac{2\pi}{2N_t} mn} \, | \, \hat{U}(n) \, |^2 \tag{82}$$

The function $\hat{U}(n)$, given by

$$\hat{U}(n) \doteq \sum_{n=0}^{2N_t - 1} e^{-i \frac{2\pi}{2N_t} nk} \, U(k) \tag{83}$$

is the discrete Fourier transform of $U(k)$, which has the length $2N_t$ and is identical with $u(k)$ for k $\in [0, N_t - 1]$. The remaining $2N_t$ data points are zero. This zero-padding is necessary to avoid spurious correlations, since $u(k)$ is a signal of *finite* length.

The application of the Fast Fourier Transform (FFT) algorithm[71] to perform the discrete forward and backward transforms reduces the complexity of the calculation of $C_{uu}(m)$ (number of multiplications) from $O\{N_t^2\}$ for the direct method according to Eq. (81) to $O\{2N_t \cdot \log_2(2N_t)\}$. To perform the FFT we use the routine DCFT of the ESSL library to obtain the maximum speed up on the IBM 3090. This routine can be replaced by an equivalent routine on other machines.

In the following we briefly mention the programs needed to calculate the dynamic structure factor and related quantities. All the programs are available to the MOTECC-users.[11] We start with the program called LISTE, which is an auxiliary program for SKWCOH and SKWINC, which calculate the coherent and incoherent dynamic structure factor respectively. LISTE generates the q-vectors needed in SKWCOH and SKWINC and dumps them on a file, which is read by SKWCOH/SKWINC. LISTE starts from the standard crystallographic parameters a, b, c and α, β, γ of the MD-cell under consideration and generates first the corresponding basis vectors $\mathbf{b}_1$, $\mathbf{b}_2$, $\mathbf{b}_3$. Then it constructs the basis vectors of the *dual basis* $\mathbf{b}^1$, $\mathbf{b}^2$, $\mathbf{b}^3$ which are defined by

$$\mathbf{b}_i \cdot \mathbf{b}^j = \delta_i^j \tag{84}$$

The q-vectors are then constructed according to

$$\mathbf{q} = 2\pi(k \cdot \mathbf{b}^1 + l \cdot \mathbf{b}^2 + m \cdot \mathbf{b}^3) \tag{85}$$

The bounds for the integer numbers k, l, m are chosen such that all vectors $\mathbf{q}_i, \mathbf{q}_j$ with inversion symmetry $\mathbf{q}_j = -\mathbf{q}_i$ are dropped, and

$$m \in \; [0, k_{\max}] \tag{86a}$$

$$l \in \begin{cases} [0, k_{\max}] & \text{if } m = 0 \\ [-k_{\max}, k_{\max}] & \text{if } m \neq 0 \end{cases} \tag{86b}$$

$$k \in \begin{cases} [0, k_{\max}] & \text{if } m = 0 \cup l = 0 \\ [-k_{\max}, k_{\max}] & \text{if } m \neq 0 \cap l \neq 0 \end{cases} \tag{86c}$$

To average the dynamic structure over all q-vectors in a shell Δq around q, which is the usual application in studying liquids, powders *etc.*, the number of q-vectors can be reduced by keeping only a small number of q-vectors per q-shell. It should be noted, that the number of q-vectors grows with third power of $k_{\max}$! As an option, the angular distribution of the q-vectors in a radial shell can be maintained or chosen to be uniformly distributed over the unit sphere. For that purpose all q-vectors, constructed according to Eq. (85), are sorted into a (q, Θ, Φ)-histogram, recording in this way their radial and angular distribution. Θ and Φ are the polar and azimuthal angle describing the direction of the q-vectors.

The program SKWINC calculates the *incoherent* intermediate scattering function and the corresponding incoherent dynamic structure factor. The intermediate scattering function is calculated according to Eq. (80) and the dynamic structure factor by a subsequent Fourier transform. The intermediate scattering $F_{inc}(\mathbf{q}, t)$ is considered to be *real*. Optionally, one can calculate the (averaged) longitudinal one particle current-current correlation function and its Fourier transform, from which one gets the density of states in the limit $q \rightarrow 0$ (see Eq. 75). The current-current correlation is calculated according to Eq. (64) by differentiating the input signal $b_\alpha(\mathbf{q}, t)$ *numerically* before autocorrelating it.

A crucial point is the definition of the input coordinates. If the MD-cell is described by three basis vectors $\mathbf{b}_1, \mathbf{b}_2, \mathbf{b}_3$, then the vector $\mathbf{x}_i$, describing the position of an arbitrary atom is written as

$$\mathbf{x} = \sum_{i=1}^{3} x^i \cdot \mathbf{b}_i \tag{87}$$

The coordinates $\{x^i\}$ are called *contravariant* coordinates. Since all particles always stay inside the MD-cell due to periodic boundary conditions, all $\{x^i\}$ stay between 0 and 1, if the origin of the coordinate system is chosen to coincide with the origin of the basis vectors. Often the origin of the coordinate system is chosen to be at the center of the MD-cell, and the basis vectors are scaled down so that the coordinates $\{x^i\}$ vary between -1 and 1.

The program SKWCOH calculates the *coherent* intermediate scattering function and the corresponding coherent dynamic structure factor. The intermediate scattering function is calculated according to Eq. (79) and the dynamic structure factor by a subsequent Fourier transform. As in the incoherent case the intermediate scattering $F_{coh}(\mathbf{q}, t)$ is considered to be *real*. Here, optionally, one can calculate the longitudinal *total* current-current correlation function and its Fourier transform. Analogous to the incoherent case the current-current correlation is calculated according to Eq. (64) by differentiating the input signal $B_\alpha(\mathbf{q}, t)$ *numerically* before autocorrelating it.

The program SKTOF calculates neutron time-of-flight spectra from the dynamic structure factor calculated by SKWCOH/SKWINC:

$$S_{cl}(q, \omega) = S_{inc,cl}(q, \omega) + S_{coh,\, cl}(q, \omega) \tag{88}$$

The so-called time-of-flight is simply the reciprocal neutron velocity and therefore a measure for energy. The differential cross section as a function of the solid angle and the time-of-flight τ can be written as follows:

$$\frac{d^2\sigma}{d\,\Omega d\tau} \propto E^{3/2} \cdot \frac{d^2\sigma}{d\,\Omega\, dE} \tag{89}$$

$$E = \frac{1}{2} m\tau^{-2} \tag{90}$$

The time-of-flight τ is usually measured in $\mu s/m$ and the energy $E = E_0 - \hbar\omega$ of the scattered neutron in meV, so that the conversion formula $\tau = 2285/\sqrt{E}$ follows. From Eq. (45) we have then immediately:

$$\frac{d^2\sigma}{d\,\Omega\, d\tau} \propto \left(\frac{E}{E_o}\right)^2 \cdot S(q, \omega) \tag{91}$$

SKTOF calculates the scattering intensity for a given scattering angle Θ according to this equation. $S(q(\Theta, \omega), \omega)$ is evaluated by interpolation using Eq. (46).

To ensure detailed balance (see Eq. 62), the following semiclassical correction formula is applied[69]

$$S(q, \omega) \simeq \frac{\beta\hbar\omega}{1 - e^{-\hbar\omega}} \cdot S_{cl}\,(q, \omega) \tag{92}$$

The important experimental data which have to be known to run SKTOF are: a) The energy E_0 of the incident neutrons [meV]; b) The scattering angle Θ [°]; c) The maximum energy transfer [meV] which, according to Eq. (46) determines the maximum energy transfer $q_{\max}$ needed in the dynamic structure factor; d) The temperature of the sample [K] which is used to calculate the detailed balance correction.

5. Periodic Boundary Conditions and Long Range Forces

Even with today's supercomputers, both the MC and the MD methods can still only be applied to samples of relatively small number of particles, on the order of a few hundreds to a few thousands depending upon the the complexity of the interaction potentials. This raises a problem: How do you simulate a macroscopic system with such a small number of particles? Fortunately, the problem can be solved because deviations from uniform density extend only to ~10 Å for most fluids at conditions far away from critical regions. In other words, atomic or molecular aspects of liquids are only important within ~10 Å. That means that we should be able to study liquid structures if we can simulate a system of this size, which translates into 100 to 1000 particles or molecules.

This, however, does not solve the problem completely because a simulation system of the above size would leave too many molecules near the surface and represent a cluster rather than a liquid. To make every molecule see the *bulk* environment, *periodic boundary conditions* were introduced by Metropolis *et al.* in 1953,[21] and have since been widely used. In this approach, the simulation system actually consists of repetitions of the base unit in all directions. However, only the base unit needs to be kept in the simulation, since all the other units are related to the base unit by translational symmetry. To facilitate our discussion, we will call this base unit the *simulation box*, containing N particles in a cube of size $L \times L \times L$.

Due to the periodic conditions, the potential energy per particle for a system interacting through a central force type potential is given by

$$V = \frac{1}{N} \sum_{\mathbf{n}} \sum_{j<i} \phi(\mathbf{r}_{ij} + L\mathbf{n}) \tag{93}$$

where the summation is to be taken over all the molecular pairs (i,j) within the simulation box, and the cell vectors $\mathbf{n}$ extend to the infinite lattice. For short range interaction forces, the summation over cell vectors would cause no problem (because $\phi(\mathbf{r})$ will be equal to zero for $\mathbf{r}$ greater than the range of the force, see Chapter 18). Unfortunately, this is not the case in most of the chemical systems that we are interested in, for example, systems with *charge-charge, charge-dipole, dipole-dipole,* and *charge-quadrupole interactions*. For these interactions, the series in Eq. (93) is only conditionally convergent. Since summation over an infinite series is impossible in practice, truncation will have to be introduced, which inevitably introduces errors. These errors has been shown to cause a steady total energy drift in MD simulations which are supposed to conserve total energy. To remedy the error introduced by the truncation, various correction schemes have been proposed.

The truncation usually is accomplished by introducing a cutoff radius R, which is often half the size of the simulation box or less. This cutoff gives rise to a situation that the interaction pairs are *not* always the pairs that are in the simulation box. Instead, a given particle i may interact with the periodic image of particle j. For example, if particle j lies outside the cutoff sphere of particle i, then the latter will interact only with the *image* of the former which lies within the cutoff sphere. Since this image is, among all the images of particle j, at the *minimum* distance from particle i, hence this method of picking up the interaction pairs is is called "minimum image" convention. The interactions with all the the particles lying outside the cutoff sphere of a given particle are not calculated directly from $\phi(\mathbf{r}_{ij})$, but are approximated by the correction schemes to be discussed next.

The oldest one, which is perfectly consistent with periodic boundary conditions, is the *Ewald-sum procedure*.[72] In this technique the summation in Eq. (93) is actually attempted by thinking every charge as made up of a Gaussian distribution. The summation is then split into two parts: One from the tail of the Gaussian distribution, the other from the rest. The former results in the complementary error function $erfc(x)$, whereas the latter is carried out in the reciprocal lattice space (Fourier $\mathbf{k}$ space). For charge-charge interactions, the resulting formula is[56]

$$NV = \sum_{\mathbf{n}} \sum_{i \leq j} \frac{erfc(\alpha |\mathbf{r}_{ij} + L\mathbf{n}|)}{|\mathbf{r}_{ij} + L\mathbf{n}|} + \frac{1}{\pi V} \sum_{\mathbf{k} \neq 0} \sum_{i \leq j} |\mathbf{k}|^{-2} e^{(-\pi^2/\alpha^2 \mathbf{k}^2)} \cos(2\pi \mathbf{k} \cdot \mathbf{r}) \quad (94)$$

The choice of the arbitrary splitting parameter α affects in opposite way the convergency of the two summations. The higher the value of α, the faster the convergency of the direct lattice term and the slower the one of the reciprocal lattice term. Practically the value of α is set such that the complementary error function decays sufficiently rapidly to truncate the direct space summation at $L/2$.

The Ewald sum procedure was originally developed for *true periodic systems* such as crystals. It is now widely adopted also for liquid simulations, which implicitly implies the periodicity of the liquid. Although not quite right, the great advantage of the method and the reason of its success lies in the fact that the conservation of the total energy is generally restored.

A second approach to the problem without the assumption of the periodicity is to use the *reaction field method*.[73] The basic physics behind this method is to treat all molecules beyond the cutoff sphere of radius R from a center charge q_i as forming a continuum with a given dielectric constant ε_F. Any charge q_j lying inside the cutoff sphere will polarize the continuum and create a reaction field at the center. From electrostatic theory, that reaction field can be shown to be

$$\mathbf{F}(r = 0) = -\frac{1 - \varepsilon_F}{\varepsilon_F + \frac{1}{2}} \frac{q_j \, \mathbf{r}_j}{R^3} \quad (95)$$

where $\mathbf{r}_j$ is the position vector of charge q_j. If instead of charges q_j's, we have an ensemble of dipoles μ_j's, then it is easily shown that the reaction field at the center is

$$\mathbf{F}(r=0) = \frac{\varepsilon_F - 1}{\varepsilon_F + \frac{1}{2}} \frac{1}{R^3} \sum_j \boldsymbol{\mu}_j \tag{96}$$

which is the Onsager's reaction field.[74]

6. Applications and Examples

In the previous sections we have outlined the computational framework for molecular dynamics simulations for pure liquids and solutions. In the following we shall discuss three applications in detail. In the first one we focus on how to construct and later verify an interaction potential, specifically a new water-water potential.[75] In the second application we discuss an extension of the MCY water-water potential[76] to include the molecular vibrations. In the third example we present a comparative study on the dynamics of a protein both *in vacuo* and in solution. This application is also given to stress that simulations of the dynamics of a protein *in vacuo* are as meaningful as studies of the dynamics of a propeller-airplane performed in a vacuum. We present, also, a discussion on the neutron scattering properties computed with the formalism of section 4.3. In the fourth application the solvent dynamics in a crystal of a protein is analyzed and compared with the bulk water dynamics.

6.1 Simulation of Liquid Water with the NCC potential

In Chapter 1 we have discussed in detail the MCY water-water potential[76] derived about 20 years ago from relatively few (about 70) *ab initio* interaction energies obtained at the CI level. The MCY potential is a two-body potential and three-[27] as well as four-body[28] corrections have been considered and found to be very important. The MCY potential has a number of shortcomings related to: 1) the limited sampling of the water-water potential energy hypersurface, especially in the repulsive regions, and 2) the analytical form selected for the potential. Concerning the latter, one of the main limitations is the use of the same value for point charges regardless of the mutual position of the two water molecules. It is recalled that the simplest way to ameliorate this approximation is to write $q_i = q_0 f(r_{AB})$, thus allowing the charges to respond to the polarization field. A more formally correct approach is to let one molecule of water be polarized by the field of the other water molecules present in the system, and vice versa. However, before considering this point any further, let us return to the energy data used to obtain a fit for the parameters in the analytical expression of the potential.

First, it is clear that the hypersurface of the water-water interaction needs to be scanned much more frequently than was done for the old MCY potential. A "reasonable" number of points lies in the 400 − 600 range. This high number brings up *practical* constraints on the desired accuracy of the quantum-mechanical computation unless one has an "infinite" amount of time on a supercomputer. The next question concerns the selection of a computational technique that we can assume as "reliable" for the level of accuracy we consider; "a must," for our example, we need at least 0.1 kcal/mole accuracy. The value 0.1 has been selected because it corresponds to about two-hundredths of the water-water binding energy (equilibrium configuration) and it is smaller than the four-body correction. Table IV of Chapter 7 indicates that such accuracy is still very hard to achieve, despite the progress in *ab initio* techniques and, even more so, the obvious progress in computer technology. For example, the MP2, MP4, and CI techniques have well known flaws. The first two are plagued by the basis set superposition error,[77] the CI approach is not size-consistent. In our opinion, the best chances to reach the accuracy discussed above are offered by those methods which have fully recognized the above limitations. These are: 1) the computation of dispersion energies *a la* Kolos,[78] 2) empirical density functional corrections as proposed, for example, in Chapter 7, 3) the Complete Basis Set methods by Petterson,[79,80] and 4) the ICF procedure by B. Liu and A.D. McLean.[81,82] It is of interest to note that the forerunner of the ICF procedure is the MCY study;[76] however, by now the approach has reached a notable systematized level. There are drawbacks for the above four techniques since there are still "problems," either explicit or "hidden," namely there is a need to fully "correct" the first, to more thoroughly recalibrate the second, to ensure consistent accuracy in different parts of the potential surface for the third, and to make the fourth more "user-friendly," being presently more of an art rather than a full-proved procedural method. This being the case, when in 1988 we took another look at the water-water interaction potential[83] it appeared reasonable to separate the problem of obtaining *truly accurate energies* for the required 400 to 600 points from the problem of *testing a new analytical form* of the potential. Indeed, the latter problem could be tested — at least to start with — by using two sets of energy data, the first one consisting of about 300 points obtained at the MP4 level, and the second one by correcting this data set with the values of the basis set superposition correction obtained at the Hartree-Fock level. These two data sets differ by about 0.5 kcal/mole at the water-water equilibrium. Now, in 1990, having finished the experimentation reported below, we can once more start all over, this time focusing on obtaining a data set of interaction energies accurate to about 0.01 kcal/mole, an elusive goal for us for the last 20 years.

The new potential is designated as NCC, a short form for Nieser-Corongiu-Clementi. It consists of two parts

$$V_{NCC} = \sum_{i,j<i} [\ V_{\text{two-body}}(i,j)\ ] + V_{pol} \tag{97}$$

where the pairwise additive part of the potential is

$$
\begin{aligned}
V_{\text{two-body}} = {} & q^2 \left(\frac{1}{R_{13}} + \frac{1}{R_{14}} + \frac{1}{R_{23}} + \frac{1}{R_{24}}\right) + \frac{4q^2}{R_{78}} \\
& -2q^2 \left(\frac{1}{R_{81}} + \frac{1}{R_{82}} + \frac{1}{R_{73}} + \frac{1}{R_{74}}\right) + A_{OO} e^{-B_{OO}R_{56}} \\
& + A_{HH}(e^{-B_{HH}R_{13}} + e^{-B_{HH}R_{14}} + e^{-B_{HH}R_{23}} + e^{-B_{HH}R_{24}}) \\
& + A_{OH}(e^{-B_{OH}R_{53}} + e^{-B_{OH}R_{54}} + e^{-B_{OH}R_{61}} + e^{-B_{OH}R_{62}}) \\
& - A'_{OH}(e^{-B'_{OH}R_{53}} + e^{-B'_{OH}R_{54}} + e^{-B'_{OH}R_{61}} + e^{-B'_{OH}R_{62}}) \\
& + A_{PH}(e^{-B_{PH}R_{73}} + e^{-B_{PH}R_{74}} + e^{-B_{PH}R_{81}} + e^{-B_{PH}R_{82}}) \\
& + A_{PO}(e^{-B_{PO}R_{76}} + e^{-B_{PO}R_{85}})
\end{aligned}
\qquad (98)
$$

where subscripts 7 and 8 correspond to the negative point charges P on the molecular C_{2v} axis of the interacting water molecules. Other subscripts are defined in Fig. 2. The geometry of the water molecule is defined as it is in the MCY potential with a OH-bond length of 0.9572 Å and a HOH-angle of 104.52°, corresponding to an isolated water molecule in the gas phase.[84] In Eq. (98), A_{PH} and A_{PO} are new terms to the MCY potential[76] required in order to correct deficiencies at short distances and to obtain a more reasonable pressure for liquid water.

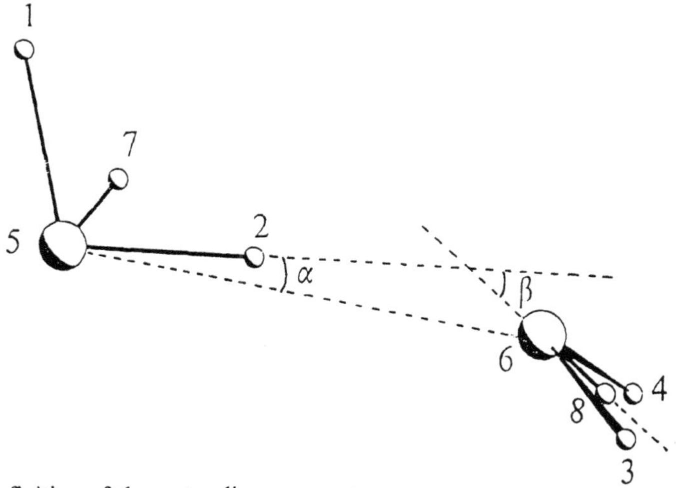

Figure 2. Definition of the water dimer geometry.

The second part of V_{NCC} is a polarization term. Previous computations, where three [27,85-87] and four-body[28] corrections were considered, have shown that the many-body corrections are necessary for accurate quantitative predictions. However, the computational time increases dramatically when the two-body MCY potential is extended with three- and four-body corrections, term by term. Therefore we require a computationally less expensive algorithm which takes advantage of the "large memory" not available on previous computers.

The need of many-body corrections can be shown on very simple physical grounds. Let us consider the electronic charge distribution for Eq. (98): it is clear that the electronic density at the lone-pairs on the oxygen atoms in gaseous H_2O must be significantly different from the electronic density in liquid water, where the lone-pairs are participating in hydrogen-bond formation. From this, it follows that the electronic distribution must be a function of the geometrical configuration of the "cluster of water molecules," solvating a given water molecule. We refer the interested reader to the study where the numerical value of the point charges was considered as a variable rather than a constant and to general volumes [88,89] and specific models[90-96] in recent literature.

In the NCC potential we chose an explicit representation of the polarization effects by inducing dipole moments on every interacting molecule. The permanent dipole moments of the water molecules are represented by three point charges per molecule as shown in Fig. 2. The polarization on one molecule is hence primarily due to the global point charge distribution of the surrounding matter. These induced dipole moments in turn cause polarization on the other molecules and thus this effect must also be included in the polarization potential. The induced polarization is, as usual, taken as a linear response to the electric field since the field is not very large.

$$\boldsymbol{\mu}_{i_\lambda}^{ind} = \tilde{\boldsymbol{\alpha}}_{i_\lambda} \mathbf{E}_{i_\lambda}^{tot} \tag{99}$$

where $\boldsymbol{\mu}_{i_\lambda}^{ind}$ is the λ-th induced moment on molecule i. The polarizability $\tilde{\boldsymbol{\alpha}}_{i_\lambda}$ is assumed to be a static property of molecule i. The electric field $\mathbf{E}_{i_\lambda}^{tot}$ at the location of the i_λ-th induced dipole moment is the sum of the electric fields caused by the surrounding point charge distribution and induced dipole moments. It can be written as [89]

$$\mathbf{E}_{i_\lambda}^{tot} = \mathbf{E}_{i_\lambda}^{q} + \mathbf{E}_{i_\lambda}^{d} \tag{100a}$$

$$\mathbf{E}_{i_\lambda}^{q} = -\sum_{j\neq i}^{N_m}\sum_{\sigma}^{N_c} \nabla_{i_\lambda}\frac{q_{j_\sigma}}{|\mathbf{R}_{i_\lambda j_\sigma}|} = \sum_{j\neq i}^{N_m}\sum_{\sigma}^{N_c} \frac{q_{j_\sigma}}{|\mathbf{R}_{i_\lambda j_\sigma}|^3}\mathbf{R}_{i_\lambda j_\sigma} \tag{100b}$$

$$\mathbf{E}_{i_\lambda}^{d} = \sum_{k\neq i}^{N_m}\sum_{\nu}^{N_d} \tilde{\mathbf{T}}_{i_\lambda k_\nu}\boldsymbol{\mu}_{k_\nu}^{ind} \tag{100c}$$

$$\tilde{\mathbf{T}}_{i_\lambda k_\nu} = -\nabla_{i_\lambda}\nabla_{i_\lambda}\frac{1}{|\mathbf{R}_{i_\lambda k_\nu}|} = \left(\frac{\tilde{\mathbf{1}}}{|\mathbf{R}_{i_\lambda k_\nu}|^3} - \frac{3\,\mathbf{R}_{i_\lambda k_\nu}\otimes\mathbf{R}_{i_\lambda k_\nu}}{|\mathbf{R}_{i_\lambda k_\nu}|^5}\right) \tag{100d}$$

where $\mathbf{E}_{i_\lambda}^{q}$ is the electric field caused by the charges q_{j_σ} on molecules j. $\mathbf{R}_{i_\lambda j_\sigma} = \mathbf{R}_{i_\lambda} - \mathbf{R}_{j_\sigma}$ is the distance vector between the induced dipole moment i_λ and charge j_σ . $\mathbf{E}_{i_\lambda}^{d}$ is the electric field due to the induced dipole moments k_ν on molecules k, and $\tilde{\mathbf{T}}_{i_\lambda k_\nu}$ is the dipole-dipole matrix. N_m is the number of molecules in the system, N_c is the number of point charges per molecule and N_d the number of induced dipole moments per molecule.

With Eqs. (99) and (100) we can set up a system of linear equations for the induced dipole moments.

$$\boldsymbol{\mu}_{i_\lambda}^{ind} = \tilde{\boldsymbol{\alpha}}_{i_\lambda}\left(\mathbf{E}_{i_\lambda}^{q} + \sum_{k\neq i}^{N_m}\sum_{\nu}^{N_d}\tilde{\mathbf{T}}_{i_\lambda k_\nu}\boldsymbol{\mu}_{k_\nu}^{ind}\right) \qquad i = 1, \ldots, N_m;\ \lambda = 1, \ldots, N_d \quad (101)$$

The induction energy of such a system of induced dipole moments is given as:

$$V_{pol} = -\frac{1}{2}\sum_{i}^{N_m}\sum_{\lambda}^{N_d}\boldsymbol{\mu}_{i_\lambda}^{ind}\mathbf{E}_{i_\lambda}^{q} \quad (102)$$

From Eq. (101) one can obtain the following identity

$$-\frac{1}{2}\sum_{i}^{N_m}\sum_{\lambda}^{N_d}\boldsymbol{\mu}_{i_\lambda}^{ind}\mathbf{E}_{i_\lambda}^{q} - \frac{1}{2}\sum_{i}^{N_m}\sum_{\lambda}^{N_d}\sum_{k\neq i}^{N_m}\sum_{\nu}^{N_d}\boldsymbol{\mu}_{i_\lambda}^{ind}\tilde{\mathbf{T}}_{i_\lambda k_\nu}\boldsymbol{\mu}_{k_\nu}^{ind} + \frac{1}{2}\sum_{i}^{N_m}\sum_{\lambda}^{N_d}\boldsymbol{\mu}_{i_\lambda}^{ind}\tilde{\boldsymbol{\alpha}}_{i_\lambda}^{-1}\boldsymbol{\mu}_{i_\lambda}^{ind} = 0 \quad (103)$$

and using this identity we rewrite (102) in the following form:

$$V_{pol} = -\sum_{i}^{N_m}\sum_{\lambda}^{N_d}\boldsymbol{\mu}_{i_\lambda}^{ind}\mathbf{E}_{i_\lambda}^{q} - \frac{1}{2}\sum_{i}^{N_m}\sum_{\lambda}^{N_d}\sum_{k\neq i}^{N_m}\sum_{\nu}^{N_d}\boldsymbol{\mu}_{i_\lambda}^{ind}\tilde{\mathbf{T}}_{i_\lambda k_\nu}\boldsymbol{\mu}_{k_\nu}^{ind} + \frac{1}{2}\sum_{i}^{N_m}\sum_{\lambda}^{N_d}\boldsymbol{\mu}_{i_\lambda}^{ind}\tilde{\boldsymbol{\alpha}}_{i_\lambda}^{-1}\boldsymbol{\mu}_{i_\lambda}^{ind} \quad (104)$$

This is, of course, only a different representation of Eq. (102) but it is a better starting point for deriving the equations for the forces. Let us briefly discuss the derivation of these equations.

The reason for expanding Eq. (102) with two additional terms to Eq. (104) is that the derivative of the polarization energy with respect to any induced dipole moment $\boldsymbol{\mu}_{k_\nu}^{ind}$ vanishes in this representation, *i.e.*

$$\frac{\partial V_{pol}}{\partial \boldsymbol{\mu}_{l_\rho}^{ind}} = -\mathbf{E}_{l_\rho}^{q} - \sum_{k\neq l}^{N_d}\sum_{\nu}^{N_d}\tilde{\mathbf{T}}_{l_\rho k_\nu}\boldsymbol{\mu}_{k_\nu}^{ind} + \tilde{\boldsymbol{\alpha}}_{l_\rho}^{-1}\boldsymbol{\mu}_{l_\rho}^{ind} = 0. \quad (105)$$

Assuming this condition, the intermolecular forces due to Eq. (104) are only a function of the derivatives of the point charge field **E**, the dipole-dipole matrix **T**, and the polarizability tensor $\boldsymbol{\alpha}$. The forces acting on the γ-th charge on molecule l are now easily available as:

$$\mathbf{F}_{l_\gamma} = \nabla_{l_\gamma} V_{pol} = \sum_{i \neq l} \sum_{\lambda} \boldsymbol{\mu}_{i_\lambda}^{ind} \nabla_{l_\gamma} \left(- \nabla_{i_\lambda} \frac{q_{l_\gamma}}{| R_{i_\lambda l_\gamma} |} \right) \tag{106}$$

and the force acting on the δ-th induced dipole moment on molecule l as:

$$\mathbf{F}_{l_\delta} = \nabla_{l_\delta} V_{pol} = \sum_{j \neq l} \sum_{\sigma} \boldsymbol{\mu}_{l_\delta}^{ind} \nabla_{l_\delta} \left(- \nabla_{l_\delta} \frac{q_{j_\sigma}}{| R_{l_\delta j_\sigma} |} \right) +$$

$$\frac{1}{2} \sum_{i \neq l} \sum_{\lambda} \boldsymbol{\mu}_{i_\lambda}^{ind} \nabla_{l_\delta} \left(- \nabla_{i_\lambda} \nabla_{i_\lambda} \frac{1}{| R_{i_\lambda l_\delta} |} \right) \boldsymbol{\mu}_{l_\delta}^{ind} + \frac{1}{2} \sum_{k \neq l} \sum_{\nu} \boldsymbol{\mu}_{l_\delta}^{ind} \nabla_{l_\delta} \left(- \nabla_{l_\delta} \nabla_{l_\delta} \frac{1}{| R_{l_\delta k_\nu} |} \right) \boldsymbol{\mu}_{k_\nu}^{ind}$$

$$- \frac{1}{2} \boldsymbol{\mu}_{l_\delta}^{ind} \nabla_{l_\delta} \tilde{\boldsymbol{\alpha}}_{l_\delta}^{-1} \boldsymbol{\mu}_{l_\delta}^{ind} \tag{107}$$

In the spirit of deriving a nonempirical potential to be used for liquid water simulations, all parameters of the NCC potential have been fitted to *ab initio* calculated data. We refer the interested reader to Ref. 83, where the *ab initio* calculations of interaction energies for various geometries of 250 water trimer and 350 water dimer configurations are discussed (a list of the interaction energies and geometries are given in Ref. 83b). We use the *ab initio* interaction energies of trimers of water molecules to fit the many-body parameters, namely in the parameterization of the locations of the induced dipole moments and the point charges, as well as the polarizability and the value q of the point charge. This set of electrostatic parameters was then used -unmodified- in the fit of the two-body potential to *ab initio* calculated interaction energies of water dimers. The potential parameters as obtained with this two-step procedure are given in Table III. The standard deviation of this fit is 0.52 kcal/mole for all configurations (attractive and repulsive) and 0.24 kcal/mole by considering only attractive configurations.

One important characterization of the water dimer potential surface is its absolute minimum, *i.e.* the most stable water dimer configuration. This configuration is predicted as an open form with a nearly linear hydrogen bond. The binding energy is computed as 5.18 kcal/mole at an intermolecular O-O distance of 2.97 Å. Table IV gives the full characterization of this configuration. The potential around this configuration is very flat, especially when changing the angle β (see Fig. 2). In comparing this data with experimental values, care has to be taken since the available experimental results span a rather broad range, particularly with respect to the binding energy.[97] Recent results[98,99] seem to settle around 5.3 kcal/mole, possibly 5.1 $\pm$ 0.3 kcal/mole.

A measure for the overall quality of the water dimer interaction hypersurface is the second virial coefficient of steam, an important property of water in the gas phase.[100] Although detailed information about the potential hypersurface cannot

Table III. Potential parameters of NCC (in atomic units).

2.8682660	A_{OO}
1.2390449	B_{OO}
0.1531258	A_{HH}
1.0133393	B_{HH}
41.5624830	A_{OH}
1.2227832	B_{OH}
40.0772130	A'_{OH}
1.2111490	B'_{OH}
14.3170660	A_{PH}
2.7567793	B_{PH}
51.1496400	A_{PO}
2.7194969	B_{PO}
0.5525372	value of point charge q
0.4481389	position of negative point charge along C_{2v} axis
0.8181759	position of induced dipole moment along O-H bond
4.1030772	α_s
1.6275863	α_b

be obtained by this quantity, it is, however, very sensitive to changes in the interaction hypersurface and is therefore a good test for the gross behavior of the potential. The calculated data are compared with experimental values in Table V. Since the NCC closely approximates the experimental data it can be assumed that the behavior of gaseous water can be represented by the potential.

Table IV. Characteristics of the most stable water dimer configuration.

5.18	Stabilization energy in kcal/mole
2.97	Intermolecular O-O distance in Å
4°	Angle α in Fig. 2
35°	Angle β in Fig. 2

Table V. Computed and experimental second virial coefficients of steam (cm^3/mole)

T	NCC	exptl.[100]
373 K	-596	-580
473 K	-239	-212
573 K	-127	-117
673 K	-77	-73

We now proceed to assess the extent to which the NCC potential can account for the available experimental data on the dynamic of the water molecules. To obtain the properties of liquid water, molecular dynamics simulations have been carried out in the microcanonical ensemble for a system of 512 water molecules at a density 0.997 g/cm^3 and a temperature 305 K . The Ewald summation[56,72] has been used for coulombic interactions. The polarization model uses a cutoff of 12.4 Å in combination with a switching function to suppress the energy drift. A sixth order Gear predictor corrector algorithm (see Eqs. 13 and 14) has been used to integrate the equations of motion. The time step was set to 0.5 femtoseconds and the simulation has been conducted for 32 picoseconds. First we will consider static properties and later we will continue with dynamic properties.

As already mentioned, the temperature (see Eq. 22) in the 32 ps simulation was 305.0 $\pm$ 10 K. The potential energy of the system (see Eq. 23) was calculated to be 10.653 $\pm$ 0.05 kcal/mole which may be compared with the evaporation energy of water of about 9.98 kcal/mole.[101] The energy is overestimated in the simulation by about 7%, but quantum corrections are expected to lower this value[102,103] by about 1.0 kcal/mole. The contribution of the polarization to the total potential energy is 3.9 kcal/mole, and the average dipole moment per water molecule is 2.8 D. The internal pressure, as calculated from the virial theorem (see Eqs. 24 and 25), is about -1180 $\pm$ 470 atm. However, very high deviations in this value can occur from small changes in the density of the system. This sensitivity, due to small changes in the interaction hypersurface, has been demonstrated in a previous work.[83] The heat capacity at constant volume, calculated without quantum correction (see Eq. 26), was 17.6 cal/(mole K); the experimental value is 17.9 cal/(mole K).[104]

The basic information on the equilibrium structure of water is represented by three radial correlation functions $g_{OO}(r)$ $g_{OH}(r)$ and $g_{HH}(r)$(see Eq. 27). Fig. 3 (top and middle insets) compares these distributions, obtained from the simulation, with experimental data. In view of experimental uncertainties [105] it appears that the computed radial distribution functions compare well with the experimental data [106,107]. However, the first neighborhood peak in the O-O radial distribution function is slightly shifted to shorter distances. This can be an indication that the two body part of the NCC potential can be further improved. However, this would require a new set of interaction energies of water dimers, using perhaps a more extended basis set, and a more accurate model for the dispersion energy.

The number of nearest neighbors, computed by integrating the radial distribution function up to the first minimum (at 3.3 Å), is predicted to be 4.3 water molecules in the first hydration shell with a preferred distance of about 2.8Å. Integration of the O-H radial distribution function predicts exactly 1 hydrogen bond between oxygen and hydrogen atoms of neighboring molecules with a preferred distance of 1.85Å.

The structure function can be calculated using Eqs. (28) and (29). Fig. 3 compares the computed structure function from X-ray (left bottom) and neutron (right bottom) scattering with the available experimental data.[106,108]

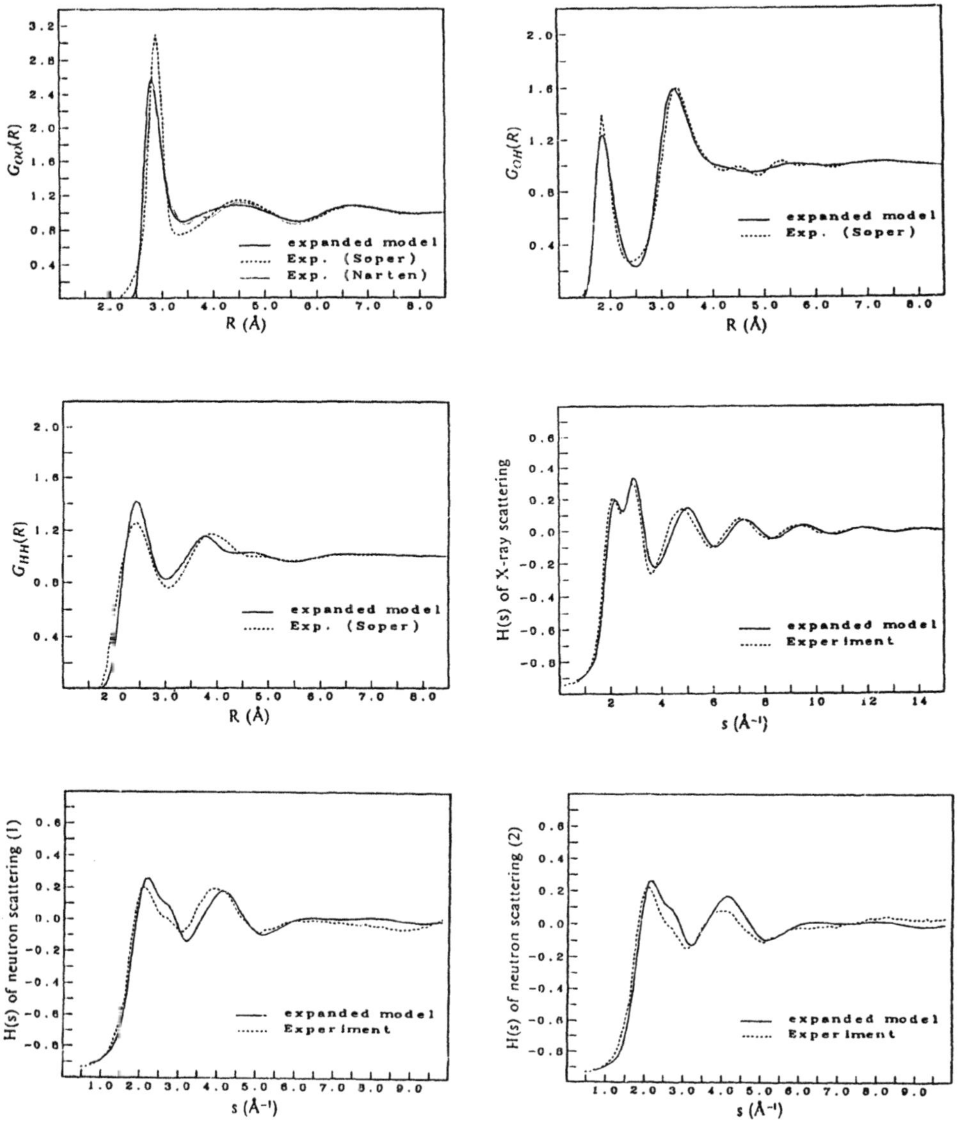

Figure 3. Top and middle insets: Comparison of the pair correlation functions obtained from NCC potential with those from experimental data. Left bottom: X-ray scattering intensity. Right bottom: Neutron beam scattering intensity.

Regarding the dynamical behavior of a single molecule in the liquid, most of the interest focuses on the translational and rotational mobility of the water molecules.

A good and experimentally accessible measure for the translational mobility is the self-diffusion coefficient, which is either computed by integrating the translational velocity autocorrelation function (see Eq. 35) or from the mean square displacements of the molecules (see Eq. 36). The velocity autocorrelation (see Eq. 34) decays rapidly and contributions after 0.6 picoseconds are almost zero. Thus setting the upper limit of the above integration to 6 picoseconds, we obtained a diffusion coefficient of $2.5 \cdot 10^{-5}$ cm^2/sec. The self-diffusion coefficient obtained from Eq. (36), using the time range between 2 and 10 ps, is $2.6 \cdot 10^{-5}$ cm^2/sec. These results are in a good agreement with the experimental value of $2.4 \cdot 10^{-5}$ cm^2/sec. [109] The determination of the diffusion coefficient of a liquid is a critical test for a potential. It has been shown that it is quite sensitive to the interaction hypersurface. [83]

A similar check for rotational mobility can be done by comparing simulated and experimental reorientation times. Experimentally, these data are indirectly available from ^{1}H-NMR relaxation measurements. In molecular ^{1}H-NMR relaxation dynamics simulations they may be estimated from the autocorrelation function of the unit vector along the intramolecular H-H axis.[58]

$$\tau_2 = \int_0^\infty < \frac{1}{2}\left(3(\mathbf{n}_{H-H}(0) \cdot \mathbf{n}_{H-H}(t))^2 - 1 \right) > dt \tag{108}$$

We determined the integral by fitting the autocorrelation function after 0.8 picoseconds to an exponential function and numerically integrating the beginning of the curve. The obtained value, $\tau_2 = 1.8$ picoseconds, has an uncertainty of probably 10%. The corresponding experimental value is 2 picoseconds.[110]

In Fig. 4, the power spectra of the translational and rotational velocity autocorrelations are displayed. These reveal the vibrational modes of single molecules in the bulk. It is interesting to note that these spectra are very similar to those obtained from the MCY two-body potential[86] and a "truncated" polarization model.[83] It appears that the effect of polarization on the vibrational modes of single molecules is rather small. The IR spectra, computed from the molecular dynamic simulation, is shown in Fig. 4. The position of the librational bands compares well with the experimental spectrum [111], which has a maximum absorption at about 700 cm^{-1}.

We now proceed to study the density correlation function and analyze the sound modes in liquid water. Based on the molecular dynamics simulation of liquid water with the ST2 potential, Rahman and Stillinger[112] were the first to predict the possibility of the existence of a high frequency sound mode with a wavelength comparable to the interatomic distance. Later, Teixeira et al[113] experimentally ver-

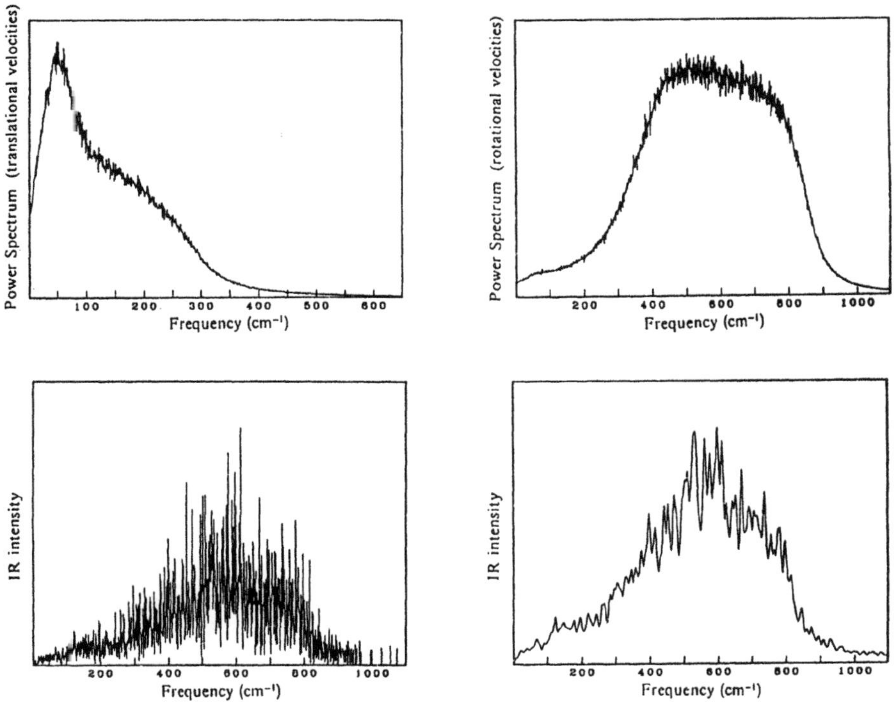

Figure 4. Power spectra for translational and rotational velocity autocorrelation functions and computed infrared spectrum.

ified its existence and argued that it is a manifestation of new collective excitation ("fast sound") propagating in the small patches of highly bonded water molecules. Rahman and Stillinger, in their spectrum of density fluctuations, identified both the ordinary sound mode and the high frequency mode; however several subsequent MD simulations [87,114,115] observed only the high frequency mode. The dynamics underlying this high frequency sound mode in liquid water remains a subject of great interest Impey et al[115] suggested that it might be due to a strong coupling between the density fluctuations and correlations in molecular orientation extending over several molecular diameters. On the basis of generalized hydrodynamics, Wojcik and Clementi[87] argued that it is a consequence of the positive dispersion of the ordinary sound mode. Ricci *et al.*,[116] in the framework of the Mori-Zwanzig formalism applied to a two component dynamical variable, found that the low frequency mode (ordinary sound) is associated with the oxygen density fluctuations, whereas the hydrogen density fluctuations are found to be governed by both modes, thus indicating that the high frequency collective motion is related to fast librations.

As discussed in section 4.3, the sound modes can be studied either from the dynamic structure factor or from a spectrum of longitudinal current correlation functions.[112,114] We will devote our attention to the analysis of the current correlation function since, as outlined in section 4.3, sound modes are more pronounced here (see Eqs. 65 to 67).

In Fig. 5 (top left) is displayed the spectral analysis of the longitudinal current correlations of the hydrogen atoms, calculated from 8192 MD configurations, for two values of q. The time interval between two consecutive configurations was 4 femtoseconds. The maximum entropy method was used to smooth the initial data. The finite size of the simulation box and the use of the periodic boundary conditions constrains the lowest wave vector accessible in the simulation (see Eqs.85 and 86). It is seen that the current correlation spectrum has three distinct peaks which correspond to the specific sound modes. For $q=0.25$, the low frequency sound propagates at a speed of 1288 m/sec and the high frequency mode travels at 3200 m/sec, while for $q=0.36$ the speed of the first sound mode is 1600 m/sec and the fast sound mode travels at a speed of 3580 m/sec. In their experiments, Texeira et al[113] found the speed of the fast sound to be approximately (3310 ± 250) m/sec. The experimental values of the low frequency sound speed vary from 1390 m/sec (Texeira and Leblond[117]) to 1490 m/sec (Matheson[118]). In addition to the low and the high frequency sound modes, the molecular dynamics simulation based on the NCC potential predicts the existence of an additional sound mode, the speed of which lies between 2290 m/sec (for $q=0.25$) and 2560 m/sec (for $q=0.36$). To investigate the dynamical origin of this additional sound mode, and to discern the q dependence of the dispersion relation of each sound mode, we are conducting much longer simulations.

The frequency spectrum of current correlations in hydrogen atoms over a much wider range of frequency is shown in Fig. 5, where the contributions arising from

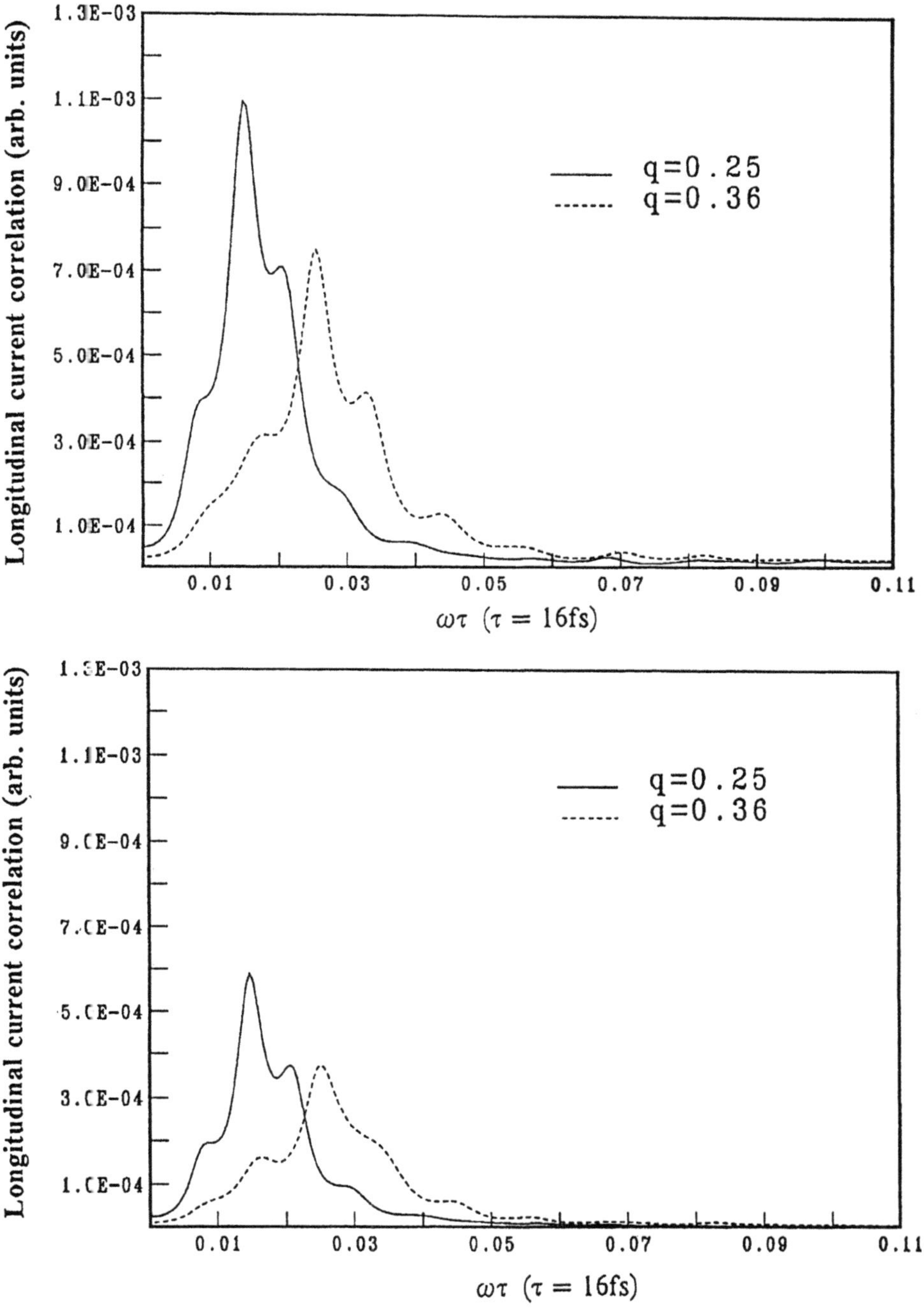

Figure 5. Temporal Fourier transform of the longitudinal current correlation function for hydrogen (top) and for oxygen (bottom).

libration motions at a higher frequency are clearly evident. In Fig. 5 we show also the spectral analysis of current correlations of oxygen atoms wherein the existence of both the low and the high frequency sound modes can be seen. Fig. 5 (top right and bottom insets) reveals that the dynamics of fast sound modes in liquid water is far more complex than thought previously. It cannot be solely due to the high frequency collective motion related to librational motion. A correct appreciation of the coupling between the density fluctuations of the atoms and the orientational dynamics of molecules might be an important step for understanding the cooperative phenomena giving rise to the fast sound mode in liquid water.

The density of states for liquid water can be obtained using Eq. (75), and is shown in Fig. 6. Two peaks are clearly seen. The first one, centered around 8 meV, is related to the flexing motion of the center of mass, while the second peak, at 57 meV, represents the libration band. The result is in qualitative agreement with the data reported by Chen and Teixeira[119] for water at a temperature −17 °*C*. In the experimental data,[119] however, the libration band is not well defined because the extrapolation $q \rightarrow 0$ is less reliable. More recently, Touken *et al*[120] reported a q dependent density of states over an energy range of 50-600 meV which covers the libration, bending and stretch vibration regions of molecular motions. Such high frequency dynamics of water can be studied only by incorporating quantum effects.

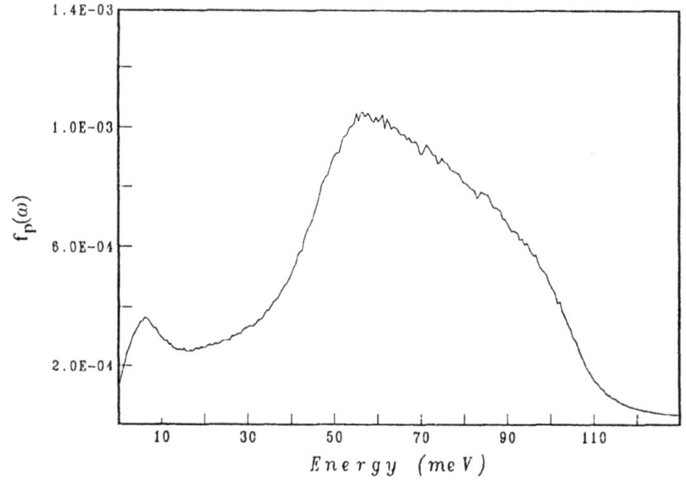

Figure 6. Density of states for liquid water.

In Fig. 7 (left) we represent the time of flight (see Eqs. 88 to 91) of cold neutrons at a scattering angle of 35° and at an incident energy of 1.236 meV. The energy transfer in the region of inelastic scattering around 300 μs/m (approximately 65 meV) corresponds to the libration mode, and the small bump around 800 μs/m can be attributed to the flexing motion of the oxygen atoms. Due to a lack of precise experimental data, we could not assess the accuracy of our time-of-flight prediction. Some experimental data on time-of-flight of cold neutrons has been described in the review article of Page[121], but it appears that these earlier experiments were not sufficiently accurate.

Lastly in Fig. 7 (right), the time variation of incoherent intermediate scattering functions (see Eq. 80) for hydrogen atoms is displayed. Since for the small values of the q vector,

$$F_{inc}(q, t) \propto e^{-Dq^2 t} \tag{109}$$

holds, this relation can be used to estimate the diffusion coefficient.[70] For $q = 0.25$ the diffusion coefficient is $2.5 \cdot 10^{-5}$ cm^2/sec which is in good agreement with the values obtained by other methods reported above.

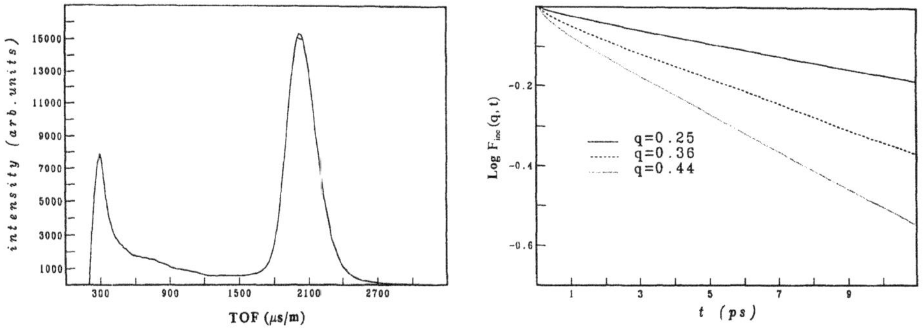

Figure 7. Left: Time-of-flight spectrum for liquid water (scattering angle = 35°, incident energy E_0 = 1.236 meV). Right: Incoherent intermediate scattering function $F_{inc}(q, t)$ for the hydrogen atoms.

In conclusion, molecular dynamics simulations based on the NCC potential confirm, once more, that the incorporation of the many-body effects due to polarization leads to considerable improvement in the prediction of the physical properties of the liquid water. For static structure factors, radial pair correlation functions, evaporation energy, heat capacity, IR spectra, diffusion coefficient, NMR relaxation time, incoherent and coherent neutron scattering, excellent agreement with the existing experimental data has been found.

6.2. MD Simulation with a Flexible Water-Water Potential

We have been discussing water-water interaction potentials in which the geometry of the individual water molecules is fixed at the experimental value. In reality, however, water molecules are continually vibrating. This can usually be neglected because vibrational motions are occurring at a much faster rate than the translational and rotational motions of the molecules. Thus most properties are affected only in a subtle way by the vibrations. In this section, we will report some results obtained by taking the vibrational motions into account.

The potential to be used for this purpose is a hybrid of two *ab initio* potentials: one for the inter- and the other for the intra-molecular part. The inter-molecular potential is the MCY potential[76], which is described by a subset of the terms in Eq. (98), whereas the intra-molecular potential is adopted from the results obtained by many body perturbation calculations.[122] The latter potential is expressed in terms of three internal coordinates of water, changes in the two OH bond lengths $\delta_i = R_i - R_e$ and the HOH bond angle $\delta_3 = R_e(\theta - \theta_e)$, up to quartic terms:

$$
\begin{aligned}
U_{intra} = {} & \frac{1}{2} f_{RR}(\delta_1^2 + \delta_2^2) + \frac{1}{2} f_{\theta\theta}\delta_3^2 + f_{RR'}\delta_1\delta_2 + f_{R\theta}(\delta_1 + \delta_2)\delta_3 \qquad (110) \\
& + \frac{1}{R_e}\Big[f_{RRR}(\delta_1^3 + \delta_2^3) + f_{\theta\theta\theta}\delta_3^3 + f_{RRR'}(\delta_1 + \delta_2)\delta_1\delta_2 \\
& + f_{RR\theta}(\delta_1^2 + \delta_2^2)\delta_3 + f_{RR'\theta}\delta_1\delta_2\delta_3 + f_{R\theta\theta}(\delta_1 + \delta_2)\delta_3^2 \\
& + f_{RRRR}(\delta_1^4 + \delta_2^4) + f_{\theta\theta\theta\theta}\delta_3^4 + f_{RRRR'}(\delta_1^2 + \delta_2^2)\delta_1\delta_2 \\
& + f_{RRR'R'}\delta_1^2\delta_2^2 + f_{RRR\theta}(\delta_1^3 + \delta_2^3)\delta_3\Big] \\
& + \frac{1}{R_e^2}\Big[f_{RRR'\theta}(\delta_1 + \delta_2)\delta_1\delta_2\delta_3 + f_{RR\theta\theta}(\delta_1^2 + \delta_2^2)\delta_3^2 \\
& + f_{RR'\theta\theta}\delta_1\delta_2\delta_3^2 + f_{R\theta\theta\theta}(\delta_1 + \delta_2)\delta_3^3\Big]
\end{aligned}
$$

The equilibrium bond length R_e and bond angle θ_e predicted from the theoretical calculations are 0.9576Å and 104.59°, respectively, to be compared with the experimental values of 0.9572Å and 104.52°. The calculated force constants are, in 10^5 dyne/cm,

$f_{RR} = 8.5120$; $f_{\theta\theta} = 0.7987$; $f_{RR'} = -0.0967$; $f_{R\theta} = 0.2732$;

$f_{RRR} = -9.497$; $f_{\theta\theta\theta} = -0.1268$; $f_{RRR'} = -0.031$; $f_{RR\theta} = -0.034$;

$f_{RR'\theta} = -0.512$; $f_{R\theta\theta} = -0.1565$;

$f_{RRRR} = 14.0$; $f_{\theta\theta\theta\theta} = -0.032$; $f_{RRRR'} = -0.05$; $f_{RRR'R'} = 0.06$; $f_{RRR\theta} = -0.2$;

$f_{RRR'\theta} = 0.1$; $f_{RR\theta\theta} = -0.08$; $f_{RR'\theta\theta} = 0.35$; $f_{R\theta\theta\theta} = 0.104$.

The method of hybridizing the two and its consequences have been discussed in detail in the original paper.[114] There it has been shown, for example, that the infrared intensities for the stretching vibrations of the water molecule should be greatly enhanced due the presence of inter-molecular binding. Experimentally, the absorption intensity of the O-H stretching increases by a factor of ten when the O-H group forms a hydrogen bond. This means that the dipole derivative with respect to the O-H bond length should increase by a factor of three, in good agreement with the prediction of the hybrid potential. This hybrid potential is referred to as the MCYL (Matsuoka-Clementi-Yoshimine-Lie) potential.

Both MCY and MCYL potentials predict that the most stable dimer form is the open form with a nearly linear hydrogen bond. The bonding energy predicted by the flexible MCYL potential is 0.07 kcal/mole lower. The bond angles of both interacting water molecules are found to decrease slightly with respect to the isolated water molecule, and the angle change should be discernible experimentally. The O-H bond, which participates in the H-bonding, is predicted to be 0.01 Å longer.

This potential was built in to a molecular dynamics computer simulation by Lie and Clementi.[114] However, due to the presence of atomic motions, we do not decompose the motion of the molecules into translational and rotational motions, as was done previously. Instead, each atom is treated as a particle subjected to forces calculated from inter-molecular, intra-molecular potentials, and reaction fields. Newtonian equations of motion are then solved individually for each atom by the Gear's predictor-corrector method (see Eqs. 13 and 14). Periodic boundary conditions are imposed. We will discuss some of the results obtained by MD simulation with 343 water molecules at 306 K. Six-th order Gear's predictor-corrector method with a time step of 0.15 femtoseconds was employed.[114] A very short time step was needed here because we are dealing with vibrational motions which have a characteristic time scale on the order of femtoseconds.

Several static and dynamical properties have been computed from the potential and compared carefully with experimental results. The heat capacity (see Eq. 26) was found to be 17.6 cal/(mole K) after quantum corrections, to be compared with the experimental value of 17.9 cal/(mole K). Quantum corrections are important here because we are dealing with motions of very high frequency. Pressure was improved a little bit from the MCY potential, but is still much too high in comparison with the experimental value.

The computer simulation produced substantial differences between the gas and the liquid phase geometries of the water molecule, which are not predicted by using potentials with fixed water geometry. The average O-H bond length is predicted to be greater in the liquid by 0.018 Å whereas bond angle smaller by ~1°. These have been confirmed indirectly by the experiments.[108]

The radial distribution functions (see Eq. 27) for O-O, O-H, and H-H atom-atom pairs were compared with experimental X-ray and neutron scattering data[106,108] and the overall fit of the MCYL potential to improved experimental data was found to be excellent, considering the experimental uncertainties.[105] Comparing with the results of MCY, the first shell water molecules are more localized in the MCYL potential.

In dynamical properties, such as the velocity auto-correlation functions (see Eq. 34), the MCYL potential introduces a vibration into the equivalent obtained from the MCY potential. By Fourier transforming the velocity auto-correlation functions of H and O atoms, good agreement was obtained with available Raman and infrared data regarding the frequency shifts from the gaseous to the liquid state of the water. As a matter of fact, the MCYL potential gives among the existing models for vibrating water the best agreement with infrared and Raman data. Not only that, Lie and Clementi predicted the existence of a combinational vibrational band at 519 meV (4188 cm^{-1}), which has subsequently been observed two years later at 525 meV by inelastic neutron scattering by Toukan, *et al.*[120]

Compared with experimental data, the hydrogen atoms in the MCYL potential seem to diffuse too slowly, so that the hydrogen bonding in the MCYL potential may be too strong. The diffusion coefficient (see Eq. 35), 2.0×10^{-5} cm^2/sec, from the computer simulation is too low compared with the experimental value of 2.4×10^{-5} cm^2/sec.

The dipole autocorrelation function and the dielectric constant (see Eqs. 30 and 31) were both computed in the simulation and the former had virtually no vibrational structure, being dominated by libration. Analysis of the dynamics of this autocorrelation function gives an estimate of about 0.03 to 0.07 ps for the lifetime of the H bond. The dielectric relaxation time was estimated as 6.1 ps, as compared with 7.4 ps estimated experimentally. The dielectric permittivity both from the MCYL and MCY potentials is too low by about a factor of three in comparison with the experimental value.

Static and dynamic structure factors and current fluctuations were also computed by Lie and Clementi, and the original article contains detailed comparisons which show overall that the MCYL model reproduced a high frequency sound mode first predicted by Rahman and Stillinger. Transverse and longitudinal current fluctuations were also computed to investigate shear waves and shear viscosity.

This concludes our first two applications examples. The next step to be taken in the simulations of water via *ab initio* potentials is rather evident: as stated before, to start with we need a large data set of *accurate* energies to recompute the potential given in Eq. (98). Once this has been accomplished, we shall return to the potential as described in Eq. (110), however coupling more extensively the two molecules and using the new potential, thus discarding MCY. We are confident that this approach will ensure much higher reliability in reproducing and predicting a broader spectrum of properties for liquid water than would the use of any empirical potential; this has been our goal since the early seventies and it appears we can finally reach it.

6.3 BPTI Simulations in Vacuo and in Solution

In this section we report on MD results and simulated neutron scattering properties for the Bovine Pancreatic Trypsin Inhibitor (BPTI) protein. For this system two simulations have been carried out: one assumes the protein *in vacuo* and the other in solution.

We recall that BPTI consists of 58 residues, ten of which carry a positive charge and four a negative one, yielding a net charge of +6 electrons, which, under experimental conditions are neutralized by counterions. In order to have a fair comparison between the *in vacuo* and the *in solution* experiments we decide to neutralize the protein by evenly distributing the extra charge among all protein's atoms (454 heavy and 438 hydrogen atoms). The initial geometry for both simulations has been generated by adding to the X-ray coordinates[123] the hydrogen atoms according to standard geometrical rules.

The intramolecular model potential used in the present two simulations is the one developed by Weiner *et al.*[20] and consists of the standard three terms:

$$E_{tot} = E_{bonded} + E_{non-bonded} + E_{H-bond}$$

where

$$E_{bonded} = \sum_{bonds} K_R(R - R_0)^2 + \sum_{angles} K_\theta(\theta - \theta_0)^2 + \sum_{dihedrals} V_n\,[1 + \cos(n\phi - \gamma)]$$

$$E_{non-bonded} = \sum_{i<j} [\frac{B_{ij}}{R_{ij}^{12}} - \frac{B_{ij}}{R_{ij}^{6}} + \frac{q_i q_j}{\varepsilon R_{ij}}]$$

$$E_{H-bond} = \sum_{H-bonds} [\frac{C_{ij}}{R_{ij}^{12}} - \frac{D_{ij}}{R_{ij}^{10}}]$$

In both simulations we used a value of unity for the dielectric constant parameter, ε, present in the electrostatic term of the *non-bonded* interactions, and in the bonded interactions we did not include the stretching term. The stretching motions, being very fast, are assumed uncoupled with the slower protein motions.

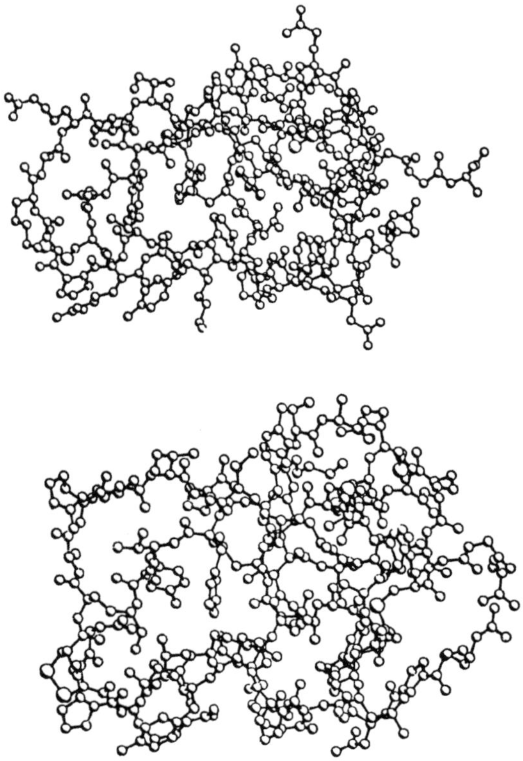

Figure 8. View of the X-ray structure of BPTI (top), and of an instantaneous conformation from the MD simulation (bottom).

For the *in vacuo* simulation we considered all possible non-bonded interactions, whereas, for the one in solution, we adopted a cutoff radius of 9Å for both the non-bonded and the intermolecular (water-water, water-protein) interactions, and we corrected the electrostatic terms with the Ewald summation technique.[56] The water-water interaction was modelled by the MCY potential[76] and the water-protein interaction by *ab initio* potentials.[124-126] In both simulations the SHAKE[53] algorithm was used for the bond constraints, since no stretching motions are allowed.

The *in vacuo* simulation was carried out with a 0.5 fs time step. Good energy conservation has been observed during the entire run. The protein has been equilibrated for 60 ps both by rescaling the velocities every 200 steps, and whenever the temperature exceeded the 310 K target value. Additional 35 ps of simulation have been carried out without velocity rescaling since the average temperature remained essentially constant around 310 K. For this part of the simulation we dumped onto disk the system history and accumulated the averages of the statistical properties. The average root mean square (RMS) displacement from the X-ray structure is relatively high, 2.7 Å, however in agreement with previously

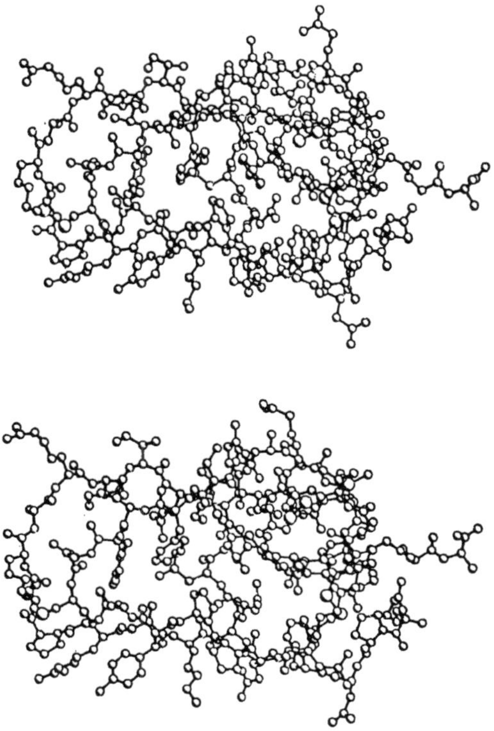

Figure 9. View of the X-ray structure of BPTI (top) and of an instantaneous conformation from the MD simulation in solution (bottom).

reported results.[127,128] In Fig. 8 we compare the X-ray structure with an instantaneous configuration from our simulation. As previously found for *in vacuo* simulations, the charged residues have the highest RMS values; for this simulation we found RMS values around 5 Å for the arginine groups. A more detailed analysis of the simulation will be published elsewhere.[129]

The *simulation in solution*[129] was carried out as a five step procedure. As a first step we equilibrated for a few picoseconds 3000 water molecules confined in an orthorombic box of size 52.8 × 40.22 × 40.22 Å^3. In the second step the protein (with the X-ray coordinates for the heavy atoms and the added hydrogen atoms) was placed in the middle of the simulation box with the smallest inertia moment parallel to the largest dimension of the box. Then, we removed from the system those water molecules with the oxygen atom within a 1.6 Å distance from any protein atom. The resulting system is composed of 2677 water molecules and 892 protein atoms all of which are at least 6 Å away from the box walls. In the third step a 1 ps simulation was performed to allow the water molecules to equilibrate in the presence of the protein (however constrained at its initial conformation). Finally (fourth step), both water and protein atoms were allowed to move. The equilibration was carried out for 33 ps; during this period the velocities were rescaled using the same procedure as for the *in vacuo* case and the mass of the hydrogen atoms was fixed to 11 u.m.a.. This choice allowed a larger than usual time step of 3 fs with still good energy conservation. In this equilibration phase, the average RMS increased very quickly during the first few picoseconds of the simulation, then started to oscillate around the value of 1.25 Å. In Fig. 9 we report the initial and the final conformation of BPTI for this simulation. Once the system was equilibrated, we started a simulation with a time step of 0.8 fs, with a value of 2 u.m.a. for the mass of the hydrogen atoms of the water molecule (D_2O). The simulation was carried out for 52 ps without velocity rescaling since the average temperature remained constant around 300 K. The stored data was then used to carry out the neutron scattering analysis described below.

Neutron Scattering Properties

It should be mentioned that the simulation has been performed in D_2O instead of H_2O because the experimental situation of inelastic neutron scattering is the one where the bulk water of the sample is replaced by D_2O to eliminate the background of the incoherent scattering of H_2O. The density of states for the system has been computed using Eq. (75). The computations has been done for both the *in vacuo* and in solution simulations. Because of the unrealistic results obtained for the *in vacuo* simulation, in the following we will discuss only on the results from the "in solution" simulation. For this one the analysis has been done twice. In the first calculation all the system atoms, including those belonging to the solvent, are taken into account, in the second one the solvent atoms are excluded by simply assigning a zero value to their scattering lengths. The results are depicted in Fig. 10 (top inset), where the solid line corresponds to the first computation (the one which includes D_2O) and the dashed line to the second one (the one without D_2O). As it is evident from the figure, inclusion of the solvent atoms enhances the intensity of the curve but does not change its shape. Experimental data in the range 0 to 150 wavenumbers have been recently published[67] for a BPTI powder. The experimental results, plotted in Fig. 10 as squares, nicely follow the computed dashed line. Experimentally it is difficult to determine which hydrogens are contributing to this region of the spectrum, whereas from a computational view point it is a simple task. As explained previously for including or not the solvent atoms, it is enough to assign a zero value to the scattering length of the atoms belonging to those particular groups that one wishes to exclude. We tried to analyze the contribution to the density of states arising from particular groups and this is reported in the bottom inset of Fig. 10. The curve labeled A) corresponds to the dashed curve of the top inset of Fig. 10: namely the computation without solvent. Curve B) reports the results obtained for the system with the CH_3 groups switched off; curve C) those where the CH_3 and CH_2 groups have been switched off; curve D) those where CH_3, CH_2 and CH groups have been switched off, and finally in curve E) all the hydrogens have been excluded. It is clear from the figure that no particular group is contributing more than others to the overall behavior, at least in this region of the spectrum.

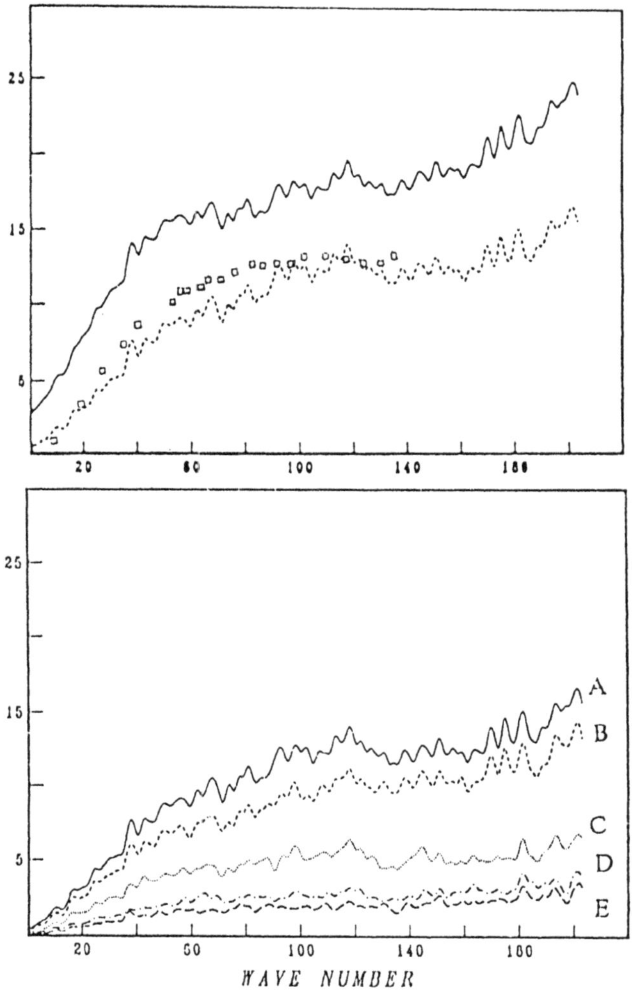

Figure 10. Density of states for BPTI in D_2O solution (arbitrary units). Inset a) Solid line: including solvent; dashed line: excluding solvent; squares: experimental values. Inset b) line A: same as dashed line in inset b; line B: excluding CH_3 groups; line C: excluding CH_3 and CH_2 groups; line D: excluding CH_3, CH_2 and CH groups; line E: excluding all hydrogens. (See text for more explanation).

In the top inset of Fig. 11 we report the computed time-of-flight spectrum (see Eqs. 88-92) at a scattering angle of 66°. As in the previous case, two computations have been done: with and without solvent. Again, in the top inset of Fig. 11 we see the same tendency as in Fig. 10, namely inclusion of the solvent (full line) enhances the intensity but does not change the shape of the curve.

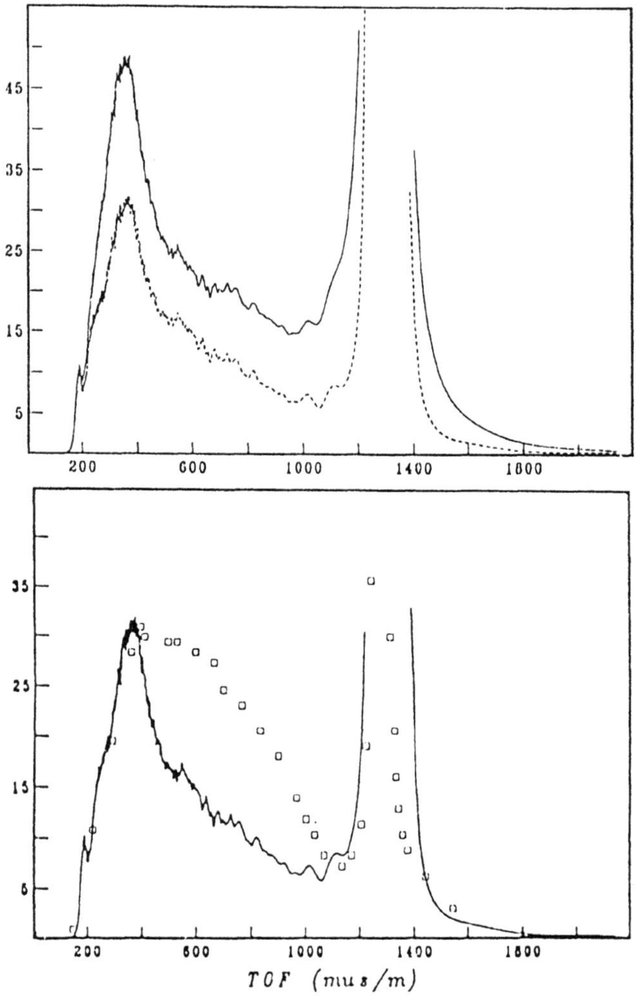

Figure 11. Time-of-flight spectrum for BPTI at $\Theta = 66°$. Top inset: Solid line: including solvent; dashed line: excluding solvent. Bottom inset: Solid line refers to the MD simulation excluding the solvent and the squares to the experimental values.

The bottom inset of Fig. 11 compares the results from the MD simulation with the available experimental data.[66] It should be mentioned that the quasi-elastic lines do not match since the MD simulation was not long enough to achieve the experimental resolution in energy/frequency. In the overall our simulated time-of-flight spectrum follows in a reasonable way the shape of the experimental one; less satisfactory is the agreement for the one computed from normal mode analysis.[66,130,131]

6.4 Water Simulation in a Crystal of Myoglobin

In the previous sections, the properties of water, as a pure liquid, have been analyzed. In the following, we will focus on the role of water as a solvent. When appropriate, comparison with the liquid water case will be made and discussed.

In recent years, due to the availability of supercomputers, computer simulations have become a powerful tool for studying structure and dynamics of biological systems. Available theoretical works on the subject have focused their attention mainly on the structural and dynamical characterization of macromolecules of biological interest, often neglecting the role of water as a solvent.

Numerous protein simulations have been done *in vacuo*, and systematically, high mean-squares displacement (MSD) values have been found.[132-134] Such findings are even more discouraging if one considers that these studies can probe only fast vibrational disorder, which is the only effective contribution to protein motions within the time generally spanned during simulations (*i.e.*, tens to hundreds of picoseconds). To the computed MSD one should add a static contribution arising from lattice disorder and a dynamical contribution due to the conformational motion of protein segments. In most simulations, however, the lattice disorder is totally neglected since the system is always assumed to be perfectly periodic under normal boundary conditions. Concerning the conformational diffusion, there are experimental evidences[135-138] that such kinds of motions generally take place in a nanosecond time scale, a time which is not yet accessible to computer simulations.

For the last decade we have called attention to the need to simulate biological systems not *in vacuo* but in their medium. Previously our interest was often in nucleic acids, but it is now extended to proteins. It is now becoming accepted that the solvent role in protein simulations is essential.[128,139-141] In the MD studies of BPTI, [129] discussed in the previous section, the agreement between experimental and computed protein mobility was dramatically improved by also including the solvent in the model. The main effect of the solvent, in MD simulations of hydrated proteins, with respect to the case *in vacuo*, is to slow down or even suppress collective motions of the side chains in the picosecond time domain. It is, therefore, reasonable to expect that a natural and straightforward approximation for investigating protein hydration at the structural level would simply consist in keeping the protein fixed. Such simulation model implies two assumptions: 1) In the picosecond time scale the protein is frozen in one particular sub-state, and no direct coupling between fast water dynamics and slow conformational diffusion can take place; 2) The vibrational modes of the protein (stretching of bonds, bendings, torsions, *etc.*), although their time scale dynamics corresponds to that of the solvent, have too small amplitudes to modify substantially the potential hypersurface on which water molecules are constrained to move.

In this section, results obtained from a molecular dynamics simulation of the water in the hydrated crystal of myoglobin[142] are presented. The structural and dynamical properties of water have been computed and compared, when available, with experimental data. A detailed analysis of the solvent has been carried out and correlation, at the structural level, between water connectivity at the protein surface and conformational diffusion of protein segments is discussed.

Myoglobin crystallizes as a monoclinic lattice (space group $P2_1$) with two proteins per unit cell in general equivalent positions (x,y,z) and (-x,y+1/2,-z). The coordinates of the 1261 heavy atoms of the protein were provided by a recent X-ray refinement.[143] Hydrogen atoms were introduced according to geometric considerations yielding a total of 2532 atoms per protein. 856 water molecules were introduced in the unit cell, yielding a crystal density of 1.2705 g/cm^3..

The protein-water interaction is represented by *ab initio* atom-atom potential functions,[124-126] and the water-water by the MCY potential.[76] The integration of the Newton-Euler equation was carried out in the (N,V,E) ensemble adopting a fifth order Gear algorithm for the translational motion. Rotational motion was treated using a sixth order integrator for the quaternions, in conjunction with a second order quaternion method for recasting the angular velocities at each time step.[51]

Long range effects due to electrostatic interaction were corrected with the Ewald sum procedure. In the present case, since we are dealing with a lattice where the solvent connectivity also shows a high level of periodicity (about half of the water molecules in the unit cell have been detected by X-ray crystallography[143]), we may confidently expect the perfect lattice approximation to be fairly reliable and physically convincing.

The simulation run was carried out for about 50 picoseconds, with a time step of 0.5 fs. The first 20 picoseconds were used for system equilibration, and the remaining 30 ps were used for statistical analysis. With this choice of the integration time step, during the last 30 picoseconds the total energy was conserved with a maximum fluctuation of 0.001 kcal/mole per ps and the velocities were never rescaled. The average temperature and average potential energy were 304.4 $\pm$ 8.15 K and -17.46 ± 0.03 kcal/mole, respectively.

Although the structure and dynamics of myoglobin has been widely studied in an enormous amount of experimental works,[144] still few experimental data are available on the hydration pattern both in the crystal and in solution. In an X-ray study by Takano,[145] 82 water molecules per protein were individually placed. Phillips[146] and Parak,[143] in two distinct X-ray refinements on deoxy and met-myoglobin, found about 160 water molecules per protein, many of them having high liquidity factors and low occupancies. In none of these studies, however, did the inferred model for the solvent yield a completely satisfactory agreement with the observed scattering intensities.

The static structure factors can be computed from the simulation data using the following equation

$$F(\mathbf{k}) = \sum_{j=1}^{NATOMS} f_{jT} \exp(i\mathbf{k} \cdot \mathbf{r}_j) + \frac{1}{nt} \sum_{i=1}^{nt} \sum_{j=1}^{NSOLV} f_j \exp(i\mathbf{k} \cdot \mathbf{r}_{ij}) . \qquad (111)$$

The first part of the equation is the contribution of the protein atoms to the overall scattered radiation. The temperature form factors f_{jT} can be computed from tabulated atomic form factors f_j and Debye-Waller factors **B** obtained from X-ray studies, according to the relations

$$f_{jT} = f_j \exp[-B(\sin\theta/\lambda)^2] \; ; \qquad 2\sin\theta/\lambda = |\mathbf{k}| \; , \qquad (112)$$

where θ and λ are, respectively, the scattering angle and wavelength of the incident beam, and **k** is a vector of the reciprocal lattice. The second part of Eq. (111) represents the contribution to the structure factors due to the solvent, *nt* being the number of time averages considered within the 30 ps time span. In Fig. 12, the computed (dashed line) and the experimental (solid line) structure amplitudes, plotted versus $\sin(\theta/\lambda)^2$, are compared and satisfactory agreement can be observed. In order to assess the water contribution to the overall scattering, we also report in the same figure (dotted line) the amplitudes obtained when only the protein atoms are considered [*i.e.*, the contribution due to the first part of Eq. (111)]. It is evident that the role of the solvent, in determining the shape of the intensity distribution at low resolution, is essential.

In Fig. 13 (right inset), radial distribution functions (RDF) and relative coordination numbers (left inset) versus distance are reported and compared with those obtained in an earlier calculation for bulk water at 298 K.[26] The curve marked with the legend 'o(w)-o(w)' refers to oxygen-oxygen water RDF in the myoglobin crystal; for 'o(w)-o(w) + p' also the heavy atoms of the protein have been included in the computation. The curve 'o(w) - p', which represents the radial distribution of protein heavy atoms with respect to a solvent molecule, can be analytically obtained by weighted difference of the two previous RDFs.

Concerning the solvent-solvent RDF (o(w)-o(w)), the position of the first peak, corresponding to the first solvation shell radius, shows no significant differences when compared to the bulk case. The enhancement of the intensity, with respect to bulk, is an artifact of the computation due to the inhomogeneity of the water density in the unit cell. For curves 'o(w)-o(w)+p' and 'o(w)-p', the first peak is shifted towards shorter distances, which is an indication of strong interaction between solvent and protein. In the right inset of Fig. 13, the coordination numbers, obtained by integrating the corresponding RDF's, are reported. The bulk water shows a well defined coordination number of five at about 3.5 Å. For the 'o(w)-o(w) + p' curve, it is hard to determine unambiguously a first shell coordination number, since the behavior is almost linear between 2.9 and 3.5 Å. However, pentacoordination is reached at 3.28 Å; at this distance the average repartition of the four coordination sites of a solvent molecule is found to be 2.8

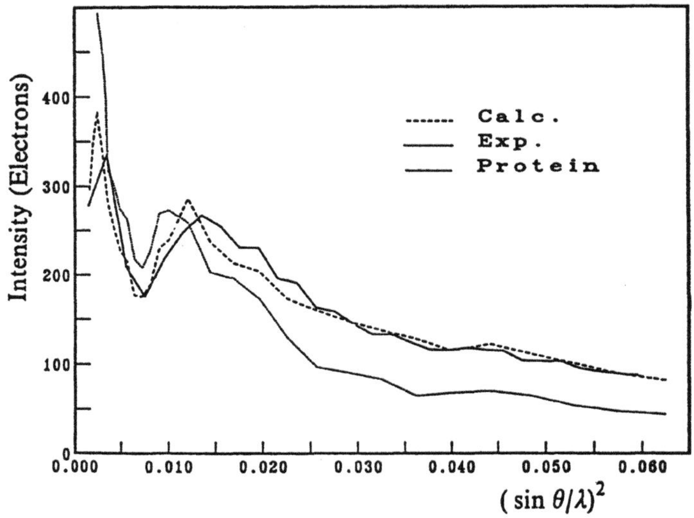

Figure 12. Comparison between computed and experimental structure factors of myoglobin in crystal. The dotted line has been obtained by considering only the protein atoms in the unit cell using experimental Debye-Waller factors.

waters (o(w) - o(w)) and 2.2 protein atoms (o(w) - p). The high coordination of water molecules to protein atoms, at short distances, indicates that there has to be a strong packing of solvent at the protein surface.

Some insights on the water diffusional regime can be obtained from the computation of the velocity autocorrelation function (VAF). In Fig. 14, the VAF for the water in the myoglobin crystal is compared with that computed for bulk water.[85] The average time for sign reverse, in the case of myoglobin-water, is faster with respect to bulk (50 fs compared to 70 fs), and the curve lies for a much longer time below zero, suggesting that diffusional motions are generally strongly hindered.

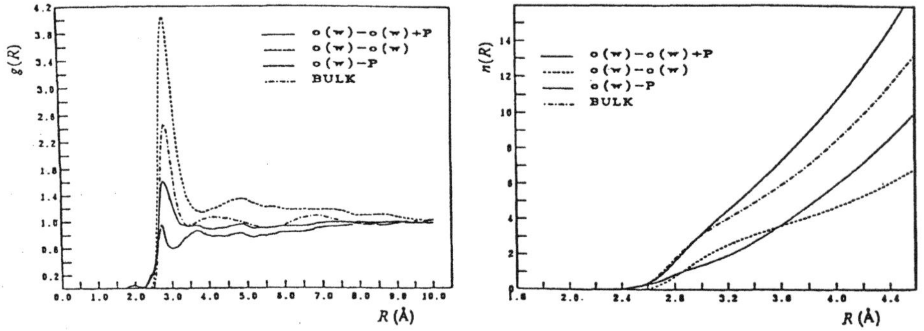

Figure 13. Water-water and water-protein radial distribution functions (left) and coordination numbers (right). See text for details.

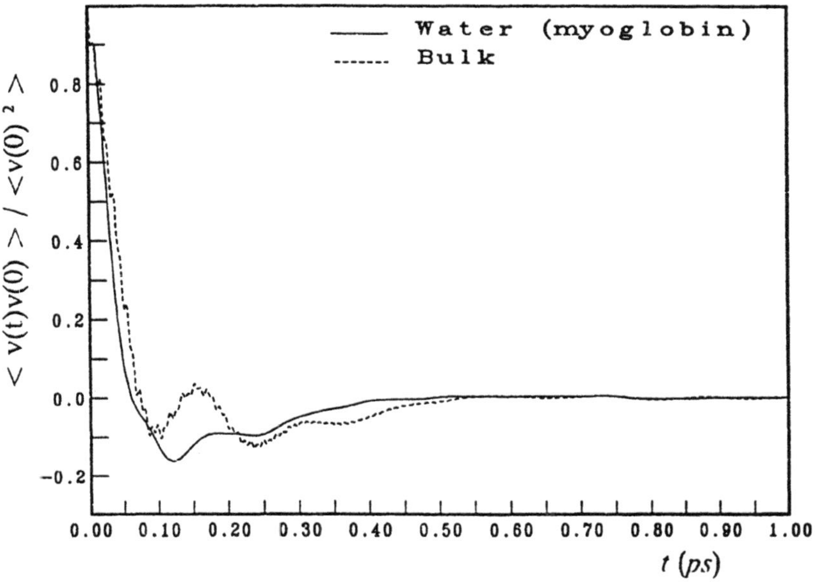

Figure 14. Velocity autocorrelation functions of bulk water and water in the myoglobin crystal.

This behavior was expected, since around 40% of the solvent molecules in the unit cell are visible in X-ray electron density maps and therefore do not diffuse.

The translational diffusional regime can also be investigated by monitoring the mean square displacements (MSD). For liquids, this quantity is a linear function

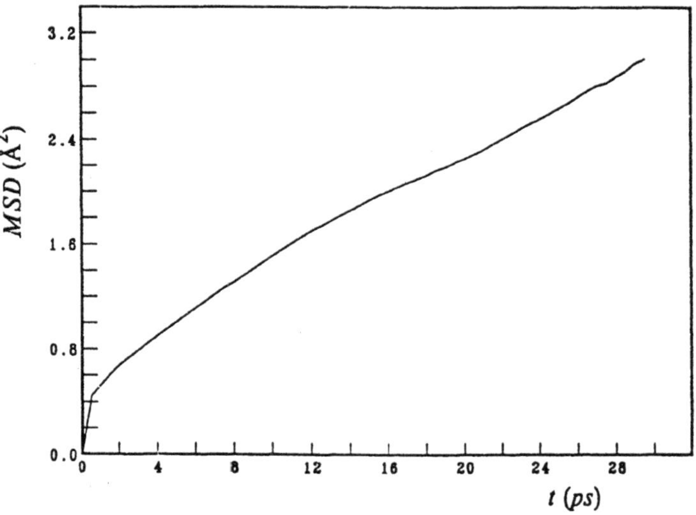

Figure 15. Mean square displacement of the solvent in myoglobin crystal as a function of time. The overall trend is typical of a liquid system.

of time and the limiting value of the slope is proportional to the self-diffusion coefficient [see Eq. (36)].

The mean square displacement plotted versus time (Fig. 15), shows a liquid-like behavior. As will be discussed in the next section, the almost linear MSD is due to the fact that a rather high percentage of solvent molecules is able to retain, even in the crystal, a high level of translational mobility.

Local structural and dynamical properties

Computation of significant local structural and dynamical properties is not straightforward since it is difficult to define sub-ensembles for which meaningful statistical averages have to be computed. We partitioned the water molecules among four classes. Those molecules which remain, during the whole simulation length, within a 3.5 Å shell from any heavy atom of charged, polar, and neutral residues, are defined to be members of class I, II and III, respectively; water molecules can be shared among classes I, II, and III. In the fourth class are gathered all the remaining molecules, namely those which either are for some time more than 3.5 Å away from any residue, or are exchanging among the three different classes defined above; members of class IV cannot be simultaneously members of any other class.

In Table VI, some general properties of each class are summarized. As expected, water-protein interaction energies are stronger for hydrophilic groups, and, although the hydrophobic residues are the most numerous, the highest level of solvation is found for charged and polar residues. About 25% of the water molecules are in class IV, they have the weakest interactions with the protein.

Water molecules in class I, II and III have low MSDs, and generally appear to be stuck at the protein surface, as it can be also observed from the typically solid-like behavior of the MSD time dependency (see Fig. 16). Water molecules in class IV show, quite surprisingly, a completely different diffusional regime, typical of a liquid system. The MSD time dependence is almost linear and the self-diffusion coefficient is about one third of that of bulk water. A similar behavior is expected to occur in solution, where the distances between water and protein atoms can be very large. In the case of the myoglobin crystal, at least 2/3 of the volume[147] of the unit cell is occupied by the two proteins and the largest distance of any solvent molecule from the protein atoms can never exceed 6.0 Å. Nevertheless, it appears that the water molecules in class IV, despite the non-negligible interaction energy with the protein (−4.2 kcal/mole), retain a high translational mobility.

The average time of solvent exchange between water molecules in the first hydration shell and the mobile layer (*e.g.*, members of class IV) can be monitored by computing, for classes I, II, and III, the time dependent correlation function

Table VI. Properties of the water at charged, polar and neutral residues.

property	class I	class II	class III	class IV
Residue type	charged	polar	neutral	
Number of residues	44	32	77	–
Number of water molecules	385	138	150	337
MSD (Å^2) after 15 ps	0.40	0.37	0.31	4.24
Water-protein energy (kcal/mole)	−23.4	−15.7	−7.3	−4.2

$$n(t) = \frac{1}{nt}\sum_{j}\sum_{i=1}^{nt} Q_j(t_i, t) \tag{113}$$

where $Q_j(t_i, t)$ is a property of molecule j. It assumes the value 1 if the molecule remains within the defined solvation shell (3.5Å) during the time interval $t_i \leq t \leq t_i + t$, and 0 otherwise. The inner sum denotes time averages over the simulation time span. For the correlation time $t = 0$, $n(t)$ is simply the average coordination number of the solvated residues; at short times the function decays exponentially and the residence time in the solvation shell is given by the exponential decay time.[148] The computed residence times increase with the residue's hydrophilicity. As it is reasonable to expect, water molecules close to charged residues exchange with the surrounding water molecules less often than those close to polar or neutral residues.

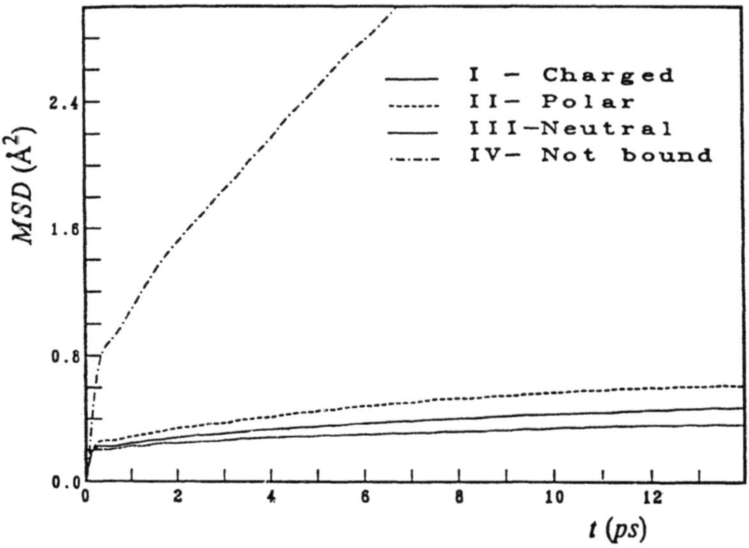

Figure 16. Mean square displacements for water molecules belonging to the four classes (see text for details).

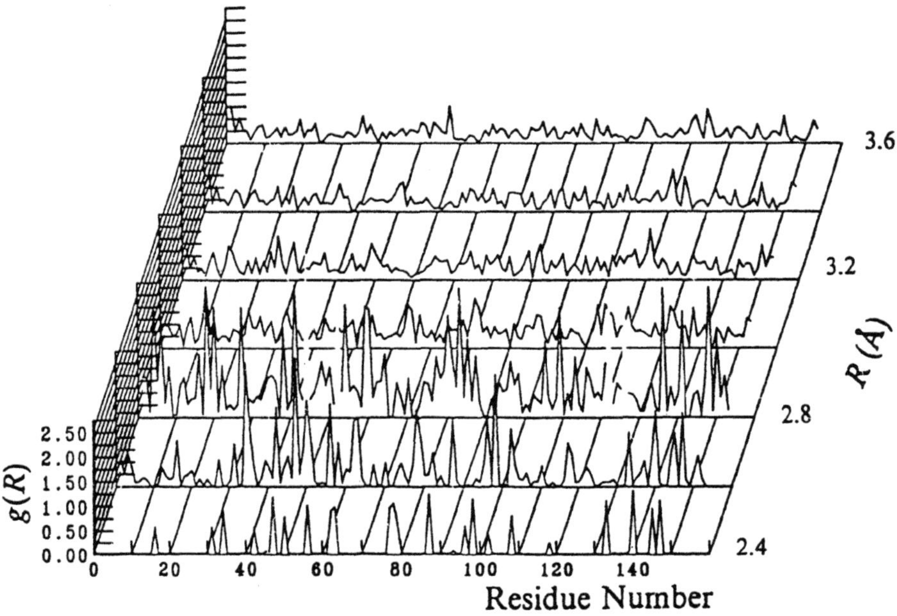

Figure 17. Residue-water radial distribution functions are plotted as a function of the residue numbers at selected distances.

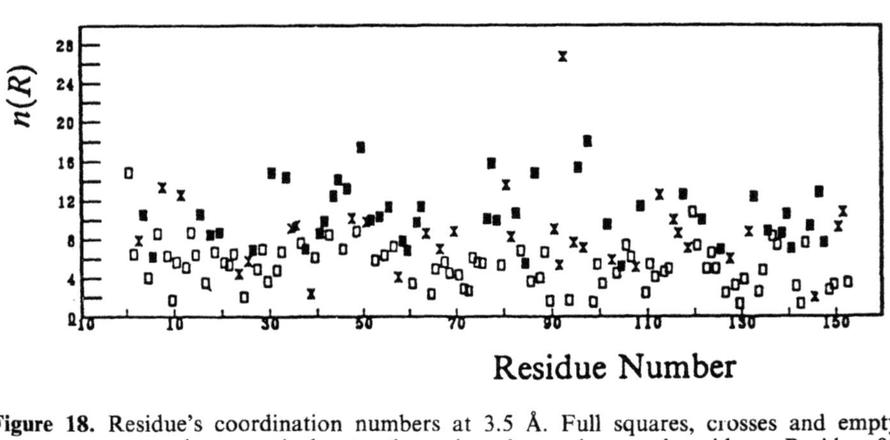

Figure 18. Residue's coordination numbers at 3.5 Å. Full squares, crosses and empty squares correspond, respectively, to charged, polar and neutral residues. Residue 93 includes the haem group.

As suggested by measurements of dielectric relaxation rate in the myoglobin crystal, the solvent structure and dynamics seem to play an important role in determining protein conformational movements. A detailed analysis on this subject can be found elsewhere,[142] here we report some schematic observations. At the structural level, we find that the residues which are embedded in regions of low conformational disorder are also poorly solvated. The number of water molecules hydrating a residue and their water-protein interaction energy grow steadily with increasing Debye-Waller factors of the solvated residues. Regions of the protein with high conformational disorder seem to be located in correspondence with the deeper minima of the water-protein potential hypersurface. At the same time computed values of MSD are generally higher when water molecules are solvating a mobile segment.

Fine details of hydration patterns can be revealed by computing residue-water oxygen RDFs. Similar calculations are of interest since few data about solvent arrangements around protein are available. The results, for solvent-residue radial distribution functions, averaged over the two proteins in the unit cell, are presented in the diagram of Fig. 17. The radial distribution functions versus residue number (x-axis) are plotted at selected distances (y-axis). It can be noticed that most of the peaks fall in the range between 2.8 and 2.9 Å, which roughly corresponds to the distance from the oxygen of an hydrogen bonded water molecule to a protein non-hydrogen atom. The sharpest peaks are generally observed for polar and charged residues, whereas gaps are mainly found in correspondence of neutral residues. Exceptions to this trend are rather frequent, especially on hydrophobic groups where often a high level of hydration is found. Fig. 18 shows the coordination number at 3.5 Å versus residue number.

High level of solvation on hydrophobic groups can be detected for example at residues Trp-7, Val-13, Leu-49, Pro-120, and at the valine terminus. Local computed MSD's for the solvent in the vicinity of neutral groups (see previous section) show that, although hydrophobic residues are in general less hydrated with respect to polar side chains, those who happen to be solvated are often surrounded by water molecules in a very stable and ordered configuration.

Appendix 17A.
Free Energy Calculations

In this appendix we address the problem of free energy calculations. As it is well known, it is very difficult to obtain entropy, and therefore free energy, from statistical mechanics computations. Various methods have been proposed to calculate *free energy changes* from molecular dynamics (MD) or Monte Carlo (MC) simulations, and are reviewed by Valleau and Torrie in Ref. 22. The use of a perturbation method, in conjunction with a coupling parameter for relatively large changes in the free energy, has been proven to be one of the most useful and encouraging methods. In the original formulation[149] the total potential energy of a given system, U, is separated into the potential energy of a reference system, U_r, and a perturbating potential U_p:

$$U = U_r + U_p$$

After some trivial manipulations of the configurational integral for the perturbed system and using the expression A = -kTlnQ for the free energy one obtains[34]

$$\Delta A = -k\,T \ln\ < e^{-U_p/kT} >_r \qquad (1A)$$

for the difference in the free energy between a reference and a perturbed system, where the average is to be taken over the reference configurational space. If the perturbing potential is much higher than kT or the configurations do not have substantial overlap with the equilibrium distribution of the reference system, the average procedure will not give reliable statistical results. In this case one can use a coupling parameter λ (0→1) to define several intermediate hybrid hypothetical systems between the reference and the fully perturbed system, in which the average in Eq. (1A) can be correctly accomplished. The total change in the free energy then will be the sum of the partial differences between two consecutive intermediate states. This method has been successfully used for the determination of the free energy of solvation for various systems from MD or MC computer simulations[150-153] and most of results agree with the available experimental data. It should be noted, however, that except for a MC study of small ion-water clusters[154] and an aqueous solution of alanine dipeptide,[155] all the interaction potentials used to calculate free energy of hydration are of empirical nature.

Here, we summarize the results[156,157] of the free energy of hydration for Li^+, Na^+, K^+, F^-, and Cl^- using the coupling parameter method and analytical potentials derived from *ab initio* calculations. The simulations were carried out in the NVT ensemble using both Metropolis[21] and preferential sampling algorithms.[158] The systems consisted of one ion and 342 water molecules confined within a cubic volume of side-length 21.74 Å. The simulated temperature was 298 K. Periodic boundary conditions were used to model the condensed phase environment. All the interactions were truncated to half of the side-length, and the configurational energy of the system was evaluated under the minimum image convention. We used the MCY pair potential for water-water interaction.[76] For ion-water interactions the analytical potential functions fitted to *ab initio* Hartree-Fock computations developed by Kistenmacher, Popkie, and Clementi[159] were used.

To calculate the free energy of hydration of ions at infinite dilution, one of the water molecules nearest to the center of the box in the starting configuration was slowly transformed into the appropriate ion. A coupling parameter λ ($0 \rightarrow 1$) was used to perform the transformation. For a given value of λ the hybrid ion-water potential was defined as

$$U_{iw}(\lambda) = (1 - \lambda) \times U_{aw} + \lambda \times U_{bw}$$

The free energy difference between initial (a, $\lambda = 0$) and final (b, $\lambda = 1$) states were then evaluated as[160]

$$\Delta A_{ab} = -kT \sum_{i=0}^{W-1} \ln \; < e^{-(U_{iw}(\lambda_{i+1}) - U_{iw}(\lambda_i))/kT} >$$

where W is the total number of windows and the canonical average is to be carried over the ensemble belonging to λ_i . The value of λ in window i+1 has been calculated as

$$\lambda_{i+1} = 0.5 \times \{ 1.0 + \tanh [(-W/2 + i) W_0] \} , \qquad i = 0, \;\; W - 1$$

where W_0 is a constant. This choice allows for a smooth interconversion of the solutes at the beginning of the simulation where most of the free energy change resides. The free energy change in each window is calculated by averaging over the M configurations generated in that window.

The calculated free energy of hydration for Li^+, Na^+, K^+, F^-, and Cl^- are presented in Table A-I and compared with available experimental data.[161,162] In order to account for hysteresis and to have a better estimate of the statistical errors, we also carried out independent backward simulations ($\lambda = 1 \rightarrow 0$), and the values reported in the table are the averages of the forward and the backward simulations. As expected from a two-body *ab initio* potential, the general agreement with the experimental data can only be judged as modest. The discrepancies are generally greater than the standard deviations, but the sign and experimental trend $Li^+ < Na^+ < K^+$ and $F^- < Cl^-$, are correctly reproduced, an indication that the two-body *ab initio* potentials are able to account for the different structural changes of water molecules around ions. The main reason for the large discrepancies is the truncation of the interaction potential at the two body level. This point has been quantified for the lithium ion by computing the free energy of hydration including also three-body *ab initio* potentials.[27,163] The average free energy of

Table A-I. Comparison of experimental and computed free energy of hydration, ΔA, in kcal/mole, for Li^+, Na^+, K^+, F^-, and Cl^-. The results are from simulations of 20.4 millions moves (3400 windows and 6000 moves in each window.)

	Monte Carlo		Expt'l	
Ion	Two-body	Three-body	Ref. 161	Ref. 162
Li^+	141.5±1.5	114.0±8.0	118.1	123.5
Na^+	109.6±1.8		90.6	98.6
K^+	76.7±2.7		73.1	80.8
F^-	99.8±2.1		110.7	103.8
Cl^-	56.7±0.6		81.5	75.8

hydration has been found to be 114 ± 8 kcal/mole; this value compares well with the experimental data reported in Table A-I.

References

1. See, for example, J. Schnitker and P. J. Rossky, J. Chem. Phys. **86**, 3471 (1986).
2. E. Clementi, *Computational Aspects for Large Chemical Systems* (Springer Verlag, Berlin, 1980).
3. G. Corongiu and E. Clementi, Gazz. Chim. It. **108**, 273 (1978).
4. J.D. Watson, N.H. Hopkins, J.W. Roberts, J.A. Steitz and A.M. Weiner, in *Molecular Biology of the Gene*, 4th ed., (The Benjamin/Cummings Publishing Company, California, 1988).
5. G. Bolis, M. Ragazzi, D. Salvaderi, D.R. Ferro and E. Clementi, Int. J. Quant. Chem.,**14**, 815 (1978).
6. L. Nilsson and M. Karplus, J. Comput. Chem., **7**, 591 (1986).
7. S.J. Weiner, P.A. Kollman, D.A. Case, U.C. Singh, C. Ghio, G. Alagona, S. Profeta Jr, and P. Weiner, J. Am. Chem. Soc., **106**, 765 (1984).
8. F.A. Momany, R.F. McGuire, A.W. Burgess and H.A. Scheraga, J. Phys. Chem., **79**, 2361 (1975).
9. R. Feldman, Atlas of Molecules (National Institutes of Health, Bethesda, MD, doc. 13, 1976).
10. A.H.J. Wang, G.J. Quigley, F.J. Kolpak, G. van der Marel, J.H. van Boom, and A. Rich, Science, **211**, 171 (1981).
11. *MOTECC-89, Input/Output Documentation*, S. Chin, E. Clementi, G. Corongiu, M. Dupuis, D. Frye, D. Logan, A. Mohanty, and V. Sonnad, Eds., (1989).
12. R.S. Mulliken, J. Chem. Phys. **1955**, **23**, 1833 (1955).

13. M. Aida, G. Corongiu, and E. Clementi, IBM Technical Report, in preparation.
14. C.L. Brooks III, M. Karplus, and B.M. Pettitt, Ad. Chem. Phys., **71**, 23 (1988).
15. I.K. Roterman, K.D. Gibson, K.D. and H.A. Scheraga, J. Biomol. Struct. Dynam., **7**, 391 (1989).
16. I.K. Roterman, M.H. Lambert, K.D. Gibson and H.A. Scheraga, J. Biomol. Struct. Dynam., **7**, 421 (1989).
17. D.A. Williams and R.R. Weller, J. Am. Chem. Soc., **105**, 4143 (1983).
18. U.A. Singh and P.A. Kollman, J. Comp. Chem., **2**, 129 (1984).
19. D.A. Pearlman and S. Kim, Biopolymers, **24**, 327 (1985).
20 S.J. Weiner, P.A. Kollman, D.T. Nguyen and D.A. Case, J. Comput. Chem., **7**, 230 (1986).
21. N. Metropolis, A. W. Rosenbluth, M. N. Rosenbluth, A. H. Teller, and E. Teller, J. Chem. Phys. 21, 1087 (1953).
22. B. J. Berne, Ed., *Statistical Mechanics,* Vol. 5 of "Modern Theroetical Chemistry Series," H. F. Schaefer, Ed. (Plenum, New York, 1977).
23. M. H. Kalos and P. A. Whitlock, *Monte Carlo Methods* (Wiley, New York, 1986).
24. D. J. Adams, Mol. Phys. **28**, 1241 (1974).
25. D. J. Adams and J. R. McDonald, J. Chem. Phys. **7**, 2761 (1974).
26. G. C. Lie, E. Clementi, and M. Yoshimine, J. Chem. Phys. **64**, 2314 (1976).
27. E. Clementi and G. Corongiu, Int. J. Quant. Chem., Symp. **10**, 31 (1983).
28. J. H. Detrich, G. Corongiu, and E. Clementi, Chem. Phys. Lett., **112**, 426 (1984).
29. E. Clementi, *Determination of Liquid Water Structure, Coordination Numbers for Ions and Solvation for Biological Molecules* (Springer Verlag, Berlin, 1976).
30. S. Romano and E. Clementi, Int. J. Quantum Chem. **14**, 839 (1978); *ibid.* **17**, 1007 (1980).
31. E. Clementi and G. Corongiu, J. Biol. Phys. **11**, 33 (1983); K. S. Kim and E. Clementi, J. Phys. Chem. **89**, 3655 (1985); *ibid.*, J. Am. Chem. Soc. **107**, 5504 (1985).
32. M. P. Allen and D. J. Tildesley, *Computer Simulation of Molecular Liquids* (Clarendon, Oxford, 1987).
33. B. J. Alder and T. E. Wainwright, Phys. Rev. **A1**, 18 (1970).
34. D. A. McQuarrie, *Statistical Mechanics* (Harper and Row, New York, 1976).
35. W. G. Hoover and W. T. Ashurst, in *Theoretical Chemistry, Advances and Perspectives,* Vol. 1, H. Eyring and D. Henderson, Eds., Academic Press, New York, p. 1 (1975).
36. M. W. Evans, P. Grigolini, and G. Pastori, *Memory Function Approaches to Stochastic Problems in Condensed Matter,* Vol. 62 of "Advances in Chemical Physics," I. Prigogine and S. A. Rice, Eds. (Wiley Interscience, New York, 1985).

37. D. L. Beveridge and W. L. Jorgensen, eds., *Computer Simulations of Chemical and Biomolecular Systems,* Annals of the New York Academy of Sciences, Vol. 482 (New York Academy of Sciences, New York, 1986).
38. J. A. McCammon and S. C. Harvey, eds., *Dynamics of Proteins and Nuclei Acids* (Cambridge Unversity Press, London, 1987).
39. S. N. Kandadai and E. Clementi, Biopolymers (in press).
40. L. Hannon, G. C. Lie, and E. Clementi, Phys. Lett. A **119**, 174 (1986).
41. D. C. Rapaport and E. Clementi, Phys. Rev. Lett. **57**, 695 (1986).
42. M. M. Mansour, A. L. Garcia, G. C. Lie, and E. Clementi, Phys. Rev. Lett. **58**, 876 (1987).
43. D. K. Bhattacharya and G. C. Lie, Phys. Rev. Lett. **62**, 897 (1989).
44. K. Singer, J. V. L. Singer and A. J. Taylor, Mol. Phys. **37**, 1239 (1979).
45. C. W. Gear, *The Numerical Integration of Ordinary Differential Equations of Various Orders,* Report ANL 7126, Argonne National Labortory (1966).
46. C. W. Gear, *Numerical Initial Value Problems in Ordinary Differential Equations* (Prentice-Hall, Englewood Cliffs, N.J., 1971).
47. L. Verlet, Phys. Rev. **159**, 98 (1967).
48. R. W. Hockney, Methods Comput. Phys. **9**, 136 (1970).
49. H. J. C. Berendsen and W. F. van Gunsteren, in *Molecular Dynamics Simulation of Statistical Mechanical Systems,* G. Ciccotti and W. G. Hood, eds. (North Holland Press, New York, 1983).
50. D. J. Evans and S. Murad, Mol. Phys. **34**, 34 (1977).
51. R. Sonnenschein, J. Comp. Chem. **59**, 348 (1985).
52. D. Fincham, CCP5 Quaterly **2**, 6 (1981).
53. J. P. Ryckaert, G. Ciccotti, and H. J. C. Berendsen, J. Comput. Phys. **23**, 327 (1977).
54. P. S. Y. Cheung, Mol. Phys. **33**, 519 (1977).
55. J.P. Hansen and I. McDonald, *Theory of simple liquids,* (Academic Press, London, 1986).
56. S. W. De Leeuw, J. W. Perram, and E. R. Smith, Proc. Roy. Soc. (London) **A 373**, 27 (1980).
57. P. H. Berens and K. R. Wilson, J. Chem. Phys. **74**, 4872 (1981).
58. R. G. Gordon, Adv. Mag. Resonance **3**, 1 (1968).
59. J. Yarwood, in *Molecular Liquids, Dynamics and Interactions,* A. J. Barnes, W. J. Orville-Thomas, and J. Yarwood, Eds., NATO ASI series Vol. 135 (Reidel, New York, 1984), p.357.
60. P. H. Berens, S. R. White, and K. R. Wilson, J. Chem. Phys. **75**, 515 (1981).
61. A. Wlodlawer, Prog. Biophys. Molec. Biol. **40**, 115 (1982).
62. H.D. Bartunik, P. Jolles, J. Berthou, and A.J. Dianoux, Biopolymers, **21**, 43 (1982).
63. H. Grimm, H. Stiller, C.F. Majkrzak, A. Rupprecht, and U. Dahlborg, Phys. Rev. Lett., **59**, 1780 (1987).
64. H. Grimm, H. Stiller, C.F. Majkrzak, A. Rupprecht, Physica B, **156**, 464 (1989).

65. L.J. Schreiner, M.M. Pintar, A.J. Dianoux, F. Volino, and A. Rupprecht, Biophys. J., **53**, 119 (1988).
66. S. Cusack, J. Smith, J. Finney, M. Karplus, and J. Trewhella, Physica **136B**, 256 (1986).
67. J. Smith, K. Kuczera, B. Tidor, W. Doster, S. Cusack, and M. Karplus, Physica B, **156-157**, 437 (1989).
68. F H. Stillinger and A. Rahman, A., in *Molecular Motions in Liquids*, J. Lacombe, Ed., Reidel, Dordrecht, Holland (1974).
69. S.W. Lovesey, in *Theory of Neutron Scattering from Condensed Matter*, Vol. 1, Clarendon Press, Oxford (1984).
70. G.R. Kneller and A. Geiger, Mol. Phys. in press (1990)
71. W.H. Press, B.P. Flannery, S.A. Teukolsky, and W.T. Vetterling, *Numerical Recipes*, Cambridge University Press, London (1986).
72. P. Ewald, Ann. Phys. **64**, 253 (1921).
73. J. A. Barker and R. O. Watts, Mol. Phys. **26**, 789 (1973).
74. L. Onsagar, J. Am. Chem. Soc. **58**, 1486 (1936).
75. U. Niesar, G. Corongiu, E. Clementi, G.R. Kneller, and D.K. Bhattacharya, J. Phys. Chem. (in press 1990). Int. J. Quant. Chem. (in press).
76. O. Matsuoka, E. Clementi, and M. Yoshimine, J. Chem. Phys. **64**, 1351 (1976).
77. J. H. van Lenthe, J. G. C. M. van Duijneveldt - van de Rijdt, and F. B. van Duijneveldt, Adv. Chem. Phys. **69**, 521 (1987).
78. K. Szalewicz, S. J. Cole, W. Kolos, and R. J. Bartlett, J. Chem. Phys. **89**, 3662 (1988).
79. G. A. Petersson, A. Bennett, T. G. Tensfeldt, M. A. Al-Laham and W. A. Shirley, J. Chem. Phys. **89**, 2193 (1988).
80. J. A. Montgomery, G. A. Petersson and N. Matsunaga, Chem. Phys. Lett **155**, 413 (1989).
81. B. Liu and A. D. McLean, J. Chem. Phys. **72**, 3418 (1980).
82. B. Liu and A.D. McLean, J. Chem. Phys., **91**, 2348 (1989).
83. a) U. Niesar, G. Corongiu, M-J. Huang, M. Dupuis, and E. Clementi, Int. J. Quant. Chem. Symp., **23**, 421 (1989); b) U. Niesar, G. Corongiu, M-J. Huang, M. Dupuis, and E. Clementi, IBM Technical Report, KGN-191 (1989).
84. W.S. Benedict, N. Gailar, and E.K. Plyler, J. Chem. Phys., **24**, 1139 (1956).
85. M. Wojcik and E. Clementi, J. Chem. Phys. **84**, 5970 (1986).
86. M. Wojcik and E. Clementi, J. Chem. Phys. **85**, 3544 (1986).
87. M. Wojcik and E. Clementi, J. Chem. Phys. **85**, 6085 (1986).
88. J.O. Hirschfelder, C.F. Curtiss, and R.B. Bird, *Molecular Theory of Gases and Liquids,* (John Wiley, New York, 1954).
89. C. J. F. Böttcher, *Theory of Electric Polarization,* Vol. 1 (Elsevier, Amsterdam, 1973).
90. F. H. Stillinger and C. W. David, J. Chem. Phys. **69**, 1473 (1978).
91. P. Barnes, J. L. Finney, J. D. Nicholas, and J. E. Quinn, Nature **282**, 459 (1979).

92. D. Levesque, J. J. Weis, and G. N. Patey, Mol. Phys. **51**, 333 (1984); J. M. Caillol, D. Levesque, J. J. Weis, P. G. Kusalik, and G. N. Patey, *ibid.* **55**, 65 (1985).
93. T.P.Lybrand and P.A. Kollman, J. Chem. Phys., **83**, 2923 (1985).
94. H.J.C. Berendsen, J.R. Grigera, and T.P. Straatsma, J. Chem. Phys., **91**, 6269 (1987).
95. J. A. C. Rullmann and P. Th. van Duijnen, Mol. Phys. **63**, 451 (1988).
96. M. Sprik and M. L. Klein, J. Chem. Phys. **89**, 7556 (1988).
97. See, for example, P. Obza, and R. Zahradnik, Chem. Rev., **88**, 871 (1988).
98. T.R. Dyke, K.M. Mack, and J.S. Muenter, J. Chem. Phys., **66**, 498 (1977).
99. L.A. Curtiss, D.J. Frurip, and M.L. Blander, J. Chem, Phys., **64**, 1120 (1982).
100. G. S. Kell, G. E. McLaurin, and E. Whalley, J. Chem. Phys. **64**, 5308 (1968).
101. N.E. Dosey, *Properties of Ordinary Water Substances*, (Reinhold, New York, 1940).
102. G. C. Lie and E. Clementi, J. Chem. Phys. **64**, 5308 (1976).
103. K. Refson, G. C. Lie, and E. Clementi, J. Chem. Phys. **87** 3634 (1987).
104. G.S. Kell, *Water: A comprehensive Treatise,* F. Franks., Ed., (Plenum, New York, 1971), Vol.1, p. 363.
105. G. C. Lie, J. Chem. Phys. **85**, 7495 (1986).
106. A. H. Narten and H. A. Levy, J. Chem. Phys. **55**, 2263 (1971).
107. A. K. Soper and R. N. Silver, Phys. Rev. Lett., **49**, 471 (1982); A. K. Soper and M. G. Phillips, Chem. Phys. **107**, 47 (1986).
108. W. E. Thiessen and A. H. Narten, J. Chem. Phys. **77**, 2656 (1982).
109. M. Krynicki, C. D. Grenn, and D. W. Sawyer, Discuss. Faraday Soc. (London) **66**, 199 (1978).
110. J. Jonas, T. DeFries, and D.J. Wilbur, J. Chem. Phys., **65**, 582 (1976).
111. G.E.Walrafen, *Water: A comprehensive Treatise*, F. Franks, Ed., (Plenum, New York, 1971), Vol.1, p. 151.
112. A. Rahman, and F.H. Stillinger, Phys. Rev. A, **10**, 368 (1974).
113. J. Teixeira, M.C. Bellissent-Funel, S.H. Chen, and B. Dorner, Phys. Rev. Lett., **54**, 2681 (1985).
114. G. C. Lie and E. Clementi, Phys. Rev. A **33**, 2679 (1986).
115. R.W. Impey, P.A. Madden, and I.R. McDonald, Mol. Phys., **46**, 513 (1982).
116. M.A. Ricci, D. Rocca, G. Ruocco, and R. Vallauri, Phys. Rev. Lett., **161**, 1958 (1988).
117. J. Teixeira, and J. Leblond, J. Phys. Lett., **39**, L83 (1978).
118. A.J. Matheson, *Molecular Acoustics*, (Wiley, New York, 1971).
119. S.H. Chen and J. Teixeira, Adv. Chem. Phys., **64**, 1 (1988).
120. K. Toukan, M.A. Ricci, S.H. Chen, C.K. Loong, D.L. Price, and J. Teixeira, Phys. Rev. A ., **37**, 2580 (1988).
121. D.I. Page, *Water: A comprehensive Treatise,* F. Franks, Ed., (Plenum, New York, 1971), Vol.1, p. 333.
122. R. J. Bartlett, I. Shavitt, and G. D. Purvis, J. Chem. Phys. **71**, 281 (1979).

123. Coordinates from Brookaven Protein Data Bank, the cited reference is A. Jack and M. Levitt, Acta Crystallogr. **A34**, 931 (1978).
124. E. Clementi, F. Cavallone, R. Scordamaglia, J. Am. Chem. Soc. **99**, 5531 (1977).
125. G. Ranghino, E. Clementi and S. Romano, Biopolymers **22**, 1449 (1983).
126. M. Ragazzi, S. Ferro, and E. Clementi, J. Chem. Phys. **70**, 1040 (1979).
127. W. F. van Gunsteren and M. Karplus, Biochemistry, **21**, 2259 (1982).
128. M. Levitt and R. Sharon, Proc. Natl. Acad. Sci. (USA) **85**, 7557 (1988).
129. M. Marchi, G.K. Kneller, G. Corongiu, and E. Clementi (to be published).
130. B. Brooks and M. Karplus, PNAS, **80**, 6571 (1983).
131. J. Smith, S. Cusack, U. Pezzeca, B. Brooks and M. Karplus, J. Chem. Phys., **85**, 3636 (1986).
132. J. A. McCammon, B. R. Gelin and M. Karplus, Nature (London), **267**, 585 (1977).
133. M. Karplus and J. A. McCammon, CRC Crit. Rev. Biochem. **9**, 293 (1981).
134. M. Levitt, J. Mol. Biol. **168**, 621 (1983).
135. H. Fraunfelder, G. Petsko and D. Tsernoglu, Nature, **280**, 558 (1979).
136. F. Parak and H. Formanek, Acta Crystallogr. **27 a**, 573 (1971).
137. F. Parak, E. W. Knapp and D. Kucheida, J. Mol. Biol. **161**, 177 (1982).
138. G. P. Singh, F. Parak, S. Hunklinger, and K. Dransfeld, Phys. Rev. Lett. **47**, 685 (1981).
139. W. F. van Gunsteren and H. J. C. Berendsen, J. Mol. Biol. **176**, 559 (1984).
140. A. J. Brunger, C. L. Brooks and M. Karplus, Proc. Natnl. Acad. Sci. (USA) **82**, 8548 (1985).
141. C. F. Wong and J. A. McCammon, Isr. J. Chem. **27**, 211 (1987).
142. P. Procacci, G. Corongiu, and E. Clementi (to be published).
143. F. Parak, private communication.
144. W. Bialek and R. F. Goldstein, Biophys. J. **48**, 1027 (1985).
145. T. Takano, J. Mol. Biol. **110**, 537 (1977).
146. S. E. V. Phillips, J. Mol. Bio. **142**, 531 (1980).
147. B. P. Schoenhorn, J. Mol. Biol. **201**, 741 (1988).
148. P. A. Madden and R. W. Impey, in *Computer Simulation of Chemical and Biomolecular Systems,* Ann. of the New York Acad. of Sci., D. L. Beveridge and W. L. Jorgensen, Eds., (1986).
149. R. W. Zwanzig, J. Chem. Phys. **22**, 1420 (1954).
150. T. P. Lybrand, I. Ghosh, and J. A. McCammon, J. Am. Chem. Soc. **107**, 7793 (1985).
151. U. C. Singh, F. K. Brown, P. A. Bash, and P. A. Kollman, J. Am. Chem. Soc. **109**, 1607 (1987).
152. P. A. Bash, U. C. Singh, R. Langridge, and P. A. Kollman, Science **236**, 564 (1987).
153. W. L. Jorgensen and C. Ravimohan, J. Chem. Phys. **83**, 3050 (1985).
154. M. R. Mruzik, F. F. Abraham, and D. E. Schreiber, J. Chem. Phys. **64**, 481 (1976).
155. M. Mezei, P. K. Mehrotra, and D. L. Beveridge, J. Am. Chem. Soc. **107**, 2239 (1985).

156. M. Migliore, G. Corongiu, E. Clementi, and G. C. Lie, J. Chem. Phys., **88**, 7766 (1988).
157. G. Corongiu, M. Migliore, and E. Clementi, J. Chem. Phys., **90**, 4629 (1989).
158. J. C. Owicki, ACS Symp. Ser. **86**, 159 (1978).
159. H. Kistenmacher, H. Popkie, and E. Clementi, J. Chem. Phys. **59**, 5842 (1973).
160. P. Cieplak, T. P. Lybrand, and P. A. Kollman, J. Chem. Phys. **86**, 6393 (1987).
161. R. Gomer and G. Tryson, J. Chem. Phys. **66**, 4413 (1977).
162. H. L. Friedman and C. V. Krishnan, in *Water: A Comprehensive Treatise,* Vol. 3, edited by F. Franks (Plenum, New York, 1973).
163. E. Clementi, H. Kistenmacher, W. Kolos, and S. Romano, Theoret. Chim. Acta, **55**, 257 (1980).

Chapter 18.

Molecular Dynamics Simulations of Fluid Flows

D.K. Bhattacharya and E. Clementi
IBM Corporation
Data Systems Division, Dept. 48B/MS 428
Neighborhood Road
Kingston, New York 12401, U.S.A.

1. Introduction

A study of nonequilibrium heat, mass and momentum transport in a fluid system is central to many disciplines of physical and engineering sciences. It is encountered in diverse industrial applications such as the design of high speed aircrafts, cooling strategies for removing heat from the integrated chips in supercomputers, oil recovery from porous rocks, dispersion of pollutants in the atmosphere, weather predictions and biomedical engineering. On a more fundamental level, problems like the cooperative behavior in liquids and polymers, energy cascading from the microscopic modes to the macroscopic modes in turbulent flows, and the emergence of an ordered structure in the seemingly incessant chaotic motion of fluid particles, remain at the heart of our quest for a precise understanding of nature.

The traditional approach of analyzing fluid dynamics is based on the methods of continuum mechanics.[1] Here the evolution of the fluid system is described in terms of a few macroscopic field variables such as mass density, velocity, and the internal energy. First the balance equations of these macroscopic variables are formulated and when these are supplemented with the linear constitutive laws (such as Fourier law, Ficks law, Newton's law), then one obtains the transport equation (Fourier equation, Ficks equation and Navier Stokes equations). Subject to appropriate boundary conditions these transport equations are solved by analytical or numerical methods, which then gives us information on the macroscopic dynamics of fluid motions.

However, in many cases of practical interest, a continuum approach to fluid flows is either inadequate or breaks down. For example:

1. In the transitional regime gas flow problems, when the mean free path of the fluid particles is comparable to the geometric dimension of the flow system, the velocity and the temperature profiles exhibit pronounced velocity and temperature slips at the boundary. The slip flow phenomena are encountered in diverse fields such as high altitude aerodynamics, gas dynamic isotopes separation techniques, aerosol dynamics, vacuum pump operation, energy transfer in molecular collision and the attainment of the milli temperature by flow coolings.[2]

2. In dense fluids the transport coefficients (such as the shear viscosity and the thermal conductivity) exhibit a complex dependence on the density and the temperature over a wide range of pressure. Since a statistical mechanical theory of dense fluids incorporating the correlations arising from N body dynamics of fluid particles is far from complete, the density dependence of the transport coefficient requires direct experimental measurements.[3]

3. Hydrodynamic phenomena at the fluid-wall interface is strongly influenced by the molecular constitution of the wall, as well as by the physical and chemical state of the interface. A complete characterization of interfacial hydrodynamics is required in many applications such as lubrication coatings, tertiary oil recovery, adhesion and the spreading of wetting liquid drops.[4]

4. In flows of polymer liquids, the deformation of the complex geometry of molecules gives rise to the nonlinear shear rate dependence of viscosity, normal stress effects, and nonequilibrium phase transition which cannot be described by the Navier Stokes equations. The rich and unusual viscoelastic behavior displayed by polymeric liquids can only be studied by investigating the microscopic dynamics associated with the topological interaction of different segments of polymers.[5]

The removal of the limitations inherent in continuum hydrodynamics requires a new approach to fluid flows, one in which we take into account the discrete molecular nature of fluid particles. The advent of high performance computers has offered a unique opportunity to study the complex fluid motion at a molecular level. Several new techniques such as Nonequilibrium Molecular Dynamic Method, Cellular Automata and the Direct Simulation Monte Carlo method have been successfully used to examine the intricate microscopic dynamics of fluid motions.[6] In this chapter we shall describe the computational algorithms for a large scale molecular dynamic simulation of fluid flows and heat transfer. We discuss some representative applications and show how molecular dynamic simulations can provide new insights into the foundations of fluid dynamics.

In their pioneering work Adler and Wainright[7] studied the fluid flows using the methods of equilibrium molecular dynamics. By analyzing the time correlations functions of the fluid particles they predicted the existence of long time tail phenomena which led to the reexamination of the kinetic theory of dense fluids. More recently Ashurst and Hoover proposed a new, Non Equilibrium Molecular Dynamic method. In this approach the fluid system is removed from the state of thermodynamic equilibrium by applying an external driving force, and then, by measuring the direct flow properties (such as shear stress and the heat flux) the transport coefficients are determined. The NEMD technique enables one to to analyze the behavior of a fluid system removed far from equilibrium and to examine the range of the validity of the classical hydrodynamics. The NEMD algorithm of Ashurst and Hoover has been extended by several authors and has been applied to a wide range of the fluid flow problems such as the nonlinear shear viscosity and viscoelasticity of simple fluids, shear induced phase transition, shock wave propagation, hydrodynamic instabilities and plastic flow[8-23] These simulations have stimulated new theories on the thermodynamics of nonequilibrium processes.[24,27] We have designed a computer program "MICRODY" for investigating the microscopic dynamics of nonequilibrium flows. The simulation strategies employed in the MICRODY code are analogous to the NEMD technique. A unique feature of this code is that it can be applied to study the nonequilibrium flow phenomena over a wide range of density, from very dense liquids to a rarefied gas. Similarly it can be applied to explore the transport phenomena in subsonic and supersonic regions and also in examining the critical influence of the imposed temperature gradient and the boundary conditions on the development of flow profiles.

The code "MICRODY" can be used as a numerical wind tunnel to test new concepts in the hydrodynamic theory of dense fluids and the transition regime gas flows. In section 2 we briefly discuss the computational algorithm for a large scale dynamic simulation of fluid flows. Subsequently in section 3 we detail several interesting simulations of fluid flows conducted at IBM, Kingston. These include dense fluid flows in a channel, the evolution of the hydrodynamic instabilities in the flow around a circular cylinder, Rayleigh-Benard convection and the transition regime gas flows. We also discuss new areas of applications.

2. Computational Algorithm for A Large Scale Molecular Dynamic Simulation

In micro-hydrodynamics, the dynamic evolution of fluid particles is studied using the molecular dynamic method. A few thousand fluid particles are enclosed in a three dimensional channel. Thermal walls placed in the z direction act as a heat reservoir. To maintain a nonequilibrium flow, an acceleration field is applied in the y direction. The fluid particles interact via a soft sphere potential.[17,20]

$$V(r) = 4\varepsilon\left[\left(\frac{r}{\sigma}\right)^{-12} - \left(\frac{r}{\sigma}\right)^{-6} + \frac{1}{4}\right], \qquad \forall\ r \le r_c = 2^{1/6}\sigma\ ;$$
$$= 0, \qquad \forall\ r > r_c$$

This potential is short ranged, *i.e.*, there is no interaction for any atomic pair that is more than r_c apart. As discussed below, the use of this kind of potential is a necessity if the computation is to be reduced to a reasonable amount of time, since the number of particles involved in this type of study is usually very large. Newton's equations of motion are solved by Gear's predictor-corrector method. The order of the Gear's algorithm appears as an input parameter. Thus for a six order Gear algorithm the evolution of the dynamical system is studied by analyzing the vector r_0 which represents a set of particle positions and 5 successive scaled time derivatives.[28] The wall-particle interactions are modelled in two different ways in the current program package. The first models the wall as a reflective boundary: when a particle hits the wall, it will be elastically reflected. This type of wall can simulate a complete slip boundary condition in the solution of the Navier-Stokes equations. The second model of the wall in the package is the diffusive wall: when a particle hits the wall, it looses completely the memory of its past and gets thermalized by the wall. The particle will be returned to the system immediately with a velocity chosen randomly from the Maxwell-Boltzmann distribution at the wall temperature, T_w. Assuming the wall is perpendicular to the x-axis, the components of the reflected velocity are given by

$$v_x = \sqrt{-2\, k_B\, T_w\, \ln \xi_1 / m}$$

$$v_y = \sqrt{-2\, k_B\, T_w\, \ln \xi_2 / m}\; \sin \xi_3$$

$$v_z = \sqrt{-2\, k_B\, T_w\, \ln \xi_2 / m}\; \cos \xi_3$$

where k_B is the Boltzmann constant, ξ_1, ξ_2, ξ_3 the random numbers between 0 and 1. Since the velocity chosen this way will average to zero in any direction perpendicular to the x-axis, this type of a wall simulates the no-slip boundary condition usually employed in solving the Navier-Stokes equations. A more sophisticated model of the wall where the atomic structure is taken into account[29,30] is currently under investigation.

The calculation of the interparticle interaction is the most time consuming part of any molecular dynamics algorithm. While designing the "MICRODY" program one important objective was to have the ability to simulate a very large number of particles. Such large scale molecular dynamic simulations are required, for example, in the study of hydrodynamic instabilities to reach the critical values of the scaling number, where bifurcations are expected to occur, and to obtain a reasonable signal to noise ratio in the statistical net to capture the nonequilibrium structure manifesting on a very small scale.[20] Since the interaction potential used in the micro-hydrodynamical simulation is short ranged, a "Link Cell" method was used for calculating the interaction forces among the fluid particles. In this approach, before starting the calculation of the interaction forces, we divide the simulation system into small cells such that each side of the cell is greater than the cutoff distance r_c. Each atom is assigned to a cell and is linked to two atoms in the same cell through the pointers. That is, cell i first points to atom k, the index vector of the atom k then points to the next atom in the same cell, m, which in turn points to the other atom n, ... etc. The calculation of the interaction forces proceeds in the following ways:

1. We choose a cell and pick up the particle appearing in the headlist (which shows the first particle in the given cell). The interaction of this particle with all other particles occupying the same cell is calculated utilizing the linked list.

2. The interaction of the chosen particle with the particle residing in the adjacent cells are calculated. A careful application of Newton's third law reduces the computations of the interaction force between the particles, however this necessitates searching the adjacent cells in a particular order.

3. After all possible interactions of the chosen particle (from a chosen cell) have been calculated, we move to the next particle (using a linked list) and then repeat the previous two steps. This process is repeated until all the particles in the chosen cells are exhausted.

4. We now move to the next cell and repeat the procedure until all cells in the simulation system have been considered.

In the "Link-Cell" approach, the computational and the memory allocation requirements for evaluating the interaction forces scales as N, which is very useful whenever the number of simulated particles is sufficiently large.

The main difference between the non-equilibrium MD and the equilibrium MD simulations described in the previous chapter is that the former requires *external* driving forces, whereas the latter does not. Several different options exist for inducing the flow. The most common practice used in NEMD studies is to realize a plane Couette flow wherein the fluid can be subjected to a uniform shear rate by manipulating the velocity of the upper plate.[8] Another widely used technique is to impose a uniform acceleration field (like a gravitational field). Here by changing the magnitude of the acceleration field the different flow regimes (subsonic or supersonic) can be achieved.[17,22] Yet another approach, due to Evans, Gillian, and Loose, consists in modifying the equation of motion of fluid particles, and have been used in studies on nonlinear thermal conductivity.[31,33] In our simulations the flow is induced via an acceleration field or by putting the fluid system in contact with a heat reservoir.

To make a connection with the continuum hydrodynamics and experiments we divide the simulation system into three dimensional boxes and calculate the average values of the velocity, number density of particles and the kinetic energy of fluid in each box at every instant of time. These average quantities are collected for a few thousand time steps, then new statistics are obtained by time averaging the dynamical information contained in each box. The advantage of this is that the data to be kept for later analysis is reduced drastically, in comparison with keeping the history of the simulation, which is almost impossible due to the huge number of particles present in the simulation system. The disadvantage is that the analysis is restricted to velocity, number of particles, and kinetic energy only.

In continuum fluid dynamics, every point is assumed to contain millions of particles averaged over seconds. This of course cannot be done in the simulations, because the number of particles considered in the simulations is usually of the order of thousands to hundreds of thousands only, and the time span is of the order of tens of picoseconds. Fortunately, experience indicates that for understanding the hydrodynamics of simple fluids it is not really necessary to carry out the averaging over experimental time and space scales. In fact, it has been shown that the dynamic simulations with 32 or 108 fluid particles averaged over several picoseconds can provide the results which are consistent with the experiments.[3] The criterion is that the averaged values should be stable in order to be meaningful. One way to check this is by taking the averages over different time spans of the simulation history and comparing them. A stable result is achieved if the fluctuations of the averages are much less than the mean of the averages (unless one is in a critical region where fluctuations can be as large as the mean). For complex molecules, stress relaxation processes occur over a much longer timescale (of the order of micro-sec onds to seconds). A molecular simulation of these systems then requires refinement of the conventional molecular dynamical algorithm.[5]

According to the ergodic hypothesis, the average of any property over an ensemble of similar systems at the same instant should be the same as the average over a single system spanning an entire history. This hypothesis is strictly true for an infinite number of systems and for an infinite duration of time, but in practice is usually interpreted more loosely. Thus we can use a larger cell size over a shorter simulation time or a smaller cell size over a longer simulation time; the results should be the same. It should be noted, however, that the cell size should not be too large so that the properties one is observing do not change appreciably over the cell. The basic structure of the MICRODY program is illustrated with a flow chart in Fig. 1.

The MICRODY code can efficiently utilize the parallel processing facilities available in a supercomputer. For example in a simulation of a nonequilibrium gas flow at $Kn=1$, the integration of the equations of motion for 13828 argon particles in a three-dimensional space for 100 time steps took 81 seconds by using one processor of an IBM 3090/400. With two processors the elapsed time was reduced to 47 seconds, thus a speed-up of a factor of 1.7 was achieved. Using a novel computer architecture, such as LCAP-3090 which can provide as many as 24 nodes, one can accelerate the sequential molecular dynamics simulation by an order of magnitude. The structure of the parallel algorithm is the following. If the number of available processors in the system is "k" then in the predictor (or corrector) step the total number of molecules are divided in k different segments and and each processor is assigned the computation of a given segment.In this step there is no overlap between the computations undertaken by the processors. For calculating the interaction force between the fluid particles, the total volume of the fluid system is divided into k parts, and each processor is assigned the computation of the interaction forces of the fluid particles occupying the volume

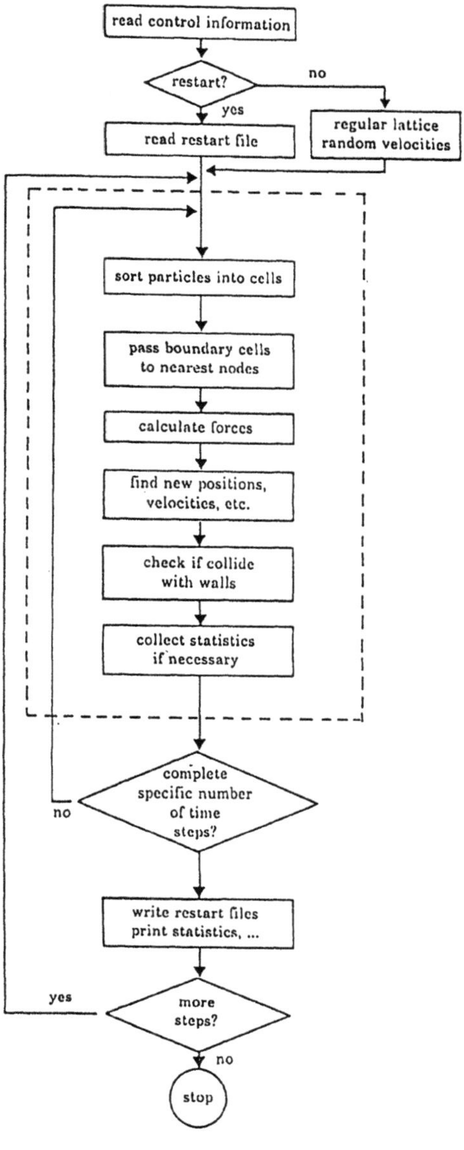

Figure 1. Basic flow chart for the MICRODY. Steps enclosed in the broken lines can be easily adapted for parallel computing.

segment assigned to the processor. Since the fluid particles in the boundary of the chosen volume segment can also interact with the fluid particles which belong to the other processor, a certain amount of the information has to exchanged between the processors. Here we might add that the parallelization of the link list molecular dynamics algorithm is different from the parallelization of the molecular dynamic algorithm for a system where the long range interactions are important. Recent years have seen some exciting developments in the microscopic simulation of nonequilbrium flows driven far from equilibrium. The implementation of molecular dynamic algorithms in massively parallel machines would further extend our search for a truly molecular basis of complex fluid motions to a new domain.

3. Applications and a Discussion of Results

One of the significant breakthroughs of the last few years has been the observation by molecular dynamics computer simulation of the hydrodynamic flow phenomena on the angstrom and picosecond scales of flowing atomic ensembles. This has closed the gap between macroscopic hydrodynamics and the microscopic atomic description of the fluid phenomena. We now proceed to briefly discuss specific features of the simulation algorithm which has made this possible.

Before we began the molecular simulations of hydrodynamics in 1985, there already existed some simulations designed to test the validities of the linear assumptions embedded in Fourier's law[34] and Newton's law[35] under extremely large external constraints. However, there seemed to be no attempts in simulating the complex hydrodynamic phenomena from the atomic scale. The first such attempt in our laboratory was carried out by Hannon, Lie, and Clementi for a 3-dimensional channel flow and the flow past a plate, with a relatively few number of particles ($\sim$1000).[18,19] We were surprised to find that despite the microscopic size of the simulation system and the magnitude of the acceleration field, the velocity and the temperature profiles developed in such a small simulation were in full agreement with the continuum hydrodynamic predictions. Encouraged by this observation, Rapaport and Clementi quickly started a systematic study of the flow past a cylinder using up to 160,000 particles, thanks to the newly assembled machines for parallel processing.[17] After that, many successful applications of the molecular simulation technique or its variant to the study of the phenomena of fluid dynamics have been reported.[21-23]

We now start describing our first attempt to simulate a micro-hydrodynamic system, a 3-dimensional channel flow, at atomic time scales and dimensions.[18] This was achieved by considering a system of particles bounded by thermal walls in one coordinate (z) and periodic in the other two (x and y). Flow is induced by an acceleration field parallel to the thermal walls. A steady state is reached when the energy of the acceleration field is balanced by its removal through the thermal walls. Assuming the fluid to be incompressible, the hydrodynamic equations for velocity and temperature reduce to

$$\nabla^2 \mathbf{v} = \frac{\rho \mathbf{g}}{\eta} \tag{3}$$

and

$$\nabla^2 T = -\frac{\eta}{2\kappa} \sum_{i,j} \left(\frac{\partial v_i}{\partial x_j} + \frac{\partial v_j}{\partial x_i} \right)^2, \qquad i,j = 1,2,3 \tag{4}$$

where ρ is the mass density, η is the viscosity, and κ the thermal conductivity. When subjected to boundary conditions

$$\begin{aligned} v_y(z = -L/2) &= v_y(z = L/2) = 0 \\ T(z = -L/2) &= T(z = L/2) = T_w \end{aligned} \tag{5}$$

these equations can be solved exactly to give

$$v_y = \frac{\rho g}{2\eta} \left[z^2 - \left(\frac{L}{2} \right)^2 \right] \tag{6}$$

$$v_x = v_y = 0 \tag{7}$$

$$T = T_w - \frac{1}{12} \frac{(\rho g)^2}{k\eta} \left[z^4 - \left(\frac{L}{2} \right)^4 \right] \tag{8}$$

Thus, macroscopic theory predicts a parabolic velocity profile and a quartic temperature profile across the channel. The transport coefficients such as the shear viscosity and the thermal conductivity can be calculated from these flow profiles. In Fig. 2 the flow profiles obtained from the molecular dynamic simulation of Argon liquid near triple point, are shown and are compared with the macroscopic hydrodynamic predictions. An excellent agreement is evident. In the case of dense fluids it is also seen that the density profile can acquire a multilayer structure. Recent experiments [36] and the molecular dynamic simulation based on slightly different algorithms also support the existence of the layering phenomena near the wall.[37,38] The shear viscosity (and the thermal conductivity) can be calculated from the simulated flow profiles. As is well known transport coefficients obtained from the molecular simulation depend, apart from the density and the temperature, on the potential used and the cutoff distance for calculating the forces. To give some idea on the nature of the results obtained with the MICRODY code we shall discuss a particular case. At a density 1.248 g/cm^3 and a temperature 306.6 Kelvin we conducted a MD simulation of 2000 fluid particles for 200,000 time steps. The obtained shear viscosity $\eta = 1.27 \times 10^{-4}$ Pa s was found to be in agreement with the experiment[39] $\eta = 1.28 \times 10^{-4}$ Pa s and the equilibrium molecular dynamic simulation [40] $\eta = 1.31 \times 10^{-4}$ Pa s. The statistical error in the molecular simulation and the experiment is around 3 percent.It is also observed that the molecular simulation based on the MICRODY code can correctly predict the temperature and the density dependence of the transport coefficients.[41]

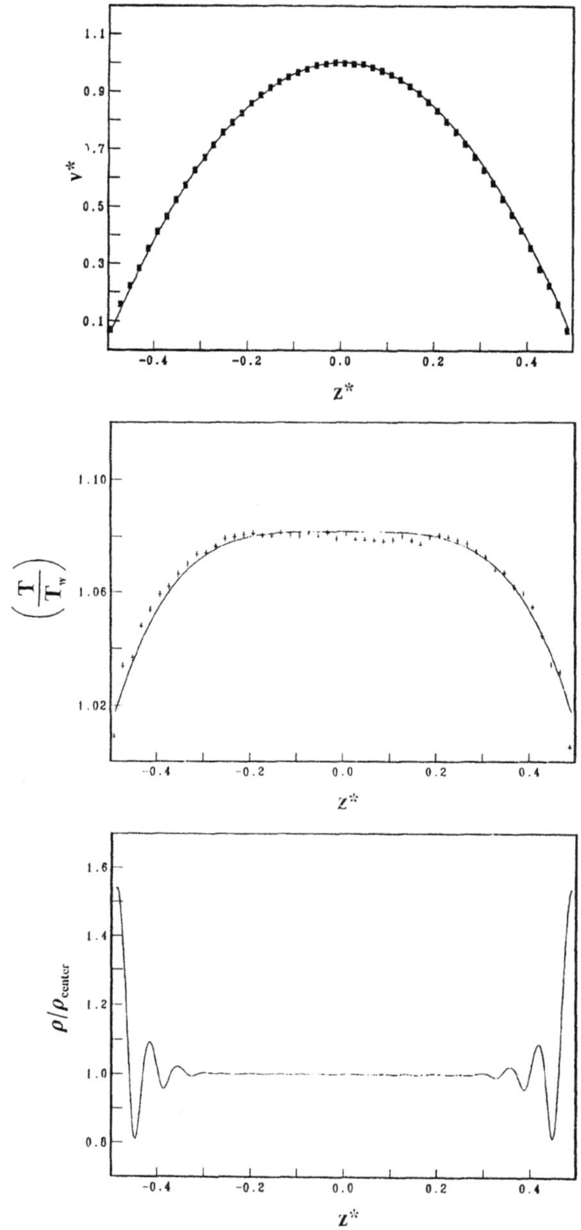

Figure 2. Dense fluid flow in a three dimensional channel. (Top) Velocity profile: the velocity has been scaled with the velocity at the center of the channel and the solid lines correspond to the macroscopic prediction. (Middle) Temperature profile: the solid lines correspond to the macroscopic prediction. (Bottom) Density profile: the density has been scaled with the density at the center.

The above simulations show that, even under extremely large non-equilbrium constraints, MD systems behave according to the predictions of the hydrodynamics equations in which the advective term, $(\mathbf{v} \cdot \nabla)\mathbf{v}$, plays no role. This is true as long as the laminar flow is stable. The next question is whether nonlinear behavior arising from this advective term can also be observed at atomic scales.In obstructed flows, the advective term converts linear momentum into angular momentum. In the region where the Reynolds number is small, linear terms in the Navier-Stokes equations dominate, and we have a laminar flow. However, when the Reynolds number exceeds some critical value, nonlinear terms dominate, the laminar flow becomes unstable, and other flow patterns begin to appear. A characteristic feature of obstructed flows is the phenomena of the eddy formation and the vortex shedding. The understanding of the forces giving rise to the vortex shedding remains one of the basic problems in fluid mechanics. Experimental as well as theoretical investigations have been undertaken to shed light on different aspects of the problem such as the generation and separation of the vortices, the stability of vortex sheets, transitions to turbulent wakes and the possibilities of manipulating or suppressing the shedding process.[42] We now consider the simulation of vortex shedding past a circular disk and a flat plate using a molecular dynamic method.

Unlike the channel flow, the observation of eddy formation and shedding in fluid flow past obstacles of various shapes depends in molecular dynamics computer simulations, on the availability of large samples, to allow the flow to propagate realistically. Experiments on fluid flows past a cylindrical obstacle reveal that the flow structure downstream of the obstruction has an intricate dependence on the Reynolds number. To model this computationally, requires at least one hundred thousand interacting particles in order to allow enough space for the various observable flow phenomena to develop. One of the first demonstrations of this by computer simulation was provided by Rapaport and Clementi, using 160,000 particles.[17]

They considered a 2-dimensional flow wherein the particles were made to flow past a disk of diameter D. At the surface of the disk the slip boundary conditions were used. Some simulations were conducted using no-slip boundary conditions at the surface and it was seen that this did not alter the nature of the flow pattern.

The boundary conditions of the simulation were such that the particles exiting downstream were reintroduced upstream with the same velocity and coordinates, and an acceleration field was applied to counteract the dissipation of the flow by collisions with the disk. The equations of motion were solved by a third order predictor-corrector. A typical run extended over 1.1 nsec, about 120,000 time steps. The computations were carried out with four FPS-264 processors, each processor being responsible for a particular sub-region of the system. A certain amount of data was exchanged between processors to allow for drift between adjacent subregions and the evaluation of interactions for particles lying within a distance r_c of the subregion boundaries.

Flow patterns were generated in the simulation by covering the area under consideration with a grid pattern, and within each grid mean flow directions were evaluated by averaging individual trajectories. The detailed evolution of flow patterns was found to depend on initial conditions, as in experimental observations. Typically, uniform flow at $t=0$ changed rapidly into Stokes flow around the obstacle. In sequence, a pair of counter-rotating eddies first develops downstream of the disk, adjacent to the downstream edge. The length of the recirculatory region grows steadily, the backward velocity amounting to about ten per cent of the overall flow speed. The fluid is compressible, because just downstream of the obstacle in the simulation there is a 25% drop in density.

The bilateral symmetry of the flow then breaks. The counterclockwise eddy grows at the expense of the clockwise (right) eddy. The latter has effectively disappeared by about $t = 310$ psec. The symmetry is then restored only to be lost again, leaving this time the clockwise eddy only. Alternate eddy formation and shedding are repeated several times and an oscillatory wake eventually appears, being the remnant of previously shed eddies. The transverse component has reversed by 840 psec, so that the eddy pattern becomes its mirror image. Two of these eddy formations and sheddings shown in Fig. 3 show the positions of the constituent atoms at two instants in time, displaying the formation of the atomic layer immediately adjacent to the cylinder. This layer is only about 1 to 2 atoms thick. The atoms in the layer are slowly drifting away from the cylinder due to their thermal motions. The recirculation that characterizes the eddies ultimately disappears, leaving only a sinuous wake, extending right up to the disk. The Reynolds number in the simulation was found to be of the same order as that observed experimentally for eddy formation, and likewise the Strouhal number, which characterizes wake oscillation. Evidence was found for a critical Reynolds number for the onset of eddy formation.

In the second simulation of flow past obstacles, the cylinder was replaced by a plate.[19] In flow past a sharp-edged plate, the critical Reynolds number should be zero, thus the non-linear terms should always dominate. Both the experimental observation and the numerical integration of the Navier-Stokes equation at macroscopic dimensions predict that the length of the vortices in the direction of the flow is directly proportional to the Reynolds number itself.

The system used for flow past a plate consisted of a flow and a bath. The flow took place in two dimensions ($L \times L$). The walls at $y = -L/2$ and $L/2$ were periodic boundaries, while those at $x = 0$ and L were in contact with the bath. The obstruction was a thermal plate situated at $x = L/3$, extending from $y = -L/6$ to $L/6$. The bath was employed as a means of alleviating the effects of the plate due to the use of a relatively small number of particles. It acted as a particle source at $x = 0$, and as a particle sink at $x = L$. The bath consisted of $N/10$ particles in a box of size $(L/10) \times L$. The walls at $y = -L/2$ and $L/2$ in the bath were periodic, but thermal walls were used for bath particles at the boundary $x = -L/10$ and $x = 0$. Particles in the bath were not accelerated. When a particle in the flow

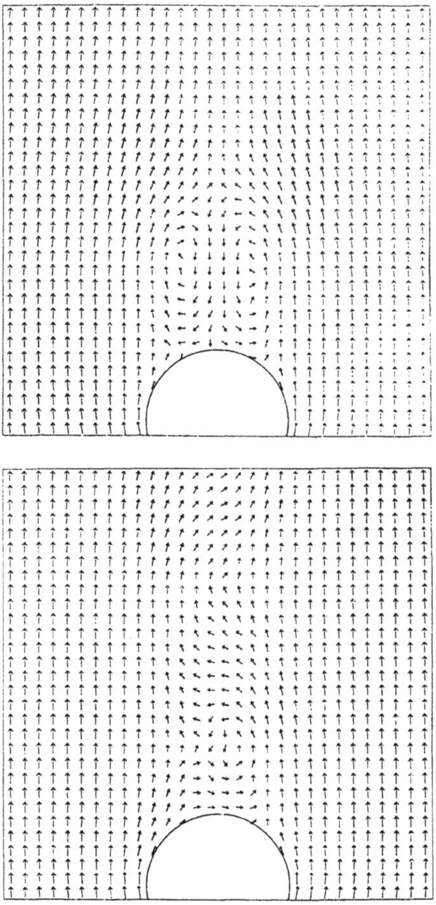

Figure 3. Flow field around the circular disk. (Top) At time t = 750 showing the position of the eddy pair relative to the location of the circular obstacle. Arrow length is related to local velocity. (Bottom) Central region of system at t = 2400 showing the single eddy is present and a portion of the oscillatory wake.

system reaches or passes the wall at x = L it is thermalized and becomes a part of the bath system. Simultaneously, the particle in the bath system nearest the wall at x = 0 becomes a member of the flow system and is subjected to an extra acceleration g. The bath system therefore maintains continuity for particle interactions and separates source behavior from sink behavior.

The particles interact through Eq. (1) and after colliding with the thermal walls are re-injected into the appropriate system with new velocities drawn from a Boltzmann distribution characterized by the wall temperature T. For particles colliding with the thermal plate this simulates no-slip boundary conditions.

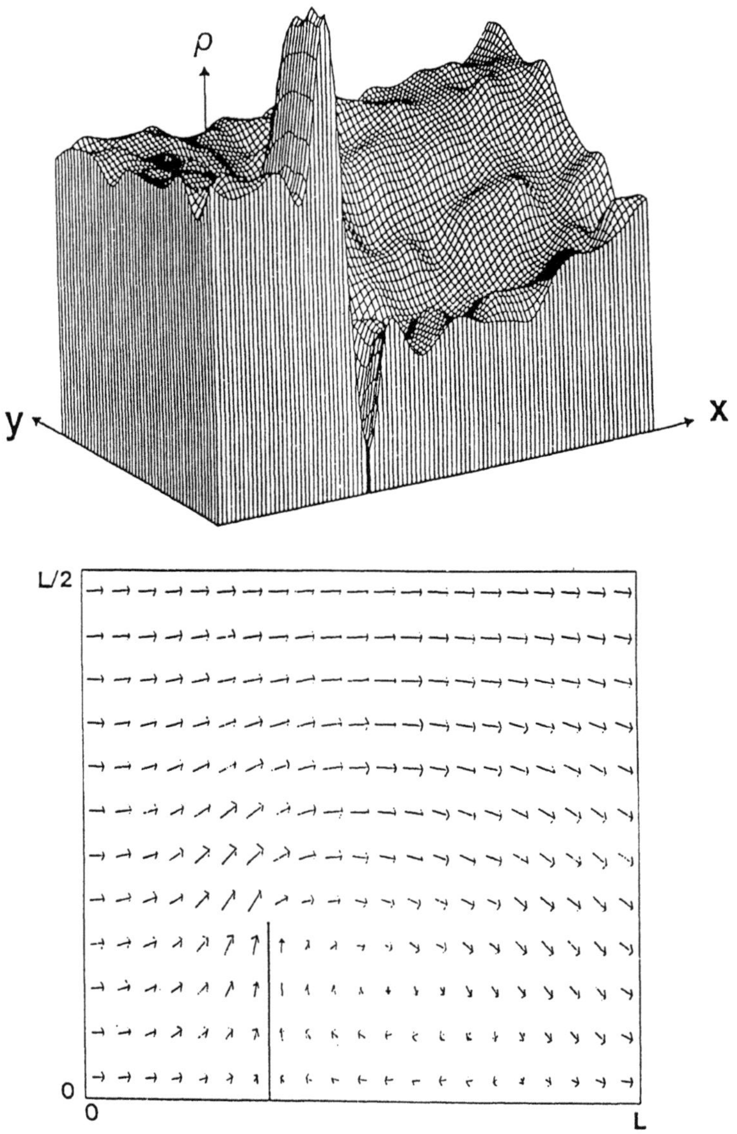

Figure 4. Obstructed flow past a plate. (Top) The velocity field: the arrow lengths are scaled as the square root of the cell velocity to maximum cell velocity. The cell centers are located at the origin of the arrows. (Bottom) The density contour: the baseline density is 1.125 g/cm^3 and the density varies from a maximum of 1.422 g/cm^3 immediately in front of the plate to 1.135 g/cm^3 immediately behind the plate.

This system was set up by Hannon, Lie, and Clementi with 10,000 particles. A fourth order Gear predictor-corrector scheme was used to integrate the equations of motion. The velocity and the density profiles are shown in Fig. 4. For statistical averages the system was divided into 500 square cells, and system behavior was sampled every 10 psec by averaging density, velocity, and temperature for each of the cells over 1000 time steps. The simulation was carried out for acceleration field strengths ranging from 0.625 to 10.0 × 10^{14} cm/sec^2. For a system with g = 2.5 × 10^{14} cm/sec^2 the appearance of two counter-rotating vortices immediately behind the obstructing plate was observed after 120 psec, with a mean velocity in the x direction of 19000 cm/sec. The distance between the plate and the center of these vortices increases with increasing velocity until the system reaches a steady state after 200 psec of simulation time. The vortex size and location, and mean velocity, remain stable over an additional 300 psec of simulation time. A drop in density from 1.422 gm/cm^3 to 1.135 gm/cm^3 was observed from immediately in front to behind the plate, so that a realistic determination of the Reynolds number was difficult to make. However the vortex size and the steady state distance between the plate and the vortex center were roughly proportional to the average mass flux, which is proportional in turn to the Reynolds number. This is in qualitative agreement with the continuum hydrodynamic prediction. A slightly different flow problem was studied by Meiberg.[23] He towed an inclined plate in a dilute gas and using the molecular dynamics and the time counter Direct Simulation Monte Carlo Method analyzed the vortex shedding. One of his remarkable conclusions was that the DSMC method may fail to reveal the vortex structure. This however has been disputed by Koura.[43] He argues that if the null collision DSMC method is used then the DSMC is capable of generating large scale vortex structure. However to accomplish this, the DSMC simulation requires substantially larger number of fluid particles (an order of magnitude more) than one considered in the MD simulations. The flow past flat plate has also been considered by Rivetet al.[44] using the lattice gas method. A detailed comparison of the flow structure obtained through different microscopic methods as well the intricate dependence of the vortex shedding process on the Mach number and the Knudsen number is currently under investigation.

We now turn our attention to analyze the Rayleigh-Benard convection. The Rayleigh-Benard Convection is the result of competition between buoyant forces, thermal dissipation, and viscous drag. A layer of fluid under the influence of a thermal gradient is set up so that the resulting density profile is inverted with respect to the acceleration field. Buoyant forces cause the lighter material to rise, viscous forces present a drag on any motion; thermal dissipation leads to equilibration of the rising material with its surroundings. The competition among these processes and the resulting behavior can be characterized by the Rayleigh number

$$Ra = \frac{\alpha \Delta T g d^3}{\nu D_T} \tag{9}$$

where α is the compressibility, ΔT the temperature difference across the fluid layer, D_T the thermal diffusivity ($\kappa/(\rho C_v)$), ν the kinematic viscosity (η/ρ) and d the depth of the fluid layer. A linear stability analysis of the hydrodynamic equations shows that at a critical value of Rayleigh number, the state wherein the heat is conducted through the layer becomes unstable and the system enters into a state in which heat is convected through the layer. Instability occurs at the minimum temperature gradient at which a balance can be steadily maintained between the kinetic energy dissipated by the viscosity and the internal energy released by the buoyancy force.[45] The nature of the top and and bottom boundaries influence the onset of convection. Once the system reaches the marginal stability curve, the convection sets in at a selected wavelength that determines the spatial periodicity in the direction perpendicular both to the thermal gradient and the acceleration field. A convection cell is defined over one wavelength, consisting of a pair of counter rotating vortices.

Hannon, Lie, and Clementi simulated this instability with 200480 particles interacting via *soft*-sphere potentials in a rectangle of length 3341 Å and depth d = 836 Å.[20] The top and bottom walls were thermalizing boundaries, the side walls were specular boundaries. The top wall was maintained at 486.5 K and the bottom at 86.5 K. This initial simulation provided a transient stream function lasting for about 500 psec, and then disappearing, a convection pattern which appeared and disappeared with time. The interaction between the onset of convection and the scale of the boundary layers probably caused the pattern to become transient. Improvements in boundary layer conditions allowed Mareschal and Kestemont,[21] and later Given and Clementi,[46] to stabilize the Rayleigh-Benard patterns for *hard*-sphere disks. One of the patterns is shown in Fig. 5. Up to now, however, it has been not possible to realize a stable Rayleigh-Benard convection using a more realistic inter-molecular potential.

The macroscopic fluid dynamic equations do not contain any information on the rapid *atomic* fluctuations. Far from equilibrium fluctuations play a crucial role in the evolution of the order or chaos in the system. According to Glansdroff and Prigogine the transition between the different steady-states is mainly due to the creation of giant fluctuations.[47] More recently Fox[48] has shown that in an unstable hydrodynamic system the dynamics of correlations is of critical importance. Near thermodynamic equilibrium, molecular simulations[48-51] have shown that the fluctuation dynamics is consistent with the Landau-Lifschitz theory.[52] However the dynamics of fluctuations at far from thermodynamics equilibrium still remains an intriguing subject. It would be perhaps worthwhile to investigate whether micro-hydrodynamic simulation can be a viable tool for studying the microscopic dynamics of a turbulent fluid system, especially the mechanism of the energy transfer from the large scale to small scale structures.

One of the principle merits of the molecular dynamics approach lies in the fact that it can be used to investigate flow phenomena in a domain where continuum hydrodynamics breaks down. One such area is that of the nonequilibrium flow and heat transfer in a rarefied gas. Experiments by Smoluchowski[53] and Knudsen[54] at the turn of the century revealed that when the gas is sufficiently dilute the velocity and the profiles differ significantly from the predictions of continuum hydrodynamics in as much as that there are pronounced velocity and temperature slips at the wall which increase with increasing rarefication. Similarly

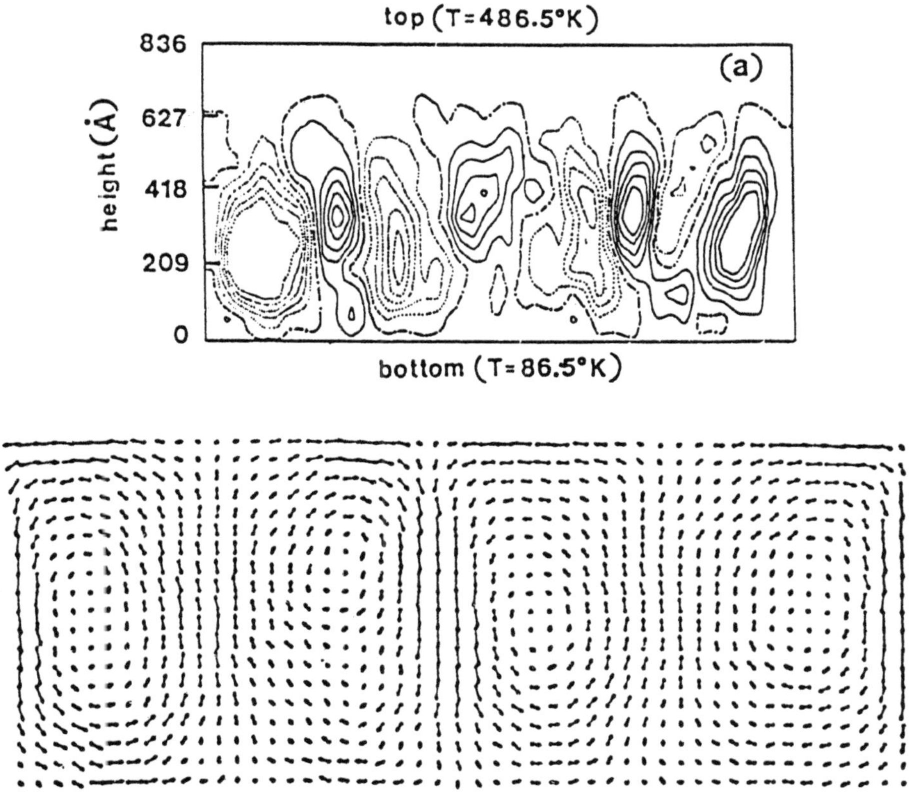

Figure 5. Rayleigh-Benard convection. (Top) Streamline contour obtained from MD simulation of 200,000 fluid particles interacting with the soft sphere potential. (Bottom) Steady state velocity field resulting from the application of cell boundary method in MD simulation of 10,200 hard disks at Ra = 5100.

the experiments of Kundt and Warburg showed that one of the basic results of the kinetic theory of gas, namely that the transport coefficients are independent of the density, is no longer valid in the rarefied gas regime.[55] They predicted that the shear viscosity and the thermal conductivity decrease monotonically with the decreasing density. Since no macroscopic transport theory has yet been developed which can explain such phenomena as observed by Kundt and Warburg, Smoluchowski, and Knudsen, a computer simulation of the flow and heat transfer in a rarefied gas is likely to be very important in the understanding of the physical principles underlying these nonequilibrium processes. Recently, Bhattacharya, Lie, and Clementi[56] have conducted a molecular dynamics simulation of the nonequilibrium flow and heat transfer in a two- and a three- dimensional channel which we will describe briefly. Further details can be found elsewhere.[57,58] In the rarefied gas flow the mean free path, λ, of the gas particles is comparable to the character-

Table I. Parameters used in the simulation of a 2-dimensional dilute gas channel flows (total number of the particles in the system is 2500).

Run No.	Length of the Box in Å	Kn	λ/σ
1	469.7	0.012	1.7
2	1183.6	0.031	10.9
3	2024.0	0.053	31.9
4	3212.9	0.085	80.3
5	4799.5	0.127	179.3
6	5494.1	0.145	235.0
7	6992.1	0.185	380.6
8	10304.1	0.272	826.6

istic length of the flow system. The degree of rarefication is characterized by the Knudsen number which is defined as

$$Kn = \frac{\lambda}{L} \tag{10}$$

where L is the characteristic length scale of the flow system, which in our simulation is the distance between the thermal walls of the channel. In the two dimensional system, we considered eight different flow configurations, varying the Knudsen number from 0.012 to 0.27 (see Table I). The temperature of the thermal walls was kept at 300 K, the acceleration field of strength $g = 5 \times 10^{13}$ cm/sec^2 was applied which produced a flow in the y-direction corresponding to a Reynolds number ~20. As the density is decreased, the dissipative effects become weak and the time to reach the steady state is increased. As a result, at the lowest density considered (Kn=0.27) we have to study the evolution of the system till one million steps. The average velocity in the x direction was found to be zero, as expected. The velocity profiles along the y direction are shown in Fig. 6. for five of the eight runs. The velocities have been scaled with a reference velocity $v_r = 0.14 \times 10^6$ cm/sec which represents the velocity at the center of the channel for Kn=0.27. For a dense fluid at Kn=0.012 the velocity at the wall is almost zero which has also been observed in several other simulations of the channel flow.[29] However, as the density decreases there is a pronounced slip at the wall. In the scaled unit the slip velocity increases from 0.05 for run 2 to 0.28 for run 8. Further, as the density decreases, the temperature jump (the difference between the wall temperature and the average temperature of gas particles next to the wall) increases rapidly. The velocity and the temperature slips found in this simulation are analogous to the ones observed by Smoluchowski and Knudsen in their pioneering experiments and are a direct consequence of the nonequilibrium arising from the infrequent collisions among the gas particles at a very low density.

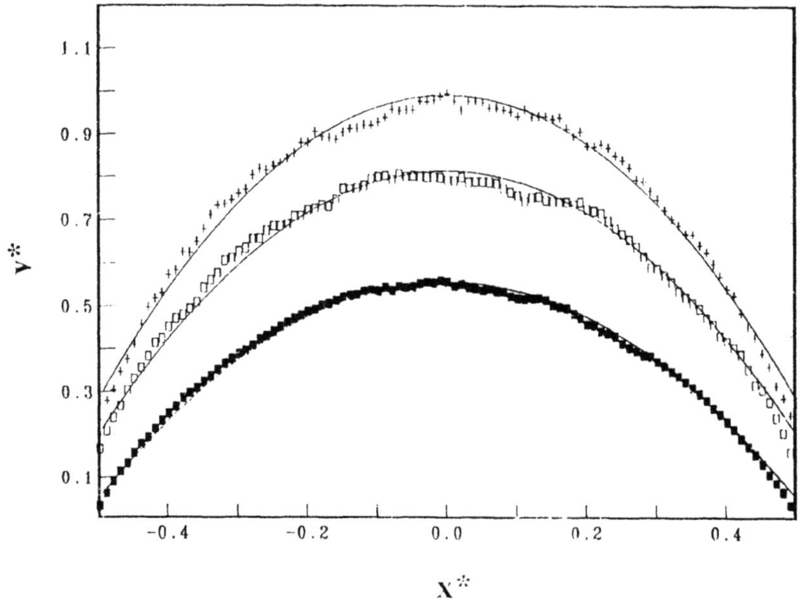

Figure 6. Velocity profiles for various values of Knudsen number. From bottom to top: Kn=0.012, 0.085, 0.14. The velocities have been scaled with a reference velocity $v_r = 1.04 \times 10^5$. The solid lines represent the corresponding macroscopic prediction.

This study also indicates that the flow profiles calculated by solving the Navier-Stokes equations with the velocity and the temperature slip conditions obtained from the simulation, also shown in Fig. 6 as solid lines, are in close agreement with the flow profiles measured in the molecular dynamics simulation.

Figure 7 displays the variation of the shear viscosity with the density. Consistent with the prediction of Kundt and Warburg, it was found that the shear viscosity of a slightly rarefied gas decreases with the decreasing density. This also indicates that the first order Chapman-Enskog solution of the Boltzmann equation showing viscosity independent of density is no longer true in the range of the Knudsen number considered in the simulation.

Maxwell's theory of the velocity slip occupies a central role in the studies on non-equilibrium rarefied gas flows.[59] According to this theory the slip velocity at the wall in the first order approximation is given by

$$v_w = \alpha \lambda \left(\frac{dv}{dx}\right)_w \tag{11}$$

where α is the slip coefficient, which depends on the momentum accommodation of particles (representing how well the gas molecules stick at the wall). Several authors[62-66] have calculated the slip coefficient by making various approximations in the Boltzmann equation and presuming that the Knudsen number is in the slip flow regime (Kn=0.01-0.1). The molecular dynamics simulation, which provides a

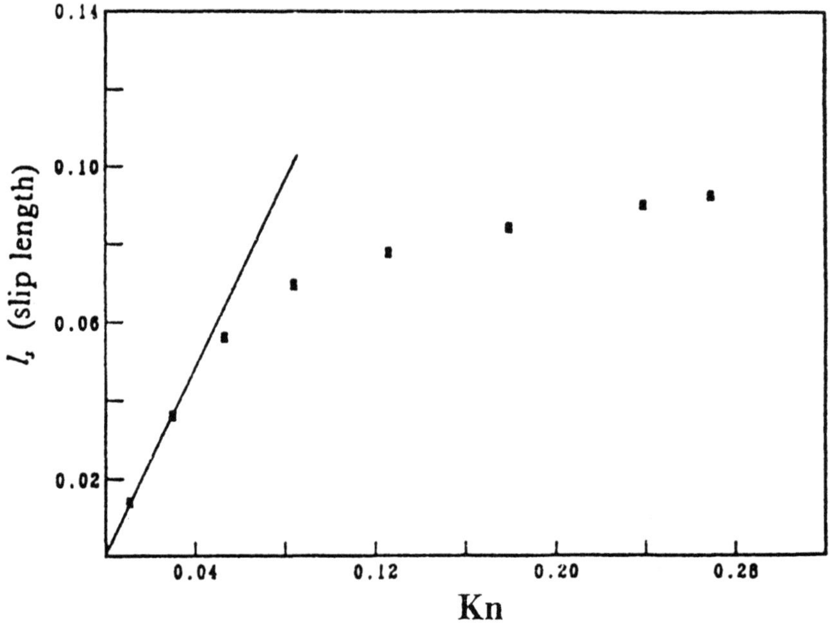

Figure 7. Variation of the shear viscosity with the Knudsen number obtained from MD simulations of dilute gas flow through a 2-dimension channel.

complete information of the flow profiles, can be used to investigate the range of validity of the Maxwell model of slip velocity. This is shown in Fig. 8 where the variation of the slip length

$$l_s = v_w/(\frac{dv}{dx})_w \tag{12}$$

is displayed.

According to Maxwell's theory the slip length is proportional to the Knudsen number.[48] For a small Knudsen number (Kn$\leq$0.05) the slip coefficient 1.16 measured from the slope of the curve is indeed constant and is in good agreement with the theoretical predictions ranging from 1.134 to 1.230 as shown in Table II. For a larger Knudsen number, however, the slope clearly varies with the density. Thus the molecular dynamics simulation of nonequilibrium rarefied gas flows leads to an important conclusion. The Maxwell model with the experimentally determined value of the slip coefficient can be used to calculate the slip velocity only if the Knudsen number is sufficiently small. The apparent breakdown of the Maxwell model at a large Knudsen number can be attributed to the higher velocity gradient and to the coupling with the temperature gradient which has been neglected in the derivation of the equation. Recent theoretical analysis of Kim et al[60] also confirm our prediction that the slip length depends on the Knudsen number.

We now proceed to analyze the simulation of the transitional regime gas flow in a three dimensional channel flow.[38] In the first simulation, designed to understand

Table II. Comparison of the slip coefficient obtained from the present MD simulations with various theoretical calculations based on the Boltzmann equations.

Authors	Slip Coefficient	Reference
Welander (1954)[a]	1.230[b]	52
Wang Chang and Uhlenbeck (1956)[a]	1.134	53
Willis (1962)[c]	1.142	54
Ziering (1963)[a]	1.16	55
Albertoni, Cercignani, and Gutusso (1963)[c]	1.146	56
Bhattacharya and Lie (1989)	1.16	46

[a] Based on approximate solutions of the Boltzmann equation for Maxwell molecules.
[b] There are some doubts as to the accuracy of this work (see Refs. 54 and 56).
[c] Based on numerical calculations of an exact analytical expression of the solution for an approximate model of the Boltzmann equation for Maxwell molecules [P. L. Bhatnagar, E. R. Gross, and M. Krook, Phys. Rev. **94**, 511 (1954)]

the heat transfer process in a very dilute gas, the dynamics of 8000 particles was studied. The upper and the lower thermal walls were 564 Å apart and kept at 288 K and 368 K, respectively. The Knudsen number

$$\mathrm{Kn} = \frac{\lambda}{\mathrm{L}} = \frac{1}{(\sqrt{2}\,\pi\, \mathrm{n}\sigma^2\mathrm{L})} \tag{13}$$

where n is the number density, σ the Lennard-Jones parameter signifying the particle diameter, and L the length of the simulation cube, for the flow system was 0.758. The statistical error in the calculated density is less than 3%. Figure 9 compares the density profile obtained through the MD approach with the moment method solution of the linearized Boltzmann equation by Gross and Ziering[62] and the experimental measurements by Tegan and Springer.[67] Even though the Knudsen number in the MD simulation and the experiment is the same, it should be noted that they are achieved in two different ways: In the simulation the gas is *dilute* and the flow system has a microscopic dimension (5.64×10^{-6} cm), whereas in the experiment the size of the flow system has a macroscopic dimension (0.13 cm) and the gas density is in the *rarefied* regime. A good agreement among the simulation, the experimental, and the theoretical results indicates that the underlying dynamics of the nonequilibrium heat transport phenomena studied in the former two methods is the same as the one embodied in the Boltzmann equation.

We now turn our attention to analyzing the velocity slip in the channel flow. We have studied seven different flow configurations, as shown in Table III. An acceleration field of strength $g = -1 \times 10^{14}$ cm/sec^2 was applied in the y direction which produced subsonic flows (Mach number varies from 0.5 to 0.99). The Knudsen number characterizing these flows are in the transition regime (Kn $\simeq$ 1). The thermal walls were kept at 300 K. The velocity slip at the wall increases with

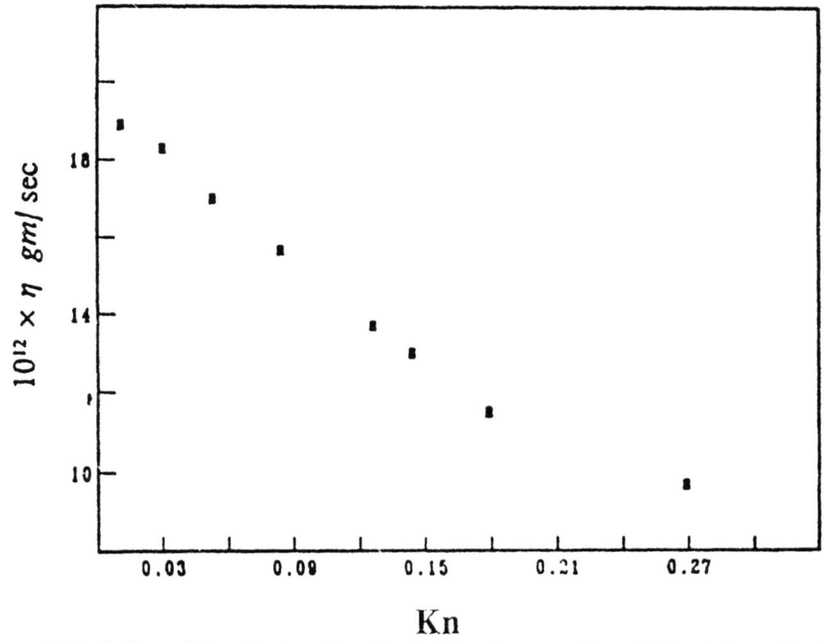

Figure 8. Variation of the slip length with the Knudsen number obtained from the simulations. The straight line is obtained by a least squares fit of the first two simulation data and the origin (which implies that we are assuming there would be no slip as Knudsen number goes to zero). The slope of the line gives the slip coefficient discussed in the text.

an increasing Knudsen number. However at all levels of the Knudsen number, the measured velocity profiles agree remarkably well with the solution of the Navier-Stokes equation with the velocity slip condition.In the transition regime it was found that the slip length varies as the logarithmic of Kn. This result indicates that the recently proposed velocity slip model by Coron,[68] using the Chapman-Enskog solution, would break down in the transition regime.

To analyze the effect of viscous heating on the velocity slip, at Kn=2.79, we increased the strength of the acceleration field and studied flows at Mach numbers (Mc) of 1.68, 2.76, and 4.05. It was found that the velocity slip is proportional to the square root of the average gas temperature, while the velocity difference between the center and the wall scales as the acceleration field, as predicted by the Navier-Stokes equations.

In the framework of linear irreversible thermodynamics,[69] the linear transport equations are derived on the assumption of local equilibrium. The assumption of local equilibrium implies that even though the dissipative system is globally in a nonequilibrium state, the irreversible system can be partitioned into small cells which remain in the state of a thermodynamic equilibrium such that the nonequilibrium entropy change is governed by the Gibbs equation. Since except for a region very close to the wall, the Navier-Stokes equations appear to be valid in our channel flows, we might anticipate that the velocity distribution is sufficiently

Table III. Parameters used in the simulation of 3-dimensional channel flow (the total number of particles is 4096).

Run No.	Length of the Box in Å	Kn	λ/σ
1	208.44	0.20	12.24
2	300.63	0.42	37.08
3	392.31	0.72	82.95
4	459.53	1.00	134.95
5	559.66	1.48	243.25
6	647.69	1.98	376.63
7	767.92	2.79	629.22

close to the local Maxwellian distribution. This is indeed true for subsonic flows as shown in Fig. 10. However in supersonic flows, also shown in Fig. 10, the presence of a high shear significantly distorts the shape of the velocity distribution function, and a deviation from the local Maxwellian function is observed.

Liu and Lees have proposed a method for solving the transition regime gas flow problem, which consists in expressing the nonequilibrium distribution function as

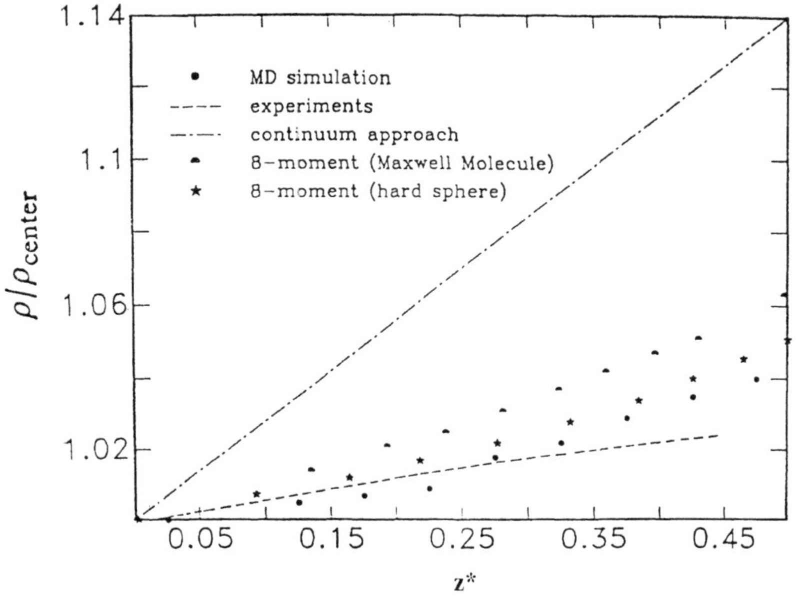

Figure 9. Comparison of the different density profiles for channel flows at Kn = 0.758. x* is a scaled coordinate across the channel (≡ x/L) with the origin at the center of the channel.

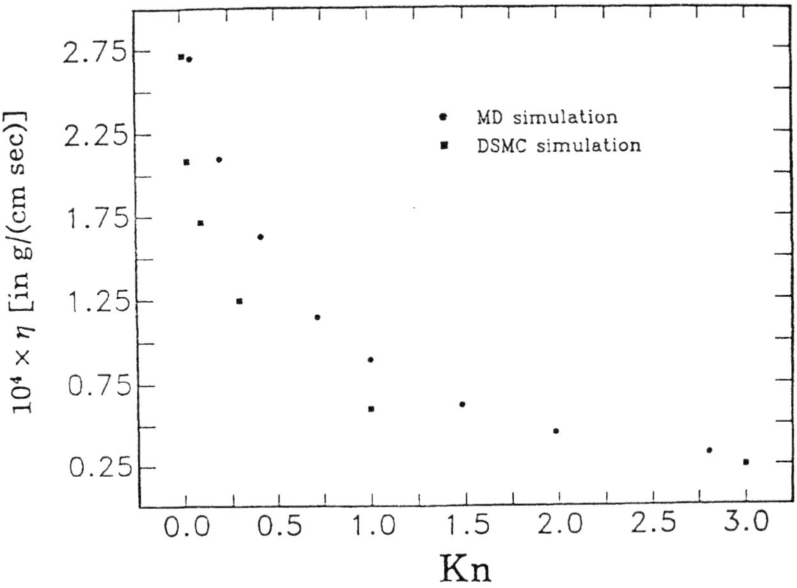

Figure 10. Nonequilibrium velocity distribution functions at Kn=2.79 for a subsonic flow (Mc=0.99) and a supersonic (Mc=4.05) flow. The solid line represents the equilibrium Maxwell velocity distribution function. v^* is a scaled dimensionaless velocity defined as $\sqrt{2k_BT/m}$. T for the subsonic and supersonic flows are 423.68 K and 1731.74 K, respectively.

a superposition of two Maxwellians.[70] It was noticed that the theory of Liu and Lees breaks down above a critical Mach number, which for the Couette flow geometry is approximately 3.1. In view of our simulation, it is clear that the limitation of the theory of Liu and Lees lies in the fact that it fails to represent, in the velocity distribution function, the nonequilibrium effects encountered in the supersonic flows.

To analyze the organization of the dissipation in the system, we have calculated the Rayleigh-Onsager dissipation function,[69] which in the framework of the linear irreversible thermodynamics represents the entropy production in the system. Consistent with the experiments of Sody and Berry,[71] we found that the dissipation in the system decreases monotonically with decreasing density, and that most of the dissipation takes place in a thin boundary layer, the thickness of which decreases with the increasing Knudsen number.

Due to numerous industrial applications, the microscopic and the macroscopic description of the transition regime flow phenomena, have been extensively investigated using methods of the kinetic theory of dilute gas and nonequilibrium thermodynamics. Grad,[72] based on his thirteen moment method solution of the Boltzmann equation, argued that the nonequilibrium transport phenomena of gas flows in the transition regime can be described by a set of generalized hydrodynamic equations, where the well known Navier-Stokes and Fourier laws are replaced by a new set of constitutive equations incorporating the nonlocal heat and stress relaxation phenomena in addition to the dissipative effects arising from spatial nonuniformities of field quantities. In recent years, in the framework of extended irreversible thermodynamics,[73] Grad's theory has been generalized to the nonlinear domain.[74] However, the full potential of the neoclassical hydrodynamics embedded in the thirteen moment method has not been realized, firstly because of the lack of appropriate boundary conditions needed to solve the heat and stress evolution equations, and secondly due to the recognition that unlike Navier-Stokes equations, these new transport equations do not admit solutions for all flow parameters.[75] A different approach, in the framework of the extended Boltzmann equation, wherein the collision dynamics of the gas surface interaction is explicitly incorporated, has been proposed by Dorfman and van Beijeren[76] and Wood[77] has proposed a theory based on the mean free path arguments. The molecular dynamics simulations can prove useful in examining the range of the validity of the generalized hydrodynamic theory derived in the framework of the Boltzmann equation or the extended irreversible thermodynamics.

The transition regime flow has also been studied with numerical stochastic simulations. The most widely used algorithm is the one proposed by Bird.[78,79] In his Direct Simulation Monte Carlo (DSMC) method, Bird simulates the physics of the transport process occurring in a rarefied hard sphere gas by uncoupling molecular motions and intermolecular collisions over small time intervals and cell sizes. The DSMC algorithm of Bird is based on the Kac master equation and is compu-

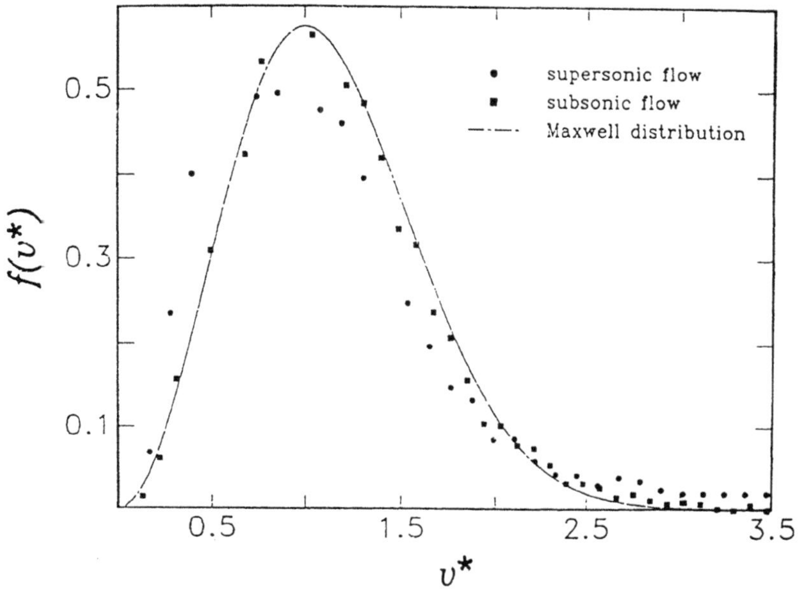

Figure 11. Comparison of the Knudsen number dependence of the shear viscosity.

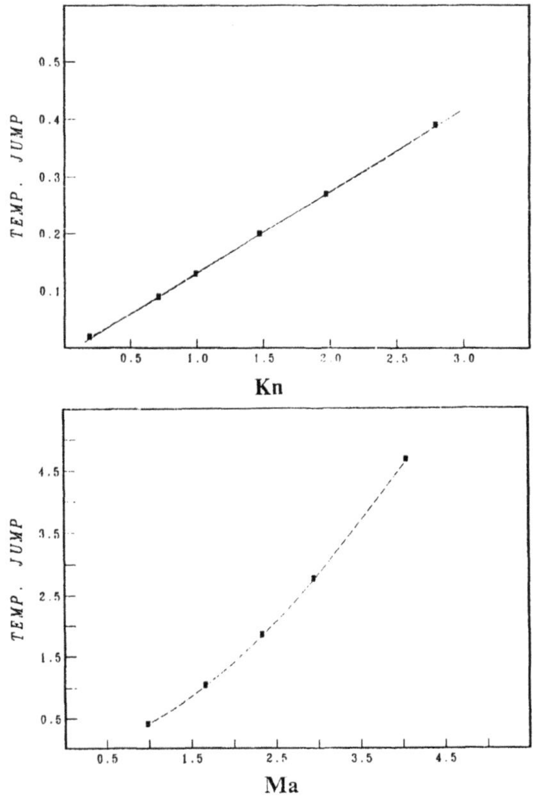

Figure 12. The variation of the temperature jump $(T_s - T_w)/T_w$ with the Knudsen number in the transition regime gas flow. The wall temperature is represented by T_w while T_s denotes the temperature of the fluid next to the wall. (Top) Knudsen number dependence at a Mach number = 0.99; (Bottom) Mach number dependence at a Knudsen number = 2.79.

tationally efficient. A slightly different DSMC method based on the Boltzmann equation has been proposed by Nanbu.[80] Meiburg has shown that the DSMC method does not conserve the angular momentum and hence may not accurately describe a high vorticity flow problem.[23] However Nanbu, Watanbe, and Igarashi argued that if cell sizes are sufficiently small, then the angular momentum is almost conserved.[81] Moreover since the vorticity is only related to the linear momentum which is conserved in the DSMC method, the method should be applicable for high vorticity problems.[81]

We now proceed to examine the comparison of the results obtained by the DSMC method and the MD method. The two methods differ in as much as the molecular dynamic method is based on the explicit evaluation of the dynamical interaction of gas particles, while the DSMC treats the collision dynamics of the fluid particles using the Monte Carlo method. We analyze the variation of shear viscosity of Argon with the Knudsen number. The viscosity has been calculated at the center of the channel from the measured velocity profile.[57] In the limit of a very low Knudsen number (Kn=0.05), we have calculated the shear viscosity at 316.7K with 8000 particles and found the result, $\eta = (2.70 \pm 0.05) \times 10^{-4}$ gm/(cm sec) to be in good agreement with the analytical result of 2.58×10^{-4} gm/(cm sec) obtained for hard-sphere particles at the same temperature.[82] Using the DSMC method Lengrand and Fadili have calculated the shear viscosity for the variable hard sphere model.[83] in plane Couette flow geometry. Figure 11 shows the Knudsen number dependence of the shear viscosity, obtained using the MD and DSMC methods. Both simulations predict that the shear viscosity of the Argon gas decreases with the Knudsen number. The difference between the two simulations is mainly due to the difference in the interaction potentials. The importance of the potential and particle collisions should diminish as the gas gets more rarefied. Thus the two methods should give the same viscosity in the limit of a high Knudsen number. Figure 11 seems to indicate that this is indeed the case.

One of the objectives of this simulation of the transition regime gas flow problem was to analyze how the temperature jump at the channel wall depends on the Knudsen number and the speed of the flow. As is well known when a high speed space vehicle (such as a space shuttle) reenters in the atmosphere it experiences a severe thermal stress due to viscous heating. To design an appropriate thermal protection system one needs an accurate estimate of the heating load at the surface.

In Fig. 12 we have shown how the temperature jump at the wall varies with the Knudsen number. It appears that in the transition regime the temperature jump at the wall increases almost linearly with the Knudsen number. The variation of the temperature jump with the Mach number is shown in Fig. 13 . Since continuum hydrodynamics can not provide any information on the temperature jump, our results outline the importance of molecular dynamic simulation in the field of high altitude aerodynamics.

Unlike the DSMC or the lattice gas dynamics method,[84] the MD method discussed above does not make any approximation in the collision dynamics of fluid particles. It can therefore be used to investigate some fundamental problems of the transition regime flows, such as the effect of the molecular constitution and the physical state of thermal walls on the structure of the nonequilibium Knudsen layer. Due to the complex many body effects associated with the scattering of the gas molecules from the lattice of the surface bounding the flow, analytical theories[85] have not been very successful in predicting how the physical and the chemical nature of the surface influences the nonequilibrium flow properties. It appears that the molecular dynamics simulation, which incorporates the dynamical interaction between the gas molecules and the molecules constituting the wall would be able to provide new insights in this area. We are currently studying this approach.

References

1. *Nonequilibrium Thermodynamics*, S.R. deGroot and P. Mazur, (North Holland, Amsterdam, 1968).
2. E. P. Muntz, D. Weaver, and D. Cambell, eds. *Rarefied Gas Dynamics,* Proceedings of the 16-th International Symposium, Progress in Astronautics and Aeronautics, (AIAA, Washington, 1989).
3. W.T.Ashurst and W.G.Hoover, Phys. Rev. A, **11**, 658, (1975).
4. D.Y. Chan and R.G.Horn, J.Chem. Phys., **83**, 5311, (1985).
5. K.Kremer, G.S.Grest and I. Carmesin, Phys. Rev. Lett., **61**, 566, (1988).
6. *Simulation of Liquids and Solids* , eds., G. Ciccotti, D. Frenkel and I.R. Mcdonald,(North Holland, Amsterdam, 1987).
7. B.J. Alder and T. E. Wainwright, Phys. Rev. A, **1**, 18, (1970).
8. *Molecular Dynamic Simulation of Statistical Mechanical System*, eds., G. Ciccoti and W. Hoover, (Clarendon, London, 1987).
9. D. J. Evans, Phys. Rev. A **23**, 1988 (1981).
10. J. P. Ryckert, A. Bellemns, G. C. Ciccotti, and G. V. Paolini, Phys. Rev. Lett. **68**, 128 (1988).
11. G. P. Morris and D. J. Evans, Phys. Rev. A **32**, 2425 (1985).
12. J. J. Erpenbeck, Phys. Rev. Lett. **52**, 1333 (1984).
13. D. M. Heys, G. P. Morris and D. J. Evans, J. Chem. Phys., **83**,4760, (1985).
14. W. Loose and S. Hess, Phys. Rev. A, **37**, 2099, (1988), Rheol. Acta, **28**, 91, (1989).
15. B.L. Holian, W.G. Hoover, B. Moran and G.K. Straub, Phys. Rev. A, **22**, 2798, (1980).
16. B. L. Holian, A. J. De Groot, W.G. Hoover and C.G. Hoover, Phys. Rev. A, **41**, 4552, (1990).
17. D. C. Rapaport and E. Clementi, Phys. Rev. Lett. **57**, 695 (1986).
18. L. Hannon, G. C. Lie, and E. Clementi, Phys. Lett. A **119**, 174 (1986).
19. L. Hannon, G. C. Lie, and E. Clementi, J. Sci. Computing **1**, 145 (1986).
20. L. Hannon, G. C. Lie, and E. Clementi, J. Stat. Phys. **51**, 965 (1988).
21. M. Mareschal and E. Kestemont, Nature **329**, 427 (1987).
22. M. Mareschal, M. Malek Mansour, A. Puhl and E. Kestmont, Phys. Rev. Letts, **61**, 2550, (1988).
23. E. Meiburg, Phys. Fluids **29**, 3107 (1986).
24. *Recent Development in Nonequilibrium Thermodynamics*, eds., J. Casas Vazquez, D. Jou and G. Lebon, (Springer, Heidelberg, 1984).
25. D.K. Bhattacharya, Acta Mech., **47**, 87, (1983); Ann. Phys., **39**, 325, (1982).
26. D.K. Bhattacharya, Warme und Stoffuberttragung, **17**, 27, (1982).
27. D.K. Bhattacharya and B. C. Eu, Phys. Rev. A, **35**,821, (1987), Mol. Phys., **59**, 1145,(1986), Phys. Rev. A, **35**, 4850, (1987).
28. *Computer Simulation of Liquids*, M.P. Allen and D.J. Tildesley, (Clarendon Press, Oxford, 1987).
29. J. Koplik, J. R. Banavar, and J. F. Willemsen, Phys. Rev. Lett. **60**, 1282 (1988).
30. U. Heinbuch and J. Fischer, Chem. Phys. Lett., **135**,587, (1987).

31. D.J. Evans, Phys. Lett., **91A**, 457, (1982).
32. M.J. Gillian and M. Dixon, J. Phys. C, **16**,869, (1990).
33. W. Loose, Phys. Rev. A, *40*, 2625, (1989).
34. C. Trozzi and G. Cicotti, Phys. Rev. A **29**, 916 (1984).
35. A. Tennenbaum, G. Cicotti, and R. Gallico, Phys. Rev. A **25**, 2778 (1982).
36. A.L. Robinson, Science, **236**, 150, (1987).
37. U. Heinbuch and J. Fischer, Phys. Rev. A, **40**, 1144, (1989).
38. I. Bitsanis, J.J. Magda, M. Tirell and H.T. Davis, J. Chem. Phys., **87**, 1733, (1987).
39. H.J. Hanley, R.D. McCarty and W.M. Haynes, J. Chem. Phys. Ref. Data, **3**, 979, (1974).
40. M. Schoen and C. Hoheisel, Mol. Phys., **56**, (1985).
41. D.K. Bhattacharya and E. Clementi, (to be submitted).
42. *Fluid Dynamics*, G.K. Batchelor, (Cambdride University Press, Cambridge, 1988).
43. K. Koura, Phys. Fluids A, **2**, 209, (1990).
44. J.P. Rivet, M. Henon, U. Frisch and D. Humieres, Europhysics Lett., **7**, 231, (1988).
45. *Hydrodynamic and Hydromagnetic Stability*, S. Chandrasekhar, (Clarendon, Oxford, 1961).
46. J. Given and E. Clementi, J. Chem. Phys., **90**, 7376, (1989).
47. *Thermodynaic theory of structure, stability and fluctuations* , P. Glansdorff and I. Prigogine, (Wiley Interscience, New York, 1971).
48. R.F. Fox, Phys. Rev. A, **41**, 2969, (1990)
49. M. M. Mansour, A. L. Garcia, G. C. Lie, and E. Clementi, Phys. Rev. Lett. **58**, 876 (1987).
50. A. L. Garcia, M. M. Mansour, G. C. Lie, and E. Clementi, J. Stat. Phys. **47**, 209 (1987).
51. A. L. Garcia, M. M. Mansour, G. C. Lie, and E. Clementi, Phys. Rev. A **36**, 4348 (1987).
52. L. D. Landau and E. M. Lifschitz, *Fluid Dynamics* (Pergamon, New York, 1959).
53. M. V. Smoluchowski, Wied Ann. Phys. **64**, 101 (1898); *ibid.*, Ann. Phys. **33**, 1559 (1910).
54. M Knudsen, Ann. Phys. **34**, 593 (1911).
55. A. Kundt and E. Warburg, Philos. Mag. **50**, 53 (1875).
56. D. K. Bhattacharya, G. C. Lie, and E. Clementi, in *Proc. of the 4th Int'l Conf. on Supercomputing,* edited by L. P. Kartashev and S. I. Kartashev (Int'l Supercomputing Inst., St. Petersburg, Florida, 1989).
57. D. K. Bhattacharya and G. C. Lie, Phys. Rev. Lett. **62**, 897 (1989).
58. D. K. Bhattacharya and G. C. Lie, *IBM Technical Report* , KGN 202, (1989).
59. J. C. Maxwell, Philos. Trans. Roy. Soc. London **170**, 231 (1867).
60. C.S. Kim, J.W. Dufty, A. Santos and J.J. Brey, Phys. Rev., *40*, 7165, (1989).
61. S. Albertoni, C. Cercignni and L. Gotusso, Phys. Fluid, **5**, 993; Systems **1**, 791 (1987).
62. E. P. Gross and S. Ziering, Phys. Fluids **2**, 701 (1959).

63. P. Welander, Arkiv. Fysik **7**, 507 (1954).
64. C. S. Wang Chang and G. E. Uhlenbeck, University of Michigan Engineering Research Institute Rept. 2457-1-T (1956).
65. D. R. Willis, Phys. Fluids **5**, 127(1962).
66. S. Ziering, Phys. Fluids **6**, 993 (1963). **5**, 993 (1963).
67. W. P. Teagan and G. S. Springer, Phys. Fluids **11**, 497 (1968).
68. F. Coron, J. Stat. Phys. **54**, 829 (1989).
69. I. Gyarmati, *Nonequilibrium Thermodynamics* (Springer Verlag, Berlin, 1970).
70. C. Y. Liu and L. Lees, in *Rarefied Gas Dynamics,* L. Talbot, ed. (Academic, New York, 1961) p.341.
71. F. Sody and A. S. Berry, Proc. Roy. Soc. (London) **83**, 254 (1910).
72. H. Grad, Commun. Pure and Appl. Math. **2**, 313 (1949).
73. I. Muller, Z. Phys. **198**, 329 (1967).
74. B. C. Eu, J. Chem. Phys. **73**, 2958 (1980); Phys. Rev. A,*40*, 6395, (1989); *A36*, 400, (1987).
75. H. Grad, in *Handbuch der Physik,* Vol. XII, G.Flugge, ed. (Springer Verlag, Berlin, 1958), p.205.
76. J. R. Dorfman and H. van Beijren, in *Statistical Mechanics,* Part **B**, **B**. J. Berne, ed., (Plenum, New York, 1977).
77. L.C. Woods, J. Fluid Mech., **93**, 585, (1979).
78. G. A. Bird, Phys. Fluid **6**, 1518 (1963); *ibid.* Ann. Rev. Fluid Mech. **18**, 243 (1978); S. M. Yen, Ann. Rev. Fluid Mech. **16**, 67 (1984); K. Koura, Phys. Fluids **13**, 1475 (1970), G. A. Bird, Phys. Fluids **30**, 364 (1987).
79. G. A. Bird, *Molecular Gas Dynamics* (Clarendon, London, 1976).
80. K. Nanbu, J. Phys. Soc. Jpn. **52**, 3382 (1983).
81. K. Nanbu, Y. Wantanbe, and S. Igarashi, J. Phys. Soc. Jpn. **57**, 2877 (1988).
82. S. Chapman and T. G. Cowling, "The Mathematical Theory of Non-Uniform Gases" (Cambridge U. P. London, 1970).
83. J. C. Lengrand and I. C. Fadili, Laboratoire D'Aerothermique du C.N.R.S., R 87-4 (1987).
84. R. Monaco, ed., *Discrete Kinetic Theory, Lattice Gas Dynamics and Foundations of Hydrodynamics* (World Scientific, Singapore, 1989).
85. F.O. Goodman and H. Y. Wachman, *Dynamics of Gas Surface Scattering* (Academic, New York, 1976).

Chapter 19.

Brownian Dynamics Simulations of a Complex Fluid System

D.K. Bhattacharya and E. Clementi
IBM Corporation
Data Systems Division, Dept. 48B/MS 428
Neighborhood Road
Kingston, New York 12401, U.S.A.

1. Introduction

In recent years there has been considerable interest in understanding the dynamical behavior of multiphase fluid systems such as colloidal dispersion, flocs and gels. These are encountered in a range of industrial and biological systems[1-9]. Industrial applications include the handling of highly concentrated coal suspensions in hydrocarbon carrier fluids, transportation of residue suspensions from separation processes, synthesis of pesticide suspension concentrates, coalesence of bubbles and droplets in foams and emulsions, detergency, condensation and power coating. Biological applications relate to the fusion of cells and vesicles, engulfment of viruses and macro-molecules by cells and enzyme-substrate reactions. A particularly interesting example pertain to the functioning of drugs such as antihistamines, antibiotics, anaesthetics and tranquillisers which possess surface active properties and are able to form aggregates in aqueous solution. The colloidal nature of these substances influence their biological activity, specially the passage of the drug through cell membrane to reach the specific binding site. One of the basic objectives in the studies of these complex systems is to determine how the macroscopic properties of the suspension such as viscosity, thermal conductivity, diffusivities, and the sedimentation rate depend on the interaction forces among suspended particles, and their size and distribution in space. A characteristic feature of these multiphase systems is the time scale separation which occurs as one form of the motion in the system is much faster than the other. The molecular dynamic simulations of such complex systems in most cases are not realizable, because it requires short time steps to treat fast motions (such as solvent particles), while very long runs are needed to allow the evolution of the slower modes (such as stress relaxation in macro-molecules, protein folding). An alternative approach wherein the dynamics of the solvent particles are omitted, and their effects are incorporated by a combination of random forces, frictional forces and hydrodynamic forces have proved very useful[10-25]. Here the dynamic evolution of the complex system is studied in the framework of the Langevin equation or the Fokker-Planck equation. Central to current studies on the stochastic dynamic simulation is the algorithm due to Ermak and McCammon[10] who first employed it to investigate the influence of the hydrodynamic effects on the dynamics of large Brownian particles whose mass and size are much larger than the host media. This algorithm has subsequently been applied in several areas such as particle congulation, polymer dynamics, suspension rheology and the gel electrophoresis phenomena. In this chapter we will describe the simulation strategies for conducting the stochastic dynamic simulation for analyzing the non-equilibrium flow properties of colloidal suspensions. In section II we briefly recall the basic theory of the Brownian dynamics description and outline the work of Ermak and McCammon. Then we discuss the modification of Ermak's algorthim for describing the Brownian dynamics of macro-molecules under shear. Section III gives an overview of the simulation program "BROWNIAN", and details a representative application of Brownian dynamics simulation techniques where we discuss the nonlinear suspension rheology of macro-molecules interacting via a repulsive Lennard-Jones potential. Lastly in section IV we comment on the generalized Brownian dynamic simulation techniques and outline new areas of applications.

2. Brownian Dynamic Simulation: Basic concepts

In 1827, the botanist Brown[27] discovered under his microscope a vigorous irregular motion of small particles originating from pollen floating on water. He also observed that very fine particles of minerals undergo similar incessant motion. Even though the idea of combining such a motion- Brownian motion with molecular motion became fairly widespread in the later half of the nineteenth century, it was the work of Einstein[28] which gave the first clear theoretical foundations of Brownian dynamics. The theory of Brownian motions was further developed by Smoluchowski, Langevin, Fokker and Planck. Two approaches have evolved[29-35]. One is based on the Fokker-Planck differential equation for the N particle phase-space probability distribution function f $(r_i p_i, t)$. The other uses 3N coupled Langevin equations to describe the motion of N particles. Although these two approaches are quite different, they are equivalent in as much as the statistical description of the N particle phase space trajectories are identical with both the methods. The solution of the Fokker-Planck equation yields directly the phase space distribution function f $(r_i p_i, t)$. On the other hand, with the set of coupled Langevin equations, one directly calculates N particle phase space trajectories and then obtains f $(r_i p_i, t)$ by averaging. We now proceed to briefly outline the mathematical framework of these theories and then discuss the algorithm of Ermak and McCammon. We are particularly interested in understanding the basic assumptions inherent in this algorithm.

2.1 Fokker- Planck Description

The system under consideration consists of N Brownian particles immersed in a viscous, incompressible solvent continuum. The particles may be subjected to configuration dependent forces due to inter-particle interaction and the external fields (such as a uniform shear or electrical field). The particles are also subjected to velocity-dependent hydrodynamic interactions. This , in general, is incorporated within the approximation of pair-wise additivity using a configuration dependent friction tensor β_{ij} . The Fokker-Planck equation for this particular system can then be written as

$$\frac{\partial f}{\partial t} + \sum_i (m_i^{-1} p_i \frac{\partial f}{\partial r_i} + F_i \frac{\partial f}{\partial p_i}) = \sum_i \sum_j \frac{\partial}{\partial p_i} \beta_{ij} \, (m_i^{-1} p_j \, f + k_B T \frac{\partial f}{\partial p_j}) \tag{1}$$

where i and j run over the particle coordinates, F_i represents the inter-particle and external forces, m_i is the mass of the particle associated with the index i, r_i, p_i are position and momentum components, respectively, k_B is the Boltzmann constant and T is the absolute temperature. The N particle phase-space distribution function $f(r_i, p_i, t)$ is the average number density of the Brownian particles at r_i, p_i at a time t which evolves from the initial distribution $f(r_i, p_i, 0)$. If the momentum variables relax to equilibrium much more rapidly than the position variables, and if we presume that all spatial gradients including the force F_j are reasonably smooth functions of the particle configurations, then the equation (1) can be reduced to

$$\frac{\partial f}{\partial t} = \sum_i \sum_j \frac{\partial}{\partial r_i} D_{ij} \, (\frac{\partial f}{\partial r_j} - \frac{1}{k_B T} F_j f) \tag{2}$$

where $f(r_i, t)$ is the N particle configuration space distribution and $D_{ij} = D_{ji}$ is the configuration dependent diffusion tensor related to β_{ij} by

$$\sum_j \beta_{ij} D_{jl} = \sum_j D_{ij} \beta_{jl} = k_B T \delta_{il} \tag{3}$$

As a consequence of the assumptions made in the above derivation, the Eq.(2) is valid for time intervals

$$\Delta t >> \frac{m_i}{\beta_{ij}} \tag{4}$$

The velocities $\{v(t + \Delta dt)\}$ are effectively uncorrelated with the velocity $v(t)$, and the mean square velocity component of each Brownian particle over time interval $\geq \Delta t$ is the equilibrium value $< v_i^2 > = k_B T / m_i$. In a Brownian dynamic simulation the particle displacements are chosen according to certain stochastic rules. The successive configurations are selected from the configuration space distribution function $f(r_i, t)$. In such a procedure the Brownian particle positions are known (

r_i^0) at the start of the given step ($\Delta t = 0$), and the corresponding initial distribution function is $f(r_i, 0) = \Pi\delta(r_i - r_i)$ where $\delta(r)$ is the Dirac delta function. Then, to the first order in Δt, the solution of the equation (2), is the multivariate Gaussian distribution function $f(r_i, \Delta t)$ which is uniquely defined by its moments

$$< \Delta r_i(\Delta t) > = \sum_j \left(\frac{\partial D_{ij}}{\partial r_j} + \frac{D_{ij}}{k_B T} F_j \right) \Delta t \tag{5}$$

and

$$< \Delta r_i (\Delta t) \Delta r_j (\Delta t) > = 2 D_{ij} \Delta t \tag{6}$$

where the diffusion tensor, the gradient of the diffusion tensor and the forces are all evaluated in the initial configuration ($\Delta t = 0$) . The configuration of the system at the end of the step is chosen from $f(r_i, \Delta t)$ and the procedure is repeated to obtain one typical Brownian trajectory. In the presence of the hydrodynamic interaction, the average displacement consists of two terms: one proportional to the gradient of the diffusion tensor and the other proportional to a weighted sum of the forces acting on all Brownian particles. The presence of the hydrodynamical interaction thus, induces a cross correlation in the displacement of different particles. Two diffusion tensors have been used to model the hydrodynamic interaction mediated by the solvent. The first is the Oseen tensor given by

$$D_{ij} = \frac{k_B T}{6\pi\eta a}\delta_{ij}, \qquad \text{on the same particle,} \tag{7}$$

$$D_{ij} = \frac{k_B T}{8\pi\eta r_{ij}}\left(I + \frac{\mathbf{r}_{ij}\,\mathbf{r}_{ij}}{r_{ij}^2}\right), i, j \qquad \text{on different particles,} \tag{8}$$

and the other is the Rotne-Prager tensor given by

$$D_{ij} = \frac{k_B T}{6\pi\eta a}\delta_{ij}, \qquad \text{on the same particle,} \tag{9}$$

$$D_{ij} = \frac{k_B T}{8\pi\eta r_{ij}}\left[\left(I + \frac{\mathbf{r}_{ij}\,\mathbf{r}_{ij}}{r_{ij}^2}\right) + \frac{2a^2}{r_{ij}^2}\left(\frac{1}{3} I - \frac{\mathbf{r}_{ij}\,\mathbf{r}_{ij}}{r_{ij}^2}\right)\right] i, j \tag{10}$$

on different particles.

Here η is the solvent viscosity, 'a' the radius of th spherical Brownian particle, δ_{ij} is the Kronecker delta, r_{ij} is the vector from the center of the Brownian particle i and the center of the Brownian particle j and I is the unit tensor. Both of these tensors have the property

$$\sum \frac{\partial D_{ij}}{\partial r_{ij}} = 0 \tag{11}$$

which speeds up the computation, as one need not compute the gradient of the diffusion tensor. A more accurate representation of the hydrodynamic interaction which includes three and four body effects has been proposed by Mazur and van Saarloos[25]. To avoid a singularity arising at certain inter-particle distances, several authors have also used the preaveraged diffusion tensor[26].

2.2 Langevin Description

In this approach the dynamic evolution of N Brownian particles are described by a set of Langevin equations:

$$m_i \dot{v}_i = -\sum_j \beta_{ij} v_j + F_j + \sum_j \alpha_{ij} \zeta_j \tag{12}$$

where the term $\Sigma\alpha_{ij}\zeta_j$ represents the randomly fluctuating force exerted on a particle by the surrounding fluid. The ζ_j are described by a Gaussian distribution with the mean and covariance

$$< \zeta_i > = 0 \tag{13}$$

and

$$< \zeta_i(t)\, \zeta_j(t') > = 2\delta_{ij}\, \delta\,(t - t') \tag{14}$$

The coefficients α_{ij} are related to the hydrodynamic friction tensor by

$$\beta_{ij} = \frac{1}{k_B T} \sum_l \alpha_{il}\, \alpha_{jl} \tag{15}$$

The term in the LHS of the Eq.(9) denotes the change in momenta $\dot{p}_i = m_i \dot{v}_i$, while the RHS represents the forces acting on the Brownian particles which are composed of three terms: a friction term which tends to decrease the particle energy, a random force which tends to increase the particle energy and a systematic force due to the interaction potential energy between the Brownian particle and the external force.

To derive the diffusive displacement equation from equation (12) we multiply it by D_{li} and after summing over i we obtain

$$\sum_j \tau_{ij}\, \dot{v}_j = -v_i + \frac{1}{k_B T} \sum_j D_{ij} \cdot F_j + \sum_j \alpha_{ij}\, \zeta_j \tag{16}$$

where

$$\tau_{ij} = \frac{D_{ij}\, m_j}{k_B T} \tag{17}$$

and

$$\sigma_{ij} = \frac{1}{k_B T} \sum_l D_{il}\, \alpha_{ij} \tag{18}$$

Now expanding τ in a Taylor series about the initial position of the Brownian particles we obtain

$$\tau_{ij} = \tau_{ij}^0 + \sum_l \frac{\partial \tau_{ij}^0}{\partial r_l} \Delta r_l + \text{-------} \tag{19}$$

where $\Delta r_l = r_l(t) - r_l(0)$ and the superscript indicates that the t=0 value is implied. Making a similar expansion for D_{ij}, σ_{ij} and F_j and substituting this in (16) and retaining the first order terms in τ_{ij} and zeroth order terms in D_{ij}, σ_{ij} and F_j , we get

$$\begin{aligned} \tau_{ii}^0 \dot{v}_i + v_i &= \tau_{ii}^0 \exp(-t/\tau_{ii}^0) \frac{d}{dt} [\exp(t/\tau_{ii}^0)\, v_i] \\ &= -\sum_j \tau_{ij}^0 \dot{v}_j - \sum_j \sum_l \frac{\partial \tau_{ij}^0}{\partial r_l} \Delta r_l \dot{v}_j \\ &\quad + \frac{1}{k_B T} \sum_j D_{ij}^0 F_j^0 + \sum_j \sigma_{ij}^0 \zeta_j, \end{aligned} \tag{20}$$

An equation for $v_i(t)$ can be obtained by multiplying Eq (20) by $(\tau_{ij}^0)^{-1}$ exp (t/τ_{il}^0) and integrating over time. A second integration over time and some straight forward mathematical manipulation leads to the following equation for the displacement of the Brownian particles:

$$\begin{aligned} \Delta r_i = & \sum_j \frac{\partial D_{ij}^0}{\partial r_j} t + \sum_j \frac{D_{ij}^0 F_j^0}{k_B T} t + \\ & \sum_j \int_0^t ds \{ 1 - \exp[-(t-s)/\tau_{ii}^0] \} \sigma_{ij}^0 \zeta_j(s) \end{aligned} \tag{21}$$

The mean displacement calculated from Eq (21) is

$$< \Delta r_i(\Delta t) > = \sum_j \left(\frac{\partial D_{ij}^0}{\partial r_j} + \frac{D_{ij}^0}{k_B T} F_j^0 \right) \Delta t \tag{22}$$

while the covariance of the displacements, to the first order in Δt is

$$< \Delta r_i (\Delta t) \Delta r_j (\Delta t) > = 2 D_{ij}^0 \Delta t \tag{23}$$

These results are seen to be in agreement with those of Eqs. 5 and 6, obtained from the diffusive Fokker Planck equation. A crucial element in this derivation is the assumption that the time step is much greater than the momentum relaxation time. In the Brownian dynamic computer simulation technique due to Ermak and McCammon the configuration space trajectories are calculated numerically. These trajectories are composed of successive displacements, each taken over a short time step ∇t. The displacements are to be randomly chosen in accordance with the multivariate distribution function whose moments are specified by Eqs (5) and (6). The system of Brownian particles are placed in an initial configuration and the particle positions after a time step ∇t are calculated using the displacement Eq. (21) which can be concisely written as

$$r_i = r_i^0 + \frac{\Delta t}{k_B T} \sum_{j=1}^{N} D_{ij}^0 \cdot F_j^0 + \Delta t \sum_{j=1}^{N} \nabla_j D_{ij}^0 + R_i (D_{ij}^0, \Delta t) \tag{24}$$

where the superscript "0" indicates that the variable is to be evaluated at the beginning of the time step. The displacement $R_i(t)$ is a random displacement with a Guassian distribution whose average value is zero and whose variance-covarience is $< R_i(\Delta t) R_j(\Delta T) > = 2 D_{ij}^0 \Delta t$. As mentioned earlier, the time step of the simulation is chosen such that $\Delta t >> m_i D_{ij}^0/(k_B T)$. The numerical accuracy limits the maximum size of the time step requiring it to be sufficiently small so that the force on the particles and the gradient of the diffusion tensor are essentially constant during ∇t.

We now proceed to discuss the generalization of the Ermak and McCammon simulation technique for studying the behavior of a colloidal suspension subjected to a constant shear in a plane Couette flow geometry[11]. In this case it is necessary to modify Eq. (24), by incorporating the displacement of the Brownian particle due to the imposed shear. We consider the specific case of a steady shear flow $(\dot{\gamma} y, 0, 0)$ of monodisperse suspension and further presume that the effect of the hydrodynamic interaction in the shear flow is negligible. Here $\dot{\gamma}$ is the imposed shear. Then the displacement equation (24) in the component form is written as

$$\begin{aligned} r_{xi} = r_{xi}^0 &+ \frac{D^0 \Delta t}{k_B T} [F_{xi}^0] + R_{xi}^0 + S_{xi}^0 + \dot{\gamma} r_{yi}^0 \Delta t + \\ & F_{zi}^0 (\dot{\gamma} \beta^{-1}) \{ \frac{\Delta^2}{2} - 2 \frac{\Delta}{\beta} - \beta^{-2} \} \end{aligned} \tag{25}$$

$$r_{yi} = r_{yi}^0 + \frac{D^0 \Delta t}{k_B T} [F_{yi}^0 + R_{yi}^0] \tag{26}$$

$$r_{zi} = r_{zi}^0 + \frac{D^0 \Delta t}{k_B T} [F_{zi}^0 + R_{zi}^0] \tag{27}$$

where

$$< R_{i\alpha}^0 (\Delta t) R_{j\beta}^0 (\Delta t) > \; = 2 D^0 \Delta t \, \delta_{ij} \, \delta_{\alpha\beta} \tag{28}$$

$$< S_i^2(t) > \; = (\frac{2 D^0}{3}) (\dot{\gamma}^2 \Delta t^3) \tag{29}$$

If the imposed shear and the time step are sufficiently small then Eq. (25) can be further simplified to

$$r_{xi} = r_{xi}^0 + \frac{D^0 \Delta t}{k_B T} [F_{xi}^0] + R_{xi}^0 + \dot{\gamma} r_{yi}^0 \Delta t \tag{30}$$

The Eqs. (25)-(27) form the basis of our Brownian dynamic simulation of a colloidal suspension. We now proceed to describe the structure of the computer program "BROWNIAN" for implementing this computational algorithm.

3. The Simulation Program "BROWNIAN": An Overview

The program BROWNIAN enables one to study the microscopic dynamics of the rheological phenomena in a colloidal suspension. Using this, one can analyze how the diffusion constant, shear viscosity, normal stresses , pressure and the structure factor of a colloidal suspension changes with the imposed shear. It also provides the computational framework for investigating how the size of the Brownian particles, their density, and the molecular interaction force changes the global macroscopic properties of the suspension.

The simulation is conducted in a cube. The shear in the x direction is imposed by using specific boundary conditions. The periodic boundary conditions are applied in the other two directions. The Brownian particles can interact either with the repulsive Lennard-Jones Potential or with the Yukawa potential. The number of the Brownian particles, particle diameter, the density, charge, cut-off distance, the solvent viscosity, temperature, time step , the imposed shear rate, and the length of the simulation are the typical input parameters. The simulation is initiated either from a regular lattice or from a previously generated restart file. The most critical and computationally demanding part of the program is the evaluation of the interaction force between the Brownian particles, which can be vectorized. The random numbers needed for calculating random displacements of the Brownian particles are calculated using the subroutine DURAND from the ESSL library. To update the particle positions, the Eqs. (26-27) and (30) are used. The computational algorithm can be effectively implemented in a parallel machine.

The configurations of the Brownian particles are periodically stored. Form the time-ordered sequence of the particle configurations, the diffusion coefficient, the radial distribution function, and the structure factor are determined. The presence of the shear in the x direction destroys the isotropy of the diffusion tensor, therefore the diffusion constants along and perpendicular to the flow direction are calculated. To monitor the convergence of the diffusion coefficients we also evaluate the mean square displacement during the simulation time, using a particular algorithm proposed by van Megen and Snook. The shear stress , normal stress difference, and the pressure are calculated by evaluating the component of the pressure tensor :

$$P_{\alpha\beta} = \sum_{i}^{N}\sum_{j}^{N}\frac{r_{\alpha ij}r_{\beta ij}(d\phi_{ij}/dr)}{2V} \tag{31}$$

where $r_{\alpha ij}$ is the α'th cartesian component of the vector between particles i and j, and V is the volume containing the N Brownian particles. The expression is analogous to the one used in the MD simulation except the fact that in the case of colloids the contributions arising from the dynamic correlation of the particle velocities are negligible and attention is focussed only on the configurational contributions.

3.1 The Nonlinear Rheology of the Colloidal Suspension

The rheology of the colloidal dispersion is of considerable practical importance and has been the subject of several theoretical investigations[36,37].

To gain some insight into the problem we conducted a BD simulations with 256 particles. The particles interact with the repulsive Lennard-Jones potential:

$$\phi_{RLJ}(r) = 4\varepsilon\left\{\left(\frac{\sigma}{r}\right)^{12} - \left(\frac{\sigma}{r}\right)^{6}\right\} + \varepsilon \quad for\, r \leq 2^{1/6}\sigma \tag{32}$$

The quantities appearing in the computations were suitably scaled. These parameters correspond to a colloidal dispersion of 10 μm silica spheres in water at room temperature[48,49]. The time step was chosen such that the maximum displacement due to the random force was 0.01 σ. The root mean square displacement due to the systematic force was typically < 1% of that due to the random force. The density of the silica particles was set to be 1.74 gcm^{-3} .

The shear rate dependence of the relative viscosity is shown in Fig.1 . The shear rate has been multiplied with the structural relaxation time

$$\tau_r = \frac{3\,\pi\,\sigma^3\,\eta_s}{4k_BT} \tag{33}$$

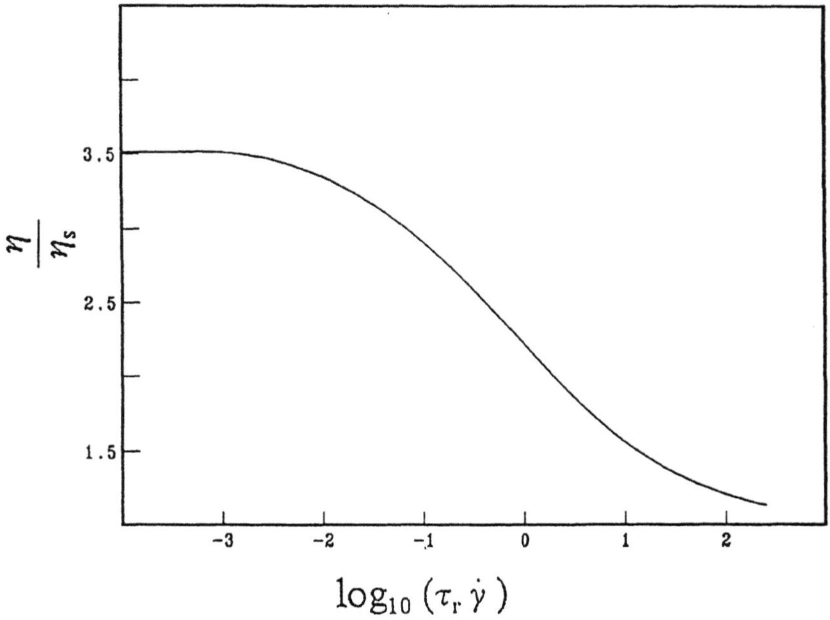

Figure 1. The variation of the reduced shear viscosity $\frac{\eta}{\eta_s}$ with the dimensionless shear rate.

and the product $\tau_r\dot{\gamma}$ now can be expressed in terms of the Peclet Number Pe in the following way :

$$Pe = \frac{\tau_r\dot{\gamma}}{2} \tag{34}$$

It is seen that for $\tau_r\dot{\gamma} \simeq 1$, the relative shear viscosity of the suspension starts decreasing. Recently Dhont et al[36-37], on the basis of their experiments and theoretical analysis predicted that the non-Newtonian viscosity of the colloidal suspension over a wide range of the shear rate varies as $(Pe)^{1/2}$. Our Brownian dynamic simulation seems to indicate that this is true as long as the particle density is sufficiently small. At a higher concentration, the exponent of the Peclet number was found to systematically increase.

The primary normal pressure difference ($P_{yy} - P_{xx}$) as well as the secondary normal pressure difference ($P_{zz} - P_{yy}$) also exhibit the shear thinning behavior, however, the shear thinning in this case begins at a much higher value of Pe ($\simeq$ 10). The shear rate dependence of the normal stress difference was found to be consistent with the experiment. The pressure as well as the infinite frequency compressional modulus are found to increase rapidly with the shear rate.

As the shear increases, the structure of the colloidal suspension changes. To study this, one can examine the pair correlation function g (r) which in an unsheared fluid, measures the average probability of finding a particle at a distance r from

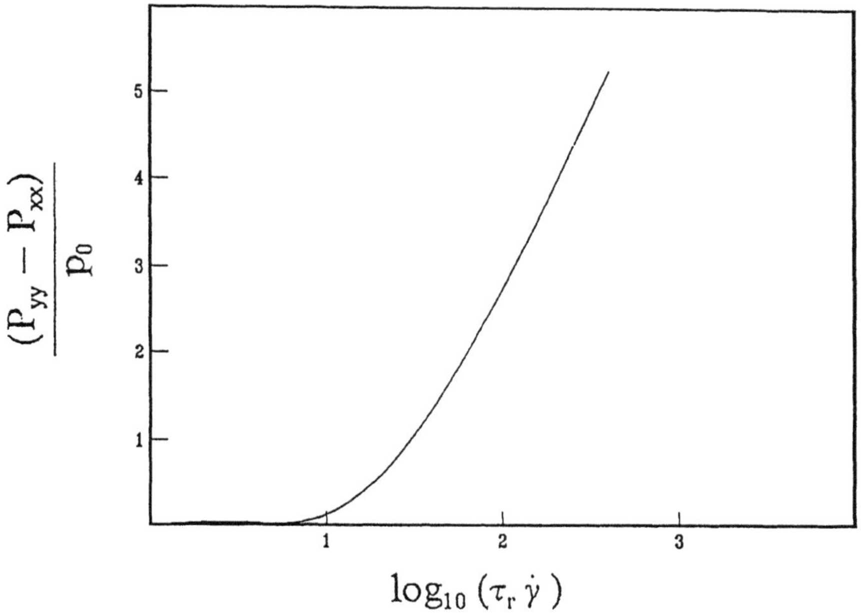

Figure 2. The variation of the primary pressure difference $(P_{yy} - P_{xx})$ with the dimensionless shear rate.

another particle. A steady state shear flow reduces the symmetry and distorts g (r). With an increasing shear rate the first peak in g(r) moves in closer separation and has a broad tail for r >> 1 while the structure beyond the first correlation shell is lost.

4. Discussion

We intend to further broaden the scope of our Brownian dynamic simulation program, both in the direction of new applications such as the phenomena of the shear induced ordering in a colloidal suspensions, as well as computationally on the more fundamental level, by incorporating the techniques of the generalized Brownian dynamics techniques[53-55] We will briefly outline these.

4.1 Shear induced phase transition in the Colloidal Suspension

At a higher shear rate, the colloidal suspension of spherical particles can exhibit some very interesting behavior.[39-42] One such phenomena is the shear induced melting [39] in which a dilute or semidilute suspension which has a long-range order (either bcc or fcc) at zero shear rate is observed to undergo a series of structural changes as the shear rate is increased. These structures include a flowing crystal, sliding layers and string and amorphous ordering. A number of theoretical and computational studies[42,51] have been recently reported for unravelling the physical processes giving rise to these structures at far from equilibrium situations. Using a non equilibrium molecular dynamic simulation technique, Erpenbeck[43] showed that at a sufficiently high shear rate a fluid undergoes a shear induced ordering and the fluid system enters into a new phase, called a string phase, which is characterized by a sudden decrease in the shear viscosity. Evans and Morris[44] argue that this phase transition like behavior may be an artifact of the specific thermostat which presumes a stable velocity profile. Loose and Hess[45] have also reported the evidence of a shear induced ordering in a simple fluid using a somewhat different computational algorithm. Evans and Heys[47] studied the behavior of the cross-correlation functions in the string phase. Bossis and Bardy[2] have noticed the transition to a layer structure in suspension flow using a Stokesian Dynamics method. More recently Heys[49] and Xue and Grest[50] found the existence of a string phase in a colloidal suspension using the Brownian dynamics techniques. Both of these results can be recreated using our simulation program. Currently we are studying the system considered by Xue and Grest.

This simulation was inspired by the experiments of Qui et al.[51] Using the forced Rayleigh scattering Qui et al measured the self diffusion of the charged polyball suspension driven far from equilibrium by an oscillating shear flow. Aside from the usual enhancement of the diffusion due to Taylor dispersion they observed an additional isotropic contribution which increases linearly with the shear rate. They also found that the anomalous contribution is absent in noninteracting samples and increases monotonically with the interparticle strength. To simulate this phenomena Xue and Grest conducted the stochastic dynamic simulation wherein the Brownian particles interact with the Yukawa potential

$$U(r) = U_0 \exp(2\kappa R)(1+\kappa R)^{-2} \exp(-\kappa r) / \varepsilon r \qquad (34)$$

where ε is the background dielectric constant, R the radius of the Brownian particle κ is the inverse screening length, $U_0 = Z^{\times 2}/\varepsilon r$, $Z^{\times}$ is the renormalized charge. The parameters chosen in the simulation were close to the one considered in the experiment. The particle radius was R = 0.091 μ and the effective charge was $Z^{\times}$ = 400. The screening length $\lambda = 5$. The number of the simulated particles was 500. The calculated diffusion coefficient in the direction perpendicular to the flow is shown in Fig. 3 . Consistent with the experiments it was found that the diffusion coefficient increases with the shear rate. However at very low shear rate, a deviation with the experimental prediction (a linear increase with the shear rate) was also noticed. It appears that this may be due to a frequency dependence or amplitude dependence of the diffusion coefficient which was not addressed in the simulation. At a higher shear rate they observed a shear induced ordering in a plane perpendicular to the flow (see Fig. 3) At a critical shear rate, which depends on the temperature, the diffusion coefficient decreases dramatically. The onset of the phase transition can be located by analyzing the anisotropic structure factor. Before the critical shear rate, the structure factor in the direction perpendicular to the flow has two peaks and the first peak is greater than the second one. However, near the transition point, as the layering phenomena begins to occur, the second peak starts growing rapidly. A study of the phase diagram reveals that

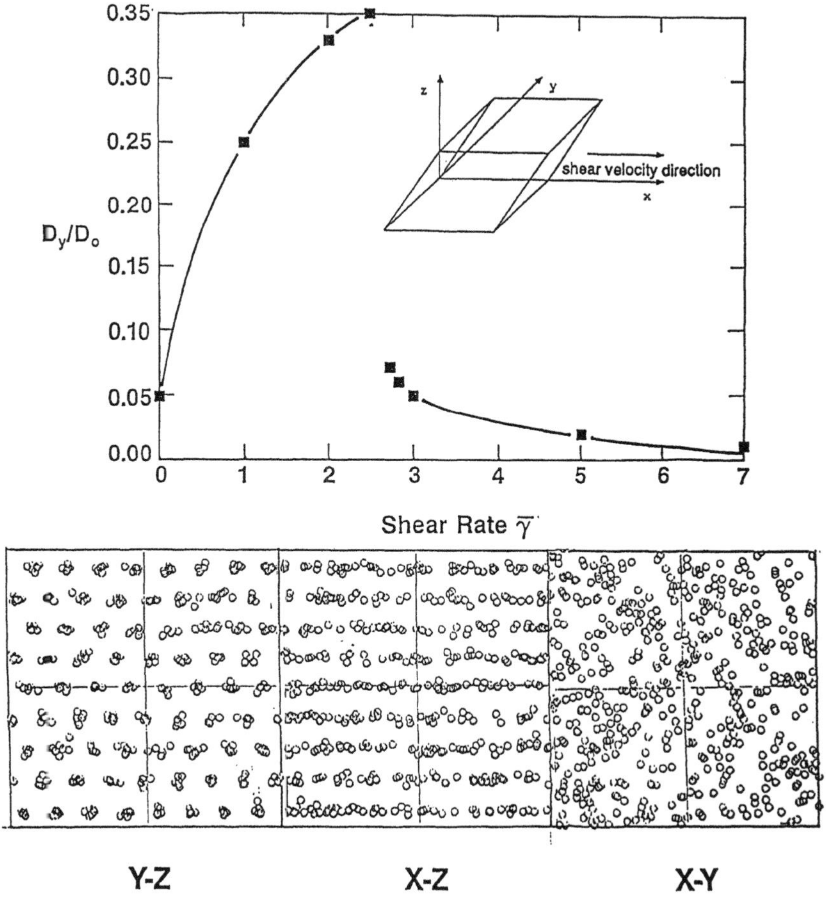

Figure 3. (Top) The shear rate dependence of the diffusion coefficient perpendicular to the flow direction. The reduced diffusion coefficient D_y/D_0 decreases dramatically after a critical shear rate ($\gamma_c = 2.7$). Here D_0 denote the diffusion constant of the suspension at a zero shear rate. (Bottom) The projection planes for a system in the string- ordered phase.

the most string order occurs in the supercooled liquid region below the melting temperature.

In our study[53] we are particularly interested in analyzing how the onset of the string phase influences the shear viscosity, shear stress and the normal stresses in the system. We are also investigating whether the hydrodynamic effects at higher concentrations delay the onset of the phase transition.

4.2 Generalized Brownian Dynamics Techniques

The present version of the BROWNIAN program was designed mainly for analyzing the microscopic dynamics of the colloidal suspensions and the computational method is based on the Ermak and McCammon algorithm. One of the assumptions inherent in the Ermak and McCammon algorithm is the fact that the nonequilibrium momentum distribution function of the Brownian particles decays very fast to it's equilibrium value , so that dynamical evolution can be effectively described by a diffusion equation in the configuration space. However if the size of the Brownian particles are comparable with those of solvent particles or if the time step is less than the momentum relaxation time then this assumption breaks down. We are now extending the computational framework of our simulation program to include the methods of the generalized Brownian Dynamic which will allow us to study the non diffusive Brownian motion. In this case the dynamics of the Brownian particle is described by the following equations[29]

$$\dot{v}(t) = -\int_0^t ds\, \beta_1\,(t-s)v(s) + + R_1(t) \tag{35}$$

$$< R_1(t) > = 0 \tag{36}$$

$$< R_1(t)R_1(s) > = \beta_1(\,|t-s|\,) \;< v^2 > \tag{37}$$

Mori[29-30] has shown that the Laplace Transform of the normalized autocorrelation function $C_v(t) = < v(t)v(0) > / < (v^2) >$ can be represented by the continued fraction

$$\tilde{C}_v(p) = \cfrac{1}{p + \cfrac{\beta_1(0)}{p + \cfrac{\beta_2(0)}{\cdot \atop {p + \cfrac{\beta_{n-1}(0)}{p + \tilde{\beta}_n(p)}}}}} \tag{38}$$

where a set of random forces $\{R_j(t), i = 1, \infty\}$ has been defined. The set $\{R_j(t), i = 1, \infty\}$ satisfies the following set of equations

$$\dot{R}(t) = -\int_0^t ds\, \beta_{(i+1)}\,(t-s)R_i(s) + + R_{(i+1)}(t) \;\; ; \;\; i = 1, \infty \tag{39}$$

and the memory functions $\{\beta_i(t),\ i = 1, \infty\}$ are proportional to the random forces auto correlation functions

$$< R_i(t)R_i(s) > = \beta_i(\,|t-s|\,) \;< R^2_{(i-1)} > \quad ; \; i = 2, \infty \tag{40}$$

The GBD technique requires the Mori continued fraction, (38) to be truncated at some suitable level n replacing β_n by the constant $\beta_n = \xi(0)$. Ciccotti et al[53] have proposed a general method of obtaining the memory functions by presuming an analytic expression for the Fourier transform of the ACF and determining the coefficients by fitting it with the molecular dynamic simulation results. A different technique for modelling the memory functions, which is based on the autoregreesive procedure, has been recently proposed by Smith and Harris.[54] The generalized Brownian techniques may be very useful for analyzing the trajectory of the Brownian particle on time scales comparable to the viscous damping time and may reveal the intricate interplay between the short range, random molecular force and the long range hydrodynamic forces on the particle.

References

1. *Physics of Complex and Supermolecular Fluids*, S. Safran and N.A. Clark, (Wiley Interscience, New York, 1987).
2. G. Bossis and J.F. Brady, in Ann. Rev. Fluid Mech., **20**, 111, (1988).
3. C.R. Wildemuth and M.C. Williams, Rheol. Acta, **23**, 627, (1984).
4. C.R. Wildemuth and M.C. Williams, Rheol. Acta, **24**, 75, (1985).
5. Q.D. Nguyen and D.V. Boger, Rheol. Acta, **24**, 427, (1985).
6. T.F. Tardos, Chem. Ind., **7**, 210, (1985).
7. T.F. Tardos and A. Zsednai, Colloids and Surfaces, **43**, 105, (1990).
8. G.C. Ansell, E. Dickison and M. Ludvigsen, J.C.S. Faraday II, **81**, 1269, (1985).
9. P. Schurtenberger, R. Scarazzini, L.J. Magid, M.E. Leser and P.L. Luisi, J. Phy. Chem., **94**, 3695, (1990).
10. D.L. Ermak and J.A. McCammon, J. Chem. Phys., **69** , 1352, (1978).
11. P.J. Dotson, J. Chem. Phys., **79**, 5730, (1983).
12. E. Helfand, Physica A, **118 A**, 123, (1983)
13. M.P. Allen, Mol. Phys., **47**, 599, (1982).
14. W.F. Gunsteren and H.J.C. Berendsen, Mol. Sim., **1**, 173, (1988).
15. K.J Gaylor, I.K. Snook, W.J. van Megen and R.O. Watts, J. Phys. Math. Gen., **13**, 2513, (1980).
16. K.J Gaylor, I.K. Snook, W.J. van Megen and R.O. Watts, J.C.S. Faraday II, **76**, 1067, (1980).
17. K.J. Gaylor, I.K. Snook and W.J. van Megen J. Chem. Phys., **75**, 1682, (1981).
18. W.J. van Megen and I.K. Snook, J.C.S. Faraday II **80**, 383, (1984).
19. S. Allison, R. Austin and M. Hogan, J. Chem. Phys., **90**, 3843, (1989).
20. J.M. Deutsch, Science, **240**, 922, (1988).
21. M. Oivera de la Cruz, D. Gersappe and E.O. Shaffer, Phys. Rev. Lett., **64**, 2324, (1990).
22. R.W. Pastor, R.M. Venable and M. Karplus, J. Chem. Phys., **89**, 1112, (1988).
23. H.J. Pearson, I.A. Valioulis and E.J. List, J. Fluid. Mech., **143**, (1984).

24. E. Dickinson and C. Elvingson, J.C.S. Faraday II, **284**, 775, (1988).
25. P. Mazur and W. van Saarloos, Physica A, **115**, 21, (1982).
26. J.J. Lopez Cascales and J.G. de la Torre, Macromolecules, **23**, 809, (1990).
27. R. Brown, Philos. Mag., **4**, 161, (1828); **6**, 161, (1829).
28. A. Einstein, Ann. Physik, **17**, 549, (1905); **19**, 371, (1906).
29. H. Mori, Prog. theor. Phys., **33**, 423, (1965).
30. H. Mori, Prog. theor. Phys., **34**, 399, (1965).
31. R. Zwanzig, Adv. Chem. Phys., **15**, 325, (1969).
32. A.J. Masters, Mol. Phys., **57**, 303, (1986).
33. R.J.A. Tough, P.N. Pusey, H.N.W. Lekkerkerker and C. van den Broeck, Mol. Phys., **59**, 595, (1986).
34. A.J. Masters, Mol. Phys., **57**, 303, (1986).
35. B.C. Eu and D.K. Bhattacharya, Phys. Rev. A, **39**, 3597, (1989).
36. D. Ronis, Phys. Rev. A, **34**, 1472, (1986).
37. J.K.G. Dhont, J.C. van der Werff and C.G. de Kruif, Physica A, **160**, 195, (1990).
38. J.C. van der Werff and C.G. de Kruif and J.K.G. Dhont, Physica A, **160**, 205, (1990).
39. B.J. Ackerson and N.A. Clark, Physica A, **118 A**, 123, (1983)
40. B.J. Ackerson and P.N. Pusey, Phys. Rev. Lett., **61**, 1033, (1988).
41. A. Chowdhury, B.J. Ackerson and N.A. Clark, Phys. Rev. Lett., **55**, 833, (1988).
42. N.J.Wagner and W. B. Russel, Phys. Fluids A , **2**, 491, (1990).
43. J. J. Erpenbeck, Phys. Rev. Lett. **52**, 1333 (1984).
44. G. P. Morris and D. J. Evans, Phys. Rev. A **32**, 2425 (1985).
45. D. M. Heys, G. P. Morris and D. J. Evans, J. Chem. Phys., **83**,4760, (1985).
46. W. Loose and S. Hess, Rheol. Acta, **28**, 91, (1989).
47. M.W. Evans and D.M. Heys, Mol. Phys. , **69**, 241, (1990).
48. G. Bossis and J.F. Brady, J. Chem. Phys., **80**, 5141, (1984).
49. D.M. Heys, Comp. Phys. Rep., **8**, 71, (1988).
50. D.M. Heys, Physica A, **27**, 47, (1988).
51. W.E. Xue and G.S. Grest, Phys. Rev. Lett., **64**, 419, (1990).
52. X. Xiu, H.D. Ou-Yang, D.J. Pine and P. Chaikin, Phys. Rev. Lett., **61**, 2554 (1988).
53. D.K. Bhattacharya and E. Clementi, (1990), to be submitted.
54. G.C. Ciccotti, M. Ferrario and J.P. Ryckaert, Mol. Phys., **46**, 875, (1982).
55. J.A. Padro, E. Guardo and G. Sese, Phys. Lett., **115**, 132, (1986).
56. D.E. Smith and C.B. Harriss, J. Chem. Phys., **92** , 1312, (1990).

Chapter 20.

Cellular Automata

R. Panda, V. Sonnad and E. Clementi
IBM Corporation
Scientific and Engineering Computations
Dept. 48B/428
Kingston, New York 12401, U.S.A.

1. Cellular Automata

Cellular automata are discrete space and time mathematical models of physical systems. A cellular automaton consists of a regular lattice, each site of which can have a finite number of states usually represented by Boolean variables. The cellular automaton (CA) evolves in time in discrete steps using deterministic or non-deterministic local rules for updating. Local rules of update require the dependence of a variable at a site at a discrete instant in time, only on the values at its nearest neighboring sites at the previous instant. All the sites are updated simultaneously which along with the locality of the updating rules, make a CA system amenable to high degrees of parallelism. The average behavior of a collection of sites of a cellular automaton is found to describe the behavior of physical systems obtained by using models with smooth continuous variables. Cellular automata were first introduced by von Neumann and Ulam[1] who also recognized the ability of CA systems in reproducing diverse physical phenomena.

2. Two-dimensional Lattice-gas Automata

The class of cellular automata used for the simulation of fluid dynamics is known as "lattice gas models." In a typical lattice gas model, all the sites are connected by links of unit length. Particles of unit mass and unit velocity hop along from site to site along the connecting links. In a lattice gas model the updating rules of a CA translate to a set of deterministic collision rules for the particles at a site. The collision rules are defined to conserve particle mass and momentum at each site. The macroscopic behavior of the lattice gas is educed by defining coarse-grained conserved quantities, e.g., density, momentum, etc. which are obtained by averaging their microscopic equivalents over subregions of the lattice. The conservation of macroscopic quantities is therefore tied to the conservation of their microscopic equivalents. One could therefore argue that by choosing a lattice with proper symmetries one could obtain the hydrodynamic behavior of a fluid.

Although Kadanoff and Swift[2] were the first to introduce a lattice gas model with fluid dynamical features, the first fully deterministic lattice gas model was introduced by Hardy, de Pazzis, and Pomeau.[3] Known in the literature as the HPP model, it has the features of discrete time, positions, and velocities. The HPP model consists of a regular, square two-dimensional lattice with unit link lengths. At each node (site) of the lattice, there can be up to four particles with unit mass and unit velocity; the velocities being in one of the four link directions. An exclusion principle is used to prevent the simultaneous occupation of any node by particles with identical velocities. The updating rule in discrete time consists of (i) propagation and (ii) collision steps. During the propagation step, each particle moves one link to the nearest node to which its velocity was pointing. The collision step prescribes that two particles moving in opposite directions at any node be replaced by two moving away from each other at right angles to the original. Any other configuration of particles at a site is left unchanged. The HPP model displays many important properties.[19] One of the crucial ones is the existence of thermodynamic equilibria. Although, there is no known ergodic theorem, the existence of thermodynamic equilibria has been demonstrated numerically.[19] The macrodynamical equations which emerge on coarse-graining differ from the Navier-Stokes equations in three respects. They fail to display Galilean invariance and isotropy, and have a crossover dimension problem.[4] The very definition of a lattice negates Galilean invariance. Without Galilean invariance thermodynamic equilibria with varying velocities cannot be related to each other by a simple transformation. The lack of Galilean invariance appears in the nonlinear terms of the momentum equations in the form of higher order corrections in velocity to the nonlinear term. However, these corrections are negligible at low Mach number, thereby ensuring incompressibility. The HPP lattice is symmetric under $\pi/2$ rotations, which is insufficient to ensure the isotropy of the momentum flux tensor. The crossover dimension is a property peculiar to two-dimensional hydrodynamics which results in the viscosity developing a logarithmic scale dependence. This difficulty disappears in three dimensions.

The model introduced by Frisch, Hasslacher and Pomeau[4] (known as the FHP model) rectifies some of the shortcomings of the HPP model. The idea that the isotropy of the momentum flux tensor is preserved when the underlying lattice is two-dimensional and invariant under the hexagonal rotation group (multiples of $\pi/3$) is incorporated in the FHP model. The FHP model is based on a triangular lattice with each node having a hexagonal neighborhood. The propagation step is the same as in the HPP model; collision step, on the other hand has a modified set of rules. There are now both two- and three-particle collisions. If we number the six links out of any node counterclockwise (Fig. 1), with an index i (a modulo 6 integer), then the allowed collisions are of the types $(i,i+3)$ and $(i,i+2,i-2)$. A 2-particle collision of type $(i,i+3)$ can result in either $(i+1,i-2)$ or $(i-1,i+2)$ with equal a posteriori probability. The 3-particle collision of type

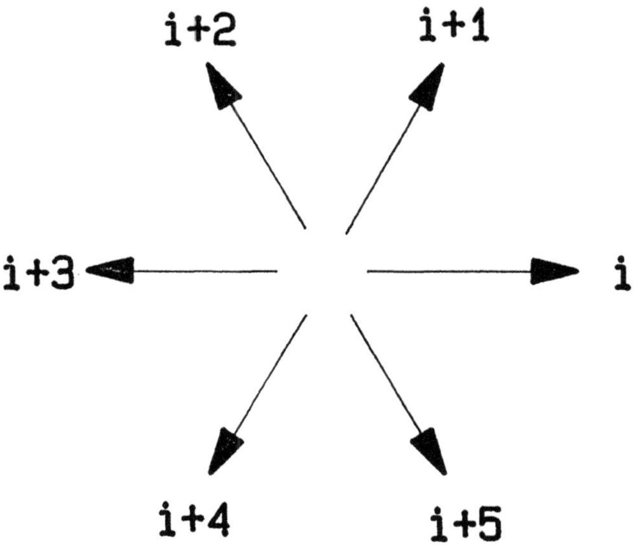

Figure 1. The six velocities of the FHP lattice.

$(i,i+2,i-2)$ leads to the combination $(i+3,i+1,i-1)$. All other combinations are not affected by these collision rules. Permissible collisions of the FHP model are shown in Fig 2. An exclusion principle similar to the one used in HPP is applied. The collision rules are designed to conserve three scalar conservation relations at each node, i.e. mass and momentum in each of two directions. Without the addition of three-particle collisions, there will be an extra spurious conservation relation. A variant on the basic FHP model is possible by adding a rest particle at each node and suitably modifying the collision rules.[4] It is possible to show[5] that a macroscopic averaging over space and time of the microscopic quantities will lead to the two-dimensional Navier-Stokes equations. These equations are valid in the limit of zero Mach number. In the limit of small Mach numbers the density fluctuations become irrelevant except in the pressure term. The nonlinear term has a multiplicative factor $g(\rho)=(\rho-3)/(\rho-6)$, where ρ is the particle density and hence can be treated as a constant. By suitably rescaling, a set of equations identical to the Navier-Stokes equations are obtained. It should be mentioned that Galilean invariance which is not valid at the lattice level is restored in the macroscopic limit. Implementation of boundary conditions is simple in lattice gas models. 'Free slip' boundary conditions can be modelled by specular reflection of particles from a rigid surface. "No-slip" boundary conditions can also be implemented by a random scattering of particles back into the flow domain from the rigid surface.

What are the advantages of lattice gas models? Frisch *et al.*[4] mention that the motivation behind using CA models for fluid simulations is to study the nature of turbulence. It is assumed that lattice gas models are not to be considered as an alternative to the standard methods of fluid simulations currently used. A standard numerical method like finite differences is used to solve the discrete

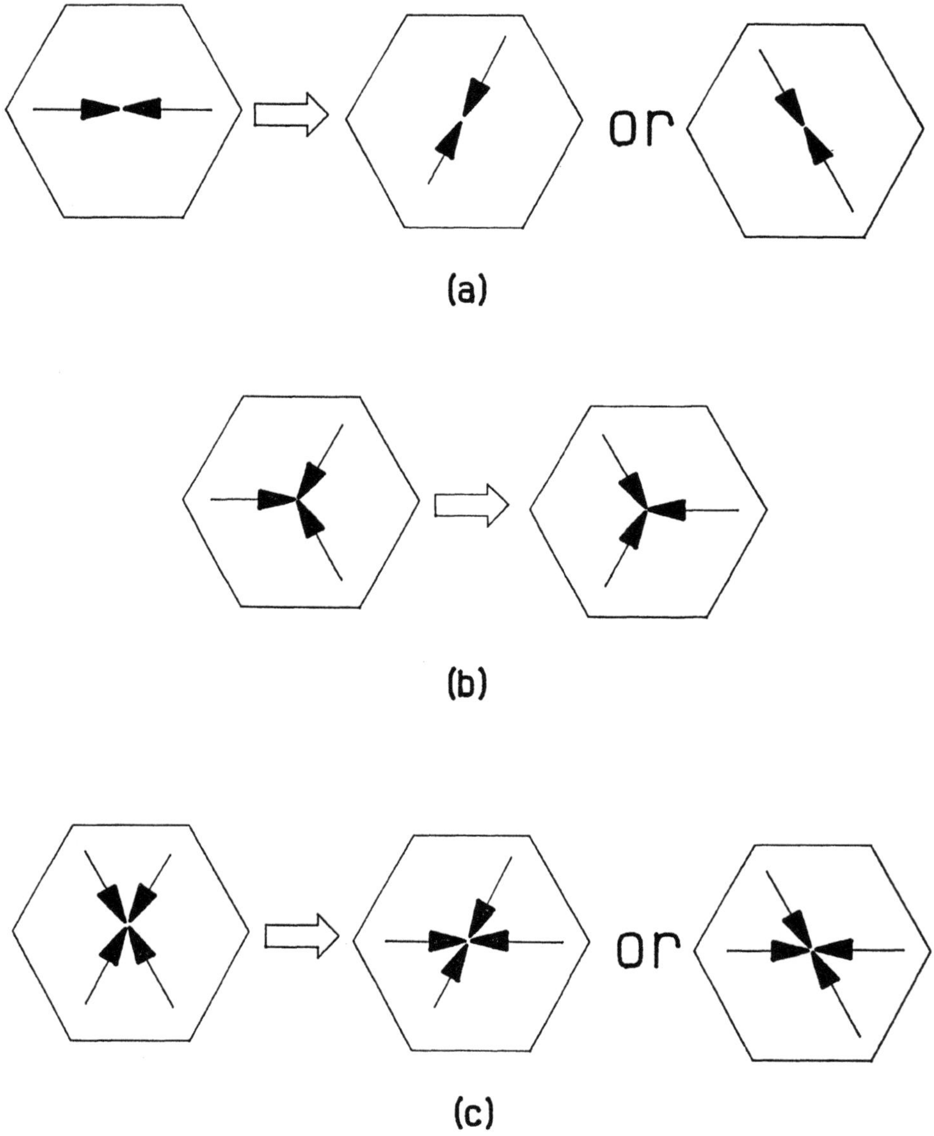

Figure 2. Collision rules of the FHP model: (a) two-particle collisions; (b) three-particle collisions; and (c) four-particle collisions.

approximations of the equations which govern the fluid behavior, namely the Navier-Stokes equations. Although, one could treat the discrete approximations as cellular automata themselves, they use floating point arithmetic which precludes 'bit-independence' in the computations. In floating-point computations all the bits are hierarchically related to one another. Furthermore, standard numerical methods are beset by problems of round-off error and stability. Lattice gas methods are free from these problems as they employ Boolean arithmetic; stability is guaranteed by theory.[5] Since in the lattice gas methods, fluid behavior arises out of the time evolution of an automaton with simple updating rules, they could be thought of as capturing and preserving the 'essence' of a fluid flow. In that sense, the study of a fluid flow through a CA simulation is similar to an experimental study.

Lattice gas methods are troubled by excessive amounts of noise. If care is not taken, excessive compressibility can lead to misleading physical behavior. Dahlburg *et al.*[6] concluded that there is a relatively narrow window in density and flow speed for obtaining true physical behavior of the fluid. As far as the memory and computing speed are concerned for simulation of 2-D and 3-D turbulent flows, Orszag and Yakhot have given very rough estimates.[7] Succi *et al.*[8] made a high-resolution (8192^2) lattice gas simulation of two dimensional turbulence and compared their results with the ones using a spectral method. For the same effective spatial resolution in both simulations, they found that the computational efficiency was much less for a lattice gas method. It does not then seem to be a practicable idea to try to simulate three dimensional turbulence using lattice gas methods at least on general purpose computers. However, on specially designed parallel computers like CAM[9] one could speed-up CA computations enormously. CA methods have been applied to study many other fluid phenomena with varying degrees of success.[10-14] It appears that CA methods are ideally suited for low-speed complex flows where an application of a regular numerical method is difficult; for example, multi-phase flows, porous flows, flows with chemical reactions, etc. where a regular numerical method will encounter problems regarding boundary conditions and moving interfaces. In lattice gas methods it is straight forward to incorporate these complexities. For example, a multi-species flow can be simulated by adding more bits to each node and appropriately 'coloring' them.

3. Simulation of 2D Fluid Flows

We describe here some of the results we obtained using the 2 dimensional lattice gas automata code KGNCA. In KGNCA code, FHP model without rest particles is used. Multispin coding technique is used which enables an efficient implementation of the lattice gas model on a general purpose computer like the IBM 3090.[16]

A rectangular computational domain (Fig 3) with NX hexagonal cells in the x direction and NY cells in the y direction is used. Six boolean variables are used to describe the state of particles at each cell. In addition, a seventh bit known as the obstacle bit is used for any obstacles that may be present inside or at the boundaries of the computational domain. The presence of the obstacle bit enables one to place arbitrary shaped solid bodies inside the flow region. The boundary conditions at the top and bottom of the flow region are in general periodic but could easily be defined as solid boundaries. When periodic boundary conditions are used, particles exiting at one end are reinjected from the other with the same momentum. No slip boundary conditions are implemented at soild boundaries by using a 'bounce-back' condition on particles impinging on solid boundaries.

The population density of particles in any direction is governed by the Fermi-Dirac statistical distribution. For small values of velocity, the Fermi-Dirac distribution correct to first order is given by&sup17.

$$N_i = \frac{\rho}{6}\left(1 + \frac{1}{2}\vec{c}_i \cdot \vec{U}\right)$$

where $\vec{c}_i$ is the unit velocity vector in the i direction and is given by,

$$\vec{c}_i = \left(\cos\frac{2\pi i}{6}, \sin\frac{2\pi i}{6}\right)$$

and N_i is the population density of particles in direction i ρ is the average density and U is the velocity. Inflow and outflow boundary conditions for the rectan-

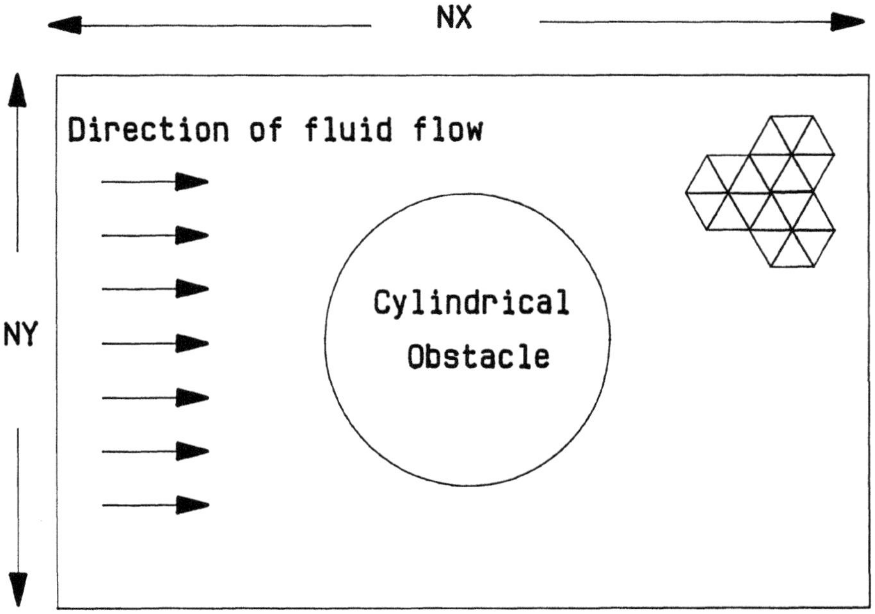

Figure 3. Rectangular computational domain of NX hexagonal cells in the X direction and NY hexagonal cells in the Y direction, with a cylindrical obstacle in the middle.

gular computational domain under consideration can be implemented using the above distribution. We allow particles to escape from the outflow boundary. To compensate for the loss of particles we inject particles according to the distribution given above. In order to maintain a flow inside the flow region, we also have to inject particles at the inflow boundary. Injection of particles at inflow and outflow boundaries leads to, on the average, 'wind-tunnel' like conditions inside the flow region.

We performed fluid flow simulation around (i) a circular cylinder and (ii) a vertical flat plate.

To simulate a fluid flow around a circular cylnder, a computational domain with NX=256 and NY=256 lattice units is taken. The computational region in the beginning is filled with a gas of average density $\rho = 2$ and a uniform velocity $U = 0.22$. A circular cylinder of diameter $D = 128$ lattice units is placed at the center of the computational domain. After the initial transients due to the impulsive starting of the flow around the cylinder subsided, a steady flow is established. Macroscopic averaging of the flow is performed over subregions of 8 × 8 hexagonal cells. The flow around the cylinder averaged over many time steps is shown in Fig 4. The effective Reynolds number and the nondimensionalized time are given by

$$R_e = \frac{g(\rho)UD}{\nu}$$

$$t_e = \frac{g(\rho)Ut}{D}$$

In the present case, g(ρ) and ν are computed to be 0.25 and 0.14 respectively. The effective Reynolds number is $R_e \simeq 50$.

For the flow around a vertical flat plate, a larger computational region with NX=512 and NY=512 is considered. The height of the plate is 160 lattice units. The velocity of the uniform flow as well as the average density are the same as before. The effective Reynolds number is $R_e \simeq 63$. The averaged flow field is shown in Fig 5. The presence of two counter-rotating vortices behind the plate can be seen in the figure.

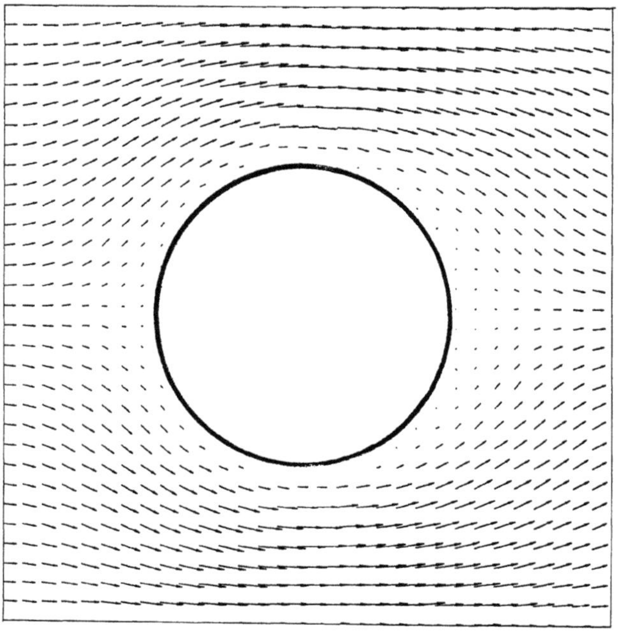

Figure 4. Simulation results from KGNCA: velocity field around a circular cylinder.

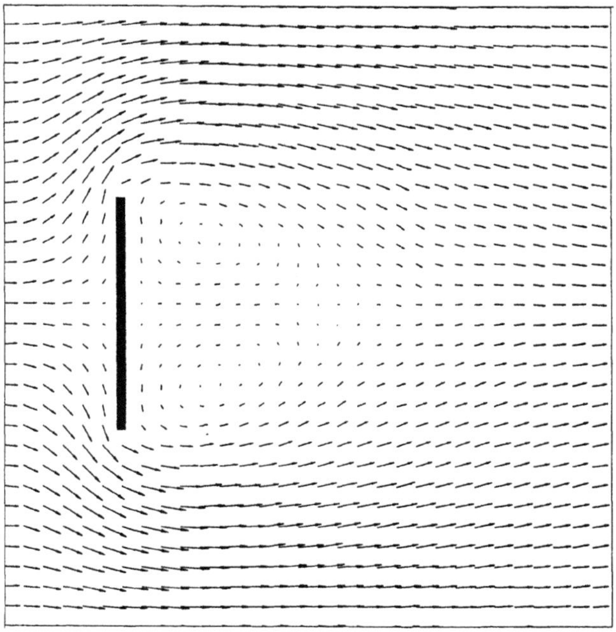

Figure 5. Simulation results from KGNCA: velocity field around a vertical flat plate.

4. Three-dimensional Lattice-gas Automata

A straightforward extension of the hexagonal lattice gas model from two to three dimesions by using any regular three dimensional lattice is not feasible for simulating fluid flows in three dimensions. The isotropy of the nonlinear and diffusion terms in the Navier-Stokes equations, which are pairwise symmetric fourth order tensors cannot be ensured by using any regular three dimensional lattice. Even when the face centered cubic (FCC) lattice which has the greatest symmetry among all space filling simplexes is used, the introduction of an isotropy breaking spurious term to the nonlinear term in the Navier- Stokes equations cannot be avoided. One of the ways of restoring isotropy in the fourth order tensor is by allowing the gas to evolve on different lattices(cubic and FCC) and over different times. By adjusting the evolution time ratio between the two lattices one could obtain an isotropic nonlinear term. However, constant transfer of information from one lattice to the other makes the idea impractical. The introduction of the isotropy breaking term can be circumvented by using a four-dimensional face-centered hypercubic(FCHC) lattice which has all the required symmetries for a three dimensional case it is sufficient to make the lattice a layer thick in the fourth dimension.[17] This gives the lattice an effective three-dimensional formalism.

It is very difficult to visualize a four dimensional lattice in three dimensions, except by analogy. In a d-dimensional close packed lattice there are 2^d nodes at the corners and $2 \times d$ nodes at centers of the faces of the lattice. For example, in the FCC lattice in three dimensions, there are $2^3 = 8$ nodes at the corners and $2 \times 3 = 5$ at centers of the faces for a total of 14. By analogy, in a FCHC lattices there are $2^4 = 16$ nodes at the corners of a hypercube while $2 \times 4 = 8$ are at centers of the 8 faces. Thus each node has $16+8=24$ nearest neighbors all of which are equidistant from the center node. The nodes of the FCHC lattice are points in four dimensional space with integer coordinates (x,y,z,w) such that $x+y+z+w$ is even. If we assign the center node to the origin of the coordinate frame, then the 24 nearest neighbors can be represented by the points,

$$(\pm 1, \pm 1, 0, 0) \quad , \quad (\pm 1, 0, \pm 1, 0)$$

$$(\pm 1, 0, 0, \pm 1) \quad , \quad (0, \pm 1, \pm 1, 0)$$

$$(0, 0, \pm 1, \pm 1) \quad , \quad (\pm 1, 0, \pm 1, 0)$$

The distance between the origin and any of the 24 nodes is $\sqrt{2}$ in 4-dimensional space.

As in the 2D case, two steps are involved in the 3D simulation of a fluid flow, namely (i) propagation and (ii) collision steps. Once the lattice with required number of symmetries is chosen, the propagation step involves movement of particles along lattices links. In the case of FCHC lattice particles can move about in 24 directions. If boolean numbers are used to represent the presence of a particle in a specific direction, a collection of 24 boolean numbers $s = (s_1, \dots , s_{24})$ will suffice to describe the state of particles at any node of the FCHC lattice at any instant. Collision at each node involves the transition of state s to another state s′ which is again a collection of 24 boolean numbers. The transition is nondeterministic with a probability $A(s,s') > = 0$ such that $\Sigma A = 1$. Of course, as in the case of 2D lattice gas automata, the collision rules must obey the the conservation of mass and momentum at each node. The condition that $\Sigma A = 1$ is known as semi-detailed balance and simply means that all states have equal probabilities before and after the collision. An additional assumption that collisions must be invariant under any isometry of the lattice is also used. An isometry of a lattice is defined as a rotation about the origin with or without a mirror symmetry. Following the above assumptions many collision rules can be defined. We are going to discuss only Henon's[18] isometric collision rules.

The isometric collision rules are based on the assumption that all collisions are isometries, i.e. the output state can be obtained from the input state by applying a proper isometry. In HPP and FHP models of 2D automata it can be seen that all collisions are isometries. Another criterion is that the isometry depend only on the momentum of the input state s For each collision with the input state having a certain momentum, one isometry out of all the possible isometries having the same momentum is chosen randomly. The choosing of the isometry, given the momentum constitutes the isometric algorithm for collisions on a FCHC lattice. In order to achieve higher Reynolds numbers the viscosity has to be minimized which means the collision rules have to be optimized to lower the shear viscosity.

An isometry of the FCHC lattice can be represented by a 4 × 4 matrix M. All the isometries of the FCHC lattice form a group designated by G of which M is an element. All the elements of G can be generated by a few isometries and hence are known as the generating set. They are known as S,P and R isometries. S -isometry: Denoted by S_i , it changes the sign of one coordinate i or inverts the axis in the 4D coordinate frame. For example S_x is given by

$$S_x = \begin{bmatrix} -1\,0\,0\,0 \\ 0\,1\,0\,0 \\ 0\,0\,1\,0 \\ 0\,0\,0\,1 \end{bmatrix}$$

P -isometry: In this isometry, any two coordinates i and $j (i \neq j)$ are permuted implying an axis interchange in 4D space. e.g. P_{12} is given by

$$P_{12} = \begin{bmatrix} 0\,1\,0\,0 \\ 1\,0\,0\,0 \\ 0\,0\,1\,0 \\ 0\,0\,0\,1 \end{bmatrix}$$

R -isometry: There are only two in this class, denoted by R_1 and R_2 . They are given by

$$R_1 = \frac{1}{2}\begin{bmatrix} 1 & 1 & 1 & -1 \\ 1 & 1 & -1 & 1 \\ 1 & -1 & 1 & 1 \\ -1 & 1 & 1 & 1 \end{bmatrix}$$

$$R_2 = \frac{1}{2}\begin{bmatrix} 1 & 1 & 1 & 1 \\ 1 & 1 & -1 & -1 \\ 1 & -1 & 1 & -1 \\ 1 & -1 & -1 & 1 \end{bmatrix}$$

Any isometry M of group G can be obtained by a composition of S,P and R isometries. The composition law is ordinary matrix multiplication. In general, M can be written as

$$M = \begin{pmatrix} I \\ S_4 \end{pmatrix} \begin{pmatrix} I \\ S_3 \end{pmatrix} \begin{pmatrix} I \\ S_2 \end{pmatrix} \begin{pmatrix} I \\ S_1 \end{pmatrix} \begin{pmatrix} I \\ P_{34} \end{pmatrix} \begin{pmatrix} I \\ P_{23} \\ P_{24} \end{pmatrix} \begin{pmatrix} I \\ P_{12} \\ P_{13} \\ P_{14} \end{pmatrix} \begin{pmatrix} I \\ R_1 \\ R_2 \end{pmatrix}$$

where in each parentheses one of the isometries is chosen, with I being an identity matrix. It can be seen from the above representation, that there are 1152 cases of M making G a group of order 1152.

Three basic steps are involved in computing collisions. In the first step an isometry N is applied which makes all the components of the momentum positive and ordered[18] This is called normalization of momentum components. The next step involves the application of an optimal isometry M which preserves the normalized momentum. Optimal isometry is one which minimizes the viscosity. There are 12 classes of optimal momentum preserving isometries. All collisions fall into one of these 12 classes. The last step is the application of the isometry N^{-1} which will reverse the effects of the isometry N. Interested reader can refer to the actual recipe for computing isometric collisions given by Henon[18] . For the isometric collision algorithm it is found that the viscosity is a minimum when the average number of particles per site is 4.

Using the above rules a 2^{24} long collision look-up table can be computed. In the practical implementation of 3D lattice gas automata using the FCHC lattice, at each time step the collision step is computed by retrieving the entry from the look-up table corresponding to the input state at the site before collision. The retrieval of an entry from the look-up table is performed using indirect addressing. In computers where indirect addressing is hardwired, implementation of the collision step will be quite efficient. One of the disadvantages of using a look-up table is that it requires a large memory for storing the look-up table. However, it is also possible to effect reduced collision tables which require much less memory.[19] Some of these algorithms are implemented without relaxing the semi-detailed balance and can be used to simulate flows with higher Reynolds numbers than with the isometric collision rules.

References

1. J. von Neumann, *Theory of Self Reproducing Automata* (Univ. of Illinois Press, 1966).
2. L. P. Kadanoff and J. Swift, Phys. Rev. **165**, 310 (1968).
3. J. Hardy, O. de Pazzis, and Y. Pomeau, Phys. Rev. A **13**, 1949 (1976).
4. U. Frisch, B. Hasslacher, and Y. Pomeau, Phys. Rev. Lett. **56**, 1505 (1986).
5. U. Frisch, D. d'Humieres, B. Hasslacher, P. Lallemand, Y. Pameau, and J. P. Rivet, Complex Systems **1**, 649 (1987).
6. J. P. Dahlburg, D. Montgomery, and G. D. Doolen, Phys. Rev. A **36**, 2471 (1987).
7. S. A. Orszag and V. Yakhot, Phys. Rev. Lett. **56**, 1691 (1986).
8. S. Succi, P. Santangelo and R. Benzi, Phys. Rev. Lett. **60**, 2738 (1988).
9. N. Margolus, T. Toffoli, and G. Vichniac, Phys. Rev. Lett. **56**, 1694 (1986).
10. D. H. Rothman, Geophys. **53**, 509 (1988).
11. D. H. Rothman and J. M. Keller, J. Stat. Phys. **52**, 1119 (1988).
12. K. Balasubramanian, F. Hayot, and W. F. Saam, Phys. Rev. A **36**, 2248 (1987).
13. C. Baudet, J. P. Hulin, P. Lallemand, and D. d'Humieres, Phys. Fluids A **1**, 507 (1989).
14. P. Clavin, P. Lallemand, Y. Pomeau, and G. Searby, J. Fluid Mech. **188**, 437 (1988).
15. F. Hayot, M. Mandal and P. Sadayappan, J. Comp. Physics **80**, 277 (1989).
16. D. d'Humieres and P. Lallemand, Complex Systems **1**, 599 (1987).
17. D. d'Humieres, P. Lallemand, and U. Frisch, Europhysics Lett. **2**, 291 (1986).
18. M. Henon, Complex Systems **1**, 475 (1987).
19. J. A. Somers and P. C. Rem, " The Construction of Efficient Collision Tables for Fluid Flow Computations with Cellular Automata", *Proceedings of the Workshop on Cellular Automata and Modeling of Complex Physical Systems* , Les Houches, France. (Springer- Verlag, Berlin, 1989)

Chapter 21.

The Equations of Fluid Flow and Their Solution by Finite Element Methods

V. Sonnad, S. Hassanzadeh, R. Panda and S. Foresti

IBM Corporation
Scientific and Engineering Computations
Dept. 48B/428
Kingston, New York 12401, U.S.A.

1. The Equations of Fluid Flow

Derivation of the Navier Stokes Equations

The Navier Stokes equations[1,2] are a set of coupled partial differential equations which describe the motion of a fluid under a wide variety of conditions of practical interest. The solution of these equations subject to different boundary conditions constitutes the major part of computational fluid dynamics. In this section we will derive the Navier Stokes equations using the simple but powerful control volume approach.

Prior to the derivation itself, we need a precise definition of a fluid. For the purpose of the derivation of the Navier Stokes equations, a fluid is defined as matter that deforms continuously under the action of a shear stress. The rate of change of strain resulting from an applied stress is given by the constitutive equations. The Navier Stokes equations in their traditional form use a linear relationship between strain rate and stress, but the derivation can easily incorporate more general constitutive equations. The major assumption underlying the derivation of the Navier Stokes equations with the control volume approach is that the fluid is a continuum, i.e., all properties of the fluid are continuous functions of position. It is interesting to note that the Navier Stokes equations can also be derived without resorting to this assumption (e.g. from the Boltzmann equations[3]). This may explain in part why the solutions to these equations often remain valid even under extreme conditions where the continuum hypothesis is violated.

The control volume approach is one of the simpler methods for the derivation of the Navier Stokes equations. In this approach, we consider an imaginary volume of infinitesimal dimensions embedded in a larger region of a fluid in motion. The basic idea of the control volume approach is to enforce conservation of a given quantity within the control volume. To illustrate the idea, consider a rectangular two-dimensional control volume of sides Δx and Δy as shown in Figure 1.

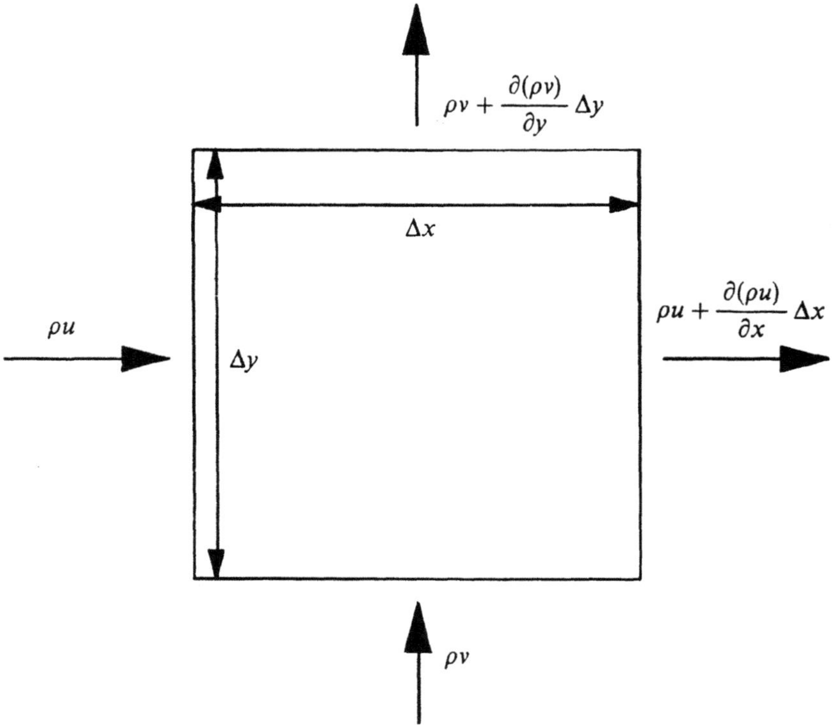

Figure 1. Control volume for the conservation of mass.

If we wish to derive the equation pertaining to the conservation of mass, we equate the net mass flux across the control volume with the rate of change of mass within the control volume.

(Mass transferred out of the control volume across all surfaces)
– (Mass transferred into the control volume across all surfaces)
= (Rate of change of mass within the control volume).

If the density of the fluid is ρ and the x and y components of velocity are u and v, respectively, the statement of conservation of mass is given by

$$-\frac{\partial(\rho u)}{\partial x}\Delta x\Delta y - \frac{\partial(\rho v)}{\partial y}\Delta x\Delta y = \frac{\partial \rho}{\partial t}\Delta x\Delta y.$$

Dividing throughout by $\Delta x \Delta y$ we get

$$\frac{\partial \rho}{\partial t} + \frac{\partial(\rho u)}{\partial x} + \frac{\partial(\rho v)}{\partial y} = 0.$$

This is the first set of partial differential equations that constitute the Navier Stokes equations, and is called the continuity equation; as mentioned before it merely expresses in differential form, the law of conservation of mass.

We next derive the equations corresponding to the conservation of momentum. Consider the x-direction momentum equation first; the statement of conservation of momentum in the x-direction is expressed as

(x-Momentum transferred out of the control volume across all surfaces)
− (x-Momentum transferred into the control volume across all surfaces)
+ (Stresses at the surface)
= (Rate of change of x-momentum within the control volume)

From Figures 2a and 2b, we see that this can be expressed as

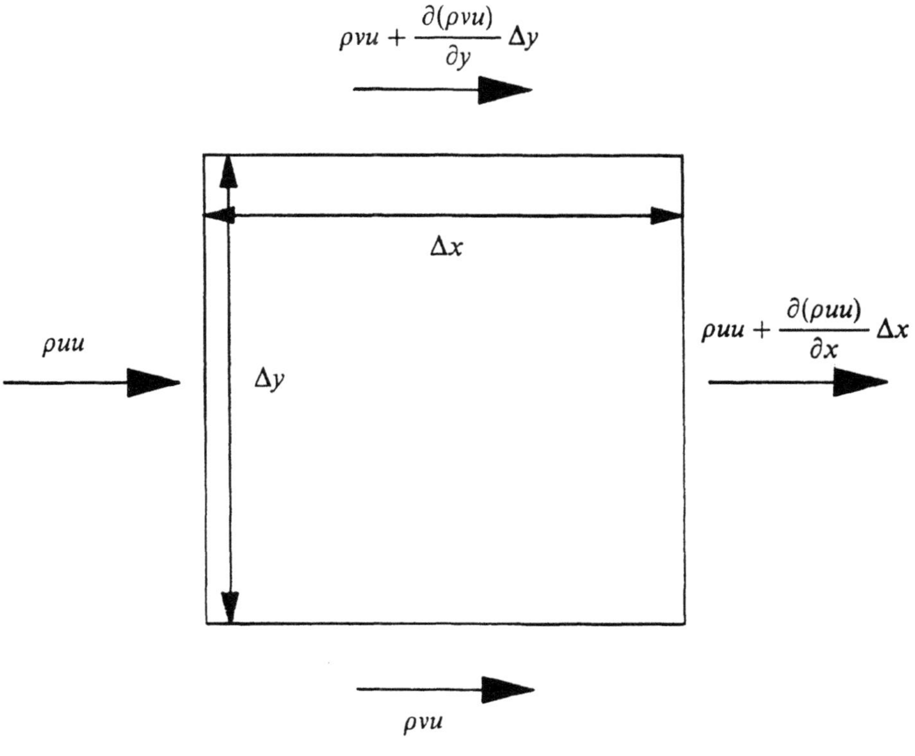

Figure 2a. Momentum flow through the control volume in the x-direction.

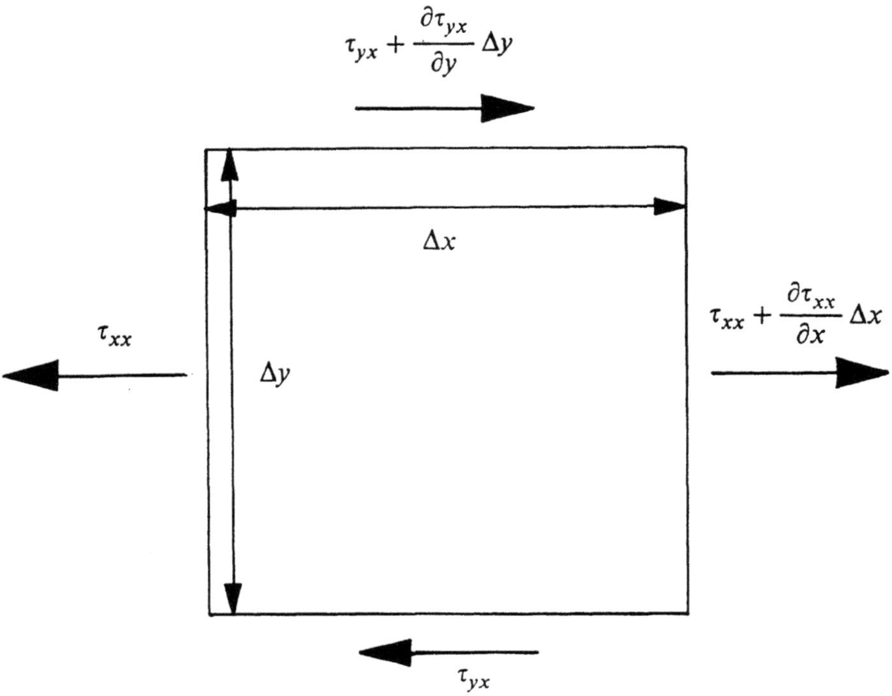

Figure 2b. Stresses at the surfaces of the control volume in the x-direction.

$$\Delta x \Delta y \frac{\partial(\rho u)}{\partial t} + \Delta x \Delta y \frac{\partial(\rho uu)}{\partial x} + \Delta x \Delta y \frac{\partial(\rho uv)}{\partial y} = \frac{\partial \tau_{xx}}{\partial x} \Delta x \Delta y + \frac{\partial \tau_{yx}}{\partial y} \Delta x \Delta y.$$

In the above equation, the quantities τ_{xx} and τ_{yx} are the stresses in the fluid at the surfaces of the control volume. We do not at this stage make any statements regarding the relation between the applied surface stresses and the velocities. This is supplied by the constitutive equations for the fluid. Again, dividing throughout by $\Delta x\ \Delta y$ the equation for the conservation of x-momentum is given as

$$\frac{\partial(\rho u)}{\partial t} + \frac{\partial(\rho uu)}{\partial x} + \frac{\partial(\rho uv)}{\partial y} = \frac{\partial \tau_{xx}}{\partial x} + \frac{\partial \tau_{yx}}{\partial y}.$$

By using identical reasoning we can derive the equation for the conservation of momentum in the y-direction:

$$\frac{\partial(\rho v)}{\partial t} + \frac{\partial(\rho vv)}{\partial y} + \frac{\partial(\rho uv)}{\partial x} = \frac{\partial \tau_{yy}}{\partial y} + \frac{\partial \tau_{xy}}{\partial x}.$$

The three-dimensional versions of these equations for the conservation of mass and momentum can be derived by considering a three-dimensional control volume; the resulting expressions are given below:

$$\frac{\partial \rho}{\partial t} + \frac{\partial(\rho u)}{\partial x} + \frac{\partial(\rho v)}{\partial y} + \frac{\partial(\rho w)}{\partial z} = 0$$

$$\frac{\partial(\rho u)}{\partial t} + \frac{\partial(\rho uu)}{\partial x} + \frac{\partial(\rho uv)}{\partial y} + \frac{\partial(\rho uw)}{\partial z} = \frac{\partial \tau_{xx}}{\partial x} + \frac{\partial \tau_{yx}}{\partial y} + \frac{\partial \tau_{zx}}{\partial z}$$

$$\frac{\partial(\rho v)}{\partial t} + \frac{\partial(\rho vu)}{\partial x} + \frac{\partial(\rho vv)}{\partial y} + \frac{\partial(\rho vw)}{\partial z} = \frac{\partial \tau_{xy}}{\partial x} + \frac{\partial \tau_{yy}}{\partial y} + \frac{\partial \tau_{zy}}{\partial z}$$

$$\frac{\partial(\rho w)}{\partial t} + \frac{\partial(\rho wu)}{\partial x} + \frac{\partial(\rho wv)}{\partial y} + \frac{\partial(\rho ww)}{\partial z} = \frac{\partial \tau_{xz}}{\partial x} + \frac{\partial \tau_{yz}}{\partial y} + \frac{\partial \tau_{zz}}{\partial z}.$$

We have at this point, derived partial differential equations for the conservation of mass and momentum in terms of velocities, density and stress. The main tool in our derivation has been the control volume approach, which is a readily understood idea, and implicitly uses a fixed Eulerian frame of reference which is the standard approach for analyzing fluid motion. The Lagrangian frame of reference which moves with the body being analyzed, is considered more appropriate for problems in solid mechanics. The power of this simple approach is evidenced by the fact that the same approach can be used to derive the basic equations in a variety of fields using only the simplest conservation concepts.

If we examine the above set of equations, we see that we have a system of four coupled partial differential equations and a total of thirteen unknowns, viz., the density ρ, three components of the velocity vector, and nine components of the stress tensor. It is however, possible to reduce the total number of unknowns that appear in the equations through the use of the constitutive equations for a fluid. The constitutive equations express the stress tensor in terms of the derivatives of the velocities (the strain rate tensor). We indicate here the major steps in the derivation of the constitutive equations for a fluid. At various stages in the derivation, there are assumptions made about the behavior and properties of the fluid based on empirical data. The result is a series of relations for the constitutive equations, that can be used across a wide range of fluid flows of practical interest.

The first step is to show that the stress tensor is symmetric. This is easily done by using the control volume approach, where we write the expression for the conservation of angular momentum. Depending upon the axes chosen for the calculation of angular momentum, it can be shown that[4]

$$\tau_{xy} = \tau_{yx}, \quad \tau_{xz} = \tau_{zx} \text{ and } \tau_{zy} = \tau_{yz}.$$

The stress tensor can therefore be written as:

$$\tau_{ij} = \begin{bmatrix} \tau_{xx} & \tau_{xy} & \tau_{xz} \\ \tau_{xy} & \tau_{yy} & \tau_{yz} \\ \tau_{xz} & \tau_{yz} & \tau_{zz} \end{bmatrix}.$$

There are six independent quantities instead of nine. In a fluid at rest, only normal stresses are exerted by the fluid, and the stress tensor is of the form

$$\tau_{ij} = -p\delta_{ij}$$

where p is the static pressure. A fluid in motion cannot be characterized by a single quantity. However it is useful to define a similar scalar for a fluid in motion – the most appropriate such quantity is the trace of the tensor and it is convenient to define the pressure in a moving fluid as

$$p = -\frac{1}{3}\tau_{ii}.$$

This quantity is not the same as the thermodynamic pressure and the relation between the two is not always a simple one.[4] It is useful to regard the stress tensor as the sum of an isotropic part $-p\delta_{ij}$ with the same form as the stress tensor in a fluid at rest, and a non-isotropic part d_{ij} called the deviatoric stress tensor due entirely to the motion of a fluid.

$$\tau_{ij} = -p\delta_{ij} + d_{ij}.$$

We now assume that the fluid is Newtonian, i.e., the stress is linearly proportional to the strain rate. We define a strain rate tensor:

$$e_{ij} = \frac{1}{2}\left(\frac{\partial V_i}{\partial x_j} + \frac{\partial V_j}{\partial x_i}\right)$$

where V_i is the component of velocity in the direction x_i.

The linear relationship between the deviatoric stress and the strain rate is

$$d_{ij} = A_{lmij}\, e_{lm}$$

where A_{lmij} is a fourth rank tensor, and would have 81 independent entries for a three-dimensional problem. However, the symmetry of the stress and strain tensor reduces the number of independent entries to 36 (six for each of the six independent quantities in the stress tensor). By imposing the condition of isotropy, the number of independent constants is reduced to two, and we can write

$$d_{ij} = 2\mu e_{ij} + \lambda e_{kk}\delta_{ij}$$

where μ is the coefficient of viscosity, and λ is the second coefficient of viscosity. For most gases, it is acceptable to assume $\lambda = -2/3\mu$ (Stokes' hypothesis[2]), and it is customary to write the expression for the stress tensor as:

$$\tau_{ij} = -p\delta_{ij} + 2\mu(e_{ij} - \frac{1}{3} e_{kk}\, \delta_{ij}).$$

The expression for the stress tensor, incorporated into the equations of conservation of mass and momentum leads to the following partial differential equations which are succinctly expressed in tensor notation.

$$\frac{\partial \rho}{\partial t} + \frac{\partial(\rho V_i)}{\partial x_i} = 0$$

$$\frac{\partial(\rho V_i)}{\partial t} + \frac{\partial(\rho V_i V_j)}{\partial x_j} = \frac{-\partial p}{\partial x_i} + \frac{\partial}{\partial x_j}\left(2\mu(e_{ij} - \frac{1}{3}e_{kk}\delta_{ij})\right).$$

These are the Navier Stokes equations of motion which express the conservation of mass and momentum. For a three-dimensional problem, we have a set of four coupled partial differential equations; there are however five unknowns to be solved for, namely V_1, V_2, V_3, ρ and p. An additional equation is required in order to define the problem completely, and this is an equation of state which links the density ρ to the pressure p. Along with the initial and boundary conditions, this set of five equations in five unknowns is a mathematically well posed problem although it may be very difficult to solve.

In the above derivation of the Navier Stokes equations, we have implicitly assumed constant temperature. If the temperature of the fluid is changing, we need an equation for the conservation of energy. This can be derived in terms of temperature by using a control volume approach. Without going into the details, we give the energy equation for an ideal gas where the effects of heating due to friction are neglected, and the flow is compressible[5]:

$$\rho C_p \frac{DT}{Dt} = \frac{\partial}{\partial x_i}\left(k\frac{\partial T}{\partial x_i}\right) + \frac{Dp}{Dt}$$

where k is the coefficient of thermal conductivity, C_p is the specific heat at constant pressure, and

$$\frac{D}{Dt} = \frac{\partial}{\partial t} + V_i\frac{\partial}{\partial x_i}.$$

Reduction to Various Specific Forms

The Navier Stokes equations express the conservation of mass and momentum in differential form. They are obtained by assuming specific forms for the constitutive equations that link the stress tensor to the strain-rate tensor. These equations can often be greatly simplified if the flow is limited to certain regimes.

If the viscosity is taken as uniform (no large temperature gradients) the Navier Stokes equation reduces to

$$\frac{\partial \rho}{\partial t} + \frac{\partial (\rho V_i)}{\partial x_i} = 0$$

$$\frac{\partial (\rho V_i)}{\partial t} + \frac{\partial}{\partial x_j}(\rho V_i V_j) = -\frac{\partial p}{\partial x_i} + \mu \left(\frac{\partial^2 V_i}{\partial x_j \partial x_j} + \frac{1}{3} \frac{\partial e_{kk}}{\partial x_i} \right).$$

Another simplification is obtained by assuming that the flow is incompressible, i.e, the density is constant throughout the flow field. (The conditions for this assumption to be valid will be discussed shortly). The equations are then given by:

$$\frac{\partial V_i}{\partial x_i} = 0$$

$$\rho \frac{DV_i}{\partial t} = -\frac{\partial p}{\partial x_i} + \mu \frac{\partial^2 V_i}{\partial x_j \partial x_j}.$$

The reduction of the Navier Stokes equations to simpler forms is greatly facilitated by the use of nondimensional numbers whose magnitudes can furnish criteria for ignoring terms in the full Navier Stokes equations. One approach in deriving useful nondimensional numbers is through arguments using order of magnitude estimates for each of the terms in the Navier Stokes equations. We illustrate this approach for the momentum equations. We assume that there is a characteristic length L, which represents the geometric dimensions of the region of interest. A characteristic velocity U is defined to represent a suitable mean of the fluid flow velocity. The steady state momentum equation is then rewritten in non-dimensional form where all the velocities are normalized by U and the all the lengths by L. By suitably regrouping terms we have the nondimensional form of the momentum equation given by

$$\overline{V}_j \frac{\partial \overline{V}_i}{\partial \overline{x}_j} = -\frac{\partial \overline{p}}{\partial \overline{x}_i} + \frac{\mu}{\rho U L} \left(\frac{\partial^2 \overline{V}_i}{\partial \overline{x}_j \partial \overline{x}_j} \right)$$

where a bar indicates a nondimensional quantity, and

$$\bar{p} = \frac{p}{\rho U^2}.$$

The nondimensional combination of terms $\rho UL/\mu$ is called the Reynolds number and is one of the most important nondimensional numbers in fluid mechanics. It is physically interpreted as the ratio of inertial forces to viscous forces. Thus, a small value for the Reynolds number implies that the flow is dominated by viscous forces, and a large value means that inertial forces dominate over viscous forces. An important consequence of putting the equations in nondimensional form is that the solution (in nondimensional form), is the same when the Reynolds number is held constant (even though U, L, and μ are varied). This principle of dynamic similarity allows data obtained from experiments conducted on model objects to be scaled up to predict flows around realistic objects. The Reynolds number also plays the role of a stability parameter governing the transition from relatively smooth laminar flow, to turbulent flow characterized by high frequency spatial and temporal fluctuations. We will postpone a detailed discussion of turbulence to section 6.

Another nondimensional number that is important in the study of fluid flow is the Mach number, which is the ratio of characteristic velocity to the acoustic velocity. Low Mach number flows can be assumed to be incompressible, i.e., the density can be assumed constant throughout the entire flow field. The Mach number separates the subsonic flow regime from the supersonic and transonic flow regimes where shocks are present in the flow. There are a variety of other nondimensional numbers such as the Froude number, the Strouhal number, the Weber number, and the Knudsen number which are useful under various physical conditions.[6]

A further simplification in the Navier Stokes equations is obtained in the limiting case of infinite Reynolds number. The viscous terms then drop out of the equations and the resulting equations are referred to as the inviscid flow or Euler equations:

$$\frac{\partial V_i}{\partial x_i} = 0$$

$$\rho \frac{DV_i}{Dt} = -\frac{\partial p}{\partial x_i}.$$

In the event that the flow is irrotational, the velocity can be expressed as the gradient of a potential

$$V_i = \frac{\partial \phi}{\partial x_i}.$$

The continuity equation is then given by

$$\frac{\partial^2 \phi}{\partial x_i \partial x_i} = 0.$$

The momentum equation can be solved by integrating along a streamline (a line that is tangent to the flow field at every point), resulting in the Bernoulli equation.

$$p + \frac{1}{2}\rho u^2 = \text{constant}.$$

The limiting case of zero Reynolds number is termed Stokes flow; the equations are linear and are given by

$$\frac{\partial V_i}{\partial x_i} = 0$$

$$\mu \frac{\partial^2 V_i}{\partial x_j \partial x_j} = \frac{\partial p}{\partial x_i}.$$

A particularly important simplification of the Navier Stokes equations results in the boundary layer equations. The practical value of these equations comes from the following considerations: the kinematic viscosities of air and water under working conditions are low, so that the Reynolds numbers are very high. A flow at high Reynolds number, at first sight, is one in which viscous forces are insignificant compared with inertial forces. If, however, the viscous forces are discarded from the Navier Stokes equations the results are not compatible with what is observed experimentally. The inviscid theory cannot explain the basics of some fundamental observations. A hypothetical fluid of zero viscosity behaves very differently from one of very low viscosity. This can be traced to the fact that a real fluid obeys the no-slip condition at a solid boundary, whereas an inviscid fluid does not.

The development of boundary layer theory represents a middle course between the assumption of inviscid flow, which neglects the effects of viscosity over the entire flow field, and the Navier Stokes equations which account for viscous effects over the entire flow field. The boundary layer assumption (which is demonstrably valid for a large number of flows of practical importance), is that the viscous effects are confined to a thin layer of fluid near the wall of a solid object. The region outside this thin layer can be considered inviscid.

An example is the two dimensional flow of an incompressible fluid over a flat plate shown in Figure 3.

In a thin region near the wall (shown diagrammatically as being of thickness δ), viscous effects dominate the flow, whereas they can be neglected outside the boundary layer. An order of magnitude analysis on the Navier Stokes equations

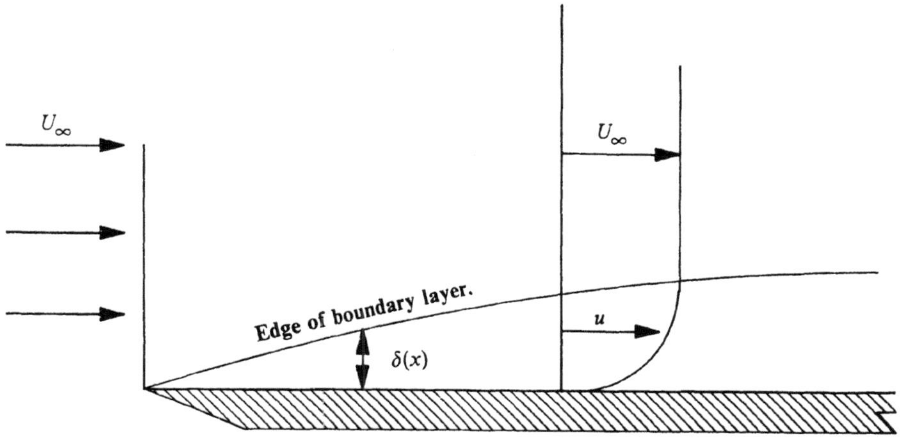

Figure 3. Boundary layer over a flat plate.

results in several terms being dropped, and the boundary layer equations for flow over a flat plate in the absence of a streamwise pressure gradient are given by:

$$\frac{\partial u}{\partial x} + \frac{\partial v}{\partial y} = 0$$

$$\rho\left(u\frac{\partial u}{\partial x} + v\frac{\partial u}{\partial y}\right) = -\frac{\partial p}{\partial x} + \mu\frac{\partial^2 u}{\partial y^2}.$$

These equations were formulated by Prandtl and solved analytically by Blasius using similarity transforms.[7] The combined work of Prandtl and Blasius on the boundary layer equations is considered an important development in modern hydrodynamics.

The Navier Stokes equations as we have derived them are expressed in terms of velocity and pressure; this is called the primitive variable form of the Navier Stokes equations. Under certain conditions, it may be desirable to use other variables to express the Navier Stokes equations. However, the approach that is now becoming prevalent in numerical simulations is to work with the primitive form of the Navier Stokes equations unless there are obvious advantages to working with alternate forms of the equations. At this point, we end our discussion of the Navier Stokes equations, and consider numerical techniques that are used in the solution of these equations.

2. The Finite Element Method

The complexity of the Navier Stokes equations allows analytic solutions for only the simplest geometries and flow conditions. For many flows of practical interest, it is necessary to use numerical techniques to solve these equations. While there are a number of techniques that are used to solve the Navier Stokes equations, we consider only the finite element method in this section. The major reason is that of all numerical methods, the finite element method is easiest able to handle arbitrary geometries which can be an important consideration in many practical problems. Another reason is that recent developments in finite element methods have greatly reduced the storage and computing requirements, making it highly competitive for fluid flow applications.

The finite element method has had an interesting history because of its origin in two distinct disciplines, one in mathematics and the other in engineering. The mathematical approach was based upon the Rayleigh Ritz method and the minimization of variational indicators. The engineering approach was based upon the discretization of complex structures into relatively simple subregions which could be analyzed easily, and then pieced together to obtain a global solution. The equivalence of these two approaches was demonstrated in the fifties, and the ad-hoc approach used by engineers was shown to have a secure mathematical basis. This led to a systematic generalization of the solution methods used by engineers who were then able to address a wider range of problems. A brief account of the historical development of the finite element method is given in Ref. 8.

While this was a major step in the development of finite element methods, the theoretical basis was still founded upon the minimization of a variational indicator. Accordingly, the approach was highly successful whenever a variational formulation was available, but was unable to deal with problems where there was no quantity to be minimized. A further step in the development of the finite element method was the use of the method of weighted residuals[9] as the theoretical basis for constructing a discretization procedure. This extends the applicability of the finite element method to many new fields, and the finite element method is now routinely applied to the solution of problems in fluid flow, convective heat transfer and semiconductor device simulation.

The Method of Weighted Residuals

We will describe the basics of the method of weighted residuals and establish its connection with the traditional finite element method. The method of weighted residuals as a basis for finite elements consists of five steps:

1. The region under consideration is divided into a number of subregions.

2. Within each region, the unknown variables are interpolated with a suitably chosen set of basis functions.

3. A residual is defined in each region by substituting the interpolating functions into the equation that is to be solved.

4. The residual in each region is then subjected to as many conditions as there are unknowns in the residual.

5. The resulting equations for each region are then assembled into a global equation and solved by suitable procedures.

We next show that specific choices in steps 1 through 5 result in the traditional finite element method. The subregions in step 1 are called elements. The shape of the elements is described by nodes. Within each element, the unknowns are approximated by low order polynomials (typically linear or quadratic polynomials). The residual is defined as the expression obtained upon substituting the approximate solution into the differential equation to be solved. The conditions placed upon the residual are the Galerkin conditions, viz., the projection of the residual on the space spanned by the basis functions is zero. This results in a set of algebraic equations for each element which are then assembled into a single global equation and solved by a suitable method.

The equations that result by following the recipe prescribed by the method of weighted residuals are identical to those that would be obtain by minimization of a variational indicator (if one is present). The advantage of the weighted residual formulation is that it can be applied intact even if there is no variational indicator that can be defined for a given problem. A more important advantage of the weighted residual framework is that, by varying the interpolating functions and the imposed constraints, it is possible to generate almost every existing approximate technique for the discretization of partial differential equations. Thus finite elements, finite differences, pseudo-spectral methods, and boundary element methods can be derived by the weighted residual method.

The H-Version of the Finite Element Method

A few more details about the method are useful in order to understand the ability of the finite element method to handle complex geometries. Explanations are provided on a two-dimensional geometry. Each element in space is described by a set of nodes. The element in physical space can assume an arbitrary shape (within some limitations). This element is then mapped onto a square computational domain (Fig. 4).

The basis functions are defined on the square transformed element. In the standard h-version of the finite element method, the basis functions are typically (but not necessarily), Lagrange polynomials that assume unit values at the nodes

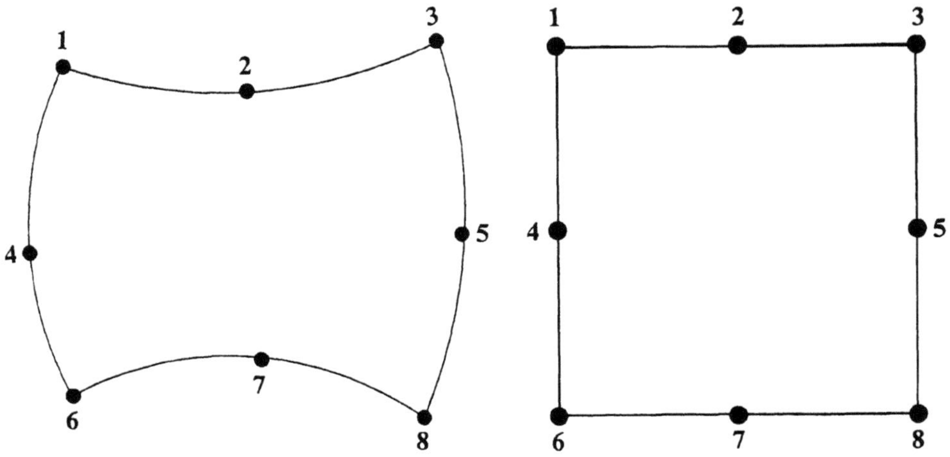

Figure 4. Transformation of an element from the physical domain (left), into the computational domain (right).

of the square element, and are zero at the other nodes. Thus, for a two-dimensional square element with four corner nodes, the basis functions are four bilinear polynomials; for a nine-noded element the basis functions are nine biquadratic polynomials. The computation of various integrals (required by the Galerkin condition), is performed in the transformed domain using numerical integration. The Jacobian of the mapping should not be singular. If the basis functions used to describe the geometry of the elements are the same as the basis functions used to describe the variation of the unknowns, the element is known as an isoparametric element. This type of element is widely used in structural analysis because the unknowns are typically displacements at the nodes; hence a change in the location of a node causes a change in the geometry of an element which must therefore be able to describe the change accurately. The isoparametric element is often unnecessary in an Eulerian description of fluid mechanics where the geometry of an element can be described by lower orders of polynomials than those required to describe the variation of velocity and pressure within an element (sub-parametric elements).

The Galerkin condition for each element results in a system of linear or nonlinear algebraic equations for that element. The individual element matrices are then assembled into a global matrix which is solved by a chosen method. The assembly process can be described very simply: assume that each element matrix is written so that it is the size of the global matrix. The only non-zero elements in this matrix are those corresponding to the degrees of freedom for that particular element. The global matrix is then the sum of the individual element matrices.

The P-Version of the Finite Element Method

The h-version of the finite element method, is at present the most widely used version. However, there are other versions of the finite element method, each with its own set of advantages and disadvantages. We describe here the p-version of the finite element method which appears to have some crucial advantages over the h-version, and relatively few disadvantages.[10]

The development of the p-version was primarily in response to the need for adaptive computing with finite elements. If there is a portion of the region under analysis where the gradients are relatively large, it is desirable to use more basis functions in that region to obtain more accuracy. In the h-version this would correspond to using more elements by decreasing the size of the elements in that region alone. This is not a trivial operation because it requires that the elements within the region of refinement maintain continuity with elements in the neighboring regions. There are several ways to attain this effect, but none of them are easy, and require elaborate programming.

The p-version of finite elements uses basis functions of higher order than those used in the h-version. However, the basis functions are not similar to those used in the h-version, and the difference is more than a matter of increasing the polynomial level. If we consider a one-dimensional element, the first two functions are the same as those used in the h-version. The subsequent ones are integrals of Legendre polynomials. The coefficients of the first two basis functions correspond to the values at the two end points; the remaining higher basis functions do not correspond to the values of the unknown at any point, but act as spectral coefficients. Thus, within a given element, the order of the polynomials can be increased to any desired level without requiring additional grid points; the basis functions are hierarchic in nature. This makes for a scheme that is particularly easy to use for adaptive computing because it is possible to increase the level of polynomials in one element without increasing the level of interpolation in the neighboring elements. Some further advantages of the p-version approach are as follows:

a) For a given number of degrees of freedom, the accuracy obtained by using the p-version is much higher than with the h-version. This is strictly true only if the underlying solution is smooth, but there are methods for dealing with singularities that can preserve the rapid convergence rate of the basis functions of the p-version.

b) Since the basis functions are hierarchic in nature, it is possible to develop highly effective solution techniques of a multilevel nature. These are described in detail in a subsequent section.

3. Solution Techniques

The discretization of continuum operators results in a system of algebraic equations for every element. The assembly of the individual element matrices results in a large global matrix that must be solved for the degrees of freedom of the given problem. We describe below the various solution techniques used in finite element analysis and point out the advantages and disadvantages of each.

Direct Solution Methods

Direct solution techniques are currently the most widely used solution techniques in finite element analysis. In this approach a matrix is written as a product of factors which are then used to solve the equations for a given right hand side. The most common factorization is the $L\,D\,L^T$ decomposition of the coefficient matrix A when this matrix is symmetric,

$$A = L\,D\,L^T$$

and the $L\,U$ decomposition when the matrix is nonsymmetric,

$$A = L\,U.$$

In order for these factorizations to be reliable, it is necessary to perform pivoting during the factorization, but the matrices resulting from finite element discretization are relatively well conditioned, and it has been traditional to use these factorization methods without pivoting. Under these conditions, the solution to a given matrix equation can be obtained in a fixed number of steps (bounds to which can be determined in advance). This has been one of the main reasons for the wide use of direct techniques in production codes.[11]

A disadvantage with direct methods is that the sparsity of the original matrix is destroyed due to fill-in during the process of factorization; the factors have significantly more non-zero elements than the original coefficient matrices. It has been observed in numerical experiments with large 3D structures that the number of non-zero elements in the factors can be as much as two orders of magnitude larger than in the original coefficient matrix.[12] These have necessitated the development of "out-of-core" solution techniques where the matrix and its factors are stored on disk and brought into main memory as needed. With the development of fast vector processors, disk access time is often very slow as compared to arithmetic operations, and out-of-core methods can greatly increase the elapsed time to factor a matrix even though the CPU time may not be very high. This has led to a renewal of interest in iterative methods for solving systems of equations encountered in finite element analysis.

Iterative Methods

Iterative methods begin with an initial guess, and compute a sequence of approximate solutions that converge to the exact solution. The accuracy of the solution does depend on the number of steps performed. They have the advantage over direct methods of preserving the sparsity of the coefficient matrix, so that they are particularly attractive for solving linear systems arising either from finite difference or finite element discretization of partial differential equations. The matrices obtained in such cases are characterized by a great number of zero entries; this reduces both the storage requirements and the computational cost per iteration, since the matrix vector product is relatively inexpensive for a sparse matrix as compared to a full matrix.

Iterative methods are usually preferred for solving large sparse systems of linear equations with a single right hand side, or for solving time dependent problems and nonlinear equations by Newton methods, in which the solution at each step provides a good starting guess for the next step. They are not considered very useful for finding inverses of matrices, or for solving systems of linear equations with a sequence of right hand sides supplied in succession. The amount of work involved in the solution of linear systems of equations using an iterative method, depends upon the convergence rate and the desired accuracy. A drawback of iterative methods is the possibility of slow or irregular convergence, the latter due to complex eigenvalues. The rate of convergence of these methods can usually be speeded by the technique of preconditioning, which will be explained in the next section.

Direct methods are able to handle symmetric and nonsymmetric matrices equally well; iterative techniques however, tend to require modifications depending upon the nature of the matrix. For symmetric positive definite matrices, (such as would be obtained by discretizing a self-adjoint operator), the iterative method of choice is the conjugate gradient method.[13] This method generates a sequence of approximate solutions using a linear combination of vectors from a Krylov space. In exact arithmetic, it is guaranteed to converge in a number of iterations equal to the order of the matrix. This however, is mainly of theoretical interest because in practice, the number of iterations is strongly affected by the condition number of the matrix.

For symmetric indefinite matrices, the conjugate gradient method is not guaranteed to converge and performs poorly in practice. The two techniques that perform relatively well on this class of problems are the conjugate residual method,[14] and the symmetric LQ method (SYMMLQ).[15]

The class of problems defined by nonsymmetric, indefinite matrices, is relatively difficult to handle with iterative methods. A host of methods have been proposed to deal with problems in this class. We mention only a few in this section. The biconjugate gradient method[16] is a generalization of the conjugate gradient method; the conjugate residual method has been generalized into GCR and ORTHOMIN.[17,18] A generalization of SYMMLQ to nonsymmetric matrices has been developed and is named USYMMLQ.[19] The generalized minimum residual method (GMRES), based on Arnoldi's method is one of the more sophisticated methods for nonsymmetric indefinite matrices.[20] A paper by Saad and Schultz[21] uses an elegant method to classify a variety of iterative methods by placing them in the framework of Petrov-Galerkin methods, and exposes underlying similarities among seemingly dissimilar methods.

Overview of Preconditioning Techniques

The spectrum of the coefficient matrix governs the rate of convergence of any iterative method. A simple way to accelerate convergence is to "precondition", that is, to replace the original problem with a new, easier problem with the same solution. The choice of the preconditioner is crucial to the success of this maneuver. It should be easy to invert and a good approximation to A. We consider some preconditioning techniques and their use in conjunction with iterative methods. Let

$$K = K_L \ K_R$$

be a nonsingular matrix. The solution to a linear system of equations

$$A \ x = b \ ,$$

can be obtained by solving the problem

$$[K_L^{-1} \ A \ K_R^{-1}] \ [K_R \ x] = K_L^{-1} \ b \ .$$

The use of such an auxiliary matrix is known as preconditioning, and the matrix K is often referred to as the preconditioner. It is possible to distinguish three different ways of applying the preconditioning[22]: the relation above defines the two sided orientation or split preconditioning. The other two ways can be considered as special cases with:

left orientation: $K_R = I$;

right orientation: $K_L = I$.

One of the more effective methods for preconditioning is based on incomplete factorization of the coefficient matrix,[23] where a lower triangular matrix L and an upper triangular matrix U are constructed that are good approximations of the LU factors of A, and are also sparse. The incomplete LU factorization (ILU) can be described as follows. Consider only those locations where A has nonzero elements; at these locations, perform operations corresponding to an LU factorization using elements only from those locations where A had nonzero elements. The sparsity structure of A is completely preserved in the resulting factors. It is possible to define sparse factorizations which allow higher levels of fill in the factors; this may lead to convergence in a lesser number of iterations although each iteration would be more expensive.

Preconditioned iterative methods have been extensively used only during the last decade. For symmetric, positive definite matrices, the conjugate gradient method with suitable preconditioning, has been shown to be highly effective in the solution of large three-dimensional problems arising in structural analysis. However, discretization of the Navier Stokes equations leads to highly nonsymmetric, indefinite matrices. To date, none of the methods described above has been sufficiently reliable to be used in a routine manner. The development of reliable iterative methods for the Navier Stokes equations is a topic of active research.

Multilevel Solution Method

The multigrid method is an iterative technique that works on different principles from the ones mentioned earlier. It exploits the ability of Gauss-Seidel and Jordan iterative methods to reduce high frequency errors very rapidly: a sequence of nested grids is constructed, and the errors are reduced at each grid; the net result is a rapid decrease in all frequencies of the error. The multigrid method is increasingly being accepted as one of the most effective computational techniques for the solution of problems arising from the discretization of partial differential equations.[24] However, for complex 3-dimensional geometries, defining this sequence of nested grids is a difficult programming problem. The p-version of finite elements offers the possibility of using multigrid concepts without the need for defining multiple grids.

The essential feature that makes the finite element method with hierarchic basis functions highly suitable in a multilevel iterative scheme is that the matrix formed in the p-version is hierarchic in nature; i.e., the matrix formed by a lower order polynomial is fully contained in the matrix formed by a higher order polynomial: the matrix is itself nested. When performing multilevel iterations, the nested matrices resulting from the p-version play a role similar to the nested grids employed in multigrid methods in the h-version.[25] Multigrid techniques use nested geometric grids to control errors at different frequencies, while the multilevel iteration scheme with hierarchic basis functions works in a more direct fashion with the frequencies.

In multilevel schemes, the work required to solve a problem to a given accuracy is kept to a minimum if there is a balance between the error at a certain level as compared to the overall error; for instance it is useless to smooth at a certain level if the components of the residual are much less than those at other levels. For this reason it is efficient to perform only a few smoothing iterations at any level. The majority of operations in the multilevel iteration scheme are taken by the smoothing operation; no calculations are required for restriction and prolongation. This is in contrast to the multigrid method with the h-version where several operations are required for restriction and prolongation. If the level is frequently changed, these operations can increase the execution time, in spite of a better rate of convergence. It is therefore possible to choose a level by level schedule with much greater freedom in the above method than with multigrid methods.

Block Gauss-Seidel schemes have been found to perform significantly better than point methods. This requires the solution of a system of equations at each iteration. This is most efficiently done by conjugate gradients using only as many iterations as are required to drop the residual below a chosen fraction of the last overall multigrid residual. The overall residual is computed at each point in the schedule where the finest level is reached.

The multilevel iterative technique is relatively recent, and has not been extensivlely used. In tests with symmetric positive definite matrices, the results have been very satisfactory, and comparable to those obtained with the best iterative solution methods. However, with nonsymmetric indefinite matrices, particularly those resulting from the Navier Stokes equations, the method is not reliable enough to be used in a routine manner.

4. Computational Solution of the Navier Stokes Equations

The discretization of the Navier Stokes equations in primitive variable form, by the p-version of the finite element method, results in a system of nonlinear algebraic equations. Since this system of equations can often be very large, it is essential that we use solution techniques that are efficient and robust. Direct methods are highly robust, but are prohibitively expensive in terms of storage and computation particularly for three-dimensional problems. Iterative methods are efficient in terms of storage and computation, but are not satisfactorily robust.

One approach to this problem is to use a formulation that does not require the solution of a system of equations as difficult as those obtained by discretization of the primitive form of the Navier Stokes equations. This can be accomplished by using an explicit time integration scheme where we simply march forward in time by using values at a previous time step; the only computation required is the updating of new values by substituting the previous values into the discretized operator. While there are stability constraints on the time step using explicit methods, we avoid entirely the difficult problem of having to solve a system of nonsymmetric algebraic equations. It is still necessary to solve a system of equations at each time step, corresponding to a consistent mass matrix, but this is symmetric and positive definite, and is very amenable to iterative solution methods.

There is however a problem with the use of an explicit time marching scheme for the incompressible Navier Stokes equations in primitive variable form. The reason is that the continuity equation can be considered as a constraint on the divergence of the velocity, rather than as a full-fledged equation coupling the pressure to the velocity. Thus, the pressure acts as a Lagrange multiplier that ensures the satisfaction of the solenoidal constraint. Since there is no time derivative of the pressure appearing in the equations, updating the velocities must be subject to the incompressibility constraint.

One approach is to replace the continuity equation with a Poisson equation for the pressure, thus decoupling the momentum and continuity equations, except on the domain boundary. This approach has been widely used in the finite difference solutions of the incompressible flow equations.[26] There are however, theoretical and practical difficulties connected with the specification of boundary conditions for the pressure Poisson equation. In the following paragraphs, we will discuss the reason for the difficulties in specifying the pressure boundary conditions, and also show how a finite element formulation can overcome these difficulties.

The pressure Poisson equation (PPE) is obtained by taking the divergence of the momentum equations:

$$\frac{\partial^2 p}{\partial x^2} + \frac{\partial^2 p}{\partial y^2} = \frac{\partial}{\partial x}\{\eta(\frac{\partial^2 u}{\partial x^2} + \frac{\partial^2 u}{\partial y^2}) - [\frac{\partial(u^2)}{\partial x} + \frac{\partial(uv)}{\partial y} + \frac{\partial u}{\partial t}]\} +$$

$$\frac{\partial}{\partial y}\{\eta(\frac{\partial^2 v}{\partial x^2} + \frac{\partial^2 v}{\partial y^2}) - [\frac{\partial(uv)}{\partial x} + \frac{\partial(v^2)}{\partial y} + \frac{\partial v}{\partial t}]\}$$

where η is the kinematic viscosity.

Since the pressure Poisson equation is an elliptic partial differential equation, it requires that at every point on the boundary, either the value of the pressure (Dirichlet conditions), or the normal derivative of the pressure (Neumann conditions) be specified. However, these are not generally determined from the physics of a situation. Earlier formulations of the PPE have either determined the normal derivative of the pressure on the boundary from the known velocity values, or have assumed homogeneous Neumann boundary conditions.[27] These measures however, are unnecessary when a finite element formulation is used for the pressure Poisson equation with integration by parts. The reasoning is relatively simple: the pressure Poisson equation is obtained by differentiating the momentum equations; weighting this with a suitable basis function and integrating by parts (in the standard manner for weighted residuals) results in a lowering of the order of differentiation. In fact the terms that result from the integration by parts are the momentum equations themselves, which are automatically satisfied. Accordingly, there is no need to specify any boundary conditions for the pressure equation that are not inherent in the physics of the problem.[28] This formulation, however, requires higher continuity condition (C^1) across the elemental boundaries. For many practical problems of interest it is possible to relax this restriction.

The discretization of the momentum and the pressure Poisson equations leads to a set of algebraic equations of the form

$$M \dot{V} = L_a V + L_v V + C P + R$$

$$F P = G U + \dot{D}.$$

In the first equation, the subscripts a and v denote the advection and viscous operators, respectively. M is the mass matrix, V is the velocity vector, C is the pressure matrix, P is the pressure vector, and R is the velocity load vector determined by the boundary conditions. In the second equation, F is the Laplacian operator, G contains derivatives of the momentum equations, and $\dot{D}$ represents the time derivative of the continuity equation.

Therefore, the problem reduces to the transient-iterative solution of these two systems of equations. We now consider a simple explicit time marching approach for these equations. The time derivatives are replaced with the backward difference equations

$$\dot{V} = \frac{V^{n+1} - V^n}{\Delta t}.$$

Once the momentum equations are integrated forward in time, the pressure at step $n+1$ is computed from the known velocity. The time integration of $\dot{D}$ is done using the same backward difference formula:

$$\dot{D} = \frac{D^{n+2} - D^{n+1}}{\Delta t}.$$

However, to enforce the divergence-free (incompressibility) constraint, D^{n+2} is set to zero. Thus, the two systems are solved iteratively in this manner.

The time integration of the momentum equations can also be achieved by a fractional time stepping (splitting) method.[29] The time-differencing scheme can be split into three steps:

$$V^1 = V^n + \Delta t \; M^{-1} \; L_a \; V^n$$

$$V^2 = V^1 + \Delta t \; M^{-1} \; L_v \; V^1$$

$$V^{n+1} = V^2 + \Delta t \; M^{-1} \; C \; P^{n+1}$$

where V^1 and V^2 are auxiliary velocity vectors, V^n and V^{n+1} are velocity vectors at time step n and n+1, respectively, and the pressure P^{n+1} is computed such that the velocity field V^{n+1} satisfies the incompressibility condition. Although in general splitting offers considerable simplifications compared to the implicit coupled schemes, it may suffer from larger time-stepping errors. However, it is possible to reduce the error of the fractional time stepping method by applying higher order local extrapolation techniques.

We next show some solutions of the Navier Stokes equations obtained by applying the above scheme to the well known problem of a driven cavity.[30] We differ only slightly from the standard problem, in that the velocity imposed at the upper surface has a parabolic profile. This is done so as to remove the singularity at the upper corners; the Reynolds number is based upon the maximum value of the imposed velocity (Fig. 5).

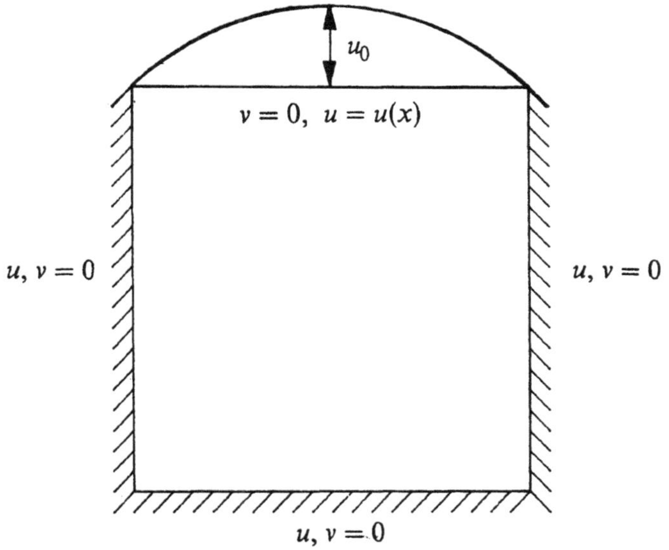

Figure 5. Driven cavity problem with imposed velocity conditions modified to remove singularities.

In Fig. 6, we compare the results obtained with this formulation for a Reynolds number of 200, with a standard formulation, where the Navier Stokes equations in primitive variable form are discretized, and the resulting system of equations solved with a direct method (LU factorization). The excellent agreement between these two very different procedures, provides a validation of the formulation explained here, and in particular, establishes the correctness of the boundary conditions for the pressure Poisson equation.

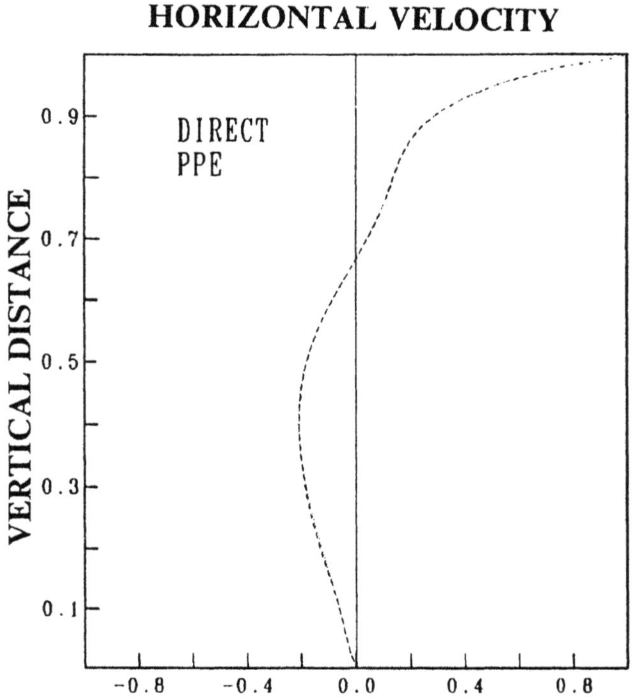

Figure 6. Comparison of results between the Pressure Poisson formulation, and the primitive variable formulation.

5. Parallel implementation of the Solution Scheme

The numerical solution of the Navier Stokes equations for large three-dimensional problems (particularly at high Reynolds numbers), is one of the most computationally demanding problems in engineering. These problems often exceed the limits of most sequential computers, and there is strong interest in the development of algorithms which can be implemented efficiently on parallel computing systems.

Explicit time integration methods, are easily and efficiently implemented in parallel with finite difference methods. When used with finite element methods, a system of equations corresponding to a consistent mass matrix needs to be solved at each time step. The solution of the Navier Stokes equations also requires the solution of a Poisson problem at each time step. The matrices corresponding to these equations are symmetric and positive definite, and are readily solved with conjugate gradient methods or multilevel iteration schemes.

The scheme for parallel implementation is particularly simple. Since the major component of the computations is based on matrix vector multiplications, we work with individual element matrices instead of assembling a global matrix. The total number of elements are broken up and assigned to processors. Each processor then performs the matrix vector multiplications independently of the other processors; there is no need for data communication at this step. After the matrix vector multiplications are completed, the resulting elemental vectors are assembled into a single global vector. This global vector is updated according to the recipe determined by the iterative technique, and is then partitioned into element vectors. These are redistributed to the processors which repeat the matrix vector products. The extraordinary simplicity of this scheme for parallel implementation also makes it ideal for massively parallel computing. The results of parallel implementation of the Poisson equation on a shared memory IBM 3090 system are shown in Fig. 7.

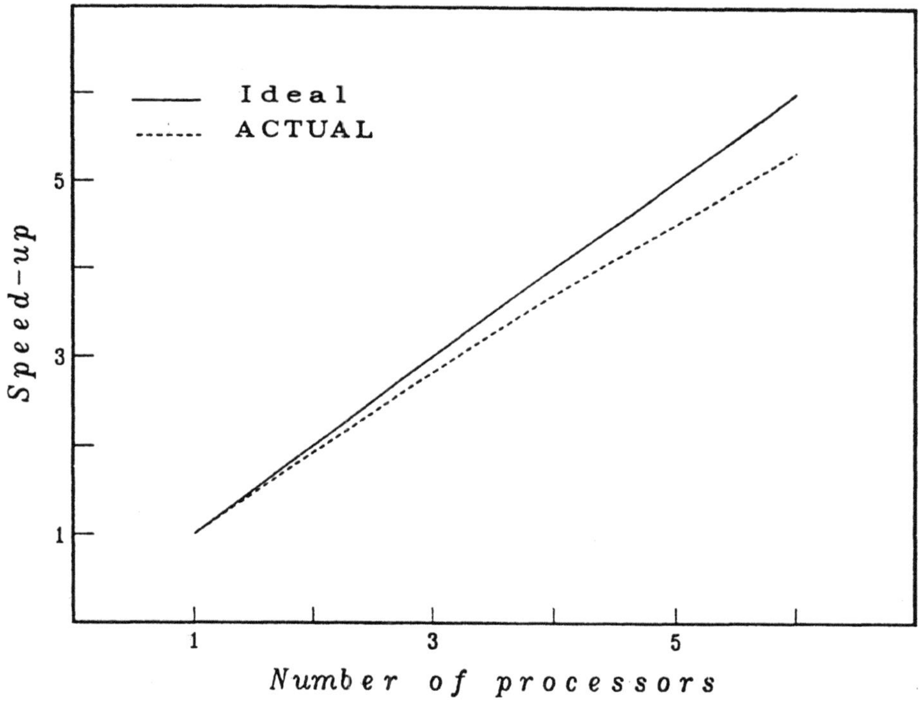

Figure 7. Speedup curve for the Poisson equation on an IBM 3090.

6. Turbulent Flow

Reynolds Equations

In the previous sections in this chapter, we have specifically avoided any discussion of turbulent flow even though most flows occurring in nature are turbulent. Turbulent flow is characterized by highly irregular, spatial and temporal fluctuations superimposed upon a mean flow. In spite of the forbidding complexity of turbulent flow it has been the subject of intense and sustained study for over a century. The major reason is that turbulence can dramatically affect the transfer of momentum and energy in a flow, and estimates that do not account for the effects of turbulence can be grossly in error.[31]

A detailed description of turbulent fluctuations would require an enormous number of degrees of freedom (impossibly large in most practical situations), and it is thus necessary to resort to averaged quantities to describe and analyze turbulent motion. We follow here the most common approach in deriving the Reynolds equations, which serve as a starting point for the analysis of fully developed turbulent flow.[7]

The Navier Stokes equations for incompressible flow can be written as:

$$\frac{\partial V_i}{\partial x_i} = 0$$

$$\rho\frac{DV_i}{Dt} = -\frac{\partial p}{\partial x_i} + \mu\frac{\partial^2 V_i}{\partial x_j \partial x_j}.$$

It is convenient to separate the flow into a mean (time averaged) motion and a superposed fluctuation; the various quantities are:

$$V_i = \overline{V}_i + V'_i \qquad p = \overline{p} + p'$$

where a bar over a quantity indicates a time average, and the primed variables are fluctutations. The time averages at a point in space are given by expressions of the form:

$$\overline{V}_i = \frac{1}{\Delta t}\int_t^{t+\Delta t} V_i dt.$$

We note here that mean values are taken over a sufficiently long interval of time so that they are independent of time. A consequence of this is that the means of the fluctuating quantities are zero.

$$\overline{V'}_i = 0; \qquad \overline{p'} = 0.$$

We can substitute for each for the field quantities V_i, p into the the Navier Stokes equations and then time average the resulting equations. If we consider first the continuity equation, we have,

$$\frac{\partial(\overline{V}_i + \overline{V'}_i)}{\partial x_i} = 0.$$

Since the means of the fluctuating quantities are all zero, the continuity equation reduces to

$$\frac{\partial \overline{V}_i}{\partial x_i} = 0.$$

After suitable algebraic manipulation,[7] the momentum equations can be written as:

$$\rho\left(\bar{u}\frac{\partial \bar{u}}{\partial x} + \bar{v}\frac{\partial \bar{u}}{\partial y} + \bar{w}\frac{\partial \bar{u}}{\partial z}\right) = -\frac{\partial \bar{p}}{\partial x} + \mu\nabla^2\bar{u} - \rho\left[\frac{\partial \overline{u'^2}}{\partial x} + \frac{\partial \overline{u'v'}}{\partial y} + \frac{\partial \overline{u'w'}}{\partial z}\right]$$

$$\rho\left(\bar{u}\frac{\partial \bar{v}}{\partial x} + \bar{v}\frac{\partial \bar{v}}{\partial y} + \bar{w}\frac{\partial \bar{v}}{\partial z}\right) = -\frac{\partial \bar{p}}{\partial y} + \mu\nabla^2\bar{v} - \rho\left[\frac{\partial \overline{u'v'}}{\partial x} + \frac{\partial \overline{v'^2}}{\partial y} + \frac{\partial \overline{v'w'}}{\partial z}\right]$$

$$\rho\left(\bar{u}\frac{\partial \bar{w}}{\partial x} + \bar{v}\frac{\partial \bar{w}}{\partial y} + \bar{w}\frac{\partial \bar{w}}{\partial z}\right) = -\frac{\partial \bar{p}}{\partial z} + \mu\nabla^2\bar{w} - \rho\left[\frac{\partial \overline{u'w'}}{\partial x} + \frac{\partial \overline{v'w'}}{\partial y} + \frac{\partial \overline{w'^2}}{\partial z}\right].$$

If we now define a virtual stress tensor (also known as the Reynolds stress tensor), as

$$\tau'_{ij} = -\overline{\rho V'_i V'_j}$$

we see that the time averaged equations can be written as

$$\rho \overline{V}_j \frac{\partial \overline{V}_i}{\partial x_j} = \frac{\partial}{\partial x_j}(\tau_{ij}).$$

These equations are called the Reynolds equations; the total stress tensor τ_{ij} is now a sum of the viscous stresses and the Reynolds stress.

$$\tau_{ij} = \bar{\tau}_{ij} + \tau'_{ij} \quad = -\bar{p}\delta_{ij} + 2\mu\bar{e}_{ij} - \rho\overline{V'_i V'_j}.$$

The overall effect of the time averaging procedure is to recast the turbulent flow equations into a form that is similar to the equations of laminar flow, except that the stress tensor now has an additional term, which is made up of time averages of fluctuating quantities. It is essential to point out here that the Reynolds stress tensor which is made up of time averages of products of fluctuations has not been defined in terms of the mean flow quantities, and the Reynolds equation in this form contains more unknowns than equations. This is a major difficulty in working with the time averaged equations for turbulent flow. Any attempts at obtaining expressions for the Reynolds stress from the Navier Stokes equations result in higher order products of fluctuations, which are themselves unknown; this is known as the closure problem of turbulence.

Empirical Relations for the Reynolds Stress Tensor

There is not as yet a rational theory that would allow for the computation of the Reynolds stress without making simplifying assumptions based on empirical considerations. We examine some of the more common means of modelling the Reynolds stress tensor.

In many flows, the normal stresses are often insignificant in comparison to the shear stresses. By analogy with laminar flow, a mixing coefficient μ_t is introduced, and it is postulated that the Reynolds stress is of the form:

$$\tau' = \mu_t \frac{\partial \bar{u}}{\partial y}.$$

Carrying this further it is possible to define an eddy viscosity

$$\nu_t = \frac{\mu_t}{\rho},$$

so that the the Reynolds shear stress is given by

$$\tau' = \rho \nu_t \frac{\partial \bar{u}}{\partial y}.$$

Prandtl defined a quantity called the mixing length which is loosely analogous to the concept of a mean free path of molecules in the kinetic theory of gases. The expression for the eddy viscosity under the mixing length hypothesis is given by[7]

$$\tau' = \rho l^2 \left| \frac{\partial \bar{u}}{\partial y} \right| \frac{\partial \bar{u}}{\partial y}.$$

The mixing length l is not a constant for a given fluid, but in some instances it is possible to establish a relation between l and some characteristic length L of the the flow. Prandtl's mixing length hypothesis represents one of the earlier attempts at solving the turbulence closure problem, and has been successfully applied to a limited class of flows. There are, however, obvious shortcomings in the model, since it fails whenever

$$\frac{\partial \bar{u}}{\partial y} = 0.$$

Numerous comparable models have been proposed which attempt to overcome the shortcomings of Prandtl's mixing length model, and are applicable to a wider range of physical flow situations. These are generally referred to as algebraic eddy viscosity models.

A different class of models are based upon the turbulence kinetic energy k and its dissipation rate ε, where,[32]

$$k = \frac{1}{2}\overline{V'_i V'_i}$$

$$\varepsilon = 2\nu\overline{e'_{ij} e'_{ij}}$$

$$e'_{ij} = \frac{1}{2}\left(\frac{\partial V'_i}{\partial x_j} + \frac{\partial V'_j}{\partial x_i}\right).$$

A relation of the form

$$\nu_t = C_M \frac{k^2}{\varepsilon}$$

is postulated for the eddy viscosity where C_M is an empirical constant. In addition, two coupled partial differential equations are derived for k and ε; these equations contain several constants that are adjusted based upon the empirical data. These models for the Reynolds stress tensor, called k-ε models, have been found to be applicable under a wider range of conditions than the algebraic eddy viscosity models.[33] Recently developed approaches, based upon renormalization group methods,[34] can derive values for the empirical constants in the k-ε models from only a few basic assumptions. The development of general models for the Reynolds stress is a topic of active research.

7. Direct Numerical Simulation of Isotropic Turbulent Flows

The nature of fluid turbulence can be studied by analytic, experimental, or numerical approaches. Although each approach has its own merits, the highly nonlinear nature of the Navier Stokes equations makes them intractable to analytic methods except in a limited set of circumstances. By comparison, experimental methods have been more successful in unraveling some of the aspects of turbulence. However they are beset by problems of high cost, accuracy and difficulty of multipoint measurements (to name a few). Numerical simulations of turbulent flows can be used to study underlying physical processes from first principles. The governing equations are integrated in time using a suitable numerical method and the solutions so obtained are assumed to be representative of a real flow occurring in the same geometry. There are many limitations to simulating turbulent flows. For a typical turbulent flow characterized by microscale Reynolds number R, computer memory and computational effort scale as $R^{3/4}$ and R^3 respectively.[35] Fluid flows encountered in practice have their Reynolds numbers in thousands, which make it impractical to handle the computations needed to simulate turbulent flows of engineering interest. If, however, the small-scale features of a turbulent flow are described by a suitably constructed model then the computational requirements can be reduced considerably.

Small-scale features of a turbulent flow are commonly assumed to be universal in character. Only the large-scale aspects are dependent on the geometry of the flow and the boundary conditions, and can be captured by standard numerical methods. Sub-grid scale models constructed to date have not been entirely satisfactory in describing the small scale features of a turbulent flow. One of the reasons is that they do not accurately describe the interaction between large and small scales of a flow. A proper understanding of the turbulent dynamics is therefore necessary for constructing efficient sub-grid scale models and direct numerical simulation can be an important research tool in that endeavor.

Direct numerical simulation of a three-dimensional isotropic turbulent flow in a simple geometry was first done by Orszag and Patterson.[36] The numerical technique used therein has paved the way for more simulations of increasing resolution over the past two decades. We will outline here, in brief, the salient features of the method. The incompressible Navier Stokes equations are solved subject to periodic boundary conditions

$$V(x + 2\pi\alpha, t) = V(x, t)$$

where V and x are vectors and α is a vector with integer components. The nonlinear term is written in a form suitable for computation called the rotation form

$$V \cdot \nabla V = V \times \omega + \frac{1}{2}\nabla(V \cdot V)$$

so that the term $\frac{1}{2}\nabla(V \cdot V)$ can be grouped with the pressure gradient term. Here ω is the vorticity vector defined by $\omega = \nabla \times V$. Spectral methods are used for the approximation of spatial derivatives of V and p in the governing equations. When the solution fields are sufficiently smooth, spectral methods provide a very accurate approximation. A Fourier series based spectral method is used to exploit the periodic boundary conditions. In this method, a truncated Fourier series

$$V(x, t) = \sum_{|m| < M} \sum_{|n| < N} \sum_{|p| < P} \tilde{V}_{mnp}(t) e^{imx} e^{iny} e^{ipz}$$

is used to approximate $V(x, t)$ and $p(x, t)$. If $V(x,t)$ and $p(x, t)$ are infinitely differentiable, then the error due to this approximation goes to zero faster than any power of the cutoffs M,N,P. Another advantage of using a Fourier series is that a Fast Fourier Transform (FFT) of the kind developed by Cooley and Tukey[37] can be used to transform from physical to Fourier space and vice versa. For reasons of accuracy, all the derivatives are evaluated in the Fourier space.

Dealiasing

The term $V \times \omega$ is evaluated in the physical space. If it were to be evaluated in the Fourier space, it would lead to the evaluation of a convolution term which can be prohibitively expensive.[38] The determination of the nonlinear term in the physical space, however, results in aliasing errors. If N is the cutoff in any one direction, the aliasing errors are due to the convolution terms from modes k > N/2. These terms are projected on to the modes of the computational domain. One way to "dealias" is to discard Fourier modes corresponding to k > N/3. Although there are several other methods[39] to dealias a computation, dealiasing by truncation is simple to implement and very effective.

Time Stepping Scheme

The time integration of the Navier Stokes equations is all done in the Fourier space except for the nonlinear term. As stated before, the nonlinear term is computed pseudospectrally. A combination of leapfrog and Crank-Nicolson schemes is used for time stepping. It is written as,

$$\frac{\hat{V}_i^{n+1} - \hat{V}_i^{n-1}}{2\Delta t} = -\hat{i}\,(k_j \hat{V}_i \hat{V}_j + k_i \hat{p})^n - \nu k^2 \frac{\hat{V}_i^{n+1} + \hat{V}_i^{n-1}}{2}$$

where $\hat{i} = \sqrt{(-1)}$, $\hat{V}_i^{n+1} = \hat{V}_i(t + \Delta t)$, $t = n\Delta t$, ν = kinematic viscosity coefficient, Δt = time step, $\hat{V}_i(k_l)$ is the Fourier transform of the velocity field $V_i(x_l)$. Leapfrog averaging is done periodically to prevent the decoupling of even and odd time steps during long-time integration.

Parallel Implementation of the Numerical Scheme

About 80% of the time is spent in transforming velocity and vorticity fields from physical to Fourier space and back. Since all the fields are three dimensional, any parallelization of the overall algorithm boils down to a parallelization of 3D FFT algorithm. A pseudospectral code has been implemented to perform 256^3 computations on an IBM 3090/600S with 512Mb of main memory and at least 1250Mb of extended memory. The extended memory can be dedicated to a single user in the VM/XA operating system. The extended memory is used for 9 large arrays, each of which is 8*256^3 bytes in size. Three arrays each are dedicated for V^n, ω^n and V^{n-1}. Blocks of data can be moved between main and extended memories in VM/XA. Most of the main memory is used as work space to perform 3D FFTs and other computations on blocks of data moved from the extended memory. We perform 3D FFTs in parallel by operating on planes of data of a 256^3 array. Parallel computation and synchronization are accomplished using the PF extensions of the VS FORTRAN compiler. Although there is only one data path between the main and extended memories to move data, by optimally staggering the computation on multiprocessors during the data transfer we have been able to obtain excellent parallel speed-ups (Fig. 8). This technique can be used for other problems which require large memories.

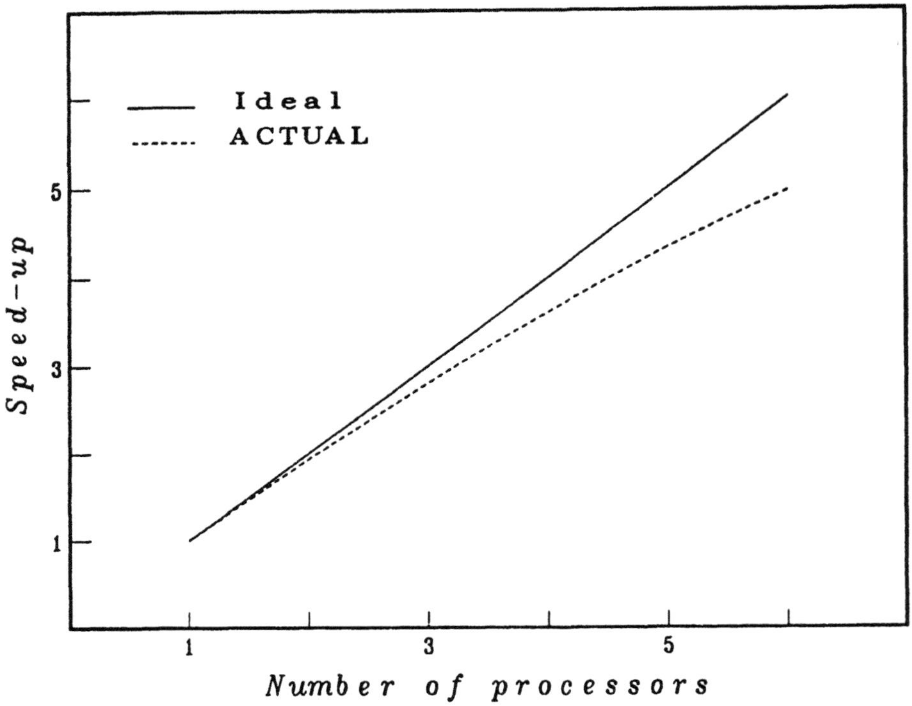

Figure 8. Speedup for a simulation of size 256×256×256.

Results

We performed a 256^3 mode freely decaying, isotropic turbulent flow simulation.[40] To date this is the largest simulation of its kind. The initial velocity field in Fourier space is a zero-mean random Gaussian field with random phases with a given energy spectrum. The energy spectrum is so chosen that only the large scales of motion contain most of the energy. Small scales of motion begin to appear with time due to the mechanism of vortex stretching. It takes many thousands of time steps before the velocity field can be characterized as fully turbulent. Turbulent velocity fields are usually characterized by a non-dimensionalized number known as the Taylor-microscale Reynolds number R_λ. The dynamics of two turbulent velocity fields can be compared provided their R_λs are the same. In the present simulation, starting with $R_\lambda = 243$ at $t = 0$, the simulation is stopped at $t = 4$ after 4000 time steps. We compared some of the important properties of the simulated velocity field with experimentally obtained results. In Figs 9 and 10 we compare normalized energy and dissipation wavenumber spectra with experimental results. Wavenumber spectra provide information content of a particular quantity like energy residing in various scales of motion of the flow. It can be seen from Figs 9 and 10 that energy and dissipation at large wavenumbers compare very well with experimetal results. The discrepancy at low wavenumbers is due to the differences in large scale structure in the initial conditions of the simulation and experiments. The real potential of direct numerical simulation can be tapped if the simulated fully developed velocity field is analyzed further to understand the dynamics of turbulence. For example, many higher order statistics of the fluid flow can be obtained which reveal much more about the turbulent interactions than wavenumber spectra of energy and dissipation. Moreover, one could also use the simulated velocity fields in understanding the topological properties of turbulent structures perhaps by using advanced computer graphics.

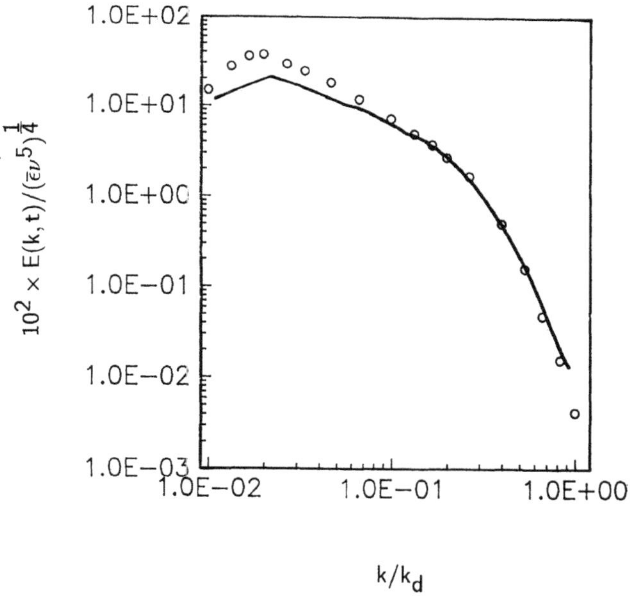

Figure 9. Energy spectrum; solid line represents the turbulent velocity field from 256^3 simulation after 4000 time steps. Open circles are Comte-Bellot and Corrsin's[41] (1971) data.

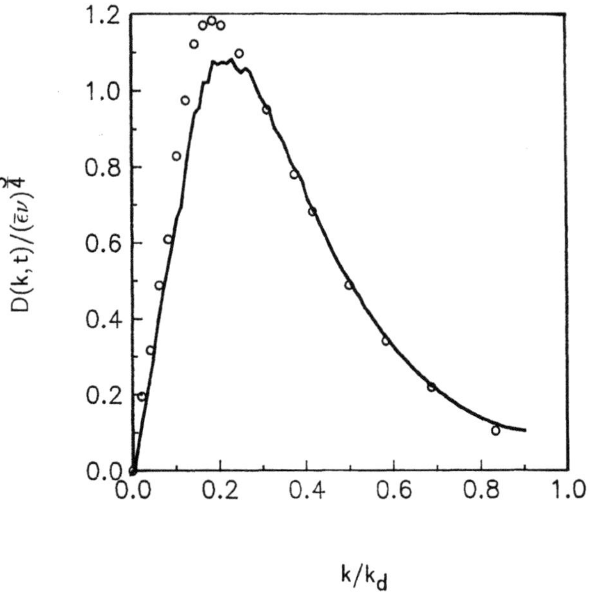

Figure 10. Dissipation spectrum; solid line represents the turbulent velocity field from 256^3 simulation after 4000 time steps. Open circles are the data on dissipation spectrum from Uberoi[42] (1963).

References

1. M. Navier, "Memoire sur les Lois Mouvement des Fluids", *Mem. de l'Acad. des Sciences*, vi, 1882.
2. G.G. Stokes, "On the Theories of Internal Friction of Fluids in Motion", *Camb. Trans.*, viii, 1845.
3. Stewart Harris, *An Introduction to the Theory of the Boltzmann Equation*, Holt, Rhinehart and Wilson, New York, 1971.
4. G. Batchelor, *An Introduction to Fluid Dynamics*, Cambridge University Press, Cambridge, 1979.
5. W. Rohsenow and H. Choi, *Heat, Mass, and Momentum Transfer*, Prentice Hall, New Jersey, 1961.
6. M.C. Potter and J.F. Foss, *Fluid Mechanics*, John Wiley and Sons, Inc., New York, 1975.
7. H. Schlicting, *Boundary Layer Theory*, McGraw Hill, New York, 1979.
8. O.C. Zienkiewicz, *The Finite Element Method*, McGraw-Hill, Inc., London, 1977.
9. S. Crandall, *Engineering Analysis: A Survey of Engineering Procedures*, McGraw Hill, New York, 1956.
10. I. Babuska, B. Szabo and I.N. Katz, "The p-version of the finite element method," *SIAM J. Numer. Anal.*, **18**, pp. 515-545, 1981.

11. K.J. Bathe, *Finite Element Procedures in Engineering Analysis*, Prentice Hall Inc., Englewood Cliffs, New Jersey, 1982.
12. F. Angeleri, V. Sonnad, and K.J. Bathe, "Studies of finite element procedures – an evaluation of preconditioned iterative solvers," *Computers and Structures*, v. 32, No. 3/4, 1989.
13. M.R. Hestenes and E. Stiefel, "Methods of conjugate gradients for solving linear systems", *Journal of Research of the National Bureau of Standards*, v. 49, pp. 409-436, December 1952.
14. R. Chandra, *Conjugate Gradient Methods for Partial Differential Equations*, Ph.D. Thesis, Department of Computer Science, Yale University, New Haven, Connecticut, 1978.
15. C.C. Paige and M.A. Saunders, "Solution of sparse indefinite systems of linear equations," *SIAM J. Numer. Anal.*, v. 12, 617-629, 1975.
16. R. Fletcher, *Conjugate Gradient Methods for Indefinite Systems*, Lectures Notes in Mathematics 506, Springer-Verlag, New York, 1976.
17. S.C. Eisenstat, "A note on the generalized conjugate gradient method," *SIAM J. Numer. Anal.*, v. 20 1983.
18. P.K.W. Vinsome, "ORTHOMIN, an iterative method for solving sparse sets of simultaneous linear equations," Paper SPE 5729, Society of Petroleum Engineers of AIME, 1976.
19. M.A. Saunders, H.D. Simon and E.L. Yip, "Two conjugate-gradient type methods for sparse unsymmetric linear equations", *Mathematics and Modeling Technical Report*, ETA-TR-18, Boeing Computer Services, Washington, 1984.
20. Y. Saad and M.H. Schultz, "GMRES: a generalized minimal residual algorithm for solving nonsymmetric linear systems," *Research Report 254*, Department of Computer Science, Yale University, New Haven, Connecticut, August 1983.
21. Y. Saad and M.H. Schultz, "Conjugate gradient like algorithms for solving nonsymmetric linear systems of equations", *Mathematics of Computation*, v 44, pp. 417-424, April 1985.
22. H.C. Elman, *Iterative Methods for Large, Sparse, Nonsymmetric Systems of Linear Equations*, Ph.D. Thesis, Department of Computer Science, Yale University, New Haven, Connecticut, April 1982.
23. J.A. Meijrink and H.A. van der Vorst, "An iterative solution method for linear systems of which the coefficient matrix is a symmetric M matrix," *Math. Comp.*, v. 31, pp. 148-162, (1977).
24. W. Hackbusch, *Multi-grid Methods and Applications*, Springer Verlag, Berlin, 1980.
25. A.W. Craig and O.C. Zienkiewicz, "A multigrid algorithm using a hierarchical finite element basis," in *Multigrid Methods for Integral and Differential Equations*, edited by D. J. Paddon and Holstein, Clarendon Press, Oxford, pp. 310-312, 1985.
26. P.J. Roache, *Computational Fluid Dynamics*, Hermosa Publishers, Albuquerque, New Mexico, p. 446, 1976.

27. P.M. Gresho and R.L. Sani, "On pressure boundary conditions for the incompressible Navier Stokes equations," *Int. J. Numer. Meth. Fluids*, **7**, pp. 1111-1145, 1987.
28. S. Hassanzadeh, V. Sonnad and S. Foresti, "Finite element implementation of boundary conditions for the pressure Poisson equation of incompressible flow," *IBM Technical Report*, KGN-184, Kingston, 1989.
29. N.N. Yanenko, *The Method of Fractional Steps*, Springer, New York, 1971.
30. S.Y. Yuan and M.D. Olson, "Review of computing methods for recirculating flow", *Journal of Computational Physics*, Vol 29, No 1, 1-19, 1978.
31. A. S. Monin and A. M. Yaglom, *Statistical Fluid Mechanics*, Vols.1 and 2, MIT Press, 1973.
32. H. Tennekes and J. L. Lumley, *A First Course in Turbulence*, MIT Press, 1972.
33. W. P. Jones and B. E. Launder, "The calculation of low Reynolds number phenomena with a two-equation model of turbulence", *Int. J. Heat Mass Transfer*, v. 16, 1119, 1973.
34. V. Yakhot and S. A. Orszag, "Renormalization group analysis of turbulence, I. Basic Theory ", *J. Sci. Comp.*, v. 1, 1986.
35. L. D. Landau and E. M. Lifshitz, *Fluid Mechanics*, Pergamon Press, Oxford 1982.
36. S. A. Orszag and G. S. Patterson, "Numerical simulation of turbulence," *Statistical Models and Turbulence* (ed. M. Rosenblatt and C. Van Atta), Lecture Notes in Physics, v. 12, Springer, 127 1972.
37. J. W. Cooley and J. W. Tukey, "An algorithm for the machine calculation of complex Fourier series", *Math. Comp.*, v. 19, 297, 1965.
38. S. A. Orszag, "Numerical simulation of incompressible flows within simple boundaries. I. Galerkin (spectral) representations," *Studies in Appl. Math.*, v. 2, No. 4, 293, 1971.
39. G. S. Patterson and S. A. Orszag, "Spectral calculations of isotropic turbulence: efficient removal of aliasing interactions," *Phys. Fluids*, v. 14, 2538 (1971).
40. R. Panda, V. Sonnad and E. Clementi, *IBM Technical Report*, KGN-205, Kingston, 1989.
41. G. Comte-Bellot and S. Corrsin, *J. Fluid Mech.*, v. 48, 273 (1971).
42. M. S. Uberoi, *Phys. Fluids*, v. 6, 1048 (1963).

Chapter 22.

The Equations of Elasticity and Their Solution by Finite Element Methods

V. Sonnad, H. Murakami, S. Foresti, S. Hassanzadeh, B. Jiang

IBM Corporation
Scientific and Engineering Computations
Dept. 48B/428
Kingston, New York 12401, U.S.A.

1. The Equations of Elasticity

The Navier equations[1] are a set of coupled partial differential equations which describe the behaviour of a loaded solid body under the assumption of small displacements, and a linear stress-strain relation. The solution of these linear equations is of great practical importance and constitutes a significant part of solid mechanics. In contrast to the equations of fluid flow which are not easily derived from a variational approach, the equations of elasticity can be readily derived from a Hamiltonian.[2] However we will not use the variational approach but will continue to use the control volume approach that was described in the previous chapter on fluid flow. This allows us to make repeated use of a single conceptual tool in deriving the equations even though the physical behaviour of a fluid in motion is very different from the motion of a solid body undergoing small displacements.

In deriving the equations of fluid flow we assumed that the control volume was fixed and the fluid flowed through the control volume, (the Eulerian approach); however in deriving the equations of elasticity, the control volume is displaced in response to applied forces (the Lagrangian approach). Because of the assumption of small displacements, the continuity equation is decoupled from the momentum equation, and can be used after the momentum equation is solved to compute changes in density if required.[3]

We can directly proceed to the derivation of the equation for the conservation of momentum. It will be seen that the procedure we follow is very similar to the procedure we use for the derivation of the equations of conservation of momentum for fluid flow. Consider first the equation for conservation of x-direction momentum; the statement for the conservation of x-momentum is

Rate of change of momentum of the control volume
= Sum of applied surface stresses.

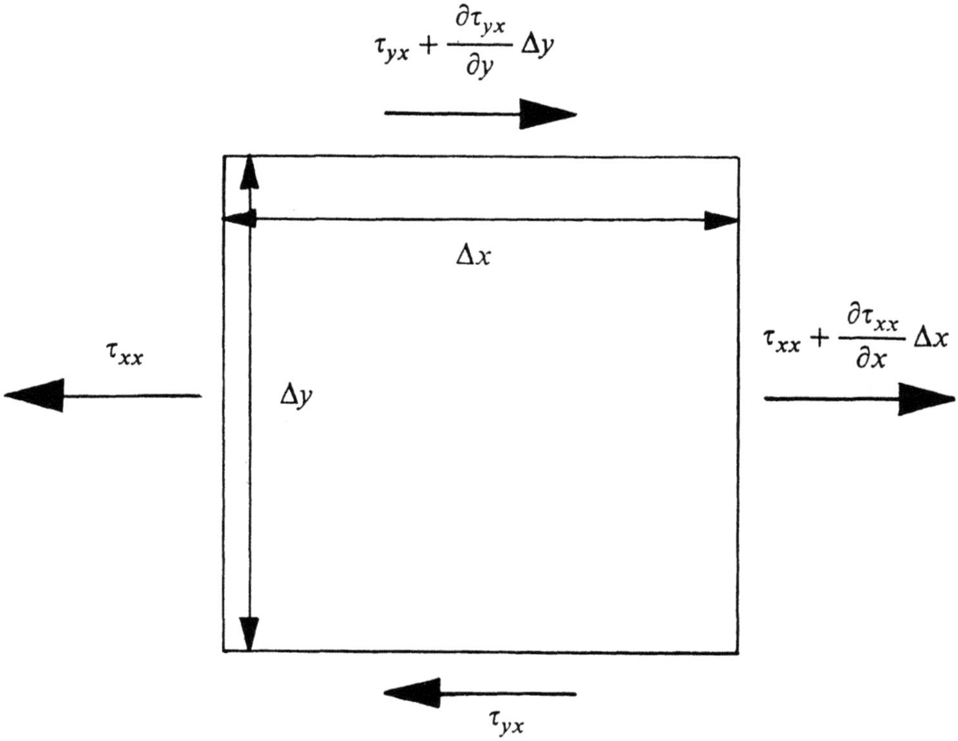

Figure 1. Stresses at the surfaces of the control volume in the x-direction.

As before, consider an infinitesimal surface element of size $\Delta x \Delta y$ with imposed surface stresses, (Figure 1). We see that the statement of conservation of momentum can be expressed in differential form as

$$\Delta x \Delta y (\rho \frac{\partial^2 u}{\partial t^2}) = \frac{\partial \tau_{xx}}{\partial x} \Delta x \Delta y + \frac{\partial \tau_{yx}}{\partial y} \Delta x \Delta y.$$

In the above equation, the quantities τ_{xx} and τ_{yx} are the stresses applied at the surfaces of the control volume; ρ is the known density; u, v, w are displacements and not velocities. At this stage, the major difference between the equation for the conservation of momentum for a solid and that of a fluid is the acceleration term. We do not at this stage make any statements regarding the relation between the applied surface stresses and the displacements. This is supplied by the constitutive equations for the solid. Again, dividing throughout by $\Delta x \Delta y$, the equation for the conservation of x-momentum is given as:

$$\rho\frac{\partial^2 u}{\partial t^2} = \frac{\partial \tau_{xx}}{\partial x} + \frac{\partial \tau_{yx}}{\partial y}.$$

By using identical reasoning, we can derive the equation for the conservation of momentum in the y-direction:

$$\rho\frac{\partial^2 v}{\partial t^2} = \frac{\partial \tau_{yy}}{\partial y} + \frac{\partial \tau_{xy}}{\partial x}.$$

The three-dimensional versions of these equations can be derived by considering a three-dimensional control volume; the resulting expressions are given below:

$$\rho\frac{\partial^2 u}{\partial t^2} = \frac{\partial \tau_{xx}}{\partial x} + \frac{\partial \tau_{yx}}{\partial y} + \frac{\partial \tau_{zx}}{\partial z}$$

$$\rho\frac{\partial^2 v}{\partial t^2} = \frac{\partial \tau_{xy}}{\partial x} + \frac{\partial \tau_{yy}}{\partial y} + \frac{\partial \tau_{zy}}{\partial z}$$

$$\rho\frac{\partial^2 w}{\partial t^2} = \frac{\partial \tau_{xz}}{\partial x} + \frac{\partial \tau_{yz}}{\partial y} + \frac{\partial \tau_{zz}}{\partial z}.$$

We have at this point, derived partial differential equations for the conservation of momentum in terms of displacements and stresses. We have purposely chosen to use a control volume approach rather than a more sophisticated variational approach so as to use the concepts explained in deriving the equations of fluid flow.

The above system of equations is a set of three coupled partial differential equations; there are a total of twelve unknowns, viz., the three components of the displacements, and nine components of the stress tensor. The number of unknowns appearing in the equations are reduced through the use of the constitutive equations for the solid which express the stress tensor in terms of the derivatives of the displacement (the strain tensor). We indicate here the major steps in the derivation of the constitutive equations for a solid where we again use a number of ideas from the derivation of the constitutive equations for fluids.

The first step is to show that the stress tensor is symmetric. This is easily done by using the control volume approach, where we write the expression for the conservation of angular momentum. Depending upon the axes chosen for the calculation of angular momentum, it can be shown that[4]

$$\tau_{xy} = \tau_{yx}, \quad \tau_{xz} = \tau_{zx} \text{ and } \tau_{zy} = \tau_{yz}.$$

The stress tensor can therefore be written as:

$$\tau_{ij} = \begin{bmatrix} \tau_{xx} & \tau_{xy} & \tau_{xz} \\ \tau_{xy} & \tau_{yy} & \tau_{yz} \\ \tau_{xz} & \tau_{yz} & \tau_{zz} \end{bmatrix}.$$

There are six independent quantities instead of nine. The concept of pressure which is so valuable in fluid mechanics does not appear directly in the equations of solid mechanics; we define a strain tensor:

$$e_{ij} = \frac{1}{2}\left(\frac{\partial V_i}{\partial x_j} + \frac{\partial V_j}{\partial x_i}\right)$$

where V_i is the component of displacement in the direction x_i. Because of the assumption of small displacements, the stress is linearly proportional to the strain, (generalized Hooke's law), and we can write a general relation between stress and strain as

$$\tau_{ij} = A_{lmij}\, e_{lm}$$

where A_{lmij} is a fourth rank tensor, and would have 81 independent entries for a three-dimensional problem. However, the symmetry of the stress and strain tensor reduces the number of independent entries to 36 (six for each of the six independent quantities in the stress tensor). By imposing the condition of isotropy, the number of independent constants is reduced to two, and we can write

$$\tau_{ij} = 2\mu e_{ij} + \lambda e_{kk}\delta_{ij}$$

where μ and λ are known as *Lame'* constants.[2] Note here the similarity of the relations between stress and strain for a solid, and between stress and strain rate for a fluid.

The expression for the stress tensor, incorporated into the equations of conservation of momentum leads to the following partial differential equations which are expressed in tensor notation.

$$\rho\frac{\partial^2 V_i}{\partial t^2} = \mu\frac{\partial^2 V_i}{\partial x_j \partial x_j} + (\lambda + \mu)\frac{\partial^2 V_j}{\partial x_j \partial x_i}$$

where V_1, V_2, V_3 are the components of the displacements, and we have implicitly assumed that λ and μ are constants. The resulting equations are the Navier equations of elasticity which express the conservation of momentum in terms of derivatives of displacements. They are a system of three partial differential equations in three unknowns; together with the initial and boundary conditions they constitute a well posed mathematical problem. As mentioned earlier, the equation of conservation of mass is decoupled from these equations, and is rarely required for most practical situations.

It is common practice in engineering to use the Young's modulus E , (ratio of stress to strain under uniaxial loading), and the Poisson's ratio v, (ratio of lateral contraction to longitudinal extension), which are related to the *Lame'* constants by the following relations:

$$E = \frac{\mu(2\mu + 3\lambda)}{\lambda + \mu}$$

$$v = \frac{\lambda}{2(\lambda + \mu)}$$

In terms of the Young's modulus and Poisson's ratio, the Navier equation takes the form:

$$\rho \frac{\partial^2 V_i}{\partial t^2} = \frac{E}{2(1+v)} \left(\frac{\partial^2 V_i}{\partial x_j \partial x_j} + \frac{1}{(1-2v)} \frac{\partial^2 V_j}{\partial x_j \partial x_i} \right)$$

When the boundary conditions are independent of time, the derivatives of the displacements with respect to time drop out and the steady state version of the Navier equation is then given by:

$$\frac{\partial^2 V_i}{\partial x_j \partial x_j} + \frac{1}{(1-2v)} \frac{\partial^2 V_j}{\partial x_j \partial x_i} = 0$$

Note here that the Young's modulus is not contained in these equations but the Poisson's ratio is part of the equations.

It is useful to introduce here some cautionary remarks regarding the applicability of the above equations. The key assumption is that of small displacements which results in relatively simple forms for the constitutive equation. A number of effects are ignored in deriving the above equations such as nonlinearity (both material and geometric), history effects (such as viscoelasticity and creep) and all manner of coupling phenomena such as thermal, magnetic, and electric effects. It is necessary to ensure that the assumptions used in the derivation of the Navier equations are valid before using them to predict the behaviour of a deforming solid.

2. Computational Techniques for Linear Static Analysis

Solution of Large 3-D Problems with Finite Elements

The equations of elasticity are in many respects simpler than the Navier Stokes' equations which are inherently nonlinear and can give rise to highly complex phenomena such as turbulence and shocks. However it is still necessary to resort to numerical methods to obtain solutions to many problems of practical interest.

For over two decades the finite element method has been the method of choice for the solution of the problems in solid mechanics (with the boundary element method now gaining popularity for the solution of certain classes of problems). As mentioned in the previous chapter, the finite element method originated in attempts to solve problems in structural analysis by discretizing a complex region into simpler regions that could be pieced together to obtain a solution over the entire region. The presence of a variational indicator allowed the finite element method to be derived by a minimization procedure; this gave the method a solid mathematical foundation and has helped establish it as the most widely used computational technique for structural analysis.

The advantages of the computational framework of the finite element method are well recognized; however there have been continuing attempts at making the finite element method easier to use and also to reduce the computing time and storage requirements in the solution of large problems. The p-version of finite elements (described in the previous chapter) was designed in order to make adaptive computations and error estimates easier than with the h-version.[10] The basic idea of the p-version approach is to increase the order of the polynomials within each element that belongs to regions where the solution needs to be refined. While it is possible to achieve excellent results with higher order Lagrange type basis functions,[11] there are significant simplifications in programming effort when Legendre type basis functions are used. A particular choice of functions uses linear combinations of Legendre polynomials leading to hierarchic basis functions. This leads to relatively well conditioned problems, and nested matrices.[12] If the underlying solution is smooth, the convergence of the p-version to the solution is rapid so that it is possible to reach a given level of accuracy with lesser degrees of freedom than with the h-version. Because of these advantages, the p-version is becoming increasingly popular in a variety of applications.

However if a direct solution method is used to solve the system of equations resulting from the use of the p-version, the memory and computing requirements can be very high for large, 3-D problems. Furthermore, the formation of the stiffness matrices can be time consuming because numerical integration of high order polynomials requires a large number of integration points and the resulting element matrices are often dense. It appears therefore that the p-version offers major advantages over the h-version from the viewpoint of simplifying the design of an adaptive code, but retains high requirements in storage and computation time for large 3-D problems when direct methods are used to factor global matrices.

In the subsequent section we describe an approach that results in significant reductions in the computation time and storage requirements for the solution of large problems with the p-version of the finite element method. It is interesting to note that all the individual components of the approach described here are well known and understood, but the combination of these ingredients has not yet been widely used for the solution of large 3-D problems in structural analysis with the p-version of finite elements.

Iterative Solution with the Conjugate Gradient Method

The first component in an approach to minimize storage is the use of an iterative technique to solve the equations obtained by the finite element discretization of the Navier equations. Since the factors of the matrix do not need to be stored, but only the matrix, this can greatly reduce the storage requirements over direct methods.

We consider here the conjugate gradient approach which can be applied to problems in static structural analysis because the matrices are symmetric and positive definite. An algorithmic description of one version of the conjugate gradient method for the solution of the system $Ax = b$ is as follows:[6] with initial vectors $x_0 = 0$ and $r_0 = b$, for i = 1,2,... until convergence, do:

$$\beta_i = \frac{(r_{i-1}, r_{i-1})}{(r_{i-2}, r_{i-2})} \qquad (\beta_1 = 0)$$

$$p_i = r_{i-1} + \beta_i p_{i-1}$$

$$\alpha_i = \frac{(r_{i-1}, r_{i-1})}{(Ap_i, p_i)}$$

$$x_i = x_{i-1} + \alpha_i p_i$$

$$r_i = r_{i-1} - \alpha_i Ap_i$$

Convergence is attained when the norm of the residual vector r_i falls below a selected value. The conjugate gradient method works by systematically eliminating components of the residual vector in an n-dimensional orthogonal space generated by successive iterations; with exact arithmetic it is guaranteed to converge in n iterations where n is the order of the matrix. Because of this property it is classified as a direct method rather than as an iterative one.[5] In practice however, it is desirable to terminate the computations when the error falls below a selected value.

The major computational component of each iteration of the conjugate gradient method is the multiplication of the matrix A by a vector p_i which is performed once during an iteration; the remaining operations involve only vectors, and have much lower operation counts than matrix vector multiplication. Since the global matrix is formed by summing up individual element matrices, the multiplication can be done without forming the global matrix. This is one of the components of the element-by-element approach popularized by Hughes.[13] A recent report gives an excellent comparison between two variations of the element-by-element approach.[14]

If we wish to reduce the storage further, the element matrices should not be stored but should be recomputed at every iteration. With the h-version this is not very expensive, and it is possible to reform the element matrices at every iteration without paying a high penalty. However, with the p-version, the reformation of the element matrix at every iteration, (particularly for 3-D problems with high levels of polynomials), can be exorbitantly expensive in solution time. Furthermore, even a single element matrix can take up a large amount of storage; e.g., for the 3-D Navier equations, a single element matrix using polynomials of order 9, requires about 4.5 million double precisison words when the symmetric part of the matrix is stored. In order to overcome this difficulty, we show in the following section an approach that can be used to minimize storage without sacrificing computational speed.

Rapid Operator Application

We show that with the use of simple and well known techniques, it is possible to get the matrix vector product without storing the element matrices, thus making the storage requirements close to minimal. The direct application of the operator (in effect, the matrix vector product) is performed with *less operations than when element matrices are formed, i.e., there is no penalty in computing time.*

The key ingredient is the use of elements that employ tensor product basis functions. Since such basis functions implicitly perform separation of variables, it is possible to group terms so that several terms are reused in the application of the operator. The savings associated with the use of tensor product basis functions are best illustrated with a simple example. Consider a cube modeled by a single element, within which we define b^3 basis functions. The variation of a quantity within this element can be expressed in the form

$$u = U_i \Psi_i(x,y,z) \qquad (i = 1 \ldots b^3)$$

where U_i are the coefficients and $\Psi_i(x,y,z)$ are the basis functions. (We use standard notation: repeated indices imply summation, and the range of summation is indicated). When the basis functions are in tensor product form, each three-dimensional basis function is expressed as a product of one-dimensional basis functions, and the above expression can be rewritten as

$$u = \phi_l(x)\phi_m(y)\phi_n(z)U_{lmn} \qquad (l,m,n = 1 \ldots b)$$

To illustrate the application of an operator employing these basis functions, assume that we wish to multiply the mass matrix by a vector, i.e, we wish to evaluate for values of $i = 1 \ldots b^3$, the expression,

$$C_i = M_{ij}U_j \qquad (j = 1 \ldots b^3)$$

where the mass matrix is given by

$$M_{ij} = \int_{-1}^{+1}\int_{-1}^{+1}\int_{-1}^{+1} \Psi_i(x,y,z)\Psi_j(x,y,z)dxdydz$$

We need for $i = 1 \ldots b^3$

$$C_i = \int_{-1}^{+1}\int_{-1}^{+1}\int_{-1}^{+1} \Psi_i(x,y,z)\Psi_j(x,y,z)\, U_j dxdydz \qquad (j = 1 \ldots b^3)$$

Using the tensor product form, we expand the above expression to get

$$C_{pqr} = \int_{-1}^{+1}\int_{-1}^{+1}\int_{-1}^{+1} \phi_p(x)\phi_q(y)\phi_r(z)\phi_l(x)\phi_m(y)\phi_n(z)\, U_{lmn} dxdydz \qquad (l,m,n = 1 \ldots b)$$

where p,q,r each assume all values between 1 and b. When we perform product type numerical integration, we replace integrals by weighted sums of the functions evaluated at the integration points. For simplicity, the number of integration points is taken equal to the number of basis functions in each of the dimensions, with an equal number of basis functions and integration points in each dimension; (this is not necessary, but simplifies the explanation).

We have thus

$$C_{pqr} = \phi_p(x^i)\phi_q(y^j)\phi_r(z^k)w(i)w(j)w(k)\phi_l(x^i)\phi_m(y^j)\phi_n(z^k)\, U_{lmn}$$

$$(i,j,k = 1 \ldots b;\ l,m,n = 1 \ldots b)$$

where the x^i, y^j, z^k are the integration points and the $w(i), w(j), w(k)$ are the weights. We note here that quantities such as $\phi_p(x^i)$ are merely the one dimensional basis functions evaluated at the integration points and are matrices which we denote by ϕ_{pi}. We therefore need to evaluate for all values of p,q,r from 1 to b, the expression,

$$C_{pqr} = \phi_{pi}\phi_{qj}\phi_{rk}w(i)w(j)w(k)\phi_{li}\phi_{mj}\phi_{nk}U_{lmn} \qquad (i,j,k = 1 \ldots b;\ l,m,n = 1 \ldots b)$$

This is done by contracting the indices from the right, one step at a time; under special circumstances it is possible to contract indices over other pairs of terms before contracting indices against U_{lmn}. Each step takes b^4 operations, and hence the total number of operations required to evaluate all C_{pqr} is $6b^4$. (An operation is defined as a a floating point multiplication and addition). By comparison, if the mass matrix were available, it would be of size $b^3 \times b^3$ and multiplication by a vector of length b^3 would take b^6 operations. For values of $b \geq 3$ we would be saving operations even though the matrix is not available.

The above derivation does not take into account the specific nature of the basis functions other than that they are in tensor product form. The basis functions for the p-version are chosen so that they are nearly orthogonal over a cube, and the evaluation of all C_{pqr} could be performed without numerical integration in order b^3 operations. However, the intention above is to illustrate the approach that is taken with more general curved elements, where no advantage can be taken of orthogonality conditions.

The procedure described above, with numerical integration of basis functions is equivalent to the following two steps:

a) Evaluate the variable u at b^3 integration points; because u is expanded in tensor product basis functions, this can be done in $3 \times b^4$ operations.

b) Sum the values of u along each of the directions of integration while weighting them with the values of the basis functions at the integration points; this is again done in $3 \times b^4$ operations.

We next show how this approach can be used for rapid application of the operator corresponding to the steady state equations of linear elasticity.

$$\frac{\partial^2 u}{\partial x^2} + \frac{\partial^2 u}{\partial y^2} + \frac{\partial^2 u}{\partial z^2} + \varepsilon\frac{\partial}{\partial x}\left(\frac{\partial u}{\partial x} + \frac{\partial v}{\partial y} + \frac{\partial w}{\partial z}\right) = 0$$

$$\frac{\partial^2 v}{\partial x^2} + \frac{\partial^2 v}{\partial y^2} + \frac{\partial^2 v}{\partial z^2} + \varepsilon\frac{\partial}{\partial y}\left(\frac{\partial u}{\partial x} + \frac{\partial v}{\partial y} + \frac{\partial w}{\partial z}\right) = 0$$

$$\frac{\partial^2 w}{\partial x^2} + \frac{\partial^2 w}{\partial y^2} + \frac{\partial^2 w}{\partial z^2} + \varepsilon\frac{\partial}{\partial z}\left(\frac{\partial u}{\partial x} + \frac{\partial v}{\partial y} + \frac{\partial w}{\partial z}\right) = 0$$

In the above equations, u, v, w are the components of the displacement along the x ,y ,z axes respectively, and ε is a constant that is related to the Poisson's ratio of the solid. In the finite element solution of these equations, we first use a standard Galerkin approach, where we require that the integral of the equations weighted by each of the basis functions is set to zero. Following standard procedure, the equations are integrated by parts resulting in a domain integral and boundary terms. We will assume that we use b^3 basis functions in tensor product form; for a single element, the resulting expressions are of the form:

$$\iiint_{Vol} \frac{\partial \phi_i}{\partial x}\frac{\partial u}{\partial x} + \frac{\partial \phi_i}{\partial y}\frac{\partial u}{\partial y} + \frac{\partial \phi_i}{\partial z}\frac{\partial u}{\partial z} + \varepsilon\frac{\partial \phi_i}{\partial x}\left(\frac{\partial u}{\partial x} + \frac{\partial v}{\partial y} + \frac{\partial w}{\partial z}\right) dxdydz$$

$$= (Boundary Terms) \qquad [1]$$

$$\iiint_{Vol} \frac{\partial \phi_i}{\partial x}\frac{\partial v}{\partial x} + \frac{\partial \phi_i}{\partial y}\frac{\partial v}{\partial y} + \frac{\partial \phi_i}{\partial z}\frac{\partial v}{\partial z} + \varepsilon\frac{\partial \phi_i}{\partial y}\left(\frac{\partial u}{\partial x} + \frac{\partial v}{\partial y} + \frac{\partial w}{\partial z}\right) dxdydz$$

$$= (Boundary Terms) \qquad [2]$$

$$\iiint_{Vol} \frac{\partial \phi_i}{\partial x}\frac{\partial w}{\partial x} + \frac{\partial \phi_i}{\partial y}\frac{\partial w}{\partial y} + \frac{\partial \phi_i}{\partial z}\frac{\partial w}{\partial z} + \varepsilon\frac{\partial \phi_i}{\partial z}\left(\frac{\partial u}{\partial x} + \frac{\partial v}{\partial y} + \frac{\partial w}{\partial z}\right) dxdydz$$

$$= (Boundary Terms) \qquad [3]$$

The above expressions hold in the Cartesian global coordinate system (x,y,z) . It is customary to map the element into a local curved coordinate system r,s,t, where the element is a cube whose sides are of length 2. Each of the above expressions must then be rewritten in the local co-ordinate system. We make use of a determinant to transform the integral from the global system to a local system; given an arbitrary function $F(x,y,z)$, the integral in the (x,y,z) coordinate system

$$\iiint_{Vol} F(x,y,z)dxdydz$$

is given in the (r,s,t) coordinate system by :

$$\int_{-1}^{+1}\int_{-1}^{+1}\int_{-1}^{+1} F(x(r,s,t),y(r,s,t),z(r,s,t)) \det |J| drdsdt$$

where J is the Jacobian matrix of the mapping from the global coordinate system into the local coordinate system. When there are derivatives present, we make use of the chain rule:

$$\frac{\partial \phi_i}{\partial x} = \frac{\partial \phi_i}{\partial r}\frac{\partial r}{\partial x} + \frac{\partial \phi_i}{\partial s}\frac{\partial s}{\partial x} + \frac{\partial \phi_i}{\partial t}\frac{\partial t}{\partial x}$$

$$\frac{\partial \phi_i}{\partial y} = \frac{\partial \phi_i}{\partial r}\frac{\partial r}{\partial y} + \frac{\partial \phi_i}{\partial s}\frac{\partial s}{\partial y} + \frac{\partial \phi_i}{\partial t}\frac{\partial t}{\partial y}$$

$$\frac{\partial \phi_i}{\partial z} = \frac{\partial \phi_i}{\partial r}\frac{\partial r}{\partial z} + \frac{\partial \phi_i}{\partial s}\frac{\partial s}{\partial z} + \frac{\partial \phi_i}{\partial t}\frac{\partial t}{\partial z}$$

Similar expressions hold for $\frac{\partial u}{\partial x}, \frac{\partial u}{\partial y}, \frac{\partial u}{\partial z}$, where it is implicitly understood that

$u = \sum_i^{b^3} \phi_i U_i$. We substitute for the chain rule expansions of $\frac{\partial \phi_i}{\partial x}, \frac{\partial \phi_i}{\partial y}, \frac{\partial \phi_i}{\partial z}$, into

Equation [1] above and then regroup as shown below:

$$\int_{-1}^{+1}\int_{-1}^{+1}\int_{-1}^{+1}\frac{\partial\phi_i}{\partial r}\left\{\frac{\partial r}{\partial x}\left[\frac{\partial u}{\partial x}+\varepsilon\left(\frac{\partial u}{\partial x}+\frac{\partial v}{\partial y}+\frac{\partial w}{\partial z}\right)\right]+\frac{\partial r}{\partial y}\frac{\partial u}{\partial y}+\frac{\partial r}{\partial z}\frac{\partial u}{\partial z}\right\}$$

$$\det|J|\,drdsdt\;+$$

$$\int_{-1}^{+1}\int_{-1}^{+1}\int_{-1}^{+1}\frac{\partial\phi_i}{\partial s}\left\{\frac{\partial s}{\partial x}\left[\frac{\partial u}{\partial x}+\varepsilon\left(\frac{\partial u}{\partial x}+\frac{\partial v}{\partial y}+\frac{\partial w}{\partial z}\right)\right]+\frac{\partial s}{\partial y}\frac{\partial u}{\partial y}+\frac{\partial s}{\partial z}\frac{\partial u}{\partial z}\right\}$$

$$\det|J|\,drdsdt\;+$$

$$\int_{-1}^{+1}\int_{-1}^{+1}\int_{-1}^{+1}\frac{\partial\phi_i}{\partial t}\left\{\frac{\partial t}{\partial x}\left[\frac{\partial u}{\partial x}+\varepsilon\left(\frac{\partial u}{\partial x}+\frac{\partial v}{\partial y}+\frac{\partial w}{\partial z}\right)\right]+\frac{\partial t}{\partial y}\frac{\partial u}{\partial y}+\frac{\partial t}{\partial z}\frac{\partial u}{\partial z}\right\}$$

$$\det|J|\,drdsdt$$

In the above expression we have deliberately not substituted for the expansions of the derivatives of any of the variables; this particular grouping is most advantageous. The application of the operator is now performed in the following steps. We give operation counts assuming that there are as many integration points as there are basis functions. We emphasize again that it is not necessary to have these equal, but it is easier to evaluate the operation counts.

1) Evaluate the Jacobian matrices, the determinants of the Jacobian matrices, and the elements of the inverse of the Jacobian matrices at each of the b^3 integration points. The operation count is heavily dependent upon the functions used to map the element from global coordinates to local coordinates, but if we assume quadratic basis functions for the element, then this is an order b^3 operation.

2) Evaluate $\frac{\partial u}{\partial r}, \frac{\partial u}{\partial s}, \frac{\partial u}{\partial t}, \frac{\partial v}{\partial r}, \frac{\partial v}{\partial s}, \frac{\partial v}{\partial t}, \frac{\partial w}{\partial r}, \frac{\partial w}{\partial s}, \frac{\partial w}{\partial t}$, at b^3 integration points. These add up to a total of $27 \times b^4$ operations.

3) Evaluate $\frac{\partial u}{\partial x}, \frac{\partial u}{\partial y}, \frac{\partial u}{\partial z}, \frac{\partial v}{\partial x}, \frac{\partial v}{\partial y}, \frac{\partial v}{\partial z}, \frac{\partial w}{\partial x}, \frac{\partial w}{\partial y}, \frac{\partial w}{\partial z}$ at b^3 integration points, using the chain rule. These are all order b^3 operations.

4)Evaluate the various quantities within the braces given in Equation [1] above, at each of the b^3 integration points; these are all order b^3 operations.

5)Weight the values so obtained by each of the b^3 basis functions and sum them up to obtain the values corresponding to the first equation. This requires $9 \times b^4$ operations.

6) Only steps 4 and 5 need to be repeated for the remaining equations 2 and 3 for an additional $18 \times b^4$ + order (b^3) operations

Thus the total number of operations required to apply the Navier operator to a vector is approximately :

$$54 \times b^4 + Order\ (b^3)$$

This estimate holds as long as the number of integration points are of order b per dimension, although the constant does not remain at 54. The value 54 is also related to the number of terms in the partial differential equation, and the ways in which they can be grouped together; a simpler equation with a lesser number of terms would have a lower constant. This approach is valid for elements that are distorted and also for nonlinear problems. We note here that when required, the element matrix can be formed in order b^7 operations, as compared to b^9 operations with the standard approach.

Because there are three unknowns u,v,w per "node", the stiffness matrix for a single element with b^3 basis functions, is of size $3b^3 \times 3b^3$, and multiplication by a vector of length $3b^3$ would take $9b^6$ operations. This compares with the value of $54 \times b^4 + Order\ (b^3)$ with rapid operator application.

Tensor product basis functions have been long utilized in mathematics (multidimensional Fourier expansions are a good example), and are well known in computer aided design.[15,16] Such basis functions have long been used in fluid mechanics, and in recent years have been elegantly exploited by Patera and his associates in the spectral element method used for fluid flow computations.[17] A few words of explanation about the differences between the spectral element method and the present approach are in order: it has become conventional to refer to a spectral element method as one that uses integration points equal to the number of basis functions; thus there is a close relation between basis functions and integration points. We follow here the traditional finite element approach where we use as many integration points as are required to evaluate the function accurately. The implications of these differences are knowledgeably addressed in a recent paper.[18] We merely note that there are advantages and disadvantages in each approach, and the choice of a method is determined by the application.

Utility of Rapid Operator Application

The discussion above shows that the use of rapid operator application in conjuction with iterative solution techniques results in significant savings in storage over traditional techniques where a global matrix is formed and factored. The storage needed to solve a problem with N unknowns is of order N, and is independent of the dimension of the physical region as well as the ordering of the unknowns. This is in contrast to direct solution techniques, based on Gaussian elimination, where the storage and solution time is very strongly affected by the bandwidth of the matrix. In reference (19) we present some comparisons between storage and solution times using factorization and iterative solution with rapid operator evaluation for large 3-D problems in structural analysis with the p-version of finite elements. For the largest problem solved, (with just under a million unknowns), the storage required with the iterative solution technique was three orders of magnitude less than that required with factorization techniques. This clearly opens up the possibility of using workstations in the solution of very large scale problems.

The solution time for the iterative approach is directly dependent upon the condition number of the matrix, and can be expected to vary widely depending upon the problem being solved, and upon the loading conditions. It can also be expected that there would be problems where the iterative approach would not converge at all. In many instances, preconditioning helps the rate of convergence. A natural approach to speed up convergence is to take advantage of the hierarchic nature of the basis functions in the p-version, and use a direct solver at a lower level as a preconditioner for an iterative solver at a higher level. This has been suggested in previous work[7, 22] and has been investigated (without rapid operator evaluation) by Morris et al;[8] preliminary theoretical results regarding the reduction of condition number of the matrix with this method of preconditioning have been obtained by Wathen.[9] The performance of such preconditioning methods is discussed in Reference (20).

The solution of large scale generalized eigenproblems in structural analysis can be achieved with minimal storage by using methods such as inverse iteration or Rayleigh quotient iteration. Results for the solution of large eigenproblems are presented in a publication by Hassanzadeh et al.[23]

Parallel Implementation of Iterative Schemes

In iterative schemes for the solution of systems of linear equations and eigenvalue problems, the major computational component is the multiplication of a matrix by a vector. We illustrate the parallel implementation of this step with reference to the conjugate gradient method described earlier, (an identical scheme could be used with eigenvalue problems).

Assume that at a given step, we have a global vector p_i which is to be multiplied by the matrix A. Let us consider the parallel implementation of this step on a shared memory machine. Since the global matrices are sums of element matrices, matrix vector multiplications can be performed element-by-element.[13] The first step is to distribute the global vector to each of the elements. This is done with a pointer vector which gives the location in the element vector for every term in the global vector. We then use the procedure of rapid operator application to each element in turn. This is a completely independent operation with no need for data communication. At the end of this operation, we have vectors corresponding to each of the elements; these are then reassembled into a global vector using a reverse pointer that gives the location in the global vector for every value in the element vector.

Because the operations on each element are completely independent, it is natural to implement this step in parallel. Once the distribution of the global vectors into the element vectors has taken place, we form a loop over the elements by assigning an element to a processor. As soon as a processor has completed the operations corresponding to its element, it indicates that it has completed its task, and is assigned a new element from those that are still unoperated on. It picks up this new task and carries it to completion, and this scheme is carried on until all the elements are operated on.

We have described above the steps corresponding to breaking up the global vector into local element vectors, and the formation of new element vectors. None of these steps require any form of synchronization between processors. In the formation of a global vector from the element vectors, it may happen that a single location in the global vector is accessed by two or more processors at once; (this corresponds in the physical discretization to several elements having a common point). These conflicts are handled by a locking mechanism which is activated when the same location in the global vector is accessed by more than one processor; the lock serializes the writing into the location. Note that the order in which the global vector is updated is not important. Once the elements are all processed, there is a synchronization step between all processors which ensure that the following step is executed with a global vector which is completely updated.

A key advantage of this approach which parcels out "processes" to processors is that there is no need to divide the physical domain into subregions which are then assigned to processors. This takes full advantage of shared memory architecture, where all data is available to all processors at all times. There are no concerns about load balancing because it is automatically adjusted at computation time; the maximum idle time would be less than the time for a single processor to complete the computations associated with one element. For the same reason, there is no problem about the level of polynomials within each element. This approach can give good results for arbitrary geometries with adaptive adjustment of levels within any region. Results using such an approach are given in Reference (21).

References

1. Navier, M., *Resume des Lecons sur L'Application de la Mecanique*, edited by B. de Saint Venant, Carilian Gouery, Paris, 1864.
2. Love, A. E. H., *A Treatise on the Mathematical Theory of Elasticity*, Cambridge University Press, Cambridge, 1927.
3. Segel, L. A., *Mathematics Applied to Continuum Mechanics*, Macmillan Publishing Company, New York, 1977.

4. Batchelor, G., *An Introduction to Fluid Dynamics*, Cambridge University Press, Cambridge, 1967.
5. Golub, G. and VanLoan, C. F., *Matrix Computations*, Johns Hopkins University Press, Maryland, 1983.
6. Strang, G., *Introduction to Applied Mathematics* , Wellesley-Cambridge Press, Massachusetts, 1986.
7. Wiberg, N. E. and Moller, P., "Formulation and solution of hierarchical finite element equations, &odqInternational Journal for Numerical Method in Engineering, Vol. 26, pp. 213-1233, 1988.
8. Morris, R. B., Tsuji, Y. and Carnevali, P., Personal Communication, 1989.
9. Wathen, A., Personal Communication, 1989.
10. Babuska, I., Szabo, B. A., and Katz, I. N., "The p-version of the finite element method, "SIAM Journal of Numerical Analysis, 18, pp. 512,1981.
11. Kato, K. and Bathe, K. J., "On the use of higher order finite elements in structural mechanics, "Report 89-4, Finite Element Research Group, Department of Mechanical Engineering, Massachusetts Institute of Technology, Cambridge, Massachusetts, October 1989.
12. Szabo, B. A., and Babuska, I., *Finite Element Analysis*, in press.
13. Winget, J., and Hughes, T., "Solution algorithms for nonlinear transient heat conduction analysis employing element-by-element iterative strategies , "Computer Methods in Applied Mechanics and Engineering, Vol 52, pp. 711-815,1985.
14. Coutinho, L. G. A., Alves, J. L. D. and Ebecken, N. E. F., "A study of implementation schemes for vectorized matrix-vector multiplication," submitted to Communications in Applied Numerical Methods.
15. Coons, S. A., "Surfaces for computer aided design of space forms," M.I.T., Project MAC, MAC-TR-41, June 1967.
16. Bezier, P. E.,"Example of an existing system in the motor industry: the Unisurf system," Proceedings of the Royal Society (London), Vol. a321, pp 207-218, 1971.
17. Patera, T. A., "A spectral element method for fluid dynamics - laminar flow in a channel expansion," Journal of Computational Physics, 54, pp. 468, 1984.
18. Maday, Y.and Ronquist, E. M., "Optimal error analysis of spectral methods with emphasis on non-constant coefficients and deformed geometries," ICASE Report No. 89-56, September 1989.
19. Sonnad, V., Foresti, S. Hassanzadeh, S. and Murakami, H., "Mininal storage solution of large 3-D problems using rapid operator application with the p-version of finite elements, "IBM Kingston Research Report, KGN-209, May 1990.
20. Foresti, S., Hassanzadeh, S., Murakami, H. and Sonnad, V., " Comparison of preconditioned iterative solution techniques with rapid operator application against direct solution methods, "IBM Kingston Research Report, KGN-216, May 1990.
21. Foresti, S., Hassanzadeh, S., Murakami, H. and Sonnad, V., " Parallel implementation of iterative solution techniques with rapid operator applica-

tion on shared memory machines, "IBM Kingston Research Report, KGN-217, May 1990.
22. Foresti, S., Brussinoh, G., Hassanzadeh, S. and Sonnad, V.," Multilevel solution method for the p-version of the finite element method," Computer Physics Communications, 53, pp. 349-355, 1989.
23. Hassanzadeh, S., Foresti, S., Murakami, H. and Sonnad, V., "Solution of very large scale generalized eigenproblems using minimal storage," IBM Kingston Research Report, KGN-219, May 1990.

Chapter 23.

Modeling of Atmospheric Pollutant Transport in Shorelines

Zaphiris D. Christidis

IBM Research
T.J. Watson Research Center
Yorktown Heights, New York 10598, U.S.A.

1. Introduction

An accurate estimation of atmospheric pollutant distributions often requires the use of mathematical transport models in combination with meteorological observations. The lack of such observations in lakeshore environments necessitates the use of weather forecasting models in order to generate physically consistent meteorological fields by dynamically forecasting wind and temperature variations. Thus, the temporal and spatial resolution of the simulated data is considerably higher than data available from observational networks. This also is advantageous because the data generated by the models represent area-averaged meteorological quantities for each grid square of the modeling domain, while data obtained by observations would only have represented the meteorology at particular points, which may not have been representative of a larger area.

This chapter describes the methodology used for the development of such hydrodynamical models and presents one that uses pseudospectral and finite differencing techniques to evaluate the modifications to wind flow and pollutant dispersion caused by the interaction of the wind field with a large lake. This approach is unique because of the use of highly accurate and efficient numerical schemes, as well as dry deposition parameterization processes which are rarely included in hydrodynamic models.

2. Model Development and Governing Equations

There are two basic frameworks for modeling atmospheric pollutant transport: Lagrangian and Eulerian. There are advantages and disadvantages inherent to each. Lagrangian models are very efficient to operate but their design limits the description of vertical variations in wind flow and pollutant concentrations and, hence, they introduce truncation errors due to the discretization of horizontal and vertical concentration gradients. Similar truncation errors are characteristic of the Eulerian models. Recently a third category of transport models, the so-called pseudospectral models,[11] has become known due to their operational accuracy and efficiency.

A fundamental approach in computing the concentrations of atmospheric pollutants is to solve a system of differential equations that govern the motions of the air and the conservation of pollutant mass. This system of equations is based on mathematical statements of conservation laws for mass, momentum and energy in the three-dimensional space, and it can be solved numerically on a horizontal and vertical grid. If molecular diffusion effects are completely ignored, these equations can be written in vector form as follows:

$$\begin{aligned} \frac{\partial \mathbf{U}}{\partial t} &= -\mathbf{U}\nabla\mathbf{U} - 2\mathbf{\Omega}\times\mathbf{U} - \frac{1}{\rho}\nabla p + \mathbf{g} + \mathbf{F} \\ \frac{\partial \rho}{\partial t} &= -\mathbf{U}\nabla\rho - \rho\nabla\mathbf{U} \\ \frac{\partial \ln\Theta^*}{\partial t} &= -\mathbf{U}\nabla\ln\Theta^* + \frac{\dot{q}}{c_pT} \\ \frac{\partial C^*_i}{\partial t} &= -\mathbf{U}\nabla C^*_i - C^*_i\nabla\mathbf{U} + Q_i \end{aligned} \tag{1}$$

where $\mathbf{U} = u^*\cdot\mathbf{i} + v^*\cdot\mathbf{j} + w^*\cdot\mathbf{k}$ is the wind velocity vector, $\mathbf{\Omega} = \hat{f}\cdot\mathbf{j}/2 + f\cdot\mathbf{k}/2$, is the vector angular velocity of the earth, $\hat{f} = 2\Omega\cos\phi$ and $f = 2\Omega\sin\phi$ are Coriolis parameters, ϕ is the latitude, $\mathbf{g} = -g\cdot\mathbf{k}$ and $\mathbf{F}$ are gravitation and friction forces, ρ, p, T and Θ^* is the density, pressure, temperature and potential temperature of the air, respectively, $\dot{q}$ is heating rate per unit mass, and C^*_i, Q_i are concentrations and sources of pollutant species.

The first equation is known as the momentum equation, the second equation is a statement of mass continuity, while the next two equations are the thermodynamic energy, and the mass conservation equation for the pollutant species i, respectively. In the atmosphere there are many pollutants that interact chemically with each other, forming new chemical constituents which in turn affect the total distribution of pollutants. Without any loss of generality, only the impact of non-reacting species as emitted from pollutant sources is examined here. Thus a pollution episode, as induced by a lake-effect phenomenon, will be described by applying the conservation of pollutant mass equation for the sum $C^* = \Sigma_i C^*_i$.

Under the assumption that the lower atmosphere is incompressible, it is convenient to eliminate ρ, T and p from the general equations by introducing the basic equation of state for an ideal gas $p = \rho RT$, the potential temperature $\Theta^* \equiv T(p_s/p)^{\eta}$, and a non-dimensional presure variable π^*, called the *Exner* functions as $\pi^* \equiv (p/p_s)^{-\eta}$ where p_s is a reference pressure, usually 1000 mb, $\eta = R/c_p$, and R, c_p are the gas constant and the specific heat of dry air at constant pressure. From the definition of Θ^* it is seen that $T = \Theta^*\pi$.

The presence of a sophisticated radiative transfer algorithm for the inclusion of radiation absorption by tropospheric species such as water vapor, carbon dioxide, etc., was omitted here. Instead, a simple parameterization scheme for heat conduction due to the radiation is considered only as a lower boundary condition for the thermodynamic energy equation, and is discussed in more detail later.

For describing phenomena induced by lakes, it is desirable first to decompose each variable according to a different horizontal scale classification, and then simplify the system of equations further by eliminating certain terms through scale analysis. Specifically, the most important atmospheric phenomena can be classified as microscale (1-20^3 meters), mesoscale (20^3-10^5 meters) and synoptic scale phenomena (10^6-10^7 meters). Thus each dependent meteorological variable A^* can be decomposed as a sum of the synoptic scale, mesoscale and microscale component ($A^* = \hat{A} + A + A'$). The microscale variables, A', cannot be resolved adequately by a mesoscale grid and they are parameterized through the so-called gradient-flux approximation.[3] In turn, the synoptic scale variables, $\hat{A}$, can be considered to vary slowly with time and therefore can be separated from the general equations. The removal of the synoptic scale variables from the model equations is useful because the remaining unknowns are the mesoscale variables (A), and represent grid-volume averaged perturbations from the synoptic state.

Scale analysis[8] shows that vertical accelerations are usually very small in a mesoscale grid. Thus, the vertical component of the momentum equation in the vertical direction reduces to the hydrostatic equation for the sum of the synoptic and mesoscale variables. It should be pointed out that the synoptic scale vertical velocity is very small, compared to the horizontal velocities, and therefore does not contribute to the transfer of heat, mass and momentum in the mesoscale regions. For this reason it is omitted from the governing equations.

To assess the effect of pollutant sources in lakeshore environments, it is assumed that synoptic scale contributions to the total pollutant distributions are zero. Also, the horizontal gradients of the synoptic wind and temperature can be considered nonexistent.[9] The synoptic pressure gradient is associated with the geostrophic wind $\mathbf{V}_g = u_g \cdot \mathbf{i} + v_g \cdot \mathbf{j}$ through the relation

$$\mathbf{V}_g = \frac{c_p\hat{\Theta}}{f}\mathbf{K} \times \nabla\hat{\pi} \qquad (2)$$

It is further assumed that the synoptic state is in hydrostatic equilibrium,

$$g = - c_p \hat{\Theta} \frac{\partial \hat{\pi}}{\partial z} \tag{3}$$

while the mesoscale potential temperature fluctuations are considered to be very small, as compared with their synoptic counterparts ($|\Theta| \ll \hat{\Theta}$).

2.1 Equations for the Mesoscale Variables

Using Eqs. (2) and (3), averaging the system (1) over a time scale compatible with the time scale of micrometeorological turbulent eddies and considering all the approximations made, the following scalar set of equations for the mesoscale variables is obtained.

$$\frac{\partial u}{\partial t} = f\left(v - v_g \frac{\Theta}{\hat{\Theta}}\right) - \hat{\Theta}\frac{\partial \pi}{\partial x} + \mathbb{Q}_T^m \hat{u} + (\mathbb{P} + \mathbb{Q}_H + \mathbb{Q}_V^m)u \tag{4}$$

$$\frac{\partial v}{\partial t} = -f\left(u - u_g \frac{\Theta}{\hat{\Theta}}\right) - \hat{\Theta}\frac{\partial \pi}{\partial y} + \mathbb{Q}_T^m \hat{v} + (\mathbb{P} + \mathbb{Q}_H + \mathbb{Q}_V^m)v \tag{5}$$

$$\frac{\partial \Theta}{\partial t} = \mathbb{Q}_T^e \hat{\Theta} + (\mathbb{P} + \mathbb{Q}_H + \mathbb{Q}_V^e)\Theta \tag{6}$$

$$\frac{\partial C}{\partial t} = Q + (\mathbb{P} + \mathbb{Q}_H + \mathbb{Q}_V^p)C \tag{7}$$

$$\frac{\partial u}{\partial x} + \frac{\partial v}{\partial y} + \frac{\partial w}{\partial z} = 0 \tag{8}$$

$$\frac{\partial \pi}{\partial z} = \frac{g}{c_p}\frac{\Theta}{\hat{\Theta}^2} \tag{9}$$

where,

$$\mathbb{P} \equiv -(\hat{u} + u)\frac{\partial}{\partial x} - (\hat{v} + v)\frac{\partial}{\partial y} - w\frac{\partial}{\partial z}, \quad \mathbb{Q}_T^q \equiv -w\frac{\partial}{\partial z} + \frac{\partial}{\partial z}\left(K_z^q \frac{\partial}{\partial z}\right),$$

$$\mathbb{Q}_h \equiv K_x \frac{\partial^2}{\partial x^2} + K_y \frac{\partial^2}{\partial y^2}, \quad \mathbb{Q}_V^q \equiv \frac{\partial}{\partial z}\left[(\hat{K}_z^q + K_z^q)\frac{\partial}{\partial z}\right].$$

Equations (4) and (5) are the x and y-momentum equations respectively, Eq. (6) is the thermodynamic energy equation for adiabatic motions ($q = 0$), while Eq. (7) is the conservation equation for pollutant mass, where Q is a term which accounts for sources or sinks of the pollutant concentration C. Equation (8) is the mass continuity equation for an incompressible atmosphere, while Eq. (9) represents the hydrostatic balance of the synoptic and mesoscale fields.

The parameters K_z^m, K_z^e, and K_z^p are eddy exchange coefficients to be defined, and they represent the parameterization of eddy transport processes in the microscale or the planetary boundary layer, while the coefficients $\hat{K}_z^m$, $\hat{K}_z^e$, and $\hat{K}_z^p$ represent eddy friction terms on the synoptic scale for the momentum, energy and pollutants respectively. These coefficients were defined in analogy with the other dependent variables, and they are functions of the synoptic scale variables only. Here it is assumed synoptic state is in stable or neutral equilibrium, and that $\hat{K}_z^m = \hat{K}_z^e = \hat{K}_z^p = \hat{K}_z$.

Unlike the vertical eddy exchange coefficients, the horizontal counterparts K_x and K_y, are not decomposed into synoptic and mesoscale components. These are chosen to be invariant with respect to the spatial variables x, y, and z. Since very little is known about their functional behavior and, from simple scale analysis, the order of magnitude of the terms in which they are involved is usually much smaller than other terms.[5] Despite this fact, these terms were retained in the equations above for the purpose of smoothing out high frequency noise from the computational schemes.

2.2 Equations for the Synoptic Scale Variables

In order to obtain the final form of the mesoscale Eqs. (4-3), the following set of equations for the synoptic state is used.

$$\frac{\partial}{\partial z}\left[\hat{K}_z \frac{\partial \hat{\mathbf{W}}}{\partial z}\right] - \imath f(\hat{\mathbf{W}} - \mathbf{W}_g) = 0 \tag{10}$$

and,

$$\frac{\partial}{\partial z}\left[\hat{K}_z \frac{\partial \hat{\Theta}}{\partial z}\right] = 0 \tag{11}$$

where $\hat{\mathbf{W}} \equiv \hat{u} + \imath v$ and $\mathbf{W}_g \equiv u_g + \imath v_g$, $(\imath^2 = -1)$. Equations (10) and (11) were obtained under the assumption that the synoptic state is in barotropic and steady state equilibrium, $(\partial/\partial x \equiv \partial/\partial y \equiv \partial/\partial t \equiv 0)$, and they describe the balance of the Coriolis, pressure gradient and eddy frictional forces. They can be solved as a system of simultaneous differential equations on a vertical grid, given certain initial and boundary conditions for $\hat{\mathbf{W}}$ and $\hat{\Theta}$. The wind and potential temperature distributions with height are dictated by the form of the synoptic eddy diffusivity coefficient, $\hat{K}_z$.

In the case where the synoptic pressure gradient $(\mathbf{W}_g)$ is constant with time, all the synoptic variables are calculated once and for all in the beginning of the simulation by numerically iterating the system of (10) and (11), until the synoptic potential temperature and wind profiles reach a steady state or equivalently when the difference of the $\|\hat{\mathbf{W}}\|$ on two successive iterations is less than 1%. The initial and boundary conditions for the synoptic variables are obtained from observations. The synoptic wind profiles calculated through Eq. (10) approach their geostrophic values close to the top of the planetary boundary layer, while close to the surface are smaller than their geostrophic values and turn towards lower pressure (Ekman spiral[3]).

3. Boundary Layer Parameterization

In parameterizing the small scale phenomena, a very good knowledge of the physics and the dynamics of the atmospheric boundary layer is required. This layer is commonly divided into two different regions. These are the constant flux layer (Prandtl Layer), close to the ground, and another deeper layer usually called Ekman layer, or the Planetary Boundary Layer (BPL). In this layer all the fluxes decrease with height. In general, over most of the boundary layer, small scale turbulence is induced by wind shear which is capable of transporting and redistributing momentum and pollutants, thermal convection and surface topography characteristics. Therefore, in order to determine the quantities $\hat{K}_z$, K_z^m, K_z^e and K_z^p that appear in the momentum energy and transport equations, it is necessary to determine the relations among the parameters responsible for small scale turbulence.

3.1 Diffusivities in the Surface Layer

A basic assumption for the boundary layer parameterization is that the turbulent flux of momentum and heat is analogous to molecular diffusion. Here it is assumed for simplicity that the pollutants, energy and heat are diffused with the same diffusion rates. Momentum is diffused in the same manner, although at a different rate than heat and pollutants. Hence the eddy diffusivity coefficient for pollutants and energy will be denoted by $\overline{K_z^h} = \hat{K}_z + K_z^h$. The momentum and heat fluxes are given by,

$$< w'u' > = -\overline{K_z^m}\frac{\partial \bar{u}}{\partial z}, \quad < w'\Theta' > = -\overline{K_z^h}\frac{\partial \overline{\Theta}}{\partial z} \tag{12}$$

where $\bar{u} = \hat{u} + u$ and $\overline{\Theta} = \hat{\Theta} + \Theta$. From Prandtl's mixing length theory, stating that $\overline{K_z^m}(= \hat{K}_z + K_z^m)$ is proportional to the vertical scale of the turbulent eddies, it follows that

$$\frac{kz}{u_*}\frac{\partial \bar{u}}{\partial z} = \Phi_m(\frac{z}{L}), \quad \frac{kz}{\Theta_*}\frac{\partial \overline{\Theta}}{\partial z} = \Phi_h(\frac{z}{L}). \tag{13}$$

The first of equations (13) is the modified logarithmic law for the Prandtl layer. The parameter k is the von Karman[1] constant = .35, u_* and Θ_* are the frictional velocity and the characteristic temperature respectively,† ρ is the density of the air, τ is a shear stress factor related to Reynolds stresses, $\Phi_m(z/L)$ is a nondimensional wind shear which basically corrects for buoyancy effects on the turbulence, $\Phi_h(z/L)$ is a dimensionless temperature gradient and L is the Monin-Obukhov length.

By combining Eqs. (12-13), the following expressions are obtained.

$$\overline{K_z^m} = ku_* z/\Phi_m(\frac{z}{L}), \quad \overline{K_z^h} = ku_* z/\Phi_h(\frac{z}{L}) \tag{14}$$

The above relations show that $\overline{K_z^m}$ and $\overline{K_z^h}$ are functions of the atmospheric stability which is specified by the quantity $z/L \equiv \zeta$. In this model the diffusivity profiles in the surface layer for stable ($\zeta > 0$), neutral ($\zeta = 0$), and unstable ($\zeta < 0$) conditions are calculated using the following relations.

$$\Phi_m \simeq \begin{cases} 1 + 4.7\zeta, & \zeta \geq 0; \\ (1 - 15\zeta)^{-1/4}, & \zeta < 0. \end{cases} \qquad \Phi_h \simeq \begin{cases} 0.74 + 4.7\zeta, & \zeta \geq 0; \\ 0.74(1 - 9\zeta)^{-1/2}, & \zeta < 0. \end{cases}$$

The friction velocity u_* and the characteristic temperature Θ_* are obtained by vertically integrating Eqs. (13) from z_0 (surface roughness) to a reference height $z_h(\sim 10\text{m})$,

$$u_* = \frac{k\overline{U}(z_h)}{\int_{z_0}^{z_h} \Phi_m(\xi/L)\xi^{-1} d\xi}, \quad \Theta_* = \frac{k[\overline{\Theta}(z_h) - \overline{\Theta}(z_0)]}{\int_{z_0}^{z_h} \Phi_h(\xi/L)\xi^{-1} d\xi} \tag{15}$$

where $\overline{U} = [(\hat{u} + u)^2 + (\hat{v} + v)^2]^{1/2}$. Here the Monin-Obukhov length is estimated using information between the Pasquill-Turner stability classes[5] and the roughness length z_0. This is achieved by evaluating a parameter β defined as,

$$\beta = \left[\left(9 + 100\frac{\partial \overline{\Theta}}{\partial z} \right)^{-2} + 0.0025\frac{\partial \overline{U}}{\partial z} \right]^{-\frac{1}{2}} \tag{16}$$

and determining L within each stability class as a function of z_0. Table I provides typical values of β and L associated with different stability regimes. In Eq. (16) $\partial\overline{\Theta}/\partial z$, $\partial\overline{U}/\partial z$ are in units of °K m^{-1} and s^{-1} respectively. The values of z_0 at grid points over land are fixed depending on the vegetation, while over water are obtained by $z_0 = 0.032u_*/g$.

---

† $u_*^2 = \tau/\rho = -<w'u'>, \quad \Theta_* = -<w'\Theta'>/u_*$

Table I. Relation of Pasquill stability classes to ranges of β, and L.

Stability and Monin-Obukhov Length Range		
Stabiity Class	β	L
A (Extremely Unstable)	$\beta \leq 7.0$	$-10\text{m} \leq L < 0\text{m}$
B (Very Unstable)	$7.0 < \beta \leq 8.0$	$-100\text{m} \leq L < -10\text{m}$
C (Slightly Unstable)	$8.0 < \beta \leq 8.75$	$-10^5\text{m} \leq L < -100\text{m}$
D (Neutral)	$8.75 < \beta \leq 9.5$	$\lvert L \rvert > 10^5\text{m}$
E (Slightly Stable)	$9.5 < \beta \leq 11.25$	$100\text{m} < L \leq 10^5\text{m}$
F (Very Stable)	$11.25 < \beta \leq 13.5$	$10\text{m} < L \leq 100\text{m}$
G (Extremely Stable)	$13.50 < \beta$	$0\text{m} < L \leq 10\text{m}$

3.2 Diffusivities in the Planetary Boundary Layer

The extent to which the described expressions for $\overline{K}_z^m$ and $\overline{K}_z^h$ are applicable depends on the depth of the surface layer and, hence, on stability. Above the surface layer, third degree Hermite interpolating polynomials[6] are used for $\overline{K}_z^m$ and $\overline{K}_z^h$. Taking into account the physical requirement that K_z^m and its vertical gradient be continuous with height in the Ekman layer,

$$\overline{K}_z^m(z) = \overline{K}_z^m(z_e) + \frac{(z - z_e)^2}{d^2}\left[\overline{K}_d^m + (z - z_s)\left(\frac{\partial \overline{K}_z^m}{\partial z}\bigg|_{z_s} + 2\frac{\overline{K}_d^m}{d}\right)\right] \tag{17}$$

where $d = z_e - z_s$ and $\overline{K}_d^m = \overline{K}_z^m(z_s) - \overline{K}_z^m(z_e)$, z_e is the top of the Ekman layer, or the effective mixing height, and z_s the top of the surface layer. A similar equation can also be written for $\overline{K}_z^h$.

The height of the Planetary Boundary Layer z_e is calculated using a rate equation.[9] This equation describes the diurnal variations of the boundary layer as a function of stability, mechanical turbulence, horizontal advection and vertical convection, and can be written as,

$$\frac{\partial z_e}{\partial t} = -\bar{u}_{z_e}\frac{\partial z_e}{\partial x} - \bar{v}_{z_e}\frac{\partial z_e}{\partial y} + w_{z_e} + Q_H z_e + F_e \tag{18}$$

where,

$$F_e = \begin{cases} \dfrac{1.8(w_*^3 + 1.1u_*^3 - 3.3u_*^2 f z_e)}{g\dfrac{z_e^2}{\overline{\Theta}_0}\dfrac{\partial \overline{\Theta}^+}{\partial z} + 9w_*^2 + 7.2u_*^2} & L < 0; \\ 0.06\dfrac{u_*^2}{z_e f}\left[1 - \left(\dfrac{3.3z_e f}{u_*}\right)^3\right] & L \geq 0. \end{cases} \tag{19}$$

The term, $\partial\overline{\Theta}^+/\partial z$ is the atmospheric lapse rate just above z_e, $\overline{\Theta}_0$ is the surface temperature, and $\bar{u}_{z_e}$, $\bar{v}_{z_e}$ and w_{z_e} are the horizontal and vertical velocity components evaluated at z_e. The convective vertical velocity scale w_*, is calculated from,

$$w_* = \begin{cases} \left(-\dfrac{g u_* \Theta_* z_e}{\overline{\Theta}_0}\right)^{1/3} & L < 0; \\ 0, & L \geq 0. \end{cases} \tag{20}$$

The height of the surface layer is approximated[9] from $z_s = \lambda z_e$ where $\lambda = 0.05$. Over the top of the Ekman layer, a value of 0.1 $m^2 s^{-1}$ is assigned to $\overline{K}_z^{(m)}$ and $\overline{K}_z^{(h)}$, while the values of $\overline{K}_z^{(m)}$ and $\overline{K}_z^{(h)}$ at the top of the surface layer are obtained by substituting $z = z_s$ in equations (13) for neutral, stable, and unstable conditions.

4. Numerical Methods

In order to obtain a detailed description of the meteorological parameters in the PBL, a variable vertical grid is introduced with small spacing Δz near the surface, and increasing with increasing z. This is achieved by mapping z into a new nondimensional height variable h as,

$$h = \tilde{h}\ln\left(\frac{z}{\tilde{z}} + 1\right) \tag{21}$$

where $\tilde{z}$ is a height amplitude parameter in meters, and $\tilde{h}$ a nondimensional constant. Thus, h is set to vary from 0 to 1, and the space and time are discretized as,

$$\begin{aligned} x_i &= x_0 + (i-1)\Delta x, \quad i = 1,2, \ldots, NX \\ y_j &= y_0 + (j-1)\Delta y, \quad j = 1,2, \ldots, NY \\ h_k &= h_0 + (k-1)\Delta h, \quad k = 1,2, \ldots, NZ \\ t_n &= t_0 + (n-1)\Delta t, \quad n = 1,2, \ldots, NT \end{aligned} \tag{22}$$

where NX, NY, NZ are the number of grid points in x, y, z respectively, and NT is the number of time steps. All grid parameters are specified in the beginning of each run by defining x_0, y_0, t_0, Δx, Δy, Δt, $\tilde{z}$ and $\tilde{h}$.

A three-level time differencing scheme (Leapfrog) in combination with a Fourier pseudospectral method[7] in x and y is used to treat the advection terms in the prognostic model equations (Eqs. 4, 5, 6 and 7). This method consists of expanding the unknown dependent variables in a series of global, orthogonal and complete set of functions, and demanding that the equations are exactly satisfied at a set of grid points in the space domain. The basis set chosen here is the Fourier functions due to their efficiency in combination with the Fast Fourier Transform (FFT). Thus, when x-gradients of the mesometeorological variables need to be evaluated at a time level n, a series of one-dimensional FFTs on grid lines parallel to the x-axis is performed, and a set of Fourier coefficients is obtained. In turn, the gradients are evaluated in the spectral domain by appropriate multiplication of the Fourier coefficients with wavenumbers. Finally, backward FFTs are performed on the calculated "spectral" gradients, transforming them back to the grid point domain with high order of accuracy. Contributions due to the horizontal diffusion terms are also calculated in the spectral domain by a simple multiplication of the Fourier coefficients with proper exponential factors.[11] The y-gradients are obtained similarly, while the Leapfrog scheme is used to advance the solution of the advective terms to the time level $n + 1$. A semi-implicit, second order in time Crank-Nicolson method, coupled with a fourth order Padé finite difference in space[2] scheme is employed to treat the vertical variability in the mesometeorological variables u, v, Θ and C. The reason for choosing the Leapfrog and Crank-Nicolson time differencing schemes was mainly due to their accuracy ($O[(\Delta t)^2]$) and their stability with respect to the processes they treat. In addition, the requirement for computer storage was minimized, as only two time levels of a particular variable needed to be stored for the performance of one time step.

The mesoscale pressure π is calculated diagnostically by integrating downwards Eq. (9), using an uncentered finite difference scheme in z, while the vertical velocity w is calculated by using a centered implicit finite differencing scheme and integrating upwards the divergence terms in Eq. (8).

Table II shows the CPU time distribution among the finite differencing (FD) and pseudospectral (PS) numerical schemes used in the model equations, for a typical integration ($NX = 64$, $NY = 1$, $NZ = 16$ and $NT = 720$). The total time required for the finite difference part is longer than the pseudospectral part of the calculation, even though the number of grid points on the x-axis to which the latter was employed, was much larger than the number of vertical levels. This difference was accounted due to the larger amount of arithmetic operations with respect to z, as the vertical dependence on all the mesometeorological variables is dominant in the model equations.

Table II. Percentages in Computer CPU Time for all Model Equations.

Percentage of CPU time among model tasks			
Task	FD	PS	Total
Total Initialization	0.1%	0.0%	0.1%
Equation for Pollutants ($\partial C/\partial t$)	6.4%	7.6%	14.0%
Energy Equation ($\partial \Theta/\partial t$)	9.8%	7.5%	17.3%
Diagnostic Hydrostatic Equation (π)	2.2%	0.0%	2.2%
x-Momentum Equation ($\partial u/\partial t$)	10.1%	13.0%	23.1%
y-Momentum Equation ($\partial v/\partial t$)	9.9%	8.0%	17.9%
Diagnostic Continuity Equation (w)	4.4%	4.9%	9.3%
Boundary Layer (z_e, u_*, Θ_*, L, K_z)	11.6%	2.6%	14.2%
Variable Inputs/Outputs (IO's)	1.9%	0.0%	1.9%
Implicit Filtering of Variables	0.2%	0.0%	0.2%
Synoptic Variable Calculation	0.0%	0.0%	0.0%
Total Run Percentages	56.5%	43.5%	100.0%

5. Initial and Boundary Conditions

At the initial time t_0 all the mesoscale components are assumed to be at rest, i.e.,

$$u = v = w = \Theta = \pi = C \equiv 0. \tag{23}$$

The lateral boundary conditions in x and y are assumed periodic for all dependent variables, while at the upper boundary ($h = 1$) all the mesoscale fluctuations are set to zero. At the lower boundary z_0 ($h = 0$), the following conditions were imposed:

$$u = v = w \equiv 0 \tag{24}$$

$$\Theta = \ell(x, y)\mathbb{R}(t) \equiv \Theta_0 \tag{25}$$

$$\pi = -\int_0^1 \frac{g\Theta}{\varphi(h)\hat{\Theta}^2}\, dh \tag{26}$$

$$S_d = K_z^{(h)}\varphi(h)\frac{\partial C}{\partial h} = -Cv_d \tag{27}$$

where $\ell(x, y)$ and $\mathbb{R}(t)$ is the spatial and temporal amplitude of the local heat forcing respectively, $\varphi(h) \equiv \partial h/\partial z = \tilde{h}\exp(-h/h)/\tilde{z}$, v_d is the so-called dry deposition velocity[10] as,

$$v_d = \begin{cases} 0.002u_* & \zeta \geq 0 \\ 0.002u_*[1 + (-300/L)^{2/3} & \zeta < 0 \end{cases} \tag{28}$$

and S_d are dry deposition rates of the pollutants C.

A two-dimensional version of the model was used here, to provide some basic understanding of how the land air is modified as it passes over a lake, and then in turn how the same air mass behaves on its subsequent passage over land. The modeling domain is centered around a lake with a pollutant source on its western shore emitting at the constant rate of 0.72 gm s^{-1} (Fig. 1). A synoptic pressure gradient parallel to the shoreline and directed toward the north is assumed ($v_g = 0$). This results in an eastward flow where the synoptic prevailing wind is perpendicular to the lake shores and directed towards increasing x. The lands on the left and right of the lake shores along the x axis are defined as upwind and downwind respectively. The length of the lake is taken to be 96 km, while the total modeling domain is 504 km. The origin of the x coordinate is at -16 km while the upwind and downwind lake shores (coasts) are at 120 and 216 km, respectively. The number of grid points in x and z is 64 and 16, while homogeneity is assumed in y ($\partial/\partial y = 0$). The grid length Δx is fixed at 8 km, while Δz is variable and corresponds to equal increments in h ($\tilde{z} = 120$ m, $\tilde{h} = 0.3$ in Eq. 21). All the numerical integrations are conducted for a total of 24 hours starting at simulation time $t = 0$, with a time step Δt of 1/30 hours. The simulation time was taken equal to the real time, so $t = 12$ would correspond to local noon. Equation (23) was used as initial condition for the mesoscale variables, while the planetary boundary layer height, z_e, was initially at 300 meters. The roughness length z_0 was set at 0.5 and 0.01 m over the land and water surfaces respectively, while over the coastal regions an average of 0.255 m was assumed. Initial vertical profiles for the synoptic wind and temperature were obtained from,

$$\hat{\mathbf{W}} = \mathbf{W}_g\{1 - \exp[-\delta(\imath + 1)z]\} \tag{29}$$

$$\hat{\Theta} = \hat{\Theta}_0 + \gamma z \tag{30}$$

where $\hat{\mathbf{W}} = \mathbf{W}_g = 7$ m s^{-1} at 1000 m, $\delta = (2\hat{K}_z/f)^{-1/2}$, $f = 10^{-4}$ s^{-1}, $\hat{K}_z = 10^{-1}$ m^2 s^{-1}, $\hat{\Theta}_0 = 290.1$°K and $\gamma = 5$°K km^{-1}. The terrain elevation was assumed 1 m over the land, zero over the lake and 0.5 m at the coastal regions. The mesoscale surface temperature fluctuations are taken from Eq. (25), with

$$\ell(x, y) = \begin{cases} \Theta_a, & \text{land} \\ \Theta_a/2, & \text{coast} \\ 0, & \text{lake} \end{cases} \tag{31}$$

and

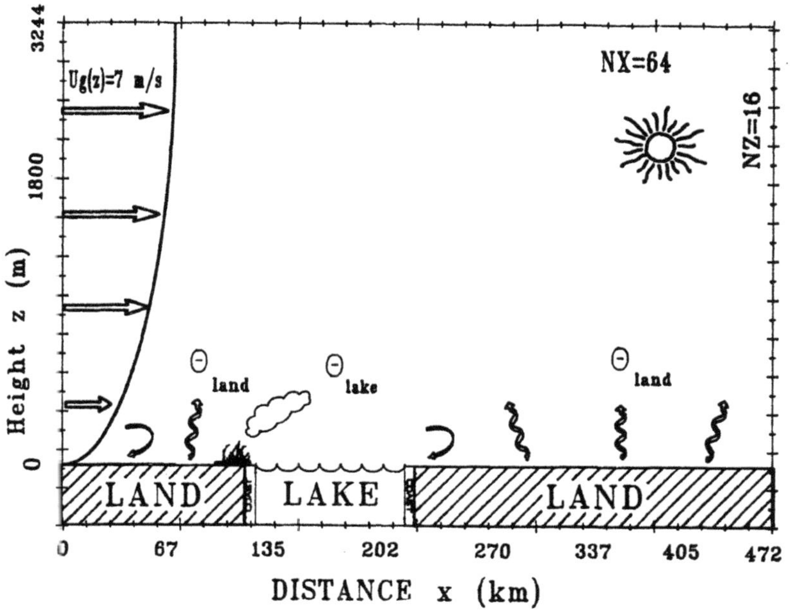

Figure 1. Coordinate system, and modeling domain.

$$\mathbb{R}(t) = \exp\left\{-\left[(t - t_l)/t_d\right]^2\right\} - \tau_h \tag{32}$$

where $\Theta_a = 11°K$ is a temperature amplitude, $t_l = 14$, $t_d = 4.9$ is a time lag and "spread" factor in hours, and $\tau_h = 0.1$ is a nondimensional bias. Thus Eq. (25) simulates typical summertime diurnal variations of land coastal and lake surface temperatures.

6. Summertime Flow Over a Lake

In the following section a series of contour plots describing the time evolution of the mesoscale plus synoptic scale predicted meterorological fields are shown. Since, initially, all the contours of the meteorological fields are parallel to the ground because of the assumption of homogeneity in the horizontal plane, the lake-breeze effects can be easily recognized from the deviations of the contours from their initial conditions. At $t = 12$, the temperature at the surface of the lake is lower than that of the land, and the air over the land is ascending due to ground heating, while there is subsidence over the lake waters. This transitional stage is shown in Fig. 2 where the vertical velocity w and the potential temperature $\overline{\Theta}$ contours (thick lines) are superimposed. Negative vertical velocities or downward motion is indicated by the light dotted contours (minimum of -3 cm s^{-1}).

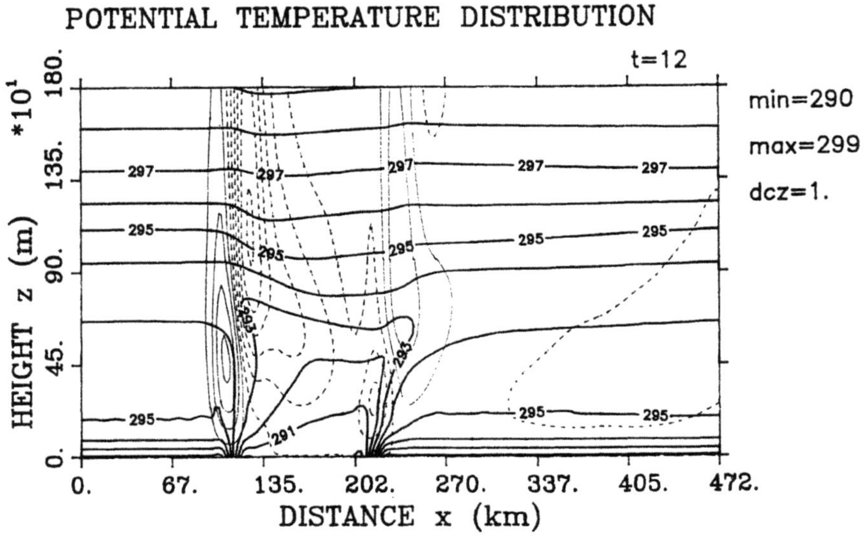

Figure 2. Simulated vertical velocity w and potential temperature $\overline{\Theta}$ at $t = 12$.

The land-lake temperature differential resulted in an increase in horizontal pressure perturbations π and therefore an increase of $\partial\pi/\partial x$ at the coastlines. This term in turn generated the onshore flows which are seen in Fig. 3, where the $\overline{u}$ component is slowed down and accelerated over the upwind and downwind shores respectively.

At $t = 15$ the land began to cool but the turbulent effects were still active as shown in Fig. 4 where $\overline{K_z^h}$ reached a maximum of 182 m^2s^{-1}. Suppressed vertical mixing above the lake is indicated by small $\overline{K_z^h}$ values. This is due to the intense stabilization of well mixed air flowing over a relatively cooler water surface. Advective effects are indicated by the slope of the isopleths on the downwind shore. Figure 5 shows that, at the lower levels, $\overline{u}$ became negative over the upwind shore with a minimum of -3 m s^{-1}, indicating that the mesoscale circulation had become opposite to the synoptic flow, while it intensified over the downwind shore. A maximum of 9 m s^{-1} was predicted above the lake and at a height of 1500 m. This was produced by the return flow replacing air that is ascending above the land at a rate of 10 cm s^{-1} due to horizontal mementum convergence (Fig. 6). The contours of $\overline{\Theta}$ were modified by the development of the high positive vertical velocities and turbulence over land, thus decreasing any vertical gradients of potential temperature due to the vertical transport and mixing of momentum and energy.

At later simulation hours the ground temperature decreased and inversions developed, resulting in stable atmospheric conditions. Furthermore, in the absence of

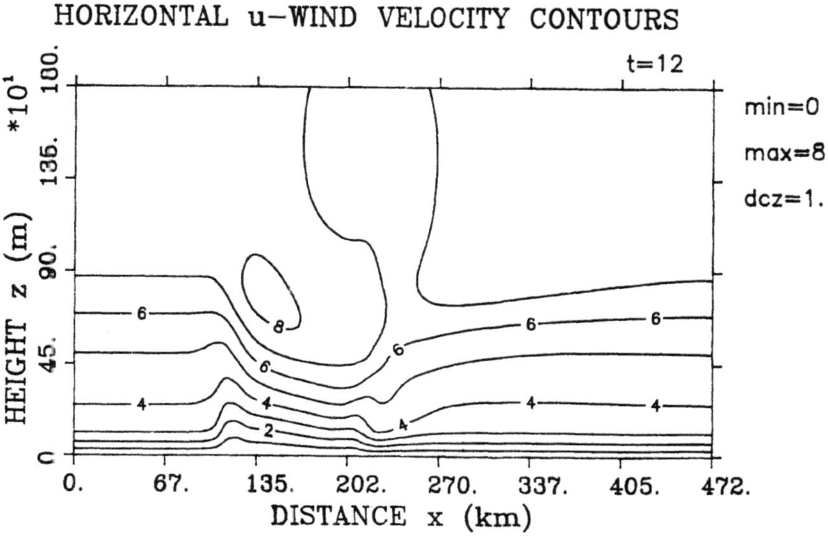

Figure 3. $\overline{u}$-wind velocity contours (m s−1) simulated at t = 12.

local forcing due to land-lake differential heating the only significant process is the horizontal transport by the synoptic wind.

Figures 7a-7d display wind vectors and pollutant concentration contours at $t = 9$, 12, 15 and 18 respectively. The concentrations are in units of μg m^{-3} while the contour interval is 50. The location of the lake is indicated by the thick line on the x-axis, while the other thick line marks the location of the boundary layer height. On the top of each figure, the solid histograms indicate schematically the amount of pollutants deposited on the ground.

In early simulation hours, pollutants were transported over the lake with little vertical mixing. The development of instabilities triggered turbulence which cased mixing of the pollutants over the land. Meanwhile v_d was increased due to increasing turbulence (Eq. 28), and more pollutants were deposited near the source. The part of the plume above the lake travelled with little dispersion and deposition until it reached the downwind shore. There, turbulent conditions were dominant and the plume moved to the ground where it was rapidly deposited. Meanwhile the boundary layer development over land caused the pollutants to mix in a larger area, while the vertical transport by the velocity field was evident at $t = 12 - 15$, from the curvature change in the concentration contours near the land-lake interface.

At $t = 18$ the area of eastward flow ($\overline{u} < 0$) propagated in the upper levels and pollutants were transported inland. The turbulent activity over the land diminished, and the pollutant deposition rates became smaller, resulting in an increase

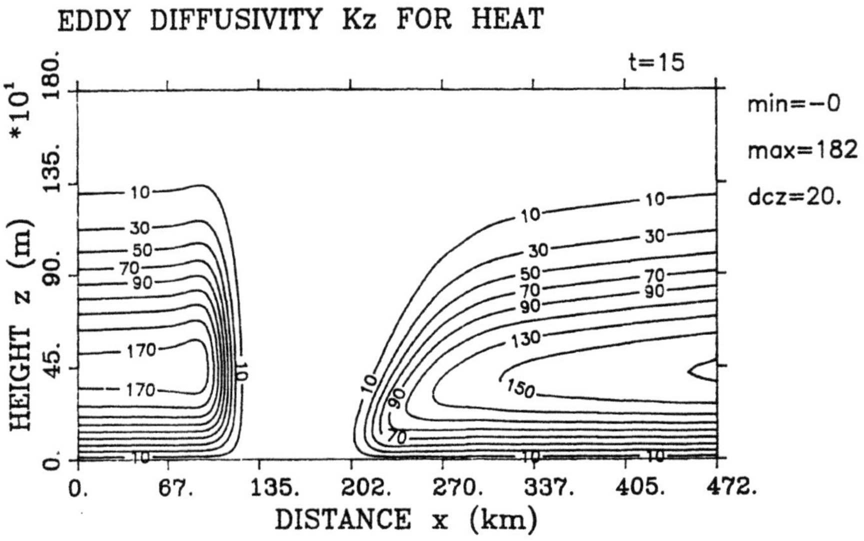

Figure 4. Simulated eddy diffusivity coefficients $\overline{K_z^h}$ ($m^2\ s^{-1}$) at t = 15.

in concentrations near the source. The development of ground inversions resulted in more stable conditions over the land and a decrease of the boundary layer height. Thus, in the absence of vertical mixing, the plume was simply transported eastward, while upwind the pollutants accumulated near the ground. Most of the pollutants were deposited near the source (total of 1000 $\mu g\ m^{-2}$ at $t = 18$), due to the stagnation of the plume by the lake-breeze flow reversal resulting in high deposition rates, while on the downwind shore there was a secondary deposition maximum that was caused by the pollutants as they were transported to the ground.

The total amount of pollutants deposited was found to be 7% of the pollutants emitted by the source. 5.2% and 0.8% was deposited on the upwind and downwind shores respectively, while 1.0% was deposited over the lake. An identical simulation was performed with the lake totally removed from the modeling domain. In this case, 8.5% of the pollutants was deposited with 5.4% and 0.1% on the upwind and downwind shores, while 2.8% was deposited in the area previously occupied by the lake.

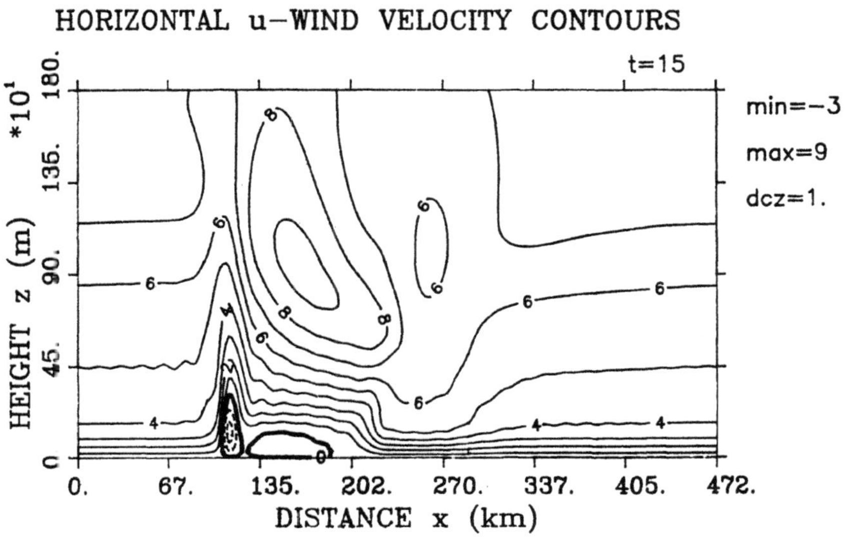

Figure 5. $\bar{u}$-wind velocity contours (m s−1) simulated at $t = 15$.

7. Summary and Conclusions

A primitive equation mesoscale model has been developed, in order to simulate weather, pollutant concentration and deposition patterns in lakeshore environments. High order numerical schemes are used, and the planetary boundary layer is parameterized with eddy diffusivity coefficients, in order to describe quantitatively the vertical transport and mixing of energy, momentum and pollutants.

Under synoptically undisturbed and homogeneous in y conditions, a westerly synoptic flow moving from a diurnally heated land to a cooler lake was simulated. During the warm air advection over the cooler lake, stable conditions were dominant over the lake while convective instabilities formed over land. In turn, strong low-level convergence patterns appeared by the shorelines with upward motions occurring over land and downward motions over the lake. On the upwind shoreline a reversal of the flow in the lower levels was present. The reversed flow propagated in the upper levels and moved inland, where it finally dissipated in the late hours. The development of the planetary boundary layer over land, caused by diurnal heating, led to intense mixing while stabilization occurred over the lake due to warm air advection over the colder water surface. The plume emitted from an upwind wource was transported with little dispersion by the synoptic wind over the lake and, around noon, was dispersed upward due to the development of turbulence over land. When the plume reached the downwind shore it dispersed to the ground and a portion was deposited due to the increased turbulence. At the same time, pollutants were deposited for the same reason on the upwind shore,

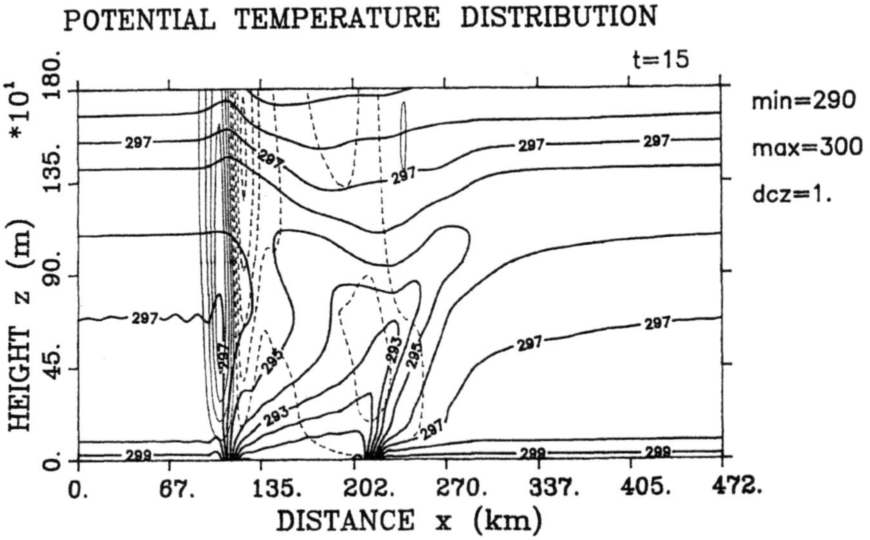

Figure 6. Simulated vertical velocity w and potential temperature $\overline{\Theta}$ at $t = 15$.

resulting in high deposition values near the source. In the late hours turbulence over land diminished and the plume was again transported with little dispersion. It was found that the presence of a lake limits pollutant deposition over its length due to stabilization.

It was demonstrated that the developed mesoscale model simulates reasonably the effects of lakes on the meteorological fields. Lyons[4] reported flow reversals in the range of -2 to -4 m s^{-1} in afternoon hours of summer days in Lake Michigan. Furthermore, the observed cloud bands on a clear day during afternoon hours coincide with the area of maximum positive vertical velocities predicted by the model, although their magnitude is underestimated as compared with observed values. However, to fully appreciate the potential and accuracy of this model, a series of improvements must be done and careful evaluation must be made in order to determine how accurately the model simulates a realistic meteorological situation. In the light of the obtained results it is recognized that the model can be significantly expanded by including features of the real atmosphere. Suggested areas of improvement can be achieved by:

1. The inclusion of the specific humidity among the model variables and replacement of the potential temperature with the equivalent potential temperature. Thus, the latent heat of evaporation and precipitation can be introduced as input of additional energy to the model.

2. The addition of a radiation scheme in order to determine the presence or formation of clouds in the model's atmosphere.

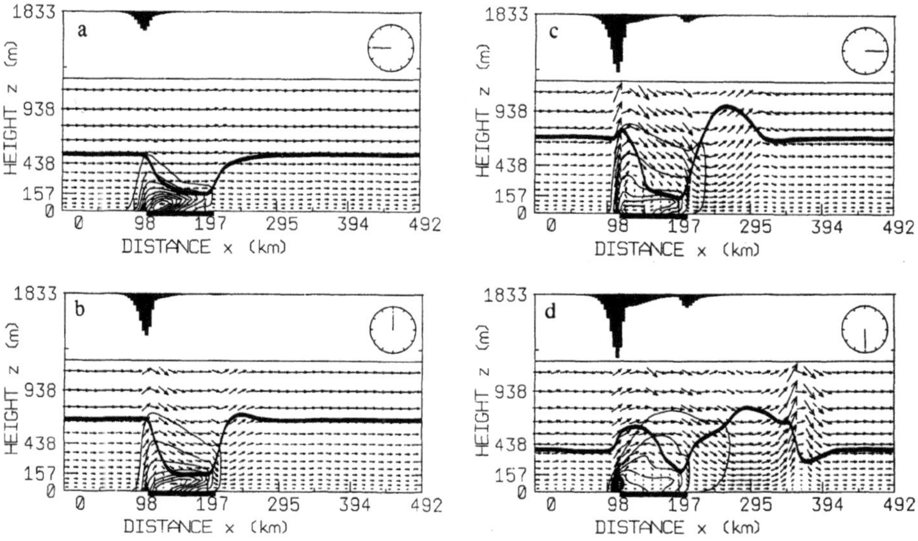

Figure 7. Vector $\bar{u}$, w wind field, boundary layer height z_e, pollutant concentration contours C and pollutant deposition patterns S_d at: a) $t = 9$, b) $t = 12$, c) $t = 15$ and d) $t = 18$.

3. The calculation of rainfall amounts and, hence, wet deposition removal processes in the equation for the conservation of pollutant mass.

4. The prediction of surface temperatures through ground physics and energy balance arguments.

5. The addition of a nonlinear chemistry module in combination with the inclusion of individual pollutant species, chemically reacting with each other.

References

1. J.A. Businger, J.C. Wyngaard, Y. Izumi and E.F. Bradley, "Flux-profile relationships in the atmospheric surface layer," J. Atmos. Sci., **28**, 181-189 (1971).
2. Z.D. Christidis, "Hydrodynamic Mesoscale Modeling of Atmospheric Transport and Pollutant Deposition in the Vicinity of a Lake," Ph.D. Thesis, Atmospheric and Oceanic Science Dept., University of Michigan, Ann Arbor (1986).
3. J.R. Holton, An Introduction to Dynamic Meteorology, Academic Press (1979)

4. W.A. Lyons, "Turbulent Diffusion and Pollutant Transport in Shoreline Environments," Lectures on Air Pollution and Environmental Impact Analyses, American Meteorological Society (1975).
5. G.J. McRae, W.R. Goodin and H.H. Seinfeld, Mathematical Modeling of Photochemical Air Pollution, Environmental Quality Laboratory Report No. 18 (1982).
6. J.J. O'Brien, "A Note on the Vertical Structure of the Eddy Exchange Coefficient in the Planetary Boundary Layer," J. Atmos. Sci., **27**, 1213-1215 (1970).
7. S.A. Orszag, "Comparison of pseudospectral and spectral approximation," Stud. Appl. Math., **51**, 253-259 (1972).
8. R.A. Pielke, "A three-dimensional numerical model of the sea breezes over South Florida," Mon. Wea. Rev., **102**, 115-139 (174).
9. R.A. Pielke, Mesoscale Meteorological Modeling, Academic Press (1984).
10. C.J. Walceck, R.A. Brost, J.S. Chang and M.L. Wesely, "SO_2 Sulfate and HNO_3 Deposition Velocities Computed Using Regional Land Use and Meteorological Data," Atmospheric Environment, **20**, 949-964a (1985).
11. Z. Zlatev, R. Berkowicz and L.P. Prahm, "Three-dimensional Advection-Diffusion Modeling for Regional Scale," Atmospheric Environment, **17**, 491-499 (1983).

Chapter 24.

Visualization Techniques and Windowing Interfaces: KGNGRAF, XWIB and REMOTE[†]

S. Chin, R. Tagliavini, J.P. Prost and M. Martins-Costa

IBM Corporation
Department 48B/428
Scientific and Engineering Computations
Kingston, New York 12401, U.S.A.

Introduction

The use of computer graphics for visualizing results of scientific and engineering applications has spread widely in the last few years. This growth has spanned a wide variety of fields, including computer aided design, computer integrated manufacturing, fluid dynamics, physics, biology, biochemistry and chemistry. Particular applications of computer graphics are quite varied within each field, however, they all have the common objective of providing a medium for manipulating and interpreting data from numerical simulation. A key ingredient required to allow this goal to be reached is the development of software and hardware tools which can be applied in a wide range of scientific fields. In particular, software must be efficient, user-friendly and, as much as possible, programmed to take advantage of the interactive aspects of the man-to-machine interface. Therefore, optimal programming of highly interactive computer graphics applications, with, if possible, a standard graphics interface, remains as one of the major tasks in advancing the use of visualization in understanding scientific and engineering research.

Increasingly more often it appears that the Programmer's Hierarchical Interactive Graphics System or PHIGS is being adopted as a standard graphics programming interface. There are still a variety of languages available to program computer graphics, but Fortran and C are the most commonly used. The PHIGS program-

† This chapter is an extension and revision of the chapter entitled, "Visualization Techniques for Science and Engineering", by S. Chin, D.P. Vercauteren, W.L. Luken, M. Re, R. Scateni, R. Tagliavini, D.J. Vanderveken and G. Baudoux in Motecc-89.

ming interface is well suited for programming graphics applications in various scientific and engineering disciplines. However, in order to fully understand the programming considerations involved in implementing particular graphics functions, it is worthwhile to have a general idea of the PHIGS graphics interface. Therefore, after a brief introduction to IBM implementation of the PHIGS standard, the IBM graPHIGS™ Application Programming Interface (API), some of the desired features of graphics programs for scientific and engineering applications will be presented, with a particular emphasis on the use of these programs to help in understanding chemical computer simulations. In particular, techniques for "interactive" computer graphics, and in particular how these techniques are implemented into our graphics system, KGNGRAF will be discussed. In addition, the rendering of "realistic" images will be discussed, along with the application of these techniques to produce computer animated videos.

Finally, as computer graphics is closely linked to and associated with the workstation environment, and since it appears likely that the future architecture of a computational environment will contain intelligent workstations networked to computational servers, we will discuss the connection among computer graphics, application programs, workstations and computational servers. This is done in the context of a user-friendly graphical interface for scientific application programs, based on X Windows and TCP/IP network protocols. This interface is designed to provide a graphical workstation platform, using a windowing environment, for the creation of input files for simulation programs, as well as facilities for submitting jobs for remote and local execution, without the necessity for the user to have an intimate knowledge of the Unix operating system and TCP/IP protocols. The two components of the interface will be discussed; first, the windows based environment for input specification to scientific application programs, XWIB; followed by a short discussion of the facilities for remote submission of jobs, REMOTE.

The graPHIGS Application Programming Interface (API)

The basic concept provided by the graPHIGS API to display and manipulate graphical data is the structure. A structure is simply an ordered list of graphical commands, called elements, that may be manipulated either individually or as an entity.[1-3] Each element may be either an output primitive such as a polyline, polymarker, polygon, triangle strip or spline surface, an attribute, such as line type, color, character font, geometrical transformation, depth cueing index, light source state, lighting calculation type, an element label (to identify particular elements within a structure), or an invocation of another structure (thereby creating hierarchies of structures).[4] Structures are stored by the API into a unique database that is common to all graphics workstations, thus making the API device independent. In addition, different types of graphics workstations can be used simultaneously within a single application program.

After structures have been defined, they are associated with a given priority level to a 2D or 3D view. A 3D view is defined as the volume of space that is visible on the graphics display. Typically, anything that is not within this viewing volume is not visible. Defining a view consists of specifying its orientation through a view transformation matrix, a clipping volume through the position of near and far clipping planes and the view reference point, a priority relative to other views, attributes of the view such as shielding, shielding color, projection type, and activity, and finally, a mapping to an output display device.[1]

As previously mentioned, graphics data are stored in a common database, independent of any workstation type. Data become eligible for display by making an association of particular portions of it with a workstation or workstations. Once an application program associates a structure with a workstation or a view, that structure becomes a root structure. The difference between associating a structure to a view or to a workstation lies in the fact that if an "intelligent" workstation is used, associating the structure to the workstation causes it to be stored in the workstation's local memory. This enables an application to achieve optimal performance from the intelligent workstation. Associating a structure to a view implicitly associates the structure to the workstation for which the view is associated. Structures may also be assigned priorities within a view, similar to views being assigned priorities within a workstation.

Once structures and views have been defined to the API, the workstation traverses the structures for image generation. Each view within a workstation is traversed in order of its assigned priority, from lowest to highest. Within each view, structures are traversed sequentially from top to bottom also according to a reverse priority.

The graPHIGS API allows an application to control the appearance of a particular view by setting the characteristics of that view. Some of the characteristics that can be set are the visibility, clipping, hidden surface removal and transparency.

At the beginning of structure traversal each element attribute and transformation takes default values. Elements within a structure are then processed sequentially, from the top of the structure to the bottom. The binding of attributes and the application of transformations to output primitives is done at traversal time. Therefore, when an output primitive is encountered, its final appearance is determined by the current attributes and transformation settings.

The graPHIGS API provides the ability to manipulate either whole structures, or individual elements. Structures may be edited at any time either inserting or deleting particular elements. This provides the possibility of changing structures without having to re-define the entire graphics image.

If the invocation of another structure is encountered during structure traversal, the API suspends the traversal of the original structure and begins traversal of the second structure. Any attributes which are bound to the first structure are also bound to the second structure. However, any attributes that are subsequently bound to the second structure are not inherited by the original structure.[1] This allows for the definition of child structures independently of parent structures.

Another useful feature of the API is the concept of filters. By defining particular elements as belonging to different classes, one can use a filter to enable or disable visibility, pickability or highlightability of selected elements within a structure. Attributes can also be stored within bundle tables. The use of those tables allows the updating of the screen without having to modify individual structures.

The graPHIGS API also contains features for programming interactive devices such as a mouse, dials and lighted programmable function keys, therefore providing flexibility in how the user is able to interact with the application program. Several modes of interaction can be used within the API: request mode, is a synchronous environment in which the application program must explicitly request information from a peripheral device; event mode, is an asynchronous environment in which the device supplies data to the application whenever an event is triggered at that device, and in sample mode, the device continually samples its value.[1]

In event mode devices operate independently of each other. A first-in-first-out central event queue (which is unique for all devices and workstations) is maintained by the API and each event is asynchronously passed to the application program. If no events are currently queued, the API may be programmed to wait for a specified time interval for the next event. Once an event is queued, which is accomplished by a user activating any of the available devices, the application retrieves data from that device by issuing a "get data command". In this fashion, applications can be programmed to wait for events, and depending upon the type of device which is activated, the application may branch to different functions.

The peripheral hardware devices commonly found on graphics workstations, including the IBM 5080/6090 Graphics System,[5] and the IBM RISC System 6000 Workstations (IBM RS/6000) include a mouse, dials, keyboard and lighted programmable function keys (LPFK). These devices can be programmed to include six different classes of logical input devices; choice (LPFK, mouse and keyboard), locator (mouse), pick (mouse), valuators (dials), stroke (mouse) and string (keyboard).

The choice device provides a non-negative integer value that represents a selection from a number of possible choices. A value of zero indicates that no choice was selected. The locator device provides the coordinates of a position on the screen. The pick device typically provides integer values representing pick and structure identifiers. The valuators provide real numbers from a given range of numbers while the stroke device provides a sequence of locator positions. String input allows an application program to input text.

A useful feature for computational chemistry that is also provided by the graPHIGS API, but which have not yet been specifically discussed is the depiction of realistic 3D surfaces, either using polygon data or through the description of a Non-Uniform Rational B-Spline (NURBS) surface. These surfaces can also be defined with particular lighting characteristics, such as reflection coefficients, transparency characteristics, colored lights. In the area of chemistry, many aspects are directly related to the physical and electronic structure of a molecule, and the realistic three dimensional representations of these structures are important in understanding the underlying science. However, it should be pointed out that not all graphics workstations have the specialized hardware to render 3D surfaces. Since in the area of chemistry optimal interactivity is a key part of any graphics application, it is realistic to only render 3D surfaces on those workstations that provide the necessary specialized hardware. In the remainder of this chapter those workstations that do not provide this specialized hardware will be referred to as 2D workstations, while those providing it will be called 3D workstations.

Introduction to KGNGRAF

For many years experimental and theoretical chemists have used mechanical models (Dreiding, CPK ..) to study and understand the structure of molecules, particularly large biological molecules. These models are built to represent the locations in three dimensional space of the atoms contained within the molecule. These three dimensional models can become quite bulky, and time consuming to build. Computer graphics are an ideal tool to supplement and even replace these mechanical models.

The use of computer graphics in chemistry has been concentrated in several main areas: molecular model building; editing and comparison; investigation of interactions between molecules; protein design and engineering and analysis of chemical and physical properties.[6-10] The basic problem common to all computational chemistry applications is the display and manipulation of three dimensional representations of molecular structures and properties associated with these structures. Some of the points which must be taken into consideration in designing graphics programs for computational chemistry is that they should be interactive, user-friendly and provide a great variety of functions that increase productivity and understanding of underlying scientific problems. These have been the basic design goals for KGNGRAF.

Many different techniques and functions are needed to accomplish these goals, and they should all be accessible using standard atomic information with as little direct input from the user as possible. For example, it should be possible to construct a molecule without any *a priori* knowledge about bond lengths, valence angles or torsion angles, or to construct a polypeptide based only on its amino acid sequence. In any case, functionality should not be specific to the desired task of the user, but should be general and applicable across all areas of chemistry.

This point of view applies not only to small molecules, but also to the larger and more complex biological molecules such as proteins and DNA.

The User Interface

The most common and logical interface to a graphics programming environment is through the graphics display itself. This in turn indicates the need for a user-friendly interface, driven by the use of peripheral input devices such as the mouse, dials and lighted programmable function keys available through the graPHIGS API. Pop-up menus and real-time interaction with the graphics objects on the screen is an absolute necessity.[11,12] In KGNGRAF this is accomplished by using the mouse to select items from a menu on the screen. A cursor on the screen is used to track the position of the mouse. The user places the mouse over any of the items in the menu and clicks on any of the buttons of the mouse to select the desired action. For other types of man-to-object interactions, other peripherals may be more desirable; for example, for real-time rotations or translations of the objects on the screen the dials are more appropriate. The use of the peripheral devices is supplemented by appropriate instructions written to a message area on the screen to direct the user towards the expected action. Interactive on-line help features are also available, and provide the user with short tutorials on each particular option within the programming environment. Action filters are also implemented to prevent the user from taking unexpected actions. In this sense, the only knowledge of the host operating system and the graphics program that is required from the user is how to log onto the computer system and start the programs. Once KGNGRAF has been initiated, all of the user interaction is accomplished directly with the graphics screen; prompts and help are available on-line as the program executes.

Display of Molecular Models

A common representation of a molecular model is to use spheres centered at each atomic position and cylinders connecting two atoms representing the presence of a chemical bond. Other typical representations are space filling or CPK models, and Dreiding or stick model representations. Each type of representation has its advantages and the most useful representation is often dependent upon the type of data and the properties being investigated. An example of a ball and stick model for a portion of the myoglobin molecule is given in Figure 1a. A CPK model is shown in Figure 1b. These models are generated and rendered in a 3-D graphics workstation, providing a realistic image through the use of shaded surfaces.

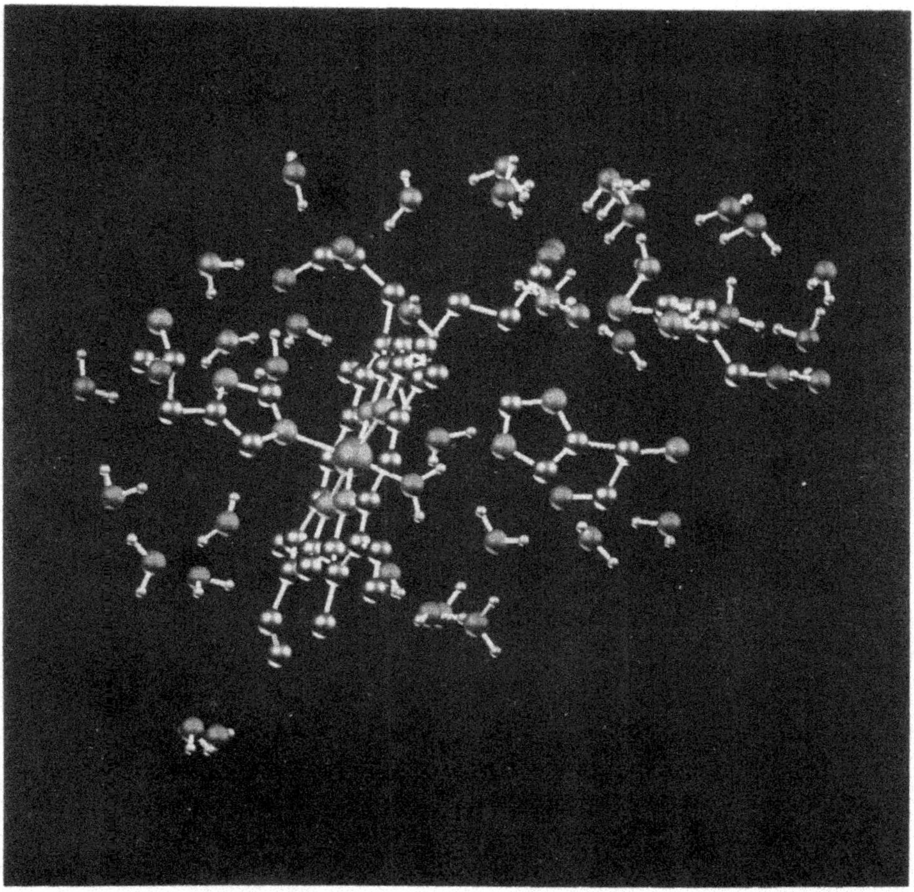

Figure 1a. Ball and stick model for a portion of the myoglobin molecule. This picture represents the area of the heme group, and some surrounding water molecules.

Most often molecular structures are presented to a graphics program in the form of either cartesian coordinates, internal coordinates or fractional coordinates, and the program must compute the atom connectivity. KGNGRAF accomplishes this by comparing the inter-atomic distance for pairs of atoms in the molecule to the sum of their covalent radii. If the distance between atoms is less than this sum within a particular tolerance, then a bond should be constructed between these two particular atoms. This algorithm is very general, and appears to give the correct bonding representation for most of the cases investigated. In addition, this type of procedure can be most easily adapted for computing hydrogen bonds (using of course a different distance criteria and taking into account the type of atoms) or any other type of inter-atomic forces that can be modelled using a distance criteria.

Stick and Ball and Stick Models

The typical ball and stick models are drawn using spheres with a radius of one-half the covalent radii of the atoms, and a cylinder to represent any bond between two atoms. This type of representation most clearly depicts both the locations of the atomic centers as well as the bonding pattern of the molecule. CPK or space filling models are drawn using spheres centered at the atomic positions, with van der Waals radii for the atoms used. This model best describes the overall spatial representation of the molecule. Dreiding or stick models are constructed by drawing sticks to represent inter-atomic bonds, each half of the bond colored according to the atom type involved in the bond. This latter model is the simplest graphical representation, and offers the fastest response to interactive manipulations by the user because only polylines are used as output primitives. Rotations and translations of these structures can then be accomplished by inserting modelling transformations within the structure. On the other hand, ball and stick models or space filling models require the use of 3D spheres, whose surfaces must be shaded and lighted according to the characteristics of the particular lighting model being used. On 3D graphics workstations, in which these features are supported in hardware, realistic images can be generated by simply making appropriate calls to graphics drivers. The hardware takes care of rendering the 3D spheres and cylinders, and since there is specialized hardware for rendering and lighting, the 3D primitive's can still be interactively manipulated. On a 2D workstation, the lighting and shading have to be computed on a local processor, making the interactive manipulation of smoothly shaded and lighted surfaces impossible.

In the case of molecules, the sphere and cylinder primitive's are sufficient for defining most molecular geometries. In KGNGRAF the particular programming implementation on 3D workstations for the sphere and cylinder primitive's are defined using NURBS surfaces. The NURBS surface is an efficient way to describe 3D surfaces, providing optimal interactive performance. The details of NURBS surfaces have been discussed elsewhere, suffice it to say here that performance advantages arise because they can be efficiently evaluated and that as coordinate transformations are performed, only the control points need to be transformed (which is much less data compared to the vertices of polygonal data). The storage required for smoothly shaded surfaces using NURBS surfaces is also greatly reduced compared to equivalent polygonal surfaces.

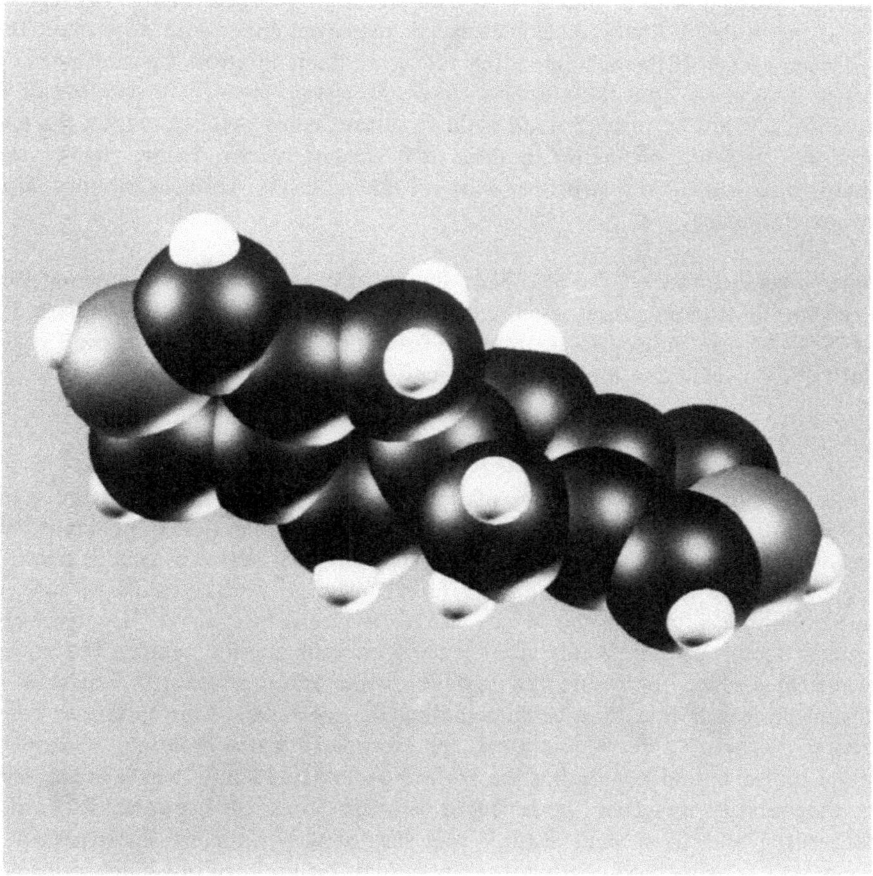

Figure 1b. CPK or space filling representation of a molecular structure.

On 2D workstations, which are primarily characterized by the lack of specialized hardware for lighting and shading computations, we are faced with the problem of representing 3D spheres and cylinders with flat polygons. In order to give the appearance that the flat polygon is a 3D primitive, the polygon must be projected onto the viewport as the rotation is performed. If this were not done, as a molecule rotates the flat polygons would eventually become perpendicular to the field of view and appear as thin lines (similar to a coins standing on their side). In order to avoid this difficulty the transformation and projection of each atom is handled by software. This sacrifices some performance, but has the advantage that the ball and stick and space filling models can be drawn and manipulated on both 2D and 3D workstations.

One point that should be noted here is that the inter-atomic bonds within a particular molecule may be divided into different classes; for example, those containing hydrogen atoms and those not containing hydrogen atoms. Through the use of the visibility filters, molecules can be displayed either with or without their hydrogen atoms. If the corresponding filters are made invisible the molecule completely disappears from the graphics screen. However, since the invocation of the structure has not been deleted, all child structures being executed within the main structure, and not belonging to these two classes, remain visible. As we shall discuss later, these child structures may include molecular surfaces, ribbons, alpha carbon traces, etc.

Other types of models have also been shown to be quite useful for investigations of chemical structure. Amongst the more popular models currently supported by KGNGRAF are molecular surfaces or envelopes, electron densities, and particularly for polypeptides, alpha carbon traces and ribbon models.

Molecular Surfaces

Molecular surfaces are commonly used in chemistry to illustrate either the spatial representation of a molecule (analogous to the space filling model), or a particular surface property of a molecule. The two most typical surfaces used in chemical applications are the van der Waals surface and the solvent accessible surface. As its name implies, the van der Waals surface is the surface of the 3D object constructed from van der Waals sized spheres at each atomic center. The solvent accessible surface represents the surface of the molecule that is accessible to solvent molecules that may be surrounding the molecule. This surface is generated by "rolling" a probe molecule, for example a water molecule, around the solute molecule and computing the surface around the solute which is accessible to the solvent molecule. It is based on the ideas of Lee and Richards,[13] Richards,[14] and Greer and Bush,[15] and has been analytically implemented by Connolly.[16] The method computes sample points on the surface which can be subsequently displayed on any display device by displaying each point as a marker on the surface. It should be noted here that the computation of the van der Waals surface is a subset of the computation of the Connolly solvent accessible surface. The van der Waals surface is computed by simply setting the radius of the probe molecule to zero. An example of a van der Waals surface, in which the spheres are colored according to atom type, is shown in Figure 2.

It is also sometimes desirable to modify molecular structures from their original conformation, therefore, neither surface is automatically computed by KGNGRAF, but must be selected for computation after the proper orientation has been chosen. Once the computation has been performed, the resultant surface can then be displayed or hidden as desired. The surface becomes a child structure of the main molecular structure and is displayed or hidden by the insertion or deletion of an execute structure element within the main structure. Molecular surfaces can also be colored according to particular properties or characteristics, such as atom type, electrostatic potential, hydrophobicity etc..

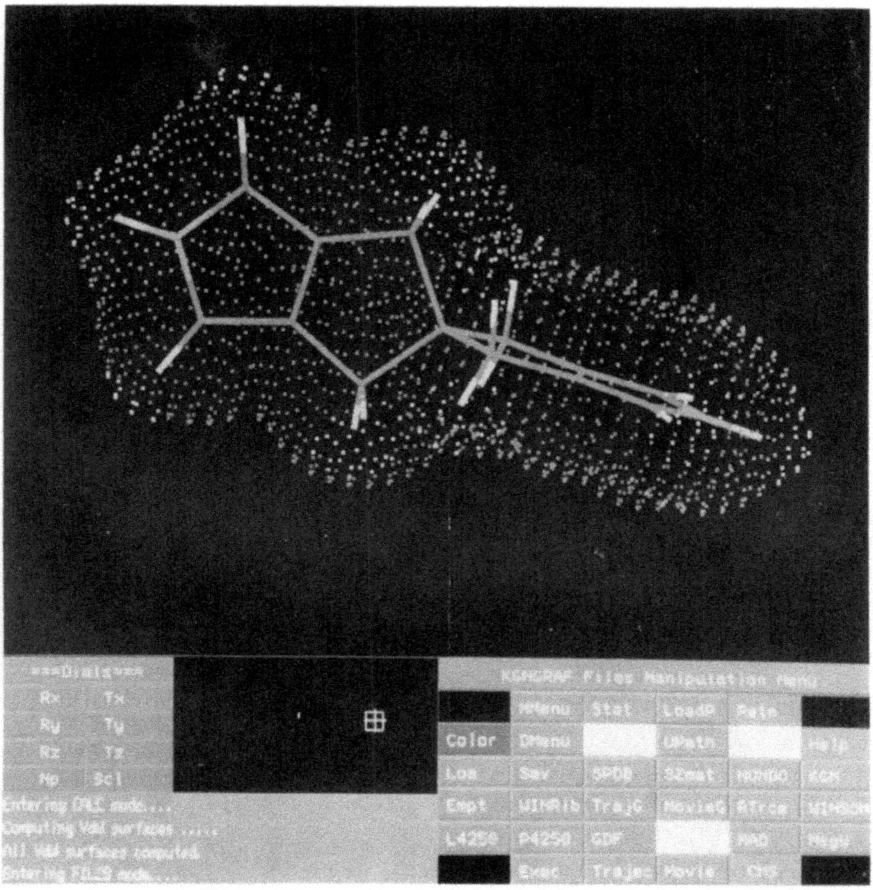

Figure 2. Van der Waals surface representation of a molecular structure.

Polypeptides

Polypeptides are a special class of molecules which have some additional representations that are particularly important in the understanding of their structures. Proteins are composed of sequences of amino acids which have a common backbone structure, but, with varying chemical groups as sidechains. A good deal of structural information can be gained by looking only at the backbone of the polypeptide. This not only simplifies the structure by reducing the number of atoms, but it also focuses attention to the secondary structures such as helices and beta sheets that are commonly found in polypeptide structures.

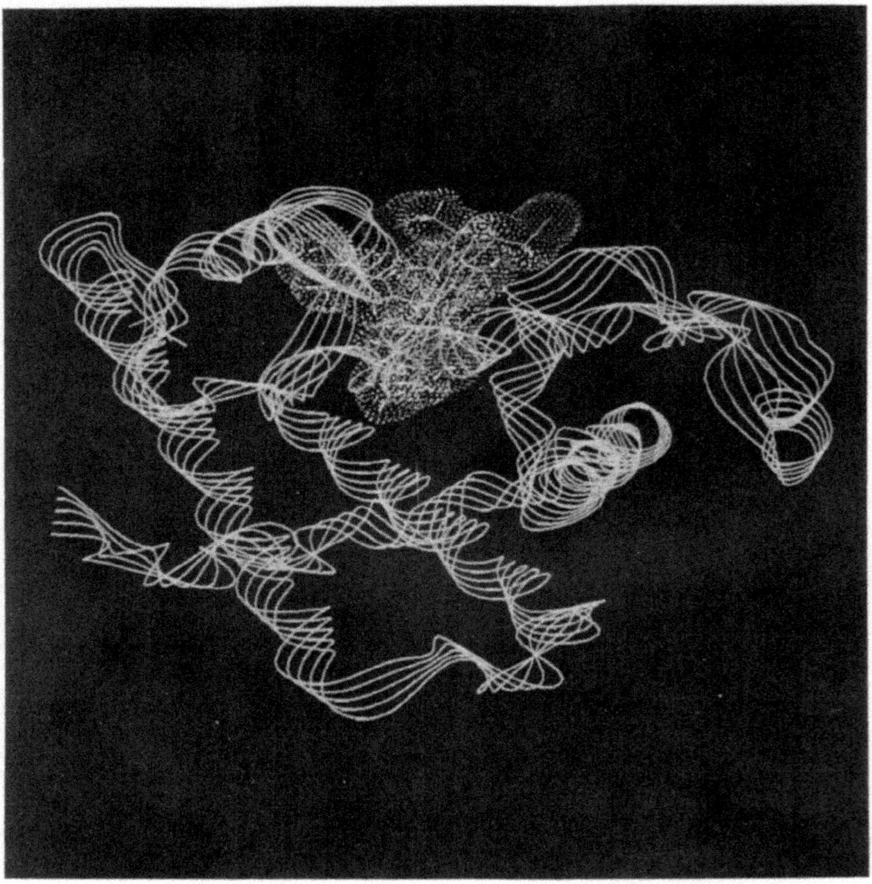

Figure 3a. Ribbon model for myoglobin with the heme group highlighted with a van der Waals surface representation.

Among the polypeptide representations that have been popularized are the alpha carbon trace and the ribbon model. The alpha carbon trace is simply a series of lines connecting the alpha carbon of each amino acid in the polypeptide. The ribbon model, popularized by Richardson,[17] depicts slightly more detailed information on the backbone structure. The polyline ribbon model used in KGNGRAF is created by tracing a number of line segments along the backbone of each amino acid residue as proposed by Carson.[18,19] The endpoints of the line segments are computed to be parallel to the C=O bond, and bisecting the C-N bond. In this fashion, when these points are connected a set of parallel lines tracing the positions of the backbone atoms of the amino acids is generated. Any number of segments can be drawn, but a good choice is usually three to five parallel segments separated by about 0.2 Å Depending upon the number of segments used, usually an equal number of segments are drawn above and below the C-N bond. In order to smooth the ribbon, a spline curve can also be drawn through the points composing the ribbon.[20]

On 2D workstations a second ribbon representation utilizes a filled polygon for each amino acid, the color being chosen according to the residue type. Each polygon is constructed from the four corners of a rectangle defined by the endpoints of the two outer line segments composing the amino acid. The polygon, like the ribbon, is flat. However, because the polygon is colored, it is much easier to identify amino acid types. Both ribbon representations are displayed as child structures of its parent molecule. Therefore, the ribbons may be superimposed on the stick model of the polypeptide (with or without hydrogen atoms), displayed alone, or displayed in combination with any other child structure. An example of a polyline ribbon model of myoglobin, with the heme group highlighted with a van der Waals surface is shown in Figure 3a.

On 3D workstations KGNGRAF represents this ribbon model with shaded surfaces, modeling the backbone trace of the protein by constructing a tube composed of 3D cylinders and spheres. For example each residue in the protein can be decomposed into segments which consist of cylinders, with the space between nonparallel cylinders being filled by spheres. In this fashion, a more realistic three dimensional image clearly exemplifying the structure of the protein can be constructed, including the effects of shading and lighting, while still maintaining interactivity. A solid ribbon model of the myoglobin protein, combined with a ball and stick representation of the heme group is given in Figure 3b.

A main advantage of the ribbon model over the alpha carbon trace is that the ribbon offers more information on the twists and turns within the backbone structure of the polypeptide. For example, in regions of alpha helical secondary structure the ribbon clearly depicts the twisting and turning of the helix, while in regions of beta sheets the ribbons are extended and flat.

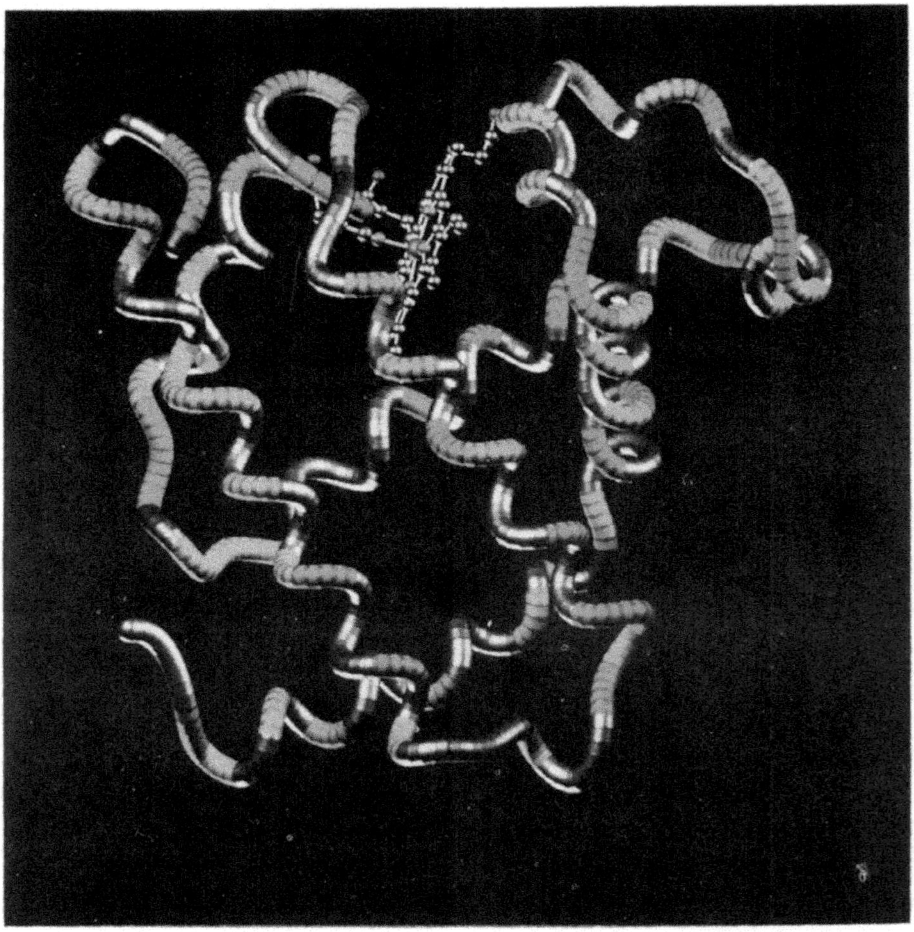

Figure 3b. Solid ribbon model for myoglobin, with the heme group represented by a ball and stick model.

Color Coding

The use of color in all of the above representations also plays an important role in elucidating information about molecular structure. In general, the default coloring scheme in KGNGRAF assigns a different color to each type of atom. However, it is also possible to use different colors to highlight different regions of a molecule, or to highlight particular properties. One example is when considering polypeptides; instead of coloring each atom according to its atom type, it is possible to color each amino acid residue by a different color, or even each amino acid residue by a different color according to a hydrophobicity scale. Another example is the coloring of a van der Waals surface according to the value of the molecular electrostatic potential computed at each point on the surface.

Different users have different preferences for choosing colors, and therefore, while default coloring schemes are generally practical for most applications, it is also useful to have a scheme for allowing users to interactively modify and choose colors. In KGNGRAF the user has complete access to the color table, both in an input file as well as within the program itself. In the latter case a palette of all color table entries is shown to the user. The user can then change any of these entries by clicking with the mouse on the particular entry and using the dials or the mouse to alter the red, green and blue components of that particular entry. All graphics output primitives that have that particular color table index associated with it will change color according to the modified entry.

To summarize briefly, with a set of cartesian, internal or fractional coordinates for a group of atoms (and a list of covalent radii for these atoms), KGNGRAF can be used to construct a number of different molecular graphics representations. Each representation has its particular advantages. In addition, by using appropriate combinations of molecular structure representation and coloring, it is possible to gain a greater understanding of molecular structures, properties and interactions. On 3D workstations, realistic images can be generated and interactively manipulated, providing added insight to understanding underlying three dimensional structures.

Display of Electron Densities and Molecular Orbitals

Two- and three-dimensional contour maps are commonly used in all areas of science and engineering. As examples, in the field of computational chemistry they are commonly used to represent iso-energy contours, (Φ-Ψ) or Ramachandran maps for polypeptides, electron density maps or molecular orbitals.[21,22] In some of these cases, the data can be represented with two-dimensional contours, while in other cases, both two- and three-dimensional contours must be used.

KGNGRAF assumes that two- and three-dimensional contours are both represented by values of a function on a grid. In the two-dimensional case, the grid is reduced to only two degrees of freedom or a single plane. In the three-dimensional case, a series of planar slices, or a 3D grid, are provided. In either case, any particular contour level can then be drawn by linear interpolation of the function values at adjacent grid points.[10,21] In two dimensions, a scan line algorithm is used to generate all line segments whose endpoints take the value of the desired contour. This can be easily expanded to three dimensions by performing the same scan line algorithm in the three sets of planes (XY, YZ, XZ). It should be noted here that this algorithm does not generate continuous line segments, but discontinuous segments. However, once these line segments have been determined, contour maps can be easily drawn by displaying all the line segments, regardless of whether they are continuous or not. After all the line segments have been drawn, by definition, the contours will be complete.

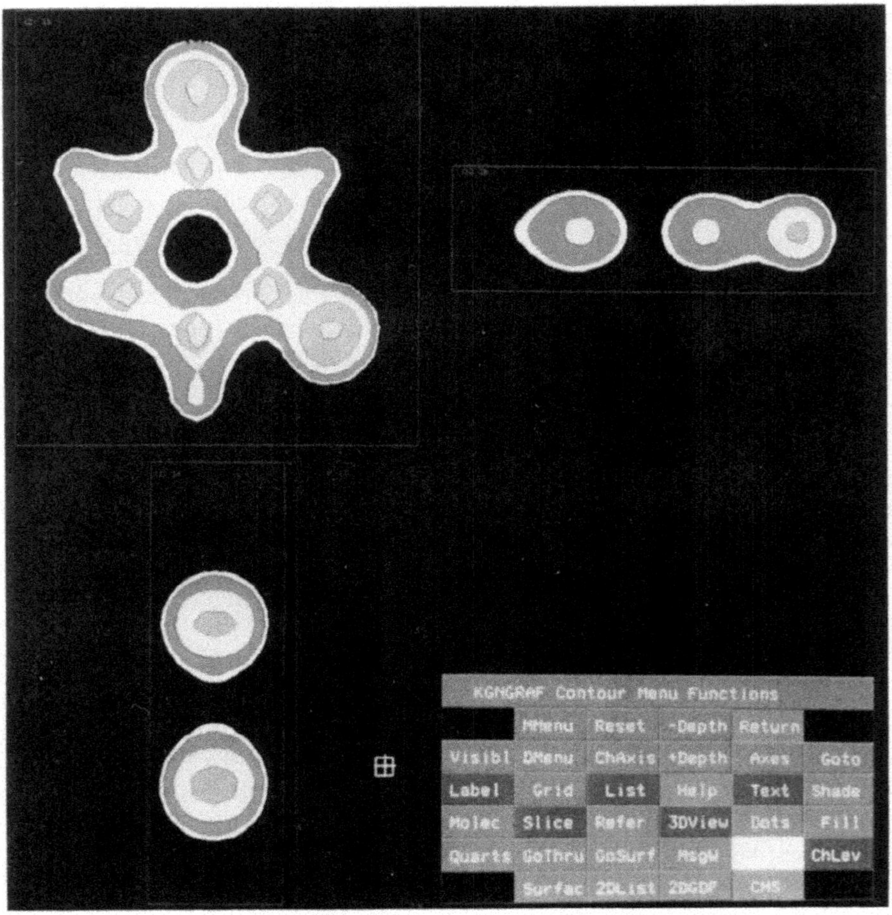

Figure 4. A view along all three axes of a filled two- dimensional contour map for the uracil molecule.

Line contours can be transformed into contours filled with a color by re-arranging the line segments so that a set of continuous line segments comprise each contour level. Once this has been accomplished, the contours can be drawn as filled polygons instead of polylines. If the contours are drawn in sequence, from the most expansive contour to the most compact contour, a two-dimensional map colored by contour level results. The color between any two adjacent contour levels is then determined by the color of the outer contour. A two-dimensional filled contour diagram for the electron density of uracil, as viewed along the X, Y and Z axes simultaneously, is shown in Figure 4. This figure also illustrates the use of multiple windows to display either different views, or properties of a molecule.

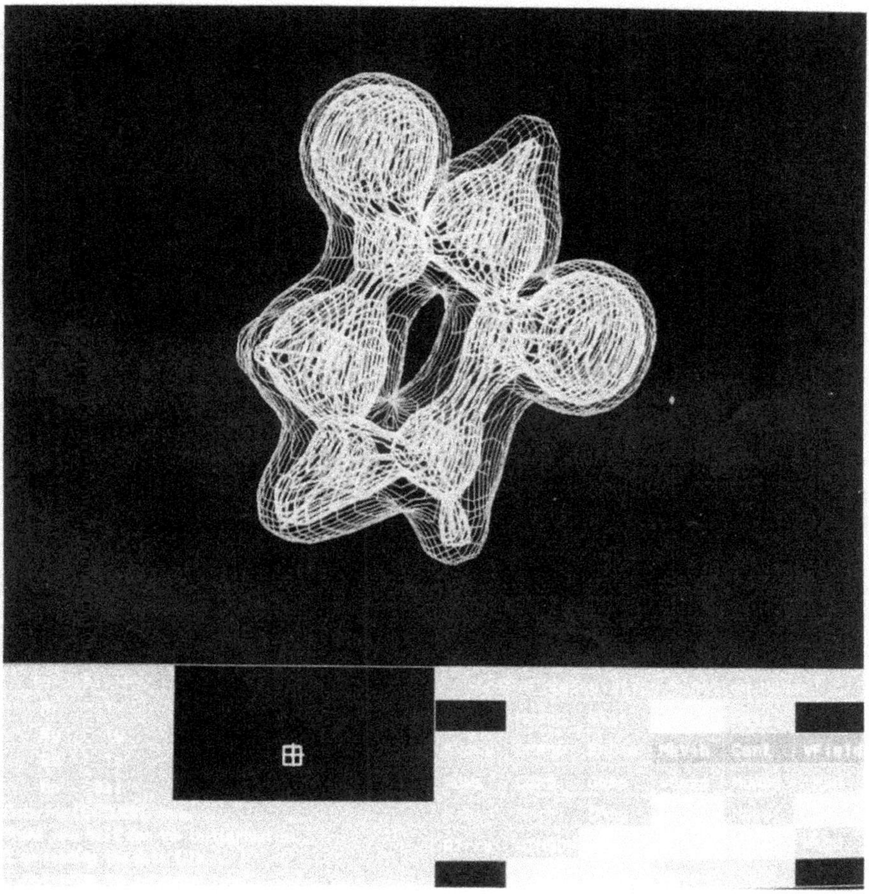

Figure 5. A representation of the slice view of a three-dimensional electron density contour map for the uracil molecule. In this representation all contour levels in all planes parallel to the plane of the molecule are computed and drawn.

The current implementation of contour plotting within KGNGRAF allows the display of both two- and three-dimensional contours. In the 2D mode, several different contour levels can be simultaneously displayed. In the 3D mode, the presence of more than a single contour value complicates the picture, and therefore the user is advised to try to keep the number of contours displayed at any given time to a minimum, though any number of contour values can be simultaneously displayed.

In both modes, a certain degree of real-time interaction with the display is provided. As previously mentioned, 2D contours are constructed from single planes in a 3D grid. However, the user can change planes by simply entering a new plane value from the terminal, or the program can be made to automatically go through all planes sequentially, from the lowest plane to the highest. It is also possible to view the contours along either the X, Y or Z axis, or along all three axes simultaneously. In addition, the user has the ability to interactively modify a contour value through the use of a dial or by entering the value directly. When this is being done, the contour level is re-computed in real-time and subsequently displayed. Of course scaling and translation, as well as a flexible labelling mechanism, is also provided. The possibility to rotate 3D contours about the X, Y or Z axis is also provided.

An additional feature that is available and complementary to the display of three-dimensional contours, is the ability to view all of the contour levels in all of the planes perpendicular to a particular axis, which we refer to as a slice representation. This representation is not a true 3D contour because only contour levels parallel to a given plane are drawn. However, due to the fact that all contours in all parallel planes are drawn, the representation does take on a 3D shape. It has the advantage over the true 3D representation in that it allows the user to view multiple contour values in a pseudo-three-dimensional representation. The user is also given the ability to selectively hide or display particular planes. This can be used for example to create a cutting plane through the middle of the molecule, and superimposing a representation of the molecule in this plane. An example of a three-dimensional contour map drawn with polylines is given in Figure 5. A similar representation, this time as a full three-dimensional solid model whose interior contours are exposed through the use of a cutting plane through the plane of the molecule, is shown in Figure 6.

The solid contour model is constructed on a 3D graphics workstation by transforming the given contour surfaces to polygonal data, and allowing the specialized hardware of the graphics workstation to render the surface. Algorithms to decompose a grid of points into polygons lying on a surface have been shown to be quite general, and easily programmable. The particular implementation used in KGNGRAF is a cell-based method for constructing polyhedral approximations, borrowed from Doi et al.[23] The details of this algorithm will be briefly summarized below.

Linear interpolation of the function inside tetrahedrons spanned by grid points is applied. Each equivalued surface within a particular tetrahedron forms a polygon. Since the cells or tetrahedrons are constructed to have adjacent edges, polygons from different cells are connected and form the polyhedral approximation to the surface. Once the vertices and the normal vectors for the polygons have been computed, this data can then be used by most 3D graphics workstations to render shaded images of surfaces.

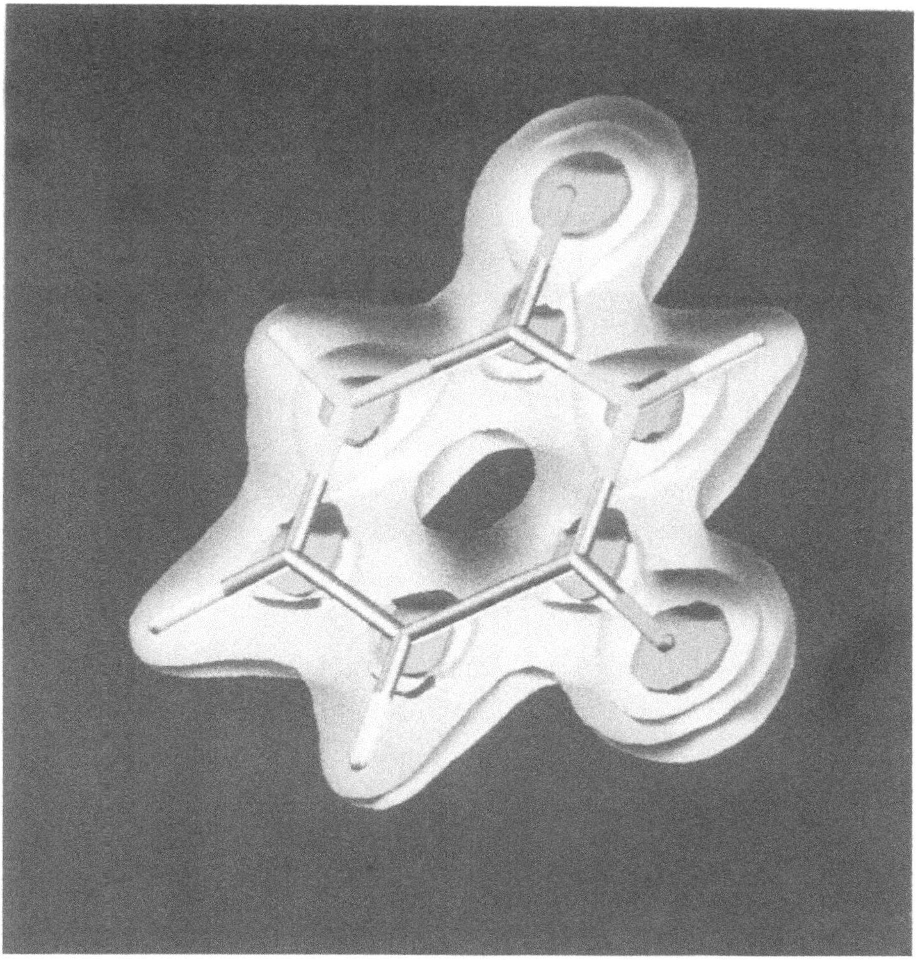

Figure 6. A three-dimensional solid model representation of the electron density of the uracil molecule. Several different density values are plotted, and a cutting plane in the plane of the molecule is used to slice the density in half.

KGNGRAF also takes advantage of transparency features of 3D graphics workstations to provide additional three dimensional information. A typical example would be a molecular structure surrounded by a transparent iso-contour surface of electron density. In this way the underlying physical structure can be seen together with the molecule's electronic structure.

A final feature that is useful for interpretation and visualization of contour data, is the ability to view the values of a function at each grid point in a particular plane. This is commonly referred to as plots of F(X,Y) as parameters of (X,Y). This type of plot exhibits the variations in the actual value of the function over a particular spatial range. By then moving through different spatial regions, animations of the variation in the function values can be generated. This is useful for visualizing potential energy surfaces or other variations in function through a spatial region. An example using the electron density of uracil is shown in Figure 7.

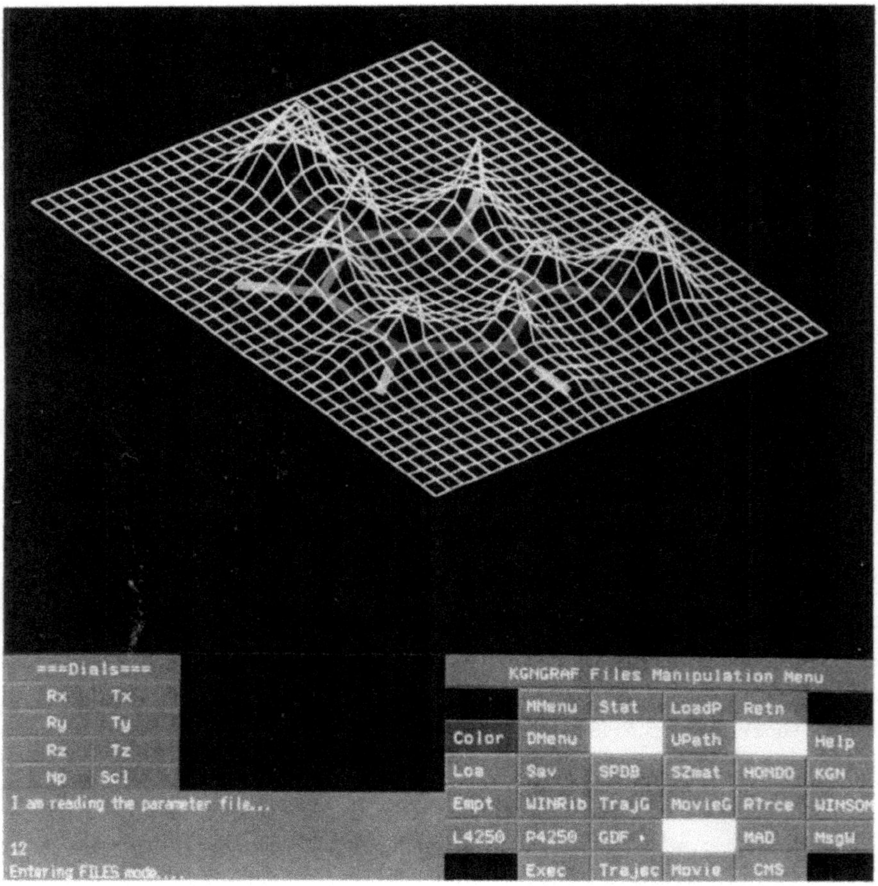

Figure 7. A surface representation of the electron density values in a given plane of the uracil molecule. The grid represents the actual values of the density in the plane.

Creation and Manipulation of Molecules

One of the advantages in using an interactive graphics system is the ability to manipulate molecular structures, either in the building process, or by altering already built molecules. By combining computer graphics techniques with geometrical model building, KGNGRAF provides the user with a much simpler and clearer understanding of the relationship between standard geometrical measurements such as bond distances, valence angles and torsions with resulting structures. It also provides simple methods for viewing a molecule from different angles and orientations, for transforming from one coordinate system to another (for example from cartesian coordinates to internal coordinates), for altering parameters and seeing the resulting structures, or even for building molecules without any prior knowledge of existing coordinates. Each of these issues will be briefly addressed below.

The global translation, rotation, scaling and zooming of a molecule is important towards understanding its three-dimensional structure. For many years chemists have been trying to understand the spatial three-dimensional relationships in a molecule by building real models of molecules using plastic balls and sticks. Holding a model and looking at it from different angles and views creates a succession of pictures that creates in the mind's eye the true shape of the molecule. A computer graphics program that lets a user rotate and translate a molecule in real time offers these same capabilities, without the problems of the models degrading with time, and with the advantage that the models can usually be easily changed through the use of programming techniques.

Other added advantages to the use of a computer graphics program, such as KGNGRAF, is that it is relatively simple to program both global transformations for rotation or translation on all molecules on the graphics screen, or to limit the transformations to only one or several particular molecules, keeping the remaining ones fixed. This is useful for investigating chemical phenomena such as interactions of two molecules or docking of molecules. In addition, the data within a computer graphics program can always be made available to the user in numerical format, offering the possibility of changing parameters as desired (for example molecular geometries).

A flexible system for performing global transformations on several molecules simultaneously, or on individual molecules (keeping the remaining molecules fixed) is important for investigating the above phenomena. To include this feature in KGNGRAF a transformation matrix is inserted at the top of each molecular structure. By selectively modifying this matrix in either all structures or only in individual structures, it is possible to rotate and/or translate several molecules at once, or one molecule individually.[24]

In general, most molecular structures are expressed either as a set of cartesian coordinates for each atom in the molecule, or as a set of internal coordinates.

The internal coordinates of a molecule are usually assumed to consist of bond distances, bond angles and torsion angles. Bond distances are defined as the distance between two bonded atoms. A bond angle is defined as the angle between three consecutively bonded atoms, while the torsion angle is the angle between the two planes defined by four bonded atoms. A regular torsion consists of four consecutively bonded atoms, while an out-of-plane wagging torsion consists of three consecutively bonded atoms with the fourth atom attached to the middle atom of the three.[25]

Internal coordinates offer a more chemically intuitive representation of molecular structure, and have been adopted for use in many programs dealing with computations of molecules. However, the definition of a complete set of internal coordinates to describe the orientations of the atoms within a molecule is not unique and it is possible to introduce redundant coordinates, or to over-specify the definition of the geometry.[25]

Graphics applications require cartesian coordinates to display the molecule. All computational methods require the use of cartesian coordinates, though the input may be specified in internal coordinates. Cartesian coordinates are also the type of data that is usually derived from X-ray crystallography, one of the major experimental techniques for investigating molecular structure.

The transformation between internal and cartesian coordinates is most commonly expressed in terms of a Z matrix. The Z matrix formalism, or similar methods, are quite common in many computational chemistry programs. Basically, the Z matrix requires a length, an angle and a torsion value to specify the position of an atom in space. These parameters must be specified in terms of atoms whose positions are already defined. Naturally, this leads to the first three atoms in the molecule being defined in a slightly differently fashion from the remaining atoms. The first atom is usually chosen to be at the origin, and no parameters are necessary to specify its position. The second atom defines the positive X-axis, and its position is defined with just a distance from the origin, or the first atom. The third atom defines the X-Y plane; a distance from either of the two previous atoms, as well as the angle between all three atoms are needed to define its position.

Using these definitions for the positions of the first three atoms, the remaining atoms can be defined in terms of their relative position to already defined atoms; hence the term internal coordinates. For the purposes of nomenclature we assign the labels I,J,K and L to the four indices defining a Z matrix element. Label I refers to the atom whose position is being defined. The distance from atom I to J is then denoted by the first entry in the Z matrix. The angle between atoms I-J-K is the second entry, and the torsion between atoms I-J-K-L is the third entry.

Given a Z matrix specification, it is possible to alter the internal coordinates of a molecule by changing the proper Z matrix elements. For modification of molecular geometries, this procedure is more chemically intuitive than cartesian coordinates. For example, it is more common for a chemist to understand that a particular bond distance for a C-C bond is too long, than to be able to determine the same information from cartesian coordinates. Interactive manipulation of Z matrix elements, with the corresponding geometrical changes reflected on the graphics screen is one method to manipulate molecular structures, and to immediately determine if the resultant structure is consistent with the expected results.

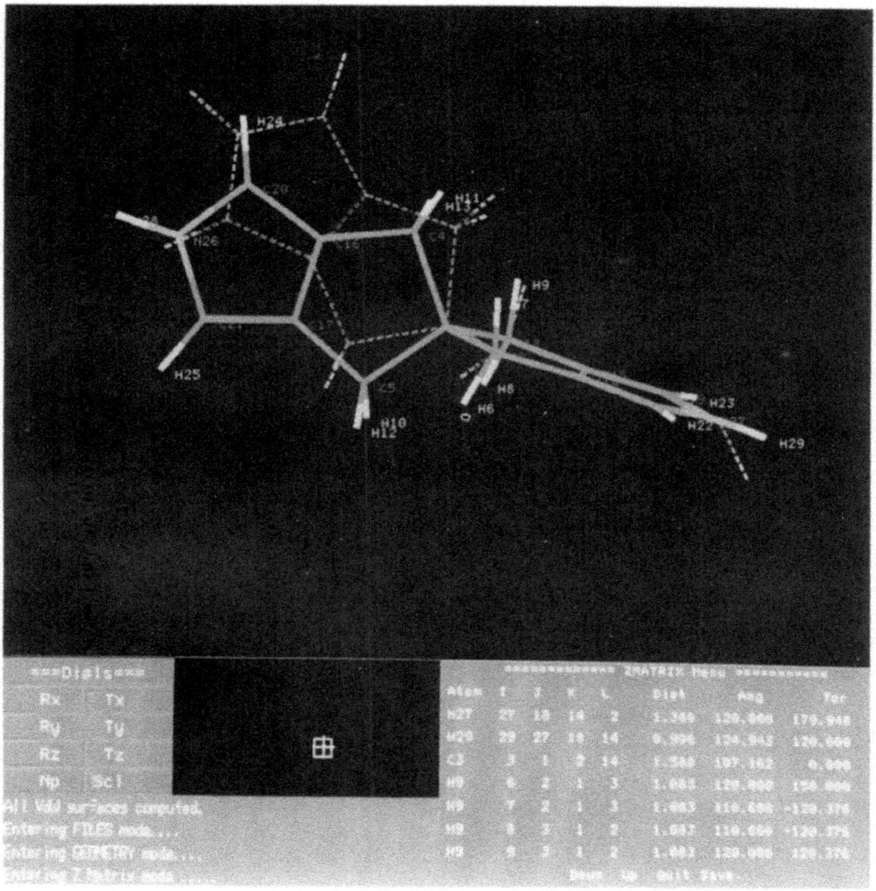

Figure 8. Example of interactive manipulation of the Z Matrix elements for a molecule. Zmatrix values are modified by changing the elements in the menu at the bottom right part of the screen. A dotted new structure is superimposed on the original structure.

Z matrix elements to describe the internal geometry of a molecule can be entered to KGNGRAF by the user in one of two ways: the first way is in a straightforward input file; the second way is a bit more complicated, and involves a transformation from user defined cartesian coordinate input to Z matrix format. As previously mentioned, this transformation is not unique, and therefore any general algorithm to build the transformation must rely on its ability to make certain assumptions. The major assumptions that must be made involve the ordering of the atoms and the choice of atoms to define a given atom's length, angle and torsion value, or J, K and L values for a particular I.

The algorithm used here makes these assumptions based on the internal connectivity table that is determined by the program. The first atom in the Z matrix is always assumed to be the first atom in the user input file and will not have its Z matrix elements individually specified. If we then assign each remaining atom in the input file a sequential number according to its position in the file, then the second entry in the Z matrix will be that atom which has the lowest number and is bonded to the first atom. The third Z matrix entry will then become the lowest numbered atom bound to either the first or second Z matrix entry.

For the remaining N-3 atoms, the atom numbers defining the bond length, angle and torsion (J, K and L values) to be used in their Z matrix definition, are determined by sequentially searching the connectivity table in ascending order. In this fashion, the majority of Z matrix elements are defined in terms of atoms which appear before themselves in the input file. Of course this will not always be the case (it is dependent upon the ordering of atoms in the input file) and in fact, it is possible that a particular Z matrix entry will refer to atom numbers that have not yet been defined in previous Z matrix entries. By the definition of the Z matrix (each atom must be defined in terms of atoms already defined), this particular atom's position cannot be determined. Therefore, this particular element must be placed as a lower entry in the Z matrix, and another element which utilizes only previously defined atoms moved into its place. It should also be noted here, that given the method adopted for assigning the (I,J,K,L) definitions, it may not be possible to replace this element with one whose (I,J,K,L) entries are all defined. In this case it then becomes necessary to modify the (I,J,K,L) entries for a particular atom to ensure that it can be defined. This is done automatically by the program, but as discussed later, can also be interactively specified by the user.

Once a Z matrix transformation has been properly defined, either in an input file or according to the above procedure, this information is passed to an interactive graphics screen which provides the user with the possibility of modifying Z matrix elements. The resultant changes in geometry are then computed and displayed on the screen along with the original geometry. The user may modify as many, or as few, of the Z matrix elements as desired. Each modification is reflected on the graphics screen, each time in comparison to the original structure. The user may then choose to exit and save the modified structure, or simply quit and return to the original structure. A sample screen from KGNGRAF, showing the interactive manipulation of the Z matrix elements is given in Figure 8.

Quite often the manipulation of the internal geometry of a molecule is limited to a small region of the molecule. In this case it is usually preferable to be able to specify a particular definition for some of the (I,J,K,L) values of some of the Z matrix definitions, for certain atoms within the molecule. Therefore, interactive modification and fixing of particular Z matrix elements is allowed. This is accomplished by giving the user the option of specifying the definition of the Z matrix elements for a particular atom in terms of the J,K and L values. This definition for this particular atom, I, is then fixed, and the resulting Z matrix elements are computed assuming these fixed value(s).

A modification of the Z matrix technique involves the rotation or stretching of bonds. This type of structure modification is quite common and ideally suited for an interactive graphics environment such as KGNGRAF. The user must choose the bond to rotate around or stretch by clicking on its two constituent atoms with the mouse. A transformation is then applied to define the chosen bond to lie along the X-axis. This is a two-step transformation: first, a rotation around the Z-axis to align the bond parallel to the X-axis, and then a rotation around the Y-axis to project the bond into the X-Y plane. A tree search on the connectivity table for the molecule is then done to determine which atoms are connected to the bond and will be modified as a result of the rotation and/or translation. The portion of the molecule to be transformed is then highlighted on the graphics screen and compared to the original structure. A dial is programmed to perform a rotation and/or translation around the new X-axis, thereby effecting a transformation of a portion of the molecule while the remainder stays fixed. If the rotation and/or translation is done around the X-axis, then the corresponding motion is a pure bond rotation or stretching. If the rotation is done around one of the remaining orthogonal axes, either angle-bending (Z-axis) or out-of-plane wagging (Y-axis) is performed.

Superposition of Molecular Structures

One aspect of chemistry involves investigating the relationships between molecular and electronic structure, and chemical and physical properties. Often, small changes in molecular structure result in larger changes in electronic structure, and large differences in chemical reactivity. For this reason, it is often useful to look at differences in structures, and to make comparisons between different structures.

In order to properly compare two slightly different molecules, it is necessary to be able to optimally superimpose the two molecules. If the two molecules are not identical, it is desirable to superimpose portions of the molecules which are similar, and to observe the differences between the two. In KGNGRAF this is accomplished by minimizing the sum of the squared distances between identified pairs of atoms. The sum of the squared distances between these pairs of atoms is then minimized by performing translations and rotations of one molecule with respect to the second molecule. By selectively identifying atoms in specific regions of a molecule, particular portions of two molecules may be superimposed. By selecting atoms throughout the molecule, two molecules can be optimally superimposed, whether they have the same number and types of atoms or not.

Molecule Building From Templates

The interactive building of molecules from a set of pre-defined templates is one method of constructing molecules without any *a priori* knowledge of bond distances, valence angles or torsions. It should be emphasized here that the geometries of molecules built in this fashion are by no means definitive, but should only be used to develop a preliminary model for further refinement and study.

A set of templates have been pre-defined in KGNGRAF. However, it is possible for a user to modify, add or delete entries according to their particular application requirements.

The construction of molecules using templates is a four-step process that can be briefly summarized as:

1. Choice of template.
2. Choice of position in molecule to add template.
3. Transformation of template coordinates to the molecular coordinate system.
4. Updating of all necessary information, such as coordinates, bond connectivity, to reflect the addition of the template to the molecule.

Templates are presented to the user in a menu format, where each template is identified by a three-letter code. Upon choosing a menu entry with the mouse, the structure of the template is displayed in a template window. This gives the user the possibility to view the structure of the template and decide if it is the one desired.

The position in the molecule is chosen by clicking the mouse on the atom to be replaced. The template is then added by deleting one atom from the molecule and one atom from the template (in order to maintain correct valencies). The template is positioned by placing the first atom in the template in the space occupied by the deleted atom in the molecule. Some attempt is made to ensure a proper bond length in the building process. This is done by creating the bond between the template and the molecule with specific distances, based on atom type and valency.

The transformation necessary to place the template in the proper orientation for the addition to the molecule is analogous to the transformation for rotation about a bond. The local coordinate system of the templates are first transformed such that they have one bond directed along the X-axis. This is the bond in the template which is used to join it to the molecule. Then the transformation to align the bond in the molecule along the X-axis is computed in the same fashion as for the rotation about a bond which is described above. If we operate on the template with the inverse of this transformation, the template will be placed in the local coordinate system of the molecule and in the proper orientation to be aligned along the given bond. The template can then be added to the molecule by updating all necessary information regarding the atoms in the molecule.

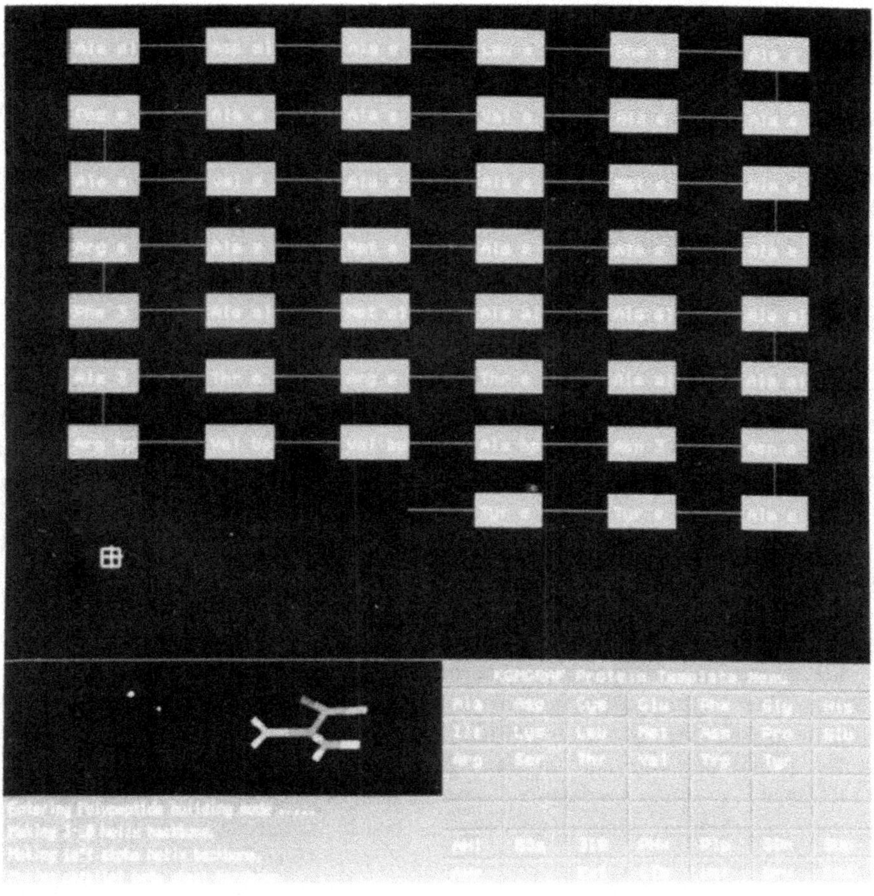

Figure 9. Example of interactive building of polypeptides. As new residues are added to the sequence the peptide editor creates a chain showing the residue names and their backbone configuration. Residues and backbone configurations can be substituted by simply clicking with the mouse.

Several of the default templates are defined only as particular valence states for an atom. For example, templates are given for a linear bond, a planar trigonal atom, a non-planar trigonal, or a tetrahedral atom. These templates are similar to the atoms commonly found in a typical set of molecular model builders. These templates are designed to be used in the same fashion as their corresponding models. The pseudo-atoms that are used in the template definition must be replaced with their appropriate atom types. This is done by either selecting some of the most common atoms from a menu (C,N,O,H), or by entering the atomic number or symbol from the keyboard.

Protein Building From Templates

Templates for the 20 naturally occurring amino acids are also given in KGNGRAF, in a separate menu. By default, the backbone configurations of the amino acids are stored in the extended form and when a polypeptide chain is constructed, it is done in the extended conformation. This default can be changed to other typical backbone conformations, such as alpha helix or beta sheet, or given by the user in terms of (Φ-Ψ) angles.[26] This is done by simply clicking with the mouse on the appropriate configuration in the menu. At the present time, the side-chain dihedral angles cannot be modified during the building process. At present this can only be done by using the Z matrix option or by rotation around a bond directly.

Polypeptide chains are built using the same formalism as adding templates to molecules (described previously). However, the situation is slightly simplified by the fact that the backbone configuration of the amino acid chain is built in a regular fashion; the nitrogen of the previous residue is always attached to the alpha carbon of the next residue. Therefore, there is no choice in how to attach the following residue – the only choice is in what particular configuration the residue should be added.

For this reason, the building of a protein molecule only involves the specification of the amino acid sequence, and the backbone configuration of each amino acid residue. The flexibility of being able to add templates at any position in the molecule, or to fuse rings or add individual atoms, which is important in building molecules, is not necessary for building proteins. The protein building portion of KGNGRAF involves the use of a graphical protein editor. As residues are selected, a graphical chain, describing the residue sequence as well as the particular backbone angles for that residue is displayed. At any time during the building process, a residue may be substituted for another residue, or the backbone dihedral angles for any residue or sequence of residues can be changed. This is accomplished through the graphical editor by simply clicking options with the mouse. At any time during the protein building process, the user can select the option to draw the protein. At this point, the program computes the cartesian coordinates of the atoms in the polypeptide, and displays them, including all hydrogen atoms. The user can then return to the protein editor to continue to build and modify the peptide. An example of using the protein editor to build a polypeptide chain is given in Figure 9.

Inquiry of Geometrical Parameters

A computer graphics environment is ideal for investigating the molecular parameters of molecular structures. The use of the graphics screen displaying the structure and the immediate feedback of the values of distances and angles, provides a user with a complete understanding of the molecular geometry. Interactive selective inquiry is possible by choosing atoms from the screen with the mouse, or if desired, it is also possible to obtain a complete list of all bond distances, valence angles and torsion angles.

KGNGRAF also provides the user with the capability to orient molecules into any desired orientation. For example, it is possible to redefine the X-axis of the molecule by selecting two atoms with the mouse. The first selected atom becomes the new origin, and the second atom defines the positive X direction. It is also possible to define the XY plane of the molecule by choosing three atoms. The first atom defines the origin, the second atom the positive X-axis, and the third atom the XY plane. The user can also put the molecule in the principal cartesian coordinate system, or perform exact rotations about the X, Y or Z axes by entering numerical values of the rotation angles.

Display of Molecular Vibrations and Molecular Spectra

Molecules are often characterized by their spectra. In particular, infrared and Raman spectra of molecules are commonly used to help identify the presence of chemical groups in a molecule. It has also become common to compute infrared and Raman spectra of molecules. These spectra are usually characterized by a set of intensities and frequencies for the normal modes of vibration. Given a set of intensities and frequencies, it is possible to construct molecular spectra by simply assuming a half-width and band shape for each normal mode. In KGNGRAF, a Lorenztian line shape with a user defined half-width describes the vibrational spectra. Both Infrared and Raman spectra can be drawn, with the possibility to interactively modify the half-width, or to zoom in on any particular region of the spectra.

It is also possible to obtain the vectors characterizing the displacements of each atom during the vibrational modes. Given this characterization, it is useful to be able to visualize the normal modes of vibration and identify their principle components (for example C-H stretching, ring breathing or C-H bending). In KGNGRAF this is accomplished by using the eigenvectors defining each normal mode to compute the displacements of the atoms in the molecule for the mode desired. Once these displacements are computed, several images of the molecule at various stages of displacement are generated, and displayed in sequence, thereby depicting visually the molecular vibration. The vibration can be changed by choosing a different vibrational frequency from a menu of frequency values. In addition, a single image of the molecule, with vectors drawn at each atom in the direction of the displacement for that atom and with the length of the vector proportional to the amount of the displacement, can be displayed for any of the vibrational frequencies. A display of one of the normal modes of vibration for the benzene molecule is shown in Figure 10. The computed IR and Raman spectra are also simultaneously displayed.

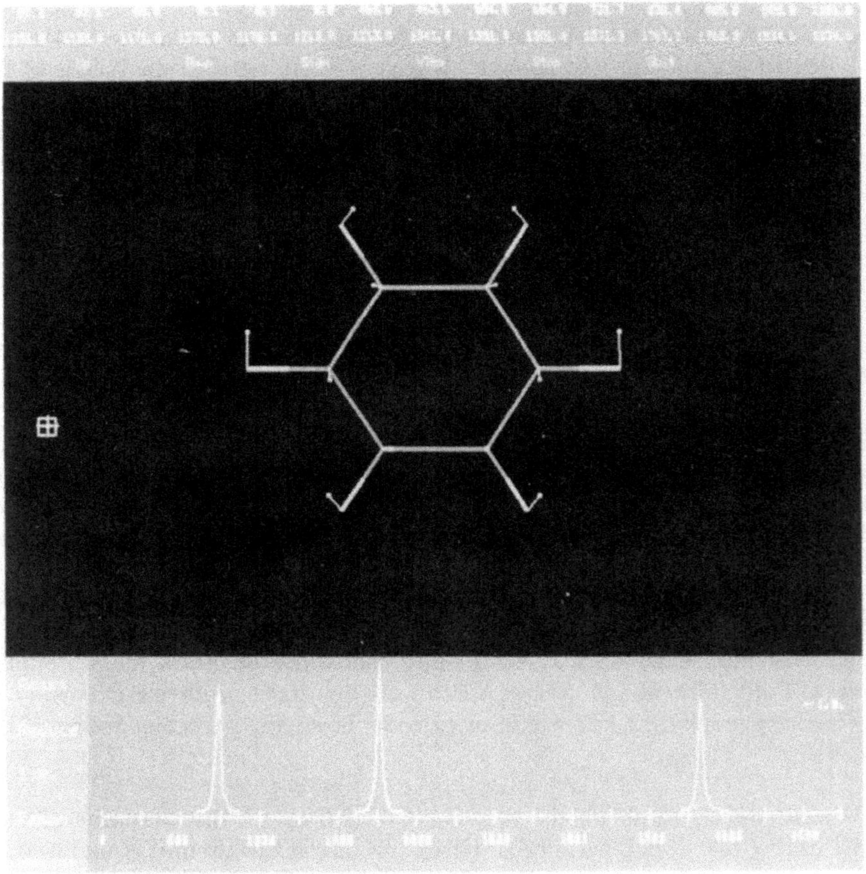

Figure 10. Display of the molecular vibration menu in KGNGRAF. In this mode, the user can view the normal mode displacements as vectors superimposed on the molecule. In addition, it is possible to animate the vibration by selecting "Vibs" with the mouse. The IR and RAMAN spectra are also shown at the bottom of the screen.

Display of Molecular Energy Diagrams

Quantum mechanical computations are used to predict the energy levels of a molecule, both the occupied and unoccupied levels. Traditional energy level diagrams order these energy levels according to energy, with the lowest energy orbitals being at the bottom and occupied first. The relative positions of the orbitals, on a scale which is proportional to the energy values themselves, is often useful in understanding the electronic structure of a molecule. It not only clearly identifies particular energy gaps between different levels, it also clearly depicts the presence or absence of degenerate levels, degenerate splitting, or coupling of energy levels. KGNGRAF draws energy level diagrams from the data typically provided by quantum chemical programs; values of orbital energies. Once drawn, the user can see only occupied or unoccupied orbitals, to zoom in on particular energy regions, to see both open and closed shell systems, and to see the occupation of each orbital. An example for the benzene molecule, as computed with HONDO is shown in Figure 11.

Interactive Animations

Animation of molecular vibrations, or molecular dynamics simulations, provides a clear and concise tool for the interpretation of large amounts of numerical data. Each frame in the animation sequence is constructed from a snapshot in time of the positions of the atoms in the molecule. The frames can then be displayed in a time ordered sequence, one after another, to create a movie. The movie illustrates the spatial relationships among the atoms in the simulations as they move and respond to the forces acting upon them. However, in addition to providing information on the spatial relationships, the movies can also be used to display the formation and breaking of hydrogen bonds or other weak inter-atomic forces. For example, it is interesting to show the formation of hydrogen bonded networks or the importance of water as a solvent. The frames can also be displayed superimposed on each other. In this way, one can easily visualize the amount of displacement that each atom undergoes during the simulation. These pictures also provide detailed information on the amplitude and direction of motion for each atom.

The animations discussed here are intended to be interactive, and used during the scientific discovery phase (as opposed to a post-processing or presentation phase). In this sense, the animations need to be generated quickly and easily so that they may be viewed as the simulation is running. Because the time to generate the images, as well as the amount of storage to hold the images must be kept to a minimum, these animations are usually constructed from relatively simple graphics representations, such as stick figures. This provides the possibility of generating and storing several hundred frames quite easily. Thus, it is possible to view results and change parameters, or input to the simulation as it progresses, which is

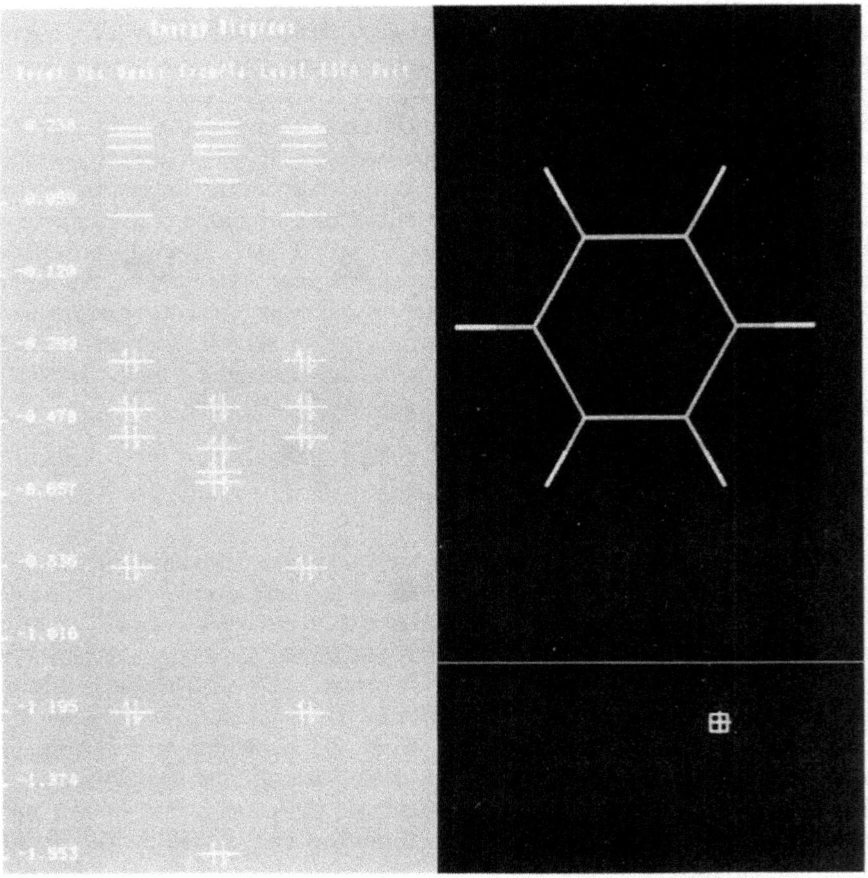

Figure 11. Display of the molecular energy levels for benzene. Core analysis can be displayed by using the option ESCA from the top menu.

the main aim of this type of movie. However, for relatively small systems, in which the accurate rendering of realistic three dimensional images is an advantage, 3D workstations can be used to generate sequences of images for space filling or ball and stick models, including the effects of lighting and shading. These images can then be displayed in sequence, providing animations with realistically rendered images. KGNGRAF provides the facilities for animations of stick figures, of ball and stick and space filling models. On 3D workstations, the ball and stick and space filling models can be animated including the full effects of lighting and shading.

It is also possible to render even more realistic three-dimensional representations, such as those incorporating shadows and reflections, and to animate these images.

However, these images can be quite time consuming to generate and the procedure is usually reserved for a post-processing mode of operation. We have developed an animation system consisting of software, for realistic rendering of molecules and hardware, for recording these images on videotape. We will leave the discussion of this system to a later section of this chapter.

Files Manipulation

Another important feature of most graphics packages, which is not specifically designed for the package, but is more of a by-product of the nature of the program, is the ability to manage and manipulate a variety of file formats. The great variety of computational chemistry programs, as well as the different number of ways to provide input to these programs, makes the creation, manipulation and maintenance of input and output files an arduous task. However, the need to display and manipulate data graphically, both before and after processing by a simulation package, is found in all areas of computational chemistry and in many ways, provides a unifying medium across all disciplines. Since most users of graphics packages will also be users of several different simulation programs, it is necessary to design flexible input and output (in the form of saving data that has been graphically manipulated and modified). Standard input and output format from the most commonly used programs within a particular research environment, should be accepted as input to a computer graphics program without the need for modification by a user. Output from the graphics program should be in the input format for the simulation programs, requiring little if any intervention by a user to run the input. While some of the issues may seem somewhat trivial, they are nevertheless important in providing a user-friendly programming and research environment. In KGNGRAF access to all the computational modules within MOTECC are provided. In this sense, access specifically refers to reading and writing input and output files directly, without any necessary modification from the user. In fact, outputs of all of the modules of MOTECC can be used directly as input for our graphics system, and the graphics system generates standard input files for all of the computational components of MOTECC. These input files can be executed without modification. However, for more flexible specification of parameters a windowing environment has been developed and this will be discussed in more detail later in this chapter.

Towards More Realistic Images

So far, most of our discussions have been centered on providing a highly interactive user-friendly environment for computer graphics. At a certain point it also becomes necessary to think about how to provide even more realistic three-dimensional computer graphics representations of three-dimensional objects on a two-dimensional screen. Several techniques are used to add realism to these two-dimensional images.[27] Among them are:

1. Hidden line removal.
2. Hidden surface removal.
3. Stereoscopic images.
4. Depth cueing.
5. Ray tracing.

Several of these techniques can be implemented in an interactive fashion through the use of specialized hardware or through the use of efficient software algorithms. Of course, as mentioned previously, the former provides higher performance, but the latter has the advantage of being more general, flexible, and applicable to a large variety of both "intelligent" and "non-intelligent" display devices. In KGNGRAF, every effort is made to support complete function in both 2D and 3D workstations. Therefore, in a sense our system is self-configuring according to workstation hardware characteristics. If a 2D workstation is being used, software implementations for hidden surface removal and depth cueing are used. On 3D workstations the program understands that the hardware implementations will not only be more efficient, but also produce more realistic images, and so the hidden surface and depth cueing are performed by the hardware.

In the discussions that follow, the software implementations for hidden surface removal and depth cueing on 2D workstations will be briefly discussed. The implementations on 3D workstations are usually hardware dependent, and will not be discussed.

Hidden surface removal and depth cueing can be efficiently implemented using a depth-sorting algorithm.[28] A depth sorting algorithm works both in the object space as well as the image space, and consists of two parts:

1. Surfaces or objects must be ordered in decreasing depth.
2. Surfaces or objects are scan-converted, starting with the surface of greatest depth and working forward.

The first step is always carried out in the object space, or the actual coordinate system of the objects in question. The second step, is carried out in image space and involves the display of pixels on the graphics screen surface.

The method for hidden surface removal is sometimes referred to as the painter's algorithm, as it is similar to the method that an artist uses to paint a picture. In the first pass, the background is painted on the screen. In the subsequent steps, the surface with the greatest depth is compared with the other surfaces in the list to determine if there are overlaps with closer surfaces. If no overlaps are present, then the surface can be painted on the screen. If there are overlaps, then additional comparisons must be made to determine whether any of the surfaces must become hidden.[29]

Of course, the algorithm cannot deal with only entire surfaces, but is complicated by the fact that it must be able to handle portions of surfaces. However, the algorithm can be modified for molecular structures and made to be quite efficient. If each atom is treated as a sphere and if the atoms are drawn in order of increasing depth value, then the hidden surfaces are automatically removed, and no surface comparisons are necessary. It should be noted here that using this technique, as rotational transformations are performed, the atoms must be re-ordered and re-drawn for each transformation. The key components of this algorithm thus are a fast sorting algorithm coupled with an efficient procedure for drawing spheres, or circles at atomic positions.

This algorithm can be easily modified for an efficient implementation of intensity or depth cueing. Again, a two-step procedure is necessary:

1. Objects must be sorted in order of decreasing depth.
2. Objects are drawn from the greatest depth and moving forward, assigning color to each object based on its depth value.

In the case of molecular structures, the objects are usually composed of lines representing chemical bonds. A more realistic three-dimensional image is obtained by dividing each bond into a number of segments, where each segment is treated as an object in the above two steps and assigned a color based on its depth level. For most molecules, dividing each bond into four segments appears to give the best compromise between smoothness of color and efficiency.

The use of stereoscopic images is also quite common in molecular graphics applications. It involves the display of two slightly rotated views simultaneously, where the views are rotated between $5-8°$ to account for the separation between the two eyes of a person. By focusing each eye only on its corresponding view and then superimposing the two views, a view with three-dimensional perspective is created. Often, it is useful to use stereoscopic viewers to decouple the eyes and to force each eye only to see its corresponding view.

Ray Tracing for Molecules

Ray tracing techniques[27,30,31] are more time-consuming than any of the techniques previously described, but these images are by far the most realistic and include the effects of shading, shadows, transparency and reflections. The realistic appearance of these images can be a valuable aid in the perception of the three-dimensional shape and the spatial distribution of the objects represented on a two-dimensional graphics screen. This is especially true for large, complicated molecular systems, or in molecular dynamics simulations composed of many independent molecules, where the molecules can appear to float in space without any apparent connections to each other, or other points in space. In these situations, the shadows and reflections of the molecules on one another can provide visual clues to the three-dimensional relationships between molecules.

The ray tracing technique can be briefly summarized[27-29, 31] by recalling that a graphics image can be considered to be composed of a large number of dots, each assigned a particular color based on surface characteristics of the objects, lighting characteristics of the screen, etc. Of course, the greater the number of dots (or pixels) the greater the resolution of the image. In principle, an infinite number of pixels could be generated over the various surfaces in an image. In practice, this number is usually limited to around 1,000,000 pixels. In the ray tracing technique, the color of each pixel is determined by tracing rays backwards from the viewing position through the pixel and back to the light source.

Starting from the viewing position, a ray is passed through each pixel in the viewing plane and then traced back into the objects defined in the image. The ray can then either strike one of the objects in the scene, or pass through the scene. If the ray strikes an object, this ray can then be reflected and refracted, generating several new child rays. The child rays are then traced until the rays end at a light source or pass out of the scene. The intensity of the corresponding pixel is then set by determining all of the contributions from the parent and all child rays. Usually the number of generations of child rays that can be generated are limited, because after several generations, the contributions become negligible. However, the amount of computation is still quite large and requires considerable time. At the present time, ray tracing is used to generate static images, which can then be animated on videotape using the animation system described in the next section.

The ray tracing algorithm that we have implemented consists of five principal processes:[31]

1. Definition of the image.
2. Calculation of ray trajectories.
3. Evaluation of picture elements.
4. Anti-aliasing.
5. Recording the image.

Each of these processes is described in more detail below. The first process and the last process are each performed only once for each image, while steps 2, 3, and 4 must be repeated for each of the pixels which form the image. The fourth process, anti-aliasing, is based on additional repetitions of processes 2 and 3.

Definition of the Image

An image is defined by three types of information:

1. The geometry of the scene.
2. The optical characteristics of the objects in the scene.
3. How the scene is illuminated and viewed.

The data required to specify this information is described in the following sections.

Geometry

The geometry of the scene is defined by the size, shape, and position of each object in the scene. The shape of an object is determined by one or more basic geometric entities called primitives. The ray tracing program used in this work, which is specialized for molecular modelling applications, recognizes only one type of primitive: the sphere.

The geometry of the scene is specified by defining each object within the scene. Each object is defined by specifying the type of primitive (i.e., sphere), the characteristic parameters (radius and center), and the optical characteristics of the object.

Optical Characteristics

The optical characteristics of each object are defined by the ability of the object to transmit and reflect light. These characteristics are determined by a number of factors including:

1. Transparency coefficients.
2. Coefficients of diffuse reflection.
3. Coefficients of specular reflection.
4. Coefficients for reflection of ambient light.

The effects of these factors are modelled by a set of corresponding parameters which are defined below, in the section on the evaluation of picture elements. Each of these factors depends on the wavelength of the transmitted or reflected light, and this wavelength dependence is responsible for the observed colors associated with each object. The current implementation treats the wavelength dependence of these factors by assigning three coefficients to each factor, corresponding to the values of the red, green, and blue characteristics of light.

If a surface is opaque, the transparency coefficients are all zero, and no refracted rays are produced by the surface. Otherwise, the surface may produce both reflected and refracted rays, and it is necessary to specify an index of refraction to calculate the paths of the refracted rays. In principle, the index of refraction also depends on the wavelength of the refracted light. However, for simplicity we assume that the index of refraction does not depend on wavelength.

Viewing and Illumination

The image produced for a particular scene depends on a number of conditions beyond the definitions of the objects in the scene. These include the location of the observer, the direction of viewing, the size of the field of view, the number of lights, and the location, color, intensity and concentration of each of the lights.

The image also depends on the number of distinct picture elements (pixels) used to cover the field of view, and the treatment of aliasing. Aliasing includes a number of artifacts, including stair steps on diagonal lines, caused by using a single point to determine the color of a pixel. Measures used to reduce or eliminate aliasing are called anti-aliasing. The anti-aliasing measures used in the current algorithm are described below.

Calculation of Ray Trajectories

Ray trajectories are calculated by following lines from the observer to the objects in the scene and from the objects to the sources of illumination. Each ray is defined by a point of origin and a unit vector which specifies the direction of the ray. A set of one or more rays is calculated for each element (pixel) of the image being rendered. These pixels form a rectangular grid, on a plane perpendicular to a line, from the eye point of the viewer, to the "gaze point" which defines the direction of view.

The first ray calculated for each pixel always has its origin at the eye point of the viewer. The direction of this ray is defined by the line from the eye point through the center of the pixel. Extension of this line through the scene leads to one of the following results:

1. The ray misses the entire scene.
2. The ray passes through the scene, but does not intersect any object within the scene.
3. The ray intersects an opaque object.
4. The ray intersects a transparent object.

Each of these conditions is illustrated in Figure 12.

If the ray misses the scene (ray 'a' in Figure 12), it is marked as a "background ray," and the ray trajectory is finished. If the ray passes through the scene, it must

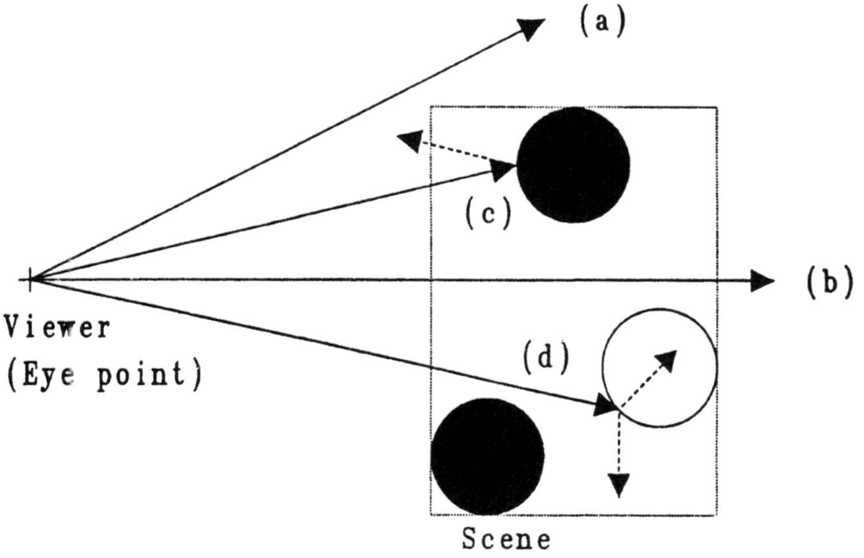

Figure 12. Examples of the possible paths that a ray may take. Initial rays originate at the viewer's eye. Ray (a) misses the entire scene, ray (b) passes through the scene without striking any object, ray (c) strikes an opaque object, and ray (d) strikes a transparent object.

be tested to determine whether it hits any object within the scene. If the ray passes through the scene without hitting any objects within the scene (ray 'b' in Figure 12), it is also marked as a "background ray," and the ray trajectory is finished.

If the ray hits one or more objects, the intersection point closest to the origin of the ray is recorded, and one or two new rays are created. The new rays each originate at the intersection point marking the end of the previous ray. If the object is not opaque, one of the new rays will be either a refracted ray, or a total internal reflection ray, depending on the index of refraction and the angle at which the ray intersects the surface. If the object is opaque, or the first ray is a refracted ray, a reflected ray is created. Each of these rays is then analyzed in the same manner until the maximum of generations of rays have been generated. The first three generations of rays arising from one initial ray are shown in Figure 13.

Evaluation of Picture Elements

The contribution of each pixel to an image is defined by the red, green, and blue intensity values of the pixel. These values are determined by the intensities of the corresponding components of one or more rays starting at the viewer and passing through the pixel. In the absence of anti-aliasing, the values for each pixel are determined by a single ray passing through the center of the pixel. With anti-aliasing, each pixel is divided into sub-pixels, as indicated in the following section.

If a ray starts from the viewer and passes through the scene without striking any objects, then the red, blue, and green intensity values of the ray are determined by the background values defined for the scene. Otherwise, the intensity values of a ray are determined by an illumination model similar to those proposed by Whitted[30] and Phong.[27-29] In this case, the intensity of each component (red, green, and blue) is given by:

$$I(\mathbf{v}) = I_d(\mathbf{v}) + I_r(\mathbf{v}) + I_a,$$

where:

I_d	is the contribution from direct illumination,
I_r	is the contribution from indirect (reflected or refracted) illumination,
I_a	is the contribution from ambient illumination, and
$\mathbf{v}$	is a unit vector in the direction of the ray,

Direct illumination represents light coming directly from a source to the point where the ray strikes the scene. Indirect illumination represents light reflected or

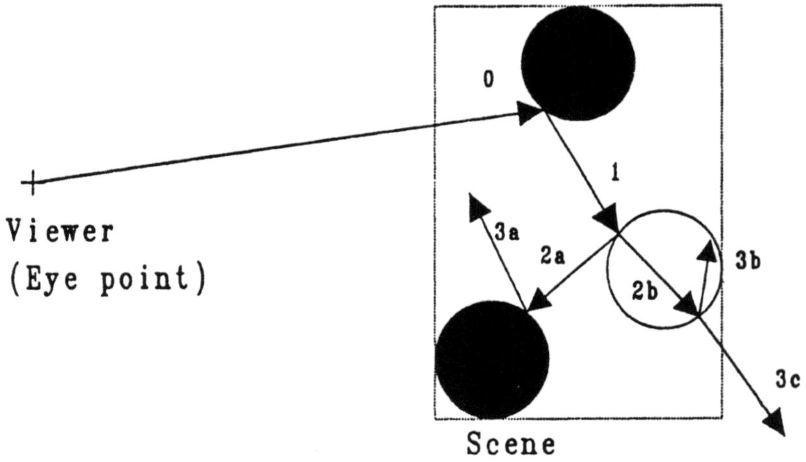

Figure 13. Generation of rays from an initial ray. The three generations of rays include ray (0) as the initial ray, ray (1) as a reflection of ray (0), rays (2a) and (2b) as reflections and refractions of ray (1), and rays (3a), (3b) and (3c) as reflections of rays (2a) and (2b), and a refraction of ray (2b), respectively.

refracted by other objects before reaching this point. Ambient illumination represents a sort of diffuse indirect illumination. Each of these contributions is described below.

Direct Illumination

The contribution of direct illumination is given by:

$$I_d(\mathbf{v}) = \sum_{i=1}^{n_l} [k_d(\mathbf{n} \cdot \mathbf{L}_i) + k_s(\mathbf{v} \cdot \mathbf{L}'_i)^{n_s}] I_i / d_i^2$$

where:

n_l	is the number of sources of illumination for which there is a direct path to the point where this ray strikes the scene,
I_i	is the intensity of the i-th source,
d_i	is the distance from the i-th source to the point where this ray strikes the scene,
k_d	is the coefficient of diffuse reflection,
k_s	is the coefficient of specular reflection,
n_s	is the specular exponent for the surface of the object where the ray strikes the scene,
$\mathbf{v}$	is a unit vector in the direction of the ray,
$\mathbf{L}_i$	is a unit vector pointing from the i-th source to the point where this ray strikes the scene, and
$\mathbf{L}'_i$	is the (reverse) reflection of $\mathbf{L}_i$ on $\mathbf{n}$:$[2\mathbf{n}(\mathbf{n} \cdot \mathbf{L}_i) - \mathbf{L}_i]$.

The relationships between the vectors $\mathbf{v}$, $\mathbf{L}_i$, and $\mathbf{L}'_i$ are shown in Figure 14. The values of k_d, k_s, and I_i depend on the color (red, green, and blue) of the ray. The value of n_s is determined by the texture of the surface, and does not depend on the color of the ray.

Light which may travel from a source of illumination, to the surface of an object, by path through a transparent surface, is ignored, because it is very difficult to determine the path(s) of such illumination. Consequently, all transparent objects cast shadows, and any surface in the interior of a transparent object will be dark unless a source of light is placed inside the transparent object.

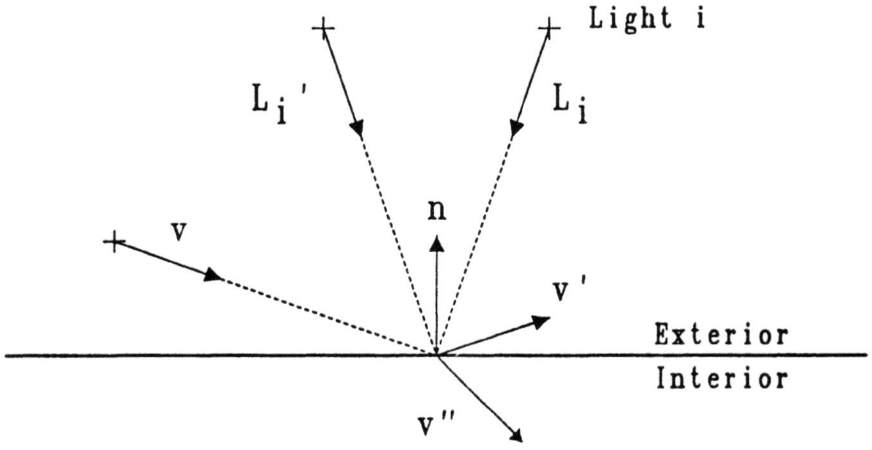

Figure 14. Geometry of the intersection of a ray with a surface with normal vector **n**. The vector **v** points away from the viewer or the previous surface, **v′** indicates the reflected ray, and **v″** indicates the refracted ray.

Indirect Illumination

Indirect illumination includes the effects of reflection and refraction. If a ray strikes an opaque object, then this contribution is given by:

$$I_r(\mathbf{v}) = k_s I(\mathbf{v}')$$

where:

k_s	is the coefficient of specular reflection,
$\mathbf{v}$	is a unit vector in the direction of the ray,
$\mathbf{v}'$	is the reflection of $\mathbf{v}$ on $\mathbf{n}$, that is $\mathbf{v} = 2\mathbf{n}(\mathbf{n} \cdot \mathbf{v})$, and
$\mathbf{n}$	is a unit vector perpendicular to the surface where the ray strikes the scene.

The relationships between these vectors are shown in Figure 14.

If the object is not opaque, and total internal reflection does not occur, then the intensity of indirect illumination is given by:

$$I_r(\mathbf{v}) = k_s I(\mathbf{v}') + k_t I(\mathbf{v}'')$$

where:

k_t is the transmission coefficient, and

$\mathbf{v}''$ is a unit vector in the direction of a refracted ray.

If total internal reflection occurs, then the intensity of indirect illumination is given by:

$$I_r(\mathbf{v}) = 0.98 I(\mathbf{v}').$$

Each of these contributions is determined by generating one or two new rays propagating in directions $\mathbf{v}'$ and/or $\mathbf{v}''$, determined by optical physics and the geometry of the intersection of the parent ray with the scene. If either of these rays strikes another object, the intensity $I(\mathrm{v}')$ or $I(\mathrm{v}'')$ is calculated in the same manner used to determine the intensity of the parent ray, $I(\mathbf{v})$. This process is repeated recursively, until either the reflected and refracted rays all leave the scene, or the maximum number of generations of new rays are generated.

If a reflected ray does not intersect an object within the scene, it does not contribute to the color of the parent ray. If a refracted ray does not intersect an object in the scene, it is assigned the background color. This allows the background to be seen through transparent objects, but prevents the background from being reflected on the surfaces of objects.

Ambient Illumination

Ambient illumination is a type of indirect illumination represented by a soft glow uniformly pervading all objects in the scene. The contribution of ambient illumination to each color (red, green, and blue) is determined by the product of the ambient light intensity and the coefficient of reflection for ambient light. The ambient light intensity is a property of the entire scene, and the coefficient of reflection for ambient light is a property of the surface at the point of intersection. This value is independent of the orientation of the surface and the direction of the ray.

The principal effects of ambient illumination are to soften shadows and to provide a minimum level of illumination to all visible portions of the scene. Excessive levels of ambient illumination tend to reduce the contrast between light and dark portions of each object, making curved surfaces appear flat. The level of ambient illumination should be kept very low to preserve the three-dimensional appearance of curved surfaces such as spheres.

Adaptive Anti-Aliasing

Anti-aliasing is the process used to eliminate the jagged or stair-step appearance of diagonal lines and the edges of curved surfaces. In the absence of anti-aliasing, one ray passes through the center of each pixel, and the color determined for this ray is assigned to the entire pixel. This causes abrupt changes in color near the edges of objects, and produces the stair-step effect seen at diagonal boundaries.

The stair-step effect (aliasing) may be minimized by dividing each picture element (pixel) into sub-pixels. The color of each sub-pixel is calculated by sending a ray through the center of the sub-pixel. The average of the colors determined for the sub-pixels is then assigned to the entire pixel. As a result, the color of each pixel represents a composite of the various objects which can be seen within the pixel. Anti-aliasing is most important at the edges of objects and other places where the colors of adjacent pixels change abruptly or rapidly. Anti-aliasing is not required in portions of an image where the colors of adjacent pixels are constant (as in the background), or slowly varying.

The anti-aliasing algorithm can be briefly summarized as follows: each pixel is compared to its eight adjacent pixels. These nine pixels form a 3×3 array, with the current pixel at the center. If the red, green, or blue color values of the current pixel differ from the corresponding values of any of the neighboring pixels by more than a fixed threshold, then the current pixel is divided into four sub-pixels and the color of each sub-pixel is calculated. If the color of any of the four sub-pixels differs from the original color of the current pixel by more than a fixed threshold, then each sub-pixel is subdivided into four smaller sub-pixels, and the color of each sub-pixel is replaced by the average of the four smaller sub-pixels. The color of the current pixel is then replaced by the average of the first four sub-pixels.

An example of the types of images that can be generated using the ray tracing technique is shown in Figure 15. This Figure shows a ray traced picture of a space filling representation of several water molecules surrounding a lithium ion. In this example the effect of a transparent shell, as well as the effects of shadows and reflections greatly enhance both the quality of the image as well as the spatial information that is available from the image.

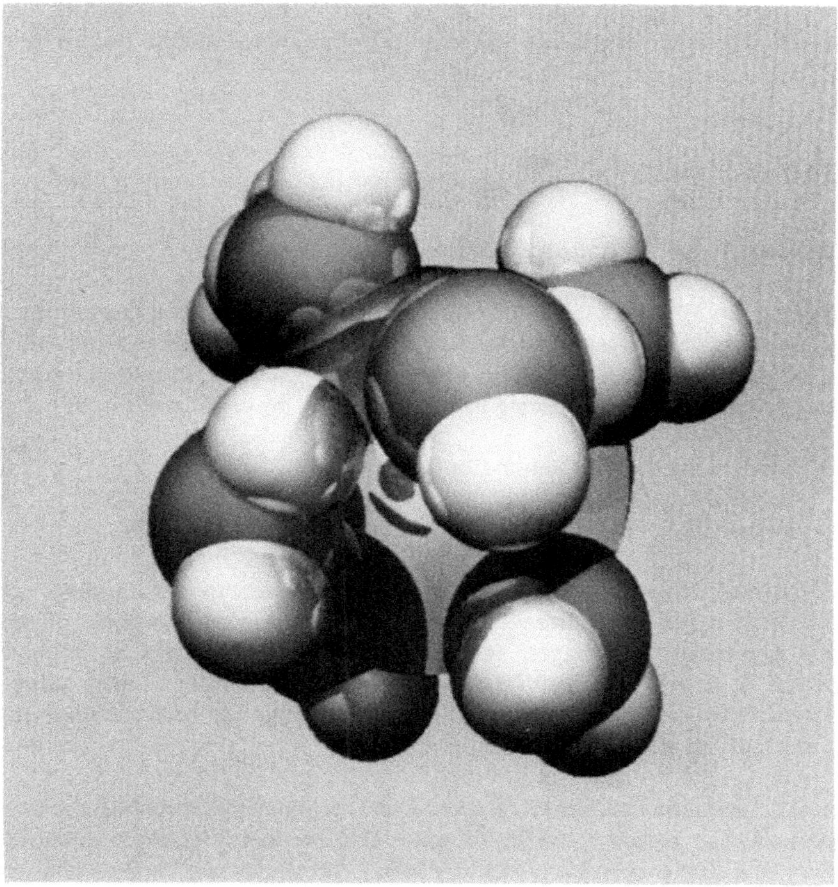

Figure 15. Ray traced representation of the hydration sphere of the first hydration shell of a lithium ion. The transparent sphere is drawn to represent the hydration shell, and all water molecules within this shell are shown.

Animations of Realistic Images

The animation of realistic three-dimensional images give higher quality images than the interactive animations previously described. These allow complex spatial motions of the atoms and molecules to be more easily seen. In addition, these animations, if recorded on videotape or some other media, provide a medium by which it is possible to exchange and present results at conferences and symposia. However, the production of computer animated videos requires special hardware and software. Our animation facility consists of the ray tracing software previously described, combined with the hardware components described below. This facility has been used to produce animated videos of molecular dynamics simulations of liquid water, ion transport through a membrane channel, hydration of

DNA, fluid flow past a cylinder and of a tidal simulation of Buzzards Bay in Massachusetts. Our experience indicates that the animations do indeed provide additional insights into the underlying science.

Animation Hardware

Selection of the Recording Medium

The animation of computer graphics based on complex scientific engineering simulations must be accomplished by recording images on a suitable medium and then displaying the recorded images at a rate of 24 to 30 images per second.[32] There are three types of recording media which may be considered:

1. Photographic media (16 mm film).
2. Magnetic recording media (video cassette recorders).
3. Optical storage media (laser video disks).

The first medium, 16 mm film, produces cinematic animation with images being displayed at a rate of 24 frames per second. This medium is capable of representing high resolution images, and the resulting films conform to a universal standard. A 16 mm film may be viewed or presented in any country using low cost commonly available projectors. Until recently, this has been the most important medium for recording animated graphics.

Cinematic animation suffers from several serious problems, including the need to process the film before it can be viewed. This processing generally requires the services of a film processing laboratory where the service can be slow and costly. For example, the turnaround time may be one week, even if there is a conveniently located laboratory. As the market for 16 mm film declines, the number of laboratories which provide this type of service is also decreasing.

As a result of the processing requirements, 16 mm films are difficult to reproduce and difficult to edit. Most often, the entire production process must be repeated in order to correct errors or to change the contents of the movie.

Magnetic and optical recording media both provide attractive alternatives to photographic movies. These media do not require processing, and the animated images may be viewed immediately after recording. These media support video animation, with the image characteristics determined by broadcast industry standards. However, the need to comply with broadcast industry standards causes two problems:

- Broadcast industry standards are not universal, and

- Image resolution is relatively low.

There are three major standards for defining how a video image is represented by an electronic signal (analog waveform): NTSC, PAL, and SECAM. The NTSC (National Television Standards Committee) standard is used in North America and Japan. PAL is used in India, China, Australia, and most of western Europe. The SECAM standard is used in France, the U.S.S.R., and most of eastern Europe. Unlike a 16 mm movie, which may be shown in any country using a common projector, a videotape which can be viewed with common equipment in the U.S.A., cannot be shown with common equipment in Europe.

The resolution and the display (refresh) frequency of the video images depends on the video standard. The NTSC standard provides a maximum of 482 lines per image, with 30 images or frames displayed each second. The PAL standard provides 25 images per second with 625 lines per frame. Some of these lines are not visible on standard television screens. Consequently, the number of lines used to represent an image is fewer than used by high resolution graphics displays (1024) or digital film recorders (2048 to 8096).

Optical storage media such as laser video disks are capable of recording and reproducing video images with high fidelity. Laser video disks must be distinguished from digital optical disks, which also use optical storage technology. A laser video disk contains a digitized recording of the analog waveform for a composite video signal. A video disk player reconstructs this signal in real time (30 frames per second) for display on a standard video monitor. A digital optical disk is the optical equivalent of a normal magnetic disk or direct access storage device (DASD). A digital optical disk stores files composed of bits, bytes, integers, characters, etc. Although a digital optical disk may be used to record a digital representations of many images, the time required to read this data and transfer it to a graphics display device usually exceeds that required for real time video display.

Most of the currently available optical storage devices are limited to read-only or write-once capability. In addition, video disk players are not as widely distributed as video cassette recorders. Consequently, we have chosen the magnetic recording technology, as represented by video cassette recorders, for the storage medium used by our animation system.

A Video Cassette Recorder Animation System

The animation hardware configuration used in our laboratory[32 on page 1089-33] is shown in Figure 16. The principal components of this system are:

1. An IBM PC-AT.
2. An analog RGB monitor.
3. An NTSC encoder.
4. A video cassette recorder (VCR).
5. A VCR controller.

6. Two NTSC video monitors.

Several components are mounted within the IBM PC-AT, or are connected only to the IBM PC-AT. The remaining components are external to the IBM PC-AT.

The Configuration of the IBM PC-AT

The IBM PC-AT provides the connection between a mainframe host computer and the animation recording hardware. This PC-AT includes the standard Intel 80286 central processing unit (CPU), 512 KByte of random access memory (RAM), color graphics adapter (CGA), hard/floppy disk controller adapter, a 1.2 MByte floppy diskette drive, and a 20 MByte fixed hard disk. In addition, the following hardware options have been installed:

1. A serial/parallel communications adapter.
2. An asynchronous serial communications adapter.
3. An IBM Personal Computer 3278/79 emulation adapter.
4. A TARGA-24 graphics display adapter.
5. 2 Mbyte of additional RAM.

TARGA stands for Truevision Advanced Raster Graphics Adapter, and Truevision is a registered trademark of AT&T Corp.[34] The TARGA-24 board produces a full color 30 Hz interlaced analog RGB graphics display with 482 lines and 512 pixels per line. This display format is electronically compatible with the NTSC video standard used in North America.

The term "full color" means the TARGA display uses 24 bits to define the color of each pixel, with 8 bits for the intensity of the red component of the color, 8 bits for the green component, and 8 bits for the blue component. This permits 256 levels of each primary color component, and provides effectively continuous shading with no discernable steps between differing colors. The image generated by the TARGA board can be displayed on an analog RGB monitor.

An IBM Enhanced Color Display is connected to a standard IBM Color Graphics Adapter. This screen is used as the text screen for normal PC-DOS functions and for terminal emulation. A 3278/79 emulation adapter is used to provide a connection to a mainframe host through a standard 3279 terminal communications port.

An asynchronous serial data port, COM1, is located on the serial/parallel communications adapter. This port provides a communications channel between the PC-AT and the Lyon-Lamb VAS-IV VCR controller. The functions of the VCR controller are described below. This adapter also provides a parallel data communications port which may be used to connect the PC-AT to a parallel printer.

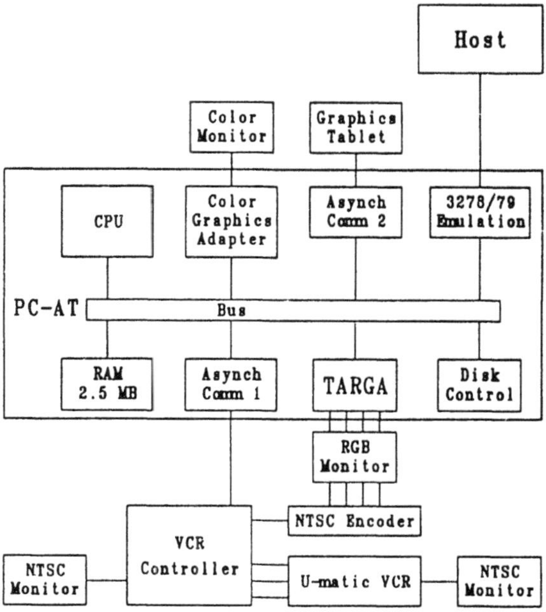

Figure 16. Schematic representation of the animation hardware configuration.

A second asynchronous serial data port, COM2, is provided by the asynchronous serial communications adapter. This port is connected to a Summagraphics model MM1201 graphics tablet. The graphics tablet may be used to control an electronic paint program called TIPS (TARGA Image Processing Software).[35] TIPS is an interactive graphics program which is useful for producing titles and diagrams, or to annotate or modify graphics images produced on the host.

A RAM expansion board provides 2 MByte of memory in addition to the basic 512 KByte of RAM.[36] The first 128 KByte of this memory are used to increase the amount of memory available to PC-DOS from 512 KByte to 640 KByte. The remaining 1920 KByte are used by TIPS to assist in certain image processing operations. This expansion RAM, the graphics tablet, the COM1 serial data port, and the TIPS software are not required for the animation of computer graphics generated on the host. These facilities are only used to create additional graphics images such as titles and diagrams.

The Animation Recording Hardware

The process of recording animated graphics requires an industrial quality VCR with editing functions, a VCR controller capable of recording single frames on this VCR, and an NTSC encoder for converting the analog RGB video signal generated by the computer into a composite video signal which can be recorded on the VCR. The hardware configuration in our laboratory uses a Lyon-Lamb ENC-VI Color Encoder and Sync Generator to perform the NTSC encoding function. In this configuration, the ENC-VI has four input cables which carry the red, green, and blue intensity signals and the sync signal produced by the TARGA board in the IBM PC-AT. These signals are used to produce an image on the RGB monitor and passed through without modification to the ENC-VI. The ENC-VI has a single output cable which carries the resulting NTSC composite video signal to a VAS-IV video animation system which functions as the VCR controller.

The VAS-IV allows an image, as defined by an NTSC video signal, to be recorded as one or more frames on the video cassette tape. As each frame is recorded, the VAS-IV also records a frame code in the vertical interval separating successive frames. This vertical interval frame code includes a scene number and a frame number. The scene number is an integer with any value from 1 to 127. This number is used to identify a contiguous sequence of related frames. A frame number in the range of 1 to 329,999 is used to identify individual frames within each scene. The frame numbers are assigned sequentially by the VAS-IV as the frames are recorded.

The VAS-IV is connected to a Sony VO-5850 VCR by three cables. One of these is a video cable which carries the NTSC video signal to be recorded from the VAS-IV to the video input connector on the VCR. The second carries the video out signal from the VCR back to the VAS-IV. This signal allows the VAS-IV to read vertical interval frame codes previously recorded on the VCR. The third cable is a remote control cable which enables the VAS-IV to control the functions of the VCR, including playing, recording, rewinding, etc.

The functions of the VAS-IV are controlled from an operator's console on the front of the VAS-IV. These functions may also be controlled through an asynchronous serial data port. This port operates at 300 baud, and it is connected to asynchronous serial data port COM1 on the IBM PC-AT. This data channel allows all functions of the VAS-IV, and therefore the VCR, to be controlled automatically by software on the PC-AT.

This hardware configuration includes a total of four video screens. The first screen is an IBM-PC color graphics monitor which serves as the PC-DOS console display and as a host terminal during terminal emulation mode. The second screen is an analog RGB monitor which displays the graphics image generated by the TARGA display adapter. The third screen is a standard NTSC video monitor connected to the VAS-IV. This screen is used to monitor the operation of the VAS-IV, including the current video input signal, vertical interval frame code, parameter entry when programming the VAS-IV, and error messages. The fourth screen is also a standard NTSC video monitor. This screen is used to monitor the operation of the VCR, including the video input signal provided when the VCR is recording frames, and the video output signal generated during playback of recorded images.

An Animation Control Program for the PC-AT

A PC-DOS program has been developed to automate the process of recording the images. This program runs on the PC-AT, and controls the video recording equipment while also controlling the connection to the mainframe host.

The first type of function provided is the user control function. This function includes parameter entry, parameter verification, program interruption, and program termination.

The status reporting function is based on a full screen display which indicates the current values of all animation parameters, the program mode, and the current program operation. This information is presented on the IBM color graphics display.

The program first receives a graphics image from the host. This image is loaded into the TARGA graphics adapter by transferring bytes of data into an array located at the absolute addresses corresponding to a 32 KByte window into the display memory on the TARGA card. Additional data must be transferred to the PC-AT I/O ports assigned to the TARGA registers which enable the TARGA memory and which control the location of the 32 KByte window within the 1 MByte address space of the TARGA memory.

The VCR control function is then used to communicate with the VAS-IV animation controller through the COM1 asynchronous communications port. This includes automated programming of the VAS-IV, commanding the VAS-IV to record the current frame, and monitoring the data transmitted by the VAS-IV to the PC-AT. These data indicate when the VAS-IV is ready to record, when recording is in progress, and when a recording has been completed. These data are also used to enable a user to interrupt the animation process by pressing key on the VAS-IV operator's console.

After the VAS-IV signals that the recording of the first image has been completed, each successive image is read from the image source, displayed on the TARGA, and then recorded on the VCR. This process continues until the program reaches the last frame in the sequence, or until the animation is interrupted.

The animation system just described has been in operation for several years and has become an integral part of the scientific process. This system has been used to produce one computer animated video entitled "The Unity of Science, Engineering and Supercomputing". This video is 45 minutes in length and contains portions relating to parallel processing, molecular dynamics of liquid water, aqueous lithium formate, hydration of DNA, ion transport through a gramicidin membrane channel, fluid flow past a cylinder and of a tidal simulation of Buzzards Bay, Massachusetts. The system has also been used for numerous other animations of molecular dynamics and fluid dynamics simulations.

Our experience with animation techniques points strongly towards two issues. First is the fact that through the use of animation, results can be easily visualized and a better understanding of the underlying physical and chemical process can be obtained. Animation adds a fourth dimension to results; not only providing a means to study time dependent phenomena effectively, but also adding the clarity derived from being able to visualize atoms and molecules in motion. A second issue is the time necessary to produce video tapes of animated computer graphics. Our procedure is currently a two step process. The simulations are performed, and then the numerical results are transformed into graphical images which are put onto video tape. Ideally, we would like to be able to produce high quality realistic images and animate these images in the same fashion as the interactive animations previously described. To accomplish this goal will require future work in algorithms (both at the application level as well as at the computer graphics level), communication between mainframe computers and graphics workstations, and computer processing speeds.

Windows Oriented Interfaces for Input Specification

Up to this point, we have been primarily demonstrating the use of computer graphics to help in understanding research problems in science and engineering. However, implicit within the previous discussions has been the notion that one major advantage which computer graphics offers is that there is a considerable improvement in the man-to-machine interface. Naturally, these improvements help a good deal in increasing the user-friendliness and productivity of a computational environment.

An area in which computer graphics can be used to greatly improve the man-to-machine interface is in specification of input parameters for computational programs. Often, most of the large scale numerical simulation packages need the introduction of a good number of input parameters that must be specified via either a line or screen editor. This requires that users have a complete and thorough knowledge of the editor being used, of the format of the input file, and of the names and possible options for all the input parameters of interest. With many sophisticated computational program packages, this type of knowledge is only obtained by continual use of the program, and with the experience of an "expert" user.

Computer graphics can help to alleviate most of these difficulties by improving the man-to-program interface through the use of an interactive windowing scheme and a graphics display. However, the windowing interface should be general and applicable across a variety of application platforms. The basic idea is that this interface should appear similar to the user whether the interest is in using quantum chemistry programs or engineering design programs or perhaps even econometric modelling programs. Therefore, it is necessary to develop a user environment consisting of a set of common tools and programs that can be easily applied across a variety of platforms.

The X Window System is fast becoming the standard windowing interface for many workstations. It is a "network-transparent" and "vendor-independent" operating environment which supports multiple overlapping windows for color and monochrome displays on a variety of vendor workstation platforms.[37-39] Using a metaphor, we may say that a typical X Window environment appears to the user as a desk with some sheets of paper on top of it. Each sheet of paper contains a certain amount of information, and as one turns the pages on the desk the information on the pages can be manipulated and modified as desired.

However, instead of the desk and sheets of paper, X Window provides a graphics screen and "windows". The windows can be viewed as small screens that are displayed on the graphics display, and which can overlap, become invisible and visible again, can accept input and display output, can be popped up and down by means of a mouse (typically menu windows) and finally, can display icons or images. Moreover, different applications or programs may run at the same time, making use of any number of windows, so that a user can perform several actions at once by simply moving the cursor from one window to another. This type of function increases productivity and functionality of the workstation environment.

X Window applications may be programmed using a low level programming interface known as Xlib, or a higher level interface composed of a set of subroutines and data structures which offer the programmer some already built facilities for programming some common features, such as menus, or dialog windows. These subroutine libraries are commonly referred to as X-Toolkits.[38, 39]

X Window is a general and powerful programming environment. General tools, or application programs, to enable this environment to be adapted to an interface for specification of input parameters to scientific and engineering application programs must still be developed. Before proceeding with the discussion of how this target has been achieved, it is worthwhile to spend a little time first describing the structure of typical input formats for large numerical simulation packages.

The first important characteristic that is observed, is that most of the input data for simulation packages have an intrinsic hierarchical structure. An example is the "namelist" data type that is one of the most common methods for data input, mainly because it belongs to the Fortran language, which is the most common programming language in scientific computing. Thinking in terms of a graph-like notation, the namelist represents the highest node in the graph, the different variables within the namelist the second level nodes, and the values of the namelist variables the third level nodes. Each of the third level nodes are connected to a second level node by a single arc in the graph, while the second level nodes are subsequently connected to the first level node, again by a single arc. A similar organization can be seen in a more complex data type called "data groups" or groups of formatted cards. In this case, the first node in the graph represents the data group, while the second level nodes are represented by the formatted data cards, which may contain multiple data elements. The third level in the graph is then the individual data elements, while the fourth level is represented by the values of these data elements. Figure 17 shows an example of this type of hierarchical data representation.

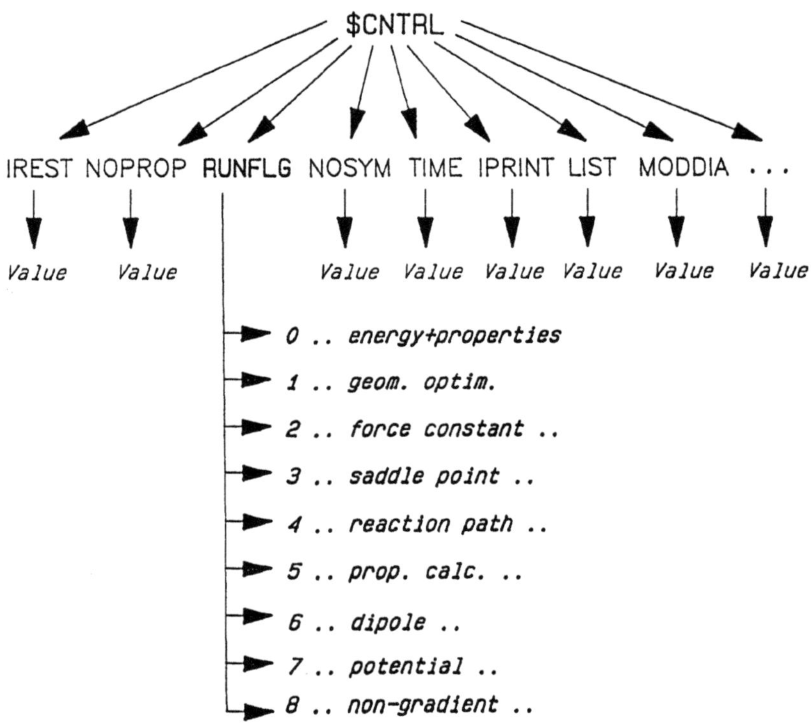

Figure 17. Example of a hierarchical data representation in a graph-like notation. The connection between different levels is always with a single arc.

In addition to the hierarchical structure present in the namelist and datagroups, there exists another higher level in the data hierarchy which appears to always be present. This level is represented by the nature of the data input, and is a result of the fact that particular computations require specific groupings of data (usually particular combinations of namelists and datagroups). This is a much more subtle hierarchy, depending for example on the type of computation to be performed. For example, in using KGNMOL for computations of interaction energies between pairs of molecules it is advisable to compute these energies including a correction for the Basis Set Superposition Error (BSSE). This is one of the options that can automatically be invoked by the KGNMOL program. However, when this option is used, a particular combination of namelist data and datagroups inputs is needed in the input file, and they must be ordered in such a way that they follow the path of the computation. This "data ordering" is actually another level in the data hierarchy, imposed by the particular computation being performed. If another type of computation is to be used, then this organizational hierarchy may not be present, or may be replaced by another organization.

In essence, what this means is that depending upon particular values of input parameters, it is often necessary to combine blocks of data in different forms; the different forms depending upon the namelists and datagroups that are needed by the computation to be performed. This type of structure can be viewed simply as another level in the data hierarchy, and can be easily represented graphically, even if it is true that this is not always easily recognizable by the application developers, or by users themselves.

In dealing with this last aspect, lies a major strength of a graphically oriented interface for input specification to large numerical application programs. For example, even though traditional program manuals are well written and programs may be well documented, it is often the case that particular relationships between input parameters are not explicitly cited in the manuals, or may not be easily recognizable from the description of the input. It may also be the case that different and varied input parameters are so numerous that the users are not aware of all the possible ways to combine them, or that they are not aware of all the necessary data structures needed for particular types of computations.

Therefore, an interface for handling these types of input specification should ideally be designed to provide an environment which graphically depicts the representation of the data structure, and also drives the user in the choice of the steps to be taken each time. This type of information can be clearly represented using a windows based input scheme, through the use of hierarchical cascade menus and dialog windows, with each level in the cascade containing a finer and finer level of information.

XWIB: An X Window Interface Builder

The current program which has been developed to create a windowing environment for input specification, is called XWIB (X Window Interface Builder). It has been implemented in the C-language and it makes use of two C-language subroutine libraries. One of them, known as Xlib, is provided with the X Window System, and contains all the functions for handling the windowing part of the program. The second library contains functions for handling the input and output flows. In particular, it includes a parser and an interpreter for a Dependency Lan-

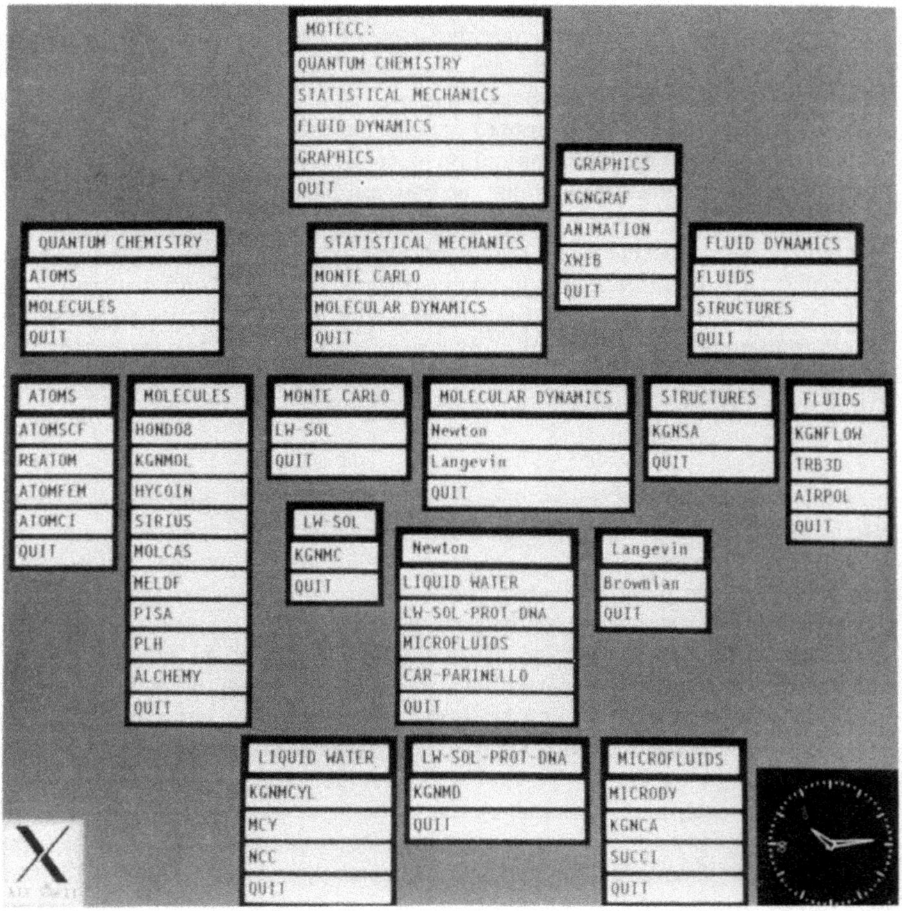

Figure 18. Example of the windowing environment for each of the components of the MOTECC package. By typing *MOTECC* at the terminal, the first window with the various fields within Motecc appears. The user can then select a field of interest, and a second level menu appears which describes the type of systems that can be computed. For example, by clicking on quantum chemistry, the systems *atoms* and *molecules* appear. The various programs in Motecc for *atoms* or *molecules* are then accessed by clicking on the respective system. The same holds true for *statistical mechanics*, *fluid dynamics*, and *graphics*.

guage developed within XWIB, and a calculator for basic arithmetic operations. The program runs on Unix-based workstations supporting a C compiler and the X Window System.

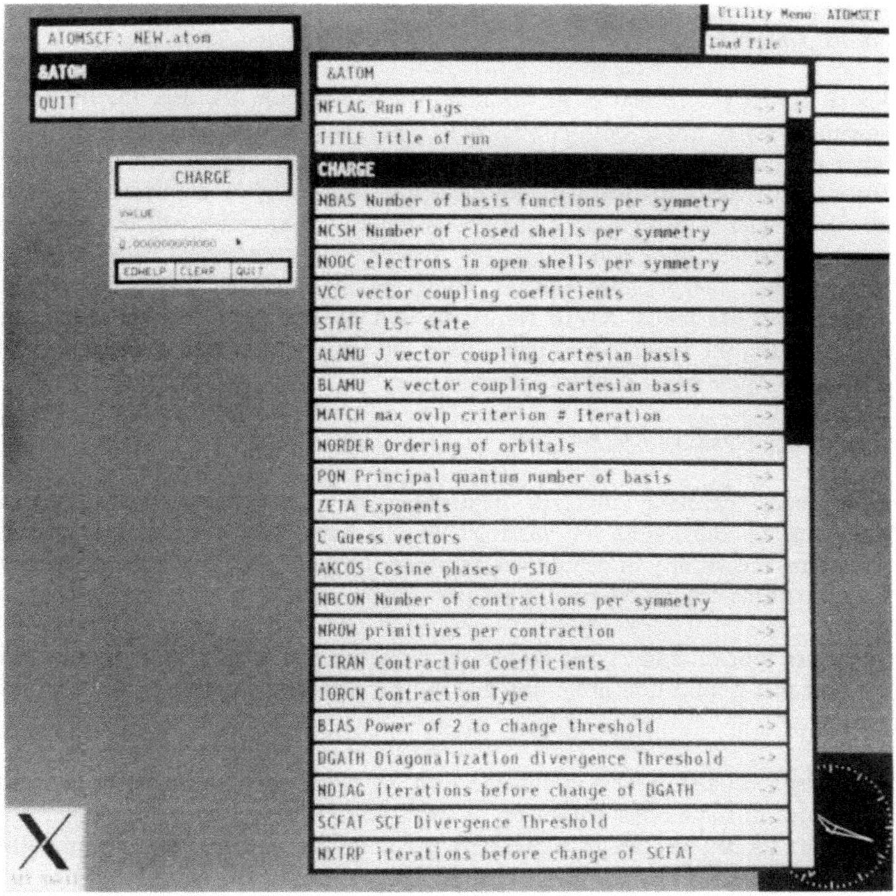

Figure 19. Example of the MOTECC windowing input specification for the Hartree-Fock atomic program ATOMSCF. The characteristics of the computation are described by specifying values for variables within the namelist &ATOM. In this particular example, the value for the charge on the atom is being specified, in the dialog window at the upper left portion of the screen.

XWIB provides a graphical representation of input parameters, and at the same time a set of facilities that include:

1. Format checking.
2. Dependency handling mechanisms.
3. Help messages.
4. On-line documentation.
5. Other utility functions.

Each of these will be briefly described below.

Format Checking

XWIB checks the format of all variables, removing from the user the burden of having to worry about both correct data types, and data formats. If a particular input parameter requires data of a particular type and length, XWIB will only allow the correctly defined variable types and lengths. For example, if a particular parameter is defined as a numeric variable, the program will not allow the user to enter any alphabetic characters. If a particular numerical value is defined to be integer, the program will not allow the user to enter a decimal number. When errors are made, XWIB warns the user by making a beeping sound, and not accepting the input. This facility frees the user from having to remember the different formats of all possible variables.

Dependency Handling Mechanisms

Dependency checking involves controlling the input specification process, according to the values of variables at the time they are specified. Before describing this facility in detail, it is necessary to summarize the types of relationships that exist among different input parameters.

There are four different types of dependencies that are commonly encountered in the scientific application programs that we have implemented using XWIB. They are:

1. The specification of a parameter, or a set of parameters is conditional to the value of some other parameter or set of parameters.
2. The format of a set of parameters is conditional to the value of some other parameters.
3. The number of entries of a particular parameter or set of parameters is conditional to the value of other parameters.
4. The flow of the input specification for a group of parameters is conditional to the values of the parameters being entered. In this case, conditional branches may occur and the order of the specification of input values is important.

A special purpose Dependency Language has been developed within XWIB to handle these dependencies and the behaviors associated with them. A complimentary language parser and interpreter have also been developed. We will briefly describe the structure of the dependency language, followed by explanations of the language parser and interpreter.

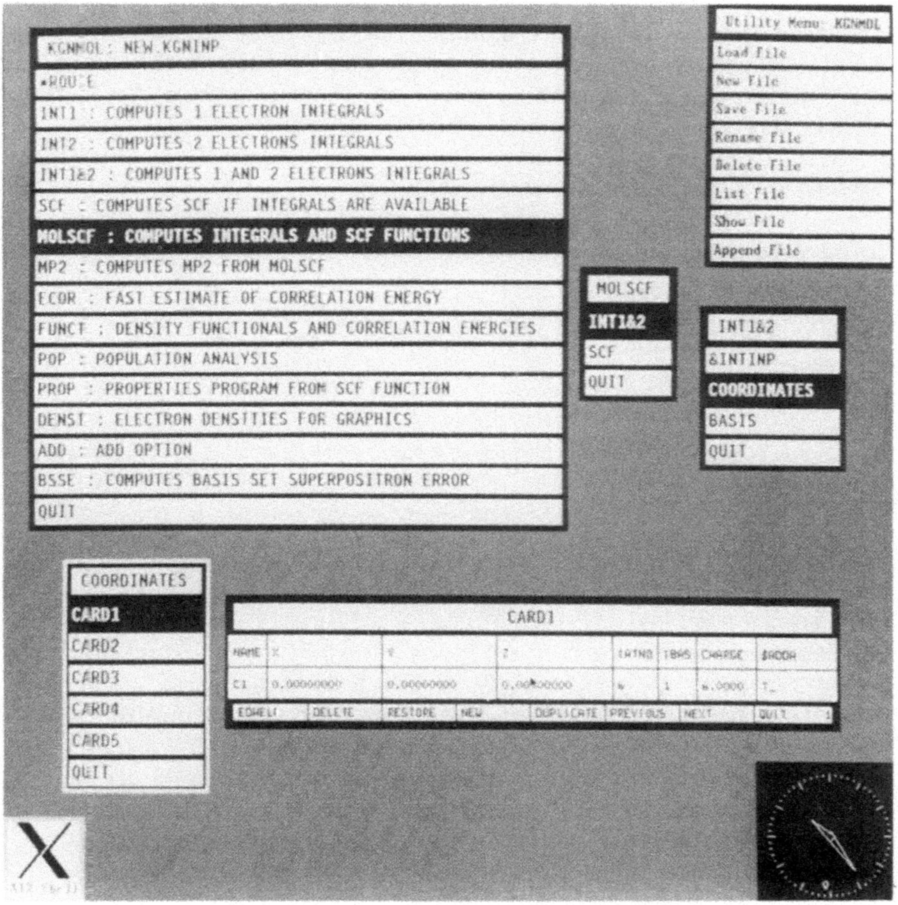

Figure 20. Example of the MOTECC windowing input specification for the quantum chemical program KGNMOL. In this example, the user is selecting an integral and SCF computation, as indicated by the selection of the route MOLSCF. This requires the user to specify parameters for INT1&2 and SCF, as indicated by the menu entitled MOLSCF in the middle right portion of the screen. By selecting INT1&2 with the mouse, the user is informed that the data needed are &INTINP, COORDINATES and BASIS. To enter the COORDINATES data, the user selects COORDINATES. This provides the user with another menu showing the format of the coordinate input, and the values for each field. The interface automatically generates the data in the correct input format, and informs the user of any format errors.

The Dependency Language

The dependencies are specified through the use of a formal language which has been developed within **XWIB**. As in every formal language, it is defined by a particular syntax. The syntax in turn is defined by an alphabet of symbols (hereby defined as operators) and a set of rules that describe how to combine these symbols in order to form legitimate statements in the language. Both parts are described below.

Dependency Construction: Operators and Statements

In the dependency language of **XWIB** there are ten different types of operators. They are:

- **Relational** operator
- **Evaluation** operator
- **Conjunction** operator
- **Disjunction** operator
- **Logical** operator
- **Assertion** operator
- **Action** operator
- **Precondition** operator
- **Conditional** operator
- **Constant** operator
- **Separator** operator

With these operators eight different types of statements can be created:

- **Evaluation** statement
- **Conjunction** statement
- **Disjunction** statement
- **Logical** statement
- **Assertion** statement
- **Action** statement
- **Precondition** statement
- **Conditional** statement

The different types of operators and the relative statements that can be built from them will be briefly described below.

A **Relational** operator can be one of: = (equal); > (greater than); > = (greater than or equal); < (lower than); < = (lower than or equal); < > (in between, extremes included); != (different). Relational operators do not generate statements by themselves, they are used as arguments in different types of statements.

An **Evaluation** operator can be one of COUNT _REP, ACT_REP, GLOB _COUNT _REP, EVAL _VAR, MAX _INT, EVAL_EXPR. Relational and Evaluation operators can be used to build **Evaluation** statements.

Evaluation statements can be used to determine the values of variables, to evaluate mathematical expressions including variables, to count the number of times a particular element has been operated on, or to count the number of repetitions of a data element.

The **Conjunction** and **Disjunction** operators are the AND and OR operators, respectively. Through the arguments of these operators, mechanisms are provided to specify the possible alternative routes or values that a particular input flow or data representation can take.

A **Precondition** operator can be one of N_COUNT _REP, N _ACT _REP, N _GLOB _COUNT, LESSTHAN, GREATHAN, EQ and NEQ. Relational operators and Evaluation statements can be combined to build Precondition statements. These operators can be used to evaluate the number of times a data element must be entered, versus the actual number of times it is needed. The traditional operators LESSTHAN, GREATHAN, EQ and NEQ perform the expected operations.

The **Logical** operators are the AND and OR operators respectively. They are used to compute the traditional AND and OR boolean functions and can only be used as components of Precondition statements.

Assertion operators consist of BLK, ASS _KEYW and FORMAT. These operators are used to specify particular formats for data entries. In special cases, blank lines or key-words play important roles in the format of input files. The assertion operators can also be used to identify these items.

There are four types of **Action** operators: SET, REPEAT, GOTO, and INCR_REP _OF. These operators are designed to perform particular actions within an input specification. By using these operators, the handling of the input flow can be specified.

Conditional operators are of the form IF --- THEN, IF --- THEN --- ELSE ---. Conditional operators have the same meaning as the traditional operators found in all programming languages.

The **Constant** operator is simply a string of alphanumeric characters or digits which is used to specify the values of variables, or path-names for menu entries.

The **Separator** operator includes ',' (comma), '(' (left parenthesis), ')' (right parenthesis), and '#' (pound sign).

The Language Parser and Interpreter

The parser is only active at the time when XWIB is begun: it reads and parses the language statements provided in an external file and loads them into a data-structure which will be continuously accessed by the interpreter. The interpreter is always active during an XWIB session, dynamically checking and handling dependencies that may arise.

The purpose of the interpreter is to control the input flow. It follows a strict policy consisting of three basic steps:

1. Checking for the validity of a particular requested action. For example, some levels in the data hierarchy can only be entered from particular points in the data entry stage.

2. Setting the environment for the specification of the data within that particular level. As an example, this is accomplished by determining the proper formats and data types, by checking any appropriate counters, and by interpreting any specified data dependencies.

3. Providing direction and assistance to the user to ensure that the correct input specification is followed. This is done by automatically inhibiting certain branches or input that are not required (according to the specified dependencies) and by prompting the user with messages concerning the next action to be followed.

   Furthermore, since the user is allowed to review and modify previously taken actions, the interpreter is designed not only to provide direction and assistance on future actions, but to also follow the user in backwards actions. In the latter case, if the user modifies data that substantially changes the previously set environment (data types, formats, counters ...) the interpreter will immediately detect the changes and will revise the input specifications accordingly. This may mean eliminating inconsistencies with previously set data. In this case the user is given the possibility of deciding what action to take with the previous data, whether to modify or destroy them.

Help Messages

Help messages to point the user towards the correct input flow are provided. This occurs when users have several options to choose from, when there is only one possible choice based on the particular input values, or when an incorrect input flow is chosen. This facility can be very useful when all the information regarding the necessary input flow cannot be easily extracted from a traditional manual.

On-Line Documentation

A complimentary facility to the help messages is the availability of on-line documentation. This provides information regarding all the levels in the data hierarchy, essentially making the manuals available at the touch of a button. An

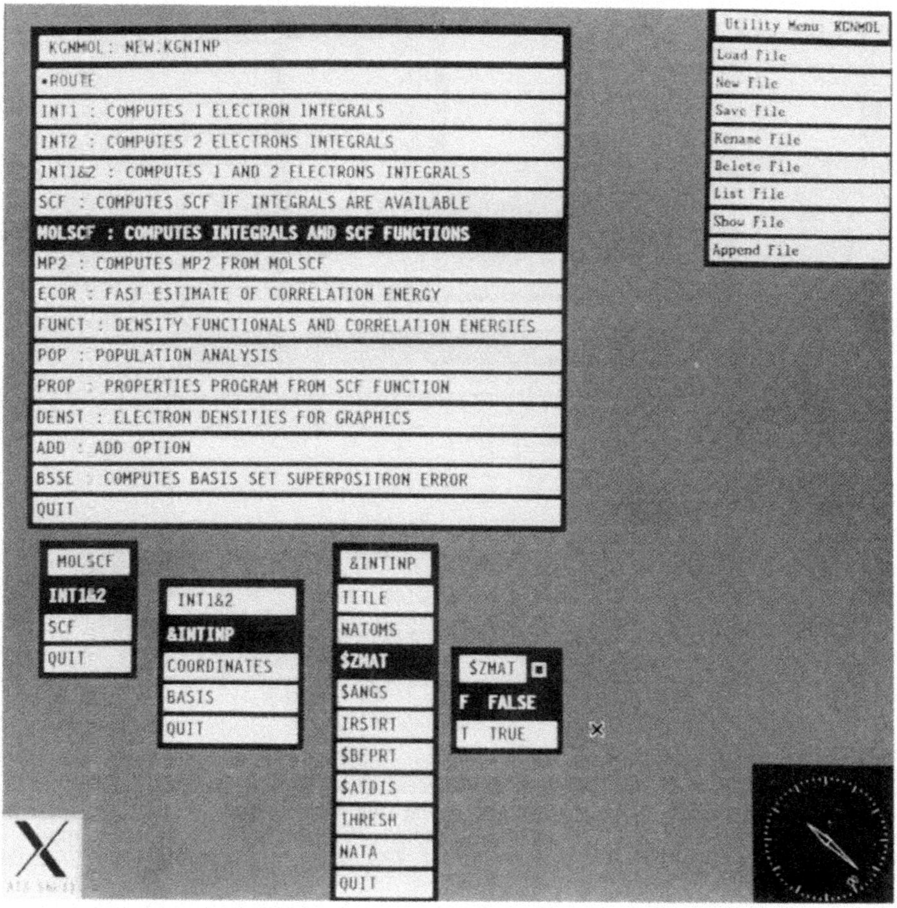

Figure 21. In this example, a MOLSCF computation from KGNMOL has been selected. This requires the specification of the data for INT1&2 and SCF, as indicated in the menu entitled MOLSCF in the lower left corner. By selecting INT1&2, the user is informed that data for &INTINP, COORDINATES and BASIS are needed. &INTINP is a namelist, containing parameters specifying some of the characteristics for the integrals computation. In this example the Z matrix format for inputting the coordinates is being set to false.

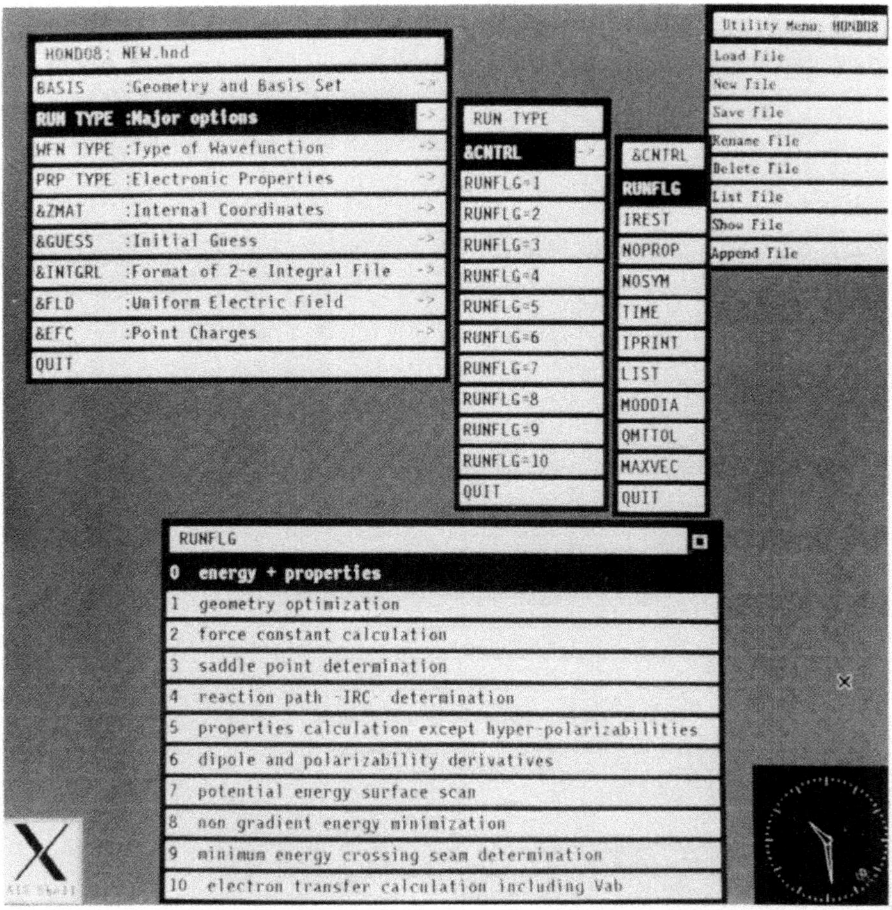

Figure 22. An example of the windowing environment for the quantum mechanical program HONDO is shown. The user has selected the option to specify the RUN TYPE. The run type in HONDO is specified using the namelist &CNTRL. By selecting &CNTRL from the RUN TYPE menu, the various parameters in this namelist appear in the window entitled &CNTRL. By selecting RUNFLG, the various values for this parameter are displayed, along with short descriptions of what each value means. It should also be noted here, that from the menu RUN TYPE, if the user already knew that a RUN TYPE with RUNFLG = 3 was desired, this could have been selected directly, and the interface would have provided a menu indicating which data need be input for this run type.

interesting aspect about this particular method of on-line documentation is that it can be structured according to the same logical organization of the parameters themselves. Therefore it is possible to organize the documentation in a hierarchical fashion exactly like the parameters of the application program. Additionally, it is possible to associate to each entry in a menu only the amount of information that is necessary to understand the meaning of the entry.

Other Utility Functions

The last facility consists of a set of "system utilities", that include the possibility to create a new file, to save, rename, delete, load, append or display the contents of a file. In the case of loading an already existing file the dependencies described to XWIB are dynamically interpreted as the file is being read, to ensure that the proper input values are assigned. Finally, a facility that is in the process of being completed, and which is described briefly in the next section, is the ability to submit jobs for remote or local execution on computational servers connected by a local area network.

So far the basic features of XWIB have been described. However, one important point that should not be overlooked is that XWIB has been designed as a "general" tool for building windows-based interfaces for different application programs. In other words the program should be considered as a general instrument capable of generating windows based interfaces for many different programs in different fields. For the moment XWIB is primarily used for numerical simulation programs in computational chemistry, however there is nothing in the design of the program which prevents it from being a completely general tool. The only requirement is that the input structure for each application program must be expressed in the language developed within XWIB. This information is contained in external data files, that are read and interpreted dynamically during the input process. In a sense, XWIB is a pre-processor, capable of taking externally provided information and building a specific input interface based on that information.

XWIB has also been programmed in a modular fashion, making extensions simple and easy. For example all the different types of windows and all the different behaviors have been coded in such a way that a precise mapping between code-name and window-behaviour-type exists. In order to add a new window and/or behavior type, a new name should be added to the mapping table, and the C-language subroutines using Xlib to handle the new window type need coding. The same can also be said for the Dependency Language: only the grammar, the parser and the interpreter have to be extended in order to handle additional constructs currently not present in the language.

In summary, XWIB has been used to implement a windows interface for specification of the input parameters for all of the programs in MOTECC-90. Examples of this interface for some specific programs within the MOTECC-90 package are given in Figures 18, 19, 20, 21 and 22.

REMOTE: Remote File Transfer and Execution

In a network environment, users wish to have the possibility to submit jobs for execution on any computer in a network. These computers may be workstations, super-minis, mini-supers, mainframes or supercomputers. The important aspect is that the submission of jobs from a workstation should be with a user-friendly graphical interface that is independent of the hardware providing the computational resource. REMOTE provides facilities for transfering files, submitting jobs, checking status of jobs, killing jobs, and retrieving outputs.

REMOTE has been implemented in the C programming language, using standard TCP/IP (Transmission Control Protocol/Internet Protocol[40-42]) protocols. It does not rely on other products, such as Remote Procedure Calls[43] (RPC) or the Network File System [44] (NFS).

REMOTE is a menu driven interface. It can be executed on any workstation running the AIX (Advanced Interactive Executive) Operating System (such as IBM PS/2, IBM RT PC, IBM RS/6000), or from any AIX/370 session. The workstation or the mainframe (which will now be referred to as the local machine) must be connected through a network, running TCP/IP, to the remote system. The remote system can be either a machine similar to the local machine (workstation or mainframe) or a mainframe running VM/XA (Virtual Machine/eXtended Architecture).

The main capabilities of the interface are to list a directory on the remote system, to delete files on the remote system, to send and receive files to/from the remote system from/to the local machine. In addition, the user can initiate, from the local machine, a remote execution, in background mode, of any non interactive program on the remote system. When the remote system is running the AIX operating system, a program is considered to be any AIX command, shell script or executable code. When the remote system is a mainframe running VM/XA, a program is defined as any CP/CMS command, any EXEC or REXX file, or any executable module. After starting a remote program, the user can monitor its execution and if desired, abort the job.

Most of these features are implemented via the remote shell command and the remote copy command provided in AIX TCP/IP.[45] For the implementation on a remote mainframe running VM/XA, both FTP (File Transfer Protocol[46]) server and REXECD (remote exec) daemon, which are both parts of VM TCP/IP[47] , are required.

These capabilities are accessed via user-friendly menus, arranged in a hierarchical fashion. The main menu gives access to one option and seven submenus; each submenu corresponds to a specific function, as shown in Figure 23.

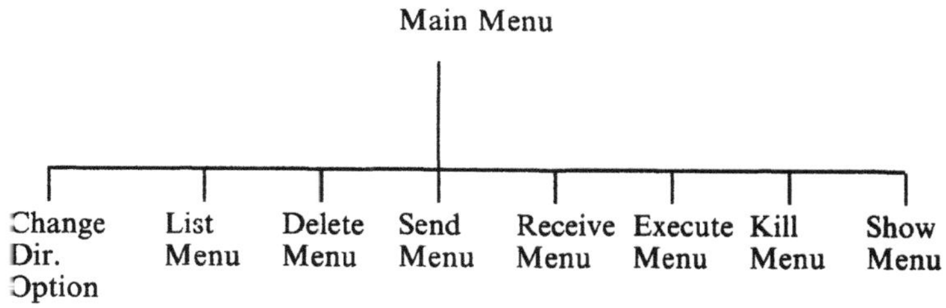

Figure 23. The sub-menus available from REMOTE. Each of these sub-menus corresponds to a particular function that has been implemented in REMOTE.

REMOTE provides two modes of operation; a basic mode for beginning users, and a more advanced mode for more expert users. In the basic mode all commands are driven by menus and knowledge of the syntax for particular commands is not needed. In the advanced mode the user is able to enter commands for any given submenu directly.

An example of how REMOTE might be used is seen below:

1. Check the existence of the necessary input files on the remote system (**List Menu**).
2. Send any missing input file to the remote system (**Send Menu**).
3. Start a background execution of the application program on the remote system (**Execute Menu**).
4. After execution begins, check the status of the job on the remote system (**Show Menu**).
5. If necessary, abort the job (**Kill Menu**).
6. Transfer the output files back to the local machine (**Receive Menu**).
7. Delete both the input and output files on the remote system (**Delete Menu**).

One added feature of REMOTE is the possibility to list the specific naming conventions of input/output files for specific applications. If this information is provided, REMOTE checks the remote machine for the existence of all needed input files before submitting a job for execution. If the necessary files do not exist, the execution will not begin. If files with the same names as the output files already exist on the remote system, the program will warn the user that these files are already existing and will be overwritten. The user then has the choice to stop the execution, or to continue.

This interface is a valuable complement to the X Windows oriented tools described in the preceding section. Undoubtedly, both contribute to providing a user-friendly workstation interface for computational application programs.

Summary

Visualization of results of scientific and engineering simulation is a quite broad field and encompasses such activities as interactive computer graphics, algorithms for realistic rendering of three dimensional computer images, animation of time dependent phenomena, manipulation and creation of models and data for computation. Another key component of graphics applications for scientific computing environments is the interface that can be provided to application programs, not only for visualization of results, but also as a front-end to the applications themselves. This means enhancing the user-friendliness of the application program interface through the use of graphics techniques such as X Windows. All of these components are important towards advancing the scientific discovery process, and are invaluable tools in performing computer simulation.

References

1. Understanding graPHIGS, IBM graPHIGS Library Manual SC33-8102, IBM Corporation, Neighborhood Road, Kingston, NY 12401.
2. Writing Applications with graPHIGS, IBM graPHIGS Library Manual SC33-8103, IBM Corporation, Neighborhood Road, Kingston, NY 12401.
3. Programmer's Reference for graPHIGS, IBM graPHIGS Library Manual SC33-8104, IBM Corporation, Neighborhood Road, Kingston, NY 12401.
4. S.S. Abi-Ezzi and A.J. Bunshaft, IEEE Computer Graphics and Applications, February 1986, page 12.
5. IBM 5080 Graphics System Operations Guide, IBM Corporation, White Plains, NY, 1986.
6. D.N.J. White and J.E. Pearson, J. Mol. Graphics, **4**, 134 (1986).
7. C.W. v.d. Lieth, J. Palm, A. Sundin, R.E. Carter and T. Liljefors, J. Mol. Graphics, **5**, 119 (1987).
8. D. Schomburg and J.Reichelt, J. Mol. Graphics, **6**, 161 (1988).
9. J. Cherfils, M.C. Vaney, I. Morize, E. Surcouf, N. Colloc'h and J.P. Mornon, J. Mol. Graphics, **6**, 155 (1988).
10. S. Chin, D.P. Vercauteren, D. Vanderveken, R. Scateni and E. Clementi, Proceedings of the Fourth International Conference on Supercomputing, Santa Clara, CA, April 30 - May 5, 1989, L.P. Kartashev and S.I. Kartashev, Eds., Volume II, page 147.
11. D. Vanderveken, Analyse, Programmation, et Applications d'un Logiciel de Graphisme MolE sub Aculaire: MGS (Molecular Graphic System), ME sub Amoire for LicenciE sub A en Sciences Chimiques, FacultE sub As Universitaires Notre Dame de la Paix, Namur, Belgium, 1987.
12. D. P. Vercauteren, Infographie MolE sub Aculaire Interactive Utilisant le Nouveau Standard Graphique PHIGS, FacultE sub As Universitaires Notre Dame de la Paix, Namur, Belgium, April 1988.
13. B. Lee and F.M. Richards, J. Mol. Biol., **55**, 379 (1971).
14. F.M. Richards, Ann. Rev. Biophys. Bioeng., **6**, 151 (1977).

15. J. Greer and B.L. Bush, Proc. Natl. Acad. Sci., USA, **75**, 303 (1978).
16. M.L. Connolly, Appl. Crystallogr., **16**, 548 (1983). M.L. Connolly, Quantum Chemistry Program Exchange (QCPE) **1**, 429 (1981).
17. J.S. Richardson, Adv. Protein Chem., **34**, 167 (1981).
18. M. Carson and C.E. Bugg, J. Mol. Graphics, **4**, 121 (1986).
19. M. Carson, J. Mol. Graphics, **5**, 103 (1987).
20. M. Re, W.L. Luken, S. Chin and E. Clementi, IBM Kingston Technical Report KGN-172, June 22, 1988.
21. P. Quarendon, Proceedings of the Fourth International Conference on Supercomputing, Santa Clara, CA, April 30 - May 5, 1989, L.P. Kartashev and S.I. Kartashev, Eds., Volume II, page 140.
22. R.J. Hinde, W. L. Luken and S. Chin, IBM Kingston Technical Report KGN-141, June 20, 1988.
23. P. Kowalczyk, W. L. Luken and S. Chin, IBM Kingston Technical Report KGN-138, June 5, 1987.
24. A. Koide, A. Doi and K. Kajioka, J. Mol. Graphics, **4**, 149 (1986).
25. E.B. Wilson, J.C. Decius and P.C. Cross, Molecular Vibrations: The Theory of Infared and Raman Vibrational Spectra, Dover Publications, New York, NY 1980.
26. M.J. Browman, L.M. Carruthers, K.L. Kashuba, F.A. Momany, M.S. Pottle, S.P. Rosen and S.M. Rumsey, Empirical Conformational Energy Program for Peptides (ECEPP) User's Guide, 1982.
27. J.D. Foley and A. Van Dam, *Fundamentals of Interactive Computer Graphics*, Addison-Wesley, Reading, Massachusetts, 1982.
28. S. Harrington, *Computer Graphics: A Programming Approach*, McGraw-Hill, New York, NY 1987.
29. D. Hearn and M.P. Baker, *Computer Graphics*, Prentice-Hall, Englewood Cliffs, NJ 1986.
30. T. Whitted, Communications of the ACM, **23**, 343 (1980).
31. W.L. Luken, N.Y. Liang, R. Caltabiano, E. Clementi, R.E. Bacon, J.M. Warren and W.F. Beausoleil, IBM Kingston Technical Report KGN-111, March 18, 1987.
32. W.L. Luken and E. Clementi, IBM Kingston Technical Report KGN-139, June 25, 1987.
33. W.L. Luken and S. Chin, IBM Kingston Technical Report KGN-140, June 12, 1987.
34. AT&T Truevision Advanced Raster Graphics Adapter, TARGA 24 and 32 User's Guide, Release 1.1, AT&T Electronic Photography and Imaging Center, Indianapolis, Indiana, October 20, 1985.
35. Truevision Image Processing Software User's Guide, TARGA 24 and 32, Release 3.0, AT&T Electronic Photography and Imaging Center, Indianapolis, Indiana, July 1986.
36. JRAM User's Manual, Tall Tree Systems, Palo Alto, California, 1986.
37. O. Jones, *Introduction to the X-Window System*, Prentice Hall Publishing, Englewood Cliffs, NJ 1989.

38. *The Hewlett Packard Widget Documentation*, Hewlett Packard Company, 1988.
39. J. McCormack, P. Asente and R.R. Swick, *X Toolkit Intrinsics - C Language Interface, X Window System X Version 11, Release 2*, Digital Equipment Corporation, 1988.
40. International Technical Support Centers: TCP/IP Tutorial and Technical Overview, IBM Corp., GG24-3376.
41. Transmission Control Protocol, DARPA Internet Program Protocol Specification, RFC 793, J. Postel.
42. Internet Protocol, DARPA Internet Program Protocol Specification, RFC 791, J. Postel.
43. RPC: Remote Procedure Call Protocol Specification, Version 2, RFC 1057.
44. Network File System (NFS) Protocol Specification, published by SUN Microsystems, Inc.
45. Interface Program for use with TCP/IP, IBM Corp., SC23-2005.
46. File Transfer Protocol (FTP), RFC 959, J. Postel.
47. TCP/IP for VM, Installation and Maintenance Manual, IBM Corp., GC09-1203.

Chapter 25.

LCAP: Loosely Coupled Array of Processors Parallel Processing Software

D. Folsom

IBM Corporation
Department 48B/428
Scientific and Engineering Computations
Kingston, New YOrk 12401, U.S.A.

1. Introduction

This chapter describes LCAP, which stands for 'Loosely Coupled Array of Processors'. LCAP is a tool for parallel processing with Fortran applications. The latest version, LCAP/3090, was developed to support parallel applications which can run across several IBM 3090 multiprocessing systems. Before proceeding, the reader may wish to review the LCAP overview in chapter 1 for a broader perspective and a brief history.

In this chapter, we will explore some basic characteristics of parallel processing and how LCAP facilities relate to them. After the introduction we explain the philosophy behind LCAP's design and usage. The remainder of the chapter describes how the important programming features in LCAP are applied to achieve parallel performance. Measurements are shown for a configuration of four 3090 systems, with a total of twenty processors.

For details on using LCAP/3090 in application programs, please refer to the LCAP chapter in the MOTECC-90 INPUT/OUTPUT documentation.

The Rationale for Parallelism

In the simplest sense, LCAP is a performance tool for numerically intensive applications. Using LCAP, applications have shown speedups of more than an order of magnitude over 'sequential' (single processor) times. Faster execution has been the primary motivation behind much of parallel processing development; it is the principle reason for LCAP's continued evolution since 1984.

This single-minded focus on performance characterizes the LCAP view of parallelism. The intent is to be able to dramatically improve execution times by replicating key portions of an application program on several processors, all operating concurrently. There should be a reasonable prospect for achieving performance gains of several times in production. If the potential speedup were only 2 times, the application would have to be very long running to justify the additional development effort.

Underlying this goal of faster execution, is the willingness to commit multiple processors to a single application. Otherwise, one pays for the higher costs for parallel processing overheads without realizing the benefits. With LCAP, we have preserved the capability to run one, extremely large application or several smaller ones; within a single multiprocessing system or across several. Experience has shown that if an application requires most of main storage, it is often detrimental not to commit all the processors to it. The resource scheduling process in LCAP was built to handle parallel requests.

Of course, we would like to achieve parallel speedups without having to restructure or rewrite entire applications. The costs would outweigh the benefits for all but a few heavily used programs. From the outset, a pragmatic approach will yield the best results. Those applications which use the most computing resources should receive the most focus, because of their potential for improving user productivity, or system efficiency. Understanding the basic constraints to parallel speedup will help in assessing each application for speedup potential. For many applications, the cost to convert to parallel under LCAP is modest (on the order of a few days to a couple of weeks).

Application Characteristics

There are three application attributes which influence the degree of speedup and the approach used to obtain parallel performance. The most obvious is the fraction of the code which must be executed sequentially (i.e. no other concurrent processing). This idea was introduced in chapter 1 (see "Amdahl's Law") as an upper bound on speedup.

Routines which establish and control the parallel environment are sequential by nature, as are those which manipulate results returned from parallel processes. Other routines may be sequential strictly for efficiency. The important point is that the sequential portion of the execution time remains small, not the portion of the code. It is not unusual for a small part of the code to consume the great majority of the compute time (e.g. the innermost of a nested set of loops). Also, the proportion of sequential time often decreases dramatically when the problem size is increased, because the parallel fraction scales up by a power of the size (Order N squared, cubed, even N to the fourth or more).

A second attribute is the amount of computation which can be done before the parallel processes need to stop and communicate results. Applications are labeled as 'coarse-grain' or 'fine-grain ' (and sometimes 'medium-grain'). If the time for a parallel compute/communicate cycle is hours or minutes, the application is considered 'coarse-grain'; if it is milli-seconds or micro-seconds, it is 'fine-grain'. (This delineation is somewhat arbitrary, but read on.)

The real significance of coarse versus fine is the portion of the total run time spend in communicating results versus computing. This parallel 'overhead' is similar to the sequential portion; it increases run time and decreases speedup. These coordination costs are unavoidable; the only ways to improve performance are to use faster communications paths or change the program to require less communication. In particular, parallel systems which have slow paths can only achieve parallel speedups for coarse-grain applications. Fortunately, many of the MOTECC applications are very coarse and do achieve good parallel speedups.

A third major application characteristic is the uniformity of the individual process run times. If the parallel computations have widely varying durations, then the run can suffer 'load imbalance'. This happens when one or more of the processors finish their computation much earlier than the others and go idle (until the others catch up). Because there are idle processors, the run takes longer than it should. One way to reduce the impact of imbalance is to dynamically assign smaller chunks of work to processors. Instead of waiting for the other processors to finish each processor immediately goes back for more work. However, this technique makes the application more fine-grain, which may lengthen the run because of higher communication times.

Configuration characteristics

There are two classical approaches to the way that parallel processor configurations communicate; through shared memories and message passing. Shared memory configurations communicate by writing information in memory locations accessible by some or all of the processors. In contrast, message passing configurations communicate by structuring data in messages, which are sent and received across communication links.

We choose this very general description because a wide variety of system configurations are possible. Indeed, as LCAP has evolved, there have been three significantly different configurations; Host to Channel attached processors, Attached processors with Bulk memories, and Channel coupled - shared memory multiprocessors. In the last two configurations, both message passing and shared memory constructs have been used together in the same applications for maximum performance. See chapter 1 for descriptions of the LCAP 1, LCAP 2 (prior to and with Bulk shared memory) and the LCAP/3090 configurations.

The mode of communication is very important in parallel programs because it directly influences both the cost of parallel overheads and the programming styles used for parallelism.

1. With message passing, processes communicate by formatted data records sent between each other. The building and storing of these records is usually part of the application program. There is also some form of message protocol which is used to allow processes to synchronize with each other and avoid deadlocks.

   A simple convention is similar to file I/O writes and reads; where message 'Sends' are asynchronous and message 'Receives' are synchronous. Once the system accepts the send request, the application continues. The application waits for a receive request, if the message has not already arrived. There may also be explicit requests to wait until a sent message has been received or until the next message is available. For two processes to exchange information, the application sequence might look like:

Process 1	Process 2
Compute	Compute
Prepare message	Prepare message
Send message to Process 2	Send message to Process 1
Receive Process 2 message	Receive Process 1 message
Save message	Save message
.	.
Repeat for other processes	Repeat for other processes
.	.
Loop to compute	Loop to compute

   With these protocols, the system assumes certain responsibilities to queue and forward messages to other processes. Often the system must buffer messages in its own storage until the recipient is ready. However, the bulk of the control is within the application. This is unavoidable. Moreover, the application must be sensitive to the costs of message transmission to get good parallel results; whether it is preferable to send many short messages or a few long messages. In some system architectures, the maximum message length may be fixed, so that longer messages have to be subdivided.

2. With shared memory, processes communicate through commonly addressed data areas. These may be arrays (often very large) or message buffers which can be in Common blocks or passed as call arguments. There are also forms of locking and synchronization to coordinate data exchanges and control program flow. Since the shared memory addresses are always accessible, each process must know when it is safe to read from and write to them. In

the following application sequence, each process writes to a separate part of the shared memory and reads the portion written by the other process:

Process 1	Process 2
Compute	Compute
Synchronize with Process 2	Synchronize with Process 1
Write data areas	Write data areas
Synchronize with Process 2	Synchronize with Process 1
Read data areas	Read data areas
Loop to compute	Loop to compute

There are often many subtleties in the management of shared data which must be dealt within the application program. The program must clearly delineate local and shared memory objects. If the hardware implementation of shared memory is external to the processor, access times can be significantly longer than local memory. Then, there is a concern for the number and length of accesses (similar to message passing). There are general questions of when to create local copies of shared data, and how to manage shared memory updates.

For both message passing and 'external' shared memories, there are two parts for the time it takes to transmit data; the startup time, or latency, and the transfer time (incremental time to transmit a character or word). For very short transfers, the latency time often dominates because it includes system software overhead. For long transfers, the transfer time determines performance. The relative times are very important to know for two reasons;

Many parallel applications communicate short sequences of data (e.g. control information and discontinuous parts of arrays). Blocking of this data into records is important to reduce communication time.

Data exchanges between processes often can not be overlapped with other execution because the data is needed to continue execution. In some cases it may be possible to segment the transfers to get partial overlap (e.g. compute,send,compute,send ... receive, compute, receive, compute), but this also introduces more overhead and synchronization time.

For fine-grain applications, the latency time must be very low because sending even a single message could exceed the compute time of a cycle.

How LCAP Simplifies Parallel Programming

From the discussion above, it should be clear that parallel applications can very easily become committed to the particular hardware configuration they were developed and tuned for. This is not just a question of program 'portability'; there is little practical experience to demonstrate that source codes can be made compatible for different parallel architectures or different vendors. The concern is more essential; to avoid 'distorting' the program's original logic and structure by the addition of many new statements for parallel control and housekeeping. (While there are several compilers that do automatic detection of parallelism, these are largely confined to fine-grain loops for shared memory multiprocessors.)

The LCAP approach to parallel programming is to preserve as much as possible of the traditional (sequential) Fortran environment for each concurrent process. Key to this is the preservation of storage areas across parallel calls. Common blocks and Fortran variables local to a process behave identically to standard Fortran (in fact they are managed by standard Fortran). Once an area has been modified inside a process, it retains its value until it is changed again under explicit program control. What is added by LCAP is the ability to copy the contents of a data area in one process to the same area in another process.

LCAP is designed to **help the programmer express parallelism simply,** without major changes to existing codes. In many applications, only a few routines need to be changed. LCAP support for expressing parallelism includes:

- A two level hierarchy, with a 'Master' process initiating and controlling the parallel 'Slave' processes. Each process has its own separate run time environment, including a local address range which can not be accessed by other processes. (Under VM/XA each process runs in its own virtual machine.)

- Parallel calls from the master process to subroutines in the slave processes. Complete flexibility is provided to execute different subroutines on some or all of the slave processes. The master executes concurrently with the slaves until it explicitly waits for one or all to complete.

- Powerful message passing primitives, during parallel calls and waits. 'Messages' are defined by listing the Fortran variables and Common blocks to be copied from the master process to the slaves as part of the call. Local images of the call arguments are created in the slave and passed to the subroutine named by the call. Arguments and Commons are copied back from the slaves to the master during wait synchronization.

- Sharing data within a single system and across systems. Shared Common blocks can be fully shared by processes running in the same system. (Only the processes running the same application have access to the same Common.) When an application runs across systems, there is a separate image of the shared Commons on each system. Parallel calls may refresh all or part of the Commons from the master's system. Parallel waits may overlay all or part of the Commons in the master's system from the slave's system.

- Configuration independence. At run time, LCAP fills a Common block with all the run time configuration data. Using this information, the application can easily determine the number of slaves, what systems they are assigned to and how to subdivide arrays for each of the slaves. There are standard LCAP parallel calls and waits for all the systems, one of systems, or a specific slave. Any of these options can be programmed with complete generality, whether the application is running on a single multiprocessor system, or several systems.

- Efficient primitives for fine-grain process synchronization inside a single system. Functions are supported to have all slaves wait for each other, allow one Slave control the others, allow each Slave to schedule itself or coordinate with adjoining neighbors, and to allow independent updates to shared arrays.

These functions are the building blocks which the programmer inserts into the application. They are specified as language extensions known as 'directives' which are translated by the LCAP Precompiler program into Fortran statements.

But LCAP is more than just a programming tool; it is a framework for parallel processing, which has proven resilient across three major architectural changes and for over a hundred different types of application programs. Other major components in LCAP include the Scheduler which controls the resources assigned to parallel runs, and a command processor which provides the user interface to initiate application runs and query on their status. In sum, it is the total complement of services in LCAP which simplifies the use of parallelism.

2. LCAP Design Philosophy

From its inception, LCAP was created as a production system. It has had to be reliable; many major research projects have run in parallel over periods of several weeks. LCAP has had to be easy to operate. It has had to be reconfigurable to bypass equipment failures or to set up special production environments. It has had to be manageable; permitting several concurrent Jobs to share parts of the total resources, and allowing application testing to have some priority over heavy production.

But most of all, LCAP has had to be usable by the scientists in our department and deliver good parallel performance. Since many of these scientists come from many countries for relatively short periods, LCAP has had to be simple to explain and straight forward to use for parallel conversions, testing and production. The essence of the LCAP design philosophy is that we try out solutions to user problems and we change them when they don't work. Fortunately, we have not had to change too many fundamental aspects of the design. We now highlight a few of the experiences which influence the design 'philosophy'.

Adapting to a Production Parallel Environment

Configurations: The first LCAP was developed as a message passing system on a 370 host with channel attached processors. This configuration proved to be a cost effective way to get supercomputer performance for the coarse-grain applications of that time (precursors to the current MOTECC). Since then, LCAP configurations have continued to couple a moderate number (no more than 32) of relatively powerful processors together to achieve supercomputer capacity. In large measure, this has been dictated by our need to use commercially available equipment and reliable interconnections.

Language: With loosely coupled systems, parallelism must be expressed at the 'coarsest' level (e.g. the outermost loop) possible in the application. With LCAP, the decomposition of the program into parallel parts is done by the application developer. The coded parallel calls, waits and data transfers between master and slaves are translated by the Precompiler into standard Fortran source code. This has made it quite easy to change to new Fortran compilers, and hopefully to provide future portability across supported systems.

Decomposition: Our experience has shown that it is not difficult for application specialists to decide on the best way to partition their programs. The goal is to minimize communication costs and maximize parallel efficiency. Depending on the application, there may be only a single interaction between the master and the slaves, or there might be many different interactions. The data flow also varies widely; from entire arrays exchanged between the master and the slaves, to only small parts of arrays passed between consecutively numbered slaves. In the examples which follow there are two techniques illustrated; static and dynamic partitioning. With static partitioning, the division of work is fixed by the program and divided among the number of slaves. With dynamic partitioning, the division of work is fluid - it is assigned as slaves complete during the run.

Hierarchy: Fast communications with a shared memory allow fine-grain applications to coordinate their own activities and data sharing. However, the master/slave hierarchy is still essential for establishing the run environment and periodically consolidating and saving results. For LCAP/3090, the master/slave hierarchy has been augmented with new support for nested parallelism. One of the slaves on each system can act as a 'leader' for the other slaves on the same system.

Scheduling: When our user group was small, processors were manually scheduled by coding addresses in the application. Very soon thereafter, the applications changed to adapt to run time information provided by a central scheduler. This run time configuration data became essential when the Bulk shared memories were partitioned among several concurrent users. More recently, run time configurations have been expanded to include 3090 system names for the master and slaves. The value, of having applications adjust to the number of slaves and the specific configuration at run time, really became apparent when we simulated performance for the LCAP scheduler design. Overall system throughput dramatically increased due to better overall load balancing.

LCAP/3090 Design Principles

Because our development group has been small, the LCAP/3090 design has also had to be simple; building on the foundations of the Fortran compiler and library, and the services in the VM/XA operating system. In order to support a loosely coupled design, each of the processes in a parallel run must be able to run on a different system. For the VM/XA system, each master and slave process runs in its own individual Virtual Machine (VM).

We have found there are several important (even if unplanned) benefits to this approach. First, error recovery is greatly simplified by having each process isolated with its own set of system resources. When a run fails, whatever the cause, LCAP can recover most or all of the assigned resources for use by other runs. There are other important performance benefits. First, each parallel process can have its own I/O files which permits parallel I/O. Second, the amount of total workload on a system can be limited by controlling the number of available slave VM's in a pool. Third, by having each process compete on an equal basis with other user VM's in the system, parallel run times are improved. Fourth, shared memory support reduces the amount of data transfer within and between systems.

LCAP software has evolved into five major components, illustrated in the figure below:

- The Precompiler which converts directives in the application into calls for the run time utilities.

- The parallel run utilities which support parallel calls/waits and the required data movements between the master and slave VM's.

- The Scheduler which assigns and initializes the resources for parallel run requests.

- The command processor which provides the operator interface to couple systems into an LCAP configuration, and the user interface to initiate runs and query on system usage.

- The communication interface used by the commands, utilities and scheduler to send messages between VM's (based on VM IUCV services).

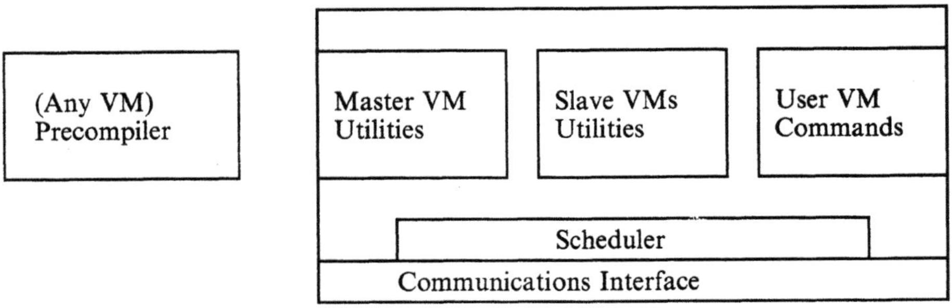

The Precompiler is a key element of the design because it isolates LCAP programs from a specific Fortran compiler. The converted program is also easier to understand and debug because all of the generated code, including calls to the parallel utilities, are visible.

The Scheduler was developed primarily to simplify the management of a parallel run; to activate Slave VM's, initialize them and ensure they terminate at the end of a run. The second role as a workload manager came later, but is now equally important.

The user commands and communication interfaces were completely redone for LCAP/3090 to support the notion of multiple systems. The basic capabilities of being able to request specific resources for a run and querying on system or run status are retained from early releases.

The parallel run utilities support the application control of parallel operations and are the subject of the rest of this chapter

3. LCAP Features for Effective Parallelism

This section discusses many of the features available with the LCAP directives (parallel language extensions) to show how different application requirements can be solved. Through use of examples, we will be able to illustrate many of the common techniques used. Although the directives are shown with proper syntax, the emphasis here is on explaining the function and how the problem is addressed. Refer to the LCAP chapter in the MOTECC-90 Input/Output Documentation for a full explanation of the directives.

Features in the Precompiler

The LCAP Precompiler parses the application program, in order to translate directives into Fortran variables and executable statements. Master and slave programs are produced in separate files. Error messages are generated for directives with syntax errors or incorrect argument values. A second input file may be used for only slave subroutines, particularly if a different routine with the same name is called by the master. The Precompiler does not have to read the entire program; only those routines with LCAP directives. However, since the Precompiler does extensive checks on data exchanged between master and slaves, it would be wise to precompile the entire program the first time.

Parallel run utilities are invoked by CALL statements inserted into the translated source. The Precompiler also generates Fortran Main program routines for both the master and slave programs; the application's Main program is called as a subroutine. These generated programs perform important initialization functions which include creating the shared Common block areas.

The Precompiler checks data declarations for all parallel call arguments and copied Common blocks. It generates error messages for those Fortran statements which would cause it to produce incorrect results. There are also warning messages when copied arrays or common blocks are not consistently defined in the master and slave routines. There is also a summary listing showing Commons declared with different sizes and the size in each routine. Just as with the Fortran compiler, these diagnostic messages are a very important aid to debugging.

Features with Parallel Calls

One of the most important parallel call features in LCAP is memory retention. Programs are loaded once during run initialization; each slave VM has its own copy of the application slave program. Unshared Common blocks and local variables have the same values they were set to during the last invocation, **the same as if they were running as a sequential subroutine.** Thus routines which use local arrays or Commons for work space will work the same in parallel. Similarly, if results are passed in work Commons between subroutines, parallel calls to these routines will work the same way within a slave. Equally important is the ability to copy a large input area once from the master to a slave and be able to access it on subsequent parallel calls.

Each slave process may run one parallel subroutine at a time. That is, once the master process has started a parallel call on a slave, it must wait before it can issue a second call to the same slave. The control flow for a parallel call and wait sequence is:

Master VM	Slave VM
Parallel Call	
Send inputs	Receive inputs
Return from call	Call Routine
. . .	. . .
Parallel Wait	Return from routine
Receive outputs	Send outputs
Return from wait	Wait for next Call

A great deal of flexibility is provided for the parallel calls. The simplest parallel call and wait has no arguments. Subroutine SUB1 runs on all slaves:

```
'C$    EXECUTE ON ALL: SUB1'
'C$    WAIT FOR ALL'
```

Note that the 'C', for Fortran, followed by '$', for LCAP, is a reminder that all the directives are Fortran comments. Other forms of the EXECUTE directive start a subroutine on a specific slave or on all the slaves on a specific system. There are also optional clauses which allow tailoring argument lists using the slave index in expressions. The WAIT directive variations are similar and are discussed below.

The directives above are not very interesting because no data is copied. Directives which cause data movement reside in the slave subroutines and are named; SLIN (for 'SLave INput'), SLOUT (OUTput) and SLIO (Input and Output). The transferred data objects named in these directives are either call arguments (scalars or arrays) or Common blocks. Entire arrays or Common blocks, as defined in the slave subroutine, are transferred to or from the master.

1). In the first example, we show a very simple routine which divides up an array into adjacent parts among the slaves. This technique is used to achieve load balancing while minimizing data transfer. Note that the program works only when the number of slaves evenly divides into the array size (e.g. 2,4,5 ...). Later examples show how to divide an array of arbitrary size among all the slaves.

```
Master:     REAL*4  ARRAY(1000)
            J = 1000/NSL
  C$        EXECUTE ON ALL, USING I: SUB1(J,ARRAY(1+J*(I-1)))
  C$        WAIT FOR ALL

Slaves:     SUBROUTINE SUB1(J,ARRAY)
            REAL*4  ARRAY(J)
  C$        SLIN J
  C$        SLIO ARRAY
  C$        ALLOCATE ARRAY(500)
```

The master partitions the ARRAY by the number of slaves (NSL). This means that every slave receives an array of size 'J'. The EXECUTE directive calculates the first ARRAY index to send each slave (LCAP utilities remember the indices when the ARRAY is returned from the slaves). In each slave, the received array starts with index 1, just as if it was passed from a normal program call!

The array size 'J' must be passed as input to the slaves, because the ARRAY is declared with an adjustable dimension. Because the Precompiler can not determine how much space to reserve for ARRAY, the ALLOCATE directive defines the space in each slave as 500 elements. This means that a minimum of 2 slaves are needed for the run. Space for copied arguments is reserved to prevent aborting a long run for lack of storage.

This example is a common form of static partitioning. Note that the technique would be awkward to use if only Common blocks could be communicated.

2). A second example will illustrate another static partitioning scheme. The difference is that each slave must process the entire ARRAY. Frequently in MOTECC, partial results (e.g. forces on a particle) have to be summed with the array in the master to get the net results. This is handled simply with a special directive named ADDING. Notice how the code fragments below have changed from the first example.

```
Master:      REAL*4  ARRAY(1000)
   C$        EXECUTE ON ALL: SUB1(ARRAY)
   C$        ADDING  ARRAY
   C$        WAIT FOR ALL

Slaves:      SUBROUTINE SUB1(ARRAY)
             REAL*4  ARRAY(1000)
   C$        SLOUT   ARRAY
```

There is no subdivision of the output array (inputs are not shown). Each slave has a complete copy which is only sent back to the master (SLOUT). It is also possible for the master to send its array image to the slaves by using SLIO, as before. There is no requirement for the ALLOCATE directive, since the array size is fixed. Then there is the new ADDING directive, which reads each slave's array image into a buffer in the master and adds the buffer to the array.

Without the ADDING directive, the master code would have to use two arrays and read each slave's results in a loop. One array would serve as a buffer, which would be added to the result array in the master. But there would also have to be a way to guarantee that only one slave at a time sent its results to the buffer. That is the function of the WAIT directive.

3). The WAIT FOR ALL is the simplest directive form. The master waits until all slaves finish their last EXECUTE call. There is no guaranteed sequence in which the slaves complete, so any SLOUT or SLIO variables in the master would contain the image of the last slave to finish.

Like the EXECUTE directive, the WAIT has several options. There is the capability to wait for a specific slave, for all the slaves on a system, for the next available slave, or for all the slaves on the next available system. These forms can be used in any reasonable combination to suit the needs of the application. The following code fragment shows how an ADDING function might be coded in the master.

```
Master:       REAL*4  ARRAY(1000),BUFFER(1000)
   C$         EXECUTE ON ALL: SUB1(BUFFER)
              DO I = 1,NSL
   C$           WAIT FOR SLAVE I
                DO J = 1,1000
                  ARRAY(J)= ARRAY(J) + BUFFER(J)
                ENDDO
              ENDDO
```

4). We finish the discussion on parallel calls with some notes on the WAIT FOR ANY directives. These directives have special uses for dynamic decomposition schemes. They are used to improve load balancing across slaves and systems when the run times are highly variable.

If all the parallel work takes the same amount of execution time, it is more efficient to do only one EXECUTE per slave with the work evenly divided. But when the parallel work takes widely varying amounts of time, the higher overhead for multiple EXECUTEs can be overcome by reducing the time that slaves wait for the others to complete. Dynamic partitioning gives each slave an initial portion of work and the remaining portions are assigned 'first come, first served' as the slaves finish. In LCAP, this is accomplished with the directive,

'C$ WAIT FOR ANY FREE SLAVE sl'

Unlike the previous forms, this WAIT directive can be used prior to an EXECUTE directive. It returns a slave identifier which is immediately available for use. If the slave was also finishing a previous EXECUTE, data transfers complete before the WAIT returns to the master program.

The need for additional 'SYSTEM' directives, became apparent when two levels of parallelism were attempted with multiple 3090 systems;

'C$ WAIT FOR ANY FREE SYSTEM sys'

'C$ WAIT FOR ANY FREE SLAVE sl, ON SYSTEM sys'

Coarse-grain chunks of work are apportioned across systems. Within each system, a 'leader' slave dynamically assigns smaller work units to individual slaves, and then returns the results to the master. As each system finishes, it is assigned more work with;

'C$ EXECUTE ON SYSTEM sys: SUB1'

Dividing up the work within a system leads us to the shared memory directives.

Features with Shared Commons

The major advantage of the parallel call directives discussed above is their generality; they can work for processes all on the same system, or on completely different systems. The major disadvantage is the performance penalty to transfer large volumes of data between systems on relatively slow links. The shared memory directives permit faster communication and finer-grain parallelism for those processes which run on the same 3090 system. Central to all the shared memory services in LCAP are shared Common blocks.

Shared Common blocks reside in a separate address range of the LCAP virtual machines, called a 'shared segment'. The shared Commons are created during run initialization and remain accessible throughout the run. The master and slave routines which update shared areas must take care not to destroy each other's data. Routines must either confine their updates to disjoint parts of Common blocks, or must do updates under some form of serial access. Note that the master, after a WAIT FOR ALL, is assured exclusive access to any shared areas.

5). For simplicity, consider first the master and all the slaves on a single system. Example 1) above, with array partitioning, can be equivalently accomplished by the following code fragments.

```
Master:
  C$          SHARED /SHR1/
              COMMON /SHR1/ ARRAY
              REAL*4      ARRAY(1000)
              INTEGER I,J(33)
              J(1)   = 1
              DO 50 I = 1,NSL
   50         J(I+1) = J(I)+(1001+NSL-J(I)-I)/(NSL+1-I)
  C$          EXECUTE ON ALL USING I: SUB1( J(I), J(I+1)-1 )

Slaves:       SUBROUTINE SUB1(J1,J2)
              COMMON /SHR1/ ARRAY
              REAL*4      ARRAY(1000)
  C$          SLIN J1,J2
```

In this example, the master and slave access the same array, because Common /SHR1/ was declared as shared. To partition the array, the master calculates and passes the starting and ending index (vector J) for each slave to process. If you display J, you will see that the DO loop evenly distributes the 1000 array elements across all slaves. Also note that J's length is set for the default maximum of 32 slaves in a run.

The vector J could also have been placed in Common /SHR1/, but the example code is simpler to use when running across systems. Since shared Commons are not addressable across systems, LCAP supports user directed copying between the master and the slaves in the same way as with unshared data. If fact, shared Commons can be copied across systems with the SLIN and SLOUT directives. But, two problems remain;

it is very inefficient to copy ARRAY to itself on the same system, and
the entire ARRAY has to be transmitted across systems, when when only a portion needs to be updated.

5a). The solution to these problems are new directives for transfer of data in shared Commons. With the SHRIN, SHROUT and SHRIO directives, scalars, arrays and partial arrays in shared Commons can be copied across systems. The same type of flexible array partitioning, used with call Arguments, is now available with shared Common blocks. To run across systems, the last example is changed by adding a SHRIO directive.

Master:
```
C$          SHARED /SHR1/
            .
            .
C$          EXECUTE ON ALL USING I: SUB1( J(I), J(I+1)-1 )
C$          SHRIO ARRAY( J(I) : J(I+1)-1 )
```

Each slave receives its slice of the array, starting with index J(I) and ending with index J(I+1)-1. This code will still work on a single system. The SHRxx utilities detect if a slave is on the same system as the master and suppress the data transfer.

6). Example 2) above can also be implemented with shared Common blocks and will also work across systems. The new directive SHRIOADD performs the ADDING function to SHRIO arrays (SHRADD is for SHROUT). The other new illustration is the conditional test to do the SHRIOADD only for the first (i.e. 'leader') slave on each system. In the loop below, SLVSYS(I) is the relative system number for slave (I) and LDRSLV(J) is the index of the first slave on system (J).

Master:

```
C$          SHARED /SHR1/
            COMMON /SHR1/ ARRAY
            REAL*4      ARRAY(1000)
            J = 1
            DO I = 1, NSL
             IF (SLVSYS(I).GT.J) J=SLVSYS(I)
             IF (I.EQ.LDRSLV(J)) THEN
C$            EXECUTE ON SLAVE I: SUB1
C$            SHRIOADD ARRAY
             ELSE
C$            EXECUTE ON SLAVE I: SUB1
             ENDIF
            ENDDO
C$          WAIT FOR ALL
```

Slaves:

```
            SUBROUTINE SUB1
            COMMON /SHR1/ ARRAY
            REAL*4      ARRAY(1000)
```

7). The prior examples are simple, but powerful, instances of the types of parallel processing that can be done with LCAP. What remains to be discussed are the other special functions in LCAP which support fine-grain applications within a single system. With shared Common blocks, the slaves in a single system do not have to rely the master to copy their data areas; they may be able to do several major steps before returning. This introduces the need to coordinate activities between slaves.

One type of coordination is the BARRIER, where all slaves in the same system wait for each other. Assume that the leader slaves in Example 6 initialized output ARRAY to zeros. The other slaves would have to know when it was safe to begin processing. This is easily accomplished by using
'C$ BARRIER'.

More typically, the coordination would occur in cycles such as the following:

Compute
Write Shared Area Changes
BARRIER
Read Shared Area Changes
BARRIER
Loop back to Compute

The two BARRIERs prevent a 'race' condition, where one slave tries to write over its changes from the last iteration - before all the other slaves have finished reading them. If, however, the slaves can alternate updates to two different data areas (e.g. double buffer), then the second BARRIER can be removed.

Other specialized types of slave synchronization can be used with the SYNCHRONIZE directive, where an array specifies which slaves must coordinate with each other. For example, the array:

0	1	0	1
1	0	1	0
0	1	0	1
1	0	1	0

creates a type of 'assembly line'; each slave produces a result used by the next slave. Slave 1, for instance, must synchronize with slaves 2 and 4 because row 1 has a 1 in columns 2 and 4. Similarly, slave 2 synchronizes with slaves 1 and 3. The array is symmetric to prevent race conditions, where one slave gets more than one cycle out of step with the others. LCAP support includes detection of deadlocks; where two or more slaves are hopelessly out of synchronization with each other and could never complete.

Another fine-grain synchronization routine, C$FADD, can be used for dynamic 'self scheduling'. By calling this routine, each slave assigns itself the next value, or range of values, from an integer variable in a shared Common. The returned value can be used to index an array or as a loop iteration.

Finally, the routines (C$ADDI4,C$ADDR4, C$ADDR8) perform coherent additions to array elements without the need for more expensive serialization routines. With these routines, slaves in the same system can perform their own, 'on the fly', ADDING function to shared arrays.

Note that LCAP does not use the traditional 'lock' or 'critical section' methods prevalent in other shared memory systems; it provides internal serialization as part of end use functions such as the BARRIER, SYNCHRONIZE, C$FADD and C$ADDxx routines. These tools are simple to use, perform efficiently within a single system and provide good protection against application deadlock.

Performance Results Attained

Performance examples are given in Chapter 1 of this volume. Several application examples running with multiple IBM 3090 clusters are shown in Fig. 10 of Chapter 1.

4. Conclusions

This chapter presents an overview of parallel processing with LCAP/3090, emphasizing the focus on multiple system configurations and the goal of supporting large parallel speedups over sequential run times. The inherent application limits to parallel speedup are reviewed, along with the two most common means of communication between multiple systems. A brief overview of LCAP's parallel facilities explains why they simplify application conversions. Essential parts of the LCAP design philosophy are explained, with some insights gained from practical experience. Several examples show how LCAP services can be used to easily control parallel applications for both static and dynamic decomposition schemes. Examples first show how to communicate across multiple systems, and then how to exploit the shared memory capability on the IBM 3090.

5. Bibliography

This section contains a list of LCAP publications grouped, by principle subject, into the following categories; general, reference/ performance, application and historical.

A) General Publications on LCAP

1. E. Clementi and D. Logan, "Parallel Processing with the Loosely Coupled Array of Processors System", IBM Research Report KGN-43, December 30, 1985,

2. M. Wojcik, G. Corongiu, J. Detrich, M. M. Mansour, E. Clementi and G. C. Lie, "Parallel Computing From Fermion Systems to Hydrodynamics: Water as an Example", Lecture Notes in Chemistry Series, "Supercomputer Simulations In Chemistry" M. Dupuis, Ed., Vol. 44, pages 224-245, Springer-Verlag, Berlin, Germany, 1986.

3. E. Clementi, S. Chin, D. Logan and J. Detrich, "Large Scale Computations On The Loosely Coupled Array of Processors", Brussels Seminar Report Part 9, Lecture Notes in Chemistry, 44, 130-164 (1986), Isreali J. Chem., 27(2), 1127 (1986), See also IBM Research Report KGN-51, May 8, 1986,

4. E. Clementi and J. Detrich, "Large Scale Parallel Computations On A Loosely Coupled Array of Processors", Elsevier Science Publishers, Amsterdam, November 29, 1986, See also IBM Research Report KGN-64, May 28, 1986,

5. E. Clementi, J. Detrich, S. Chin, G. Corongiu, D. Folsom, D. Logan, and R. Caltabiano, A. Gnudi, A. Cornevali, J. Helin, P. Palamidese, M. Russo, "Large Scale Computations on a Vector, Scalar and Parallel Supercomputer",

in Structure & Dynamics of Nucleic Acids, Proteins and Membranes, E. Clementi and S. Chin, Eds., Plenum Publishing Co., New York, 403-449 (1986); Parallel Computing, 5(1&2), 13-44 (1987), See also IBM Research Report KGN-77, August 15, 1986,

6. E. Clementi, S. Chin, G. Corongiu, J. Detrich, M. Dupuis, D. Folsom, G. C. Lie, D. Logan, D. Meck, V. Sonnad and R. Caltabiano, A. Carnevali, L. Hannon, J. Helin, D. C. Rapaport, M. Russo and M. Wojcik, "Large-Scale Computations on a Scalar, Vector and Parallel Supercomputer and Global Ab Initio Simulations With a Study on Liquid Water as an Example", IBM Research Report KGN-86, November 14, 1986,

7. E. Clementi, D. Logan and V. Sonnad, "Parallel Solution of Fundamental Algorithms Using a Loosely Coupled Array of Processors", Numerical Algorithms for Modern Parallel Computer Architectures, Martin Schultz, Ed., Springer Verlag, (1988), See also IBM Research Report KGN-105, March 30, 1987,

8. E. Clementi, in the Chapter "Supercomputers for Chemical Research and Development", pps. 27-80, in Computer Applications in Chemical Research and Education, Josef Brandt and Ivar K. Ugi Eds., Verlag Gmbh, Heidelberg, Germany, 1989,

9. E. Clementi, S. Chin, Z. Christidis, C. Corongiu, J. Detrich, M. Dupuis, D. Folsom, G. J. B. Hurst, G. C. Lie, D. Logan, J. Perez, M. Re, L. Rosenzweig, J. Saarinen, V. Sonnad and H. O. Villar, "Invited Papers Related to LCAP/3090 Presented by the Scientific Engineering Computations Department at the Boston Third International Conference on Supercomputing", IBM Research Report KGN-173, February 18, 1988,

    - E. Clementi, "Global Simulations on Scalar, Vector and Parallel LCAP-Type Supercomputers",

    - D. Logan, J. Saarinen and E. Clementi, "LCAP/3090-Genesis and Evolution of a Parallel Processing System",

    - J. H. Detrich, D. G. Folsom and L. J. Rosenzweig, "LCAP/3090 at IBM Kingston: Evolution of Software to Support Parallel Execution",

    - Z. D. Christidis and V. Sonnad, "Parallel Implementation of a Pseudospectral Method on a Loosely Coupled Array of Processors",

    - M. Dupuis, H. Villar, G. Hurst and E. Clementi, "Quantum Supercomputer Simulations for Electronic Devices at the

Molecular Level",

- J. Perez, M. Re, S. Chin and E. Clementi, "Determination of Protein Three Dimensional Structure: Computations and Graphics on LCAP",

10. E. Clementi, S. Chin, G. Corongiu, J. H. Detrich, M. Dupuis, D. Folsom, G. C. Lie, D. Logan and V. Sonnad, "Supercomputing and Super Computers: for Science and Engineering in General and For Chemistry and Biosciences in Particular", in Biological and Artificial Intelligence Systems, E. Clementi and S. Chin, Eds., pp. 319-424, ESCOM Science Publishers, B.V., Leiden, the Netherlands, (1988), Int'l J. Quant. Chem., 35, 3-89 (1989), See also T. Theophanides(ed.), Spectroscopy of Inorganic Bioactivators, Theory and Applications - Chemistry, Physics, Biology and Medicine, 1-112, See also IBM Research Report KGN-121, July 28, 1988,

11. E. Clementi, D. Logan and J. Saarinen, "LCAP/3090 Parallel Processing for Large Scale Scientific and Engineering Problems", IBM Research Report KGN-153, August 26, 1988, IBM Systems Journal, 27, 475 (1988),

12. E. Clementi, D. Logan and V. Sonnad, "Solution of Large Scale Engineering Problems Using a Loosely Coupled Array of Processors", in Numerical Algorithms for Modern Parallel Computer Architectures, M. Schultz, Ed., pp. 11-27, Springer-Verlag, New York, (1988),

13. E. Clementi and D. Logan, "Parallel Processing with the Loosely Coupled Array of Processors System", Special Purpose Computers, B. J. Alder, Ed., 135-183, Academic Press, (1988),

14. E. Clementi, D. K. Bhattacharya, S. Chin, G. Corongiu, M. Dupuis, K. Dyall, D. Folsom, S. Foresti, S. Hassanzadeh, G. C. Lie, R. Lindh, B. Liu, D. R. Logan, P. Mougenot, P. Procacci, R. Scateni, D. J. Schneider, V. Sonnad, J. M. Tyler, D. Vanderveken and d. P. Vercauteren, "Invited Papers Presented by the Scientific Engineering Computations Department at the Santa Clara Fourth International Conference on Supercomputing", IBM Research Report KGN-185, March 7, 1989,

B) Reference/Performance Publications on LCAP

1. L. Brochard, "Communication and Control Costs on a Parallel Multiprocessor", IBM Research Report KGN-71, May 20, 1987,

2. J. Helin, A. Perez-Ambite and J. Detrich, "Computation on the Loosely Coupled Array of Processors System: A Flexible Batch Processing Environment for the VM/CMS Operating System", IBM Research Report KGN-93, February 5, 1987,

3. R. Caltabiano, M. Russo, A. Carnevali, J. Detrich and D. Folsom "Parallel Computation on the Loosely Coupled Array of Processors: Shared Common Blocks and Synchronization", IBM Research Report KGN-103, February 17, 1987,

4. R. Caltabiano, A. Carnevali and J. Detrich, "Directives for the Use of Shared Bulk Memories: A Precompiler Extension", IBM Research Report KGN-110, March 3, 1987,

5. S. Cauberghs and J. Detrich, "Computation on the Loosely Coupled Array of Processors System: An Automatic and Interactive Backup Facility for the VM/CMS Environment", IBM Research Report KGN-125, June 3, 1987,

6. M. Russo, A. Perez-Ambite, R. Caltabiano, J. Detrich and D. Folsom, "An Approach To Parallel Scheduling For The LCAP System", IBM Research Report KGN-135, June 17, 1987,

7. M. Russo, A. Perez-Ambite, S. Cauberghs, J. Detrich and D. Folsom, "Simulation Of Parallel Scheduling For The LCAP System", IBM Research Report KGN-136, June 24, 1987,

8. S. Cauberghs and J. Detrich, "Computation on the Loosely Coupled Array of Processors System: Parallel Scheduler for the lCAP System: Operator and User Guide", IBM Research Report KGN-168, January 13, 1988,

9. G. Corongiu, J. Detrich and E. Clementi, "Study of Communication and Synchronization Overhead on the Loosely Array of Processors Systems", IBM Research Report KGN-170, March 29, 1988,

10. J. Prost, J. Detrich and M. Becker, "A Simulator of the Parallel Scheduler for the LCAP System", IBM Research Report KGN-54, May 16, 1988,

11. J. Prost, J. Detrich and M. Becker, "Cost Characterization of the Precompiler Communication and Synchronization Directives on the LCAP System", IBM Research Report KGN-73, May 19, 1988,

12. J. Prost, D. Folsom, J. Detrich, and Monique Becker, "Modeling the LCAP/FPS264 Parallel Archecture for Evaluating Its Performance", IBM Research Report KGN-178, January 24, 1989,

13. D. Folsom, M. Klonowski, N. Pitsianis, A. Trannoy and S. Veronese, "LCAP/3090 User Guide", IBM Research Report KGN-193, December 31, 1989,

C) Application Publications on LCAP

1. L. Domingo and E. Clementi, "Parallel Computation of the Migration of Seismic Data on the lCAP", IBM Research Report KGN-17, January 17, 1985,

2. M. Wojcik, "MD Simulation of MCY Water Using LCAP", IBM Research Report KGN-28, August 21, 1985,

3. L. Brochard, "Parallelization of Transonic Flow on lCAP", See IBM Research Report KGN-37, September 26, 1985,

4. P. E. Bowen-Jenkins and M. Dupuis, "Parallel Computation of Molecular Electronic Similarity", IBM Research Report KGN-46, November 6, 1985,

5. Luigi Brochard, "Parallelization of Compressible Turbulence Flows on LCAP and TCAP", See IBM Research Report KGN-44, January 6, 1986,

6. J. E. Watts, M. Dupuis and H. Villar, "Implementation of Large Fortran Programs On The Loosely Coupled Array of Processors: Parallel Computation of Molecular SCF Energy Gradients", Theo. Chim. Acta, 71(91), (1987), See also IBM Research Report KGN-78, August 29, 1986,

7. A. Capotondi, V. Sonnad and S. Chin, "Parallel Resolution of the Shallow Water Equations Using An Explicit Finite Difference Algorithm", Computer Physics Communications, 52, 195-205, (1989), See also IBM Research Report KGN-57, September 8, 1986,

8. E. Clementi, D. Logan, V. Sonnad, Z. Christidis and A. Capotondi, "Solving Engineering Problems in Parallel With a Loosely Coupled Array of Processors", Computers in Mechanical Engineering, 5(3), November 1986,

9. R. King and V. Sonnad, "Implementation of an Element-by-Element Solution Algorithm for the Finite Element Method on a Coarse Grained Parallel Computer", Computer Methods in Applied Mechanics and Engineering, 65(1), November 1987,

10. Z. Christidis, V. Sonnad and D. Logan, "Parallel Implementation of a 2-D Fast Fourier Transform on a Loosely Coupled Array of Processors", IBM Research Report KGN-68, January 26, 1987,

11. J. D. Watts and M. Dupuis, "Towards Efficient Parallel Computation of Correlated Wave Functions. Implementation of the Two-Electron Integral Transformation on the lCAP Parallel Supercomputer", IBM Research Report KGN-100, January 31, 1987,

12. J. D. Watts and M. Dupuis, "Parallel Computation of the Moller-Plesset Second-Order Contribution to the Electronic Correlaton Energy", J. Comp. Chem., 9(2), 158-170, (1988), See also IBM Research Report KGN-106, February 10, 1987,

13. W. Luken, N. Liang, R. Caltabiano, E. Clementi, E. Bacon, J. M. Warren and W. F. Beausoleil, "Application of Parallel Processing to Molecular Modeling Graphics", IBM Research Report KGN-111, March 18, 1987,

14. R. Herbin, W. D. Gropp, D. E. Keyes and V. Sonnad, "A Domain Decomposition Technique on a Loosely Coupled Array of Processors", IBM Research Report KGN-124, May 21, 1987,

15. A. Capotondi, R. Signell, R. Beardsley and V. Sonnad, "Tide-Induced Residual Circulation Simulated on a Parallel Computer", IBM Research Report KGN-132, June 15, 1987,

16. Hsiao-Ming Hsu and V. Sonnad, "Parallelization Of A Numerical Model For Ocean Circulation", IBM Research Report KGN-134, July 8, 1987,

17. Z. D. Christidis and V. Sonnad, "Parallel Implementation Of A Pseudospectral Method On A Loosely Coupled Array Of Processors", IBM Research Report KGN-143, July 15, 1987,

18. R. Herbin, S. Gerbi and V. Sonnad, "Parallel Implementation Of A Multigrid Method On The LCAP Supercomputer", Applied Mathematics & Computation, 27, 281-312, (1988), See also IBM Research Report KGN-144, July 24, 1987,

19. R. Herbin, S. Gerbi and V. Sonnad, "Parallel Implementation Of A Multigrid Method On The Experimental LCAP Supercomputer: Part II", IBM Research Report KGN-155, September 14, 1987,

20. J. C. Antoranz, V. Sonnad and D. Logan, "Use Of A Loosely Coupled Array Of Processors To Determine The Fractal Dimension Of A Strange Attractor", IBM Research Report, KGN-157, September 17, 1987,

21. M. Mareschal, E. Kestemont and S. Chin, "Monte-Carlo Molecular Dynamics Experiments on the Loosely Coupled Array of Processors", IBM Research Report KGN-56, October 7, 1987,

22. R. Donati and V. Sonnad, "Coupling a Finite Element Frontal Solver Code to a Parallel Conjugate Gradient Solver", IBM Research Report KGN-98, October 8, 1987,

23. A. Milani, M. Carpino and D. Logan, "Parallel Computation of Planet-Crossing Asteroid Orbits", IBM Research Report KGN-161, November 17, 1987,

24. S. Foresti, G. Brussino S. Hassanzadeh, and V. Sonnad, "Multilevel Solution Method for the P-Version of Finite Elements: Parallel Implementation and Comparison With Other Solution Methods", IBM Research Report KGN-137, July 8, 1988, Computer Physics Communications, 53,349 (1989).

D) Historical Publications on LCAP (Early Implementation)

1. A. Carnevali, L. Domingo, and E. Clementi, "A Precompiler for Parallel Programs in lCAP Environment (1CAPAR): User's Notes", IBM Research Report KGN-24, March 12, 1985,

2. L. Domingo and S. Chin, "Parallel Computation on the lCAP Computer System: Tools and Guidelines", IBM Research Report KGN-25, September 15, 1985,

3. S. Chin, L. Domingo, A. Carnevali, R. Caltabiano and J. Detrich, "Parallel Computation on the Loosely Coupled Array of Processors: A Guide to the Precompiler", IBM Research Report KGN-42, November 25, 1985,

4. Jukka Helin and John Detrich, "Parallel Computation On The Loosely Coupled Array of Processors: Scheduler For The VM System", IBM Research Report KGN-58, August 14, 1986,

6. Acknowledgments

The LCAP/3090 system was developed by a small group of capable, highly motivated people. In the beginning of the project, most of the design work and prototype coding was done by Jukka Saarinen and Thierry Delbecque. Their long hours and dedication created the building blocks. Antoine Trannoy was the principle author of the Precompiler and synchronization support; Pierre Ferron developed the Precompiler lexical analyzer. Laura Evans' experience was valuable in all areas, for debugging, organizing the libraries and devising the operational controls. Marian Klonowski, Nikos Pitsianis and Silvia Veronese worked as a team on the scheduler and LCAP command processors; their willingness to experiment and revise are evident in the usability of the final programs. Philippe Mouthon contributed to an efficient set of synchronization routines.

Chapter 26.

Parallel Algorithms

Doug Logan

IBM Corporation
Scientific and Engineering Computations
Department 48B/428
Kingston, New York 12401, U.S.A.

This chapter is meant to be a short introduction to parallel algorithms. It will focus on parallel implementations of several important algorithms in matrix computations. The objective will be to develop their structures based on a number of fundamental matrix operators. Parallel performance will be assessed in terms of their implementation on lCAP systems. In particular, we will discuss the solution of linear systems of equations, the algebraic eigenvalue problem, and generation of orthogonal basis sets.

Solving sets of linear equations is, for example, central to solving elliptic partial differential equations. Solvers can be optimized to correspond to the characteristics of the system to be solved, i.e. whether dense, sparse, symmetric and whether approximate answers are suitable. Eigenvalue problems arise in many disciplines, and have been discussed in previous chapters that describe self-consistent field methods for determining eigenstates of many body systems in the Hartree Fock approximation. The algorithms selected here depend again on whether the matrix is sparse, dense, symmetric and also on how many and which eigenvalues and/or eigenvectors are required. Lastly, the generation of orthogonal basis sets is often an important component of larger algorithms, and is key to optimization problems, such as least squares solutions to over or under determined systems of equations.

While we limit ourselves to a description of several algorithms in these areas, it is important to develop a library of parallel algorithms that covers a much larger area. This is because algorithms can be thought of as macro instructions of the parallel machine. Optimizing these instructions can have a far reaching effect on all applications' performance. This is particularly true of engineering applications where, from an algorithmic perspective, fields as different as fluid dynamics and materials science to magnet design, use the same tools to solve their problems.

This chapter is organized as follows. In section 1 we state general parallel processing performance principles and discuss some of the particular issues that affect algorithms. In section 2 we examine four basic matrix operators, that together, form much of the framework under which most algorithms in linear algebra are expressed. Devising parallel algorithms for matrix computations will often amount to understanding how these operators may be applied in parallel. In section 3 we discuss two methods of solving linear sets of equations, one for dense and one for sparse systems. In section 4 we present two eigenvalue solvers that calculate all eigenvalues and eigenvectors. In section 5 we discuss orthogonal basis set generation in the context of linear least squares problems. Section 6 concludes with several observations on generalizing this operator based approach to other parallel algorithms and performance limiting issues.

1. General Principles

Algorithms, like larger scale applications, must satisfy the three following constraints in order to be effectively executed in parallel. First, a large fraction of the algorithm must be parallelizable, i.e. a sequential component limits the asymptotic potential speedup in inverse proportion to its relative size. Second, interprocessor communication must be minimized as this represents an effective sequential component. Lastly, it is essential to distribute the work as evenly as possible among processors. A full discussion of these performance issues may be found in Chapter 1. Here we elaborate on the specific properties of numeric algorithms that may distinguish them from large applications.

In general, the parallel grain size of numeric algorithms is much smaller than, say, large Monte Carlo sampling codes. This means that communication (including synchronization) between processors is often the overriding issue in whether or not an algorithm will exhibit good speedup and for what problem size. Algorithms tend to have good load balancing properties. Their rigid computational structures allow a concise analysis of the amount of parallel work and the optimal number of processors to use. However, it is often the case that both decrease as the algorithm progresses through its various steps. We will observe this behavior with several of our examples.

The goal of minimizing communication is equivalent to finding the best possible mapping of algorithm to machine. This means different things for different architectures. For distributed systems, such as hypercubes, it is achieved by a mapping that only requires nearest neighbor communication. This means that global synchronizations, sums or reduction operations that are explicit in many algorithms must be reformulated or cleverly executed to avoid large performance penalties. In contrast, shared memory machines are less sensitive to the assignment of tasks to processors because, by definition, all processors are nearest neighbors. However, the size of such systems are limited by increased memory contention, the degree of memory interleaving/partitioning and multistage network (if used) latency, and cache (if used) coherency effects and updating policy. In this chapter we limit ourselves to a discussion of several algorithms on shared memory systems, in particular the lCAP systems.

Another issue that affects not only communication, but also vector performance, is that of devising "block" algorithms. Most algorithms are conventionally cast in the form where a sequence of simple matrix operators are successively applied on a system until "completion." The predominant operations that result are saxpys or saxpy like operations (expressions of the form $y = ax + y$ or $y = ay + x$ where x and y are vectors and a a constant). However, it is often advantageous to reformulate the algorithm such that products of these operators are applied, in that the predominant operations are now matrix multiplies. The latter are executed at higher MFLOP rates because the contents of vector registers may be reused thereby reducing vector stores and fetches. More arithmetic is carried out per unit time. Thus on a sequential vector machine the preferred algorithms are those cast in block form. For a thorough discussion of these issues, Refs. 1 and 2 may be consulted.

Added to this benefit, in the context of parallel implementation, is that block algorithms may increase the parallel grain size of the tasks and lessen the number of synchronizations required. A common strategy is to assign one processor the task of computing the next product of matrix operators, while other processors apply the previous product using a domain decomposition. To accommodate load balancing, the first processor may also apply the first product to a smaller domain before computing the next product. These techniques will be seen in several of the examples.

A recurring observation, noted in the following, will be: for a fixed number of processors, parallel performance improves as problem size, n, increases. This is a fundamental tenet for parallel numeric algorithms. It follows because, where parallel grain sizes generally increase in proportion to n^x $(x \geq 1)$, sequential components and/or communication overhead typically scale as n^{x-1}. Thus, for large problems, the degrading factors may become relatively insignificant.

2. Basic Matrix Operators

Most matrix computations can be described in terms of the application of a small set of basic operators. All do the same sort of thing. They operate on a vector or matrix and introduce zeros either explicitly or indirectly. That is, they distill the information of a full system into a more meaningful system with few nonzero entries. In the following, we examine four basic operators and their block representations. They are: the Gram Schmidt projection operator, the Gauss transform, the Givens transform, and the Householder transform.

It is important to note that, with the exception of the Givens transform, all have the same structure, i.e. a simple rank one correction to the identity. Thus, it should come as no surprise that parallel algorithms employing these operators will have similar structures. The Givens operator is a simple rank two correction to the identity, and this will be shown to give rise to slightly more complex parallel applications. While these operators are formally matrices, they are preferentially applied in factored form so that operation counts are $O(n)$ saxpys rather than $O(n^2)$ matrix vector multiplies.

2.1 Gram Schmidt Projection

Given two linearly independent vectors x and y, this operator applied to one of them makes the new pair orthogonal, i.e. their inner product is zero. If we represent the operator as P_x then we we have:

$$y' = P_x y$$

where:

$$x^T y' = 0$$

and P_x is constructed as:

$$P_x = I - \frac{xx^T}{x^T x}$$

where I is the identity matrix. Given this operator, we can orthogonalize a system of linearly independent vectors.

Properly organized, this process describes the Modified Gram Schmidt algorithm for solving the OB (Orthonormal Basis) problem. Products of these operators in factored form may be applied as a block to the unorthogonalized system. Standard domain decomposition, using segmentation over columns, provides an efficient means of accomplishing this in parallel. We will examine in more detail the Householder QR factorization algorithm for the OB problem because, although somewhat slower, it is less sensitive to near rank deficiency and is in more general usage. However, the Gram Schmidt operator is a fundamental construct that is used in many algorithms and is introduced here for completeness, and to point out that its parallel applications can be easily effected.

2.2 The Gauss Transform

The Gauss Transform is constructed to operate on a vector such that a given range of elements is zeroed. Given a vector x with n elements a Gauss transform, M_k, that zeros elements x_{k+1} to x_n can be constructed as:

$$M_k = I - \alpha e_k^T$$

where vector α has elements $\alpha_i = 0$ for $i \leq k$ and

$$\alpha_i = \frac{x_i}{x_k} \quad for\ i > k$$

This operator has the appearance of what is called a spike matrix (the identity plus one lower spike at diagonal position k).

The operator is not orthogonal ($M_k^{-1} \neq M_k^T$) but is useful primarily because of other properties of its inverse. It is trivially shown that M_k^{-1} is computed by changing the sign on each entry α_i. An important consequence is that products of inverses, applied in sequence from k large to small, results in a lower triangular matrix L with the same entries as each individual inverse. Thus, as may be anticipated, this operator forms the basis of LU factorization for the solution of linear systems of equations.

Products of Gauss transforms can be stored in factored form (or in L form if it is anticipated that a triangular solve will follow) and applied as a block to vectors or columns of a matrix. This is called a rank n update for a block of n transforms.

2.3 The Givens Rotation

The Givens rotation, or plane rotation, is an operator that rotates a vector in a plane defined by two elementary directions e_i and e_j. It is constructed from the identity by assigning values to elements $I_{i,i}$ $I_{i,j}$ $I_{j,i}$ and $I_{j,j}$ as:

$$I_{i,i} = I_{j,j} = \cos(\theta)$$

$$I_{i,j} = -I_{j,i} = \sin(\theta)$$

where $\cos(\theta)$ and $\sin(\theta)$ are the cosine and sine of some angle θ. The angle θ can be calculated to effect different objectives. If we wish to zero the jth element in a vector by a rotation in the i,j^{th} plane, θ is uniquely defined by a linear equation in the values of the ith and j^{th} elements.

The Givens rotation is an orthogonal operator, i.e. if represented as G then

$$G\,G^T = G^T G = I$$

where G^T is its transpose. Orthogonal operators are of fundamental importance in matrix computations because they do not affect the 2 norms of the systems they operate on. More importantly, they preserve the eigenvalues of a matrix when used to effect orthogonal transformations of the form:

$$B = G^T A\,G$$

B and A have the same eigenvalues. Furthermore, if A is symmetric then so is B. Thus orthogonal transformations also preserve symmetry. These properties are of fundamental importance as we will see.

An orthogonal transformation, using a Givens rotation, can be constructed to zero a transpose pair of elements in a symmetric matrix. This is called a Jacobi transformation and is the basic operation in the Jacobi method for solving the symmetric eigenvalue problem. The appropriate value of the angle θ for each pair is defined by a quadratic equation involving the values of the elements in the (i,i^{th}) and (j,j^{th}) diagonal and (i,j^{th}) off diagonal positions. The solution corresponding to the smaller angle is always selected for reasons of numerical stability.

Products of Givens rotations can be used to develop block variations of the algorithms that employ them. Such products can either be left in factored form or explicitly formed as block matrices. We will see both variants in several algorithms to be discussed.

2.4 The Householder Transform

The Householder transform, like the Givens rotation, is an orthogonal operator but with structure and zeroing abilities similar to the Gauss transform. It can be used to zero a range of elements of a vector, as contrasted with the plane rotation, where only one element can be annihilated at a time. It has the structure of a rank one correction to the identity:

$$H = I - \frac{u\,u^T}{h}$$

Application of this operator to an n element vector x , Hx, can be used to zero elements in positions $k+1$ to n. The value of the kth element is set to:

$$x_k = -\,sign(x_k)\,|x_{k,n}|$$

where $|x_{k,n}|$ is the 2 norm of the partial vector over the range of x_k to x_n and function $sign(x_k)$ returns the sign of the original value of x_k.

The vector u is constructed according to the rules:

$$u_i = 0 \quad \text{for } i < k$$

$$u_i = x_k + sign(x_k)\;|x_{k,n}| \quad \text{for } i = k$$

$$u_i = x_i \quad \text{for } i > k$$

and h is a normalization constant calculated as:

$$h = \frac{u^T u}{2}$$

This operator can be constructed to operate in alternative ways. For example, reversing the direction on the two inequalities defining the Householder vector u, and replacing the partial 2 norm $|x_{k,n}|$ with $|x_{1,k}|$, defines a transform that zeros elements in positions 1 to $k-1$ and sets value of the kth element to:

$$x_k = -sign(x_k)\,|x_{1,k}|$$

The Householder matrix H is both symmetric and orthogonal:

$$H = H^T \text{ and } HH = I$$

The orthogonality property is important for the same reasons as discussed in section 2.3.

There is a lot of flexibility associated with products of this operator. They can be left in factored form, explicitly created or represented in block factored form. The last instance is referred to as the WY factorization,[3] where assuming we have a sequence of matrices $H_k\,H_{k+1}\,....H_l$, defined as above, we can represent their product as:

$$H_k\,H_{k+1}....H_l = I + WY^T$$

where W is full and of size $R^{k,l-k}$ and Y is of similar size and is lower trapezoidal. This factorization can be used to implement block Householder operations.

3. Linear System Solvers

Solving systems of linear equations is one of the fundamental operations in applied mathematics, spanning disciplines as different as econometric modelling to airplane design. Its importance underlies the specification of equilibrium, where one must satisfy a set of linear constraints on a number of independent variables. Provided that the set of constraints is not redundant and equals in number the number of variables, we define the problem to be full rank and nonsingular, i.e. there exists a unique solution. It is the solution of this full rank problem that we address in this section.

Representing the independent variables as a vector x, we require the solution to the problem in matrix form:

$$A\,x = b$$

How best to proceed depends on several factors. Foremost is the density of A. If A is full and nonsymmetric, Gaussian elimination is generally the preferred algorithm. If full and symmetric, a variation of Gaussian elimination, called Cholesky factorization, is optimal. Both of these methods are direct solvers; they must be carried out completely in order to calculate the values of the unknowns. If, however, A is sparse, then depending on its size and the degree and structure of the sparsity, one of a number of iterative non direct solvers may be advantageous.

Solvers in this class can be terminated when an appropriate convergence criteria is satisfied, indicating a solution to some degree of accuracy.

In the following we discuss one solver from each class and their parallel implementation and performance on the lCAP systems. The analysis of parallel content permits these algorithms to be adapted to other parallel systems.

3.1 LU Factorization

Gaussian elimination, or LU factorization, is the standard method for solving a dense nonsymmetric set of linear equations. Its execution speed has become a quasi standard for comparing the performance of different computers. It admits a high degree of vectorization and parallelization is thus effective at differentiating between competing high performance systems. For a discussion of vector performance guidelines in implementing this algorithm see Ref. 1.

Its development is straightforward. Given:

$$A\,x = b$$

factor A into a lower triangular matrix, L, times an upper triangular matrix, U:

$$A = L\,U$$

Defining the product Ux as a vector y, solve for y :

$$y = L^{-1}\,b$$

then solve for the solution x:

$$x = U^{-1}\,y$$

The LU factorization is accomplished through a sequence of Gauss transformations discussed in section 2.2. The operation count of this step is proportional to n^3, where n is the size of the system. The triangular solvers for y and x each have an operation count proportional to n^2. Thus, for large n, the factorization is the performance governing step. We now address the parallel and vector implementation of this step; the more problematic but less critical triangular solvers will be briefly discussed in section 5.

The algorithm (without "pivoting") proceeds as follows:

for $j = 1$ to $n - 1$
1. calculate M_j from column A_j
   store the spike $-\alpha_j$ over A_j
2. update columns A_k $k = j + 1$ to n
   $A_k = M_j A_k$

On completion, A is overwritten by the strictly lower component of L and the upper triangular U. The diagonal portion of L is the identity. The triangular solvers are now ready to be executed.

Step 2 represents independent (hence parallelizable) updates to columns A_{j+1} to A_n. With p processors, each can apply the current Gauss transform, M_j, over equal or nearly equal domains of the remaining columns of A. Moreover, step 1, the generation of the next Gauss transform can be calculated as soon as column A_{j+1} has been updated. If the processor that updates this column performs this additional task then the partitioning of the total number of remaining columns between processors can be adjusted so that the parallel computation remains load balanced. The only significant (depending on size n) sequential computation is the initial generation of M_1 (which could also be computed in parallel, but with often prohibitively small parallel grain size). This sequence of rank 1 operations describes a parallel algorithm with $n-1$ synchronizations and $n-1$ parallel updates.

However, as the factorization proceeds, the number of columns to update in parallel decreases until there are less columns than processors. As the amount of work per parallel update decreases, synchronization overhead becomes relatively more costly. The latter effect can be somewhat alleviated (but not done away with) by employing a rank n or block variation of the algorithm. Here, one processor now generates $m > 1$ successive Gauss transforms, after which each remaining column is repetitively updated m times. This has several advantages: the number of synchronizations decrease from approximately $n-1$ to $\frac{n}{m}$, parallel grain sizes increase by a factor of m, and vector performance is improved. Load balancing is achieved by allocating less updates to the processor that generates, from its updated columns, the next product of transforms. The optimal blocking factor, m, domain decomposition and their dependence on the remaining problem size can be found by measuring the times to calculate Gauss transforms, perform updates, and synchronize. In general, the greater the synchronization time, the more desirable is a large m. Towards the end, parallel grain size and synchronization costs dictate that less processors are used, until only one completes the factorization.

This strategy will work with high efficiency for large problems, i.e. $n >> p$. However, one additional procedure may be needed to provide numerical stability and minimize roundoff error, i.e. pivoting. Pivoting is the process of selecting the element about which to generate the next Gauss Transform. Full pivoting chooses the largest remaining element and permutes the current row and column with those belonging to this element. Partial pivoting restricts the search for the maximum size element to the current column and performs a permutation of current row with the row of this element. Following the permutation(s), the calculation of the transform and updating proceed as usual. It is commonly thought that partial pivoting provides sufficient stability.

Using pivoting procedures results in the LU factorization of a system PA where P is a permutation matrix. Bearing in mind that the final objective is to obtain a solution to $Ax = b$, the triangular solvers that follow must be applied to the right hand side Pb.

Partial pivoting is accommodated in the parallel algorithm by assigning to the processor generating the next product of m transforms, the additional task of maxima searches. A permutation vector is updated indicating which rows require permuting prior to the update process. Again, one can determine heuristically the optimal blocking factors and domain decompositions to load balance work across processors.

Shown in Fig. 1 is the speedup curve for the LU factorization of a double precision matrix of size $n = 1000$, employing partial pivoting, $m = 20$, and a simple domain decomposition strategy on the lCAP system. As usual, performance improves for larger problems, and vice versa.

3.2 Preconditioned Conjugate Gradient Methods

This algorithm is principally applied to the problem $Ax = b$ where A is large, symmetric, sparse and positive definite. Such properties are met in finite element or difference discretizations of elliptic partial differential equations, describing such phenomena as heat flow over structures with suitable boundary conditions. While the solution could proceed by Cholesky factorization, this is often impractical because sparsity is not maintained in factored form. The amount of fill in, can for large problems, exceed available memory. In this case, iterative methods which preserve sparsity, such as the conjugate gradient method, may be preferred providing there is an acceptable rate of convergence.

There are a number of other general iterative techniques. In order of increasing convergence rate, they include the Jacobi, Gauss Seidel and SOR algorithms; each based on a specific splitting of $A = M - N$ and iterative solutions to x as $M x_{k+1} = N x_k + b$, where $x_{k+1} \rightarrow x$ as k becomes large. Their convergence rate is governed by the spectral radius of the iteration matrix $M^{-1}N$. These techniques may be parallelized (see Refs. 5–10 for example). However, for the symmetric, sparse positive definite problem, the preconditioned conjugate gradient algorithm is often superior.

While introduced as an iterative technique, this is formally a direct method, i.e., for a system of size n, in exact arithmetic, the solution is obtained in exactly n steps. However, because roundoff error may severely degrade this convergence, the algorithm is properly viewed as an iterative method. While a full description is outside the scope of this chapter, we give a brief overview of its development. For more detail, Ref. 17 may be consulted.

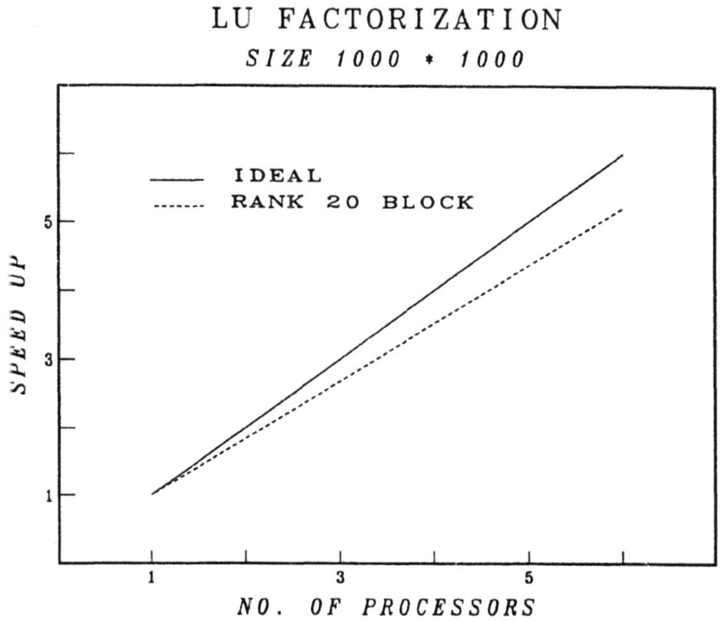

Figure 1. Speedup curve for LU factorization on the ICAP system.

The algorithm can be described as a minimization procedure, developing the solution vector $x = A^{-1}b$, that minimizes the functional:

$$\phi(x) = \frac{1}{2} x^T A x - x^T b$$

Starting with an initial trial vector x_0, a solution is progressively improved upon by defining a sequence of search directions p_k, by which we update:

$$x_{k+1} = x_k + \alpha_{k+1} p_{k+1}$$

Convergence is monitored by the 2 norm of the residual $Ax_{k+1} - b$.

To converge in n steps, the conjugate gradient vectors, p_k, must be linearly independent and A orthogonal:

$$p_j^T A p_i = 0 \quad j \neq i$$

The classical conjugate gradient algorithm, developed by Hestenes and Stiefel,[4] specifies how to efficiently generate these vectors.

n step convergence is rarely achieved because roundoff error tends to destroy A orthogonality. This results in indefinite convergence properties. The extent of this degradation is intimately tied to the condition number of A. That is, the better conditioned, the better A orthogonality is retained and the faster will be the convergence. This suggests the approach of preconditioning the problem to solve an equivalent, well conditioned problem, that will converge quickly. An equivalent problem is defined as:

$$A'x' = b'$$

where

$$A' = C^{-1} A C^{-1}$$

$$x' = C x$$

$$b' = C^{-1} b$$

where C is a symmetric non singular matrix chosen to make the condition number of A' small.

These ideas are tied together to describe the preconditioned conjugate algorithm, which we now state without further elaboration. Starting from a best guess of the final solution, x_0 (if no best guess exists, select $x_0 = 0$), and residual vector $r_0 = Ax_0 - b$, the following iterations are repeated until the residual's 2 norm is acceptably small:

for $k = 1, \ldots\ldots$
  if $|r_{k-1}| \leq \varepsilon$
    set $x = x_{k-1}$ and end
  else

1. $z_{k-1} = M^{-1} r_{k-1}$

2. $\beta_k = \dfrac{z_{k-1}^T r_{k-1}}{z_{k-2}^T r_{k-2}} \qquad \beta_1 = 0$

3. $p_k = z_{k-1} + \beta_k p_{k-1} \qquad p_1 = z_0$

4. $\alpha_k = \dfrac{z_{k-1}^T r_{k-1}}{p_k^T A p_k}$

5. $x_k = x_{k-1} + \alpha_k p_k$

6. $r_k = r_{k-1} - \alpha_k A p_k$

$|r_{k-1}|$ represents the 2 norm of the residual vector r_{k-1} and ε is a limiting value of the accuracy desired. The preconditioning matrix, M ($M = C^2$), is critical in determining the rate of convergence. Ideally, it should be chosen such that $M^{-1}r_{k-1}$ is easy to compute and $M^{-1}A$ approaches the identity. Unfortunately, these goals are usually contradictory and some compromise is needed.

Implementing this algorithm in parallel merits the following considerations. Step 1, solving for vector z_{k-1}, can be parallelized with ease or difficulty depending on

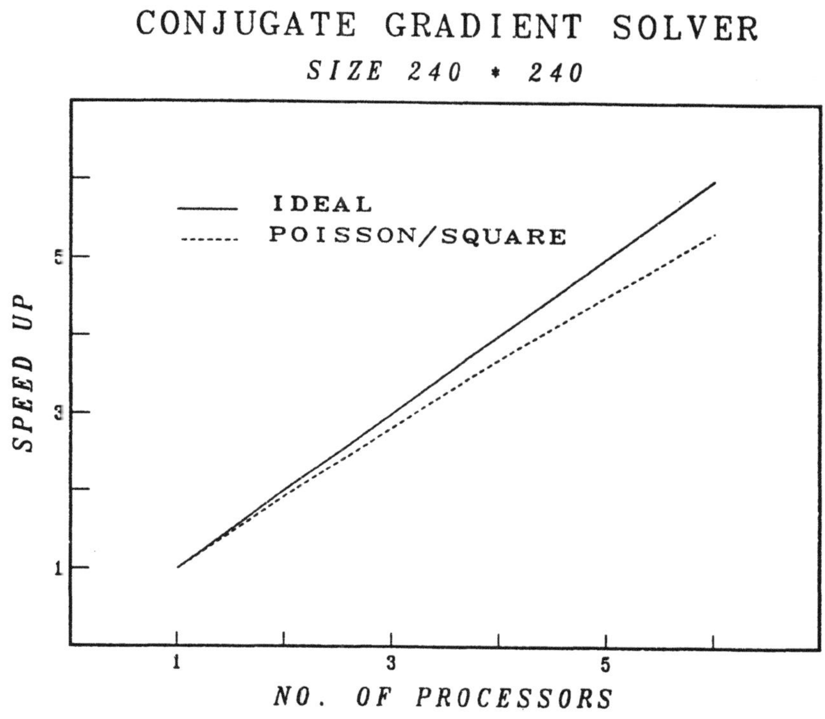

Figure 2. Speedup curve for the Conjugate Gradient Solver on lCAP.

the structure of preconditioner M. If M can be organized in block diagonal form, to correspond to the number of processors, then sub-domains of the vector z can be effectively computed in parallel. The simplest example of this is choosing M as the diagonal of A . This preconditioning is called diagonal scaling. If, however, A is not diagonally dominant, this will not be effective. A more general preconditioner is the "incomplete" Cholesky factorization[11] of A, with that decomposition compelled to retain the same sparsity structure as A. In this case we require efficient parallel implementations of sparse triangular solvers. These are often difficult to achieve.

The remaining steps, while parallelizable, have small operation counts and require a number of costly synchronizations. Thus step 2 requires the parallel inner product of $z_{k-1}^T r_{k-1}$. After a synchronization, β_k is calculated. The saxpy for p_k is next computed in parallel, followed by another synchronization. Step 4 is a matrix vector multiplication Ap_k to form a vector w. Each processor can compute equal parts of w and accumulate partial contributions to inner product $p_k^T w$. Following another synchronization, parameter α_k is calculated. The two saxpys of steps 5 and 6 are next computed in parallel, as well as partial contributions to the convergence criteria $|r_k|$. Following another synchronization, this process is repeated if convergence has not been reached.

This is a challenging algorithm to perform well in parallel. There are a minimum of four synchronizations per iteration (depending on how $M^{-1}r_{k-1}$ is done) and all intervening parallel operations have complexity of $O(n)$, with the exception of the sparse matrix vector multiply (which, because it is sparse, will be small also).

Test cases of solving Poisson's equation on a square, using simple diagonal scaling as a preconditioner, reveal that of the order of 90% of linear speedup can be achieved with a six processor lCAP system for problem sizes of $n \simeq 240$ (see Fig. 2). Much of this success results from the favorable structure of A that obtains from a p version finite element discretization of the Laplacian operator.[37] However, this may not be a discriminating problem, in that the conjugate gradient algorithm is not usually the algorithm of choice for the Poisson equation (there are a host of faster spectral[12,13] or multigrid[14] type Poisson solvers that can also be parallelized). For more relevant problems, the success of the conjugate gradient method depends on the preconditioner, and its parallel implementation on how effectively the inverted preconditioner can be applied. Some of these issues are discussed in greater detail in Refs. 15 and 16. Optimal parallel performance may be achieved by employing a block version of this algorithm, wherein blocks of conjugate gradient vectors are developed during each iteration, thus increasing parallel grain size and limiting the degrading effect of many synchronizations. A full treatment of this approach may be found in Ref. 33.

4. Real Symmetric Eigenvalue Solvers

As may be appreciated from the previous chapters, the eigenvalue problem is fundamental in computational chemistry and physics. We require, for a system of size n , the eigenvectors, x_i, and eigenvalues, λ_i, which satisfy:

$$A\, x_i = \lambda_i\, x_i \qquad i = 1,n$$

For $n \geq 5$, all eigenvalue solvers are iterative because the underlying problem is necessarily iterative. That is, the problem is formally equivalent to solving for the n roots of an nth order polynomial. Moreover, all eigenvalue solvers employ the same iterative procedure; the use of sequences of similarity transforms that progressively bring A to upper triangular form and hopefully diagonal form:

$$V_k^{-1} V_{k-1}^{-1} \ldots\ldots V_1^{-1}\; A\; V_1 \ldots\ldots V_{k-1} V_k \;\rightarrow\; D$$

If this process is successful, then the entries on the diagonal represent the eigenvalues, and if D is strictly diagonal, then the product of operators the eigenvectors:

$$X = |x_1 \ldots\ldots x_n| = V_1 \ldots\ldots V_{k-1} V_k$$

$$\lambda = |d_1 \ldots\ldots d_n|$$

However, when A is symmetric real (or Hermitian) then life becomes much simpler. A solution exists and all eigenvalues are real and all eigenvectors orthogonal. The last property is important as it means that orthogonal operators may be used in the similarity transformations. Since orthogonal transformations preserve symmetry, the process can be organized in a well structured manner.

Because of the prevalence of the symmetric real problem, we examine the parallel implementation of two eigenvalue solvers for this case. The first relies on orthogonal transformations using Givens rotations and is called the Jacobi method. The second relies on orthogonal transformations using Householder and Givens operators and is called the symmetric, implicitly shifted, QR algorithm. In general, the second method is superior but under certain conditions the first may enjoy certain advantages.

4.1 Jacobi Diagonalization

The basic operation in this method is the application of a succession of Givens transformations that progressively zero all off diagonal elements. The conventional EISPACK cyclic by row algorithm[18,19] can be described as follows:

Do until all $a_{i,j} \le \varepsilon$
for $i = 1, n-1$
for $j = i+1, n$
if $a_{i,j} > \varepsilon$ then do
$A = G_{i,j}^T A G_{i,j}$

$G_{i,j}$ is the Givens rotation constructed to annihilate off diagonal pair $a_{i,j}$ and $a_{j,i}$, as described in section 2.3. The double loop over all elements $a_{i,j}$ is called one full Jacobi sweep. Convergence may be improved if the threshold criteria, ε, is lowered on successive full sweeps until it is assigned the required precision (usually machine precision, although one can require less, if desired, using the Gershgorin circle theorem[17] to assign uncertainties to the eigenvalues). The eigenvectors, if needed, can be accumulated as the product of the Givens rotations.

This method works because an orthogonal transformation preserves the Frobenius norm of any matrix, where this norm is defined as the sum of the squares of all entries $a_{i,j}$. A Givens transformation will, however, redistribute the contribution from the off diagonal element transpose pair to the diagonal. Ultimately then, if this sequence of transformations is repetitively applied, diagonal form is attained. More than one full Jacobi sweep is generally required because off diagonal pairs previously annihilated can be reintroduced by later transformations.

As has been pointed out many times (see Ref. 20 and references therein), this algorithm can be reorganized to allow a high degree of parallel computation. The central idea is that for a problem size of n, where there are a total of $\frac{n}{2}(n-1)$

symmetric off diagonal elements, one can find sequences of n/2 transpose pairs which can be annihilated in parallel. These n/2 pairs necessarily have unique column and row indices. This parallelization strategy is evident given that an orthogonal transformation using a Givens rotation, $G_{i,j}$, effects only columns and rows with indices i and j. The parallel algorithm thus requires a parallel ordering scheme, which is a specification of $n-1$ groups of n/2 transpose pairs (whose zeroing we refer to as a minisweep), such that one full Jacobi sweep is defined.

There are a combinatorially large number of parallel ordering schemes. A optimal scheme should satisfy the following criteria. First, it should be simple to compute. Second, performing each minisweep in parallel requires a simple and economic task assignment policy. This means different things for different parallel computers. For a distributed system using message passing, it dictates that passing groups of Givens rotations and partial updates to A, from one processor to another, can be done efficiently. On a shared memory machine this criteria is moot. Lastly, the ordering scheme should have provable convergence properties, i.e. it should be "equivalent" to a cyclic by row (column) ordering.[34]

Given an ordering scheme, a parallel Jacobi method proceeds through $n-1$ minisweeps, during each of which $\frac{n}{2}$ parallel transforms are partitioned among p processors. During a minisweep, each processor updates its set of diagonal elements, and then with right and left applications of its Givens rotations, updates its pairs of columns and rows respectively. The row updating requires a synchronization to avoid incorrect updates. That is, all processors update columns, synchronize, and then and only then update rows. Following another synchronization, this procedure is repeated for the next minisweep; and so on until one complete Jacobi sweep has been performed. The process continues until all elements satisfy the lowest threshold criteria for one full Jacobi sweep.

An example of obtaining good parallel performance with this method is described in Ref. 20. However, several objections may be made. First, twice as much work is being performed than is necessary. Because orthogonal transformations preserve symmetry only the upper triangular form (or lower triangular form) need be updated. This also implies that twice as much memory as needed is being used. Second, while the updating process is rich in saxpy like operations, better performance can be obtained if the algorithm is expressed in terms of matrix multiplies. These objectives are met by a block Jacobi algorithm.

There are a number of ways of defining a block Jacobi algorithm that differ in how, and the degree with which, "zeroing" is accomplished. One version, discussed here, follows a simple reformulation of the element by element algorithm. For p processors, the essential point is to reorganize A in upper block triangular form with block sizes equal to n/2pm, where m is some integer m$\geq$ 1. The basic operation of applying a Givens transformation to zero a single off diagonal pair is replaced by determining the product of Givens rotations that obtain in using the usual cyclic be row algorithm to zero elements within a given block. Note that the block is not "zeroed" as it would be in an exact block analogue of the Jacobi algorithm. However, its contribution to the Frobenius norm has been significantly reduced, and by a much smaller number of operations than would be required if full zeroing was accomplished. Some experience has been obtained with block Jacobi SVD algorithms that suggests the present strategy is superior.[35]

With this blocking factor a full Jacobi sweep can be defined by $2p - 1$ block mini-sweeps, each with m block transforms for each processor to perform in parallel. A block transformation can be written as:

$$A = J_{i,j}^T A J_{i,j}$$

where $J_{i,j}$ is a two by two block rotation that is constructed by application of the serial algorithm over block $A_{i,j}$:

$$J_{i,j} = \prod_{k,l=1}^{b} G_{k,l}$$

Subscripts k and l cycle through all elements in $A_{i,j}$ and b = n/2pm is the size of the block. An analogous block parallel ordering scheme is specified such that only blocks with unique block indices are concurrently updated.

This is illustrated with the parallel ordering scheme of Brent and Luk,[24] 3 processors, and a blocking factor with $m = 1$. This defines $2p - 1 = 5$ block mini-sweeps, during which each processor performs $m = 1$ parallel block updates. The first minisweep assigns the following jobs to processors labelled $p1$, $p2$ and $p3$:

$p1$	$p2$	$p3$	
1	3	5	*block label* 1
2	4	6	*block label* 2

Block labels 1 and 2 refer to the block row and column indices that make sense if operating on the upper triangular form. After each processor has updated its block $A_{i,j}$ and corresponding diagonal blocks $A_{i,i}$ and $A_{j,j}$, the remaining blocks in these block columns and rows are updated. Writing this operation as:

$$A = J_{p3}^T J_{p2}^T J_{p1}^T A J_{p1} J_{p2} J_{p3}$$

we proceed, exploiting symmetry to update only those blocks on the upper triangular form, by assigning nonconflicting pairs of blocks to update in parallel. A job list of such parallel updates is constructed for each minisweep, going through the block rotations calculated by processors $p1$, $p2$ and $p3$, and reused on successive full Jacobi sweeps. The number of block pair updates per minisweep is $(2pm - 2)p$. They may be partitioned to achieve ideal load balancing. The eigenvectors, initialized to the identity, are efficiently updated in parallel with each processor applying its block rotations to its own unique block columns of eigenvectors.

The next minisweep assignments are computed by holding block label 1 fixed for processor $p1$ and rotating all other block labels right circular. This leads to the minisweep number 2 assignments:

$p1$	$p2$	$p3$	
1	2	3	*block label* 1
4	6	5	*block label* 2

Repeating this process 3 more times generates the last three minisweep assignments. One full Jacobi sweep is completed when the diagonal blocks are similarly processed. Rather than establish a separate minisweep for these, it is preferable to process them as a preface to the first minisweep, using their 1 by 1 block rotations as initial blocks from which to construct the first minisweep's 2 by 2 block rotations. This saves one synchronization and minimizes the number of operations in the block column and row updating. If, at the end of the full Jacobi sweep, convergence has not been attained, a new Jacobi sweep is reinitiated, starting again with the first minisweep assignments.

The algorithm, as described, was coded using lCAP parallel task constructs, shared memory to hold the blocked matrix and block rotations, and synchronization by shared variables. It was tested on large random, double precision, matrices and found to converge in the same number of sweeps as the sequential algorithm, and with the same answers. The parallel version, executed with one processor, was typically several times faster than the best organized sequential cyclic by row EISPACK algorithm. This may be attributed to the updating process, which uses matrix multiplies to achieve superior vector performance. In Fig. 3 is shown the speedup curve for 1 to 4 processors on an IBM 3090 model 400 for a problem size of $n = 800$ and $m = 4, 3, 2, 1$ respectively (same blocking structure in all cases). With 4 processors, approximately 90% of linear speedup was observed. The deviation from ideal speedup is primarily due to synchronization overhead. As usual, performance improves for larger problems.

4.2 The Symmetric QR Algorithm

As frequently observed (see Ref. 36 for example), this method is typically an order of magnitude or so faster than the Jacobi algorithm, for large systems ($n \simeq 1000$). In fact, it is hard to justify the Jacobi method unless there are many parallel processors available, or A is very diagonally dominant, or very accurate eigenvectors are required. Otherwise, this method is recommended when all eigenvalues and eigenvectors are needed. When only one or several eigenvalues/eigenvectors are needed, Lanczos variants or orthogonal iteration methods are preferable.

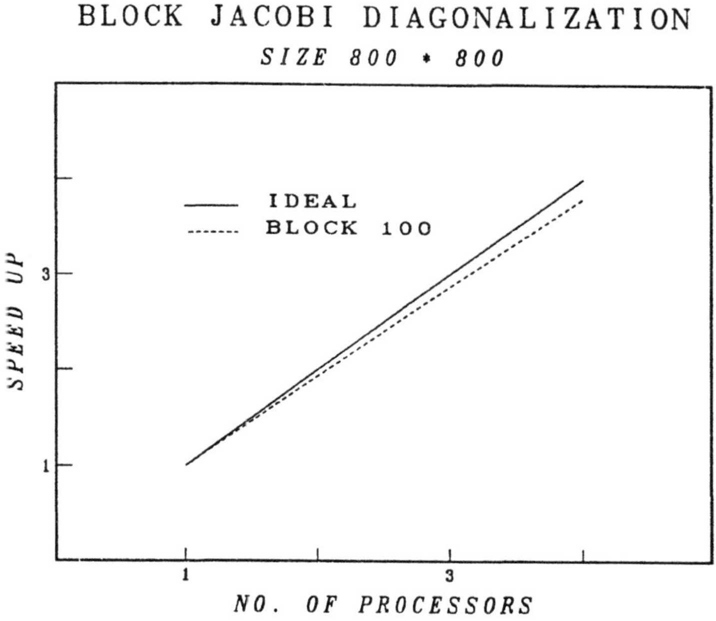

Figure 3. Speedup curve for block Jacobi Diagonalization on ICAP.

To develop all eigenvalues and eigenvectors, the algorithm proceeds in four steps. First, A is reduced to tridiagonal form using a sequence of Householder transformations:

$$T = H_3.....H_n \, A \, H_n.....H_3$$

If A were nonsymmetric, this procedure describes an algorithm for reduction to Hessenberg form, from which the nonsymmetric eigenvalue problem is best approached, with steps similar to the following. For the symmetric case, given that orthogonal transformations preserve symmetry, only the upper (lower) triangular form of A is used and operated upon.

Second, initial approximations to the eigenvectors are accumulated as the product of the Householder transforms:

$$V = H_n.....H_3$$

The third step iteratively reduces the tridiagonal to final diagonal form. It is the QR or QL factorization (Q orthogonal, R upper triangular, L lower triangular) of T followed by an orthogonal transformation of T with Q :

$$T = Q\,L$$

$$T = Q^T \, T \, Q \, (= LQ) \qquad T \rightarrow D$$

The last step, like the second, is performed if all eigenvectors are required. It continues, after each QR (QL) step, the accumulation of the orthogonal transforms:

$$V = VQ$$

The operation count of these four steps, carried out to the final diagonalization, reveals that the first, second, and fourth are $O(n^3)$ and the third $O(n^2)$. The $O(n^3)$ steps are easily parallelized, resulting in a parallel algorithm that is efficient for large n. It is this limited objective that we will satisfy. The third step can also be parallelized using a "tearing" procedure described in Refs. 21 and 22. This may be necessary if n is small, or a large number of processors are used, or the eigenvectors are not developed. In these cases, the fraction of the code left in sequential mode becomes more significant.

The tridiagonalization of A can be equivalently started from the upper left corner and proceed downwards, or from the lower right corner and proceed upwards. We describe the latter procedure with each Householder operator H_j constructed to zero elements of column j in row indices 1 through $j-2$ (see section 2.4). Thus step 1 is:

for $j = n$ to 3

determine H_j from *column* A_j

update $A = H_j A H_j$

Each iteration of this procedure is recursive, i.e. generation of H_{j-1} requires the prior full orthogonal transformation, $H_j A H_j$. However, a transformation can be parallelized because each is simply the product of several matrices. To do this efficiently, taking advantage of the structure of H_j and the symmetry of A, requires this product be calculated in three stages.

With the current transform defined by a Householder vector u_j and normalization constant h_j, $H_j = I - \frac{u_j u_j^T}{h_j}$, a vector q and coefficient k are calculated:

$$q = A \frac{u_j}{h_j}$$

$$k = \frac{u_j^T q}{2h_j}$$

Next, update q by a saxpy:

$$q = q - k u_j$$

The transformation $H_j A H_j$ is then computed as a rank 2 update to A:

$$A = A - qu_j^T - u_j q^T$$

Each of these steps is parallelized. The first is a matrix vector multiply and an accumulation of an inner product. To get good vector performance for the former it is important to exploit the symmetry of A. The partitioning of work between processors follows a segmentation over the columns of A. After a synchronization, the saxpy is executed in parallel. After another synchronization, the rank 2 update on A is done in parallel. Again, only the upper triangular part is updated and the work is distributed among processors on equal or nearly equal areas (by columns).

Given the similar structures of the Householder and Gauss transforms, the parallel structure of the tridiagonalization procedure is similar to the parallel rank 1 LU factorization, discussed in section 3. u_{j-1} and h_{j-1} can be concurrently computed during the rank two update at the jth stage, such that the total work is load balanced. This is analogous, in the previous case, to precomputing the next Gauss transform. Also, the same decrease in parallel grain size occurs as tridiagonalization proceeds, as occurred in LU decomposition, as that factorization progressed. Unfortunately, unlike LU factorization (or QR factorization described in section 5), it is difficult to cast this process in a more efficient block form (using products of Householder transforms) because a full transformation H_jAH_j is required before obtaining column (row) A_{j-1}, from which H_{j-1} can be computed.

The second step, multiplication of the Householder matrices, is less restrictive. The parallel development of this product is evident if written, post multiplying by identity, as:

$$V = H_n H_{n-1} \ldots\ldots H_3 \; I$$

Considering the development of the kth eigenvector, starting from the kth column of I, e_k, reveals that:

$$H_j \, e_k = e_k \quad \text{for all } j < k + 1$$

Therefore, each independent eigenvector x_k can be efficiently computed in parallel, as:

for $j = \max(k+1, 3)$ to n
$x_k = H_j \, x_k$

with x_k initialized to e_k. Noting that the amount of work to develop the kth eigenvector is $O((n-k)^2)$, leads to a simple load balanced assignment of sets of eigenvectors for each processor to calculate.

The third step is executed sequentially, justified by its smaller operation count and algorithmic complexity. A brief description of this step is given, in order to appreciate the parallelization of the fourth step (the continued updating of the eigenvectors). For a more detailed description, consult Ref. 17.

The rationale for convergence to diagonal form, by repeatedly factoring T to $Q\,L$ and performing the orthogonal transformation Q^TTQ, is provided by the power methods for developing single eigenvalues. It can be shown that diagonalization proceeds in developing eigenvalues in sequence from largest to smallest. Thus sequences of transformations are applied to smaller size tridiagonal systems as each new eigenvalue is produced. The power method indicates that faster convergence may be obtained if a good estimate of an eigenvalue is known. Calling this estimate a shift, μ, the QR algorithm is preferentially applied to the system $T - \mu I$. The Wilkinson shift, which is usually favored, is found by solving the eigenvalue problem for the two by two problem of the current lowest tridiagonal sub matrix. That eigenvalue closest in magnitude to the lowest tridiagonal element is selected. Thus diagonalization under this method progresses upwards from the lower right-hand corner of the tridiagonal. Good numerical stability is retained by performing the factorization of $T - \mu I$ and orthogonal transformation implicitly, using a number of tricks legitimized by what is known as the Implicit Q theorem.

The fourth step continues the accumulation of eigenvectors. Following each Q^TTQ transformation of step 3, they are updated as $V = V\,Q$, where Q is a product of Givens rotations:

$$Q = G_{k,k-1}\,G_{k-1,k-2}\cdots\cdots G_{2,1}$$

Subscript k corresponds to current k^{th} eigenvalue being developed for the current tridiagonal of size k.

To do the updating in parallel with p processors, V is segmented in p groups by rows and the same update applied by each processor to its own domain. As is evident, only the first k eigenvectors are updated, given that a right hand side application of a Givens rotation, $G_{i,j}$, affects only columns i and j. As diagonalization proceeds, this amount of parallel work decreases significantly. Towards the end, synchronization costs become comparable to the benefits of the parallel execution.

This algorithm, coded as described, using lCAP software and shared common blocks, was tested on large random, double precision, matrices. It was found to give equivalent solutions to those obtained by the ESSL symmetric, implicitly shifted QR library routine.[23] For large size problems ($n = 1000 - 1500$), using 4 processors, a speedup of approximately 80% of ideal was observed (see Fig. 4). In absolute performance, this was roughly three times faster than the ESSL routine.

Parallel performance may be improved by parallelizing the third step and

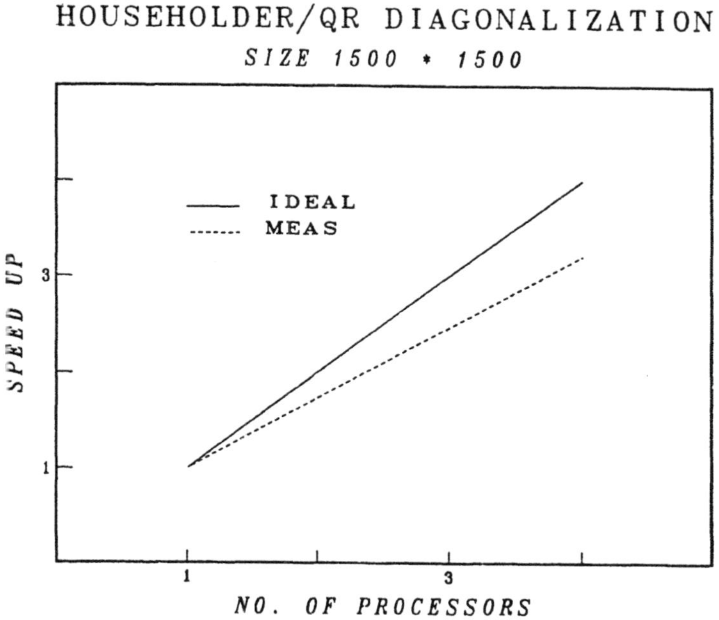

Figure 4. Speedup curve for symmetric QR Diagonalization on lCAP.

tional improvements are constrained by the cost of synchronizations bracketing parallel kernels whose grain sizes decrease monotonically as the algorithm proceeds. In absolute performance, assembler coding of the important kernels should pay additional dividends.

5. Orthogonal Basis Set Generation

The orthogonal basis set problem can be stated as: given a set of n vectors $x_i \in R^m \quad i = 1,n,$ where $m \geq n$, determine a set of orthogonal vectors $q_i \in R^m \quad i = 1,k$ such that:

$$span(q_1, \ldots.., q_k) = span(x_1, \ldots..,x_n)$$

Thus we require an orthogonal basis $Q^{m,k}$, for the space defined by $X^{m,n}$. If X is full rank, i.e. all vectors x_i are linearly independent, then $k = n$. In the following we limit ourselves to this condition.

To motivate orthogonal basis generation, we illustrate their use in solving for optimal solutions to over determined systems of linear equations: $Ax = b$ where $A \in R^{m,n} \quad m > n$, i.e. there are more equations than unknowns. An optimal solution will be defined in the linear least squares sense: find x such that the 2 norm of the residual, $|Ax - b|_2$, is minimized.

Historically, the oldest least squares procedure relies on solving the Normal equations, defined as:

$$x = (A^TA)^{-1} A^Tb$$

This entails forming a symmetric system $B = A^TA$, its Cholesky factorization, a new right hand side A^Tb, and applying two triangular solvers. This algorithm may be parallelized effectively for large problems as most of the operation count is in the $O(n^3)$ first two steps. Matrix multiplication is trivially parallelized and Cholesky factorization is a special case of LU factorization, described in section 3.1. Of the remaining $O(n^2)$ steps, matrix vector multiplication is readily parallelized, the triangular solvers less so.

However, if A is poorly conditioned or near rank deficient, this method becomes numerically unstable. A better method, where near rank deficiency is handled with more stability, introduces the subject of this section. If an orthogonal basis Q can be found by a factorization:

$$A = QR$$

then the full rank least squares solution is given by:

$$x = R^{-1}Q^Tb$$

As mentioned in section 2.1, one method of generating orthogonal basis (and achieving the QR factorization required here) is the Modified Gram Schmidt algorithm. However, this factorization can be more effectively generated and applied by using either Householder or Givens transformations. Of the two, the Householder method is faster, although for very structured problems the Givens method may be superior, given the selectivity of the Givens operator. Both methods may be parallelized (see for example Ref. 25), and can take advantage of blocking to give high performance. Because of its greater general usage, the Householder method will be described.

A third method of solving the general least squares problem, and of determining rank, is the most expensive but also the most exact. It is based on the Singular Value Decomposition (SVD), a full treatment of which may be found in Ref. 17, and parallel implementations in Refs. 20, 24, 34 and 35. It is the only recommended approach for solving the rank deficient problem.

QR factorization by Householder transformations is accomplished by the following $O(n^3)$ algorithm:

for $j = 1, n-1$
  calculate H_j to *zero elements* $a_{k,j}$ $k = j+1,m$ of *column* A_j

do $k = j+1,\ n$
$A_k = H_j\ A_k$

On completion, A is over written by triangular form R. The orthogonal basis may be stored in factored form by storing the Householder vectors u_j in the lower triangular part of A and the Householder normalization constants h_j in another array. If the non factored form of Q is required, then the product of Householder matrices may be computed:

$$Q = H_1 \ldots\ldots H_{n-1}$$

Parallelization is achieved by applying the current a Householder transform, H_j, in parallel to all remaining columns. This algorithm is essentially identical in structure to LU factorization, because the structure of the Gauss and Householder operators are identical. The same parallelization strategies that were used to load balance and maximize the performance of the updating (precomputation of the next transform, blocking by construction of products of transforms, etc.) may be used. Similarly, as the factorization proceeds, the parallel grain sizes diminish and this must be accommodated by selecting optimal blocking factors and number of processors to employ.

While structurally similar, the block QR factorization differs from its LU counterpart in forming the operator products. Optimal updating performance may be achieved by using the WY representation[3] for products of Householder transforms:

$$H_k \ldots H_{k-m} = I + W\ Y^T$$

where W is full, Y is lower trapezoidal and both are of size k by $k-m$. The WY factorization permits the updating to proceed by matrix multiplication. If at the end, the full form of Q is desired, operations analogous to those discussed in section 4.2 permit this to be done in parallel.

In the linear least squares problem, the system may be reduced in parallel to the stage $Rx = c$ where $c = Q^Tb$ without having to explicitly form Q. This is done by modifying the previous algorithm (or its block version) to include an update on b:

for $j = 1,\ n-1$
calculate H_j to *zero elements* $a_{k,j}$ $k = j+1, m$ of *column* A_j
$b = H_j b$
do $k = j+1,\ n$
$A_k = H_j\ A_k$

This is comparable to the QR factorization of a system of size m by $n+1$, with b representing the extra column, and the above strategies may be used to do this effectively in parallel. The dense triangular solver for $Rx = c$, may be performed sequentially (justified by its smaller $O(n^2)$ operation count) or attempted in parallel if synchronization costs are not prohibitive. A "column sweep" parallel triangular solver[29] is based on partitioning the update of the right hand side as each new element of x is developed. Thus, its parallel grain size is $O(n)$, and decreases as the solution proceeds. Some advantage is gained by proceeding in a block fashion to minimize the number of synchronizations.

Given the similarity of LU and QR factorizations, their parallel performances are found to be similar, when Q is left in factored form. When Q is explicitly formed, the efficient parallelization of computing products of Householder matrices, improves the latter's. When applied to the least squares problem, inclusion of the triangular solver degrades performance. As usual, larger problems are favorable.

6. Conclusions

We have attempted to show that parallel algorithms are best examined in terms of the application of basic operators. Their structures and optimization strategies are readily identified. For matrix computations, four operators, that form much of the basis of linear algebra, have been shown effectively applied in parallel for a number of algorithms.

This principle is extendable. Consider, for example, initial value problems in partial differential equations. Two popular solution techniques are the Alternating Direction Implicit (ADI) method[27] and the (pseudo) spectral method.[26] In the former, the basic operator is often a tridiagonal linear system solver; in the latter, the one dimensional Fourier transform. In both cases the parallel structure of the key computational kernels may be similarly expressed. For example, an n dimensional problem using the pseudo spectral method, requires at each time step an n dimensional Fourier transform on some data structure $A(N_1, N_2, \ldots, N_n)$. This can be written:

$$\textit{for dimensions } j = 1,2,3,\ldots\ldots n$$

$$\textit{for } k = 1 \text{ to } N \qquad N = \prod_{i=1}^{n} N_i \quad i \neq j$$

$$A_k = F_{Nj} A_k$$

The operator, F_{Nj}, is the one dimensional Fourier transform for an array of size N_j (the size of the data structure in dimension j), and can be applied in parallel, using a simple domain decomposition, to independent arrays A_k (subscript k represents all data projected onto dimension j). A synchronization is required before beginning each new dimension. For the ADI method a similar parallel structure is obtained where the operator is a tridiagonal linear system solver. For either operator, an optimized library routine may be used.

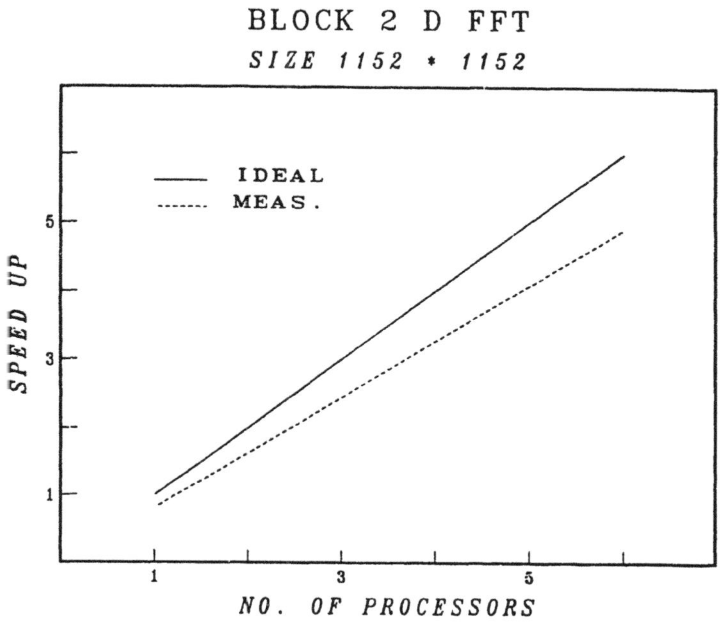

Figure 5. Speedup curve for a 2 D FFT on lCAP.

Vector and cache performance are usually optimized by successively transposing the data structure so that operators work on contiguous arrays. Additional cache performance may be obtained by blocking the data structure. Shown in Fig. 5 is the parallel speedup curve for a two dimensional Fourier transform of a complex, double precision matrix of size 1152. The speedup is with respect to the single processor ESSL optimized two dimensional transform.[31] The parallel algorithm, coded with lCAP software, employed the ESSL one dimensional transform.[32]

The one dimensional Fourier transform may be similarly parallelized[28] by an implicit representation of the one dimensional array in a two dimensional form. This is possible if its size can be factored into a product of powers of small prime numbers. Tridiagonal solvers can be parallelized,[30] but with very small parallel grain sizes.

The performance of parallel algorithms are very dependent on problem size. As problem size decreases, overall performance often becomes increasingly dependent on the parallel performance of an $O(n^2)$ operation (triangular solvers, diagonalization of a tridiagonal system), or worse, an $O(n)$ operation (inner product, saxpy, tridiagonal solver). If the $O(n^2)$ operations require $O(n)$ synchronizations, or the $O(n)$ operations even the minimum one synchronization, communication overhead may become significant. This may result in parallel speed down as the number of processors increase and parallel grain sizes decrease.

However, it is for large problems that parallel processing is directed. Thus, the performance of the few, but representative, parallel algorithms cited here is encouraging. A larger survey of parallel and vector implementations of algorithms, for solving partial differential equations, may be found in Ref. 30.

References

1. J.J. Dongarra, F. Gustavson and A. Karp, (1984), "Implementing Linear Algebra Algorithms on a Vector Pipeline Machine," SIAM Rev. 26, pp. 91-112.
2. J.J. Dongarra and D.C. Sorensen, (1985), "A Look at the Evolution of Mathematical Software for Dense Matrix Problems over the Past Fifteen Years," in: Numerical Algorithms for Modern Parallel Computer Architectures, ed. M. Shultz, Springer-Verlag, N.Y.
3. C. Bischoff and C. Van Loan, (1985), "The WY Representation of Products of Householder Matrices," Technical Report No. TR85-681, Dept. of Comp. Sci., Cornell Univ., Ithaca, N.Y.
4. M. Hestenes and E. Stiefel, (1952), "Methods of Conjugate Gradient for Solving Linear Systems," J. Res. Nat. Bur. Standards Sect. B, 49, pp. 409-436.
5. L. Adams and J. Ortega, (1982), "A Multi-color SOR Method for Parallel Computation," Proc. Int. Conf. Par. Proc., pp.53-56.
6. L. Adams and H. Jordan, (1986), "Is SOR Color-blind ?", ICASE Rep. No. 84-14, NASA Langley Research Center, Hampton, VA.
7. D.J. Evans, (1984), "Parallel SOR Iterative Methods," Parallel Computing, Vol. 1, No. 1, pp. 3-19.
8. M. Dubois and F. Briggs, (1982), "Performance of Synchronized Iterative Processes in Multiprocessor Systems," IEEE Trans. Softw. Eng. Se. 8, pp. 419-431.
9. N.M. Missirlis, (1987), "Scheduling Parallel Iterative Methods on Multiprocessor Systems," Parallel Computing, Vol. 5, No. 3, pp. 295-303.
10. W. Hackbusch and U. Trottenberg, eds., (1982), **Multigrid Methods,** Springer-Verlag, Berlin.
11. J. Meijerink and H. van der Vorst, (1981), "Guidelines for the Usage of Incomplete Decompositions in Solving Sets of Linear Equations as They Occur in Practical Problems," J. Comp. Phys., 14, pp. 134-155.
12. P. Swarztrauber, (1977), "The Methods of Cyclic Reduction, Fourier Analysis and the FACR Algorithm for the Discrete Solution of Poisson's Equation on a Rectangle," SIAM Rev., 19, pp. 490-501.
13. W.L. Briggs and T. Turnbull, (1988), "Fast Poisson Solvers for MIMD Computers," Parallel Computing, Vol. 6, No. 3, pp. 265-275.
14. C. Grosch, (1979), "Performance Analysis of Poisson Solvers on Array Computers," in: Infotech State of the Art Report: Supercomputers, Infotech, Maidenhead, England, C. Jesshope and R. Hockney, eds., Vol. 2, pp. 147-181.
15. 15. G. Meurant, (1987), "Multitasking the Conjugate Gradient Method on

16. M.K. Seager, (1986), "Parallelizing Comjugate Gradient for the Cray X-MP," Parallel Computing, Vol. 3, No. 1, pp. 35-49.
17. G.H. Golub and C.F. Van Loan, (1983), Matrix Computations, Johns Hopkins Univ. Press.
18. Matrix Eigensystems Routines - EISPACK Guide, 2nd ed., Vol. 6 of Lecture Notes in Computer Science, B.T. Smith, et. al., Springer-Verlag, 1976.
19. W.H. Press, B.P. Flannery, S.A. Teukolsky, and W.T. Vetterling, (1986), Numerical Recipes: The Art of Scientific Computing, Cambridge Univ. Press.
20. M. Berry and A. Sameh, (1986), "Multiprocessor Jacobi Algorithms for Dense Symmetric Eigenvalue and Singular Value Decompositions," Proc. Int Conf. Par. Proc., pp. 433-440.
21. J.J.M. Cuppen, (1981), "A Divide and Conquer Method for the Symmetric Tridiagonal Eigenproblem," Numerische Mathematik 36, pp. 177-195.
22. J.J. Dongarra and D.C. Sorensen, (1986), "A Fully Parallel Algorithm for the Symmetric Eigenvalue Problem," Argonne Natl. Lab. Rep. MCS-JM-62.
23. Subroutine DSLEV, IBM Engineering and Scientific Subroutine Library, Product No. 5668-863.
24. R.P. Brent and F.T. Luk, (1985), "The Solution of the Singular Value and Symmetric Eigenproblems on Multiprocessor Arrays," SIAM J. Sci. Stat. Comput., Vol. 6, pp. 69-84.
25. J.J. Dongarra, A.H. Sameh and D.C. Sorensen, (1986), "Implementation of Some Concurrent Algorithms for Matrix Factorization," Parallel Computing, Vol. 3, No. 1, pp. 25-35.
26. R. Voigt, D. Gottlieb and M. Hussaini, eds., (1984), Spectral Methods for Partial Differential Equations, SIAM Press, Philadelphia.
27. J. Ericksen, (1972), "Iterative and Direct Methods for Solving Poisson's Equation and their Adaptability to ILLIAC IV," Center for Advanced Computation Document 60, Univ. Illinois, Urbana-Champaign.
28. P.N. Swarztrauber, (1987), "Multiprocessor FFTs," Parallel Computing, Vol. 5, No 1, pp. 197-211.
29. D. Kuck, (1976), "Parallel Processing of Ordinary Programs," Advances in Computers 15, Academic Press, N.Y. pp. 119-179.
30. J.M. Ortega and R.G. Voigt, (1985), Solution of Partial Differential Equations on Vector and Parallel Computers, SIAM Press, Philadelphia.
31. Subroutine SCFT2, IBM Engineering and Scientific Subroutine Library, Product No. 5668-863.
32. Subroutine SCFT, IBM Engineering and Scientific Subroutine Library, Product No. 5668-863.
33. D.P. O'Leary, (1987), "Parallel Implementation of the Block Conjugate Gradient Algorithm," Parallel Computing, Vol. 5, No 1, pp. 127-139.
34. F.T. Luk and H. Park, (1988), "A Proof of Convergence for Two Parallel SVD Algorithms," IEEE Trans. Comp., Vol. 38, No. 6, pp. 806-811.
35. C. Bischoff, (1987), "Computing the Singular Value Decomposition on a Distributed System of Vector Processors," Technical Report, Dept. of Comp. Sci., Cornell Univ., Ithaca, N.Y.

36. A.H. Karp and J. Greenstadt, (1987), "An Improved Parallel Jacobi Method for Diagonalizing a Symmetric Matrix," Parallel Computing, Vol.5, No. 3, pp. 281-294.
37. S. Foresti, G. Brussino and V. Sonnad, (1988), "Multilvel Solution Method for the P-version of Finite Elements: Parallel Implementation and Comparison with Other Solution Methods," Technical Report KGN-137, IBM-Kingston, New York.

Index

A

B

C

D

E

F

J

K

L

M

N

O

P

Q

R

S

T

U

V

W

X

Y

Z

Numerics

www.ingramcontent.com/pod-product-compliance
Ingram Content Group UK Ltd.
Pitfield, Milton Keynes, MK11 3LW, UK
UKHW021901220926
475870UK00004B/128

* 9 7 8 9 4 0 1 0 7 4 9 5 7 *